THE MATHEMATICAL PAPERS OF SIR WILLIAM ROWAN HAMILTON

Dunsink Observatory by W. B. Taylor

Cunningham Memoir No. XVI

THE MATHEMATICAL PAPERS OF SIR WILLIAM ROWAN HAMILTON

VOL. IV

GEOMETRY, ANALYSIS, ASTRONOMY,
PROBABILITY AND FINITE DIFFERENCES,
MISCELLANEOUS

EDITED FOR THE
ROYAL IRISH ACADEMY

BY

B. K. P. SCAIFE, PH.D., D.SC. (ENG.), M.RI.A.

Fellow Emeritus,
Trinity College, Dublin

Shaftesbury Road, Cambridge CB2 8EA, United Kingdom

One Liberty Plaza, 20th Floor, New York, NY 10006, USA

477 Williamstown Road, Port Melbourne, VIC 3207, Australia

314–321, 3rd Floor, Plot 3, Splendor Forum, Jasola District Centre, New Delhi – 110025, India

103 Penang Road, #05–06/07, Visioncrest Commercial, Singapore 238467

Cambridge University Press is part of Cambridge University Press & Assessment,
a department of the University of Cambridge.

We share the University's mission to contribute to society through the pursuit of
education, learning and research at the highest international levels of excellence.

www.cambridge.org
Information on this title: www.cambridge.org/9781009414883

First published 2000
First paperback edition 2023

A catalogue record for this publication is available from the British Library

ISBN 978-1-009-41488-3 Paperback

CONTENTS

* In the case of papers communicated, the date immediately following the title is the date of communication.

PART III. ANALYSIS

PART IV. ASTRONOMY

PART V. PROBABILITY AND FINITE DIFFERENCES

PART VI. MISCELLANEOUS

PREFACE

This fourth volume of the Mathematical Papers of Sir William Rowan Hamilton completes the project begun, in 1925*, by the instigators and first Editors: Arthur William Conway (1875–1950) and John Lighton Synge (1897–1995). It contains Hamilton's published papers on geometry, analysis, astronomy, probability and finite differences, and a miscellany of publications including several addresses. There are also three previously unpublished manuscripts, namely: the, unfortunately incomplete, Third Part of the Systems of Rays, the earlier parts of which were published in Volume I; two letters to Augustus De Morgan, one devoted to definite integrals and the other to a third order differential equation; and a very long letter and postscript addressed to Andrew Searle Hart on Anharmonic Coordinates.

At the end of this volume will be found a list of Hamilton's papers in approximate chronological order. As well as an index to this volume, there is also a combined index for all four volumes.

The frontispiece is a view of Dunsink Observatory, where Hamilton lived and worked from the time of his appointment, in 1827, as Andrews Professor of Astronomy and Royal Astronomer of Ireland until his death in 1865. It is a reproduction of an aquatinted etching published in 1820 by William Benjamin Sarsfield Taylor (1781–1850) and was intended to be part of his History of the University of Dublin†.

The Royal Irish Academy acknowledges with gratitude the generous financial assistance towards the publication of this volume from the National University of Ireland; Trinity College, Dublin; University College, Dublin; University College, Cork; University College, Galway; The Queen's University of Belfast; The Dublin Institute for Advanced Studies.

A warm expression of thanks must be paid to the staffs of the Departments of Manuscripts and of Early Printed Books in the Library of Trinity College, Dublin, and to the staff of the Royal Irish Academy, for their willing and expert assistance at all times.

For help in the preparation of this volume especial thanks are due to Dr. Ian Elliott, of Dunsink Observatory, who suggested the frontispiece and who advised on astronomical matters, and in Trinity College to Dr. B. P. McArdle, Professor Petros Florides, Miss Hannah O'Connor, and my brother Professor W. Garrett Scaife. Prof. T. J. Gallagher of University College, Dublin, and my daughter, Lucy, gave invaluable help in proof reading.

The care with which the Cambridge University Press has produced this work is most gratefully acknowledged.

B. K. P. S

Department of Electronic and Electrical Engineering
Trinity College, Dublin
January 2000

* See *Selected papers of Arthur William Conway*, ed. James McConnell, pp. 8–9, Dublin Institute for Advanced Studies, Dublin: 1953.

† The history was not published until 1845 and did not contain this particular etching: W. B. S. Taylor *History of the University of Dublin*, T. Cadell, London, and J. Cumming, Dublin: 1845.

The odd spelling of 'Dunsinok' for Dunsink is puzzling and has not been explained. Dunsink is the accepted English version of the Gaelic name 'Dún Sinche' ('Fort of Sinneach').

Part I
MANUSCRIPTS

I.

SYSTEMS OF RAYS

*PART THIRD**

On extraordinary systems, and systems of rays in general

XXIII. On plane Systems of Rays

110. If the aberrations be measured from a focus, at which the radius of curvature of the caustic vanishes, the expression (M)″ for the aberrations, or rather for their first terms, vanish also, and the convergence is much more accurate than at other points of the caustics; on which account we may call these foci the *principal foci* and the corresponding rays the *axes of the system.* Resuming the general formulæ (H)″, (K)″, and putting for simplicity $\mu=0$, $i=\theta$, as well as $\psi''=0$, that is taking the axis of the system for the axis of (x) we find the following values for the aberrations from a principal focus,

$$l=\tfrac{1}{6}\psi'''\theta^2, \quad \lambda=\tfrac{1}{6}\psi'''\theta^3. \tag{P)''$$

Besides, by (G)″, the arc of the caustic, comprised between the two rays, has for expression

$$S=\tfrac{1}{2}\psi'''\theta^2; \tag{Q)''$$

we have therefore, by elimination of ψ''',

$$l=\tfrac{1}{3}S, \quad \lambda=\tfrac{1}{3}S\theta. \tag{R)''$$

111. The coefficient ψ''', which thus enters into the expressions (P)″, (Q)″, for the aberrations on the arc, measured from a principal focus, has a remarkable geometrical meaning, and is equal to three halves of the parameter of a parabola, which cuts perpendicularly the rays near the axis of the system. For, if we take the principal focus for origin, and the axis of the system for the axis of (x), conditions which give $\mu=0$, $\psi=0$, $\psi'=0$, $\psi''=0$, the focus of a near ray has for coordinates,

$$X=\tfrac{1}{2}\psi'''.i^2, \quad Y=\tfrac{1}{3}\psi'''.i^3,$$

and the approximate equation of the caustic is

* [This part is now printed for the first time from the manuscript found among Hamilton's papers after the publication, in 1931, of Vol. I of these papers (MS 1492/312 in the Library of Trinity College, Dublin). The manuscript is undated and incomplete. According to the published table of contents (see Vol. I, pp. 1–9) this part should begin with paragraph **107** and finish with paragraph '**161** to the end', but the MS begins with the last two words of the paragraph **109** on the 57th sheet and continues to the end of paragraph **159** on the 79th sheet. Sheets 61, 62, and 63 are missing; these contain the end of paragraph **119** to the beginning of paragraph **128**.]

$$X^3 = \tfrac{9}{8}\psi'''.Y^2, \tag{S)''$$

which shews that unless ψ''' be nothing or infinite, the caustic has, at the principal focus, a cusp like the cusp of a semicubical parabola, and may be considered as the evolute of the following common parabola

$$y^2 = \tfrac{2}{3}\psi'''.(x + \tfrac{1}{3}\psi'''), \qquad \text{(T)}''$$

to which therefore the rays near the axis are perpendicular, and of which the parameter is equal to two thirds of ψ'''.

112. We have just seen that the rays near the axis are cut perpendicularly by the common parabola (T)″; in general, the rays of a plane system are perpendicular to a series of rectangular trajectories, which have for equation $dx + \mu dy = 0$, that is, by integration,

$$x\sqrt{1+\mu^2} + \int \frac{\mu d\psi}{\sqrt{1+\mu^2}} = T, \qquad \text{(U)}''$$

(μ) being considered as a given function of (x, y) deduced from the equations of the ray, and (T) being the arbitrary constant, which may be called the parameter of the trajectory. These curves possess many interesting properties, a few of which I shall mention. In the first place, they have the caustic for their common evolute, and the plane zone bounded by any two of them is of the same breadth throughout, namely the difference of their parameters. A trajectory, where it meets the caustic, has in general a cusp like the cusp of a semicubical parabola, and may be considered as the evolute of a common parabola; of which the directrix bisects the radius of curvature of the caustic, and of which the parameter is equal to two thirds of that radius. The osculating parabola to a trajectory, at the point where it crosses a given ray, has its axis inclined in general to that ray, at an angle whose tangent is the third part of the radius of curvature of the caustic, at the focus of the ray, divided by the radius of curvature of the trajectory, that is by the distance of the focus from the point of osculation. But if the ray be an axis of the system, then this angle vanishes, and the axis of the parabola coincides with the axis of the system. The parabola (T)″ is a particular case of these osculating parabolas, but it has this peculiarity, that whereas in general there is only contact of the third order between a parabola and the curve to which it osculates, the parabola (T)″ has contact of the fourth order with the trajectory to which it belongs.

113. If the axes of the coordinates be chosen as in [**111.**], the equation of a ray, near the axis of the system, will be

$$y = ix - \tfrac{1}{6}\psi'''.i^3, \qquad \text{(V)}''$$

in which we may consider (i) as equal to the angle that the near ray makes with the axis. If among all these near rays, we consider only those which make with the axis, at either side, angles not exceeding some given small value θ, we shall have $i = \theta.\sin\xi$, (ξ) being an auxiliary angle, introduced for the sake of confining (i) within the limits assigned; and the entire space over which the near rays are perpendicularly diffused, at any given distance (x) from the principal focus, is the difference of the extreme values of (y) corresponding to the given value of (x). In this manner it may be proved that the least linear space into which the given parcel can be collected, has for expression

$$\Sigma = \tfrac{1}{12}\psi'''.\theta^3, \qquad (W)''$$

and corresponds to

$$x = \tfrac{1}{8}\psi'''.\theta^2; \qquad (X)''$$

that is, it is equal to half the lateral aberration of the extreme rays, and its distance from the principal focus is three quarters of their longitudinal aberration. It may also be shewn, in a similar manner, that the ends of this linear space Σ, are situated on the two branches of the caustic curve, at the points where those branches are crossed by the extreme rays that touch the alternate branches.

114. We have seen, that when the function ψ is given, that is when we know the nature of the system, we can deduce the equation and the properties of the caustic curve, by means of the formula (D)″,

$$X = \psi', \qquad Y = \mu\psi' - \psi.$$

If then it were required, reciprocally, to determine the nature of the function ψ, the equation of the caustic being given, we should have to integrate the following equation of the first order,

$$\mu\psi' - \psi = f(\psi'), \qquad (Y)''$$

$y = f(x)$ being the given equation of the caustic. Differentiating (Y)″, we find

$$\{\mu - f'(\psi')\}.d\psi' = 0; \qquad (Z)''$$

the first factor

$$\mu - f'(\psi') = 0$$

belongs to a singular primitive of (Y)″, and contains the true solution of the question: the other factor

$$d\psi' = 0, \qquad \text{or}, \qquad \psi'' = 0 \qquad (A')''$$

belongs to the complete integral, and represents a set of systems, in which the rays do not touch the given caustic, but converge to some one point upon it.

115. In all the applications of the preceding theory, to the plane systems produced by ordinary reflection or refraction, or by any other optical law, we may consider the rays as emanating from the last reflecting or refracting curve, according to a given law; and if we represent by a, b, the coordinates of this reflecting or refracting curve, the equation of a ray is of the form

$$\frac{y - b}{x - a} = f(\nu, b'), \qquad (B')''$$

in which ν is the tangent of the angle that the incident ray makes with the axis of (x); $b' = (db/da)$, the corresponding quantity for the tangent to the reflecting or refracting curve;

and f is a known function, depending on the law of reflection or refraction. Comparing this equation (B′)″, with the form (A)″ which we have before employed, we find

$$\mu = f(\nu, b'), \qquad \psi = a\mu - b, \tag{C′}''$$

and therefore

$$d\mu = \frac{df}{d\nu}\,d\nu + \frac{df}{db'}\,db', \qquad d\psi = ad\mu + \mu da - db, \qquad \psi' = a + \frac{\mu - b'}{\mu'}, \tag{D′}''$$

if we put for abridgment $d\mu = \mu' da$. Substituting this value for ψ', in the formulæ (D)″, we find the focal coordinates

$$X = a + \frac{\mu - b'}{\mu'}, \qquad Y = b + \mu.\frac{\mu - b'}{\mu'}; \tag{E′}''$$

and therefore, for the focal length,

$$F = \surd\{(X - a)^2 + (Y - b)^2\} = \frac{(1 + \mu^2)^{\frac{1}{2}}(\mu - b')}{\mu'}. \tag{F′}''$$

Hence it follows that the reciprocal of the focal length, being given by the formula

$$\frac{1}{F} = \frac{\dfrac{df}{d\nu}.\nu' + \dfrac{df}{db'}.b''}{(1 + \mu^2)^{\frac{1}{2}}.(\mu - b')}, \tag{G′}''$$

consists of two parts; the one independent of the curvature of the reflecting or refracting curve, and varying inversely as the distance of that curve from the focus of the incident ray; the other independent of that distance, and varying inversely as the radius of curvature: a result remarkable for its generality, since it is independent of the form of the function f, and therefore holds, not only for the plane systems produced by ordinary reflection and refraction, but also for the extraordinary systems, produced by refraction at a chrystal of revolution, the axis of revolution being supposed to coincide with the axis of chrystallization, and the incident rays being contained in the plane of one of the meridians.

116. As the caustic of a given curve, is found by the formulæ

$$X = a + \frac{\mu - b'}{\mu'}, \qquad Y = b + \mu\frac{\mu - b'}{\mu'},$$

so, reciprocally, if it were required to find the curves corresponding to a given caustic, we should have to integrate the equation

$$Y = f(X)$$

$X\,Y$ having the same meanings as before, and f denoting a given function. Differentiating this equation we find either

$$dX = 0$$

or

$$\mu - f'(X) = 0;$$

the latter factor, which belongs to the singular primitive, contains the true solution of the question; the other factor, which belongs to the complete integral, represents a class of focal curves which would cause the rays to pass through some one point upon the given caustic. This point in which a ray of a plane system touches the caustic, is the focus of a curve of this kind which has contact of the second order with the given reflecting or refracting curve; the principal foci belong to focal curves which have contact of the third order.

XXIV. On Developable Systems

117. It was shewn in the IVth Section, that systems of rays may be divided into classes, according to the number of their elements of position. It was also shewn that the rays of a system of the first class, that is a system with but one such element, are contained upon a surface or pencil as their locus; in the preceding section we have considered those systems of the first class, in which this locus of the rays is plane; let us now pass to the case where it is a curve surface, developable, or undevelopable, and let us call the system corresponding, *developable* and *undevelopable systems.*

118. The equations of a ray being put under the form $x = \alpha + \mu z$ $y = \beta + \nu z$, in which α β do not now represent the cosines of the angles that it makes with the axes, but the coordinates of the point in which it intersects the horizontal plane; if the ray belong to a given system of the first class, we may consider the coefficients α β ν μ as given functions of some one element of position, which we shall represent by γ. Denoting the derived functions corresponding by α' β' μ' ν' α'' ... the partial differentials of the pencil are

$$p = \frac{dz}{dx} = \frac{\beta' + \nu' z}{\mu(\beta' + \nu' z) - \nu(\alpha' + \mu' z)}$$

$$q = \frac{dz}{dy} = \frac{-(\alpha' + \mu' z)}{\mu(\beta' + \nu' z) - \nu(\alpha' + \mu' z)}$$

$$r = \frac{d^2 z}{dx^2} = \frac{\nu^2 E + 2(\mu'\beta' - \nu'\alpha')\nu(\beta' + y' z)}{\{\mu(\beta' + \nu' z) - \nu(\alpha' + \mu' z)\}^3}$$

$$s = \frac{d^2 z}{dxdy} = -\frac{\mu\nu E(\mu'\beta' - \nu'\alpha')\{\mu(\beta' + \nu' z) + \nu(\alpha' + \mu' z)\}}{\{\mu(\beta' + \nu' z) - \nu(\alpha' + \mu' z)\}^3}$$

$$t = \frac{d^2 z}{dy^2} = \frac{\mu^2 E + 2(\mu'\beta' - \nu'\alpha')\mu(\alpha' + \mu' z)}{\{\mu(\beta' + \nu' z) - \nu(\alpha' + \mu' z)\}^3}$$

in which $E = (\alpha' + \mu' z)\ (\beta'' + \nu'' z) - (\beta' + \nu' z)(\alpha'' + \mu'' z)$; as appears by differentiating the equations of the ray, and eliminating the differentials of γ. Hence

$$rt - s^2 = \frac{-(\mu'\beta' - \nu'\alpha')^2}{\{\mu(\beta' + \nu' z) - \nu(\alpha' + \mu' z)\}^4}$$

and the condition for the pencil being a developable surface is

$$\mu'\beta' - \nu'\alpha' = 0.$$

When this condition is satisfied, the rays are in general tangents to a caustic curve, whose equations are had by eliminating γ between the following expressions

$$X = \alpha + \mu Z, \quad Y = \beta + \nu Z, \quad Z = -\frac{\alpha'}{\mu'} = -\frac{\beta'}{\nu'}:$$

which expressions, when we assign any particular value to γ, that is when we consider any particular ray, determine the focus of that ray, or the point where it touches the caustic. The rays which are very near to a given ray, may be considered as intersecting that ray in the corresponding focus, and as being contained in the tangent plane to the pencil; this plane, which is the osculating plane of the caustic, has for equation

$$y - \beta - \nu z = \frac{\nu'}{\mu'}(x - \alpha - \mu z).$$

119. This supposition, of a ray being intersected in its own focus by all the near rays, and of these near rays being contained in the osculating plane of the caustic, being only approximately true, it is important to investigate the errors which it leads to, this is to calculate the aberrations, lateral and longitudinal. Denoting by (i) the small increment which the element of position receives, in passing from a given ray to a near ray, the equations of that near ray will be

$$x = \alpha + \alpha' i + \tfrac{1}{2}\alpha'' i^2 + \cdots + (\mu + \mu' i + \cdots) z,$$

$$y = \beta + \beta' i + \tfrac{1}{2}\beta'' i^2 + \cdots + (\nu + \nu' i + \cdots) z;$$

besides by the expressions for Z, we have

$$\alpha' = -\mu' Z, \quad \beta' = -\nu' Z,$$

$$\alpha'' = -(\mu'' Z + \mu' Z'), \quad \beta'' = -(\nu'' Z + \nu' Z'),$$

$$\alpha''' = -(\mu''' Z + 2\mu'' Z' + \mu' Z''), \quad \beta''' = -(\nu''' Z + 2\nu'' Z' + \nu' Z''):$$

and if, for simplicity, we take the given focus for origin; the ray for axis of z, and the osculating plane for the plane of xz, conditions which give

$$Z = 0, \quad \alpha = 0, \quad \beta = 0, \quad \mu = 0, \quad \nu = 0, \quad \nu' = 0, \quad \alpha' = 0, \quad \beta' = 0,$$

$$\alpha'' = -\mu' Z', \quad \beta'' = 0, \quad \alpha''' = -(2\mu'' Z' + \mu' Z''), \quad \beta''' = -2\nu'' Z',$$

the equations of the near ray may be thus written

$$x = (\mu' i + \tfrac{1}{2}\mu'' i^2 + \cdots) z - \tfrac{1}{2}\mu' Z' i^2 - \tfrac{1}{6}(2\mu'' Z' + \mu' Z'') i^3 \cdots,$$

$$y = (\tfrac{1}{2}\nu'' i^2 + \cdots) z - \tfrac{1}{3}\nu'' Z' i^3 \cdots$$

and if we neglect the cube of (i), the lateral aberration measured from the focus is

$$\lambda = \tfrac{1}{2}\mu' Z' i^2 = \tfrac{1}{2}\frac{Z'}{\mu'}.\theta^2,$$

θ being the angle which the near ray makes with the given ray. With respect to the longitudinal aberration, it does not exist; in the same sense as the plane systems, because the near ray does not in general intersect the given ray. But if we investigate the point on the given ray, which is nearest to the near ray, this point may be considered as the virtual intersection, and its distance from its limiting position may be called the

[There is a gap in the manuscript at this point. Sheets 61, 62, and 63 are missing.]

XXV. On Undevelopable Systems

[Paragraphs **125–128** are missing.]

129. We have shewn how to calculate, for any given undevelopable system, the directrix of the pencil, and the breadth of the generating rectangle; for the directrix is the *arrêt de rebroussement* of the envelope of the limiting planes, and the breadth of the rectangle is the interval (I), between the tangent of the directrix, and the ray to which it is parallel. Reciprocally if we know the directrix, and the breadth of the rectangle corresponding to any given point on that curve, we can deduce the equations of the ray, and all the other properties of the system. In this manner we find, for the ray corresponding to any given point (abc) on the directrix, the following equations

$$x - a = a'.(z - c) + I.a''.\left(\frac{1 + a'^2 + b'^2}{a''^2 + b''^2 + (b'a'' - a'b'')^2}\right)^{\frac{1}{2}},$$

$$y - b = b'.(z - c) + I.b''.\left(\frac{1 + a'^2 + b'^2}{a''^2 + b''^2 + (b'a'' - a'b'')^2}\right)^{\frac{1}{2}},$$

in which the accents denote derived functions for c; and eliminating c, we shall have the equation of the pencil.

130. If in the equations of the preceding paragraph, we consider the three quantities a, b, I, as arbitrary functions of c, then that system of equations represents all surfaces composed of straight lines; if we consider a, b, as arbitrary functions of c, and I as a constant quantity, then the same system of equations represents a particular class of straight-lined surfaces, which may be called *isoplatal surfaces*, because the breadth of their generating rectangle is constant; and this class includes developable surfaces, for which that breadth is nothing. Finally, if we consider a, b, as given functions of c, and I as an arbitrary function, then the system represents a class of pencils which have all the same given directrix; eliminating c, we find the two following equations for those pencils, one with an arbitrary function ψ, the other in partial differentials of the first order,

$$y - b'.z = \psi(x - a'z), \qquad a'p + b'q = 1,$$

in which a' b' are to be considered as known functions of x y z, deduced from the following equation, in which there is nothing arbitrary,

$$a''.\{y - b - b'(z - c)\} = b''.\{x - a - a'(x - c)\}.$$

131. The normals to a developable pencil, for all points of any given ray, have for their locus a plane passing through that ray; but when the pencil is undevelopable, the locus of these normals is not in general a plane but an irregular hyperboloid, having for equation

$$(\mu\mu' + \nu\nu')\{z + \mu(x - \alpha) + \nu(y - \beta)\}^2 - \{\alpha'(x - \alpha) + \beta'(y - \beta)\}(1 + \mu^2 + \nu^2)^2$$

$$- \{\mu'(x - \alpha) + \nu'(y - \beta) - (\mu\alpha' + \nu\beta')\}(1 + \mu^2 + \nu^2)\{z + \mu(x - \alpha) + \nu(y - \beta)\} = 0.$$

If then we eliminate γ between this equation, and its derived [equation], we shall have the joint equation of the two surfaces of centres of the pencil; or, if we assign any particular value to γ, we shall find the locus of the centres of curvature for all the points of the corresponding ray. Differentiating therefore the equation of the hyperboloid, and putting for simplicity

$$\alpha = 0, \quad \beta = 0, \quad \mu = 0, \quad \nu = 0, \quad \nu' = 0, \quad \alpha' = 0, \quad \beta' = \mu' u,$$

that is taking the ray for axis z, the virtual focus for origin, and the limiting plane for the plane of xz, we find the two following equations for the curve of the centres,

$$xz + uy = 0, \quad \mu'^2(z^2 - x^2 + u^2) = x(\alpha'' + \mu'' z) + y(\beta'' + \nu'' z);$$

from which it follows that the two radii of curvature of the pencil, at any given distance, $z = \delta$, from the virtual focus, are given by the following formula,

$$\rho = \frac{(\delta^2 + u^2)^{1/2}}{2u^2} . [n(\delta - \delta')(\delta - \delta'') \pm \sqrt{\{n^2(\delta - \delta')^2(\delta - \delta'')^2 + 4u^2(\delta^2 + u^2)\}}]$$

in which (u) is the coefficient of undevelopability;

$$n = \frac{\nu''}{\mu'^2};$$

and δ', δ'' are the roots of this equation

$$\nu''\delta^2 + (\beta'' - \mu'' u)\delta - \alpha'' u = 0.$$

We see then that the two radii of curvature of an undevelopable pencil have opposite signs, that is are turned in opposite directions; the geometric mean between them has for expression

$$\sqrt{\rho'\rho''} = \frac{\delta^2 + u^2}{u},$$

which at the virtual focus reduces itself to the coefficient of undevelopability; and they are equal to one another in length, at two distinct points upon the ray, namely those points whose focal distances are δ' δ'': we shall therefore call these points, the *points of equal and opposite curvature,* and the two curves upon the pencil, which are their loci, we shall call the *lines of equal and opposite curvature.*

132. If the ray be one of those axes condsidered in paragraph **126.**, the hyperboloid of normals reduces itself to a plane, and the curve of centres becomes

$$x = 0, \quad \mu'^2 z^2 = y(\beta'' + \nu'' z);$$

it is therefore in this case a common hyperbola touching the ray at the origin, that is at the focus, and having one of its asymptotes normal to the pencil at the point

$$z = -\frac{\beta''}{\nu''}.$$

At this point both radii of curvature of the pencil are infinite, so that we may call it the *point of evanescent curvature*; in general, at a ray of the kind that we are now considering, one radius of curvature is infinite, while the other, the ordinate of the hyperbola of centres, has for expression

$$\rho = \frac{\mu'^2}{\nu''}\frac{\delta^2}{\delta - \epsilon},$$

ϵ being the focal distance of the point of evanescent curvature. At this point ρ changes sign; the pencil undergoes an inflexion, and is cut by its tangent plane. We may remark also, that the points in which the ray is crossed by the lines of equal and opposite curvature, being given by an equation which becomes

$$\delta^2 - \epsilon\delta = 0,$$

are the focus and the point of evanescent curvature. Finally, since $u = 0$ gives $I = 0$, I being the breadth of the generating rectangle, the ray that we are now considering touches the directrix of the pencil, and since it may be shewn that the point of contact is determined by the equation

$$z = \frac{\theta''}{\nu''} = -\epsilon$$

this point is as far from the focus, at one side, as the point of evanescent curvature is at the other.

133. We have seen that the two radii of curvature of an undevelopable pencil are always turned in opposite directions. Now upon every surface which satisfies this condition, there exist two series of lines determined by the differential equation

$$rdx^2 + 2sdxdy + tdy^2 = 0,$$

($r\ s\ t$ being partial differentials of the second order) which possess this remarkable property, that the tangents to the two lines of this kind, passing through any given point on the surface, coincide with the directions in which the surface is cut by its tangent plane at that point. On an undevelopable pencil, one set of these lines are evidently the rays themselves; in general they may be called the *lines of inflexion* on the surfaces, and they are connected with the curvatures by many interesting relations; for example, the acute angle between them is bisected by the tangent to the greater circle of curvature, and the obtuse angle by the tangent to the lesser; also the square of the tangent of half the acute angle is equal to the lesser radius of curvature divided by the greater, and the square of the tangent of half the obtuse angle is equal to the greater radius divided by the lesser; so that the only surfaces upon which these lines of inflexion are constantly perpendicular to one another, are those which have their radii of curvature equal and opposite, namely, the surfaces which La Grange has shewn to have the least possible area, for any given perimeter; and on a given undevelopable pencil, the only points at which a ray is crossed perpendicularly by its conjugate line of inflexion, are the points where the same ray meets the lines of equal and opposite curvature; finally the lines of inflexion have this other distinguishing property, that when they are curves, their osculating plane is always a tangent plane to the surface to which they belong.

134. After the preceding remarks upon undevelopable pencils, considered as curve surfaces, let us now investigate the aberration of a ray of such a pencil, at a small but finite distance from a given ray. By calculations similar to those of the preceding section, we find the following general expression for the lateral aberration

$$\lambda = \sqrt{(\delta^2 + u^2)}.\theta,$$

θ being the angle between the two rays, δ the distance from the virtual focus, and u the coefficient of undevelopability. If the given ray be one of those which touch the virtual caustic, and if the aberration be measured from its focus, we have then $u = 0$, $\delta = 0$, and by new calculations we find

$$\lambda = \tfrac{1}{2}\frac{R}{\cos^2 V}.\theta^2,$$

R being the radius of curvature of the virtual caustic, and V the angle at which the osculating plane of that curve is inclined to the tangent plane of the pencil, that is here, to the osculating plane of the directrix. With respect to the longitudinal aberration, if we define and calculate it by the same reasonings as in developable systems, we find

$$l = \tfrac{1}{2}\frac{u}{\tan v}\theta$$

v being the angle at which the ray crosses the virtual caustic.

135. The coefficient

$$\left(\frac{u}{\tan v}\right)$$

in this formula for the longitudinal aberration, is the radius of curvature of a remarkable curve, to which we may be conducted by the following reasonings. We have seen that the coefficient of undevelopability (u) is equal to the least distance between two infinitely near rays, divided by the angle between them; we have therefore

$$u = \frac{dS.\sin v}{d\theta}, \quad \frac{u}{\tan v} = \frac{\cos v.dS}{d\theta},$$

dS being the element of the arc of the virtual caustic, and $d\theta$ the angle between the rays; if then we can find a curve such that the arc comprised between two consecutive tangents, which make with one another the same angle as that made by the two consecutive rays, may be equal to $(\cos v.dS)$, that is to the projection of the element of the virtual caustic upon the given ray, the radius of curvature (R') of this curve will be equal to the coefficient

$$\frac{u}{\tan v},$$

and the formula for the longitudinal aberration may be thus written

$$l = \tfrac{1}{2}R'.\theta.$$

Now, there are two different ways of finding a plane curve which shall satisfy this condition; first by making a plane roll round the developable envelope of the limiting planes of the pencil, collecting the rays in its progress, till it makes them all tangents to one plane curve; and secondly by making each ray successively a momentary axis of revolution round which all the rays that precede it in the system, beginning with some given ray, describe little stripes of one-branched hyperboloids, in such a manner that they are all at last brought to be parallel to a given plane, and therefore horizontal tangents to a vertical cylinder, if we suppose the given plane horizontal; the base of this cylinder will be the same plane curve as that before obtained, and will, like it, satisfy the required condition. Finally we may remark, that the two constructions here indicated, include as particular cases the known manner of unrolling a developable surface, and may be called the *virtual developments* of the pencil.

XXVI. On Systems of the Second Class

136. In a system of the second class, that is a system with two arbitrary constants, if we represent a ray, as in the two preceding sections, by equations of the form

$$x = \alpha + \mu z, \qquad y = \beta + \nu z,$$

we may in general consider any two of the coefficients, for example α, β, as being given functions of the two others, and put their differentials under the form

$$d\alpha = \alpha' d\mu + \alpha_, d\nu, \; d\beta = \beta' d\mu + \beta_, d\nu.$$

We may also consider [**19.**] the coefficients μ, ν, as being themselves given functions of x, y, z; the derived functions corresponding are,

$$\frac{d\mu}{dx} = \frac{z+\beta_,}{\sigma}, \; \frac{d\mu}{dy} = -\frac{\alpha_,}{\sigma}, \; \frac{d\mu}{dz} = -\left(\mu.\frac{d\mu}{dx} + \nu.\frac{d\mu}{dy}\right),$$

$$\frac{d\nu}{dx} = \frac{-\beta'}{\sigma}, \; \frac{d\nu}{dy} = \frac{z+\alpha'}{\sigma}, \; \frac{d\nu}{dz} = -\left(\mu.\frac{d\nu}{dx} + \nu.\frac{d\nu}{dy}\right),$$

in which, $\sigma = (z+\alpha')(z+\beta_,) - \alpha_,\beta'$. By means of these formulæ, we can find the condition for the system being rectangular, that is for the rays being cut perpendicularly by a series of surfaces; for if this be the case, the differential equation of those surfaces,

$$\mu dx + \nu dy + dz = 0,$$

will be integrable, and we shall have

$$\frac{d\mu}{dy} - \nu.\frac{d\mu}{dz} = \frac{d\nu}{dx} - \mu.\frac{d\nu}{dz},$$

that is

$$\mu\nu\beta_, - (1+\nu^2)\alpha_, = \mu\nu\alpha' - (1+\mu^2)\beta'.$$

When this condition is satisfied we shall have

$$\mu d\alpha + \nu d\beta = \sqrt{(1+\mu^2+\nu^2)}.d\varphi,$$

ϕ being a function of μ, ν, considered as independent variables; and the rays will be cut perpendicularly by a series of surfaces having for equation

$$z\surd(1+\mu^2+\nu^2)+\varphi=T,$$

T being an arbitrary constant which may be called the parameter of the surface; we may also remark, that the shell of space, bounded by any two of these surfaces, is of the same thickness throughout, equal to the difference of their parameters.

137. Establishing any arbitrary relation between μ, ν, the rays which satisfy that relation have a surface or pencil for their locus; these pencils may all be represented by the partial differential equation

$$\mu p+\nu q=1,$$

or by the functional equation

$$\nu=\psi(\mu)$$

and the arbitrary function may be determined by the condition that the rays shall pass through a given curve, or envelope a given surface. Among all these pencils there is only a certain series developable; the differential equation of the series is, by XXIV.,

$$d\mu . d\beta - d\nu . d\alpha = 0,$$

that is, in our present notation,

$$\alpha_{,}d\nu^2+(\alpha'-\beta_{,})d\nu d\mu-\beta' d\mu^2=0.$$

Denoting by a, a' the roots of this quadratic, so that

$$2\alpha_{,}a=\beta-\alpha'-\surd\{(\beta_{,}-\alpha')^2+4\alpha_{,}\beta'\},$$

$$2\alpha_{,}a'=\beta_{,}-\alpha'+\surd\{(\beta_{,}-\alpha')^2+4\alpha_{,}\beta'\},$$

the tangent planes to the two developable pencils passing through any given ray have for equations

$$y-\beta-\nu z=a(x-\alpha-\mu z),$$

$$y-\beta-\nu z=a'(x-\alpha-\mu z),$$

and the corresponding foci of the ray, that is the points in which it touches the *arrêts de rebroussement* of these two pencils, have for their vertical ordinates

I^{st} $\quad Z=-\tfrac{1}{2}(\alpha'+\beta_{,})+\tfrac{1}{2}\surd\{(\beta_{,}-\alpha')^2+4\alpha_{,}\beta'\},$

$\mathrm{II}^{\mathrm{nd}}$ $\quad Z'=-\tfrac{1}{2}(\alpha'+\beta_{,})-\tfrac{1}{2}\surd\{(\beta_{,}-\alpha')^2+4\alpha_{,}\beta'\};$

these focal ordinates are therefore roots of the quadratic

$$Z^2+Z(\alpha'+\beta_{,})+\alpha'\beta_{,}-\alpha_{,}\beta'=0,$$

and the joint equation of the two caustic surfaces is

$$\sigma=0,$$

σ having the same meaning as in the preceding paragraph. It may also be shewn that this equation is a singular primitive of the partial differential equation

$$\mu p + \nu q = 1,$$

of which the complete integral $\nu = \psi(\mu)$ represents all the pencils of the system.

138. To find the intersection of the two caustic surfaces, we have the condition

$$(\beta_{,} - \alpha')^2 + 4\alpha_{,}\beta' = 0;$$

when the system is rectangular, this condition resolves itself into the three following,

$$\beta_{,} - \alpha' = 0, \quad \alpha_{,} = 0, \quad \beta' = 0,$$

which are however equivalent to but two distinct equations, in consequence of the condition of rectangularity; and thus the intersection sought, reduces itself in rectangular systems to a finite number of *principal foci*, together with a corresponding number of rays, each of which may be shewn to be intersected in its own focus by all the rays infinitely near it, and which may therefore be called the *axes of the system*. We have already arrived at this result, by different reasonings, in the case of the systems produced by ordinary reflection and refraction, which as we have seen are all rectangular; and though the systems produced by extraordinary refraction are not in general rectangular, yet we shall see that for these systems also, there are certain particular rays, or axes of the system, which satisfy the same conditions

$$\beta_{,} - \alpha' = 0, \quad \alpha_{,} = 0, \quad \beta' = 0.$$

If in any particular system, these three conditions be identically satisfied, by the nature of the functions α, β, then it is easy to shew that those functions must be of the form

$$\alpha = a - \mu . c, \quad \beta = b - \nu . c,$$

and therefore that the rays all pass through some one point $(a\ b\ c)$.

139. Besides the two foci of a ray, which correspond to the two developable pencils, and which are points of intersection with rays infinitely near; a given ray has in general an infinite number of virtual foci, which correspond to the undevelopable pencils, and which are nearest points to rays infinitely near. These virtual foci are however, all ranged upon a finite portion of the ray, which has its middle point at the middle of the interval between the foci of the developable pencils, and its length equal to that interval divided by the sine of the angle between the same developable pencils. The extreme virtual foci thus determined, form two surfaces, which have for their joint equation

$$\{\mu\nu\beta_{,} - (1+\nu^2)\alpha_{,} - \mu\nu\alpha' + (1+\mu^2)\beta'\}^2 = 4\sigma(1+\mu^2+\nu^2),$$

σ having the same meaning as in **136.**: these may be called the *virtual caustic surfaces*; they coincide with the ordinary caustic surfaces, where the first member of their equation vanishes, that is when the system is rectangular; in general they have the same *diametral surface*

$$2z + \alpha' + \beta_{,} = 0;$$

and their intersection with one another reduces itself to a finite number of isolated points, which are determined by the following equations,

$$\frac{\alpha' - \beta,}{2\mu\nu} = \frac{\beta'}{1+\nu^2} = \frac{-\alpha,}{1+\mu^2}$$

and which may be called the *principal virtual foci.* When the system is rectangular, these points coincide with the foci considered in **138**.

140. With respect to the law, according to which the virtual focus varies between its extreme limits, it may be shewn that these extreme positions correspond to two undevelopable pencils, or rather to two sets of such pencils whose limiting planes are perpendicular to one another, and are symmetrically situated with respect to the tangent planes of the developable pencils, with which they coincide when the system is rectangular; and that if we take the ray for the axis of z and denote by (η) the angle which the limiting plane of any other undevelopable pencil, makes with that corresponding to one of the extreme virtual foci, and focal ordinate of this pencil will be

$$z = z_1 \cos^2\eta + z_2 \sin^2\eta,$$

z_1, z_2, denoting the extreme values. Moreover if we take the planes corresponding to the extreme virtual foci, for the planes of xz, yz, and place the origin at the middle point of the interval between those foci, the length of which interval we shall denote by V; then, the equations of a near ray may be thus written

$$x = (z - V).i + V\cos P.k, \quad y = (z + V).k - V\cos P.i,$$

$i.\ k$ being the small increments received by μ, ν, in passing from the given ray to the near ray, and P the angle between the developable pencils; the general equation of thin pencils is therefore $k = \psi(i)$, that is

$$\frac{y(z-V) + x.V\cos P}{z^2 - V^2\sin^2 P} = \psi\left(\frac{x(z-V) - yV\cos P}{z^2 - V^2\sin^2 P}\right)$$

and it may be shewn as in IX. that the area of a perpendicular section is proportional to the product of the distances from the foci of the developable pencils. It may also be shewn that if we determine the form of the function ψ, by the condition that the rays shall pass through a little circle having for equations

$$z = \delta, \quad x^2 + y^2 = e^2;$$

for example through the circumference of the pupil of the eye, the radius of this pupil being (e), and the eye being placed at a distance (δ) from the middle point between the foci; every section of the pencil made by a plane perpendicular to the given ray, (supposed to coincide with optic axis), is a little ellipse, whose semi-axis major is

$$a = \frac{e.\{V(\delta - z) + \sqrt{V^2(\delta - z)^2 + (\delta^2 - V^2\sin^2 P)(z^2 - V^2\sin^2 P)}\}}{\delta^2 - V^2\sin^2 P},$$

of which the minimum value

$$a = \frac{e.V.\sin^2 P}{\delta - \cos P.\sqrt{(\delta^2 - V^2\sin^2 P)}}$$

corresponds to

$$z = \frac{V^2.\sin^2 P}{\delta - \cos P.\sqrt{(\delta^2 - V^2\sin^2 P)}}.$$

141. These results include, as particular cases, those which we had before found for systems of reflected and refracted rays. We might, in a similar manner proceed to generalise the calculations and reasonings into which we have entered, respecting the aberrations and density in those systems; but it is more interesting to examine the general properties of systems of rays, proceeding from any given surface according to any given law. Suppose then that from every point of a given curve surface

$$F(x\ y\ z) = 0,$$

proceeds a ray having for equations

$$x' - x = \mu(z' - z) \quad y' - y = \nu(z' - z),$$

μ, ν being given functions of x, y, z, dz/dx, dz/dy, that is, ultimately, of x and y; the differentials of μ, ν, z, will be of the form

$$d\mu = \mu' dx + \mu_, dy,$$

$$d\nu = \nu' dx + \nu_, dy,$$

$$dz = \mu dx + q dy;$$

and because $\alpha = x - \mu z$ [and] $\beta = y - \nu z$, we have

$$(z + \alpha')(\mu'\nu_, - \mu_,\nu') = (1 - \mu p).\nu_, + \mu q \nu',$$

$$-\alpha_,(\mu'\nu_, - \mu_,\nu') = (1 - \mu p).\mu_, + \mu q \mu',$$

$$-\beta'(\mu'\nu_, - \mu_,\nu') = (1 - \nu q).\nu' + \nu p \nu,$$

$$(z + \beta_,)(\mu'\nu_, - \mu_,\nu') = (1 - \nu q)\mu' + \nu p \mu_,,$$

α', $\alpha_,$, β', $\beta_,$ having the same meanings as in the preceding paragraphs of the present section. By means of these relations, we can translate all the formulæ that we have found for systems of the second class in general into the notation of emanating systems. We find, for example, that the differential equation of the curves in which the developable pencils intersect the surface from which the rays emanate, is

$$d\mu.(dy - \nu dz) = d\nu.(dx - \mu dz);$$

the condition for these developable pencils cutting one another at right angles, is

$$\mu(\nu + q).\mu' + (1 - \mu p + \nu^2).\mu_, = (1 - \nu q + \mu^2)\nu' + \nu(\mu + p)\nu_,;$$

the equation which determines the two foci, is

$$\{\mu'(z' - z) + 1 - \mu p\}\{\nu_,(z' - z) + 1 - \nu q\} = \{\mu_,(z' - z) - \mu q\}.\{\nu'(z' - z) - \nu p\}:$$

and the condition for these two foci coinciding, that is for the intersection of the caustic surfaces, is

$$\{(1 - \nu q)\mu' + \nu p\mu_, - \mu q\nu' - (1 - \mu p)\nu_,\}^2 + 4\{(1 - \mu p)\mu_, + \mu q\mu'\}.\{(1 - \nu q)\nu' + \nu p\nu_,\} = 0.$$

We may remark that the condition for the rectangularity of an emanating system, is the condition for the formula

$$\frac{\mu dx + \nu dy + dz}{\sqrt{(1 + \mu^2 + \nu^2)}}$$

being an exact differential.

142. We have seen, in the two preceding parts of this essay, that if upon the plane passing through a given reflected or refracted ray and through any given direction upon the reflecting or refracting surface, we project the ray reflected or refracted from the consecutive point upon the given direction, the projection will cross the given ray in a point whose distance (h) from the mirror is determined by the formula

$$\frac{1}{h} = \frac{\cos^2\psi}{\rho_1} + \frac{\sin^2\psi}{\rho_2},$$

ρ_1, ρ_2 being the distances of the foci of the developable pencils from the mirror, and ψ the angle which the plane of projection makes with one of the tangent planes to those developable pencils. In general, it may be shewn that

$$\frac{1}{\varepsilon} = \frac{\cos^2\eta}{\varepsilon_1} + \frac{\sin^2\eta}{\varepsilon_2},$$

ε being the distance of the analogous point from the surface from which the rays emanate; η the angle which the plane of projection makes [with] the plane corresponding to one of the extreme virtual foci **140.**: and ε_1, ε_2 the extreme values of ε, which are connected with ρ_1, ρ_2, that is with the distances of the foci of the developable pencils, by the following relations

$$\frac{1}{\varepsilon_1} + \frac{1}{\varepsilon_2} = \frac{1}{\rho_1} + \frac{1}{\rho_2}, \quad \frac{1}{\varepsilon_1} - \frac{1}{\varepsilon_2} = \frac{1}{\sin P}\left(\frac{1}{\rho_1} - \frac{1}{\rho_2}\right),$$

P being the angle between the developable pencils. We may call the point thus determined, the *focus by projection*; and we see that the extreme foci by projection cannot coincide with one another, except at the intersection of the caustic surfaces, when the system is rectangular; or, more generally, at the principal virtual foci (see **139.**).

143. In the two former parts of this essay, we have arrived at various theorems respecting the osculating foci of reflected and refracted systems; that is, the foci of surfaces which osculate to the given reflector or refractor, and which would themselves reflect or refract the given incident rays to some one point or focus. Those theorems, abstracts as they appear, are however only particular cases of certain more general properties of emanating systems, which we come now to consider. Representing by (x, y, z, p, q, r, s, t), the coordinates and partial differentials, first and second order, of the given surface from which the rays emanate; and by $(x', y', z', p', q', r', s', t')$ the corresponding quantities for a *focal surface*, that is for a surface which would cause the rays to pass through some one point (a, b, c); the coefficients (μ, ν) which enter into the equation of the ray being given functions of the coordinates of the surface and the partial differentials of the first order, their differentials will be, for the given surface,

$$d\mu = \frac{d\mu}{dx}.dx + \frac{d\mu}{dy}.dy + \frac{d\mu}{dz}.dz + \frac{d\mu}{dp}.dp + \frac{d\mu}{dq}.dq,$$

$$d\nu = \frac{d\nu}{dx}.dx + \frac{d\nu}{dy}.dy + \frac{d\nu}{dz}.dz + \frac{d\nu}{dp}.dp + \frac{d\nu}{dq}.dq,$$

and for the focal surface

$$d\mu = \frac{d\mu}{dx'}.dx' + \frac{d\mu}{dy'}.dy' + \frac{d\mu}{dz'}.dz' + \frac{d\mu}{dp'}.dp' + \frac{d\mu}{dq'}.dq',$$

$$d\nu = \frac{d\nu}{dx'}.dx' + \frac{d\nu}{dy'}.dy' + \frac{d\nu}{dz'}.dz' + \frac{d\nu}{dp'}.dp' + \frac{d\nu}{dq'}.dq'.$$

But we have also, by the notation of **141.**

$$d\mu = \mu' dx + \mu_{,} dy \qquad d\nu = \nu' dx + \nu_{,} dy$$

for the given surface; and by the nature of the focal surface, we have, for that surface,

$$d\mu = \frac{\mu dz' - dx'}{c - z'}, \qquad d\nu = \frac{\nu dz' - dy'}{c - z'}.$$

Comparing these equations, and supposing that the given surface is touched by the focal surface, an hypothesis which gives

$$x' = x, \quad y' = y, \quad z' = z, \quad p' = p, \quad q' = q,$$

and renders μ, ν the same for both surfaces, we find

$$\mu' = M' + \frac{d\mu}{dp}.r + \frac{d\mu}{dq}.s,$$

$$\mu_{,} = M_{,} + \frac{d\mu}{dp}.s + \frac{d\mu}{dq}.t,$$

$$\nu' = N' + \frac{d\nu}{dp}.r + \frac{d\nu}{dq}.s,$$

$$\nu_{,} = N_{,} + \frac{d\nu}{dp}.s + \frac{d\nu}{dq}.t,$$

$$\frac{\mu p - 1}{c - z} = M' + \frac{d\mu}{dp}r' + \frac{d\mu}{dq}.s',$$

$$\frac{\mu q}{c - z} = M_{,} + \frac{d\mu}{dp}.s' + \frac{d\mu}{dq}.t',$$

$$\frac{\nu p}{c - z} = N' + \frac{d\nu}{dp}.r' + \frac{d\nu}{dq}.s',$$

$$\frac{\nu q - 1}{c - z} = N_{,} + \frac{d\nu}{dp}.s' + \frac{d\nu}{dq}.t',$$

if we put for abridgement

$$M' = \frac{d\mu}{dx} + \frac{d\mu}{dz}.p \quad M_{,} = \frac{d\mu}{dy} + \frac{d\mu}{dz}.q,$$

$$N' = \frac{d\nu}{dx} + \frac{d\nu}{dz}.p \quad N_{,} = \frac{d\nu}{dy} + \frac{d\nu}{dz}.q.$$

If now we wish to determine the focal surface by the condition of osculating to the given surface in any given direction, for which $dy = \tau dx$, we are to employ the following formula

$$r' - r + 2(s' - s)\tau + (t' - t)\tau^2 = 0,$$

r', s', t' being considered as given functions of the focal ordinate c, determined by the preceding equations. And if among all the osculating focal surfaces thus determined, we wish to find those for which this ordinate is a maximum or a minimum, together with the corresponding directions of osculation; we shall have to employ the two following equations

$$r' - r + (s' - s).\tau = 0 \quad s' - s + (t' - t).\tau = 0,$$

that is

$$dp' - dp = 0, \quad dq' - dq = 0,$$

or finally

$$\frac{\mu dz - dx}{c - z} = \mu' dx + \mu_{,} dy, \quad \frac{\nu dz - dy}{c - z} = \nu' dx + \nu_{,} dy;$$

formulæ which give by elimination

$$(\mu dz - dx)(\nu' dx + \nu_{,} dy) = (\nu dz - dy)(\mu' dx + \mu_{,} dy),$$

$$\{\mu'(c - z) + 1 - \mu\beta\}\{\nu_{,}(c - z) + 1 - \nu q\} = \{\mu_{,}(c - z) - \mu q\}\{\nu'(c - z) - \nu p\}.$$

Comparing these results with the formulæ **141.** we see that the foci of the greatest and least osculating focal surfaces, are the points in which the ray touches the two caustic surfaces; and that the directions of osculation corresponding, are the directions of the curves in which the developable pencils intersect the given surface from which the rays proceed.

144. The preceding theory of osculating focal surfaces conducts to many important consequences respecting reflected and refracted systems, ordinary and extraordinary; for in every such system, the rays may be considered as emanating from the last reflecting or refracting surface, according to a given law, which depends both upon the law of reflexion or refraction, and also upon the nature of the incident system: and it will be shewn that it is always possible to assign such a form to the last reflector or refractor, as to make the rays converge to any given focus $(a\ b\ c)$; that is in the notation that we have just employed, the differential equation obtained by eliminating p', q' between the three following

$$a - x' = \mu(c - z'), \quad b - y' = \nu(c - z'), \quad dz' = p'dx' + q'dy',$$

is always integrable, and the integral represents a series of focal surfaces. With respect to the nature of this series, we have already assigned their integral equation, for the case of ordinary reflexion and refraction; and we shall assign it, in subsequent sections of this essay, for the case of extraordinary reflexion and refraction, by chrystals with one or with two axes. The integral equation of this series may therefore, in all cases, be found by elimination alone; and supposing it known, we may determine the four constants which it involves, (namely the three coordinates of the focus, and the parameter introduced by integration,) by supposing the focal surface to osculate to the given reflector or refractor, at any given point, in any given direction. And if, among all the focal surfaces thus osculating at any given point, we seek the greatest and the least, by means of the formulæ

$$dp' = dp, \quad dq' = dq,$$

it follows from the theory of the preceding paragraph, that we shall find the two foci of the ray, and the directions of the two lines of reflexion and refraction, by calculations which are generally more simple than those which would be required in differentiating the equations of the ray.

145. Another application of this theory, relates to the surfaces of constant action. We have seen, in the two former parts of this essay, that the rays of an ordinary system are normals to this series of surfaces; and it will be shewn, that in extraordinary systems, the rays may be considered as proceeding from the corresponding series of surfaces according to a law of such a nature that if the rays were to converge to any one point, the surfaces would become spheroids of certain known forms, having that point for centre. Hence it follows, by our general theory of osculating focal surfaces, that as in ordinary systems, the surfaces of constant action have the centres of their greatest and least osculating spheres contained upon the two caustic surfaces; so, in extraordinary systems, the two caustic surfaces contain the centres of the greatest and least spheroids, which osculate at any given point to the surfaces of constant action.

146. A third application relates to the axes of an extraordinary system. We have seen, in the two first parts, that in ordinary systems, there are in general one or more such axes, each of which is intersected by all the consecutive rays, in a point which is the focus of a focal reflector or refractor, that has contact of the second order with the given reflector or refractor. We can now extend this theorem to extraordinary systems also; for if we determine the focus, and the point of osculation, by the following equations

$$x' = x, \quad y' = y, \quad z' = z, \quad p' = p, \quad q' = q, \quad r' = r, \quad s' = s, \quad t' = t,$$

which express the conditions for contact of the second order, we shall have by the equations of [**143.**],

$$\mu' = \frac{\mu p - 1}{c - z}, \quad \mu_{,} = \frac{\mu q}{c - z}, \quad \nu' = \frac{\nu p}{c - z}, \quad \nu_{,} = \frac{\nu q - 1}{c - z},$$

conditions which express that the ray is an axis of the system, that is that it is intersected by all the rays infinitely near it, in one and the same focus; namely the focus of the osculating surface. It may also be shewn, by similar reasonings, that this focus is the common centre of a series of spheroids, which have contact of the second order with the surfaces of constant action.

147. A fourth application relates to the determination of the surface of a chrystal, by means of one of its two caustic surfaces. We have given, in Section XIX., the solution of the corresponding questions, for the case of ordinary reflectors and refractors; it will be sufficient therefore to point out, here, the principal steps of the reasoning.

From what we have already shewn, it follows that the equation of the caustic surfaces (a, b, c), may be obtained by eliminating x', y', z' between the equations of the chrystal and of the ray, combined with an equation of the form

$$(r' - r)(t' - t) - (s' - s)^2 = 0,$$

in which r, s, t are the partial differentials of the second order of the given chrystal, and r', s', t' the corresponding quantities for a focal chrystal which would refract the rays to the point (a, b, c). If then we are given the equation of one of the caustic surfaces

$$f(a\ b\ c) = 0,$$

and are required to find the equation of the chrystal, we have to integrate the equation

$$(r' - r)(t' - t) - (s' - s)^2 = 0,$$

considering (r, s, t) as partial differentials of the second order belonging to the unknown chrystal, and r', s', t' as expressions composed in a known manner of (x, y, z, p, q, a, b, c), or simply of (x, y, z, p, q), since (a, b, c) are themselves known functions of (x, y, z, p, q), deduced from the equations

$$f(a\ b\ c) = 0, \quad a - x = \mu(c - z), \quad b - y = \nu(c - z),$$

in which μ, ν are given functions of x, y, z, q depending on the law of extraordinary refraction and on the nature of the incident system. The complete integral, with two arbitrary functions, represents the envelope of a series of focal chrystals, which have their foci on the given caustic surface; there is also a singular primitive of the first order, which contains the true solution, representing a series of chrystals, each of which would refract the given incident rays so as to make them touch the given caustic surface.

148. We may also generalise the investigations of the second part, respecting the caustics of a given curve, and the surfaces of circular profile; and shew that a given curve may have an infinite number of caustics by extraordinary refraction, according to the infinite number of chrystals, on which it may be a line of extraordinary refraction, that is the base of a developable pencil. The locus of these curves, is the envelope of a certain series of cones, the equation of which may be calculated and, of which we shall assign the equation in a subsequent section. In the mean time we may remark, that although we have contented ourselves with applying to the case of extraordinary refraction, the theorems of the present section, they possess a much greater degree of generality, and may be applied to all other optical laws which may hereafter be discovered: provided that those laws shall be of such a nature as to render integrable the equation of the focal surfaces **143.**, that is the equation which results from elimination of p', q' between the three following,

$$a - x' = \mu(c - z'), \qquad b - y' = \nu(c - z'), \qquad dz' = pdx' + qdy'.$$

The conditions for the integrability are two; they are obtained by eliminating r', s', t' between the four equations **143.** which result from the two following

$$(c - z')d\mu = \mu dz' - dx', \qquad (c - z')d\nu = \nu dz' - dy',$$

considering c as arbitrary: and it will be shewn that they are necessarily satisfied, not only for the laws of ordinary and extraordinary reflexion and refraction, but also for all other laws which are included in the principle of least action.

XXVII. Systems of the Third Class

149. Before we proceed to examine specially the properties of extraordinary systems, let us make a few remarks upon systems of classes higher than the second (IV), that is systems in which the position of the rays depends on more than two arbitrary quantities. We have an example of such a system, in the case of heterogeneous rays which after issuing from a luminous point have been any number of times refracted; for besides the two arbitrary coordinates of the first point of incidence, there is also a third arbitrary element depending on the colour of the ray: a system of this kind is therefore a system of the third class. It is possible also to conceive systems, in which the position of a ray shall depend on more than three arbitrary elements, these elements being restricted to remain within certain limiting values; but as all the important questions of optics may be reduced to the consideration of systems of the first and second classes, of which we have already treated, I shall confine myself to a brief view of the properties of systems of the third class.

Representing therefore, as before, a ray by the two equations

$$x = \alpha + \mu z, \qquad y = \beta + \nu z,$$

let us suppose that the four coefficients $(\alpha, \beta, \mu, \nu)$ are connected by some one given relation

$$u = 0 = \varphi(\alpha\ \beta\ \mu\ \nu);$$

it is evident that then the system will be of the third class, because three constants remain arbitrary, and the properties of the system will depend upon the form of the function φ. Through any given point of space, x, y, z pass in general an infinite number of rays, composing a cone, which has for equation $u = 0$, α, β, μ, ν being changed into their values

$$\mu = \frac{X - x}{Z - z}, \qquad \nu = \frac{Y - y}{Z - z}, \qquad \alpha = x - \mu z, \qquad \beta = y - \nu z,$$

and X, Y, Z representing the coordinates of the cone. In the same manner, if we consider a new point of space, infinitely near the former, and having for coordinates $x+dx$, $y+dy$, $z+dz$, the rays passing through this point compose another cone, having for equation

$$u+\frac{du}{dx}.dx+\frac{du}{dy}.dy+\frac{du}{dz}.dz=0;$$

which may also be thus written,

$$u+\left(\frac{du}{d\mu}-Z\frac{du}{d\alpha}\right)d\mu+\left(\frac{du}{d\nu}-Z\frac{du}{d\beta}\right)d\nu=0.$$

The intersection of these two cones is a curve having for equations

$$u=0,\quad \frac{du}{d\mu}-Z\frac{du}{d\alpha}+\left(\frac{du}{d\nu}-Z\frac{du}{d\beta}\right).\frac{d\nu}{d\mu}=0;$$

and although the nature and position of this curve depend on the value of $\frac{d\nu}{d\mu}$, that is on the direction in which we pass from the point x, y, z to the infinitely near point $x+dx$, $y+dy$, $z+dz$; yet the curve of intersection, and therefore the second cone, always passes through the point (X, Y, Z) which satisfies the three following conditions

$$u=0,\quad \frac{du}{d\mu}-Z\frac{du}{d\alpha}=0,\quad \frac{du}{d\nu}-Z\frac{du}{d\beta}=0.$$

This point I shall therefore call the *focus* of the first cone, and the ray passing through it I shall call a *focal ray*; the values of μ, ν corresponding to this ray are determined by the two equations

$$u=0,\quad \frac{du}{d\mu}\frac{du}{d\beta}=\frac{du}{d\nu}\frac{du}{d\alpha}.$$

150. We have just seen that in a system of the third class, any assigned point of space is in general the centre of a cone of rays, which is intersected by all the infinitely near cones in one and the same point of focus. It may easily be shewn that these foci have a curve surface for their locus, the equation of which in X, Y, Z is had by eliminating μ, ν between the three following equations in $(\alpha, \beta, \mu, \nu)$

$$u=0,\quad \frac{du}{d\mu}-Z\frac{du}{d\alpha}=0,\quad \frac{du}{d\nu}-Z\frac{du}{d\beta}=0,$$

after changing α, β to their values $\alpha=X-\mu Z$, $\beta=Y-\nu Z$; and since we should obtain the same three equations, if we sought the maximum or minimum of Z, corresponding to given values of X, Y, and considered as a function of μ, ν, (that is the highest or lowest point upon a given vertical ordinate, which has a ray of the system passing through it,) I shall call the locus of the conic foci the *limiting surface* of the system. This surface is touched by all the cones and other pencils of the system, and also by all the focal rays; if the condition

$$\frac{du}{d\mu}\frac{du}{d\beta}=\frac{du}{d\nu}\frac{du}{d\alpha}$$

be identically satisfied, for all the rays of the system, then those rays are all tangents to the limiting surface. Reciprocally if the rays of a system of the third class are all tangents to any one surface, or all pass through any one curve, the condition which has just been given is then identically satisfied.

151. Many other remarks might be made upon the properties of the focal rays and of the limiting surface; but I believe it more important to notice here the investigations which Malus* has prefixed to his *Traité d'optique*, which according to my division relate to systems of the third class. Malus supposes that from every point of space x', y', z' proceeds a ray of light having for equations

$$\frac{x-x'}{m}=\frac{y-y'}{n}=\frac{z-z'}{o}$$

(m, n, o) being functions of (x', y', z'), the forms of which depend upon the nature of the system. The ray from an infinitely near point has for equations

$$\frac{x-x'-dx'}{m+dm}=\frac{y-y'-dy'}{n+dn}=\frac{z-z'-dz'}{o+do}$$

and in order that it should intersect the former ray, we have the following equation of condition

$$(ndz'-ody')dm+(odx'-mdz')dn+(mdy'-ndx')do=0,$$

which may be put under the form

$$\alpha dx'^2+\beta dy'^2+\gamma dz'^2+\delta\, dx'dy'+\varepsilon dx'dz'+\zeta dy'dz'=0;$$

and therefore the directions in which we can pass from the given point x', y', z' to the infinitely near point $x'+dx'$, $y'+dy'$, $z'+dz'$, are contained upon a conic locus, which has for equation

$$\alpha(x-x')^2+\beta(y-y')^2+\gamma(z-z')^2+\delta(x-x')(y-y')+\varepsilon(x-x')(z-z') \\ +\zeta(y-y')(z-z')=0.$$

This cone being of the second degree, is in general cut in two directions by a plane passing through its centre; and it is in this manner that Malus proves, that when rays issue from a given surface according to a given law, there exist in general two series of lines upon that surface, analogous to the lines of curvature, the rays from which compose developable surfaces, and are tangents to two series of caustic curves, which are contained upon two caustic surfaces: results to which we have already arrived by different reasonings.

XXVIII. On Extraordinary Systems, Produced by Single-Axed Chrystals

152. When a ray of light is refracted by extraordinary refraction, in passing from an unchrystallised medium into a chrystal with one axis, we know by experience that if we describe a spheroid of revolution, having its centre at the point of incidence, its axis of revolution parallel to the axis of the chrystal, and its polar and equatorial diameters equal to two

* [Étienne Louis Malus (1775–1812), his *Traité d'optique* appeared in 1807. See Vol. I p. 463.]

quantities $\frac{2}{m}, \frac{2}{m'}$, which depend on the nature of the medium and on the colour of the ray; the tangent plane to this spheroid, at the point where it meets the extraordinary ray, is perpendicular to the plane of incidence, and passes through a point taken on the prolongation of the projection of the incident ray, upon the tangent plane to the chrystal, at a distance from the centre of the spheroid is equal to the reciprocal of the sine of incidence. Representing therefore by X, Y, Z the coordinate of the point last mentioned, thus taken, we shall have the three following conditions

$$Z = pX + qY,$$
$$(X + pZ)(\beta' + q\gamma') = (Y + qZ)(\alpha' + p\gamma'),$$
$$\alpha' X + \beta' Y + \gamma' Z = -1,$$

if we take the point of incidence for origin, denoting by p, q the partial differentials, first order, of the chrystal at that point, and by α', β', γ' the cosines of the angles which the incident ray makes with the axes of coordinates. Hence

$$X = -\frac{\alpha' + p\gamma' - q(\beta' p - \alpha' q)}{(\alpha' + p\gamma')^2 + (\beta' + q\gamma')^2 + (\beta' p - \alpha' q)^2},$$
$$Y = -\frac{\beta' + q\gamma' + p(\beta' p - \alpha' q)}{(\alpha' + p\gamma')^2 + (\beta' + q\gamma')^2 + (\beta' p - \alpha' q)^2},$$
$$Z = -\frac{p(\alpha' + p\gamma') + q(\beta' + q\gamma')}{(\alpha' + p\gamma')^2 + (\beta' + q\gamma')^2 + (\beta' p - \alpha' q)^2};$$

also the equation of the spheroid is

$$m^2 z^2 + m'^2 (x^2 + y^2) = 1 \tag{A}$$

if we take the axis of the chrystal parallel to the axis of z; and if we represent by $\frac{\alpha}{v}, \frac{\beta}{v}, \frac{\gamma}{v}$ the coordinates of the point in which the extraordinary ray meets the spheroid, α, β, γ being the cosines of the angles which this ray makes with the axes, we shall have

$$v^2 = m^2 + (m'^2 - m^2)(1 - \gamma^2) \tag{B}$$

and the equation of the tangent plane to the chrystal will be

$$m^2 \gamma z + m'^2 (\alpha x + \beta y) = v.$$

And since this plane is to pass through the point X, Y, Z, and to be perpendicular to the plane of incidence which has for equation

$$(x + pz)(\beta' + q\gamma') = (y + qz)(\alpha' + p\gamma'),$$

we have the two following conditions

$$m^2 \gamma (\beta' p - \alpha' q) + m'^2 \{\alpha(\beta' + q\gamma') - \beta(\alpha' + p\gamma')\} = 0,$$
$$m^2 \gamma Z + m'^2 (\alpha X + \beta Y) = v,$$

which may be put under this simpler form

$$\left.\begin{aligned} m'^2\alpha + m^2 p\gamma + v(\alpha' + p\gamma') = 0, \\ m'^2\beta + m^2 q\gamma + v(\beta' + q\gamma') = 0. \end{aligned}\right\} \qquad \text{(C)}$$

These two formulæ, which we shall presently combine into one, by Laplace's* principle of least action, are the analytic representation of the law of Huyghens,† for the extraordinary refraction produced by single-axed chrystals.

153. Adding the two preceding formulæ, after multiplying them respectively by $(\delta x, \delta y)$ we find

$$m'^2(\alpha\delta x + \beta\delta y) + m^2\gamma\delta z + v(\alpha'\delta x + \beta'\delta y + \gamma'\delta z) = 0$$

δx, δy, δz being variations of the coordinates of the chrystal. Let a', b', c', represent a point upon the incident ray, ρ' its distance from the chrystal, and a, b, c, ρ the corresponding quantities for a point on the refracted ray; we shall have

$$\alpha'\delta x + \beta'\delta y + \gamma'\delta z = -\delta\rho', \quad \alpha\delta x + \beta\delta y = -\gamma\delta z - \delta\rho,$$

$$\gamma\delta z = -\gamma\delta(\rho\gamma) = -\gamma^2\delta\rho - \rho\gamma\delta\gamma$$

and the formula may be thus written

$$\{m'^2 + (m^2 - m'^2)\gamma^2\}\delta\rho + \rho(m^2 - m'^2)\gamma\delta\gamma + v\delta\rho' = 0$$

or finally

$$\delta(v\rho + \rho') = 0 \qquad \text{(D)}$$

because

$$m'^2 + (m^2 - m'^2)\gamma^2 = v^2, \quad (m^2 - m'^2)\gamma\delta\gamma = v\delta v$$

v being, as before, the reciprocal of the radius vector of the spheroid. If then we take this reciprocal for the measure of the velocity of the extraordinary ray, that of the incident ray being unity, the law of extraordinary refraction is included in the principle of least action. And as in the systems produced by ordinary reflexion and refraction we found that there existed a certain characteristic function from which all the properties of the system might be deduced, namely the action of the light considered as depending on the coordinates of the point to which it is measured: so also in the case of extraordinary refraction, the properties of the system may all be deduced by uniform methods from the form of the same characteristic function; because its partial differentials of the first order, although not proportional to the cosines of the angles which the ray makes with the axes, as in the case of ordinary systems, are yet connected with those cosines by the following fixed relations

$$v\frac{dV}{da} = m'^2\alpha, \quad v\frac{dV}{db} = m'^2\beta, \quad v\frac{dV}{dc} = m^2\gamma, \qquad \text{(E)}$$

V being the action, measured to the point a, b, c; the velocity of the incident rays being still supposed equal to unity, and the axis of the chrystal being taken for the axis of (z).

To prove these formulæ (E), let us observe that when light issues in all directions from a

* [Pierre-Simon Laplace (1749–1827).]
† [Christian Huyghens (Christiaan Huygens) (1629–1695).]

luminous point, and after undergoing any number of ordinary reflexions and refractions is finally refracted at the surface of a chrystal; the whole variation of the action corresponding to a variation in the point to which it is measured, is the same (by the principle of least action) as if the last point of incidence had remained univaried. Hence,

$$\upsilon dV = \upsilon.\delta'(\upsilon\rho) = \upsilon^2.\delta'\rho + \rho.\upsilon\delta'\upsilon = \upsilon^2\delta'\rho + (m^2 - m'^2)\gamma\rho\delta'\gamma$$
$$= m'^2\delta'\rho + (m^2 - m'^2)\gamma\delta'(\rho\gamma),$$

δ' denoting the variation arising from the change in the point, a, b, c, to which the action is measured; we have also

$$\delta'\rho = \alpha da + \beta db + \gamma dc, \qquad \delta'(\rho\gamma) = \delta'(c - z) = dc$$

and therefore finally

$$\upsilon dV = m'^2(\alpha da + \beta db) + m^2\gamma dc.$$

154. It appears from the preceding paragraph, that the surfaces of constant action of the extraordinary system have for their differential equation

$$m'^2(\alpha da + \beta db) + m^2\gamma dc = 0 \tag{F}$$

and that therefore the rays may be considered as proceeding from them according to a law expressed by the two following formulæ

$$\mu = -\frac{m^2}{m'^2}.P, \qquad \nu = -\frac{m^2}{m'^2}.Q, \tag{G}$$

μ, ν, denoting the ratios $\frac{\alpha}{\gamma}$, $\frac{\beta}{\gamma}$, and P, Q being partial differentials of the surface of constant action. It follows also, that if we construct a series of spheroids similar to the spheroid (A), having their axes parallel to the axis of the chrystal, and touching the surface of constant action, the centres of these spheroids will be upon the extraordinary ray. Hence we may infer that if the rays of the extraordinary system converge to any one point, the surfaces of constant action are spheroids having that point for their common centre; and reciprocally, that we can always find a series of focal chrystals, which shall refract to a given point the rays of a given system; namely by so choosing the surface of the chrystal that the action, measured to the given point, may be equal to any constant quantity. These theorems, combined with the general properties of emanating systems, conduct to many interesting consequences, most of which we have anticipated in Section XXVI., and upon which therefore we shall not at present delay. Neither shall we stop to investigate the formulæ that determine the pencils of the system, the caustic curves and surfaces, the axes, images, foci and aberrations and density; because these formulæ are computed on the same principles, and nearly by the same methods, as those which we have had occasion to employ in the two preceding parts of this essay.

155. The Equation (F), of the preceding paragraph, shews that the extraordinary ray is not in general perpendicular to the surfaces of constant action, but inclined to the perpendicular at an angle ε, such that

$$v^2 \tan \varepsilon = (m^2 - v^2)^{\frac{1}{2}}(v^2 - m'^2)^{\frac{1}{2}}. \tag{H}$$

It is therefore interesting to investigate the condition for the rectangularity of an extraordinary system, that is the condition for the rays being cut perpendicularly by any series of surfaces. Employing for this purpose the methods of section XXVI., we find the following formula

$$\frac{(m^2\gamma + v\gamma')(\alpha r + \beta s) + v(\alpha A' + \beta B')}{(m^2\gamma + v\gamma')(\alpha s + \beta t) + v(\alpha\beta' + \beta C')} = \frac{\alpha\gamma - (\alpha^2 + \beta^2)p}{\beta\gamma - (\alpha^2 + \beta^2)q}, \tag{I}$$

in which (p, q, r, s, t) are partial differentials of the chrystal, and A', B', C' represent for abridgement

$$\frac{d^2\rho'}{dx^2} + 2p\frac{d^2\rho'}{dxdz} + p^2\frac{d^2\rho'}{dz^2} = A',$$

$$\frac{d^2\rho'}{dxdy} + q\frac{d^2\rho'}{dxdz} + p\frac{d^2\rho'}{dydz} + pq\frac{d^2\rho'}{dz^2} = B',$$

$$\frac{d^2\rho'}{dy^2} + 2q\frac{d^2\rho'}{dydz} + q^2\frac{d^2\rho'}{dz^2} = C',$$

ρ' being the distance of the point of incidence from a surface which cuts perpendicularly the incident rays produced, so that

$$\alpha' = \frac{d\rho'}{dx} \quad \beta' = \frac{d\rho'}{dy} \quad \gamma' = \frac{d\rho'}{dz}.$$

Suppose, for example, that the face of the chrystal is a plane perpendicular to the axis; we shall then have

$$p = 0, \quad q = 0, \quad r = 0, \quad s = 0, \quad t = 0, \quad A' = \frac{d^2\rho'}{dx^2}, \quad B' = \frac{d^2\rho'}{dxdy}, \quad C' = \frac{d^2\rho}{dy^2},$$

and because by (C)

$$\alpha = -\frac{v}{m'^2}.\alpha', \quad \beta' = -\frac{v}{m'^2}.\beta',$$

the condition (I) becomes

$$\frac{d\rho'}{dy}\left(\frac{d\rho'}{dx}.\frac{d^2\rho'}{dx^2} + \frac{d\rho'}{dy}\frac{d^2\rho'}{dxdy}\right) = \frac{d\rho'}{dx}\left(\frac{d\rho'}{dx}\frac{d^2\rho'}{dxdy} + \frac{d\rho'}{dy}\frac{d^2\rho'}{dy^2}\right), \tag{K}$$

a partial differential equation of the second order, of which the integral is the result of elimination of (a) between the two following

$$\rho' = \psi\{(x-a)^2 + (y-b)^2\}, \quad x - a + (y-b).\frac{db}{da} = 0, \tag{L}$$

(b) being considered as an arbitrary function of (a) and ψ denoting another arbitrary function. In this case therefore, that is when the face of the chrystal is a plane perpendicular to the axis, the extraordinary system will not be rectangular unless the distance of the point of incidence from a surface which cuts the incident rays perpendicularly, be a function of the perpendicular distance of the same point of incidence, from some arbitrary curve traced upon the face of the chrystal; a condition which is manifestly satisfied when the incident rays diverge from a luminous point, the arbitrary curve becoming in this instance a circle. In general, this condition (I) may be put under this other form

$$Q(PR + QS) = P(PS + QT), \tag{M}$$

P, Q, R, S, T being the partial differentials of the surfaces of constant action; and integrating this equation (M), which is of the same form as (K), we find that if the extraordinary system be rectangular, the normals to each surface of constant action are tangents to a cylindric surface, whose generating line is parallel to the axis of the chrystal.

156. The preceding remarks are sufficient to shew the manner in which we ought to proceed, in order to investigate the properties of the systems produced by reflexion at the interior surface of a single-axed chrystal or by the passage from one such chrystal into another. I shall therefore conclude this section by shewing that when the extraordinary rays recover their ordinary state, by emerging into an unchrystallised medium, or in any other manner, they become again perpendicular to the surfaces of constant action; a theorem which enables us to extend the reasonings of the two preceding parts of this Essay, to the systems produced by combinations of mirrors, lenses and chrystals.

For this purpose it is sufficient to observe, that when homogeneous rays issuing from a luminous point have been any number of times modified, by reflexion and refraction, ordinary and extraordinary, the whole variation of the action arises from the variation of the point to which it is measured; and therefore, if the final velocity of the ray be independent of its direction, this variation has for expression

$$\delta \int v d\rho = v(\alpha \delta x + \beta \delta y + \gamma \delta z)$$

as in **106.**; from which, it follows that the partial differentials of the action are proportional to the cosines of the angles which the ray makes with the axes, and that the rays are cut perpendicularly by the surfaces for which the action is constant.

XXIX. On Other Extraordinary Systems

157. We come now to make some remarks upon the properties of other extraordinary systems, beginning with those produced by the extraordinary refraction of chrystals with two rectangular axes. Brewster* has discovered that the increment of the square of the velocity, in a chrystal of this kind, is represented by the diagonal of a parallelogram, whose sides are the increments produced by each axis separately, according to the law of Huyghens, and whose angle is double of the angle formed by the planes which pass through those axes and through the extraordinary ray. Denoting this increment by i, and putting i_1, i_2 to denote the two separate increments of which it is composed, we have by the law of Huyghens

$$i_1 = k(1 - \alpha^2), \quad i_2 = k'(1 - \beta^2),$$

* [David Brewster (1781–1868).]

k, k' being constant coefficients, and (α, β) the cosines of the angles which the ray makes with the axes of the chrystal; axes which we shall take for those of (x) and (y). The equations of the extraordinary ray being put under the form

$$\frac{x}{\alpha} = \frac{y}{\beta} = \frac{z}{\gamma},$$

the planes which pass through it and through the axes have for equations

$$\text{I}^{\text{st}}\ \frac{y}{\beta} = \frac{z}{\gamma} \qquad \text{II}^{\text{nd}}\ \frac{x}{\alpha} = \frac{z}{\gamma};$$

and if we denote by P the angle which these planes form with one another, we shall have

$$\cos^2 P = \frac{\alpha^2\beta^2}{(\beta^2 + \gamma^2)(\alpha^2 + \gamma^2)}, \qquad \cos 2P = \frac{\alpha^2\beta^2 - \gamma^2}{\alpha^2\beta^2 + \gamma^2},$$

and since the law of Brewster gives

$$i^2 = i_1^2 + i_2^2 + 2 i_1 i_2 \cos 2P,$$

we have finally

$$i^2 = k^2(1 - \alpha^2)^2 + k'^2(1 - \beta^2)^2 + 2kk'(\alpha^2\beta^2 - \gamma^2). \tag{A}$$

Denoting therefore by (v') the velocity of the incident, and by (v) the velocity of the extraordinary ray, we have

$$v^2 = v'^2 + \sqrt{\{k^2(1 - \alpha^2)^2 + k'^2(1 - \beta^2)^2 + 2kk'(\alpha^2\beta^2 - \gamma^2)\}},$$

and substituting this value in the equation of least action

$$\delta(v'\rho' + v\rho) = 0 \tag{B}$$

we shall have the relations which exist between the directions of the incident and the refracted rays. And putting $i = 0$ in the formula (A), we find the position of the apparent or resultant axes, which, before the discoveries of Brewster, were thought to be the real axes of the chrystal.

158. In the extraordinary systems produced by double-axed chrystals, as well as in those other reflected and refracted systems of which we have already treated, the action may be considered as the characteristic function, from which all the properties of the system may be deduced; its partial differentials of the first order being connected with the cosines of the angles which the ray makes with the axes, by the following relations,

$$\begin{aligned}
\frac{iv}{\gamma}\frac{dV}{dc} &= iv^2 - (k\alpha^2 + k'\beta^2)^2 + (k - k')(k\alpha^2 - k'\beta^2), \\
\frac{iv}{k}\left(\frac{1}{\alpha}\frac{dV}{da} - \frac{1}{\gamma}.\frac{dV}{dc}\right) &= k\alpha^2 + k'\beta^2 - (k - k'), \\
\frac{iv}{k'}\left(\frac{1}{\beta}\frac{dV}{db} - \frac{1}{\gamma}\frac{dV}{dc}\right) &= k\alpha^2 + k'\beta^2 + k - k';
\end{aligned} \tag{C}$$

which when the two axes of the chrystal are of equal intensity, so that

$$k = k' = m'^2 - m^2, \quad k\alpha^2 + k'\beta^2 = i = v^2 - m^2,$$

reduce themselves to the corresponding formulæ (E) of the preceding section. And as in that section, we found that the surfaces of constant action were touched by the spheroids of Huyghens, so here the corresponding surfaces are touched by the spheroids of Brewster; that is by spheroids which have their centres on the extraordinary rays, and their radius vector inversely as the velocity. When the corresponding rays converge to any one point, the surfaces of constant action become a series of concentric spheroids; and we can always choose the surface of the double-axed chrystal so as to satisfy this condition, by making the action equal to any constant quantity, in the same manner as in the analogous questions representing other focal surfaces. For the consequences which follow from these principles, respecting the general properties of double-axed chrystals, and of the systems which they produce, we must refer to the theory which has been given, for systems of the second class (XXVI.).

159. When light passes through a chrystallised medium of continually varying nature; we must consider the velocity as depending not only on the direction of the ray but also on the coordinates. Putting therefore

$$v = f(x\ y\ z\ \alpha\ \beta\ \gamma),$$

we are to express, according to the principle of least action, that the variation of the integral

$$\int v ds = \int v.\sqrt{dx^2 + dy^2 + dz^2}$$

is nothing, the limits being fixed. Now

$$\delta \int v ds = \int \delta(v ds) = \int (\delta v.ds + v\delta ds);$$

also,

$$\delta v.ds = \left(\frac{dv}{dx}\delta x + \frac{dv}{dy}\delta y + \frac{dv}{dz}\delta z\right)ds + \frac{dv}{d\alpha} ds.\delta\frac{dx}{ds} + \frac{dv}{d\beta} ds.\delta\frac{dy}{ds} + \frac{dv}{d\gamma} ds.\delta\frac{dz}{ds},$$

and

$$\delta ds = \alpha\delta dx + \beta\delta dy + \gamma\delta dz;$$

if then we put

$$\left.\begin{aligned} X &= \frac{dv}{dx} + \alpha\left(v - \alpha\frac{dv}{d\alpha} - \beta\frac{dv}{d\beta} - \gamma\frac{dv}{d\gamma}\right), \\ Y &= \frac{dv}{d\beta} + \beta\left(v - \alpha\frac{dv}{d\alpha} - \beta\frac{dv}{d\beta} - \gamma\frac{dv}{d\gamma}\right), \\ Z &= \frac{dv}{d\gamma} + \gamma\left(v - \alpha\frac{dv}{d\alpha} - \beta\frac{dv}{d\beta} - \gamma\frac{dv}{d\gamma}\right), \end{aligned}\right\} \qquad \text{(D)}$$

we shall have

$$\delta(v ds) = X\delta dx + Y\delta dy + Z\delta dz + \left(\frac{dv}{dx}\delta x + \frac{dv}{dy}\delta y + \frac{dv}{dz}\delta z\right)ds,$$

$$\delta\int(vds) = X\delta x + Y\delta y + Z\delta z - (X'\delta x' + Y'\delta y' + Z'\delta z') + \int\left(\frac{dv}{dx}.ds - dX\right)\delta x$$

$$+ \int\left(\frac{dv}{dy}ds - dY\right)\delta y + \int\left(\frac{dv}{dz}ds - dZ\right)\delta z,$$

the quantities X', Y', Z', $\delta x'$, $\delta y'$, $\delta z'$ being referred to the first limit of the integral. Hence it follows that the equations of the ray are the three following

$$\frac{dv}{dx}.ds = dX, \quad \frac{dv}{dy}.ds = dY, \quad \frac{dv}{dz}.ds = dZ, \tag{E}$$

which are however equivalent to but two distinct equations, as appears by adding them multiplied respectively by α, β, γ; and if the first limit be fixed, that is if we make the integral begin at the luminous point from which the rays originally proceed, the partial differentials of the action, considered as a function of the coordinates of the point to which it is measured, are

$$\frac{dV}{dx} = X, \quad \frac{dV}{dy} = Y, \quad \frac{dV}{dz} = Z. \tag{F}$$

These partial differentials are therefore, for any given point of the medium, connected by fixed relations with the cosines of the angles which the tangent to the curved ray makes with the axes of coordinates; a theorem which enables us to consider the action, in these systems also, as the characteristic function from which all the other properties may be deduced. With respect to the manner of making this deduction, the extent to which we have already proceeded obliges us to refer to the examples which have been given in the foregoing sections, and to the general remarks of the following one, with which we shall conclude the essay.

Conclusion

It has been already stated, that the object of this Essay is to investigate, in the most general manner, (the *consequences of the law of least action*, and) the *properties of systems of rays*. We have, in the preceding sections, endeavoured to effect this object, for the cases that are most likely to occur; we have established principles respecting the systems produced by combinations of mirrors, lenses, and crystals, and have shewn that the properties of every such system may be deduced from the form of one characteristic function. We have, also, pointed out some analogous principles, respecting the systems of curved rays produced by varying mediums, and respecting systems of rays in general. It remains therefore, in this concluding section, to draw together these principles into one view, and to present the reasonings by which they have been established, under a more simple and general form; that so we may not complete our theory, with regard to those laws and systems which have been hitherto discovered; but also may be prepared to extend that theory, to the examination of those new laws and systems, which the progress of optical science may hereafter require to be considered.

II.

TWO LETTERS TO AUGUSTUS DE MORGAN (1858)*

---

1. On Definite Integrals and Diverging Series

Observatory – Feb. 15th 1858

My dear De Morgan

Although I indulge myself by taking this large sheet of paper, I hope that I shall not fill it: but wish to tell you something about a few of my recent investigations, which have little *apparent* affinity to those more *general researches* respecting definite integrals, whereof I wrote to you not long ago: notwithstanding that they have all a certain nucleus of relation. The last results, if that be not too grand a name to give them, may possibly interest you a little, because they have a connection with the theory of *Diverging Series*: for the very welcome duplicate of your Paper on which subject, I trust that you have ere now received my acknowledgment. And I am the more willing to lose no time in stating to you something about my last calculations, because I have lately been requested, or summoned, to supply some Dublin Printers with an account of those more general Theorems, above alluded to, which relate to Definite Integrals. Besides, I have had the fortune to discover, a few hours ago, a certain *linear differential equation*, of the *third order*, with *variable coefficients* (between only *two* variables), which I think that I shall be able to integrate, at least *in series*, though not proceeding solely according to powers, ascending or descending, but introducing *logarithms also*: which linear equation appears to me to be likely to be found important, at no distant time, in several *physical* inquiries. It has, at least, a connection with the mathematical theories of heat, and light and vibration; for it is connected with that very useful Transcendental function, the *ft* of my lately printed paper, which is one of the links between all those theories: and the function which satisfies the equation represents, with a proper selection of constants, what may be called the *square of the amplitude of a vibration*, in a certain class of questions. It throws a strong illumination on some old unpublished researches of mine; but since, I did not happen to perceive it, 18 years ago, I could almost wish that it had hidden itself a little longer from my view. For it

* [These letters are now printed for the first time. They were sent in instalments, over a period of months, to Augustus De Morgan (1806–1871) who was professor of Mathematics at University College, London University. The first letter started on 15 February 1858 and ended on 22 May 1858; the second letter began on 15 July 1858 and finished on 15 August 1858. Copies of the letters, as sent, are MSS 1493/972 and 1493/997 in the Library of Trinity College, Dublin. Hamilton had them both copied, with the help of his assistant Charles Thompson, into his Notebook D of 1858 (first letter begins on p. 151) and Notebook E of 1858 (second letter begins on p. 301) (Trinity College Dublin MSS 1492/144 and 1492/145).]

seems quite possible that this triordinal equation* *may* conduct to finite expressions, or at least to new and manageable series, which shall dispense with my working out, as I had begun to do, some methods for *approximating* (at least *probably*) to the values of certain Diverging Series, whereof only a few of the first terms have been actually computed; but which there is reason to believe to belong to that great Class, respecting which you made very important remarks: namely the class of ''Alternating Series'', in which each + or −, after any one calculated term, determines at least the algebraical *sign* of the *correction* which is to be made, in order to change the *sum of the terms so far*, into the *true theoretical value* of the *function* whereof the Series is *one development.* But if any of my recent labours of calculation should in this way come to be theoretically superceded, let us take courage. I need not tell De Morgan, that the last thing an inventive mathematician need ever fear to run short of, is a stock of *difficulties*!

* One form of the triordinal but linear equation, to which I refer, is the following:

$$(xD)^3 y + x^2 Dxy = 0. \tag{a}$$

One particular integral is,

$$y = a \sum_{m=0}^{m=\infty} \left[\frac{1}{2}\right]^m ([0]^{-m})^3 \left(\frac{x}{2}\right)^{2m}; \qquad \text{(b)}^\dagger$$

where a is an arbitrary constant. Another particular integral, with another constant, is:

$$y = \frac{b}{x} \sum_{m=0}^{m=\infty} [0]^{-m} \left(\left[\frac{1}{2}\right]^m\right)^3 \left(\frac{x}{2}\right)^{-2m}. \tag{c}$$

These *two series*, ascending and descending, seem to *exhaust* all the particular integrals *expressible by powers alone*; but I have found a *third particular integral*, not less simple than either of them, which is of the form,

$$y = c(B_x \log x + A_x), \tag{d}$$

where B_x is the series (b), and A_x is another ascending series; and which (speaking theoretically) *completes the integration of the triordinal equation* (a). I think that I see how to get a *fourth* particular integral, which shall involve $(\log x)^2$, and even that I shall want that form; but if (as I suspect) it *exists*, it cannot be altogether *independent* of the *other three.* On this I must write again.

† [In which the symbol (introduced by Alexandre-Théophile Vandermonde (1735–1796, 'Mémoires surdes irrationelles de différens ordres avec une application au cercle', pp. 489–498, *Histoire de l'Académie Royale des Sciences*, Part I, 1772.)

$$[x]^n = x(x-1)(x-2)\cdots(x-n+1), \tag{α}$$

so that

$$[x]^n = [x]^m [x-m]^{n-m}, \tag{β}$$

$$[x]^n = ([x]^{n+m})/([x-n]^m), \tag{γ}$$

$$[0]^{-n} = \frac{1}{[n]^n} = \frac{1}{1.2.3\cdots n} = \frac{1}{n!}, \tag{δ}$$

and

$$(1+x)^n = \sum_{m=0}^{m=\infty} [n]^m [0]^{-m} x^m. \tag{ε}$$

See Vol. I, p. 468, and this volume, pp 134.]

I. If I now resume the function

$$T = 2t\epsilon^{+t^2}\int_t^\infty \epsilon^{-t^2}\,dt, \tag{1}$$

you conceive that it is precisely because this function T is one *so very well known*, that it may be conveniently *used* to *test* the success, or failure, of any proposed *method* for *summing* diverging series, or for *approximating* to the numerical *value* of the sum, when the value of the variable is given. We know, in fact, that if

$$q = \frac{1}{2t^2}, \tag{2}$$

then T can be developed in the diverging series,

$$T = 1 - 1.q + 1.3.q^2 - 1.3.5.q^3 + \&c; \tag{3}$$

which relatively to the variable t, is a *descending* one. We know also that the same function can be *otherwise developed*, for I should not choose to call *this* an *en*velopment, –

$$T = \frac{1}{1+}\frac{q}{1+}\frac{2q}{1+}\frac{3q}{1+}\frac{4q}{1+}\ \&c. \tag{4}$$

The function T can also be developed, in a third known way, according to ascending powers of t; for it is well known that

$$T = \pi^{\frac{1}{2}}t\epsilon^{t^2} - 2t\epsilon^{t^2}\int_0^t \epsilon^{-t^2}\,dt$$

$$= \pi^{\frac{1}{2}}t\left(1 + \frac{t^2}{1.2} + \frac{t^4}{1.2.3.4} + \frac{t^6}{1.2.3.4.5.6} + \cdots\right) - 2t^2\left(1 + \frac{2t^2}{3} + \frac{4t^4}{3.5} + \frac{8t^6}{3.5.7} + \cdots\right). \tag{5}$$

And finally, the *numerical values* of the connected integral

$$2^{-1}t^{-1}\epsilon^{-t^2}T = G = \int_t^\infty \epsilon^{-t^2}\,dt, \tag{6}$$

have been *tabulated*, by Kramp*; whose book I have unluckily not seen, but from which I understood you to state that you reprinted Tables I & II at the end of your Treatise on the theory of Probabilities in the Encyclopædia Metropolitana†. The function, here called T, may therefore be accounted to be almost as well known, *as regards its values*, as the functions sine and cosine, although I am not aware that many, or any, *properties* of it are *known*, which could be expressed as *functional equations*. As regards values, it is easy for instance to compute, from Kramp's† Table (as reprinted by you), that

$$\text{for } t = 1, \qquad LT = \bar{1}{\cdot}8795960; \qquad T = 0{\cdot}75787; \tag{7}$$ ‡

$$\text{for } t = 2, \qquad LT = \bar{1}{\cdot}9568185; \qquad T = 0{\cdot}90535; \tag{8}$$

* [Christian Kramp (1760–1826), *Analyse des réfractions*, pp. 195–206, Strasbourg: 1799.]

† [*Encyclopædia Metropolitana*, eds. E. Smedley, H. J. Rose, and H. J. Rose, Vol. II, pp. 393–490, London: 1845.]

‡ [$L \equiv \log_{10}$.]

$$\text{for } t = 3, \qquad LT = \bar{1}{\cdot}9785541; \qquad T = 0{\cdot}95182. \tag{9}$$

II. Before proceeding to test any new method for even *approximately summing*, or in any other way attempting to envelope, without definite integrals, the diverging series (3) for T, I shall just remark, that on representing the finitely continued fraction

$$T_n = \frac{1}{1+}\frac{q}{1+}\frac{2q}{1+}\cdots\frac{nq}{1+0}, \tag{10}$$

by the expression

$$T_n = \frac{N_n}{M_n}, \tag{11}$$

when M_n and N_n are polynomial functions (rational and integral) of q, the known equations in differences,

$$M_n - M_{n-1} = nqM_{n-2}, \qquad N_n - N_{n-1} = nqN_{n-2}, \tag{12}$$

do not merely enable us to calculate expressions, algebraical or arithmetical, for all the numerators N, and all the denominators M, *successively*, from the known initial pairs,

$$M_0 = 1, \quad M_1 = 1 + q; \quad N_0 = 1, \quad N_1 = 1; \tag{13}$$

but also give, by *finite integration*, the following *Rule* for forming at *once* the polynomial M_n, or the *denominator* of T_n, when the index n is given:

Develop by the binomial theorem, the power

$$(1 \pm q^{\frac{1}{2}})^{n+1},$$

under the form,

$$(1 \pm q^{\frac{1}{2}})^{n+1} = 1 + a_n q + b_n q^2 + c_n q^3 + \&\text{c.} \pm \&\text{c.}; \tag{14}$$

then

$$M_n = 1 + 1.a_n q + 1.3.b_n q^2 + 1.3.5.c_n q^3 + \&\text{c.}, \tag{15}$$

without the $\pm$ terms. For example

$$(1 \pm q^{\frac{1}{2}})^8 = 1 + 28q + 70q^2 + 28q^3 + q^4 \pm \&\text{c.}; \tag{16}$$

$$(1 \pm q^{\frac{1}{2}})^9 = 1 + 36q + 126q^2 + 84q^3 + 9q^4 \pm \&\text{c.}; \tag{17}$$

$$(1 \pm q^{\frac{1}{2}})^{10} = 1 + 45q + 210q^2 + 210q^3 + 45q^4 + q^5 \pm \&\text{c.}; \tag{18}$$

whence, by the rule (15)

$$M_7 = 1 + 28q + 210q^2 + 420q^3 + 105q^4; \tag{19}$$

$$M_8 = 1 + 36q + 378q^2 + 1260q^3 + 945q^4; \tag{20}$$

$$M_9 = 1 + 45q + 630q^2 + 3150q^3 + 4725q^4 + 945q^5; \tag{21}$$

with the verification, by (12), $M_9 - M_8 = 9qM_7$.

As regards the *numerator*, N_n, although, like the denominator, M_n, it can easily be computed, algebraically, or arithmetically, as I suppose that it usually is, by the help of the equation of differences which it satisfies, through successive *steps* from its initial pair of values; (and although I have integrated the equation just referred to;) yet it may be allowed me here to make the remark, – a sufficiently obvious one, indeed, – that whenever we know, by any process, for any given value of n, the coefficients of the polynome M_n, we can easily find the corresponding coefficients of the other polynome N_n, by forming the product TM_n, where T is the known series (3), and suppressing all the terms whose exponents are greater than n. Verifications will offer themselves, in the evanescence of the coefficients of certain lower powers of q. For example, in the product TM_7, the coefficients of q^4, q^5, q^6, q^7 all vanish; in the product TM_8, the coefficients of q^5, q^6, q^7, q^8 are each zero; in the product TM_9, the coefficients of q^5, q^6, q^7, q^8, q^9 disappear.

And in this way, among others, we may find that

$$N_7 = 1 + 27q + 185q^2 + 279q^3; \tag{22}$$

$$N_8 = 1 + 35q + 345q^2 + 975q^3 + 384q^4; \tag{23}$$

$$N_9 = 1 + 44q + 588q^2 + 2640q^3 + 2895q^4; \tag{24}$$

with the verification,

$$N_9 - N_8 = 9qN_7. \tag{25}$$

III. If we assume as a numerical example

$$t = 1, \qquad q = \tfrac{1}{2}, \tag{26}$$

so that we have here

$$T = 2\epsilon \int_1^\infty \epsilon^{-t^2}\, dt, \tag{27}$$

and if we denote by A_n and B_n the whole numbers which take respectively the place of M_n and N_n, when those are multiplied by suitable powers of 2, we easily find in this way the values (without passing through earlier stages,)

$$A_7 = 16M_7 = 2025; \qquad B_7 = 16N_7 = 1530; \tag{28}$$

$$A_8 = 16M_8 = 5281; \qquad B_8 = 16N_8 = 4010; \tag{29}$$

$$A_9 = 32M_9 = 28787; \qquad B_9 = 32N_9 = 21790; \tag{30}$$

with the verifications,

$$A_9 - 2A_8 = 9A_7; \qquad B_9 - 2B_8 = 9B_7. \tag{31}$$

Again

$$\left.\begin{aligned} A_{10} - A_9 &= 10A_8; & B_{10} - B_9 &= 10B_8; \\ A_{11} - 2A_{10} &= 11A_9; & B_{11} - 2B_{10} &= 11B_9; \\ A_{12} - A_{11} &= 12A_{10}; & B_{12} - B_{11} &= 12B_{10}; \end{aligned}\right\} \tag{32}$$

whence

$$\left.\begin{array}{ll} A_{10} = 81597; & B_{10} = 61890; \\ A_{11} = 479851; & B_{11} = 363470; \\ A_{12} = 1459015; & B_{12} = 1106150. \end{array}\right\} \quad (33)$$

Hence

$$\left.\begin{array}{llll} T_7 = \dfrac{N_7}{M_7} = \dfrac{B_7}{A_7} & = \dfrac{1530}{2025} & = \dfrac{34}{45} = 0{\cdot}75556; \\ T_8 = \quad = \dfrac{B_8}{A_8} & = \dfrac{4010}{5281} & = 0{\cdot}75933; \\ T_9 = \quad = \dfrac{B_9}{A_9} & = \dfrac{21790}{28787} & = 0{\cdot}75694; \\ T_{10} = \quad = \dfrac{B_{10}}{A_{10}} & = \dfrac{61890}{81597} & = 0{\cdot}75848; \\ T_{11} = \quad = \dfrac{B_{11}}{A_{11}} & = \dfrac{363470}{479851} & = 0{\cdot}75746; \\ T_{12} = \quad = \dfrac{\frac{2}{10}B_{12}}{\frac{2}{10}A_{12}} & = \dfrac{221230}{291803} & = 0{\cdot}75815. \end{array}\right\} \quad (34)$$

We have the inequalities,

$$T > T_7, \quad T < T_8, \quad T > T_9, \quad T < T_{10}, \quad T > T_{11}, \quad T < T_{12}; \quad (35)$$

which are, as we see, all consistent with the order of T deduced from Kramp's Table; namely with $T = 0{\cdot}75787$. But it was for my purpose desirable to deduce these limits by a *special* calculation, independent of any conceivable errors in the construction, transcription, or impression of that Table; although I am very willing to believe that such errors are few or none. For I wished to be *quite sure*, for reasons which will afterwards appear, that the two following inequalities are satisfied by T, in the case $t = 1$:

$$T < 0{\cdot}75984; \qquad T < 0{\cdot}75839. \quad (36)$$

The inequality $T < T_8$ proves the first; and $T < T_{12}$ proves the second. The first presented itself to me thus, as a thing to be examined: was it true that

$$T < 3^{-\frac{1}{4}}? \quad \text{or that } T^{-4} > 3? \quad (37)$$

The tabular value of LG gave LT as in (7); whence

$$4LT = \bar{1}{\cdot}5183840; \quad 4L\frac{1}{T} = 0{\cdot}4816160; \quad T^{-4} = 3{\cdot}03121; \quad (38)$$

but I thought it *more completely satisfactory* to infer the inequalities (37) by the simple *arithmetical* remark, that

$$T_8^{-4} = \left\{\frac{5281}{4010}\right\}^4 = \frac{777\,794\,145\,659\,521}{258\,569\,616\,010\,000} > 3; \quad (39)$$

while it was known from theory, that

$$T^{-4} > T_8^{-4}. \tag{40}$$

IV. The method of continued fractions, applied to the function T, gives the expression:

$$T = \frac{1}{1+}\frac{q}{1+}\frac{2q}{1+}\frac{3q}{1+}\frac{4q}{1+}\cdots\frac{(n-1)\,q}{1+n\theta_n q};\ \theta_n > 0,\ < 1; \tag{41}$$

at least if $n > 1$, for there appears to be a sort of discontinuity at the very commencement of the fractional development, and it seems safer to write, as a separate formula,

$$T = \frac{1}{1+\theta_1 q};\ \theta_1 > 0,\ < 1. \tag{42}$$

Suppose now *that we had only been given the first few terms of the diverging series for* T; let us say, the six terms,

$$T = 1 - q + 3q^2 - 15q^3 + 105q^4 - 945q^5 + \cdots, \tag{43}$$

where the final $+\cdots$ means nothing more than the *next* (*or seventh*) *term has a positive coefficient.* We are not supposed to have even *perceived* that if T be written as

$$T = 1 - aq + bq^2 - cq^3 + dq^4 - eq^5 + \cdots, \tag{44}$$

then not only $a = 1$, but also

$$b = 3a, \qquad c = 5b, \qquad d = 7c, \qquad e = 9d. \tag{45}$$

Or, if *this* be assuming too vast a want of sagacity, let it at least be here supposed that the five coefficients a, b, c, d, e, have in some laborious way been *separately* (or successively) calculated, and that we have nothing beyond a *guess* to go upon, as to whether the *law* (45) will continue. But let it be admitted that we know, or think that we have good reasons for believing, that the series for T is an *alternating* one, of the kind signalised by you as important; so that we have at *least* the *six inequalities,*

$$\left.\begin{array}{l} T < 1, \quad > 1 - q, \quad < 1 - q + 3q^2, \quad > 1 - q + 3q^2 - 15q^3, \\ < 1 - q + 3q^2 - 15q^3 + 105q^4, \quad > 1 - q + 3q^2 - 15q^3 + 105q^4 - 945q^5; \end{array}\right\} \tag{46}$$

if q be any real quantity > 0; to which I find it necessary to append the following *other supposed inequality,*

$$T > 0, \quad \text{if} \quad q > 0, \quad < \infty. \tag{47}$$

How far, with only these suppositions (or at most with only a few others entirely analogous to these) *can we recover the expression* (41)?

V. Consider generally the function

$$X = (1 + \alpha x(1 + \alpha' x(1 + \alpha'' x(1 + \cdots)^{-\frac{1}{\nu}})^{-\frac{1}{\nu}})^{-\frac{1}{\nu}})^{-\frac{1}{\nu}}; \tag{48}$$

and suppose that it is to be otherwise developed in a series of the form,

$$X = 1 - ax + bx^2 - cx^2 + dx^4 - ex^5 + \cdots; \tag{49}$$

where the coefficients a, b, c, d, e, ... are certain algebraic functions of the other coefficients

$\alpha, \alpha', \alpha'', \alpha''', \alpha^{iv}, \ldots$ or $\alpha, \beta, \gamma, \delta, \varepsilon, \ldots$ which functions we are now to determine. Raising the series (49) to the power of which the exponent is $-\nu$, we get an expression which may be thus denoted,

$$X^{-\nu} = 1 + \nu a x X'; \tag{50}$$

where X' is a new series, of the form

$$X' = 1 - a'x + b'x^2 - c'x^3 + d'x^4 - \cdots; \tag{51}$$

$a', b', c', d', \ldots$ being algebraical functions of a, b, c, d, e, and of ν: namely

$$\begin{aligned} a' &= \frac{b}{a} - \frac{\nu+1}{1}\frac{a}{2}; \\ b' &= \frac{c}{a} - \frac{\nu+1}{1} b + \frac{\nu+1}{1}\frac{\nu+2}{2}\frac{a^3}{3}; \\ c' &= \&c. \end{aligned} \tag{52}$$

But also, by (48),

$$X^{-\nu} = 1 + \alpha x X', \tag{53}$$

where

$$X' = (1 + \alpha' x(1 + \alpha'' x(1 + \cdots)^{-\frac{1}{\nu}})^{-\frac{1}{\nu}})^{-\frac{1}{\nu}}. \tag{54}$$

Comparing, we are led to infer that

$$\alpha = \nu a, \ a = \nu^{-1}\alpha; \tag{55}$$

and also that the two developments, (51), (54), for X', are to be regarded as transformations of each other. A *law of accentuation* of the letters is thus suggested; and we find that if we write

$$\left.\begin{aligned} a'' &= \frac{b'}{a'} - \frac{\nu+1}{1}\frac{a'}{2}, \\ b'' &= \frac{c'}{a'} - \frac{\nu+1}{1}\frac{b'}{2} + \frac{\nu+1}{1}\frac{\nu+2}{2}\frac{a'^2}{3}, \\ &\&c., \end{aligned}\right\} \tag{56}$$

$$a''' = \frac{b''}{a''} - \frac{\nu+1}{1}\frac{a''}{2}, \quad \&c., \quad \&c. \tag{57}$$

then,

$$X'^{-\nu} = 1 + \nu a' x X'', \quad X''^{-\nu} = 1 + \nu a'' x X''', \quad \&c., \tag{58}$$

where

$$X'' = (1 + \alpha'' x(1 + \cdots)^{-\frac{1}{\nu}})^{-\frac{1}{\nu}}, \quad X''' = \&c., \tag{59}$$

and

$$\alpha' = \nu a', \quad \alpha'' = \nu a'', \quad \&c., \quad \text{or } a' = \nu^{-1}\alpha', \quad a'' = \nu^{-1}\alpha'', \quad \&c. \tag{60}$$

In this manner a system of equations can be formed, for connecting $a, b, c, d, e, \ldots$ with $\alpha, \beta,$

$\gamma, \delta, \varepsilon, \ldots$ and deducing successively the former from the latter, or conversely the latter from the former, with the help of the auxiliary quantities: $a', b', c', d', \ldots, a'', b'', c'', \ldots, a''', b''', \ldots, a^{iv}, \ldots$.

VI. For example, if $\nu = 1$, the equations between the constants are:

$$\left.\begin{aligned} a' = \frac{b}{a} - a;\ a'' = \frac{b'}{a'} - a';\ a''' = \frac{b''}{a''} - a'';\ a^{iv} = \frac{b'''}{a'''} - a''';\\ b' = \frac{c}{a} - 2b + a^2;\ b'' = \frac{c'}{a'} - 2b' + a'^2;\ b''' = \frac{c''}{a''} - 2b'' + a''^2;\\ c' = \frac{d}{a} - \left(2c + \frac{b^2}{a}\right) + 3ab - a^3;\ c'' = \frac{d'}{a'} - \left(2c' + \frac{b'^2}{a'}\right) + 3a'b' - a'^3;\\ d' = \frac{e}{a} - 2\left(d + \frac{bc}{c}\right) + 3(ac + b^2) - 4a^2b + a^4; \end{aligned}\right\} \tag{61}$$

together with

$$\alpha = a, \quad \beta = a', \quad \gamma = a'', \quad \delta = a''', \quad \varepsilon = a^{iv}, \tag{62}$$

if we write $\beta, \gamma, \delta, \varepsilon$, for $\alpha', \alpha'', \alpha''', \alpha^{iv}$, and if we neglect x^6 in X.

Thus the polynome (49) may be thrown into the form of the continued fraction,

$$X = \frac{1}{1+}\frac{\alpha x}{1+}\frac{\beta x}{1+}\frac{\gamma x}{1+}\frac{\delta x}{1+}\frac{\varepsilon x}{1+\cdots}, \tag{63}$$

and the coefficients $\alpha, \beta, \gamma, \delta, \varepsilon$ computed, if a, b, c, d, e be given or vice versa. For instance, if, as in (45), we assume that

$$a = 1, \quad b = 3, \quad c = 15, \quad d = 105, \quad e = 945, \tag{64}$$

then

$$\left.\begin{matrix} a' = 2, & b' = 10, & c' = 74, & d' = 706; & a'' = 3, & b'' = 21, & c'' = 207;\\ a''' = 4, & b''' = 36; & a^{iv} = 5; & \alpha = 1, & \beta = 2, & \gamma = 3, \quad \delta = 4, & \varepsilon = 5; \end{matrix}\right\} \tag{65}$$

and, *so far at least*, we recover the known expression for T, for we find

$$1 - x + 3x^2 - 15x^3 + 105x^4 - 945x^5 \cdots = \frac{1}{1+}\frac{1x}{1+}\frac{2x}{1+}\frac{3x}{1+}\frac{4x}{1+}\frac{5x}{1+\cdots}; \tag{66}$$

the remaining constants of the fractional development being unknown *until* we know the remaining coefficients of the diverging series, or at least their *law*, which for the moment we are not supposed to do.

VII. The process may of course be varied, and the following seems to be a more elegant form of it. Since $\nu = 1$, the equation (53) gives here

$$1 = (1 + \alpha x X')X = (1 + \alpha x - \alpha a' x^2 + \alpha b' x^3 - \cdots)(1 - ax + bx^2 - \cdots); \tag{67}$$

whence

$$\left.\begin{array}{l}\alpha^{-1}a = 1; \quad \alpha^{-1}b = a' + a; \quad \alpha^{-1}c = b' + aa' + b; \\ \alpha^{-1}d = c' + ab' + ba' + c; \quad \alpha^{-1}e = d' + ac' + bb' + ca' + d;\end{array}\right\} \tag{68}$$

of which equations each may be *accented* throughout, and α', α'', α''', α^{iv} then changed to β, γ, δ, ε, if we think fit. Thus

$$\left.\begin{array}{l}\beta^{-1}a' = 1, \quad \beta^{-1}b' = a'' + a', \quad \beta^{-1}c' = b'' + a'a'' + b', \quad \beta^{-1}d' = c'' + a'b'' + b'a'' + c'; \\ \gamma^{-1}a'' = 1, \quad \gamma^{-1}b'' = a''' + a'', \quad \gamma^{-1}c'' = b''' + a''a''' + b'' \\ \delta^{-1}a''' = 1, \quad \delta^{-1}b''' = a^{iv} + a'''; \quad \varepsilon^{-1}a^{iv} = 1.\end{array}\right\} \tag{69}$$

These 15 equations, (68), (69), are merely *transformations* of the 15 equations, (61), (62), but they are more *symmetric*, and better fitted for the easy *elimination* of the auxiliary constants, so as to give expressions for a, b, c, d, e in terms of α, β, γ, δ, ε, involving *nothing foreign*, and taking rather simple shapes. In this way, by eliminating a', b', c', d', a'', b'', c'', a''', b''', a^{iv}, I find the expressions,

$$\left.\begin{array}{l}a = \alpha; \quad b = \alpha\beta + \alpha^2; \quad c = \alpha\beta\gamma + \alpha(\beta + \alpha)^2; \\ d = \alpha\beta\gamma\delta + \alpha\beta(\gamma + \beta + \alpha)^2 + \alpha^2(\beta + \alpha)^2; \\ e = \alpha\beta\gamma\delta\varepsilon + \alpha\beta\gamma(\delta + \gamma + \beta + \alpha)^2 + \alpha\{\beta\gamma + (\beta + \alpha)^2\}^2;\end{array}\right\} \tag{70}$$

which may be verified by observing that they reproduce the numerical values, 1, 3, 15, 105, 945, for a, b, c, d, e, when α, β, γ, δ, ε have the values 1, 2, 3, 4, 5. It is a consequence from the process employed that all the subsequent coefficients, f, g, h, ..., of the series (49), must in like manner admit of being expressed by polynomial functions of α, β, γ, δ, ε, ζ, η, θ, ..., of the dimensions 6, 7, 8, ..., and with all their coefficients equal to positive whole numbers.* And of course, it is about as easy to deduce α, β, γ, δ, ε from a, b, c, d, e, by the equations (70) as to deduce the latter from the former.

VIII. *If only the five coefficients a, b, c, d, e, be given,* with the values (45) or (64), we can, as above, *only deduce the five related constants* α, β, γ, δ, ε, with the known values (65). We have *no right to assume even that the next constant* ζ *is positive*; for although we are supposed to know that the *next coefficient* f *is positive*, and may even *presume that it is greater than* 945, we cannot say before-hand whether it will be found to be *sufficiently great to render* $\zeta > 0$, when the *sixth equation of the series* (70) comes to be resolved for ζ. To form that sixth equation, on the plan of (68), the formula

$$\alpha^{-1}f = e' + ad' + bc' + cb' + dd' + e; \tag{71}$$

that is, by (65),

$$f = e' + 2233; \tag{72}$$

*

$$f = \alpha\beta\gamma\delta\varepsilon\zeta + \alpha\beta\gamma\delta(\epsilon + \delta + \gamma + \beta + \alpha)^2 + \alpha\beta\{\gamma(\delta + \gamma + 2\beta + \alpha) + (\beta + \alpha)^2\}^2 + \alpha^2\{\beta\gamma + (\beta + \alpha)^2\}^2;$$

if $\alpha, \beta, \gamma, \delta, \varepsilon = 1, 2, 3, 4, 5$, respectively, $f - 120\zeta = (24 + 18 + 1).\,225 = 9675$.

where, by accenting the fifth equation (70), we have

$$e' = \beta\gamma\delta\varepsilon\zeta + \beta\gamma\delta(\varepsilon + \delta + \gamma + \beta)^2 + \beta\{\gamma\delta + (\beta + \alpha)^2\}^2; \tag{73}$$

&c. Therefore, again by (65),

$$e' = 120\zeta + 7462; \tag{74}$$

so that, finally the relation between f and ζ is the following:

$$f = 120\zeta + 9675. \tag{75}$$

Unless, then, we know that

$$f > 9675, \tag{76}$$

we cannot be sure that $\zeta > 0$. *In point of fact,* the series for T is such that in it the coefficient of q^2 is $f = 11 \times 945 = 10\,395$; giving

$$\zeta = \frac{10\,395 - 9675}{120} = 670. \tag{77}$$

But it might have been supposed that the divergent character of the series would be sufficiently *saved* by our *guessing* a *less value* than 9675 for f; and in *that* case, if the guess were right, the constant ζ would be *negative.*

IX. *Even then, if we suppose,* which is certainly supposing a good deal, that *all the series* X', X'', X''', X^{iv}, X^{v}, ... successively derived from X, or from T, by the process of paragraph V., *belong to the same great and general class as the original series itself,* we cannot be sure that the series

$$X^{v} = 1 - a^{v}x \ldots, \tag{78}$$

where $a^{v} = \zeta$, has the property of *being less than* 1; *nor even that of being greater than* 0; so long as we are left uncertain, whether ζ (or a^{v}) is a *positive* or a *negative* number. Let us, however, here admit that the value of the (*presumably divergent*) *series* denoted by X^{v} is *likely to be greater than* 0, *on account of its first term being positive* (= 1), and let us count up all the results which, at this stage, we might fairly be induced to call ***probable***, respecting the *value of the function,* which we conceive to be represented by the series X, for real and positive values of x.

X. I think, then, that we might, not unfairly, *presume* that, for any proposed real and positive value of x, – *at least,* if such value of x were *not too large,* – we should have the following *chain of inequalities.* We have just now *supposed* that

$$x > 0, \ X^{v} > 0; \tag{79}$$

and we have proved that α, β, γ, δ, ε are each > 0, because they have the values 1, 2, 3, 4, 5: though the next constant ζ, for anything *yet known, may perhaps be negative;* namely by the unknown coefficient f of x^6 in X, though positive being *less than the limit* 9675. We may then infer that because

$$X^{iv} = (1 + \varepsilon x X^{v})^{-1},$$

therefore

$$X^{iv} > 0, < 1; \tag{80}$$

but we are not entitled to assume that

$$X^{iv} < (1+\varepsilon x)^{-1}. \tag{80'}$$

Again since

$$X''' = (1+\delta x X^{iv})^{-1};$$

therefore

$$X''' < 1, > (1+\delta x)^{-1}; \tag{81}$$

although it remains uncertain whether

$$X''' < (1+\delta x(1+\varepsilon x)^{-1})^{-1}. \tag{81'}$$

Again since

$$X'' = (1+\gamma x X''')^{-1},$$

therefore

$$X'' > (1+\gamma x)^{-1}, \quad < (1+\gamma x(1+\delta x)^{-1})^{-1}; \tag{82}$$

$$X' = (1+\beta x X'')^{-1}, \quad < (1+\beta x(1+\gamma x)^{-1})^{-1}, \quad > (1+\beta x(1+\gamma x(1+\delta x)^{-1})^{-1})^{-1}; \tag{83}$$

and finally, the equation

$$X = (1+\alpha x X')^{-1}$$

gives the two limits

$$X > (1+\alpha x(1+\beta x(1+\gamma x)^{-1})^{-1})^{-1}; \qquad X < (1+\alpha x(1+\beta x(1+\gamma x(1+\delta x)^{-1})^{-1})^{-1})^{-1}; \tag{84}$$

which may be written thus,

$$X > \frac{1}{1+}\frac{\alpha x}{1+}\frac{\beta x}{1+}\frac{\gamma x}{1+0}, \quad < \frac{1}{1+}\frac{\alpha x}{1+}\frac{\beta x}{1+}\frac{\gamma x}{1+\delta x}; \tag{85}$$

and which give (α, β, γ, δ, and x being all positive) the following expression for X:

$$X = \frac{1}{1+}\frac{\alpha x}{1+}\frac{\beta x}{1+}\frac{\gamma x}{1+\theta\delta x}; \quad \theta > 0, \quad < 1. \tag{86}$$

Thus with our values of α, β, γ, δ, if we write q for x, and T for X, we have indeed, the (probable) limits $T > T_3$, $T < T_4$; that is,

$$T > \frac{1+5q}{1+6q+3q^2}, \quad T < \frac{1+9q+8q^2}{1+10q+15q^2}; \tag{87}$$

or, for the case $q = 1/2$,

$$T > \frac{14}{19}, \quad < \frac{30}{39}; \quad \text{or } T > 0{\cdot}73684, \quad T < 0{\cdot}76923. \tag{88}$$

But whether T be greater or less than the next fraction, namely than

$$T_5 = \frac{1+14q+33q^2}{1+15q+45q^2+15q^3} = \frac{130}{173} = 0{\cdot}75145, \tag{89}$$

we should not (as I conceive) *be entitled to assert anything, as even probable, with our present suppositions.* We might, indeed, *write*

$$X = \frac{1}{1+}\frac{\alpha x}{1+}\frac{\beta x}{1+}\frac{\gamma x}{1+}\frac{\delta x}{1+\vartheta\varepsilon x}; \tag{90}$$

where ϑ should represent the series $X^v = 1 \ldots$, *whereof only the first term* 1 *is known*; or, in the case above considered, might form the *expression*,

$$T = \frac{1}{1+}\frac{\frac{1}{2}}{1+}\frac{\frac{2}{2}}{1+}\frac{\frac{3}{2}}{1+}\frac{\frac{4}{2}}{1+}\frac{5\vartheta}{2}; \tag{91}$$

but in our uncertainty whether ϑ, although *presumed* to be *positive, is less or greater than* 1, we should get *no more of (even probable) information* from this last expression for T, than from the following case of the *simpler form* (86),

$$T = \frac{1}{1+}\frac{\frac{1}{2}}{1+}\frac{\frac{2}{2}}{1+}\frac{\frac{3}{2}}{1+}\frac{4\theta}{2}, \qquad \theta > 0, \quad < 1. \tag{92}$$

It is unnecessary to repeat, that all this depends on the ***hypothesis***, *that we know nothing of the law of the coefficients a, b, c, d, e, f*, ... of the series for X, *or for* T; and in particular, that *even* if we have *guessed the law*, (as for the present series, from a few of its first terms, we could scarcely fail to do,) yet we have *no such assurance of the correctness of that guess*, as to pronounce with any confidence that *the sixth coefficient f must exceed the limit* 9675. In other respects, you see that I am *seeking* to give here the *fullest scope* to principles which *favour diverging series*, and that, in fact, the *supposed doubt* is precisely this, *whether the coefficients increase* ***fast enough***? Is the series ***sufficiently divergent***, for the *safe application* of the *Method of Converging Fractions*? (Of course, ***we know that it is so***: but then *we* are *in full possession of the law* and know that it must continue.)

XI. That method of converging fractions which must also be called (I think) the "Method of Reciprocals" may be presented in the following form. Let φx denote the function,

$$\varphi x = x^{-1} - 1; \tag{93}$$

the operating by $\alpha^{-1}x^{-1}\varphi$ on the series X, of which the first term is unity, if the constant α be suitably chosen, we deduce from it another series X', with its first term also $= 1$; and so proceeding, we have the succession of equations

$$\alpha^{-1}x^{-1}\varphi X = X'; \quad \beta^{-1}x^{-1}\varphi X' = X''; \quad \gamma^{-1}x^{-1}\varphi X'' = X'''; \\ \delta^{-1}x^{-1}\varphi X''' = X^{iv}; \quad \varepsilon^{-1}x^{-1}\varphi X^{iv} = X^{v}; \tag{94}$$

whence

$$\left.\begin{aligned} &X'' = \beta^{-1}x^{-1}\varphi\alpha^{-1}x^{-1}\varphi X; \quad X''' = \gamma^{-1}x^{-1}\varphi\beta^{-1}x^{-1}\varphi\alpha^{-1}x^{-1}\varphi X; \\ &X^{iv} = \delta^{-1}x^{-1}\varphi\gamma^{-1}x^{-1}\varphi\beta^{-1}x^{-1}\varphi\alpha^{-1}x^{-1}\varphi X; \\ &X^{v} = \varepsilon^{-1}x^{-1}\varphi\delta^{-1}x^{-1}\varphi\gamma^{-1}x^{-1}\varphi\beta^{-1}x^{-1}\varphi\alpha^{-1}x^{-1}\varphi X; \end{aligned}\right\} \tag{95}$$

&c., each symbol φ governing the entire system of symbols to its right. If then we write

$$X^{iv} = \theta, \qquad X^{v} = \vartheta, \tag{96}$$

we have the transformation, in which

$$\varphi^{-1} x = (1 + x)^{-1}: \tag{97}$$

$$X = \varphi^{-1}\alpha x \varphi^{-1}\beta x \varphi^{-1}\gamma x \varphi^{-1}\theta\delta x; \quad X = \varphi^{-1}\alpha x \varphi^{-1}\beta x \varphi^{-1}\gamma x \varphi^{-1}\delta x \varphi^{-1}\vartheta\varepsilon x; \tag{98}$$

where each symbol φ^{-1} governs in like manner all that follows it. And if, as above, we have computed the values of the five constants, α, β, γ, δ, ε, and find them all to be positive, we may *presume* that θ lies between the limits 0 and 1; and may employ the first of the two expressions (98), so as to deduce from it, *with a reasonable probability*, the two following limits for X, at least *if x be not assumed too great*:

$$X > \varphi^{-1}\alpha x \varphi^{-1}\beta x \varphi^{-1}\gamma x; \quad X < \varphi^{-1}\alpha x \varphi^{-1}\beta x \varphi^{-1}\gamma x \varphi^{-1}\delta x; \tag{99}$$

the inverse function $\varphi^{-1}x$ being here one which decreases continually from 1 to 0, as x increases from 0 to ∞. But because we are yet entirely ignorant whether ζ is positive or negative, we cannot say whether the series denoted by X^{v} or by ϑ, is greater than, or less than one; the second expression (98) is therefore here (not false but) useless: and we have *no right to establish*, as *even probably true, and even for small values of x, this other and analogous inequality*,

$$X > \varphi^{-1}\alpha x \varphi^{-1}\beta x \varphi^{-1}\gamma x \varphi^{-1}\delta x \varphi^{-1}\varepsilon x. \tag{100}$$

Perhaps ϕ might be written for the inverse function φ^{-1}, as you propose to write γ instead of $\log^{-1}$.

XII. Instead of assuming the *form* (93), *for the auxiliary function φx*, I was lately led to try the effect of assuming this *other form*,

$$\varphi x = l'x = l\frac{1}{x}; \tag{101}$$*

which gave, as its inverse,

$$\varphi^{-1}x = l'^{-1}x = \epsilon^{-x}; \tag{102}$$

that is, with your notation γ, and with my symbol ϕ for φ^{-1},

$$\phi x = \gamma(-x); \quad \text{or simply, } \phi x = \gamma - \quad x, \tag{103}$$

the $-$ being here treated as a *factor*, or as an operator. As I wrote to you lately on some of the results of the assumption of *this* auxiliary function, φ, it may suffice here to sketch, in the briefest way, a few of the steps of the proofs which I employed. As analogous to the equations of paragraph VI., I had,

* $[l = \log_\epsilon]$

$$\left.\begin{aligned}
&a' = \frac{b}{a} - \frac{a}{2}, \quad a'' = \frac{b'}{a'} - \frac{a'}{2}, \quad a''' = \frac{b''}{a''} - \frac{a''}{2}, \quad a^{iv} = \frac{b'''}{a'''} - \frac{a'''}{2};\\
&b' = \frac{c}{a} - b + \frac{a^2}{3}, \quad b'' = \frac{c'}{a'} - b' + \frac{a'^2}{3}, \quad b''' = \frac{c''}{a''} - b'' + \frac{a''^2}{3};\\
&c'' = \frac{d}{a} - \left(c + \frac{b^2}{2a}\right) + ab - \frac{a^3}{4}, \quad c'' = \frac{d'}{a'} - \left(c' + \frac{b'^2}{2a}\right) + a'b' - \frac{a'^3}{4};\\
&d' = \frac{e}{a} - \left(d + \frac{bc}{a}\right) + (ac + b^2) - a^2 b + \frac{a^4}{5};
\end{aligned}\right\} \tag{104}$$

along with

$$\alpha = a, \beta = a', \gamma = a'', \delta = a''', \varepsilon = a^{iv}. \tag{105}$$

$$\left[e' = \frac{f}{a} - \left(e + \frac{bd}{a} + \frac{c^2}{2a}\right) + \left(ad + 2bc + \frac{b^3}{3a}\right) - a\left(ac + \frac{3b^2}{2}\right) + a^3 b - \frac{a^5}{6}.\right]$$

As transformations, of the same kind as those already given for another form of φ, in paragraph VII., it was easy to deduce the following:

$$\left.\begin{aligned}
&2\alpha^{-1}b = 2a' + a,\ 2\beta^{-1}b' = 2a'' + a',\ 2\gamma^{-1}b'' = 2a''' + a'',\ 2\delta^{-1}b''' = 2a^{iv} + a''';\\
&3\alpha^{-1}c = 3b' + 2aa' + b,\ 3\beta^{-1}c' = 3b'' + 2a'a'' + b',\ 3\gamma^{-1}c'' = 3b''' + 2a''a''' + b'';\\
&4\alpha^{-1}d = 4c' + 3ab' + 2ba' + c,\ 4\beta^{-1}d' = 4c'' + 3a'b'' + 2b'a'' + c';\\
&5\alpha^{-1}e = 5d' + 4ac' + 3bb' + 2ca' + d;
\end{aligned}\right\} \tag{106}$$

together with the equation (105), or with

$$\alpha^{-1}a = 1, \quad \beta^{-1}a' = 1, \quad \gamma^{-1}a'' = 1, \quad \delta^{-1}a''' = 1, \quad \varepsilon^{-1}a^{iv} = 1. \tag{107}$$

Eliminating a', b', c', d', a'', b'', c'', a''', b''', a^{iv}, we obtain expressions analogous to those marked (70), namely the following:

$$\left.\begin{aligned}
a - \alpha = 0;\ 2(b - \alpha\beta) = \alpha^2;\ 2.3(c - \alpha\beta\gamma) &= \alpha(3\beta^2 + 6\alpha\beta + \alpha^2)\\
&= 3\alpha(\beta + \alpha)^2 - 2\alpha^3;\\
2.3.4(d - \alpha\beta\gamma\delta) &= 12\alpha\beta\gamma(\gamma + 2\beta + 2\alpha) + 4\alpha\beta(\beta^2 + 6\alpha\beta + 3\alpha^2) + \alpha^4\\
&= 12\alpha\beta(\gamma + \beta + \alpha)^2 + \alpha(\alpha^3 - 8\beta^3);\\
\text{and}\quad & \\
2.3.4.5(e - \alpha\beta\gamma\delta\varepsilon) &= 60\alpha\beta\gamma\delta^2 + 120\alpha\beta\gamma\delta(\gamma + \beta + \alpha)\\
&\quad + 20\alpha\beta\gamma^2(\gamma + 6\beta + 3\alpha)\\
&\quad + 60\alpha\beta\gamma(\beta^2 + 4\alpha\beta + \alpha^2)\\
&\quad + \alpha(5\beta^4 + 80\alpha\beta^3 + 90\alpha^2\beta^2 + 20\alpha^3\beta + \alpha^4).
\end{aligned}\right\} \tag{108}$$

XIII. Assuming, as a particular example, the values (45) or (64) for a, b, c, d, e, I found the following values, for the other quantities above mentioned:

$$\left.\begin{aligned}
&a' = \frac{5}{2}; \quad b' = \frac{37}{3}; \quad c' = \frac{353}{4}; \quad d' = \frac{4081}{5}\\
&a'' = \frac{221}{60}; \quad b'' = \frac{501}{20}; \quad c'' = \frac{1\,690\,091}{7200};\\
&a''' = \frac{131\,519}{26\,520}; \quad b''' = \frac{103\,112\,711}{2\,366\,800};\\
&a^{iv} = \frac{26\,082\,306\,193}{4\,185\,460\,656};
\end{aligned}\right\} \tag{109}$$

$$\left.\begin{aligned}
&\alpha = 1; \quad \beta = \frac{5}{2}; \quad \gamma = \frac{221}{60}; \quad \delta = \frac{131\,519}{26\,520};\\
&\varepsilon = \frac{26\,082\,306\,193}{4\,185\,460\,656};
\end{aligned}\right\} \tag{110}$$

where it is to be noted that *these* values of α, β, γ, δ, are exactly the *doubles* of the quantities denoted by the same symbols in my letter of February 10th, for a reason which will easily appear. There was no difficulty in hence deducing the formula, rule and type of that recent letter; except that the numerator of ε requires to be corrected as above since ε was found to be *positive* it seemed to be reasonable to expect that the inequalities (99), or these,

$$X > \phi\alpha x\phi\beta x\phi\gamma x, \qquad < \phi\alpha x\phi\beta x\phi\gamma x\phi\delta x, \tag{111}$$

would turn out to hold good, unless x were too large a number; while it remained doubtful whether

$$X \gtreqqless \phi\alpha x\phi\beta x\phi\gamma x\phi\delta x\phi\varepsilon x, \tag{112}$$

XIV. To convert these formulæ into numbers, we must consider how to calculate the ordinary logarithm, L, of the expression of the form $X = \phi\alpha x X'$, where ϕ has the signification (103), so that

$$\phi x = \epsilon^{-x}, \tag{113}$$

It is clear that this last formula of definition gives

$$L\phi x = -xL\epsilon = xL\frac{1}{\epsilon}; \tag{114}$$

so that if we agree, at least for the present, to write (as I did lately)

$$L' = -L, \qquad \text{or } L'x = L\frac{1}{x}, \tag{115}$$

we shall have generally

$$L'\phi x = xL\epsilon; \tag{116}$$

and therefore, in particular,

$$L'X = L'\phi\alpha x X' = \alpha x X' L\epsilon. \tag{117}$$

(I take the trouble *just now*, of writing ϵ to denote the natural base, because I have been *using both* ϵ and ε, to denote other things in this letter.) Taking again the ordinary logarithms of both numbers, and observing that all the factors in the expression (117) are positive, for all cases to which it is proposed to apply the method, we may go on to write,

$$LL'X = LL'\phi\alpha xX' = LX' + x_1 = x_1 - L'X', \tag{118}$$

if we make

$$x_1 = Lx + L\alpha + LL\epsilon = \alpha_1 + Lx, \tag{119}$$

where

$$\alpha_1 = L\alpha + LL\epsilon = L\alpha + \bar{1}{\cdot}6377843. \tag{120}$$

In like manner

$$LL'X' = LL'\phi\beta xX'' = x_2 - L'X'', \tag{121}$$

where

$$x_2 = \alpha_2 + Lx, \tag{122}$$

and

$$\alpha_2 = L\beta + L^2\varepsilon = L\beta + \bar{1}{\cdot}6377843. \tag{123}$$

And so we may go on. When a, b, c, d, e, and $\therefore$ $\alpha, \beta, \gamma, \delta, \varepsilon$, have the values recently assigned, we find easily that the values of $\alpha_1, \ldots, \alpha_5$ are:

$$\alpha_1 = \bar{1}{\cdot}6377843;\ \alpha_2 = 0{\cdot}0357243;\ \alpha_3 = 0{\cdot}2040254;\ \alpha_4 = 0{\cdot}3331993;\ \alpha_5 = 0{\cdot}4323870. \tag{124}$$

If we choose to write, with reference to the known definite integral, which was cited at the commencement of this Letter,

$$x = (q =)2^{-1}t^{-2}, \tag{125}$$

we shall then have

$$Lx = \bar{1}{\cdot}6989700 - 2Lt; \tag{126}$$

and may write

$$x_n = t_n = a_n - 2Lt, \tag{127}$$

where

$$a_n = \alpha_n + \bar{1}{\cdot}6989700; \tag{128}$$

the symbol t_n serving merely to connect this letter with a former one. For example, confining ourselves to five decimal places, which are quite enough, the values found above for $\alpha, \beta, \gamma, \delta, \varepsilon$ give the five following constants:

$$a_1 = \bar{1}{\cdot}33675; \quad a_2 = \bar{1}{\cdot}73469; \quad a_3 = \bar{1}{\cdot}90300; \quad a_4 = 0{\cdot}03217; \quad a_5 = 0{\cdot}13136; \tag{129}$$

and there, accordingly, were the logarithmic constants proposed in my memorandum of February 9th, to be added algebraically to the common term, $-2Lt$, except that the value then assigned to what is here called a_5 was somewhat different, because an error (of no great importance in its effects) had been committed in the calculation of the numerator of the fraction denoted in the present Letter by ε (110): such error having since been discovered by the *double system of equations* between the constants above described, and having been traced to the inadvertent substitution of $-625/16$, instead of $-625/64$, for $-a'^4/4$, as one of the terms of the expression for $a'c''$; whence c'', and thence b''' and a^{iv}, as derived from it, came to have not quite accurate values assigned to them. From the checks since employed, I think that we may now rely on the correctness of the fractions (109), and therefore on that of the set (110); which latter (except $\alpha = 1$) are merely selected from among the former. But to show how small was the practical effect of the detected mistake, and therefore of the introduced correction, I may remark that on recomputing the approximate values of the function X (or T),

$$X_3 = \phi\alpha x\phi\beta x\phi\gamma x, \qquad X_4 = \phi\alpha x\phi\beta x\phi\gamma x\phi\delta x, \qquad X_5 = \phi\alpha x\phi\beta x\phi\gamma x\phi\delta x\phi\varepsilon x, \tag{130}$$

I find now, for the case $t = 2$, or $x = \frac{1}{8}$,

$$LX_3 = \bar{1}{\cdot}95543; \qquad LX_4 = \bar{1}{\cdot}95746; \qquad LX_5 = \bar{1}{\cdot}95666; \tag{131}$$

giving

$$X_3 = 0{\cdot}90246; \qquad X_4 = 0{\cdot}90669; \qquad X_5 = 0{\cdot}90461. \tag{132}$$

The correction of the fraction ε in (110), or of the logarithmic constant a_5, has made *no change* in the *two first* of the three logarithms (131); and has not importantly altered the *third*; for the value adopted for what is here called X_5, in my note referred to, was 0·9048. If, on the plan of that recent note, we assume it to be likely that

$$X = \frac{X_4^2 - X_3 X_5}{2X_4 - X_3 - X_5}, \quad \text{or that } \frac{1}{X_4 - X} = \frac{1}{X_4 - X_3} + \frac{1}{X_4 - X_5}, \quad \text{nearly,} \tag{133}$$

the approximations (132), whereof only the last has been modified give, nearly,

$$X = X_4 - 0{\cdot}00139 = 0{\cdot}90530; \tag{134}$$

which coincides almost exactly with the theoretical values computed with the help of your reprint of Kramp's Table, namely,

$$4\epsilon^4 \int_2^\infty \epsilon^{-t^2}\, dt = 0{\cdot}90534. \tag{135}$$

It is not worth while to try whether 7 places would bring us any nearer. [An earlier value of] a_4 was 0·032 40, instead of 0·032 17, [was obtained] in consequence of my having taken out the logarithm of 13 159 instead of that of 131 5<u>19</u>. It is curious how this error (of dropping the 1 before the 9 of 131 519) has *haunted* me. It led me into at least half a dozen puzzles, but I trust that I have now (by means of checks) completely rectified every thing.

XV. Still, as the calculation will cost *me* but little trouble (the type being prepared), and as it need not cost *you* any at all, so that the worst effect will be the adding slightly to the length of this Letter, – and as even the *last* mentioned values of LX_5 and X_5 were rather hastily computed, – I shall here go on to perform, and to write down *in full*, the work for finding, with 7 decimal places of logarithms, the values of X_1, X_2, X_3, X_4, X_5, for the case $x = \frac{1}{8}$, on the plan of the preceding paragraph.

Type of the Calculation* (136)

Constants: $\alpha_1 = \bar{1}\cdot6377843$; $\alpha_2 = 0\cdot0357243$; $\alpha_3 = 0\cdot2040254$; $\alpha_4 = 0\cdot3331993$; $\alpha_5 = 0\cdot4323870$;

$x = 0\cdot125$; $Lx = \bar{1}\cdot0969100$ $= \bar{1}\cdot0969100$ $= \bar{1}\cdot0969100$ $= \bar{1}\cdot0969100$ $= \bar{1}\cdot0969100$.

$L0\cdot0542868 =$ $\underline{\underline{x_1 = \bar{2}\cdot7346943}}$ $\left(= L^2\dfrac{1}{X_1} = Lx + \alpha_1\right)$

$\underline{\underline{X_1 = 0\cdot882497;}}$ $LX_1 = \bar{1}\cdot9457132$.

$L0\cdot1357170 =$ $\underline{\underline{x_2 = \bar{1}\cdot1326343}}$ $\left(= Lx + \alpha_2 = L\left(L^2\dfrac{1}{X_1} - L^2\dfrac{1}{X_2}\right).\right)$

$L0\cdot0397171$ $= \bar{2}\cdot5989773$ $[= x_1 - L^{-1}x_2.]$

$\underline{\underline{X_2 = 0\cdot912605;}}$ $LX_2 = \bar{1}\cdot9602829$. $L0\cdot1999565 =$ $\underline{\underline{x_3 = \bar{1}\cdot3009354}}$

$L0\cdot0856402$ $= \bar{2}\cdot9326778$ $[= x_2 - L^{-1}x_3.]$

$L0\cdot0445712$ $= \bar{2}\cdot6493541\,[= x_1 - L^{-1}(x_2 - L^{-1}x_3).]$ $L0\cdot2692212 =$ $\underline{\underline{x_4 = \bar{1}\cdot4301093}}$

$\underline{\underline{X_3 = 0\cdot902462;}}$ $LX_3 = \bar{1}\cdot9554288$. $L0\cdot1075757$ $= \bar{1}\cdot0317142$ $[= x_3 - L^{-1}x_4.]$

$L0\cdot1059397$ $= \bar{1}\cdot0250586\,[= x_2 - L^{-1}(x_3 - L^{-1}x_4)]$. $L0\cdot3382963 =$ $\underline{\underline{x_5 = \bar{1}\cdot5292973}}$

$L0\cdot0425358$ $= \bar{2}\cdot6287546$ $L0\cdot1235415$ $= \bar{1}\cdot0918130$ Assuming

$\underline{\underline{X_4 = 0\cdot906701;}}$ $LX_4 = \bar{1}\cdot9574842$. $L0\cdot1504506$ $= \bar{1}\cdot1773939$ $\dfrac{X_5 - X_4}{X_4 - X_3} = \dfrac{X_6 - X_5}{X_5 - X_4} =$ &c,

$L0\cdot0959806$ $= \bar{2}\cdot9821837$ $L(X_4 - X_3) = L0\cdot004239$ $= \bar{3}\cdot62726$

$L0\cdot0435225$ $= \bar{2}\cdot6387137$ $(X_4 - X_5)^2 = L^{-1}\bar{6}\cdot62646$ $L(X_4 - X_5) = L0\cdot002057$ $= \underline{\bar{3}\cdot31323}$

$\bar{6}\cdot94049$

$\underline{\underline{X_5 = 0\cdot904644;}}$ $LX_5 = \bar{1}\cdot9564775$ $2X_4 - X_3 - X_5 = L^{-1}\underline{\bar{3}\cdot79906}$ $\underline{\bar{3}\cdot79906}$

in this example $(x = \frac{1}{8})$, $L0\cdot000672 = \bar{4}\cdot82740$ $L0\cdot001385 = \bar{3}\cdot14143$

$X_5 = 0\cdot904644$ $X_4 = 0\cdot906701$

true values are by equation (8), $\therefore X_\infty = 0\cdot905316$; $X_\infty = 0\cdot905316$, (137)

$X = 0\cdot905354$; $LX = \bar{1}\cdot9568185$ because $X_\infty = X_5 + \dfrac{(X_4 - X_5)^2}{2X_4 - X_3 - X_5} = X_4 - \dfrac{(X_4 - X_3)(X_4 - X_5)}{2X_4 - X_3 - X_5}$, (138)

because $X = 4\epsilon^4\displaystyle\int_2^\infty \epsilon^{-t^2}\,\mathrm{d}t$.

* [The following relations, (which follow from equations (118), (121), and (130)) have been used:

$$X_1 = L^{-1}(-L^{-1}x_1); \quad X_2 = L^{-1}[-L^{-1}(x_1 - L^{-1}x_2)];$$

$$X_3 = L^{-1}\{-L^{-1}[x_1 - L^{-1}(x_2 - L^{-1}x_3)]\};$$

$$X_4 = L^{-1}(-L^{-1}\{x_1 - L^{-1}[x_2 - L^{-1}(x_3 - L^{-1}x_4)]\});$$

$$X_5 = L^{-1}\{-L^{-1}(x_1 - L^{-1}\{x_2 - L^{-1}[x_3 - L^{-1}(x_4 - L^{-1}x_5)]\})\}.$$

These equations, together with a preliminary version of the calculation (136), are contained in a short letter from Hamilton to De Morgan, dated 10 February 1858 (Trinity College, Dublin, MS 1493/971).]

on the hypothesis of a geometric progression of the differences, $X_4 - X_3$, $X_5 - X_4$, $X_6 - X_5$, etc.; which indeed is a very precarious one, or rather is certainly incorrect in rigour of theory, but which has a degree of plausibility as a rule of approximation, and in several instances, as here, gives results pretty near the truth. The method of converging fractions gives, as the approximations answering to X_3, X_4, X_5, for the same case,

$$\left(q = x = \frac{1}{8}, \quad \text{or } t = 2,\right)$$

the following closer limits, $X > T_3$, $< T_4$, $> T_5$; where

$$T_3 = \frac{104}{115}, \quad T_4 = \frac{48}{53}; \quad T_5 = \frac{1672}{1847}; \text{ that is,}$$

$$T_3 = 0{\cdot}904348; \quad T_4 = 0{\cdot}905660; \quad T_5 = 0{\cdot}905252; \tag{139}$$

whence on the plan (137) it might be inferred that the value of X was nearly

$$= 0{\cdot}905349. \tag{140}$$

XVI. It must be admitted that, in this example, the exponential method, or "method of continued and converging exponentials", gives with more trouble a less rapidly converging series, or system of limits, X_1, X_2, X_3, X_4, X_5, for the function X or T, than the corresponding limits T_1, T_2, T_3, T_4, T_5, which are obtained by the known and usual *method of reciprocals*, or of "continued and converging fractions": although it happens that the aproximate values,

$$X = \frac{X_4^2 - X_3 X_5}{2X_4 - X_3 - X_5}, \quad T = \frac{T_4^2 - T_3 T_5}{2T_4 - T_3 - T_5}, \tag{141}$$

are here about equally near to the true value of X or of T. Again, by calculating with the same *exponential type* (136), but with only five places of decimals in the logarithms, I find for the case $x = \frac{1}{2}$, the values;

$$X_1 = 0{\cdot}6065; \quad X_2 = 0{\cdot}8665; \quad X_3 = 0{\cdot}6636; \quad X_4 = 0{\cdot}8426; \quad X_5 = 0{\cdot}6740; \tag{142}$$

and therefore, nearly, on the present plan,

$$X = \frac{X_4^2 - X_3 X_5}{2X_4 - X_3 - X_5} = X_5 + \frac{(X_4 - X_5)^2}{2X_4 - X_3 - X_5} = 0{\cdot}7560; \tag{143}$$

whereas the method of *converging fractions* gives for this case, when only as many terms of the diverging series are supposed to be known,

$$T_1 = 0{\cdot}66667; \quad T_2 = 0{\cdot}80000; \quad T_3 = 0{\cdot}73684; \quad T_4 = 0{\cdot}76923; \quad T_5 = 0{\cdot}75145;$$

with the resulting aproximation

$$T = T_5 + \frac{(T_4 - T_5)^2}{2T_4 - T_3 - T_5} = 0{\cdot}75775; \tag{144}$$

the true value being here (because $t = 1$), by (7),

$$X = T = 2\epsilon^2 \int_t^\infty \epsilon^{-t^2}\, dt = 0{\cdot}75787. \tag{145}$$

And here again the *method of converging fractions has an advantage, over the method of converging exponentials,* as I believe that it will *always* be found to have, when applied to the *particular diverging series* (43), and *generally, to any alternating series, which diverges sufficiently fast.* For

$$t = 3, \qquad x = q = \frac{1}{2t^2} = \frac{1}{18}, \tag{146}$$

the convergence is rapid enough, which ever method* of approximation we employ: but *this* may be considered to be a case *too easy* and *favourable,* and one which need not interfere with the belief above expressed: besides that, even in this case, the believed *advantage* partly shows itself. But to explain my *reason* for so believing, and what I imagine to be the *ground* of the *preference* (independently of its simplicity) which I admit that the *usual* method *here* deserves, I must enter into a few more details, respecting the laws of the connexion between the coefficients a, b, c, d, e of the series, and the constants α, β, γ, δ, ε which are deduced from those coefficients, by one or by another process, according as we adopt one of the two foregoing methods, or the other.

XVII. Suppose that we only knew *four terms* of our *series* (43), or had only

$$X = 1 - x + 3x^2 - 15x^3 + \cdots \tag{147}$$

and that we could not at all tell, as yet what might be the value of the following coefficient d, in the 5th term, $+dx^4$; except that we may (let it be supposed) *presume* that this coefficient d is not only >0, but >15. Then there are many (perhaps infinitely many) ways of setting about to transform this series for X into an expression of the form,

$$X = \phi\alpha x\phi\beta x\phi\gamma x\phi\delta x \cdots; \tag{148}$$

whereof *two* only (but those perhaps the most essential) have been hitherto considered in the present Letter. In the Ist, which is also the *usual* way, the *direct function,* or that with which we *operate on the given series,* and on others successively derived from it, has the form,

$$\varphi x = x^{-1} - 1; \tag{93}$$

so that the *inverse* function corresponding is,

$$\varphi^{-1} x = \phi x = (1 + x)^{-1} \cdots \tag{97}$$

On *this* plan, the supposed values, 1, 3, 15, of the 3 first coefficients a, b, c, give, as we have seen, the corresponding values, 1, 2, 3, for the 3 first constants; α, β, γ; and then the 4th of the equations (70) enables us to infer that the 4th coefficient d of the diverging series for X, and the 4th constant δ of the continued fraction which is a transformation of that series, are connected by the linear relation, analogous to (75),

* The type gives here, (if we use only 5 figure logarithms,) $X_3 = 0{\cdot}95159$; $X_4 = 0{\cdot}95188$; $X_5 = 0{\cdot}95179$; $X = 0{\cdot}95181$, nearly; fractions give $T_3 = 0{\cdot}951724$; $T_4 = 0{\cdot}951830$; $T_5 = 0{\cdot}951810$; $T = 0{\cdot}951813$, nearly the true value being by (9), $X = T = 6\epsilon^9 \int_3^\infty \epsilon^{-t^2}\,\mathrm{d}t = L^{-1}\bar{1}{\cdot}9785541 = 0{\cdot}951818_4$. I find that seven figures of logarithms gives

$$X_3 = 0{\cdot}951600; \qquad X_4 = 0{\cdot}951870; \qquad X_5 = 0{\cdot}951797;$$

(by the exponential method) and therefore, $X = 0{\cdot}951813$, nearly. (I find that $T_5 = 3654/3839$.)

$$d = b\delta + 81. \tag{149}$$

In order, then, that the constant δ of the fraction may be >0, it is necessary that the coefficient d of the series should be greater than the limit 81. If for instance, in any application of the method of continued fractions, we should meet with the following succession of initial terms,

$$X = 1 - x + 3x^2 - 15x^3 + 60x^4 - ex^5 + \cdots \tag{150}$$

we must transform the series, so far, into a continued fraction of the form,

$$X = \frac{1}{1+}\frac{x}{1+}\frac{2x}{1+}\frac{3x}{1+}\frac{-\frac{7}{2}x}{1+\varepsilon x X^v}; \tag{151}$$

X^v denoting here (as in previous paragraphs) a series of which the 1st term is $= 1$; and the coefficient ε being as yet unknown. And here we should *not* be entitled to establish the inequality,

$$X > \phi\alpha x\phi\beta x\phi\gamma x; \tag{111}$$

or, more particularly, we could *not* expect to be able to write,

$$X > T_3, \quad \text{if } T_3 = \frac{1+5x}{1+6x+3x^2}; \quad \text{compare (87);}$$

but should, on the contrary, have ground for believing the *opposite inequality* ($X < T_3$) to be *probable.* The *method of converging fractions would* ∴ *fail, for this case* (150), *and at this stage of the approximation*; *in the sense* that it would conduct to a quantity T_3, which was *not a limit of an opposite character, as compared with the next preceding limit, or aproximate value* T_2. In fact, if we develope in series the fractional expressions for these two successive approximations, T_2 and T_3 we find:

$$T_2 = \frac{1}{1+}\frac{x}{1+}\frac{2x}{1+0} = \frac{1+2x}{1+3x} = 1 - x + 3x^2 - 9x^3 + \cdots; \tag{152}$$

$$T_3 = \frac{1}{1+}\frac{x}{1+}\frac{2x}{1+3x} = \frac{1+5x}{1+6x+3x^2} = 1 - x + 3x^2 - 15x^3 + 81x^4 \cdots; \tag{153}$$

on comparing which with the *true* series,

$$T = 1 - x + 3x^2 - 15x^3 + 105x^4 \cdots \tag{3}$$

we see, indeed, that $T < T_2$, $T > T_3$, as before, because $15 > 9$, and $105 > 81$; *but,* on comparing these with the *assumed* (or *hypothetical*) *series* (150), for X, we find that while X is *still* $< T_2$, because we have still $15 > 9$, yet it must *now be presumed, so far as comparison of coefficients goes,* that X is *also* $< T_3$, because $60 < 81$. – What my *argument insists on,* is simply this: that if we had merely been given, or had found, the *three coefficients,* 1, 3, 15, of $-x^1$, $+x^2$, $-x^3$, and knew as yet *no proof of any law,* (even if such *law* were *suspected,*) which could *oblige the coefficient of* x^4 *to be as great as it really is,* ($105 = 7 \times 15$,) we *could not be sure, from the mere divergent character of the series, that this unknown coefficient of* x^4, *although probably greater than* 15, *would turn out to be greater than* 81; and therefore, that although we should be able to derive successively from the *three* given (or known) coefficients, 1, 3, 15, of the *diverging series,* the *three constants,* 1, 2, 3, of the *continued fraction;* and could thus *form the finite expressions* for the *three approximate values*, T_1, T_2, T_3, $\left(\text{where } T_1 = \frac{1}{1+x};\right)$ we should indeed be *entitled to presume* that (at least *if x be not to large,*) $X > T_1$, and $X < T_2$; but should *not be at all entitled to call it even probable* (as distinguished from being improbable), that it would turn out that $X > T_3$: the function X being *as yet merely known, by the few initial terms which are expressed,* in the formula (147).

XVIII. Let us now try the effect of the IInd. Method, mentioned in this Letter, by applying it to reduce the *same given beginning*, (147), of the diverging series for *X*, into an expression of the *same general functional form*, (148), but with a *different assumption* of the *direct function* φ, (and ∴ also of the inverse function ϕ,) and with *different values of the constants* α, β, γ. In other words, assuming now the forms,

$$\varphi x = l\frac{1}{x}, \ (101), \qquad \text{and } \phi x = \varepsilon^{-x}, \tag{113}$$

we are to determine the *three constants*, α, β, γ, and to assign the *relation*, analogous to (149), between the 4*th constant* δ, and the 4th coefficient d of the diverging series, by the condition that

$$\epsilon^{-\alpha x.}\epsilon^{-\beta x.}\epsilon^{-\gamma x.}\epsilon^{-\delta x.\&c.} = 1 - ax + bx^2 - cx^3 + dx^4 - \&c.; \tag{154}$$

when

$$a = 1, \quad b = 3, \quad c = 15,$$

but d is supposed to be unknown. In passing, I may just ask whether you think that this *notation*,

$$\epsilon^{-\alpha x.}\epsilon^{-\beta x.\&c.}$$

or more fully

$$\epsilon^{-\alpha x \times}\epsilon^{-\beta x \times \&c.}$$

might not with advantage replace what you justly call the "sprawling" form,

$$\epsilon^{-\alpha x \epsilon^{-\beta x \epsilon^{-\gamma x \epsilon^{-\delta x}}}} \tag{155}$$

which you so amusingly converted into a *ladder*, with your humble servant walking up its steps? – At all events, we have here, as in (110) the values,

$$\alpha = 1, \quad \beta = \frac{5}{2}, \quad \gamma = \frac{221}{60},$$

which may be deduced from *a, b, c,* by the three first equations (108), or by other processes already given or indicated: and thus the 4th of those equations (108) becomes,

$$120(24d - 221\delta) = 170881. \tag{156}$$

That is to say, if d and δ be connected by this linear relation (156), we shall have the transformation,

$$X = 1 - x + 3x^2 - 15x^3 + dx^4 - \&c. = \epsilon^{-x.}\epsilon^{-\frac{5x}{2}.}\epsilon^{-\frac{221x}{60}.}\epsilon^{-\delta x.\&c.}; \tag{157}$$

and the *constant* δ *will be negative, if the unknown coefficient be less than a certain limit,* which is nearly $= 59:33$; for

$$d = \frac{221\delta}{24} + \frac{170881}{2880}. \tag{158}$$

But, for the same reason, δ will be *positive,* if d *exceed* the *limit* thus assigned; that is, more fully

$$\delta > 0, \quad \text{if } 2880d > 170881. \tag{159}$$

For example, if

$$d = 60, \quad \text{then } \delta = +\frac{1919}{26520}; \tag{160}$$

so that

$$\epsilon^{-x} \cdot \epsilon^{-\frac{5x}{2}} \cdot \epsilon^{-\frac{221x}{60}} \cdot \epsilon^{-\frac{1919x}{26520}} \cdot \epsilon^{-\varepsilon x.\&c.} = 1 - x + 3x^2 - 15x^3 + 60x^4 - ex^{-5}\&c. \tag{161}$$

which equals the series (150), so far as that series has been written down: where e and ε are still unknown coefficients, or constants, and ϵ still denotes (though clumsily) the natural base of logarithms. Again, by (156) or (158), if $\delta = 0$, then

$$d = \frac{170881}{2880} = 59{\cdot}33368, \text{ nearly:} \tag{162}$$

therefore

$$X_3 = \epsilon^{-x} \cdot \epsilon^{-\frac{5x}{2}} \cdot \epsilon^{-\frac{221x}{60}} = 1 - x + 3x^2 - 15x^3 + \frac{170881}{2880}x^4 - ex^5 + \cdots, \tag{163}$$

where the coefficient of $+x^4$ is < 60. In fact, without necessarily depending on the investigations contained in earlier paragraphs of this Letter, we have

$$\begin{aligned} X_3 &= \epsilon^{-\alpha x} \cdot \epsilon^{-\beta x} \cdot \epsilon^{-\gamma x} = \epsilon^{-\alpha x} \cdot \epsilon^{-(\beta x - \beta\gamma x^2 + \&c.)} = \epsilon^{-Ax + Bx^2 - Cx^3 + Dx^4 \ldots} \\ &= 1 - ax + bx^2 - cx^3 + dx^4 - \&c., \end{aligned} \tag{164}$$

$$\text{when } A = \alpha, \quad B = \alpha\beta, \quad C = \alpha\beta\left(\gamma + \frac{\beta}{2}\right), \quad D = \alpha\beta\left(\frac{\gamma^2}{2} + \beta\gamma + \frac{\beta^2}{6}\right), \tag{165}$$

and

$$a = A, \quad b = B + \frac{1}{2}A^2, \quad c = C + AB + \frac{1}{6}A^3, \quad d = D + AC + \frac{1}{2}B^2 + \frac{1}{2}A^2B + \frac{1}{24}A^4 \tag{166}$$

whence the series (163), so far as it has been written, may be deduced anew; and also this other comencement of a diverging series,

$$X_2 = \epsilon^{-x} \cdot \epsilon^{-\frac{5x}{2}} = 1 - x + 3x^2 - cx^3 + dx^4 - \&c., \tag{167}$$

where

$$c = \frac{139}{24}, \quad \text{and therefore } c < 15. \tag{168}$$

And because

$$X_1 = \epsilon^{-x} = 1 - x + bx^2 - \cdots, \quad \text{where } b = \frac{1}{2} < 3, \tag{169}$$

we see, upon the whole, so far as coefficients enable us to judge, that if X still denote a diverging series, of which the initial terms are those assigned in the equation (150), and if X_1, X_2, X_3, X_4 retain their recent significations, then

$$X > X_1, \quad X < X_2, \quad X > X_3; \tag{170}$$

but whether $X >$ or $< X_4$, we are not yet prepared to say.

XIX. Suppose, however, in the next place, that we are *given the coefficient d of* $+x^4$ in X, and *know* that it is $= 105$; so that the series to be discussed is *known to begin* as follows,

$$X = 1 - x + 3x^2 - 15x^3 + 105x^4 - ex^5 + \&c.; \tag{171}$$

where the coefficient e has a value not yet assigned. We now (by this new supposition) *know more* than we did when we had *only* the terms written in (147); and ∴ may expect to be able to deduce *closer limits* of approximation, for the sought value of the function X. We can infer, for instance, that if we adopt the "Method of Continued Fractions", the constant which I have called δ has now a *positive value*, namely the value 4 but that if we adopt, instead, the "Method of Continued Exponents", then the corresponding coefficient, denoted above by the same symbol δ, has this *other* positive value

$$\delta = +\frac{131\,519}{26\,520},$$

assigned in (110).

If we now proceed to consider the *linear relations*, which exist in the two methods respectively, between the constants which (for the present) I continue to denote by e and ε, we find that they are the following. In the first method,

$$e = 24\varepsilon + 825; \tag{172}$$

and consequently, (while $\varepsilon = 5$, if $e = 945$, as in the *true* series for T,)

$$q < 0, \quad \text{if } e < 825. \tag{173}$$

In the Second Method, there is this other relation, more complex, arithmetically speaking, but quite analogous in point of theory, and the coefficients of which have, as I think, been examined with sufficient care:

$$31824(2880e - 131519\varepsilon) = 60\,529\,892\,207. \tag{174}$$

I do not recommend this process, as one which is remarkably exempt from labour; but it shows that (in this method)

$$\varepsilon > 0, \text{ if } e > \frac{60\,529\,892\,207}{91\,653\,120}; \tag{175}$$

or (taking the next greater integer), that

$$\varepsilon > 0, \text{ if } e \geqslant 661. \tag{176}$$

We see, then, here as in former comparisons of the two methods, that when positive constants

α, β, γ, ... have been assigned, or found, up to a certain point, it requires a *less divergence* in the series for X to secure a *positive character* for the *next* of those constants, (as here for what I am, for the moment, calling ε,) in the Second Method, than in the Ist. It is, therefore, *not quite so likely, à priori,* when only the coefficients, *up to some given point* are known, and have been *found* to give, *so far* in both methods, results (in the way of approximation) *alternately less and greater* than the truth, that this *condition of alternation* will *continue* to be satisfied, if the *First* Method be adopted, as it is if the *Second* Method be used. *But,* for precisely the *same reason, whenever* the Method of Continued Fractions *gives* such *alternating approximations,* it gives them *closer* to the *true* value sought, than does (I confess) the Method of Continued Exponentials. At least, it does so, in all cases which I have *tried*: and I see no reason for expecting that a different result will *ever* be experienced. – The *old* method is *here* the *good* one.

XX. When what I may call (for shortness) the "Exponential Method" occurred to me, not long ago, in connexion with some of those Diverging Series which arise out of the Researches, in part of a physical character, that were alluded to at the commencement of this Letter; and before I had thought of applying that Method to the function T, of equation (1), as a test; I tried it on the diverging but *geometric series;*

$$X = 1 - x + x^2 - x^3 + x^4 - x^5 + \&c., \tag{177}$$

where x was supposed to be > 1. In this application, if we use the formulæ of paragraph XII., we have the given values,

$$a = b = c = d = e = 1; \tag{178}$$

and easily deduce from them the constants,

$$\alpha = 1, \quad \beta = \frac{1}{2}, \quad \gamma = \frac{5}{12}, \quad \delta = \frac{47}{120}, \quad \varepsilon = \frac{12917}{33840}; \tag{179}$$

with the approximate expression corresponding:

$$X = (1+x)^{-1} = \epsilon^{-x.}\epsilon^{-\beta x.}\epsilon^{-\gamma x.}\epsilon^{-\delta x.}\epsilon^{-\theta\varepsilon x}; \quad \theta > 0, < 1. \tag{180}$$

When x was $> 0, < 1$, I found that this gave a set of fairly rapid successive approximations; but *even when x was* $\geqslant 1$, the *alternate character* was *still preserved;* and, on the whole, the *degree of approach* obtained, to the true value, was not to be altogether despised. By substituting the values (179) for α, β, γ, δ, ε, the constant logarithms, analogous to those marked (124), are found to be the following:

$$\left.\begin{array}{llll} \alpha_1 = \bar{1}{\cdot}6377843; & \alpha_2 = \bar{1}{\cdot}3367543; & \alpha_3 = \bar{1}{\cdot}2575721; & \alpha_4 = \bar{1}{\cdot}2307010; \\ \alpha_5 = \bar{1}{\cdot}2195156; & & & \end{array}\right\} \tag{181}$$

and we have merely to write these numbers in the first line of the Type (136), of paragraph XV., and then proceed as in that Type. For the case $x = \frac{9}{10}$, it is thus found that

$$X_3 = 0{\cdot}5165530; \quad X_4 = 0{\cdot}5289111; \quad X_5 = 0{\cdot}5356387; \quad X = 0{\cdot}5263238 \text{ nearly}; \tag{182}$$

the true value being here, by a converging geometric series,

$$X = \frac{10}{19} = 0{\cdot}5263258 = 1 - \frac{9}{10} + \left(\frac{9}{10}\right)^2 - \text{\&c.} \tag{183}$$

For $x = 1$, or for the *neutral* series,

$$X = 1 - 1^1 + 1^2 - 1^3 + 1^4 - 1^5 + \cdots, \tag{184}$$

the type gives

$$X_3 = 0{\cdot}4871434; \quad X_4 = 0{\cdot}5037234; \quad X_5 = 0{\cdot}4989456; \quad X = 0{\cdot}5000144 \text{ nearly}, \tag{185}$$

the *theoretical* value being here $X = \frac{1}{2}$, and the *approximate* X of (185), like that of (182), being deduced from X_3, X_4, X_5, by the formula (138). Even for $x = 2$, or for the *diverging* series,

$$X = 1 - 2^1 + 2^2 - 2^3 + 2^4 - 2^5 + \text{\&c.}, \tag{186}$$

the same type (136) gives with the help of the same formula (138),

$$X_3 = 0{\cdot}27388, \quad X_4 = 0{\cdot}36428, \quad X_5 = 0{\cdot}31933; \quad X = 0{\cdot}33426, \text{ nearly}; \tag{187}$$

the usual theoretical value being in this case,

$$X = 0{\cdot}33333. \tag{188}$$

And I *believe*, without pretending to have yet *proved* it, that *however large* a (positive) value may be assigned to x, we should *in all cases* find the *inequalities* subsisting,

$$X = (1 + x)^{-1} > X_1, \quad < X_2, \quad > X_3, \quad < X_4, \quad > X_5, \quad \text{\&c.} \tag{189}$$

XXI. In other words, although I have taken pains in this Letter to prove that a certain degree of rapidity of divergence is *requisite*, if the coefficients of the series *begin* by diverging; and that, at all events, when the coefficients *up to a certain point* are given, the next *following* coefficient must be at least equal to a certain *minor limit*, in order that the constants of the Exponential Development of the function represented by the series may all continue to be *positive* yet I *believe* that all such conditions will be found to be *satisfied*, however far the examination may be pursued, for the geometric series, or function,

$$X = (1 + x)^{-1} = 1 - x + x^2 - x^3 + x^4 - x^5 + \text{\&c.}; \tag{177}$$

and also for several other important functions, as for instance, the one to which most of the foregoing investigations relate; namely the function (1), represented by the series,

$$T = 1 - q + 3q^2 - 15q^3 + 105q^4 - 945q^5 + \text{\&c.} \tag{3}$$

For various particular values of the variable x, (or q,) and with the omission of powers of that variable which exceeded the fifth, this fact of the condition being satisfied has been proved by the success, such as it has been, of what may be called the "Exponential Method of Approximation" to the Value of the Series: *including*, as what is, for our present purpose, an essential *element* of such success, the *alternating character* of the approximate values which the Method gives. But because, when we take for the variable a *value too large* to render the *method useful*, as giving (within any *reasonable* limits of labour), an *arithmetical approach* to the truth, it has appeared to me worth while to examine, in another way, and by a *different type* of calculation, for values of x which thus render (not only the original series itself divergent, but also) the exponential development *too slowly convergent*, whether at least certain *conditions of inequality*, of a kind already alluded to, continue *still* to be satisfied: at least when we content ourselves with considering the *five first exponential constants*, for each of the two functions above mentioned, because those constants alone have been as yet determined.

XXII. It is, then, to be shown, by an examination of what may seem to be a sufficient range of instances, that if we assign to the real variable x any value > 0, the *function*

$$X = (1 + x)^{-1}, \tag{178}$$

to take it first, as being the most elementary in its form, – satisfies the five following *inequalities*: (compare equation (94):)

$$\frac{1}{X'} = \frac{\alpha x}{\varphi X} > 1; \quad \frac{1}{X''} = \frac{\beta x}{\varphi X'} > 1; \quad \cdots \frac{1}{X^{v}} = \frac{\varepsilon x}{\varphi X^{iv}} > 1; \tag{190}$$

where the function φx denotes the (natural) *logarithm of the reciprocal* of x, so that

$$\varphi x = l\frac{1}{x}; \tag{101}$$

while $\alpha, \beta, \ldots, \varepsilon$ are the five constants and positive fractions,

$$\alpha = 1, \quad \beta = \frac{1}{2}, \quad \gamma = \frac{5}{12}, \quad \delta = \frac{47}{120}, \quad \varepsilon = \frac{12917}{33840}, \tag{179}$$

which depend partly on the form of the function φx, and partly on that of X. Or because (101) gives

$$\frac{1}{\varphi x} = \frac{L\epsilon}{L\frac{1}{x}}, \quad L\frac{1}{\varphi x} = L^2\epsilon - L^2\frac{1}{x}, \tag{191}$$

we are to show that these other five inequalities subsist:

$$\left.\begin{array}{l} L\frac{1}{X'} = \alpha_1 + Lx - L^2\frac{1}{X} > 0; \quad L\frac{1}{X''} = \alpha_2 + Lx - L^2\frac{1}{X'} > 0; \\ \qquad \cdots; L\frac{1}{X^{v}} = \alpha_5 + Lx - L^2\frac{1}{X^{iv}} > 0; \end{array}\right\} \tag{192}$$

where

$$\alpha_1 = L\alpha + L^2\epsilon, \ (120), \quad \alpha_2 = L\beta + L^2\epsilon, \ \&c.; \tag{123}$$

so that these five logarithmic constants α_1, α_2 &c. have here the values (181). Writing, therefore, on the plan of the equations (119), (122), for the present question,

$$\left.\begin{array}{l} x_1 = Lx + \bar{1}{\cdot}6377843; \quad x_2 = Lx + \bar{1}{\cdot}3367543; \quad x_3 = Lx + \bar{1}{\cdot}2575721; \\ \qquad x_4 = Lx + \bar{1}{\cdot}2307010; \quad x_5 = Lx + \bar{1}{\cdot}2195156; \end{array}\right\} \tag{193}$$

we are to show, for variously selected (but positive) values of x, that

$$\begin{array}{l} L\frac{1}{X'} = x_1 - L^2(1 + x) > 0; \quad L\frac{1}{X''} = x_2 - L^2\frac{1}{X'} > 0; \\ \qquad \cdots; L\frac{1}{X^{v}} = x_5 - L^2\frac{1}{X^{iv}} > 0. \end{array} \tag{194}$$

These formulæ will enable us at the same time, in each particular example, to *assign the fractions,* X', X'', ..., X^{v}, (analogous to θ in well known formulae,) which enter as final or *correcting factors* into those *finite exponential expressions* whereto the Method conducts. But if we do not care for more than a mere verification of the *Theorem,* that each such factor is >0, <1, we may eliminate X', ..., X^{v}, and write simply, with the signification (193) of $x_1, \ldots, x_5$, and as the inequalities to be numerically verified, the following:

$$\left.\begin{array}{l} x_1 > L^2(1+x); \quad x_2 > L(x_1 - L^2(1+x)); \quad x_3 > L(x_2 - L(x_1 - L^2(1+x))); \\ x_4 > L(x_3 - L(x_2 - L(x_1 - L^2(1+x)))); \\ x_5 > L(x_4 - L(x_3 - L(x_2 - L(x_1 - L^2(L+x))))). \end{array}\right\} \qquad (195)$$

I proceed to give the calculation *in full,* for such verification of the *Theorem,* in the case where $x = 100$; in which case the Method, regarded as one of *approximation,* would be practically *useless,* on account of the slow convergence of its results.

XXIII. I find it convenient here to employ the following *Type*:

$$\left.\begin{array}{lll}
x = 100;\ L(1+x) = L101 = & & \left\{\begin{array}{l} L\dfrac{1}{X} = \underline{2{\cdot}004\,321\,4} \\ \qquad = L^{-1}0{\cdot}301\,967\,3; \end{array}\right. \\
X = 0{\cdot}009\,901\,0 = L^{-1}\bar{3}{\cdot}995\,678\,6; & & \\
Lx = 2{\cdot}000\,000\,0; & \alpha_1 = \bar{1}{\cdot}637\,784\,3; & x_1 = \underline{1{\cdot}637\,784\,3;} \\
X' = 0{\cdot}046\,151\,2 = L^{-1}\bar{2}{\cdot}664\,183\,0; & & \left\{\begin{array}{l} L\dfrac{1}{X'} = \underline{1{\cdot}335\,817\,0} \\ \qquad = L^{-1}0{\cdot}125\,747\,0; \end{array}\right. \\
Lx = 2; & \alpha_2 = \bar{1}{\cdot}336\,754\,3; & x_2 = \underline{1{\cdot}336\,754\,3;} \\
X'' = 0{\cdot}061\,516\,7 = L^{-1}\bar{2}{\cdot}788\,992\,7; & & \left\{\begin{array}{l} L\dfrac{1}{X'} = \underline{1{\cdot}211\,007\,3} \\ \qquad = L^{-1}0{\cdot}083\,146\,7; \end{array}\right. \\
Lx = 2; & \alpha_3 = \bar{1}{\cdot}257\,572\,1; & x_3 = 1{\cdot}257\,572\,1; \\
X''' = 0{\cdot}066\,922\,9 = L^{-1}\bar{2}{\cdot}825\,574\,6; & & \left\{\begin{array}{l} L\dfrac{1}{X'''} = \underline{1{\cdot}174\,425\,74} \\ \qquad = L^{-1}0{\cdot}069\,825\,4; \end{array}\right. \\
Lx = 2; & \alpha_4 = \bar{1}{\cdot}230\,701\,0; & x_4 = 1{\cdot}230\,701\,0; \\
X^{iv} = 0{\cdot}069\,043\,8 = L^{-1}\bar{2}{\cdot}839\,124\,4; & & \left\{\begin{array}{l} L\dfrac{1}{X^{iv}} = 1{\cdot}160\,875\,6 \\ \qquad = L^{-1}0{\cdot}064\,785\,7; \end{array}\right. \\
Lx = 2; & \alpha_5 = \bar{1}{\cdot}219\,515\,6; & x_5 = 1{\cdot}219\,515\,6; \\
X^{v} = 0{\cdot}070\,043\,9 = L^{-1}\bar{2}{\cdot}845\,370\,1; & & L\dfrac{1}{X^{v}} = 1{\cdot}154\,729\,9
\end{array}\right\} \qquad (196)$$

The inequalities,

$$X' > 0, < 1; \quad X'' > 0, < 1; \quad X''' > 0, < 1; \quad X^{iv} > 0, < 1;$$
$$X^{v} > 0, < 1, \tag{197}$$

are therefore, in this example, satisfied.

XXIV. The same type, (196), gives me for $x = 10$, $X = \frac{1}{11}$, the values

$$\left.\begin{array}{lll} X' = 0{\cdot}239\,789\,5; & X'' = 0{\cdot}285\,598\,7; & X''' = 0{\cdot}300\,760\,9; \\ & X^{iv} = 0{\cdot}306\,750\,5; & X^{v} = 0{\cdot}309\,587\,6; \end{array}\right\} \tag{198}$$

which again satisfies the inequalities (197). We may observe that these values of which we may call the *"fractional factors"*, (again analogous to the usual θ,) X', &c., are here also *increasing* among themselves, from X' to X^{v}, like those for $x = 100$, in (196); but that they are also *greater* than *those* corresponding factors. And generally it appears to me likely that *each* of the 5 *factors,* $X', \ldots, X^{v}$, *increases from* 0 *to* 1 while x *decreases from* ∞ *to* 0. This seems to be connected with the possibility of the *development* of *each* factor, X', &c., in the form of a *series,* of which the *first term* is $= 1$, while the next term of each appears to have a *negative coefficient.* At least, I have *proved* theoretically, or *algebraically,* that this *negative character* belongs to the coefficient of x^1 in each of the *four series,* for X', X'', X''', X^{iv}; and the comparison of the *arithmetical values* of X^{v} seems to establish the *same* result, for the coefficient of the series for X^{v} *also.*

XXIV. The same type (196) gives for $x = 2$, $X = 1/3$, these other values:

$$\left.\begin{array}{lll} X' = 0{\cdot}549\,306\,2; & X'' = 0{\cdot}599\,099\,3; & X''' = 0{\cdot}614\,795\,0; \\ & X^{iv} = 0{\cdot}621\,202\,8; & X^{v} = 0{\cdot}624\,024\,9; \end{array}\right\} \tag{199}$$

which satisfy still the same inequalities as before, and follow the same general laws of progress. And similar remarks apply to the case

$$\left.\begin{array}{lll} x = 1,\ X = \dfrac{1}{2}, \text{ for which } & X' = 0{\cdot}693\,147\,3; & X'' = 0{\cdot}733\,025\,7; \\ X''' = 0{\cdot}745\,380\,7; & X^{iv} = 0{\cdot}750\,281\,3; & X^{v} = 0{\cdot}752\,687\,7; \end{array}\right\} \tag{200}$$

and to the case where $x = 9/10$, $X = 10/19$, with the fractional factors,

$$\left.\begin{array}{lll} X' = 0{\cdot}713\,181\,0; & X'' = 0{\cdot}751\,186\,7; & X''' = 0{\cdot}762\,937\,7; \\ & X^{iv} = 0{\cdot}767\,599\,3; & X^{v} = 0{\cdot}769\,894\,4. \end{array}\right\} \tag{201}$$

Indeed, the mere fact of the requisite *inequalities* ($X^{(n)} > 0, < 1$) being satisfied for *these three cases,*

$$\left(x = 2, \quad x = 1, \quad x = \frac{9}{10},\right)$$

might have been inferred from the results of the calculations of *approximation,* which had been previously performed with a *different type* (136), and of which some account was given in paragraph XX. of this Letter.

XXV. When *smaller positive values* are assigned to *x*, the *Method* naturally *succeeds* still *better*, as giving still *more rapid approximation*; but it does not follow that the *Theorem* can always be *more easily;* or (in extreme cases) *at all verified*, with our usual Tables of Logarithms: a circumstance which deserves to be remarked, because the neglect to notice it might lead to false conclusions. On trying, for instance, with the type (196), and with just about the same degree of care, but still with logarithms of *seven figures only*, (they happened to be Hutton's, edited by O. Gregory, 1838,*) what were the factors X' &c. for the case

$$x = \frac{1}{10}, \quad X = \frac{10}{11}, \tag{202}$$

I *seemed* to find the values,

$$\begin{aligned} L\frac{1}{X'} = 0{\cdot}020\,860\,5; \quad & L\frac{1}{X''} = 0{\cdot}017\,429\,6; \quad L\frac{1}{X'''} = 0{\cdot}016\,284\,7 \\ & L\frac{1}{X^{iv}} = 0{\cdot}018\,921\,2; \quad L\frac{1}{X^{v}} = \bar{1}{\cdot}942\,566\,9; \end{aligned} \tag{203}$$

and it was already *suspicious*, that $L\frac{1}{X^{iv}}$ appeared to be *greater* than $L\frac{1}{X'''}$; but *monstrous*, and *intolerable*, that $L\frac{1}{X^{v}}$ should be actually *negative*! But as no *blunder* was detected, on revisal, I proceeded to calculate X_5, by the *other type*, (136), as the best *single approximation* which our 5 *constants*, α_1 to α_5, (the only ones as yet determined,) would here give: though it might perhaps be improved, on the plan (138), by combining it with X_3 and X_4. The resulting number was,

$$X_5 = L^{-1}\bar{1}{\cdot}958\,607\,3 = 0{\cdot}909\,090\,9 = \frac{10}{11}; \tag{204}$$

which is here (to the 7† places used) the *exact theoretical value of* X!! And now I saw, on a little consideration, that the *least little error* of the 7 figure tables – such as that committed by taking $\log 11 = 1{\cdot}041\,392\,7$, instead of $1{\cdot}041\,392\,685\ldots$, – came to be *magnified* with a frightful rapidity, in applying the type (196) to the case $x = \frac{1}{10}$, so as quite to interfere with our power of *verifying the Theorem*, although it was so easy and satisfactory to *apply the Method* here.

XXVI. Some reasonings, partly mental, which it is not worth while writing to develop here, lead me to think that, instead of being < 0, $L\frac{1}{X^{v}}$ is nearly $= +0.015$, but probably a little greater, for this last case of $x = \frac{1}{10}$; and accordingly, if we *asume* that

$$L\frac{1}{X^{v}} = 0{\cdot}015\,000\,0, \tag{205}$$

* [See note on p. 166.]

† It is to be noted, generally, that I do not pretend to be *sure* of the *last* decimal figure set down: nor, always, of the one before it.

we find, by (193), (194), combined with the value

$$Lx = -1 = \overline{1}\cdot 000\,000\,0, \tag{206}$$

$$\left.\begin{aligned} L^2\frac{1}{X^{iv}} &= x_5 - L\frac{1}{X^v} = \overline{2}\cdot 219\,515\,6 - 0\cdot 015\,000\,0 = \overline{2}\cdot 204\,515\,6 = L0\cdot 016\,014\,6;\\ L^2\frac{1}{X'''} &= x_4 - L\frac{1}{X^{iv}} = \overline{2}\cdot 230\,701\,0 - 0\cdot 016\,014\,6 = \overline{2}\cdot 214\,686\,4 = L0\cdot 016\,394\,1;\\ L^2\frac{1}{X''} &= x_3 - L\frac{1}{X'''} = \overline{2}\cdot 257\,572\,1 - 0\cdot 016\,394\,1 = \overline{2}\cdot 241\,178\,0 = L0\cdot 017\,425\,2;\\ L^2\frac{1}{X'} &= x_2 - L\frac{1}{X''} = \overline{2}\cdot 336\,754\,3 - 0\cdot 017\,425\,2 = \overline{2}\cdot 319\,329\,1 = L0\cdot 020\,860\,7;\\ L^2\frac{1}{X} &= x_1 - L\frac{1}{X'} = \overline{2}\cdot 637\,784\,3 - 0\cdot 020\,860\,7 = \overline{2}\cdot 616\,923\,6 = L0\cdot 041\,392\,7; \end{aligned}\right\} \tag{207}$$

whence

$$L\frac{1}{X} = 0\cdot 041\,392\,7 = L\frac{11}{10}; \tag{208}$$

giving again

$$X = \frac{10}{11} = (1+x)^{-1},$$

as we know that X ought to be.

XXVII. The assumption $X = X_5$, which had been found to be sensibly correct, corresponds, in the theory of this Letter, to the values

$$X^v = 1, \quad L\frac{1}{X^v} = 0; \tag{209}$$

we see, then, that if our object be merely to calculate, approximately, the *value of the function* X, (supposed to be known *only* by its development in *series*, and by the *six first terms* thereof,) it is not of much importance whether we attribute to $L\frac{1}{X^v}$ the *positive* value (205), or the *null* value (209), or the *negative* value (203). The *Method* is therefore *here* a *good* one; but the verification of the *Theorem* is *precarious;* since it would be difficult, or perhaps impossible, from the foregoing calculations *alone*, to pronounce whether the factor X^v is (in this case of $x = \frac{1}{10}$) < 1, or > 1. But, as above hinted, I have reasons for thinking that the assumptions and results of paragraph XXVI. are in this case *nearly* correct, or at least are not *very far* from the truth; and thus that, without fear of any remarkable error, at least as affecting the theory, we may write here, instead of (203), the following equations, in which the value of $L\frac{1}{X^v}$ has been a little increased from that assumed above, and may still require some slight correction, but the others are more to be relied on:

$$\left.\begin{array}{lll} L\dfrac{1}{X'} = 0{\cdot}020\,86; & L\dfrac{1}{X''} = 0{\cdot}017\,43; & L\dfrac{1}{X'''} = 0{\cdot}016\,39; \\ & L\dfrac{1}{X^{iv}} = 0{\cdot}016\,00; & L\dfrac{1}{X^{v}} = 0{\cdot}015\,50; \end{array}\right\} \tag{210}$$

with these corresponding values of the factors, $X', \ldots, X^{v}$:

$$\left.\begin{array}{lll} X' = 0{\cdot}953\,10; & X'' = 0{\cdot}960\,67; & X''' = 0{\cdot}962\,95; \\ & X^{iv} = 0{\cdot}963\,84; & X^{v} = 0{\cdot}964\,94. \end{array}\right\} \tag{211}$$

One confirmation of the four first values (210) is, that these four logarithms are (as they ought to be, on account of the smallness of x) nearly proportional to the four last constants, $\beta, \gamma, \delta, \varepsilon$, of (179).

XXVIII. Before quite leaving the subject, of the development by a Continued Exponential, of the function $X = (1 + x)^{-1}$, I shall mention another instance, in which not only does the "Method of Successive Approximation" succeed with sufficient accuracy, but also the "Theorem of Alternate Inequalities" can be verified by the usual Tables: namely the case when $x = \frac{1}{2}$. I find, in this case, by using the type (136), the converging approximate values,

$$\left.\begin{array}{lll} X_1 = 0{\cdot}606\,530\,7; & X_2 = 0{\cdot}677\,463\,0; & X_3 = 0{\cdot}664\,882\,0; \\ & X_4 = 0{\cdot}666\,956\,6; & X_5 = 0{\cdot}666\,620\,3; \end{array}\right\} \tag{212}$$

whence, by the formula (138), we derive this improved approximation,

$$X = X_5 + \frac{(X_4 - X_5)^2}{2X_4 - X_3 - X_5} = 0{\cdot}666\,666\,4, \quad \text{nearly;} \tag{213}$$

the true value being here

$$X = \frac{2}{3} = 0{\cdot}666\,666\,7.$$

And if setting out with this last value for X, we compute, by the type (196), the factors X', $\ldots, X^{v}$ we find that they satisfy the required inequalities; their values being, in this example,

$$\left.\begin{array}{lll} X' = 0{\cdot}810\,93; & X'' = 0{\cdot}838\,29; & X''' = 0{\cdot}846\,67; \\ & X^{iv} = 0{\cdot}849\,94; & X^{v} = 0{\cdot}851\,92; \end{array}\right\} \tag{214}$$

and their progression following the same general laws as before.

XXIX. It will be remembered that these *factors*, X', &c., are certain *functions* of x, which can be developed in certain ascending *series*, namely, in the notation of this Letter, (compare equation (51),)

$$\left.\begin{array}{l} X' = 1 - a'x + b'x^2 - c'x^3 + d'x^4 - \cdots; \\ X'' = 1 - a''x + b''x^2 - c''x^3 + \cdots; \\ X''' = 1 - a'''x + b'''x^2 - \cdots; \\ X^{iv} = 1 - a^{iv}x + \cdots; \\ X^{v} = 1 - \cdots; \text{ \&c.} \end{array}\right\} \tag{215}$$

For

$$X = (1+x)^{-1} = \phi\alpha x\phi\beta x\phi\gamma x \cdots, \quad \text{with} \quad \phi x = \epsilon^{-x},$$

I find that the *coefficients* of these series, so far as they are written in (215), have the values:

$$\left.\begin{array}{llll} a' = \frac{1}{2}, & b' = \frac{1}{3}, & c' = \frac{1}{4}, & d' = \frac{1}{5}; \\ a'' = \frac{5}{12}, & b'' = \frac{1}{4}, & c'' = \frac{251}{1440}; & \\ a''' = \frac{47}{120}, & b''' = \frac{2443}{10\,800}; & a^{iv} = \frac{12\,917}{33\,840}; & \end{array}\right\} \tag{216}$$

where a', a'', a''', a^{iv} are the same fractions as those denoted by β, γ, δ, ε in (179). I have not determined the fractional expression for the coefficient a^{v}, or ζ, which belongs to the same forms of the two functions, X and ϕx; but a comparison of the computed values of X^{v} seems to make it certain that this coefficient is *positive,* like the others; and even that its *value* cannot differ *much* from that of the fraction $\frac{1}{3}$, although (no doubt) its numerator and denominator are numbers which may be spoken of as large. I judge, also, from the same sort of comparison of numerical values, that if we write, according to the same analogy, but more fully than in (215),

$$X^{v} = 1 - a^{v}a + b^{v}x^2 - \cdots, \tag{217}$$

the coefficient b^{v}, as well as a^{v}, will be found to be *positive;* and I *think* that I see, *nearly,* what its *value* will turn out to be. But it would be a waste of time to delay, for the purpose of verifying *such* estimations: and I pass to some examples of the calculated values of the factors, or series, X', &c., for that *other form* of the function X, which I began by considering, in this Letter.

XXX. After the full details which have been given, respecting the *plan* of calculation adopted, it may be quite enough to mention a few of the numerical *results* obtained. I find, then, that if we write

$$X = \left(\frac{2}{x}\right)^{\frac{1}{2}} \epsilon^{\frac{1}{2x}} \int_{(2x)^{-\frac{1}{2}}}^{\infty} \epsilon^{-t^2}\, dt, \tag{218}$$

so that

$$X = 1 - 1x + 1.3x^2 - 1.3.5x^3 + 1.3.5.7x^4 - 1.3.3.7.9x^5 + \text{\&c.}, \tag{219}$$

as in earlier paragraphs, the *law* being now supposed to be *known*; (on which account there is the less harm in my indulging the last series with a new *number of reference,* although it had occurred before: – at worst, the doing so can produce no confusion, nor embarrassment, although I admit that a greater *economy* of such numbers would have had a *better appearance* throughout:) and if we write, on the plan of paragraph XIV., instead of (193), the equations,

$$\left.\begin{array}{lll} x_1 = Lx + \bar{1}{\cdot}637\,784\,3; & x_2 = Lx + 0{\cdot}035\,724\,3; & x_3 = Lx + 0{\cdot}204\,025\,4; \\ & x_4 = Lx + 0{\cdot}333\,199\,3; & x_5 = Lx + 0{\cdot}432\,387\,0, \end{array}\right\} \tag{220}$$

where the constants added to Lx are those denoted by $\alpha_1, \ldots, \alpha_5$ in (124); and employed in the type (136); if also, subject to this selection of values of those constants, we determine successively $X', \ldots, X^v$ by the formulæ (192), remembering that $\frac{1}{x}$ is not now equal to $(1 + x)$, but to the reciprocal of the expression (218), so that we have still the relations,

$$\left.\begin{array}{lll} L\dfrac{1}{X'} = x_1 - L^2\dfrac{1}{X}; & L\dfrac{1}{X''} = x_2 - L^2\dfrac{1}{X'}; & L\dfrac{1}{X'''} = x_3 - L^2\dfrac{1}{X''}; \\ & L\dfrac{1}{X^{iv}} = x_4 - L^2\dfrac{1}{X'''}; & L\dfrac{1}{X^v} = x_5 - L^2\dfrac{1}{X^{iv}}, \end{array}\right\} \tag{221}$$

with the values (220) for $x_1, \ldots, x_5$; then the *five inequalities,*

$$L\frac{1}{X'} > 0, \quad L\frac{1}{X''} > 0, \quad L\frac{1}{X'''} > 0, \quad L\frac{1}{X^{iv}} > 0, \quad L\frac{1}{X^v} > 0, \tag{222}$$

continue to subsist, at least in all the cases which I have tried.

XXXI. Thus, for the case where

$$x = \frac{1}{18}, \qquad X = 6\epsilon^9 \int_3^\infty \epsilon^{-t^2}\, dt = L^{-1}\bar{1}{\cdot}978\,554\,1$$
$$= 0{\cdot}951\,818\,4, \tag{223}$$

I find the five following positive values:

$$\left.\begin{array}{lll} L\dfrac{1}{X'} = 0{\cdot}051\,167\,5; & L\dfrac{1}{X''} = 0{\cdot}071\,457\,6; & L\dfrac{1}{X'''} = 0{\cdot}094\,704\,4; \\ & L\dfrac{1}{X^{iv}} = 0{\cdot}101\,556\,7; & L\dfrac{1}{X^v} = 0{\cdot}170\,405\,9; \end{array}\right\} \tag{224}$$

giving the five following factors, $X', \ldots, X^v$, to seven decimal places, such as they were taken out:*

$$\left.\begin{array}{lll} X' = 0{\cdot}888\,858\,2; & X'' = 0{\cdot}848\,286\,2; & X''' = 0{\cdot}804\,073\,2; \\ X^{iv} = 0{\cdot}791\,486\,2; & X^v = 0{\cdot}675\,451\,5 & \end{array}\right\} \tag{225}$$

whereof each is seen to be > 0, but less than 1. Again for the case where $x = \frac{1}{8}$,

* I think that I have declared that I do not answer for all the decimal figures set down, although it does not seem worth while to abridge their extent, by way of precaution. I give them just as they came, and shan't be surprised if corrections shall be found; or rather should be surprised at the contrary: though I think that the *theory* is safe.

$$X = 4\epsilon^4 \int_2^\infty \epsilon^{-t^2}\, dt = L^{-1}\bar{1}{\cdot}956\,818\,5 = 0{\cdot}905\,354\,2, \tag{226}$$

I find these other values,

$$\left.\begin{array}{lll} L\dfrac{1}{X'} = 0{\cdot}099\,396\,5; & L\dfrac{1}{X''} = 0{\cdot}135\,263\,2; & L\dfrac{1}{X'''} = 0{\cdot}169\,755\,8; \\ & L\dfrac{1}{X^{iv}} = 0{\cdot}200\,284\,7; & L\dfrac{1}{X^{v}} = 0{\cdot}227\,649\,5; \end{array}\right\} \tag{227}$$

$$\left.\begin{array}{lll} X' = 0{\cdot}795\,432\,7; & X'' = 0{\cdot}732\,380\,5; & X''' = 0{\cdot}676\,463\,3; \\ & X^{iv} = 0{\cdot}630\,543\,9; & X^{v} = 0{\cdot}592\,039\,3; \end{array}\right\} \tag{228}$$

which satisfy the same inequalities. For the case,

$$x = \frac{1}{2}, \qquad X = 2\epsilon \int_1^\infty \epsilon^{-t^2}\, dt = L^{-1}\bar{1}{\cdot}879\,596\,0 = 0{\cdot}757\,872\,3, \tag{229}$$

it is found that

$$\left.\begin{array}{lll} L\dfrac{1}{X'} = 0{\cdot}256\,113\,4; & L\dfrac{1}{X''} = 0{\cdot}326\,262\,0; & L\dfrac{1}{X'''} = 0{\cdot}389\,428\,9; \\ & L\dfrac{1}{X^{iv}} = 0{\cdot}441\,741\,1; & L\dfrac{1}{X^{v}} = 0{\cdot}486\,189\,2; \end{array}\right\} \tag{230}$$

$$\left.\begin{array}{lll} X' = 0{\cdot}554\,480\,9; & X'' = 0{\cdot}471\,778\,3; & X''' = 0{\cdot}407\,916\,3; \\ & X^{iv} = 0{\cdot}361\,625\,3; & X^{v} = 0{\cdot}326\,445\,5. \end{array}\right\} \tag{231}$$

And finally for the case, which may be regarded as almost an *extreme* one, when

$$x = 50, \qquad X = \tfrac{1}{5}\epsilon^{\frac{1}{100}} \int_{\frac{1}{10}}^\infty \epsilon^{-t^2}\, dt = L^{-1}\bar{1}{\cdot}201\,104\,4 = 0{\cdot}158\,892\,8, \tag{232}$$

the same system of formulæ conducts to the values,

$$\left.\begin{array}{lll} L\dfrac{1}{X'} = 1{\cdot}434\,264\,3; & L\dfrac{1}{X''} = 1{\cdot}578\,063\,1; & L\dfrac{1}{X'''} = 1{\cdot}704\,871\,0; \\ & L\dfrac{1}{X^{iv}} = 1{\cdot}800\,477\,7; & L\dfrac{1}{X^{v}} = 1{\cdot}875\,969\,2; \end{array}\right\} \tag{233}$$

$$\left.\begin{array}{lll} X' = 0{\cdot}036\,790\,5; & X'' = 0{\cdot}026\,420\,2; & X''' = 0{\cdot}019\,730\,1; \\ & X^{iv} = 0{\cdot}015\,831\,5; & X^{v} = 0{\cdot}013\,305\,5; \end{array}\right\} \tag{234}$$

which still satisfy the same inequalities. In all those cases, therefore, the *theorem* (of alternate inequalities) is verified, on account of the factors $X', \ldots, X^v$ being always included between the limits 0 and 1. As regards the *method* (of successive approximations), it has been found to succeed *well* for the *first* case, (223); (see note to paragraph XVI;) since it then gave

$$X = 0{\cdot}951\,81, \quad \text{nearly, instead of} \quad X = 0{\cdot}951\,818. \tag{235}$$

In the *second* case, (226), it gave

$$X = 0{\cdot}905\,316, \quad \text{nearly, instead of} \quad X = 0{\cdot}905\,354. \tag{236}$$

In the *third* case, (229), it gave

$$X = 0{\cdot}7560, \quad \text{nearly, instead of} \quad X = 0{\cdot}757\,87. \tag{237}$$

The success was therefore respectable, in the second case; and *not quite to be despised* in the third. For the *fourth* case (232), I find that the same method gives numbers which are *quite useless*, as approximations; namely, to several decimal places,

$$X_1 = 0; \quad X_2 = 1; \quad X_3 = 0; \quad X_4 = 1; \quad X_5 = 0; \tag{238}$$

so it would merely enable us to infer that X lies between 0 and 1. But the *theorem* continues to be verified; although the *series* to which it is here applied is one of *excessive divergence.**

XXXII. We have hitherto considered only *two Transformand Functions*, namely,

$$X = 1 - x + 3x^2 - 15x^3 + 105x^4 - 945x^5 + \&c., \quad \text{and}$$
$$X = 1 - x + x^2 - x^3 + x^4 - x^5 + \&c.;$$

and in order to *develop* them *otherwise*, namely under the *general form*,

$$X = \phi\alpha x\phi\beta x\phi\gamma x\phi\delta x\phi\varepsilon x \ldots,$$

we have used only *two Auxiliary Functions*, namely,

$$\phi x = (1 + x)^{-1}, \quad \text{and } \phi x = \epsilon^{-x};$$

which may perhaps be called *Inverse Transformers*, if we give the name of *Direct Transformers* to the two other Functions

$$\varphi x = x^{-1} - 1, \quad \text{and } \varphi x = l\frac{1}{x},$$

whereof the two foregoing forms are the respective inverses, and with which we have *operated* on the *given series* X, in order to *deduce* from it the *constants*, α, β, γ, δ, ε. I wish now to say a few words respecting a *third transformand*; and afterwards to make some remarks upon a *third transformer*: with hints about indefinitely *varying both*. For the moment I take the series, (mentioned in a separate note; lately,)

* [At this point Hamilton attached the following note:

Observatory March 23rd 1858

My Dear De Morgan,

I send on 8th Sheet, which winds up all that I wished to say, respecting the *two Transformands*,

$$X = 1 - x + 3x^2 - 15x^3 + \&c., \text{ and}$$
$$X = 1 - x + x^2 - x^3 + \&c.,$$

considered as combined with the *two Transformers*,

$$\phi x = (1 + x)^{-1}, \quad \text{and} \quad \phi x = \epsilon^{-x}.$$

But I must write about a *third Transformand, X,* and a *third Transformer,* ϕx. *Then,* perhaps, the perturbed spirit which evoked my Letter of February 15th may be allayed: – and I shall be free to *begin* a *new* one!]

$$X = \frac{1^{-1}x^0}{\Gamma 1} - \frac{2^0 x^1}{\Gamma 2} + \frac{3^1 x^2}{\Gamma 3} - \frac{4^2 x^3}{\Gamma 4} + \frac{5^3 x^4}{\Gamma 5} - \frac{6^4 x^5}{\Gamma 6} \,\&\text{c.}; \qquad (239)^*$$

which I have not met with in any book, nor do I attach much importance to it, but it will serve for illustration. Here,

$$a = 1, \quad b = \frac{3}{2}, \quad c = \frac{8}{3}, \quad d = \frac{125}{24}, \quad e = \frac{54}{5}; \qquad (240)$$

whence, if we adopt the first method of transformation, namely the method of reciprocals, we derive (on the plan of paragraph VI.) this other series,

$$X' = \frac{X^{-1} - 1}{ax} = 1 - a'x + b'x^2 - c'x^3 + d'x^4 - \&\text{c.}, \qquad (241)$$

when

$$a' = \frac{1}{2}, \quad b' = \frac{2}{3}, \quad c' = \frac{9}{8}, \quad d' = \frac{32}{15}; \qquad (242)$$

so that we have, so far, (but I have verified the law for several other successive terms, and see no reason to doubt that it continues indefinitely,)

$$X^{-1} = \frac{-(-1)^{-1}x^0}{\Gamma 1} + \frac{0^0 x^1}{\Gamma 2} - \frac{1^1 x^2}{\Gamma 3} + \frac{2^2 x^3}{\Gamma 4} - \frac{3^3 x^4}{\Gamma 5} + \frac{4^4 x^5}{\Gamma 6} \,\&\text{c.}, \qquad (243)$$

where 0^0 is to be interpreted as equal to 1. Pursuing we have

$$X'' = \frac{X'^{-1} - 1}{a'x} = 1 - a''x + b''x^2 - c''x^3 + \&\text{c.}, \qquad (244)$$

where

$$a'' = \frac{5}{6}, \quad b'' = \frac{7}{6}, \quad c'' = \frac{721}{360}. \qquad (245)$$

Again,

$$X''' = \frac{X''^{-1}}{a''x} = 1 - a'''x + b'''x^2 - \&\text{c.}, \qquad (246)$$

$$a''' = \frac{17}{30}, \quad b''' = \frac{172}{225}. \qquad (247)$$

And finally, we may write, on the same plan,

$$X^{iv} = \frac{X'''^{-1} - 1}{a'''x} = 1 - a^{iv}x + \&\text{c.}, \qquad X^{v} = \frac{X^{iv^{-1}} - 1}{a^{iv}x} = 1 - \&\text{c.}, \qquad (248)$$

with the numerical value,

$$a^{iv} = \frac{133}{170}. \qquad (249)$$

* [Γn denotes the gamma function; $\Gamma(n+1) = n\Gamma(n)$.]

If then we write,

$$\alpha = 1, \quad \beta = \frac{1}{2}, \quad \gamma = \frac{5}{6}, \quad \delta = \frac{17}{30}, \quad \varepsilon = \frac{133}{170}, \tag{250}$$

we shall have for the series (239), the transformation, (compare (90),)

$$X = \frac{1}{1+}\frac{\alpha x}{1+}\frac{\beta x}{1+}\frac{\gamma x}{1+}\frac{\delta x}{1 + \varepsilon x X^{v}}; \tag{251}$$

in which the five constants, $\alpha, \beta, \gamma, \delta, \varepsilon$, have the positive values (250), but nothing has as yet been *proved* respecting the value of the function X^{v}; not even that it is always > 0, for $x > 0$; much less that it is < 1. If, however, we *presume* that this function X^{v} is *likely* to have a *positive* value, because it can be developed in a series whereof the first term is $= 1$, – a presumption precarious enough in general, but which I think will *here* be found correct; and if we form the fractional expressions,

$$\left.\begin{aligned}
&X_1 = \frac{1}{1 + \alpha x}, \quad X_2 = \frac{1 + \beta x}{1 + (\alpha + \beta) x}, \quad X_3 = \frac{1 + (\beta + \gamma) x}{1 + (\alpha + \beta + \gamma) x + \alpha\gamma x^2}, \\
&X_4 = \frac{1 + (\beta + \gamma + \delta) x + \beta\delta x^2}{1 + (\alpha + \beta + \gamma + \delta) x + (\alpha\gamma + \alpha\delta + \beta\delta) x^2}, \\
&X_5 = \frac{1 + (\beta + \gamma + \delta + \varepsilon) x + (\beta\delta + \beta\varepsilon + \gamma\varepsilon) x^2}{1 + (\alpha + \beta + \gamma + \delta + \varepsilon) x + (\alpha\gamma + \alpha\delta + \alpha\varepsilon + \beta\delta + \beta\varepsilon + \gamma\varepsilon) x^2 + \alpha\gamma\varepsilon x^3}
\end{aligned}\right\} \tag{252}$$

then, in consistency, expect to find that the following inequalities are satisfied,

$$X < 1, \quad > X_1, \quad < X_2, \quad > X_3, \quad < X_4; \tag{253}$$

but whether $X >=< X_5$ we are not yet prepared to say, because we have not yet determined the algebraic sign of the next following constant, ζ, or a^{v}. Or, substituting for $\alpha, \beta, \gamma, \delta, \varepsilon$ their values (250), we have the converging fractions,

$$\left.\begin{aligned}
&X_1 = \frac{1}{1 + x}; \quad X_2 = \frac{2 + x}{2 + 3x}; \quad X_3 = \frac{6 + 8x}{6 + 14x + 5x^2}; \\
&X_4 = \frac{60 + 114x + 17x^2}{60 + 174x + 101x^2}; \quad X_5 = \frac{1020 + 2736x + 1353x^2}{1020 + 3756x + 3579x^2 + 665x^3};
\end{aligned}\right\} \tag{254}$$

which in fact may be developed as follows,

$$\left.\begin{aligned}
&X_1 = 1 - x + x^2 - \cdots; \quad X_2 = 1 - x + \frac{3x^2}{2} - \frac{9x^3}{4} + \cdots; \\
&X_3 = 1 - x + \frac{3x^2}{2} - \frac{8x^3}{3} + \frac{179x^4}{36} - \cdots; \\
&X_4 = 1 - x + \frac{3x^2}{2} - \frac{8x^3}{3} + \frac{125x^4}{24} - \frac{7643x^5}{720} + \cdots; \\
&X_5 = 1 - x + \frac{3x^2}{2} - \frac{8x^3}{3} + \frac{125x^4}{24} - \frac{54x^5}{5} + \frac{2843701x^6}{122400} - \cdots.
\end{aligned}\right\} \tag{255}$$

If then we only know the coefficients of the series (239) for X, *so far* as they have been actually written down so that we are only informed that

$$X = 1 - x + \frac{3x^2}{2} - \frac{8x^3}{3} + \frac{125x^4}{24} - \frac{54x^5}{5} + fx^6 - \&c., \tag{256}$$

where f may be *presumed* to be *positive,* and even to be $>54/5$; we can indeed infer the inequalities (253), at least *for small and positive values of* x; but cannot conclude that, even for such values, $X > X_5$; until we know that

$$f > \frac{2\,843\,701}{122\,400}. \tag{257}$$

This result is exactly analogous to one of paragraph VIII., in which it was found to be necessary, for the *continued success* of the method of reciprocals, or at least for that method's continuing to give *alternating limits,* or approximate values alternately less and greater than a certain *other minor limit;* which was, in that former application, $= 9675$. But as we had, *then,* by the *law then existing,*

$$f = 11.945 = 10\,395 > 9675, \qquad \text{(compare [(75) and] (77),)}$$

so we have *now,* by the *present law* of the series (239) for X, the value,

$$f = \frac{7^5}{\Gamma 7} = \frac{16\,807}{720} > \frac{2\,843\,701}{122\,400}; \quad \& \text{ therefore } X > X_5, \tag{258}$$

at least if x be not too large. But I *think* that this inequality, $X > X_5$, will be found to hold good, *for all positive values of* x; if X be that function of x from which, by processes not worth dwelling upon, the series (239) was deduced: namely that real and positive quantity which satisfies the equation,

$$lX + xX = 0; \quad \text{or,} \quad \frac{1}{X}\, l \frac{1}{X} = x. \tag{259}$$

XXXIII. It may be noted, that because the general term of the series (239) for X is

$$\frac{(n+1)^{n-1}(-x)^n}{1.2\ldots n} = \frac{(n+1)^n(-x)^n}{\Gamma(n+2)} \tag{260}$$

so that we may write

$$-xX = \sum_{m=1}^{m=\infty} \frac{m^{m-1}(-x)^m}{\Gamma(m+1)}; \tag{261}$$

therefore a *distant term* of this series for $-xX$ is nearly equal (by a well known property of the function Γ[viz: $n! \sim \sqrt{2\pi n}(n/\epsilon)^n$, due to James Stirling (1692–1770).]) to the expression,

$$(-\epsilon x)^m (2m^3\pi)^{-\frac{1}{2}}; \tag{262}$$

whence it appears to follow that the series in question becomes *divergent,* if

$$x > \epsilon^{-1};$$

but is on the contrary *convergent,* if

$$x > 0, \quad x \leqq \epsilon^{-1}. \tag{263}$$

Compare the remarks in Cauchy's Cours d'Analyse,* on the conditions of Convergence of Series – which I have not very carefully studied. The Corollary in his page 137 appears to show that the series (239) converges, *even if x be negative,* provided that

$$x \geqslant -\epsilon^{-1} \quad (x < 0). \tag{264}$$

In fact, Cauchy seems to make out that the series,

$$1 + \frac{1}{2^{\mu}} + \frac{1}{3^{\mu}} + \frac{1}{4^{\mu}} + \&c.; \tag{265}$$

converges if $\mu > 1$, but *diverges* in the contrary case: and

$$\mu = \frac{3}{2}, \tag{266}$$

when we apply this principle to the *asymptotic series* of which the general term is (261), supposing

$$x = -\epsilon^{-1}; \quad \text{and therefore by (259),} \quad X = \epsilon \tag{267}$$

It appears then to follow that the series,

$$\epsilon = 1 + \epsilon^{-1} + \frac{3^1\epsilon^{-2}}{1.2} + \frac{4^2\epsilon^{-3}}{1.2.3} + \frac{5^3\epsilon^{-4}}{1.2.3.4} + \frac{6^4\epsilon^{-5}}{1.2.3.4.5} + \&c., \tag{268}$$

ought to be found to converge, though slowly, to the limit written here; but it is evident that this other series is a divergent one,

$$\epsilon^{-1} = 1 - \epsilon + \frac{3^1\epsilon^{2}}{1.2} - \frac{4^2\epsilon^{3}}{1.2.3} + \frac{5^3\epsilon^{4}}{1.2.3.4} - \frac{6^4\epsilon^{5}}{1.2.3.4.5} + \&c.; \tag{269}$$

and indeed the recent analysis seems to show that if a positive quantity, x, which exceeds ϵ^{-1}, be substituted for ϵ in the second member of the equation (269), so as to reproduce the development (239), this resulting series will diverge. The series (243), for X^{-1}, appears to be just about *as* convergent, or *as* divergent, as (239), of which it is the reciprocal, for its general term is,

$$\frac{-(m-1)^{m-1}(-x)^m}{\Gamma(m+1)} = -\epsilon^{-1}(-\epsilon x)^m (2m^3\pi)^{-\frac{1}{2}}, \quad \text{nearly,} \tag{270}$$

where m is a large positive number. The following series therefore would seem to be a convergent one, though the convergence is but slow:

$$\epsilon^{-1} = 1 - \epsilon^{-1} - \frac{\epsilon^{-2}}{1.2} - \frac{2^2\epsilon^{-3}}{1.2.3} - \frac{3^3\epsilon^{-4}}{1.2.3.4} - \frac{4^4\epsilon^{-5}}{1.2.3.4.5} - \&c. \tag{271}$$

In general, for any positive quantity X, we have (it seems) these two series, converging or diverging, as the case may be:

* [Augustin Louis Cauchy (1789–1857), *Cours d'analyse de l'école royale polytechnique,* Paris: 1821 (Reprinted in: *Œuvres complètes d'Augustin Cauchy,* 2nd Ser., Vol. I, Paris: 1905.).]

$$X = 1 + \frac{2^0 X^{-1} lX}{1} + \frac{3^1 (X^{-1} lX)^2}{1.2} + \frac{4^2 (X^{-1} lX)^3}{1.2.3} + \frac{5^3 (X^{-1} lX)^4}{1.2.3.4} + \&c.; \qquad (272)^*$$

$$X^{-1} = 1 - \frac{X^{-1} lX}{1} - \frac{1^1 (X^{-1} lX)^2}{1.2} - \frac{2^2 (X^{-1} lX)^3}{1.2.3} - \frac{3^3 (X^{-1} lX)^4}{1.2.3.4} - \&c. \qquad (273)$$

It is true that (after happening to perceive them), I have not done more than *verify* these *laws*, for about 7 or 8 terms of *each separate series*; but I cannot doubt that they *continue* to hold good: especially as I observe that in *combination* with *each other* they satisfy the *differential equation,*

$$\frac{d}{dx}\frac{1}{X} + \frac{xd}{dx}.xX = 1, \qquad (274)$$

which is a consequence of the equation (259); so that *each series* can be *deduced from the other.* In fact, the *complete integral* of the equation (274) is,

$$lX + xX = \text{constant}; \qquad (275)$$

and the constant is here $= 0$, because the series for X reduces itself to 1, when x becomes $= 0$.

XXXIV. After this discussion on the *form* of the series (239), let us pay a moment's attention to its *value,* for the case $\alpha = \frac{1}{8}$, which may serve as a convenient example. The *converging fractions* (254) become here,

$$X_1 = \frac{8}{9} = 0{\cdot}888\,888\,9; \quad X_2 = \frac{17}{19} = 0{\cdot}894\,736\,8; \quad X_3 = \frac{448}{501} = 0{\cdot}894\,211\,6;$$

$$X_4 = \frac{4769}{5333} = 0{\cdot}894\,243\,4; \quad X_5 = \frac{708\,168}{791\,921} = 0{\cdot}894\,240\,7; \quad \text{giving } X = 0{\cdot}894\,240\,9, \text{ nearly}; \qquad (276)$$

and in fact the equation,

$$\frac{1}{X}\,l\frac{1}{X} = (x =)\,\frac{1}{8}, \quad \text{gives} \quad X = 0{\cdot}894\,240\,9; \qquad (277)$$

so that the convergence is here sufficiently rapid, and the inequalities (253) (258) are satisfied. *Continued exponentials* give, here, the very simple and elegant *expression,*

$$X = \epsilon^{-x.}\epsilon^{-x.}\epsilon^{-x.}\&c., \qquad (278)$$

to be interpreted on the same plan as the expression (154); but there is *still a practical inferiority* in this *second method,* as compared with the *first*: since the *corresponding aproximations* which it gives are *somewhat less close,* being these,

$$\left.\begin{array}{lll} X_1 = 0{\cdot}882\,497; & X_2 = 0{\cdot}895\,555; & X_3 = 0{\cdot}894\,094; \\ & X_4 = 0{\cdot}894\,257; & X_5 = 0{\cdot}894\,239; \end{array}\right\} \qquad (279)$$

* Note, Thursday April 15, 1858: De Morgan has recently called my attention to an anticipation of the series (272) by Murphy†. [Letter of De Morgan to Hamilton, April 11, 1858, Trinity College Dublin MS 1493/987.]

† [Robert Murphy, 1806–1843; *A treatise on the theory of algebraic equations,* p. 82, London: 1839.]

from which, however, we might infer, on the plan (138), that

$$X = 0{\cdot}894\,241, \text{ nearly.} \tag{280}$$

Indeed, in this case, the original series (239) converges sufficiently fast to give without much trouble the few first decimals of X.

XXXV. If we take x even so great as $\frac{1}{2}$, then, (since $\frac{1}{2} > \epsilon^{-1}$,) the series (239) will *diverge*; its first terms becoming now,

$$\begin{aligned} X &= 1 - \frac{a}{2} + \frac{b}{4} - \frac{c}{8} + \frac{d}{16} - \frac{e}{32} + \cdots = 1 - \frac{1}{2} + \frac{3}{8} - \frac{1}{3} + \frac{125}{384} - \frac{27}{80} + \cdots \\ &= 1 - 0{\cdot}500\,00 + 0{\cdot}375\,00 - 0{\cdot}333\,33 + 0{\cdot}325\,52 - 0{\cdot}337\,50 \cdots; \end{aligned} \tag{281}$$

whence we could only infer, without some such *transformation* as we have been discussing, that the real and positive root X of the equation,

$$\frac{1}{X}\, l \frac{1}{X} = \frac{1}{2}, \tag{282}$$

considered as developed in this *diverging series*, was subject to the two following inequalities,

$$\begin{aligned} &X > \frac{13}{24}, \quad < \frac{111}{128}; \quad \text{or} \\ &X > 0{\cdot}541\,67, \quad < 0{\cdot}867\,19; \end{aligned} \tag{283}$$

and in fact this root of (282) is found by trials to be, nearly,

$$X = 0{\cdot}703467. \tag{284}$$

Converging fractions give here, by making $x = \frac{1}{2}$ in (254),

$$\left. \begin{aligned} X_1 = \frac{2}{3}, \quad X_2 = \frac{5}{7}, \quad X_3 = \frac{40}{57} = 0{\cdot}70175, \quad X_4 &= \frac{485}{689} = 0{\cdot}703\,92, \\ X_5 &= \frac{21\,810}{31\,007} = 0{\cdot}703\,39; \end{aligned} \right\} \tag{285}$$

whence, on the plan (138), we might infer that

$$X = 0{\cdot}703\,49, \quad \text{nearly.} \tag{286}$$

Converging exponentials give the corresponding, but *less close*, approximations:

$$X_3 = \epsilon^{-\frac{1}{2}\cdot\epsilon^{-\frac{1}{2}\cdot\epsilon^{-\frac{1}{2}}}} = 0{\cdot}691\,286; \quad X_4 = 0{\cdot}707\,765; \quad X_5 = 0{\cdot}701\,957; \tag{287}$$

from which, however, on the same plan (138), the value (284) might very nearly be recovered. But the *first* method (that of fractions) preserves still its *advantage* over the *second*.

XXXVI. When we take $x = 1$, the series (239), of course, diverges faster than before; *so fast*, indeed, as to furnish no useful *guess* at the true value of X. But the five converging fractions (254) give, still, the formation which is not to be disdained: even when we abstain, as we have done, from determining the fractions which follow them. In fact, they give the successive approximations,

$$X_1 = \frac{1}{2}, \quad X_2 = \frac{3}{5}, \quad X_3 = \frac{14}{25}, \quad X_4 = \frac{191}{335}, \quad X_5 = \frac{5109}{9020}; \tag{288}$$

or, expanding the three last fractions decimally,

$$X_3 = 0{\cdot}56000; \quad X_4 = 0{\cdot}57015; \quad X_5 = 0{\cdot}56641;$$

$$\text{whence} \quad X = 0{\cdot}56742, \quad \text{nearly.} \tag{289}$$

The *true value* is here the real and positive root of this transcendental equation,

$$lX + X = 0, \quad \text{or}$$

$$\frac{1}{X} l \frac{1}{X} = 1; \quad \text{namely (by trials),} \quad X = 0{\cdot}5671433. \tag{290}$$

The *exponential method* (as in all the other cases hitherto considered) approaches to the mark *more slowly*; although it *would* infallibly *reach** that mark, after an infinite time allowed, if we had the patience to shoot with it, *over and over again,* each time acquiring thereby a slightly better ground for taking aim. In fact, if we now write,

$$X_1 = \epsilon^{-1}; \quad X_2 = \epsilon^{-1\cdot}\epsilon^{-1}; \quad X_3 = \epsilon^{-1\cdot}\epsilon^{-1\cdot}\epsilon^{-1}; \text{ \&c.}; \tag{291}$$

so that

$$X_{n+1} = \epsilon^{-X_n}, \quad \text{and} \quad X_0 = 1; \tag{292}$$

then, *rigorously,* as the *limit* of an *indefinitely continued* process, we may write the equation, whereof the *theory* may perhaps be developed somewhat farther on:

$$X = X_\infty. \tag{293}$$

Numerical calculation gives, with these last meanings of X_1, X_2, &c., the values:

$$\left.\begin{aligned} X_1 = 0{\cdot}367\,879; \quad X_2 = 0{\cdot}692\,200; \quad X_3 = 0{\cdot}500\,474; \\ X_4 = 0{\cdot}606\,243; \quad X_5 = 0{\cdot}545\,396; \end{aligned}\right\} \tag{294}$$

whence, on a plan already often referred to, it might be inferred that

$$X = 0{\cdot}567617, \quad \text{nearly;} \tag{295}$$

and in fact this value is not much less accurate than the value (289). I have had the patience to push this method of approximation farther, the type employed being easy enough; it seems that after about twenty steps we might perhaps be content to stop: for I find, as the result of calculations performed with seven decimal places throughout,

$$X_{18} = 0{\cdot}567\,156\,9; \quad X_{19} = 0{\cdot}567\,135\,5; \quad X_{20} = 0{\cdot}567\,147\,7, \text{ nearly;} \tag{296}$$

which differs little from the value (290) of X; being, however as it ought to be, a little in excess. But the method of converging exponentials is still inferior, in practice, to the method of converging fractions: although, if we applied the plan (138) to the values (296), we should very exactly recover the value (290), under the term

* For the proof of this assertion, I must refer to some subsequent paragraphs.

$$X = X_{20} - \frac{(X_{20} - X_{19})^2}{X_{20} - 2X_{19} + X_{18}} = 0{\cdot}567\,143\,3. \tag{297}$$

XXXVII. I scarcely intended to say anything more about the "*Third Transformand*" of this Letter, namely the Function or Series (239). But before I pass to the consideration of what I have already alluded to, as a "*Third Transformer*", or third form of an *auxiliary function*, a word or two may be added respecting the case,

$$x = \epsilon, \qquad X = \epsilon^{-1} = 0{\cdot}367\,88; \tag{298}$$

in which case we are conducted to the highly divergent series (269). The *exponential method* would here be *practically useless; so slow* would be the convergence of its successive approximations be: but still *it would give convergence,* from the very outset, and therefore, *theoretically speaking, it would not fail.* In fact we have now instead of (292), the relations,

$$X_{n+1} = \epsilon^{-\epsilon X_n}, \qquad X_0 = 1; \tag{299}$$

whence,

$$X_1 = \epsilon^{-\epsilon} < \epsilon^{-1}; \quad \epsilon X_1 < 1; \quad X_2 > \epsilon^{-1}; \quad X_3 < \epsilon^{-1}; \quad X_4 > \epsilon^{-1}; \quad X_5 < \epsilon^{-1}; \quad \text{\&c.} \tag{300}$$

But also,

$$X_1 > 0; \quad \text{therefore } X_2 < 1; \; X_3 > \epsilon^{-\epsilon}, \quad \text{therefore } > X_1; \; X_4 < \epsilon^{-\epsilon X_1} \quad \text{therefore} < X_2; \text{ \&c.}; \tag{301}$$

thus, $X_1, X_3, X_5, \ldots$ form an *increasing series,* >0, but never quite *ascending to* ϵ^{-1}; and $X_2, X_4, X_6, \ldots$ form a *descending series* <1, but never quite *descending to* ϵ^{-1}; or, in symbols,

$$\epsilon^{-1} > X_{2n+1} > X_{2n-1} > \cdots > X_3 > X_1 > 0; \tag{302}$$

and

$$\varepsilon^{-1} < X_{2n} < X_{2n-2} < \cdots < X_4 < X_2 < 1. \tag{303}$$

We are therefore already entitled to infer that X_{2n+1} and X_{2n} tend, *each separately,* to certain positive *limits,* whereof the former *cannot exceed* ϵ^{-1}, and the latter *cannot fall short* of it; or in symbols, that

$$\lim_{n=\infty} X_{2n-1} = A > 0, \not> \epsilon^{-1}; \qquad \lim_{n=\infty} X_{2n} = B < 1, \not< \epsilon^{-1}; \tag{304}$$

where the marks $\not>$, $\not<$, are the *contradictories* of $>$, $<$, or, signify, respectively, "not less than", and "not greater than". (I sometimes write, in like manner, the mark $\neq$, to signify "not equal to".) If, then, we could, at this stage, assent that X_n tended to *any one fixed limit, independent of the odd or even character of the index n;* or, in symbols, if we were already certain that

$$A = B, \qquad \text{or that } \lim_{n=\infty} X_{2n-1} = \lim_{n=\infty} X_{2n}, \tag{305}$$

n here denoting *any whole number*; we might at once conclude that

$$X_\infty = \lim_{n=\infty} X_n = A = B = \epsilon^{-1}. \tag{306}$$

Accordingly, if we suppose that X_∞ has any *one fixed value,* which is *independent* of the *odd* or *even* character of the whole number denoted here by ∞, the first equation (299) will give the condition,

$$X_\infty = \epsilon^{-\epsilon X_\infty}; \tag{307}$$

but if we write, for a moment,

$$\psi y = y\epsilon^{\epsilon y}, \tag{308}$$

the function ψy increases, constantly and continuously, with y, from 0 to ∞; it passes therefore once, and only once, during this increase, through any one given real and positive value; or in other words, the *equation,*

$$y\epsilon^{\epsilon y} = c, \tag{309}$$

has one, and only one, real and positive root, y, for any given real and positive value of c: and *all its other roots are imaginary.* But, with the form (308),

$$\psi(\epsilon^{-1}) = 1; \tag{310}$$

the equation (307) has therefore *no other real root,* except the root (306).

XXXVIII. It is, however, *conceivable,* before examination, that the *two limits, A and B,* in (304), may be *unequal;* in which case, either the *former* limit must be *less* than ϵ^{-1}, or the *latter* limit must be *greater* than ϵ^{-1}, or else *both* these inequalities must be satisfied together. And, in fact, if we go *only a short way* in the *actual calculation* of the *numerical values* of X_1, X_2, X_3, &c., we shall not *perceive any marked indication* of the lately supposed *convergence,* of X_{2n-1} and X_{2n}, *to any common limit* X_∞. I find, for example, by calculations similar to those lately mentioned, that

$$\left.\begin{array}{lll} X_1 = 0{\cdot}065\,988\,0, & X_2 = 0{\cdot}835\,793\,1, & X_3 = 0{\cdot}103\,113\,9, \\ & X_4 = 0{\cdot}755\,562\,6, & X_5 = 0{\cdot}128\,242\,5; \end{array}\right\} \tag{311}$$

and even after pursuing the process (with fewer decimals) *to a stage much more advanced,* say to about X_{60}, there is still *observed to remain a very decided excess* of X_{2n} over $X_{2n\pm1}$; although this *excess* is *observed* to *diminish* continually, as by the theory it ought to do. In fact, it is found that we have, nearly, the values (calculated with five decimals only,)

$$X_{59} = 0{\cdot}263\,89; \quad X_{60} = 0{\cdot}488\,06; \quad X_{61} = 0{\cdot}265\,36. \tag{312}$$

One might, therefore, perhaps, *doubt,* for a while, whether this *diminution,* though *continual,* would *ever quite bring down, even for infinite values of n,* the *difference* here spoken of to *zero;* and whether we might not, perhaps, have, on the contrary, the final and *limiting inequality,*

$$\lim_{n=\infty} X_{2n} > \lim_{n=\infty} X_{2n-1}? \quad \text{or, briefly,} \quad B > A? \tag{313}$$

XXXIX. Retaining, for a moment longer, the *two symbols, A and B,* to denote the *two limits,* without *yet* deciding whether those limits are *equal,* or *unequal,* we easily deduce from (299) the two equations,

$$B = \epsilon^{-\epsilon A}; \quad A = \epsilon^{-\epsilon B}; \tag{314}$$

which may also be written thus,

$$lB + \epsilon A = 0; \qquad lA + \epsilon B = 0. \tag{315}$$

If then we write,

$$\chi y = \epsilon^{\epsilon y} l \frac{1}{y}, \tag{316}$$

a very easy process of elimination shows that A and B must both be roots of the transcendental equation,

$$\chi y = \epsilon; \tag{317}$$

whereof the value,

$$y = X = \epsilon^{-1}, \tag{298}$$

is also obviously a root; so that we have,

$$\chi A = \chi B = \chi \frac{1}{\epsilon} = \epsilon. \tag{318}$$

The relations,

$$A = B = \epsilon^{-1}, \tag{306}$$

must therefore exist, *unless* the equation (317), in y, admits *at least two* real, positive, and *unequal roots.* But it is easy to prove that

$$\left. \begin{aligned} \chi' y &= \frac{d\chi y}{dy} \\ &= \epsilon^{\epsilon y}\left(\epsilon l \frac{1}{y} - \frac{1}{y}\right) < 0, \quad \text{if} \quad y > 0, \quad y \gtrless \epsilon^{-1}; \end{aligned} \right\} \tag{319}$$

while for the particular case $y = \epsilon^{-1}$; we have

$$\chi' y = \chi' \frac{1}{\epsilon} = 0. \tag{320}$$

Hence, the function χy decreases constantly and continuously, from an indefinitely large positive to an indefinitely large negative value, while y increases from an indefinitely small to an indefinitely large positive number; this function χy being, however, for a moment *stationary* in this decrease, where $y = \epsilon^{-1}$; at which stage we have seen that it attains the value, $\chi(\epsilon^{-1}) = \epsilon$. It follows hence that *the transcendental equation,*

$$\chi y = c, \tag{321}$$

when c denotes *any given real quantity,* whether positive or negative, or zero, *has, in general, only one real and positive root,* y; but that in the *singular case,* where $c = \epsilon$, that is, in the case (317), the equation may be said to have *two** *real and positive roots,* which are however (in this case) *equal to each other,* and to ϵ^{-1}. The values (306) are therefore thus proved to be correct; and the *convergence* of the *two systems* of increasing values $X_1, X_3, X_5, \ldots,$ and of decreasing

* Rather *three* real, positive and equal roots. See p. 88.

values X_2, X_4, X_6, ..., to *one common limit*, $X_\infty = X = \epsilon^{-1}$, is thus *completely established.* This *theoretical convergence* is, however, here, *in practice*, excessively *slow*: for a rough calculation gives me the (still) very sensibly unequal values,

$$X_{99} = 0{\cdot}285; \qquad X_{100} = 0{\cdot}460. \tag{322}$$

The converging *fractions* (254) give far more rapidly the following much less rude approximations, to what is here the true value (298), of the divergent series (269):

$$\left.\begin{aligned} X_1 = 0{\cdot}26894; \quad X_2 = 0{\cdot}464\,63; \quad X_3 = 0{\cdot}342\,54; \\ X_4 = 0{\cdot}387\,33; \quad X_5 = 0{\cdot}361\,63. \end{aligned}\right\} \tag{323}$$

XL. More generally, in applying the exponential method to transform the series (239), we have, instead of (292) or (299), the equation

$$X_{n+1} = \epsilon^{-xX_n}, \qquad X_0 = 1; \tag{324}$$

in which we at present suppose that $x > 0$, in order that the terms of the transformand series may be alternately positive and negative. By an analysis similar to that employed in recent paragraphs, but made a little more general, we can show that the transcendental equation

$$X = \epsilon^{-xX}, \tag{325}$$

which is only another form of

$$lX + xX = 0, \qquad \text{or} \qquad \frac{1}{X}\, l \frac{1}{X} = x, \tag{259}$$

has *one*, and *only* one, real and positive root X, for any given real and positive value of x. In fact, when x is thus given, and real, and > 0, the function

$$\psi X = X\epsilon^{xX}, \tag{326}$$

which is slightly extended from the form (308), increases constantly and continuously with X, from 0 to ∞; it passes therefore once but *only once*, through any assigned stage of real and positive value; and the equation (325) under the form

$$\psi X = 1, \tag{327}$$

is thus seen to have, as above asserted, one real positive root X, and only one. Or we may observe that (x being still > 0) the function,

$$l\psi X = lX + xX, \tag{328}$$

increases constantly and continuously from $-\infty$ to $+\infty$, while X increases from 0 to ∞; this function therefore passes once, but once only through any assigned stage of real value, and in particular through the value 0, during this increase of X. Or again we must consider this other function

$$\varphi X = \frac{1}{X}\, l \frac{1}{X}; \tag{329}$$

of which the differential coefficient relatively to X is

$$\varphi' X = X^{-2}(lX - 1); \tag{330}$$

hence

$$\varphi' X < 0, \quad \text{if} \quad X > 0, < \epsilon, \quad \text{but} \quad \varphi' X > 0, \quad \text{if} \quad X > \epsilon; \tag{331}$$

therefore φX decreases from ∞ to 0, and passes once, but only once, during such decrease, through any one assigned stage x of real and positive value, while X increases from 0 to 1. Or finally we might remark that while the one member, X, of the equation (325), increases from 0 to 1, the other member, ϵ^{-xX}, decreases from 1 to a quantity < 1, namely to ϵ^{-x}. In any of these ways we may easily see, not only (as above stated) that the equation (325) has always a real and positive and *unique* root X, if $x > 0$, but also that this root

$$X < 1. \tag{332}$$

Such, then, must be the value of X_∞, *if it have any determined value;* that is if, in the first equation (324), we are at liberty to consider X_n as tending *to any one fixed limit,* (while the whole number n increases,) which limit is *independent of the odd or even character of* n. But the *conditions of such independence* remain to be investigated.

XLI. Meanwhile, this seems to be a natural occasion for observing, that since, by (331), the function φX in (329) decreases from 0 to $-\epsilon^{-1}$, while X increases from 1 to ϵ, and afterwards increases from $-\epsilon^{-1}$ to 0 again, while X continues to increase from ϵ to ∞, each process of decrease or of increase being constant and continuous, therefore each of these three forms of one common transcendental equation,

$$\varphi X = \frac{1}{X} l \frac{1}{X} = -x, \qquad lX - xX = 0, \qquad X = \epsilon^{xX}, \tag{333}$$

$$\left.\begin{array}{l}\text{has } \textit{two real and unequal and positive roots}, \text{ one} > 1, < \epsilon, \\ \text{and the other} > \epsilon, \text{ if } x > 0, < \epsilon^{-1};\end{array}\right\} \tag{334}$$

but that the same equation, under any of these forms (333), has *two real and equal roots,* each being

$$X = \epsilon, \quad \text{if } x = \epsilon^{-1}; \tag{335}$$

and finally that each equation (333) has *no real root X,*

$$\text{if } x > \epsilon^{-1}. \tag{336}$$

The *critical stage* of *equal roots,* (335), corresponds to the case (267), and to the *barely convergent series* (268), of which the value has been seen to be ϵ. And we now see that the series,

$$X = 1 + \frac{2^0 x}{\Gamma 2} + \frac{3^1 x^2}{\Gamma 3} + \frac{4^2 x^3}{\Gamma 4} + \frac{5^3 x^4}{\Gamma 5} + \frac{6^4 x^5}{\Gamma 6} + \&\text{c.}, \tag{337}$$

which is formed from (239) by changing the sign of x, and may be obtained by suitable processes from any one of the forms (333), may be said, in language of your own, to "escape from imaginariness, by becoming divergent", when x comes to exceed the limit ϵ^{-1}. At the same time we see that *this series* (337), *under the conditions* (334), (compare paragraph XXXIII.,) *converges to a limit* X, $> 1, < \epsilon$; *namely to the lesser of the two real and unequal roots* of the equation (333).

XLII. Hence if in this series (337), we change x to lX/X, and so recover the series,

$$X = 1 + \frac{2^0 X^{-1} lX}{1} + \frac{3^1 (X^{-1} lX)^2}{1.2} + \frac{4^2 (X^{-1} lX)^3}{1.2.3} + \frac{5^3 (X^{-1} lX)^4}{1.2.3.4} + \&\text{c.}, \tag{272}$$

the second member will indeed be the *algebraical development* of the first, when X is treated as a function of $X^{-1} lX$, and developed according to *ascending powers* thereof; in such a manner that the equation

$$\epsilon^z = 1 + \frac{2^0 z\epsilon^{-z}}{1} + \frac{3^1 z^2 \epsilon^{-2z}}{1.2} + \frac{4^2 z^3 \epsilon^{-3z}}{1.2.3} + \frac{5^3 z^4 \epsilon^{-4z}}{1.2.3.4} + \&\text{c.}, \tag{338}$$

is an *algebraical identity*, so far as agreement between the *coefficients of powers of* z is concerned; for instance, after multiplying the coefficient of z^n in ϵ^z by $1.2.3 \ldots n$, or differentiating the last equation n times, and then making $z = 0$, we have the true *arithmetical relations,*

$$1 = 1 = 2^0 = -2.2^0 + 3^1 = +3.2^0 - 3.2^1.3^1 + 4^2 = -4.2^0 + 6.2^2.3^1 - 4.3^1.4^2 + 5^3$$

$$= +5.1^4.2^0 - 10.2^3.3^1 + 10.3^2.4^2 - 5.4^1.5^3 + 6^4 = \&\text{c.}; \tag{339}$$

yet we must not conclude, *unconditionally*, that the *arithmetical values* of the *two members* of the formula (272), or (338), are in *all* cases *equal to each other*. In fact, with our recent meaning (329) of φ, the function,

$$-\varphi(\epsilon^z) = z\epsilon^{-z}, \tag{340}$$

increases from 0 to ϵ^{-1}, while z increases from 0 to 1, and afterwards decreases from ϵ^{-1} to 0, while z increases from 1 to ∞. If then we *substitute for* z *any assumed real value* $c > 1$, in the series (338), *that series will converge to a real limit* $C < \epsilon$, but > 1, for reasons already stated; so that we may write,

$$C = 1 + \frac{2^0 c\epsilon^{-c}}{1} + \frac{3^1 c^2 \epsilon^{-2c}}{1.2} + \frac{4^2 c^3 \epsilon^{-3c}}{1.2.3} + \&\text{c.} > 1, < \epsilon, \quad \text{if } c > 1; \tag{341}$$

but this arithmetical value, C, of the second member of the equation (338), obtained from the *convergent series* (341), will *not be equal to the arithmetical value* ϵ^c *of the first member of the same equation* (338), obtained by the *same substitution* of the value c for z: since we shall have, on the contrary the *inequality,*

$$C < \epsilon^c, \quad \text{if,} \quad c > 1. \tag{342}$$

To obtain a *finite equation,* which the *sum* C of the series (341) shall satisfy, let us write, for a moment,

$$a = c\epsilon^{-c}; \tag{343}$$

then

$$C = 1 + \frac{2^0 a^1}{1} + \frac{3^1 a^2}{1.2} + \frac{4^2 a^3}{1.2.3} + \&\text{c.}, \tag{344}$$

so that C is what X in (337) becomes, when x is changed to a; and the conditions (334) are satisfied. Hence, by what was lately shown, *C is the lesser root of the equation obtained from* (333), by changing x and X to $c\epsilon^{-c}$ and C; and thus, not withstanding the *in*equality (342), we have the *equation,*

$$C^{-1}\,lC = c\epsilon^{-c}; \tag{345}$$

whereof the *greater root* may be denoted by

$$C' = \epsilon^{c}. \tag{346}$$

In like manner, if we substitute for X in the *series* (272), any value $A > \epsilon$, the series will *converge* to a *limit*, and will therefore have an *arithmetical sum*, say B, which will be $> 1, < \epsilon$; so that

$$B = 1 + \frac{2^0 A^{-1}\,lA}{1} + \frac{3^1 (A^{-1}\,lA)^2}{1.2} + \frac{4^2 (A^{-1}\,lA)^3}{1.2.3} + \&\text{c.}, > 1, < \epsilon, \quad \text{if } A > \epsilon; \tag{347}$$

but this sum B will not be equal to A, since we shall have, on the contrary,

$$B < A \tag{348}$$

it will however be true that A and B are connected, by the relation

$$B^{-1}\,lB = A^{-1}\,lA. \tag{349}$$

For example, if we assume

$$c = 5, \quad a = 5\epsilon^{-5}, \quad A = \epsilon^5, \tag{350}$$

and substitute these values in the series (341), (344), (347), we obtain the convergent series

$$(B =)\ C = 1 + \frac{2^0 5^1 \epsilon^{-5}}{1} + \frac{3^1 5^2 \epsilon^{-10}}{1.2} + \frac{4^2 5^3 \epsilon^{-15}}{1.2.3} + \&\text{c.}, \tag{351}$$

of which the first terms are decimally,

$$1 + 0{\cdot}033\,689\,7 + 0{\cdot}001\,702\,5 + 0{\cdot}000\,102\,0 + 0{\cdot}000\,006\,7 + 0{\cdot}000\,000\,5; \tag{352}$$

so that the *sum* may thus be seen to be nearly

$$C = 1{\cdot}035\,501\,4; \tag{353}$$

which is very decidedly *less* than ϵ^5, being indeed (as it ought to be) less than ϵ itself. And if we resolve, by trials, the equation,

$$(C^{-1}\,lC =)(B^{-1}\,lB =) 5\epsilon^{-5}, \tag{354}$$

rejecting the greater root, $C' = \epsilon^5$, we find that the *lesser root* is,

$$C = L^{-1} 0{\cdot}015\,150\,7; \tag{355}$$

that is, again,

$$C = 1{\cdot}035\,501\,4.$$

XLIII. We might also approximate to this lesser root, and therefore to the sum of the series (351), by an application of the *exponential method* of the present Letter. Making

$$x = 5\epsilon^{-5} = L^{-1}\bar{2}{\cdot}527\,497\,6 = 0{\cdot}033\,689\,7, \tag{356}$$

and writing now, instead of (324), the equations,

$$X_{n+1} = \epsilon^{xX_n}, \quad X_0 = 1, \tag{357}$$

we easily obtain the successive approximations following:

$$\left.\begin{array}{lll} LX_1 = 0{\cdot}014\,631\,3; & LX_2 = 0{\cdot}015\,132\,6; & LX_3 = 0{\cdot}015\,150\,1; \\ & LX_4 = 0{\cdot}015\,150\,7; & LX_5 = 0{\cdot}015\,150\,7; \\ X_1 = 1{\cdot}034\,263\,8; & X_2 = 1{\cdot}035\,458\,2; & X_3 = 1{\cdot}035\,500\,0; \\ & X_4 = 1{\cdot}035\,501\,4; & X_5 = 1{\cdot}035\,501\,4; \end{array}\right\} \tag{358}$$

at least so far as logarithms with 7 decimal places enable us to judge. Converging fractions give, with a little more trouble, for the same value (356), of *x*, the following approximation, which are somewhat more accurate, at least in their first stages:

$$X_1 = \frac{1}{1-\alpha x}; \quad X_2 = \frac{1}{1-}\frac{\alpha x}{1-\beta x}; \quad X_3 = \frac{1}{1-}\frac{\alpha x}{1-}\frac{\beta x}{1-\gamma x}; \text{ \&c.,} \tag{359}$$

with the values (250) of $\alpha, \beta, \gamma, \delta, \varepsilon$; that is, after the substitutions indicated,

$$\left.\begin{array}{lll} LX_1 = 0{\cdot}014\,883\,4; & LX_2 = 0{\cdot}015\,142\,9; & LX_3 = 0{\cdot}015\,150\,6; \\ LX_4 = 0{\cdot}015\,150\,7; & LX_5 = 0{\cdot}015\,150\,7. & \end{array}\right\} \tag{360}$$

Almost the same logarithms are obtained (as they ought to be) by changing x to $-5\epsilon^{-5}$, in the fractional expressions (254). It is to be noted that, whichever of the two methods we employ, the approximate values successively obtained form an *increasing series,* and *not* an *alternating* one, which is a consequence of our having *changed the sign of x.*

XLIV. Before discussing the *conditions,* alluded to at the end of paragraph XL., for the *convergence of the continued exponential,*

$$X_n = (\epsilon^{-x.})^n 1, \tag{361}$$

to any one fixed limit, X_∞, when $x > 0$, and the whole number n increases indefinitely, I shall mention a mode of *proving* the correctness of the two *series,* (272), (273), at least so far as their *coefficients* are concerned, which had not (as I confessed) occurred to me, when I gave those series as having been found through only an *induction,* though one which was carried pretty far, and might (as I conceived) be relied on. Lagrange's* Theorem (as given in page 170 of your *Differential and Integral Calculus*)† shows (if usual things be admitted), that

$$\left.\begin{array}{l} \psi u = \psi z + \varphi z.\psi' z.x + \dfrac{d}{dz}((\varphi z)^2.\psi' z).\dfrac{x^2}{z} \\ \qquad + \left(\dfrac{d}{dz}\right)^2((\varphi z)^3.\psi' z).\dfrac{x^3}{2.3} + \text{\&c., if } u = z + x\varphi u. \end{array}\right\} \tag{362}$$

Make, then, in particular,

$$\varphi u = -\epsilon^u, \quad \text{and} \quad \psi u = \epsilon^{ru}, \tag{363}$$

so that the finite equation connecting *x, z, u,* is

* [Joseph Louis Lagrange (1736–1813).]
† [A. De Morgan, *The Differential and Integral Calculus,* London: 1842.]

$$u = z - x\epsilon^{u}, \tag{364}$$

and the development of ϵ^{ru} according to ascending powers of x is the thing sought. The Theorem gives us this development,

$$\epsilon^{ru} = \epsilon^{rz} - rx\epsilon^{(r+1)z} + \frac{r(r+2)^1}{2}x^2\epsilon^{(r+2)z} - \frac{r(r+3)^2}{2.3}x^3\epsilon^{(r+3)z}\&c., \tag{365}$$

and the equation (354) gives,

$$-x\epsilon^{z} = (u-z)\epsilon^{z-u} = y\epsilon^{-y}, \quad \text{if} \quad y = u - z; \tag{366}$$

we have, therefore, the *identity*,

$$\left.\begin{array}{l} r^{-1}\epsilon^{ry} = r^{-1} + y\epsilon^{-y} + \dfrac{(r+2)^1}{2}y^2\epsilon^{-2y} \\ \qquad + \dfrac{(r+3)^2}{2.3}y^3\epsilon^{-3y} + \&c. = \displaystyle\sum_{m=0}^{m=\infty}\frac{(r+m)^{m-1}}{\Gamma(m+1)}y^m\epsilon^{-my}; \end{array}\right\} \tag{367}$$

or, changing ϵ^{y} to X,

$$r^{-1}X^r = \sum_{m=0}^{m=\infty}\frac{(r+m)^{m-1}}{\Gamma(m+1)}\left(\frac{lX}{X}\right)^m; \tag{368}$$

which includes the two series referred to. By changing r^{-1} to α, and ry to hD, the series (367) becomes,

$$\epsilon^{hD} = \sum_{m=0}^{m=\infty}\frac{(md+1)^{m-1}}{\Gamma(m+1)}h^m D^m \epsilon^{-m\alpha hD}; \tag{369}$$

hence (subject to exceptions for divergence &c.), we have the *general transformation,*

$$f(x+h) = \sum_{m=0}^{m=\infty}\frac{(m\alpha+1)^{m-1}}{\Gamma(m+1)}h^m\varphi^m f(x - m\alpha h); \tag{370}$$

where α is an *arbitrary constant,* which may even have an imaginary value. And if in particular, we assume $\alpha = 1$, $\alpha = -1$, changing also in the last case h to $-h$, we obtain the two following series:

$$f(x+h) = fx + hDf(x-h) + \frac{3^1h^2}{2}D^2f(x-2h) + \frac{4^2h^3}{2.3}D^3f(x-3h) + \&c.; \tag{371}$$

$$f(x-h) = fx - hDf(x-h) - \frac{1^1h^2}{2}D^2f(x-2h) - \frac{2^2h^3}{2.3}D^3f(x-3h) - \&c. \tag{372}$$

These general transformations pretend only to the same kind and degree of correctness, as that which would now be conceded to the series of Taylor* and Lagrange. In fact, a recent paragraph (XLII.) of the present Letter proves that if $fx = \epsilon^x$, and if $h > \epsilon$, the development in the second member of (371) converges to a limit different from $f(x+h)$.

* [Brook Taylor (1685–1731).]

XLV. Resuming now the investigation begun in paragraph XL., and therefore returning to the equations

$$X_{n+1} = \epsilon^{-xXn}, \qquad X_0 = 1, \tag{324}$$

with the supposition that x is *positive*; we see, first, by a slight extension of the analysis employed in paragraph XXXVII., that if X denote the real and *unique* root of the equation (259), or (326), so that (see (325), (332),)

$$X = \epsilon^{-xX} > 0, \quad < 1, \quad (x > 0,)$$

then, unlimitedly many inequalities of the same general forms as those numbered (302) and (303) exist: so that

$$1 > X_2 > X_4 > X_5 > \cdots > X_{2n-2} > X_{2n} > X_{2n+1} > X_{2n-1} > \cdots > X_3 > X_1 > 0. \tag{373}$$

As an extension of the formulæ (304), we may therefore write now,

$$\lim_{n=\infty} X_{2n-1} = A > 0, \not> X; \qquad \lim_{n=\infty} X_{2n} = B < 1, \not< X; \tag{374}$$

and the question (313) recurs, can we have $B > A$?

XLVI. It is evident that A and B must satisfy conditions analogous to those marked (314), namely the following,

$$B = \epsilon^{-xA}, \qquad A = \epsilon^{-xB}; \tag{375}$$

and therefore that whether they be equal or unequal to each other, they must be *roots* of the transcendental equation in y,

$$\chi y = \epsilon^{+xy} l \frac{1}{y} = x; \tag{376}$$

which is extended from (316) (317), and of which the quantity X is *also* seen to be a root. The question therefore reduces itself to this: can the last written equation have real and *unequal* roots? Can we have

$$\chi A = \chi X = \chi B = x, \tag{377}$$

without the real and positive quantities A and B being, *each* of them, $= X$? In paragraph XXXIX., it was shown that for the *case* when x was $= \epsilon$, and $X = \epsilon^{-1}$, this latter value for X was the *only real root* of the transcendental equation (317), which then took the place of (376); or at least that there was *no other real root* of the equation, which was *either greater or less* than this one. Indeed in consequence of our then having $\chi' \frac{1}{\epsilon} = 0$, (320), to which I may now add that

$$\chi'' \frac{1}{\epsilon} = 0, \quad \text{if} \quad x = \epsilon, \tag{378}$$

because, by (376),

$$\chi' y = \epsilon^{xy} \left(xl \frac{1}{y} - \frac{1}{y} \right), \tag{379}$$

and

$$\chi''y = \epsilon^{xy}\left(x^2 l\frac{1}{y} - \frac{2x}{y} + \frac{1}{y^2}\right), \tag{380}$$

while yet

$$\chi'''y = \epsilon^{xy}\left(x^3 l\frac{1}{y} - \frac{3x^2}{y} + \frac{3x}{y^2} - \frac{2}{y^3}\right), \tag{381}$$

so that

$$\chi'''\frac{1}{\epsilon} = -\epsilon^4 < 0, \quad \text{if} \quad x = \epsilon, \tag{382}$$

the equation which replaced (376), for this value ϵ of x, namely the equation

$$\epsilon^{\epsilon y} l\frac{1}{y} = \epsilon, \tag{383}$$

may be said to have *three real, positive, and equal roots* in y, each $= \epsilon^{-1}$; *but not to have any fourth real root,* whether *equal* or *unequal* to these. (In XXXIX., I omitted to notice the existence of the *third* real and *equal root* of the equation (317), or (383); but the fact that $\chi''\frac{1}{\epsilon}$ was $= 0$, and that $\chi'''\frac{1}{\epsilon}$ was < 0, for the case of χy there considered, might have been inferred from the remark there made, that the function χy in (316) *decreased, constantly and continually,* from $+\infty$ to $-\infty$, while y increased from 0 to ∞, being however for a moment *stationary* at the value ϵ, when y was $= \epsilon^{-1}$.)

XLVII. But we are now to consider those *other* cases, in which x, though still > 0, is either $<$ or $> \epsilon$. In general, by (379),

$$\epsilon^{-xy}\chi'y = xl\frac{1}{y} - \frac{1}{y}; \tag{384}$$

which function when x has any given real and positive value, increases from $-\infty$ to the maximum value,

$$\epsilon^{-1}\chi'\frac{1}{x} = xl\frac{x}{\epsilon} \tag{385}$$

while y increases from 0 to x^{-1}, and then decreases again to $-\infty$, while y increases to ∞: each increase or decrease being constant and continuous. It follows then, that for all real and positive values of y

$$\chi'y < 0, \quad \text{if} \quad x > 0, < \epsilon; \tag{386}$$

and consequently that the function χy, in (376), decreases, constantly and continuously, from $+\infty$ to $-\infty$, without even a moment's stationariness, while y increases from 0 to ∞. This function χy passes therefore *once,* but *only once* through *any one given* stage, such as x, of *real values*; therefore the equation $\chi y = x$, or (376), has *one real and positive root, y,* there being in this case *not even any other equal root*: and we conclude, as an extension of the result (306), that we are justified in writing,

$$A = B = X = X_\infty = (\epsilon^{-x.})^\infty 1 = \lim_{n=\infty} X_n, \quad \text{if} \quad x > 0, < \epsilon. \tag{387}$$

Hence the *Exponential Method* will give, *when x is thus* >0 *but* $<\epsilon$, a succession of approximate values, X_1, X_2, X_3, ... *alternately in defect and in excess, but always converging to the limit X*; that is, to the real, positive, and unique root of the equation $X = \epsilon^{-xX}$, (325), *even when the series,*

$$X = \frac{1^{-1} x^0}{\Gamma 1} - \frac{2^0 x^1}{\Gamma 2} + \frac{3^1 x^2}{\Gamma 3} - \frac{4^2 x^3}{\Gamma 4} + \&\text{c.}, \tag{239}$$

diverges, by our having $x > \epsilon^{-1}$. Of such *alternation,* combined with such *convergence,* (more or less rapid according to the value assumed for *x*,) we have already had several examples: namely, in XXXIV., for the case $x = \frac{1}{8}$, where the *series* (239) *converged,* although not quickly, and X_5, as given by the exponential method, was already almost equal to X; in XXXV., where x was $= \frac{1}{2}$, so that the series (239) *diverged,* but the converging exponentials, X_3, X_4, X_5, though not very close approximations themselves, would have given, if *combined* on the plan (138), a value for X very near to the true one; and finally in XXXVI., where x being assumed $= 1$ produced a pretty *rapid divergence* in the *series* (239), but the *exponentials* still *continued to converge,* though not so fast as before, and X_{18}, X_{19}, X_{20} gave X with sufficient exactness. – The extreme case, $x = \epsilon$, for which we had *still, in theory,* $X = X_\infty = A = B$, but in which the exponential method is *practically useless,* has been also sufficiently discussed in other recent paragraphs. I may however just add, to show more fully the *excessive slowness* of the convergence of the exponentials in this case, that calculations* conducted with 4 decimals, as a sequel to those referred to in (312), gave here,

$$\left.\begin{aligned} X_{149} = (\epsilon^{-\epsilon.})^{149} 1 = 0{\cdot}2987; \quad X_{150} &= (\epsilon^{-\epsilon.})^{150} 1 = 0{\cdot}4440; \\ X_{151} &= (\epsilon^{-\epsilon.})^{151} 1 = 0{\cdot}2991; \end{aligned}\right\} \tag{388}$$

the true value being, in this case, as stated in (298),

$$X = X_\infty = (\epsilon^{-\epsilon.})^\infty 1 = \epsilon^{-1} = 0{\cdot}3679. \tag{389}$$

XLVIII. It remains to consider the case where $x > \epsilon$. In this case, by (384), (385), the function $\epsilon^{-xy}\chi' y$ increases from $-\infty$ *to a positive maximum,* namely to $xl\frac{x}{\epsilon}$, while y increases from 0 to $\frac{1}{x}$; and the same function afterwards decreases from this positive maximum to $-\infty$ again, while y continues to increase from $\frac{1}{x}$ to ∞; it therefore passes *twice,* but *only twice,* through the value 0, in such a manner that the equation,

$$yl\frac{1}{y} = \frac{1}{x}, \quad \text{or} \quad \chi' y = 0, \tag{390}$$

has two real and positive roots; which we may call y_1 and y_2. (In fact, I see that this agrees with a remark in page 132, of your *Differential and Integral Calculus.*) Thus,

* [The details are recorded on pp. 113–115 of Notebook D of 1858, Trinity College Dublin MS 1492/144.]

$$\left.\begin{array}{l}\chi y < 0, \quad \text{if} \quad y > 0, < y_1; \quad \chi' y_1 = 0; \quad \chi' y > 0, \quad \text{if} \quad y > y_1, < y_2; \\ \chi' y_2 = 0; \quad \chi y < 0, \quad \text{if} \quad y > y_2.\end{array}\right\} \tag{391}$$

Hence, χy decreases from $+\infty$ to χy_1; increases to χy_2; and then decreases to $-\infty$. We see also, already, that

$$y_1 > 0, < x^{-1}, \quad \text{but that} \quad y_2 > x^{-1}; \tag{392}$$

indeed we have *now*

$$\chi' \frac{1}{\epsilon} > 0, \quad \text{and therefore} \quad y_2 > \epsilon^{-1}. \tag{393}$$

The limits for χy_1, χy_2 are perhaps a little less obvious; yet there is not much difficulty in proving that

$$\chi y_1 < x, \quad \text{and that} \quad \chi y_2 > x. \tag{394}$$

For this purpose it is enough to prove that χy becomes equal to x, or that the equation (376) is satisfied, for a value of y which lies between y_1 and y_2; or for which $\chi' y$ is positive: and X is such a value. In fact, we have already seen (377) that X is a root of the equation (376); so that $\chi X = x$; but it is also a root of (325), or of (259), so that

$$xX = l\frac{1}{X} > 1, \quad X > x^{-1}, \quad \text{because} \quad x > \epsilon, \quad X < \epsilon^{-1}, \tag{395}$$

since we have seen that $X = \epsilon^{-1}$ when $x = \epsilon$, and that X decreases thence forward to 0, while x increases to ∞; (for we said, in XL., that the differential coefficient $\frac{dx}{dX} = \varphi' X$ was negative, if X was > 0, but $< \epsilon$;) hence, by (379),

$$\chi' X = x^2 - X^{-2} > 0; \tag{396}$$

and therefore by (391),

$$X > y_1, < y_2. \tag{397}$$

The inequalities (394) therefore subsist; because the function χy, while constantly and continuously *increasing*, from χy_1 to χy_2, passes through the stage of value, χX, or x. For the same reason, *the equation* (376), *or* $\chi y = x$, *has three real and unequal positive roots, y*, but *only three such roots*, whereof the middle one is $Y_2 = X$, and the least and greatest may be denoted by Y_1, Y_3; so that

$$\chi Y_1 = \chi X = \chi Y_3 = x, \tag{398}$$

and

$$Y_3 > X > Y_1. \tag{399}$$

(In fact, the function χy must have passed *once*, but *only once*, through the value x, while it was decreasing from $+\infty$ to the value $\chi y_1 < x$, during the increase of y from 0 to y_1; and it must pass again, *once more*, but *not oftener*, through the same stage of value, x, while decreasing from $\chi y_2 > x$ to $-\infty$, during the increase of y from y_2 to ∞.) There is, therefore, no absurdity in supposing that we *may* have, notwithstanding (377), the equations

$$A = Y_1, \quad B = Y_3; \tag{400}$$

but it is not yet proved, that we *have,* in fact, these equations.

XLIX. We see, however, already, that if

$$a = \epsilon^{-x} > 0, < \epsilon^{-\epsilon}, \tag{401}$$

then the equation

$$a^{a^y} = y, \tag{402}$$

has *these real, positive, and unequal roots, y,* and *only three*; whereof the *second* is also the *unique* real and positive root, X of this *other* and simpler equation

$$a^X = X. \tag{403}$$

Making, more generally,

$$a^y = z, \tag{404}$$

and supposing merely that y is *some* root of the equation (402), we have, by that equation,

$$a^z = y; \tag{405}$$

hence, by elimination of y,

$$a^{a^z} = z; \tag{406}$$

so that z must be *either* the *same root,* or else *another* (real and positive) *root,* of the *same equation* (402). But it cannot be the *same* root, unless it coincide with what has been already seen to be the *unique* root, X, of the equation (403). If then the *two other* (real and positive) *roots* of (402) be denoted, as lately, by Y_1, Y_2, we must have the *relations,*

$$a^{Y_1} = Y_3, \quad a^{Y_3} = Y_1; \tag{407}$$

which exactly *correspond* to the equations (375); but which *do not yet prove* that these *least and greatest real roots,* Y_1, Y_3, of the equation (402), or (376), with the two *limits,* A and B, (374) of those *alternate approximations,* X_{2n-1} and X_{2n}, which err respectively by defect and by excess, as compared with the sought value of X, and are given by the Exponential Method; or by the formulæ,

$$X_{n+1} = a^{Xn}, \quad X_0 = 1. \tag{408}$$

L. But it is easy to *complete the proof* of the two equations (400). In fact, the function a^y, under the conditions (401), decreases, while the (positive) variable y increases; if, then in (404), we assume $y < Y_1$, we shall have, by (407), $z > Y_3$; and in like manner, if $y > Y_3$, then $z < Y_1$. It might be enough to have made this remark; yet it may be added, that

$$\left.\begin{array}{lll} \text{if } y > 0, < Y_1, & \text{then } a^{a^y} > y, < Y_1; & \text{and that} \\ \text{if } y > Y_3, & \text{then } a^{a^y} < y, > Y_3; & \end{array}\right\} \tag{409}$$

inequalities which may be proved, among other ways, by remarking that, in the first of these two cases (409), and with the signification (376) of χy we have

$$\chi y > x, \tag{410}$$

because the function χy *has not yet decreased* from $+\infty$ to the value $\chi Y_1 = x$, (398); therefore, in the case here considered,

$$\epsilon^{xy} l\frac{1}{y} > x, \quad l\frac{1}{y} > x\epsilon^{-xy}, \quad \frac{1}{y} > \epsilon^{x\epsilon^{-xy}}, \quad a^{a^y} = \epsilon^{-x\epsilon^{-xy}} > y; \tag{411}$$

and similarly for the second case (409). The equations (400) are therefore, *fully proved.* It may, however, be interesting to *exemplify* their correctness, by a few *numerical applications* of the method.

LI. As such an application, let

$$x = 3, \quad a = \epsilon^{-3} = L^{-1}\bar{2}\text{·}697\,116\,6 = 0\text{·}049\,787\,1; \tag{412}$$

then the unique root Y_2 of the equation (403), or of

$$\epsilon^{-3x} = X, \tag{413}$$

is found to be, nearly,

$$Y_2 = X = L^{-1}\bar{1}\text{·}544\,030\,4 = 0\text{·}349\,969\,7; \tag{414}$$

and the *two other* real roots, Y_1, Y_3, or A, B, of the equation (402), or of

$$\epsilon^{-3\epsilon^{3y}} = y, \tag{415}$$

are found, by trials, to be nearly the following numbers:

$$\left.\begin{aligned} A = Y_1 = L^{-1}\bar{1}\text{·}133\,921\,5 = 0\text{·}136\,119\,9; \\ B = Y_3 = L^{-1}\bar{1}\text{·}822\,615\,7 = 0\text{·}664\,739\,8; \end{aligned}\right\} \tag{416}$$

with the relations, included in (407),

$$\epsilon^{-3Y_1} = Y_3, \quad \epsilon^{-3Y_3} = Y_1; \quad \text{or} \quad \epsilon^{-3A} = B, \quad \epsilon^{-3B} = A. \tag{417}$$

If, in the next place, by (408), (412), we write,

$$X_{n+1} = \epsilon^{-3X_n} = a^{X_n}, \quad X_0 = 1, \tag{418}$$

we find (nearly) the numerical values, (with, as usual, some doubtful figures,)

$$\left.\begin{aligned} &X_1 = a = 0\text{·}049\,787\,1;\ X_2 = a^a = 0\text{·}861\,257\,8;\ X_3 = a^{a^a} = 0\text{·}075\,488\,4; \\ &X_4 = a^{a^{a^a}} = 0\text{·}797\,346\,6;\ X_5 = a^{a^{a^{a^a}}} = 0\text{·}091\,442\,9;\ X_6 = a^{a^{a^{a^{a^a}}}} = 0\text{·}760\,082\,1 \end{aligned}\right\} \tag{419}$$

among which it may be noticed that X_1, X_3, X_5 are increasing, though *not very* rapidly, towards the theoretical limit A; and that X_2, X_4, X_6 are decreasing, though rather slowly, towards the other limit B. Pursuing somewhat farther, but with fewer decimals, the same train of calculation, we find, nearly

$$X_{17} = 0\text{·}125\,89; \quad X_{18} = 0\text{·}685\,46; \quad X_{19} = 0\text{·}127\,91; \quad X_{20} = 0\text{·}681\,30; \tag{420}$$

the same slow convergence towards the *two alternate limits* being seen to continue. With four decimal places, a rough calculation gives these other values:

$$X_{37} = 0{\cdot}1348; \quad X_{38} = 0{\cdot}6673; \quad X_{39} = 0{\cdot}1351; \quad X_{40} = 0{\cdot}6668; \tag{421}$$

and now we are pretty near to the limits, A and B; but the convergence is even slower than before.

LII. I have made a few other verifications of the theory, set forth in paragraphs XLVIII., XLIX., L., that the equation (402) has 3 real, positive, and *unequal* roots, A, X, B, under the conditions (401); and may state, generally, that the *exponential method* appears to give a *more rapid convergence,* than that above exemplified, towards the extreme and *alternate limits,* A and B, when x, in the equation

$$\epsilon^{-x\epsilon^{-xy}} = y,$$

is assumed *greater than* 3, being *already* greater than ϵ; or when a, in the equation

$$a^{a^y} = y,$$

is assumed less than ϵ^{-3}, or than $0{\cdot}049\,787\,1$, being already smaller than $\epsilon^{-\epsilon}$, or than $0{\cdot}065\,988\,0$. On these other *verifications* I do not wish to delay you, nor myself; but as regards the whole *Theory* of the present Letter, it seems not unimportant to remark, that on trying how the *method of converging fractions* would *work,* as applied to the recent question, or to the case where $x = 3$, the results were found to be by (254), the following:

$$\left.\begin{aligned} &X_1 = \frac{1}{4}; \quad X_2 = \frac{5}{11}; \quad X_3 = \frac{10}{31} = 0{\cdot}322\,580\,6; \\ &X_4 = \frac{185}{497} = 0{\cdot}372\,233\,4; \quad X_5 = \frac{7135}{20\,818} = 0{\cdot}342\,732\,3; \end{aligned}\right\} \tag{422}$$

whence, on the plan (138), it might be inferred, as at least a less rude approximation to X than any one of these three last fractions, that

$$X = 0{\cdot}353\,73, \text{ nearly.} \tag{423}$$

At all events, there is an evident convergence, by alternate deficit and excess, to *some* value not very different from this last, or from 0.35; and accordingly we saw, in (414), that the *only real root* of the equation $\epsilon^{-3X} = X$, or the *middle real root* of the equation

$$\epsilon^{-3\epsilon^{-3y}} = y,$$

is $X = 0{\cdot}349\,97$, nearly. I therefore cannot doubt that it is *to this root* that the *fractions converge,* in this application of the "method of reciprocals:" although it might almost be worth while to carry the calculation *one step further* for the sake of an additional verification. Meanwhile we see clearly, that ***the fractions*** (422) *have no tendency whatever to either of the two alternate limits, A and B,* to which we saw, in LI., that the values given by the "method of exponentials" tend.

LIII. Since writing the foregoing paragraph, I have made the *additional step* of calculation, which was suggested towards its end: Continuing a little farther the process of XXXIII., I find that because the coefficients of x^6, in the series (239) for X, and of x^5 in the series (241) for X', are respectively,

$$f = \frac{7^5}{\Gamma 7} = \frac{16\,807}{720}, \quad \text{and} \quad e' = \frac{5^5}{\Gamma 7} = \frac{625}{144}, \tag{424}$$

or simply from the latter of these two values, combined with those of a', b', c', d' already deduced in (242), we can not only recover, by a process formerly explained, the values (245), (247), (249), of a'', b'', c'', a''', b''', a^{iv}, but can also deduce these others:

$$d'' = \frac{1379}{360}; \quad c''' = \frac{6997}{5400}; \quad b^{iv} = \frac{917}{850}; \quad \text{and} \quad \zeta = a^{v} = \frac{1927}{3230}. \tag{425}$$

We have then this *sixth converging fraction,*

$$X_6 = \frac{1}{1+}\frac{\alpha x}{1+}\frac{\beta x}{1+}\frac{\gamma x}{1+}\frac{\delta x}{1+}\frac{\varepsilon x}{1+\zeta x}; \tag{426}$$

where, by (250) and (425), the constants have the values,

$$\alpha = 1, \quad \beta = \frac{1}{2}, \quad \gamma = \frac{5}{6}, \quad \delta = \frac{17}{30}, \quad \varepsilon = \frac{133}{170}, \quad \zeta = \frac{1927}{3230}; \tag{427}$$

and it seems not uninteresting to notice, that the following *chain* of inequalities subsists,

$$\alpha > \gamma > \varepsilon > \frac{2}{3} > \zeta > \delta > \beta > 0. \tag{428}$$

I *suspect* that, in the present question, a *law* of this sort would be found to *continue* to hold good; or in other words, that *the constants* $\alpha, \beta, \gamma, \delta, \epsilon, \zeta, \ldots$ *of the development of the function* X, *in the form of a continued fraction,* would be found to *continue to converge to some one fixed limit,* either exactly or nearly equal to the fraction $\frac{2}{3}$, being *still* (as above) *alternately in excess and in deficit.*

LIV. Waiving, however, *this* question of *suspected convergence of the constants,* I observe that if the numerators of the 5 fractions (254) be called $N_1, \ldots, N_5$, and the denominators $M_1, \ldots, M_5$, and if we also write,

$$\left.\begin{array}{lllll} N_1 = \nu_1, & N_2 = 2\nu_2, & N_3 = 6\nu_3, & N_4 = 60\nu_4, & N_5 = 1020\nu_5, \\ M_1 = \mu_1, & M_2 = 2\mu_2, & M_3 = 6\mu_3, & M_4 = 60\mu_4, & M_5 = 1020\mu_5, \end{array}\right\} \tag{429}$$

then $\nu_1, \ldots, \nu_5$ will be the numerators, and $\mu_1, \ldots, \mu_5$ will be denominators of the more *algebraically expressed* fractions (252); they will therefore be connected by the relations,

$$\left.\begin{array}{l} \nu_1 = 1,\ \nu_2 = \nu_1 + \beta x,\ \nu_3 = \nu_2 + \gamma x \nu_1,\ \nu_4 = \nu_3 + \delta x \nu_2,\ \nu_5 = \nu_4 + \varepsilon x \nu_3; \\ \mu_1 = 1 + \alpha x,\ \mu_2 = \mu_1 + \beta x,\ \mu_3 = \mu_2 + \gamma x \mu_1,\ \mu_4 = \mu_3 + \delta x \mu_2,\ \mu_5 = \mu_4 + \varepsilon x \mu_3; \end{array}\right\} \tag{430}$$

whereof the analogy would become still more obvious, if we agreed to write,

$$\left.\begin{array}{l} \nu_0 = 1, \quad \nu_{-1} = 0, \quad \mu_0 = 1, \quad \mu_{-1} = 1, \quad \text{or} \\ X_0 = \dfrac{\nu_0}{\mu_0} = \dfrac{1}{1}, \quad X_{-1} = \dfrac{\nu_{-1}}{\mu_{-1}} = \dfrac{0}{1}; \end{array}\right\} \tag{431}$$

for then we shall have,

$$\nu_1 = \nu_0 + \alpha x \nu_{-1}, \quad \nu_2 = \nu_1 + \beta x \nu_0; \ \mu_1 = \mu_0 + \alpha x \mu_{-1}, \quad \mu_2 = \mu_1 + \beta x \mu_0. \tag{432}$$

Combining (427), (429), (430), we easily deduce the relations,

$$N_1 = 1,\ N_2 = 2N_1 + x,\ N_3 = 3N_2 + 5xN_1,\ N_4 = 10N_3 + 17xN_2,\ N_5 = 17N_4 + 133xN_3; \tag{433}$$

$$\left.\begin{aligned} &M_1 = 1 + x,\ M_2 = 2M_1 + x,\ M_3 = 3M_2 + 5xM_1,\ M_4 = 10M_3 + 17xM_2,\\ &M_5 = 17M_4 + 133xM_3; \end{aligned}\right\} \tag{434}$$

which may seem to verify the expressions (254), for the fractions,

$$X_1 = \frac{N_1}{M_1}, \quad X_2 = \frac{N_2}{M_2}, \quad \ldots, \quad X_5 = \frac{N_5}{M_5}, \tag{435}$$

and might have been employed to form these; though I think that in point of fact I deduced those expressions in some other way, perhaps by direct substitution of the constants (250) in the algebraical fractions (252). We have now

$$X_6 = \frac{N_6}{M_6}, \tag{436}$$

where if we make

$$N_6 = 193\,800\nu_6, \qquad M_6 = 193\,800\mu_6, \tag{437}$$

we may employ the relations

$$\nu_6 = \nu_5 + \zeta x\nu_4, \qquad \mu_6 = \mu_5 + \zeta x\mu_4, \tag{438}$$

with the value (425) or (427) of the constant ζ; thus,

$$N_6 = 190N_5 + 1927xN_4, \qquad M_6 = 190M_5 + 1927xM_4; \tag{439}$$

and finally, by substitution of the polynomes already computed, we find this *sixth converging fraction,* of the series begun in (254):

$$X_6 = \frac{193\,800 + 635\,460x + 476\,748x^2 + 32\,759x^3}{193\,800 + 829\,260x + 1\,015\,308x^2 + 320\,977x^3}; \tag{440}$$

of which the development in series, according to ascending powers of x, would necessarily have its *first 7 terms* including the term $+fx^6$, *coincident with the corresponding terms of the series* (239); so that, *at least if x be small,* whether positive or negative, it *must* give a very *good approximation to the sum of that series.* But I believe that *even when the series diverges,* if it does not diverge *very fast indeed* this sixth fraction (440) will give a *fair approach* to the unique real root X of the equation $lX + xX = 0$; where it is supposed that x is given and positive. For $x = 3$, we have seen that this root is, $X = 0{\cdot}34997$, nearly; and the fraction (440) gives, in this case,

$$X_6 = \frac{2\,425\,135}{6\,828\,577} = 0{\cdot}355\,145\,0. \tag{441}$$

Thus X_6, like X_2 and X_4, is somewhat greater than the true value of X; and it may be a little improved, by subtracting from it

$$\frac{(X_6 - X_5)^2}{X_6 - 2X_5 + X_4} = 0{\cdot}003\,676\,0; \tag{442}$$

which leaves the remainder,

$$X = 0{\cdot}351\,47, \text{ nearly;} \tag{443}$$

as a better approximation than (423).

LV. It may be added that if we compute X_6 from (440), for the case $x = 1$, and combine it on the same plan with the values (289) of X_4, X_5, we find,

$$X_6 = \frac{1\,338\,767}{2\,359\,345} = 0{\cdot}567\,43; \qquad X = 0{\cdot}567\,21, \text{ nearly;} \tag{444}$$

the true value of X, considered as the root of $lX + X = 0$, being, nearly, $X = 0{\cdot}567\,14$, by (290). The real root of $lX + \frac{1}{2}X = 0$, or of the equation (282), is by (284), $X = 0{\cdot}703\,47$, nearly; or to seven decimals, it is a little more exactly,

$$X = 0{\cdot}703\,467\,4; \tag{445}$$

and if we make $x = \frac{1}{2}$ in the fraction (440), and combine the result with the values (285), or with the following which are slightly more accurate,

$$X_4 = \frac{485}{689} = 0{\cdot}703\,918\,7, \qquad X_5 = \frac{21\,810}{31\,007} = 0{\cdot}703\,389\,6, \tag{446}$$

then for this value of x, which is still large enough to render (as we saw by (281),) the series (239) for X divergent, we find

$$X_6 = \frac{507\,849\,5}{721\,903\,3} = {\cdot}703\,486\,9; \qquad X = {\cdot}703\,471\,8 \quad \text{nearly.} \tag{447}$$

For $x = \frac{1}{8}$, in which case the series (239) converges, having still alternately positive and negative terms, the true value of the real root X of the equation $lX + \frac{1}{8}X = 0$ has been seen, in (277), to be, so far as 7 decimals seemed able to give it, $X = 0{\cdot}894\,240\,9$; and the fractional expression (440) may be said to reproduce this root, by giving

$$X_6 = \frac{143\,741\,783}{160\,741\,681} = 0{\cdot}894\,240\,9. \tag{448}$$

Finally for the case

$$x = -5\epsilon^{-5} = -L^{-1}\overline{2}{\cdot}527\,497\,6 = -0{\cdot}033\,689\,7, \tag{449}$$

(compare (356),) the development (239) becomes the converging series (351), or (352), with all its terms positive, and with a *sum*

$$C = L^{-1}0{\cdot}015\,150\,7 = 1{\cdot}035\,501\,4,$$

nearly, which sum has been seen to be the *lesser of the two real roots of the equation* $lC - 5\epsilon^{-5}C = 0$, (compare (354),) the *greater root* being $C' = \epsilon^5$; and if we substitute for x the negative value (449), in the fractional expression (440), that expression becomes, nearly,

$$X_6 = \frac{172\,931{\cdot}38}{167\,002{\cdot}55} = L^{-1}0{\cdot}015\,150\,7 = C; \tag{450}$$

which value may also be completed from the equivalent expression (426), with the constants (427); so that the fraction X_6 is here *very nearly equal to the lesser real root* of the equation $lX + xX = 0$, as I believe that it will *always nearly* be, when $x < 0, > -\epsilon^{-1}$, so as to allow of the series having all its terms of one common sign, and converging to a sum $X, > 1, < \epsilon$; at least if x be *not too near the limiting value,* $-\epsilon^{-1}$, the fractions (254) and (440) become, nearly,

$$X_1 = 1{\cdot}581\,98; \quad X_2 = 1{\cdot}820\,83; \quad X_3 = 2{\cdot}002\,77;$$
$$X_4 = 2{\cdot}108\,38; \quad X_5 = 2{\cdot}196\,51; \quad X_6 = 2{\cdot}255\,90; \tag{451}$$

and we see that they form an increasing series, converging slowly towards the *only real root* ϵ of the equation $lX - \epsilon^{-1}X = 0$; at least their march in (451) is quite consistent with this theoretical convergence. When $x < -\epsilon^{-1}$, the equation $lX + xX = 0$ has *only imaginary roots,* by elementary principles already referred to; and in such a case, I presume that the *fractions would not converge to any limit.* If we assume, as an example,

$$x = -\frac{1}{2}, \tag{452}$$

the fractions then become,

$$X_1 = 2; \quad X_2 = 3; \quad X_3 = 8; \quad X_4 = \frac{29}{-7}; \quad X_5 = \frac{-78}{-371}; \quad X_6 = \frac{-70\,703}{-57\,001}; \tag{453}$$

whereof one is even *negative*: and indeed it seems that *any one* of the fractions X_n (besides its power of vanishing with its numerator) can be made to become infinite, and in so doing to reverse its *sign,* by our assuming a *real and negative value* for x, if this value is suitably chosen. Such is at least the case with each of the *six* fractions X_n, of which the denominators M_n have been computed, in (254) and (440): for each of the *six equations,*

$$M_1 = 0, \quad M_2 = 0, \quad M_3 = 0, \quad M_4 = 0, \quad M_5 = 0, \tag{454}$$

has *at least one real and negative root,* the *quadratics* of this series having *two* such roots, whereof each is $< -\epsilon^{-1}$: an inequality which seems *likely* to be satisfied by *all the real roots* of *all the equations of the series*

$$M_n = 0. \tag{455}$$

Indeed, a little reflexion leads me to *believe,* that *the roots of every such equation,* (although *all negative*) will *all* be *real*; at least if the constants α, β, γ, &c. of the *continued fraction* shall *all* be found to be *positive.* But this must be reserved for discussion in another sheet.

LVI. Writing $\alpha_1, \alpha_2, \alpha_3$, &c., instead of α, β, γ, &c., and forming thus the continued fraction,

$$X_n = \frac{1}{1+}\frac{\alpha_1 x}{1+}\frac{\alpha_2 x}{1+}\cdots\frac{\alpha_{n-1}x}{1+\alpha_n x}, \tag{456}$$

in which it is here assumed that

$$\alpha_1 > 0, \quad \alpha_2 > 0, \quad \ldots, \quad \alpha_n > 0; \tag{457}$$

we have, by usual principles, the transformation,

$$X_n = \frac{\nu_n}{\mu_n}, \tag{458}$$

where μ_n and ν_n are rational and integer polynomes; such that if we write

$$\nu_0 = 1, \quad \nu_{-1} = 0, \quad \mu_0 = 1, \quad \mu_{-1} = 1,$$

as in (431), then, for all integer values of $n > 0$, we may write,

$$\mu_n = \mu_{n-1} + \alpha_n x \mu_{n-2}; \qquad \nu_n = \nu_{n-1} + \alpha_n x \nu_{n-2}; \tag{459}$$

and may thus determine, to any proposed extent, the denominators μ_n, and the numerators ν_n, of the algebraical fraction (252). Attending at present only to the denominators μ, it is easy to see that the polynome μ_n is of the dimension $\frac{n}{2}$, if n be even; or

$$\frac{n+1}{2}, \text{ if } \quad n \text{ be odd.}^* \tag{460}$$

Such, then, must be the number ρ_n of the *real or imaginary* roots (but negative if real) of the equation,

$$\mu_n = 0; \tag{461}$$

which roots may be called, for shortness, "roots of μ_n", and may be denoted, (in order of decrease from 0, if they be real,) by the symbols,

$$x_n^{(0)}, \quad x_n^{(1)}, \quad x_n^{(2)}, \quad \ldots, \quad x_n^{(\rho_n - 1)}; \tag{462}$$

the first few of these symbols being also, if we choose, abridged to

$$x_n, \quad x'_n, \quad x''_n, \quad \ldots. \tag{463}$$

But I propose to prove (what is very likely to be known), that *all these roots are real;* and that *zero* and *the roots of any one equation* μ_{n-1}, *are limits of the same number of roots of each of the two next following equations* μ_n, μ_{n+1}; *although* μ_{n+1}, always, and μ_n also if n be odd, has one additional root, which lies between the lesser root of μ_{n-1}, and the limit $-\infty$. You will tell me, at your leisure, whether this is merely a "discovery of the Mediterranean Sea". Meantime I may amuse myself, without of necessity boring you, (since you have the power to *skip,*) by noting down the chief steps of the proof, such as it has occurred to me.

LVII. Beginning at the beginning, it is obvious that

$$x_1 = -\alpha_1^{-1}; \tag{464}$$

because x_1 is the root of the linear equation,

$$1 + \alpha_1 x = 0. \tag{465}$$

But

$$\mu_2 = \mu_1 + \alpha_2 x = 1 + (\alpha_1 + \alpha_2)x; \quad \text{therefore} \quad x_2 = -(\alpha_1 + \alpha_2)^{-1} > x_1. \tag{466}$$

Without actually resolving the two quadratic equations, μ_3, μ_4, since we have, by (459),

$$\mu_3 = \mu_2 + \alpha_3 x \mu_1, \qquad \mu_4 = \mu_3 + \alpha_4 x \mu_2, \tag{467}$$

we see with respect to the first, that when

* The dimension of ν_n is $\frac{n}{2}$ if n be even; but is $\frac{(n-1)}{2}$ if n be odd.

$$x = x_1, \quad \mu_1 = 0, \quad \mu_2 < 0, \quad \mu_3 < 0; \tag{468}$$

and that when

$$x = x_2, \quad \mu_2 = 0, \quad \mu_1 > 0, \quad \mu_3 < 0; \tag{469}$$

whence μ_3 has real roots, x_3, x'_3, whereof

$$x_3 > x_2, \quad \text{and} \quad x'_3 < x_1; \tag{470}$$

because the function μ_3 is positive $(= 1)$ when $x = 0$, and is again positive $(= \infty)$ when $x = -\infty$. The three equations μ_1, μ_2, μ_3, give therefore the *chain* of inequalities,

$$0 > x_3 > x_2 > x_1 > x'_3. \tag{471}$$

In like manner, when $x = x_2$, $\mu_2 = 0$, $\mu_3 < 0$, as before; therefore

$$\mu_4 < 0, \quad \text{and} \quad x_4 > x_2, \quad x'_4 < x_2; \tag{472}$$

the second quadratic equation $\mu_4 = 0$ having thus two real roots, which are separated by the limit x_2. To compare these roots of μ_4 with the roots of the former quadratic μ_3, we may observe that when $x = x_3$, then

$$\mu_3 = 0, \quad \mu_2 > 0, \quad \mu_4 < 0; \tag{473}$$

but that when $x = x'_3$, then

$$\mu_3 = 0, \quad \mu_2 < 0, \quad \mu_4 > 0; \tag{474}$$

whence x_4 must lie between 0 and x_3, and x'_4 between x_2 and x'_3; so that the three equations, μ_2, μ_3, μ_4, supply the chain

$$0 > x_4 > x_3 > x_2 > x'_4 > x'_3; \tag{475}$$

in which it may be noticed that no place has been as yet assigned to the root x_1 of the *first* equation μ_1. We have only shown that x_1, like x'_4 lies between x_2 and x'_3; but have not determined whether the (algebraically) lesser root, x'_4, of μ_4, is $>$, $=$, or $<$ the unique root x_1 of μ_1. And in fact this question *cannot* be determined, without our making a *new* supposition. For, where $x = x_1$, then

$$\mu_1 = 0, \quad \mu_3 = \mu_2 = \alpha_2 x_1 = -\frac{\alpha_2}{\alpha_1}, \quad \mu_4 = (1 + \alpha_2 x_1)\mu_2 = \frac{\alpha_2(\alpha_4 - \alpha_1)}{\alpha_1^2}; \tag{476}$$

so that the sign of μ_4, for $x = x_1$, will be the same as that of the difference, $\alpha_4 - \alpha_1$, of two of the constants of the continued fraction, which constants as yet, have only been supposed to be each > 0. In the example last considered, we had, by (427),

$$\alpha_1 = \alpha = 1, \quad \text{and} \quad \alpha_4 = \delta = \frac{17}{30};$$

therefore, in *this* case,

$$\alpha_4 < \alpha_1, \quad \text{and} \quad \mu_4 < 0, \quad \text{when} \quad x = x_1; \tag{477}$$

but we saw, in (474), that $x = x'_3$ gives $\mu_4 > 0$; consequently, *here*, x'_4 must lie between x_1 and x'_3; and the *complete chain* of inequalities, so far, must be that assigned by the following formula; if $\alpha_4 < \alpha_1$, then

$$0 > x_4 > x_3 > x_2 > x_1 > x_4' > x_3'. \tag{478}$$

Accordingly if we seek the roots of the four first equations (454), under the particular forms given by the fractions (254), namely, the equations,

$$\left.\begin{array}{l} 0 = M_1 = 1 + x; \quad 0 = M_2 = 2 + 3x; \quad 0 = M_3 = 6 + 14x + 5x^2; \\ \qquad\qquad\qquad\qquad\qquad\qquad\qquad 0 = M_4 = 60 + 174x + 101x^2; \end{array}\right\} \tag{479}$$

we find that they are, nearly,

$$\left.\begin{array}{l} x_1 = -1; \quad x_2 = -0{\cdot}667; \quad x_3 = -0{\cdot}528; \\ x_3' = -2{\cdot}272; \quad x_4 = -0{\cdot}477; \quad x_4' = -1{\cdot}246; \end{array}\right\} \tag{480}$$

and all the inequalities of the chain (478) are seen to be satisfied. But *it may* happen that α_4 shall, in some other example, be *greater* (instead of being *less*) than α_1; and if so, then in *that* case μ_4 will, by (476), be >0, when $x = x_1$; but we saw that $\mu_4 < 0$, for $x = x_2$; therefore x_4' *now* lies between x_2 and x_1, and the chain becomes,

$$\text{if} \quad \alpha_4 > \alpha_1, \quad \text{then} \quad 0 > x_4 > x_3 > x_2 > x_4' > x_1 > x_3'. \tag{481}$$

For instance, when the function T, which was just considered in this Letter, is transformed in the well known way, (whereof Laplace appears to claim the discovery *,) into the continued fraction (4), we have the constants $\alpha_1 = 1$, $\alpha_4 = 4 > \alpha_1$, and the inequalities (481) must subsist. And accordingly, if we solve the four equations,

$$\left.\begin{array}{l} 0 = M_1 = 1 + x; \quad 0 = M_2 = 1 + 3x; \quad 0 = M_3 = 1 + 6x + 3x^2; \\ \qquad\qquad\qquad\qquad\qquad\qquad\qquad 0 = M_4 = 1 + 10x + 15x^2, \end{array}\right\} \tag{482}$$

(compare the foot of page 34 [426?] of your *Theory of Probabilities,*) obtained by equating to zero the denominators of the four first of the converging fractions, we find, nearly the roots,

$$\left.\begin{array}{l} x_1 = -1; \quad x_2 = -{\cdot}3333; \quad x_3 = -{\cdot}1835; \quad x_3' = 1{\cdot}8165; \\ \qquad\qquad\qquad\qquad\qquad x_4 = -{\cdot}1225; \quad x_4' = -{\cdot}5442; \end{array}\right\} \tag{483}$$

which satisfy the chain (481), but not (478).

LVIII. In general, let us suppose it to have been proved that, for some one given value of ν, (as for the value 2 in (473), (474),) the ν roots of $\mu_{2\nu-1}$ are real and unequal, (being of course also negative,) and that their substitution for x (in an order decreasing from 0) renders the function $\mu_{2\nu}$ alternately negative and positive; so that, for instance, when

$$x = x_{2\nu-1}^{(\nu-1)}, \quad \text{then} \quad (-1)^\nu \mu_{2\nu} > 0. \tag{484}$$

Then, remembering that $x = 0$ renders $\mu_n = 1$, we see that the ν roots of $\mu_{2\nu}$ are real and unequal, and lie within the ν intervals comprised between the $\nu + 1$ limits,

* By his reference, in [Pierre Simon Laplace, 1749–1827] *Théorie analytique des probabilités*, Livre Ier, Article 27, [p. 104, 3rd edn, Paris: 1820] to the *Méc. Cél.*† "comme je l'ai fait ..., ou j'ai trouvé".

† [P. S. Laplace, *Traité de Mécanique céleste*, Livre Xme, pp. 255–6. Paris: 1805.]

$$0, \quad x_{2\nu-1}, \quad x'_{2\nu-1}, \quad \ldots, \quad x^{(\nu-1)}_{2\nu-1};$$

whence it may easily be inferred that these ν roots of $\mu_{2\nu}$ render $\mu_{2\nu-1}$ alternately positive and negative; so that in particular, if we consider the last or νth root,

$$x = x^{(\nu-1)}_{2\nu} \quad \text{gives} \quad (-1)^\nu \mu_{2\nu-1} < 0. \tag{485}$$

But we have, by (459),

$$\mu_{2\nu+1} = \mu_{2\nu} + \alpha_{2\nu+1} x \mu_{2\nu-1}; \tag{486}$$

where $\alpha_{2\nu+1} > 0$, by (457), but $x < 0$, for each of the roots just now mentioned. Consequently, whether we substitute (in algebraically decreasing order) the ν roots of $\mu_{2\nu-1}$, or the ν roots of $\mu_{2\nu}$, in the function $\mu_{2\nu+1}$, that function will take alternately negative and positive values; and in particular, whether we make

$$x = x^{(\nu-1)}_{2\nu-1}, \quad \text{or} \quad x = x^{(\nu-1)}_{2\nu},$$

we shall have

$$(-1)^\nu \mu_{2\nu+1} > 0. \tag{487}$$

Already therefore we see that there are at least ν roots, and therefore that there are $\nu+1$ such roots, of the equation $\mu_{2\nu+1}$; because this last equation is of the dimension $\nu+1$. But the substitution

$$x = -\infty \text{ gives } (-1)^\nu \mu_{2\nu+1} = -\infty; \tag{488}$$

and therefore, whether we take the $\nu+2$ limits,

$$0, \quad x_{2\nu-1}, \quad x'_{2\nu-1}, \quad \ldots, \quad x^{\nu-1}_{2\nu-1}, \quad -\infty,$$

or the $\nu+2$ other limits,

$$0, \quad x_{2\nu}, \quad x'_{2\nu}, \quad \ldots, \quad x^{(\nu-1)}_{2\nu}, \quad -\infty,$$

we shall in each case have $\nu+1$ intervals, within which, respectively, the $\nu+1$ roots of $\nu_{2\nu+1}$ (now seen to be all real, unequal, and negative,) must be comprised. Combining the above results, we have the *chain:*

$$0 > x_{2\nu+1} > x_{2\nu} > x_{2\nu-1} > x'_{2\nu+1} > x'_{2\nu} > x'_{2\nu-1} > x''_{2\nu+1} > \cdots$$
$$> x^{(\nu-1)}_{2\nu+1} > x^{(\nu-1)}_{2\nu} > x^{(\nu-1)}_{2\nu-1} > x^{(\nu)}_{2\nu+1}; \tag{489}$$

which gives, in particular, for the relative arrangement of the roots of the two quadratic equations, μ_3, μ_4, and of the cubic equation μ_5,

$$0 > x_5 > x_4 > x_3 > x'_5 > x'_4 > x'_3 > x''_5. \tag{490}$$

This chain, for instance, must be satisfied by the 3 roots, x_5, x'_5, x''_5, of the cubic equation,

$$0 = M_5 = 1020 + 3756x + 3679x^2 + 665x^3, \tag{491}$$

obtained by supposing the denominator of the 5th fraction (254) to vanish, when those roots are compared with the 4 roots (480) of the two quadratic equations (479); and the same chain of inequalities (490) would doubtless be found to connect the 3 roots of this other cubic equation,

$$0 = 1 + 15x + 45x^2 + 15x^3, \tag{492}$$

(compare (89)), with the roots (483) of the two quadratic equations (482), which arise from the making null the denominators of certain celebrated fractions, connected with a continually occurring definite integral, to which this Letter partly relates.

LIX. Again, since

$$\mu_{2\nu+2} = \mu_{2\nu+1} + a_{2\nu+2}x\mu_{2\nu}, \tag{493}$$

and since, by the chain (489), the successive substitution of the $\nu + 1$ roots of $\mu_{2\nu+1}$ renders each of the two preceding functions $\mu_{2\nu}$ and $\mu_{2\nu-1}$, alternately positive and negative, while we have seen that the substitution of the ν roots of $\nu_{2\nu}$ renders $\mu_{2\nu+1}$ alternately negative and positive, it follows that whether we substitute successively these ν roots of $\mu_{2\nu}$, or those $\nu + 1$ roots of $\mu_{2\nu+1}$, the function $\mu_{2\nu+2}$ (which is of the dimension $\nu + 1$) will become alternately negative and positive; it has therefore $\nu + 1$ real and unequal roots, and it can have no more; and we have this other chain,

$$\begin{aligned} 0 > x_{2\nu+2} > x_{2\nu+1} > x_{2\nu} > x'_{2\nu+2} > x'_{2\nu+1} > x'_{2\nu} > x''_{2\nu+2} > \cdots \\ > x^{(\nu-1)}_{2\nu+2} > x^{(\nu-1)}_{2\nu+1} > x^{(\nu-1)}_{2\nu} > x^{(\nu)}_{2\nu+2} > x^{(\nu)}_{2\nu+1}. \end{aligned} \tag{494}$$

For example, having proved the chain (490), we may infer that

$$0 > x_6 > x_5 > x_4 > x'_5 > x'_4 > x''_6 > x''_5; \tag{495}$$

and this Letter must connect the roots of the quadratic equation μ_4, and those of the two cubic equations, μ_5, μ_6. It determines for instance, (as would doubtless be found on trial) the relative order (of decrease from 0) of the two roots (480) of the second quadratic equation (479), the three roots of the cubic equation (491), and the three roots x_6, x'_6, x''_6 of this other cubic,

$$0 = M_6 = 193\,800 + 829\,260x + 1\,015\,308x^2 + 320\,977x^3, \tag{496}$$

obtained by equating to zero the denominator of the sixth converging fraction (440). And in like manner, it cannot fail to assign the mutual arrangement of the roots (483) of the second quadratic (482), the roots of the cubic (492), and the roots x_6, x'_6, x''_6 of this other cubic, obtained from the sixth of Laplace's converging fractions,

$$0 = 1 + 21x + 105x^2 + 105x^3. \tag{497}$$

We see, moreover, that the laws expressed by the two chains of inequalities, (489) and (494), must continue to hold good, for all greater values of ν; so that, for example, if we still denote by M_7, M_8, M_9 the functions so called in (19), (20), (21), but write x for q, then the 4 roots $x_7, \ldots, x'''_7$ of the biquadratic equation $M_8 = 0$, and with the 5 roots $x_9, \ldots, x^{iv}_9$ of the quintic equation $M_9 = 0$, by the arrangement,

$$0 > x_9 > x_8 > x_7 > x'_9 > x'_8 > x'_7 > x''_9 > x''_8 > x''_7 > x'''_9 > x'''_8 > x'''_7 > x^{iv}_9. \tag{498}$$

All this may be perfectly well known, but I have made it out for myself as above.

LX. It may be remarked that no place has been assigned in the chain (489) for any of the roots of any equation earlier than $\mu_{2\nu-1}$; nor in (494) for those of any equation before $\mu_{2\nu}$; so that generally each of these chains assigns only the arrangement of the roots of *three successive equations*, of the form $\mu_n = 0$. And in fact we saw, in LVII., that the place where x_1 ought to be inserted, with respect to x'_4, in the chain (475), depended on the relative magnitude of the two constants α_1, α_4 of the continued fraction. In like manner I find that

$$x_2 \gtreqless x'_5, \quad \text{when} \quad \alpha_1 + \alpha_2 \gtreqless \alpha_5; \tag{499}$$

and that

$$x_1 \gtreqless x'_5, \quad \text{according as} \quad \alpha_1 \gtreqless \alpha_4 + \alpha_5. \tag{500}$$

It would be tedious to pursue the investigation of these *particular conditions*; but just as *examples* of the application of those above assigned, I may remark that with the system of constants (427),

$$\alpha_1 > \alpha_5 - \alpha_2, \quad \alpha_1 < \alpha_5 + \alpha_4;$$

therefore, by (499), (500), $x_2 > x'_5$, $x_1 < x'_5$; but also $\alpha_1 > \alpha_4$, and therefore $x_1 > x'_4$, as in (478); the chain (490) comes therefore to be thus completed, for the 9 roots of the 5 equations (479), (491):

$$0 > x_5 > x_4 > x_3 > x_2 > x'_5 > x_1 > x'_4 > x'_3 > x''_5. \tag{501}$$

On the other hand, the constants of *Laplace's Fractions* ($\alpha_1 = 1$, $\alpha_2 = 2$, $\alpha_3, \ldots, \alpha_n = n$) are such that $\alpha_1 < \alpha_4$, $\alpha_1 < \alpha_5 - \alpha_2$; therefore in *this* case, $x_1 < x'_4$, as in (481), and $x_2 < x'_5$; but $x_2 > x'_4$, by (475); and $x_1 > x'_3$ by (472); thus the places of x_1 and of x_2 in (490) are known, and that chain becomes, for the 9 roots of the 5 equations (482), (492):

$$0 > x_5 > x_4 > x_3 > x'_5 > x_2 > x'_4 > x_1 > x'_3 > x''_5. \tag{502}$$

It is clear that results of the same general character must hold good, with respect to the equations of the form

$$\nu_n = 0, \quad \text{or} \quad N_n = 0, \tag{503}$$

so far as regards the reality and inequality of *all* the roots of *each*, and the existence of *chains* of inequalities, *analogous* to (489) and (494), and connecting the roots of any *three* successive equations. – Is all this old?

LXI.a* After this digression, on the nature and the limits of the roots of certain algebraical equations, suggested by the converging fractions which express approximately the values of certain functions, I wish to make a few remarks, chiefly in the way of recapitulation, on what I find that I must now call "Murphy's Series". Although, when I wrote it as number (239) of

* [The next three paragraphs, labelled LXI.a, LXII.a, and LXIII.a (unfinished), appear only in Notebook D (Trinity College Dublin MS 1492/144, pp. 83–88) but not in the copy of the Letter (Trinity College, Dublin MS 1493/972). Each of these pages has a vertical line down the centre and carries the caption: *Page not adopted.* Revised versions of these paragraphs follow. It appears that the content of these paragraphs (written on 26 April 1858) was strongly influenced by De Morgan's letter to Hamilton of 11 April 1858 (Trinity College Dublin MS 1493/987) drawing attention to the work of Murphy. See footnote to equation (272).]

this Letter, I had not happened to see it in print. Rewriting it in the form before employed, but marking it now with the letter (M), we have

$$X = \frac{1^{-1}x^0}{\Gamma 1} - \frac{2^0 x^1}{\Gamma 2} + \frac{3^1 x^2}{\Gamma 3} - \frac{4^2 x^3}{\Gamma 4} + \text{\&c.} \qquad (239) = (M)$$

as the development, and to ascending powers of x, of the real root X of the equation

$$lX + xX = 0, \quad \text{or} \quad x = \frac{1}{X}\, l \frac{1}{X}, \qquad (259)$$

where that real root is unique and admits of such a development. This series was shown in XXXIII. to converge, if x^2 were not $> \epsilon^{-2}$; its *value* then admitting of being expressed by the *continued exponential*,

$$X = (\epsilon^{-x.})^{\infty} 1$$

where $\epsilon^{-x.}y$ is interpreted as equal to ϵ^{-xy}. Accordingly, I observe, (since you called my attention to his "Theory of Equations",) that Murphy writes (in his page 81), "if $a < \frac{1}{\epsilon}$ then

$$\epsilon^{a.\epsilon^{a.\epsilon^{a.\epsilon^{\cdot^{\cdot^{(ad.inf.)}}}}}} = 1 + \frac{2a}{1.2} + \frac{3^2 a^2}{1.2.3} + \frac{4^3 a^3}{1.2.3.4} + \text{\&c.;}\text{''}$$

adding that "if we put $a = \frac{1}{\epsilon}$ we find the left member of this equation to be merely ϵ; which is obvious, by supposing the series of indices to terminate at any distance, however remote. Hence

$$1 = \frac{1}{\epsilon} + \frac{2}{1.2.\epsilon^2} + \frac{3^2}{1.2.3\epsilon^3} + \frac{4^3}{1.2.3.4\epsilon^4} + \text{\&c.''}$$

I am not sure that this *inference* is *quite* so *obvious*, as he says that it is; at least if his *ladder** denote exactly the same thing as my own: but at all events, I had perceived the *result*, which in this Letter was written as

$$\epsilon = 1 + \epsilon^{-1} + \frac{3^1 \epsilon^{-2}}{1.2} + \frac{4^2 \epsilon^{-3}}{1.2.3} + \frac{5^3 \epsilon^{-4}}{1.2.3.4} \text{\&c.,} \qquad (268)$$

before I saw it in his pages. Now it is clear to me that Murphy had considered the effects of the substitution of a *negative* number for a in his series: but by all means let *him* have all the credit – and if I should ever publish even a *tithe*, or any smaller decimal fraction, of what I have been abusing your good-humour by writing to you, let *me* have the common sense not to pester general readers, by telling them that I had happened to perceive and discuss, for myself, if not for you, the series above marked (M).

* In fact, I admit that *each* of the expressions, $\epsilon^{a\epsilon}$, $\epsilon^{a\epsilon^{a\epsilon}}$, etc., *ad infinitum*, is 'obviously' equal to ϵ, when $a = \frac{1}{\epsilon}$; but this was not, as I conceive, the thing to be proved. What was to be shown (and has been or will be shown) in this Letter, is relatively to *this Letter*, that (when a is my $-\frac{1}{\epsilon}$) these exponential expressions ϵ^a, $\epsilon^{a\epsilon^a}$, $\epsilon^{a\epsilon^{a\epsilon^a}}$, etc., *converge*, though slowly, to the *limit* ϵ. In short I adopt in this place the *result*, but not the *logic* of Murphy.

LXII.a As long as that Series is *convergent*, whether what I have called x be positive or negative, the "exponential method" (perhaps not, *as such*, anticipated by Murphy, although he has certainly printed *one ladder*) *succeeds*; that is to say, if we write

$$X_n = (\epsilon^{-x\cdot})^n 1; \quad \text{then} \quad X_\infty = X;$$

X being the unique real root of the transcendental equation above written. But my discussion has shown, that even when x, being > 0, is also $> \epsilon^{-1}$, and so renders the *series divergent*, still if it be *not* $> \epsilon$, and so allow this *other* transcendental equation, (perhaps mine,)

$$\epsilon^{-x\epsilon^{-xX}} = X, \quad \text{or} \quad (\epsilon^{-x\cdot})^2 X = X,$$

to have *only one real root*, X, which in that case is *also* still the unique real root of the former equation, (undoubtedly in substance Murphy's) $\epsilon^{-xX} = X$, then the "exponential method" *still theoretically succeeds*, by giving a series or succession of values, X_n, which *converges to the limit* X; being *less* than that limit when the whole number n is *odd*, but *greater* than the same limit x, when the index n is *even*: although this convergence is *very slow*, and the *method* consequently *useless* (in *practice*); when x is taken *equal to*, or *very little less than*, this *new limit*, ϵ, which Murphy does not seem to have noticed. For example, whethter the "∞" denote an *odd* or an *even infinity*, we have, in *each* case, by the present Letter, definitely, or *unambiguously*, in point of *theory*

$$X_\infty = (\epsilon^{-\epsilon\cdot})^\infty 1 = \epsilon^{-1}; \tag{389}$$

while the *series* (M) becomes a *very divergent* one, when my x is replaced, in it, by $+\epsilon$; or when Murphy's a is made $= -\epsilon$, since it then becomes the series (269). But nobody, perhaps would have *patience* actually to *work out*, by *arithmetic*, including the use of logarithmic *tables*, the *value* (389) of this *last limit*, X_∞ even to (say) *five* decimal places; at least, if my numbers (388) be as I think that they are, nearly correct,* so far as they go: namely,

$$X_{149} = 0{\cdot}2987, \quad X_{150} = 0{\cdot}4440, \; X_{151} = 0{\cdot}2991; \tag{388}$$

the *distinction* between the *results*, for odd and even values of the *index*, n, continuing still to be so very well marked; although it is *proved*, that *this distinction must ultimately vanish*.

LXIII.a Still greater chance of being new, it seems that those subsequent results of the present Letter may have, according to which, when Murphy's "a", or my "$-x$", is replaced by a real value which is (algebraically) less than $-\epsilon$, (for example by the negative number, -3) then the *sum of the diverging series*, lately called (M), comes to admit of *three mutually distinct, but separately definite, interpretations*. If we write (to use the example just now referred to),

$$X = \frac{1^{-1}3^0}{\Gamma 1} - \frac{2^0 3^1}{\Gamma 2} + \frac{3^1 3^2}{\Gamma 3} - \frac{4^2 3^3}{\Gamma 4} + \frac{5^3 3^4}{\Gamma 5} - \frac{6^4 3^5}{\Gamma 6} + \&\text{c.},$$

then, as I need not tell you, from a simply *arithmetical* point of view, the *sum*, X, has *no meaning at all*. But if we inquire *from what algebraical process, applied to some function* X, might have arisen

* At the end of the last sheet of this Letter (in the future existence of what sheet you must be pardoned if you do not believe), I intended to give a short list of numerical errata *some* perhaps the fault of the copyist, but which I have happened to observe while glancing over the copy prepared.

a series, proceeding according to ascending powers of a variable, x, which should *come to coincide, term for term,* with the diverging series last written, when the symbol x was replaced by 3; and *what the arithmetical value of this function X becomes,* when x was so replaced: *then,* it is only just (as you have led me to see) to admit that Murphy – though apparently with a *leaning to reject diverging series,* – has yet supplied *one answer.* For he has shown, what in substance amounts to this: – that the series last written *might* have been formed, but seeking to resolve, or at least to find *one root,* (which turns out to be the *only real one,*) of the following equation in X:

$$\epsilon^{-3X} = X; \tag{413}$$

the number 3 being first replaced by the symbol x, and x being (at first) treated as small. The *real root of* (413) has been found to be, nearly

$$X = 0{\cdot}349\,97; \tag{414}$$

and to this limit, accordingly, it has been shown (in LIV.) that the "Method* of Continued Fractions", as used with the formulæ of this Letter, gives a not despicable approximation, derived *from the series itself,* or rather from its seven first terms; namely, $X = 0{\cdot}351\,47$, (443). But I am *not yet aware* – being however prepared to *thank you* very cordially, if you shall *inform* me – whether he, or any one else, has published the remarks:

Ist, that the *same series* (M) may be obtained by seeking to develop X, according to ascending powers of x, from the transcendental equation (above cited, as *perhaps my own*),

$$\epsilon^{-x\epsilon^{-xX}} = X, \quad \text{or} \quad a^{a^X} = x, \quad \text{when} \quad a = \epsilon^{-x};$$

the letter a having *here* a different signification from that in which it is used by Murphy;
IInd, that this equation has 3 *real and unequal roots, but only three,* if $x > \epsilon$, or if $a > 0, < \epsilon^{-\epsilon}$;
IIrd, that of *these* three real roots, the middle one is precisely the *unique* real root of the equation which I am willing (or rather bound) to call Murphy's – namely, in our recent notation,

$$\epsilon^{-xX} = X \qquad \text{or} \qquad a^X = X;$$

IVth, that *to this root,* X, we should *never* (under the *supposed condition* of inequality, $x > \epsilon$) *make any approach,* by continuing, *ever so far,* the process which is suggested by the notation,

$$X_\infty = (\epsilon^{-x\cdot})^\infty 1 = (a^\cdot)^\infty 1,$$

when $a^\cdot b$ is to be interpreted as denoting the exponential a^b;
Vth, that (even under the same condition of inequality) we should approach to the (middle) root X just mentioned; by the *method of converging fractions applied to the diverging series,* as in the lately cited instance (443), in which x was equal to 3;
VIth, that *to the least real root,* say A, of what (for the present) I call *my* transcendental equation (as distinguished from Murphy's) or to the *least real number* A which satisfies the condition

$$\epsilon^{-x\epsilon^{-xA}} = A, \quad \text{when} \quad x > \epsilon,$$

* I have not yet read enough of Murphy's *Book on Equations,* to know whether it occurred to him to use that *method* of transforming what I have called *his* series, (M).

we should indefinitely approximate, by using the formula,

$$A = \lim_{n=\infty} (\epsilon^{-x.})^{2n+1} 1;$$

VIIth, that *to the greatest real root, B,* of the same transcendental equation in *X*, or to the *greatest real number B* which satisfies the condition

$$\epsilon^{-x\epsilon^{-xB}} = B, \quad \text{when} \quad x > \epsilon,$$

we should make an indefinite approach, by using this other formula,

$$B = \lim_{n=\infty} (\epsilon^{-x.})^{2n} 1;$$

VIIIth, that *both* these formulæ, for the *two extreme roots A* and *B, as well* as the *fractions* which converge to the *intermediate root X*, might have been *suggested* (and in fact *were* so *to me*) by applying *to the series itself* the *methods* developed in the present Letter;
IXth, that the two extreme roots, *A* and *B*, are *connected with each other*, by the relations,

$$A = \epsilon^{-xB}, \; B = \epsilon^{-xA};$$

Xth, what perhaps I have not formally proved, but what easily follows from principles already stated, that when we have thus *two real and unequal numbers*, A and $B > A$, which satisfy the two equations, $A = a^B$, and $B = a^A$, where $a > 0$, $< \epsilon^{-\epsilon}$, then although there is *one real and intermediate number*, X, $> A$, $< B$, such that $a^X = X$, yet if we take *any other real number*, say Y, and determine Z by the equation $Z = a^Y$, the number Z thus determined will be *unequal* to Y; Z being $> B$, if $Y < A$; $Z < A$, if $Y > B$; $Z < B$, $> X$, if $Y > A$, $< X$; and $Z > A$, $< X$, if $Y < B$, $> X$.

LXI. After this digression, on the nature and limits of the roots of certain algebraic equations, suggested by the converging fractions which express approximately the values of certain functions, I wish to make a few remarks, *partly* in the way of *recapitulation*, on the finite or infinite but real values, of the combined exponential

$$b^{b^{b^{\cdot^{\cdot^{\cdot^{c}}}}}} = (b^{\cdot})^{\infty} c; \tag{504}$$

where b and c are supposed to be real and b is positive; all *logarithms of negatives* being for the present *excluded* from our view; and the function

$$b^c = b^{\cdot} c \tag{505}$$

receiving an *arithmetical* signification, so as to denote *definitely one real and positive number*. The following are my results, (perhaps all anticipated,) respecting the values in question. Some are obvious at sight; others have been proved in this Letter; and the rest can easily be deduced from principles of the same general class.

(1) If $b > \epsilon^{\frac{1}{\epsilon}}$, then $(b^{\cdot})^{\infty} c = \infty$. (506)
(2) If $b = \epsilon^{\frac{1}{\epsilon}}$, and $c > \epsilon$, $(b^{\cdot})^{\infty} c = \infty$. (507)
(3) If $b = \epsilon^{\frac{1}{\epsilon}}$, and $c \ngtr \epsilon$, $(b^{\cdot})^{\infty} c = \epsilon$. (508)
(4) If $b < \epsilon^{\frac{1}{\epsilon}}$, but > 1, and $c > y'$, $(b^{\cdot})^{\infty} c = \infty$; (509)
y' being the greater of the two real roots of the equation
$b^y = y$. (510)

(5) If $b < \epsilon^{\frac{1}{\epsilon}}, > 1$, and $c = y'$, $(b^{\cdot})^{\infty} c = y'$. (511)

(6) If $b < \epsilon^{\frac{1}{\epsilon}}, > 1$, and $c < y'$, $(b^{\cdot})^{\infty} c = y$; (512)

y being the lesser of the two real roots of the equation (510).

(7) If $b = 1$, $(b^{\cdot})^{\infty} c = 1$. (513)

(8) If $b < 1$, but $> \epsilon^{-\epsilon}$, then $(b^{\cdot})^{\infty} c = y$; (514)

y being here the only real root of the equation (510).

(9) If $b = \epsilon^{-\epsilon}$, the equation $b^y = y$ (510) has still only one real root, which is now $= \epsilon^{-1}$; and $(b^{\cdot})^{\infty} c = (\epsilon^{-\epsilon^{\cdot}})^{\infty} c = \epsilon^{-1}$. (515)

(10) If $b < \epsilon^{-\epsilon}$ but > 0, the equation (510) has still only one real root, y; and we have (still) $(b^{\cdot})^{\infty} y = y$. (516)

(11) If $b < \epsilon^{-\epsilon}, > 0$, the following other transcendental equation,

$$(b^{\cdot})^2 y = b^{b^y} = y, \tag{517}$$

has three real and unequal roots, whereof the middle one is the root y of (510), while the least and greatest may be denoted respectively by A and B; and then if $c > y$,

$$(b^{\cdot})^{2\infty} c = B, \quad (b^{\cdot})^{2\infty+1} c = A. \tag{518}$$

(12) But, finally, under the same conditions, $b < \epsilon^{-\epsilon}$, $b > 0$, and with the same significations of A, y, B, as the 3 real roots of the equation (517),

$$\text{if } c < y, \text{ then } (b^{\cdot})^{2\infty} c = A, \ (b^{\cdot})^{2\infty+1} c = B. \tag{519}$$

And thus *all possible cases* of values of the continued exponential (504), (for any real c, and for any real and positive b, with arithmetical interpretations) appear to be *exhausted* – A *case* of number (3) has been remarked by Murphy in his *Equations* (page 81), – for calling my attention to which work I have again to thank you; namely the case (sufficiently "obvious" indeed, as he himself pronounces it to be), where

$$c = \epsilon; \tag{520}$$

so that the formula becomes, Limit of the Exponentials,

$$\epsilon^{a\epsilon}, \epsilon^{a\epsilon^{a\epsilon}}, \ldots$$

or (as he writes it)

$$\epsilon^{a.\epsilon^{a.\epsilon^{a.\epsilon^{\cdot^{\cdot^{(ad\ inf.)}}}}}} = \epsilon, \text{ when } a = \epsilon^{-1}.$$

LXII. Since I have again mentioned Murphy, let me try to do to him some further justice, by pointing out an *approach* which I observe that he made, in his *Equations*, to one of the Theorems of the present Letter; namely to that which is expressed by the formula*

* The notation here used, $\sum\limits_{m=0}^{m=\infty}$, is certainly cumbersome to write, and perhaps expensive to print. I have seen one like this, $\sum_{m\ 0}^{m=\infty}$. Have *you* any favourite notation of this sort of sum? I formerly accustomed myself to write $\sum_{(m)0}^{\infty}$, and am still open to suggestions – my habits have not yet quite hardened me.

$$f(x+h) = \sum_{m=0}^{m=\infty} \frac{(m\alpha+1)^{m-1}}{\Gamma(m+1)} h^m D^m f(x - m\alpha h). \tag{370}$$

It is very possible that I may yet find for myself, or have it pointed out to me by you, that this Theorem has been *completely* anticipated by him: but, *as yet*, I can only say that I admit him to have *at least* come *very near* to seeing it. At this moment I cannot lay my hand upon his Book, but from my own "E.1858" I extract (with some trifling alteration) the following Memorandum.* "Murphy derives the series,

$$x^n = c^n + nac^{n+1} + \frac{n}{2}(n+2)\,a^2 c^{n+2} + \frac{n}{2}\frac{(n+3)^2}{3}\,a^3 c^{n+3} + \&c.,$$

from $x = c\epsilon^{ax}$, by considering $x^n = fx$ as the coefficient of $\frac{1}{x}$ in the development of

$$-f'x.\left(1 - \frac{c\epsilon^{ax}}{x}\right) = -nx^{n-1}\, l\left(1 - \frac{c\epsilon^{ax}}{x}\right);$$

(writing l as the characteristic of natural logarithms;) or in

$$n\left\{\frac{1}{n}\frac{c^n\epsilon^{nax}}{x} + \frac{1}{n+1}\frac{c^{n+1}\epsilon^{(n+1)ax}}{x^2} + \&c.\right\}.$$

Change x to cX; then $ac = \frac{lX}{X}$; and it is found, from Murphy's cited series, that

$$X^n = 1 + n\frac{lX}{X} + \frac{n}{2}(n+2)\left(\frac{lX}{X}\right)^2 + \frac{n}{2}\frac{(n+3)^2}{3}\left(\frac{lX}{X}\right)^3 + \&c.;$$

a result which I do not observe that *he* has *expressly* drawn, but which exactly agrees with one of my own deductions from Lagrange's Theorem. I see another mode of deducing the same series from Murphy's principles. – But it seems to me that if *he had* perceived it, he would at once have *gone on* to derive from it that development for $f(x+h)$, which I lately communicated to Charles Graves†, and to De Morgan, and which has the peculiarity of involving a *foreign and arbitrary constant.* For as soon as he obtained the equation, answering to the particular $n = 1$, namely

$$a = 1 + \frac{2A}{2} + \frac{(3A)^2}{2.3} + \frac{(4A)^3}{2.3.4} + \&c.,$$

where $A = lA/a$, he seems to have *immediately* seen that it admitted of what he calls the "remarkable extension", namely

$$f(x+k) = fx + \frac{2k}{2} f'(x-k) + \frac{(3k)^2}{2.3} f''(x-2k) + \&c.$$

And it would scarcely have occupied *more room* in his Treatise on *Equations*, to have written the *more general series*, which I happened to perceive for myself,

* [Notebook E (1858), p. 13. Trinity College, Dublin MS 1492/145.]

† [Charles Graves (1812–1899) was Erasmus Smith's professor of mathematics at Trinity College, Dublin, from 1843 until 1862.]

$$f(x+h) = fx + hDf(x-\alpha h) + \frac{(1+2\alpha)^1}{1.2}h^2D^2f(x-2\alpha h)$$
$$+\frac{(1+3\alpha)^2}{1.2.3}h^3D^3f(x-3\alpha h) + \&c.$$

– But he may possibly have printed this last series in the *Philosophical Transactions* [*of the Royal Society of London*] or in the Cambridge Memoirs [*Transactions of the Cambridge Philosophical Society*]. In the meantime I am most willing to give him *all* credit in the matter – and consider it as a sufficient repayment for my trouble in writing the present Letter – which trouble, most certainly, *you* never *asked* me to take – that my attention has been called, by it and you, to what you describe in your *Differential and Integral Calculus* (page 328), as "One of the most general and interesting contributions which analysis has received for many years". I also admit, most cheerfully, that Murphy printed *one* "*ladder*": although I do not *yet* see that he had anticipated the "Exponential Method" of the present Letter, nor the series of Theorems respecting $(b^{\cdot})^{\infty}c$, which is given in the foregoing paragraph (LXI.).

LXIII. You can scarcely be more glad than myself to have *done* with what I must now call "Murphy's Series" – though you know that I merely brought it in (when I thought it my own), for the sake of *illustration* of some general methods. Still a remark or two, in the nature of "wind-up", may be endured, before wholly leaving that *part* of the whole subject. Rewriting the series, in the form before adopted in this Letter, but with the signature (M), as

$$X = \frac{1^{-1}x^0}{\Gamma 1} - \frac{2^0x^1}{\Gamma 2} + \frac{3^1x^2}{\Gamma 3} - \frac{4^2x^3}{\Gamma 4} + \&c., \qquad (239) = (M)$$

I admit, (I.), that Murphy preceded me in perceiving (not only the *existence* but) the *convergence* of this series, when $x^2 \not> \epsilon^2$. (He used terms all positive, but let me not nibble at a point of that sort) – I admit, (II.), that he saw the series to be, *when convergent,* the development of at least *one root* of the equation which I wrote as

$$lX + xX = 0, \quad \text{or} \quad x = \frac{1}{X}l\frac{1}{X}; \qquad (259)$$

and have no doubt, (III.) that he was well aware that *this* was the *lesser of the two real roots,* of the last equation considered as resolved for X, when x was given, and negative, but $> -\epsilon^{-1}$. He probably saw, (IV.), that, in your words, or something like them, the series "escapes from *imaginariness,* by becoming *divergent*", when $x > \epsilon^{-1}$. He perceived, (V.), that the series, *when* convergent, must be regarded as the development of a *continued exponential.* But I am not yet aware that it even *occurred* to him, or that he thought it at all *worth his while* to inquire, (VI.), *what value ought* to be, *or might be,* attributed to the series, when x, being positive, (in my form of the development) *exceeds the limit* ϵ^{-1}, and so causes the *series* to *diverge,* while yet the *equation* from which it was derived continues to have a *single real root. As yet,* the results of this Letter upon *that* subject appear to me to be new; but you know how very rash it would be to make *assertion* of any such novelty. Still, to pursue our *recapitulation,* I remark, (VII.), that when $x > \epsilon^{-1}$, *provided that x does not exceed a certain greater limit,* ϵ, the *series* indeed *diverges,* but the *only real root* X of the *equation* $lX + xX = 0$ may *still* be obtained, with any required approximation, by the *continued exponential,* $X = (\epsilon^{-x\cdot})^{\infty}1$, which might, by one of the methods of the present Letter, have been obtained *from the series itself.* But (VIII.), when x *is near* the *limit*, ϵ, the *convergence* of this *exponential* becomes extremely *slow*; since it was found that if $X_n = (\epsilon^{-\epsilon \cdot})^n 1$, then

$$X_{149} = 0{\cdot}2987, \quad X_{150} = 0{\cdot}4440, \quad X_{151} = 0{\cdot}2991, \tag{388}$$

nearly, the *alternate values* showing, even at this advanced stage of the process, (which thus becomes *here* practically *useless*,) a *well-marked difference* from each other, which difference however would necessarily *at last disappear*; so that the process would *ultimately* give (it has been proved) the equation $(\epsilon^{-\epsilon \cdot})^\infty = \epsilon^{-1}$ (389) – (IX.) The exponential Method would thereby assign what *may* (or *must*)* be considered as *the value*, namely ϵ^{-1}, of the very *divergent series*,

$$\epsilon^{-1} = 1 - \epsilon + \frac{3^1\epsilon^2}{2} - \frac{4^2\epsilon^3}{2.3} + \frac{5^3\epsilon^4}{2.3.4} - \&c.; \tag{269}$$

which Murphy has abstained from considering, apparently as not choosing to deal with any series except converging ones, but which it has been part of the object of this Letter to discuss. In like manner, this *less divergent* series,

$$X = 1 - 2^{-1} + \frac{3^1 2^{-2}}{2} - \frac{4^2 2^{-3}}{2.3} + \frac{5^3 2^{-4}}{2.3.4} - \&c., \tag{281}$$

may be considered as representing this other continued exponential,

$$X = (\epsilon^{-\frac{1}{2}\cdot})^\infty 1 = 0{\cdot}703\,467\,4,$$

nearly, by (445), namely the only real root X of the equation

$$\frac{1}{X} l \frac{1}{X} = \frac{1}{2}, \tag{282}$$

and similarly in other cases. It has been shown, (X.), that the *same value* X for the *same diverging series*, under the *same conditions*, $x > \epsilon^{-1}$, $\not> \epsilon$, may be approached to, and with *greater rapidity*, by the use of *converging fractions*, obtained *from the series itself.* We found (XI.), that when $x > \epsilon$, the equation $lX + xX = 0$ has *still* one and only *one real root* X, to which the *fractions still converge*; although the *series* is now *still more divergent* than before. For example, the seven first terms of the highly divergent series,

$$X = 1 - 3 + \frac{3^1 3^2}{2} - \frac{4^2 3^3}{2.3} + \frac{5^3 3^4}{2.3.4} - \&c., \tag{521}$$

the *Method of Continued Fractions*, combined with a plan of improving an approximation by *combining* the few last results, which has somewhere been used by Legendre,† has given us the approximate value, $X = 0{\cdot}351\,47$; (443): while the real root of the equation $lX + 3X = 0$, or $\epsilon^{-3X} = X$, (413), is $X = 0{\cdot}349\,97$ nearly, by (414); so that the agreement of results even in this unfavourable case, is not to be entirely despised. But (XII.), when x is thus $> \epsilon$, the *Method*

* At least I think so, if a diverging series be admitted to have *any value*; on account of the *agreement* of the *results* here obtained, by *different processes* of transformation.

† [Adrien Marie Legendre (1752–1833).]

of Continued Exponentials gives *no approach to the value of X, however far* the process may be carried; for example it has been found that, nearly,

$$(\epsilon^{-3\cdot})^{37}1 = 0{\cdot}1348;\ (\epsilon^{-3\cdot})^{38}1 = 0{\cdot}6673;\ (\epsilon^{-3\cdot})^{39}1 = 0{\cdot}1351;\ (\epsilon^{-3\cdot})^{40}1 = 0{\cdot}6668; \qquad (421)$$

the convergence being *here* towards *two alternate limits,* A and B, which are respectively the *least and greatest real roots* of the transcendental equation,

$$(\epsilon^{-3\cdot})^{2}X = \epsilon^{-3\epsilon^{-3X}} = X, \qquad (415)$$

namely, $A = 0{\cdot}136\,12$, and $B = 0{\cdot}664\,74$, nearly, by (416); while the same equation (415) has *another real root, intermediate* between these roots A and B, namely the root $X = 0{\cdot}349\,97$ of the former equation $\epsilon^{-3X} = X$ (413). – This is perhaps the *most remarkable result of the present Letter* (which may not be saying much for it); and it may be worth while to stop for a moment to consider its *geometrical signification.*

LXIV. Suppose that we met the equation,

$$\epsilon^{-x\epsilon^{-xy}} = y, \qquad (522)$$

and chose to regard it as the *equation of a curve.* Abstracting from all *imaginary points,* and from all *logarithms of negatives, the shape* would be found such as is rudely sketched, in the annexed Figure 1. For I can prove that the *real roots* y of the equation (522) are precisely the *same as those of the simpler equation*

$$\epsilon^{-xy} = y, \qquad (523)$$

if x be less than ϵ; but that *as soon as* x *becomes* equal to, or greater than ϵ, then *two new real roots start into existence,* for the equation (522), *which do not belong* to the less complex equation (523), except that when x is *exactly* equal to ϵ, then the two new roots of (522) are *equal* to each other, and to the root $y = \epsilon^{-1}$, of the equation (523). (Figure 1.) Now the curve represented by this *last* equation, or by this other form of it,

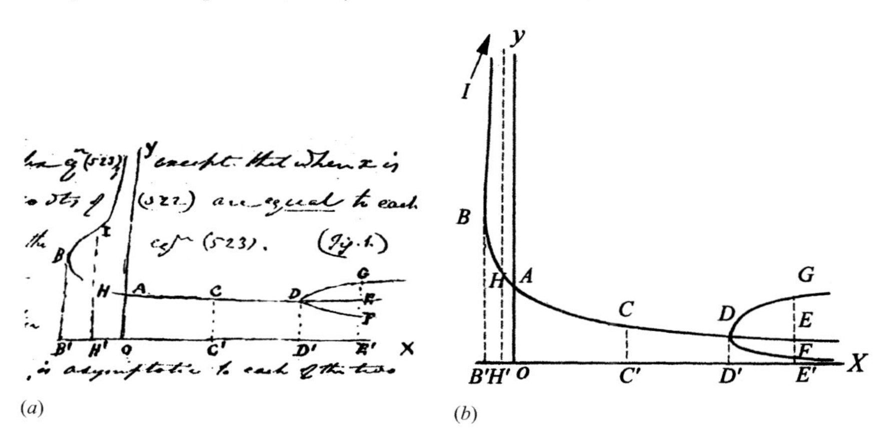

Fig. 1 (*a*) This shows the curves as sketched in Hamilton's notebook D (p. 95 Trinity College, Dublin, MS 1492/144); (*b*) accurately drawn versions of the same curves.

$$x = \frac{1}{y}\, l \frac{1}{y}, \tag{524}$$

is asymptotic to each of the two positive semiaxes *OX*, *OY* of coordinates; crossing the axis of y at a point *A* for which $x = 0$, $y = 1$; and extending behind that axis, to a point *B*, for which

$$x = \epsilon^{-1}, \qquad y = \epsilon, \tag{525}$$

and at which it is *touched* by the ordinate *BB'*: in such a manner that for every point H', between 0 and B', an ordinate *cuts* the curve in 2 *real points*, *H* and *I*; but for every point C', D', E', ..., on the positive side of *O*, *an ordinate intersects the curve* (523) *in only one real point*, *C*, *D*, *E*, ... The cuve (522) *contains all these real points of* (523); but *it undergoes a trifurcation* at a point *D*, for which

$$x = +\epsilon, \qquad y = \epsilon^{-1}. \tag{526}$$

Each curve, for real points not too far from A, namely for the real range from *B* to *C*, if *C* be such that $OC' = \epsilon^{-1}$, has its ordinate, $B'B$, or $H'H$, or *OA*, or $C'C$, respresented by the *convergent series* of Murphy, or here by

$$y = 1 - x + \frac{3x^2}{2} - \frac{4^2 x^3}{2.3} + \&\text{c.}; \tag{M}$$

and the *value* of this ordinate, or of this series, (for $x^2 \ngtr \epsilon^{-2}$) may be *approximately found*, either by a *continued exponential*, or by a *continued fraction*, as in this Letter. For the range from *C* to *D*, these *methods still give the common ordinate of the* 2 *curves*, although now the *series diverges*. But *beyond the point of trifurcation D*, the 3 real ordinates of the curve (522) are found *approximately*, and *separately*, as 3 *distinct limits*, by methods already explained, as *so many different transformations of the diverging series*; namely the *middle ordinate*, y or $E'E$, which is *also* that of the *simpler curve* (523), as the limit of the *converging fractions* deduced *from the series itself*; the *least ordinate*, *A* [see equation (518)], or $E'F$, as the limit of the *increasing exponentials of odd orders*, $(\epsilon^{-x\cdot})^{2n+1}1$; and finally the *greatest ordinate B* [see equation (518)], or $E'G$, as the limit of the *decreasing exponentials of even orders*, $(\epsilon^{-x\cdot})^{2n}1$. The branch *DF*, like *DE*, is *asymptotic to the axis OX*; and the branch *DG* is *asymptotic to a parallel* to that axis, drawn through the point *A* – And now at last I take (for the present) my farewell of this curve and series. But something still remains to be said, on the subject of those *general transformations, to which allusions have been made in this Letter.*

LXV. Let us therefore now suppose that $f_1 x$, $f_2 x$, ... are functions such that

$$f_1 x = 1 + 1_1 x + 1_2 x^2 + \cdots, \qquad f_2 x = 1 + 2_1 x + 2_2 x^2 + \cdots,$$
$$f_3 x = 1 + 3_1 x + 3_2 x^2 + \cdots, \ \&\text{c.}, \tag{527}$$

where m_n is a constant coefficient, independent of x. Let us write also,

$$F_1 x = f_1 x f_2 x f_3 x f_4 x \ldots, \qquad F_2 x = f_2 x f_3 x f_4 x \ldots, \qquad F_3 x = f_3 x f_4 x \ldots, \ \&\text{c.} \tag{528}$$

where each functional characteristic is conceived to govern the whole system of symbols which follow it, so that, for instance, we have, more fully,

$$F_1 x = f_1 (x f_2 (x f_3 (x f_4 (x \ldots)))). \tag{529}$$

Then each of these functions $F_r x$ may in general be developed in a series according to ascending powers of x, which may be thus denoted:

$$F_r x = 1 + (r,\,1)\,x + (r,\,2)\,x^2 + (r,\,3)\,x^3 + \text{\&c.}; \tag{530}$$

and because we have the relation,

$$F_r x = f_r(xF_{r+1}\,x); \tag{531}$$

we may write also,

$$F_1 x = 1 + r_1\,xF_{r+1}\,x + r_2\,x^2(F_{r+1}\,x)^2 + r_3\,x^3(F_{r+1}\,x)^3 + \text{\&c.} \tag{532}$$

Hence,

$$\left.\begin{aligned}
&(r,\,1) = r_1; \quad (r,\,2) = r_1(r+1,\,1) + r_2; \quad (r,\,3) = r_1(r+1,\,2) + 2r_2(r+1,\,1)\,r_3;\\
&(r,\,4) = r_1(r+1,\,3) + r_2\{2(r+1,\,2) + (r+1,\,1)^2\} + 3r_3(r+1,\,1) + r_4;\\
&(r,\,5) = r_1(r+1,\,4) + r_2\{2(r+1,\,3) + 2(r+1,\,1)(r+1,\,2)\}\\
&\qquad + r_3\{3(r+1,\,2) + 3(r+1,\,1)^2\} + 4r_4(r+1,\,1) + r_5;\\
&(r,\,6) = r_1(r+1,\,5) + r_2\{2(r+1,\,4) + 2(r+1,\,1)(r+1,\,3)\\
&\qquad + (r+1,\,2)^2\} + r_3\{3(r+1,\,3) + 6(r+1,\,1)(r+1,\,2) + (r+1,\,1)^3\}\\
&\qquad + r_4\{4(r+1,\,2) + 6(r+1,\,1)^2\} + 5r_5(r+1,\,1) + r_6; \quad \text{\&c.:}
\end{aligned}\right\} \tag{533}$$

where the calculus of derivations* may be used, to assist in determining the coefficients and exponents. In this manner it may be shown that

$$\left.\begin{aligned}
&(1,\,1) = 1_1; \quad (1,\,2) = 1_1 2_1 + 1_2; \quad (1,\,3) = 1_1 2_1 3_1 + 1_1 2_2 + 2.1_2 2_1 + 1_3;\\
&(1,\,4) = 1_1 2_1 3_1 4_1 + 1_1 2_1 3_2 + 2.1_1 2_2 3_1 + 1_1 2_3 + 2.1_2 2_1 3_1 + 1_2(2.2_2 + 2_1^2) + 3.1_3 2_1 + 1_4;\\
&(r,\,5) = 1_1 2_1 3_1 4_1 5_1 + 1_1 2_1 3_1 4_2 + 2.1_1 2_1 3_2 4_1 + 2.1_1 2_2 3_1 4_1 + 2.1_2 2_1 3_1 4_1\\
&\qquad + 1_1 2_1 3_3 + 1_1 2_2(2.3_2 + 3_1^2) + 2.1_2 2_1 3_2 + 3.1_1 2_3 3_1 + 2.1_2(2.2_2 + 2_1^2)3_1\\
&\qquad + 3.1_3 2_1 3_1 + 1_1 2_4 + 2.1_2(2_3 + 2_1 2_2) + 3.1_3(2_2 + 2_1^2) + 4.1_4 2_1 + 1_5;\\
&(1,\,6) = 1_1 2_1 3_1 4_1 5_1 6_1 + 1_1 2_1 3_1 4_1 5_2 + 2.1_1 2_1 3_1 4_2 5_1 + 2.1_1 2_1 3_2 4_1 5_1 + 2.1_1 2_2 3_1 4_1 5_1\\
&\qquad + 2.1_2 2_1 3_1 4_1 5_1 + 1_1 2_1 3_1 4_3 + 1_1 2_1 3_2(2.4_2 + 4_1^2) + 2.1_1 2_2 3_1 4_2\\
&\qquad + 2.1_2 2_1 3_1 4_2 + 3.1_1 2_1 3_3 4_1 + 2.1_1 2_2(2.3_2 + 3_1^2)4_1 + 4.1_2 2_1 3_2 4_1\\
&\qquad + 3.1_1 2_3 3_1 4_1 + 2.1_2(2.2_2 + 2_1^2)3_1 4_1 + 3.1_3 2_1 3_1 4_1 + 1_1 2_1 3_4\\
&\qquad + 2.1_1 2_2(3_3 + 3_1 3_2) + 2.1_2 2_1 3_3 + 3.1_1 2_3(3_2 + 3_1^2)\\
&\qquad + 1_2(2.2_2 + 2_1^2)(2.3_2 + 3_1^2) + 3.1_3 2_1 3_2 + 4.1_1 2_4 3_1 + 6.1_2(2_3 + 2_1 2_2)3_1\\
&\qquad + 6.1_3(2_2 + 2_1^2)3_1 + 4.1_4 2_1 3_1 + 1_1 2_5 + 1_2(2.2_4 + 2.2_1 2_3 + 2_2^2)\\
&\qquad + 1_3(3.2_3 + 6.2_1 2_2 + 2_1^3) + 1_4(4.2_2 + 6.2_1^2) + 5.1_5 2_1 + 1_6;
\end{aligned}\right\} \tag{534}$$

* [See: A. De Morgan, *The differential and integral calculus*, pp. 328—335, London: 1842.]

where it is worth noticing that, with the incorporations indicated by the parentheses, of which it is easy to see the reason, the *number of terms goes on doubling*, being

$$\left.\begin{array}{llll} = 1, \text{ for } (1, 1); & = 2, \text{ for } (1, 2); & = 2^2, \text{ for } (1, 3); & = 2^3, \text{ for } (1, 4); \\ = 2^4, \text{ for } (1, 5); & \text{and} = 2^5, \text{ for } (1, 6). & & \end{array}\right\} \tag{535}$$

As a verification, if we assume

$$m_n = m^n, \tag{536}$$

we have the particular forms,

$$f_1 x = \frac{1}{1-x}, \; f_2 x = \frac{1}{1-2x}, \ldots, \; f_m x = \frac{1}{1-mx}; \tag{537}$$

thus, we must have, by a transformation which Laplace appears to claim, and which has before been cited in this Letter, (with merely a change of certain signs,)

$$F_1 x = \frac{1}{1-}\frac{1x}{1-}\frac{2x}{1-}\frac{3x}{1-}\text{\&c.} = 1 + 1x + 1.3x^2 + 1.3.5x^3 + \text{\&c.}: \tag{538}$$

we ought therefore to have, with the assumption (536),

$$(1, n) = (2n-1)(2n-3)\ldots 5.3.1 \tag{539}$$

and accordingly we find now the particular numerical values following:

$$\left.\begin{array}{l} (1, 1) = 1; \; (2, 1) = 2 + 1 = 3; \; (1, 3) = 6 + 4 + 4 + 1 = 15 = 3.5; \\ (1, 4) = 24 + 18 + 24 + 8 + 12 + 12 + 6 + 1 = 105 = 3.5.7; \\ (1, 5) = 120 + 96 + 144 + 96 + 48 + 54 + 108 + 36 + 72 + 72 + 18 + 16 \\ \qquad + 32 + 24 + 8 + 1 = 945 = 3.5.7.9; \\ (1, 6) = 720 + 600 + 960 + 720 + 480 + 240 + 384 + 864 + 384 + 192 + 648 \\ \qquad + 864 + 288 + 288 + 288 + 72 + 162 + 432 + 108 + 432 + 324 + 54 + 192 \\ \qquad + 288 + 144 + 24 + 32 + 80 + 80 + 40 + 10 + 1 = 10\,395 = 3.5.7.9.11. \end{array}\right\} \tag{540}$$

In general let there be two mutually inverse functions, of the respective forms,

$$\begin{gathered} \varphi x = b_1(x-1) + b_2(x-1)^2 + \text{\&c.}, \quad \text{and} \\ \varphi^{-1} x = \phi x = 1 + c_1 x + c_2 x^2 + \text{\&c.}; \end{gathered} \tag{541}$$

and let

$$X = \phi\alpha_1 x\phi\alpha_2 x\phi\alpha_3 x\phi\alpha_4 x\phi\alpha_5 x\phi\alpha_6 x \ldots = 1 + a_1 x + a_2 x^2 + \text{\&c.} \tag{542}$$

Then, by making

$$m_n = c_n \alpha_m^n, \quad \text{and} \quad (1, n) = a_n, \tag{543}$$

the expressions (534) – respecting the general composition of which, some interesting remarks might be made – become:

$$\left.\begin{aligned}
a_1 &= c_1\alpha_1; \quad a_2 = c_1^2\alpha_1\alpha_2 + c_2\alpha_1^2;\\
a_3 &= c_1^3\alpha_1\alpha_2\alpha_3 + c_1c_2\alpha_1\alpha_2^2 + 2c_1c_2\alpha_1^2\alpha_2 + c_3\alpha_1^3;\\
a_4 &= c_1^4\alpha_1\alpha_2\alpha_3\alpha_4 + c_1^2c_2\alpha_1\alpha_2(\alpha_3 + 2\alpha_2 + 2\alpha_1)\alpha_3 + c_1c_3\alpha_1\alpha_2^3\\
&\quad + c_2(2c_2 + c_1^2)\alpha_1^2\alpha_2^2 + 3c_1c_3\alpha_1^3\alpha_2 + c_4\alpha_1^4\\
&= c_1^4\alpha_1\alpha_2\alpha_3\alpha_4 + c_1^2c_2\alpha_1\alpha_2\{\alpha_3^2 + 2(\alpha_2+\alpha_1)\alpha_3 + \alpha_1\alpha_2\}\\
&\quad + c_1c_3\alpha_1\alpha_2(\alpha_2^2 + 3\alpha_1^2) + 2c_2^2\alpha_1^2\alpha_2^2 + c_4\alpha_1^4;\\
a_5 &= c_1^5\alpha_1\alpha_2\alpha_3\alpha_4\alpha_5 + c_1^3c_2\alpha_1\alpha_2\alpha_3\{\alpha_4^2 + 2(\alpha_3+\alpha_2+\alpha_1)\alpha_4 + \alpha_2(\alpha_3 + 2\alpha_1)\}\\
&\quad + c_1^2c_3\alpha_1\alpha_2(\alpha_3^3 + 3\alpha_2^2\alpha_3 + 3\alpha_1^2\alpha_3 + 3\alpha_1^2\alpha_2) + 2c_1c_2^2\alpha_1\alpha_2(\alpha_2\alpha_3^2 + \alpha_1\alpha_3^2\\
&\quad + 2\alpha_1\alpha_2\alpha_3 + \alpha_1\alpha_2^2) + c_1c_4\alpha_1\alpha_2(\alpha_2^3 + 4\alpha_1^3) + c_2c_3\alpha_1^2\alpha_2^2(2\alpha_2 + 3\alpha_1) + c_5\alpha_1^5;\\
a_6 &= c_1^6\alpha_1\alpha_2\alpha_3\alpha_4\alpha_5\alpha_6 + c_1^4c_2\alpha_1\alpha_2\alpha_3\{\alpha_4(\alpha_5^2 + 2\alpha_4\alpha_5 + 2\alpha_3\alpha_5 + 2\alpha_2\alpha_5\\
&\quad + 2\alpha_1\alpha_5 + \alpha_3\alpha_4 + 2\alpha_2\alpha_3 + 2\alpha_1\alpha_2) + \alpha_1\alpha_2\alpha_3\}\\
&\quad + c_1^3c_3\alpha_1\alpha_2\{\alpha_3\alpha_4^3 + 3\alpha_3\alpha_4(\alpha_3^2 + \alpha_2^2 + \alpha_1^2) + 3\alpha_2\alpha_3(\alpha_2\alpha_3 + 2\alpha_1^2)\\
&\quad + \alpha_1^2\alpha_2^2\} + 2c_1^2c_2^2\alpha_1\alpha_2\alpha_3\{(\alpha_3+\alpha_2+\alpha_1)\alpha_4^2 + 2(\alpha_2\alpha_3 + \alpha_1\alpha_3 + \alpha_1\alpha_2)\alpha_4\\
&\quad + \alpha_2(\alpha_3^2 + 2\alpha_1\alpha_3 + 3\alpha_1\alpha_2)\} + c_1^2c_4\alpha_1\alpha_2\{\alpha_3^4 + 4\alpha_3(\alpha_2^3 + \alpha_1^3) + 6\alpha_1^3\alpha_2\}\\
&\quad + c_1c_2c_3\alpha_1\alpha_2\{\alpha_3(2\alpha_2\alpha_3^2 + 2\alpha_1\alpha_3^2 + 3\alpha_2^2\alpha_3 + 3\alpha_1^2\alpha_3 + 6\alpha_1\alpha_2^2 + 6\alpha_1^2\alpha_2)\\
&\quad + 2\alpha_1\alpha_2^2(\alpha_2 + 3\alpha_1)\} + c_1c_5\alpha_1\alpha_2(\alpha_2^4 + 5\alpha_1^4) + c_2^3\alpha_1^2\alpha_2^2(4\alpha_3^2 + \alpha_2^2)\\
&\quad + 2c_2c_4\alpha_1^2\alpha_2^2(\alpha_2^2 + 2\alpha_1^2) + 3c_3^2\alpha_1^3\alpha_2^3 + c_6\alpha_1^6.
\end{aligned}\right\} \tag{544}$$

If we make, as a verification and application

$$c_n = (-c)^n, \tag{545}$$

so that the second equation (541) shall give

$$\phi x = (1 + cx)^{-1} \tag{546}$$

then

$$\left.\begin{aligned}
&-c^{-1}a_1 = \alpha_1; \quad +c^{-2}a_2 = \alpha_1(\alpha_2+\alpha_1); \quad -c^{-3}a_3 = \alpha_1\alpha_2\alpha_3 + a_1(\alpha_2+\alpha_2)^2;\\
&+c^{-4}a_4 = \alpha_1\alpha_2\alpha_3\alpha_4 + \alpha_1\alpha_2(\alpha_3+\alpha_2+\alpha_1)^2 + \alpha_1^2(\alpha_2+\alpha_1)^2;\\
&-c^{-5}a_5 = \alpha_1\alpha_2\alpha_3\alpha_4\alpha_5 + \alpha_1\alpha_2\alpha_3(\alpha_4+\alpha_3+\alpha_2+\alpha_1)^2 + \{\alpha_2\alpha_3 + (\alpha_2+\alpha_1)^2\};
\end{aligned}\right\} \tag{547}$$

and these functions of $\alpha_1, \alpha_2, \alpha_3, \ldots$, agree precisely with those functions of $\alpha, \beta, \gamma, \ldots$, which were called $a, b, c, \ldots$, in (70), and which were the coefficients of $-x, +x^2, -x^3, \ldots$, in the development of the continued fraction,

$$\frac{1}{1+}\frac{\alpha x}{1+}\frac{\beta x}{1+}\frac{\gamma x}{1+\ldots}.$$

The recent expression (544) for a_6 may have its terms otherwise arranged, as follows:

$$\left.\begin{aligned}
a_6 = {} & c_1^6\alpha_1\alpha_2\alpha_3\alpha_4\alpha_5\alpha_6 + c_1^4c_2\alpha_1\alpha_2\alpha_3\alpha_4\alpha_5^2 + 2c_1^4c_2\alpha_1\alpha_2\alpha_3\alpha_4\alpha_5(\alpha_4+\alpha_3+\alpha_2+\alpha_1)\\
& + c_1^3c_3\alpha_1\alpha_2\alpha_3\alpha_4^3 + c_1^2c_2(2c_2+c_1^2)\alpha_1\alpha_2\alpha_3^2\alpha_4^2 + 2c_1^2c_2^2\alpha_1\alpha_2(\alpha_1+\alpha_2)\alpha_3\alpha_4^2\\
& + 3c_1^3c_3\alpha_1\alpha_2\alpha_3^3\alpha_4 + 2c_1^2c_2(2c_2+c_1^2)\alpha_1\alpha_2^2\alpha_3^2\alpha_4 + 4c_1^2c_2^2\alpha_1^2\alpha_2\alpha_3^2\alpha_4\\
& + 3c_1^3c_3\alpha_1\alpha_2^3\alpha_3\alpha_4 + 2c_1^2c_2(2c_2+c_1^2)\alpha_1^2\alpha_2^2\alpha_3\alpha_4 + 3c_1^3c_3\alpha_1^3\alpha_2\alpha_3\alpha_4\\
& + c_1^2c_4\alpha_1\alpha_2\alpha_3^4 + 2c_1c_2(c_3+c_1c_2)\alpha_1\alpha_2^2\alpha_3^3 + 2c_1c_2c_3\alpha_1^2\alpha_2\alpha_3^3\\
& + 3c_1c_3(c_2+c_1^2)\alpha_1\alpha_2^3\alpha_3^2 + c_2(2c_2+c_1^2)^2\alpha_1^2\alpha_2^2\alpha_3^2 + 3c_1c_2c_3\alpha_1^3\alpha_2\alpha_3^2\\
& + 4c_1^2c_4\alpha_1\alpha_2^4\alpha_3 + 6c_1c_2(c_3+c_1c_2)\alpha_1^2\alpha_2^3\alpha_3 + 6c_1c_3(c_2+c_1^2)\alpha_1^3\alpha_2^2\alpha_3 + 4c_1^2c_4\alpha_1^4\alpha_2\alpha_3\\
& + c_1c_5\alpha_1\alpha_2^5 + c_2(2c_4+2c_1c_3+c_2^2)\alpha_1^2\alpha_2^4 + c_3(3c_3+6c_1c_2+c_1^3)\alpha_1^3\alpha_2^3\\
& + 2c_4(2c_2+3c_1^2)\alpha_1^4\alpha_2^2 + 5c_5\alpha_1^5\alpha_2 + c_6\alpha_1^6;
\end{aligned}\right\} \quad (548)$$

which when we make the substitutions (545), gives, for what would have been called the coefficient f in (70), the expression:

$$\left.\begin{aligned}
+c^{-6}a_6 = {} & \alpha_1\alpha_2\alpha_3\alpha_4\alpha_5\alpha_6 + \alpha_1\alpha_2\alpha_3\alpha_4\alpha_5^2 + 2\alpha_1\alpha_2\alpha_3\alpha_4\alpha_5(\alpha_4+\alpha_3+\alpha_2+\alpha_1)\\
& + \alpha_1\alpha_2\alpha_3\alpha_4^2(\alpha_4+3\alpha_3+2\alpha_2+2\alpha_1) + \alpha_1\alpha_2\alpha_3^2\alpha_4(3\alpha_3+6\alpha_2+4\alpha_1)\\
& + 3\alpha_1\alpha_2\alpha_3\alpha_4(\alpha_2+\alpha_1)^2 + \alpha_1\alpha_2\alpha_3^3(\alpha_3+4\alpha_2+2\alpha_1)\\
& + 3\alpha_1\alpha_2\alpha_3^2(2\alpha_2^2+3\alpha_1\alpha_2+\alpha_1^2) + 4\alpha_1\alpha_2\alpha_3(\alpha_2+\alpha_1)^3 + \alpha_1(\alpha_2+\alpha_1)^5\\
= {} & \alpha_1\alpha_2\alpha_3\alpha_4\alpha_5\alpha_6 + \alpha_1\alpha_2\alpha_3\alpha_4(\alpha_5+\alpha_4+\alpha_3+\alpha_2+\alpha_1)^2\\
& + \alpha_1\alpha_2\{\alpha_3(\alpha_4+\alpha_3+2\alpha_2+\alpha_1)+(\alpha_2+\alpha_1)^2\}^2 + \alpha_1^2\{\alpha_2\alpha_3+(\alpha_2+\alpha_1)^2\}^2.
\end{aligned}\right\} \quad (549)$$

Accordingly when we make $\alpha_n = n$, this last expression becomes

$$720 + (24+18+1).225 = 45(16+215) = 45.231 = 1.3.5.7.9.11,$$

as otherwise found before, in (540). We also see that, in the notation of (70), the sixth equation of that system would be

$$\begin{aligned}
f = {} & \alpha\beta\gamma\delta\epsilon\zeta + \alpha\beta\gamma\delta(\epsilon+\delta+\gamma+\beta+\alpha)^2 + \alpha\beta\{\gamma(\delta+\gamma+2\beta+\alpha)+(\beta+\alpha)^2\}^2\\
& + \alpha^2\{\beta\gamma+(\beta+\alpha)^2\}^2;
\end{aligned} \quad (550)$$

whence by making $\alpha = 1, \beta = 2, \gamma = 3, \delta = 4, \epsilon = 5$, the linear equation (75), or

$$f = 120\zeta + 9675,$$

may be deduced *at sight*, by observing that $43.225 = 9675$. As another bit of verification we may assume the values

$$\begin{gathered}
\alpha = \beta = \gamma = \delta = \epsilon = \zeta = 1, \quad \text{or}\\
\alpha_1 = \alpha_2 = \alpha_3 = \alpha_4 = \alpha_5 = \alpha_6 = 1;
\end{gathered} \quad (551)$$

and then if we still adopt the expression (545) for c_n, we shall have

$$X = \frac{1}{1+}\frac{cx}{1+}\frac{cx}{1+}\cdots \&c., \quad (552)$$

which may be written as

$$X = \left(\frac{1}{1+cx.}\right)^{\infty} 1, \tag{553}$$

and gives

$$X = \frac{1}{1+cxX}, \quad cxX^2 + X = 1,$$

$$X = \frac{(1+4cx)^{\frac{1}{2}} - 1}{2cx} = 1 - cx + 2c^2x^2 - 5c^3x^3 + 14c^4x^4 - 42c^5x^5 + 132c^6x^6 - \cdots; \tag{554}$$

and accordingly if α_n be made equal to unity in (547), (549), the resulting values of $a_1, \ldots, a_6$ are

$$a_1 = -c, \quad a_2 = 2c^2, \quad a_3 = -5c^3, \quad a_4 = 14c^4, \quad a_5 = -42c^5, \quad a_6 = 132c^6. \tag{555}$$

LXVI. When the coefficients $c_1, c_2, \ldots,$ of the series (541) for the *transforming function* ϕ, are either immediately given, or have been deduced by reversion of series, or otherwise from the coefficients $b_1, b_2, \ldots,$ of the series for the inverse function ϕx, they can be substituted in the expressions (544); and then the coefficients $a_1, a_2, \ldots,$ of the series (542), for the *transformand function* X, come to be expressed as *known and explicit functions, rational and integral,* of the constants $\alpha_1, \alpha_2, \ldots,$ of the *transformed functions,* or expression, $X = \phi\alpha_1 x\phi\alpha_2 x\phi\alpha_3 x \ldots,$ to which this Letter mainly relates; as for example, in the recent equations, (547), (549). And conversely, from the *form* of the equations thus obtained, it is evident that we can *determine, successively,* and *unambiguously,* without having to resolve any equation higher than the *first degree,* the constants $(\alpha_1, \alpha_2, \ldots)$ of the *transformee,* when those of the *transformand,* $(a_1, a_2, \ldots)$ and of the *transformer,* $(c_1, c_2, \ldots)$ are *given*; at least if c_1 has any value different from 0. But although it is satisfactory and useful to have the expressions (544), yet I think that in practice it may often be convenient to employ another *method,* for this converse deduction of $\alpha_1, \alpha_2, \ldots,$ from $a_1, a_2, \ldots,$, and from $c_1, c_2, \ldots$; or more immediately, from $a_1, a_2, \ldots,$ and $b_1, b_2, \ldots$. This method uses the *Calculus of Derivations* and is merely a slight extension of the processes employed in paragraphs VI. and XII. of the Letter. Writing

$$X' = \phi\alpha_2 x\phi\alpha_3 x \ldots = 1 + a_1' x + a_2' x^2 + \cdots, \tag{556}$$

we have, by (561) and (562),

$$\alpha_1 xX' = \varphi X = b_1(a_1x + a_2x^2 + a_3x^3 + \cdots) + b_2(a_1x + a_2x^2 + \cdots)^2 + \cdots, \tag{557}$$

whence

$$\left.\begin{aligned} \alpha_1 &= b_1 a_1, \qquad \alpha_1 a_1' = b_1 a_2 + b_2 a_1^2, \cdots \\ b_1 a_1 a_n' &= b_1 D^n a_1 + b_2 D^{n-1} a_1^2 + b_3 D^{n-2} a_1^3 + \cdots + b_{n+1} a_1^{n+1}; \end{aligned}\right\} \tag{558}$$

where $D^n a_1 = a_{n+1}$, and generally $D^n a_1^m$ is interpreted on the plan of your *Differential and Integral Calculus** (p. 328 *et seq.*), for I have not Arbogast's book† at hand, though it is somewhere in the house*, and I read a good part of it long ago; so that, for instance

* [See footnote to paragraph XLIV.]

† [Louis François Antoine Arbogast (1759–1803), *Du calcul des dérivations,* Strasbourg: 1800.]

$$\left.\begin{array}{l} Da_1^2 = 2a_1 a_2, \quad D^2 a_1^2 = 2a_1 a_3 + a_2^2, \\ D^3 a_1^2 = 2a_1 a_4 + 2a_2 a_3, \quad D^4 a_1^2 = 2a_1 a_5 + 2a_2 a_4 + a_3^2, \\ \&\text{c.} \end{array}\right\} \tag{559}$$

On the same plan if we write

$$X'' = \phi\alpha_3 x \phi\alpha_4 x \ldots = 1 + a_1'' x + a_2'' x^2 + \cdots, \tag{560}$$

we have

$$\alpha_2 x X'' = \phi X', \ \alpha_2 = b_1 a_1', \ b_1 a_1' a_n'' = b_1 D^n a_1' + b_2 D^{n-1} a_1'^2 + b_3 D^{n-2} a_1'^3 + \&\text{c.}; \tag{561}$$

where

$$Da_1'^2 = 2a_1' a_2', \quad \&\text{c.}$$

Thus, when b_1, b_2, &c., are given, α_1 is found from a_1; α_2, through a_1', from a_1 and a_2; α_3, through a_1'', from a_1' and a_2', and so from a_1, a_2, and a_3; and so on, for we have generally

$$\alpha_n = b_1 a_1^{(n-1)}; \tag{562}$$

the accents at the top serving only to attach the coefficients $a^{(m)}$ to the series $X^{(m)}$; and *one* such *series* being *deduced* (rather than in any *technical* and *usual* sense *derived*) *from the preceding series* of the same system, by the general relation,

$$\alpha_n x X^{(n)} = \varphi X^{(n-1)}. \tag{563}$$

As a system of verifications, we may operate on this last equation by ϕ, and so change it to what indeed follows at once from the notation used,

$$X^{(n-1)} = \phi\alpha_n x X^{(n)}; \tag{564}$$

for this will give

$$\begin{aligned} a_1^{(n-1)} &= \alpha_n c_1, \quad \text{and generally} \\ a_m^{(n-1)} &= c_1\alpha_n D^{m-1} a_0^{(n)} + c_2\alpha_n^2 D^{m-2} a_0^{(n)2} + c_3\alpha_n^3 D^{m-3} a_0^{(n)3} + \cdots + c_m\alpha_n^m a_0^{(n)m}; \end{aligned} \tag{565}$$

where the derivations D are so interpreted that (for example)

$$Da_0^{(n)2} = 2a_0^{(n)} a_1^n, \quad D^2 a_0^{(n)2} = 2a_0^{(n)} a_2^{(n)} + a_1^{(n)2}, \quad \&\text{c.}; \tag{566}$$

and $a_0^{(n)}$ is made $= 1$, after all these derivations have been performed. Thus

$$\begin{aligned} a_4^{(n-1)} &= c_1\alpha_n D^3 a_0^{(n)} + c_2\alpha_n^2 D^2 a_0^{(n)2} + c_3\alpha_n^3 Da_0^{(n)3} + c_4\alpha_n^4 a_0^{(n)4} \\ &= c_1\alpha_n a_3^{(n)} + c_2\alpha_n^2 (2a_2^{(n)} + a_1^{(n)2}) + 3c_3\alpha_n^3 a_1^{(n)} + c_4\alpha_n^4; \quad \&\text{c.} \end{aligned} \tag{567}$$

Finally, the relations between the two systems of constants, b and c, of the two mutually inverse or reciprocal functions φ and ϕ, may be expressed by the formula,

* 15 May 1858 – I have since found my Arbogast, and read, with the greatest pleasure, the first thirty pages, or thereabouts, at one stretch, verifying every step, as I went along, by mental calculation.

$$D^{n-1}1 = b_1 D^{n-1} c_1 + b_2 D^{n-2} c_1^2 + b_3 D^{n-3} c_1^3 + \cdots + b_n c_1^n. \tag{568}$$

LXVII. Before, I began this Letter, I had considered (to some small extent) the case alluded to in paragraph V., where the *transforming functions* were,

$$\varphi x = x^{-y} - 1,\ \phi x = \varphi^{-1} x = (1+x)^{-\frac{1}{\nu}}; \tag{569}$$

ν being any given exponent. In this case, with the recent meaning of b_n, we have the expression

$$b_n = \frac{(-1)^n \nu(\nu+1)(\nu+2)\ \ldots\ (\nu+n-1)}{1.2.3\ \ldots\ n}. \tag{570}$$

The formula (558) becomes therefore here, compare the equations (52):

$$a_1 a'_n = D^n a_1 - \frac{\nu+1}{2} D^{n-1} a_1^2 + \frac{\nu+1}{2}\frac{\nu+2}{3} D^{n-2} a_1^3 + \cdots + \frac{\nu+1}{2}\frac{\nu+2}{3} \cdots \frac{\nu+n}{n+1} a_1^{n+1} \tag{571}$$

in which every a may receive an additional accent, as in (561), and so on for ever. Hence if we know the unaccented coefficients a, as far as a_n, we can first deduce the singly accented coefficients a', as far as a'_{n-1}; then a'', as far as a''_{n-2}, &c.; till we come to the coefficient $a_1^{(n-1)}$; after which the first n constants α_n will be known, by the formula (562), which here becomes,

$$\alpha_n = -\nu a_1^{(n-1)}. \tag{572}$$

(Compare (55) and (60).) – The *method of reciprocals*, which leads to the *continued fractions*, already discussed at some length, was, as remarked in paragraph VI., merely that particular *sub-case*, for which $\nu = 1$. The *case next in simplicity*, and which perhaps will be found to be quite devoid of *utility*, appears to me to be that for which

$$\nu = 2; \tag{573}$$

so that the *transformation* to be considered here is (compare equation (48)) expressed by the formula:

$$X = 1 + a_1 x + a_2 x^2 + \&c. = \phi\alpha_1 x \phi\alpha_2 x \ldots = (1 + \alpha_1 x(1 + \alpha_2 x(1 + \cdots)^{-\frac{1}{2}})^{-\frac{1}{2}})^{-\frac{1}{2}}. \tag{574}$$

The equations of the forms (571) are, for the present transformations,

$$\left.\begin{aligned}
&a_1 a'_1 = a_2 - \frac{3}{2} a_1^2;\ a_1 a'_2 = a_3 - 3a_1 a_2 + 2a_1^3;\ a_1 a'_3 = a_4 - 3\left(a_1 a_3 + \frac{a_2^2}{2}\right) + 6a_1^2 a_2 - \frac{5}{2} a_1^4;\\
&a_1 a'_4 = a_5 - 3(a_1 a_4 + a_2 a_3) + 6(a_1^2 a_3 + a_1 a_2^2) - 10a_1^3 a_2 + 3a_1^5;\\
&a_1 a'_5 = a_6 - 3\left(a_1 a_5 + a_2 a_4 + \frac{1}{2} a_3^2\right) + 6\left(a_1^2 a_4 + 2a_1 a_2 a_3 + \frac{1}{3} a_2^3\right) - 10\left(a_1^3 a_3 + \frac{3}{2} a_1^2 a_2^2\right)\\
&\qquad + 15 a_1^4 a_2 - \frac{7}{2} a_1^6;
\end{aligned}\right\} \tag{575}$$

the last or right hand terms being those of the development of

$$-\frac{1}{2}(1+a_1)^{-2},$$

and others being afterwards *derived* from these, with the greatest possible facility, by the Rules of the Calculus of Derivations. In like manner, by accenting,

$$a_1' a_1'' = a_2' - \frac{3}{2} a_1'^2, \quad a_1' a_2'' = a_3' - 3a_1' a_2' + 2a_1'^3, \quad \text{\&c.}; \tag{576}$$

and when a', a_1'', ..., have thus been computed from a_1, a_2, a_3, ..., we have

$$\alpha_1 = -2a_1, \quad \alpha_2 = -2a_1', \quad \text{\&c.} \tag{577}$$

And the numerical calculations may be checked in various ways, on principles already stated or suggested.

LXVIII. It seems to be worth while, however, to write down the following system of formulæ, which may indeed be deduced by elimination from (575) but may also be otherwise and independently found; and which may serve, not merely as a check on the numerical results obtained from the last cited equations, in any application of this "*method of inverse square roots*" but also as an independent instrument of calculation, for successively determining a_1', a_2', ..., when a_1, a_2, a_3, ... are given. The system to which I refer is included in the more general system (567), and may be deduced there from, by simply introducing the present values of the coefficients c_n; or by writing,

$$(-1)^n c_n = \frac{1.3.5.7\cdots(2n-1)}{2.4.6.8\cdots 2n}, \tag{578}$$

for in this manner we obtain the following equations, with the help of (577):

$$\left.\begin{aligned}
&a_2 = a_1 a_1' + \frac{3}{2} a_1^2; \; a_3 = a_1 a_2' + 3a_1^2 a_1' + \frac{5}{2} a_1^3; \; a_4 = a_1 a_3' + 3a_1^2\left(a_2^1 + \tfrac{1}{2}a_1'^2\right) + \frac{15}{2} a_1^3 a_1' + \frac{35}{8} a_1^4;\\
&a_5 = a_1 a_4' + 3a_1^2(a_3' + a_1' a_2') + \frac{15}{2} a_1^3(a_2^1 + a_1'^2) + \frac{35}{2} a_1^4 a_1' + \frac{63}{8} a_1^5;\\
&a_6 = a_1 a_5' + 3a_1^2\left(a_4' + a_1' a_3' + \tfrac{1}{2}a_2'^2\right) + \frac{15}{2} a_1^3\left(a_3' + 2a_1' a_2' + \frac{1}{3} a_1'^3\right)\\
&\qquad + \frac{35}{2} a_1^4\left(a_2' + \frac{3}{2} a_1'^2\right) + \frac{315}{8} a_1^5 a_1' + \frac{231}{16} a_1^6.
\end{aligned}\right\} \tag{579}$$

I think that it will be scarcely *possible* for any *arithmetical mistake* to escape notice, in the calculation of the values of a_1', ..., a_5', from those of a_1, ..., a_6, in any particular question, if these *two systems of equations*, (575) and (579), be *separately used* for that purpose: so different are the auxiliary or *intermediate members* which present themselves, in one and in the other process; and so different is the play of the *signs*. Such, at least, is my impression from my own little experience of such calculations; mistakes of mine having been detected, by the foregoing system of checks; but an entire confidence being reposed by me in the final values, when the results of the two processes agree. – I forget whether I made much *use* of the following other system of expressions, when I was applying the "method of inverse square roots", a few months ago, before I began to write the present Letter: if

$$X = 1 + a_1 x + a_2 x^2 + \cdots = (1 + \alpha_1 x(1 + \alpha_2 x(1 + \cdots)^{-\frac{1}{2}})^{-\frac{1}{2}})^{-\frac{1}{2}} \tag{574}$$

then

$$\left.\begin{aligned}
&a_1 = -\tfrac{1}{2}\alpha_1; \quad a_2 = \frac{1}{4}\alpha_1\left(\alpha_2 + \frac{3}{2}\alpha_1\right); \quad a_3 = -\frac{1}{8}\alpha_1\alpha_2\alpha_3 - \frac{\alpha_1}{16}(3\alpha_2^2 + 6\alpha_1\alpha_2 + 5\alpha_1^2);\\
&a_4 = \frac{1}{16}\alpha_1\alpha_2\alpha_3\alpha_4 + \frac{1}{32}\alpha_1\alpha_2(3\alpha_3^2 + 6\alpha_2\alpha_3 + 6\alpha_1\alpha_3 + 5\alpha_2^2 + 12\alpha_1\alpha_2 + 15\alpha_1^2) + \frac{35}{128}\alpha_1^4;\\
&a_5 = -\frac{1}{32}\alpha_1\alpha_2\alpha_3\alpha_4\alpha_5 - \frac{1}{64}\alpha_1\alpha_2\alpha_3(3\alpha_4^2 + 6\alpha_3\alpha_4 + 6\alpha_2\alpha_4 + 5\alpha_3^2 + 12\alpha_2\alpha_3 + 15\alpha_2^2)\\
&\qquad - \frac{35}{256}\alpha_1\alpha_2^4 - \frac{3}{64}\alpha_1^2\alpha_2\alpha_3(2\alpha_4 + 3\alpha_3 + 8\alpha_2 + 5\alpha_1) - \frac{3}{8}\alpha_1^2\alpha_2^3\\
&\qquad - \frac{75}{128}\alpha_2^3\alpha_2^2 - \frac{35}{64}\alpha_1^4\alpha_2 - \frac{63}{256}\alpha_1^5;\\
&a_6 = \frac{1}{64}\alpha_1\alpha_2\alpha_3\alpha_4\alpha_5\alpha_6 + \text{(a long polynome)} + \frac{231}{1024}\alpha_1^6.
\end{aligned}\right\} \tag{580}$$

The arrangement of the terms in a_5 might be improved, but it is not worth our while to delay upon it. I must, however, remark that the system (580) is *analogous* to those before marked as (70) and (108); and that it is, like them, *included* in the *more general system* of such equations, which has been numbered as (544). It is clear, also, that these last equations (580) *may* be used, to *determine* successively and linearly, the constants $\alpha_1, \ldots, \alpha_4$, when the coefficients $a_1, \ldots, a_4$ are given; and also to *verify* the values of those constants, if they have been determined by any *other* process, such as those which have been already described.

LXIX. If now we introduce the values,

$$a_1 = 1, \ a_2 = 3, \ a_3 = 15, \ a_4 = 105, \ a_5 = 945, \tag{581}$$

without at first assuming any value for a_6, we find, successively,

$$\left.\begin{aligned}
&a_1' = \frac{3}{2}, \quad a_2' = 8, \quad a_3' = 62, \quad a_4' = 612; \quad a_1'' = \frac{37}{12}, \quad a_2'' = \frac{131}{6}, \quad a_3'' = \frac{3545}{16};\\
&a_1''' = \frac{727}{296}, \quad a_2''' = \frac{67\,591}{2664}; \quad a_1^{iv} = \frac{25\,743\,589}{3\,873\,456}; \quad \alpha_1 = -2a_1 = -2;\\
&\alpha_2 = -2a_1' = -3; \quad \alpha_3 = -\frac{37}{6}; \quad \alpha_4 = -\frac{727}{148}; \quad \alpha_5 = -\frac{25\,743\,589}{1\,936\,728};
\end{aligned}\right\} \tag{582}$$

that is to say, the series

$$X = 1 + x + 3x^2 + 15x^3 + 105x^4 + 945x^5 + \&c., \tag{583}$$

which has occurred before in this Letter, where the &c. merely means *some polynome* or series, of the *6th and higher dimensions*, may be transformed into this other expression not hitherto written down,

$$X = \left(1 - 2x\left(1 - 3x\left(1 - \frac{37}{6}x\left(1 - \frac{727}{148}x\left(1 - \frac{25\,743\,589}{1\,936\,728}xX^{v}\right)^{-\frac{1}{2}}\right)^{-\frac{1}{2}}\right)^{-\frac{1}{2}}\right)^{-\frac{1}{2}}\right)^{-\frac{1}{2}}; \tag{584}$$

where we only know of X^{v}, that it is a polynome or series, *whose first term is unity*. Or, changing the sign of x, we have the transformation,

$$\begin{aligned} X &= \left(1 + 2x\left(1 + 3x\left(1 + \frac{37}{6}x\left(1 + \frac{727}{148}x\left(1 + \frac{25\,743\,589}{1\,906\,728}xX^{v}\right)^{-\frac{1}{2}}\right)^{-\frac{1}{2}}\right)^{-\frac{1}{2}}\right)^{-\frac{1}{2}}\right)^{-\frac{1}{2}} \\ &= 1 - 1x + 1.3x^2 - 1.3.5x^3 + 1.3.5.7x^4 - 1.3.5.7.9x^5 + a_6x^6 - a_7x^7 + \&\text{c.}; \end{aligned} \tag{585}$$

where

$$X^{v} = 1 - a_1^{v}x + \&\text{c.} \tag{586}$$

and the coefficients a_6, a_7, ..., of the one development are connected with the coefficients a_1^{v}, a_2^{v}, ..., of the other, by laws of the kind which has been already considered in this Letter. And if we write,

$$\left.\begin{aligned} &X_1 = (1 + 2x)^{-\frac{1}{2}}; \quad X_2 = (1 + 2x(1 + 3x)^{-\frac{1}{2}})^{-\frac{1}{2}}; \\ &X_3 = \left(1 + 2x\left(1 + 3x\left(1 + \frac{37}{6}x\right)^{-\frac{1}{2}}\right)^{-\frac{1}{2}}\right)^{-\frac{1}{2}}; \\ &X_4 = \left(1 + 2x\left(1 + 3x\left(1 + \frac{37}{6}x\left(1 + \frac{727}{148}x\right)^{-\frac{1}{2}}\right)^{-\frac{1}{2}}\right)^{-\frac{1}{2}}\right)^{-\frac{1}{2}} \\ &X_5 = \left(1 + 2x\left(1 + 3x\left(1 + \frac{37}{6}x\left(1 + \frac{727}{148}x\left(1 + \frac{25\,743\,589}{1\,936\,728}x\right)^{-\frac{1}{2}}\right)^{-\frac{1}{2}}\right)^{-\frac{1}{2}}\right)^{-\frac{1}{2}}\right)^{-\frac{1}{2}}; \end{aligned}\right\} \tag{587}$$

then with the *last* series, $1 - x + 3x^2 - \&\text{c.}$, which has *alternate signs*, and for values of x which are *positive*, and *not too large*, we may establish the *inequalities*, $X < 1, > X_1, X_2, > X_3, < X_4$, as in (253); or more fully, we may write the *chain* of inequalities,

$$1 > X_2 > X_4 > X > X_3 > X_1 > 0. \tag{588}$$

But whether the function X, in (585), is *greater or less than* X_5, *even for positive and moderate values* of x we cannot yet decide, until we are given the coefficient a_6, and have deduced from it *at least the algebraical sign* of the connected coefficient a_1^{v}, so as to determine *whether* X^{v} *is less or greater than unity*.

LXX. We may also write, with the last adopted sign of x, the series of equations,

$$\left.\begin{aligned} &X^{-2} = 1 + 2xX'; \; X'^{-2} = 1 + 3xX''; \; X''^{-2} = 1 + \frac{37}{6}xX'''; \\ &X'''^{-2} = 1 + \frac{727}{148}xX^{iv}; \; X^{iv^{-2}} = 1 + \frac{25\,743\,589}{1\,936\,728}xX^{v}; \end{aligned}\right\} \tag{589}$$

and

$$\left.\begin{aligned}
X' &= 1 - a'_1 x + a'_2 x^2 - a'_3 x^3 + a'_4 x^4 - a'_5 x^5 + \cdots; \\
X'' &= 1 - a''_1 x + a''_2 x^2 - a''_3 x^3 + a''_4 x^4 - \cdots; \\
X''' &= 1 - a'''_1 x + a'''_2 x^2 - a'''_3 x^3 + \cdots; \\
X^{iv} &= 1 - a^{iv}_1 x + a^{iv}_2 x^2 - \cdots; \\
X^{v} &= 1 - a^{v}_1 x + \cdots \text{ as above};
\end{aligned}\right\} \tag{590}$$

where a'_1, a'_2, a'_3, a'_4, a''_1, a''_2, a''_3, a'''_1, a'''_2, a^{iv}_1, have still the values (582); and a'_5, a''_4, a'''_3, a^{iv}_2, a^{v}_1, will admit of being successfully calculated, as soon as a_6 shall be known; a_1, a_2, a_4, a_5, retaining still the values of (581). And for such successive calculation we may employ, at pleasure, either of the two following systems of formulæ, which may therefore serve as checks upon each other, and which are *collected* for convenience here, although they have substantially been given already, in (575) and (579):

(*I.*)

$$\left.\begin{aligned}
a_1 a'_5 &= a_6 - 3\left(a_1 a_5 + a_2 a_4 + \frac{1}{2} a_3^2\right) + 6\left(a_1^2 a_4 + 2 a_1 a_2 a_3 + \frac{1}{3} a_2^3\right) \\
&\quad - 10\left(a_2^3 a_3 + \frac{3}{2} a_1^2 a_2^2\right) + 15 a_1^4 a_2 - \frac{7}{2} a_1^6; \\
a'_1 a''_4 &= a'_5 - 3(a'_1 a'_4 + a'_2 a'_3) + 6(a_1'^2 a'_3 + a'_1 a_1'^2) - 10 a_1'^3 a'_2 + 3 a_1'^5; \\
a''_1 a'''_3 &= a''_4 - 3\left(a''_1 a''_3 + \frac{1}{2} a_2''^2\right) + 6 a_1''^2 a''_2 - \frac{5}{2} a_1''^4; \\
a'''_1 a^{iv}_2 &= a'''_3 - 3 a'''_1 \alpha'''_2 + 2 a_1'''^3; \qquad a^{iv}_1 a^{v}_1 = a^{iv}_2 - \frac{3}{2} a_1^{iv^2};
\end{aligned}\right\} \tag{591}$$

(*II.*)

$$\left.\begin{aligned}
a_6 &= a_1 a'_5 + 3 a_1^2\left(a'_4 + a'_1 a'_3 + \frac{1}{2} a_2'^2\right) + \frac{15}{2} a_1^3\left(a'_3 + 2 a'_1 a'_2 + \frac{1}{3} a_1'^3\right) + \frac{35}{2} a_1^4\left(a'_2 + \frac{3}{2} a_1'^2\right) \\
a'_5 &= a'_1 a''_4 + 3 a_1'^2 (a''_3 + a''_1 a''_2) + \frac{15}{2} a_1'^3 (a''_2 + a_1''^2) + \frac{35}{2} a_1'^4 a''_1 + \frac{63}{8} a_1'^5; \\
a''_4 &= a''_1 a'''_3 + 3 a''_1\left(a''''_2 + \frac{1}{2} a_1'''^2\right) + \frac{15}{2} a_1''^3 a'''_1 + \frac{35}{8} a_1''^4; \\
a'''_3 &= a'''_1 a^{iv}_2 + 3 a_1'''^2 a^{iv}_1 + \frac{5}{2} a_1'''^3; \\
a^{iv}_2 &= a^{iv}_1 a^{v}_1 + \frac{3}{2} a_1^{iv^2}.
\end{aligned}\right\} \tag{592}$$

Each system conducts us, with the values (581) for a_1, a_2, a_3, a_4, a_5, and therefore with the values (582) for a'_1, a'_2, a'_3, a'_4, a''_1, a''_2, a''_3, a'''_1, a'''_2, a^{iv}_1, to the following *chain of linear equations*, connecting successively

$$\left.\begin{array}{l} a_5', a_4'', a_3''', a_2^{iv}, \text{ and } a_1^{v} \text{ with } a_6: \\ a_5' = a_6 - 3137;\ 48a_4'' = 32a_5' - 98\,439;\ 127\,872a_3''' = 41\,472a_4'' - 72\,370\,301; \\ 95\,545\,248a_2^{iv} = 38\,901\,504a_3''' - 6\,119\,799\,487; \\ 199\,433\,318\,547\,168a_1^{v} = 30\,007\,322\,767\,872a_2^{iv} - 1\,988\,197\,123\,802\,763; \end{array}\right\} \quad (593)$$

in which I think that the coefficients, though large, may be relied on. Eliminating, successively, a_5', a_4'', a_3''', and a_2^{iv}, we find these other equations:

$$\left.\begin{array}{l} 48a_4'' = 32a_6 - 198\,823; \\ 127\,872a_3''' = 27\,648a_6 - 244\,153\,373; \\ 859\,907\,232a_2^{iv} = 757\,00\,224a_6 - 723\,570\,130\,657; \\ \text{and finally,} \\ 199\,433\,318\,547\,168a_1^{v} = 2\,641\,635\,016\,704a_6 - 27\,237\,900\,403\,209\,435\ \ldots. \end{array}\right\} \quad (594)$$

It appears, then, that if we confine ourselves to integer numbers as limits,

$$\left.\begin{array}{l} a_5' < 0, \quad \text{if } a_6 < 3137; \quad a_4'' < 0, \quad \text{if } a_6 < 6213; \quad a_3''' < 0, \quad \text{if } a_6 < 8830; \\ a_2^{iv} < 0, \quad \text{if } a_6 < 9558; \quad \text{and finally, } a_1^{v} < 0, \quad \text{if } a_6 < 10\,311: \end{array}\right\} \quad (595)$$

whereby it is not implied that the *conditioned* inequalities would in all cases *not* be satisfied, if the *conditioning* inequalities should cease to exist. In fact, we should have, in addition to the result (595), these others:

$$\left.\begin{array}{l} a_5' = 0, \quad \text{if } a_6 = 3137; \quad a_4'' < 0, \quad \text{if } a_6 = 6213; \quad a_3''' < 0, \quad \text{if } a_6 = 8830; \\ a_2^{iv} < 0, \quad \text{if } a_6 = 9558; \quad \text{and } a_1^{v} < 0, \quad \text{if } a_6 = 10\,311. \end{array}\right\} \quad (596)$$

On the other hand, using still integer limits,

$$\left.\begin{array}{l} a_5' > 0, \quad \text{if } a_6 > 3137; \quad a_4'' > 0, \quad \text{if } a_6 \geq 6214; \quad a_3''' > 0, \quad \text{if } a_6 \geq 8831; \\ a_2^{iv} > 0, \quad \text{if } a_6 \geq 9559; \quad \text{and finally, } a_1^{v} > 0, \quad \text{if } a_5 \geq 10\,312. \end{array}\right\} \quad (597)$$

Thus, in particular, when we assume

$$a_6 = 11.9.7.5.3.1 = 10\,395, \quad (598)$$

(compare (540),) then

$$a_5' > 0, \quad a_4'' > 0, \quad a_3''' > 0, \quad a_2^{iv} > 0, \quad a_2^{v} > 0; \quad (599)$$

although the condition for the existence of this *last* inequality is only *barely* satisfied. And, on computing the fractional values of these coefficients, we find them to be, for this last value of a_6, (namely for that which presents itself in a series of Laplace, with which we started in this Letter,) the following:

$$\left.\begin{array}{l} a_5' = 7258; \quad a_4'' = \dfrac{133\,817}{48}; \quad a_3''' = \dfrac{43\,247\,587}{127\,872}; \quad a_2^{iv} = \dfrac{63\,333\,697\,823}{859\,907\,232}; \\ \text{and finally } a_1^{v} = \dfrac{221\,895\,595\,428\,645}{199\,433\,318\,547\,168}. \end{array}\right\} \quad (600)$$

LXXI. *Recapitulating* the results obtained by the *present method of transformation,* as applied to the *last mentioned series,* we may say that if

$$X = \left(\frac{2}{x}\right)^{\frac{1}{2}} \epsilon^{\frac{1}{2x}} \int_{(2x)^{-\frac{1}{2}}}^{\infty} \epsilon^{-t^2}\, dt = 1 - 1.x + 1.3x^2 - 1.3.5x^3 + 1.3.5.7x^4$$
$$- 1.3.5.7.9x^5 + 1.3.5.7.9.11x^6 - \text{\&c., } \textit{ad infinitum}, \qquad (601)$$

then

$$X^{-2} = 1 + 2xX', \text{ when } X' = 1 - \frac{3}{2}x + 8x^2 - 62x^3 + 612x^4 - 7258x^5 + \text{\&c.}; \qquad (602)$$

$$X'^{-2} = 1 + 3xX'', \; X'' = 1 - \frac{37}{12}x + \frac{131}{b6}x^2 - \frac{3545}{16}x^3 + \frac{133\,817}{48}x^4 - \text{\&c.}; \qquad (603)$$

$$X''^{-2} = 1 + \frac{37}{6}xX''', \; X''' = 1 - \frac{727}{296}x + \frac{67\,591}{2664}x^2 - \frac{43\,247\,587}{127\,872}x^3 + \text{\&c.}; \qquad (604)$$

$$X'''^{-2} = 1 + \frac{737}{148}xX^{iv}, \; X^{iv} = 1 - \frac{25\,743\,589}{3\,873\,456}x + \frac{63\,333\,697\,823}{859\,907\,232}x^2 - \text{\&c.}; \qquad (605)$$

and

$$X^{iv-2} = 1 + \frac{25\,743\,589}{1\,936\,728}xX^{v}, \; X^{v} = 1 - \frac{221\,895\,595\,428\,645}{199\,433\,318\,547\,168}x \text{ \&c.}; \qquad (606)$$

where each of the "&c."s, in (602) to (606), remains as yet uncalculated, even the algebraic *signs* of the unwritten terms being (in rigour) as yet unknown, whatever *guesses* may be made respecting them: because, although the *law* of the celebrated series (601) *is known,* we have hitherto *used* only the terms which have been lately written down, – nor am I likely to think it worth while to take even *one more term* of the series into account, so as to deduce any others of these frightful fractions from it! And it is unnecessary to make the admission, which however I cheerfully do make, that although, perhaps, some little refinement may have been brought into play, in the *manner* of calculating and verifying the *fractions* in these series, (602), ..., (606), yet there is *no difficulty* in the calculation of *them,* which *might* not have been overcome by the mere *patience* of a *beginner,* who was just acquainted with the *rules of vulgar arithmetic,* and with a *few* of the *early* rules of *algebra;* not even the *general binomial theorem* being *required.*

LXXII. But now that the fractional coefficients *have* been, so far, computed, can any *use* of them be made? I think that there can – though well inclined to admit that, perhaps, "le jeu ne vaut pas la chandelle". – I shall begin with an Example, which had been considered by me *before* this Letter was commenced, and to which you will find an *allusion,* (unintelligible, doubtless, hitherto,) so early as in paragraph III. Let then x be supposed to be equal to $\frac{1}{2}$, in the series for X, which thus acquires the value, $X = L^{-1}1{\cdot}879\,596\,0 = 0{\cdot}757\,87$, nearly as has been seen in the course of this Letter (compare (7)). I did not know, at the time, what the fractional coefficient of x in X^v, (606), might be; not even its algebraic *sign. Assuming,* therefore, as the best *guess* in my power to make, that x^v was *equal to unity,* or that the series for X^v reduced itself to its *first term,* I had, successively, on the *hypothesis* that every one of the preceding quantities of the form $X^{(n)}$ was *positive*, and with the corresponding *interpretation* of the *radicals* $(\ )^{-\frac{1}{2}}$,

$$X^{iv} = (1 + \alpha_5 x)^{-\frac{1}{2}} = L^{-1}\bar{1}{\cdot}558\,28, \text{ nearly, if } x = \tfrac{1}{2}, \text{ and } \alpha_5 = \frac{25\,743\,589}{1\,936\,728}; \tag{607}$$

$$X''' = (1 + \alpha_4 x X^{iv})^{-\frac{1}{2}} = (1 + 0{\cdot}888\,22)^{-\frac{1}{2}} = L^{-1}\bar{1}{\cdot}861\,97, \text{ where } \alpha_4 = \frac{72}{148}; \tag{608}$$

$$X'' = (1 + \alpha_3 x X''')^{-\frac{1}{2}} = (1 + 2{\cdot}2438)^{-\frac{1}{2}} = L^{-1}\bar{1}{\cdot}744\,47, \text{ where } \alpha_3 = \frac{37}{6}; \tag{609}$$

$$X' = (1 + \alpha_2 x X'')^{-\frac{1}{2}} = (1 + 0{\cdot}832\,84)^{-\frac{1}{2}} = L^{-1}\bar{1}{\cdot}868\,44, \text{ where } \alpha_2 = 3; \tag{610}$$

and finally X (or rather X_5) $= (1 + \alpha_1 x X')^{-\frac{1}{2}} = (1 + 0{\cdot}738\,65)^{-\frac{1}{2}}$

$$= L^{-1}\bar{1}{\cdot}879\,89 = 0{\cdot}758\,39 \text{ nearly, with } \alpha_1 = 2. \tag{611}$$

(I pick out these logarithms, &c., from a page of more than 3 months old, without at present seeking to correct or to examine them; but the *same approximate value*, to 5 decimals, namely $X = 0{\cdot}75839$ was obtained, about the same time, by a different logarithmic process – introducing cosines and tangents, from the same data. –) It struck me as remarkable *then*, and *still* appears to me worthy of remark, that the approximate X, or rather rigorously defined X_5, see (587) namely the approximate value 0·75839, – deduced in *this way*, by *positive inverse square roots*, is *much nearer to the true value*, ($X = 0{\cdot}75787$,), *not only* than that obtained by my *exponential method*, which value was $X_5 = 0{\cdot}6748$, (142), but *even* than that of the *method of reciprocals*, which gave by (89), X_5 (or T_5) $= 0{\cdot}75145$. But it *surprised* me to see that the *error* (such as it was) of the "Method of Square Roots" was in the *wrong direction* or at least in an *unusual one*: since it gave (compare (36)) X_5 *greater* than the true X, instead of giving one *less* than it.

LXXIII. It will serve to throw light on this result, if we pursue an opposite course, or adopt an inverted order calculation, and deduce successively X', X'', X''', X^{iv}, X^{v}, from the value of X considered as known, by means of the formulæ

$$\left.\begin{array}{c} X' = \dfrac{X^{-2} - 1}{\alpha_1 x}, \quad X'' = \dfrac{X'^{-2} - 1}{\alpha_2 x}, \quad X''' = \dfrac{X''^{-2} - 1}{\alpha_3 x}, \quad X^{iv} = \dfrac{X'''^{-2} - 1}{\alpha_4 x} \\ X^{v} = \dfrac{X^{iv-2} - 1}{\alpha_5 x}, \end{array}\right\} \tag{612}$$

with the recent values of $\alpha_1, \alpha_2, \alpha_3, \alpha_4, \alpha_5$; to which we may add the analogous formula,

$$X^{vi} = \frac{X^{v-2} - 1}{\alpha_6 x}, \quad \text{if } \alpha_6 = \frac{22\,189\,595\,428\,645}{99\,716\,659\,273\,584}; \tag{613}$$

so that we have, at least nearly, these 5 constant logarithms, for the question,

$$\left.\begin{array}{l} L\alpha_1 = 0{\cdot}301\,030\,0,\ L\alpha_4 = 0{\cdot}477\,121\,3,\ L\alpha_3 = 0{\cdot}790\,050\,4, \\ L\alpha_4 = 0{\cdot}691\,272\,7,\ L\alpha_5 = 1{\cdot}123\,600\,4; \end{array}\right\} \tag{614}$$

with which may be combined, if we choose, this 6th logarithmic constant,

$$L\alpha_6 = 0{\cdot}347\,381\,0. \tag{615}$$

If we suppose

$$X = \cos\xi, \quad X' = \cos\xi', \quad X'' = \cos\xi'', \quad X''' = \cos\xi''', \quad X^{iv} = \cos\xi^{iv}, \quad \ldots, \tag{616}$$

and if each X, at least as far as $X^{(n)}$, is >0, <1, then ξ, ξ', ξ'', ... as far as $\xi^{(n)}$, will be real, and may be regarded as being in the first positive quadrant; and after calculating, with the help of Kramp's Table, or otherwise, the numerical value of X, and therefore that of ξ, we can go on to deduce from it ξ', ξ'', &c., in succession, as long as they continue to be real, by the formulæ,

$$\alpha_1 x \cos\xi' = (\tan\xi)^2, \quad \alpha_2 x \cos\xi'' = (\tan\xi')^2, \quad \&c. \tag{617}$$

In this manner, by supposing

$$x = \frac{1}{2}, \quad \text{and therefore } X = 2\epsilon\int_1^\infty \epsilon^{-t^2}\,dt = L^{-1}\bar{1}{\cdot}879\,596\,0 = \cos 40°43'23'', \tag{618}$$

I have found,

$$\xi = 40°43'23''; \ \xi' = 42°10'48''; \ \xi'' = 56°48'50''; \ \xi''' = 40°41'42''; \text{ and } \xi^{iv} = 72°28'29''; \tag{619}$$

but ξ^v would be imaginary, because its cosine is equal to

$$X^v = L^{-1}0{\cdot}178\,691\,0 = 1{\cdot}508\,87 > 1. \tag{620}$$

Now it is from this circumstance that the angle ξ^v is here *imaginary*, or that its theoretical *cosine*, X^v, is in the present question *greater than unity*, that the fact of the approximation X_5 being *in excess* (and not in *defect*) depends: as easily appears, if we apply the principles of the present Letter. It is true, indeed, that the *series* (606) for X^v begins with the terms $1 - \frac{1}{2}\alpha_6 x$, where α_6 has been found to be *positive*, (613); but this series may be *presumed* to be a very *divergent* one, and it is not surprising, *on consideration*, that the terms which *follow* those here written should *raise its value*, above that of the 1st term, 1; because *no proof* has been given that *this* development (606) belongs to that great *class of alternating series*, which you and others with justice consider to possess a *prerogative*, and to be by eminence important. Still, I confess that I did at first *expect* – or at least was under the *impression* – until numerical calculation, in the recent example, proved the contrary, – that X^v would be found to be *always less than unity*, if x were *positive and real*, and if the coefficient α_6 were *positive.* And hence, when I perceived that the approximate value X_5 was *slightly in excess*, (after X_1, X_2, X_3, X_4 had comported themselves as usual, so as to be alternately less and greater than X,) I not only *inferred* that X^v was >1, (which it really *is*), for $x = \frac{1}{2}$, but also was led to *anticipate* that α_6 would be found to be *negative*; or that the *series* for X^v must *begin* with $1 + \cdots$, and *not* with $1 - \cdots$. Yet I see no reason to suspect that any arithmetical error has crept into the calculation of the value (613) for α_6; and therefore I must *now* condemn that *anticipation*, as having been erroneous.

LXXIV. It still appeared (and appears) to me, notwithstanding, that from the way in which the *series* for X^v *begins*, the function which that series represents must be both *positive* and *less than unity*, if x be *positive*, and *not too large*; and that the same thing may be asserted, respecting each of the preceding functions of the same group: so that the *chain* of inequalities, (588), ought to admit of being *enlarged*, by the addition of *another link*, as follows:

$$1 > X_2 > X_4 > X > X_5 > X_3 > X_1 > 0, \tag{621}$$

if we do not exceed some positive limit, which I have not attempted to determine. Or, writing not only

$$\left.\begin{array}{ll} X = \cos\xi, \quad \text{but also} & X_1 = \cos\xi_1, \\ X_2 = \cos\xi_2, & X_3 = \cos\xi_3, \ldots \end{array}\right\} \tag{622}$$

I still expected (and expect) that, for sufficiently *moderate and positive values of x*, there ought to be found this other *chain*:

$$\frac{\pi}{2} > \xi_1 > \xi_3 > \xi_5 > \xi > \xi_4 > \xi_2 > 0; \tag{623}$$

where ξ_1, ξ_2, ..., ξ_5, and (if necessary) ξ_6, may be computed from the usual Tables of Logarithmic Sines, by a very easy and uniform set of processes, which admits of being formulated thus:

$$\tan\xi_1 = (\alpha_1 x)^{\frac{1}{2}}; \tag{624}$$

$$\tan T_2 = (\alpha_2 x)^{\frac{1}{2}}, \ \tan\xi_2 = (\alpha_1 x \cos T_2)^{\frac{1}{2}}; \tag{625}$$

$$\tan T_3 = (\alpha_3 x)^{\frac{1}{2}}, \ \tan T_3' = (\alpha_2 x \cos T_3)^{\frac{1}{2}}, \ \tan\xi_3 = (\alpha_1 x \cos T_3')^{\frac{1}{2}}; \tag{626}$$

$$\tan T_4 = (\alpha_4 x)^{\frac{1}{2}}, \ \tan T_4' = (\alpha_3 x \cos T_4)^{\frac{1}{2}}, \ \tan T_4'' = (\alpha_2 x \cos T_4')^{\frac{1}{2}}, \ \tan\xi_4 = (\alpha_1 x \cos T_4'')^{\frac{1}{2}}; \tag{627}$$

$$\begin{gathered} \tan T_5 = (\alpha_5 x)^{\frac{1}{2}}, \ \tan T_5' = (\alpha_4 x \cos T_5)^{\frac{1}{2}}, \ \tan T_5'' = (\alpha_3 x \cos T_5')^{\frac{1}{2}}, \\ \tan T_5''' = (\alpha_2 x \cos T_5'')^{\frac{1}{2}}, \ \tan\xi_5 = (\alpha_1 x \cos T_5''')^{\frac{1}{2}}; \end{gathered} \tag{628}$$

&c. – If I shall go on to give an Example, and a Type, it will of course not be that I can suppose *you* to stand in any manner in *need* of any such illustration, but simply, or chiefly, because it may be a comfort to *myself*, hereafter, to have such an example, and such a type, *collected* along with others of the same general kind, in the copy that is to be presented for me of this Letter.

LXXV. Let therefore x be now assumed equal to $\frac{1}{18}$, so that the definite integral, to be by the present method studied, is (compare a note to paragraph XVI.), (see also equation (9),)

$$X = 6\epsilon^9 \int_3^\infty \epsilon^{-t^2}\, dt = L^{-1}\bar{1}{\cdot}978\,554\,1 = \cos 17°51'29''{\cdot}5 = 0{\cdot}951\,818. \tag{629}$$

Here

$$\begin{gathered} La_1 x = \bar{1}{\cdot}045\,757\,5; \ La_2 x = \bar{1}{\cdot}221\,848\,8; \ La_3 x = \bar{1}{\cdot}534\,777\,9; \ La_4 x = \bar{1}{\cdot}436\,000\,2; \\ La_5 x = \bar{1}{\cdot}868\,327\,9; \end{gathered} \tag{630}$$

and the calculations for the determination of ξ_3, ξ_4, ξ_5 may proceed as follows.

Type (numbered for references as) ... (631)

[which is displayed on p. 131].

And *here* we see that $\xi_3 > \xi_5 > \xi > \xi_4$, because

$$17°51'54''\cdot6 > 17°51'32''\cdot6 > 17°51'29''\cdot5 > 17°51'26''\cdot6. \quad (632)$$

In fact we find that, in *this* case, the angles ξ', ξ'', ξ''' ξ^{iv}, ξ^{v} are all *real*, although ξ^{vi} is *imaginary* or in other words, X', X'', X''', X^{iv}, and X^{v} are here each >0, and <1, but X^{vi} is here >1, as appears by the little Type annexed, which I number as (633).

Type. (633)	$2L\tan\xi = \overline{1}\cdot016\,212\,4$
	(sub.) $La_1 x = \overline{1}\cdot045\,757\,5$
$\xi' = 20°53'43''\cdot9$	$= (L\cos)^{-1} = \overline{1}\cdot970\,454\,9$
	$2L\tan\xi' = \overline{1}\cdot163\,611\,4$
	(sub.) $La_2 x = \overline{1}\cdot221\,848\,8$
$\xi'' = 29°0'48''\cdot5$	$(L\cos)^{-1} = \overline{1}\cdot941\,762\,6$
	$2L\tan\xi'' = \overline{1}\cdot487\,985\,5$
	(sub.) $La_3 x = \overline{1}\cdot534\,777\,9$
$\xi''' = 26°7'19''\cdot5$	$= (L\cos)^{-1} = \overline{1}\cdot953\,207\,6$
	$2L\tan\xi''' = \overline{1}\cdot381\,053\,1$
	(sub.) $La_4 x = \overline{1}\cdot436\,000\,2$
$\xi^{iv} = 28°13'4''\cdot3$	$(L\cos)^{-1} = \overline{1}\cdot945\,052\,9$
	$2L\tan\xi^{iv} = \overline{1}\cdot459\,296\,0$
	(sub.) $La_5 x = \overline{1}\cdot868\,327\,9$
$\xi^{v} = 67°3'3''\cdot2$	$(L\cos)^{-1} = \overline{1}\cdot590\,968\,1$
	$2L\tan\xi^{v} = 0\cdot746\,443\,0$
	(sub.) $La_6 x = \overline{1}\cdot092\,108\,5$
$X^{vi} = 45\cdot11643$	$= L^{-1}1\cdot654\,334\,5$
numbered for reference as (633)	

and in which (as usual) I do not pledge myself for all the figures that are set down: for (if I have made no important mistake) X^{vi} even exceeds 45. Hence I should expect to find that X_6 *errs by defect*; or that, in the present case $(x = \frac{1}{8})$, we must have $\xi_6 > \xi$. And accordingly, on calculating ξ_6 in the same way as ξ_3, ξ_4, ξ_5, (631), I find the following auxilliary angles

$$\left.\begin{array}{lll} T_6 = 19°22'18''\cdot8, & T_6' = 39°51'1''\cdot2, & T_6'' = 24°35'40''\cdot4, \\ T_6''' = 29°10'2''\cdot2, & T_6^{iv} = 20°52'52''\cdot6; & \end{array}\right\} \quad (634)$$

and finally,

$$\xi_6 = 17°51'32''\cdot4 > 17°51'29''\cdot5. \quad (635)$$

In this case, ξ_6 seems to be almost exactly the same as ξ_5, being however *a little less* than it. But I give the numbers merely as *illustrations*; and with the same view add that in the present case of the present method,

$$X_3 = \cos\xi_3 = 0\cdot951\,781; \quad X_4 = \cos\xi_4 = 0\cdot951\,823; \quad X_5 = \cos\xi_5 = 0\cdot951\,814; \quad (636)$$

the corresponding approximations in the method of continued fractions, being

$$X_3 = \frac{138}{145} = 0\cdot951\,724; \quad X_4 = \frac{494}{519} = 0\cdot951\,830; \quad X_5 = \frac{3654}{3839} = 0\cdot951\,810; \quad (637)$$

Table II.I.

$(\xi = \cos^{-1} X = 17°51'29''{\cdot}5)$		$L\alpha_5 x = \bar{1}{\cdot}8683279$
		$T_5 = 40°40'25''{\cdot}4 = (L\tan)^{-1}\bar{1}{\cdot}9341639\vert 5$
		$L\cos T_5 = \bar{1}{\cdot}8799173$
	$L\alpha_4 x = \bar{1}{\cdot}4360002$	$L\alpha_4 x = \bar{1}{\cdot}4360002$
		$2)\bar{1}{\cdot}3159175$
	$T_4 = 27°34'56''{\cdot}6 = (L\tan)^{-1}\bar{1}{\cdot}7180001$	$T_5' = 24°27'46''{\cdot}5 = (L\tan)^{-1}\bar{1}{\cdot}6579587\vert 5$
	$L\cos T_4 = \bar{1}{\cdot}9476032$	$L\cos T_5' = \bar{1}{\cdot}9591509$
$L\alpha_3 x = \bar{1}{\cdot}5347779$	$L\alpha_3 x = \bar{1}{\cdot}5347779$	$L\alpha_3 x = \bar{1}{\cdot}5347779$
	$2)\bar{1}{\cdot}4823811$	$2)\bar{1}{\cdot}4939288$
$T_3 = 38°20'27''{\cdot}7 = (L\tan)^{-1}\bar{1}{\cdot}7673889\vert 5$	$T_4' = 28°51'25''{\cdot}0 = (L\tan)^{-1}\bar{1}{\cdot}7411905\vert 5$	$T_5'' = 29°10'48''{\cdot}2 = (L\tan)^{-1}\bar{1}{\cdot}7469644$
$L\cos T_3 = \bar{1}{\cdot}9360280$	$L\cos T_4' = \bar{1}{\cdot}9424186$	$L\cos T_5'' = \bar{1}{\cdot}9410600$
$L\alpha_2 x = \bar{1}{\cdot}2218488$	$L\alpha_2 x = \bar{1}{\cdot}2218488$	$L\alpha_2 x = \bar{1}{\cdot}2218488$
$2)\bar{1}{\cdot}1578768$	$2)\bar{1}{\cdot}1642674$	$2)\bar{1}{\cdot}1629088$
$T_3' = 20°46'11''{\cdot}3 = (L\tan)^{-1}\bar{1}{\cdot}5789384$	$T_4'' = 20°54'35''{\cdot}8 = (L\tan)^{-1}\bar{1}{\cdot}5821337$	$T_5''' = 20°52'48''{\cdot}3 = (L\tan)^{-1}\bar{1}{\cdot}5814544$
$L\cos T_3' = \bar{1}{\cdot}9708175$	$L\cos T_4'' = \bar{1}{\cdot}9704132$	$L\cos T_5''' = \bar{1}{\cdot}9704996$
$L\alpha_1 x = \bar{1}{\cdot}0457575$	$L\alpha_1 x = \bar{1}{\cdot}0457575$	$L\alpha_1 x = \bar{1}{\cdot}0457575$
$2)\bar{1}{\cdot}0165750$	$2)\bar{1}{\cdot}0161707$	$2)\bar{1}{\cdot}0162571$
$\xi_3 = 17°51'54''{\cdot}6 = (L\tan)^{-1}\bar{1}{\cdot}5082875$	$\xi_4 = 17°51'26''{\cdot}6 = (L\tan)^{-1}\bar{1}{\cdot}5080853\vert 5$	$\xi_5 = 17°51'32''{\cdot}6 = (L\tan)^{-1}\bar{1}{\cdot}5081285\vert 5$

while the true theoretical value is nearly,

$$X = 0{\cdot}951\,818, \tag{629}$$

as above stated; so that the *method of square roots* gives *here* results which are *decidedly more exact* (though not very *importantly* so in *practice*) that those given by the *method of reciprocals*: although this last mentioned method was seen (in the lately cited note to paragraph XVI.) to give results superior in accuracy to those furnished by the *methods of exponentials* namely than the following:

$$X_5 = 0{\cdot}951\,600; \quad X_4 = 0{\cdot}951\,870; \quad X_5 = 0{\cdot}951\,797. \tag{638}$$

LXXVI. I shall only add, as regards the "method of square roots", than when I took, as another example, $x = \frac{1}{8}$,

$$X = 4\epsilon^4 \int_2^\infty \epsilon^{-t^2}\,dt = \bar{1}{\cdot}956\,818\,5 = \cos\xi = 0{\cdot}905\,354, \; \xi = 25°7'44''{\cdot}3 \tag{639}$$

(compare equation (8)) I found (by a rather hasty calculation) the values,

$$\xi' = 28°21'15'', \; \xi'' = 39°2'48'', \; \xi''' = 31°24'4'', \; \xi^{iv} = 52°38'14'' \tag{640}$$

so that *these* 4 angles, like ξ itself, are *real*; but ξ^v is an *imaginary* angle,

$$X^v = \cos\xi^v = L^{-1}0{\cdot}013\,838\,8 > 1; \tag{641}$$

whence I inferred that the approximate value X_5 should *err in excess*; or that we should *here* find

$$\xi_5 > \xi. \tag{642}$$

Accordingly on computing, with a *little* more care, the values of $T_5, \ldots, \xi_5$, by the type (631), I found the values

$$T_5 = 52°11'45''{\cdot}7; \quad T_5 = 31°11'43''{\cdot}4; \quad T_5'' = 39°1'39''{\cdot}2;$$
$$T_5''' = 28°21'26''{\cdot}6; \tag{643}$$

and finally,

$$\xi_5 = 25°7'42''{\cdot}8 < 27°7'44''{\cdot}3; \quad X_5 = \cos\xi_5 = 0{\cdot}905\,357 > 0{\cdot}905\,354. \tag{644}$$

It is remarkable, however, *how trifling these little errors, in the present method, are,* as compared with those which remained, at the corresponding stages of the applications of the two other methods of this Letter, to the same case, $x = \frac{1}{8}$, or to the definite integral (639), or (8). Thus, the *method of reciprocals* has given by (139), $X_5 = \dfrac{1672}{1847} = 0{\cdot}90\,464\,4$.

LXXVII. It appears, then, that of the *three methods of transformation of a diverging series,* which have been selected for special study and discussion in this long and now almost concluded Letter, the "Method of Inverse Square Roots" possesses some important *advantages,* as regards *rapidity of approximation,* over the other two although it *may fail, at a certain stage,* to produce results *alternately less and greater than the true theoretical value of the function,* from which series is conceived to have been developed. The *rationale* of this *advantage,* in one respect, and comparative *disadvantage* in *another* respect, of the *third method* of this Letter, as compared with the *two other* methods, and also of the *first method* as compared with the *second*, appears to consist in the circumstance that of the *three elementary series*,

$$\left.\begin{array}{ll}(I.), & \phi x = (1+x)^{-1} = 1 - x + x^2 - x^3 + \&c.; \\ (II.), & \phi x = \epsilon^{-x} = 1 - x + \dfrac{x^2}{2} - \dfrac{x^3}{2.3} + \&c.; \\ \text{and} & \\ (III.), & \phi x = (1+2x)^{-\frac{1}{2}} = 1 - x + \dfrac{3x^2}{2} - \dfrac{3.5x^3}{2.3} + \&c.,\end{array}\right\} \quad (645)$$

the Ist, when $x > 1$ is *more divergent* than the IInd, but *less so* than the IIIrd. We use a *more divergent tool* in one transformation than in the other, but we *may overdo* the business, by taking *too powerful* an instrument into our hands. It occurred to me, however, from such *general* considerations as those to which I have been now alluding, and before I began to write the present Letter, that the following *fourth form*, of an *auxilliary* and *elementary series*, analogous to those lately mentioned, namely this one:

$$(IV.) \quad \phi x = (1+4x)^{-\frac{1}{4}} = 1 - x + \frac{5x^2}{2} - \frac{5.9x^3}{2.3} + \&c. \qquad (646)$$

might possibly be found to be *still more powerful*, and useful in *some* important cases of *transformation of divergent developments*, although it might require, for such *utility*, or at least for continuing to exhibit an *alternating* character in its results, a *still greater degree of rapidity in the increase of the coefficients of the transformand*. And such, precisely, continues to be my *impression*, although I have not expended much *labour*, (nor indeed any that is worth mentioning,) upon the examination of *this part* of the whole question and subject. I may just add that papers, not since examined, which were written before this Letter was begun appeared to give me, for a transformation of the recent function *X*, (so well known in connexion with Laplace, &c.,) the expression:

$$X = \left(1 + 4x\left(1 + 2x\left(1 + \frac{35}{4}x\left(1 - \frac{817}{14}xX^{iv}\right)^{-\frac{1}{4}}\right)^{-\frac{1}{4}}\right)^{-\frac{1}{4}}\right)^{-\frac{1}{4}}; \qquad (647)$$

where X^{iv} denotes a series of which the Ist term is unity. You conceive, of course, that I should have remarks to make on this last expression, of the same general *kind* as those which have been already made. But the only remark which at present, and in the conclusion of this Letter, I shall allow myself to add, is this, to show the great *caution* with which such diverging tools as these ought to be used: that if we assume $x = \frac{1}{2}$, we have *not* $X > (1+4x)^{\frac{1}{4}}$, for *all* positive values of *x*; since we saw, very early in this Letter, (37), that $X < 3^{\frac{1}{4}}$ when $x = \frac{1}{2}$. – Such *caution*, combined with such *encouragement*, for the use of *Diverging Series*, may induce you (perhaps) to forgive me, for having (potentially but I suppose *not actually*) occupied so much of your time as would be required for *even glancing* through this Epistle, begun* on the 15th of February, but only finished on the 22nd of May. I am yours faithfully,

W. R. Hamilton.

* I find an *entry* in my Book C. 1858, [Trinity College, Dublin MS 1492/143]. p. 119 *dated* February 15th 1858, "Would this appear a paradox to De Morgan?" All this long *Letter* has been designed, while preserving the date of that entry, to do full justice to its spirit. . . .

2. On The Solution of a Third Order Differential Equation

Observatory, July 15th, 1858

My dear De Morgan,

When I began, on the 15th of February last, that terribly long Letter which you were so good humoured as to allow me to continue, by instalments, until it came at length to a point at which I was content to close it. I stated that I had just then been led to perceive that certain *linear equations of the third order*, between two variables and with variable coefficients, had important connexions with some mathematico-physical investigations; and that I hoped to be able to integrate the equation in question, at least with the help of series, though not proceeding according to *powers* alone, but introducing *logarithms* (or a logarithm) also. Indeed, I find a memorandum that in a note* to the First Sheet of that long Letter, I remarked that the equation,

$$(xD)^3 y + x^2 Dxy = 0, \qquad (a)$$

had for *one* particular integral the *ascending series*,

$$y = a \sum_{m=0}^{m=\infty} \left[-\frac{1}{2}\right]^m ([0]^{-m})^3 \left(\frac{x}{2}\right)^{2m}; \qquad (b)$$

where a is an arbitrary constant, and $[-1/2]^m$, $[0]^{-m}$ are factorials, to be interpreted as in the notation of Vandermonde,* which I learned long ago from the third Volume of La Croix,† (long missing from my library,) and in which notation the symbol $[n]^m$ denotes the same thing as

$$\frac{\Gamma(n+1)}{\Gamma(n-m+1)};$$

so that, if m be a positive whole number,

$$\begin{cases} [n]^m = n(n-1)(n-2)\cdots(n-m+1); \\ [n]^{-m} = (n+1)^{-1}(n+2)^{-1}(n+3)^{-1}\cdots(n+m)^{-1}. \end{cases}$$

I remarked also that *another* particular integral, with another arbitrary constant b, was given by the *descending series*,

$$y = \frac{b}{x} \sum_{m=0}^{m=\infty} [0]^{-m} \left(\left[-\frac{1}{2}\right]^m\right)^3 \left(\frac{x}{2}\right)^{-2m}; \qquad (c)$$

and that although these two series, (b) and (c), appeared to *exhaust* all the particular integrals of the *triordinal* equation (a), which are expressible by *powers alone*, yet I had found a *third particular integral*, of the form

$$y = c(A_x + B_x \log x); \qquad (d)$$

* [See p. 35 of this volume.]

† [Sylvestre François Lacroix (1765–1843), *Traité des différences et des séries* (faisant suite au *Traité du calcul différentiel et du calcul intégral*), p. 74. Paris: 1800.]

where c is a third arbitrary constant, B_x is the ascending series (b), and A_x is a new ascending series. I stated at the same time, that although the equation of the *third order* (a) was thus *completely integrated*, by the combination of these *three particular integrals*, (b), (c), (d), yet I thought that I saw how to obtain a *fourth particular integral*, (with a *fourth arbitrary constant*,) which should involve the *square of the logarithm of* x; and even that I should *want* such *fourth form*, for the purposes of my investigation. And finally, in the same note, – of which I repeat the substance now, to save you the trouble of a reference, – even in case that you have preserved and can find the sheet of which I speak, – I observed that if (as I suspected) the *fourth* particular integral *existed*, it *could not be independent of the other three*: but that on this whole subject I must write again. I now propose to redeem that promise, or to fulfil that expressed intention: and trust that if you read this *new* long Letter, – for such I fear that it must be, – you will consider me to have at last completely overcome the difficulties, first, of *assigning* the *fourth form* of integral, and second, of *connecting* it with the *other three.*

I. The manner in which $\frac{x}{2}$ enters into the lately cited series, (b) and (c), may serve to suggest the convenience of assuming, as the fundamental Equation of the *present* Letter, the following:

$$(xD)^3\,y + 4x^2 Dxy = 0; \tag{1}$$

where $D = \frac{d}{dx}$ and the introduction of the coefficient 4 makes, of course, no essential change, in the nature of the question to be discussed. If we make $\theta = \log x$, the equation becomes,

$$D_\theta^3\,y + 4\epsilon^\theta D_\theta \epsilon^\theta\,y = 0; \tag{2}$$

and many other transformations may be made, whereof I have found some to be useful. For example, we have the symbolic equations,

$$(xD)^3 = x^3D^3 + 3x^2D^2 + xD, \quad \text{and } Dx = xD + 1;$$

writing therefore

$$Dy = y', \quad D^2y = y'', \quad D^3y = y''',$$

we have

$$x^2y''' + 3xy'' + (1 + 4x^2)\,y' + 4xy = 0. \tag{3}$$

Let δ be a symbol of operation, such that

$$\delta = (xD)^3 + 4x^2Dx = D_\theta^3 + 4\epsilon^\theta D_\theta \epsilon^\theta; \tag{4}$$

then the triordinal equation may be thus briefly written:

$$\delta\,y = 0. \tag{5}$$

The operation δ is such that

$$\delta\,x^\nu = \nu^3 x^\nu + 4(\nu + 1)\,x^{\nu+2}; \tag{6}$$

if then we assume that y can be expressed by a *series*, such as

$$y = \sum a_\nu x^\nu, \tag{7}$$

the differential equation will give this equation in finite differences,

$$\nu^3 a_\nu + 4(\nu - 1)\, a_{\nu-2} = 0. \tag{8}$$

This last equation gives,

$$a_1 = 0, \quad a_3 = 0, \quad a_5 = 0, \quad \&\text{c.}$$

unless a_1 be infinite; it gives also

$$a_2 = 0, \quad a_4 = 0, \quad a_6 = 0, \quad \&\text{c.},$$

unless a_0 be infinite; if then we *exclude infinite coefficients, and retain only whole exponents,* we must suppose the series (7) to involve only powers whose exponents *are positive and even,* including 0; or only those whose exponents are *negative* and *odd*; or else a mixture of *both.* The first supposition gives the particular integral,

$$y = a_0 \sum_{m=0}^{m=\infty} \overline{-\tfrac{1}{2}}\Big\rceil^{m} \left(\overline{0}\rceil^{-m}\right)^3 x^{2m}; \tag{9}$$

where I write for shortness,

$$\overline{n}\rceil^{m} \text{ instead of } [n]^m, \text{ or of } \frac{\Gamma(n+1)}{\Gamma(n - m + 1)};$$

so that

$$\overline{-\tfrac{1}{2}}\Big\rceil^{m} = -\tfrac{1}{2}.-\frac{3}{2}.-\frac{5}{2}.\cdots.\frac{1-2m}{2} = \left(-\frac{1}{2}\right)^m .1.3.5.7.\cdots.(2m-1);$$

and

$$\overline{0}\rceil^{-m} = \frac{\Gamma 1}{\Gamma(m+1)} = \frac{1}{1.2.3.4.\cdots.m}.$$

The second supposition gives this other particular integral,

$$a_{-1} x^{-1} \sum_{m=0}^{m=\infty} \left(\overline{-\tfrac{1}{2}}\Big\rceil^{m}\right)^3 \overline{0}\rceil^{-m} x^{-2m}; \tag{10}$$

and the third supposition gives merely the *sum* of the last two values for y. If we admitted *fractional exponents,* (or incommensurable ones,) the series (7) would then, of necessity, extend *indefinitely both ways,* ascending and descending; and I wish for the present to *avoid* the consideration of such *mixed series.* It appears, then, that if we choose that our series (7) for y should *only ascend,* we *must* adopt the form (9), which answers to the series (*b*) of my lately cited note; and that if we wish the series *only to descend,* it *must* be of the form (10), which corresponds to the development (*c*) of the same note.

II. But the form (2) of the triordinal equation (1) suggests the assumption of this other expression for y:

$$y = A_x + \theta B_x + \theta^2 C_x; \tag{11}$$

where $\theta = \log x$, but A_x, B_x, C_x are three functions of x, which do not expressly involve this logarithm. In fact, with the definition (4) of the symbol δ, we shall have

$$\left.\begin{aligned} \delta\theta B_x - \theta\delta B_x &= 3D_\theta^2 B_x + 4x^2 B_x = (3D_\theta^2 + 4x^2)\,B_x;\\ \delta\theta^2 C_x - \theta^2\delta C_x &= 2\theta(3D_\theta^2 + 4x^2)\,C_x + 6D_\theta C_x; \end{aligned}\right\} \tag{12}$$

therefore

$$\delta y - (\delta A_x + \theta\delta B_x + \theta^2\delta C_x) = (3D_\theta^2 + 4x^2)\,B_x + 6D_\theta C_x + 2\theta(3D_\theta^2 + 4x^2)\,C_x; \tag{13}$$

and the differential equation $\delta y = 0$ will be satisfied, if we satisfy the systems of the three following equations:

$$\delta C_x = 0; \quad \delta B_x + 2(3D_\theta^2 + 4x^2)\,C_x = 0; \quad \delta A_x + (3D_\theta^2 + 4x^2)\,B_x + 6D_\theta C_x = 0 \tag{14}$$

in which,

$$D_\theta = xD, \quad \text{and } D_\theta^2 = (xD)^2 = x^2D^2 + xD.$$

We shall therefore at least *satisfy* the triordinal equation $\delta y = 0$ by the expression (11) for y, if we suppose, first, that $C_x = 0$; second, that B_x is any particular integral of the equation (1), for example the ascending series (9), so that $\delta B_x = 0$; and third, that A_x satisfies this other differential equation,

$$\delta A_x + (3D_\theta^2 + 4x^2)\,B_x = 0. \tag{15}$$

We shall thus have

$$B_x = a_0 + a_2x^2 + a_4x^4 + \&\text{c.}, \tag{16}$$

with the equation in differences (8), to connect the coefficients $a_2, a_4, \ldots$ with a_0 or to determine their ratios to that initial and arbitrary constant; which equation (8), because $\nu = 2m$, becomes, in this application,

$$2m^3a_{2m} + (2m-1)\,a_{2m-2} = 0. \tag{17}$$

To satisfy (15), without introducing into A_x a part equal to $B_x\times$ a constant, which would be useless, because we already know that such a part will be a particular integral of $\delta y = 0$, and may be added, at the end of the process, to any other integral of that equation, we may assume

$$A_x = \alpha_1a_2x^2 + \alpha_2a_4x^4 + \alpha_3a_6x^6 + \&\text{c.}, \tag{18}$$

where $\alpha_1, \alpha_2, \alpha_3, \ldots,$ are constants to be determined by (15). The operation δ will give, by (6),

$$\delta\alpha_m a_{2m}x^{2m} = 4\alpha_m a_{2m}\{2m^3x^{2m} + (2m+1)\,x^{2m+2}\}; \tag{19}$$

also,

$$(3D^2 + 4x^2)\,a_{2m}x^{2m} = 12m^2a_{2m}x^{2m} + 4a_{2m}x^{2m+2}; \tag{20}$$

therefore collecting the coefficients of x^{2m} in (15), changing $(2m-1)\,a_{2m-2}$ to $-2m^3a_{2m}$ by (17), and dividing by $8m^3a_{2m}$, we find this new equation in differences,

$$\alpha_m - \alpha_{m-1} = \frac{1}{2m-1} - \frac{3}{2m}. \tag{21}$$

Thus,

$$\alpha_0 = 0, \quad \alpha_1 = \frac{1}{1} - \frac{3}{2} = -\frac{1}{2}; \quad \alpha_2 = -\frac{1}{2} + \frac{1}{3} - \frac{3}{4} = \frac{-11}{12}, \quad \alpha_3 = \frac{-11}{12} + \frac{1}{5} - \frac{3}{6} = \frac{73}{60}, \text{ \&c.} \tag{22}$$

but

$$a_2 = -\frac{1}{2} a_0, \quad a_4 = \frac{1.3}{1^3.2^3}\left(-\frac{1}{2}\right)^2 a_0 = \frac{3a_0}{32}, \quad a_6 = \frac{1.3.5}{1^3.2^3.3^3}\left(-\frac{1}{2}\right)^3 a_0 - \frac{-5a_0}{576}, \text{ \&c.}; \tag{23}$$

therefore,

$$\alpha_1 a_2 = \frac{1}{4} a_0, \quad \alpha_2 a_4 = -\frac{11}{128} a_0, \quad \alpha_3 a_6 = +\frac{73}{6912} a_0, \quad \text{\&c.}; \tag{24}$$

and finally the series for A_x and B_x are the following:

$$A_x = a_0 \left\{ \frac{x^2}{4} - \frac{11x^4}{128} + \frac{73x^6}{6912} - \text{\&c.} \right\}; \tag{25}$$

$$B_x = a_0 \left\{ 1 - \frac{x^2}{2} + \frac{3x^4}{32} - \frac{5x^6}{576} + \text{\&c.} \right\}. \tag{26}$$

These series, (except that x has since been changed to $2x$,) were those alluded to in my cited note, as entering into the *third particular integral* (d) of the equation of the 3rd order (a); and (with our present x) the corresponding integral of the triordinal equation (1) may be thus written:

$$y = \frac{c}{a_0}(A_x + B_x \log x) = c\left(\frac{x^2}{4} - \frac{11x^4}{128} + \frac{73x^6}{6912} - \cdots\right) + c\theta\left(1 - \frac{x^2}{2} + \frac{3x^4}{32} - \frac{5x^6}{576} + \cdots\right) \tag{27}$$

where c is an arbitrary constant. As a verification, this last expression (27) for y gives,

$$\begin{aligned}
4x^2 Dxy = 4(x^3 Dy + x^2 y) &= 4c\left(\frac{x^4}{2} - \frac{11x^6}{32}\left[+\frac{73x^8}{1152}\right]^* - \cdots\right) \\
&+ 4c\left(\frac{x^4}{4} - \frac{11x^6}{128}\left[+\frac{73x^8}{6912}\right]^* - \cdots\right) + 4c\left(x^2 - \frac{x^4}{2} + \frac{3x^6}{32}\left[-\frac{5x^8}{576}\right]^* + \cdots\right) \\
&+ 4c\theta\left(-x^4 + \frac{3x^6}{8}\left[-\frac{5x^2}{96}\right]^* + \cdots\right) + 4c\theta\left(x^2 - \frac{x^4}{2} + \frac{3x^6}{32}\left[-\frac{5x^8}{576}\right]^* + \cdots\right) \\
&= c\left(4x^2 + x^4 - \frac{43}{32}x^6 \cdots\right) + c\theta\left(4x^2 - 6x^4 + \frac{15}{8}x^6 \cdots\right);
\end{aligned}$$

this then is one part of δy; and the other part is its negative, namely,

* The terms within square brackets, though written down, have not been used in this verification.

$$(xD)^3 y = c\left(2x^2 - \frac{11x^4}{2} + \frac{73x^6}{32} - \cdots\right) + 3x\left(-2x^2 + \frac{3x^4}{2} - \frac{5x^6}{16} + \cdots\right)$$

$$+ c\theta\left(-4x^2 + 6x^4 - \frac{15}{8}x^6 + \cdots\right) = c\left(-4x^2 - x^4 + \frac{43}{32}x^6 \cdots\right) + c\theta\left(-4x^2 + 6x^4 - \frac{15}{8}x^6 \cdots\right);$$

so that we have, in fact, $\delta y = 0$, as we described, if the values (25) and (26) be substituted for A_x and B_x, and if we neglect $2x^8$ and θx^8. The neglected term would be made to vanish like the ones here attended to, if the developments (25) and (26) we carried on.

III. Let us now consider that *fourth particular integral*, which is of the *trinomial form* (11); the coefficient C_x, of the *square of log x*, being not now supposed to vanish. It is evident, from the analysis of the preceding paragraph II. that we shall satisfy the *two first* of the three equations of condition (14), if we *now* assume, for C_x and B_x, the expressions:

$$C_x = \frac{c}{a_0}(a_0 + a_2 x^2 + a_4 x^4 + \cdots) = c \sum_{m=0}^{m=\infty} \overline{-\frac{1}{2}}\Big|^{m} (\overline{0}|^{-m})^3 x^{2m}; \tag{28}$$

$$B_x = \frac{2c}{a_0}(\alpha_1 a_2 x^2 + \alpha_2 a_4 x^4 + \cdots) = \frac{-c}{a_0}\left(a_2 x^2 + \frac{11}{6} a_4 x^4 + \frac{73}{30} a_6 x^6 + \cdots\right); \tag{29}$$

or, developing as far as terms of the 6th dimension inclusive,

$$C_x = c\left(1 - \frac{x^2}{2} + \frac{3x^4}{32} - \frac{5x^6}{576} + \cdots\right); \quad B_x = c\left(\frac{x^2}{2} - \frac{11x^4}{64} + \frac{73x^6}{3456} - \cdots\right); \tag{30}$$

where c is an arbitrary constant. The *third* equation of condition (14) may then be satisfied, by assuming for A_x a series of the form,

$$A_x = \frac{c}{a_0}(\beta_2 a_4 x^4 + \beta_3 a_6 x^6 + \cdots); \tag{31}$$

in which $\beta_2, \beta_3, \ldots,$ are constants yet to be determined, and terms below the 4th dimension are suppressed, because such terms (with subsequent terms connected with them) may be conceived to arise* from former particular integrals. In fact, with our last development (30) for C_x and B_x, we have a new series of the form

$$(3D_\theta^2 + 4x^2) B_x + 6D_\theta C_x = \frac{c}{a_0}(b_4 x^4 + b_6 x^6 + \cdots) \tag{32}$$

the coefficient of x^2 vanishing; and after assuming the series (31) for A_x we shall satisfy the third equation (14): if we so choose the factors β_2, β_3, &c., as to fulfill the condition

$$\delta(\beta_2 a_4 x^4 + \beta_3 a_6 x^6 + \cdots) + b_4 x^4 + b_6 x^6 + \cdots = 0. \tag{33}$$

By an analysis similar to that of paragraph II., it is found that the coefficient of x^{2m} in

$$\delta(\beta_2 a_4 x^4 + \&c.) \text{ is equal to } 8m^3 a_{2m}(\beta_m - \beta_{m-1});$$

therefore

* Perhaps this logic is a little obscure, but I have convinced myself of the correctness of the conclusion. Besides we only want to satisfy the third of equations (14).

$$8m^3 a_{2m}(\beta_m - \beta_{m-1}) + b_{2m} = 0; \tag{34}$$

with the initial values $\beta_0 = 0$, $\beta_1 = 0$. As regards the coefficient b_{2m}, the part $(3D_\theta^2 + 4x^2)B_x$, of the left hand member contributes, by (20) and (29), the part,

$$8m^3\left(\frac{3\alpha_m}{m} - \frac{2\alpha_{m-1}}{2m-1}\right)a_{2m}$$

to the value of b_{2m}; while the other part, $6D_\theta C_x$, of the same left member, contributes this other part, $12ma_{2m}$, to the same value of b_{2m}. Collecting these parts of b_{2m}, substituting them in (34), and dividing by $8m^3 a_{2m}$, we find,

$$\beta_m - \beta_{m-1} = \frac{2\alpha_{m-1}}{2m-1} - \frac{3\alpha_m}{m} - \frac{3}{2m^2}; \tag{35}$$

where α_m retains its recent signification. Hence, with the help of the values (22) of α_m, we find $\Delta\beta_0 = 2\alpha_0 - 3\alpha_1 - \frac{3}{2} = 0$; but this is a mere varification, for we knew that $\beta_1 = \beta_0 = 0$ but proceeding,

$$\beta_2 = \Delta\beta_1 = \frac{2\alpha_1}{3} - \frac{3\alpha_2}{2} - \frac{3}{8} = -\frac{1}{3} + \frac{11}{8} - \frac{3}{8} = \frac{2}{3};$$

$$\beta_3 = \frac{2}{3} + \Delta\beta_2 = \frac{2}{3} + \frac{2\alpha_2}{5} - \alpha_3 - \frac{1}{6} = \frac{2}{3} - \frac{11}{30} + \frac{73}{60} - \frac{1}{6} = \frac{27}{20}; \text{ \&c.}; \tag{36}$$

and we had, by (23),

$$a_4 = \frac{3a_0}{32}, \quad a_6 = -\frac{5a_0}{576}; \tag{37}$$

therefore

$$\beta_2 a_4 a_0^{-1} = +\frac{1}{16}; \quad \beta_3 a_6 a_0^{-1} = \frac{-3}{256}; \ldots$$

therefore by (31), developing still only as far as x^6 inclusive,

$$A_x = c\left(\frac{x^4}{16} - \frac{3x^6}{256} + \cdots\right). \tag{38}$$

Our *fourth* particular integral of the triordinal equation (1) is therefore the following:

$$y = c(\theta^2 U + 2\theta V + W); \tag{39}$$

where c is still an arbitrary constant; θ still equals $\log x$, U, V, W are 3 ascending series, which *we know how to continue indefinitely*, and which begin as follows:

$$U = 1 - \frac{x^2}{2} + \frac{3x^4}{32} - \frac{5x^6}{576} + \cdots; \; V = \frac{x^2}{4} - \frac{11x^4}{128} + \frac{73x^6}{6912} - \cdots, \; W = \frac{x^4}{16} - \frac{3x^6}{256} + \cdots. \tag{40}^*$$

Another particular integral (the *third*), was of the form,

$$y = 2b(\theta U + V); \tag{41}$$

* (July 19) The next term of U is $+\frac{35x^8}{73\,728}$.

b being another constant. And *another* particular integral (the *first*) was,

$$y = aU; \tag{42}$$

namely the series (9), with a for a_0. The *complete integral* of the triordinal equation (1) may therefore be thus written:

$$y = (a + 2b\theta + c\theta^2)\,U + 2(b + c\theta)\,V + cW. \tag{43}$$

But *how to bring under this general form that other particular integral* (the *second*), which is expressed by the *descending series* (10), was (or appeared to me to be) a *difficulty*, which it has cost me some pains to surmount. (It will be proved that

$$V^2 = UW;$$

that

$$V = U\int_0^x \frac{dx}{x}\,(U^{-1} - 1);$$

and that

$$U = \frac{4}{\pi^2}\left(\int_0^{\frac{\pi}{2}} d\omega \cos(x\cos\omega)\right)^2.)$$

IV. The *difficulty* being thus, (for it is not essentially altered by supposing that the coefficient a_{-1} in the development (10) is equal to unity,) to see *how* the *particular integral*,

$$y = \sum_{m=0}^{m=\infty}\left(\left\lceil -\frac{1}{2}\right\rceil^m\right)^3 \overline{0}\rceil^{-m} x^{-2m-1} = x^{-1} - \frac{x^{-3}}{8} + \frac{27x^{-5}}{128} - \frac{1125x^{-7}}{1024} + \cdots \tag{44}$$

of the *triordinal equation* (1), namely

$$(xD)^3\,y + 4x^2Dxy = 0,$$

(whereof there is no reason for supposing that it is a *singular solution*,) can be consistent with, and included under, the *complete integral*, with *three arbitrary constants*, expressed by our recent formula,

$$y = (a + 2b\theta + c\theta^2)\,U + 2(b + c\theta)\,V + cW; \tag{43}$$

it appears to be natural to look out for some *intermediate integrals*, which may assist in establishing *relations between the 3 constants a, b, c*; and so, perhaps, in discovering, ultimately the *value* of *each* of those three. Among several attempts which I have made, the following seems to have been, on the whole, the most successful. Adopting the form (2), as thus modified,

$$D_\theta^3\,y + 4xD_\theta\,xy = 0; \tag{45}$$

multiplying by $2y$; and observing that

$$2yD_\theta^3\,y = D_\theta\{2yD_\theta^2\,y - (D_\theta\,y)^2\}; \tag{46}$$

we find, by an easy *first integration*,

$$2\,yD_\theta^2\,y - (D_\theta\,y)^2 + 4x^2\,y^2 = 4\,h^2; \tag{47}$$

where h is an arbitrary constant, real or imaginary. On substituting for y the descending series (44), we see at once that $4\,h^2 = 4$, and that therefore we may write,

$$h = 1. \tag{48}$$

It is a *little* more troublesome to substitute the expression (43) for y, in the first member of the equation (47): but if we confine ourselves, as we may, to the terms which do not involve x^2, we may reduce that expression (43) to $a + 2b\theta + c\theta^2$, and suppress the term $4x^2\,y^2$ in that left member of (47): which then becomes,

$$4c(a + 2b\theta + c\theta^2) - 4(b + c\theta)^2 = 4\,h^2,$$

or briefly,

$$ac - b^2 = h^2. \tag{49}$$

Comparing, therefore, the two expressions for the constant h^2; in the result (47) of a first integration of the triordinal equation, we find that the 3 constants a, b, c of the complete integral (43), must be *connected* with each other by the relation,

$$ac - b^2 = 1; \tag{50}$$

in order that this complete or *general* integral may be *reduced* to represent, or to coincide, with the *particular* integral represented by the descending series (44).

V. It is not so easy, or at least it seems to be less obvious, to see how we are to proceed to a *second successive integration* of the triordinal equation (1). But if we observe that the supposition $y = aU$, or simply $y = U$, must satisfy the first integral equation (47), provided that we suppose $h = 0$, in virtue of (43) and (49), we shall see that the function U, or aU, must be a particular solution of the following differential equation of the *2nd order*,

$$2\,UD_\theta^2\,U - (D_\theta\,U)^2 + 4x^2\,U^2 = 0. \tag{51}$$

Eliminating x^2 between (47) and (51), we find,

$$4\,h^2\,U^2 = 2\,Uy(\,UD_\theta^2\,y - yD_\theta^2\,U) + (\,yD_\theta\,U)^2 - (\,UD_\theta\,y)^2. \tag{52}$$

Multiply by

$$U^{-2}\,y^{-2}(\,yD_\theta\,U - UD_\theta\,y);$$

the equation becomes integrable, and gives, as the *sought second integral, of the first order of* (1),

$$4\,h^2\,Uy^{-1} + U^{-1}\,y^{-1}(\,UD_\theta\,y - yD_\theta\,U)^2 = 4\,k^2; \tag{53}$$

where k is a new arbitrary constant: that is,

$$4\,h^2\,U^2 + (\,UD_\theta\,y - yD_\theta\,U)^2 = 4\,k^2\,Uy; \tag{54}$$

or, by another easy transformation,

$$h^2 + \left(\frac{U}{2}\,D_\theta\,\frac{y}{U}\right)^2 = k^2\,\frac{y}{U}. \tag{55}$$

Changing $\frac{y}{U}$ to $a+2b\theta+c\theta^2$, and U to 1, while $h^2=ac-b^2$, we obtain the equation:

$$k^2 = c; \tag{56}$$

because we have $(ac-b^2)+(b+c\theta)^2 = k^2(a+2b\theta+c\theta^2)$. But some other considerations must be introduced, before we can obtain the value of k^2 *from the descending series* (44), and so *determine the constant c.*

VI. In the equation (51), make

$$U = v^2; \tag{57}$$

it then becomes

$$(D_\theta^2 + x^2)v = 0; \tag{58)*}$$

or, if we return to xD instead of its equivalent symbol D_θ,

$$(D^2 + x^{-1}D + 1)v = 0. \tag{59}$$

This differential equation of the 2nd order presented itself to *Fourier,* in the study of the *propagation of heat in a cylinder,* and it has since occurred to others, myself included, in other important connexions. The only value of v, consistent with this equation (59), which admits of being developed according to *ascending* powers of x, and which, like $\sqrt{U}$, reduces itself to unity when $x=0$, may, as Fourier pointed out, be expressed, at pleasure, either by the series,

$$v = 1 - \left(\frac{x}{2}\right)^2 + \left(\frac{x^3}{2.4}\right)^2 - \left(\frac{x^3}{2.4.6}\right)^2 + \&\text{c.} = \sum_{m=0}^{m=\infty} (\overline{0|}^{-m})^2 \left(\frac{-x^2}{h}\right)^m; \tag{60}$$

or else by the *definite integral,*

$$v = \frac{2}{\pi}\int_0^{\frac{\pi}{2}} d\omega \cos(x\cos\omega). \tag{61}$$

It follows that the series lately found for U, namely the first of the 3 developments (40), must be equal to the *square of Fourier's Series,* denoted just now by v; or that we must have the *identity*:

$$\sum_{m=0}^{m=\infty} \left.-\frac{1}{2}\right]^m (\overline{0|}^{-m})^3 x^{2m} = \left(\sum_{m=0}^{m=\infty} (\overline{0|}^{-m})^2 \left(\frac{-x^2}{4}\right)^m\right)^2; \tag{62}$$

that is, more fully,

$$1 - \frac{1}{1^3}\cdot\frac{x^2}{2} + \frac{1.3}{1^3.2^3}\left(\frac{x^2}{2}\right)^2 - \frac{1.3.5}{1^3.2^3.3^3}\left(\frac{x^2}{2}\right)^3 + \cdots = \left\{1 - \left(\frac{x}{2}\right)^2 + \left(\frac{x^2}{2.4}\right)^2 - \left(\frac{x^3}{2.4.6}\right)^2 + \cdots\right\}, \tag{63}$$

a relation between these two series which perhaps is new, but which can be verified with ease, to any required extent. It follows also that the series U admits of being thus expressed;

* $D_\theta^2 = (xD)^2 = xDxD = x^2D^2 + xD.$

$$U = \frac{4}{\pi^2}\left(\int_0^{\frac{\pi}{2}} d\omega \cos(x\cos\omega)\right)^2. \tag{64}$$

Now it was shown by *Poisson**, that for very large values (real and positive) of x, the definite integral above cited from Fourier took nearly the value of the following approximate expression:

$$v = \left(\frac{2}{\pi x}\right)^{\frac{1}{2}} \cos\left(x - \frac{\pi}{4}\right); \tag{65}$$

our function $U = v^2$ becomes therefore, for large values of x,

$$U = \frac{2}{\pi x}\cos^2\left(x - \frac{\pi}{4}\right), \quad \text{nearly.} \tag{66}$$

But the descending series (44), for y, reduces itself, at the same time, to

$$y = \frac{1}{x}, \quad \text{nearly;} \tag{67}$$

whence

$$\frac{y}{U} = \frac{\pi}{2}\sec^2\left(x - \frac{\pi}{4}\right), \quad D_\theta \frac{y}{U} = xD\frac{y}{U} = \pi\tan\left(x - \frac{\pi}{4}\right)\sec^2\left(x - \frac{\pi}{4}\right),$$
$$\frac{U}{2} D_\theta \frac{y}{U} = \tan\left(x - \frac{\pi}{4}\right);$$

and finally, because $h^2 = 1$, by (48), the equations (55) and (56) give,

$$c = k^2 = \frac{2}{\pi}. \tag{68}$$

VII. It is not difficult, after proceeding so far, to accomplish a *third* and *last integration* of the triordinal equation (1), which shall introduce a *third arbitrary constant*, l, to be combined with h and k. In fact, if we substitute, in our *second successive integral* equation (54), the value of $4h^2U^2$ which is given by (52), and then divided by $2U^2y$, we find this integrable equation,

$$2k^2U^{-1} = U^{-1}(UD_\theta^2 y - yD_\theta^2 U) - U^{-2}D_\theta U(UD_\theta y - yD_\theta y); \tag{69}$$

that is,

$$\left.\begin{aligned} &2k^2U^{-1} = D_\theta . U^{-1}(UD_\theta y - yD_\theta U) = D_\theta UD_\theta \frac{y}{U}, \\ \text{or} \\ &\frac{U}{2}D_\theta \frac{y}{U} = (k^2 D_\theta^{-1} U^{-1}) = k^2 \int \frac{dx}{xU}. \end{aligned}\right\} \tag{70}$$

It is true that even if we suppose the constant, say l, under this sign of integration $\int$ to be known, as well as the constant k, this last equation (70) will *still* be a *differential one*, of the *first*

* [Siméon Denis Poisson (1781–1840), 'Sur la distribution de la chaleur dans les corps solides, second mémoir', pp. 248–403, *Journal de l'École Royale Polytechnique*, Vol. XII, Cahier 19, 1823.]

order; as it *ought* to be, in order to permit the introduction of a *third constant*, such as h, into the *final* integral equation. We might indeed *write*,

$$y = 2k^2 UD_\theta^{-1} U^{-1} D_\theta^{-1} U^{-1}; \tag{71}$$

and such, in fact, I suppose to be the very *simplest* and *most elegant form*, (except that we might replace $2k^2$ by a *single letter*,) under which it is *possible* to *express the complete integral of the proposed triordinal equation*: of which equation it may be remembered that *one* fundamental form was,

$$D_\theta^3 y + 4\epsilon^\theta D_\theta \epsilon^\theta y = 0. \tag{2}$$

And it is important to observe that this equation (2) will be satisfied by the expression (71) for y, without it being *necessary* to assume for U the value (64), provided that U satisfies (51); or that

$$2UD_\theta^2 U - (D_\theta U)^2 + 4\epsilon^{2\theta} U^2 = 0. \tag{72}$$

In fact this last equation gives, by differentiation,

$$D_\theta^3 U + 4\epsilon^\theta D_\theta \epsilon^\theta U = 0, \quad \text{or} \quad \delta U = 0, \tag{73}$$

a result which agrees with the analysis of paragraph III.; to satisfy therefore the equation $\delta y = 0$, (5) we may conveniently assume for y, an expression of the form,

$$y = UD_\theta^{-1} z; \tag{74}$$

which will give

$$D_\theta^3 y = D_\theta^3 U . D_\theta^{-1} x + 3D_\theta^2 U . z + 3D_\theta U . D_\theta z + UD_\theta^2 z;$$

$$4\epsilon^\theta D_\theta \epsilon^\theta y = 4\epsilon^{2\theta}(D_\theta + 1)\, y = 4\epsilon^{2\theta}(D_\theta + 1)\, U . D_\theta^{-1} z + 4\epsilon^{2\theta} Uz;$$

therefore

$$(U\delta y - U\delta U . D_\theta^{-1} z =)\, U^2 D_\theta^2 z + 3UD_\theta U . D_\theta z + (3UD_\theta^2 U + 4\epsilon^{2\theta} U^2) . z = 0; \tag{75}$$

therefore substituting for $4\epsilon^{2\theta} U^2$ its value given by (72), the original equation comes to be transformed to the following:

$$0 = U^2 D_\theta^2 z + 3UD_\theta U . D_\theta z + D_\theta UD_\theta U . z = D_\theta (U^2 . D_\theta z + UD_\theta U . z);$$

that is

$$0 = D_\theta UD_\theta Uz = (D_\theta U)^2 z; \tag{76}$$

therefore the integral is, evidently

$$z = \text{constant} \times U^{-1} D_\theta^{-1} U^{-1}, \tag{77}$$

thus reproducing (71), without any restriction on U, beyond what is expressed by the *biordinal equation* (72). – Or take it thus, for the sake of exercise, and of variety: let us propose to *eliminate* U, between the 2 equations (71) and (72), and so to find *what differential equation the function y must satisfy*; *if* it depend on U by the law expressed in the former of those 2 equations, and if U be any integral of the latter of the same pair. We have, now

$$2k^2 = \text{constant} = (UD_\theta)^2 \frac{y}{U} = UD_\theta U^{-1}(UD_\theta y - yD_\theta U)$$

$$= UD_\theta^2 y - UD_\theta . yU^{-1} D_\theta U = UD_\theta^2 y - D_\theta yD_\theta U + yU^{-1}(D_\theta U)^2 - yD_\theta^2 U \tag{78}$$

therefore by differentiation,

$$0 = UD_\theta^3 y + ((D_\theta U)^2 - 2UD_\theta^2 U).D_\theta y + \{UD_\theta . U^{-1}(D_\theta U)^2 - UD_\theta^3 U\}.y; \tag{79}$$

but

$$(D_\theta U)^2 - 2UD_\theta^2 U = 4\epsilon^{2\theta} U^2, \text{ by (72)};$$

and

$$UD_\theta . U^{-1}(D_\theta U)^2 - UD_\theta^3 U = 2UD_\theta(D_\theta^2 U + 2\epsilon^{2\theta} U) - UD_\theta^3 U$$

$$= U\{D_\theta^3 U + 4\epsilon^{2\theta}(D_\theta + 2U)\} = 4\epsilon^{2\theta} U^2, \text{ by (73)};$$

therefore

$$D_\theta^3 y + 4\epsilon^{2\theta}(D_\theta y + y) = 0, \tag{80}$$

or $\delta y = 0$, as before.

VIII. But simple and elegant as the **form** (71) of the *complete integral* of the equation $\delta y = 0$ may be admitted to be, I think that for the *applications* which I wish to make, a certain *other* form of that complete or final integral possesses some important advantages. To obtain this other form, we have merely to substitute the expression (70) for $\frac{U}{2} D_\theta \frac{y}{U}$, in the equation (55); which equation thus becomes

$$\frac{y}{U} = \left(\frac{h}{k}\right)^2 + k^2 \left(\int \frac{dx}{xU}\right)^2; \tag{81}$$

a third arbitrary constant, l, in addition to h and k, being still conceived as entering, under the sign of integration $\int$. As a verification, if we differentiate this last equation (81), we obtain

$$D_\theta \frac{y}{U} = \frac{2k^2}{U} \int \frac{dx}{xU}; \tag{82}$$

therefore

$$U^{-1}(UD_\theta y - yD_\theta U) = 2k^2 \int \frac{dx}{xU} = 2k^2 D_\theta^{-1} U^{-1}; \tag{83}$$

and therefore by another differentiation,

$$(UD_\theta^2 y - yD_\theta^2 U) - U^{-1} D_\theta U.(UD_\theta y - yD_\theta U) = 2k^2; \tag{84}$$

which coincides with the equation (78). Conversely, from the biordinal equation (78), we may pass through a multiplication by U^{-1}, and an integration relatively to θ; to the equation (83); and thence, by a process exactly similar, to the complete integral (81). It must, however, be observed that we have no right, in the *general* question, to *assume* that h and k are *real*; so that we can only say, *in general*, and until we come to limit ourselves by some *application*, that the final integral of $\delta y = 0$ may be reduced to the form

$$y = AU + BU\left(\int \frac{dx}{xU}\right)^2; \tag{85}$$

where A and B are *any constants, real or imaginary*; and U is any integral of the biordinal equation (72). – It is worthy of remark, that while the *part* AU of this last expression for y is thus an integral of that equation (72), the *other part*,

$$BU\left(\int \frac{dx}{xU}\right)^2,$$

is also an integral of that equation of the 2nd order: although, on account of its *non-linear* character, the biordinal equation here referred to is *not* (in general) satisfied by the *sum* of those two parts, or by the expression y *itself*, which has been obtained as the integral of a *higher* equation. To justify the assertion just now made, respecting *both* parts of y in (85) being integrals of (72), whenever the *first* of them is such, let U be changed in (72) to $w^2 U$, and let the resulting equation be reduced by the help of (72) itself; it is found that

$$D_\theta U D_\theta w = 0; \ldots \tag{86}$$

therefore &c.. Indeed, from what was shown at the commencement of paragraph VI., it would have been sufficient to remark, that if v and vw *both* satisfy the equation (58), then,

$$0 = 2v D_\theta v D_\theta w + v^2 D_\theta^2 w = D_\theta v^2 D_\theta w; \tag{87}$$

and conversely that if v be any integral of (58), and if $v^2 D_\theta w$ be constant, then vw is *also* an integral. Accordingly, it has been remarked by Fourier*, that the complete† integral of the equation

$$gu + \frac{d^2u}{dx^2} + \frac{1}{x}\frac{du}{dx} = 0, \qquad (F_0)^{\ddagger}$$

is

$$u = \left(A + B\int \frac{dx}{x(\int \cos(x\sqrt{g}.\sin r)dr)^2}\right)\int \cos(x\sqrt{g}.\sin r)dr, \qquad (F_1)^{\ddagger}$$

A and B being arbitrary constants. But I am not aware that it occurred to Fourier to consider *that other function*, which I am at present denoting by y; and which, to imitate as closely as possible his just now cited notation, might be expressed as follows:

* [Jean Baptiste Joseph Fourier (1786–1830).]

† J.B.J. Fourier, *Théorie analytique de la Chaleurs*, p. 378, Paris: 1822 [*The analytical theory of heat* (trans. by A. Freeman), ch. VI, p. 298, Cambridge: 1878. Reprinted New York: 1955.]

Fourier seems to have treated this *complete* integral of his Equation for Heat in a Cylinder as a mere matter of *Mathematical* curiosity. I am not *sure* that it has not physical bearings. At all events, my own train of research, resumed of late without any recollection of that page of Fourier's Théorie, has been encouraged by a *hope* that *some such* bearings may exist.

‡ These markings are my own.

$$y = \left\{ \left(A + B \int \frac{dx}{x\left(\int \cos(x\sqrt{g}.\sin r)\,dr\right)^2} \right)^2 + C^2 \right\} . \left(\int \cos(x\sqrt{g}.\sin r)\,dr \right)^2 . \qquad (F_2)$$

(It is scarcely worth while mentioning, that I had not had access to Fourier's work since the year 1840, or thereabouts, till a few months ago; but that I had made out, for purposes of my own, a formula lately, similar to that marked (F_2), just above, before I succeeded in re-borrowing, for I have not had an opportunity of purchasing, – the cited work of Fourier.)

IX. In the expression (85), or in the equation (81), it is a *sufficiently general* procedure, and I am content, to assume for U the value (64); which it has been shown how to develop in an ascending series, proceeding by even powers of x, the terms being alternately positive and negative, and the first term being unity. It follows that U^{-1} may (with this selection of value) be developed, in a series of the form,

$$U^{-1} = 1 + u_1 x^2 + u_2 x^4 + u_3 x^6 + \text{ \&c.}; \qquad (88)$$

the values of the first coefficients being,

$$u_1 = \frac{1}{2}; \quad u_2 = \frac{5}{32}; \quad u_3 = \frac{23}{576}; \quad u_4 = \frac{677}{73728}; \quad \text{\&c.} \qquad (89)$$

which are, *so far*, all positive; and though I have not *proved* that they *must* be such *for ever*, I think it *likely* that they will be found to be so. Hence,

$$D_\theta^{-1} U^{-1} = \int \frac{dx}{xU} = \text{constant} + \log x + \frac{1}{2} u_1 x^2 + \frac{1}{4} u_2 x^4 + \frac{1}{6} u_3 x^6 + \text{\&c.}; \qquad (90)$$

that is

$$D_\theta^{-1} U^{-1} = \chi - \lambda, \qquad (91)$$

where

$$\lambda = \log \frac{\gamma}{x}, \qquad (92)$$

and

$$\chi = \int_0^x \frac{dx}{x}(U^{-1} - 1); \qquad (93)$$

γ being an arbitrary constant, such that $-\log\gamma$ is equal to the lately mentioned constant l, introduced by $\int \frac{dx}{xU}$. We may write also

$$-\lambda = l + \theta; \qquad (94)$$

$$\int \frac{dx}{xU} = \theta + \chi + l; \qquad (95)$$

and therefore by (81)

$$\frac{y}{U} = \left(\frac{h}{k}\right)^2 + k^2(\theta + \chi + l)^2. \tag{96}$$

Comparing this last *form* (96) of the *complete integral* of the triordinal equation (1) with the form (43), we not only recover the two *relations*, $k^2 = c$, (56), and $h^2 = ac - b^2$, (50), between the *constants*, h, k, l, of the *one* form, and the constants a, b, c of the *other*, but also obtained this *third* relation,

$$l = \frac{b}{c}. \tag{97}$$

At the same time, we are conducted to the following relations between the new *function*, χ, and the old functions, U, V, W:

$$V = U.\chi; \tag{98}$$

and

$$W = U.\chi^2; \tag{99}$$

which give, by elimination of χ, the simple and remarkable equation,

$$UW = V^2. \tag{100}$$

The properties (64), (98), (100), of the functions U, V, W, with the signification (93) of χ, were mentioned by anticipation before (at the end of paragraph III.); they can be verified, to any proposed extent, by actual substitution of the 5 series, for U, V, W, χ, and v; but the following other proof of the equation (98) may not be without interest and instruction.

X. The series for V may be considered to have been derived from that for U, by the equation

$$\delta V + (3D_\theta^2 + 4x^2)U = 0, \tag{101}$$

which results from the substitution, by (25), (26), and (40), of $a_0 V$ for A_x, and of $a_0 U$ for B_x, in the equation (15); combined with the rejection of any absolute term, or coefficient of x^0 in this sought series V, for reasons stated in paragraph II. *Assume*

$$V = UD_\theta^{-1}v, \tag{102}$$

where v denotes a new series, ascending by even powers of x, and beginning with the second power. Then, because $\delta U = 0$, (73), we have this equation of the second order, of which v must be a particular integral,

$$(3D_\theta^2 + 4x^2)U.(v+1) + 3D_\theta U.D_\theta v + UD_\theta^2 v = 0. \tag{103}$$

Multiplying by $2U$, and observing that, by (51),*

$$2U(3D_\theta^2 + 4x^2)U = D_\theta^2 U^2, \tag{104}$$

we have this easily integrable equation,

$$D_\theta^2 U^2.(v+1) + 3D_\theta U^2.D_\theta v + 2U^2 D_\theta^2 v = 0; \tag{105}$$

* As to the *notation* employed, you will easily perceive that, at present, I wish $D_\theta U^2$, $D_\theta^2 U^2$, to denote $D_\theta(U^2)$, $D_\theta^2(U^2)$, and *not* to denote $(D_\theta U)^2$, $(D_\theta^2 U)^2$; and similarly in other cases.

which gives, first,

$$D_\theta U^2.(v+1) + 2U^2 D_\theta v = \text{constant} = 0; \tag{106}$$

and secondly, by another integration, after dividing by $2U$,

$$U.(v+1) = \text{constant} = 1, \tag{107}$$

so that finally,

$$v = U^{-1} - 1, \tag{108}$$

and

$$V = UD_\theta^{-1}(U^{-1} - 1) = U\int_0^x \frac{dx}{x}(U^{-1} - 1), \tag{109}$$

or

$$V = U\chi$$

as in (98).

XI. By a similar analysis about equally simple, the expression $W = U\lambda^2$, (99), can be deduced from the equation

$$\delta W + 2(3D_\theta^2 + 4x^2)V + 6D_\theta U = 0, \tag{110}$$

which is obtained from the third equation of condition (14), by making, in (30), (38), and (40),

$$C_x = cU,\ B_x = 2cV,\ A_x = cW; \tag{111}$$

and by assuming that W is a series proceeding according to ascending and even powers of x, beginning with the 4th; so that if we write

$$W = Uw, \tag{112}$$

w becomes another even and ascending series beginning with the same 4th power. From the properties of U, V, χ, already employed, we shall have, on the one hand the transformation

$$U\delta W = D_\theta UD_\theta UD_\theta w; \tag{113}$$

and, on the other hand,

$$\begin{aligned} &2U(3D_\theta^2 + 4x^2)V + 6UD_\theta U \\ &\quad = \chi D_\theta^2 U^2 + 6D_\theta\chi.D_\theta U^2 + 6D_\theta^2\chi.U^2 + 3D_\theta U^2 \\ &\quad = \chi D_\theta^2 U^2 + 12(1-U)D_\theta U - 6D_\theta U + 6UD_\theta U \\ &\quad = \chi D_\theta^2 U^2 + 6(1-U)D_\theta U \\ &\quad = D_\theta\{\chi D_\theta U^2 - 2(1-U)^2\}; \end{aligned} \tag{114}$$

a *first* integration of the *differential equation* (110) gives, therefore,

$$(UD_\theta)^2 w - 2(1-U)^2 + \chi D_\theta U^2 = \text{constant} = 0, \tag{115}$$

that is,

$$D_\theta U D_\theta w = -2\chi D_\theta U + 2(1-U) D_\theta \chi = 2D_\theta \chi (1-U).$$

A *second* integration gives, after division by U,

$$UD_\theta w - 2\chi(1-U) = \text{constant} = 0; \tag{116}$$

$$\text{(and consequently } D_\theta w = 2\chi D_\theta \chi = D_\theta \chi^2\text{)}$$

and a *third* integration gives after a similar division,

$$w - \chi^2 = \text{constant} = 0; \tag{117}$$

so that $W = U\chi^2$, as above. In determining the three *constants*, in these last equations, regard has been had to the *known form* of the series for W, which is here only a *particular* (and not a complete or *general*) integral of the differential equation of the third order (110).

XII. In this manner, after perceiving the chief properties of the simplest, namely U, of the three functions, U, V, W in paragraph III., we might have been led *without actually forming the developments* of which the first terms are given in the equations (40); to infer the existence of those relations between them, which were stated by anticipation at the end of the same (cited) paragraph; and so to transform the expression (43) into this other:

$$y = \frac{ac-b^2}{c} U + cU\left(\theta + \frac{V}{U} + \frac{b}{c}\right)^2 = U\left\{\frac{ac-b^2}{c} + c(D_\theta^{-1} U^{-1})^2\right\}; \tag{118}$$

which agrees with (81), and with (96), because we have

$$c = k^2, \ (56); \qquad ac - b^2 = h^2, \ (49); \qquad \text{and } \frac{V}{U} = \chi, \ (98).$$

As regards the function U itself, if we had happened to perceive the *identity*,

$$yD_\theta^3 y = 2D_\theta y^{\frac{3}{2}} D_\theta^2 y^{\frac{1}{2}}, \tag{119}$$

(in which, as usual, (when no point is inserted,) the symbol D_θ governs all those that follow it,) we should have seen at once that the triordinal equation (2), which is the main subject of this whole Letter, may be thus written:

$$D_\theta y^{\frac{3}{2}} (D_\theta^2 + \epsilon^{2\theta}) y^{\frac{1}{2}} = 0; \tag{120}$$

and then it would have been extremely natural to *begin* with the *lower* equation,

$$(D_\theta^2 + x^2)v = 0, \ (58) \quad \text{or}$$

$$(D^2 + x^{-1}D + 1)v = 0, \ (59)$$

which has already been cited from Fourier: for it would have been obvious that if v be any particular integral of this last biordinal equation, then $U = v^2$ (57) must at the same time be a particular integral of my triordinal. But there is *only one ascending series* for v, proceeding according to *powers alone*, which satisfies Fourier's equation (59); namely his series (60), which he enveloped as a definite integral (61). There is therefore a *motive* for paying great attention, in this inquiry, to the function above called U; which we have expressed, in (64), as the *square of a definite integral*; and have found to be the *only integral* of the triordinal equation

(1), which admits of being developed in *ascending powers of x alone*: except that, of course, the series v in the one question, or U in the other, may be multiplied by an arbitrary constant, without ceasing to satisfy the equation.

XIII. The complete integral of the triordinal is scarcely, even yet, under a *form* sufficiently suited to my purpose. But it may easily be further transformed. In fact the form (81) *suggests* the assumption of a new *auxiliary angle*, or other quantity ω, real or imaginary such that

$$\left(\frac{h}{k}\right)^2 = \frac{y}{U}\sin^2\omega; \tag{121}$$

and therefore, at the same time,

$$\left(k^2\left(\int\frac{dx}{xU}\right)^2 =\right) k^2(D_\theta^{-1}U^{-1})^2 = \frac{y}{U}\cos^2\omega. \tag{122}$$

(I adopt *here*, without quite liking it, the *usual* notation of $\sin^2\omega$ for $(\sin\omega)^2$. At least, it saves parentheses; and the symbol $\sin\omega^2$ which perhaps I commonly prefer, has really in *these* investigations, some risk of being mistaken for $\sin(\omega^2)$.) Eliminating y between the two last equations, we find

$$\cot\omega = -k^2h^{-1}D_\theta^{-1}U^{-1}; \tag{123}$$

the negative sign being preferred, because it conducts to the relation,

$$yD_\theta\omega = h, \tag{124}$$

whereas the other sign for $\cot\omega$ would have given the less simple formula,

$$yD_\theta\omega = -h.$$

The *system of these two last equations,* (123) and (124), may be regarded as a *new form of the complete integral* required. The auxiliary quantity ω is any solution of the *new biordinal equation*

$$D_\theta UD_\theta\cot\omega = 0; \quad \text{or,} \quad (UD_\theta)^2\cot\omega = 0. \tag{125}$$

In the *completed integration* of *this* equation, *two constants* must be conceived to be introduced; and a *third constant,* namely h, is brought in by the equation (124). And it appears to me that the *clue* to the *arithmetical determination* of that *third constant* which has not *yet* had its value assigned, but which is to be conceived as entering under the sign $\int$ in the equation,

$$y\sum_{m=0}^{m=\infty}\left(\left.-\frac{1}{2}\right|^m\right)^3 \overline{0|}^{-m}x^{-2m-1} = \frac{\pi}{2}U + \frac{2}{\pi}U\left(\int\frac{dx}{xU}\right)^2, \tag{126}$$

where

$$U = \frac{4}{\pi^2}\left(\int_0^{\frac{\pi}{2}} d\alpha\cos(x\cos\alpha)\right)^2, \tag{64}$$

is to be found in the use of this *auxiliary angle* ω, and in the study and *comparison* of the two recent formulæ, (124) and (125).

XIV. Before my own triordinal had occurred to me, I thought of solving the related biordinal of Fourier in nearly the following manner. To satisfy the equation $(D^2 + x^{-1}D + 1)v = 0$, (59), I assumed

$$v = a\rho \sin \omega, \tag{127}$$

where a was an arbitrary constant, and ρ, ω were two variables, between which it was permitted to establish a relation. In this manner, writing

$$D\rho = \rho', \quad D\omega = \omega', \quad D^2\rho = \rho'', \text{ \&c.}$$

I had

$$a^{-1}Dv = D\rho \sin \omega = \rho' \sin \omega + \omega' \rho \cos \omega;$$

$$a^{-1}D^2v = \rho'' \sin \omega + 2\omega'\rho' \cos \omega + \rho\omega'' \cos \omega - \rho\omega'^2 \sin \omega,$$

and finally, (59) was *satisfied*, if I made,

$$\rho'' + x^{-1}\rho' + \rho = \rho\omega'^2, \tag{128}$$

and,

$$2\omega'\rho' + \rho\omega'' + x^{-1}\omega'\rho = 0. \tag{129}$$

On multiplying the last equation by $x\rho$, and integrating, the result was,

$$x\rho^2\omega' = h; \tag{130}$$

where h is an arbitrary constant. Multiplying (128) by $x^2\rho^3$, and eliminating ω' by (130), I had

$$x^2\rho^3\rho'' + x\rho^3\rho' + x^2\rho^4 = h^2; \tag{131}$$

that is,

$$\rho^3\{x(x\rho')' + x^2\rho\} = h^2; \tag{132}$$

which, if I had made $x = \epsilon^\theta$, would have at once become,

$$\rho^3(D_\theta^2 + x^2)\rho = h^2; \tag{133}$$

and, by a single differentiation, would have given, under a form only very slightly different, the triordinal equation (120), and therefore ultimately the fundamental equation (1), if I had thought of writing,

$$y = \rho^2; \tag{134}$$

which would also have reproduced, through (130), the recent equation $yD_\theta\omega = h$ (124). You see then, that I was *very nearly* on my recent track, – so far as regards the *arithmetical constant* which enters into the expression for

$$\cot \omega = -D^{-1}\frac{\omega'}{\sin^2 \omega} = \text{\&c.},$$

– *before* I *began* to write to you that long *arithmetical Letter*, which was dated February 15th, 1858: and, in fact, I had *fully arrived* at the *point* at which *arithmetical labour must still begin*, until some more decided *theoretical advance* shall have been made, than that which has recently occurred to me. But on anything which may seem to relate to (even private) *history*, it seems better to *postpone* any account, (beyond perhaps some short remark in passing,) till after the (abridged) statement of my own (rather) recent calculations, and of the numerical result to which they have led.

XV. In the proposed application of the Theory of the present Letter, as regards the integration of the triordinal equation, &c., we have $h = 1$, (48), and $y =$ the descending and diverging series (44). Changing therefore $D_\theta\omega$ to $xD\omega$, or to $x\omega'$, in the important formula (124), that formula becomes,

$$xy\omega' = 1; \tag{135}$$

or more fully,

$$1 = \omega' \sum_{m=0}^{m=\infty} \left(\left\lceil -\frac{1}{2} \right\rceil^m \right)^3 \overline{0}\rceil^{-m} x^{-2m} = \omega' \left(1 - \frac{1}{1.(8x)^2)^1} + \frac{1^3.3^3}{1.2(8x^2)^2} - \frac{1^3.3^3.5^3}{1.2.3(8x^2)^3} + \&\text{c.} \right). \tag{136}$$

Thus

$$\omega' = (xy)^{-1} = \left(\sum_{m=0}^{m=\infty} \left\lceil -\frac{1}{2} \right\rceil^m \overline{0}\rceil^{-m} x^{-2m} \right)^{-1} = \left(1 - \frac{x^{-2}}{8} + \frac{27x^{-4}}{128} - \frac{1125x^{-6}}{1024} + \cdots \right)^{-1}; \tag{137}$$

that is, developing this reciprocal,

$$\omega' = 1 + \left(\frac{x^{-2}}{8} - \frac{27x^{-4}}{128} + \frac{1125x^{-6}}{1024} - \frac{385\,875x^{-8}}{32\,768} \right)$$
$$+ \left(\frac{x^{-4}}{64} - \frac{27x^{-6}}{512} + \frac{5229x^{-8}}{16\,384} - \cdots \right) + \left(\frac{x^{-6}}{512} - \frac{81x^{-8}}{8192} + \cdots \right) + \left(\frac{x^{-8}}{4096} - \cdots \right) + \cdots \tag{138}$$

or, collecting terms which have the same exponents,

$$\omega' = 1 + \frac{x^{-2}}{8} - \frac{25x^{-4}}{128} + \frac{1073x^{-6}}{1024} - \frac{375\,733x^{-8}}{32\,768} + \&\text{c.} \tag{139}$$

Intergrating, we have this other series:

$$\omega = \omega_0 + x - s, \tag{140}$$

where ω_0 is a constant, not yet determined, and

$$s = \int_x^\infty dx(1 - x^{-1}y^{-1}) = \frac{x^{-1}}{8} - \frac{25x^{-3}}{384} + \frac{1073x^{-5}}{5120} - \frac{375\,733x^{-7}}{229\,376} + \&\text{c.}; \tag{141}$$

or multiplying by 1 [radian] = 206,264″·806,

$$s = \frac{25\,783''\!\cdot\!1}{x} - \frac{13\,428''\!\cdot\!7}{x^3} + \frac{43\,227''\!\cdot\!0}{x^5} - \frac{337\,875''\!\cdot\!3}{x^7} + \ \&\text{c.} \tag{142}$$

When x is even *so small as* 1, *this descending series diverges rapidly*, and becomes almost totally *useless*, unless it be treated by some such *methods of transformation* as those which I have sketched in my Letter of February 15, and which I have since applied (with others only *hinted* in that Letter) to transform the series for the *tangent* of this angle s and so to make what has turned out to be a curiously near approach to the value of s for $x = 1$. But when x is moderately large, the series (141) or (142) for s then converges sufficiently fast, at least in its early terms, to give with ease an approximation as close as can be useful, for any practical purpose. For example,

$$s = +2578''\cdot 31 - 13''\cdot 43 + 0''\cdot 43 - 0''\cdot 03 = +2565''\cdot 28 = 42'45''\cdot 3, \text{ for } x = 10; \tag{143}$$

and

$$s = +1841''\cdot 65 - 4''\cdot 89 + 0''\cdot 07 - 0''\cdot 00 = +1836''\cdot 83 = +30'36''\cdot 8, \text{ for } x = 14; \tag{144}$$

and I suppose that we may rely on these two values of s, as correct to the tenth of a second.

XVI. As regards the constant ω_0, it may be determined in the following way. When x is large enough to render s a very small angle, we have, nearly, by (140),

$$\omega = x + \omega_0; \tag{145}$$

and therefore by (121), and by the values (48) and (68) of h and k^2, we have nearly

$$\frac{\pi}{2} \cdot \frac{U}{y} = \sin^2 (x + \omega_0). \tag{146}$$

But also, under the same circumstances, we have, nearly, by (66) and (67),

$$\frac{\pi}{2} \cdot \frac{U}{y} = \sin^2 \left(x + \frac{\pi}{4} \right). \tag{147}$$

The comparison of these two ultimate or *limiting forms* gives, *rigorously*, the value:

$$\omega_0 = \frac{\pi}{4}. \tag{148}$$

The final and unambiguous (or determinate) expression for ω is therefore,

$$\omega = x + \frac{\pi}{4} - s; \tag{149}$$

where s is to be computed as above, and can be so to a great degree of accuracy, by an *initially convergent series*, unless x be taken too small. For example,

$$\omega + s = 10 + \frac{\pi}{4} = 617°57'28''\cdot 06, \text{ and } \omega = 617°57'28''\cdot 1 - 42'45''\cdot 3 = 3\pi + 77°14'42''\cdot 8,$$
$$\text{for } x = 10; \tag{150}$$

$$\omega + s = 14 + \frac{\pi}{4} = 847°8'27''\cdot 28, \ \omega = 847°8'27''\cdot 3 - 30'36''\cdot 8 = 4\pi + 126°37'50''\cdot 5, \text{ for } x = 14; \tag{151}$$

The angle ω may thus be considered as known, very exactly, when x is moderately large; consequently we can take out, from the usual Tables, under the same conditions, to about *seven* decimal places, the values of each of *two* of the terms of the first member of the following equation, slightly transformed from (123),

$$\frac{\pi}{2} \cot \omega + \log x + \chi = -l; \tag{152}$$

where l has the same meaning as before, denoting an as yet unknown *constant*; χ is an *integral* already considered; and the *logarithm* of x (often in this Letter denoted by θ) is the *natural* one.* *At this stage* I had arrived, in January or February last, before I began my long Letter of the latter month, and before I thought of the triordinal equation (1). That is, I had the *formula* just marked (152), the *ascending series* for χ, and the *descending series* for ω, and *wished* to determine the *constant*.

But I was embarrassed by the difficulty of finding a value for x, which should admit of my computing, with such means as I then possessed, and with as much accuracy as I desired, the values of *both* ω *and* χ; the former being given, as above, by a *descending series*, and the latter by an *ascending* one: namely, by paragraph IX.,

$$\chi = \frac{x^2}{4} + \frac{5x^4}{128} + \frac{23x^6}{3456} + \frac{677x^8}{589\,824} + \text{ \&c.}; \tag{153}$$

which *converges* moderately fast, when x is tolerably *small*, but *not very rapidly*, when x is even *so great as* 1; and which, besides, is subject to the *very grave inconvenience*, that *all its terms are positive*, at least so far as the coefficients have been computed; and I see no reason to suspect that the signs will afterwards change. Indeed, it was easy to see that the definite integral (93), of which this series (153) is a development, must attain an *infinite value*, as often as x passes through the stage of being one of the *roots of the transcendental equation*,

$$v = \frac{2}{\pi}\int_0^{\frac{\pi}{2}} d\alpha \cos(x\cos\alpha) = 0; \tag{154}$$

which equation is important in physics. In fact, for each such value of x we have

$$U = v^2 = 0; \tag{155}$$

we have also,

$$\cot\omega = \pm\infty, \tag{156}$$

and therefore, must, by (152), have then,

$$\chi = \pm\infty. \tag{157}$$

Nor is it as yet proved, that the *constant*, $-l$, does not then undergo *discontinuity*. Nevertheless, for $x = 1$, the 4 written terms (153), with an estimated† correction for the remainder of the series gave

$$\chi = 0{\cdot}250\,00 + 0{\cdot}039\,06 + 0{\cdot}006\,65 + 0{\cdot}0014 + (\text{say})0{\cdot}000\,23 = 0{\cdot}297\,08; \tag{158}$$

and from the *diverging series* for tan s, *before* I had thought of a method (not yet explained) of transforming χ, so as to compute *its* value for moderately *large values of* x, I *estimated*, by the "method‡ of inverse cube roots",

* I shall here mention, by anticipation, that I have lately found $\left(\frac{\pi}{2}\right)^{\frac{1}{2}}\cot\omega + \left(\frac{2}{\pi}\right)^{\frac{1}{2}}(\theta+\chi) = \frac{37}{400}$, *at least very nearly*; but I *think, rigorously*.

† Each term seeming to be here *about* one *sixth* of the term before it, I add as the correction for the remainder, about one *fifth* part of the *last* computed term.

‡ This method is merely a modification of the "*method* of *inverse square roots*", which was explained near the close of my Letter of February the 15th.

$$s = 5°43', \quad \text{giving } \omega = 96°35'; \tag{159}$$

this method of transforming the diverging series gave therefore, for $x = l$

$$\frac{\pi}{2} \cot \omega = -L^{-1}\bar{1}{\cdot}258\,36 = -0{\cdot}181\,28; \tag{160}$$

also $\theta = \log 1 = 0$; therefore this process, depending as it did on *two distinct and independent estimations*, gave:

$$\text{constant} = -l = +0{\cdot}1158. \tag{161}$$

And here my *good luck* was really remarkable: for a process *far more* to be relied on has given me since,

$$\frac{\pi}{2} \cot \omega + \theta + \chi = \text{constant} = 0{\cdot}115\,931, \tag{162}$$

$$= \frac{37}{400} \times \left(\frac{\pi}{2}\right)^{1/2} [= 0{\cdot}159\,315\,58] \text{ at least very nearly.}$$

XVII. The *more reliable process*, here referred to, was of the following kind. Although it has been seen that the series (153) for χ becomes infinite when the condition (155) is satisfied, by x becoming a root of the equation

$$v = \frac{2}{\pi} \int_0^{\frac{\pi}{2}} d\alpha \cos(x \cos \alpha) = 0; \tag{154}$$

and although I think that it becomes divergent, as soon as x passes through the stage of the least positive root of that transcendental equation, which root I have with great labour ascertained to be, nearly*

$$x = +2{\cdot}404\,825\,56; \tag{163}$$

yet you may have noticed that the said series (153) for χ, *with all its terms positive*, is, by (98), the *quotient of two other series*, U and V, which have, by (40), *their terms alternately positive and negative*: a "progressing" series, in a phraseology of your own, being thus equal to the quotient of two "*alternating*" ones. And it is rather the *product-series* $V = U\chi$, (98), than the *quotient-series*,

* Note – Professor Stokes has adopted the value 2·4050 for this root, as a result of interpolation from other values of the integral, which have been tabulated by Mr. Airy† – but this value is certainly too great, and indeed I think that you may rely on all the figures of the root, which I have above set down. Stokes wanted the root for a physical purpose, and his estimate was quite near enough for the occasion. It was necessary, however, for a theoretical object of my own, that I should endeavour to attain a much greater degree of accuracy. You are probably acquainted with the Paper of Professor Stokes, in the *Cambridge Philosophical Transactions* to which I allude, but on which I cannot at this moment lay my hand.‡ It may, and I hope will, turn up, before the present Letter is finished – and I shall leave a line or two for reference (Aug 10).

I can only say, in general, the Stoke's Paper relates to the Numerical Calculation of certain Definite Integrals, and that he referred, in it, to my own old Memoir on Fluctuating Functions.

† [George Biddell Airy (1801–1892), 'On the diffraction of an annular aperture', *Philosophical Magazine*, Ser. 3, Vol. 18, pp. 1–10, 1841.]

‡ [See note on p. 172].

$$\chi = \frac{V}{U}, \tag{164}$$

or the equivalent *integral* (93) which enters *immediately* into the form (43) of the complex integral of the triordinal equation (1). That *form* (43) contains however another function, W, expressed by *another alternating series*, of the same set (40), and connected with U and χ by the relation, $W = U.\chi^2$; (99) while U was, by (57), or (64), the *square of the definite integral* (61), or v; so that

$$\chi = \frac{W^{\frac{1}{2}}}{v}, \tag{165}$$

where $W^{\frac{1}{2}}$ is still *another* alternating series, of which I find that the *law* is *simpler* than that of V, or of W, and almost as simple, and as well adapted to numerical calculation, as that of the series U, or even of *its* square-root, v. In fact, on proceeding to form this *new product-series*,

$$\begin{aligned}
W^{\frac{1}{2}} &= v\chi = U^{\frac{1}{2}}\int_0^x \frac{dx}{x}(U^{-1}-1) \\
&= \left\{1-\left(\frac{x}{2}\right)^2+\left(\frac{x^2}{2.4}\right)^2-\left(\frac{x^3}{2.4.6}\right)^2+\cdots\right\}\left\{\frac{x^2}{4}+\frac{5x^4}{128}+\frac{23x^6}{3456}+\frac{677x^8}{589\,824}+\cdots\right\} \\
&= \frac{x^2}{4}\left(1-\frac{x^2}{4}+\frac{x^4}{64}-\frac{x^6}{2304}\right)+\frac{5x^4}{128}\left(1-\frac{x^2}{4}+\frac{x^4}{64}-\cdots\right) \\
&\quad +\frac{23x^2}{3456}\left(1-\frac{x^2}{4}+\cdots\right)+\frac{677x^8}{589\,824}(1-\cdots)+\cdots \\
&= \frac{x^2}{4}+\frac{x^4}{128}(-8+5)+\frac{x^6}{13\,824}(54-135+92)+\frac{x^8}{1\,769\,472}(-192+1080-2944+2031) \\
&\quad +\cdots = \frac{x^2}{4}-\frac{3x^4}{128}+\frac{11x^6}{13\,824}-\frac{25x^8}{1\,769\,472}+\cdots = \frac{x^2}{4}-\frac{3}{2}.\frac{x^4}{64}+\frac{11}{4}.\frac{x^6}{2304}-\frac{25}{12}.\frac{x^8}{147\,456}+\cdots \\
&= 1\left(\frac{x}{2}\right)^2-\left(1+\frac{1}{2}\right)\left(\frac{x^2}{2.4}\right)^2+\left(1+\frac{1}{2}+\frac{1}{3}\right)\left(\frac{x^3}{2.4.6}\right)^2-\left(1+\frac{1}{2}+\frac{1}{3}+\frac{1}{4}\right)\left(\frac{x^4}{2.4.6.8}\right)^2+\cdots
\end{aligned} \tag{166}$$

a *sufficiently simple law* becomes apparent; and I can prove that it *must continue*; because, in the notation of this Letter, not only v, but also $v(\chi+\theta)$ satisfies the differential equation (58); in such a manner that we have the relation,

$$0 = (D_\theta^2 + x^2)v\chi + 2D_\theta v. \tag{167}$$

XVIII. Indeed, I have *just now noticed*, for the first time, a series of very *similar form*, in Carmichael's Book* on the Calculus of operations, which has *more* than an *accidental* connexion with this part of my subject, and which I must pause for a minute or two to describe, of course for my own sake, rather than for yours, since I think that you have got the

* [Robert Bell Carmichael (1828–1861), *A treatise on the calculus of operations*, London: 1855.]

Book. In his page 49, he cites from Gregory*, who cited it from Fourier, the differential equation of the 2nd order,

$$xD^2 y + Dy + y = 0, \tag{F_4}$$

which I thus mark as (F_4), because it is intimately related to the equation (F_0) of paragraph VIII. of this Letter: and I shall continue to number a few equations similarly on the present occasion. Carmichael transforms (in his page 53) the equation (F_4) to

$$(xD)^2 y + xy = 0, \tag{F_5}$$

and presents its integral under the symbolical form,

$$y = \left\{1 - \frac{1}{(xD)^2} x + \frac{1}{(xD)^2} x \frac{1}{(xD)^2} x - \cdots\right\}(C_1 \log x + C_2); \tag{F_6}$$

or, as I might perhaps prefer to write it,

$$y = \{1 + (xD)^{-2} x\}^{-1} (xD)^{-2} 0. \tag{F_7}$$

Hence he obtains the final transformation, in page 54,

$$\begin{aligned} y = (C_1 \log x + C_2)&\left\{1 - \frac{x}{1^2} + \frac{x^2}{1^2.2^2} - \frac{x^3}{1^2.2^2.3^2} + \&\text{c.}\right\} \\ + 2C_1&\left\{\frac{1}{1^2} x - \left(\frac{1}{1} + \frac{1}{2}\right) \frac{1}{1^2.2^2} x^2 + \left(\frac{1}{1} + \frac{1}{2} + \frac{1}{3}\right) \frac{1}{1^2.2^2.3^2} x^3 - \&\text{c.}\right\}. \end{aligned} \tag{F_8}$$

(I find that I have transposed some factors, but not so as to affect the sense.) On turning to Gregory's Examples, (at page 339, not page 343, as cited by Carmichael, at least in my copy, dated Cambridge, 1841,) I find a more elementary method, but what seems to me a less correct result: and in short, Gregory had not caught the *law* of the series in the 2nd line of (F_8), although he found a few terms of it, and was aware (page 311) that the series on the first line was (as Fourier showed) the expansion of the definite integral,

$$\frac{1}{\pi} \int_0^\pi d\theta \cos(2 \sin \theta\, x^{\frac{1}{2}}), \tag{F_9}$$

to which this Letter partly relates. Change x to $\frac{x^2}{4}$; the definite integral (F_9) becomes what I have also called v; $\log x$ becomes $2\theta - \log 4$; the series (see above) in the 2nd line of (F_8) becomes identical with the one which in my development (166) has been assigned as representing what I have called $v\chi$, or $W^{\frac{1}{2}}$; and the complete (although until today unknown) agreement of my results on *this* point with those of Carmichael is established by the corresponding transformation of his differential equation (F_5), which is changed to the following,

$$(D_\theta^2 + x^2)\, y = 0; \tag{168}$$

agreeing thus with my equation (58), except that y replaces v; and having, for its complete integral, whether by his method or by mine,

* [Duncan Farquharson Gregory (1813–1844), *Examples of the processes of the differential and integral calculus*, Cambridge: 1841.]

$$y = av + bv(\chi + \theta), \tag{169}$$

where a and b are arbitrary constants. – Gregory, I find, writes, instead of (F_8), for the integral of Fourier's equation (F_4),

$$y = 2A\left(x - \frac{3x^2}{2^3} + \frac{11x^3}{2^3.3^3} - \frac{51x^4}{2^3.3^3.4^3} + \&\text{c.}\right)$$
$$+ \left(1 - \frac{x}{1^2} + \frac{x^2}{1^2.2^2} - \frac{x^3}{1^2.2^2.3^2} + \frac{x^4}{1^2.2^2.3^2.4^2} - \&\text{c.}\right)\log cx; \tag{F_{10}}$$

where the 51, as I conceive, ought to be 50, at least if the result is to agree with Carmichael's and mine. But you know that the integration of my *own triordinal equation* (1), and *not* that of *Fourier's biordinal equation,* (F_0), which had indeed been sufficiently accomplished by himself in assigning the formula (F_1), has been the *chief object* of the present Letter: and it was only *today* (July 31st) that I thought of opening, or at least that I did open, for comparison, the books of Gregory and of Carmichael.

XIX. Resuming my own investigation, let me now write

$$v\chi = W^{\frac{1}{2}} = w; \tag{170}$$

so that, although the symbol w may have just been used for a moment, in passing, for another purpose, as in (86) and (87), or in (112), &c., yet we have *now* the series,

$$w = v\chi = 1v_1 - (1+\tfrac{1}{2})v_2 + (1+\tfrac{1}{2}+\tfrac{1}{3})v_3 - (1+\tfrac{1}{2}+\tfrac{1}{3}+\tfrac{1}{4})v_4 + \&\text{c.} \tag{171}$$

if

$$v_n = \left(\frac{x^n}{2.4.6.\cdots.(2n)}\right)^2 = 2^{-2n}x^{2n}([0]^{-n})^2, \tag{172}$$

and

$$v = \frac{2}{\pi}\int_0^{\frac{\pi}{2}} d\alpha \cos(x\cos\alpha) = v_0 - v_1 + v_2 - v_3 + v_4 - \&\text{c.} \tag{173}$$

In this manner, we shall have this other formula for computing w:

$$w = \frac{1}{1}(v_0 - v) + \frac{1}{2}(v_0 - v_1 - v) + \frac{1}{3}(v_0 - v_1 + v_2 - v) + \frac{1}{4}(v_0 - v_1 + v_2 - v_3 - v) + \&\text{c.}; \tag{174}$$

which is, I think, more convenient in practice than (171); and then χ will be given as the *quotient of two alternating series,*

$$\chi = \frac{w}{v}. \tag{175}$$

Taking, for example, $x = 1$, I find, in this way, for v, the terms,

$$\left.\begin{aligned}
v_0 &= 1{\cdot}000\,000\,000\,000 \qquad -v_1 = -{\cdot}250\,000\,000\,0000\\
v_2 &= {\cdot}015\,625\,000\,000 \qquad -v_3 = -{\cdot}000\,434\,027\,778\\
v_4 &= {\cdot}000\,006\,781\,684 \qquad -v_5 = -{\cdot}000\,000\,067\,817\\
v_6 &= {\cdot}000\,000\,000\,471 \qquad -v_7 = -{\cdot}000\,000\,000\,002\\
&\text{and therefore}\\
v &= (+1{\cdot}015\,631\,782\,155 - 0{\cdot}250\,434\,095\,597)\\
&= \frac{2}{\pi}\int_0^{\frac{\pi}{2}} d\alpha \cos(\cos\alpha) = +0{\cdot}765\,197\,686\,558 = +L^{-1}\bar{1}{\cdot}883\,773\,7, \text{ nearly.}
\end{aligned}\right\} \tag{176}$$

Writing also

$$v = v_0 - v_1 + v_2 - \cdots + (-1)^{n-1} v_{n-1} + (-1)^n n w_n, \tag{177}$$

we shall have, in general,

$$w = +w_1 - w_2 + w_3 - w_4 + \&\text{c.}; \tag{178}$$

and in the present example, by successively subtracting, with their signs, the terms v_0, $-v_1$, v_2, ... from their sum, we find the numerical values:

$$\left.\begin{aligned}
-1w_1 &= v - v_0 = -0{\cdot}234\,802\,313\,442; & +2w_2 &= -1w_1 + v_1 = +0{\cdot}015\,197\,686\,558;\\
-3w_3 &= 2w_2 - v_2 = -0{\cdot}000\,427\,313\,442; & +4w_4 &= -3w_3 + v_3 = +{\cdot}000\,006\,714\,336;\\
-5w_5 &= 4w_4 - v_4 = -0{\cdot}000\,000\,067\,348; & +6w_6 &= -5w_5 + v_5 = +{\cdot}000\,000\,000\,469;\\
-7w_7 &= 6w_6 - v_6 = -0{\cdot}000\,000\,000\,002; & +8w_8 &= -7w_7 + v_7 = +{\cdot}000\,000\,000\,000;
\end{aligned}\right\} \tag{179}$$

where the final 0 is a verification of the correctness of all the subtractions. Dividing next by -1, -2, -3, -4, &c., we obtain the following terms of w:

$$\left.\begin{aligned}
w_1 &= +0{\cdot}234\,802\,313\,442; & -w_2 &= -0{\cdot}007\,598\,843\,279;\\
w_3 &= \quad +{\cdot}000\,142\,437\,814; & -w_4 &= \quad -{\cdot}000\,001\,678\,584;\\
w_5 &= \quad +{\cdot}000\,000\,013\,470; & -w_6 &= \quad -{\cdot}000\,000\,000\,078;\\
&\text{therefore}\\
w &= +0{\cdot}234\,944\,764\,726 & &\qquad - 0{\cdot}007\,600\,521\,941\\
&= +0{\cdot}227\,344\,242\,785 = +L^{-1}\bar{1}{\cdot}356\,684\,0, \text{ nearly;}
\end{aligned}\right\} \tag{180}$$

whence, by logarithmic division,

$$\chi = \frac{w}{v} = +L^{-1}\bar{1}{\cdot}472\,910\,3 = +0{\cdot}297\,10|52; \tag{181}$$

where the last figures may be doubtful, and might be more accurately obtained by arithmetical division, from the fraction,

$$\chi = \frac{10^{12}w}{10^{12}v} = \frac{227\,344\,242\,785}{765\,197\,686\,558} \cdots. \tag{182}$$

However, the value (181) of χ agrees well enough with the former value, $\chi = +{\cdot}297\,08$, (158) which was deduced from the ascending series (153), with the help of a certain allowance, or estimation, for the terms not computed, in the case $x = 1$. When x is larger, the operations are of course more laborious. For

$$x = 14, \tag{183}$$

I find, by 29 terms of v,

$$v = \left.\begin{matrix} +64\,709{\cdot}366\,887\,062\,333 \\ -64\,709{\cdot}195\,813\,586\,219 \end{matrix}\right\} = \left\{\begin{matrix} +0{\cdot}171\,073\,476|114 \\ = +L^{-1}\bar{1}{\cdot}233\,182\,6 \end{matrix}\right\}; \tag{184}$$

and by 28 terms of w,*

$$w = \left.\begin{matrix} -4\,779{\cdot}565\,492\,127\,139 \\ +4\,779{\cdot}333\,645\,842\,994 \end{matrix}\right\} = \left\{\begin{matrix} -0{\cdot}231\,846\,284\,145 \\ = -L^{-1}\bar{1}{\cdot}365\,200\,1 \end{matrix}\right\}; \tag{185}$$

whence, by logarithmic division

$$\chi = \frac{w}{v} = -L^{-1}0{\cdot}132\,017\,5 = -1{\cdot}355\,2|44; \tag{186}$$

or more exactly, by common arithmetic,

$$\chi = \frac{w}{v} = -\frac{231\,846\,284}{171\,073\,476} = -1{\cdot}355\,243\,9|0. \tag{187}$$

For $x = 10$, I find

$$v = -0{\cdot}245\,935\,764\,4;^{\dagger}$$

$$w = +0{\cdot}625\,224\,384\,2; \quad \chi = -2{\cdot}542\,226, \quad \text{nearly.} \tag{188}$$

XX. Although I have extracted the last results from one of my manuscript books, and suppose them to be more than sufficiently exact for any possible *practical* application, yet I am tempted by an impulse of curiosity, connected with a *theoretical* question to which I alluded in a recent note (of July 26th) from Edgeworthstown, to resume the whole calculation of χ for $x = 10$, with a few additional figures at each step, and to write it out on the present page, in order that the chief points of the investigation may be preserved, in a collected form, for myself at least, in the copy which is to be made for me of what remains of this Letter.

We have now, by (172),

$$v_0 = 1, \text{ and } v_n = \left(\frac{5}{n}\right)^2 v_{n-1}. \tag{189}$$

Hence,

* Although the negative part of w happens here to be written above the other, it was obtained from $-w_2$, $-w_4$, &c., and the positive part from $+w_1$, $+w_3$, &c.

† Aug. 18. Stokes found, $-0{\cdot}245\,94$. Airy, he says, had found $-{\cdot}2450$.

$$\left.\begin{array}{r|l}
v_0 = 1{\cdot}000\,000\,000\,000\,000 & v_1 = 25{\cdot}000\,000\,000\,000\,000 \\
v_2 = 156{\cdot}250\,000\,000\,000\,000 & v_3 = 434{\cdot}027\,777\,777\,777\,778 \\
v_4 = 678{\cdot}168\,402\,777\,777\,778 & v_5 = 678{\cdot}168\,402\,777\,777\,778 \\
v_6 = 470{\cdot}950\,279\,706\,790\,123 & v_7 = 240{\cdot}280\,754\,952\,443\,940 \\
v_8 = 93{\cdot}859\,669\,903\,298\,414 & v_9 = 28{\cdot}969\,033\,920\,771\,115 \\
v_{10} = 7{\cdot}242\,258\,480\,192\,779 & v_{11} = 1{\cdot}496\,334\,396\,734\,045 \\
v_{12} = 0{\cdot}259\,780\,277\,210\,772 & v_{13} = 0{\cdot}038\,429\,035\,090\,351 \\
v_{14} = 0{\cdot}004\,901\,662\,639\,075 & v_{15} = 0{\cdot}000\,544\,629\,182\,119 \\
v_{16} = 0{\cdot}000\,053\,186\,443\,566 & v_{17} = 0{\cdot}000\,004\,600\,903\,423 \\
v_{18} = 0{\cdot}000\,000\,355\,007\,980 & v_{19} = 0{\cdot}000\,000\,024\,585\,040 \\
v_{20} = 0{\cdot}000\,000\,001\,536\,565 & v_{21} = 0{\cdot}000\,000\,000\,087\,107 \\
v_{22} = 0{\cdot}000\,000\,000\,004\,499 & v_{23} = 0{\cdot}000\,000\,000\,000\,213 \\
v_{24} = 0{\cdot}000\,000\,000\,000\,009 & v_{25} = 0{\cdot}000\,000\,000\,000\,000
\end{array}\right\} \qquad (190)$$

and therefore

$$\begin{aligned} v &= +1407{\cdot}735\,346\,901\,560 - 1407{\cdot}981\,282\,115\,352\,909 \\ &= -0{\cdot}245\,935\,764\,451\,349 = \frac{2}{\pi}\int_0^{\frac{\pi}{2}} d\alpha \cos(10\cos\alpha), \end{aligned} \qquad (191)$$

(or sin: it is immaterial which). I suppose that we shall be pretty safe in writing,

$$v = -0{\cdot}245\,935\,764\,451\,35 = \text{nearly} - L^{-1}\bar{1}{\cdot}390\,821\,7. \qquad (192)$$

Retaining however 15 places of figures, and subtracting successively from v its terms, v_0, $-v_1$, &c., we find:

$$\left.\begin{array}{r|l}
-1\,w_1 = -1{\cdot}245\,935\,164\,451\,349 & +2\,w_2 = +23{\cdot}754\,064\,235\,548\,651 \\
-3\,w_3 = -132{\cdot}495\,935\,764\,451\,349 & +4\,w_4 = +301{\cdot}531\,842\,013\,326\,429 \\
-5\,w_5 = -376{\cdot}636\,560\,764\,451\,349 & +6\,w_6 = +301{\cdot}531\,842\,013\,326\,429 \\
-7\,w_7 = -169{\cdot}418\,437\,693\,463\,694 & +8\,w_8 = +70{\cdot}862\,317\,258\,980\,246 \\
-9\,w_9 = -32{\cdot}997\,352\,644\,318\,168 & +10\,w_{10} = +5{\cdot}971\,681\,276\,452\,947 \\
-11\,w_{11} = -1{\cdot}270\,577\,203\,739\,832 & +12\,w_{12} = +0{\cdot}225\,757\,192\,994\,213 \\
-13\,w_{13} = -0{\cdot}034\,023\,084\,216\,559 & +14\,w_{14} = +0{\cdot}004\,405\,950\,873\,792 \\
-15\,w_{15} = -0{\cdot}000\,495\,711\,765\,283 & +16\,w_{16} = +0{\cdot}000\,048\,917\,416\,836 \\
-17\,w_{17} = -0{\cdot}000\,004\,269\,026\,730 & +18\,w_{18} = +0{\cdot}000\,000\,331\,876\,693 \\
-19\,w_{19} = -0{\cdot}000\,000\,023\,131\,287 & +20\,w_{20} = +0{\cdot}000\,000\,001\,453\,753 \\
-21\,w_{21} = -0{\cdot}000\,000\,000\,082\,812 & +22\,w_{22} = +0{\cdot}000\,000\,000\,004\,295 \\
-23\,w_{23} = -0{\cdot}000\,000\,000\,000\,204 & +24\,w_{24} = +0{\cdot}000\,000\,000\,000\,009 \\
-25\,w_{25} = -0{\cdot}000\,000\,000\,000\,000 &
\end{array}\right\} \qquad (193)$$

with the final 0 as a verification. Dividing by -1, -2, &c. the terms of w are thus found to be:

$$\left.\begin{array}{r|r} w_1 = +1{\cdot}245\,935\,764\,451\,349 & -w_2 = -11{\cdot}877\,032\,117\,774\,325 \\ w_3 = +44{\cdot}165\,311\,921\,483\,783 & -w_4 = -75{\cdot}382\,960\,503\,331\,607 \\ w_5 = +75{\cdot}327\,312\,152\,890\,270 & -w_6 = -50{\cdot}255\,307\,002\,221\,072 \\ w_7 = +24{\cdot}202\,633\,956\,209\,099 & -w_8 = -8{\cdot}857\,789\,657\,372\,531 \\ w_9 = +2{\cdot}555\,261\,404\,924\,241 & -w_{10} = -0{\cdot}597\,168\,127\,645\,295 \\ w_{11} = +0{\cdot}115\,507\,018\,521\,803 & -w_{12} = -0{\cdot}018\,813\,099\,416\,184 \\ w_{13} = +0{\cdot}002\,617\,160\,324\,351 & -w_{14} = -0{\cdot}000\,314\,710\,776\,699 \\ w_{15} = +0{\cdot}000\,033\,047\,451\,019 & -w_{16} = -0{\cdot}000\,003\,057\,338\,552 \\ w_{17} = +0{\cdot}000\,000\,251\,119\,219 & -w_{18} = -0{\cdot}000\,000\,018\,437\,594 \\ w_{19} = +0{\cdot}000\,000\,001\,217\,436 & -w_{20} = -0{\cdot}000\,000\,000\,072\,688 \\ w_{21} = +0{\cdot}000\,000\,000\,003\,943 & -w_{22} = -0{\cdot}000\,000\,000\,000\,195 \\ w_{23} = +0{\cdot}000\,000\,000\,000\,009 & -w_{24} = -0{\cdot}000\,000\,000\,000\,000 \\ \hline \end{array}\right\} \tag{194}$$

giving

$$\left.\begin{aligned} w &= +147{\cdot}614\,612\,678\,596\,522 - 146{\cdot}989\,388\,294\,386\,742 \\ &= +0{\cdot}625\,224\,384\,209\,780 = \text{nearly} + L^{-1}\bar{1}{\cdot}796\,032\,59 \end{aligned}\right\} \tag{195}$$

whence (for $x = 10$),

$$\begin{aligned} \chi = \frac{w}{v} = -\frac{625\,224\,384\,209\,780}{245\,935\,764\,451\,349} &= (\text{nearly}) - L^{-1}0{\cdot}405\,214\,2 \\ &= (\text{nearly}) - 2{\cdot}542\,226, \text{ as in (188).} \end{aligned} \tag{196}$$

More exactly, by arithmetical division,

$$\chi = -2{\cdot}542\,226\,363\,882\,354; \tag{197}$$

and I think that this value of χ, for $x = 10$, may be relied on, to at least 13 figures after the point. Adding to it the known logarithm,

$$\log_e 10 = \theta = +2{\cdot}302\,585\,092\,994\,405, \tag{198}$$

we obtain

$$\chi + \theta = -0{\cdot}239\,641\,270\,887\,947, \tag{199}$$

whence, at least very nearly,

$$\chi + \theta = -L^{-1}\bar{1}{\cdot}379\,561\,4, \tag{200}$$

for our present case of $x = 10$.

XXI. Admitting then that we can safely rely upon the value

$$\chi + \theta = -0{\cdot}239\,641\,270\,88, \tag{201}$$

as being at least *very near* to the truth, for the case of $x = 10$, we have in the next place to inquire to what degree of accuracy we can compute, without (or even, if it be necessary, with) excessive labour, the *other part* namely $\frac{\pi}{2}\cot\omega$, of the expression (compare (152) and (162))

$$\frac{\pi}{2}\cot\omega + \chi + \theta = \text{constant}. \tag{202}$$

It is true that this question may seem to have been already answered, by our having found, in (150), the value for this case,

$$\omega = 3\pi + 77°14'42''{\cdot}8;$$

or a little more exactly, by (143), taking account of hundredths of seconds of arc,

$$\omega = 45° + 572°57'28''{\cdot}06 - 42'58''{\cdot}31 + 13''{\cdot}43 - 0''{\cdot}43 + 0''{\cdot}03 = 3\pi + 77°14'42''{\cdot}78; \tag{203}$$

which is certainly not wrong by so much as $0''{\cdot}1$, but *may*, so far as we *yet* know, be in error by about $0''{\cdot}02$, or $0''{\cdot}03$, from the nature of the process employed. Adopting it, however, as a temporary determination, we have, by Taylor's Tables,*

$$\frac{\pi}{2}\cot\omega = \frac{\pi}{2}\tan 12°45'17''{\cdot}22 = L^{-1}(0{\cdot}196\,119\,9 + \bar{1}{\cdot}354\,808\,7)$$
$$= L^{-1}\bar{1}{\cdot}550\,928\,6 = +0{\cdot}355\,572\,9, \text{ nearly}, \tag{204}$$

adding to which our recent value (201), reduced to 7 decimals, namely, $\chi + \theta = -0{\cdot}239\,641\,3$, we find for the constant, in the case $x = 10$, the approximate value,

$$\text{constant} = +0{\cdot}115\,931\,6; \tag{205}$$

where the 7th decimal must, at this stage of the process, be reckoned as a doubtful figure. It is, however, already worthy of note, that the constant thus found does not *greatly* differ from the rough approximation (161), namely $+0{\cdot}1158$, which was deduced by a far *ruder method*, and with a much freer use of *estimation*, for the case $x = 1$; whence we may fairly infer that *the constant in question is not a discontinuous one*. For, in the passage from the value $x = 1$ to the value $x = 10$, the variable x has passed, not only through the stage (163), at which x was equal to the *first*, (or least) positive *root* of the transcendental equation (154), but also through the *second* and the *third* roots of that equation ($v = 0$), which the researches of Fourier have made immortal. Thus, the *auxiliary angle* ω, which I am disposed (compare my expression (127)) to call the ***Phase*** of *Fourier's Definite Integral* (in this Letter denoted by

$$v = \frac{2}{\pi}\int_0^{\frac{\pi}{2}} d\alpha \cos(x\cos\alpha),)$$

because that Integral vanishes when the angle ω is any whole multiple of π, and has passed not only through π, itself, but also through 2π, and through 3π, while x has increased from 1 to 10; the terms, $\frac{\pi}{2}\cot\omega$, and χ, in the trinomial expression of our constant, becoming *both* infinite, at *each* such stage of the progress of x, or of the connected angle ω, but their

* [Michael Taylor (1756–1789), *Tables of logarithms and of sines and tangents*, London: 1792.]

algebraic *sum* being found to have remained *finite*, and *continuous*. This important conclusion is confirmed, when we advance to the case $x = 14$, whereby we *pass the fourth root* of the transcendental equation (154); for we have then, to the same order of approximation as for the case $x = 10$, by (144) and (151),

$$\omega = 4\pi + 126°37'50''{\cdot}45; \tag{206}$$

therefore, by Taylor's Tables,

$$\frac{\pi}{2}\cot\omega = -\frac{\pi}{2}\tan 36°37'50''{\cdot}45$$

$$= -L^{-1}0{\cdot}067\,399\,0 = -1{\cdot}167\,882\,0; \tag{207}$$

but also, by (187), for the case $x = 14$, we have, very nearly,

$$\chi = -1{\cdot}355\,243\,9; \tag{208}$$

and Hutton* (in his Table of Hyperbolic Logarithms) gives,

$$\theta = \log 14 = +2{\cdot}639\,057\,3; \tag{209}$$

whence by addition of the three last terms, we find, for $x = 14$, the value:

$$\text{constant} = \frac{\pi}{2}\cot\omega + \chi + \theta = 0{\cdot}115\,731\,4; \tag{210}$$

the seventh decimal figure still being doubtful.

XXII. Having arrived at this degree of approximation, I might well have been *content* to stop *there*, and to regard the solution of the main object of the present Letter as having been already, for all practical purposes, *accomplished*. And, in fact, my *most recent* calculations have given me, to *seven decimals*, a value which lies *between* the two (almost coincident) approximations, (205) and (210): namely the following value, on the *improvement* of which I may however still return,

$$\text{constant} = \frac{\pi}{2}\cot\omega + \chi + \theta = +0{\cdot}115\,931\,5. \tag{211}$$

But here, unluckily, came into play the *inconvenience* of possessing a certain share of what some people, in their politeness, might call *inventive* power – or at least some moderate degree of mathematical *sagacity*. Perceiving that $\sqrt{\pi}$, and still some more that $\sqrt{\frac{\pi}{2}}$, played an important part, as a *familiar* though *transcendental constant*, in a large portion of the whole investigation, I *suspected* that the constant of my own theory *might* (perhaps) be *related*, in some simple way, to the said square root, $\sqrt{\frac{\pi}{2}}$. This *thought* having occurred to me, it was not difficult to see that

$$\frac{37}{400}\sqrt{\frac{\pi}{2}} = 0{\cdot}115\,931\,558, \text{ nearly}; \tag{212}$$

* [Charles Hutton (1737–1823), *Mathematical Tables* first published in London in 1785; the seventh, published in 1830, and later editions were edited by Olinthus Gilbert Gregory (1774–1841).]

and then I think that I may fairly count on being *forgiven*, if I confess again, what you already know, from shorter notes of mine, – mere pilot balloons, – that I was, *for a while*, disposed to *believe* that the following equation was perfectly *rigorous*, as in fact I *find* it to be wonderfully *approximate*:

$$\sqrt{\frac{\pi}{2}}\cot\omega + \sqrt{\frac{2}{\pi}}(\chi+\theta) = \frac{37}{400} = 0{\cdot}092\,500. \tag{213}$$

My latest investigation makes this *last* constant to be *less* than the fraction $\frac{37}{400}$ by somewhat *about* one *thirty-millionth part of unity*; *so* that the value (213) *must be retained*, if we wish to write it correctly to the *nearest figure* in the 7th *place* of decimals; the error of my first approximation would bear well to be thus represented, for the benefit of readers in general. Suppose that somebody stated to a friend, that he knew for certain that Baron Rothschild had just that day netted, on some great loan transaction, a Million Sterling (not a French million). I beg your pardon, replies the friend, I am more correctly informed. The Baron *did*, no doubt, *receive* £1,000,000 from the Treasury: but I happened to be present when he handed out no less a sum than *seven shillings and six pence afterwards, as the charge for a receipt stamp* – Was I *lucky, or unlucky*, to come, so soon, *so near* the truth, without entirely attaining it? (The 7/6 appears to me, at present, to be an allowance slightly *too large*. See the *adopted constant*, in a subsequent paragraph.)

XXIII. It became, therefore, necessary for my purpose, that I should investigate, *more minutely*, the value of the *constant* in question: because if the equation (213) had turned out to be *rigorously accurate*, an *indication* would thereby have been given of (what I should have considered as being) a *very curious theoretical result*: deserving to be carefully examined, as perhaps *pointing* to some *quite new part* of the *whole Doctrine* of Definite Integrals. – And I still *think* that the *result points* to *something* of interest, which is as yet unknown to me.

XXIV. You see for my purpose it was necessary that I should render myself quite *sure*, not merely of the 7th, but even of the 8th decimal figure, in the expression of my constant; or that I should reduce its error *below* 0·000 000 005, or at all events *below* 0·000 000 01. The value of $\chi+\theta$ is already known, for $x=10$, *beyond* this limit of accuracy, and indeed I think to *fully twelve places*, as in the equation (201); and it can easily be found, with the help of the calculations already made, to almost as great a degree of accuracy as this or at any rate to about *ten* places, for the case $x=14$; in which case we have already determined that two *ascending* series give,

$$v = \frac{2}{\pi}\int_0^{\frac{\pi}{2}} d\alpha \cos(14\cos\alpha) = 0{\cdot}171\,073\,476\,114 \tag{184}$$

and

$$w = -0{\cdot}231\,846\,284\,145; \tag{185}$$

whence, by a somewhat more careful *division* than that mentioned in (186), or in (187) and with 12 figures written, after the point, although the 2 last may be doubtful,

$$\chi = \frac{w}{v} = -\frac{231\,846\,284\,145}{171\,073\,476\,114} = -1{\cdot}355\,243\,895\,2|63, \text{ for } x = 14. \tag{214}$$

But also,

$$\theta = \log 14 = L14 \times \log 10 = 1{\cdot}146\,128\,035\,678 \times 2{\cdot}302\,585\,092\,994 = +2{\cdot}639\,057\,329\,6|14, \tag{215}$$

whence, with an accuracy of perhaps about ten figures,

$$\chi + \theta = +1{\cdot}283\,813\,434\,3|51, \text{ for } x = 14. \tag{216}$$

(The values (208), (209) would have given, $\chi + \theta = 1{\cdot}283\,813\,4$.) We may therefore regard $\chi + \theta$ as *sufficiently well known*, for each of these 2 values of x; but the values of $\frac{\pi}{2}\cot\omega$ have not yet been computed, to the degree of accuracy which is at present aimed at. We know, however, to what may be called (for our purpose) *perfect* exactness, the value of the factor $\frac{\pi}{2}$; and although the Tables of Taylor will not give tangents to 8 or 9 decimals, yet we can compute such tangents for ourselves, by an endurable amount of industry. The important thing, then, and the only remaining (theoretical) difficulty, is to determine, more accurately than we have yet done, the values of what I call the *Phase* of the Integral, or the *auxiliary angle* ω. And these values must be computed, *at least to thousandths of seconds,* and not merely to *tenths,* or even to *hundredths,* as above. For, when ω is nearly equal to the value (203), as it is for $x = 10$, an increase of a *single tenth* of a second in this angle ω produces a decrease of about $0{\cdot}000\,000\,8$ in the value of $\frac{\pi}{2}\cot\omega$; and therefore an increase of ω by $0''{\cdot}01$ decreases $\frac{\pi}{2}\cot\omega$ by about $0{\cdot}000\,000\,08$, that is, by almost 1 in the 7th place of decimals: I must therefore aim at knowing the value of ω, if possible, without an error of $0''{\cdot}001$, in order that I may avoid an incorrectness of 1 in the 8th place, in the resulting product. And when ω is nearly equal to its other value (206), as it is when $x = 14$, an increase of a *single thousandth of a second* in it decreases $\frac{\pi}{2}\cot\omega$ by *rather more than* 1 *in the 8th place* of decimals: it is therefore *even more necessary,* in this case, for our purpose, to *avoid so great an error* as $0''.001$ in the determination of ω. For these reasons I am compelled to resume and improve the method, which has been employed in this Letter, for the calculation of that angle, or phase.

XXV. The descending series (44), for that particular integral y of the fundamental triordinal equation (1), to which the present Letter mainly relates, may be thus written:

$$y = x^{-1}(1 - \overline{0}|^{-1}\alpha_1 t + \overline{0}|^{-2}\alpha_2 t^2 - \overline{0}|^{-3}\alpha_3 t^3 + \&c.); \tag{217}$$

where

$$t = \frac{1}{8x^2}; \tag{218}$$

$$\overline{0}|^{-m} = \frac{1}{1.2.3.\cdots m}, \text{ as before; and } \alpha_m = 1^3 3^3 5^3 7^3 \cdots (2m-1)^3, \tag{219}$$

so that

$$\alpha_1 = 1, \quad \alpha_2 = 27, \quad \alpha_3 = 3375, \quad \alpha_4 = 1\,157\,625,$$
$$\alpha_5 = 843\,908\,625, \quad \alpha_6 = 1\,123\,242\,379\,875. \tag{220}$$

Hence, by (135),

$$\begin{aligned} \omega' = (xy)^{-1} &= (1 - \overline{0}|^{-1}\alpha_1 t + \overline{0}|^{-2}\alpha_2 t^2 - \&c.)^{-1} \\ &= 1 + \overline{0}|^{-1}\beta_1 t - \overline{0}|^{-2}\beta_2 t^2 + \overline{0}|^{-3}\beta_3 t^3 - \&c., \end{aligned} \tag{221}$$

with the relations,

$$\alpha_1 = \beta_1, \quad \alpha_2 = 2\alpha_1\beta_1 + \beta_2, \quad \alpha_3 = 3\alpha_2\beta_1 + 3\alpha_1\beta_2 + \beta_3, \ \&c., \tag{222}$$

which gives

$$\beta_1 = 1, \quad \beta_2 = 25, \quad \beta_3 = 3219, \quad \beta_4 = 1\,127\,199,$$
$$\beta_5 = 830\,771\,625, \quad \beta_6 = 1\,112\,086\,390\,905. \tag{223}$$

Restoring for t its value (218), and then integrating the series (211) with respect to x, and adding the constant $\frac{\pi}{4}$, we find, as in (149),

$$\omega = \frac{\pi}{4} + x - s,$$

with the following descending series for s, carried somewhat farther than (141) or (142):

$$\left.\begin{aligned} s &= \frac{\beta_1 x^{-1}}{8} - \frac{\beta_2 x^{-3}}{3.2.8^2} + \frac{\beta_3 x^{-5}}{5.2.3.8^3} + \frac{\beta_4 x^{-7}}{7.2.3.4.8^4} + \frac{\beta_5 x^{-9}}{9.2.3.4.5.8^5} - \frac{\beta_6 x^{-11}}{11.2.3.4.5.6.8^6} + \&c. \\ &= \frac{1}{2}\left\{(4x)^{-1} - \frac{25}{3}(4x)^{-3} + \frac{2146}{5}(4x)^{-5} - \frac{375\,733}{7}(4x)^{-7} + \frac{110\,769\,550}{9}(4x)^{-9} \right. \\ &\quad \left. - \frac{49\,426\,061\,818}{11}(4x)^{-11} + \&c.\right\} = 2578''{\cdot}310\,078\left(\frac{10}{x}\right)^1 - 13''{\cdot}428\,698\left(\frac{10}{x}\right)^3 \\ &\quad + 0''{\cdot}432\,270\left(\frac{10}{x}\right)^5 - 0''{\cdot}033\,788\left(\frac{10}{x}\right)^7 + 0''{\cdot}004\,842\left(\frac{10}{x}\right)^9 - 0''{\cdot}001\,105\left(\frac{10}{x}\right)^{11} + \&c.\}; \end{aligned}\right\} \tag{224}$$

the two last terms being new, and the coefficients of the four others being now computed to greater exactness than before. For $x = 10$, the initial convergence of this ultimately diverging series is scarcely so rapid as we require: that convergence is also inconveniently decreasing. However, if we sum separately the first five and the first six terms, as just now set down, we find the two near limits,

$$s < 2565''{\cdot}2847, \ \text{ and } s > 2565''{\cdot}2836, \ \text{ for } x = 10; \tag{225}$$

and accordingly our former approximation to s, for this value of x, was $s = 2565''{\cdot}28$, (143). – Expanding $\frac{\pi}{4} + x$ to the 4th decimal of a second of arc

$$\omega + s = \frac{\pi}{4} + x = 45^\circ + 2\,062\,648''\cdot 0625; \tag{226}$$

therefore

$$\omega - 45^\circ > 2060082''\cdot 7778,\ \omega - 45^\circ < 2060082''\cdot 7789; \tag{227}$$

that is,

$$\omega - 3\pi > 77^\circ 14' 42''\cdot 7778, \text{ but } < 77^\circ 14' 42''\cdot 7789, \text{ for } x = 10; \tag{228}$$

and accordingly, for this value of x, we lately found that

$$\omega = 3\pi + 77^\circ 14' 42''\cdot 78, \quad \text{nearly}. \tag{203}$$

Hence, for $x = 10$, we have

$$\frac{\pi}{2}\cot\omega < \frac{\pi}{2}\tan 12^\circ 45' 17''\cdot 2222, \quad > \frac{\pi}{2}\tan 12^\circ 45' 17''\cdot 2211. \tag{229}$$

The arc of which the tangent is here taken is only a little greater than

$$\frac{2}{9} = 12^\circ 43' 56''\cdot 6236 \tag{230}$$

in such a manner that we must have, *very nearly*,

$$\frac{\pi}{2}\cot\omega = \frac{\pi}{2}\frac{t+T}{1-tT}, \tag{231}$$

$$\text{if } t = \tan\frac{2}{9}, \quad \text{and } T = \tan 80''\cdot 598. \tag{232}$$

By the known series for sine and cosine,

$$t = \frac{\sin\frac{2}{9}}{\cos\frac{2}{9}} = \frac{\cdot 220\,397\,743\,456\,123}{\cdot 975\,410\,085\,389\,448} = 0\cdot 225\,953\,931\,333|533; \tag{233}$$

where I suppose that at least 12 or 13 figures are correct, the values of sin and cos having been verified by their giving,

$$\left.\begin{array}{l} \sin^2\frac{2}{9} = \cdot 048\,575\,165\,320\,551 \\ \cos^2\frac{2}{9} = \cdot 951\,424\,834\,679\,450 \end{array}\right\} \text{sum} = 1\cdot 000\,000\,000\,000\,001. \tag{234}$$

The other tangent T is so small, that for *it* we may conveniently employ the single ascending series, without computing sine and cosine: and thus we have

$$T = \tan(\alpha = 80''\cdot 598) = \cdot 000\,390\,750\,130\,70 = \alpha + \frac{\alpha^3}{3} + \cdots$$

$$= \left\{\begin{array}{l} \cdot 000\,390\,750\,131 \\ +\cdot 000\,000\,000\,020 \end{array}\right\} = \cdot 000\,390\,750\,151: \tag{235}$$

wherefore, by (231) and (233), at least very nearly,

$$\cot\omega = \tan 12°45'17''\cdot 2216 = \frac{t+T}{1-tT} = 0\cdot 226\,364\,667\,569; \tag{236}$$

and therefore

$$\frac{\pi}{2}\cot\omega = +0\cdot 355\,572\,788\,334. \tag{237}$$

(We had found, by logarithms,

$$\frac{\pi}{2}\cot\omega = \text{nearly } 0\cdot 3555729.)$$

Adding $\chi + \theta = -0\cdot 239\,641\,270\,888$, (201), we have thus, *for* $x = 10$, the *improved approximate value*:

$$\text{costant} = \frac{\pi}{2}\cot\omega + \chi + \theta = +0\cdot 115\,931\,517\,446. \tag{238}$$

For $x = 14$, the series for s converges *much* more rapidly, in its initial terms; and we thus find,

$$\begin{aligned}\omega &= 45° + 2\,887\,707''\cdot 287\,459\,3 - s = 4\pi + 127°8'27''\cdot 287\,459\,3\\ &\quad - 30'41''\cdot 650\,055\,7 + 4''\cdot 893\,840\,4 - 0''\cdot 080\,373\,9 + 0''\cdot 003\,205\,3\\ &\quad - 0''\cdot 000\,234\,4 + 0''\cdot 000\,027\,3 - \&c. + \&c.\\ &= 4\pi + 126°37'50''\cdot 453\,86\ldots,\end{aligned} \tag{239}$$

so that we may be pretty sure of *at least 4 decimals of a second*, and may safely write,

$$\omega = 4\pi + 126°37'50''\cdot 4539. \tag{240}$$

However it is almost as easy to calculate $\frac{\pi}{2}\cot\omega$ from (239), as from (240), as follows: Let

$$t = \tan 36° = \sqrt{5 - 2\sqrt{5}} = 0\cdot 726\,542\,528\,005; \tag{241}$$

$$T = \tan 37'50''\cdot 453\,86 = 0\cdot 011\,007\,915\,529; \tag{242}$$

then

$$\frac{\pi}{2}\cot\omega = -\frac{\pi}{2}\frac{t+T}{1-tT} = -1\cdot 167\,881\,918\,679. \tag{243}$$

Adding to this,

$$\chi + \theta = +1\cdot 283\,813\,434\,351, \tag{216}$$

$$\text{constant} = \frac{\pi}{2}\cot\omega + \chi + \theta = +0\cdot 115\,931\,515\,6|72. \tag{244}$$

And now, *at last*, I am conducted to a value of the constant, with which I am disposed to be *content*, as with one *likely* to have 9 (nay, almost *ten*) figures correct. In fact, if we suppose, *as we may*, by (228), that

$$\omega = 3\pi + 77°14'42''\cdot 7786 \text{ for } x = 10, \tag{245}$$

then the value (238) is slightly but sufficiently diminished, and becomes consistent with the following, which I **adopt** as **final**:

$$\textbf{constant} = +0{\cdot}115\,931\,516. \tag{246}$$

But we see by (212) that this constant (246) is *not exactly equal* to the value, namely

$$\frac{37}{400}\left(\frac{\pi}{2}\right)^{\frac{1}{2}},$$

or 0·115 931 558, with which I had for a while *supposed* it to coincide, and to which in fact it makes a very remarkable degree of approach: its defect from the conjectured value being less than 0·000 000 4 [times] the whole, and nearly [equal to] that whole [multiplied by] 3 and [divided by] 8 millions; or (as I said) to 7 shillings and 6 pence in a million of pounds sterling. (The result may be thus illustrated: Let the sum on which we are to operate be one million of Pounds Sterling; my fraction $\frac{37}{400}$ would give a result [equal to] £92,000; and the *correction of this fraction,* [equal to] $-1/30\,000\,000$, would diminish this by about eight pence.)

XXVI. Conceiving that I have now completely established, as true to (at least) the 7th decimal, the equation

$$\left(\frac{\pi}{2}\right)^{\frac{1}{2}}\cot\omega + \left(\frac{2}{\pi}\right)^{\frac{1}{2}}.(\chi+\theta) = +0{\cdot}092\,500\,0 = \frac{37}{400}, \tag{213}$$

where ω, χ, and θ have the significations already explained, I proceed to make a few applications of this formula. And first let me connect it with some of the investigations contained in the Memoir by Professor Stokes, to which I could only refer from memory, when writing my fourth Sheet, but which has been found today, after the Fifth Sheet was posted. The Paper is probably in your Library, but for my own convenience I shall here copy its Title, and a small part of its contents. It is entitled; "On the Numerical Calculation of a Class of Definite Integrals and Infinite Series"*. From the *Transactions of the Cambridge Philosophical Society*. Volume IX. Part I. By G. G. Stokes, M.A., Fellow of Pembroke College, and Lucasian Professor of Mathematics in the University of Cambridge. Engraving. Cambridge: printed at the Pitt Press, by John W. Parker, Printer to the University, M.DCCC.L. The atrocious custom, to which I own that I have too often been a consenting party, of giving to authors their private copies *with private paging*, renders it impossible for a reader to make a *reference* which shall be intelligible, or at least shall be sure to be so, to any one else who may have access to the public copies only, by mentioning the *page* alone; but in *this* case, the leading *formula* (as well as *articles*) being *numbered*, we can get along well enough: besides that I suppose Stokes is likely to have sent *you* a separate copy. Turn then to his "Second Example," articles 19 to 22. In that Example, Stokes considers the integral

$$u = \frac{2}{\pi}\int_0^{\frac{\pi}{2}} \cos(x\cos\theta)d\theta = 1 - \frac{x^2}{2^2} + \frac{x^4}{2^2 4^2} - \frac{x^6}{2^2 4^2 6^2}\cdots,$$

which, as he remarks, "occurs in a great many physical investigations", and which has in this Letter been denoted by v. He says: "If we perform the operation $x\frac{d}{dx}$ twice in succession on

* [Reprinted in: George Gabriel Stokes (1819–1903), *Mathematical and physical papers,* Vol. II, pp. 329–357, Cambridge: 1883.]

the series, we get the original series multiplied by $-x^2$, whence

$$\frac{d^2u}{dx^2}+\frac{1}{x}\frac{du}{dx}+u=0.\text{''} \qquad St(47)$$

(Compare the equations (58) (59) of this Letter.) As the complete integral of this differential equation, he assigns the form,

$$u=Ax^{-\frac{1}{2}}(R\cos x+S\sin x)+Bx^{-\frac{1}{2}}(R\sin x-S\cos x), \qquad St(49)$$

where

$$R=1-\frac{1^2.3^2}{1.2(8x)^2}+\frac{1^2.3^2.5^2.7^2}{1.2.3.4(8x)^4}\cdots,\qquad S=\frac{1^2}{1.8x}-\frac{1^2.3^2.5^2}{1.2.3(8x)^3} \qquad St(50)$$

while A and B are arbitrary constants. For very large values of x, he obtains the same approximate results as Poisson, for the particular integral in question, namely

$$u=(\pi x)^{-\frac{1}{2}}(\cos x+\sin x);$$

(compare equation (65) of this Letter;) whence he infers that

$$A=B=\pi^{-\frac{1}{2}},\text{ and finally that } u=\left(\frac{2}{\pi x}\right)^{\frac{1}{2}}R\cos\left(x-\frac{\pi}{4}\right)+\left(\frac{2}{\pi x}\right)^{\frac{1}{2}}S\sin\left(x-\frac{\pi}{4}\right); \qquad St(52)$$

being pleased to observe, in a Note, that "this expression for u, or rather an expression differing from it in nothing but notation and arrangement, has been already obtained in a different manner by Sir William Rowan Hamilton, in a memoir On Fluctuating Functions. See *Transactions of the Royal Irish Academy*, Volume XIX. page 313".* He refers, of course, to a formula which (with a slightly modified notation) I shall now number with the equations of this Letter, as follows:

$$(u,\text{ or})v=\frac{2}{\pi}\int_0^{\frac{\pi}{2}}d\alpha\cos(x\cos\alpha)$$

$$=\left(\frac{\pi x}{2}\right)^{-\frac{1}{2}}\sum_{m=0}^{m=\infty}\overline{0}|^{-m}\left(\left[-\frac{1}{2}\right]^{m}\right)^2(2x)^{-m}\cos\left(x-\frac{\pi}{4}-\frac{m\pi}{2}\right). \qquad (247)$$

XXVII. I am content to adopt Stokes's *symbols*, R and S, for the coefficients of the cosine and sine of $x-\frac{\pi}{4}$, in the expansion of

$$\left(\frac{2x}{\pi}\right)^{\frac{1}{2}}\int_0^{\frac{\pi}{2}}d\alpha\cos(x\cos\alpha),$$

as given by this formula of mine, in finding which I was assisted by a hint in one of Poisson's Memoirs; and then, taking nothing on the present occasion from Professor Stokes, *except those two letters*, R and S, I suppose myself to be entitled to write, and to number, as follows:

* [See p. 585 of this volume.]

$$\left(\frac{\pi x}{2}\right)^{\frac{1}{2}} v = \left(\frac{2x}{\pi}\right)^{\frac{1}{2}} \int_0^{\frac{\pi}{2}} d\alpha \cos(x \cos \alpha) = (R \sin - S \cos)\left(x + \frac{\pi}{4}\right); \tag{248}$$

$$R - \sqrt{-1}S = \sum_{m=0}^{m=\infty} \overline{0}|^{-m} \left(\left[-\frac{1}{2}\right]^m\right)^2 (2x\sqrt{-1})^{-m}. \tag{249}$$

Stokes gives, as an example, for the case $x = 10$, the values,

$$\text{``}R = 1 - \cdot 000\,07 + \cdot 000\,01 = \cdot 999\,31; \quad S = \cdot 012\,50 - \cdot 000\,10 = \cdot 012\,40;$$

$$\text{Angle } x - \frac{\pi}{4} = 527^\circ \cdot 957\,80 = 3 \times 180^\circ - 12^\circ 2' 32''; \quad \text{whence } u = - \cdot 245\,94,$$

which agrees with the number $(-\cdot 2460)$ obtained by Mr. Airy by a far more laborious process, namely by calculating from the original Series.'' The only liberty which I have taken in the transcription has been, that I have written the decimal point a little higher up, than it is written in his Paper. You may remember, or on turning back will find, that I lately deduced, from 25 terms of the ascending series, for the case $x = 10$, the value, more exact than Stokes required for his purpose, but so far as his goes agreeing with his result, and not professing to be itself corrected to *all* its 15 decimals:

$$v = \left.\begin{matrix} +1407 \cdot 735\,346\,350\,901\,560 \\ -1407 \cdot 981\,282\,115\,352\,907 \end{matrix}\right\} = 0 \cdot 245\,935\,764\,451\,349 = \frac{2}{\pi} \int_0^{\frac{\pi}{2}} d\alpha \cos(10 \cos \alpha). \tag{191}$$

And possibly you may recollect, that in my Essay on Fluctuating Functions, to which Professor Stokes has referred, and of which you were (I think) at one time pleased to desire that I should send you a copy, – and I trust that the request (if made) was acted on, – I published the analogous results, cited in a rather recent little paper of mine, in the Philosophical Magazine for last November, 1857*, but which I shall now, for reference sake, incorporate with the present Letter:

$$\frac{2}{\pi} \int_0^{\frac{\pi}{2}} d\alpha \cos(40 \cos \alpha) = \left.\begin{matrix} +7447\,387\,396\,709\,949 \cdot 965\,7957 \\ -7447\,387\,396\,709\,949 \cdot 958\,4289 \end{matrix}\right\} = +0 \cdot 007\,366\,8, \tag{250}$$

from the *ascending* series, by *sixty* terms employed; and from the *descending* series, by only *three* terms,

$$\frac{2}{\pi} \int_0^{\frac{\pi}{2}} d\alpha \cos(40 \cos \alpha) = \left(1 - \frac{9}{204\,800}\right) \frac{\cos 86^\circ 49' 52''}{\sqrt{20\pi}} + \frac{1}{320} \frac{\sin 86^\circ 49' 52''}{\sqrt{20\pi}}$$

$$= 0 \cdot 006\,973\,6 + 0 \cdot 000\,393\,6 = +0 \cdot 007\,367\,2. \tag{251}$$

XXVIII. Professor Stokes proceeds to introduce two new quantities, which he denotes by M and ψ, and which are such that

$$R = M \cos \psi \qquad S = M \sin \psi.$$

(He does not actually print these very equations, but it is obvious that they were in his mind.) He goes on to say:- ``whence we get for calculating u for a given value of x

* [See p. 652 of this volume.]

$$M = 1 - \frac{1}{16}x^{-2} + \frac{53}{512}x^{-4}, \qquad \tan\psi = \frac{1}{8}x^{-1} - \frac{33}{512}x^{-3} + \frac{3417}{16\,384}x^{-5},$$

$$u = \left(\frac{2}{\pi x}\right)^{\frac{1}{2}} M\cos\left(x - \frac{\pi}{4} - \psi\right).\text{''} \qquad St(53)$$

If I were writing for the mathematical public at large, and I gladly admit that these Letters to De Morgan constitute in themselves a *quasi-publication*, very useful to myself, as forcing me to express myself with some degree of method and arrangement, – I might shrink from the egotism of saying, what I cannot formally prove, and what it would be worth nobody's while not even my own, that I should prove if I could, – but what I still choose to assert to you, on manuscript evidence satisfactory to me, – that I possessed the transformation last cited, ten years before Stokes printed it. Admitting, however, that, by all laws of evidence, he is entitled to call his deductions (St(53)) from my formula (St(52)) *his own*, I suppose that I may be allowed to *adopt* them: and to *adopt* them to the system of this Letter by a slight modification of the symbols, and a slight extension of the developments. The auxiliary angle ψ of Stokes is precisely the s of these pages; permit me therefore to write, now,

$$R = r\cos s, \qquad S = r\sin s; \tag{252}$$

where R and S have the same significations as in Stokes's "Second Example"; so that (though not in his notations)

$$R = \sum_{m=0}^{m=\infty} \overline{0}\rceil^{-2}\left(\overline{\left.-\frac{1}{2}\right|}^{2m}\right)^2(-4x^2)^{-m}; \qquad S = \frac{1}{2x}\sum_{m=0}^{m=\infty} \overline{0}\rceil^{-(2m+1)}\left(\overline{\left.-\frac{1}{2}\right|}^{2m+1}\right)^2(-4x^2)^{-m}. \tag{253}$$

Although the *angle* which I here denote by s, namely

$$s = \tan^{-1}\frac{S}{R}, \tag{254}$$

and which Stokes has denoted by ψ, has been *itself* developed in this Letter to a (much) greater degree of accuracy, (compare (224)), on principles which do not appear to have occurred to Professor Stokes, and on which I may perhaps return, yet for the *tangent* of this angle I am here content to write down an expansion, which has only *one term more* than that cited from the Professor's Paper:

$$(\tan\psi,\text{ or) } \tan s = \frac{S}{R} = (8x)^{-1} - 33(8x)^{-3} + 6834(8x)^{-5} - 342\,731\,7(8x)^{-7} + \&\text{c.} \tag{255}$$

The "one term more", it must be said in passing, I was able to turn to some important account, by one of my methods of "Transformation of Diverging Series", even for the very unfavourable case $x = 1$, *before* I had made out the processes of the *present* Letter, whereby the value of the angle in question can be computed (as I conceive) *with certainty*, for every real and positive value of x, to decidedly *less than the thousandth part of a second*, if the *improved* value (246) of my constant be adopted; and with an error which can *scarcely ever* amount to the *two-hundredth part of a second*, if we content ourselves with the approximate formula (213), and deduce s from ω, by the relation (149): the *semicircle*, to which ω (given at first by its *cotangent*) belongs, being determined by a comparatively slight and rude and easy examination of the *roots* of the transcendental equation $v = 0$: (154) of which equation Stokes has tabulated the *first twelve roots*, with (as I presume) an accuracy sufficient for his purpose, although I judge that the *first root*, say x_1, as by him deduced by interpolation from a Table of Mr Airy, is wrong by almost two units in the fourth place of decimals: Professor Stokes adopting for this root the value 2·4050, while I have found it to be, much more nearly, $x_1 = +2{\cdot}40482556$, (163) – As regards the quantity which I above, in (252), call r, and he calls M namely the (Argandian) *modulus* of

$$R + \sqrt{-1}S, \quad \text{or} \quad (M =) r = (R^2 + S^2)^{\frac{1}{2}}, \tag{256}$$

I shall just write down my own result, which goes a *couple of terms farther* than that of Stokes:

$$(M =) r = 1 - \frac{1}{2^4 x^2} + \frac{53}{2^9 x^4} - \frac{4447}{2^{13} x^6} + \frac{3\,066\,403}{2^{19} x^8} - \&\text{c.} \tag{257}$$

Of course, I am not so *foolish* as to entertain even the *slightest doubt*, but that Stokes could *easily* have deduced these additional *terms* from my series, – if it ought not rather to be called Poisson's, since I admit that I took the hint from him, although Poisson does not seem to have assigned the *general term*, – if Professor Stokes had judged it to be *useful*, *for his purpose*, to have done so. But I confess that I *do* doubt whether he saw the *laws* which regulate the series for what he denotes by M, or rather the *square of that series*, and which connect the reciprocal *of that square*, with the differential coefficient of the angle ψ, or s.

XXIX. For, with every disposition in the world to look *up* to Professor Stokes, – with whom I have now a strange and Irish feeling of being almost in some degree *connected*, since he has married that charming Irish girl, whom I knew a little from her infancy, – Mary, the Daughter of my old friend, the Reverend Doctor Thomas Romney Robinson of Armagh, – I must, in justice to myself, or rather to the subject of the present Letter, go on to point out a few things, which did not, perhaps, lie exactly in Stokes's *way*, and which he is most excusable for not having *dwelt* upon: but which it is still a little *curious* to me that he did not in some way *notice*, or *allude* to, if he happened to be aware of their existence, at a time when his subject brought him *so near* them.

For the moment, then, calling those series *mine*, which Stokes has lettered R and S, but has in print acknowledged to have been substantially anticipated by my Paper on Fluctuating Functions, – which Paper I remember Jacobi holding in his hand at Manchester in 1842, when he was pleased to style me, in his address to Section A, of the British Association, "le Lagrange de vôtre pays," – I venture to remark that Professor Stokes does not seem to have been aware of some of those *different relations* between them, respecting which I now presume on your good-humour, to give to them, through you, that sort of "quasi-publication", to which I have already alluded: although I have no wish to *print*, until I can much more closely condense.

XXX. In order that the descending series,

$$v = bx^{-\frac{1}{2}}(R \sin - S \cos)(x + c), \tag{258}$$

may express (as Stokes was quite aware that it does) *one form of the complete integral* of the linear and biordinal equation of Fourier,

$$(D^2 + x^{-1} D + 1) v = 0, \tag{59}$$

while b and c are arbitrary constants, it is necessary and sufficient that the functions R and S should be connected with each other by the two *relations*:

$$D^2 R + 2DS + \frac{R}{4x^2} = 0; \qquad D^2 S - 2DR + \frac{S}{4x^2} = 0, \tag{259}$$

which I have not noticed in any book, but which are easily verified, as being *satisfied* by the two series (253), or by Stokes's expressions ($St(50)$) for R and S. Multiplying the first of these equations (259) by S, and the second by $-R$, adding, and integrating, we get,

$$SDR - RDS + R^2 + S^2 = \text{constant} = 1; \tag{260}$$

another relation which I suppose to be new. Introducing the expressions (252), we find

$$r^{-2} = 1 - s'; \tag{261}$$

or, in the notation of Stokes's Paper,

$$\frac{d\psi}{dx} = 1 - M^{-2};$$

a relation of which, however, I do not see any indication that he was at all aware. (It is a *case* of a *much more general** *relation*, which I perceived, but did not publish, in 1840, and to which I may perhaps return, in the present, or in some future Letter.) – Substitute (252) in the 1st equation of the 2nd order,

$$r^3 r'' + r^4 \left(1 + \frac{1}{4x^2}\right) = 1: \tag{262}$$

so that in Stokes's notation, but (as it seems) without his knowing it,

$$M^3 \frac{d^2 M}{dx^2} + M^4 \left(1 + \frac{1}{4x^2}\right) = 1.$$

Divide the differential of (262) by $\frac{r^2}{2}$; we find that

$$2(rr''' + 3r'r'') + 8rr'\left(1 + \frac{1}{4x^2}\right) - \frac{r^2}{x^3} = 0; \tag{263}$$

that is

$$(D^3 + (4 + x^{-2}) D - x^{-3}) r^2 = 0;$$

or finally,

$$(xD)^3 x^{-1} r^2 + 4x^2 Dr^2 = 0, \tag{264}$$

when Dr^2 denotes $D(r^2)$, not $(Dr)^2$. Comparing with the fundamental and triordinal (but linear) equation (1), of this Letter, we see that

$$r^2 = xy, \tag{265}$$

* I have not time just now to write a note about this.

with the value (44) for y, there being *only one descending series* for $x^{-1}r^2$, or for y, which satisfies the triordinal equation, and gives $xy = 1$ when $x = \infty$ hence, by the present Letter, the square of Stokes's M or of my r, is:

$$r^2 = \sum_{m=0}^{m=\infty} \overline{0}|^{-m}\left(\overline{-\frac{1}{2}}|^{m}\right)^3 x^{-2m} = 1 - \frac{1^3}{1(8x^2)} + \frac{1^3 3^3}{1.2(8x^2)^2} - \frac{1^3.3^3.5^3}{1.2.3(8x^2)^3} + \cdots, \tag{266}$$

and this result which I suppose to be new. Finally, if we make $\omega = x + \frac{\pi}{4} - s$ *as in* (149), equal to what I have called the *Phase of Fourier's Integral,* that definite integral becomes,

$$v = \frac{2}{\pi}\int_0^{\frac{\pi}{2}} d\alpha \cos(x\cos\alpha) = \left(\frac{2}{\pi x}\right)^{\frac{1}{2}} r\sin\omega; \tag{267}$$

and the formula*

$$\left(\frac{\pi}{2}\right)^{\frac{1}{2}}\cot\omega + \left(\frac{2}{\pi}\right)^{\frac{1}{2}}(\chi + \theta) = \frac{37}{400}, \tag{213}$$

enables me to find the phase for small as well as for great values of x: which it appears to have been quite *impossible* to do, by the method of Stokes's Paper. I am enabled, for instance, to enunciate this ***Theorem.*** "*The Phase* ω *tends to* 0, when x tends to 0"; or this: "While x, being real and positive, tends to 0, *the angle s tends to* 45°, *and the diverging series* (255) *tends to* 1."

* Note added on 1 June 1860:
If v be developed in the *ascending* series

$$v = v_0 - v_1 + v_2 - v_3 + \text{\&c.}$$

where

$$v_0 = 1,\ v_1 = \frac{x^2}{4},\ \ldots,\ v_n = 2^{-2n}x^{2n}(\overline{0}|^{-n})^2,$$

and if

$$w = w_1 - w_2 + w_3 - w_4 + \text{\&c.},$$

where

$$-w_1 = v - v_0,\ -w_2 = v - v_0 + v_1,\ -w_3 = v - v_0 + v_1 - v_2,\ \text{\&c.},$$

then

$$\chi = \frac{w}{v};\ \text{and } \theta = \log x.$$

Thus $\cot\omega$ is *definitely found,* by the equation (213), for all real and positive values of x. As to the *semicircle* in which the *phase* ω lies, it is the 1st *positive semicircle,* ($\omega > 0, < \pi$) if $x > 0$, $x <$ *least positive root* of equation $v = 0$; it becomes [equal to] π for this root; is in 2nd positive semicircle, until x [equals] *second* positive root of same transcendental equation; and so on; being, as I find, [equal to] 0, when $x = 0$, which is one of the chief theoretical results of this Letter, as regards Fourier's Integral.

III.

LETTER TO HART ON ANHARMONIC COORDINATES (1860)*

Observatory, February 27th 1860

My dear Dr Hart,

1. Although I lately wrote in reply to your note about the pyramids, and suppose that you may not consider that subject as completely disposed of, yet I believe that I have your permission to address to you a letter, which may become a long one, on the subject of *Anharmonic Coordinates*: especially as Salmon† and others, to whom I lately spoke on that subject, did not appear to have heard of such things, nor can I find any equivalent expression, nor any such *coordinate system*, in the *Barycentric Calculus of Möbius*;‡ although it was by combining some points of the calculus with quaternions that I happened to form the *conception*. Möbius, in fact, expressly defined that his coordinates of a *point* P, in the plane of a given *fundamental triangle* ABC, are the *weights* (or at least the *ratio* of the weights,) with which the corners of that triangle must be *loaded*, in order that the *point* P may become their *centre of gravity*: and similarly for a point P in space referred to a fundamental pyramid $ABCD$. He certainly has some investigations, in which, after selecting some one point, say D, in the plane of a fundamental triangle, and denoting *its* weights, or its *barycentric coordinates*, by a, b, c, so that in his notation we have the *congruence*, $D = aA + bB + cC$, he passes then to the consideration of *another* point P in the same plane, which satisfies this *other* congruence $P = xaA + ybB + zcC$. But I do not find, perhaps my search has not been sufficiently close, and I shall be glad, if so, to be set right, – that he gives any *name* to *these new coefficients* x, y, z; or that he in any way proposes to make them the main *elements of a system*: with the corresponding use of *quotients of quotients of weights* for *space*, as when

$$E \equiv aA + bB + cC + dD, \quad \text{and} \quad P \equiv xaA + ybB + zcC + wdD,$$

where $ABCD$ is the fundamental pyramid. Even his use of Greek letters to denote these

* [This long letter, which now appears in print for the first time, was addressed to Dr Andrew Searle Hart (1811–1890) who was, in 1860, a senior fellow and bursar of Trinity College, Dublin. The present text is based on two, incomplete, copies contained in Trinity College MSS 1493/1170 (articles 1–203, 213–357, 367–382, 393–490) and 1493/1171 (articles (169–438). A short publication, based on this letter, will be found on pp. 507–515 of this volume; see the first footnote to article 490 on p. 427.]

† [George Salmon (1819–1904), fellow of Trinity College, Dublin, and, in 1860, assistant to the Erasmus Smith's professor of mathematics, Charles Graves (1812–1899), and Donegal lecturer.]

‡ [August Ferdinand Möbius (1790–1868), *Der barycentrische Calcul*, Leipzig: 1827.]

quotients is to me some argument that he regarded them as only of a secondary or *derivative* character, and of merely *occasional* use and application, while his true *coordinates* (which he denotes by italic letters) are *still* the *weights themselves*, or numbers proportional thereto. But I shall leave the remainder of this page for a few extracts from his book, which is now just beside me. And I shall pass, in page 3, to an account of *my own way* of looking at the whole subject, which appears to me to be much more simple and elementary; and to show that the Anharmonic Coordinates to which I allude ought really to be introduced, quite early, into the Elements of at least the Higher Geometry, including the application of Algebra: but not *necessarily* requiring an employment of Quaternions, although I find that the methods easily combine. I shall continue to *number paragraphs* [later referred to as articles], namely as 1, 2, 3, . . . , leaving the Roman numerals I., II., . . . , to serve [but not exclusively] as references to my former letters.

2. Extracts from the Barycentric Calculus of Möbius.*

3. To begin at the beginning, let us take the case of Cartesian coordinates, in a given plane. Assuming, as usual, an origin O, and two axes, OX, OY, which may be conceived to be continued to meet the line XY at infinity, we have thus what may be called a *fundamental triangle*: and it seems to be taken for granted that this is sufficient, for the determination or *construction* of any point P in the plane, as soon as its *numerical coordinates*, x and y, are given: at least, I have never happened to see any use made of any *fourth point of reference*, as among the *data* of the *coordinate system*. But suppose that $P = (1, 1)$, or that it is required to construct the point which has each coordinate equal to positive unity. How is this to be done? Of course, the answer would be, that we must employ some assumed *unit of length*, – or it may be, *two such units*, one for one axis, and the other for the other, – and must also select one of the *four regions* about the origin, as that which is wholly *positive*. Very well: but might it not as well be said at once, that we *select a fourth point* U in the given plane, (X and Y being treated as

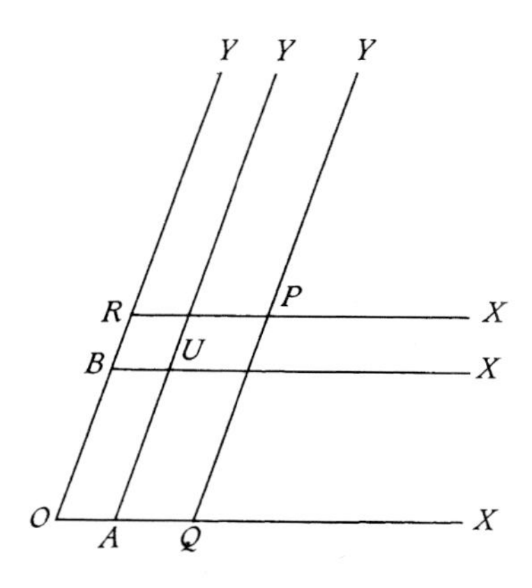

Fig. 1

* [This section has been left blank.]

given,) of which the coordinates are 1 and 1, and which may therefore be briefly called the *unit-point*? At all events, we see that such a *selection* is *virtually made*, in the process *usually* employed. And when it is made *avowedly*, we need no *additional convention* respecting *units*, or positive direction. – Indeed, I conceive that it is then unnecessary so much as to say that any one direction is positive. I would prefer to speak of *positive or negative quotients of segments*, their *relative directions*, only being considered; and then I think that the grand and received *convention*, respecting *signs of segments*, won't fall away as *useless*. We can surely *admit* that *the quotient* $AB : AB = +1$, and that *this other quotient* $AB : BA = -1$, without ever *thinking* of the question, whether AB *itself* be a *positive* or a *negative line*: and in fact, you are aware that I admit *no such lines, in quaternions*. – But be that as it may, I proceed to lay down my own *definitions*, of *coordinates* within the plane.

4. Having assumed a *triangle* OXY, and a *point* U in its plane, let two other fixed points A and B be determined, by the conditions [Figure 2]

$$A = YU^{\cdot}OX, \qquad B = XU^{\cdot}OY;^{*}$$

that is to say, let A be the intersection of the two given lines YU, OX, and let B be the intersection of the other given lines XU, OY. Let P be any variable or *arbitrary point* in the given plane, and let Q, R be its *projections* on the two given *axes*, OX, OY, obtained by drawing lines *through the given points* Y and X; so that, in symbols, $Q = YP^{\cdot}OX$, and $R = XP^{\cdot}OY$. Finally, let x, y, z be *numbers* (positive or negative, as the case may be), such that

$$\frac{x}{z} = \frac{OQ}{QX} \cdot \frac{XA}{AO}; \qquad \frac{y}{z} = \frac{OR}{RY} \cdot \frac{YB}{BO}.$$

Then I define that x, y, z, (or any numbers proportional to them) are the *three anharmonic coordinates of the point* P, with respect to the given (or assumed) triangle ABC, and the given point U, which may (as in 3.) be called the *unit-point*, because by the definition, its three coordinates are *equal* to each other, and may therefore be taken as 1, 1, 1, or more fully as

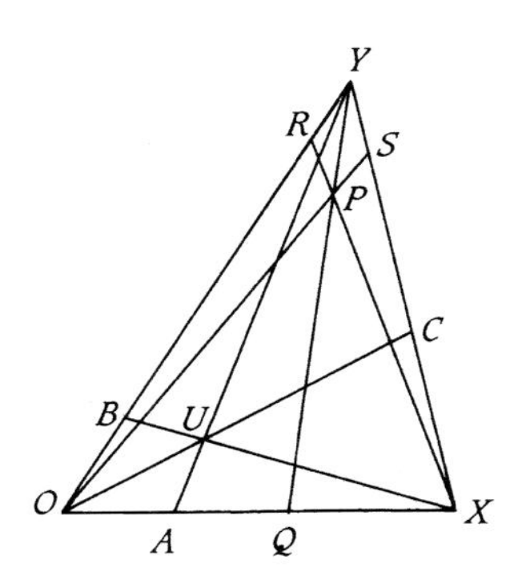

Fig. 2

* [The $^{\cdot}$ here denotes intersection.]

each $= +1$. And it is evident that when the line XY goes off to infinity, as in Figure 1, and when we assume $z = 1$, we thus fall back on the *simple ratios*, or quotients,

$$x = \frac{OQ}{OA}, \quad y = \frac{OR}{OB};$$

which are of the Cartesian kind, if we further suppose that the two given lines OA and OB are *equally long*. *In general*, however, $OAUB$ may be *any plane quadrilateral*; and the points X and Y may be deduced from it, as the *intersections of its opposite sides*, by the formulae, $X = OA^{\cdot}BU$, $Y = OB^{\cdot}AU$; so that we might speak of the point P as being *referred to this quadrilateral $OAUB$*, of which the corners, expressed by their coordinates, are as follows:

$$O = (0, 0, 1), \quad A = (1, 0, 1), \quad B = (0, 1, 1), \quad U = (1, 1, 1);$$

at least if we choose, as we may, that z shall be $= 1$ for each of them.

5. The usage respecting the symbol $(ABCD)$, considered as denoting an *anharmonic ratio*, or *quotient*, appearing to be not quite fixed as regards the *order* of the letters, or of the points, it is proper that I should state precisely what significance I attach to that symbol. If, then A, B, C, D be any collinear *group* of four points, I write,

$$(ABCD) = \frac{AB}{BC} \cdot \frac{CD}{DA} = \frac{AB}{CB} : \frac{AD}{CD}, \qquad \&c.,$$

so that

$$\begin{aligned}(ABCD) &= (BCDA)^{-1} = (CDAB) = (DABC)^{-1} = (DCBA) \\ &= (CBAD)^{-1} = (BADC) = (ADCB)^{-1} = \text{say, } a;\end{aligned}$$

and if this anharmonic be thus denoted by a, we have

$$\begin{aligned}(ACBD) &= (CBDA)^{-1} = (BDAC) = (DACB)^{-1} = (DBCA) \\ &= (BCAD)^{-1} = (CADB) = (ADBC)^{-1} = 1 - a,\end{aligned}$$

because

$$\frac{AB}{BC} . CD + \frac{AC}{CB} . BD = DA;$$

whence also,

$$\begin{aligned}(ACDB) &= (CDBA)^{-1} = (DBAC) = (BACD)^{-1} = (BDCA) \\ &= (DCAB)^{-1} = (CABD) = (ABDC)^{-1} = 1 - a^{-1}.\end{aligned}$$

These known relations between the 24 anharmonics of a group were quite *familiar* to Möbius, who seems to have invented, for himself, the whole doctrine of anharmonic ratio* (*Doppelschnittsverhältniss* = *ratio bissectionalis*†), without even knowing (as historical) the property of a *pencil* cited from Pappus by Chasles,‡ as I mentioned in a former letter (par.). I write

* [See reference to Möbius in article 1, pp. 243–263.]
† [The same references, p. 244 §182.]
‡ [Michel Chasles, 1793–1880.]

them here to have them ready, and partly because the symbol $(ABCD)$ is not always *interpreted* as above: it was used differently by Möbius, for example. In the present notation, the equation $(ABCD) = -1$, or $(ACBD) = 2$, or $(ADBC) = \frac{1}{2}$, &c., expresses of course that D is *harmonically conjugate* to B, with respect to A and C; a relation which Möbius would have expressed by $(ACBD) = -1$, writing *conjugates together*, whereas I prefer to *separate* them. – I may just remark, in passing, that even if $ABCD$ be a gauche* *quadrilateral*, I still write

$$(ABCD) = \frac{AB}{BC} \cdot \frac{CD}{DA},$$

as above; and that I then say that this *product of two quotients, interpreted as quaternions*, is the ***Anharmonic Quaternion*** *of that quadrilateral.* In like manner, I write

$$(ABCDEF) = \frac{AB}{BC} \cdot \frac{CD}{DA} \cdot \frac{EF}{FA}.$$

and call this the *Evolutionary Function*, or briefly, the *evolutionary*, of the system of the *six points* A, B, C, D, E, F; extending this also, by quaternions, to space, and therefore to gauche hexagons.

6. With the same *arrangements* of the letters, if the four points A, B, C, D be collinear, and if four other points A', B', C', D', not necessarily situated on any one right line, be on the four right lines drawn to the former points from any origin O, I naturally write [Figure 3]

$$(O.A'B'C'D') = (ABCD) = \frac{AB}{BC} \cdot \frac{CD}{DA};$$

and I interpret similarly the symbol $(\alpha\beta\gamma\delta)$, for the *anharmonic of the pencil* of the *four rays* α, β, γ, δ, or OA, OB, OC, OD or OA', OB', OC', OD' from any common *vertex* O. Thus,

$$(\alpha\beta\gamma\delta)(\alpha\delta\gamma\beta) = 1 = (\alpha\beta\gamma\delta) + (\alpha\gamma\beta\delta), \text{ etc.}$$

With these notations, the definition (4.) of the anharmonic coordinates of a point P may be expressed as follows:

$$\frac{x}{z} = (OQXA) = (Y.OPXU)$$

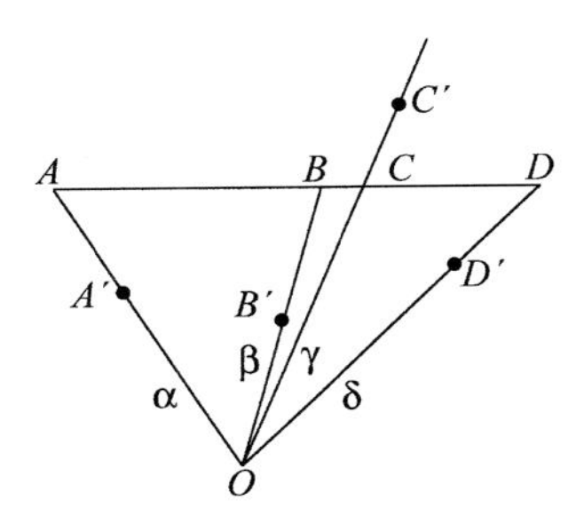

Fig. 3

* [That is, 'not planar'.]

$$\frac{y}{z} = (ORYB) = (X.OPYU);$$

and thus (as in Figure 4), by considering anharmonics of pencils, we need not introduce *expressly* the *projected points ABQR*.

7. If, however, we choose to introduce (as in Figure 2) two *other* projected points, on the third side XY of the triangle, namely

$$C = OU^{\cdot}XY, \quad \text{and} \quad S = OP^{\cdot}XY,$$

then the well known *equations of six segments*,

$$\frac{OA}{AX} \cdot \frac{XC}{CY} \cdot \frac{YB}{BO} = 1, \qquad \frac{OQ}{QX} \cdot \frac{XS}{SY} \cdot \frac{YR}{RO} = 1,$$

show that we have

$$\frac{y}{x} = (ORYB)(OAXQ) = \frac{OR}{RY} \cdot \frac{YB}{BO} \cdot \frac{OA}{AX} \cdot \frac{XO}{QO} = \frac{XS}{SY} \cdot \frac{YC}{CX}$$

or briefly that

$$\frac{y}{x} = (XSYC) = (O.XPYU).$$

Of course we have also,

$$\frac{x}{y} = (O.YPXU); \quad \frac{z}{x} = (Y.XPOU); \quad \frac{z}{y} = (X.YPOU).$$

8. Already we may see several examples of ***anharmonic equations of lines***. Thus, the *sides of the triangle OXY* have for their respective equations,

$$OY, \ x = 0; \qquad OX, \ y = 0; \qquad XY, \ z = 0;$$

and the equations of the three lines drawn to its corners from the point U, are

$$OUC, \ y = x; \qquad YUA, \ x = z; \qquad XUB, \ y = z;$$

while any other right lines through O, Y, X are represented by equations of the slightly more general form,

$$\frac{y}{x} = \text{constant}, \quad \frac{x}{z} = \text{constant}, \quad \frac{y}{z} = \text{constant}.$$

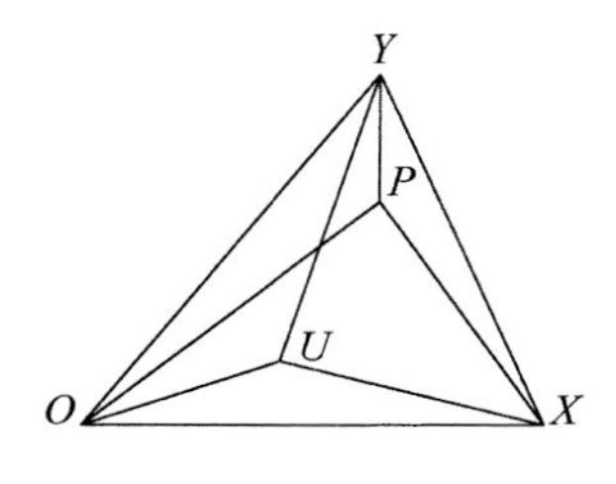

Fig. 4

9. To generalise these results, we may proceed as follows. Let *LMN* be any rectilinear transversal [Λ], cutting the sides *OX, OY, XY,* in the points *L, M, N,* respectively, and passing through the point *P* [see Figure 5]. By a very well known theorem,* respecting the *six segments* made by such transversal, we shall have the equation,

$$\frac{OL}{LX} \cdot \frac{XN}{NY} \cdot \frac{YM}{MO} = -1.$$

But we also have the analogous equation,

$$\frac{OA'}{A'X} \cdot \frac{XC'}{C'Y} \cdot \frac{YB'}{B'O} = -1,$$

if, A', B', C', be the harmonic conjugates of *A, B, C,* with respect to the three sides *OX, XY, YO.* (I am not aiming yet, at any great *symmetry* of notation, but am content to use these three letters, *O, X, Y,* as reminding us of Cartesian coordinates.) Dividing, therefore, the former of these two equations by the latter, we arrive at this other equation:

$$(OLXA')\,(XNYC')\,(YMOB') = 1;$$

which allows us to assume,

$$\frac{m}{n} = (YMOB'), \quad \frac{n}{l} = (OLXA'), \quad \frac{l}{m} = (XNYC'),$$

or

$$\frac{n}{m} = (OMYB'), \quad \frac{l}{n} = (XLOA'), \quad \frac{m}{l} = (YNXC');$$

where *l, m, n,* are three new numbers, of which the ratios depend upon, and conversely characterise, the *position* of the *transversal* line [Λ]; and which (or any numbers proportional to them) I propose to call the ***Anharmonic Coordinates*** *of that* ***Line.*** I also propose to denote the line [Λ], above considered by the **symbol** [*l, m, n*]; for example, the three points A', B', C', are situated on one right line [Υ], of which the *symbol* is [1, 1, 1], and which I therefore

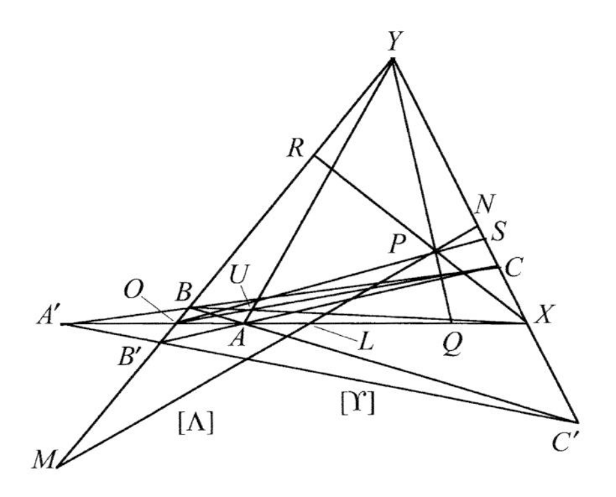

Fig. 5

* [Theorem of Menelaus; see, for example, D. Pedoe *Geometry*, p. 28, Cambridge University Press, Cambridge: 1970.]

propose to call the ***Unit-Line***. It follows that, in a known sense, (compare Salmon, *Higher Plane Curves*, page 149*) the *Unit-Point* U, and the ***Unit-Line*** $[\Upsilon]$ are related to each other, and to the given triangle OXY, as Pole and Polar.

10. I next *combine* the coordinates of *point* and *line*, as follows. Multiplying the three anharmonics

$$\frac{l}{n} = (LXA'O), \quad \frac{x}{z} = (OQXA), \quad -1 = (OAXA'),$$

we get the product,

$$-\frac{l}{n}\frac{x}{z} = (LXQO) = (OQXL);$$

where $Q = (x, 0, z)$ is derived as before from $P = (x, y, z)$, namely from a point of the line $[\Lambda]$, by drawing a line through Y to AX. Multiplying, in like manner, the anharmonics,

$$\frac{m}{n} = (MYB'O), \quad \frac{y}{z} = (ORYB), \quad -1 = (OBYB'),$$

we get this other product,

$$-\frac{m}{n}\frac{y}{z} = (MYRO) = (ORYM) = (OXQL);$$

the last transformation being effected by means of the pencil through P. But (by article 5.),

$$(OQXL) + (OXQL) = 1;$$

therefore

$$lx + my + nz = 0:$$

that is to say, "if the anharmonic coordinates x, y, z of any point P be multiplied by the anharmonic coordinates l, m, n of any right *line* $[\Lambda]$ through that point, *the sum of the products is zero*". And according as we conceive, Ist, a *variable point* P to move *along* a *given right line* $[\Lambda]$, or IInd, a *variable right line* $[\Lambda]$ to *turn round a given point* P, (so as always to pass *through* that point,) we may regard this homogeneous equation of the first degree,

$$lx + my + nz = 0,$$

as being either I, the anharmonic and *local equation of the line*, or II, the *anharmonic and tangential equation of this point*. For example, the *local* equation of the *unit-line* $[\Upsilon]$ is,

$$x + y + z = 0;$$

and the *tangential* equation of the *unit-point*, U, is,

$$l + m + n = 0.$$

And already we may see that *any homogeneous equation* of the pth *degree*,

$$f(x, y, z) = 0$$

* [G. Salmon, *A treatise on the higher plane curves*, Hodges and Smith, Dublin: 1852; hereinafter referred to as: *HPC*.]

between the anharmonic coordinates of a point, is the *local equation* of a *curve* of the pth *order*: because it is met by any given right line $[l,\ m,\ n]$ in p points, real or imaginary.

11. Consider, in particular, this equation of the *second degree,*

$$xy = z^2,$$

which, (by what has just been said) must represent a *conic.* The line OX or (by article 8.) $y = 0$, meets it at two coincident points, or *touches* it at the point X; and in like manner the line OY, or $x = 0$, touches the conic at Y, so that O is the *pole* of the *chord* XY, with reference to the curve [see Figure 6]. The equation shows that the unit-point U is on the conic; but (by article 8.) the local equation of the line OU is $y = x$; the *other* intersection of this line with the curve is therefore $U' = (1,\ 1,\ -1) = (-1,\ -1,\ 1)$; and if $C = OU^{\cdot}XY$, (as before) then the equation of XY being $z = 0$, we have $C = (1,\ 1,\ 0)$.

But in general, if there be any four collinear points,

$$P_0 = (x_0,\ y_0,\ z_0), \quad P_1 = (x_1,\ y_1,\ z_1), \quad P_2 = \&\text{c.}, \quad P_3 = \&\text{c.},$$

so that

$$x_2 = tx_0 + ux_1, \quad y_2 = ty_0 + uy_1, \quad z_2 = tz_0 + uz_1,$$
$$x_3 = t'x_0 + u'x_1, \quad y_3 = t'y_0 + u'y_1, \quad z_3 = t'z_0 + u'z_1,$$

we have the anharmonic,*

$$(P_0 P_2 P_1 P_3) = \frac{t'u}{u't}.$$

* [In his Notebook 160 (Trinity College MS 1492/160) Hamilton, on 6th March 1860, comments: 'The very important *Theorem* of article 11. . . . deserves a clear and separate proof of a purely geometrical character; for I have a proof by *quaternions*.' Such a proof will be found on p. 33 of his *Elements of quaternions*, Longmans, Green, & Co., London: 1866. The expression for $(P_0 P_2 P_1 P_3)$ may be obtained as follows.

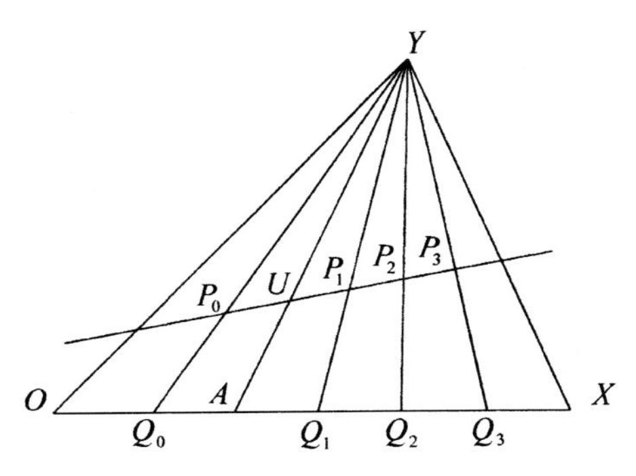

Fig. A

In Figure A the points P_0, P_1, P_2, P_3, lie on the same straight line; the line YA passes through the unit-point; the lines YP_0, YP_1, YP_2, YP_3, when projected, intersect the side OX at the points Q_0, Q_1, Q_2, Q_3, respectively. Therefore, the anharmonic $(P_0 P_2 P_1 P_3)$ is equal to the anharmonic $(Q_0 Q_2 Q_1 Q_3)$. By definition

continued overleaf

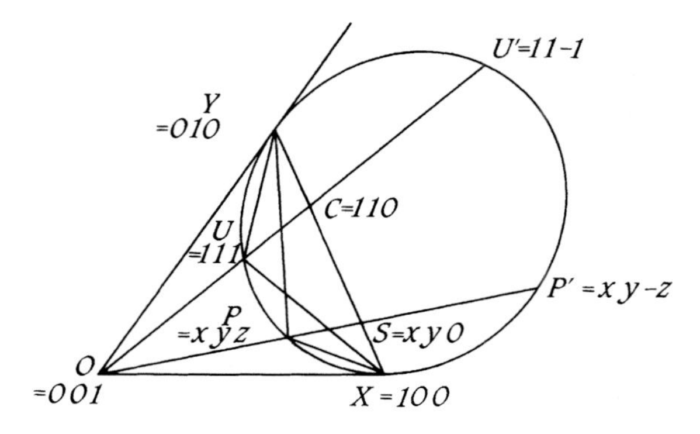

Fig. 6

$$\frac{(OQ_1Q_2X)\,(OQ_0Q_3X)}{(OQ_0Q_2X)\,(OQ_1Q_3X)} = \frac{OQ_1}{Q_1Q_2}\frac{Q_2X}{XO}\frac{Q_0Q_2}{OQ_0}\frac{XO}{Q_0X}\frac{OQ_0}{Q_0Q_3}\frac{Q_3X}{XO}\frac{Q_1Q_3}{OQ_1}\frac{XO}{Q_3X}$$
$$= \frac{Q_0Q_2}{Q_1Q_2}\frac{Q_1Q_3}{Q_0Q_3} = \frac{Q_0Q_2}{Q_2Q_1}\frac{Q_1Q_3}{Q_3Q_0} = (Q_0Q_2Q_1Q_3);$$

by the equations in article 5,

$$(OQ_1Q_2X) = [1-(OQ_2XQ_1)]^{-1},$$

and since

$$(OQ_2XQ_1) = \frac{OQ_2}{Q_2X}\frac{XQ_1}{Q_1O} = \frac{OQ_2}{Q_2X}\frac{XA}{AO}\frac{XQ_1}{Q_1O}\frac{AO}{XA}$$
$$= \frac{OQ_2}{Q_2X}\frac{XA}{AO}\frac{Q_1X}{OQ_1}\frac{AO}{XA} = \frac{(OQ_2XA)}{(OQ_1XA)},$$

we obtain

$$(OQ_1Q_2X) = \left[1-\frac{(OQ_2XA)}{(OQ_1XA)}\right]^{-1},$$

which means that

$$(Q_0Q_2Q_1Q_3) = \frac{[(OQ_0XA)-(OQ_2XA)][(OQ_1XA)-(OQ_3XA)]}{[(OQ_1XA)-(OQ_2XA)][(OQ_0XA)-(OQ_3XA)]}.$$

By article 6.

$$(OQ_0XA) = \frac{x_0}{z_0}, \quad (OQ_1XA) = \frac{x_1}{z_1}, \quad (OQ_2XA) = \frac{x_2}{z_2}, \quad (OQ_3XA) = \frac{x_3}{z_3},$$

which leads to the equation

$$(Q_0Q_2Q_1Q_3) = \frac{(x_0z_2-x_2z_0)(x_1z_3-x_3z_1)}{(x_1z_2-x_2z_1)(x_0z_3-x_3z_0)},$$

and when x_2, x_3, z_2, and z_3, are expressed in terms of x_0, x_1, z_0, and z_1, we finally obtain the relation:

$$(P_0P_2P_1P_3) = (Q_0Q_2Q_1Q_3) = \frac{t'u}{tu'}.\Big]$$

Taking then O, C, U, U' for the four points P_0, P_1, P_2, P_3, so that $t = 1$, $u = 1$, $t' = -1$, $u' = 1$, we have $(OUCU') = -1$, that is to say, "the point C of the *polar*, is *harmonically conjugate* to the *pole* O, with respect to the intercepted *chord* UU'" as was known, I suppose to Apollonius.

12. The equation of the conic may be written thus:

$$\frac{y}{z} = \frac{z}{x}; \quad \text{or, } (X.OPYU) = (Y.XPOU),$$

(see articles 6 and 7). That is to say, (XO and YO being tangents,) "the anharmonic of the *pencil* of which the rays pass through the four points X, P, Y, U of the conic, has the same value, whether we place the *vertex* of that pencil at the point X, or at the point Y, of the conic". Of course, this is merely a particular case of a well known theorem: yet it would be sufficient to give the *anharmonic equation*, $xy = z^2$, from which all other properties of the general conic may be deduced.

13. For example, let OP meet the curve again in P'; while it meets XY in S, as before. Then

$$P = (x, y, z), \qquad S = (x, y, 0), \qquad P' = (x, y, -z);$$

and because $O = (0, 0, 1)$, we have

$$(OPSP') = -1;$$

or P, P' are conjugates, with respect to O and S. This, however, is merely a little exercise of calculation, the well known theorem which it expresses being sufficiently represented by the formula

$$(OUCU') = -1, \text{ of article 11.}$$

14. Let $V = (a, b, c)$ be a new point upon the conic [see Figure 7], so that $ab = c^2$. Let the chords PU, PV meet the chord XY in the points T, W; so that

$$W = (cx - az, cy - bz, 0), \quad \text{and} \quad T = (x - z, y - z, 0).$$

Then because

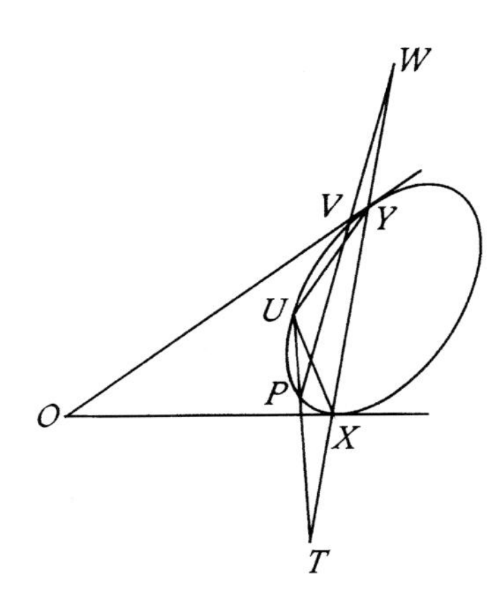

Fig. 7

$$X = (1, 0, 0), \quad \text{and} \quad Y = (0, 1, 0),$$

we have

$$(P.XYVU) = (XWYT) = \frac{(x - y)(cy - bz)}{(y - z)(cx - az)} = \frac{c}{a} = \frac{b}{c};$$

or in words, "the anharmonic of a pencil whose four rays pass through any four given points X, V, Y, U of the conic, and whose vertex P is on that curve, has a value independent of the position of that vertex, or does not change while P moves along the conic": which is the well known property, whereof that in article 12 was a case.

15. We see, at the same time, – what, to say the least, I was not *familiar* with, if I ever met it before, – that this *constant* value of the anharmonic $(P.XVYU)$, is just the common value of $\frac{z}{x}$ and of $\frac{y}{z}$ for the point V; that is, by article 12,

$$(P.XVYU) = (X.OVYU) = (Y.XVOU).$$

However, on a moment's reflection, I see that this comes merely to placing P at X, and at Y, alternately. But I make the little calculation in the foregoing article, without *thinking* of this easy verification.

16. In general, let $f(x, y, z) = 0$ be (as in article 10) the local equation of a curve of the pth order, f being a homogeneous function, rational and whole, and of the pth dimension. Writing

$$df = D_x f.dx + D_y f.dy + D_z f.dz;$$

we have

$$0 = xD_x f + yD_y f + zD_z f;$$

and therefore

$$0 = x' D_x f + y' D_y f + z' D_z f,$$

if

$$x' - x : y' - y : z' - z = dx : dy : dz;$$

that is if the point (x', y', z') be on the tangent to the curve, f, drawn at the point (x, y, z). The anharmonic *coordinates* of this tangent are *therefore* the following:

$$l = D_x f, \quad m = D_y f, \quad n = D_z f;$$

or the tangent itself has for its *symbol* (compare article 9.)

$$[D_x f, D_y f, D_z f].$$

17. For example, when $f = xy - z^2$, as in article 11, we have

$$l = y, \quad m = x, \quad n = -2z;$$

so that the relation, between a point of contact (x, y, z), and a point (x', y', z') upon the tangent at that point, is expressed by the equation,

$$yx' + xy' - 2zz' = 0;$$

which may be called, more generally, the ***Equation of Conjugation***, with respect to the conic $xy = z^2$, as serving to express that the two points (x, y, z) (x', y', z') are *conjugate*, relatively thereto. Thus, if the line at infinity be $[\lambda, \mu, \nu]$ so that the equation of this line is

$$\lambda x + \mu y + \nu z = 0,$$

then the *centre of the conic* is the point $(2\mu, 2\lambda, -\nu)$; and generally, the point $(2\mu, 2\lambda, -\nu)$ is the *pole* of the *line* $[\lambda, \mu, \nu]$ *with respect to the given conic*, $xy = z^2$. For example, the line XY is $z = 0$, or $[0, 0, 1]$; its pole is therefore the point $(0, 0, -1)$, or $(0, 0, 1)$, or O as before. (The pole of $[\lambda, \mu, \nu]$ *with respect to the given triangle OXY*, is the *point* $(\lambda^{-1}, \mu^{-1}, \nu^{-1})$.)

18. The four tangents at X, V, Y, U have for their respective symbols, $[0, 1, 0]$, $[b, a, -2c]$, $[1, 0, 0]$, $[1, 1, -2]$; and they are met by the fifth tangent $[y, x, -2z]$, which is drawn to the conic at P, in the five points $(2z, 0, y)$, $(2az - 2cx, 2cy - 2bz, ay - bx)$, $(0, 2z, x)$, $(2z - 2x, 2y - 2z, y - x)$. Calling these four points X_1, V_1, Y_1, U_1, and comparing the coordinates of V_1 and U_1, with those of X_1 and U_1, as the coordinates of P_2 and P_3, were compared with those of P_0 and P_1 in article 11., we find that

$$t = az - cx, \qquad u = cy - bz, \qquad t' = z - x, \qquad u' = y - z;$$

whence

$$(X_1 V_1 Y_1 U_1) = \frac{t'u}{tu'} = \frac{(z - x)(cy - bx)}{(az - cx)(y - z)} = \frac{c}{a} = \frac{a}{b},$$

as in 14; so that "if four fixed tangents to a conic be cut by any fifth tangent, the anharmonic of the *group* of the four points of section is constant, and is equal to the anharmonic of the *pencil* of four rays, drawn to the four fixed points of contact from any fifth point of the conic": a well known theorem which, as we see, is here obtained, with scarcely any trouble of calculation – I forget whether any particular *symbol* is commonly used, to denote this constant anharmonic; perhaps it might be denoted by the symbol $((XVYU))$, in the case which we have here been considering.

19. We have not much occasion, in the present method, to distinguish between the different sorts of conics; yet it is easy to do so, by considering the line at infinity. Let this line be, as in article 17, $[\lambda, \mu, \nu]$, or $\lambda x + \mu y + \nu z$, where λ, μ, ν are supposed to be given: or in other words let λ^{-1}, μ^{-1}, ν^{-1} be the *given coordinates* of the *mean point* or (in the simplest sense) *the centre of gravity* G, of the *given triangle*, OXY. Then the points of the conic which are at infinity are given by the quadratic,

$$(\lambda x + \mu y)^2 - \nu^2 xy = 0;$$

so that the curve is an *ellipse*, a *parabola*, or a *hyperbola*, according as $\nu^2 <, =,$ or $> 4\lambda\mu$. For example, if $G = (2, 2, 1)$, so that C bisects XY, U bisects OC, then $\lambda = 1$, $\mu = 1$, $\nu = 2$, and the line at infinity is [1, 1, 2], which satisfies the condition $\nu^2 = 4\lambda\mu$, so that the conic is now a parabola; [see Figure 8].

20. The same *condition of parabolic form* may be obtained from the ***tangential equation*** *of the conic*, namely

$$n^2 = 4\,lm;$$

when [l, m, n] is a *variable tangent*, of which the conic is the *envelope*. For then the equation $\nu^2 = 4\lambda\mu$ expresses that the curve has a *tangent at infinity*. Or this last equation may be considered as expressing that the *point* $(2\mu, 2\lambda, -\nu)$ is *on the line* $[\lambda, \mu, \nu]$; and therefore that *the centre* (article 17) of the conic is *infinitely distant.*

21. In general, if the line OUC pass through the mean point G, so that C bisects XY, then $\lambda = \mu$, and

$$\frac{\nu}{\lambda} = \frac{\nu}{\mu}(OGCU) = 2\frac{CU}{UO};$$

if then $CU = UO$, the curve is a parabola, as above, but if $CU < UO$, (U being *on* the *finite* bisector OC,) then the conic is an *ellipse*; and if $CU > UO$ (U being still *on* OC,) the curve is a *hyperbola*; because $(\nu^2/4\lambda\mu)$ is in these three cases, $= 1$, < 1, and > 1, respectively [see Figure 9].

22. The *simplest arrangement* possible, of the four given points, appears to be that in which the triangle OXY is *equilateral* and U its *mean point* (or its *centre*). In this case, the *conic* $xy = z^2$

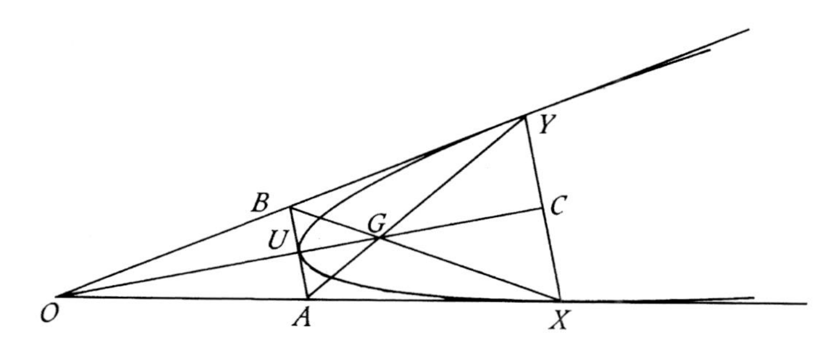

Fig. 8

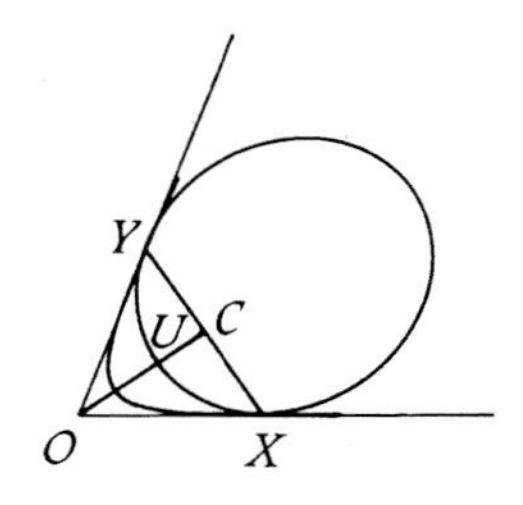

Fig. 9

is easily seen to become a *circle*; of which the *centre* is, by article 17., the point $K' = (2, 2, -1)$,because the *line at infinity* is now the *unit-line* [1, 1, 1] or Υ, as it always is when the *unit-point* (1, 1, 1), or *U*, is the *mean* point of the triangle *OXY* [see Figure 10].

Let *I*, *J* be the *circular points at infinity*; the coordinates of each must satisfy the two equations,

$$xy = z^2, \quad x + y + z = 0;$$

they are therefore the following,

$$I = (1, \theta, \theta^2), \quad J = (1, \theta^2, \theta),$$

where θ is an imaginary cube-root of unity: and *every other conic*,

$$f(x, y, z) = ax^2 + by^2 + cz^2 + 2a'yz + 2b'zx + 2c'xy = 0,$$

which passes through these two points, *I*, *J*, or of which the coefficients satisfy the two equations,

$$f(1, \theta, \theta^2) = 0, \quad f(1, \theta^2, \theta) = 0,$$

or

$$a + 2a' = b + 2b' = c + 2c',$$

is necessarily a *circle*. The *general equation of a circle*, referred to the present system of coordinates, is therefore of the form,

$$a(x^2 - yz) + b(y^2 - zx) + c(z^2 - xy) + e(yz + zx + xy) = 0,$$

when *a*, *b*, *c*, *e*, are any constants. For example,

$$x^2 = yz \text{ is the circle } OUY, \text{ with centre } K'' = (-1, 2, 2);$$

$$y^2 = zx \text{ is the circle } OUX, \text{ with centre } K''' = (2, -1, 2);$$

and $yz + zx + xy = 0$ is the circumscribed circle *OXY*, with *U* for centre: and these *four real circles* have all the same *imaginary intersections*, *I* and *J*, situated upon the unit-line, or at infinity. By making $a = b = c = e$, we get a fifth, but an *imaginary circle*,

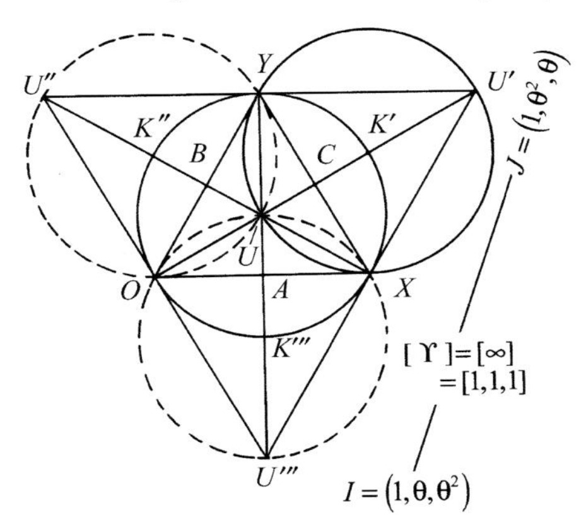

Fig. 10

$$x^2 + y^2 + z^2 = 0;$$

which has this characteristic property, that the *three corners* of the triangle OXY are, with respect to it, the poles of the opposite sides. In general

$$xx' + yy' + zz' = 0,$$

is for this last circle, the *equation of conjugation* (article 17), which connects the coordinates of any two conjugate points, P and P'.

And the *osculating circle* to *any given conic,*

$$0 = Ax^2 + By^2 + Cz^2 + 2A'yz + 2B'zx + 2C'xy,$$

at any given point (x, y, z) is obtained by comparing this last equation with the first and second differentials thereof. – I do not [know] whether the imaginary *cube-roots of unity* have, in any other systems of coordinates, been shown to have a connexion with the *circular points at infinity.*

23. It may not be quite useless to verify that the points I, J, *thus determined*, coincide with the *usual* circular points &c. Their characteristic property being that $x^3 = y^3 = z^3$ for each, *without* our having $x = y = z$, we must, by article 7, have, in particular, the equation

$$(O.XIYU)^3 = 1; \quad (XI'YC) = 1,$$

if I' be the (imaginary) intersection of OI with XY. But with the present arrangement of the given points, $YC = CX$; therefore

$$\left(\frac{X'I}{I'Y}\right)^3 = 1, \quad \left(\frac{XI'}{I'Y}\right)^2 + \frac{XI'}{I'Y} + 1 = 0, \quad XI' = \theta I'Y;$$

but

$$XI' = XC + CI', \quad I'Y = I'C + CY = XC - CI';$$

therefore

$$(1 + \theta)\, CI' = (\theta - 1)\, XC.CI' = (\theta - \theta^2)\, XC = \pm\sqrt{-3}.XC = \pm\sqrt{-1}.OC,$$

or,

$$\tan UOI = \pm\sqrt{-1},$$

which is the known and distinguishing property of the *circular points, usually so called.*

24. In general, let OXY be any triangle, inscribed in any conic [see Figure 11]. Draw tangents at the corners YA', XB', OC', meeting the respectively opposite sides in the points A', B', C', which will therefore be on one right line. From these points draw the other tangents, $A'A''$, $B'B''$, $C'C''$, and join YA'', XB'', OC''; these three last chords will intersect each other in one point U; and they will cross the sides OX, OY, XY in the points A, B, C, which are evidently the harmonic conjugates of A', B', C'. In short the line $A'B'C'$ will be the *polar line* of the *point* U, and the curve $OA''XC''YB''$ will be the polar conic of that point, with respect to the triangle OXY. (*HPC*, p. 149). All this being admitted, let OXY be taken for the *anharmonic triangle* of article 4, of which the sides OY, OX, XY have for their respective equations (by

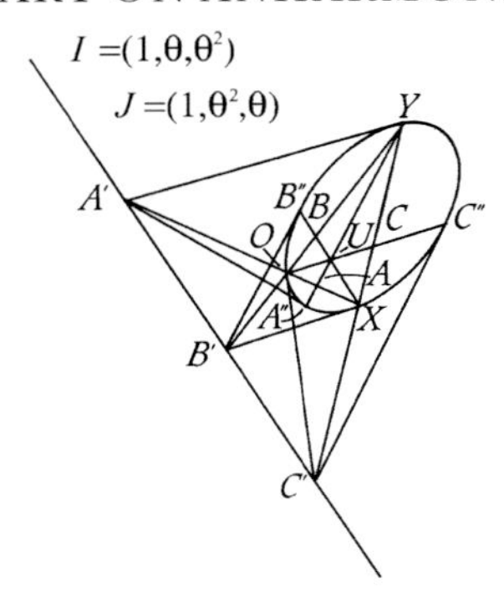

Fig. 11

article 8), $x = 0$, $y = 0$, $z = 0$; and let the point U be taken for the anharmonic unit-point, $x = y = z$; its *polar line* $A'\ B'\ C'$ will become the *anharmonic unit line*, $x + y + z = 0$; and its *polar conic* will have for its *anharmonic equation*,

$$yz + zx + xy = 0,$$

or

$$x^{-1} + y^{-1} + z^{-1} = 0;$$

on which account I am disposed to call it, with reference to the present system of coordinates, the *Unit-Conic*, or briefly the ***Unit-Curve*** of the system. In the case of Figure 10, this unit curve was the *circumscribed circle OXY*, and in general, it is that particular circumscribed *conic*, which *touches* (as above) the three lines $A'Y$, $B'X$, $C'O$; A', B', C' being points upon the unit line, determined as in Figure 5; and the *unit-point U* being the *pole* of the *unit-line*, with respect to the *unit-curve*.

25. If now we seek the *intersections*, *I* and *J*, of the *real* unit-*line* $A'B'C'$ with the *real* unit-*curve*, or with the conic *OXY*, we find that these intersections are *always imaginary*; and that they may *still*, as in article 22, be represented by the symbols,

$$I = (1,\ \theta,\ \theta^2), \quad J = (1,\ \theta^2,\ \theta),$$

although they are *not now* (in general) the *circular points at infinity*, because we do not *now* assume the particular *arrangements* of Figure 10, or of articles 22, 23. And the equation

$$a(x^2 - yz) + b(y^2 - zx) + c(z^2 - xy) + e(yz + zx + xy) = 0,$$

represents now (in general) *not a circle, but a conic*, which had the *same pair of imaginary intersections I, J with the unit-line, as the unit-curve.* To this new *system of conics* belongs therefore the curve $xy = z^2$, which was considered in several former articles; and also the two other real curves, $yz = x^2$, $zx = y^2$, which likewise pass through the point *U*, and of which the first is *touched* at *O* and *Y*, but the second at *O* and *X*, by *sides of the* ***unit-triangle***: for so I think that we may conveniently call the triangle *OXY*.

The points *I*, *J* may perhaps be called the ***Imaginary Unit-Points***: *U* being then called, by contrast, the ***Real Unit-Point***. And the *imaginary conic*, $x^2 + y^2 + z^2 = 0$, (compare article 22) which passes through the two imaginary unit-points, and with respect to which each corner of the unit-triangle is the pole of the opposite side, may, in consequence of these relations, and of the simple form of its equation, be called the ***Imaginary Unit-Curve.*** More generally, I am disposed to say that all the conics represented (as above) by a *homogeneous and linear equation* between the *four quadratic functions,*

$$x^2 - yz, \quad y^2 - zx, \quad x^2 - xy, \quad yz + zx + xy,$$

are curves of the ***Unit-System.***

26. Another general form for the anharmonic equation of such curves is the following:

$$(\alpha x + \beta y + \gamma z)(x + y + z) = yz + zx + xy$$

where α, β, γ are any three constants. Let

$$(\alpha' x + \beta' y + \gamma' z)(x + y + z) = yz + zx + xy$$

be the equation of a second curve of the same system; then of their four intersections, two are *always imaginary,* and are on the unit-line as a *common chord,* namely the points I and J; and the other two, whether real or imaginary, are situated on the *second common chord,*

$$(\alpha' - \alpha)x + (\beta' - \beta)y + (\gamma' - \gamma)z = 0$$

which is *always real,* if the curves be such, and may be said to be their anharmonic ***radical axis.*** The point for which

$$\alpha x + \beta y + \gamma z = \alpha' x + \beta' y + \gamma' z = \alpha'' x + \beta'' y + \gamma'' z,$$

may be called the (anharmonic) ***radical centre*** of the *three* curves of the unit-system, which have $\alpha, \beta, \gamma, \alpha', \beta', \gamma'$, and $\alpha'', \beta'', \gamma''$ for their constants.

27. If

$$u = f(x, y, z) = Ax^2 + \&c. + 2A'yz \ \&c., \quad \text{and} \quad l = Ax + C'y + B'z, \ \&c.$$

so that

$$lx + my + nz = f(xyz) \text{ and } l = \frac{1}{2} D_x f, \ \&c.$$

then the equation

$$f(x, y, z).f(x', y', z') = (lx' + my' + nz')^2,$$

considered as an equation in x', y', z', represents the *pair of tangents* drawn to the curve $f(x, y, z) = 0$, from the point (x, y, z). If this pair passes through I and J, x', y', z' must admit of being equated either to $1, \theta, \theta^2$ or to $1, \theta^2, \theta$. Let therefore

$$f(1, \theta, \theta^2) = a + b\theta^2 + c\theta,$$

so that

$$a = A + 2A', \quad b = B + 2B', \quad c = C + 2C',$$

let also

$$\lambda = l^2 + 2mn, \qquad \mu = m^2 + 2nl,$$

so that

$$(l + m\theta + n\theta^2)^2 = \lambda + \mu\theta^2 + \nu\theta;$$

then the system of the *two real equations of the 2nd degree,*

$$au - \lambda = bu - \mu = cu - \nu,$$

will determine *four points* (x, y, z), such that the *two (imaginary) tangents* from *any one* of them, to *the conic* $u = 0$, *will pass through the two imaginary unit-points,* I and J. I therefore propose to call these four points the ***Four Anharmonic Foci*** *of that conic.* They become the *four ordinary foci,* when I and J become (as in articles 22, 23) *the two circular points at infinity.*

28. With the recent significations of $a, b, c, \lambda, \mu, \nu$, and u, if α, β, γ, be any three constants, such that

$$\alpha + \beta + \gamma = 0,$$

the equation

$$\alpha(au - \lambda) + \beta(bu - \mu) + \gamma(cu - \nu) = 0$$

represents a conic, passing through the four anharmonic foci. And if we take

$$\alpha = b - c, \qquad \beta = c - a, \qquad \gamma = a - b,$$

this *new conic* breaks up into a pair of *real right lines.* For its equation is then,

$$(b - c)(l^2 + 2mn) + (c - a)(m^2 + 2nl) + (a - b)(n^2 + 2lm) = 0,$$

when (by article 27) a, b, c are known and real constants, and l, m, n are known and real functions of x, y, z, homogeneous and of the first degree: but one of the many possible transformations of this last equation is the following,

$$\{(b - c)l + (a - b)m + (c - a)n + h(m - n)\}$$
$$\times \{(b - c)l + (a - b)m + (c - a)n + h(n - m)\} = 0,$$

when

$$h^2 = a^2 + b^2 + c^2 - bc - ca - ab > 0,$$

so that h is a *real constant,* depending on the given and *real conic,* u. The two distinct and real right lines,

$$(b - c)l + (a - b)m + (c - a)n + h(m - n) = 0,$$
$$(b - c)l + (a - b)m + (c - a)n + h(n - m) = 0,$$

on which the four anharmonic foci are thus situated, I propose to call the ***Two Anharmonic Axes of the conic,*** $u = 0$; and the real point of intersection of these two axes, namely the point for which $l = m = n$, *and which is* therefore the *pole of the unit-line,* with respect to that conic u, may perhaps be not inconveniently called the ***Anharmonic Centre*** *of that conic.* – In the case of article 22, or of Figure 10, when the *unit-line* goes off to *infinity,* and the *unit-curve* becomes a *circle,* these *anharmonic axes* and *centre* become the *usual* axes and centre of the conic u.

29. As an *example,* let the conic of which we wish to determine the foci, &c., have for its (local) equation,

$$u = x^2 + y^2 - z^2 = 0;$$

so that

$$A = B = -C = 1, \quad A' = B' = C' = 0, \quad a = b = -c = 1$$

and

$$l = x, \quad m = y, \quad n = -z.$$

We have now $h^2 = 4$, and we may write $h = 2$; after which the equations of the *anharmonic axes of the conic* become (by article 28),

$$l + m - 2n = 0, \quad l - m = 0;$$

or, substituting for l, m, n, their values, and transposing,

$$x - y = 0, \quad x + y + 2z = 0;$$

so that the *anharmonic centre* of the conic is the point $(1, 1, -1)$, of which the *polar,* with respect to the same conic, is the unit line $[1, 1, 1]$, as by the general theory it ought to be. *One* of the conics through the *four anharmonic foci is*

$$au - \lambda = bu - \mu, \quad \text{or} \quad \lambda - \mu = 0;$$

but this is merely the *pair of axes,*

$$(l - m)(l + m - 2n) = 0.$$

If however we take this other conic through those foci,

$$0 = (a + b - 2c)u + 2\nu - \lambda - \mu$$

or

$$0 = 4u + 2n^2 - l^2 - m^2 + 4lm - 2n(l + m)$$

or

$$0 = 3(x^2 + y^2) - 2z^2 + 4xy + 2yz + 2zx,$$

we are conducted, by the equations of the axes, to the following quadratics:

$$\begin{cases} \text{Ist, } 5x^2 + 2xz - z^2 = 0, \text{ for the axis } y = x; \text{ and} \\ \text{IInd, } x^2 + 2xz + 3z^2 = 0, \text{ for the axis } x + y + 2z = 0. \end{cases}$$

The Ist gives the *two real foci,* $(1, 1, 1 \pm \sqrt{6})$; and the IInd gives the *two imaginary foci,* $(-1 \pm \sqrt{-2}, -1 \pm \sqrt{-2}, 1)$. Let F, F' be the two real foci; they are on the line OUC, of several former figures, and we have the two anharmonics,

$$(CFOU) = \left(\frac{z}{x} =\right) 1 + \sqrt{6}, \quad (CF'OU) = 1 - \sqrt{6};$$

whence (by article 5),

$$(COFU) = -\sqrt{6}, \quad (COF'U) = +\sqrt{6}.$$

But, if the unit-line be *at infinity*, *U trisects OC* and $\frac{CO}{UC} = -3$; therefore, in this case,

$$\frac{FU}{OF} = +\sqrt{\frac{2}{3}}, \quad \frac{F'U}{OF'} = -\sqrt{\frac{2}{3}};$$

whence also

$$\frac{OF}{OU} = \frac{\sqrt{3}}{\sqrt{3}+\sqrt{2}} = 3 - \sqrt{6}, \quad \text{and} \quad \frac{OF'}{OU} = 3 + \sqrt{6};$$

the anharmonic *centre* $K = (1, 1, -1)$ of the conic being also such, that $OK = 3OU = 2OC$ because $(CKOU) = -1$; and the curve being a *hyperbola*, with $x + z = 0$ and $y + z = 0$, or YA' and XB', for its *asymptotes*; and with $x - z = 0$, and $y - z = 0$, or YA and XB for *tangents* at finite distances. Its *anharmonic summits*, on the axis $y = z$, are

$$V = (1, 1, \sqrt{2}), \quad V' = (1, 1, -\sqrt{2});$$

whence

$$(CVOU) = \sqrt{2}, \quad \frac{CV}{VO} = \sqrt{\frac{1}{2}}, \quad \frac{OV}{OC} = \frac{\sqrt{2}}{1+\sqrt{2}} = 2 - \sqrt{2},$$

$$\frac{OV}{OU} = 3 - \frac{3}{\sqrt{2}}, \quad \frac{VK}{OU} = \frac{3}{\sqrt{2}}; \quad \text{but} \quad \frac{FK}{OU} = \sqrt{6};$$

therefore

$$\frac{KF}{KV} = \frac{2}{\sqrt{3}}$$

[which] may be called the *anharmonic excentricity* of the hyperbola.

30. To fix more closely the conceptions, and to verify the results by comparison with other methods, let O be taken for the origin $(0, 0)$ of a system of *Cartesian* and *rectangular coordinates*; and let U be the point $(1, 0)$ on the axis of x; so that the points $ABCXYK$ are now denoted as follows:

$$A = \left(\frac{3}{4}, -\frac{\sqrt{3}}{4}\right); \quad B = \left(\frac{3}{4}, \frac{\sqrt{3}}{4}\right); \quad C = \left(\frac{3}{2}, 0\right);$$

$$X = \left(\frac{3}{2}, -\frac{\sqrt{3}}{2}\right); \quad Y = \left(\frac{3}{2}, \frac{\sqrt{3}}{2}\right); \quad K = (3, 0);$$

the triangle OXY being now again supposed (as in article 22) to be *equilateral*, and U to be its *mean point*, while $XOYK$ is a parallelogram [see Figure 12].

The equations of the right lines KA', KB' are now

$$x - 3 = \mp y\sqrt{3};$$

and therefore the *equation of the hyperbola*, which has these lines for its asymptotes, and which has

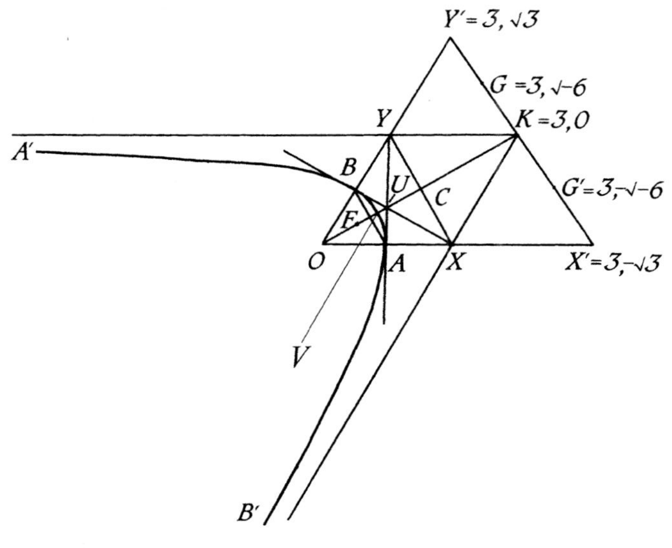

Fig. 12

$$V = \left(3 - \frac{3}{\sqrt{2}},\ 0\right)$$

for a point upon it, is

$$(x-3)^2 - 3y^2 = \frac{9}{2}.$$

The usual methods show next that the point V is a *summit*: and that the lines UA, UB, of which the joint equation is

$$3(x-1)^2 - y^2 = 0,$$

are *tangents* to the curve; as, by article 29, they ought to be. In the general equation

$$0 = \left(x^2 - 3y^2 - 6x + \frac{9}{2}\right)\left(x'^2 - 3y'^2 - 6x' + \frac{9}{2}\right) - \left(xx' - 3yy' - 3x - 3x' + \frac{9}{2}\right)^2,$$

represents the *pair of tangents* to the hyperbola, drawn from any given point (x', y'); and if this point be a *focus*, this pair must pass through the *circular points at infinity*, for which we have *now* $x = \infty$, $y = \pm x\sqrt{-1}$; substituting therefore these last values, we get the two equations,

$$x'^2 - y'^2 - 6x' + 3 = 0, \quad \text{and } y'(x'-3) = 0,$$

as the conditions by which the *four foci* are to be determined: these four are therefore the two *real points*, $F = (3 - \sqrt{6}, 0)$, $F' = (3 + \sqrt{6}, 0)$, and the two *imaginary* points, $G = (3, \sqrt{-6})$,

$G' = (3, -\sqrt{-6})$. The *real foci, thus determined,* are seen at once to *coincide* with those which were otherwise found in the preceding article; because they give

$$\frac{OF}{OU} = 3 - \sqrt{6}, \quad \frac{OF'}{OU} = 3 + \sqrt{6},$$

as before.

And to verify the agreement of the *imaginary foci,* as found by these two systems of coordinates, we may prolong OX, OY to meet the conjugate axis GG' of the hyperbola, in the points $X'(3, 3 - \sqrt{3})$, $Y' = (3, 3 + \sqrt{3})$; and thus shall have the *imaginary anharmonic*

$$(O.XAYU) = (X'AY'K) = \frac{\sqrt{3} + \sqrt{-6}}{\sqrt{3} - \sqrt{-6}} = \frac{1 + \sqrt{-2}}{1 - \sqrt{-2}}$$

which exactly agrees with the imaginary value of $\frac{y}{x}$, for the first imaginary focus in article 29, although found by so different an analysis. Finally, the *real excentricity,* $\frac{KF}{KV}$, is again seen to be

$$\frac{\sqrt{6}}{3.\sqrt{\frac{1}{2}}} = \frac{2}{\sqrt{3}},$$

as in that article.

31. Resuming our *anharmonic coordinates,* and making no particular supposition respecting the *shape* of the given triangle, OXY, or respecting the *position* of the given point U; writing also for greater symmetry, the equation (article 26) of the curves of the *unit-system* under the form

$$(\alpha x + \beta y + \gamma z)(x + y + z) + \delta(yz + zx + xy) = 0$$

we see that the (anharmonic) *centre* (a, b, c) of any such curve, considered as being determined by the equations $l = m = n$, must satisfy the conditions

$$(\alpha + \delta)(a + b + c) - \delta a = (\beta + \delta)(a + b + c) - \delta b = (\gamma + \delta)(a + b + c) - \delta c$$

so that we may write,

$$\delta = a + b + c, \quad \alpha = a + e, \quad \beta = b + e, \quad \gamma = c + e,$$

where e is a new constant. The equation of the curve thus becomes,

$$0 = (ax + by + cz)(x + y + z) + (a + b + c)(yz + zx + xy) + e(x + y + z)^2;$$

and we see that "*any two* (anharmonically) *concentric curves of the unit-system have double contact with each other, at the two imaginary unit-points, I* and *J, in which they intersect the unit-line*". – It is easy also to prove that for any curve of this system "the *four* (anharmonic) *foci close up, into the* (anharmonic) *centre*"; in fact, the two equations of the second degree (article 27), by which those foci are determined become, in this case, $\lambda = \mu = \nu$, and can only be satisfied by supposing $l = m = n$.

32. Supposing next that α, β, γ are any three new constants, of which (as in article 28) the sum is zero, so that $\alpha + \beta + \gamma = 0$, and that the point (α, β, γ) is situated upon the unit-line, we may propose to join this point to the (anharmonic) centre (a, b, c) of the recent conic, and to find the points (x, y, z) in which the joining line will intersect that curve. For this purpose, we may write

$$x = a + t\alpha, \quad y = b + t\beta, \quad z = c + t\gamma,$$

and substitute these expressions for x, y, z, in the last equation of the conic, so as to obtain a quadratic in t, which must evidently be a pure one, on account of the existing *harmonic* relations; and in fact the coefficient of $(a + b + c)t$, in the result of this substitution, is

$$(a\alpha + b\beta + c\gamma) + (b\gamma + c\beta) + (c\alpha + a\gamma) + (a\beta + b\alpha) = (a + b + c)(\alpha + \beta + \gamma) = 0.$$

If we make $k = a + b + c + e$, so that the equation of the curve becomes

$$0 = a(y^2 + yz + z^2) + b(z^2 + zx + x^2) + c(x^2 + xy + y^2) - k(x + y + z)^2,$$

the complete result of the substitution is easily found to be

$$\tfrac{1}{2}(\alpha^2 + \beta^2 + \gamma^2)t^2 = k(a + b + c) - (bc + ca + ab);$$

division by $a + b + c$ being performed, and $\alpha\beta + \beta\gamma + \gamma\alpha$ being changed to its equivalent, $-\frac{1}{2}(\alpha^2 + \beta^2 + \gamma^2)$. According, then, as

$$k(a + b + c) >, \quad =, \quad \text{or} < bc + ca + ab,$$

the two values of t are real and unequal, or equal and null, or unequal but imaginary; and therefore the curve will, in these three respective cases, be *real* and *finite*, or *evanescent*, or *imaginary*. For example, if $a = b = c = 1$, so that the centre of the curve is at the unit point, the (finite) reality, evanescence, or imaginariness of that curve will depend on the constant k being $>$, $=$, or < 1. Accordingly, when $k = 2$, we get the *real unit-curve* (article 24), namely the conic for which $yz + zx + xy = 0$; when $k = \frac{1}{2}$, we get the *imaginary unit-curve* (article 25), $x^2 + y^2 + z^2 = 0$; and when $k = 1$, we get what may be called the ***Evanescent Unit-Curve***,

$$x^2 + y^2 + z^2 + -xy - yz - zx = 0,$$

which has *no real point, except the unit-point.*

33. These *evanescent curves of the unit-system,* which *correspond to infinitesimal circles,* and *become* such, when the triangle OXY is made an equilateral one by projection, and the point U is made its mean point, – may deserve a somewhat closer attention. Eliminating the constant k from the last written equation of the curve, by the condition

$$k(a + b + c) = bc + ca + ab,$$

we are conducted without difficulty to an equation, which breaks up into two *linear* (but imaginary) *factors*, and may be written thus:

$$W.W' = 0$$

when

$$W = (bz - cy) + \theta(cx - az) + \theta^2(ay - bx),$$

$$W' = (bz - cy) + \theta^2(cx - az) + \theta(ay - bx),$$

and

$$\theta^2 + \theta + 1 = 0,$$

as before.

This *evanescent curve* may therefore be considered as a *pair of imaginary lines,*

$$W = 0, \quad W' = 0,$$

which are drawn *from the given centre* (a, b, c) or K to the *two imaginary unit-points, I* and *J*, or $(1, \theta, \theta^2)$ and $(1, \theta^2, \theta)$.

34. Curves of the unit-system occur so often in these researches, that we want a shorter name for them. To call them ***Anharmonic Circles***, (because they won't be *projected into circles*, under conditions above assigned,) might be too great an innovation; suppose then that we agree to name them *generally*, ***Unit-Curves***: calling, then, for distinction, the three *chief* curves of this system, which have been above considered, namely,

$$yz + zx + xy = 0, \quad x^2 + y^2 + z^2 = 0, \quad x^2 + y^2 + z^2 - yz - zx - xy = 0,$$

the ***principal real unit-curve***, the ***principal imaginary unit-curve*** and the ***principal evanescent unit-curve***, respectively. We may then say (by article 33), that "the evanescent unit-curve, with *any* given point K as centre, is equivalent to the pair of imaginary right lines, KI, KJ."

35. If an evanescent unit-curve have its *centre* at a *focus* of a conic, the two imaginary rectilinear *branches* of the first are two of the *tangents* of the second; the two curves must therefore be considered to have **double contact** with each other; namely, at the *two imaginary points*, where the conic is cut by that *real right line*, which is, with respect to it, the *polar of its focus*: and which I therefore propose to call its ***Anharmonic Directrix***, corresponding to that focus. – I remember *staring*, ten or twelve years ago, when Townsend* told me, in what I call *Brinkley's Garden*,† that a plane conic has double contact with each of its foci, considered, of course, as infinitesimal *circles*; and he admitted that my i, j, k, which were then comparatively recent, had nothing in them more *paradoxical*, on the first appearance. You see that I have made a step or two forward since then, aided (no doubt) *mainly* by Salmon's books, – though I know *something* of foreign ones also, – in what is called "Modern Geometry": but I have the deepest feeling of my inferiority, in that respect, to persons who have made the subject their special study.

36. If V be any linear and homogeneous function of the coordinates x, y, z, then the equation

$$WW' - V^2 = 0$$

represents a conic, which has the real intersection K of the two imaginary right lines $W = 0$, $W' = 0$ for one focus, and the right line $V = 0$ for the directrix corresponding. – An analogous theorem exists, of the kind here considered, for *two foci on one axis*, which answers

* [Richard Townsend (1821–1884), fellow of Trinity College, Dublin.]
† [John Brinkley (1763–1835) was Hamilton's predecessor at the Dunsink Observatory as Andrews professor of astronomy and as Astronomer Royal for Ireland.]

to the usual *sum-or-difference* property of the conic sections: but you will dispense with my writing it down, at least at present, for it is just possible that I may return to that part of the subject, before the present long letter is finished.

37. You conceive that several other things of the same sort, in this (perhaps new) system of coordinate analysis, which however I can *connect* with *quaternions*, or rather which I am *now detaching from them*, – have occurred to me. But to vary the illustrations, let us next investigate the *tangential equation* (compare article 20) of the general *unit-curve*, taking the form (article 32),

$$u = a(y^2 + yz + z^2) + b(z^2 + zx + x^2) + c(x^2 + xy + y^2) - k(x + y + z)^2 = 0,$$

for the *local equation* of that curve. Differentiating, we have (by article 16)

$$l = D_x u = b(z + 2x) + c(y + 2x) - 2k(x + y + z),\ m = \&c.,\ n = \&c.;$$

between which three linear equations, and $lx + my + nz = 0$, we are to eliminate x, y, z. Making, for a moment,

$$x + y + z = t,\ ax + by + cz = v, \quad \text{and} \quad a + b + c = g,$$

$$w = gt - v - 2kt = a(y + z) + b(z + x) + c(x + y) - 2k(x + y + z),$$

we have

$$l = w + gx - at,$$

$$m = w + gy - bt,$$

$$n = w + gz - ct.$$

Hence

$$l + m + n = 3w;$$

$$al + bm + cn = \{g(w + v) - (a^2 + b^2 + c^2)t = (g^2 - a^2 - b^2 - c^2 - 2kg)t\}$$

$$= +2(bc + ca + ab - gk)t;$$

$$l^2 + m^2 + n^2 = w(l + m + n) - (al + bm + cn)t;$$

and now we have only to eliminate t and w between these three last equations. Writing, for abridgement,

$$\Lambda = l^2 + m^2 + n^2 - lm - mn - nl = (l + \theta m + \theta^2 n)(l + \theta^2 m + \theta n)$$

the result of this elimination is easily seen to be

$$4\{k(a + b + c) - (bc + ca + ab)\}\Lambda = 3(al + lm + cn)^2;$$

so that, Λ can never be negative, the *condition of reality* of the general unit curve (article 32) is again found to be expressed by the inequality,

$$k(a + b + c) > bc + ca + ab.$$

As an example, the constants of the *principal real unit-curve* (article 33) are $a = b = c = 1$, $k = 2$; so that while the *local* equation of this curve is, as before,

$$yz + zx + xy = 0,$$

the *tangential* equation of the same curve is,

$$4\Lambda = (l + m + n)^2,$$

or,

$$l^2 + m^2 + n^2 - 2mn - 2nl - 2lm = 0.$$

Accordingly, this last form expresses, that the two tangents from any corner of the unit triangle coincide, and have the positions assigned in (article 24); thus, the tangential equation of the point O is $n = 0$, which gives $(l - m)^2 = 0$; and the *unique* tangent, $[1, 1, 0]$, or $x + y = 0$, is the line OC', as before. – In general, the form of the tangential equation, assigned above, for an unit-curve, may be verified by observing that it shows the *points I, J* to be *on* the curve, and the *tangents* at those points to *meet* in K.

38. Writing, for abridgement,

$$k(a + b + c) - (bc + ca + ab) = 3\kappa^2,$$

the general tangential equation (article 37) of an unit-curve becomes,

$$4\kappa^2\Lambda = (al + bm + cn)^2$$

a, b, c being still the coordinates of the centre K, and Λ being still the same quadratic and essentially positive function of the coordinates l, m, n of the tangent, as before. The tangential equation of a second unit-curve is, in like manner,

$$4\kappa'^2\Lambda = (a'l + b'm + c'n)^2;$$

and if we combine these two equations, in order to determine the *common tangents* to these two curves, we find that these *four tangents* pass, two by two, through two ***Anharmonic Centres of Similitude***, namely through the two points,

$$S = (\kappa a' - \kappa' a,\ \kappa b' - \kappa' b,\ \kappa c' - \kappa' c),$$

$$S' = (\kappa a' + \kappa' a,\ \kappa b' + \kappa' b,\ \kappa c' + \kappa' c),$$

If either of these two points be on the unit-line, so that

$$\kappa(a' + b' + c') = \pm\kappa'(a + b + c),$$

then the two curves may be said to be ***Anharmonically Equal***, because they would be *projected into equal circles*, by throwing off the unit line to ∞, &c. For example, if we make $a' = b' = c' = \kappa' = 1$, the second curve becomes, by article 37, the principal real unit-curve; if then $3\kappa = \pm(a + b + c)$, or if the first curve have for its tangential equation,

$$4(a + b + c)^2\Lambda = 9(al + bm + cn)^2,$$

it will be, in the foregoing sense, *anharmonically equal* to that principal unit-curve.

39. Without *yet* requiring any *general method* for deducing the *local equation* of a curve from its *tangential equation*, we can easily here obtain the local equation which corresponds to any given values of the four constants a, b, c, κ, by eliminating κ from the equation in article 32, with the help of the relation (article 38) between k and κ: a process which gives easily,

$$WW' = 3\kappa^2(x + y + z)^2,$$

as the *local form* of the tangential equation (article 38), namely,

$$4\kappa^2\Lambda = (al + bm + cn)^2.$$

For example, when $\kappa = 0$, the local equation is $WW' = 0$, as in article 33; and the tangential equation becomes $(al + bm + cn)^2 = 0$, implying that "every tangent to an evanescent unit-curve passes through the centre of that curve". In general, the *local form*, assigned in the present article, expresses, (when compared with the equation $WW' = V^2$ in article 36,) that "an unit-curve is a conic which has its *centre for a focus*, and the *unit-line* for a directrix": whence we might see again, that such a curve would be projected into a *circle*, under conditions already assigned. And if we make $3\kappa = (a + b + c)r$, so that the curve is represented, at pleasure, by either of the two equations,

$$3WW' = r^2(a + b + c)^2(x + y + z)^2,$$

$$4r^2(a + b + c)^2\Lambda = 9(al + bm + cn)^2,$$

then the *new constant*, r, which is equal to *unity* for the principal real unit-curve, and which has a *common value* for all *anharmonically equal* unit-curves, may be called the ***Anharmonic Radius*** of the unit curve to which it belongs.

40. To justify more completely this designation of that constant r, we may resume the analysis of article 32, eliminating the constant k, and introducing r in its stead. We then obtain

$$\tfrac{3}{2}t^2(\alpha^2 + \beta^2 + \gamma^2) = r^2(a + b + c)^2,$$

with the signification of t in that article; hence the same right line, KL, or $(a, b, c)(\alpha, \beta, \gamma)$, drawn from the centre to the unit-line, which meets [see Figure 13] a given unit-curve (a, b, c, r) in the point $P = (a + t\alpha, b + t\beta, c + t\gamma)$, meets a *concentric* unit-curve (a, b, c, r') in the point $P' = \left(a + \frac{r'}{r}t\alpha,\ b + \frac{r'}{r}t\beta,\ c + \frac{r'}{r}t\gamma\right)$; and therefore, by article 11, we have the anharmonic $(KP'LP) = r'/r$: so that when L is thrown off to infinity, we have the simple ratio,

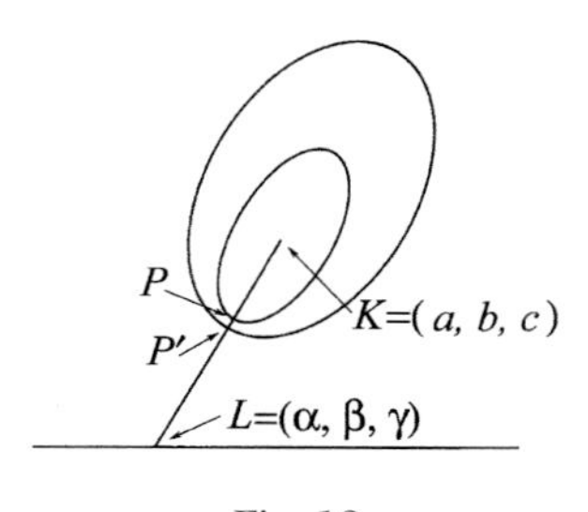

Fig. 13

$$\frac{KP'}{KP} = \frac{r'}{r}.$$

41. If we develop the product WW' in article 33, we find

$$WW' = (b^2 + bc + c^2)x^2 + (c^2 + ca + a^2)y^2 + (a^2 + ab + b^2)z^2 + (a^2 - bc)yz$$
$$+ (b^2 - ca)zx + (c^2 - ab)xy - (bc + ca + ab)(yz + zx + xy)$$
$$= \varphi(x, y, z, a, b, c) = \varphi(a, b, c, x, y, z) > 0,$$

except in the case $x : y : z = a : b : c$, for which we have $(a, b, c, a, b, c) = 0$. And we see that the function,

$$(r =)\psi(x, y, z, a, b, c) = \frac{\pm\sqrt{3WW'}}{(a + b + c)(x + y + z)},$$

which in like manner does not change its value when x, y, z are interchanged with a, b, c, may naturally be said to be an ***Anharmonic Distance***, *between the* ***two points***, $P = (x, y, z)$, and $K = (a, b, c)$. In fact, when we project, as before, OXY into an equilateral triangle, and U into its mean point, and take for unity the length of the radius UO of the circumscribed circle (Figure 10), then the *numerical value* of r or ψ will *remain unchanged by such projection*, as *depending solely on anharmonic quotients*; but, *in the new state of the figure*, it will *represent* the *length of the projected line*, $K'P'$, if K' be the projection of the point K, and P' of P. And similarly, this other function,

$$(r =)\chi(l, m, n, a, b, c) = \frac{\pm 3(al + bm + cn)}{2(a + b + c)\Lambda^{\frac{1}{2}}},$$

may be said to represent the *Anharmonic Distance of the Point* (a, b, c) *from the Line* $[l, m, n]$; or the *Length of the* ***Anharmonic Perpendicular***, *let fall from that point on that line*. For here again the *numerical value* of r, or of the function χ, will remain *unaltered by projection*; but will *come* to denote the *length* of the *perpendicular* of the usual kind, let fall *from the projected point, on the projected line*.

42. As an example, let

$$a = b = c = 1, \quad y = z = 0, \quad x = 1, \quad l = m = 0, \quad n = 1;$$

so that

$$K = U, \quad P = X, \quad \text{and the line } [l, m, n] \text{ is } z = 0, \quad \text{or } XY.$$

The formulæ of article 41 give $W.W' = 3$, and $r = \varphi = 1$, for the anharmonic *distance* from K to P, or from U to X. They give also, (because we have now $\Lambda = 1$) $r = \chi = \frac{1}{2}$, for the anharmonic *perpendicular* from U on XY. Accordingly, when OXY is projected into an equilateral triangle, &c., as in Figure 10, the *length of UX* becomes *equal* to that of UO; but the perpendicular UC on XY has a length only *half* as great.

43. It comes naturally to be noticed in passing, that if we take, as constants, the values $a = b = c = 1$, $r = \frac{1}{2}$, we are conducted to the local and tangential equations following,

$$x^2 + y^2 + z^2 = 2(yz + zx + xy), \quad mn + nl + lm = 0,$$

whereof the latter may be written thus,

$$l^{-1} + m^{-1} + n^{-1} = 0,$$

and which represent what we may call the *inscribed unit-curve*, touching the sides of the unit-triangle in the points A, B, C of former figures, and having its centre at the unit point U [see Figure 14]. When the triangle is made equilateral by projection, and the unit-line thrown off to infinity, this curve becomes of course the *inscribed circle* with a *radius UA* equal to *half* the *unit-radius UO*, of the *circumscribed circle OXY*; which may be considered as an illustration of the result (article 39) respecting the *anharmonic radius* of an unit-curve. In general, (compare *HPC* p. 153,) the curve represented as above by the tangential equation,

$$\frac{1}{l} + \frac{1}{m} + \frac{1}{n} = 0,$$

is the polar conic of the unit-line, with respect to the unit-triangle.

44. We have hitherto only passed *from local to tangential equations* of curves; but it is in general as easy to return, or to pass in the opposite order, from their tangential to their local equations. Let the given tangential equation be

$$0 = F(l, m, n),$$

and its differential

$$0 = dF = D_l F . dl + D_m F . dm + D_n F . dn;$$

comparing this with the equation

$$0 = xdD_x F + ydD_y F + zdD_z F,$$

or

$$0 = xdl + ydm + zdn,$$

which results from article 16, we see that

$$x : y : z = D_l F : D_m F : D_n F;$$

so that the symbol, $(D_l F, D_m F, D_n F)$ represents the *point of contact*, of the *variable right line* $[l, m, n]$, with its envelope, $F = 0$. The same result may be obtained, without any *previous* consideration of the *local* equation $f = 0$, by seeking the *intersection* (x, y, z) of a *given* line $[l, m, n]$ with a *consecutive* line of the same system; or the condition of *equal roots* in that

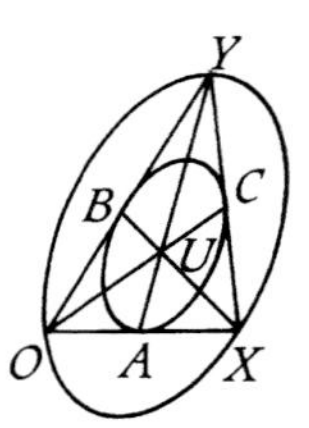

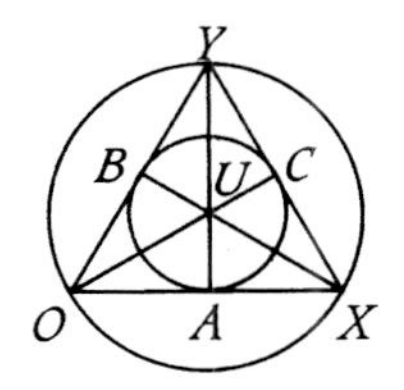

Fig. 14

homogeneous equation in l, m, which results, from the elimination of n, between the given equation $F=0$, and the linear equation $lx+my+nz=0$. With the help, then, of this last equation we can pass either from the tangential equation $F=0$, to the local equation under the form $f(D_lF, D_mF, D_nF)=0$, or from the local equation $f=0$, to the tangential equation $F(D_xf, D_yf, D_zf)=0$, as exemplified in former articles.

45. Examples are, of course, unnecessary: the *work* being exactly of the same kind, as in calculations with *trilinear* coordinates. It may, however, just be noticed, as we pass, that from the recent *tangential form* (article 43),

$$l^{-1}+m^{-1}+n^{-1}=0,$$

of the principal *inscribed* unit-curve ABC, we pass at once to the following *local form* of the same curve,

$$x^{\frac{1}{2}}+y^{\frac{1}{2}}+z^{\frac{1}{2}}=0;$$

and that, in like manner, from the local equation (article 24),

$$x^{-1}+y^{-1}+z^{-1}=0$$

of the *circumscribed* unit curve OXY, we pass with equal ease to the corresponding tangential equation of the curve,

$$l^{\frac{1}{2}}+m^{\frac{1}{2}}+n^{\frac{1}{2}}=0;$$

where by elimination of the radicals, we get, for these two curves,

$$x^2+y^2+z^2-2xy-2yz-2zx=0,$$

and

$$l^2+m^2+n^2-2lm-2mn-2nl=0,$$

as in articles 43 and 37.

46. Differentiating the last general form (article 39), of the tangential equation of any unit-curve with its centre at (a, b, c), we have (by article 44) for the coordinates of the point contact of such a curve with a line $[l, m, n]$ which satisfies that equation of it or for the coordinates of the *foci of the anharmonic perpendicular* from (a, b, c) on $[l, m, n]$ the expressions:

$$x=2r^2(a+b+c)^2D_l\Lambda-9a(al+bm+cn),$$
$$y=2r^2(a+b+c)^2D_m\Lambda-9b(al+bm+cn),$$
$$z=2r^2(a+b+c)^2D_n\Lambda-9c(al+bm+cn).$$

Let $[l';\ m',\ n']$ be the right line $(a, b, c)(x, y, z)$, or KP, which is drawn from the centre K to the point of contact P; we shall have $al'+bm'+cn'=0$, and $xl'+ym'+zn'=0$; where the coordinates of the two lines must be connected by the equation:

$$0=l'D_l\Lambda+m'D_m\Lambda+n'D_n\Lambda;$$

or, expanding,

$$2(ll' + mm' + nn') = l'(m + n) + m'(n + l) + n'(l + m).$$

Such, then, is the *equation of condition* of what may be called ***Anharmonic Perpendicularity*** considered as a *relation between two lines* which *become perpendicular* to each other (in the usual sense), *by projection*. It enables us to *erect* or *let fall* an *anharmonic perpendicular*, *at* or *from* a given point, to or on a given right line; and so to deal, if we choose, *anharmonically*, with problems respecting *normals* and *evolutes*, &c.

47. ***Anharmonic Parallelism*** may be said to exist between two lines which *meet upon the unit-line*; the *parallels* to $[l, m, n]$ are therefore in this system represented by symbols of the form, $[l + h, m + h, n + h]$, where h is any constant. The *anharmonic direction* of a line $[l, m, n]$, – or of the line into which it is to be projected, – can therefore depend only on the *ratios of the differences of the coordinates* l, m, n of that line. Accordingly, if we make

$$l_1 = 2l - m - n, \quad m_1 = 2m - n - - l, \quad n_1 = 2n - l - m,$$
$$l_1' = 2l' - m' - n', \quad m_1' = 2m' - n' - l', \quad n_1' - 2n' - l' - m',$$

so that $[l_1, m_1, n_1]$ is (anharmonically) parallel to $[l, m, n]$ and $[l_1', m_1', n_1']$ to $[l', m', n']$, the recent *equation of perpendicularity* may be thus written:

$$l_1 l_1' + m_1 m_1' + n_1 n_1' = 0;$$

the new lines $[l_1, m_1, n_1]$ and $[l_1', l_1', l_1']$, passing through the unit-point, because their coordinates satisfy the conditions,

$$l_1 + m_1 + n_1 = 0, \quad l_1' + m_1' + n_1' = 0.$$

48. These new coordinates, l_1, m_1, n_1 have for the sum of their squares,

$$l_1^2 + m_1^2 + n_1^2 = 6\Lambda;$$

if then we make

$$\lambda = l_1(l_1^2 + m_1^2 + n_1^2)^{-\frac{1}{2}},$$
$$\mu = m_1(l_1^2 + m_1^2 + n_1^2)^{-\frac{1}{2}},$$
$$\nu = n_1(l_1^2 + m_1^2 + n_1^2)^{-\frac{1}{2}}$$

we shall have

$$l_1 = \lambda\sqrt{6\Lambda}, \quad m_1 = \mu\sqrt{6\Lambda}, \quad n_1 = \sqrt{6\Lambda},$$

and

$$3l = \lambda\sqrt{6\Lambda} + l + m + n, \quad 3m = \mu\sqrt{6\Lambda} + l + m + n, \quad 3n = \nu\sqrt{6\Lambda} + l + m + n.$$

I call these new quantities, λ, μ, ν, which are connected by the two relations

$$\lambda + \mu + \nu = 0, \lambda^2 + \mu^2 + \nu^2 = 1,$$

and of which the geometrical signification will afterwards appear, the ***Unit-Point Coordinates of the Line*** $[l, m, n]$, or of *any line* which is (article 47) *anharmonically parallel* thereto; in fact, they do not change, in passing from a line to a parallel; and the line $[\lambda, \mu, \nu]$, (which has the same position as the recent line $[l_1, m_1, n_1]$) is a line through the unit-point.

49. I am disposed to write also,

$$\xi = \frac{x}{x+y+z}, \quad \eta = \frac{y}{x+y+z}, \quad \zeta = \frac{z}{x+y+z},$$

and

$$\alpha = \frac{a}{a+b+c}, \quad \beta = \frac{b}{a+b+c}, \quad \gamma = \frac{c}{a+b+c};$$

and to call these new coordinates, which satisfy the two equations,

$$\xi + \eta + \zeta = 1, \quad \alpha + \beta + \gamma = 1,$$

the ***Unit-Coordinates*** *of the* ***Points***,

$$P = (x, y, z), \quad \text{and } K = (a, b, c),$$

for reasons which will be seen hereafter. Adopting in the mean time these names and symbols, we see (by articles 41 and 48) that the *anharmonic perpendicular* p, from (x, y, z) on $[l, m, n]$, may be expressed as follows:

$$p = \frac{3(\alpha l + \beta m + \gamma n)}{2\sqrt{\Lambda}} = \sqrt{\frac{3}{2}}(\alpha\lambda + \beta\mu + \gamma\nu) + \frac{l+m+n}{2\sqrt{\Lambda}};$$

where the first part of the expression represents the perpendicular from (α, β, γ) on $[\lambda, \mu, \nu]$, and the second part represents the perpendicular from $(1, 1, 1)$ on $[l, m, n]$. And it is not difficult to prove that the expression (article 41) for the *anharmonic distance*, r, between the two points K and P, becomes in unit-line coordinates of those points, the following:

$$r = \sqrt{\frac{3}{2}}\{(\xi-\alpha)^2 + (\eta-\beta)^2 + (\zeta-\gamma)^2\}^{\frac{1}{2}}.$$

50. Suppose now that the line $[l, m, n]$, passes, with any arbitrary direction, through the unit-point, or that $l + m + n = 0$, and that the point P coincides with that unit-point, U, so that

$$\xi = \eta = \zeta = \tfrac{1}{3};$$

while K shall still remain an arbitrary point, with any unit-line coordinates α, β, γ of which the sum is zero. Let λ', μ', ν' be the unit-point coordinates of the line UK, so that

$$\alpha\lambda' + \beta\mu' + \gamma\nu' = 0$$

as well as

$$\lambda' + \mu' + \nu' = 0, \quad \text{and} \quad \lambda'^2 + \mu'^2 + \nu'^2 = 1.$$

We shall then have

$$\frac{\beta-\gamma}{\lambda'}=\frac{\gamma-\alpha}{\mu'}=\frac{\alpha-\beta}{\nu'}=\sqrt{\{(\alpha-\beta)^2+(\beta-\gamma)^2+(\gamma-\alpha)^2\}}$$

$$=\sqrt{3}\sqrt{(\alpha^2+\beta^2+\gamma^2-\tfrac{1}{3})}=\sqrt{3}\sqrt{\{(\alpha-\tfrac{1}{3})^2+(\beta-\tfrac{1}{3})^2+(\gamma-\tfrac{1}{3})^2\}}=\sqrt{2},$$

where r is (by article 49) the anharmonic distance UK of the point (α, β, γ) from the unit-point $(\frac{1}{3}, \frac{1}{3}, \frac{1}{3})$; so that we have the expressions

$$\lambda'=\frac{\beta-\gamma}{r\sqrt{2}}, \quad \mu'=\frac{\gamma-\alpha}{r\sqrt{2}}, \quad \nu'=\frac{\alpha-\beta}{r\sqrt{2}}$$

which enables us to *distinguish* between the line $[\lambda', \mu', \nu']$ itself considered as having a *given direction*, and the line $[-\lambda', -\mu', -\nu']$ of which the direction is the *opposite*. Conversely, we have, for the unit-line coordinates (article 49) of a point K upon this line, the expressions

$$\alpha=\frac{1}{3}+\frac{r\sqrt{2}}{3}(\nu'-\mu'), \quad \beta=\frac{1}{3}+\frac{r\sqrt{2}}{3}(\lambda'-\mu'), \quad \gamma=\frac{1}{3}+\frac{r\sqrt{2}}{3}(\mu'-\lambda').$$

51. Substituting these last values for α, β, γ, in the expression (article 49) for the perpendicular p let fall from K on the line $[\lambda, \mu, \nu]$, we are conducted to the formula:

$$\frac{p\sqrt{3}}{r}=\lambda(\nu'-\mu')+\mu(\lambda'-\nu')+\nu(\mu'-\lambda')$$

$$=(\mu\lambda'-\lambda\mu')+(\nu\mu'-\mu\nu')+(\lambda\nu'-\nu\lambda').$$

But the three parts of this last expression are equal, because

$$(\mu\lambda'-\lambda\mu')-(\nu\mu'-\mu\nu')=\mu(\nu'+\lambda')-\mu'(\nu+\lambda)=-\mu\mu'+\mu'\mu=0, \text{ \&c.};$$

hence,

$$\frac{p}{r}=(\mu\lambda'-\lambda\mu')\sqrt{3}=(\nu\mu'-\mu\nu')\sqrt{3}=(\lambda\nu'-\nu\lambda')\sqrt{3}.$$

Hence also,

$$\left(\frac{p}{r}\right)^2=(\mu\lambda'-\lambda\mu')^2+(\nu\mu'-\mu\nu')^2+(\lambda\nu'-\nu\lambda')^2$$

$$=(\lambda^2+\mu^2+\gamma^2)(\lambda'^2+\mu'^2+\nu'^2)-(\lambda\lambda'+\mu\mu'+\nu\nu')^2$$

$$=1-(\lambda\lambda'+\mu\mu'+\nu\nu')^2;$$

and therefore

$$\frac{r^2-p^2}{r^2}=(\lambda\lambda'+\mu\mu'+\nu\nu')^2.$$

I am therefore conducted to say that the ***Sine of the Anharmonic Angle*** φ *between the* ***two lines*** $[\lambda, \mu, \nu]$, $[\lambda', \mu', \nu']$, is

$$\sin\varphi=(\mu\lambda'-\lambda\mu')\sqrt{3};$$

and that the ***Cosine*** of the same *anharmonic angle* is

$$\cos\varphi = \lambda\lambda' + \mu\mu' + \nu\nu'.$$

Accordingly, we have (by articles 47 and 48) the equation,

$$\lambda\lambda' + \mu\mu' + \nu\nu' = 0,$$

as the *condition of anharmonic perpendicularity.*

52. If we place K at O, so that

$$\alpha = 0, \quad \beta = 0, \quad \gamma = 1, \quad r = 1,$$

the expressions (article 50) for λ', μ', ν' become,

$$\lambda' = \frac{-1}{\sqrt{2}}, \quad \mu' = \frac{1}{\sqrt{2}}, \quad \nu' = 0$$

so that the recent sine and cosine become

$$\sin\varphi = -(\lambda+\mu)\sqrt{\frac{3}{2}}, \quad \cos\varphi = \frac{\mu-\lambda}{\sqrt{2}};$$

and therefore the unit-point coordinates λ, μ, ν of any line $[l, m, n]$ can be expressed in terms of the angle φ, as follows:

$$\lambda = \sqrt{\frac{2}{3}}\sin(\varphi - 120°), \quad \mu = \sqrt{\frac{2}{3}}\sin(\varphi + 120°), \quad \nu = \sqrt{\frac{2}{3}}\sin\varphi;$$

with the equations of verification,

$$\lambda + \mu + \nu = 0, \quad \lambda^2 + \mu^2 + \nu^2 = 1.$$

And if we make in like manner,

$$\lambda_1 = \sqrt{\frac{2}{3}}\sin(\varphi_1 - 120°), \quad \mu_1 = \sqrt{\frac{2}{3}}\sin(\varphi_1 + 120°), \quad \nu_1 = \sqrt{\frac{2}{3}}\sin\varphi_1;$$

then

$$(\mu\lambda_1 - \lambda\mu_1)\sqrt{3} = \sin(\varphi - \varphi_1), \quad \text{and} \quad \lambda\lambda_1 + \mu\mu_1 + \nu\nu_1 = \cos(\varphi - \varphi_1).$$

53. These results enable us to establish easily *connexions* between *anharmonic* and *rectangular* or *polar coordinates,* of the *usual* kinds, and so to *pass,* at pleasure, from either to the other. Let ζ, η, ξ be the *unit-line* anharmonic coordinates (article 49) of a *point* P; and let λ, μ, ν be the unit-point anharmonic coordinates (article 48) of the *line* UP, which is drawn to that point P from the unit-point. Adopting then the expressions (article 52), which I shall here write again in order to have all the necessary *Elements of Transformation* in *one view* before us,

$$\lambda = \sqrt{\frac{2}{3}}.\sin\left(\varphi - \frac{2\pi}{3}\right),$$

$$\mu = \sqrt{\frac{2}{3}}.\sin\left(\varphi + \frac{2\pi}{3}\right),$$

$$\nu = \sqrt{\frac{2}{3}}.\sin\varphi,$$

or making

$$\cos\varphi = \frac{\mu - 2}{\sqrt{2}}, \qquad \sin\varphi = -(\mu + \lambda)\sqrt{\frac{3}{2}} = \nu\sqrt{\frac{3}{2}},$$

we see that the *directions* of this line *UP* depends on the value of the *auxiliary angle* φ; which, when thus used, I propose to call the ***Anharmonic Polar Angle***, of the *line UP*, or of the *point P* upon it. And changing, in the expression (article 49) for r, each of the coordinates α, β, γ, (instead of ζ, η, ξ, as in article 50) to $\frac{1}{3}$, so that it becomes ($\xi + \eta + \zeta$ being equal to 1,)

$$\begin{aligned} r &= \sqrt{\frac{3}{2}}\cdot\sqrt{\left\{\left(\xi - \frac{1}{3}\right)^2 + \left(\eta - \frac{1}{3}\right)^2 + \left(\zeta - \frac{1}{3}\right)^2\right\}} \\ &= \sqrt{\frac{1}{2}}\cdot\sqrt{\{(\eta - \zeta)^2 + (\zeta - \xi)^2 + (\xi - \eta)^2\}} \\ &= \sqrt{\{\xi^2 + \eta^2 + \zeta^2 - \eta\zeta - \zeta\xi - \xi\eta\}} \\ &= (\xi + \theta\eta + \theta^2\zeta)^{\frac{1}{2}}(\xi + \theta^2\eta + \theta\zeta)^{\frac{1}{2}}, \end{aligned}$$

whence (compare article 50) we have the expressions

$$\lambda = \frac{\eta - \zeta}{r\sqrt{2}}, \qquad \mu = \frac{\zeta - \xi}{r\sqrt{2}}, \qquad \nu = \frac{\xi - \eta}{r\sqrt{2}},$$

which satisfy the three conditions

$$\lambda + \mu + \nu = 0, \quad \lambda\xi + \mu\eta + \nu\zeta = 0, \quad \lambda^2 + \mu^2 + \nu^2 = 1.$$

I call this quantity r the ***Anharmonic Radius Vector*** of the point (ξ, η, ζ), or P: in fact, it represents what has been called the *anharmonic distance* (article 41), of that point *P from the unit-point U*, which *latter* point I propose to call the ***Anharmonic Origin***, on account of its *central position*, and *properties*, when the system of the four given points *OXYU* is *projected* into what I call the ***Canonical Arrangement***, illustrated by Figure 10 of article 22.

54. Finally, if we write

$$X = r\cos\varphi = (2\zeta - \xi - \eta)\cos\frac{\pi}{3},$$

$$Y = r\sin\varphi = (\xi - \eta)\sin\frac{\pi}{3}$$

which will give

$$r = \nu(X^2 + Y^2), \qquad \varphi = \tan^{-1}\frac{Y}{X},$$

and

$$\xi = \tfrac{1}{3}(1 - X + Y\sqrt{3}),$$
$$\eta = \tfrac{1}{3}(1 - X - Y\sqrt{3}),$$
$$\zeta = \tfrac{1}{3}(1 + 2X),$$
$$\lambda = \frac{-Y - X\sqrt{3}}{r\sqrt{6}},$$
$$\mu = \frac{-Y + X\sqrt{3}}{r\sqrt{6}},$$
$$\nu = \frac{2Y}{r\sqrt{6}},$$

then I call these quantities, X and Y, the ***Anharmonic Rectangular Coordinates*** of the point P, or (ξ, η, ζ); of which point, r and φ are the ***Anharmonic Polar Coordinates.***

55. An important consequence of this theory is, that if we have in any manner been led to form an anharmonic and local equation of any curve, under the form $f(x, y, z) = 0$, when the function f is *homogeneous*, then, because

$$\xi : \eta : \zeta = x : y : z \text{ (by article 49)}$$

we may *substitute* these *unit-line* coordinates, ξ, η, ζ, *or their triples*, for x, y, z, in that equation of the curve: which thus becomes,

$$0 = f(1 - X + Y\sqrt{3}, 1 - X - Y\sqrt{3}, 1 + 2X);$$

when X and Y may be *interpreted* as *Cartesian* and *rectangular coordinates*: because, while *their own numerical values* (compare article 41) remain entirely *unaltered by projection,* they *become such usual coordinates,* when the *system of the four given points, O, X, Y, U* is *projected* (as already said) into the canonical arrangement (article 22).

56. In this new or *projected system,* of *ordinary* and *rectangular* coordinates, the *origin* (compare article 53) is at the projected point U [see Figure 15]; the *positive semiaxis* of X has the *direction* of the line UO, and the *length* of that line is taken for the *unit* of length; the *radius vector* r becomes (compare article 53) the quotient,

$$\frac{\text{length of } UP}{\text{length of } UO},$$

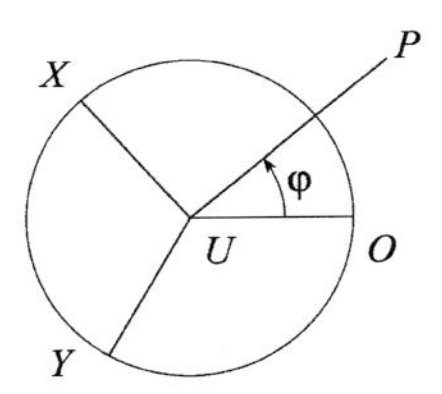

Fig. 15

or simply the numerical expression of the distance of P from U; the *polar angle* φ, or OUP, is *measured positively* from UO towards UX, becoming [equal to] 120° when it has attained this last position, but increasing to +240°, or decreasing to −120°, when UP becomes UY: also the three points O, X, Y are *equally distant* from U, so that OXY is (as in Figure 10) an *equilateral triangle*, with U for *the centre* (or *mean point*) thereof. Thus, if we denote the point P, in this projected figure by the new symbol $\{X, Y\}$, we shall have

$$U = \{0, 0\}; \quad O = \{1, 0\}; \quad X = \left\{-\frac{1}{2}, \frac{\sqrt{3}}{2}\right\}; \quad Y = \left\{-\frac{1}{2}, -\frac{\sqrt{3}}{2}\right\}.$$

57. As a very simple *example* of such transformations, let XY be any chord of any conic, and O its pole, while U and P are any two points of the curve; also let YU, YP meet OX in A and Q, and let XU, XP meet OY in B and R [see Figure 16]. Then, with the *earliest definitions* in this Letter, we saw (articles 11, &c.) that the conic is represented by the local equation $xy = z^2$; implying, as the *immediate consequence* of those definitions, or as the *first interpretation* derived from them, that $(OQXA).(ORYB) = 1$; or that $(ORYB) = (XQOA)$, or that $(X.XPYU) = (Y.XPYU)$ or, in words, that the *two pencils*, $X.XPYU$ and $Y.XPYU$, have *equal anharmonic ratios*. Now let the four points, O, X, Y, U be *canonically projected*, as above, and let it be required to assign the *equation of the projected curve*, in *Cartesian* and *rectangular coordinates*, X and Y. *Substituting* as in article 55, we have at once, for this sought equation, the following form:

$$(1 - X + Y\sqrt{3})(1 - X - Y\sqrt{3}) = (1 + 2X)^2;$$

or

$$X^2 + Y^2 + 2X = 0;$$

the projection is therefore a *circle*; namely the circle XUY of Figure 10: whence we might *infer* that "*every conic can be projected into a circle*", if we were to adopt the foregoing *anharmonic property*, as furnishing the *definition of a conic*.

58. Again, take the form (article 24),

$$x^{-1} + y^{-1} + z^{-1} = 0,$$

of the principal real unit-curve, or of the *polar conic of the unit-point U, with respect to the unit-triangle OXY*. Substituting for x, y, z, as above, we have *immediately*

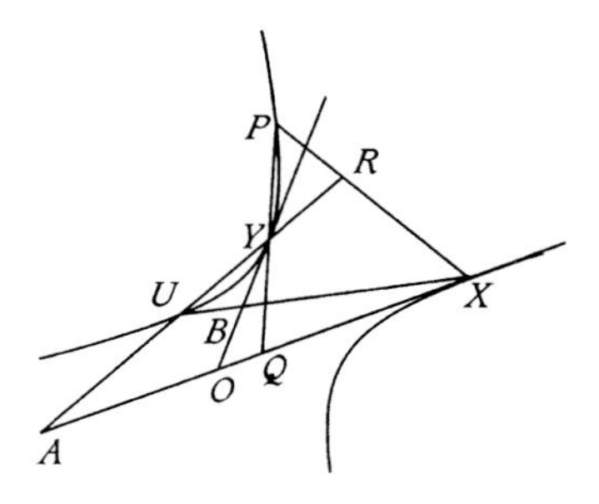

Fig. 16

$$\frac{1}{1 - X + Y\sqrt{3}} + \frac{1}{1 - X - Y\sqrt{3}} + \frac{1}{1 + 2X} = 0,$$

or

$$X^2 + Y^2 = 1;$$

the canonical projection is therefore again a *circle*, namely the *circumscribed* circle *OXY* of Figure 10.

59. More generally, if we make the same substitutions for x, y, z, in the general form (article 26) for the anharmonic equation of *any curve of the unit-system, the result is,*

$$0 = X^2 + Y^2 + (2\gamma - \alpha - \beta)X + \sqrt{3}(\alpha - \beta)Y + \alpha + \beta + \gamma - 1;$$

where α, β, γ are constants; *all unit-curves,* that is (by article 25) "*all curves which pass through the two imaginary unit-points* I and $(1, \theta, \theta^2)$ and J, or $(1, \theta^2, \theta)$ *are therefore*" seen anew to be "*canonically projected into circles*".

60. Although I feel that I owe you an apology for entering into so many details, when doubtless a mere *hint* might have sufficed to let you see my meaning; yet, as I expressed, near the commencement of this letter, an opinion that this ***Anharmonic Method*** might advantageously be introduced into *elementary teaching,* I shall go on for a while, with your hoped-for permission, to show how easily one or two problems of conic sections may be solved by it, through the facility which is conferred by the power of *selecting any four points, O, X, Y, U,* or *any four lines,* such as the sides of the unit-triangle and the conic line, as *data,* in any construction.

61. Suppose, then that we are (for the moment) in the position of persons who have *never heard of conic sections,* but are *familiar* with the elementary geometrical properties of anharmonic ratios; and also, that we have in some manner learned the ***Fundamental Theorem*** of this letter, expressed by the formula of article 10,

$$lx + my + nz = 0$$

which *connects* the *anharmonic coordinates* of *point* and *line.* This *formula* is a *sufficient basis,* whereon to build *all* investigations respecting *local* and *tangential equations of plane curves*; of *any order,* and of any *class*: with *all* their applications, to *poles and polars, inflexions, nodes, cusps, double tangents*; and generally to whatever depends, ultimately, on *collinearity of points,* or upon *concurrence of lines.* We shall find, indeed, that all which relates to *distances, normals, foci, evolutes,* and generally whatever involves in any way the *conception of a circle, requires* (as has been already seen in part) *modification*; and conversely receives (as it appears to me) extension, in the present view.

62. Let there be now proposed the following

Problem:- "To find the *locus* of the *vertex* P of a *pencil,* $P.XOYU$, of which the *rays* pass through *four given points* , X, O, Y, U, (*no three* of these points being *collinear,*) and which has a *given anharmonic ratio,* $= -\frac{b}{a}$, where a and b are given."

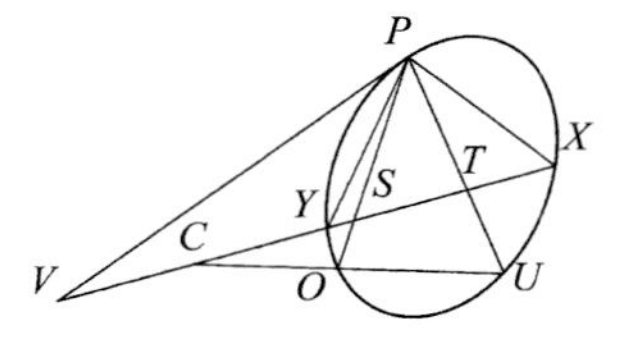

Fig. 17

***Solution*:-** The anharmonic coordinates of the five points being as in this letter, so that

$$O = (0, 0, 1),\ X = (1, 0, 0),\ Y = (0, 1, 0),$$
$$U = (1, 1, 1),\ P = (x, y, z),$$

and therefore

$$S = OP^{\cdot}XY = (x, y, 0)$$

and

$$T = UP^{\cdot}XY = (x - z, y - z, 0),$$

we have, on the plan of article 11,

$$-\frac{b}{a} = (P.XOYU) = (XSYT) = \frac{(x - z)y}{(y - z)x};$$

making therefore for symmetry, $a + b + c = 0$, the *local equation* sought is the following:

$$ayz + bzx + cxy = 0.$$

And because this equation is of the *second dimension* we learn that the sought *locus* is a *curve of the second order,* namely, one which is *cut,* by an *arbitrary transversal,* in *two points,* real or imaginary: while this curve *passes* (as its local equation shows) *through the four given points, O, X, Y, U.*

63. ***Problem*:-** "To find the *tangential equation* of the curve, which is thus the locus of *P*."

***Solution*:-** Writing the local equation (article 62) as follows:

$$0 = -u = \frac{a}{x} + \frac{b}{y} + \frac{c}{z},$$

we find by differentiation (article 16), the expressions,

$$l = D_x u = \frac{a}{x^2}, \quad m = D_y u = \frac{b}{y^2}; \quad n = D_z u = \frac{c}{z^2};$$

whence

$$(al)^{\frac{1}{2}} + (bm)^{\frac{1}{2}} + (cn)^{\frac{1}{2}} = 0,$$

or

$$(al)^2 + (bm)^2 + (cn)^2 - 2al.bm - 2bm.cn - 2cn.al = 0$$

or finally,

$$MN + NL + LM = 0$$

or

$$L^{-1} + M^{-1} + N^{-1} = 0;$$

where

$$L = bm + cn - al, \qquad M = cn + al - bm, \qquad N = al + bm - cn.$$

64. ***Problem***:- "To *determine the tangent at the vertex* P of the pencil; or more fully, to assign the anharmonic *coordinates*, l, m, n, of that tangent, PV, considered as a *line*; and *also*, the anharmonic coordinates, x', y', z', of the point V, in which that tangent crosses the given chord XY."

Solution: The *symbol* of the *tangent* is,

$$[l, m, n] = [ax^{-2}, by^{-2}, cz^{-2}];$$

the *equation* of that *line* PV is,

$$\frac{ax'}{x^2} + \frac{by'}{y^2} + \frac{cz'}{z^2} = 0;$$

and the *symbol* of that *point* V is,

$$V = (bx^2 - ay^2, 0).$$

65. ***Problem***:- To *interpret* the last result.

Solution:- By article 7, &c., if $C = OU^{\cdot}XY$,

$$\frac{y}{x} = (O.XPYU) = (XSYC), \quad \text{and} \quad \frac{y'}{x'} = (XVYC);$$

also

$$-\frac{b}{a} = (XSYC), \text{ by article 62,} \quad \text{and} \quad -\frac{b}{a}.\frac{y'}{x'} = \left(\frac{y}{x}\right)^2, \text{ by article 64,}$$

hence one interpretation is, that

$$(XSYT).(XVYC) = (XSYC)^2.$$

But we can simplify, by observing that

$$\frac{(XVYC)}{(XSYC)} = (XVYS), \quad \text{and} \quad \frac{(XSYC)}{(XSYT)} = (XTYC);$$

whence the equation to be interpreted becomes,

$$(XVYS) = (XTYC), \quad \text{or} \quad (P.XPYO) = (U.XPYO);$$

a result with which, of course, we are familiar, but which is *here* deduced without the supposition of any *previous* knowledge of the doctrine of conic sections.

66. ***Problem*:-** "To deduce the anharmonic property of *tangents* to the locus."
***Solution*:-** Differentiating the form (article 62) of the *local* equation, namely,

$$ayz + bzx + cxy = 0$$

we find that the general tangent PV may be denoted by the symbol

$$[l, m, n] = [bz + cy, cx + az, ay + bx];$$

whence, in particular, the four fixed tangents at X, O, Y, U are,

$$[0, c, b], [b, a, 0], [c, 0, a], [b + c, c + a, a + b].$$

But any two lines, $[l, m, n]$, and $[\lambda, \mu, \nu]$, intersect in the point $(\mu n - \nu m, \nu l - \lambda n, \lambda m - \mu l)$; forming therefore thus the symbols of the four points X_1, O_1, Y_1, U_1, in which the variable tangent PV cuts the four fixed tangents at O, Y, U, and comparing the coordinates of O_1, and U_1 with those of X_1 and Y_1, on the plan of article 11, we find, after a few reductions, and with the recent significations (article 63) of L, M, N, the anharmonic,

$$(X_1 O_1 Y_1 U_1) = \frac{bM(cN - aL)}{aL(cN - aM)} = -\frac{b}{a} = (P.XOYU);$$

so that (as is well known), "the anharmonic ratio of the *group* of *intersections* of *four fixed tangents* with a *variable tangent*, is the same as the anharmonic ratio of the *pencil*, of the *chords* drawn to the *four fixed points of contact*, from the *variable point of contact*". (Compare article 18.)

67. As one more ***Problem***, of this *elementary* kind, let us take the following: "To find the *envelope* of the *line*, which moves so as to be always cut by *four given lines*, (of which *no three* are concurrent,) in a *given anharmonic ratio*".
***Solution*:-** Let the four given lines be the *sides* OY, OX, XY, of the unit-triangle, and the *unit-line* A', B', C' of Figure 5; the intersections of $[l, m, n]$ with these are,

$$M = (0, -n, m),\ L = (n, 0, -l),\ N = (-m, l, 0), \quad \text{and} \quad V = (n - m, l - n, m - l)$$

therefore

$$(LMNV) = \frac{l(n - m)}{m(n - l)} = \text{constant} = (\text{say}) \frac{-a}{b}, \quad \text{with } a + b + c = 0;$$

"the envelope is therefore a *curve of the second class, touching the four given lines*," because its tangential *equation* is,

$$amn + bnl + clm = 0; \quad \text{or} \quad \frac{a}{l} + \frac{b}{m} + \frac{c}{n} = 0;$$

it is therefore also of the *second order*, because its *local equation* is

$$(ax)^{\frac{1}{2}} + (by)^{\frac{1}{2}} + (cz)^{\frac{1}{2}} = 0,$$

or

$$X^{-1} + Y^{-1} + Z^{-1} = 0,$$

where

$$X = by + cz - ax, \qquad Y = cz + ax - by, \qquad Z = ax + by - cz.$$

68. As another example of an *envelope*, which is even easier than the last in *calculation*, but is in *principle* perhaps less elementary, or at least assumes a greater degree of previous acquaintance with the theory, let us take the case of *two unit-curves* (article 24) with *given anharmonic centres*, $K = (a, b, c)$, and $K' = (a', b', c')$, and with a *given product* R of their *anharmonic radii*, r and r'; and let us investigate the tangential equation of the curve which is *always touched* by one of their common tangents. The tangential equation of the two unit curves being, by article 39,

$$4r^2(a + b + c)^2\Lambda = 9(al + bm + cn)^2,$$

$$4r'^2 = (a' + b' + c')^2\Lambda = 9(a'l + b'm + c'n)^2,$$

while

$$rr' = R.$$

The coordinates, l, m, n of a common tangent must be such that

$$\pm 4R(a + b + c)(a' + b' + c')\Lambda = 9(al + bm + cn)(a'l + b'm + c'n);$$

such then is the required *equation of the envelope*; and we see that this curve consists of a *system of two biconfocal conics*, one answering to the upper, and the other to the lower sign; for when the sign is given, the curve is of the *second class*, and has the two given points K and K' for foci. In fact, the tangents from K (for example) satisfy the equation $al + bm + cn = 0$; they therefore satisfy also $\Lambda = 0$, or

$$(l + \theta m + \theta^2 n)(l + \theta^2 m + \theta n) = 0,$$

and consequently they pass through the two imaginary unit-points $(1, \theta, \theta^2)$ and (l, θ^2, θ).

69. If we introduce unit-line coordinates $\alpha, \beta, \gamma, \alpha', \beta', \gamma'$ (article 49) of the two given points K, K', we write $\pm\frac{4}{9}R = \delta$ the recent equation of the envelope becomes, more briefly

$$(\alpha l + \beta m + \gamma n)(\alpha' l + \beta' m + \gamma' n) = \delta.\Lambda;$$

and we see that this *form may* represent any conic, with (α, β, γ) $(\alpha', \beta', \gamma')$ for two foci. In general, when we have the tangential equation of a conic, under the form (article 44),

$$F(l, m, n) = 0$$

the partial derivatives $D_l F$, $D_m F$, $D_n F$ are the *coordinates of the pole* of the *arbitrary line* $[l, m, n]$, taken with respect to the conic; so that the coordinates of the (anharmonic) *centre* (article 28) are found by making $l = m = n$ in the expression of these derivatives. But $D_l\Lambda$, $D_m\Lambda$, $D_n\Lambda$ all *vanish* for equal values l, m, n; and

$$\alpha + \beta + \gamma = 1, \quad \alpha' + \beta' + \gamma' = 1$$

if then we denote by $\alpha'', \beta'', \gamma''$, the unit-line coordinates of the centre of the recent conic, so that $\alpha'' + \beta'' + \gamma'' = 1$, we have

$$\alpha'' = \frac{\alpha + \alpha'}{2}, \quad \beta'' = \frac{\beta + \beta'}{2}, \quad \gamma'' = \frac{\gamma + \gamma'}{2};$$

whence it may be inferred that "the *two foci K, K′* are *on one axis* (article 28) of the conic, and the *interval* between them is *harmonically divided* by the *centre* and the *unit-line*".

70. The equation in (article 68) may be interpreted, and might have been more rapidly deduced, by the help of the expression in (article 41), for the *anharmonic perpendicular,* say, p, let fall from any given point (a, b, c) on any given line $[l, m, n]$; namely by the formula,

$$p = \pm \frac{3(al + bm + cn)}{2(a + b + c)\Lambda^{\frac{1}{2}}};$$

for thus the equation becomes simply,

$$pp' = \pm R.$$

The product of the two anharmonic perpendiculars, let fall from the two real anharmonic foci of a conic, (situated on a common axis,) upon any tangent to that curve is therefore constant". In other words, if two unit-curves, be described so as to *touch that tangent,* and to have the *foci* for *their centres,* their *anharmonic radii* (article 39) will have a *constant product.*

71. Employing *unit-line coordinates* of points, and the formula of (article 49) to express an *anharmonic distance,* the equation

$$r \pm r' = s\sqrt{6},$$

in which r and r' denote the distance of (ξ, η, ζ) from (α, β, γ) and $(\alpha', \beta', \gamma')$ and s is any constant, will become,

$$\{(\xi - \alpha)^2 + (\eta - \beta)^2 + (\zeta - \gamma)^2\}^{\frac{1}{2}} \pm \{(\xi - \alpha')^2 + (\eta - \beta')^2 + (\zeta - \gamma')^2\}^{\frac{1}{2}} = 2s;$$

and may without difficulty be shown to represent a system of two conics, with the two given points for common foci. This was what I alluded to in (article 36), as the *anharmonic analogue* (and *extension*) of the usual *sum or difference property* of conics. – But I am sure that you must be quite tired of reading what I am almost tired of writing, on a subject which *looks* so rudimentary; and which *is* such, except in so far as the ***Conception*** employed may introduce an element of *novelty.* Even on this *last* point, however, I await the judgment of others, yourself included. And I shall say no more about *conics* in this Letter.

72. But as the quantities ξ, η, ζ which I have called the *Anharmonic Unit-Line Coordinates* of a point P, are found to play an important part in *simplifying* several formulæ – especially expressions for the *anharmonic analogue* of *distance* – it may be worth while to show that they are not merely *functions of anharmonic quotients,* as in the equations

$$\xi = \frac{x}{x + y + z}, \quad \&c.,$$

by which they have been (in article 49) *defined,* but have also *independent geometrical significations,* and can be *expressed as anharmonic, themselves.* With this view, however, I shall consider first the following more general

Problem:- "given any six constants, l, m, n, l', m', n', to interpret the function

$$\psi(x, y, z) = \frac{l'x + m'y + n'z}{lx + my + nz},$$

as representing the anharmonic of a pencil.''

73. Let K be the point (a, b, c) for which this function becomes

$$\psi(a, b, c) = \frac{0}{0},$$

or of which the coordinates satisfy the two equations,

$$la + mb + nc = 0, \qquad l'a + m'b + n'c = 0;$$

then these coordinates will also satisfy this third equation,

$$(l' - kl)a + (m' - km)b + (n' - kn)c = 0,$$

or

$$\psi(x, y, z) = k = \text{constant}$$

will be the *equation of a right line through K*. Let L, M, N be three points anywhere taken, upon the three lines KL, KM, KN for which the constant k is, respectively, equal to: zero, unity, and infinity; then it will be found that for any point P, or (x, y, z), the value of the function is

$$\psi(x, y, z) = (K.PLMN).$$

74. To prove this *Theorem*, which contains the solution of the foregoing general problem, we may suppose that the four points, $L = (a', b', c')$, $M = (a'', b'', c'')$, $N = (a''', b''', c''')$, and $P = (x, y, z)$ are situated on one common transversal of the pencil, so that

$$0 = \begin{vmatrix} a' & b' & c' \\ a'' & b'' & c'' \\ a''' & b''' & c''' \end{vmatrix} \quad \text{and} \quad 0 = \begin{vmatrix} a' & b' & c' \\ x & y & z \\ a''' & b''' & c''' \end{vmatrix};$$

in which case we may write briefly,

$$(M) = t(N) + u(L), \qquad (P) = t'(N) + u'(L),$$

or more fully

$$a'' = ta''' + ua', \qquad b'' = tb''' + ub', \qquad c'' = tc''' + uc',$$
$$x = t'a''' + u'a', \qquad y = t'b''' + u'b', \qquad z = t'c''' + u'c';$$

whence the anharmonic of the lately mentioned pencil will be (by article 11).

$$(K.PLMN) = (K.NMLP) = \frac{t'u}{tu'};$$

and we are to show that this is equal to $\psi(x, y, z)$, under the assigned conditions of construction. Those conditions give, as what may be called the *equations of the three fixed rays*, KL, KM, KN, the following:

$$l'a' + m'b' + n'c' = 0; \qquad l'a'' + m'b'' + n'c'' = la'' + mb'' + nc'';$$
$$la''' + mb''' + nc''' = 0;$$

hence

$$t(l'a''' + m'b''' + n'c''') = u(la' + mb' + nc'),$$
$$l'x + m'y + n'z = t'(l'a''' + m'b''' + n'c'''),$$
$$lx + my + nz = u'(la' + mb' + nc'),$$

and the anharmonic becomes

$$\frac{t'u}{tu'} = \frac{l'x + m'y + n'z}{lx + my + nz} = \psi(x, y, z)$$

as asserted. – We derive, at the same time, this slightly more general result:

$$\frac{\psi(x', y', z')}{\psi(x, y, z)} = \frac{l'x' + m'y' + n'z'}{lx' + my' + nz'} \cdot \frac{lx + my + nz}{l'x + m'y + n'z} = (K.NPLP');$$

in which KL and KN, are still the rays which make the function (x, y, z) vanish and become infinite, but P and P' are arbitrary points of the given plane.

75. Applying now the theorem of article 73, to the intepretation of the symbol ξ [see article 72] or of the quotient

$$\frac{x}{x + y + z} = \psi(x, y, z),$$

the denominator shows that KN is the unit-line, and the numerator shows that KL coincides in position with the side OY of the unit-triangle: K is therefore the point B' of Figure 5, which thus becomes the vertex of the pencil: and the ray KM passes through the point X, for which we have $\psi(1, 0, 0) = 1$; in fact the equation of the line $B'X'$ is

$$y + z = 0, \quad \text{or } \xi = 1.$$

Hence for any point P, we have, by the theorem,

$$\xi = (B'.POXA');$$

and similarly,

$$\eta = (A'.POYB'), \quad \text{and} \quad \zeta = (C'.PXOA').$$

Or we may write,

$$\xi = (B'.PYXC') \quad \text{or} \quad \xi = (B'.POXC')\&\text{c}.$$

As verification, when P is anywhere on the line $B'OY$, each of these expressions for ξ vanishes; when P is on $B'X$, each becomes $= 1$, as above; and when P is on the unit line $A'B'C'$, then each of them becomes infinite.

76. If then any triangle OXY have its sides OX, OY, XY cut by any transversal in the points A', B', C', and if P be any point in its plane, we have the *following geometrical theorem*, which can be very simply otherwise proved, by projecting the transversal to infinity:

$$(B'.POXA') + (A'.POYB') + (C'.PXOA') = 1;$$

the three parts, or terms, of the first member being those anharmonics which I have above denoted by ξ, η, ζ; and my *motives* for calling which (in article 49) the ***Unit-Line Coordinates*** of the point P, may now be more fully understood.

77. The *geometrical notation* of this whole Letter might have been made much *more symmetric*, by *abandoning, from the outset, all reference* (even in thought) to *Cartesian coordinates*, and by using consequently *other letters* for the points: for example, by substituting, as in Figure 18, the letters O, A, B, C, A', B', A'', B'', C'', in the places which were occupied by U, X, Y, O, B', A', B, A, C, while C' retains its position [see Figure 5]. Calling then ABC the *unit-triangle*, and $A'B'C'$ the *unit-line*, as before, I might have *begun* with **unit-line coordinates** ξ, η, ζ of any point P, *defining* them by the equation,

$$\xi = (A'.PCAB'), \quad \eta = (B'.PABC'), \quad \zeta = (C'.PBCA');$$

or, which comes to the same thing, by these other equations,

$$\xi = (RCAB') = (S'BAC'), \quad \eta = (SABC') = (Q'XBA'),$$
$$\zeta = (QBCA') = (R'ACB'),$$

if we make, as in Figure 19,

$$Q = C'P^{\cdot}BC, \quad R = A'P^{\cdot}CA, \quad S = B'P^{\cdot}AB,$$

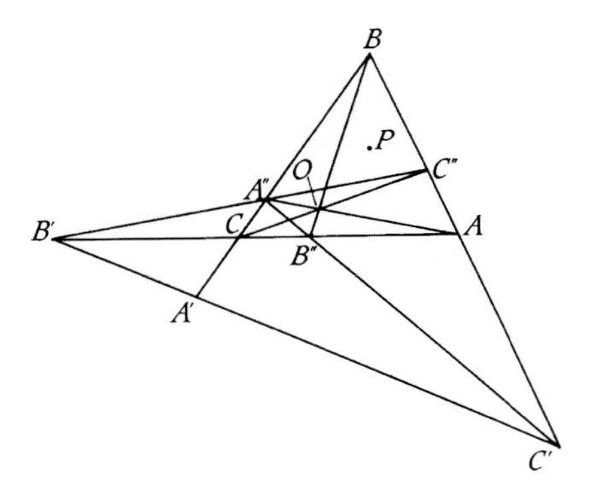

Fig. 18

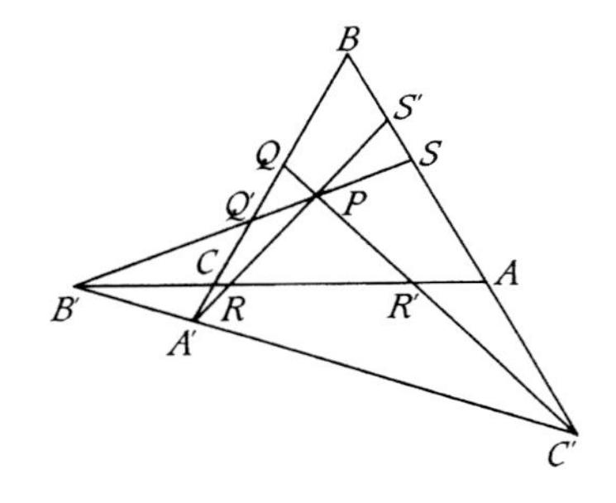

Fig. 19

$$Q' = B'P^{\cdot}BC, \qquad R' = C'P^{\cdot}CA, \qquad S' = A'P^{\cdot}AB.$$

When the unit-line is thrown off to infinity, these anharmonics ξ, η, ζ become, more simply, as in Figure 20, the quotients of segments,

$$\xi = \frac{CR}{CA} = \frac{S'B}{AB}, \qquad \eta = \frac{AS}{AB} = \frac{Q'C}{BC}, \qquad \zeta = \frac{BQ}{BC} = \frac{R'A}{CA};$$

but, by similar triangles,

$$\frac{R'A}{CA} = \frac{PS}{CA} = \frac{SS'}{AB}; \qquad \text{and } S'B + AS + SS' = AB;$$

and generally, we have the relation

$$\xi + \eta + \zeta = 1,$$

which (as before) connects the unit-line coordinates, but is now obtained by (perhaps) the most elementary process possible: so that the theorem of article 76 is anew obtained, under this form, that "if a triangle ABC be cut by a transversal $A'B'C'$, and if P be any point in its planes, then

$$(A'.PCAB') + (B'.PABC') + (C'.PBCA') = 1."$$

78. Again, if we make, as in Figure 20

$$Q'' = AP^{\cdot}BC, \qquad R'' = BP^{\cdot}CA, \qquad S'' = CP^{\cdot}AB,$$

and observe that A'', B'', C'' are, in this projected figure, the middle points of the sides of the triangle, we find that

$$\frac{\xi}{\eta} = \frac{S'B}{AS} = \frac{PQ}{R'P} = \frac{S''B}{AS''} = (S''BC''A) = (C.PBOA),$$

where O is the mean point of the triangle; and similarly for other quotients of ξ, η, ζ. If then we return to the general state of the construction, as represented in Figure 18, and denote by x, y, z *any three quantities* which are *proportional to the unit-line coordinates* ξ, η, ζ, but which are *not connected among themselves* by the condition that their *sum* shall be *unity, nor by any other relation* whatever; we shall have the values,

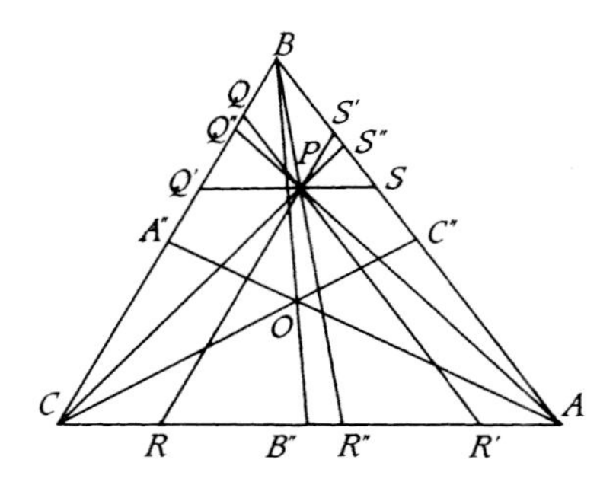

Fig. 20

$$\frac{y}{z} = (A.PCOB), \quad \frac{z}{x} = (B.PAOC), \quad \frac{x}{y} = (C.PBOA);$$

$$\frac{z}{x} = (A.PBOC), \quad \frac{x}{z} = (B.PCOA), \quad \frac{y}{z} = (C.PAOB);$$

which may with advantage, so far as *notation* is concerned, replace the corresponding formula of earlier articles (articles 6 and 7). And I think that *these* anharmonic coordinates x, y, z, when thus *defined* by reference to the unit-point O, or when thus proved to have essential and simple *relations* to that *point*, instead of having them to the line $A'B'C'$, may conveniently be called, for distinction's sake, the ***Unit-Point Coordinates*** of P.

79. Continuing thus to improve a little the symmetry of our formulæ, let Figure 5 be replaced by the annexed Figure 21, in which LMN is a *second transversal* of the triangle ABC. I would then define that the ***Unit-Line Coordinates*** of this *line LMN*, are any quantities l, m, n which satisfy the two first, and therefore all the others, of the six following equations (compare article 9):

$$\frac{m}{n} = (BLCA'), \quad \frac{n}{l} = (CMAB'), \quad \frac{l}{m} = (ANBC');$$

$$\frac{n}{m} = (CLBA'), \quad \frac{l}{n} = (AMCB'), \quad \frac{m}{l} = (BNAC').$$

And then I would go on to *prove*, (on the plan of article 10,) what I have elsewhere called (see article 10) the *Fundamental Theorem* of this *Anharmonic Method*: namely, that "if (x, y, z), or P, be a point situated anywhere upon this line $[l, m, n]$, or LMN, then the *coordinates of point and line* are *connected by the following equation*:

$$lx + my + nz = 0\text{''};$$

which (as in article 10) may be regarded either as the local equation of the *line*, or as the *tangential equation* of the point. After this, all would proceed exactly as before, only with a more symmetrical geometrical notation (article 75).

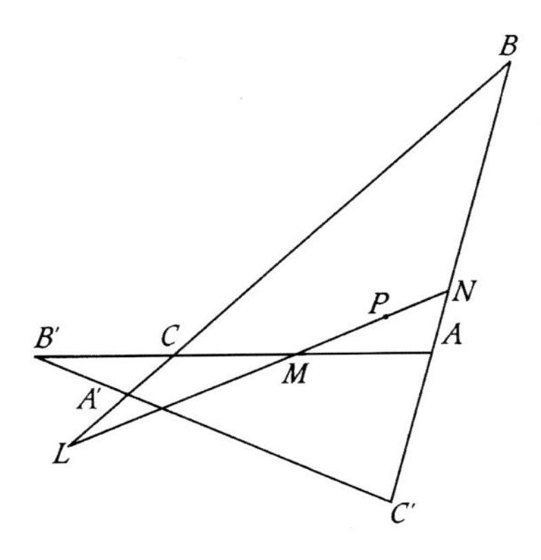

Fig. 21

80. Elementary, however, as was the proof given in article 10, of the important theorem just now cited, that proof may be made perhaps a little clearer and more simple, by employing unit-line coordinates for *both* point and line, or by proving that

$$l\xi + m\eta + n\zeta = 0;$$

and by throwing off the unit-line to infinity [see Figure 22]. In this manner we have (compare article 77),

$$-\frac{l}{n} = \frac{AM}{MC}, \quad -\frac{m}{n} = \frac{BL}{LC}, \quad \frac{\xi}{\zeta} = \frac{Q'P}{PS}, \quad \frac{\eta}{\zeta} = \frac{PR}{S'P};$$

therefore

$$-\frac{l\xi}{n\zeta} = \frac{AM}{PS} \cdot \frac{Q'P}{MC} = \frac{NM}{PN} \cdot \frac{LP}{ML} = (LPNM),$$

$$-\frac{m\eta}{n\zeta} = \frac{BL}{S'P} \cdot \frac{PR}{LC} = \frac{NL}{NP} \cdot \frac{PM}{LM} = (LNPM),$$

but

$$(LPNM) + (LNPM) = 1 \text{ (by article 5)};$$

therefore

$$-\frac{l\xi}{n\zeta} - \frac{m\eta}{n\zeta} = 1, \quad \text{or,} \quad l\xi + m\eta + n\zeta = 0;$$

and consequently,

$$lx + my + nz = 0, \quad \text{as before.}$$

81. When l, m, n are given constants, and ξ, η, ζ are the unit-line coordinates of a point P which is not situated upon the line $[l, m, n]$, then the expression $l\xi + m\eta + n\zeta$ has a *value* different from *zero*; and the theorem of article 73, which is unaffected by the recent change of geometrical notation, enables us to *interpret* this expression as representing the *anharmonic of a pencil.* Let K be the point in which the given line $[l, m, n]$ meets the unit-line; so that the *unit-point coordinates* (article 78) of this *point* K may be written thus

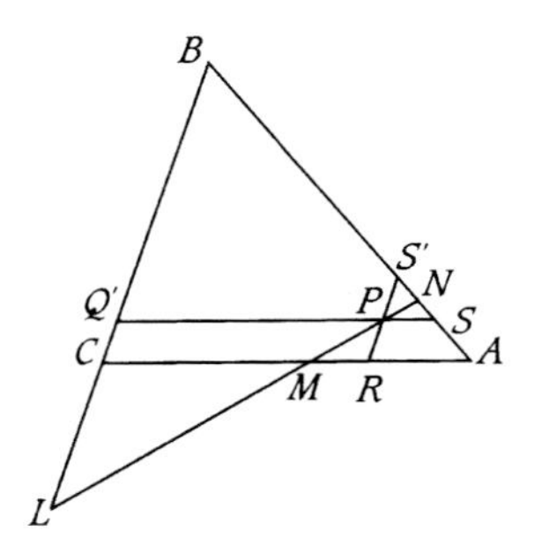

Fig. 22

$$a = m - n, \quad b = n - l, \quad c = l - m,$$

although the *unit-line coordinates* (article 77) *are infinite.* Let L be any other point on the given line $[l, m, n]$, and N any other point on the unit-line $[1, 1, 1]$; finally, let M be any point for which the proposed expression, $l\xi + m\eta + n\zeta$, becomes $= -1$. Then, generally, by the theorem (article 73), for any point P, of which the unit-line coordinates are ξ, η, ζ, we have the equation,

$$l\xi + m\eta + n\zeta = (K.PLMN)$$

and if P' have ξ', η', ζ'' for its corresponding coordinates, then, by (article 74),

$$\frac{l\xi + n\eta + n\zeta}{l\xi' + m\eta' + n\zeta'} = (K.PLP'N).$$

82. As an example, let us take the expressions (article 54),

$$X = (2\zeta - \xi - \eta)\cos\frac{\pi}{3}, \quad Y = (\xi - \eta)\sin\frac{\pi}{3},$$

and seek, on this plan, their anharmonic interpretations. For the first of these expressions the given line is OC', so that we must place K at C', and may place L at O, and N at A', also $X = 1$ for the point C, which may therefore be taken for M: so that we may write

$$X = (C'.POCA') = (QOCC'''), \quad \text{if} \quad Q = C'P^{\cdot}OC, \quad \text{and} \quad C''' = C'A'^{\cdot}OC.$$

For the second expression the given line is OC, so that we must take $K = OC^{\cdot}A'B' = C'''$; also OC intersects AB in C'', and for the point A we have $\xi = 1, \eta = 0, Y = \sin\frac{\pi}{3}$; if then we write $R = KP^{\cdot}AB$, (Q and R being here not the same points as in article 77,) and place N on AB at C', we shall have

$$\xi - \eta = (C'''.POAC'), \quad \text{and} \quad Y = (RC''AC')\sin\frac{\pi}{3}.$$

83. If we now *project canonically* (articles 22, 55, &c.) the unit-triangle ABC into an *equilateral* triangle, and the unit-point O into its mean point (or centre), as in Figure 23, the points C''', and C', in these expressions for the functions X and Y, will go off into infinity, and Q and R will become the feet of the perpendiculars from P on OC and AB; also C'' will bisect AB, and we shall have

$$C''A = OD.\sin\frac{\pi}{3},$$

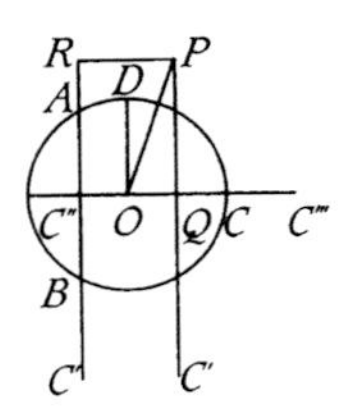

Fig. 23

if OD be the radius of the circumscribed circle ABC, erected towards the same side of OC as OA. The recent anharmonic expressions become, therefore

$$X = \frac{OQ}{OC}, \qquad Y = \frac{C''R}{OD};$$

and we see anew that *these two functions* are simply (in this canonical projection) the *two rectangular* and *Cartesian* coordinates of the point P, referred to the *unit-point* O as their origin, with the radius OC for the *unit of length*, and at the same time for the *positive semiaxis of* x; while OD is the positive semiaxis of y: as was otherwise found in articles 54 and 56, with a less symmetric geometrical *notation*, and as *the result of an altogether different train* of reasoning, and of calculation.

84. The interpretations of these two functions, X and Y, having been thus in a new way established, if we write, as in (article 54),

$$r\cos\varphi = X = (2\zeta - \xi - \eta)\cos\frac{\pi}{3},$$

$$r\sin\varphi = Y = (\xi - \eta)\sin\frac{\pi}{3},$$

we shall again have, as in (article 53)

$$r^2 = X^2 + Y^2 = \xi^2 + \eta^2 + \zeta^2 - \xi\eta - \eta\zeta - \zeta\xi,$$

as an expression for the square of the anharmonic radius vector of a point P, while φ is (as before) the (anharmonic) *polar angle* of that point; also the square of the (anharmonic) *distance* R between any *two* points, of which the unit-line coordinates are ξ, η, ζ, and ξ', η', ζ', is

$$R^2 = (X' - X)^2 + (Y' - Y)^2 = r'^2 + r^2 - 2rr'\cos(\varphi' - \varphi)$$

where

$$rr'\cos(\varphi' - \varphi) = XX' + YY' = \xi\xi' + \eta\eta' + \zeta\zeta' - \tfrac{1}{2}(\xi\eta' + \eta\xi') - \tfrac{1}{2}(\eta\zeta' + \zeta\eta') - \tfrac{1}{2}(\zeta\xi' + \xi\zeta')$$

with reductions on which it is not worth while to delay, except that we may here just notice the very simple *symmetric* formula (compare article 49),

$$\frac{2}{3}R^2 = (\xi - \xi')^2 + (\eta - \eta')^2 + (\zeta - \zeta')^2$$

which may be proved, among other ways, by substituting in the right hand member the values (article 54),

$$\xi = \tfrac{1}{3}(1 - X + Y\sqrt{3}),\ \eta = \tfrac{1}{3}(1 - X - Y\sqrt{3}),\ \zeta = \tfrac{1}{3}(1 + 2X),$$

with the corresponding values of ξ', η', ζ'.

85. Let $f(x, y, z) = 0$, or $f(\xi, \eta, \zeta) = 0$, be any *homogeneous* and local equation of a curve, which we may always suppose it to be; because if the equation be not *given* as homogeneous in ξ, η, ζ we can always *render* it such, without altering its degree, by the help of the relation $\xi + \eta + \zeta = 1$. Then as in (article 55), we have, at once,

$$f(1 - X + Y\sqrt{3},\ 1 - X - Y\sqrt{3},\ 1 + 2X) = 0,$$

as the equation in *Cartesian* and *rectangular coordinates*, of what the curve becomes, by the *canonical projection* which has already been sufficiently described. Such substitutions having been, in former articles 57, 58 and 59 exemplified for *conics*. I shall now venture to give an example or two, for the case of *cubic curves*: although conscious how vastly inferior my knowledge of that subject is to yours.

86. Taken then the following equation of the third degree

$$27\xi\eta\zeta = k,$$

when k is any real constant, and which is equivalent to the homogeneous form,

$$27xyz = k(x + y + z)^3.$$

The substitutions in question give, immediately,

$$(1 - X + Y\sqrt{3})(1 - X - Y\sqrt{3})(1 + 2X) = k;$$

or expanding *and making* $X = r\cos\varphi,\ Y = r\sin\varphi$

$$1 - 3r^2 + 2r^3\cos 3\varphi = k;$$

such, then, are the equations in rectangular and polar coordinates, of the canonically projected curve.

87. The last *homogeneous form* (article 86) has been discussed by others, and must be considered as *well known*, although *I* know it only from Salmon's Book (*HPC*, pp. 136 and 172) but I shall be curious to learn, whether any one has assigned a short and easy process, like that given above, for *transforming* it to *common coordinates*. Any remark on its interpretation, that I may go on to make, will of course be purely for *my own* satisfaction; and because it is really *easier* for me, when the pen is in my hand, to write with some fullness than to be concise. If I had taken time to write these sheets *over again*, instead of sending them to you as fast as they were written, I could easily have compressed them into *less than half*, perhaps in to a *third part*, of their extent. As it is, you will just suppose me to be writing in a private book, the pages of which I allow you to look over, without the slightest hope of teaching you anything.

88. Two principal cases may be distinguished [see Figure 24]: Ist, the case where $k > 0$, but $< \infty$; and IInd, the case $k \ngtr 0$, $> -\infty$. In the Ist case, there exist *three infinite branches* of the curve, lying in the three *vertically opposite* angles $B'AC'$, $C'BA$, $A'CB'$, of the triangle ABC; but besides *these* (always real) branches, there exists a *real oval, interior* to those triangles, if $k < 1$; which reduces itself to a *conjugate point, at the centre* O of the same triangle, if $k = 1$; and becomes an *imaginary oval,* if $k > 1$; all this appearing easily from the considerations of the *cubic equation* in r. In the IInd case, if $k = 0$, the cubic *curve* degenerates into a *system of three right lines,* namely, the 3 sides of the triangle; but if $k < 0$, it consists again of *three infinite but curvilinear branches,* which are however situated now in the three infinite trapezoids, $C'BCB'$, $A'CAB'$, $B'ABA'$, formed by prolonging two of the sides *beyond the third* [see Figure 25]. In each case, (at least if we set aside the *non-curvilinear subcase* $k = 0$), the *three sides* of the

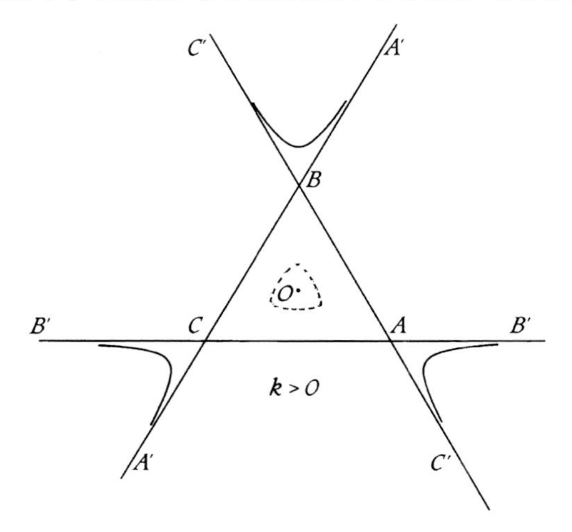

Fig. 24

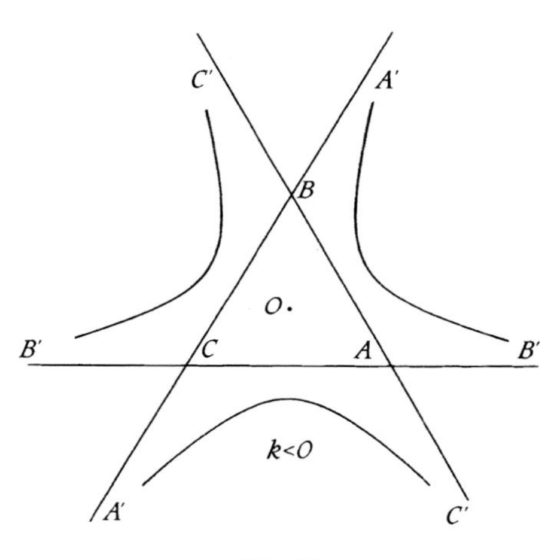

Fig. 25

triangle are the *three asymptotes*; and the points at infinity, in which they meet the curve, are the *three real points of inflexion*. The *six imaginary points* of inflexion are the *intersections* of the *cubic* with the *infinitely small circle* at O. (If $k = \pm\infty$ the curve degenerates into *three right lines* which however in this case are *all at infinity*.) And it appears that *cubics of the sixth class* may *generally* be *projected* into *one or other* of these forms: namely by projecting the *triangle of tangents* drawn *at the three real and distinct points of inflexion*, into an *equilateral triangle*; and at the same time projecting *to infinity* the *line* which *joins those points*. Indeed, it seems to be sufficient to accomplish this *later* projection, or to consider only the *triangle of asymptotes*, when these are *real* lines, and their points at infinity are points of *inflexion* on the curve. For then I do not see how the general *shapes* can change, any otherwise than as the shape of an *ellipse* differs from that of a *circle*. But we should not have *equations quite so simple*, in *rectangular*, or in *polar coordinates*.

89. I grant that all this construction (article 88) comes *virtually to the very simple proposition,* which I suppose is perfectly well known, although I cannot refer to it:- "The *General Cubic is some projection of the locus of a point P, from which the three perpendiculars* (p, p', p''), *let fall on the three sides of a given equilateral triangle* (*ABC*), *have their product equal to a given constant*"; *each* such perpendicular being regarded as *positive,* when let fall from the opposite *vertex,* but is passing from positive to *negative,* when the point comes to *cross the side.*

90. Indeed, on turning again to Salmon (*HPC,* p. 136), I see a *Theorem,* respecting the *interpretation of the equation* $ACE - B^3 = 0$, containing (as he observes) *nine* constants: which theorem, if *we admit* that the *General Cubic* has *three real and distinct points of inflexion, the tangents* at which form a *real* and *non-evanescent triangle,* (as *not generally* meeting in *one point*), may certainly be said to *include* the foregoing proposition; although I happened to deduce it, and the constructions in (article 88), from the *Polar Equation* assigned in (article 86).

91. The theorem (p. 171, article 184 of *HPC*), respecting the *position of the conjugate point,* when such a point exists, is reproduced by the analysis of our recent article 88. – By *starting* with the supposition that the *unit-line* contains *one real,* and *two imaginary* points of inflexion, (instead of *three real* points,) I was led to the *form of equation,*

$$U = (x + y + z)^3 + 6kz(x^2 + xy + y^2) = 0;$$

which gave, by Salmon's formula, the Hessian,

$$H = k^2 U + 4k^2 (y - z)(z - x)(x + y + z);$$

and accordingly, the *biquadratic* (*HPC,* p. 187) for $\frac{\lambda}{\mu}$, in the system of three right lines through the nine points of inflexion, represented by the equation

$$\lambda U + \mu H = 0,$$

becomes here (if I have made no slip in the calculation of S and T),

$$0 = (\lambda + k^2\mu)\{3\lambda^3 - 3k^2\lambda^2\mu - 3k^3(8 + 5k)\lambda\mu^2 - k^4(4 + 3k)^2\mu^3\}$$

the *cubic factor* equated to 0, giving generally *one real root,* and *two imaginary roots.* – But whether I shall enter on any account of my *interpretation* of *these* equations is doubtful, as I wish to *finish soon.*

92. I may however jot down a few formulæ for reference, in connexion with this form,

$$U = (x + y + z)^3 + 6kz(x^2 + xy + y)^2,$$

or briefly

$$U = s^3 + 6kzp,$$

if

$$s = x + y + z, \quad p = x^2 + xy + y^2;$$

which form is by no means proposed as the base, but merely as one which, for a particular purpose, I lately found it convenient to discuss. Differentiating U, we have

$$l = \tfrac{1}{3}D_x U = s^2 + 2kz(2x + y),$$

$$m = \tfrac{1}{3}D_y U = s^2 + 2kz(x + 2y),$$

$$n = \tfrac{1}{3}D_z U = s^2 + 2kp,$$

$$l' = \tfrac{1}{6}D_x^2 U = s + 2kz,$$

$$m' = \tfrac{1}{6}D_y^2 U = s + 2kz,$$

$$n' = \tfrac{1}{6}D_z^2 U = s,$$

$$l'' = \tfrac{1}{6}D_y D_z U = s + k(x + 2y),$$

$$m'' = \tfrac{1}{6}D_z D_x U = s + k(2x + y),$$

$$n'' = \tfrac{1}{6}D_x D_y U = s + kz;$$

whence

$$H = l'l''^2 + m'm''^2 + n'n''^2 - l'm'n' - 2l''m''n'', \ (\textit{HPC}, \text{ p. } 71)$$

$$= k^2 s\{(x - y)^2 + 6(x + y)z - 3z^2\} + 6k^2 zp$$

$$= k^2\{U + 4s(y - z)(z - x)\}$$

as in our recent article 91. – The S, T, R came out, for this form of U,

$$S = 3k^3(4 + 3k); \quad T = 72k^4(2 + 6k + 3k^2); \quad R = -2^8 3^3 k^8(3 + 2k);$$

whence the biquadratic in $\frac{\lambda}{\mu}$, namely (*HPC*, p. 187)

$$27\lambda^4 - 18S\lambda^2\mu^2 - T\lambda\mu^3 - S^2\mu^4 = 0$$

because

$$0 = 3\lambda^4 - 6k^3(4 + 3k)\lambda^2\mu^2 - 8k^4(2 + 6k + 3k^2)\lambda\mu^3 - k^6(4 + 3k)^2\mu^4;$$

or, as in, (article 91),

$$0 = (\lambda + k^2\mu)\{3\lambda^3 - 3k^2\lambda^2\mu - 3k^3(8 + 5k)\lambda\mu^2 - k^4(4 + 3k)^2\mu^3\}$$

where the factor $\lambda + k^2\mu$ corresponds to the decomposition, assigned above, of the combination $H - k^2U$ into three linear factors, s, $y - z$, $z - x$, answering to a system of three real right lines, on which the nine points of inflexion are contained.

93. The line

$$s = 0 \quad \text{or} \quad x + y + z = 0$$

is what has been called, in this Letter (article 10) the *unit-line*; it cuts the cubic U in one real point of inflexion, namely in $C' = (1, -1, 0)$; and in two imaginary points of inflexion, namely,

$$I = (1, \theta, \theta^2) \quad J = (1, \theta^2, \theta),$$

which have been called (in article 25) the two *imaginary unit-points*. The tangents to the curve, at these three points of inflexion, are

$$z = 0, \quad \text{or} \quad C'AB; \quad y - \theta x = 0, \quad \text{or} \quad IC; \text{ and } y - \theta^2 x = 0, \quad \text{or} \quad JC;$$

as may be perceived on inspection of the given equation $U = 0$, and may be confirmed by observing that, (see article 92,)

$$x = 1, \quad y = -1, \quad z = 0 \quad \text{give} \quad l = 0, \quad m = 0, \quad n = 2k;$$

$$x = 1, \quad y = \theta, \quad z = \theta^2 \quad \text{give} \quad l = 2k\theta^2(2+\theta), \quad m = 2k\theta^2(1+2\theta), \quad n = 0;$$

and

$$x = 1, \quad y = \theta^2, \quad z = \theta \quad \text{give} \quad l = 2k\theta(2+\theta^2), \quad m = 2k\theta(1+2\theta^2), \quad n = 0;$$

so that the tangents at C', I, J have for their equations,

$$2kz = 0, \quad 2k\theta^2(2+\theta)x + 2k\theta^2(1+2\theta)y = 0,$$
$$2k\theta(2+\theta^2)x + 2k\theta(1+2\theta^2)y = 0;$$

or, because

$$\frac{2+\theta}{1+2\theta} = -\theta \quad \text{and} \quad \frac{2+\theta^2}{1+2\theta^2} = -\theta^2,$$
$$z = 0, \quad y - \theta x = 0, \quad y - \theta^2 x = 0, \quad \text{as above.}$$

The point C, or $(0, 0, 1)$, in which the two imaginary tangents intersect, is real and is one of the corners of what we have called the *unit-triangle*; and the real tangent at C' is the opposite side of that triangle.

And if we draw, as in former figures, the line CO to the unit-point O, meeting AB in C'', and $A'B'C'$ in $C^{\backprime}$, and having $y = x$ for its equation, this line will be the *harmonic polar* (*HPC*, p. 140) of the point of inflexion C', with reference to the given cubic. Thus the real point C, or $(1, 1, -2)$, is the harmonic conjugate of C', or $(1, -1, 0)$, not only with respect to the two real points $A' = (0, 1, -1)$, and $B' = (-1, 0, 1)$, but also with respect to the two imaginary points I and J, so that

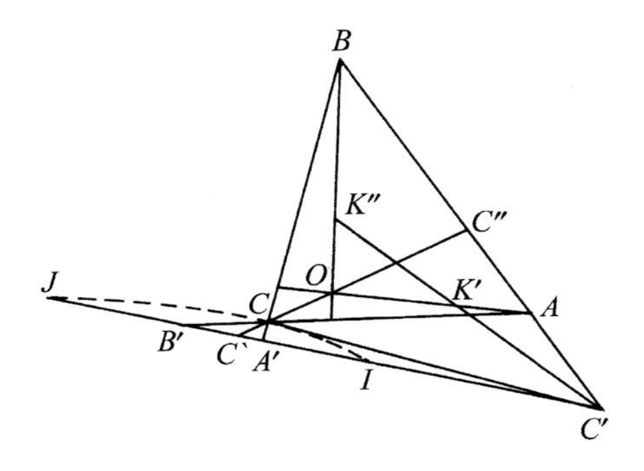

Fig. 26

$$\frac{2}{C'C'} = \frac{1}{C'A'} + \frac{1}{C'B'} = \frac{1}{C'I} + \frac{1}{C'J};$$

and the imaginary tangents at I and J *are conjugate rays of a pencil in involution*, whereof the sides CA, CB are also conjugate rays, CC' and CO being the *double rays*.

94. The constructions in the foregoing article are independent of any use, or even knowledge, of the *Hessian*, H; but we may now go on to employ the equation

$$k^{-2}H - U = 4s(y-z)(z-x),$$

in order to discover the six remaining points of inflexion of U: wherof we see that three are on the real right line $y - z = 0$, or OA; and three others on the other real line, $z - x = 0$, or OB. And *now* it is evident that at least *this* cubic U has *three real points of inflexion*; namely the point C' on the line $s = 0$, as before, and *two new points*, which we may call K' and K'', situated on the right lines OA, OB. The imaginary points on OA may be called I' and J'; and those on OB, I'' and J''.

95. Writing

$$k' = \sqrt[3]{k}, \qquad k'' = \sqrt[3]{(12+8k)}$$

(where I intend that *real* cube-roots should be taken,) and

$$w = k'k''(k' + k''), \qquad w' = k'k''(\theta k' + \theta^2 k''), \qquad w'' = k'k''(\theta^2 k' + \theta k''),$$

the quantities w, w', w'' are the one real and the two imaginary roots of the following cubic equation:

$$w^3 + 12k(3+2k)w - 4k(3+2k)(12+7k) = 0.$$

Eliminating z from $U = 0$, by the equation $y - z = 0$ of OA, we obtain this other cubic equation, to determine the directions of the three lines, CK', CI', CJ', which are drawn from the point C to the three points K', I', J' on inflexion on the line OA:

$$\begin{aligned} 0 &= (x+2y)^3 + 6ky(x^2 + xy + y^2) \\ &= x^3 + 6(1+k)x^2y + 6(2+k)xy^2 + 2(4+3k)y^3; \end{aligned}$$

in which if we make

$$(8+6k)y + (4+2k)x = wx,$$

we are conducted to the recent cubic in w, the roots of which have been assigned. The real point of inflexion in OA is therefore

$$K' = (8+6k,\ w-4-2k,\ w-4-2k);$$

and the two imaginary points of inflexion, on the same line OA, are,

$$\begin{aligned} I' &= (8+6k,\ w'-4-2k,\ w'-4-2k), \\ J' &= (8+6k,\ w''-4-2k,\ w''-4-2k). \end{aligned}$$

In like manner the real point of inflexion on OB is

$$K'' = (w - 4 - 2k,\ 8 + 6k,\ w - 4 - 2k);$$

and the two imaginary points of inflexion, on the same line, are,

$$I'' = (w' - 4 - 2k,\ 8 + 6k,\ w' - 4 - 2k),$$

$$J'' = (w'' - 4 - 2k,\ 8 + 6k,\ w'' - 4 - 2k).$$

96. As verifications, we may try whether the 9 points are collinear, 3 by 3. One such collinearity $(C'IJ)$ we started with; and two others, $(I'J'K')$ and $(I''J''K'')$, are immediate results of the calculation. The three real points of inflexion, C', K', K'', are collinear, because their coordinates satisfy the common equation,

$$2(w + k)z = (w - 4 - 2k)(x + y + z),$$

which represents a real right line. The two imaginary lines,

$$2(w' + k)z = (w' - 4 - 2k)(x + y + z),$$
$$2(w'' + k)z = (w'' - 4 - 2k)(x + y + z),$$

connect in like manner the two sets of three points, C', I', I'', and C', J', J''; and thereby exhibit two other collinearities. But besides the three lines, one real and two imaginary, of which the equations have just been written, and the three real lines, above alluded to, of which the equations are

$$x + y + z = 0, \qquad y - z = 0, \qquad x - z = 0,$$

namely the three linear factors of $k^2 U - H$, we ought to be able to assign *six other lines*, to make up the known number of twelve.

97. For this purpose, I modify a little the foregoing expressions for the points I', J', K', I'', J'', K'', as follows. Writing, more briefly,

$$K' = (t,\ v,\ v), \qquad K'' = (v,\ t,\ v),$$

where t is a real, which may still be assumed equal to $8 + 6k$, while $\frac{v}{t}$ is the real root of the cubic equation,

$$(2v + t)^3 + 6kv(v^2 + tv + t^2) = 0$$

and supposing that $\frac{v'}{t}$, $\frac{v''}{t}$ are the two imaginary roots of the same equation, for it will, in general, have two such roots: then by developing and reducing the equation,

$$(v' - v)^{-1}\{v(v^2 + tv + t^2)(2v' + t)^3 - v'(v'^2 + tv' + t^2)(2v + t)^3\} = 0,$$

I am conducted to a quadratic, which may be written thus:

$$0 = \left(\frac{v'}{t} - \frac{\theta^2 t - \theta v}{t + 2\theta^2 v}\right)\left(\frac{v'}{t} - \frac{\theta t - \theta^2 v}{t + 2\theta v}\right).$$

Hence the following new symbols may be employed to represent the points I', ..., J'':

$$I' = (t + 2\theta^2 v,\ \theta^2 t - \theta v,\ \theta^2 t - \theta v);$$

$$J' = (t + 2\theta v,\ \theta t - \theta^2 v,\ \theta t - \theta^2 v);$$

$$I'' = (\theta^2 t - \theta v,\ t + 2\theta^2 v,\ \theta^2 t - \theta v);$$

$$J'' = (\theta t - \theta^2 v,\ t + 2\theta v,\ \theta t - \theta^2 v);$$

t and v having the same real values in all these formulæ. The six former collineations are again put in evidence; thus besides the original equation $x + y + z = 0$, which is satisfied by the three points C', I, J, we have $y = z$ for each of the points K', I', J', and $x = z$ for each of K'', I'', J''; the equation $(t + 2v)z = vs$, (where s is still equal to $x + y + z$,) represents the real right line $C'K'K''$; and the two imaginary lines $C'I'I''$, $C'J'J''$, have for their respective equations,

$$(\theta^2 - \theta)(t + 2v)z = (\theta^2 t - \theta v)s,\ (\theta - \theta^2)(t + 2v)z = (\theta t - \theta^2 v)s.$$

But the *six other* relations of collinearity, for which we were in search, come also now into view. Thus the equation,

$$0 = (\theta^2 - \theta)vx + (v - \theta^2 t)y + (\theta t - v)z, \quad \text{or} \quad 0 = \begin{vmatrix} x, & y, & z \\ t, & v, & v \\ 1, & \theta, & \theta^2 \end{vmatrix},$$

is satisfied, whether we suppose $x = t,\ y = v,\ z = v$, as for K'; or $x = 1,\ y = \theta,\ z = \theta^2$, as for I; or

$$x = z = \theta t - \theta^2 v \quad \text{and} \quad y = t + 2\theta v$$

as for J'': these three points of inflexion K', I, J'', whereof the first is real, and the two others are imaginary, are therefore situated on that one imaginary right line, of which the equation has just been written. In like manner, K', J, I'' are on this other imaginary line,

$$0 = (\theta - \theta^2)vx + (v - \theta t)y + (\theta^2 t - v)z, \quad \text{or} \quad \begin{vmatrix} x, & y, & z \\ t, & v, & v \\ 1, & \theta^2, & \theta \end{vmatrix};$$

K'', I, I' are on

$$0 = \begin{vmatrix} x, & y, & z \\ v, & t, & v \\ 1, & \theta, & \theta^2 \end{vmatrix} = \theta(\theta t - v)x + (1 - \theta^2)vy + (\theta v - t)z;$$

and K'', J, J' are on

$$0 = \begin{vmatrix} x, & y, & z \\ v, & t, & v \\ 1, & \theta^2, & \theta \end{vmatrix} = \theta(t - \theta v)x + (1 - \theta)vy + (\theta^2 v - t)z;$$

so that besides the real line $C'K'K''$, connecting the three real points, we have already found three other real lines, and six imaginary lines, passing each through *one* of those three real points, and connecting them with imaginary points. Finally, if we write

$$I' = (t', v', v'), \quad J' = (t'', v'', v''), \quad I'' = (v', t', v'), \quad J'' = (v'', t'', v''),$$

the equations

$$0 = \begin{vmatrix} x & y & z \\ t' & v' & v' \\ v'' & t'' & v'' \end{vmatrix} = v'(v'' - t'')x + v''(v' - t')y + (t't'' - v'v'')z,$$

$$0 = \begin{vmatrix} x & y & z \\ t'' & v'' & v'' \\ v' & t' & v' \end{vmatrix} = v''(v' - t')x + v'(v'' - t'')y + (t't'' - v'v'')z,$$

or, after expansion and division by $t - v$, the equations,

$$0 = \theta(v - \theta t)x + (t - \theta v)y + (\theta^2 - 1)vz,$$

$$0 = \theta^2(v - \theta^2 t)x + (t - \theta^2 v)y + (\theta - 1)vz,$$

or still more briefly

$$0 = v'x - \theta^2 v''y + (1 - \theta)^2 vz, \quad \text{and} \quad 0 = v''x - \theta v'y + (1 - \theta)vz,$$

represent two imaginary lines, namely $JI'J''$, and $IJ'I''$, which connect the six imaginary points *among themselves*, and are the two lines that remained to be found. – It is to be noted that these two lines intersect each other in a *real point* $O' = (v, v, t)$, which is *on the harmonic polar* (article 93) of C'; and that the product of their equations, namely

$$0 = (t^2 + tv + v^2)(x^2 + y^2) + (t^2 - 2tv - 2v^2)xy - 3tv(x + y)z + 3v^2z^2,$$

or

$$0 = (t^2 + tv + v^2)(x - y)^2 + 3(tx - vz)(ty - vz),$$

represents an *imaginary* or *infinitesimal conic*, which has the recently determined O' for its *only real point*, but which is to be considered as *passing through* the six *imaginary points of inflexion*. (Compare article 88.)

98. If we multiply the equation of this conic, by the equation

$$0 = vx + vy - (t + v)z,$$

of the real right line $C'K'K''$ which connects the *3 real points* of inflexion, we obtain a new equation of the third degree, which may be written thus:

$$0 = (t^2 + tv + v^2)U + (t + 2v)^2 s(y - z)(z - x);$$

and which represents a locus, containing *all* the nine points of inflexion. Accordingly, (in articles 91 and 92) we found that

$$4s(y - z)(z - x) = k^{-2}H - U,$$

so that the last equation is of the form $\lambda U + \mu H = 0$, as it ought to be; and I have verified that the resulting value of $\frac{\lambda}{\mu}$, namely $\frac{3t^2k^2}{(t + 2v)^2}$, is the one real root of the cubic (article 91),

$$0 = 3\lambda^2 - 3k^2\lambda^2\mu - 3k^3(8 + 5k)\lambda\mu^2 - k^4(4 + 3k)^2\mu^3,$$

where k is treated as a function of t and v, determined by the first equation in article 97, or by the formula,

$$6k = \frac{-(t+2v)^3}{v(t^2+tv+v^2)}.$$

99. Writing, for greater symmetry, K instead of C', to denote the first real point of inflexion, which is at once suggested by the mere form of U, without any trouble of calculation, we may collect the twelve collineations into four groups, as follows:

$$\text{I.}\quad \begin{cases} (KK'K'') = v(x+y) - (t+v)z = 0; \text{ (equation of right line through } K,\ K',\ K'') \\ (IJ'I'') = v''x - \theta v'y + (1-\theta)vz = 0; \text{ (equation \&c. through) } I,\ J',\ I'') \\ (JI''J'') = v'z - \theta^2 v''y + (1-\theta^2)vz = 0; \text{ (equation \&c.;)} \end{cases}$$

$$\text{II.}\quad \begin{cases} (KIJ = x+y+z = 0; \\ (K'I'J') = y - z = 0; \\ (K''I''J'') = z - x = 0; \end{cases}$$

$$\text{III.}\quad \begin{cases} (KI'I'') = v'(x+y) - (t'+v')z = 0; \\ (K'IJ'') = (1-\theta)vx + v''y - \theta v'z = 0; \\ (K''JJ') = v''x + (1-\theta)vy - \theta v'z = 0; \end{cases}$$

$$\text{IV.}\quad \begin{cases} (KJ'J'') = v''(x+y) - (t''+v'')z = 0; \\ (K'JI'') = (1-\theta^2)vx + v'y - \theta^2 v''z = 0; \\ (K''II') = v'x + (1-\theta^2)vy - \theta^2 v''z = 0; \end{cases}$$

it being understood that each of these twelve linear equations in x, y, z holds only for the set of three points of inflexion to which it belongs; and t', v', t'', v'', still depending on t and v, by the formulæ (article 97),

$$t' = t + 2\theta^2 v; \quad v' = \theta^2 t - \theta v;$$

$$t'' = t + 2\theta v; \quad v'' = \theta t - \theta^2 v.$$

100. Multiplying the three equations of Group I, we find, by article 99,

$$(KK'K'').(IJ'I'').(JI'J'') = vv'v''U + v(t+2v)^2 s(y-z)(z-x) = \lambda_0 U + \mu_0 H$$

where

$$\frac{\lambda_0}{\mu_0} = 3\left(\frac{kt}{t+2v}\right)^2; \quad \text{and} \quad s = x+y+z. \text{ (Compare 98.)}$$

Multiplying the three equations of Group II, we have, at once, the product (article 91),

$$(KIJ).(K'I'J').(K''I''J'') = s(y-z)(z-x) = \lambda_1 U + \mu_1 H; \quad \frac{\lambda_1}{\mu_1} = -k^2.$$

Multiplying the equations of Group III, we find, after reduction,

$$(KI'I'').(K'IJ'').(K''JI') = (1-\theta)vv'v''U - \theta^2 v'(t+2v)^2 s(y-z)(z-x)$$

$$= \lambda_2 U + \mu_2 H;$$

and in like manner, the product of the three equations of Group IV is,

$$(KJ'J'').(K'JI'').(K''II') = (1-\theta^2)vv'v''U - \theta v''(t+2v)^2 s(y-z)(z-x)$$
$$= \lambda_3 U + \mu_3 H;$$

where

$$\frac{\lambda_2}{\mu_2} = -\left(\frac{kt'}{t+2v}\right)^2, \quad \text{and} \quad \frac{\lambda_3}{\mu_3} = -\left(\frac{kt''}{t+2v}\right)^2.$$

I have verified that while $\frac{\lambda_0}{\mu_0}$ and $\frac{\lambda_1}{\mu_1}$ are (as already seen, or stated) the *real positive root* and the *real negative root*, respectively, $\frac{\lambda_2}{\mu_2}$ and $\frac{\lambda_3}{\mu_3}$ are the *two imaginary roots*, of the *biquadratic equation*, already cited from Salmon, namely,

$$27\lambda^4 - 18S\lambda^2\mu^2 - T\lambda\mu^3 - S^2\mu^4 = 0.$$

101. The following method has occurred to me, – but it is probably not new to you, – of proving that this important biquadratic has *generally* two roots *imaginary*. Write, for a moment $\frac{\lambda}{\mu} = t$, and consider the function,

$$f(t) = 27t^3 - 18St - S^2t^{-1}.$$

The derivative of this function is,

$$f'(t) = 81t^2 - 18S + S^2t^{-2} = (9t - St^{-1})^2 > 0 \text{ unless } 9t^2 = S;$$

and even for such a value of t, the function would only be for a moment stationary, in its positive progress, from

$$f(-\infty) = -\infty, \quad \text{to} \quad f(0_-) = +\infty;$$

where by 0_- I denote an *infinitely small* but *negative* value; or again in its (*still positive*) progress, from

$$f(0_+) = -\infty \quad \text{to} \quad f(+\infty) = +\infty,$$

where 0_+ denotes a *positive infinitesimal.* Hence, while the real variable t advances from $-\infty$ to $+\infty$, the function $f(t)$ *passes twice,* but *not more than twice,* through *any assigned and real value,* T, whether that *given* value be positive or negative. Consequently, *the biquadratic equation,*

$$0 = tf(t) - tT, \quad \text{or} \quad 27t^4 - 18St^2 - Tt - S^2 = 0,$$

has, generally, two real and unequal roots, and *not more than two*; it must therefore in general have *two imaginary roots*; the *coefficients* S and T being always supposed to be real. – Determinants, no doubt, might be employed, and in part I have made a verification by them. – It may be noticed that the case $f'(t) = 0$, above hinted at, will not give *equal roots* in the *biquadratic* equation, unless we have at the same time

$$T = f(t) = \mp 8S^{\frac{3}{2}}, \quad \text{and therefore} \quad R = 64S^3 - T^2 = 0.$$

102. In some such way, it is probably well known that *two* of the four roots of that biquadratic are generally imaginary: but there may be a *chance* of novelty in the *form* of those *four* roots to which the foregoing discussion has conducted me; and which had offered themselves also to my observation, in discussing, with a similar view, the form of U in article 86. These forms are:

$$\frac{\lambda_0}{\mu_0} = +\alpha^2; \quad \frac{\lambda_1}{\mu_1} = -\frac{(\alpha+\beta)^2}{3}; \quad \frac{\lambda_2}{\mu_2} = -\frac{(\alpha+\theta\beta)^2}{3}; \quad \frac{\lambda_3}{\mu_3} = -\frac{(\alpha+\theta^2\beta)^2}{3};$$

where α and β are real quantities, of which the first may be taken as positive. (As compared with the formulæ of 100, θ and θ^2 are here interchanged.) To test the *generality* of these forms, – which are to *me* quite *new*, – I form the product

$$0 = (t - 3\alpha^2)\{t + (\alpha+\beta)^2\}\{t + (\alpha+\theta\beta)^2\}\{t + (\alpha+\theta^2\beta^2)\},$$

and find it to be,

$$0 = t^4 - 6\alpha(\alpha^3+\beta^3)\,t^2 - (8\alpha^6 - 20\alpha^3\beta^3 - \beta^6)\,t - 3\alpha^2(\alpha^3+\beta^3)^2;$$

comparing which with the biquadratic

$$0 = t^4 - 6St^2 - Tt - 3S^2, \quad \left(\text{where } t = \frac{3\lambda}{\mu}\right),$$

it only remains to prove that the *two equations*

$$\alpha(\alpha^3+\beta^3) = S \quad \text{and} \quad 8\alpha^6 - 20\alpha^3\beta^3 - \beta^6 = T,$$

can be jointly satisfied by *real values* of α and β, *whatever real values* (including *signs*) may be proposed for S and T: and this I conceive that I can do.

103. For this purpose I first eliminate β by assuming

$$\beta = \alpha\sqrt[3]{(\gamma-1)}$$

which reduces the two equations to be satisfied, to the forms:

$$\alpha^4\gamma = S; \quad \alpha^6(27 - 18\gamma - \gamma^2) = T.$$

And now we see that it is sufficient to prove that the *new biquadratic equation,*

$$(\gamma^2 + 18\gamma - 27)^2 = \frac{T^2}{S^3}\gamma^3 = 0,$$

has at least one real root, γ, which satisfies the condition,

$$\frac{\gamma^2 + 18\gamma - 27}{T} < 0;$$

S and T having any given values, supposed at first to be different from zero.

104. Consider the *biquadratic curve,*

$$(x^2 + 18x - 27)^2 - yx^3 = 0;$$

which is easily found to give

$$(x-9)^2(x^2+18x-27) - y'x^4 = 0,$$

if $y' = \dfrac{dy}{dx}$; or, slightly changing the notation,

$$y_x = x^{-3}(x-x_1)^2(x-x_2)^2,$$
$$y'_x = x^{-4}(x-9)^2(x-x_1)(x-x_2),$$
$$x_1 = -9-6\sqrt{3}, \quad x_2 = -9+6\sqrt{3}.$$

Attributing (or conceiving as attributed) to x all real values in succession, from negative to positive infinity, and examining the correspondent march of y, or y_x, we have,

$$y_x < 0, \quad y'_x > 0, \quad \text{if} \quad x < x_1;$$
$$y_x < 0, \quad y''_x < 0, \quad \text{if} \quad x > x_1, \quad < 0;$$
$$y_x > 0, \quad y'_x < 0, \quad \text{if} \quad x > 0, \quad < x_2;$$
$$y_x > 0, \quad y'_x > 0, \quad \text{if} \quad x > x_2, \quad < 9, \quad \text{or} \quad \text{if } x > 9;$$

with the following particular values:

$$y_{-\infty} = -\infty; \quad y_{x_1} = 0; \quad y_{0_-} = -\infty; \quad y_{0_+} = +\infty; \quad y_{x_2} = 0; \quad y_9 = 64; \quad y_{+\infty} = +\infty;$$
$$y'_{-\infty} = +1; \quad y'_{x_1} = 0; \quad y'_{0_-} = -\infty; \quad y'_{0_+} = -\infty; \quad y'_{x_2} = 0; \quad y'_9 = 0; \quad y'_{+\infty} = +1;$$

to which it may be added that not only y'_x but also y''_x or $\dfrac{d^2y}{dx^2}$ vanishes for $x = 9$, so that we may write,

$$y''_9 = 0.$$

105. The curve of which the equation may be written thus,

$$y = x + 36 + \frac{270}{x} - \frac{972}{x^2} + \frac{729}{x^3},$$

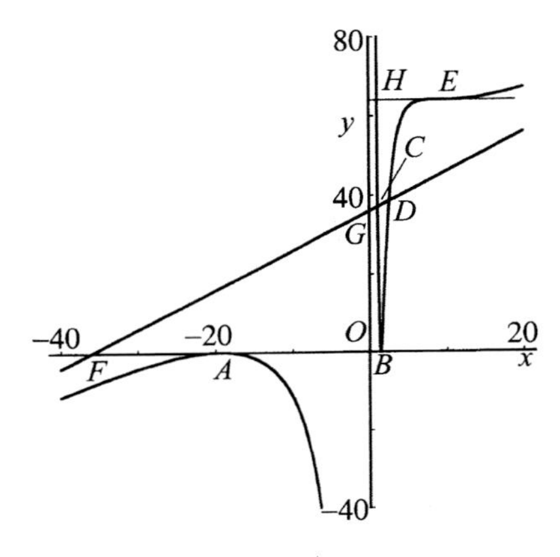

Fig. 27

and which has therefore the right line $y = x + 36$ for an asymptote, must consequently have a *shape* somewhat like that which is sketched in figure 27; being *below* this asymptote for large negative values of x, but *above* it for large positive values, and *crossing* it in two real and finitely distant points, $(x_1', x_1' + 36)$, and $(x_2', x_2' + 36)$, when x_1', x_2' are the two roots of the quadratic,

$$10x'^2 - 36x' + 27 = 0;$$

so that

$$x_1' = \frac{3}{10}(6 - \sqrt{6}), \quad x_2' = \frac{3}{10}(6 + \sqrt{6})$$

or nearly

$$x_1' = +1{\cdot}07, \quad x_2' = +2{\cdot}53;$$

while we had

$$x_2 = -9 + 6\sqrt{3} = +1{\cdot}39, \text{ nearly;}$$

so that

$$x_2 > x_1', \quad < x_2'.$$

The curve, after leaving the above mentioned asymptote at $(-\infty, -\infty)$, ascends until it comes to touch the axis of x, at a point $A = (x_1, 0)$, when $x_1 = -19{\cdot}39$, nearly; it then descends again, until it touches the axis of y, which is another asymptote, at $(0, -\infty)$. But this is, in modern Geometry, the same point as $(0, +\infty)$; and accordingly there arises immediately a new infinite branch, which descends until it comes to touch the axis of x, at the point $B = (x_2, 0)$, having crossed the former asymptote on its way, at a point $C = (x_1', 36 + x_1')$. It then ascends for ever afterwards, crossing that asymptote FG, – when $F = (-36, 0)$, and $G = (0, 36)$, – in a point $D = (x_2', 36 + x_2')$, after undergoing (as it seems) an inflexion between B and D; but its velocity of ascent may be said to vanish for a moment, at a point of *horizontal inflexion*, $E = (9, 64)$.

106. The foregoing *construction*, of this *auxiliary quartic*, makes it evident, *even to the eye*, that *every right line*, which coincides with neither the axis AB of x, nor with the parallel tangent of inflexion EH, but is *parallel* to each of those two lines, *cuts the curve in two real and distinct points*, and in *no more than two*: so that *if any real value*, excepting 0 and 64, *be substituted for the ordinate* y, *in the biquadratic equation* of article 104, namely in

$$0 = (x^2 + 18x - 27)^2 - yx^3$$

that equation must have two of its roots real and unequal, and *coincident* points at E, and again in a *fourth real point* H; and accordingly, by supposing $y = 64$, the biquadratic becomes,

$$\begin{aligned} 0 &= (x^2 + 18x - 27)^2 - 64x^3 \\ &= (x - 9)^3(x - 1); \end{aligned}$$

so that the point $H = (1, 64)$.

107. Of course, it was *not necessary* to have thus *constructed the curve,* in order to deduce these results, respecting the biquadratic *equation.* We might have simply considered the *function* y_x of article 104, and have observed that by the formula of that article,

$$\begin{cases} y_x \text{ increases (constantly and continuously) from } -\infty \text{ to } 0, \\ \quad \text{while } x \text{ increases from } -\infty \text{ to } x_1; \\ y_x \text{ decreases (\&c.) from } 0 \text{ to } -\infty, \text{ while } x \text{ increases from } x_1 \text{ to } 0_-; \\ y_x \text{ decreases (\&c.) from } +\infty \text{ to } 0, \text{ while } x \text{ increases from } 0, \text{ to } x_2; \\ y_x \text{ increases (\&c.) from } 0 \text{ to } +\infty, \text{ while } x \text{ increases from } x_2 \text{ to } +\infty; \end{cases}$$

its *rate* of increase vanishing, however, for a moment, in the last of these four intervals, for $x = 9$, $y = 64$. For this would have been sufficient to prove, that if (without any reference to geometry) we regard y as an arbitrary but real constant in the biquadratic equation in article 104, this equation in x will have the following solutions:

I. two real and unequal roots, and two imaginary roots if $y \gtrless 0$, and $\lessgtr 64$;or
II. two distinct pairs of equal and real roots, x_1, x_1, and x_2, x_2, if $y = 0$; or
III. four real roots, 1, 9, 9, 9, whereof the three are equal to each other if $y = 64$.

108. Returning now to the equations of article 103, we see that if (T^2/S^3) have *any given and real value,* distinct from 0 and from 64, *two real and unequal values* (but *not more*) can be found for γ, which shall each satisfy the biquadratic equation of *that* article. There was however a certain *condition of inequality* to be *also* satisfied, if possible, by γ; and we now see that it *is* possible to satisfy that condition likewise. For the preceding analysis shows that *one* of the two real values of γ falls *within,* and that the *other* of those two values falls *without,* the interval between the two roots of the quadratic equation,

$$\gamma^2 + 18\gamma - 27 = 0;$$

according, then, as the *given* T is *positive or negative,* we take the *former* or the *latter* of these two values of γ: and then, in each case, the condition in article 103,

$$T^{-1}(\gamma^2 + 18\gamma - 27) < 0,$$

will be satisfied. And if we make α the real and positive value of

$$\left(\frac{T}{27 - 18\gamma - \gamma}\right)^{\frac{1}{6}}$$

and β the real value of

$$\alpha(\gamma - 1)^{\frac{1}{3}},$$

we have $(S/\gamma) = \alpha^4$, and are led back to the final equations of article 102, namely,

$$\alpha(\alpha^3 + \beta^3) = S, \quad 8\alpha^6 - 20\alpha^3\beta^3 - \beta^6 = T;$$

which we now see to admit of being satisfied by *one,* but by *only one,* system of *real values* of α, β, with the *condition* $\alpha > 0$, if S, T, and $R(= 64S^3 - T^2)$ are *all real,* and *different from zero.*

109. As an *example* of this determination of the real constants α, β, γ, when S and T are given, let us suppose that U has the known *canonical form,*

$$U = x^3 + y^3 + z^3 + 6exyz,$$

whence (*HPC*, p. 183, §198)

$$H = e^2(x^3 + y^3 + z^3) - (1 + 2e^3)xyz,$$

$$S = e^4 - e, \ T = 1 - 20e^3 - 8e^6.$$

We are then to satisfy, by real values of α, β, γ, the equations,

$$\alpha^4 + \alpha\beta^3 = e^4 - e; \quad 8\alpha^6 - 20\alpha^3\beta^3 - \beta^6 = 1 - 20e^3 - 8e^6; \quad \gamma = \frac{\alpha^3 + \beta^3}{\alpha^3};$$

for which purpose we are led to consider the resulting biquadratic,

$$0 = e^3(e^3 - 1)^3(\gamma^2 + 18\gamma - 27)^2 - (8e^6 + 20e^3 - 1)^2\gamma^3.$$

But the same biquadratic in γ would have been obtained, if the second equation to be satisfied had been

$$8\alpha^6 - 20\alpha^3\beta^3 - \beta^6 = 8e^6 + 20e^3 - 1;$$

and in *that* case we might have assumed

$$\alpha = e, \quad \beta = -1, \quad \gamma = 1 - e^{-3};$$

this last value of γ must therefore be a root of the biquadratic, but a root that is to be rejected, as answering to a wrong sign of T. Dividing then by $(e^3\gamma + 1 - e^3)$, we are led to the cubic equation,

$$0 = (e^3 - 1)^3\gamma^3 - 27e^3(e^6 + 16e^3 + 10)\gamma^2 + 243e^3(e^3 - 1)(e^3 - 4)\gamma - 729e^3(e^3 - 1)^2;$$

which I find to break up into two rational factors, one linear, namely,

$$0 = (e - 1)^3\gamma - 9e(e^2 + e + 1),$$

and the other quadratic, namely,

$$0 = (e^2 + e + 1)^3\gamma^2 - 9e(e - 1)(2e^4 + 4e^3 - 6e^2 - 8e - 1)\gamma + 81e^2(e - 1)(e^3 - 1).$$

The roots of the quadratic factor must, by the theory, be imaginary, and are here to be set aside. Confining ourselves therefore to the linear factor, we have the value,

$$\gamma = \frac{9e(e^2 + e + 1)}{(e - 1)^3};$$

whence

$$\frac{(e - 1)^6}{27}(\gamma^2 + 18\gamma - 27) = 3e^2(e^2 + e + 1)^2 + 6(e - 1)^3e(e^2 + e + 1) - (e - 1)^6$$

$$= 8e^6 + 20e^3 - 1 = -T;$$

so that the condition,

$$T^{-1}(\gamma^2 + 18\gamma - 27) < 0,$$

of article 103, is satisfied by *this* value of γ, but *not* by the rejected value $(1 - e^{-3})$, which would have given, if we denote *that* value by γ',

$$T^{-1}(\gamma'^2 + 18\gamma' - 27) = T^{-1}(-8 - 20e^{-3} + e^{-6}) = e^{-6} > 0.$$

Making next, by article 108,

$$\alpha^{-6} = T^{-1}(27 - 18\gamma - \gamma^2) = 27(e - 1)^{-6},$$

we infer that, since α is to be real we must take

$$\alpha = \frac{e - 1}{\pm\sqrt{3}};$$

and then by the same article,

$$\frac{\beta}{\alpha} = \sqrt[3]{\gamma - 1} = \frac{2e + 1}{e - 1},$$

so that

$$\beta = \frac{2e + 1}{\pm\sqrt{3}}.$$

Or, taking the sign $+$ for each radical whether e be $>$ or < 1, (since it is not really important whether α be positive or negative, provided that $\frac{\beta}{\alpha}$ has the proper sign,) we may write definitely,

$$\alpha = \frac{e - 1}{\sqrt{3}}, \quad \beta\frac{2e + 1}{\sqrt{3}},$$

as sufficiently representing, in the present example, the real system of values of α and β, determined by the two given equations,

$$\alpha^4 + \alpha\beta^3 = S, \quad 8\alpha^6 - 20\alpha^3\beta^3 - \beta^6 = T;$$

their satisfying which may be verified by actual substitution.

110. When real constants, α and β, have in any manner been found so as to satisfy the last two equations, for any given values of S and T, of which neither vanishes, and which are not connected by the relation $R = 0$, it follows from article 102 that the biquadratic equation,

$$0 = \left(\frac{3\lambda}{\mu}\right)^4 - 6S\left(\frac{3\lambda}{\mu}\right)^2 - T\left(\frac{3\lambda}{\mu}\right) - 3S^2,$$

has its four roots expressible as follows:

$$\frac{3\lambda_0}{\mu_0} = +3\alpha^2; \quad \frac{3\lambda_1}{\mu_1} = -(\alpha + \beta)^2; \quad \frac{3\lambda_2}{\mu_2} = -(\alpha + \theta\beta)^2; \quad \frac{3\lambda_3}{\mu_3} = -(\alpha + \theta^2\beta)^2.$$

In other words, the four following equations represent each a system of three right lines, passing through the 9 points of inflexion of U:

$$\begin{cases} \text{I.} & \alpha^2 U + H = 0; \\ \text{II.} & (\alpha+\beta)^2 U - 3H = 0; \\ \text{III.} & (\alpha+\theta\beta)^2 U - 3H = 0; \\ \text{IV.} & (\alpha+\theta^2\beta)^2 U - 3H = 0. \end{cases}$$

And from comparison of these general forms with those in articles 99, 100, and with the corresponding forms derived from the case in article 86, where

$$U = k(x+y+z)^3 - 27xyz,$$

it appears that equation I, represents *generally a system of one real right line*, containing the *three real points of inflexion* of *U, and of an imaginary* or *infinitesimal conic*, which is equivalent to a *pair of imaginary right lines*, meeting each other in *one real point* (but *not* in a point of inflexion,) and containing the *six imaginary points of inflexion* of the cubic curve (cf., articles 88 and 97). On the other hand, equation II. represents generally a *system of three real right lines*, whereof *each* passes through *one real point of inflexion* and through *two imaginary points* of inflexion of *U*. Finally, each of the two remaining equations, III. and IV., represents a *system of three imaginary right lines*, whereof each still passes through one real point, and through two imaginary points of inflexion.

I do not know whether the *distinction*, thus drawn, between the geometrical significations of the *two* (*positive* and *negative* but) *real roots of the biquadratic equation* in $\frac{\lambda}{\mu}$, which expresses the condition necessary for $\lambda U + \mu H$ being a product of the three linear factors, (*HPC*, p. 187, §201) has been published, or perceived by anyone.

111. To verify it for the *canonical form* of *U* in article 109, I observe that, for this form, the equations I. and II. of article 110 become, by article 109,

$$\text{I.}\quad (e-1)^2 U + 3H = 0; \qquad \text{II.}\quad e^2 U - H = 0;$$

that is, here, after dividing by $4e^2 - 2e + 1$, and by $8e^3 + 1$, respectively,

$$\text{I.}\quad x^3 + y^3 + z^3 - 3xyz = 0; \qquad \text{II.}\quad xyz = 0.$$

The latter is evidently a system of *three real* right *lines*, eash passing through *one real point* of inflexion. The form may be written thus;

$$\text{I.}\quad (x+y+z)(x+\theta^2 y+\theta z)(x+\theta y+\theta^2 z) = 0;$$

or,

$$\text{I.}\quad (x+y+z)(x^2+y^2+z^2 - yz - zx - xy) = 0;$$

it consists therefore of *one real right line*, passing through the *three real points* of inflexion, and of an *imaginary* conic through the *six other points*. All which agrees with the general theory enunciated in article 110.

112. We may as well verify equations III. and IV. of that article, by means of the same canonical form of *U*. It is easily found that the values in article 109 of α and β in terms of *e*, give

$$-(\alpha+\theta\beta)^2 = (e-\theta^2)^2, \quad -(\alpha+\theta^2\beta)^2 = (e-\theta)^2;$$

the two remaining but imaginary systems of right lines are therefore here,

$$\text{III.}\quad 0 = (e - \theta^2)^2 U + 3H; \qquad \text{IV.}\quad 0 = (e - \theta)^2 U + 3H.$$

Accordingly, there is no difficulty in seeing that

$$\frac{(e - \theta^2)^2 U + 3H}{4e^2 - 2\theta^2 e + \theta} = x^3 + y^3 + z^3 - 3\theta^2 xyz,$$

and that

$$\frac{(e - \theta)^2 U + 3H}{4e^2 - 2\theta^2 e + \theta^2} = x^3 + y^3 + z^3 - 3\theta xyz;$$

so that these two remaining systems are,

$$\text{III.}\quad 0 = (x + \theta y + \theta z)(x + \theta^2 y + z)(x + y + \theta^2 z),$$

and

$$\text{IV.}\quad 0 = (x + \theta^2 y + \theta^2 z)(x + \theta y + z)(x + y + \theta z).$$

And in fact, if we denote the nine points of inflexion (for the canonical U), as follows:

$$\begin{cases} K' = (0,\ 1,\ -1), & I' = (0,\ 1,\ -\theta), & J' = (0,\ 1,\ -\theta^2), \\ K'' = (-1,\ 0,\ 1), & I'' = (-\theta,\ 0,\ 1), & J'' = (-\theta^2,\ 0,\ 1), \\ K''' = (1,\ -1,\ 0), & I''' = (1,\ -\theta,\ 0), & J''' = (1,\ -\theta^2,\ 0), \end{cases}$$

we find twelve formulæ of collineation between them, which may be briefly written thus, on nearly the same plane as in article 99:

$$\begin{cases} \text{I.} & (K'K''K''') = x + y + z; & (I'I''I''') = x + \theta^2 y + \theta z; & (J'J''J''') = x + \theta y + \theta^2 z; \\ \text{II.} & (K'I'J') = x; & (K''I''J'') = y; & (K'''I'''J''') = z; \\ \text{III.} & (K'I''J''') = x + \theta y + \theta z; & (K''J'I''') = x + \theta^2 y + z; & (K'''I'J'') = x + y + \theta^2 z; \\ \text{IV.} & (K'J''I''') = x + \theta^2 y + \theta^2 z; & (K''I'J''') = x + \theta y + z; & (K'''J'I''') = x + y + \theta z. \end{cases}$$

113. Suppose that in this, or in some other way, we had found that for the canonical form of U in article 109, the biquadratic in $\frac{3\lambda}{\mu}$, or t, becomes,

$$\begin{aligned} 0 &= t^4 - 6St^2 - Tt - 3S^2 \\ &= t^4 - 6(e^4 - e)\,t^2 + (8e^6 + 20e^3 - 1)\,t - 3(e^4 - e)^2 \\ &= (t + 3e^2)\{t - (e - 1)^2\}\{t - (e - \theta)^2\}\{t - (e - \theta^2)^2\}, \end{aligned}$$

without any *previous* knowledge of the values of α and β; we might have proceeded to *find* (instead of *using*) those values, by comparing this last equation with the corresponding equation of article 102,

$$0 = (t - 3\alpha^2)\{t + (\alpha + \beta)^2\}\{t + (\alpha + \theta\beta)^2\}\{t + (\alpha + \theta^2\beta)^2\}.$$

Thus we should have,

$$3\alpha^2 = (e-1)^2; \quad (\alpha+\beta)^2 = 3\alpha^2;$$

and *either*

$$(\alpha + \theta\beta)^2 = -(e-\theta)^2,$$

or

$$(\alpha + \theta\beta)^2 = -(e-\theta^2)^2;$$

but a little examination would show that the latter alternative was to be adopted, and would conduct to the same values as in article 109, namely,

$$\alpha = \frac{e-1}{\sqrt{3}}, \quad \beta = \frac{2e+1}{\sqrt{3}}.$$

114. Again, if we start with the following modification of the form in article 86,

$$U = (c^3-1)(x+y+z)^3 + 27xyz,$$

which gives

$$27U + 4H = 108(c^3-1)(x^3+y^3+z^3-3xyz),$$

we have the 9 following points of inflexion:

$$\begin{cases} K' = (0,\,1,\,-1), & K'' = (-1,\,0,\,1), & K''' = (1,\,-1,\,0); \\ I' = (c-1,\,\theta c-1,\,\theta^2 c-1), & I'' = (\theta^2 c-1,\,c-1,\,\theta c-1), & I''' = (\theta c-1,\,\theta^2 c-1,\,c-1); \\ J' = (c-1,\,\theta^2 c-1,\,\theta c-1), & J'' = (\theta c-1,\,c-1,\,\theta^2 c-1), & J''' = (\theta^2 c-1,\,\theta c-1,\,c-1); \end{cases}$$

with the 4 groups of right lines (s being equal to $x+y+z$):

$$\begin{cases} \text{I.} & (K'K''K''') = x+y+z, & (I'I''I''') = x+\theta y+\theta^2 z, & (J'J''J''') = x+\theta^2 y+\theta z, \\ \text{II.} & (K'I'J') = 3x+(c-1)s, & (K''I''J'') = 3y+(c-1)s, & (K'''I'''J''') = 3z+(c-1)s; \\ \text{III.} & (K'I'''J'') = 3x+(\theta c-1)s, & (K''I'J''') = 3y+(\theta c-1)s, & (K'''I''J') = 3z+(\theta c-1)s; \\ \text{IV.} & (K'I''J''') = 3x+(\theta^2 c-1)s, & (K''I'''J') = 3y+(\theta^2 c-1)s, & (K'''I'J'') = 3z+(\theta^2 c-1)s; \end{cases}$$

and the four products:

$$\begin{cases} \text{I.} & K'K''K'''.I'I''I'''.J'J''J''' = \dfrac{27U+4H}{108(c^3-1)}; \\ \text{II.} & K'I'J'.K''I''J''.K'''I'''J''' = \dfrac{9(2c+1)^2U-4H}{36(c^2+c+1)}; \\ \text{III.} & K'I''J''.K''I'J'''.K'''I''J' = \dfrac{9(2\theta c+1)^2U-4H}{36(\theta^2c^2+\theta c+1)}; \\ \text{IV.} & K'I''J'''.K''I'''J'.K'''I'J'' = \dfrac{9(2\theta^2 c+1)^2U-4H}{36(\theta c^2+\theta^2 c+1)}. \end{cases}$$

The four values of $\dfrac{\lambda}{\mu}$ are, therefore, in this Example,

$$\frac{\lambda_0}{\mu_0} = +\frac{27}{4}; \quad \frac{\lambda_1}{\mu_1} = -\frac{9}{4}(2c+1)^2; \quad \frac{\lambda_2}{\mu_2} = -\frac{9}{4}(2\theta c+1)^2; \quad \frac{\lambda_3}{\mu_3} = -\frac{9}{4}(2\theta^2 c+1)^2;$$

and we have thus a verification (above alluded to) of the Theorem of article 110; for we see that the *real* and *positive value* of $\frac{\lambda}{\mu}$ answers here again (as in the cases before considered) to a system of *one real* and *two imaginary right lines*, whereof the former connects the *three real points* of inflexion, while the latter pass through the *six imaginary points*; and that, on the other hand, the *real and negative value* of $\frac{\lambda}{\mu}$ corresponds (in this, as in other instances) to a system of *three real right lines*, whereof *each* passes through *one real point* of inflexion, and through *two imaginary points*. (– Have you elsewhere met with this *distinction*, of article 110, between the geometrical significations of the *two real roots* of the biquadratic $\frac{\lambda}{\mu}$?).

115. Comparing the recent values of the four roots of that biquadratic, with the forms in article 102, namely with

$$\frac{\lambda_0}{\mu_0} = \alpha^2, \quad \frac{\lambda_1}{\mu_1} = -\frac{1}{3}(\alpha+\beta)^2, \quad \frac{\lambda_2}{\mu_2} = -\frac{1}{3}(\alpha+\theta\beta)^2, \quad \frac{\lambda_3}{\mu_3} = -\frac{1}{3}(\alpha+\theta^2\beta)^2,$$

we see that, in the present Example,

$$\alpha = \frac{3\sqrt{3}}{2}, \quad \beta = 3\sqrt{3}.c;$$

whence (by the cited article),

$$S = \alpha(\alpha^3+\beta^3) = 2^{-4}3^6(1+8c^3);$$

$$T = 8\alpha^6 - 20\alpha^3\beta^3 - \beta^6 = 2^{-3}3^9(1-20c^3-8c^6).$$

Accordingly, if we take the form,

$$U = a(x+y+z)^3 + 6(d-a)xyz,$$

and seek the values of S and T in terms of a and d, by the general formula of *HPC* (p. 184, §199) after correcting the signs of the last groups of terms in S, in conformity with the remark made in the Note* to page 113 of the "Lessons" on "Modern Higher Algebra", – which remark I may be excused for saying that I made for myself in 1857, while engaged in a correspondence, chiefly on Quaternions, with our friend Salmon, who appeared to be a little surprised at the time, and to have supposed that nobody but Cayley† would have detected the necessity of the correction, – but since the summer of that year, until quite recently, I have not been thinking of Cubics at all, – we find,

$$S = d^4 - 6a^2d^2 + 8a^3d - 3a^4 = (d-a)^3\{(d-a)+4a\};$$

$$T = -8(d-a)^4(d^2+4ad+a^2) = -8(d-a)^4\{(d-a)^2+6a(d-a)+6a^2\}.$$

Making then, more particularly,

$$U = a(x+y+z)^3 + 27xyz, \quad d-a = \frac{9}{2},$$

* [Salmon G., *Lessons introductory to the Modern Higher Algebra*, p. 113, 1st edn. Hodges Smith: Dublin 1859, referred to hereinafter as *LHA*.]

† [Arthur Cayley, 1821–1895.]

we have

$$4S = \left(\frac{27}{2}\right)^2 (9 + 8a), \qquad T = -\left(\frac{27}{2}\right)^3 (27 + 36a + 8a^2).$$

And if we finally write $(c^3 - 1)$ instead of a, we obtain,

$$4S = \left(\frac{27}{2}\right)^2 (1 + 8c^3), \qquad T = \left(\frac{27}{2}\right)^3 (1 - 20c^3 - 8c^6);$$

values agreeing perfectly with those found above, by a totally different method, with the assistance of the *two constants*, α, and β: which if not really new, are at least such to me.

116. I shall employ those new (or supposed new) constants, α, β, to assist in resolving the following Problem:-

"To assign the coefficients, a and d, of the Canonical Form of the Cubic, thus written,

$$U = a(x^3 + y^3 + z^3) + 6dxyz,$$

in terms of the constant c, so as to make the S and T of *this* form *coincide*, respectively, with the S and T of the form in article 114, namely,

$$U = (c^3 - 1)(x + y + z)^3 + 27xyz."$$

In other words, we are to assign real values of a and d, which shall satisfy, the two algebraic equations, (cf. *HPC*, p. 183, §198, and the last article, 409, of the present Letter,)

$$(S =)\ d^4 - a^3 d = 2^{-4} 3^6 (1 + 8c^3);$$

$$(T =)\ a^6 - 20a^3 d^3 - 8d^6 = 2^{-3} 3^9 (1 - 20c^3 - 8c^6);$$

c being still supposed to have a real value.

117. For this purpose I observe that the second form of U in article 115, gives $\frac{\beta}{\alpha} = 2c$; and that the second form gives, by the results of article 109,

$$\frac{\beta}{\alpha} = \frac{2e + 1}{e - 1},$$

where $e = \frac{d}{a}$; comparing, therefore, we have the equation

$$2c = \left(\frac{\beta}{\alpha} =\right) \frac{2d + a}{d - a};$$

where

$$d = \frac{(2c + 1)\,a}{2(c - 1)}.$$

Substituting this value for d, in the two last equations of article 116, and observing that

$$(2c + 1)^4 - 8(c - 1)^3 (2c + 1) = 9(1 + 8c^3),$$

and

$$8(c-1)^6 - 20(c-1)^3(2c+1)^3 - (2c+1)^6 = 27(1 - 20c^3 - 8c^6)$$

we find that these equations become,

$$a^4 = 3^4(c-1)^4, \quad a^6 = 3^6(c-1)^6;$$

they are therefore both satisfied by our supposing

$$a = 3(c-1), \quad d = 3(c+\tfrac{1}{2}),$$

although of course the signs of both a and d may be changed together.

Hence "the function,

$$U = 3(c-1)(x^3 + y^3 + z^3) + 9(2c+1)xyz,$$

which may also be thus written,

$$U = 3(c-1)\,ss's'' + 27cxyz,$$

where

$$s = x + y + z, \quad s' = x + \theta y + \theta^2 z, \quad s'' = x + \theta^2 y + \theta z,$$

while c is still an *arbitrary constant*, has *the same* S, *and the same* T, namely,

$$S = 2^{-4}3^6(1 + 8c^3), \quad \text{and} \quad T = 2^{-3}3^9(1 - 20c^3 - 8c^6),$$

as the function $U = (c^3 - 1)s^3 + 27xyz$", or more fully,

$$U = (c^3 - 1)(x + y + z)^3 + 27xyz:$$

and the Problem of article 116 has been resolved.

118. Whatever values the constants α and β may have, the expressions in article 102 for S and T, in terms of those constants, may easily be decomposed into *linear factors*, as follows. First, we have evidently,

$$S = \alpha(\alpha^3 + \beta^3) = \alpha(\alpha + \beta)(\alpha + \theta\beta)(\alpha + \theta^2\beta);$$

or briefly,

$$S = \alpha\Pi_\iota(\alpha + \iota\beta),$$

where ι is *any cube root of* $+1$, and

$$\Pi_\iota f(\iota) = f(1)f(\theta)f(\theta^2).$$

Secondly, if we resolve the equation $T = 0$, as a quadratic in $\dfrac{8\alpha^3}{\beta^3}$, we get the two roots,

$$\frac{8\alpha^3}{\beta^3} = 10 \pm \sqrt{108} = (1 \pm \sqrt{3})^3;$$

writing therefore

$$\alpha_1 = \frac{1 - \sqrt{3}}{2}, \quad \alpha_2 = \frac{1 + \sqrt{3}}{2},$$

we have

$$T = 8\alpha^6 - 20\alpha^3\beta^3 - \beta^6 = 8(\alpha^3 - \alpha_1^3\beta^3)(\alpha^3 - \alpha_2^3\beta^3)$$
$$= 8(\alpha - \alpha_1\beta)(\alpha - \theta\alpha_1\beta)(\alpha - \theta^2\alpha_1\beta)(\alpha - \alpha_2\beta)(\alpha - \theta\alpha_2\beta)(\alpha - \theta^2\alpha_2\beta)$$
$$= 8\Pi_\iota(\alpha - \iota\alpha_1\beta)(\alpha - \iota\alpha_2\beta);$$

and the required decomposition has been effected.

119. Since

$$2(\alpha - \iota\alpha_1\beta)(\alpha - \iota\alpha_2\beta) = 2\alpha^2 - 2\iota\alpha\beta - \iota^2\beta^2,$$

we have

$$T^2 = \Pi_\iota(2\alpha^2 - 2\iota\alpha\beta - \iota^2\beta^2)^2 = \Pi_\iota(4S - \iota B) = 64S^3 - B^3,$$

where

$$B = \beta(8\alpha^3 - \beta^3);$$

but the known

$$R = 64S^3 - T^2;$$

we have therefore

$$\sqrt[3]{R} = \beta(8\alpha^3 - \beta^3) = \beta\Pi_\iota(2\alpha - \iota\beta).$$

Thus, making $\alpha = (3/2)\sqrt{3}$, $\beta = (3\sqrt{3})c$, as in article 115, we have, for *each* of the two functions U of article 117, the *common value*,

$$R = 3^6 c(1 - c^3).$$

120. Let us now apply a similar analysis to the solution of this other Problem:-

"To assign real values of the coefficients g and h, in the form

$$U = 9(gs^3 + hzp),$$

where (as in article 92)

$$s = x + y + z, \quad \text{and} \quad p = x^2 + xy + y^2,$$

so that the resulting S and T may have the recent values (in articles 115, 116, 117),

$$S = \frac{1}{4}\left(\frac{27}{2}\right)^2 (1 + 8c^3), \quad \text{and} \quad T = \left(\frac{27}{2}\right)^3 (1 - 20c^3 - 8c^6)."$$

In other words (cf. article 92) we are to find two real, and if possible rational functions, g and h, of the real constant c, which shall satisfy the two algebraic equations

$$h^3(h + 8g) = 1 + 8c^3 h^4(h^2 + 12gh + 24g^2) = 1 - 20c^3 - 8c^6.$$

121. To solve this problem on the same general plan as before, I compare (after interchanging θ and θ^2, as remarked in article 102) the expressions of article 100 for the four roots of the biquadratic in $\left(\dfrac{\lambda}{\mu}\right)$, namely,

$$\frac{\lambda_0}{\mu_0} = \frac{3k^2 t^2}{(t+2v)^2}, \quad \frac{\lambda_1}{\mu_1} = -k^2, \quad \frac{\lambda_2}{\mu_2} = -k^2\left(\frac{t+2\theta v}{t+2v}\right)^2, \quad \frac{\lambda_3}{\mu_3} = -k^2\left(\frac{t+2\theta^2 v}{t+2v}\right)^2,$$

in which $k = \left(\frac{h}{6g}\right)$, and $\left(\frac{v}{t}\right)$ is the real root of the cubic equation

$$U_{t,v,v} = 0,$$

if $U_{x,y,z}$ be written for U, with the general forms of article 102,

$$\frac{\lambda_0}{\mu_0} = \alpha^2, \quad \frac{\lambda_1}{\mu_1} = -\frac{1}{3}(\alpha+\beta)^2, \quad \frac{\lambda_2}{\mu_2} = -\frac{1}{3}(\alpha+\theta\beta)^2, \quad \frac{\lambda_3}{\mu_3} = -\frac{1}{3}(\alpha+\theta^2\beta)^2;$$

and thus obtain, for the present form of U, the following values of the constants α and β:

$$\alpha = \frac{kt\sqrt{3}}{t+2v}; \quad \beta = \frac{2kv\sqrt{3}}{t+2v}.$$

Comparing these with the values of article 115, namely

$$\alpha = \frac{3\sqrt{3}}{2}, \quad \beta = 3c\sqrt{3},$$

we find that we are to make

$$\frac{2kt}{t+2v} = 3, \quad \frac{2kv}{t+2v} = 3c;$$

and therefore $v = ct$; so that the constants of U must be such as to allow the equation

$$U_{1,c,c} = 0$$

to subsist, and therefore we must have

$$\frac{h}{g} = -\frac{(1+2c)^3}{c(1+c+c^2)}, \quad \text{or} \quad \frac{g}{h} = -\frac{c(1+c+c^2)}{(1+2c)^3}.$$

But we have, *identically*,

$$(1+2c)^3 - 8c(1+c+c^2) = 1 - 2c + 4c^2 = \frac{1+8c^3}{1+2c};$$

and

$$(1+2c)^6 - 12c(1+c+c^2)(1+2c)^3 + 24c^2(1+c+c^2)^2 = 1 - 20c^3 - 8c^6;$$

the two last equations of article 120 become, therefore

$$h^4 = (1+2c)^4, \quad h^6 = (1+2c)^6,$$

and agree in giving $h^2 = (1+2c)^2$. We may then take, as the solution of the problem of article 120, the values,

$$h = -(1+2c), \quad g = \frac{c(1+c+c^2)}{(1+2c)^2},$$

giving

$$U = \frac{9c(1+c+c^2)}{(1+2c)^2}s^3 - 9(1+2c)zp;$$

although, no doubt, the signs of g, h, U might all be changed together.

122. Collecting recent results, we see that the *three* following *distinct forms* of the *cubic function*, namely,

$$\begin{cases} \text{I.} & U = 3(c-1)ss's'' + 27cxyz; \\ \text{II.} & U = (c^3-1)s^3 + 27xyz; \\ \text{III.} & U = 9c(1+c+c^2)(1+2c)^{-2}s^3 - 9(1+2c)zp, \end{cases}$$

(in which we still employ the abridgments,

$$s = x+y+z, \quad s' = x+\theta y+\theta^2 z, \quad s'' = x+\theta^2 y+\theta z, \quad \text{and} \quad p = x^2+xy+z^2,$$

while c is still an arbitrary constant,) have *all* the *same values* of S, T, and of course R: since, for *each* of these three functions, we have the expressions,

$$4S = \left(\frac{27}{2}\right)^2(1+8c^3); \quad T = \left(\frac{27}{2}\right)^3(1-20c^3-8c^6); \quad \sqrt[3]{R} = 3^6c(1-c^3);$$

which can all, by article 118, be decomposed into linear factors. Thus, in the notation of the last cited article, we may write,

$$s = 2^{-4}3^6\Pi_\iota(1+2\iota c); \quad T = -3^9\Pi_\iota(\alpha_1+\iota c)(\alpha_2+\iota c); \quad \sqrt[3]{R} = 3^6c\Pi_\iota(1-\iota c).$$

123. A few words may here be said, on those particular *cases* ($S=0$, $T=0$, $R=0$), which in several of the preceding articles we have set aside, when deducing the values of the constants α, β which answered to given values of S and T.

I. The case $S=0$, $T>0$, or <0, presents no special difficulty. We have only to consider the *ordinate*, $y = \left(\frac{T^2}{S^3}\right)$, of the biquadratic curve in Figure 27, as here becoming *infinite*; and therefore (by the nature of that curve) as corresponding *either* to an *infinite*, *or* to a *null value* of the *abscissa*, $x=\gamma$. The *former* supposition gives $\alpha=0$, and the *latter* gives $\alpha=-\beta$; with *each*, therefore, we satisfy, by article 102, the proposed condition $S=0$. But $\alpha=0$ gives $T=-\beta^6<0$, while $\alpha=-\beta$ gives $T=+27\beta^6>0$; the *distinction* between the *two real values of* γ, or between the two real points in which the curve in article 104 is cut by a given parallel to the axis of x, *depends* therefore *still*, as in the general method or process of article 108, *on the given sign of* T. Thus,

(1) if $S=0$, $T<0$, we take $\alpha=0, \beta=(-T)^{\frac{1}{6}}$; but

(2) if $S=0$, $T>0$, we then take $\alpha=-\beta=-\left(\frac{T}{27}\right)^{\frac{1}{6}}$.

124. In the first subcase, the biquadratic in $\left(\frac{\lambda}{\mu}\right)$ becomes,

$$0 = 27\left(\frac{\lambda}{\mu}\right)^4 + \beta^6\frac{\lambda}{\mu},$$

and its roots may be thus expressed,

$$\frac{\lambda_0}{\mu_0}=0,\quad \frac{\lambda_1}{\mu_1}=-\frac{\beta^2}{3},\quad \frac{\lambda_2}{\mu_2}=-\frac{(\theta\beta)^2}{3},\quad \frac{\lambda_3}{\mu_3}=-\frac{(\theta^2\beta)^2}{3};$$

in the second subcase, the biquadratic is

$$0=\left(\frac{\lambda}{\mu}\right)^4-\beta^6\left(\frac{\lambda}{\mu}\right),$$

and because

$$\theta=-\tfrac{1}{3}(\theta-1)^2,\quad \text{and}\quad \theta^2=-\tfrac{1}{3}(\theta^2-1)^2,$$

its roots may be written thus,

$$\frac{\lambda_0}{\mu_0}=\beta^2,\quad \frac{\lambda_1}{\mu_1}=0,\quad \frac{\lambda_2}{\mu_2}=-\tfrac{1}{3}(\theta\beta-\beta)^2,\quad \frac{\lambda_3}{\mu_3}=-\tfrac{1}{3}(\theta^2\beta-\beta)^2;$$

but, in each of these two subcases, the expressions for the four roots agree perfectly with the general forms of article 102. We can also verify here our general Theorem of article 110, respecting the *geometrical distinction between the two real roots* of the biquadratic equation in $\left(\frac{\lambda}{\mu}\right)$. For we must evidently consider the *null* root, in subcase (1), as the *limit* of the *positive* root; but in subcase (2), as the limit of the *negative* root. The theorem to which we refer leads us therefore to conclude, that when $S=0$, and $T<0$, the Hessian H is the product of *one real linear factor*, and of *two imaginary ones*; but that when $S=0$, $T>0$, H is, on the contrary, a product of linear factors, which are all *three real*. Now this pair of contrasted conclusions agrees precisely with what might have been otherwise inferred, from the principles of Salmon's book, on *Higher Plane Curves*. For he remarks (*HPC*, p. 184), that $S=0$ is the condition for the *Hessian H* breaking up into *three right lines*; and also that, if we set aside the case when U itself so breaks up, the equation $S=0$ expresses also the condition for the *given equation* $U=0$ being reducible to a *sum of three cubes*. Now there are *two ways*, and there seem to be *only two*, in which this *last* property can hold good; the given coefficients of the equation being always supposed to be *real*: namely, the *cubes* may be *all three real*, or *two* of them may be *imaginary*. But the *former* of these two alternatives answers to the form,

$$U=a^3x^3+b^3y^3+c^3z^3;$$

which gives (cf. *HPC*, p. 183),

$$S=0,\quad T=a^6b^6c^6>0,\quad H=-a^3b^3c^3xyz;$$

it belongs therefore to our recent subcase (2), and accordingly we see that the *three* linear factors of the Hessian are *all real here*: as was above expected, from the general Theorem in article 110 of the present Letter. On the other hand, if we take, in conformity with the *latter* alternative, respecting the composition of U,

$$U=a^3x^3+b^3y^3+c^3z^3+6abcxyz,$$

so that

$$3U = (ax + by + cz)^3 + (ax + \theta by + \theta^2 cz)^3 + (ax + \theta^2 by + \theta cz)^3,$$

we have *still* $S = 0$, but we have *now* $T = -27a^6b^6c^6 < 0$; we are therefore here in the subcase (1) of the present article, and accordingly we find (by the last cited page),

$$H = a^2b^2c^2(a^3x^3 + b^3y^3 + c^3z^3 - 3abcxyz)$$
$$= a^2b^2c^2(ax + by + cz)(ax + \theta by + \theta^2 cz)(ax + \theta^2 by + \theta cz);$$

two linear factors of H are therefore here *imaginary*, as by the general theorem of this Letter, above referred to, they ought to be.

125. Let us next consider briefly the case,

II. $T = 0$, $S > 0$, or < 0. In this case, the ordinate $y = \left(\frac{T^2}{S^3}\right)$ of the auxiliary quartic constructed in Figure 27, becomes equal to zero, instead of infinity; and because the axis of x has double contact with that curve, the abscissa $x = \gamma$ has four real values, equal however two by two, namely those which were denoted in article 107 by x_1, x_1, and x_2, x_2; where, article 104,

$$x_1 = -9 - 6\sqrt{3}, \quad x_2 = -9 + 6\sqrt{3},$$

so that these *two distinct and real values* of x, or of γ, are the two roots, one negative and the other positive, of the quadratic equation,

$$\gamma^2 + 18\gamma - 27 = 0.$$

But *both* these values of γ give $T = 0$; they cannot therefore be here *distinguished*, or *separated*, by any reference to the *sign of T*, which reference has hitherto been our resource. Fortunately, however, a *new* resource, which had *not* been previously available, starts up here, exactly when it is wanted. In general, *any parallel to the axis of x cuts the curve* of Figure 27 in *two distinct and real points*, which are *at one common side of the axis of y*; so that they give one *common sign* to the abscissa γ, and therefore also to $S = \alpha^4\gamma$ (see article 103). The *sign of S* is therefore *generally useless*, for any purpose of *separation between the two real roots* γ, of the biquadratic equation of article 103; and in fact, without thinking of any *curve*, the cited equation, namely,

$$0 = (\gamma^2 + 18\gamma - 27)^2 - \frac{T^2}{S^3}\gamma,$$

shows immediately that, *if* $\left(\frac{T^2}{S^3}\right) > 0$, or < 0, *each real root* γ *must have the same sign as S.* Nor could we, *from this biquadratic alone, assign any reason* for *preferring one root* to another, of this *quadratic* equation above written. But neither could we, in former cases, consider one real root of the biquadratic as deserving any preference to the other, until we went *back* to the equations involving α and β, from which the equation in γ was a result. In the present case, the *two points of contact, of the curve with the axis of x*, are *at opposite sides of the axis of y*; they give, therefore, *opposite signs* to γ, and consequently to S: which corresponds analytically to the *negative product* of the *two roots* of the *quadratic.*

The *sign of S is therefore useful*, and in fact *decisive, here*, as regards the *choice of the root* γ. Accordingly, we now have the two subcases:

(1) if $T=0$, $S<0$, then $\gamma=-9-6\sqrt{3}$;
(2) if $T=0$, $S>0$, then $\gamma=-9+6\sqrt{3}$.

In the first subcase, the formula of article 103,

$$\beta=\alpha\sqrt[3]{\gamma-1},$$

gives $\beta=-\alpha(1+\sqrt{3})$; in the second subcase, $\beta=-\alpha(1-\sqrt{3})$.

And because in each we have, by article 103, $\alpha^4\gamma=S$, we may write definitely,

(1) if $T=0$, $S<0$, then

$$\alpha=\left(\frac{-S}{9+6\sqrt{3}}\right)^{\frac{1}{4}},\quad \beta=-(1+\sqrt{3})\alpha;$$

but

(2) if $T=0$, $S>0$, then

$$\alpha=\left(\frac{+S}{-9+6\sqrt{3}}\right)^{\frac{1}{4}},\quad \beta=-(1-\sqrt{3})\alpha.$$

126. These values give, by the general formulæ of article 102,

$$\begin{cases}\dfrac{\lambda_0}{\mu_0}=-\dfrac{\lambda_1}{\mu_1}=\dfrac{1}{3}[(3\mp 2\sqrt{3})\,S]^{\frac{1}{2}},\\[2ex] \dfrac{\lambda_2}{\mu_2}=-\dfrac{\lambda_3}{\mu_3}=\dfrac{1}{3}[(3\pm 2\sqrt{3})\,S]^{\frac{1}{2}};\end{cases}$$

the upper signs being taken when $S<0$, and the lower signs when $S>0$. Hence for each of these two subcases, $\left(\dfrac{\lambda}{\mu}\right)$ must be a root of the equation,

$$\left(\left(\frac{3\lambda}{\mu}\right)^2-3S\right)^2=12S^2;$$

and accordingly, under the form

$$27\lambda^4-18S\lambda^2\mu^2-S^2\mu^4=0,$$

this is precisely what the general equation, already cited (*HPC*, p. 187) namely

$$27\lambda^4-18S\lambda^2\mu^2-T\lambda\mu^3-S^2\mu^4=0,$$

becomes for the case $T=0$.

127. If we write a new modification of the known canonical form of U,

$$U=ss's''+3\varepsilon xyz,$$

where s, s', s'' denote the same three linear functions as in articles 117 or 122, and ε is a new arbitrary constant, then it is not difficult to prove that

$$4H=(\varepsilon-1)^2ss's''-(\varepsilon-3)^2\varepsilon xyz;$$

also

$$16S = (\varepsilon - 1)(\varepsilon - 3)(\varepsilon^2 + 3),$$

and

$$8T = (3 - \varepsilon^2)\{(3 - 3\varepsilon + \varepsilon^2)^2 + 3\varepsilon^2\}.$$

The only real values of ε, which render $T = 0$, are therefore $\varepsilon = +\sqrt{3}$, and $\varepsilon = -\sqrt{3}$; whereof the first gives $S < 0$, but the second gives $S > 0$. Hence, for these values of ε, we have the particular forms,

$$U = ss's'' \pm 3\sqrt{3}xyz, \quad (\varepsilon + 1)^2 H = ss's'' \mp 2\sqrt{3}xyz;$$

the upper signs answering to the first subcase, and the lower signs to the second subcase, in article 119. We thus confirm, then, the known theorem (*HPC*, p. 184), that *when* $T = 0$, *the second** *Hessian is the original curve*; since a second change of sign of the radical $\sqrt{3}$ would bring us here from H to U again. And whether T be, or be not, thus equal to zero, we see that

$$(\varepsilon - 3)^2 U + 12H = \{(\varepsilon - 3)^2 + 3(\varepsilon - 1)^2\} ss's'';$$

$$(\varepsilon - 1)^2 U - 4H = \{(\varepsilon - 3)^2 + 3(\varepsilon - 1)^2\} \varepsilon xyz;$$

so that we thus obtain a new verification of the (*perhaps new*) *theorem* of article 110, that "if $\lambda U + \mu H$ be a product of three linear factors, the coefficients λ and μ being *real*, then according as these two coefficients *agree* or *differ* in *sign*, either (1) *two* of the *factors* are *imaginary*, (as here s', s'',) or else (2) *all* the *three* factors (as here x, y, z) are *real*". – By changing ε to $\pm\sqrt{3}$, we must recover the real values (article 126) of $\frac{\lambda}{\mu}$ for the two subcases of article 119 under the forms

$$\frac{\lambda_0}{\mu_0} = -\frac{\lambda_1}{\mu_1} = \frac{(\varepsilon - 1)^2}{4} = \frac{2 \mp \sqrt{3}}{2};$$

and accordingly we have, (for $\varepsilon = \pm\sqrt{3}$,)

$$S = \frac{3}{4}(3 - 2\varepsilon) = \frac{3}{4}(3 \mp 2\sqrt{3}),$$

and

$$(3 \mp 2\sqrt{3})^2 = 3(2 \mp \sqrt{3})^2,$$

so that

$$\frac{1}{3}((3 \mp 2\sqrt{3})S)^{\frac{1}{2}} = \frac{2 \mp \sqrt{3}}{2}.$$

128. Another important though only particular case is the following

III. $R = 0$, $T > 0$, or $T < 0$; whence also $S > 0$.

In this case, the ordinate of the auxiliary quartic of article 104 is $y = \frac{T^2}{S^3} = 64$; we are therefore led to consider, as at the end of article 106, that particular parallel, EH, to the axis

* The word "second" has dropped out of the cited page, but the context leaves no ambiguity.

of x, which is the horizontal tangent (see Figure 27) of inflexion at the point E, and which therefore meets the curve in *three* coincident points *there*, cutting it *again* at the point H: where

$$E = (9, 64), \quad \text{and} \quad H = (1, 64).$$

In other words, (as in articles 106 and 107) the biquadratic which determines the abscissæ x, or γ, answering to the ordinate $y = 64$, is

$$0 = (x-9)^3(x-1);$$

its roots are, therefore, *all real*, but *three* of them are *equal*; and they are

$$1, 9, 9, 9.$$

The *distinction*, between the *two real* and *distinct values* of the abscissa, comes *again* to *depend on the sign of* T, but *not* on that of S, which is here *given*. In the Figure the two intersections E, H, of the above mentioned tangent of inflexion with the quartic curve, are both *at one common side of the axis of* y; but they are *separated* by the *parallel to that axis* of which the equation is $\gamma = x = x_2 = -9 + 6\sqrt{3}$; namely by the ordinate at the point b, at which (for the second time) the function T vanishes. Accordingly, the formulæ of article 103,

$$S = \alpha^4\gamma, \quad T = \alpha^6(27 - 18\gamma - \gamma^2),$$

give

$$S = 9\alpha^4, \quad T = -216\alpha^6, \quad R = 64S^3 - T^2 = (3^4 2^6 - (3^2 2^3)^2)\alpha^{12} = 0,$$

for the point of inflexion E; but

$$S = \alpha^4, \quad T = 8\alpha^6, \quad R = 0,$$

for the other point of intersection H. Combining therefore these last results with the formula

$$\beta = \alpha\sqrt[3]{\gamma - 1}$$

of article 103, we find the following values of α and β, for the subcases of III:

(1) if $R = 0$, $T < 0$, then

$$\alpha = \frac{(-T)^{\frac{1}{6}}}{\sqrt{6}}, \quad \beta = 2\alpha;$$

(2) if $R = 0$, $T > 0$, then

$$\alpha = \frac{T^{\frac{1}{6}}}{\sqrt{2}}, \quad \beta = 0.$$

129. If we proceed to calculate hence the value of $\left(\frac{\lambda}{\mu}\right)$, by the general formulæ of article 102, applied to the recent subcase (1), we find for that subcase,

$$\frac{\lambda_0}{\mu_0} = \alpha^2 = \tfrac{1}{6}\sqrt[3]{-T}; \quad \frac{\lambda_1}{\mu_1} = -\tfrac{1}{3}(\alpha + \beta)^2 = -3\alpha^2 = -\tfrac{1}{2}\sqrt[3]{-T}.$$

These are the *two roots* which are *generally real*, the first being *generally positive*, and the second *generally negative*, although as a *limit* (cf. article 124) we have found examples of one or the other *vanishing*. The *two other roots* of the biquadratic in $\left(\frac{\lambda}{\mu}\right)$ have been seen, in article 101, to be *generally imaginary*; but in the present instance they are *both real*, being also *equal* to each other, and to the *positive* root $\left(\frac{\lambda_0}{\mu_0}\right)$. For the formulæ of article 102 give now,

$$\frac{\lambda_2}{\mu_2} = -\tfrac{1}{3}(\alpha+\theta\beta)^2 = -\frac{\alpha^2}{3}(1+2\theta)^2 = +\alpha^2 = \tfrac{1}{6}\sqrt[3]{-T};$$

$$\frac{\lambda_3}{\mu_3} = -\tfrac{1}{3}(\alpha+\theta^2\beta)^2 = -\frac{\alpha^2}{3}(1+2\theta^2)^2 = +\alpha^2 = \tfrac{1}{6}\sqrt[3]{-T}.$$

The four roots $\left(\frac{\lambda}{\mu}\right)$ have therefore been assigned for the subcase (1) of III.; that is, for $R=0$, $T<0$; and we see that *three* of them are *equal*, and *positive*; while the *fourth* (or rather the *second*, in the foregoing order of enumeration) is *negative*, being equal to the *negative triple* of each of the other three; so that (as always) the *sum* of the *four* roots is *zero*.

130. In the subcase (2), $R=0$, $T>0$, the same general formulæ of article 102 give, by article 128,

$$\frac{\lambda_0}{\mu_0} = \alpha^2 = \tfrac{1}{2}\sqrt[3]{T}; \quad \frac{\lambda_1}{\mu_1} = -\tfrac{1}{3}(\alpha+\beta)^2 = \frac{-\alpha^2}{3} = \tfrac{1}{6}\sqrt[3]{T};$$

$$\frac{\lambda_2}{\mu_2} = -\tfrac{1}{3}(\alpha+\theta\beta)^2 = -\frac{\alpha^2}{3}; \quad \frac{\lambda_3}{\mu_3} = -\tfrac{1}{3}(\alpha+\theta^2\beta)^2 = -\frac{\alpha^2}{3};$$

so that here the *two* latter roots, which are *generally imaginary*, become *real*, and *equal* each to the *negative* root of the equation. Whether we suppose T to be less than, or greater than, zero, we see that in *each* subcase the biquadratic must admit of being written thus:

$$\begin{aligned}0 &= \left(\frac{\lambda}{\mu} - \frac{1}{2}T^{\frac{1}{3}}\right)\left(\frac{\lambda}{\mu} + \frac{1}{6}T^{\frac{1}{3}}\right)^3 \\ &= \left(\frac{\lambda}{\mu}\right)^4 - \frac{1}{6}T^{\frac{2}{3}}\left(\frac{\lambda}{\mu}\right)^2 - \frac{1}{27}T\frac{\lambda}{\mu} - \frac{1}{432}T^{\frac{4}{3}} \\ &= \left(\frac{\lambda}{\mu}\right)^4 - \frac{2}{3}S\left(\frac{\lambda}{\mu}\right)^2 - \frac{1}{27}T\frac{\lambda}{\mu} - \frac{1}{27}S^2,\end{aligned}$$

because $R=0$ gives $T^{\frac{2}{3}} = 4S$; and accordingly this last form agrees with that which is given by Salmon.

131. The *distinction* between the two subcases of article 128 may be illustrated by the following Example. Let the three coefficients c_1, c_2, c_3 vanish, in Salmon's general form (*HPC*, p. 99) for the cubic function U; that function then becomes,

$$U = a_1x^3 + 3a_2x^2y + 3b_1xy^2 + b_2y^3 + 3z(a_3x^2 + 2dxy + b_3y^2).$$

At the same time, the general formulæ for S and T, – the *correction of signs* in the first (cf. article 115 of this Letter) not coming *here* into play, – become,

$$S = (d^2 - a_3 b_3)^2, \quad T = -8(d^2 - a_3 b_3)^3;$$

so that the condition $T^2 = 64S^3$, or $R = 0$, is satisfied. And we see that we have the *first* or the *second* subcase of article 128, $T < 0$ or $T > 0$, according as $(d^2 - a_3 b_3) > 0$, or < 0; that is, according as the *quadratic*,

$$a_3 x^2 + 2dxy + b_3 y^2 = 0,$$

has *real* or *imaginary roots*; or finally, according as the *point* (0, 0, 1), – which is called by Salmon (but not always by me) the *origin*, – is a *node*, or a *conjugate point*, of the *cubic curve*, $U = 0$: the *two tangents* to that curve, at that point, being *real* if $T < 0$, but *imaginary* if $T > 0$.

132. No essential generality is lost, when we suppose for simplicity that $d = 0$; which gives $S = a_3{}^2 b_3{}^2$, $T = 8a_3{}^3 b_3{}^3$; the *distinction* between the two cases, $T < 0$, and $T > 0$, being now simply this, that the coefficients of the two terms of $a_3 x^2 + b_3 y^2$, whereby $3z$ is the multiplier in *U*, have *opposite signs* in the first subcase (namely for a *node*), but have *similar* signs in the second subcase (namely, when there is a *conjugate point*). But when we thus make c_1, c_2, c_3, and d all vanish, the expressions (*HPC*, p. 182) for the coefficients of the Hessian become:

$$\mathrm{a}_1 = b_1 a_3^2; \quad \mathrm{b}_2 = a_2 b_3^2; \quad \mathrm{c}_3 = 0; \quad 3\mathrm{a}_2 = a_3(a_3 b_2 - 2a_2 b_3); \quad 3\mathrm{b}_1 = b_3(b_3 a_1 - 2a_3 b_1);$$

$$3\mathrm{a}_3 = a_3^2 b_3; \quad 3\mathrm{c}_1 = 0; \quad 3\mathrm{b}_3 = a_3 b_3^2; \quad 3\mathrm{c}_2 = 0; \quad 6\mathrm{d} = 0;$$

so that

$$H = b_1 a_3^2 x^3 + a_3(a_3 b_2 - 2a_2 b_3)x^2 y + b_3(b_3 a_1 - 2b_1 a_3)xy^2 + a_2 b_3^2 y^3 + a_3 b_3(a_3 x^2 + b_3 y^2)z.$$

At the same time, in the first subcase, $T < 0$, we have by article 129,

$$\frac{\lambda_0}{\mu_0} = \frac{\lambda_2}{\mu_2} = \frac{\lambda_3}{\mu_3} = -\tfrac{1}{3}a_3 b_3; \quad \frac{\lambda_1}{\mu_1} = a_3 b_3;$$

but in the second subcase, $T > 0$, we have, by article 130,

$$\frac{\lambda_0}{\mu_0} = a_3 b_3; \quad \frac{\lambda_1}{\mu_1} = \frac{\lambda_2}{\mu_2} = \frac{\lambda_3}{\mu_3} = -\tfrac{1}{3}a_3 b_3.$$

Whichever sign T may have, we ought therefore to find that the two combinations, $a_3 b_3 U - 3H$, and $a_3 b_3 U + H$, are products of three linear factors; and the Theorem of article 110 leads us to expect, that if $a_3 b_3 < 0$, $T < 0$, the first of these combinations should have *only one real factor*, while the *second* should have *all* its *three* factors *real*; but that, on the contrary, if $a_3 b_3 > 0$, $T > 0$, then the three factors of $a_3 b_3 U - 3H$ *should all* be *real*, whereas *two* of the factors of $a_3 b_3 U + H$ should here become *imaginary*.

133. It did not, however, seem to me easy, at first sight, to *verify* this expectation, without any *restriction* being laid upon the *constants*, a_1, a_2, b_1, b_2, which enter into *U* and *H*, but not into *S*, nor into *T*; except that the *given coefficients*, to which these constants belong, are *always supposed to be real* (see article 124). And I was even inclined to console myself, for what I thought *might* turn out to be a *failure* of the *general* theorem of article 110, in this *singular case* (or case of singularity), $R = 0$, by observing that because *S* and *T* thus *take no cognizance (here) of the four constants* a_1, a_2, b_1, b_2, – *attending*, as it were, only to the *immediate neighbourhood* of the *singular point,* whether *nodal* or *conjugate,* – therefore, *perhaps,* that theorem could not *fairly* be *required to extend to this case* since it was established on a study of the *relations* generally existing, between the roots $\left(\frac{\lambda}{\mu}\right)$, and the *constants S and T,* without *any other constants* being supposed to be *given*; although I had found it convenient to introduce *two auxiliary constants,* α and β, which have been proved (articles 102 to 108) to be *real functions of S and T,* and have largely figured in several recent articles of this Letter. But happily I soon ascertained, as follows, that this species of *consolation* was *unnecessary*: the Theorem still holding good, without any exception or modification whatever.

134. Writing the first combination of article 132 as follows:

$$a_3 b_3 U - 3H = Ax^3 + 3Bx^2 y + 3Cxy^2 + Dy^3,$$

where

$$A = a_3(a_1 b_3 - 3a_3 b_1), \quad B = -a_3(a_3 b_2 - 3a_2 b_3),$$

$$D = b_3(a_3 b_2 - 3a_2 b_3), \quad C = -b_3(a_1 b_3 - 3a_3 b_1),$$

we see at once that this combination is a product of *some three* linear factors, because z has been eliminated; but it is not *obvious* whether all are *real,* or *two imaginary.* Calculating, however, the *discriminant,*

$$\Delta = (AD - BC)^2 - 4(B^2 - AC)(C^2 - BD),$$

which is greater than zero if there be two imaginary roots, but less than zero if all three be real and unequal, (cf. *HPC,* p. 296,) we find that

$$AD - BC = 0;$$

$$a_3^2(B^2 - AC) = a_3(a_3 B^2 + b_3 A^2), \quad a_3^2(C^2 - BD) = b_3(a_3 B^2 + b_3 A^2);$$

therefore

$$\Delta = -4a_3^{-3} b_3(a_3 B^2 + b_3 A^2)^2.$$

Hence, my first expectation is confirmed; for we see that in the *first* (or *nodal*) subcase, $T < 0$, $a_3 b_3 < 0$, we have $\Delta > 0$, and there are *two imaginary factors* of $a_3 b_3 U - 3H$; but that in the *second* (or *conjugate*) subcase, $T > 0$, $a_3 b_3 > 0$, we have $\Delta < 0$, and *all* the *factors* of $a_3 b_3 U - 3H$ are *real.*

135. The *other combination,* $a_3 b_3 U + H$, *although involving* z, is still more easily discussed. For we soon find that

$$a_3 b_3 U + H = (a_3 x^2 + b_3 y^2)\{4a_3 b_3 z + (a_1 b_3 + a_3 b_1)x + (a_2 b_3 + a_3 b_2)y\};$$

where the *linear* factor $\{4a_3 b_3 z + \cdots\}$ is *always real*; and the *quadratic factor,* $(a_3 x^2 + b_3 y^2)$, when equated to zero, gives *real* or *imaginary roots,* according as $a_3 b_3$ is less than, or greater than zero, and therefore according as $T < 0$, or > 0.

The ***Theorem*** receives therefore, again, for the *singular case* $R = 0$, a critical and scarcely hoped for confirmation.

136. To complete the discussion of the equations at the end of article 102, or the determination of real values of α and β, answering to any given and real values of S and T, a few words must be said on the remaining *case*,

IV. $S = 0$, $T = 0$; that is the case of a *cusp* (*HPC*, p. 193). Here the *ordinate* (S^2/T^3) of the auxiliary quartic of article 104 becomes indeterminate; and therefore the *abscissa* γ cannot be definitely found. But the equations of article 102 show, here, without any reference to that auxiliary curve, or to its equation, that we are, in the present case, to take, simply,

$$\alpha = 0, \quad \beta = 0,$$

as the only real values which satisfy the given conditions. The four values of $\left(\frac{\lambda}{\mu}\right)$, when expressed by the equations of article 102 in terms of α and β, all vanish here; and accordingly the biquadratic reduces itself to $\lambda^4 = 0$, in the case when $S = 0$, and $T = 0$. In this case, then, the *only* combination of the form $\lambda U + \mu H$, which breaks up into linear factors, is the Hessian H *itself*; but the Theorem of article 110 fails here to *distinguish* between the cases of real and imaginary factors, because the zero, which is at once equal to $\left(\frac{\lambda_0}{\mu_0}\right)$ and to $\left(\frac{\lambda_1}{\mu_1}\right)$, is here the *common limit* of the *real positive* and of the *real negative root* of the biquadratic. Hence I conceive that we may safely in this case infer, that the *imaginary factors of H become real*, by becoming *equal* to each other; and therefore the Hessian ought then to be found to be the product of *one* real and linear function, multiplied by the *square* of *another* such factor. Accordingly if we make $a_3 b_3 = 0$, as well as $d = 0$, we shall have the case of a cusp; and at the same time the Hessian will become,

$$H = a_3^2 x^2 (b_1 x + b_2 y) + b_3^2 y^2 (a_1 x + a_2 y),$$

which evidently admits of the decomposition into three real factors, whereof two are equal, whether we suppose that b_3 or that a_3 vanishes.

137. It will now, I think, be instructive, to resume the *three* general forms of article 122, of which each involves only *one* arbitrary constant c, and which have been *adjusted* so as to give *all* one *common set of values of S and T*; and to consider *what particular forms* they assume, in the three particular *cases* I., II., III., of articles 123, 125, 128; for we cannot expect to meet the case IV. of article 136, or to satisfy the *two* equations $S = 0$, $T = 0$, by any *one* value of the still disposable constant c.

138. Form I. of article 122 was,

$$U = 3(c-1)ss's'' + 27cxyz;$$

and since it, like the two others, gave $S = 2^{-4}3^6(1+8c^3)$, it is evident that if we had written that form as

$$U = 3(c-a)ss's'' + 27cxyz,$$

we should have found $S = 2^{-4}3^6(a^4 + 8ac^3 + 0c^4)$, because S must always be conceived to rise as high as the fourth dimension, with respect to the coefficients of U. Hence when we come to consider the real values of c, which render $S = 0$, in order to adapt the three forms of article 122 to the case I. of article 123, we must not merely employ the value $c = -(\frac{1}{2})$, which renders $1 + 8c^3 = 0$, but must also take, as an alternative, the value $c = \infty$. The first value gives $T = +2^{-6}3^{12}$; the second value gives $T = -\infty$; we are, then, to take the latter for the subcase (1) of article 123, and the former for the subcase (2). But the reasoning will become much clearer, if we modify the forms of article 122, by *dividing each U by c*; which will of course have the effect of dividing S and $\sqrt[3]{R}$ each by c^4, and T by c^6.

139. Consider therefore these *three new forms*:

$$\begin{cases} \text{I.} \quad U = 3(1 - c^{-1})\, ss's'' + 27\, xyz; \\ \text{II.} \quad U = (c^2 - c^{-1})\, s^3 + 27 c^{-1}\, xyz; \\ \text{III.} \quad U = 9\dfrac{(1 + c + c^2)}{(1 + 2c)^2}\, s^3 - 9\dfrac{1 + 2c}{c}\, zp; \end{cases}$$

where s, s', s'', p have the same significations as in article 122. For *each* of these forms of U, we must, by the theory already explained, have the common set of values following:

$$S = 2^{-4}3^6(c^{-4} + 8c^{-1}); \quad T = 2^{-3}3^9(c^{-6} - 20c^{-3} - 8); \quad \sqrt[3]{R} = 3^6(c^{-3} - 1);$$

which may be decomposed into linear factors, on the plan of article 122.

140. If now it be required to assign cubics, included under the three last forms, which shall satisfy the two conditions $S = 0$, $T < 0$, we see that we are to take c infinite, and the three required cubics are, or *appear* to be, for we shall find cause to *reject* the second:

$$\begin{cases} \text{I.} \quad x^3 + y^3 + z^3 + 6xyz = 0; \text{ or } s^3 + s'^3 + s''^3 = 0; \text{ (cf. article 124)}; \\ \text{II.} \quad (x + y + z)^3 = 0; \text{ or } s^3 = 0; \\ \text{III.} \quad (x + y + z)^3 - 8z(x^2 + xy + y^2) = 0; \text{ or } s^3 - 8zp = 0. \end{cases}$$

As a verification of this last form, we may observe that it, like the first, is the *sum of three cubes*, one real and two imaginary: namely,

$$(x + y - z)^3 + \left(z + \frac{y - x}{\theta - \theta^2}\right)^3 + \left(z + \frac{y - x}{\theta^2 - \theta}\right)^3 = s^3 - 8zp.$$

But we said that the second form must be rejected. In fact, although it is true that $U = as^3 + 6(d - a)xyz$ gives generally $T = -8(d - a)^4(d^2 + 4ad + a^2)$, and therefore (by making $a = c^2 - c^{-1}$, $d - a = \frac{9}{2}c^{-1}$,) $T = 2^{-3}3^9(c^{-6} - 20c^{-3} - 8)$, as in article 139, if $U = (c^2 - c^{-1})s^3 + 27c^{-1}xyz$; and although *this* value of T tends to the finite limit $T = -3^9 < 0$, as c tends to ∞; yet when we *again divide U*, namely by c^2, in order to reduce the second cubic to $s^3 = 0$, as above, we virtually divide the last expression for T by c^{12}, and so *reduce it from a negative to zero*. Accordingly if we operate directly on $U = s^3$, we get $T = 0$, as may be seen, among other ways, by making $d = a$ in a formula of the present article. We conclude, then, that *no actual cubic curve* (with *real and finite coefficients*), *of the second form* of article 139, (nor even a system of *three coincident right lines,*) *can satisfy the two conditions,*

$$S = 0, \quad T < 0;$$

but that these conditions are both satisfied by the two cubics, of the *first* and *third forms* respectively,

$$\text{I.}\quad s^3 + s'^3 + s''^3 = 0, \qquad \text{and} \qquad \text{III.}\quad s^3 - 8zp = 0;$$

whereof *each* is the *sum of three cubes, one real,* and *two imaginary* (cf. *HPC,* p. 184).

141. For $S = 0$, $T > 0$, there is less difficulty; and *now the two first forms* of article 139, or of article 122, are available. We simply change c to $-\frac{1}{2}$, in each of them, and divide by numerical coefficients; and thus obtain the two following cubics, which both satisfy the proposed conditions, and are of the required forms:

$$\text{I.}\quad x^3 + y^3 + z^3 = 0; \qquad \text{II.}\quad s^3 - 24xyz = 0;$$

where both I. and II. are *sums of three real cubes,* since

$$(x + y - z)^3 + (y + z - x)^3 + (z + x - y)^3 = (x + y + z)^3 - 24xyz.$$

But when we endeavour to render the *cubic* III. *actual,* or to get rid of *infinite coefficients* in its equation, by multiplying its U by $(1 + 2c)^2$, we thereby multiply the $T = +2^{-6}3^{12}$ of article 138, by $(1 + 2c)^{12}$, or by 0^{12}, and so *reduce* it from being *positive* to be *null*; so that the equation $s^3 = 0$, which might seem to belong here to form III., is to be *rejected,* as it was then regarded as belonging to form II., in the analysis of article 140.

142. For $T = 0$, $S < 0$, we take (cf. article 122), $c = -\alpha_2 = -\{(1 + \sqrt{3})/2\}$; and therefore $c^{-1} = 1 - \sqrt{3} = 2\alpha_1$, so that form I. becomes,

$$ss's'' + (3\sqrt{3})xyz = 0,$$

as in article 125. At the same time,

$$c^2 = \frac{2 + \sqrt{3}}{2}, \qquad c^2 - c^{-1} = \frac{3\sqrt{3}}{2},$$

and form II. of article 139 becomes,

$$s^3 - 18xyz + (6\sqrt{3})xyz = 0;$$

also

$$1 + 2c = -\sqrt{3}, \quad 1 + c + c^2 = \frac{3}{2}, \quad \frac{1 + c + c^2}{(1 + 2c)^2} = \frac{1}{2}, \quad \frac{1 + 2c}{-c} = (\sqrt{3}) - 3;$$

therefore the form III. becomes here,

$$s^3 - 6zp + (2\sqrt{3})zp = 0.$$

Collecting, then, these last results, we have, for $T = 0$, the following cubic curves, of the three forms proposed:

$$\begin{cases} \text{I.} & ss's'' \pm (3\sqrt{3})xyz = 0; \\ \text{II.} & s^3 - 18xyz \pm (6\sqrt{3})xyz = 0; \\ \text{III.} & s^3 - 6zp \pm (2\sqrt{3})zp = 0; \end{cases}$$

the upper signs answering to $S < 0$, and the lower sign to $S > 0$. As a verification, the *second*

Hessian of each must admit of being shown (cf. article 127) to coincide with the original curve.

143. Finally, for $R=0$, $T<0$, if we set aside the cases in which the cubic U breaks up into three lines, whether all real, or two of them imaginary, we take the value $c=1$, and the form III. and find III.: $s^3-9zp=0$, which is a curve with a *cusp*, (cf. article 131,) at the point $(1, 1, 1)$; but for $R=0$, $T>0$, we take the value $c=0$, and the form II., namely $s^3-27xyz=0$; and this cubic has a *conjugate point*, (cf. again article 131,) at the same *unit-point* $(1, 1, 1)$, or O, as earlier and otherwise seen in article 88.

144. It was early observed, in article 61, that the fundamental formula of article 10,

$$lx+my+nz=0,$$

which *connects* the anharmonic coordinates of a *point* (x, y, z) with those of a *right line* $[l, m, n]$ whereon that point is situated, is a *sufficient basis* for all investigations that depend ultimately on the conceptions of *collinearity of points*, and of *concurrence of right lines*: and I think that I have somewhere remarked that the *mechanism* of all *such* researches, considered merely as a branch of *calculation*, cannot be expected to differ, in any essential respect, from the *usual* mechanism of *trilinear coordinates*: although the *interpretations* of the elementary *symbols* will be different, and *may* suggest, sometimes, improvements or modifications even in the *work*. Accordingly, for at least *nine* of the last sheets, and in fact ever since the little investigations of articles 86 and 88, respecting the form $ks^3=27xyz$ of the cubic (when $s=x+y+z$,) I may *seem* to have quite *dropped* the consideration of *anharmonic* coordinates; although the *name* has been, for uniformity's sake, retained, in the *heading* of all the sheets of this long Letter. But besides that we may always *think* of xyz, in all these later investigations, as *meaning* anharmonic coordinates, I shall now go on, – as it is indeed high time, – to mention one or two ***Problems***, which are perhaps more immediately *suggested* by the *conception* of *such* coordinates than by the notion of trilinears; and for the solution of which, the Method of this Letter appears to offer some facilities. A brief *recapitulation* of the first *principles* of that *Method* may however occupy, not uselessly, the remaining pages of the present sheet.

145. Allow me therefore to state, – selecting rather than *merely repeating* from former parts of this Letter, – that with the improved geometrical notation of article 77, and with a corresponding modification of the early and very simple Figure 4, my *fundamental conception* of the anharmonic coordinates of a *point*, when disengaged from everything foreign, or superfluous, or merely illustrative, admits of being enunciated as follows (cf. article 78). Let ABC be any given triangle, Figure 4 bis, [which is a slightly altered version of Figure 4] and O,

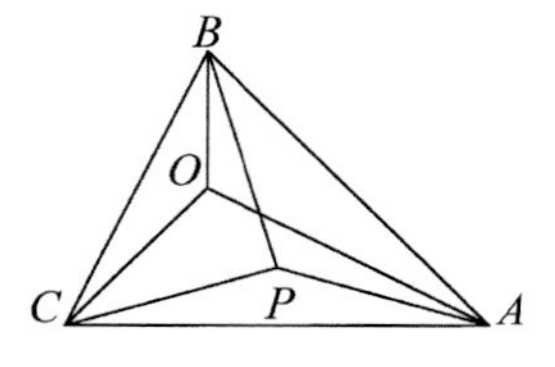

Fig. 4 *bis*

P any two points in its plane, whereof O is treated as given or constant, but P as arbitrary or variable. Then, whereas, by a well known theorem of segments, we have the constant product,

$$(A.PCOB).(B.PAOC).(C.PBOA) = +1$$

I write

$$\frac{y}{z} = (A.PCOB), \quad \frac{z}{x} = (B.PAOC), \quad \frac{x}{y} = (C.PBOA);$$

and *define* that x, y, z, or any other quantities proportional to them, are the *Anharmonic Coordinates of the point P*, taken with respect to the given triangle ABC, and to the given point O. Accordingly I *denote* the variable point P by the *symbol* (x, y, z); and because the corresponding symbols of the four given points are thus,

$$A = (1, 0, 0), \quad B = (0, 1, 0), \quad C = (0, 0, 1), \quad O = (1, 1, 1),$$

I call (as in former articles) the triangle ABC the *Unit-Triangle*, and the point O the *Unit-Point* of the System.

146. Permit me also to *restate*, – chiefly with a view to disengage it from anything irrelevant, and to present it in what appears to be its last degree of simplicity, – my *definition* of the anharmonic coordinates of a right *line*, taken here as being, in their *conception*, independent of the coordinates of a *point*, although they must afterwards be connected therewith.

I take, then, again any triangle ABC, which I treat as a given one, and I cut its sides by any two transversals, $A'B'C'$ and LMN, whereof I (see Figure 21) consider the former as given, but the latter as arbitrary; and then whereas, by another well known theorem of segments, we have this other constant product,

$$(LBA'C).(MCB'A).(NAC'B) = +1,$$

(the $+$ arising from a combination of two $-$'s,) I write (cf. article 79),

$$\frac{m}{n} = (LBA'C), \quad \frac{n}{l} = (MCB'A), \quad \frac{l}{m} = (NAC'B),$$

and define that l, m, n, or any quantities thereto proportional, are the *anharmonic coordinates of the line LMN*, with respect to the given triangle ABC, and to the given transversal $A'B'C'$. I also *denote* (as before) this variable line by the *symbol* $[l, m, n]$; and because we have thus,

$$[1, 0, 0], \quad [0, 1, 0], \quad [0, 0, 1],$$

as the symbols of the *three sides BC, CA, AB* of the triangle ABC, considered here merely as having *position*, we derive a *new* and independent *motive* for calling that *given triangle* (as above) the *Unit-Triangle* of this, as well as of the former construction. At the same time, the symbol of the line $A'B'C'$ is, on the same plan, $[1, 1, 1]$; whence it is natural to call that *given transversal*, as in preceding articles, the *Unit-Line*.

147. It is important to observe that *in this statement* of my fundamental conceptions and definitions, the principle of *Geometrical Duality* is fully recognised: *no prerogative* whatever [can exist], of simplicity or anything else, belonging to *point-coordinates*, or to *line-coordinates*, as if either were more fundamental than the other, when they are thus *independently defined.* It is just *as simple*, – but not a whit *more* so, – to consider the *sides* of a triangle as *cut* by *two right lines*; so as to form a *group* of *four points* on each, as to consider the *corners* of the same triangle as *connected* with *two points*, so as to form, with each corner as vertex, a *pencil* of *four rays*; it is then *as natural*, – but not more so, – to study the *anharmonic functions* of those *three groups*, as of those *three pencils*.

148. *No theorem* however emerges, of any novelty or interest, until we *combine* these two elementary constructions, by supposing that the *point P* is situated *upon the line LMN*, and that the *unit-point O* is the *pole* of the *unit-line* $A'B'C'$, with respect to the *unit-triangle ABC*. *Then*, however, with the help of the six auxiliary points, (see [a slightly altered and relabelled version of Figure 5] Figure 5 bis)

$$A'' = OA^{\cdot}BC, \quad B'' = OB^{\cdot}CA, \quad C'' = OC^{\cdot}AB,$$
$$Q = PA^{\cdot}BC, \quad R = PB^{\cdot}CA, \quad S = PC^{\cdot}AB,$$

and of the three harmonic equations,

$$(BA'CA'') = (CB'AB'') = (AC'BC'') = -1,$$

or even with only the two first of these three harmonic groups, and the four auxiliary points, A'', B'', Q, R, combined with the pencil of four rays through P, we easily prove, on the plan of article 10, that

$$-\frac{l}{n} = (MAB''C), \quad \frac{x}{z} = (RCB''A), \quad \frac{-lx}{nz} = (MARC),$$

and that

$$\frac{-m}{n} = (LBA''C), \quad \frac{y}{z} = (QCA''B), \quad \frac{-my}{nz} = (LBQC) = (MRAC) = 1 - (MARC);$$

whence

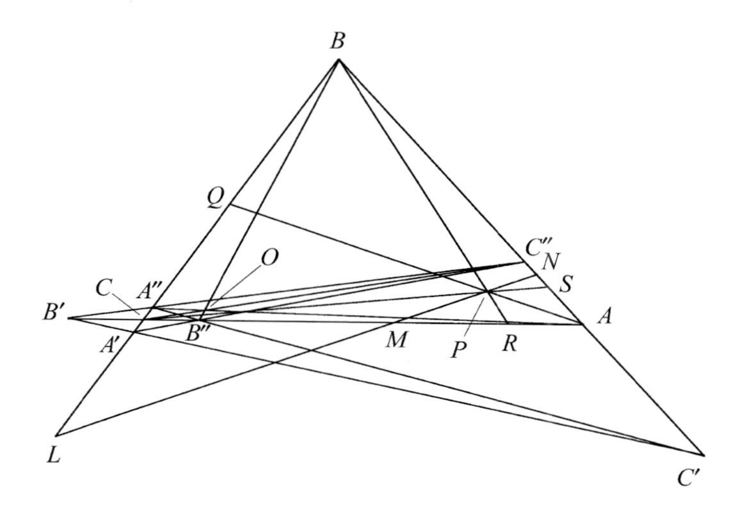

Fig. 5 *bis*

$$lx + my + nz = 0,$$

as before. The same important and fertile Theorem may also be proved, in this comparatively symmetric geometrical notation, (substituted for that of *OXY*, with which mode of denoting the given triangle this Letter began,) by projecting the given transversal $A'B'C'$ to infinity, as was done in article 80, with the help of the simple Figure 22, but perhaps, upon the whole, less elegantly.

149. After this *recapitulation*, which I thought not useless, of some first *principles* of this whole theory, I proceed to a few general *problems*, which they suggest, and which can be understood and resolved without any reference to *curves*. And first I shall consider the following Problem, which is evidently suggested by those principles:-
"To express the anharmonic of a pencil,

$$(P_0 . P_1 P_2 P_3 P_4),$$

in terms of the anharmonic coordinates of its vertex,

$$P_0 = (x_0, \ y_0, \ z_0),$$

and of any four points,

$$P_1 = (x_1, \ y_1, \ z_1), \ldots, \ P_4 = (x_4, \ y_4, \ z_4),$$

which are situated on its four rays respectively".

150. The most obvious process appears to be the following. Assuming

$$P'_2 = P_0 P_2 \cdot P_1 P_3 = (x'_2, \ y'_2, \ z'_2),$$
$$P'_4 = P_0 P_4 \cdot P_1 P_3 = (x'_4, \ y'_4, \ z'_4);$$

we shall thus reduce the sought anharmonic of a *pencil* to the anharmonic of a *group*, and we shall have the transformation,

$$(P_0 . P_1 P_2 P_3 P_4) = (P_1 P'_2 P_3 P'_4) = \frac{t'u}{tu'},$$

by the already often employed theorem of article 11; if, with an abridged notation similar to that of article 74, we write

$$(P'_2) = t(P_1) + u(P_3), \quad (P'_4) = t'(P_1) + u'(P_3),$$

or more fully,

$$x'_2 = tx_1 + ux_3, \quad y'_2 = ty_1 + uy_3, \quad z'_2 = tz_1 + wz_3,$$
$$x'_4 = t'x_1 + u'x_3, \quad y'_4 = t'y_1 + u'y_3, \quad z'_4 = t'z_1 + u'z_3;$$

as we are allowed to do, on account of the two equations,

$$0 = \begin{vmatrix} x_1, & y_1, & z_1 \\ x'_2, & y'_2, & z'_2 \\ x_3, & y_3, & z_3 \end{vmatrix}, \quad 0 = \begin{vmatrix} x_1, & y_1, & z_1 \\ x'_4, & y'_4, & z'_4 \\ x_3, & y_3, & z_3 \end{vmatrix},$$

which serve, in combination with these two other equations of the same kinds,

$$0 = \begin{vmatrix} x_0, & y_0, & z_0 \\ x_2, & y_2, & z_2 \\ x_2', & y_2', & z_2' \end{vmatrix}, \quad 0 = \begin{vmatrix} x_0, & y_0, & z_0 \\ x_4, & y_4, & z_4 \\ x_4', & y_4', & z_4' \end{vmatrix},$$

to determine the coordinates of the two auxiliary points of intersection, P_2', and P_4', in terms of the coordinates of the five given points, $P_0, P_1, \ldots, P_4$.

151. It remains then only to express the coefficients t, u, and t', u', of their ratios, in terms of those given coordinates. But for this purpose, it is only necessary to substitute, in the latter pair of determinant equations, the expressions

$$x_2' = tx_1 + ux_3, \text{ \&c.,} \quad \text{and} \quad x_4' = t'x_1 + u'x_3, \text{ \&c.,}$$

which satisfy the former pair. For we thus obtain,

$$0 = -t \begin{vmatrix} x_0, & y_0, & z_0 \\ x_1, & y_1, & z_1 \\ x_2, & y_2, & z_2 \end{vmatrix} + u \begin{vmatrix} x_0, & y_0, & z_0 \\ x_2, & y_2, & z_2 \\ x_3, & y_3, & z_3 \end{vmatrix};$$

$$0 = t' \begin{vmatrix} x_0, & y_0, & z_0 \\ x_4, & y_4, & z_4 \\ x_1, & y_1, & z_1 \end{vmatrix} - u' \begin{vmatrix} x_0, & y_0, & z_0 \\ x_3, & y_3, & z_3 \\ x_4, & y_4, & z_4 \end{vmatrix};$$

whence, by article 150, the required anharmonic in article 149 may be thus expressed, as the quotient of two products of determinants:

$$(P_0.P_1P_2P_3P_4) = \begin{vmatrix} x_0, & y_0, & z_0 \\ x_1, & y_1, & z_1 \\ x_2, & y_2, & z_2 \end{vmatrix} \begin{vmatrix} x_0, & y_0, & z_0 \\ x_3, & y_3, & z_3 \\ x_4, & y_4, & z_4 \end{vmatrix} : \begin{vmatrix} x_0, & y_0, & z_0 \\ x_2, & y_2, & z_2 \\ x_3, & y_3, & y_3 \end{vmatrix} \begin{vmatrix} x_0, & y_0, & z_0 \\ x_4, & y_4, & z_4 \\ x_1, & y_1, & z_1 \end{vmatrix}.$$

152. Verifications offer themselves at sight. Thus we see that the function of the fifteen coordinates with second members vanishes if either the two points P_1, P_2, or the two points P_3, P_4, be collinear with the point P_0; that is, if either the two first rays, or the two last rays, of the pencil in the first member coincide in position with each other. Again, if either the two middle rays, or the two extreme rays, of the pencil coincide in position, the anharmonic in the first member becomes infinite; and so does that quotient of products of determinants, which is the second member of the equation. Again, the anharmonic in the first member, and the quotient in the second member, become each equal to unity, when either the first ray coincides with the third, or the second with the fourth; for if we make, for example,

$$(P_3) = a(P_0) + b(P_1),$$

we shall have

$$\begin{vmatrix} x_0, & y_0, & z_0 \\ x_3, & y_3, & z_3 \\ x_4, & y_4, & z_4 \end{vmatrix} = -b \begin{vmatrix} x_0, & y_0, & z_0 \\ x_4, & y_4, & z_4 \\ x_1, & y_1, & z_1 \end{vmatrix}, \quad \text{and} \quad \begin{vmatrix} x_0, & y_0, & z_0 \\ x_2, & y_2, & z_2 \\ x_3, & y_3, & z_3 \end{vmatrix} = -b \begin{vmatrix} x_0, & y_0, & z_0 \\ x_1, & y_1, & z_1 \\ x_2, & y_2, & z_2 \end{vmatrix}$$

and similarly if we made

$$(P_4) = a'(P_0) + b'(P_2).$$

Finally, if we place any one of the four points P_1, P_2, P_3, P_4 at the vertex P_0 of the pencil, the anharmonic of that pencil becomes indeterminate; and, at the same time, the quotient, to which it is equated in the formula, takes the form $\frac{0}{0}$.

153. If we attribute a given or constant value k to the anharmonic of the pencil, so as to have

$$(P_0.P_1P_2P_3P_4) = k = \text{constant},$$

then, by the formula of article 151, the 15 coordinates of the five points are connected by the following equation, which is freed from fractions, and is homogeneous with respect to each set of coordinates:

$$\begin{vmatrix} x_0, & y_0, & z_0 \\ x_1, & y_1, & z_1 \\ x_2, & y_2, & z_2 \end{vmatrix} \begin{vmatrix} x_0, & y_0, & z_0 \\ x_3, & y_3, & z_3 \\ x_4, & y_4, & z_4 \end{vmatrix} = k \begin{vmatrix} x_0, & y_0, & z_0 \\ x_2, & y_2, & z_2 \\ x_3, & y_3, & z_3 \end{vmatrix} \begin{vmatrix} x_0, & y_0, & z_0 \\ x_4, & y_4, & z_4 \\ x_1, & y_1, & z_1 \end{vmatrix}.$$

If *any four* of the five points, *including* P_0, be *given*, this equation assigns a *right line through* P_0, as the *locus of the remaining point*; but if P_0 *alone* be undetermined, while the *other four* points are *given*, then because the equation is of the *second dimension* in x_0, y_0, z_0, and is satisfied when these coordinates are replaced by x_1, y_1, z_1; or by x_2, y_2, z_2, or [by] &c., we see that, in this case, *the locus of the vertex* P_0 *of the pencil is a conic, which passes through the four given points,* P_1, P_2, P_3, P_4. But these results are otherwise well known from geometry: they may therefore be considered as additional verifications of the correctness of the expression in article 151, for the anharmonic of the pencil in article 149.

154. As an example, let the vertex P_0, be the first real point (see article 111) of inflexion, K' or A', of the cubic curve in the canonical form of article 109,

$$U = x^3 + y^3 + z^3 + 6exyz = 0;$$

taking also the second real point K'' or B', of inflexion on the first ray P_1 of the pencil, and obliging the three other rays to pass in order through the three imaginary points of inflexion I', I'', I''', the line connecting which is *itself* imaginary (see article 111). We have then the anharmonic,

$$(K'.K''I'I''I''') = \begin{vmatrix} 0, & 1, & -1 \\ -1, & 0, & 1 \\ 0, & 1, & -\theta \end{vmatrix} \begin{vmatrix} 0, & 1, & -1 \\ -\theta, & 0, & 1 \\ 1, & -\theta, & 0 \end{vmatrix} : \begin{vmatrix} 0, & 1, & -1 \\ 0, & 1, & -\theta \\ -\theta, & 0, & 1 \end{vmatrix} \begin{vmatrix} 0, & 1, & -1 \\ 1, & -\theta, & 0 \\ -1, & 0, & 1 \end{vmatrix}$$

$$= \frac{(1-\theta)(1-\theta^2)}{(\theta^2-\theta)(\theta-1)} = -\theta.$$

If then we denote the four rays of this pencil, or the *four chords of inflexion* from K', whereof the two first are real, and the two last are imaginary, by the symbols K_1, K_2, K_3, K_4, we may write, concisely,

$$(K_1K_2K_3K_4) = -\theta;$$

whence also,

$$(K_1 K_3 K_2 K_4) = 1 + \theta = -\theta^2 = (-\theta)^{-1} = (K_1 K_4 K_3 K_2);$$

so that we have here

$$(K_1 K_2 K_3 K_4) = (K_1 K_4 K_2 K_3) = (K_1 K_3 K_4 K_2) = -\theta,$$

and

$$(K_1 K_3 K_2 K_4) = (K_1 K_2 K_4 K_3) = (K_1 K_4 K_3 K_2) = -\theta^2,$$

the *anharmonic function* of *this pencil* having thus only *two distinct values* (both imaginary), *in whatever order the four rays may be taken* whereas, *in general,* a pencil has *six distinct values* of its anharmonic function, such as h, $1 - h$; $h - 1$, $1 - h^{-1}$; and $(1 - h)^{-1}$, $1 - (1 - h)^{-1}$ $(= h/(h - 1))$, when the rays are variously arranged; and even a *harmonic pencil* has *three* distinct values (-1, 2, and $\frac{1}{2}$). I am therefore disposed to call the *pencil of chords,* above considered, a ***Di-anharmonic Pencil***; and to extend *this name* to every pencil which has *only two values of its anharmonic*: those two values being then necessarily $-\theta$ and $-\theta^2$, because $1 - h = h^{-1}$.

155. The particular problem of the last article might have been resolved with even less trouble of calculation, by our availing ourselves of the *collinearity* (see article 112) of the *three* points of (imaginary) inflexion, I', I'', I''', and using the (imaginary) intersection, say for a moment I^0, of the (imaginary) right line on which they are situated, with the first (and real) ray $K'K''$ of the pencil. That intersection is easily seen from the expressions already assigned, to be precisely the point which in some early articles of this Letter was denoted by J, and was called one of the *two imaginary unit-points* (I and J, articles 25 &c.,) namely the point $I^0 = J = (1, \theta^2, \theta)$; in fact, the coordinates of this point satisfy at once the equation $s = 0$ of the real line $K'K''$, and the equation $s'' = 0$ of the imaginary line $I'I''I'''$. In this manner it is easily seen that

$$(I') = \theta(I^0) + (I''), \quad \text{and} \quad (I''') = -\theta^2(I^0) + (I''),$$

a notation already explained; whence

$$(K'.K''I'I''I''') = (I^0 I'I''I''') = -\theta,$$

as above. But we shall soon consider the whole subject of the determination of such anharmonics of pencils, in an entirely different way.

156. Meantime, as another example of the process last employed, let us take the form of the cubic in article 114, namely:

$$U = (c^3 - 1)(x + y + z)^3 + 27xyz = 0,$$

and seek the value of the anharmonic of the pencil which has still $(K'.K''I'I''I''')$ for its symbol, and which is still composed of *four chords of inflexion,* two real and two imaginary, with the real point of inflexion K' for vertex as before, although (by article 114) the expressions for the three collinear and imaginary points of inflexion I', I'', I''' involve now the arbitrary constant c; being (by the cited article),

$$I' = (c - 1, \theta c - 1, \theta^2 c - 1),$$

$$I'' = (\theta^2 c - 1,\ c - 1,\ \theta c - 1),$$
$$I''' = (\theta c - 1,\ \theta^2 c - 1,\ c - 1);$$

while the imaginary line that conects them has now $s' = 0$ for its equation. Combining this with the equation of the first real ray of the pencil, which is still $s = 0$, we are conducted to the following auxiliary point, of intersection of the two last lines,

$$I^0 = K'K''.I'I'' = (1,\ \theta,\ \theta^2) = I,$$

namely the other of our old imaginary unit-points; hence

$$(I') = (1 - \theta^2)c(I^0) + (I''), \qquad (I''') = (\theta - \theta^2)c(I^0) + (I''),$$

so that the arbitrary constant c disappears from the expression for the sought anharmonic of the pencil, which is thus found to be,

$$(K'.K''I'I''I''') = (I^0 I'I''I''') = \frac{(\theta - \theta^2)c}{(1 - \theta^2)c} = -\theta^2;$$

the *pencil* is therefore again a *di-anaharmonic* one, in the sense of article 154, although we happen to have started here with an *order of the rays*, and with a *selection* of the cube root of unity, which gives $-\theta^2$, instead of $-\theta$, for the particular value resulting. Of course, by varying the order, we should have here,

$$(K'.K''I''I'I''') = 1 + \theta^2 = -\theta, \text{ \&c.}$$

157. Although, after solving the recent problem for the *canonical* form of U, it may be regarded as merely an exercise of calculation to resolve it for any *other* form, since the geometrical result must be the same, yet, *as* such an exercise, I will take, for a moment, the equation of article 92:

$$U = s^3 + 6kzp = 0,$$

with the resulting expressions, given in article 97, for the nine points of inflexion; in which however I shall write, as in article 121, c for $\frac{v}{t}$. We shall thus have, for two of the real points of inflexion, the values,

$$P_0 = K = (1,\ -1,\ 0), \qquad P_1 = K' = (1,\ c,\ c);$$

and for three of the imaginary points, on one imaginary line,

$$P_2 = I = (1,\ \theta,\ \theta^2); \qquad P_3 = J' = (1 + 2c\theta,\ \theta - c\theta^2,\ \theta - c\theta^2);$$
$$P_4 = I'' = (\theta^2 - c\theta,\ 1 + 2c\theta^2,\ \theta^2 - c\theta).$$

There is no difficulty in determining the point of intersection,

$$I^0 = KK'.IJ' = (1 + (1 - \theta^2)c + 2\theta c^2, \quad \theta + (\theta - \theta^2)c + 2c^2, \quad -\theta^2 c(1 + 2c));$$

nor in deducing the formulæ,

$$(J') = t(I) + u(I''), \qquad t = (c - 1)\theta, \qquad u = -1,$$
$$(I^0) = t'(I) + u'(I''), \qquad 3t' = (1 - \theta)(1 - c)(1 + 2c), \qquad 3u' = (\theta - 1)(1 + 2c)$$

whence the arbitrary constant c again disappears, in the resulting expression for the anharmonic,

$$(K.IJ'I''K') = (IJ'I''I^0) = \frac{t'u}{tu'} = -\theta^2,$$

which gives

$$(K.K'IJ'I'') = -\theta.$$

But perhaps in this case it is simpler to use the general formula of article 151, observing that because $P_0 = (1, -1, 0)$, we have

$$\begin{vmatrix} x_0, & y_0, & z_0 \\ x_m, & y_m, & z_m \\ x_n, & y_n, & z_n \end{vmatrix} = (y_m z_n - y_n z_m) - (z_m x_n - x_m z_n) = s_m z_n - s_n z_m,$$

if we write $s_m = x_m + y_m + z_m$, $s_n = x_n + y_n + z_n$. We have here,

$$z_1 = c, \quad z_2 = \theta^2, \quad z_3 = \theta(1 - \theta c), \quad z_4 = \theta(\theta - c),$$

$$s_1 = 1 + 2c, \quad s_2 = 0, \quad s_3 = (\theta - \theta^2)(1 + 2c), \quad s_4 = (\theta^2 - \theta)(1 + 2c);$$

therefore

$$\begin{aligned}(K.K'IJ'I'') &= (P_0.P_1P_2P_3P_4) \\ &= \frac{s_1z_2 - s_2z_1}{s_2z_3 - s_3z_2} \cdot \frac{s_3z_4 - s_4z_3}{s_4z_1 - s_1z_4} \\ &= \frac{\theta^2(1+2c)}{(\theta-1)(1+2c)} \cdot \frac{(\theta-\theta^2)(c-1)(1+2c)}{\theta^2(c-1)(1+2c)} = \frac{\theta^2}{\theta-1} \cdot \frac{1-\theta}{\theta} = -\theta,\end{aligned}$$

as before. This pencil, therefore, as was to be expected from what was established for the canonical form of the cubic, is found to be *also di-anharmonic* (see article 154).

158. It may be worth examining, whether anything depends on our having taken, in the three preceding examples, a *real* point of inflexion for the *vertex* of the *pencil of four chords.* Employing for this purpose the canonical form, let the pencil be now $(I'.K'K''K'''I'')$; where the vertex is *imaginary*, and only the *first ray* is real. It is here convenient to consider, as in article 155, the imaginary intersection, $J = I'I''\cdot K'K'' = (1, \theta^2, \theta)$, and thus we find, $(K'') = -(K') - (K''')$, $(J) = -\theta(K') + (K''')$; whence $(I'.K'K''K'''I'') = (K'K''K'''J) = -\theta$; and the pencil is *still di-anharmonic.*

159. We arrive then at this *Theorem*, which probably is known, although I do not see any mention of it in Salmon's Book:

"If any one of the points of inflexion of a cubic be made the vertex of a pencil of four chords, which pass through the eight other points, the anharmonic function of this pencil, in whatever order the rays may be arranged, is always equal to one of the imaginary cube-roots of negative unity."

In other words, *every such pencil is di-anharmonic.*

160. As another example of the determination of the anharmonic function of a pencil, which has an important connexion with the theory of cubic curves, let us take this problem:- To determine the value of the anharmonic of the pencil,

$$(K'.OI'I''I''');$$

where $K'I'$, $K'I''$, $K'K'''$ are the same three chords of inflexion as in article 154, but O is that real point (1, 1, 1), which we have called in many former articles, (article 78, etc.) the anharmonic *unit-point*; so that the two first rays of this pencil are real, and the two last imaginary. The general expression of article 151 becomes here,

$$(K'.OI'I''I''') = \begin{vmatrix} 0, & 1, & -1 \\ 1, & 1, & 1 \\ 0, & 1, & -\theta \end{vmatrix} \begin{vmatrix} 0, & 1, & -1 \\ -\theta, & 0, & 1 \\ 1, & -\theta, & 0 \end{vmatrix} : \begin{vmatrix} 0, & 1, & -1 \\ 0, & 1, & -\theta \\ -\theta, & 0, & 1 \end{vmatrix} \begin{vmatrix} 0, & 1, & -1 \\ 1, & -\theta, & 0 \\ 1, & 1, & 1 \end{vmatrix} = -\theta^2;$$

so that *this pencil also* belongs to the di-anharmonic class. We might here have availed ourselves of the facility given by the *collinearity* of the four points O, I', I'', I''', since the real point O is the intersection of the two imaginary chords of inflexion ($s'' = 0$, $s' = 0$), which connect, three by three, the six imaginary points (cf. article 112); for thus we should have found,

$$(I') = (O) + \theta^2(I''), \quad (I''') = -\theta(O) + \theta(I''),$$

and therefore $(K'.OI'I''I''') = (OI'I''I''') = -\theta^2$, as before. As a verification, interchanging θ and θ^2 we have $(K'.OJ'J''J''') = -\theta$; therefore $(K'.OJ'J'''J'') = -\theta^2 = (K'.OI'I''I''')$; and accordingly these two last pencils coincide, the real and two imaginary chords, $I'J'$, $I''J'''$, $I'''J''$, converging, by article 112, to the real point of inflexion K'.

161. The *five collinear points*

$$O = (1, 1, 1), \quad J = (1, \theta^2, \theta), \quad I' = (0, 1, -\theta),$$
$$I'' = (-\theta, 0, 1), \quad I''' = (1, -\theta, 0),$$

have evidently their symbols connected by the *ten* following *linear equations*, answering to their ten *ternary combinations*, of which some have been already employed:

$$\begin{cases} (O) = (I'') - \theta^2(I'''), \quad (O) = (I''') - \theta^2(I'), \quad (O) = (I') - \theta^2(I''); \\ \theta^2(J) = (I'') - (I'''), \quad (J) = (I''') - (I'), \quad \theta(J) = (I') - (I''); \\ (1-\theta^2)(I') = (O) - (J), \quad (1-\theta^2)(I'') = (O) - \theta(J), \quad (1-\theta^2)(I''') = (O) - \theta^2(J); \\ (I') + \theta(I'') + \theta^2(I''') = 0; \end{cases}$$

respecting which equations it is important to observe, that *no one of them*, taken *simply*, gives *any information* respecting the *relative position* of the *three points* involved, beyond the mere fact of their *collinearity*: but that because the five *symbols*, (O), (J), (I'), (I''), (I'''), are taken to represent, in all of these equations, the *same systems of anharmonic coordinates*, – the symbol (O), for example, meaning the system of coordinates (1, 1, 1) throughout, and *not* that system affected with any factor, such as (t, t, t), where t is different from unity – therefore any two of the ten equations, from which one *common symbol* is excluded, are adapted to give the value of the *anharmonic of a group*, formed by *rejecting one* of the 5 points, or by taking their *quarternary combinations.*

162. In this manner, we derive the five *anharmonics* of groups of which some have occurred before:-

$$\text{Retaining } O \text{ and } J, \quad (OI'''JI'') = (OI'JI''') = (OI''JI') = +\theta$$

$$'' \quad '' \quad O, \text{ but not } J, \ (OI'I''I''') = -\theta^2;$$

$$'' \quad '' \quad J, \text{ but not } O \ (JI'I''I''') = -\theta.$$

The *two last* are therefore seen again to be *di-anharmonic groups*, so that we may write,

$$(OI'I''I''') = (OI''I'''I') = (OI'''I'I'') = (I'I''I'''J) = (I''I'''I'J) = (I'''I'I''J) = -\theta^2;$$
$$(I'I''I'''O) = (I''I'''I'O) = (I'''I'I''O) = (JI'I''I''') = (JI''I'''I') = (JI'''I'I'') = -\theta.$$

But the *three first groups*, – each formed by excluding one of the 3 imaginary points of inflexion, I', I'', I''', on the given line $s''=0$, and by comparing the *two others* with the two auxiliary points, O and J, in which that line is cut by the *two other lines*, $s'=0$ and $s=0$, or $J'J''J'''$ and $K'K''K'''$, – have each six different values (cf. article 154) *of its anharmonic function*, according to the *order* in which the four points of that group are arranged. Thus, besides the 3 anharmonic equations for those 3 groups, which have been written in the present article, we have (by article 5) these 15 others:-

$$(OI''JI''') = (OI'''JI') = (OI'JI'') = +\theta^2;$$
$$(OJI'''I'') = (OJI'I''') = (OJI''I') = 1-\theta;$$
$$(OJI''I''') = (OJI'''I') = (OJI'I'') = 1-\theta^2;$$
$$(OI''I'''J) = (OI'''I'J) = (OI'I''J) = (1-\theta)^{-1};$$
$$(OI'''I''J) = (OI'I'''J) = (OI''I'J) = (1-\theta^2)^{-1}.$$

163. The two first of the 18 anharmonic equations, which thus involve *both O* and *J, suggest* the following Theorem:-

"If *any two*, of any three collinear points of inflexion, (as here any two of I', I'', I''') be combined *as second and third*, or as first and fourth with the two points (as here with O and J) in which that *line* is *cut* by the *two other chords*, (here $J'J''J'''$ and $K'K''K'''$), containing each *three other points* of inflexion, *the anharmonic of the group is an imaginary cube root of positive unity.*" I only say, for the moment, that they "suggest" this Theorem: for I do not think it unreasonable to desire a *proof* or at least a *verification*, such as can easily be supplied, of the Theorem *still* holding good, when we pass from a line through *three imaginary points* of inflexion, to a line containing three points of which *one at least* is *real.*

164. Consider then the five collinear points,

$$I = (1, \theta, \theta^2), \quad J = (1, \theta^2, \theta), \quad K' = (0, 1, -1), \quad K'' = (-1, 0, 1), \quad K''' = (1, -1, 0);$$

where the *three last* are points of *real* inflexion, while the two first points are respectively on the lines $s'=0$, $s''=0$, or $J'J''J'''$ and $I'I''I'''$, which connect, 3 by 3, the 6 imaginary points of inflexion of the cube, expressed still by its canonical form (article 109 &c.). We have now these ten linear equations, analogous to those of article 161:

$$\theta^2(I) = \theta(K'') - (K'''), \quad \theta(I) = \theta(K''') - (K'), \quad (I) = \theta(K') - (K'');$$

$$\theta(J) = \theta^2(K'') - (K'''), \quad \theta^2(J) = \theta^2(K''') - (K'), \quad (J) = \theta^2(K') - (K'');$$

$$(\theta - \theta^2)(K') = (I) - (J), \quad (\theta - \theta^2)(K'') = \theta^2(I) - \theta(J), \quad (\theta - \theta^2)(K''') = \theta(I) - \theta^2(J);$$

$$(K') + (K'') + (K''') = 0;$$

whence we have not only the *two di-anharmonic groups*,

$$(IK'K''K''') = (K'K''K'''J) = -\theta,$$

whereof the latter was employed in article 158, but also these *three other groups*, which are of the *usual*, or *hex-anharmonic* kind, – as having each *six unequal values* for its anharmonic function, –

$$(IK''JK''') = (IK'''JK') = (IK'JK'') = +\theta;$$

which confirms the theorem in article 163. Any further verification of that theorem appears to be now unnecessary.

165. It is an immediate consequence of that Theorem of article 163, that the *cube* of any one of the anharmonics, which are formed according to its enunciation, must be equal to *positive unity*. Substituting, then, for such an anharmonic, its value as defined in article 5, we deduce from article 162 the equations

$$\left(\frac{OI'}{JI'}\right)^3 = \left(\frac{OI''}{JI''}\right)^3 = \left(\frac{OI'''}{JI'''}\right)^3;$$

and from article 163 the analogous equations,

$$\left(\frac{IK'}{JK'}\right)^3 = \left(\frac{IK''}{JK''}\right)^3 = \left(\frac{IK'''}{JK'''}\right)^3.$$

Indeed, I have (quite recently) reason to believe that *Hesse** arrived at results of the *same form* as *these*, but by an entirely different analysis, and without using *anharmonics* at all, but rather the known properties of *transversals* of a *triangle*, which I have not had occasion (for this purpose) to employ.

It must also be observed, that the *three quotients of segments*, of which the *cubes* are here equated, *are not anharmonic functions*, and therefore *are* (generally) altered by projection: although we see that, if so, their *cubes* must all be altered *in one common ratio*, in order that the *equality* of those cubes may be undisturbed. Indeed it is an immediate consequence of the fundamental property of the anharmonic of a group, that the *quotients themselves* alter (if at all), in one common ratio, by projection: because, for example,

$$\frac{OI'}{JI'} : \frac{OI''}{JI''} = \text{the anharmonic, } (OI'JI'').$$

* [Ludwig Otto Hesse, 1811–1874.]

166. In words, we may enunciate the result of article 165 as follows:-

"If the *interval*, on *any one* of the *12 chords* of inflexion, which is comprised *between the two points, not* of inflexion, where *that chord is cut* by two of the eleven other chords, be divided into *two segments* by *any one* of the *three points of inflexion* upon that chord assumed; then the *cube of the quotient* of those two segments has a *common value, independent* of the *particular* point of inflexion which may be *selected* for the section."

167. A great variety of other results, which it might be too pompous to call Theorems, may be deduced from the foregoing formulæ: expecially if we agree to introduce certain new and auxiliary points, defined by the harmonic relations,

$$(OI''O_1I''') = (OI'''O_2I') = (OI'O_3I'') = -1;$$

$$(I'''I'I''I_1') = (I'I''I'''I_1'') = (I''I'''I'I_1''') = -1;$$

with others of the same kind, which I am not sure that I shall think it necessary or useful to write down. One result, however, may be here briefly mentioned:- "*If any two* of the 3 collinear points of inflexion, I', I'', I''', be treated as *conjugate points* of an involution, while the two points O, J are made *another conjugate pair* the remaining point of inflexion $I^{(n)}$ will be *one* of the two *double points* of the involution, its harmonic conjugate $I_1^{(n)}$ being the *other* double point".

168. (April 18th, 1860) The discursive character of this long Letter allows me here to leave for awhile the discussion of such relations between these, between the points of inflexion of a cubic, and other points derived from them, and to take up a totally different question, which is however connected quite as closely with the title of these sheets: since it will require, or at least employ, the method of anharmonic coordinates, as applied to the discussion, – and in a slight degree the extension, – of a very elegant *Theorem* of plane Geometry, which you communicated to me, in a note that arrived yesterday.

169. You tell me, then, that if a *circle* be described so as to pass *through the middle points* of the three sides of any triangle, it will not only pass (which you say is well known, and can in fact, as I see be very easily proved) *through the feet of the three perpendiculars* from the corners of the triangle, but will also *touch* the *inscribed circle*, and each of the three *exscribed circles.* This theorem was wholly new to me, and I thought it well worth trying to deduce a proof of it, by the method of this Letter, but with a somewhat generalized enunciation, which that method naturally *suggested*: although, indeed, the same extension has probably occurred to yourself, from principles sufficiently known.

170. Instead of the *middle points* of the sides of the triangle ABC, I take then (compare, if you choose, the figure of article 148) the points A'', B'', C'', which are the harmonic conjugates of those other points A', B', C', where the sides are cut by a *transversal*, assumed by me as the *unit-line* (article 146, &c.). And instead of the *circular points at infinity*, I assume now (more generally than in some early articles) *any two points Q, Q', real or imaginary, upon the unit-line*: of which points I thus suppose the coordinates xyz to satisfy the system of the two equations

$$a + y + z = 0, \qquad ax^2 + by^2 + cz^2 = 0,$$

the coefficients a, b, c, being three arbitrary constants. Your *bisecting circle* becomes thus the *conic* through the *five points* A'', B'', C'', Q, Q'; and its equation is easily found to be

$$0 = U = 2(ax^2 + by^2 + cz^2) - (x + y + z)(ax + by + cz),$$

if we remember that the points A'', B'', C'', are (0, 1, 1), (1, 0, 1), (1, 1, 0).

171. Such being then the *conic* which replaces your first or *bisecting circle*, I substitute next for your *inscribed* (or exscribed) *circle another conic, touched by the three sides* of the triangle ABC, and passing *through the same two points* Q, Q', of the unit-line. The three tangencies give this form of its equation,

$$O = V = \alpha^2 x^2 + \beta^2 y^2 + \gamma^2 z^2 - 2\beta\gamma\, yz - 2\gamma\alpha\, zx - 2\alpha\beta\, xy;$$

because the unit line, $x + y + z = 0$, is to be a *common chord* QQ' of this and of the former conic, we must suppose that the two equations are connected by a relation of the following form:

$$U = 2V + (x + y + z)(\lambda x + \mu y + \nu z);$$

when $\lambda x + \mu y + \nu z = 0$ is *another common chord*, passing through the *two other common points*, or points of intersection of the two conics, which we may call P and P'. Comparing coefficients, in the two expressions for U we have

$$a = \lambda + 2\alpha^2; \quad b = \mu + 2\beta^2; \quad c = \nu + 2\gamma^2;$$
$$-(b + c) = \mu + \nu - 4\beta\gamma; \quad -(c + a) = \nu + \lambda - 4\gamma\alpha; \quad -(a + b) = \lambda + \mu - 4\alpha\beta;$$

whence, if we write for conciseness $\delta = \alpha + \beta + \gamma$, we derive

$$a = \alpha\delta - \beta\gamma, \quad b = \beta\delta - \gamma\alpha, \quad c = \gamma\delta - \alpha\beta,$$

and

$$\lambda = (\gamma - \alpha)(\alpha - \beta), \quad \mu = (\alpha - \beta)(\beta - \gamma), \quad \nu = (\beta - \gamma)(\gamma - \alpha)$$

indeed these last values, of λ, μ, ν, or the equivalent formulæ,

$$\mu + \nu = -(\beta - \gamma)^2, \quad \nu + \lambda = -(\gamma - \alpha)^2, \quad \lambda + \mu = -(\alpha - \beta)^2,$$

may be more rapidly obtained by observing that the second expression for U must vanish, when x, y, z are replaced by the coordinates of any one of the three points, A'', B'', C''. It may be noticed in passing that

$$b + c = (\beta + \gamma)^2, \quad c + a = (\gamma + \alpha)^2, \quad a + b = (\alpha + \beta)^2.$$

172. Let us now calculate those two intersections, above named P and P', of the two conics U and V which are *not* upon the unit line QQ'; and which therefore satisfy the equation of the *other* common chord, $\lambda x + \mu y + \nu z = 0$. Eliminating z, by this linear equation, from the equation $V = 0$, we are conducted to the following quadratic,

$$0 = \{(\gamma\lambda + \alpha\nu)x - (\beta\nu + \gamma\mu)y\}^2 + 4\gamma(\alpha\mu\nu + \beta\nu\lambda + \gamma\lambda\mu)xy;$$

but

$$\alpha\mu\nu + \beta\nu\lambda + \gamma\lambda\mu = (\alpha - \beta)(\beta - \gamma)(\gamma - \alpha)\{\alpha(\beta - \gamma) + \beta(\gamma - \alpha) + \gamma(\alpha - \beta)\} = 0;$$

the quadratic is therefore an *exact square*, as by your (extended) Theorem it ought to be, and the points P, P' *coincide*: the coordinates of the point of contact P, which is thus proved to exist, for the two conics U and V, admitting of being written thus,

$$x = -(\beta\nu + \gamma\mu) = \alpha(\beta - \gamma)^2, \quad y = \beta(\gamma - \alpha)^2, \quad z = \gamma(\alpha - \beta)^2.$$

And although it was *not necessary*, the foregoing *analysis* being quite *sufficient* – yet as a sort of *synthesis* or at least as an *a posteriori* proof, I have verified that when the last values of x, y, z are substituted in the expression,

$$D_xU = 2ax - (a + b)y - (a + c)z, \quad D_yU = \&\text{c.}, \quad D_zU = \&\text{c.},$$

and

$$D_xV = 2\alpha(\alpha x - \beta y - \gamma y), \quad D_yV = \&\text{c.}, \quad D_zV = \&\text{c.},$$

the results are

$$D_xU = (\alpha + \beta)(\beta + \gamma)(\gamma + \alpha)\lambda, \quad D_yU = (\alpha + \beta)(\beta + \gamma)(\gamma + \alpha)\mu,$$
$$D_zU = (\alpha + \beta)(\beta + \gamma)(\gamma + \alpha)\nu,$$
$$D_xV = 4\alpha\beta\gamma\lambda, \quad D_yV = 4\alpha\beta\gamma\mu, \quad D_zV = 4\alpha\beta\gamma\nu$$

being thus, for each function, proportional to λ, μ, ν; we have also

$$\lambda x + \mu y + \nu z = \lambda\alpha(\beta - \gamma)^2 + \&\text{c.} = (\alpha - \beta)(\beta - \gamma)(\gamma - \alpha)\{\alpha(\beta - \gamma) + \beta(\gamma - \alpha) + \gamma(\alpha - \beta)\}$$
$$= 0$$

so that $xD_xU + yD_yU + zD_zU = 0$, $xD_xV + yD_yV + zD_zV = 0$, or briefly $U = 0$, $V = 0$, for these values of x, y, z; whence we derive a verification, or rather a proof, that the two conics U and V *do in fact touch each other* at the point P, or (x, y, z) which has been above assigned.

173. Your Theorem is therefore by my method proved, with this slightly extended enunciation:- "If a conic V be touched by the three sides of a triangle ABC, and if through the two (real or imaginary) points, Q, Q', in which that conic is cut by any transversal $A'B'C'$, and through the harmonically conjugate points $A''B''C''$ upon the sides, another conic (U) be described, then these two conics *touch* each other;" namely, in the point P, determined as above.

174. I grant that the foregoing *mechanism of calculation* (compare many similar admissions made before) *might* have occurred to anyone, who knew any *bilinear coordinates*. But when we come to *interpret the results*, *then* I conceive that the *geometrical advantage* of my method comes into play. We have, for instance, here, by article 172, the equation,

$$\frac{\alpha y}{\beta x} = \left(\frac{\gamma - \alpha}{\gamma - \beta}\right)^2;$$

which I now propose to *interpret*, so as to derive from it a *construction* for the point of contact P, of which the *existence* was discovered by you. Let A_1, B_1, C_1 be the points in which the inscribed (or exscribed) conic (V) touches the 3 sides of the triangle; so that

$$A_1 = (0, \beta^{-1}, \gamma^{-1}), \quad B_1 = (\alpha^{-1}, 0, \gamma^{-1}), \quad C_1 = (\alpha^{-1}, \beta^{-1}, 0).$$

Then $P = (x, y, z)$ being still the point of contact, and Q, R, S being points derived from it as in the figure of article 148, my general formulæ give

$$\frac{\alpha}{\beta} = (C_1 AC''B), \quad \frac{\beta}{\gamma} = (A_1 BA''C), \quad \frac{\gamma}{\alpha} = (B_1 CB''A),$$

$$\frac{y}{x} = (SAC''B), \quad 1 - \frac{\beta}{\gamma} = (A_1 A''BC), \quad 1 - \frac{\alpha}{\gamma} = (B_1 B''AC),$$

and therefore

$$(SAC''B)(C_1 AC''B) = (B_1 B''AC)^2 (A_1 CBA'')^2;$$

thus S cuts the side AB in a known ratio, and similarly for Q and R.

175. When we confine ourselves to the *circles* of your enunciation, a few easy reductions give me the very simple geometrical proportion of rectangles and squares; $SAC_1 : SBC_1 = \overline{B''B_1}^2 : \overline{A''A_1}^2$; or with the usual trigonometrical significations of a, b, c, s, and for the case of the *inscribed* circle $\frac{AS}{SB} = \frac{s-b}{s-a} \cdot \left(\frac{c-a}{c-b}\right)^2$. Accordingly it is clear that this quotient, $\frac{AS}{SB}$, is equal to zero, unity, or infinity, in the three respective cases, $c = a$, $a = b$, $b = c$; because, in these three cases, the point of contact P coincides evidently with B_1, C_1, A_1; that is, here, with B'', C'', A'': the projections of which on AB, by lines from C, are A, C'', B.

176. A *simpler construction* may however be derived, by considering rather the *coordinates of the common tangent* $[\lambda, \mu, \nu]$, which may be conceived to be the line LMN in the figure of article 148, than the *coordinates of the point of contact* (x, y, z), or P. We have, by article 171

$$\lambda : \mu : \nu = (\beta - \gamma)^{-1} : (\gamma - \alpha)^{-1} : (\alpha - \beta)^{-1};$$

also if we write λ, μ, ν, for l, m, n in article 146, we have

$$\frac{\mu}{\nu} = (LBA'C), \quad \frac{\nu}{\lambda} = (MCB'A), \quad \frac{\lambda}{\mu} = (NAC'B),$$

or

$$-\frac{\mu}{\nu} = (LBA''C), \quad \frac{-\nu}{\lambda} = (MCB''A), \quad \frac{-\lambda}{\mu} = (NAC''B);$$

but, by what we have just now written,

$$\frac{-\mu}{\nu} = \frac{\alpha - \beta}{\alpha - \gamma}, \quad -\frac{\nu}{\lambda} = \frac{\beta - \gamma}{\beta - \alpha}, \quad -\frac{\lambda}{\mu} = \frac{\gamma - \alpha}{\gamma - \beta}$$

and, by article 174,

$$\frac{\gamma - \alpha}{\gamma - \beta} = (B_1 B''AC)(A_1 CBA''), \text{ \&c.},$$

therefore finally the 3 points L, M, N, in which your tangent cuts the sides of the triangle ABC, are given by the three anharmonic equations,

$$\begin{cases} (LBA''C) = (B_1ACB'').(C_1C''BA); \\ (MCB''A) = (C_1BAC'').(A_1A''CB); \\ (NAC''B) = (A_1CBA'').(B_1B''AC); \end{cases}$$

where, as a verification, it may be proved that the product of the six anharmonics, in the three right hand members is equal to negative unity, as it ought to be, in order to allow the product of the three anharmonics in the left hand members to be equal to the same given value, as the transversal requires.

177. When by throwing off the line $A'B'C'$ to infinity, my enunciation, in article 173, of your Theorem takes the form;

"*If through the middle points A'', B'', C'' of the sides of a triangle circumscribed to a given conic* (V), *a homothetic conic* (U) *be made to pass, these two conics will touch one another*",

the recent equations for the determination of the common tangent LMN, at the point of contact P of the two conics, are simplified to the following:

$$\frac{LB}{CL} = \frac{B_1A}{B''B_1} \cdot \frac{C_1C''}{AC_1}; \quad \frac{MC}{AM} = \frac{C_1B}{C''C_1} \cdot \frac{A_1A''}{BA_1}; \quad \frac{NA}{BN} = \frac{A_1C}{A''A_1} \cdot \frac{B_1B''}{CB_1};$$

of which, as a verification, the product is still equal to negative unity; while A_1, B_1, C_1 are (as before) the points in which the conic (V) is touched by the three sides of the triangle. And if we suppose that this conic is the *inscribed circle*, while the length of the *sides* are denoted (as usual) by a, b, c, then a few easy reductions give, more simply,

$$\frac{LB}{CL} = \frac{a-b}{a-c}; \quad \frac{MC}{AM} = \frac{b-c}{b-a}; \quad \frac{NA}{BN} = \frac{c-a}{c-b};$$

so that "*the tangent LMN cuts the greatest and least sides internally, but cuts the mean side externally*", becoming of course *indeterminate*, when the triangle is equilateral, because the two conics then *coincide*. Verifications for the case of an isosceles triangle, as in article 175, are obvious.

178. If we consider that *exscribed circle* (V') which touches the side BC *itself*, in a point A_1', and the sides AC, AB *prolonged*, in points B_1', C_1', and denote by $L'M'N'$ the tangent to *this* circle, at the new point P' where (by your Theorem) it touches the same bisecting circle (U) as before, we easily find, on the same plan, that

$$\frac{L'B}{CL'} = \frac{a+b}{a+c}; \quad \frac{M'C}{AM'} = \frac{b-c}{b+a}; \quad \frac{N'A}{BN'} = \frac{c+a}{c-b};$$

so that this tangent cuts *internally* the side BC, and the greater of the two sides, AB, CA; but cuts the lesser of those two sides *externally*. In like manner, your common tangents to the bisecting circle (U), and to the *two other exscribed circles* (V'') and (V'''), are on the present plan determined by the following ratios of section:

$$\frac{L''B}{CL''} = \frac{a+b}{a-c}; \quad \frac{M''C}{AM''} = \frac{b+c}{b+a}; \quad \frac{N''A}{BN''} = \frac{c-a}{c+b};$$

$$\frac{L'''B}{CL'''} = \frac{a-b}{a+c}; \quad \frac{M'''C}{AM'''} = \frac{b+c}{b-a}; \quad \frac{N'''A}{BN'''} = \frac{c+a}{c+b}.$$

– I do not think that we need *wish* for any *simpler construction* of your *four tangents,* than those given by these ratios of segments; and I doubt whether *trilinear coordinates* would have *suggested* constructions *so simple.* Besides, it is to be remembered that my method gives formulæ scarcely less simple (see article 177), for the case of *two homothetic conics*; and even (by article 176), through a few *anharmonics,* for the more general case of article 173, where the two conics *cut,* in two real or imaginary points Q, Q', on a *finitely distant line* $A'B'C'$.

179. (April 21st, 1860) – Since the foregoing pages were written, I have received from you a note, dated the 19th of this month, from which I collect that you have found yourself anticipated, in the Theorem that the *Bisecting Circle* (U) is touched by each of the four circles, inscribed or exscribed, which, or their analogous *Four Conics,* I have above marked as (V) (V') (V'') (V'''). And although I have not yet attempted to study your said Note, I see enough, from a mere *glance* at it, to lead me to suppose that you, or the French, had anticipated by *ratios and section,* of the sides of the triangle, made of the common tangents, at the four points of contact. *Ainsi soit il,* I cordially add: for I wish now to pass or return to other subjects which are to me more interesting at present.

180. I have already remarked, – either in some article to which I have not just now turned back, of this long Letter, or in some separate and recent Note, – that "when we replace the bisecting and inscribed *circles,* (U) and (V), by two *homothetic conics* (article 177), the two right lines drawn from the second intersection of the conic (U) with any one of the three sides of the given triangle, so as to be parallel to the two (real or imaginary) *asymptotes* of that or of the other conic, (since these are supposed parallel to each other,) are *harmonically conjugate* with respect to the intersected side, and the line drawn from the intersection to the vertex". A little more generally, let the conic (U) meet the sides BC, CA, AB, not only in the three former points A'', B'', C'', but also in the three points of second intersection, A''', B''', C''', while it still meets the unit line $A'B'C'$ in the two real or imaginary points Q and Q' (see article 170); I say that we have then the three following harmonic equations of pencils,

$$(A'''.AQA''Q') = (B'''.BQB''Q') = (C'''.CQC''Q') = -1$$

or the three corresponding harmonic equations of groups, upon the unit line.

$$(A'_1QA'Q') = (B'_1QB'Q') = (C'_1QC'Q') = -1$$

if A'_1, B'_1, C'_1 be the points $AA'''\cdot B'C'$, &c., in which the right lines drawn from the second intersection with (U) to the opposite vertices of the triangle, intersect that unit-line; whence it will follow as may be just remarked in passing, that the *three pairs of points,* A', A'_1; B', B'_1; and C', C'_1, form an *involution,* of which Q and Q' are the two (real or imaginary) *double points.*

181. The easiest or at all events the most *obvious* way, of proving this theorem by the formulæ of the present Letter, is perhaps the following. The coordinates of A', B', C' being already known, since $A' = (0, 1, -1)$, &c., let us investigate those of A'_1, B'_1, C'_1; and first of A''', B''', C'''. Cutting for this purpose the conic U by the side BC, or making $x = 0$ in the equation of article 170, namely:

$$2(ax^2 + by^2 + cz^2) - (x + y + z)(ax + by + cz) = 0,$$

we get the quadratic, $(y - z)(by - cz) = 0$; rejecting therefore the point A'', we have $A''' = (0, c, b)$; and similarly, $B''' = (c, 0, a)$, $C''' = (b, a, 0)$. The equations of the right lines AA''', BB''', CC''' are thus seen to be,

$$by = cz, \quad cz = az, \quad ax = by,$$

respectively; cutting these then by the unit line $x + y + z = 0$, and making for conciseness

$$a' = -b - c, \quad b' = -c - a, \quad c' = -a - b,$$

we find

$$A'_1 = (a', c, b), \quad B'_1 = (c, b', a), \quad C'_1 = (b, a, c').$$

Any two points, Q and Q' upon the unit line, may be thus symbolized,

$$(Q) = t(A') + (A'_1), \quad (Q') = t'(A') + (A'_1);$$

or thus

$$Q = (a', c + t, b - t); \quad Q' = (a', c + t', b - t');$$

with the resulting anharmonic equation $(A'QA'_1Q') = \frac{t'}{t}$. In the present case, A and Q' are both on the auxiliary conic of article 170,

$$ax^2 + by^2 + cz^2 = 0;$$

hence t and t' are the roots of the quadratic $ab + bc + ca + t^2 = 0$; so that $t' = -t$, and the harmonic relation is proved. – But variations of this proof or rather investigations quite different in their form, may instructively be made to replace it.

182. The following variation, for instance, may be noted, although it cannot pretend to be important. The vertex of the first pencil (article 180) being $A''' = (0, c, b)$, while $A = (1, 0, 0)$, and $A'' = (0, 1, 1)$, if (with new meanings of $\alpha, \beta, \gamma, \alpha', \beta', \gamma'$) we write

$$Q = (\alpha, \beta, \gamma), \quad Q' = (\alpha', \beta', \gamma'),$$

we shall have, by article 151,

$$(A'''.AQA''Q') = \begin{vmatrix} 0, & c, & b \\ 1, & 0, & 0 \\ \alpha, & \beta, & \gamma \end{vmatrix} \begin{vmatrix} 0, & c, & b \\ 0, & 1, & 1 \\ \alpha', & \beta', & \gamma' \end{vmatrix} : \begin{vmatrix} 0, & c, & b \\ \alpha, & \beta, & \gamma \\ 0, & 1, & 1 \end{vmatrix} \begin{vmatrix} 0, & c, & b \\ \alpha', & \beta', & \gamma' \\ 1, & 0, & 0 \end{vmatrix} = \frac{\alpha'}{\alpha} \cdot \frac{b\beta - c\gamma}{b\beta' - c\gamma'};$$

But, with the present significations of $\alpha, \beta, \gamma, \alpha', \beta', \gamma'$,

$$0 = \alpha + \beta + \gamma = \alpha' + \beta' + \gamma'$$
$$= a\alpha^2 + b\beta^2 + c\gamma^2 = a\alpha'^2 + b\beta'^2 + c\gamma'^2;$$

therefore

$$\beta\gamma' - \gamma\beta' = \gamma\alpha' - \alpha\gamma' = \alpha\beta' - \beta\alpha',$$

and

$$a: b: c = \beta^2\gamma'^2 - \gamma^2\beta'^2 : \gamma^2\alpha'^2 - \alpha^2\gamma'^2 : \alpha^2\beta'^2 - \beta^2\alpha'^2$$
$$= \beta\gamma' + \gamma\beta' : \gamma\alpha' + \alpha\gamma' : \alpha\beta' + \beta\alpha';$$

so that

$$(A'''.AQA''Q') = -1,$$

and the pencil is harmonic, as before.

183. Another mode of viewing the question is the following. The equation of the *four rays* of the pencil here considered are, respectively,

$$by - cz = 0; \quad tx + by - cz = 0; \quad x = 0; \quad -tx + by - cz = 0;$$

where, as in article 181, $t^2 + ab + bc + ca = 0$. Hence multiplying these four equations, we get what may be called the *equation of the pencil*, under the form,

$$0 = (ab + bc + ca)x^3(by - cz) + x(by - cz)^3$$

And it may perhaps be considered as evident, that every equation of the form

$$0 = BX^3Y + DXY^3$$

where B and D are any constants, and X, Y are any homogeneous rational and integral functions, of the first dimension in x, y, z, represents a harmonic pencil, with its vertex at the point for which $X = 0$, $Y = 0$; where the present pencil is *again* seen to be a harmonic one. But this suggests a much *more general investigation*, to which we shall now proceed; *dismissing* henceforth, for the purpose of the present Letter, all further consideration of the Theorem cited in article 169, and of every collateral theorem or *construction*, as being doubtless far better treated by yourself, and by the (unnamed) French writer or writers to whom you have alluded, in a note received by me not long ago.

184. I take therefore now the following *General Problem*, analogous to that proposed in article 149, and resolved in articles 150, and 151:

"Let Λ_0, or $[l_0, m_0, n_0]$, be a given transversal, cutting in P_1, P_2, P_3, P_4 or in $(x_1, y_1, z_1), \ldots, (x_4, y_4, z_4)$, the four given right lines $\Lambda_1, \ldots, \Lambda_4$, or $[l_1, m_1, n_1], \ldots, [l_4, m_4, n_4]$, which (for the sake of greater generality) we shall (at first) suppose *not* to *concur* in any common point, and therefore *not* to form a *pencil*, although they may still be said to compose a *system*; it is required to express the *anharmonic of the group*, of the *four points of section P*, which anharmonic we shall denote thus,

$$h = (P_1P_2P_3P_4) = [\Lambda_0.\Lambda_1\Lambda_2\Lambda_3\Lambda_4],$$

in terms of the 15 anharmonic coordinates, $l_0, m_0, n_0, l_1, m_1, n_1, \ldots, l_4, m_4, n_4$, of the five right lines Λ."

185. For this purpose I observe first that the coordinates of P_r may be thus expressed

$$x_r = \begin{vmatrix} m_0, & n_0 \\ m_r, & n_r \end{vmatrix}, \quad y_r = \begin{vmatrix} n_0, & l_0 \\ n_r, & l_r \end{vmatrix}, \quad z_r = \begin{vmatrix} l_0, & m_0 \\ l_r, & m_r \end{vmatrix};$$

so that we may write

$$(P_r) = \left(\begin{vmatrix} m_0, & n_0 \\ m_r, & n_r \end{vmatrix}, \begin{vmatrix} n_0, & l_0 \\ n_r, & l_r \end{vmatrix}, \begin{vmatrix} l_0, & m_0 \\ l_r, & m_r \end{vmatrix} \right).$$

But if we take any three of the four points of section, the collinearity of these three points enables us to foresee that we must have a symbolic linear equation of the form,

$$0 = p_{s,t}(P_r) + p_{t,r}(P_s) + p_{r,s}(P_t);$$

which is to be considered as including the three following,

$$0 = p_{s,t}x_r + p_{t,r}x_s + p_{r,s}x_t, \quad 0 = \ldots, \quad 0 = p_{s,t}z_r + p_{t,r}z_s + p_{r,s}z_t.$$

Accordingly it is not difficult to perceive that these equations are satisfied, when we make

$$p_{s,t} = \begin{vmatrix} l_0, & m_0, & n_0 \\ l_s, & m_s, & n_s \\ l_t, & m_t, & n_t \end{vmatrix}; \quad p_{t,r} = \&\text{c.}; \quad p_{r,s} = \&\text{c.}$$

And from the two symbolical equations,

$$0 = p_{2,3}(P_1) + p_{3,1}(P_2) + p_{1,2}(P_3),$$
$$0 = p_{3,4}(P_1) + p_{1,3}(P_4) + p_{4,1}(P_3),$$

we derive by a principle already often employed in this Letter, the sought anharmonic under the form

$$h = (P_1 P_2 P_3 P_4) = \frac{p_{1,2}\, p_{3,4}}{p_{2,3}\, p_{4,1}};$$

or more fully,

$$[\Lambda_0.\Lambda_1\Lambda_2\Lambda_3\Lambda_4] = \begin{vmatrix} l_0, & m_0, & n_0 \\ l_1, & m_1, & n_1 \\ l_2, & m_2, & n_2 \end{vmatrix} \begin{vmatrix} l_0, & m_0, & n_0 \\ l_3, & m_3, & n_3 \\ l_4, & m_4, & n_4 \end{vmatrix} : \begin{vmatrix} l_0, & m_0, & n_0 \\ l_2, & m_2, & n_2 \\ l_3, & m_3, & n_3 \end{vmatrix} \begin{vmatrix} l_0, & m_0, & n_0 \\ l_4, & m_4, & n_4 \\ l_1, & m_1, & n_1 \end{vmatrix};$$

so that the solution of the recent Problem (article 184) is expressed by a formula *exactly analogous* to the formula of solution, in article 151, for the corresponding problem of article 149, in which *five points,* instead of *five lines,* were given. In fact, I had *expected* to find the formula last deduced, as the *analogue* of the formula (of article 151), in virtue of that principle of *Geometrical Duality,* which is so fully recognised in, and incorporated with, my Method of *Anharmonic Coordinates.*

186. *Verifications,* analogous to those of articles 152 and 153, offer themselves with ease; among which I shall only mention that if we assign a *constant value* to the *anharmonic* h, and fixed positions to the *four* (non-concurrent) *right lines* $\Lambda_1, \ldots, \Lambda_4$, the *variable transversal* Λ_0 is then seen to *envelope a conic,* which *touches the four given lines.* As an *example,* by making those four lines Λ_1, Λ_2, Λ_3, Λ_4 coincide respectively, with the three sides of the unit-triangle ABC, and with the unit-line $A'B'C'$, while the suffix 0 may be suppressed, we have $h = [\Lambda.\Lambda_1\Lambda_2\Lambda_3\Lambda_4] = \dfrac{n(m-l)}{l(m-n)}$; so that by making $h = -1$ and letting off the unit-line to infinity, we obtain the following *tangential equation* of what may be called the *inscribed and harmonic parobola,*

$$l^{-1} + n^{-1} = 2\,m^{-1};$$

which gives the local equation

$$x^{\frac{1}{2}} + z^{\frac{1}{2}} = (-2\,y)^{\frac{1}{2}}$$

or

$$0 = (x - z)^2 + 4\,y(x + y + z).$$

In fact, it is evident on inspection that this conic is touched by each of the four lines, $x = 0$, $y = 0$, $z = 0$, $x + y + z = 0$, namely in the four points $(0,\ 1,\ -2)$, $(1,\ 0,\ 1)$, $(1,\ -2,\ 1)$. And although it is almost too elementary to mention, yet I may just add, what I saw while writing the last sentence, that we have thus the solution of the following little problem [see Figure 28]: "To find the envelope of a transversal *LMN* of a given triangle *ABC*, so drawn as to be bisected at *M*." The curve is the parabola, of which the anharmonic and local equation has now been written; it touches the side *CA*, at the middle point B'' of that side; it has BB'' prolonged for a diameter; and it is touched at *Q* and *S* by the sides *BC* and *BA* prolonged so as to become the doubles of their given tangents; the tangents from *B* being thus both bisected by the side *AC*. Of course, I suppose that all this is given in quite elementary books.

187. But to return to things more general. When the four lines $\Lambda_1, \ldots, \Lambda_4$ happen to be *concurrent*, and so to form a *pencil*, the position of the transversal Λ_0 is, of course, *unimportant*, the coordinates l_0, m_0, n_0 of the *line* Λ_0 must therefore in this case disappear from the result, which consequently may be written in any of the three forms following:

$$h = [\Lambda_1\Lambda_2\Lambda_3\Lambda_4] = \left|\begin{matrix} m_1, & n_1 \\ m_2, & n_2 \end{matrix}\right|\left|\begin{matrix} m_3, & n_3 \\ m_4, & n_4 \end{matrix}\right| : \left|\begin{matrix} m_2, & n_2 \\ m_3, & n_3 \end{matrix}\right|\left|\begin{matrix} m_4, & n_4 \\ m_1, & n_1 \end{matrix}\right|;$$

$$h = \left|\begin{matrix} n_1, & l_1 \\ n_2, & l_2 \end{matrix}\right|\left|\begin{matrix} n_3, & l_3 \\ n_4, & l_4 \end{matrix}\right| : \left|\begin{matrix} n_2, & l_2 \\ n_3, & l_3 \end{matrix}\right|\left|\begin{matrix} n_4, & l_4 \\ n_1, & l_1 \end{matrix}\right|;$$

$$h = \left|\begin{matrix} l_1, & m_1 \\ l_2, & m_2 \end{matrix}\right|\left|\begin{matrix} l_3, & m_3 \\ l_4, & m_4 \end{matrix}\right| : \left|\begin{matrix} l_2, & m_2 \\ l_3, & m_3 \end{matrix}\right|\left|\begin{matrix} l_4, & m_1 \\ l_1, & m_1 \end{matrix}\right|;$$

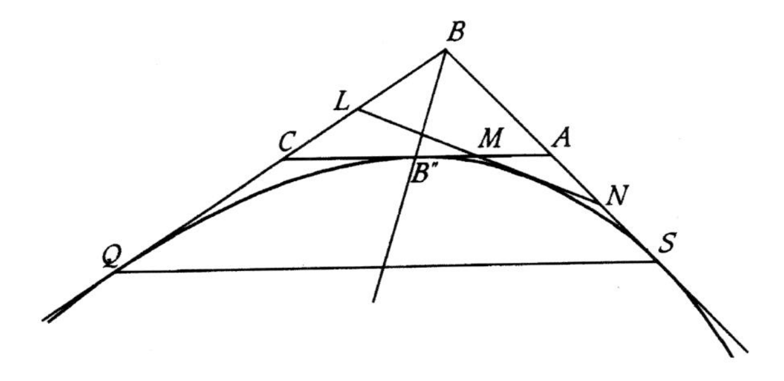

Fig. 28

and accordingly the *consistency* of these different expressions for the anharmonic h of the pencil is easily verified by means of the now existent determinant equations,

$$0 = \begin{vmatrix} l_1, & m_1, & n_1 \\ l_2, & m_2, & n_2 \\ l_3, & m_3, & n_3 \end{vmatrix}, \quad 0 = \begin{vmatrix} l_3, & m_3, & n_3 \\ l_4, & m_4, & n_4 \\ l_1, & m_1, & n_1 \end{vmatrix};$$

which give

$$\begin{vmatrix} m_1, & n_1 \\ m_2, & n_2 \end{vmatrix} : \begin{vmatrix} n_1, & l_1 \\ n_2, & l_2 \end{vmatrix} : \begin{vmatrix} l_1, & m_1 \\ l_2, & m_2 \end{vmatrix} = \begin{vmatrix} m_2, & n_2 \\ m_3, & n_3 \end{vmatrix} : \begin{vmatrix} n_2, & l_2 \\ n_3, & l_3 \end{vmatrix} : \begin{vmatrix} l_2, & m_2 \\ l_3, & m_3 \end{vmatrix},$$

and

$$\begin{vmatrix} m_3, & n_3 \\ m_4, & n_4 \end{vmatrix} : \begin{vmatrix} n_3, & l_3 \\ n_4, & l_4 \end{vmatrix} : \begin{vmatrix} l_3, & m_3 \\ l_4, & m_4 \end{vmatrix} = \begin{vmatrix} m_4, & n_4 \\ m_1, & n_1 \end{vmatrix} : \begin{vmatrix} n_4, & l_4 \\ n_1, & l_1 \end{vmatrix} : \begin{vmatrix} l_4, & m_4 \\ l_1, & m_1 \end{vmatrix};$$

in fact, each of these four last sets of determinants, taken separately maybe expressed as representing the coordinates of the vertex of the pencil.

188. Quite similarly, if the *four points* $P_1, \ldots, P_4$ in the formula of article 151, be *collinear*, the *position* of the *vertex* P_0 of the pencil is immaterial, and its coordinates x_0, y_0, z_0 must disappear from the expression of the anharmonic; which thus assumes any one of the three following forms,

$$h = ((x_1, y_1, z_1)(x_2, y_2, z_2)(x_3, y_3, z_3)(x_4, y_4, z_4))$$

$$= (P_1 P_2 P_3 P_4) = \begin{vmatrix} y_1, & z_1 \\ y_2, & z_2 \end{vmatrix} \begin{vmatrix} y_3, & z_3 \\ y_4, & z_4 \end{vmatrix} : \begin{vmatrix} y_2, & z_2 \\ y_3, & z_3 \end{vmatrix} \begin{vmatrix} y_4, & z_4 \\ y_1, & z_1 \end{vmatrix};$$

$$h = \begin{vmatrix} z_1, & x_1 \\ z_2, & x_2 \end{vmatrix} \begin{vmatrix} z_3, & x_3 \\ z_4, & x_4 \end{vmatrix} : \begin{vmatrix} z_2, & x_2 \\ z_3, & x_3 \end{vmatrix} \begin{vmatrix} z_4, & x_4 \\ z_1, & x_1 \end{vmatrix};$$

$$h = \begin{vmatrix} x_1, & y_1 \\ x_2, & y_2 \end{vmatrix} \begin{vmatrix} x_3, & y_3 \\ x_4, & y_4 \end{vmatrix} : \begin{vmatrix} x_2, & y_2 \\ x_3, & y_3 \end{vmatrix} \begin{vmatrix} x_4, & y_4 \\ x_1, & y_1 \end{vmatrix};$$

and the mutual compatibility of these three expressions for h maybe proved by the two determinant equations

$$0 = \begin{vmatrix} x_1, & y_1, & z_1 \\ x_2, & y_2, & z_2 \\ x_3, & y_3, & z_3 \end{vmatrix}, \quad 0 = \begin{vmatrix} x_3, & y_3, & z_3 \\ x_4, & y_4, & z_4 \\ x_1, & y_1, & z_1 \end{vmatrix},$$

which give the proportions,

$$\begin{vmatrix} y_1, & z_1 \\ y_2, & z_2 \end{vmatrix} : \begin{vmatrix} z_1, & x_1 \\ z_2, & x_2 \end{vmatrix} : \begin{vmatrix} x_1, & y_1 \\ x_2, & y_2 \end{vmatrix} = \begin{vmatrix} y_2, & z_2 \\ y_3, & z_3 \end{vmatrix} : \begin{vmatrix} z_2, & x_2 \\ z_3, & x_3 \end{vmatrix} : \begin{vmatrix} x_2, & y_2 \\ x_3, & y_3 \end{vmatrix},$$

and

$$\begin{vmatrix} y_3, & z_3 \\ y_4, & z_4 \end{vmatrix} : \begin{vmatrix} z_3, & x_3 \\ z_4, & x_4 \end{vmatrix} : \begin{vmatrix} x_3, & y_3 \\ x_4, & y_4 \end{vmatrix} = \begin{vmatrix} y_4, & z_4 \\ y_1, & z_1 \end{vmatrix} : \begin{vmatrix} z_4, & x_4 \\ z_1, & x_1 \end{vmatrix} : \begin{vmatrix} x_4, & y_4 \\ x_1, & y_1 \end{vmatrix};$$

in fact, *any one* of these four last sets of determinants separately taken, maybe considered as the anharmonic *coordinates of the line*, on which, by supposition, the four points $P_1, \ldots, P_4$ are situated.

189. As the formulæ of the foregoing article enable us to calculate the *anharmonic of a group*, in a *direct* manner, from the *coordinates of the four points* of that group, without seeking the values of any *auxiliary coefficients*, such as those denoted by t, u, and t', u', in many former investigations, it may not be quite uninteresting to give an *example* or two of their application to questions already discussed. In fact, if this Letter were to be rewritten, they ought to be introduced quite early, and treated as among the fundamental *elements* of the *Method*.

190. Our first Example may be taken from article 11 itself, in which those auxiliary coefficients t, u, t', u' were first introduced. We were to calculate

$$h = (OUCU') = ((0,\, 0,\, 1)(1,\, 1,\, 1)(1,\, 1,\, 0)(1,\, 1,\, -1));$$

for which the first of our recent expressions (article 188) gives,

$$h = \begin{vmatrix} 0, & 1 \\ 1, & 1 \end{vmatrix} \begin{vmatrix} 1, & 0 \\ 1, & -1 \end{vmatrix} : \begin{vmatrix} 1, & 1 \\ 1, & 0 \end{vmatrix} \begin{vmatrix} 1, & -1 \\ 0, & 1 \end{vmatrix} = 1 : -1 = -1;$$

the group is therefore harmonic as in the cited article. The second expression of article 188 gives, in like manner

$$h = \begin{vmatrix} 1, & 0 \\ 1, & 1 \end{vmatrix} \begin{vmatrix} 0, & 1 \\ -1, & 1 \end{vmatrix} : \begin{vmatrix} 1, & 1 \\ 0, & 1 \end{vmatrix} \begin{vmatrix} -1, & 1 \\ 1, & 0 \end{vmatrix} = 1 : -1 = -1;$$

but the third expression of article 188 conducts to the (not false but) useless form,

$$h = \begin{vmatrix} 0, & 0 \\ 1, & 1 \end{vmatrix} \begin{vmatrix} 1, & 1 \\ 1, & 1 \end{vmatrix} : \begin{vmatrix} 1, & 1 \\ 1, & 1 \end{vmatrix} \begin{vmatrix} 1, & 1 \\ 0, & 0 \end{vmatrix} = \frac{0}{0}.$$

In general, if any one of the four points of a given group be at a corner of the unit triangle, (as the point O was, in the notation of article 11,) so that *two* of its coordinates *vanish*, we are then to reject, as useless, that one of the three expressions of article 188, which combines those *two* evanescent coordinates.

191. A Second Example, of slightly greater complexity, may be taken from article 18; in which we were to calculate this other anharmonic,

$$h = (X_1\, V_1\, Y_1\, U_1)$$

when

$$X_1 = (2z,\, 0,\, y), \quad V_1 = (2az - 2cx,\, 2cy - 2bz,\, ay - bx),$$
$$Y_1 = (0,\, 2z,\, x), \quad U_1 = (2z - 2x,\, 2y - 2z,\, y - x).$$

In this case

$$\begin{vmatrix} y_1, & z_1 \\ y_2, & z_2 \end{vmatrix} = \begin{vmatrix} 0, & y \\ 2cy - 2bz, & ay - bx \end{vmatrix} = -2y(cy - bz);$$

$$\begin{vmatrix} y_2, & z_2 \\ y_3, & z_3 \end{vmatrix} = \begin{vmatrix} 2cy - 2bz, & ay - bx \\ 2z, & x \end{vmatrix} = -2y(az - cx);$$

$$\begin{vmatrix} y_3, & z_3 \\ y_4, & z_4 \end{vmatrix} = \begin{vmatrix} 2z, & x \\ 2y - 2z, & y - x \end{vmatrix} = 2y(z - x);$$

$$\begin{vmatrix} y_4, & z_4 \\ y_1, & z_1 \end{vmatrix} = \begin{vmatrix} 2y - 2z, & y - x \\ 0, & y \end{vmatrix} = 2y(y - z);$$

so that the first expression of article 188 becomes here,

$$h = \frac{(cy - bz)(z - x)}{(az - cx)(y - z)},$$

as in the cited article. The subsequent reduction here given, of this anharmonic to the form $\frac{c}{a}$, or $\frac{b}{c}$, depended on certain supposed relations between the quantities ($2y = z^2$, $ab = c^2$), which had geometrical meanings assigned to them, but are foreign to our present purpose.

192. As a third Example, which will introduce imaginaries, let us take the group of article 155 no one of the four points of which is real. Here,

$$h = (I^0 I' I'' I''') = ((1, \theta^2, 0)(0, 1, -\theta)(-\theta, 0, 1)(1, -\theta, 0));$$

and if, for the sake of variety, we employ the third expression of article 188, we find

$$h = \begin{vmatrix} 1, & \theta^2 \\ 0, & 1 \end{vmatrix} \begin{vmatrix} -\theta, & 0 \\ 1, & -\theta \end{vmatrix} : \begin{vmatrix} 0, & 1 \\ -\theta, & 0 \end{vmatrix} \begin{vmatrix} 1, & -\theta \\ 1, & \theta^2 \end{vmatrix} = \theta^2 : -\theta = -\theta,$$

as in the cited article; this imaginary group being therefore a *di*-anharmonic one, as was otherwise found before. The first expression of article 188 would have given, in like manner, $h = 1 : -\theta^2 = -\theta$; and the second expression $h = \theta : -1 = -\theta$. But it seems useless to multiply such examples of the use of the general expressions of article 188, for the anharmonic of a *group*; nor shall we think it necessary to give any instances, so particular as these, of the corresponding application of the analogous and equally general formulæ of article 187, for the anharmonic of a *pencil*.

193. The following application of the last mentioned formula appears however to deserve attention. Let the point C, or $(0, 0, 1)$, be assumed as the vertex of the pencil; then $n_1 = n_2 = n_3 = n_4 = 0$, and the two first of the expressions of article 187 for h become illusory; but the last of them is applicable, and gives

$$h = [\Lambda_1 \Lambda_2 \Lambda_3 \Lambda_4] = \frac{(l_1 m_2 - m_1 l_2)(l_3 m_4 - m_3 l_4)}{(l_2 m_3 - m_2 l_3)(l_4 m_1 - l_1 m_4)}.$$

Under the same condition, the *equations of the four rays* become

$$l_1 x + m_1 y = 0, \quad l_2 x + m_2 y = 0, \quad l_3 x + m_3 y = 0, \quad l_4 x + m_4 y = 0;$$

if then the *equation of the pencil*, obtained by multiplying together the equations of its *rays*, be in any manner known to be

$$Ax^4 + 4Bx^3 y + 6Cx^2 y^2 + 4Dxy^3 + Ey^4 = 0$$

when A, B, C, D, E are given or known coefficients, we are sure that the four ratios of m to l must be the roots of the corresponding equation,

$$Am^4 - 4Blm^3 + 6Cl^2 m^2 - 4Dl^3 m + El^4 = 0.$$

Or, if we *make* for abridgment,

$$m_1 = \alpha l_1, \quad m_2 = \beta l_2, \quad m_3 = \gamma l_3, \quad m_4 = \delta l_4,$$

then $\alpha, \beta, \gamma, \delta$ are the four roots of the biquadratic equation,

$$A\alpha^4 - 4B\alpha^3 + 6C\alpha^2 - 4D\alpha + E = 0.$$

At the same time, *one value* of the *anharmonic* (h) *of the pencil*, answering to *one arrangement* of its *rays*, – or rather to *four* such arrangements, because we have (cf. article 5) the four *equal* anharmonics,

$$[\Lambda_1\Lambda_2\Lambda_3\Lambda_4] = [\Lambda_2\Lambda_1\Lambda_4\Lambda_3] = [\Lambda_3\Lambda_4\Lambda_1\Lambda_2] = [\Lambda_4\Lambda_3\Lambda_2\Lambda_1],$$

$$h_1 = [\Lambda_1\Lambda_2\Lambda_3\Lambda_4] = \frac{(\beta - \alpha)(\delta - \gamma)}{(\gamma - \beta)(\alpha - \delta)};$$

the *five other values* of h, answering to the *twenty other* arrangements of the rays, being (cf. again, article 5)

$$h_2 = [\Lambda_1\Lambda_3\Lambda_2\Lambda_4] = \frac{(\gamma - \alpha)(\delta - \beta)}{(\beta - \gamma)(\alpha - \delta)} = 1 - h_1;$$

$$h_3 = [\Lambda_1\Lambda_4\Lambda_3\Lambda_2] = \frac{(\delta - \alpha)(\beta - \gamma)}{(\gamma - \delta)(\alpha - \beta)} = h_1^{-1};$$

$$h_4 = [\Lambda_1\Lambda_3\Lambda_4\Lambda_2] = \frac{(\gamma - \alpha)(\beta - \delta)}{(\delta - \gamma)(\alpha - \beta)} = 1 - h_1^{-1};$$

$$h_5 = [\Lambda_1\Lambda_4\Lambda_2\Lambda_3] = \frac{(\delta - \alpha)(\gamma - \beta)}{(\beta - \delta)(\alpha - \gamma)} = (1 - h_1)^{-1};$$

$$h_6 = [\Lambda_1\Lambda_2\Lambda_4\Lambda_3] = \frac{(\beta - \alpha)(\gamma - \delta)}{(\delta - \beta)(\alpha - \gamma)} = 1 - (1 - h_1)^{-1} = -h_1(1 - h_1)^{-1}.$$

And it is now required *to form that equation of the sixth degree,* of which the *roots* shall be *these six values,* (generally unequal), namely, $h_1, \ldots, h_6$, of the anharmonic h of the pencil, that pencil being supposed to be only *given by its equation,* $Ax^4 + 4Bx^3 + \&c. = 0$; or to *determine the coefficients of the resultant sextic in h, as rational functions of the coefficients A, B, C, D, E, of the given biquadratic equation* of the said pencil.

194. In this investigation, I do not find that I can much assist myself, by taking for a guide the analysis given [by Salmon (*HPC*, p. 191 §205)]. The author of that article forms first, in a known way, the *cubic* equation of which the roots are the three combinations,

$$\alpha\beta + \gamma\delta, \quad \alpha\gamma + \beta\delta, \quad \alpha\delta + \beta\gamma;$$

and then he forms a *sextic*, reducible (with the help of an ambiguous square root) to a *cubic form*, of which the roots are the *differences* of the roots of the foregoing cubic equation. He then asserts (*in loco citato*) that "the anharmonic functions in question are the ratios of the roots of this equation"; namely, of that *sextic* which, with the help of a $\pm$ sign, is thus presented under a *cubic form*: or more fully, to copy from his p. 191,

$$t^3 - 12t \pm 2\sqrt{\left(\frac{R}{S^3}\right)} = 0.$$

But, if you allow me here to repeat part of what I have said in a recent and separate note, – a *general sextic* would conduct to an equation of the *30th degree* in h, if h were employed to denote the quotient of one of its six roots divided by another. And even when we assume that the *coefficients of odd powers vanish* in the *given sextic* (as here in the equation

$$t^6 - 24t^4 + 144t^2 - \frac{4R}{S^3} = 0,$$

which results immediately from Salmon's cited page) still the resultant equation in h, – after being divided by the foreign factor $(h+1)^6$, and after the (demonstrably possible) extraction of a square root of the quotient has been accomplished, – is only depressed as low *as the 12th degree*, although in this last *reduction*, as in earlier forms *no odd powers of h will appear.* If the equation last written be called *Salmon's sextic*, in relation to our present object then I find that the resultant equation of the *twelth* degree breaks up into *two rational factors*; which are each of the *sixth* degree; and I am pretty certain, – indeed I feel *quite sure*, in my own mind, although I have not set about to find any formal and general *proof* of it, – that the same sort of *rational decomposition*, into *two distinct sextics*, must be possible universally: that is to say, whatever may be the values of the given coefficients $A, \ldots, E$. But how are we to decide; *which of the two sextic factors* is to be *retained*, to the *exclusion* of the *other*? Or how are we to know, whether *some* of the six values of h may not belong to *one* of those two sextic factors; and the other values of that anharmonic h to the *other*?

195. For these and other reasons, I decided on attacking the Problem of article 193 by a totally *different*, and yet I think by an easier analysis; namely, by forming that *new auxiliary cubic*, (without any square or other radical entering into its composition,) of which the three roots η_1, η_2, η_3 shall be *the three values of the combination*, $\eta = h(1-h)$. Writing then

$$\eta_1 = h_1 h_2 = h_1(1 - h_1) = \frac{(\beta - \alpha)(\delta - \gamma).(\gamma - \alpha)(\delta - \beta)}{-(\beta - \gamma)^2.(\alpha - \delta)^2},$$

$$\eta_2 = h_3 h_4 = -h_1^{-2}(1 - h_1) = \frac{(\delta - \alpha)(\beta - \gamma).(\gamma - \alpha)(\beta - \delta)}{-(\delta - \gamma)^2.(\alpha - \beta)^2},$$

$$\eta_3 = h_5 h_6 = -h_1(1 - h_1)^{-2} = \frac{(\delta - \alpha)(\gamma - \beta).(\beta - \alpha)(\gamma - \delta)}{-(\delta - \beta)^2.(\alpha - \gamma)^2},$$

my proofs may be said to consist in forming the cubic equation, which has for roots these three (partially symmetric) functions of the roots $\alpha, \beta, \gamma, \delta$ of the given biquadratic equation – the first of these for example, not changing value, when β and γ, or when δ and α are interchanged. But for this purpose I assist myself by the expressions here given, for the three new roots η_1, η_2, η_3, in terms of the single quantity h_1.

196. In fact I have thus not only the relation,

$$\eta_1\eta_2\eta_3 = 1,$$

which is evident on mere inspection, and might have been foreseen from the geometrical principle, that to each anharmonic of a group or pencil a reciprocal anharmonic corresponds, but also the less obvious formula of relation,

$$\eta_1^{-1} + \eta_2^{-1} + \eta_3^{-1} = 3;$$

or, in combination with the former,

$$\eta_2\eta_3 + \eta_3\eta_1 + \eta_1\eta_2 = 3;$$

because

$$\eta_1^{-1} + \eta_2^{-1} = \eta_1^{-1}(1 - h_1^3) = h_1^{-1}(1 + h_1 + h_1^2) = 3 + h_1^{-1}(1 - h_1)^2 = 3 - \eta_3^{-1}.$$

If then we make

$$\eta_1 + \eta_2 + \eta_3 = 3 - \varepsilon$$

where ε is a new constant, the *cubic* in η will take the *very simple form,*

$$(\eta - 1)^3 + \varepsilon\eta^2 = 0;$$

or because $\eta = h(1 - h)$, the sextic in h will assume this form of corresponding simplicity,

$$(h^2 - h + 1)^3 - \varepsilon(h^2 - h)^2 = 0;$$

in which it only remains to express the constant ε, as a rational function of the coefficients A, B, C, D, E, of the proposed biquadratic.

197. As I have not seen this very simple *form* of the sought sextic equation in print, it may not be amiss to offer here a brief confirmation of its correctness. Writing, for this purpose, $h = \dfrac{(\beta - \alpha)(\delta - \gamma)}{(\gamma - \beta)(\alpha - \delta)}$, we wish to calculate the coefficients an an equation of the 6th degree, which shall have for its 6 roots the 6 values of this function h of the 4 roots $\alpha, \beta, \gamma, \delta$ of a given biquadratic equation. By interchanging β and δ, we change h to h^{-1}, by interchanging, on the other hand, β and γ, we change h to $1 - h$. If then a rational function h can be found, such that $\varphi(h) = \varphi(h^{-1}) = \varphi(1-h)$, *by the mere form of the function,* we shall have $\varphi(h_1) = \varphi(h_2) = \varphi(h_3) = \varphi(h_4) = \varphi(h_5) = \varphi(h_6) =$ some constant $\varepsilon =$ a rational function of the coefficients of the biquadratic; but the function $\varphi(h) = \dfrac{(h^2 - h + 1)^3}{(h^2 - h)}$ has the properties required; and $\varphi(h) = \varepsilon$ when cleared of fractions is an equation of the sixth degree: it is therefore the sought sextic.

198. For the calculation of ε, in terms of $A, \ldots, E$, two different processes present themselves. The most obvious is to calculate it as a symmetric function of the roots $\alpha, \beta, \gamma, \delta$, by substituting the expressions in article 195 for η_1, η_2, η_3 in the formula of article 196, $\varepsilon = 3 - \eta_1 - \eta_2 - \eta_3$; but it is also allowed to substitute the value in article 197 for h, in the expression $\varepsilon = \varphi(h)$, which must then be found to reduce either to such a symmetric function of the roots, or to a rational function of the coefficients, of the given biquadratic equation.

199. Adopting the former method, and denoting for the present the product of the squares of the differences of the roots of the given equation by Δ, so that

$$\Delta = (\alpha-\beta)^2(\alpha-\gamma)^2(\alpha-\delta)^2(\beta-\gamma)^2(\beta-\delta)^2(\gamma-\delta)^2,$$

we have

$$(\varepsilon - 3)\Delta = (\alpha-\beta)^3(\alpha-\gamma)^3(\beta-\delta)^3(\gamma-\delta)^3 + (\alpha-\gamma)^3(\alpha-\delta)^3(\beta-\gamma)^3(\beta-\delta)^3 + (\alpha-\beta)^3(\alpha-\delta)^3(\gamma-\beta)^3(\gamma-\delta)^3.$$

This symmetric function or the function ε which depends upon it, occurs, I suppose, somewhere in books; but as I do not remember meeting it, I have been obliged to strike out modes of treating it, or of calculating ε for myself. One such mode is the following. Writing for abridgment,

$$\alpha\beta + \gamma\delta = z, \quad \alpha\gamma + \delta\beta = z', \quad \alpha\delta + \beta\gamma = z'',$$

we have

$$\Delta = (z - z')^2(z' - z'')^2(z'' - z)^2,$$

and

$$(\varepsilon - 3)\Delta = (z - z'')^3(z' - z'')^3 + (z' - z)^3(z'' - z)^3 + (z'' - z')^3(z - z')^3.$$

Hence,

$$\Delta + (z - z'')^3(z' - z'')^3 = G(z' - z'')^2(z'' - z)^2,$$
$$\Delta + (z' - z)^3(z'' - z)^3 = G(z'' - z)^2(z - z')^2,$$
$$\Delta + (z'' - z')^3(z - z')^3 = G(z - z')^2(z' - z'')^2,$$

if we write

$$G = z^2 + z'^2 + z''^2 - z'z'' - z''z - zz';$$

but this gives

$$(z' - z'')^2(z'' - z)^2 + (z'' - z)^2(z - z')^2 + (z - z')^2(z' - z'')^2 = G^2,$$

so that $\varepsilon = \frac{G^3}{\Delta}$; Δ is a well known function of the coefficients of the biquadratic, we have only to express G as another rational function of them. But this is easily done: for we soon find that

$$A^2 G = 12(AE - 4BD + 3C^2) = 144S,$$

with the same meaning of S as in *HPC* p. 191 and since we have also, with the meaning there assigned to R, $A^6\Delta = 2^8 3^3 R$, we may write finally $\varepsilon = \frac{2^4 3^3 S^3}{R}$. The sextic equation of which the anharmonics h of the pencil are the roots, is therefore by article 196, (as was stated in a recent note,)

$$(h^2 - h + 1)^3 - \frac{432 S^3}{R}(h^2 - h)^2 = 0.$$

200. The same result may be obtained, with even less trouble of calculation, by the second method of article 198; namely by substituting the value

$$h = \frac{(\beta - \alpha)(\delta - \gamma)}{(\gamma - \beta)(\alpha - \delta)},$$

which gives

$$1 - h = \frac{(\alpha - \gamma)(\delta - \beta)}{(\gamma - \beta)(\alpha - \delta)},$$

in the functional expression

$$h = \frac{(h^2 - h + 1)^3}{(h^2 - h)^2},$$

which must then be found to be equal to a constant ε. For we should thus have, at once, $\varepsilon\Delta = G^3$, when

$$\begin{aligned} G &= (\beta - \alpha)^2(\delta - \gamma)^2 - (\beta - \alpha)(\delta - \gamma)(\alpha - \delta)(\gamma - \beta) + (\alpha - \delta)^2(\gamma - \beta)^2 \\ &= \Sigma\alpha^2\beta^2 - \Sigma\alpha^2\beta\gamma + 6\alpha\beta\gamma\delta \\ &= (\Sigma\alpha\beta)^2 - 3\Sigma\alpha\Sigma\alpha\beta\gamma + 12\alpha\beta\gamma\delta; \end{aligned}$$

[where $\sum \alpha^2\beta^2$ stands for $\alpha^2(\beta^2 + \gamma^2 + \delta^2) + \beta^2(\alpha^2 + \gamma^2 + \delta^2) + \gamma^2(\alpha^2 + \beta^2 + \delta^2)$, and so on; see *LHA*, p. 77,] or

$$\begin{aligned} A^2 G &= (6C)^2 - 3(4B)(4D) + 12AE \\ &= 12(AE - 4BD + 3C^2) = 144S, \end{aligned}$$

as before; the rest of the proof being unaltered. Of course it would be easy to modify these expressions in articles 199 and 200, for the symmetric function G of the roots, so as to exhibit their agreement (numerical coefficients excepted) with the analogous expressions given by Salmon for S. But I do not know whether it has occurred to him to observe, by the use of imaginary cube roots of unity, that this function S, which is of the fourth *dimension* relating to

α, β, γ, δ, may be *decomposed into quadratic factors.* In fact, the first expression of the present article for G may be thus written:

$$G = \{(\beta - \alpha)(\delta - \gamma) + \theta(\alpha - \delta)(\gamma - \beta)\}\{(\beta - \alpha)(\delta - \gamma) + \theta^2(\alpha - \delta)(\gamma - \beta)\};$$

or thus

$$G = (-\theta z - \theta^2 z' - z'')(-\theta^2 z - \theta^2 z' - z'')$$
$$= (+\theta^2 z' + \theta^2 z'')(z + \theta^2 z' + \theta z'')$$

which gives accordingly, as in article 199,

$$G = z^2 + z'^2 + z''^2 - z'z'' - z''z - zz';$$

and therefore, (cf. *HPC*, note in p. 297, and *LHA* p. 100)

$$\frac{288S}{A^2} = 2G = (z' - z'')^2 + (z'' - z)^2 + (z - z')^2$$
$$= (\alpha - \beta)^2(\gamma - \delta)^2 + (\alpha - \gamma)^2(\delta - \beta)^2 + (\alpha - \delta)^2(\beta - \gamma)^2.$$

As a verification, by changing $(\alpha - \gamma)(\delta - \beta)$ to its equivalent, namely, $(\gamma - \beta)(\alpha - \delta) - (\beta - \alpha)(\delta - \gamma)$, and halving, we return to the first value of G assigned in the present article.

201. The constant T may be usefully introduced, as follows. When the pencil is a *harmonic* one, so that h has one of the three values, -1, $+2$, $+\frac{1}{2}$, the constant $\varepsilon = \varphi(h)$ takes the particular value $\frac{27}{4}$; and accordingly we have the identity,

$$4(h^2 - h + 1)^3 - 27(h^2 - h)^2 = (h + 1)^2(h - 2)^2(2h - 1)^2;$$

so that

$$4\varepsilon - 27 = \left\{\frac{(h + 1)(h - 2)(2h - 1)}{h(h - 1)}\right\}^2$$

and this last function of h must be a symmetric function of α, β, γ, δ. Writing for abridgment

$$\zeta = z' - z'', \quad \zeta' = z'' - z, \quad \zeta'' = z - z'$$

so that

$$\zeta + \zeta' + \zeta'' = 0, \quad \zeta^2\zeta'^2\zeta''^2 = \Delta$$

and

$$\zeta^2 + \zeta'^2 + \zeta''^2 = 2G, \quad \zeta'^2\zeta''^2 + \zeta''^2\zeta^2 + \zeta^2\zeta'^2 = G^2$$

ζ^2 being thus (as it is easy to verify) a root of the cubic equation

$$(\zeta^2)^3 - 2G(\zeta^2)^2 + G^2(\zeta^2) - \Delta = 0,$$

to which we may add that (as appears from article 199)

$$\zeta'^3\zeta''^3 + \zeta''^3\zeta^3 + \zeta^3\zeta'^3 = 3\Delta - G^3,$$

and that

$$\frac{\zeta^2}{\zeta'\zeta''}+\frac{\zeta'^2}{\zeta''\zeta}+\frac{\zeta''^2}{\zeta\zeta'}=3,$$

we have $h=-\frac{\zeta}{\zeta''}$, and therefore

$$\zeta''(h-2)=\zeta'-\zeta'',\quad \zeta''(h+1)=\zeta''-\zeta,\quad \zeta''(1-2h)=\zeta-\zeta';$$

also

$$h-1=\frac{\zeta'}{\zeta''},\quad \zeta''^3h(1-h)=\zeta\zeta'\zeta''.$$

Hence

$$(4\varepsilon-27)\Delta=\{(\zeta'-\zeta'')(\zeta''-\zeta)(\zeta-\zeta')\}^2;$$

where not only the second member itself, but the product of which it is the square, is a symmetric function of the roots of the biquadratic.

In fact we have

$$\begin{aligned}A(\zeta'-\zeta'')&=A(z'+z''-2z)=6C-3Az=3(2C-Az);\\ A(\zeta''-\zeta)&=A(z''+z-2z')\qquad\qquad=3(2C-Az')\\ A(\zeta-\zeta')&=A(z+z'-2z'')\qquad\qquad=3(2C-Az'')\end{aligned}$$

because $\frac{6C}{A}=\Sigma\alpha\beta=z+z'+z''$. [See note about Σ in article 200.] Also

$$z'z''+z''z+zz'=\Sigma\alpha^2\beta\gamma=\Sigma\alpha.\Sigma\alpha\beta\gamma-4\alpha\beta\gamma\delta=4A^{-2}(4BD-AE);$$

and

$$\begin{aligned}zz'z''=\alpha\beta\gamma\delta\Sigma\alpha^2+\Sigma\alpha^2\beta^2\gamma^2&=(\Sigma\alpha\beta\gamma)^2-4\alpha\beta\gamma\delta\Sigma\alpha\beta+\alpha\beta\gamma\delta(\Sigma\alpha)^2\\ &=8A^{-3}(2AD^2-3ACE+2EB^2);\end{aligned}$$

so that z is a root of the cubic equation,

$$A^3z^3-6A^2Cz^2+4A(4BO-AE)z-8(2AD^2-3ACE+2EB^2)=0:$$

which indeed is given by Salmon (*HPC* p. 191), apparently from Lacroix.* Changing then Az to $2C$ in the first member of this equation, and dividing by 8, we find,

$$\begin{aligned}\frac{A^3}{216}(\zeta'-\zeta'')(\zeta''-\zeta)(\zeta-\zeta')&=(C-\tfrac{1}{2}Az)(C-\tfrac{1}{2}Az')(C-\tfrac{1}{2}Az'')\\ &=-2C^3+C(4BD-AE)-(2AD^2-3ACE+2EB^2)\\ &=2(-C^3+ACE-AD^2+2BCD-EB^2)=2T\end{aligned}$$

with the signification of T in the last cited page, and in the page preceding; we have therefore

* [See footnote on p. 134 of this volume.]

$$\begin{aligned}(\zeta' - \zeta'')(\zeta'' - \zeta)(\zeta - \zeta') &= \{(\alpha - \gamma)(\delta - \beta) - (\alpha - \delta)(\beta - \gamma)\}\\ &\quad \times \{(\alpha - \delta)(\beta - \gamma) - (\alpha - \beta)(\gamma - \delta)\}\\ &\quad \times \{(\alpha - \beta)(\gamma - \delta) - (\alpha - \gamma)(\delta - \beta)\}\\ &= 432 A^{-3} T = 2^4 3^3 A^{-3} T;\end{aligned}$$

which agrees with the note (*HPC* p. 297, and *LHA* p. 100), except that I have been obliged to supply the numerical coefficient. Hence, $4\varepsilon - 27 = \dfrac{2^8 3^6 T^2}{A^6 \Delta} = \dfrac{27 T^2}{R}$, which might have been at once deduced from the value $\varepsilon = \dfrac{432 S^3}{R}$ in article 199, if I had chosen to use, as known, the relation $R = 64 S^3 - T^2$, instead of *proving* that relation anew, as has here been virtually done. And thus is obtained the following new form of the sextic in h:

$$\left\{\frac{(h+1)(h-2)(2h-1)}{h(h-1)}\right\}^2 = \frac{27 T^2}{R};$$

on eliminating R, and reintroducing S,

$$\left(\frac{\varepsilon}{4\varepsilon - 27}\right) = \frac{(h^2 - h + 1)^3}{(h+1)^2 (h-2)^2 (2h-1)^2} = \frac{16 S^3}{T^2}.$$

202. These various forms of *my sextic* in h being admitted, it becomes an interesting question to examine whether, and how far, they are reconcileable with Salmon's Analysis. In his article 205, as remarked in article 194 of this Letter, he arrives virtually at the equation of the sixth degree;

$$t^2 (t^2 - 12)^2 - \frac{4R}{S^3} = 0,$$

as one of which the ratios of the roots are the sought anharmonic ratios; or rather, as I prefer to state it, *among* the quotients of the roots of which, those anharmonics of the pencil must all be found. Writing then ht, as a value for a second root; subtracting, and dividing not merely by $(h-1)t$, but by $(h^2 - 1)t^2$, which will get rid of *some* of the foreign factors; we have this second equation,

$$t^4 (h^4 + h^2 + 1) - 24 t^2 (h^2 + 1) + 144 = 0;$$

between which, and *Salmon's sextic*, the quantity t is to be eliminated. Or, making

$$t^2 = \frac{12}{v}, \quad \text{and} \quad \varepsilon = \frac{432 S^3}{R},$$

where ε is *now* used as a mere *arbitrary abbreviation*, which will however be useful in *comparison of processes*, we are to eliminate v between the two equations

$$v^3 - \varepsilon (v - 1)^2 = 0, \quad \text{and} \quad v^2 - 2(h^2 + 1) v + h^4 + h^2 + 1 = 0.$$

The result of this elimination may at once be written as follows,

$$\{v'^3 - \varepsilon (v' - 1)^2\}\{v''^3 - \varepsilon (v'' - 1)^2\} = 0;$$

where

$$v' + v'' = 2(h^2 + 1), \quad \text{and} \quad v'v'' = h^4 + h^2 + 1.$$

Substituting and developing, we have the following *equation of the twelvth degree* (cf. article 194 of this Letter):

$$(h^4 + h^2 + 1)^3 - 2\varepsilon h^2(h^8 + h^6 - 2h^4 + h^2 + 1) + \varepsilon^2 h^4(h^2 - 1)^2 = 0$$

which at first looks rather unmanageable. But if we observe that

$$(h^4 + h^2 + 1)^3 = (h^2 + h + 1)^3(h^2 - h + 1)^3, \quad h^4(h^2 - 1)^2 = (h^2 + h)^2 \times (h^2 - h)^2$$

and that

$$2h^2(h^8 + h^6 - 2h^4 + h^2 + 1) = h^2(h^3 - 1)^2(h^2 + h + 1) + h^2(h^3 + 1)^2(h^2 - h + 1)$$
$$= (h^2 - h)^2(h^2 + h + 1)^3 + (h^2 + h)^2(h^2 - h + 1)^3,$$

we shall perceive that *decomposition into two rational and sextic factors* (article 194), of which I spoke before. These factors are the following:

I. $(h^2 + h + 1)^3 - \varepsilon(h^2 + h)^2 = 0;$
II. $(h^2 - h + 1)^3 - \varepsilon(h^2 - h)^2 = 0;$

where $\varepsilon = \dfrac{432S^3}{R}$, as above. In fact, these correspond to the two separate roots of the quadratic in v, namely

$$v' = h^2 + h + 1, \quad v'' = h^2 - h + 1.$$

But *my direct analysis* (article 199), which *introduced no foreign factor*, conducted to the *second* of these two sextics; *that* sextic therefore is to be *retained*, and the *other* is to be *rejected* as *foreign*. (Compare a separate Note, dated [ ].)

203. If we had supposed, – what I own that Salmon's text seems to *suggest* – that the sought *anharmonics h* are the *quotients of the roots* of the *cubic equation,*

$$t^3 - 12t + 2r = 0,$$

where r is some *one* value of the radical $\sqrt{\dfrac{R}{S^3}}$, so that $r^2 = \dfrac{432}{\varepsilon}$, we should have had to eliminate t with the help of

$$t^2(h^2 + h + 1) - 12 = 0,$$

and the result would have been the *foreign sextic* I. Hence my *completion,* – if I may not call it *correction* – of Salmon's Rule is as follows: "*Determine the six quotients of the three roots of the* ***cubic,***

$$t^3 - 12t + 2\sqrt{\frac{R}{S^3}} = 0,$$

attributing some ***one sign*** *to the radical; those quotients, with their* ***signs changed,*** *will be the sought* ***anharmonics*** *h of the pencil*". For example, when $T = 0$, the roots of

$$t^3 - 12t + 16 = 0$$

are 2, 2, and -4; and the *negatives* of their *quotients*, namely -1, $+2$, $+\frac{1}{2}$, but not the quotients *themselves*, are the values of h for the pencil. And as Salmon calls *such* a cubic $(T = 0)$ *Harmonic*, so I am disposed to give the name of ***Di-anharmonic Cubics*** (cf. article 154) to those for which $S = 0$, and therefore, $h = -\theta$ or $h = -\theta^2$.

204. We may now perhaps dismiss the Subject, of the determination of the anharmonics of a *pencil* which is only given by the joint *equation of the System* of its four rays, as having been quite fully enough discussed, for the purposes of the present Letter, although, if I were not desirous to approach to a termination of these Sheets, I might have other connected remarks to make, especially as regards the application of the Theorem in article 73, or of its extension in article 74, to equations in which anharmonic coordinates of *points* and *lines* are mixed. But at present I prefer to pass to the important, and indeed, (in this Calculus) *fundamental* problem, of the *transformation* of such *coordinates*, through the adoption of *new points and lines of reference.*

205. Suppose, then that we assume, as a *new unit-triangle* and *new unit-point*, any arbitrary but given triangle, A_1, B_1, C_1, and any arbitrary but given point, O_1, in the given plane; and let the *old* coordinates of these four *given* points be,

$$O_1 = (x_0,\ y_0,\ z_0); \quad A_1 = (x_1,\ y_1,\ z_1); \quad B_1 = (x_2,\ y_2,\ z_2); \quad C_1 = (x_3,\ y_3,\ z_3).$$

The old coordinates of a *variable* point P being x, y, z, it is required to find its *new* coordinates x_1, y_1, z_1, or some quantities proportional to these, in terms of the 15 coordinates,

$$x,\ y,\ z; \quad x_0,\ y_0,\ z_0; \quad x_1,\ y_1,\ z_1; \quad x_2,\ y_2,\ z_2; \quad x_3,\ y_3,\ z_3.$$

206. Applying here the *definitions* of article 145, I have first the equation:

$$\frac{y_1}{z_1} = (A_1.PC_1\,O_1\,B_1); \quad \frac{z_1}{x_1} = (B_1.PA_1\,O_1\,C_1); \quad \frac{x_1}{y_1} = (C_1.PB_1\,O_1\,A_1).$$

Writing, then,

$$O_1 = P_0, \quad A_1 = P_1, \quad B_1 = P_2, \quad C_1 = P_3,$$

the general formula of article 151 supplies us with the expressions following:

$$\frac{y_1}{z_1} = (P_1.PP_3\,P_0\,P_2), \text{ \&c.};$$

that is to say,

$$
\begin{aligned}
\frac{y_1}{z_1} &= \begin{vmatrix} x_1, & y_1, & z_1 \\ x, & y, & z \\ x_3, & y_3, & z_3 \end{vmatrix} \begin{vmatrix} x_1, & y_1, & z_1 \\ x_0, & y_0, & z_0 \\ x_2, & y_2, & z_2 \end{vmatrix} : \begin{vmatrix} x_1, & y_1, & z_1 \\ x_3, & y_3, & z_3 \\ x_0, & y_0, & z_0 \end{vmatrix} \begin{vmatrix} x_1, & y_1, & z_1 \\ x_2, & y_2, & z_2 \\ x, & y, & z \end{vmatrix} \\
&= \begin{vmatrix} x, & y, & z \\ x_3, & y_3, & z_3 \\ x_1, & y_1, & z_1 \end{vmatrix} \begin{vmatrix} x_0, & y_0, & z_0 \\ x_1, & y_1, & z_1 \\ x_2, & y_2, & z_2 \end{vmatrix} : \begin{vmatrix} x, & y, & z \\ x_1, & y_1, & z_1 \\ x_2, & y_2, & z_2 \end{vmatrix} \begin{vmatrix} x_0, & y_0, & z_0 \\ x_3, & y_3, & z_3 \\ x_1, & y_1, & z_1 \end{vmatrix} \\
&= \frac{x \begin{vmatrix} y_3, & z_3 \\ y_1, & z_1 \end{vmatrix} + y \begin{vmatrix} z_3, & x_3 \\ z_1, & x_1 \end{vmatrix} + z \begin{vmatrix} x_3, & y_3 \\ x_1, & y_1 \end{vmatrix}}{x_0 \begin{vmatrix} y_3, & z_3 \\ y_1, & z_1 \end{vmatrix} + y_0 \begin{vmatrix} z_3, & x_3 \\ z_1, & x_1 \end{vmatrix} + z_0 \begin{vmatrix} x_3, & y_3 \\ x_1, & y_1 \end{vmatrix}} : \\
&\quad \frac{x \begin{vmatrix} y_1, & z_1 \\ y_2, & z_2 \end{vmatrix} + y \begin{vmatrix} z_1, & x_1 \\ z_2, & x_2 \end{vmatrix} + z \begin{vmatrix} x_1, & y_1 \\ x_2, & y_2 \end{vmatrix}}{x_0 \begin{vmatrix} y_1, & z_1 \\ y_2, & z_2 \end{vmatrix} + y_0 \begin{vmatrix} z_1, & x_1 \\ z_2, & x_2 \end{vmatrix} + z_0 \begin{vmatrix} x_1, & y_1 \\ x_2, & y_2 \end{vmatrix}};
\end{aligned}
$$

with analogous expressions for the two other quotients of the new anharmonic coordinates, cyclically taken, namely,

$$\frac{z_1}{x_1}, \quad \text{and} \quad \frac{x_1}{y_1}.$$

Hence we may write, as the formulae of transformation required, the following:

$$x_1 = \begin{vmatrix} x, & y, & z \\ x_2, & y_2, & z_2 \\ x_3, & y_3, & z_3 \end{vmatrix} : \begin{vmatrix} x_0, & y_0, & z_0 \\ x_2, & y_2, & z_2 \\ x_3, & y_3, & z_3 \end{vmatrix};$$

$$y_1 = \begin{vmatrix} x, & y, & z \\ x_3, & y_3, & z_3 \\ x_1, & y_1, & z_1 \end{vmatrix} : \begin{vmatrix} x_0, & y_0, & z_0 \\ x_3, & y_3, & z_3 \\ x_1, & y_1, & z_1 \end{vmatrix};$$

$$z_1 = \begin{vmatrix} x, & y, & z \\ x_1, & y_1, & z_1 \\ x_2, & y_2, & z_2 \end{vmatrix} : \begin{vmatrix} x_0, & y_0, & z_0 \\ x_1, & y_1, & z_1 \\ x_2, & y_2, & z_2 \end{vmatrix};$$

or more fully,

$$
\begin{cases}
x_1 = \dfrac{\alpha x + \beta y + \gamma z}{\alpha x_0 + \beta y_0 + \gamma z_0}; \\[2ex]
y_1 = \dfrac{\alpha' y + \beta' z + \gamma' x}{\alpha' y_0 + \beta' z_0 + \gamma' x_0}; \\[2ex]
z_1 = \dfrac{\alpha'' z + \beta'' x + \gamma'' y}{\alpha'' z_0 + \beta'' x_0 + \gamma'' y_0};
\end{cases}
$$

where

$$\alpha = \begin{vmatrix} y_2, & z_2 \\ y_3, & z_3 \end{vmatrix}; \quad \beta = \begin{vmatrix} z_2, & x_2 \\ z_3, & x_3 \end{vmatrix}; \quad \gamma = \begin{vmatrix} x_2, & y_2 \\ x_3, & y_3 \end{vmatrix};$$

$$\alpha' = \begin{vmatrix} z_3, & x_3 \\ z_1, & x_1 \end{vmatrix}; \quad \beta' = \begin{vmatrix} x_3, & y_3 \\ x_1, & y_1 \end{vmatrix}; \quad \gamma' = \begin{vmatrix} y_3, & z_3 \\ y_1, & z_1 \end{vmatrix};$$

$$\alpha'' = \begin{vmatrix} x_1, & y_1 \\ x_2, & y_2 \end{vmatrix}; \quad \beta'' = \begin{vmatrix} y_1, & z_1 \\ y_2, & z_2 \end{vmatrix}; \quad \gamma'' = \begin{vmatrix} z_1, & x_1 \\ z_2, & x_2 \end{vmatrix}.$$

207. It is, then, permitted to write generally, in this species of anharmonic transformation of coordinates, the formulæ:

$$\begin{cases} x_1 = ax + by + cz; \\ y_1 = a'y + b'z + c'x; \\ z_1 = a''z + b''x + c''y; \end{cases}$$

in which the *nine coefficients*,

$$a, b, c; \quad a', b', c'; \quad a'', b'', c'',$$

are *all arbitrary*, although only their *eight ratios* are important. In fact, let us suppose that the *nine values* of their coefficients are assigned; and that we wish to *interpret* such *data* of transformation, as having reference to the *choice* of a new unit triangle $A_1\ B_1\ C_1$ and of a new unit point O_1. Since

$$\begin{cases} a : b : c = \alpha : \beta : \gamma, \\ a' : b' : c' = \alpha' : \beta' : \gamma', \\ a'' : b'' : c'' = \alpha'' : \beta'' : \gamma'', \end{cases}$$

the assigned *ratios* of a, b, c will give as the *position* of the *new side* $B_1 C_1$, for any point P on which side we have the equation,

$$0 = \begin{vmatrix} x, & y, & z \\ x_2, & y_2, & z_2 \\ x_3, & y_3, & z_3 \end{vmatrix} = \alpha b + \beta y + \gamma z,$$

and therefore, also,

$$ax + by + cz = 0.$$

In like manner, the ratios of a', b', c' will give us the position of the side $C_1 A_1$; and those of a'', b'', c'' will give the position of $A_1 B_1$. *Six ratios* of the 9 assigned coefficients, $a, \ldots, c''$, are therefore sufficient, but not more than sufficient, to determine the *position of the new unit triangle*, including of course the positions of its three corners. And when we are *also* given the *two other ratios, a, a', a''*, we have *three* (necessarily *concurrent*) *right lines*; as *loci for the new unit point*, O_1; because our formulæ of transformation give,

$$ax_0 + by_0 + cz_0 = a'y_0 + b'z_0 + c'x_0 = a''z_0 + b''x_0 + c''y_0 = 1.$$

208. It follows, then, that ***eight*** *distinct* and *independent* ***coefficients*** (or *constants*) of *transformation* are introduced by this *Anharmonic Method*, so even when it is *confined*, as we at present confine it, to the ***plane*** and when only *ratios* are retained. And it is easy to foresee that there will be, on the same or on similar principles, no fewer than ***fifteen*** such *independent* constants, or ratios for ***space***. For we may already expect to find, in applications of the Method to the Geometry of ***Three*** Dimensions, – on which I am not likely to enter in this Letter, – if the *four* anharmonic coordinates of a point P be denoted by w, x, y, z, (a notation to which however I do not wish to be considered as pledged,) that we shall have *four formulæ*, such as the following:

$$\begin{cases} w_1 = aw + bx + cy + dz; \\ x_1 = a'x + b'y + c'z + d'w; \\ y_1 = a''y + b''z + c''w + d''x; \\ z_1 = a'''z + b'''w + c'''x + d'''y; \end{cases}$$

where *all the sixteen coefficients*, $a, \ldots, d'''$, are *arbitrary*, although only their *fifteen ratios* are important; as depending upon, and conversely serving to determine, the ***five points***, or the ***five planes***, of *reference*, in that new System of what I have on that account proposed (in some former letter) to denominate "***Quinquipunctual***", or "***Quinquiplanar Coordinates***." – *Mais, revenons à nos moutons*: let us stick for the present, to the *plane*, and be content with *quadripunctual*, or with *quadrilinear coordinates*.

209. As an example of such transformation of anharmonic coordinates within the plane, let us take the equation of article 114 of a cubic curve,

$$U = (c^3 - 1)\,s^3 + 27xyz = 0;$$

where $s = x + y + z$, and c is an arbitrary constant. The old or given unit-triangle is here a *triangle of tangents of inflexions*; and we now propose to substitute for it a *triangle of chords* of inflexion; *retaining*, still, as *unit-line*, that real right line $s = 0$, which connects the *three real points* of inflexion. The new coordinates of an arbitrary point P are soon found to be, in this example, the following linear functions of the old ones:

$$x_1 = \frac{(c-1)\,s + 3x}{3c}; \quad y_1 = \frac{(c-1)\,s + 3y}{3c}; \quad z_1 = \frac{(c-1)\,s + 3z}{3c};$$

whence

$$s_1 = x_1 + y_1 + z_1 = s,$$

and therefore,

$$3x = 3cx_1 + (1-c)\,s_1; \quad 3y = 3cy_1 + (1-c)\,s_1; \quad 3z = 3cz_1 + (1-c)\,s_1.$$

It follows that the transformed equation of the curve is,

$$\frac{U}{3c^2} = \frac{(c^3-1)s_1^3 + \{(1-c)s_1 + 3cx_1\}\{(1-c)s_1 + 3cy_1\}\{(1-c)s_1 + 3cz_1\}}{3c^2}$$

$$= \frac{(c^3-1) + (1-c)^3 + 3c(1-c)^2}{3c^2} s_1^3 + 3(1-c)s_1(x_1y_1 + y_1z_1 + z_1x_1) + 9cx_1y_1z_1$$

$$= (c-1)s_1(s_1^2 - 3x_1y_1 - 3y_1z_1 - 3z_1x_1) + 9cx_1y_1z_1$$

$$= (c-1)(x_1 + y_1 + z_1)(x_1^3 + y_1^3 + z_1^3 - y_1z_1 - z_1x_1 - x_1z_1) + 9cx_1y_1z_1$$

$$= (c-1)(x_1^3 + y_1^3 + z_1^3) + 3(2c+1)x_1y_1z_1 = 0.$$

But this, when we suppress the lower accents, and multiply by the number 3, is precisely that modification of the canonical form of the cubic, which was introduced in article 117, as giving the *same* S and the *same* T, as the curve

$$(c^3-1)s^3 + 27xyz = 0,$$

with which in the present article, and in article 114, we set out this last curve, and the curve

$$(c-1)ss's'' + 9cxyz = 0,$$

where (as in article 117),

$$s' = x + \theta y + \theta^2 z, \qquad s'' = x + \theta^2 y + \theta z,$$

are therefore absolutely the *same cubic*, only *referred to different triangles.*

210. Perhaps it may not be too much of a digression here, if I enter into some account of what *led* me to fix my attention, for a while, on *another form* (article 91) of the equation of a cubic; or on the equivalent but later form (article 121). But I must exhibit briefly the *reduction* of the earlier form, namely

$$U = s^3 + 6kzp = 0,$$

(articles 91, 92, &c.) when $s = x + y + z$, as before; $p = x^2 + xy + y^2$; and k is an arbitrary constant, to the known *canonical form*

$$x^3 + y^3 + z^3 + 6mxyz = 0,$$

by a suitable transformation of coordinates.

211. The three real points of inflexion are here (by articles 97, 98, and 121),

$$K = (1, -1, 0), \qquad K' = (1, c, c), \qquad K'' = (c, 1, c),$$

where c is the real root of the cubic equation

$$(2c+1)^3 + 6kc(c^2 + c + 1) = 0;$$

so that we may eliminate k (as in article 121), and denote

$$c(c^2 + c + 1)U = c(c^2 + c + 1)s^3 - (2c+1)^3 zp = 0,$$

as a form for the equation of the cubic curve now under discussion. Its three real points of inflexion are on the right line which has for its equation in the older given coordinates,

$$KK'K'': \ c(x+y) - (1+c)z = 0;$$

and we propose to take this line for the *new unit line*, so as to have

$$c(x+y) - (1+c)^2 = \nu(x_1 + y_1 + z_1),$$

where ν is some constant coefficient. Again by equation II of article 98, the three real right lines, of which each passes through one real and through two imaginary points of inflexion, are for the present curve the following:

$$KIJ: \ x+y+z=0; \quad K'I'J': \ y-z=0; \quad K''I''J'': \ z-x=0;$$

and we shall take these lines for the *sides* $A'B'$, $B'C'$, $C'A'$, of the *new unit-triangle*, writing thus

$$y - z = \kappa x_1, \quad z - x = \lambda y_1, \quad x+y+z = \mu z_1,$$

where κ, λ, μ are constants. These four assumed equations between old and new coordinates become compatible, when we select the four new constants as follows:

$$\kappa = -\lambda = 3(c-1); \quad \mu = 3(2c+1); \quad \nu = (c-1)(2c+1);$$

and then we have formulæ of transformation,

$$x = (c-1)(-x_1 + 2y_1) + (2c+1)z_1;$$
$$y = (c-1)(2x_1 - y_1) + (2c+1)z_1;$$
$$z = -(c-1)(x_1 + y_1) + (2c+1)z_1;$$

which give

$$s^3 = 27(2c+1)^3 z_1^3;$$

$$\frac{p}{3} = \frac{x^2 + y^2 + (x+y)^2}{6} = (c-1)^2(x_1^2 - x_1 y_1 + y_1^2) + (c-1)(2c+1)(x_1+y_1)z_1 + (2c+1)^2 z_1^2;$$

$$\frac{pz}{3} = -(c-1)^3(x_1^3 + y_1^3) - 3(c-1)^2(2c+1)x_1 y_1 z_1 + (2c+1)^3 z_1^3.$$

But

$$9c(c^2+c+1) - (2c+1)^3 = (c-1)^3;$$

therefore

$$\frac{c(c^2+c+1)}{3(c-1)^2(2c+1)^3} U = (c-1)(x_1^3 + y_1^3 + z_1^3) + 3(2c+1)x_1 y_1 z_1 = 0;$$

and the reduction to the canonical form has been accomplished.

212. We see, then, the *three cubic forms* of article 122, namely,

$$\begin{cases} \text{I.} & U = 3(c-1)\,ss's'' + 27cxyz, \\ \text{II.} & U = (c^3-1)s^3 + 27xyz, \\ \text{III.} & U = 9c(1+c+c^2)(1+2c)^{-2}s^3 - 9(1+2c)zp, \end{cases}$$

with the recent meanings of p, s, s', s'', and with *one common value of the constant* c, have *not only*, as in the cited article, *one common set of values of* S, T, *and* $\sqrt[3]{R}$, namely,

$$\begin{cases} S = 2^{-4}3^{6}(1+8c^{3}), \\ T = 2^{-3}3^{9}(1-20c^{3}-8c^{6}), \\ \sqrt[3]{R} = 3^{6}c(1-c^{3}), \end{cases}$$

(which are all decomposable, as in article 122, into linear factors), but also, when equated to zero, *represents absolutely the same cubic curve*, only *compared with different systems* of *points and lines of reference.* Indeed, I suppose that wherever *both* S *and* T have the *same values, including signs*, for *any two cubic curves*, not affected with any of the *singularities*, which answer to the case $R = 0$; or (which comes to the same thing), when $S' = e^{4}S$, and $T' = e^{6}T$, if S and T belong to the *one* cubic, and S', T' to the *other*, while e is any *real* constant, then either curve may be *identified* with the other, by mere transformation of coordinates, as in the examples given above. But I *think* that we are *not at liberty* to *change the sign of* T; and therefore that $\frac{T^{2}}{S^{3}}$ is *not* sufficient to *determine the curve*, in *all* its *essential properties*: although I grant that it determines the *six anharmonics of the pencil of tangents.* (Compare the *two subcases* of article 123, &c.) And when there is a *double point*, ($R = 0$,) whether *nodal* or *conjugate*, we seem to have found, in article 133, that then the functions S and T *take cognizance* (so to speak) *only* of the course of the curve in the *immediate neighbourhood* of that *singular point*: and consequently that *their* values are in this case *insufficient* to determine the *whole* of the cubic. In fact, we saw (in article 131, &c.) that if we suppose c_1, c_2, c_3, to vanish, so that

$$U = a_1x^{3} + 3a_2x^{2}y + 3b_1xy^{2} + b_2y^{3} + 3z(a_3x^{2} + 2dxy + b_3y^{2}),$$

then although the *four coefficients* a_1, a_2, b_1, b_2, thus enter into the *given cubic function* U, (and also into its *Hessian* H,) *they do not enter at all into the composition of* S *and* T; which latter functions of the constants were found to be expressed by the formulæ,

$$S = (d^{2} - a_3b_3)^{2}; \quad T = -8(d^{2} - a_3b_3)^{3}.$$

Thus, S and T would *in this case* remain *unaltered*, (their signs being, as above, included in their *values*,) if we were to *reduce the cubic* to a *system of three right lines*, namely the line $z = 0$, and the two (real or imaginary) *tangents* to the curve at the singular point (0, 0, 1). Leaving, however, at least for the present, the consideration of such *singularities*, I proceed to answer the question at the beginning of article 210, or to show *what led me* to discuss, with some detail, the *form* $U = s^{3} + 6kzp$, of article 91, &c.

213. You may remember that I had my doubts, after I had begun to write this Letter, whether Salmon held it to be proved, that a cubic curve has *always three real points of inflexion*; allowance being made for the possible *absorption* of such points, by a *double* point of the curve: which allowance I did not sufficiently make at first, in endeavouring to interpret some parts of the phraseology of *HPC*. At this moment, I have not beside me the notes in which you referred me to pages of that very able and important Work, sufficing not only to clarify the Author's own *conviction* on the subject, but also to supply *geometrical grounds* for that conviction; on which grounds I am quite willing to accept, as satisfactory to a candid inquirer, or (let us say) to a *docile student*: such as I *wish* to be, in whatever degree it is *permitted* to be one, although haunted and disturbed perpetually, by aspirations after knowledge more complete.

214. I admit that Section III., of Chapter III., – and especially the *Figures* in page 145 of *HPC*, – may be considered to leave *no reasonable doubt*, on the mind of a *geometrical* student, as to the existence of *three real inflexions* on any *cubic of the sixth class*. Yet there seems to be something unsatisfactory, – as, in a note of yours (not just now at my hand), I thought that you appeared to admit, – in our being obliged to appeal *so much* to considerations of ***Shape***, – and in fact to what our *eye* can take in, – when we are dealing *virtually* with this question of pure ***Algebra***: Whether that *Equation of the Ninth Degree*, which results from elimination between the two equations,

$$U = 0, \quad H = 0,$$

of the original cubic and of its Hessian, *can ever have eight imaginary roots*? For it is, substantially, *this* Question, and not any that relates to *visible Flexures* of the Curve, which we are called upon to investigate. And, charmed as I always am, with any help which Algebra can derive from Geometry, I confess that, *as an algebraist*, I cannot help myself from even *still* feeling, that we are expected to place *too much reliance* on our own *power, or want of power*, of *imagining Geometrical Form*, when we are asked to draw so important, so general, and so algebraical a conclusion, as that which I have just now referred to, from the reasonings and the diagrams of the Section above cited.

215. Quite recently, – at least long since this Letter was begun, – I have happened to see a Note (numbered as XII.), to the Second Edition ([Paris] 1854) of Serret's ''*Cours d'Algebra Supérieure*''*, which note is entirely devoted to an account of Hesse's *own* Analysis respecting the Nine Points of Inflexion of a Curve of the Third Degree, and the Equation of the Ninth Degree which determines them. It is only in a very cursory way that I have yet attempted to read the cited Note: but its mere Title, – ''Sur la Résolution Algébrique de l'Équation du Neuvième Degré, à laquelle conduit la Recherche des Points d'Inflexion des Courbes du Troisième Degré,'' – is sufficient to show that it must have a very important bearing upon the present Subject. Serret says, in his page 539: ''L'analyse de M. Hesse est assez remarkable pour que je croie devoir la reproduire ici:'' I suppose we may consider the twenty following octavo pages in French to be, at least substantially, translated from the German of Hesse, which I cannot at present consult. Now, just at the middle of the Note in question (page 549), I find the following passage:

''On voit donc qu'une courbe réelle du troisième degré ne peut avoir plus de trois points d'inflexion réels, lesquels sont toujours en ligne droite, d'après le théorème I. Je dis, en outre, qu'il y a effectivement des courbes du troisième degré qui ont trois points d'inflexion réels. Par exemple, la courbe dont $\left(\frac{x}{y}, \frac{y}{z}\right)$ désignent les coordonnées rectilignes et qui a pour équation

* [Joseph Alfred Serret, 1819–1885.]

$$y = \frac{x^3 - xz^2}{3x^2 + z^2},$$

est rencontrée par l'arc des abscisses en trois points d'inflexion réels.''

216. I suppose that we may admit that when Hesse wrote the Memoir thus translated, or in some form reproduced, *he* was not aware of any *proof* of the *general reality* of so many as *three* roots, of his equation of the ninth degree; since otherwise he was not likely to have taken the pains to give so simple looking an *example* (the correctness of which I have not thought it necessary to verify) of there being *sometimes three* such *real* roots. My own *reading* on these subjects, as I believe that you well know, or can guess, is indeed of very small extent; but I still think that an algebraical proof, – even if mixed with *some geometry*, – of the *impossibility of eight imaginary roots* in the equation of the ninth degree, which (as above) results from the given cubic and its Hessian, is a *desideratum* in this theory. And my *object* in discussing that (only partially symmetric) *form* of the cubic U, which was introduced in article 91 of this Letter, was to contribute *something* towards the supplying of a *want*, which was then supposed by me to exist.

217. I therefore *started* with the much more easily proved and indeed unquestioned theorem, that there always exists *at least one real point of inflexion* on any real cubic: assuming also, as a thing admitted, that *at least one real right line* can be drawn through that point, so as to contain *two other points of inflexion*, real or imaginary. In fact, the *biquadratic* equation in $\frac{\lambda}{\mu}$ has, by its known form, *always two real roots*, one positive and the other negative; but it was sufficient for my purpose to know that it had always *at least one real root*: and therefore that there was always *at least one real and linear combination*, of the form

$$\lambda\mu U + \mu H = 0,$$

representing a *system of three chords of inflexion*, whereof *one chord at least* must be *real.* I took *that real chord* for my (anharmonic) *unit-line;* the *real point of inflexion* upon it for the point $C' = (1, -1, 0)$; and the *tangent* at that point for the *side AB*, or $z = 0$, or $[0, 0, 1]$ of the *unit-triangle.* Assuming also for argument's sake, that the *two other points* of inflexion on the *same* chord were *imaginary,* I denoted them by the symbols $(1, \theta, \theta^2)$ and $(1, \theta^2, \theta)$, as my method allowed me to do; and chose the *real intersection* of the two *imaginary tangents* at those points for the *corner C* or $(0, 0, 1)$ of the unit-triangle. This *led* me to the *form*

$$U = s^3 + 6kzp = 0,$$

of article 92; and because this succeeded in proving that there were, *with these suppositions, two other real points of inflexion,* I *inferred,* – in a way satisfactory to myself at least, – that there were always three such *real points.*

218. The calculations might, with little increase of trouble, have been presented more generally as follows. The side AB, or $z = 0$, of the unit-triangle being still supposed to touch the curve at the point C', or $(1, -1, 0)$, where it meets the unit-line, and this real point of contact C' it being still supposed to be a point of inflexion, let the unit-line $A'B'C'$, or $x + y + z = 0$, or $s = 0$, be still supposed to meet the curve in two other points of inflexion, respecting which we shall only *now* assume that the *real intersection* of the two (*real or imaginary*) *tangents* at them is made, as before, the corner C, or the point $(0, 0, 1)$, of the unit-triangle. Then the *equation of the pair* of those two tangents to the cubic will be a quadratic of the form,

$$v = ax^2 + 2bxy + cy^2 = 0,$$

in which the *coefficients* a, b, c are *real*, but may satisfy *either* of the two *inequalities*,

$$\text{I.}\quad b^2 - ac > 0, \qquad \text{or} \qquad \text{II.}\quad b^2 - ac < 0.$$

In the Ist case, the quadratic has two *real roots*, answering to two real points of inflexion of the unit-line, *besides* the assumed point of inflexion C', so that in *this* case there is no question about the existence of *three real points* of inflexion. In the IInd case, – which *includes* that of articles 91 &c., – the quadratic has *imaginary roots*, and the *two other* points of inflexion on the unit-line (additional to the given real point C') are *imaginary points*; although the *tangents* to the curve at those points still meet, as before, in the real corner C of the unit-triangle. But in *each* case, the *equation of the cubic* takes the *form*, which we shall now proceed to discuss,

$$U = s^3 + 3vz = 0;$$

where

$$s = x + y = z, \qquad \text{and} \qquad v = ax^2 + 2bxy + cy^2,$$

as above, while a, b, c are arbitrary constants.

219. Introducing, for the sake of homogeneity of constants, another coefficient e, or writing

$$U = es^3 + 3vz,$$

differentiation gives, on the plan of the expressions,

$$l = \tfrac{1}{3}D_xU = es^2 + 2(ax + by)z,$$
$$m = \tfrac{1}{3}D_yU = es^2 + 2(bx + cy)z,$$
$$n = \tfrac{1}{3}D_zU = es^2 + v;$$

$$l' = \tfrac{1}{6}D_x^2U = es + az,$$
$$m' = \tfrac{1}{6}D_y^2U = es + cz,$$
$$n' = \tfrac{1}{6}D_z^2U = es;$$

$$l'' = \tfrac{1}{6}D_yD_zU = es + bx + cy,$$
$$m'' = \tfrac{1}{6}D_zD_xU = es + ax + by,$$
$$n'' = \tfrac{1}{6}D_xD_yU = es + bz;$$

whence (writing out for my own convenience the calculations at full length) we have the combinations,

$$l'l''^2 = e^3s^3 + e^2s^2(2bx + 2cy + az) + es(bx + cy)(bx + cy + 2az) + az(bx + cy)^2,$$

$$m'm''^2 = e^3s^3 + e^2s^2(2ax + 2by + cz) + es(ax + by)(ax + by + 2cz) + cz(ax + by)^2,$$

$$n'n''^2 = c^3s^3 + 2e^2s^2bz + esb^2z^2,$$

$$-l'm'n' = -e^3s^3 - e^2s^2(a + c)z - esacz^2,$$

$$\begin{aligned}-2l''m''n'' = {} & -2e^3s^3 - 2e^2s^2\{(a + b)x + (b + c)y + bz\} \\ & - 2es\{(ax + by)(bx + cy) + (ax + by)bz + (bx + y)bz\} \\ & - 2bz(ax + by)(bx + cy);\end{aligned}$$

so that the Hessian for the present form of U, is

$$\begin{aligned}H &= l'l''^2 + m'm''^2 + n'n''^2 - l'm'n' - 2l''m''n'' \\ &= csX + zZ,\end{aligned}$$

where

$$\begin{aligned}X &= \{(ax + by) - (bx + cy)\}^2 + 2(c - b)(ax + by)z \\ &\quad + 2(a - b)(bx + cy)z + (b^2 - ac)z^2, \\ &= (b^2 - ac)(z - x - y)^2 + (ac - b^2)(x + y)^2 + \{(ax + by) - (bx + cy)\}^2, \\ &= (b^2 - ac)(x + y - z)^2 + (a - 2b + c)(ax^2 + 2bxy + cy^2); \\ Z &= c(ax + by)^2 + a(bx + xy)^2 - 2b(ax + by)(bx + xy), \\ &= (ac - b^2)(ax^2 + 2bxy + cy^2);\end{aligned}$$

and therefore, finally, since

$$ax^2 + 2bxy + cy^2 = v,$$

$$H = (b^2 - ac)es(x + y - z)^2 + \{(a - 2b + c)es + (ac - b^2)z\}v.$$

220. We have therefore the obvious combination,

$$(b^2 - ac)U + 3H = esY;$$

where

$$\begin{aligned}Y &= (b^2 - ac)\{(x + y + z)^2 + 3(x + y - z)^2\} \\ &\quad + 3(a - 2b + c)(ax^2 + 2by + cy^2) \\ &= (b^2 - ac)(x + y - 2z)^2 + 3\{(a - b)x + (b - c)y\}^2\end{aligned}$$

which is equal to a product of two linear factors; which are evidently *real*, when the factors of v are *imaginary*; although they are, on the other hand, *imaginary*, when those of v are real. It follows, then, that "*the system of the two new chords of inflexion*, represented by the equation $Y = 0$, *is real or imaginary*, represented by the equation $v = 0$, is *imaginary or real*"; that is, according as the *second*, or the *first*, of the two *inequalities* in article 218, is satisfied: and it may be noticed that *in each case*, and for *all values* of the *constants a, b, c, e*, the *two chords* of inflexion ($Y = 0$) *intersect* each other in a *real point*, which is situated *on the given right line*,

$$x + y - 2z = 0, \quad \text{or} \quad OC';$$

this intersection having been (as in article 94) the unit-point O itself, when we had the relation

$$a = 2b = c,$$

as in the less general form of article 91. Our analysis therefore enables us to conclude, that "*when two of the points of inflexion on the given chord* $s = 0$ *are imaginary, there exist generally two real points of inflexion, distinct from the given real point* C' *and situated on those two other real chords, which are then represented jointly by the equation* $Y = 0$." And I must say that *my own conviction*, of the *general existence of three real points of inflexion on a cubic, rests mainly*, at present, *on this argument.*

221. The *six points* of inflexion, whether some real or all imaginary, which are the intersections of the pair of chords $Y = 0$ with the given cubic $U = 0$, ought to be arranged upon a new System of *three chords*, through the given point of inflexion C'. Accordingly, if we write

$$W = (a - 2b + c)es^3 + 4(ac - b^2)z\{z^2 - (x + y)z + (x + y)^2\},$$

the elimination of v between the two equations,

$$U = 0, \quad Y = 0,$$

will conduct to the equation

$$W = 0,$$

which is of cubic form with respect to $\frac{x + y}{z}$, and therefore represents a system of three right lines through that given point C': our analysis receiving thus a verification.

222. If we substitute for $x + y$ its value $s - z$, and write, for conciseness,

$$\frac{9(a - 2b + c)e}{4(ac - b^2)} = g,$$

we have

$$\frac{9W}{4(ac - b^2)} = gs^3 + 3s^2(3z) - 3s(3z)^2 + (3z)^3;$$

this cubic function equated to zero gives therefore the values of $\frac{3z}{s}$ for the three chords of inflexion through C', drawn to the points for which $Y = 0$; and if we combine with these the given chord $s = 0$, we shall have *all the nine points of inflexion*, answering to the System of the two equations, $U = 0$, $H = 0$, as contained upon the pencil of four chords *through* C', (the vertex C' being included,) which *pencil* is represented by the *biquadratic* equation

$$0 = gs^4 + 3(3z)s^3 - 3(3z)^2s^2 + (3z)^3s;$$

or

$$0 = As^4 + 4Bs^3(3z) + 6Cs^2(3z)^2 + 4Ds(3z)^3 + E(3z)^4,$$

where

$$A = 4g, \quad B = 3, \quad C = -2, \quad D = 1, \quad E = 0.$$

Hence, for this pencil,

$$12S = AE - 4BD + 3C^2 = 0,$$

whatever the values of the constants a, b, c, e may be; and therefore, by articles 200, &c., "*This pencil of four chords of inflexion is di-anharmonic*": which agrees with my Theorem of article 159, and is a new verification of the analysis.

223. Writing $\alpha^3 - 1$ instead of g, and supposing for simplicity that α is real, so that

$$\alpha = \sqrt[3]{\left\{1 + \frac{9(a - 2b + c)e}{4(ac - b^2)}\right\}},$$

the cubic equation $W = 0$ gives for the *three new chords* the three *separate* equations,

$$\frac{3z}{s} = 1 - \alpha; \quad \frac{3z'}{s'} = 1 - \theta\alpha; \quad \frac{3z''}{s''} = 1 - \theta^2\alpha;$$

one being thus a *real right line*, and the two others being *imaginary* lines, although they all *pass* through the given and real point of inflexion C'. Hence the *anharmonic* of the *pencil* of the *four* chords, including the *given* unit-line, and treating *it* as the *first* of the four, while the three others follow in the *order* written above, is (by the principles of article 184, &c.) the following:

$$h = \begin{vmatrix} 1, & 0 \\ 1 - \alpha, & 1 \end{vmatrix} \begin{vmatrix} 1 - \theta\alpha, & 1 \\ 1 - \theta^2\alpha, & 1 \end{vmatrix} : \begin{vmatrix} 1 - \alpha, & 1 \\ 1 - \theta\alpha, & 1 \end{vmatrix} \begin{vmatrix} 1 - \theta^2\alpha, & 1 \\ 1, & 0 \end{vmatrix} = \frac{(\theta^2 - \theta)\alpha}{(1 - \theta)\alpha} = -\theta$$

and thus the *di-anharmonic property* of the pencil (compare article 159) is proved anew.

224. It may be remembered here that, for any point P, we have $\dfrac{3z}{s} = (C'.PAOA')$, and therefore $1 - \dfrac{3z}{s} = (C'.POAA')$; if then P, P', P'' be any points on the three new chords, we have the relations,

$$(C'.P'OAA') = \theta(C'.POAA'),$$
$$(C'.P''OAA'') = \theta^2(C'.POAA');$$

and consequently,

$$(C'.POAA')^3 = (C'.P'OAA')^3 = (C'.P''OAA')^3 = \alpha^3.$$

If we make $e = 1$, and $a = 2b = c = 2k$, so as to return to the particular form (articles 91 and 92)

$$U = s^3 + 6kzp,$$

we have

$$a^3 = 1 + \frac{3}{2k};$$

so that the cube roots of this last constant are then the values of the anharmonic $(C'.POAA')$ for the System of three new chords through C'. But we had, in article 96, for the new chord $C'K'K''$, an equation which may be written thus,

$$\frac{3z}{s} = \frac{3(w-4-2k)}{2(w+k)}, \quad \text{or} \quad 1 - \frac{3z}{s} = \frac{4(3+2k) - w}{2(w+k)},$$

with corresponding equations for the two other chords, $C'I'I''$, and $C'J'J''$, formed by changing w to w' and w'', that is, to the two other roots of a certain cubic equation (article 95); that cubic equation in w ought therefore to admit of being written under the form,

$$\left\{\frac{4(3+2k) - w}{2(w+k)}\right\}^3 = 1 + \frac{3}{2k},$$

or,

$$k\{w - 4(3+2k)\}^3 + 4(3+2k)(w+k)^3 = c;$$

and accordingly when we expand these cubes, and divide the result by $3(4+3k)$, we get the equation

$$w^3 + 12k(3+2k)w - 4k(3+2k)(12+7k) = 0,$$

exactly as in article 95.

225. The examination of the roots w, w', w'' of that Cubic equation supplies another interesting verification of the theory of the di-anharmonic property of the pencil of four chords, which has only recently occurred to me. A given chord being still the unit-line, $s = 0$, but the three other chords through C' be the lines $z = \tau s$, $z = \tau' s$, $z = \tau'' s$; the anharmonic of the pencil is then

$$h = \begin{vmatrix} 1, & 0 \\ \tau, & 1 \end{vmatrix} \begin{vmatrix} \tau', & 1 \\ \tau'', & 1 \end{vmatrix} : \begin{vmatrix} \tau, & 1 \\ \tau', & 1 \end{vmatrix} \begin{vmatrix} \tau'', & 1 \\ 1, & 0 \end{vmatrix} = \frac{\tau' - \tau''}{\tau' - \tau} = -\iota,$$

where

$$\iota = \theta^{\pm 1},$$

by the property referred to; an *ambiguous exponent* being thus used, to allow of variety of arrangement of the chords, or rays of the pencil. In each case, $\iota^2 + \iota + 1 = 0$, and $\tau + \iota\tau' + \iota^2\tau'' = 0$; wherefore the relation

$$\tau^2 + \tau'^2 + \tau''^2 - \tau'\tau'' - \tau''\tau - \tau\tau' = 0,$$

or

$$(\tau + \tau' + \tau'')^2 = 3(\tau'\tau'' + \tau''\tau + \tau\tau'),$$

must hold good for *all* such arrangements. If then we suppose that τ, τ', τ'' are the three roots of the following cubic,

$$a\tau^3 - 3b\tau^2 + 3c\tau - d = 0,$$

(where a, b, c are new constants,) the coefficients a, b, c of this new cubic must satisfy the *equation of condition*

$$b^2 - ac = 0.$$

226. It is worth observing that this equation between the new coefficients can be deduced, in a quite different way, as follows, from the same di-anharmonic theorem or property.

The *equation of the pencil* of the *four* chords being

$$0 = as(\tau s - z)(\tau' s - z)(\tau'' s - z) = s(ds^3 - 3czs^2 + 3bz^2 s - az^3),$$

or

$$0 = As^4 + 4Bs^3 z + 6Cs^2 z^2 + hDsz^3 + Ez^4,$$

when

$$A = 9, \quad B = -3c, \quad C = 2b, \quad D = -a, \quad E = 0,$$

we must have, by the theorem

$$0 = 12S = AE - 4BD + 3C^2 = 12(b^2 - ac)$$

or $b^2 = ac$, as before. And it is now proposed to prove that this relation does in fact exist, between the coefficients of the cubic of which the roots are τ, τ', τ''; or (which will do as well), any linear functions of those quantities, such as $\lambda\tau + \mu$, $\lambda\tau' + \mu$, $\lambda\tau'' + \mu$; when

$$2\tau = \frac{2z}{s} = \frac{w - 4 - 2k}{w + k} = 1 - \frac{4 + 3k}{w + k},$$

$$2\tau' = 1 - \frac{4 + 3k}{w' + k}, \qquad 2\tau'' = 1 - \frac{4 + 3k}{w'' + k}.$$

227. It is sufficient, then, to make

$$w + k = v^{-1}$$

and to try whether, on eliminating w by this from the cubic equation in article 95, which was lately cited in article 224, the resulting equation

$$av^3 - 3bv^2 + 3cv - d = 0,$$

or

$$d(w + k)^3 - 3c(w + k)^2 + 3b(w + k) - a = 0,$$

has its coefficients a, b, c connected by the recent relation $b^2 = ac$; whatever value the remaining coefficient d may happen to have. Now it is easy to perform the elimination here required, which is found to give the coefficients:

$$a = 9k(4 + 3k)^2, \quad b = 3k(4 + 3k), \quad c = k, \quad d = 1;$$

in fact these give $3kd - 3c = 0$, $3k^2 d - 6kc + 3b = 12k(3 + 2k)$,

$$k^3 d - 3k^2 c + 3bk - a = -4k(3 + 2k)(12 + 7k),$$

whence, by substituting them in the last cubic in $w + k$, we are led back to the cubic (article

95) in w. And it is evident that a, b, c have the required relation; so that this expectation of theory is fulfilled.

228. Again, since the calculations of the last article show that the three roots, v, v', v'', of the cubic in v, or in $(w+k)^{-1}$, are connected by the relation,

$$0 = v^2 + v'^2 + v''^2 - v'v'' - v''v - vv'$$

$$= (v + \theta v' + \theta^2 v'')(v + \theta^2 v' + \theta v''),$$

it follows that we must have *one or other* of the two factor equations

I. $\quad 0 = v + \theta v' + \theta^2 v''$,

or

II. $\quad 0 = v + \theta^2 v' + \theta v''$;

that is, either

I. $\quad 0 = \dfrac{1}{w+k} + \dfrac{\theta}{w'+k} + \dfrac{\theta^2}{w''+k}$,

or

II. $\quad 0 = \dfrac{1}{w+k} + \dfrac{\theta^2}{w'+k} + \dfrac{\theta}{w''+k}$;

which it is necessary here to distinguish, because the expressions for w' and w'', in article 95, already involve θ and θ^2. Nor did I at once see how to decide between these two last alternatives; until the following analysis occurred to me.

229. Writing, as their *common expression*, the formula

$$0 = \frac{1}{w+k} + \frac{\iota}{w'+k} + \frac{\iota^2}{w''+k},$$

when as before, $\iota = \theta^{\pm 1}$, and clearing of fractions, I had

$$0 = (w'+k)(w''+k) + \iota(w''+k)(w+k) + \iota^2(w+k)(w'+k)$$

$$= w'w'' + \iota w''w + \iota^2 ww' - k(w + \iota w' + \iota^2 w'').$$

But the three roots w, w', w'' were assigned in article 95 under the forms

$$w = w_1 + w_2, \qquad w' = \theta w_1 + \theta^2 w_2, \qquad w'' = \theta^2 w_1 + \theta w_2;$$

hence

$$w'w'' + \iota w''w + \iota^2 ww' = (1 + \iota\theta^2 + \iota^2\theta)\, w_1^2 + (1 + \iota\theta + \iota^2\theta^2)\, w_2^2,$$

and

$$w + \iota w' + \iota^2 w'' = (1 + \iota\theta^2 + \iota^2\theta)\, w_2 + (1 + \iota\theta + \iota^2\theta^2)\, w_1;$$

we are therefore to prove that the ambiguous exponent of θ in ι can be so selected as to give

$$0 = (1 + \iota\theta^2 + \iota^2\theta)(w_1^2 - kw_2) + (1 + \iota\theta + \iota^2\theta^2)(w_2^2 - kw_1);$$

which can only be by our having *either*

$$w_1^2 = kw_2, \quad \text{or} \quad w_2^2 = kw_1;$$

the first alternative answering to $\iota = \theta$, and the second to $\iota = \theta^2$. Hence, on comparing the values in article 95 of the constants w_1 and w_2, namely $w_1 = k'^2 k''$, $w_2 = k' k''^2$, we see that we are to try whether first, $k'^3 = k$; or, second, $k''^3 = k$. And because, on referring to article 95, we find that the first, but not the second, of these two last equations to be good, we infer finally that we are to take the value $\iota = \theta^{+1} = \theta$; or that the roots w, w', w'' of the cubic in article 95 are connected by the equation

I. $\quad 0 = (w + k)^{-1} + \theta(w' + k)^{-1} + \theta^2(w'' + k)^{-1},$

but *not* by the equation II. of article 228; at least if we still make θ and θ^2 enter into w' and w'', in the same way as we did in that article 95, which has been so often lately referred to.

230. As another verification, though not an important one, it ought to be found (cf. article 223) that the product of the three values of $\dfrac{3z}{s}$ in article 224, is equal to $1 - \alpha^3$; or that

$$9k(w - 4 - 2k)(w' - 4 - 2k)(w'' - 4 - 2k) + 4(w + k)(w' + k)(w'' + k) = 0;$$

where w, w', w'' are still the roots of the same cubic of article 95 or writing for abridgement $4 + 3k = \delta$, and (as in article 227), $w + k = v^{-1}$, we ought to have

$$0 = 9k(1 - \delta v)(1 - \delta v')(1 - \delta v'') + 4;$$

v, v', v'' being the roots of the cubic in v, resulting from elimination of w. But this latter cubic has been found (article 227) to be

$$0 = k\delta^2 v^3 - 9k\delta v^2 + 3kv - 1,$$

or

$$0 = (3\delta v - 1)^3 + 1 - \frac{3\delta}{k};$$

whence

$$\delta(v + v' + v'') = 1,$$
$$3\delta^2(v'v'' + v''v + vv') = 1,$$
$$9k\delta^3 vv'v'' = \delta;$$

therefore

$$9k(1 - \delta v)(1 - \delta v')(1 - \delta v'') + 4 = 9k - 9k + 3k - \delta + 4 = 0,$$

and the verification succeeds. (Of course, the process may be varied.)

231. Since I have been thus led to speak again of some of the earlier formulæ of this Letter respecting cubics, I shall just observe here that the equations of article 99 (for example), for the *twelve chords of inflexion* of the curve (articles 91 and 92)

$$U = S^3 + 6kzp = 0,$$

as well as the coordinates of the 9 *points* of inflexion themselves, for the same cubic, assigned in article 97, be considerably simplified, by introducing the constant c of article 121, namely the real root of the cubic equation of article 211,

$$(2c+1)^3 + 6kc(c^2+c+1) = 0.$$

Writing

$$t' = 1 + 2\theta^2 c, \qquad v' = \theta^2 - \theta c,$$

$$t'' = 1 + 2\theta c, \qquad v'' = \theta - \theta^2 c,$$

so that

$$v'' - \theta v' = \theta - 1, \qquad v' - \theta^2 v'' = \theta^2 - 1,$$

$$v'' - \theta^2 v' = (1-\theta^2)\,c, \qquad v' - \theta v'' = (1-\theta)\,c,$$

$$\theta^2 v'' - t' = (1-\theta^2)\,c, \qquad \theta v' - t'' = (1-\theta)\,c,$$

so we have the following

Table of Coordinates of the Nine Points of Inflexion

$K = (1, -1, 0)$;	$K' = (1, c, c)$;	$K'' = (c, 1, c)$;
$I = (1, \theta, \theta^2)$;	$J' = (t'', v'', v'')$;	$I'' = (v', t', v')$;
$J = (1, \theta^2, \theta)$;	$I' = (t', v', v')$;	$J'' = (v'', t'', v'')$.

232. We have also the following

Table of Coordinates of the Twelve Chords of Inflexion

I.	$[c, c, -1, -c]$;	$[v'', -\theta v', c-\theta c]$;	$[v', -\theta^2 v'', c-\theta^2 c]$;
II.	$[1, 1, 1]$;	$[0, 1, -1]$;	$[-1, 0, 1]$;
III.	$[v', v', -t'-v']$;	$[c-\theta c, v'', -\theta v']$;	$[v'', c-\theta c, -\theta v']$;
IV.	$[v'', v'', -t''-v'']$;	$[c-\theta^2 c, v', -\theta^2 v'']$;	$[v', c-\theta^2 c, -\theta^2 v'']$.

The notation $[l, m, n]$ having been already explained (articles 9, 79, 146, &c.) it is only necessary to add that in constructing the foregoing *Table of Coordinates of Chords*, I have followed the arrangement in article 99; namely

I. $KK'K''$, $IJ'I''$, $JI'J''$;
II. KIJ, $K'I'J'$, $K''I''J''$;
III. $KI'I''$, $K'IJ''$, $K''JJ'$;
IV. $KJ'J''$, $K'JI''$, $K''II'$.

233. Combining the equations thus grouped, and tabulating in order, the coordinates of the intersection of the sides 2 and 3 of the triangle I as (I, 1) etc., we soon construct the following

Table of Coordinates of the Twelve Corners of Triangles of Chords

	1.	$(c, c, 1)$;
I.	2.	$(1+\theta^2 c, \theta^2 + c, -\theta c)$;
	3.	$(1+\theta c, \theta + c, -\theta^2 c)$.

II.	1.	$(1, 1, 1)$;
	2.	$(1, -2, 1)$;
	3.	$(-2, 1, 1)$.
III.	1.	$(\theta - c, \theta - c, \theta^2 + 2\theta c)$;
	2.	$(1 - 2c + \theta c, -2 - \theta^2 c, \theta - c)$;
	3.	$(-2 - \theta^2 c, 1 - 2c + \theta c, \theta - c)$.
IV.	1.	$(\theta^2 - c, \theta^2 - c, \theta + 2\theta^2 c)$;
	2.	$(1 - 2c + \theta^2 c, -2 - \theta c, \theta^2 - c)$;
	3.	$(-2 - \theta c, 1 - 2c + \theta^2 c, \theta^2 - c)$.

(Division of all the coordinates by $2c + 1$ has been several times here employed.)

234. These Tables are not *completely* symmetrical; nor could it be expected, from the partially unsymmetrical arrangement with which we geometrically started, in deducing the here adopted *form* of article 91, that they should be so. But if we now employ the formulæ of article 211 of transformation of anharmonic coordinates, we arrive at the following *new* Tables, in which we write

$$x_1 = \frac{y - z}{c - 1}, \quad y_1 = \frac{x - z}{c - 1}, \quad z_1 = \frac{x + y + z}{2c + 1},$$

$$(x, y, z) = \left(\frac{y - z}{c - 1}, \frac{x - z}{c - 1}, \frac{x + y + z}{2c + 1}\right)_1$$

changing also cyclically the arrangements of the points from 1, 2, 3 to 3, 1, 2 and multiplying or dividing, where convenient, by any common factor:

Table of Transformed Coordinates of Points of Inflexion

$K' = (0, 1, -1)_1$;	$K'' = (-1, 0, 1)_1$;	$K = (1, -1, 0)_1$;
$J' = (0, 1, -\theta)_1$;	$I'' = (-\theta, 0, 1)_1$;	$I = (1, -\theta, 0)_1$;
$I' = (0, 1, -\theta^2)_1$;	$J'' = (-\theta^2, 0, 1)_1$;	$J = (1, -\theta^2, 0)_1$.

235. *Table of Transformed Coordinates of Chords of Inflexion*

(Arrangement cyclically altered from that of article 232, in each of the three latter groups of chords.)

I_1.	$[1, 1, 1]_1$;	$[1, \theta^2, \theta]_1$;	$[1, \theta, \theta^2]_1$;
II_1.	$[1, 0, 0]_1$;	$[0, 1, 0]_1$;	$[0, 0, 1]_1$;
III_1.	$[\theta, 1, 1]_1$;	$[1, \theta, 1]_1$;	$[1, 1, \theta]_1$;
IV_1.	$[\theta^2, 1, 1]_1$;	$[1, \theta^2, 1]_1$;	$[1, 1, \theta^2]_1$.

– I am not certain whether I shall devote any space to the *general theory* of such *transformations of coordinates of lines*, before this Letter ends; but in the present instance the meaning and validity of the resulting expressions are evident. Thus, the three symbols of the group I_1 imply that the nine points of inflexion may be arranged on the three chords $(K'K''K, J'I''I, I'J''J)$, which have for their transformed equations

$$x_1 + y_1 + z_1 = 0, \quad x_1 + \theta^2 y_1 + \theta z_1 = 0, \quad x_1 + \theta y_1 + \theta^2 z_1 = 0;$$

and similarly for the three other groups.

236. *Table of Transformed Coordinates of Corners of Chord Triangles*

I_1.	$(1, 1, 1)_1$;	$[1, \theta, \theta^2)_1$;	$(1, \theta^2, \theta)_1$;
II_1.	$(1, 0, 0)_1$;	$(0, 1, 0)_1$;	$(0, 0, 1)_1$;
III_1.	$(\theta^2, 1, 1)_1$;	$(1, \theta^2, 1)_1$;	$(1, 1, \theta^2)_1$;
IV_1.	$(\theta, 1, 1)_1$;	$(1, \theta, 1)_1$;	$(1, 1, \theta)_1$.

This last Table may be constructed from the one in article 235, by seeking the intersection of chords 2 and 3, chords 3 and 1, and chords 1 and 2, respectively, in each of the four groups which have been tabulated in that article; or it may be deduced, by the transformations in article 234, from the corresponding Table of article 233, with cyclical changes of arrangement in the three latter groups. For example, from the corner 2 of Triangle I, in the last cited Table, of which the coordinates are $x = 1 + \theta^2 c$, $y = \theta^2 + c$, $z = -\theta c$, we derive immediately the transformed coordinates,

$$x_1 = \frac{y - z}{c - 1} = -\theta^2, \quad y_1 = \frac{x - z}{c - 1} = -1, \quad z_1 = \frac{x + y + z}{2c + 1} = -\theta;$$

which may then be changed, as in the Table of the present article, to the proportional expressions, 1, θ, θ^2. Or we may determine the new coordinates x, y, z, of this corner by the system of the two linear equations,

$$x_1 + \theta y_1 + \theta^2 z_1 = 0, \quad x_1 + y_1 + z_1 = 0,$$

which represent the two chords whereof the transformed symbols are $[1, \theta, \theta^2]_1$ and $[1, 1, 1]_1$, in the 1st Triangle of article 235. I see, indeed, that *some* of the formulæ of recent articles had been in substance given before, (or at least formulæ connected with them) for instance in articles 112, &c., and 211; but it may not have been quite useless to show *how completely the constant c disappears by this transformation of coordinates* (article 211), *from all expressions which involve only the nine points of inflexion*, or the *intersections of the chords* connecting them, &c.

237. (May 7th, 1860) Although I may have something more to say respecting points of inflexion, and the various forms which the cubic equation may assume, when considered relatively to those points, and to the tangents at them, as indeed I have suggested in some separate notes, written during the progress of this Letter, yet it will not perhaps, be considered an unpardonable violation of system, if I now *resume* the consideration of that Theorem of Pure Geometry, which you mentioned to me a few weeks ago, and which has recently been brought under my notice again, by a reperusal of your note of April 25th: notwithstanding my having designed to content myself with such (rather long) discussion as I had given to it, in articles 170 to 183, written before the last mentioned note arrived; which articles I shall suppose to be within your reach, and of which therefore I shall here *retain the notations* without any new explanation of them.

238. In articles 174, and 175, a construction was given for the point S, in which the right line CP, drawn from the corner C of the given triangle to the point of contact P of two conics or circles, of the question, met (as in the figure of article 148) the opposite side AB; but I admit that this construction, or the ratio which it assigned between the segments AS, SB of that side, was rather complicated. In articles 176 and 177, a *much simpler construction*, and one more *geometrically relevant* (so to speak) was given; namely one for the point (L or M or N) in which the *common tangent LMN* to the two conics, or circles drawn at the point P where they *touch one another*, intersects any one of the three sides of the given triangle ABC; and I still think (as in article 170) that any one of the three formulæ (article 177)

$$\frac{LB}{CL} = \frac{a-b}{a-c}, \quad \frac{MC}{AM} = \frac{b-c}{b-a}, \quad \frac{NA}{BN} = \frac{c-a}{a-b},$$

which belong to the case of contact with the *inscribed* circle, is not only *sufficient to determine*, but also *determines with sufficient simplicity*, the point of contact P, by assigning *a point on one of the sides* from which a *tangent* to the inscribed circle is to be drawn. Perhaps, therefore, I should have been *content* with this construction; if it had not been for your subsequently received, but above cited, note of the 25th of last month [April]; and for the elegant and simple determinations given in it, for the point above called P, on which my eye has later fallen again, and which I wish to prove, and in some degree also to extend, by the Method of Anharmonic Coordinates.

239. Your little figure (of April 25th) was perfectly clear, and therefore was, no doubt, more elegant than one which involves so many lines and lettered points, as those which I have just now amused myself, by introducing into the annexed Diagram (Figure 29). It represents, as you see, a system of two circles; one with its centre at I, which is inscribed in the triangle ABC, touching the sides thereof at the points A_1, B_1, C_1; and the other, with centre K, bisecting those sides in A'', B'', C'', and intersecting them again in the feet A''', B''', C''' of the three perpendiculars from the points A, B, C, which are exhibited as meeting one another in a point ϖ. The two circles are seen to touch at a point P, their common tangent LMN at that point being *parallel* to the mean side CA, and *trisecting* each of the extreme sides, AB, BC; because it has been assumed, in the construction of this Diagram (Figure 29), that the three sides of the triangle are, as regards their relative lengths,

$$a = 4, \quad b = 3, \quad c = 2;$$

whence, by my formulæ of article 178, cited in article 238, we have the three ratios of section

$$\frac{LB}{CL} = +\frac{1}{2}; \quad \frac{MC}{AM} = -1; \quad \frac{NA}{BN} = +2.$$

And so far, I have made no use of the information contained in your cited note on this subject.

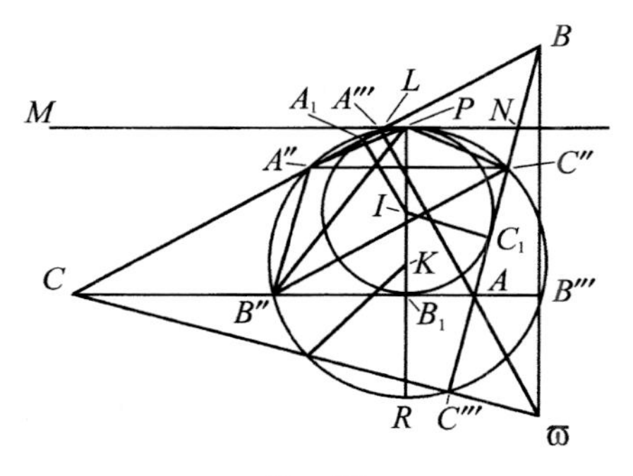

Fig. 29

240. You inform me, however, in that Note, and I understand the reasoning by which you prove it, although my memory is not *familiar* with all the elementary theorems to which you refer, that if, *on that arc* of the bisecting circle (K) which lies *between the mean chord* $C''A''$ *and the mean* corner B (my own letters being here used), a point P be taken, whose distances $C''P$ and $A''B$ from the extremities of that chord shall be proportional to the tangents $C''C_1$ and $A''A_1$, which are drawn from the same points C'' and A'' to the inscribed circle (I), we shall then have the more full *proportion of chords and tangents*

$$\overline{A''P} : \overline{B''P} : \overline{C''P} = \overline{A''A_1} : \overline{B''B_1} : \overline{C''C_1};$$

whence you infer (I.), that the two circles touch, in the point P thus determined. And you add, without proof, an elegant construction (II.) which may be thus described, for *graphically determining that point of contact*:- Bisect, say in R, the arc $B''B'''$ of the circle (K), which is remote from B; the right line RB_1, drawn to the point of contact B_1 of this side CA with the circle (I) will pass through the point P.

241. To investigate these things in my own way, I form first the three following *anharmonics of pencils*:

$$h = (A''.B''CC''P); \qquad h' = (B''.C''AA''P); \qquad h'' = (C''.A''BB''P);$$

where I begin by supposing that P is *any point* (x, y, z) of the plane. The formula of article 151 gives, next,

$$h = \begin{vmatrix} 0, & 1, & 1 \\ 1, & 0, & 1 \\ 0, & 0, & 1 \end{vmatrix} \begin{vmatrix} 0, & 1, & 1 \\ 1, & 1, & 0 \\ x, & y, & z \end{vmatrix} : \begin{vmatrix} 0, & 1, & 1 \\ 0, & 0, & 1 \\ 1, & 1, & 0 \end{vmatrix} \begin{vmatrix} 0, & 1, & 1 \\ x, & y, & z \\ 1, & 0, & 1 \end{vmatrix} = -\frac{z + x - y}{x + y - z};$$

and similarly,

$$h' = -\frac{x + y - z}{y + z - x}, \quad \text{and} \quad h'' = -\frac{y + z - x}{z + x - y};$$

so that $hh'h'' = -1$.

But we had (in article 172), for the point P of contact of the two conics (U) and (V) the coordinates,

$$x = \alpha(\beta - \gamma)^2; \quad y = \beta(\gamma - \alpha)^2; \quad z = \gamma(\alpha - \beta)^2;$$

whence the three recent anharmonics become, for that point

$$h = \frac{\alpha - \beta}{\alpha + \beta} \cdot \frac{\alpha + \gamma}{\alpha - \gamma}; \quad h' = \frac{\beta - \gamma}{\beta + \gamma} \cdot \frac{\beta + \alpha}{\beta - \alpha}; \quad h'' = \frac{\gamma - \alpha}{\gamma + \alpha} \cdot \frac{\gamma + \beta}{\gamma - \beta}.$$

Again, because (by article 174) the contacts of the conic (V) with the sides are

$$A_1 = (0, \gamma, \beta), \quad B_1 = (\gamma, 0, \alpha), \quad C_1 = (\beta, \alpha, 0)$$

we have (cf. article 174)

$$\frac{\beta}{\alpha} = (C_1 BC''A) = -(C_1 BC'A);$$

and therefore

$$1-\frac{\beta}{\alpha}=(C_1C''BA),\quad \left(1+\frac{\beta}{\alpha}\right)^{-1}=(C_1ABC');$$

so that

$$\frac{\alpha-\beta}{\alpha+\beta}=(C_1C''BA).(C_1ABC')=(C_1C''BC')=-(C_1C''AC'),$$

and similarly for the rest. Hence, on the conic (U), we have the three following anharmonics of groups (cf. end of article 18):-

$$h=((B''A'''C''P))=-(C_1C''AC'):(B_1B''CB')=(B_1B'AB'').(C_1C''AC');$$

$$h'=((C''B'''A''P))=-(A_1A''BA'):(C_1C''AC')=(C_1C'BC'').(A_1A''BA');$$

$$h''=((A''C'''B''P))=-(B_1B''CB'):(A_1A''BA')=(A_1A'CA'').(B_1B''CB');$$

of which, as a verification, the product is negative unity. Already the expressions thus obtained – any one of which suffices to determine the point P, – are not of great complexity; but they become of course, much simpler, when we throw off the line $A'B'C'$ to infinity: giving then

$$h=\frac{AB''}{B_1B''}.\frac{C_1C''}{AC''};\quad h'=\frac{BC''}{C_1C''}.\frac{A_1A''}{BA''};\quad h''=\frac{CA''}{A_1A''}.\frac{B_1B''}{CB''};$$

(U) and (V) being now two homothetic conics (article 177), whereof the former *bisects* and the latter *touches*, the sides of the given triangle ABC. Supposing, finally, that these *conics* become *circles*, and that the latter is *inscribed*, we have the *anharmonics of circular groups*,

$$h=\frac{b}{c}.\frac{a-b}{a-c},\quad h'=\frac{c}{a}.\frac{b-c}{b-a},\quad h''=\frac{a}{b}.\frac{c-a}{c-b},$$

which are easily seen to contain the first or the metric part of your construction. It may be remarked that when $a>b>c$, we have not only $h>0$, but also $h<1$; whence it may be inferred that P falls (as in Figure 29) *between* A''' *and* C''; which indeed is geometrically evident.

242. If we now proceed to consider, – and to *transform*, – the second or graphic construction cited (in article 240) from your Note, and which I understand to be *your own*, it becomes necessary to lay down rules for dealing with questions of *bisection of angles*, in that extended Calculus, to which the present Letter relates. How am I to *express*, – even as a thing to be *proved*, – that the chord PB_1, of the inscribed circle *bisects* the *angle* $B''PB'''$? And how *extend* this theorem, so as to meet the more general case of two mutually *tangent conics*?

243. For this purpose, it will be useful to consider first the analogous *extension of the relation of* ***perpendicularity*** *of lines*, of which we have already had an *example* in article 180; where the elementary relation, mentioned by you in your first note on the subject, namely that the *second intersection* of the side with the bisecting circle is the *foot of the perpendicular* let fall from the opposite corner of the triangle, or (with my symbols) that each of the *angles* $A''A'''A$, $B''B'''B$, $C''C'''C$ (as in Figure 29) is *right*, was replaced by the *harmonic* property of the *pencil*, which had its *vertex* at that point of second intersection, and had the two perpendicular lines for *conjugate rays*, which the *other pair* of conjugates passed through the two (real or imaginary) points Q, Q', in which (by articles 170 and 171) the *two conics* (U) and (V) *intersected each other on the unit-line*: these two latter points, Q and Q', being in fact what I substitute, for the *circular points at infinity*, in my treatment of this whole question. In short, I stated in article 180 and proved in articles 181 and 182, the harmonic equation,

$$(A'''.AQA''Q') = -1,$$

as my *extension* of the known theorem that AA''' is, in the simple case of the *bisecting circle*, *perpendicular* to $A'''A''$, or to BC.

244. Let it therefore be now proposed to investigate *generally* the relation which must exist between the nine anharmonic coordinates of any three points

$$P = (x,\ y,\ z), \qquad P_1 = (x_1,\ y_1,\ z_1), \qquad P_2 = (x_2,\ y_2,\ z_2),$$

in order that we may have the *harmonic relation*,

$$(P.P_1QP_2Q') = -1$$

when Q and Q' are still (as in article 170) the two real *or* imaginary points, – for it is important to observe that they *may* here be *real*, – in which the *given conic* $ax^2 + by^2 + cz^2 = 0$, is cut by the *given unit-line*

$$x + y + z = 0.$$

For we shall thus obtain the condition of perpendicularity of PP_1 to PP_2, in that *limiting* case, when the *conic* becomes a *circle*, and the *unit-line* goes off *to infinity*.

245. Writing

$$Q = (x',\ y',\ z'), \qquad Q' = (x'',\ y'',\ z'')$$

so that $\frac{x'}{y'}$ and $\frac{x''}{y''}$ are the roots of the quadratic equation

$$(a + c)x^2 + 2cxy + (b + c)y^2 = 0;$$

for which roots expressions were given in article 181, but respecting which it is sufficient here to observe that by the quadratic, we have the proportion,

$$x'x'' : x'y'' + y'x'' : y'y'' = b + c : -2c : a + c,$$

which, by the linear equation, we have also

$$x' + y' + z' = 0, \qquad x'' + y'' + z'' = 0;$$

the general theory of article 151 gives the following expression for the anharmonic function of the pencil in question

$$h = (P.P_1 Q P_2 Q') = \begin{vmatrix} x, & y, & z \\ x_1, & y_1, & z_1 \\ x', & y', & z' \end{vmatrix} \begin{vmatrix} x, & y, & z \\ x_2, & y_2, & z_2 \\ x'', & y'', & z'' \end{vmatrix} : \begin{vmatrix} x, & y, & z \\ x', & y', & z' \\ x_2, & y_2, & z_2 \end{vmatrix} \begin{vmatrix} x, & y, & z \\ x'', & y'', & z'' \\ x_1, & y_1, & z_1 \end{vmatrix}$$

$$= \begin{vmatrix} x', & y', & z' \\ x, & y, & z \\ x_1, & y_1, & z_1 \end{vmatrix} \begin{vmatrix} x'', & y'', & z'' \\ x, & y, & z \\ x_2, & y_2, & z_2 \end{vmatrix} : \begin{vmatrix} x'', & y'', & z'' \\ x, & y, & z \\ x_1, & y_1, & z_1 \end{vmatrix} \begin{vmatrix} x', & y', & z' \\ x, & y, & z \\ x_2, & y_2, & z_2 \end{vmatrix}$$

or

$$h = \frac{(a_1 x' + b_1 y' + c_1 z')(a_2 x'' + b_2 y'' + c_2 z'')}{(a_1 x'' + b_1 y'' + c_1 z'')(a_2 x' + b_2 y' + c_2 z')},$$

if we write for conciseness,

$$a_1 = \begin{vmatrix} y, & z \\ y_1, & z_1 \end{vmatrix}, \quad b_1 = \begin{vmatrix} z, & x \\ z_1, & x_1 \end{vmatrix}, \quad c_1 = \begin{vmatrix} x, & y \\ x_1, & y_1 \end{vmatrix},$$

$$a_2 = \begin{vmatrix} y, & z \\ y_2, & z_2 \end{vmatrix}, \quad b_2 = \begin{vmatrix} z, & x \\ z_2, & x_2 \end{vmatrix}, \quad c_2 = \begin{vmatrix} x, & y \\ x_2, & y_2 \end{vmatrix}.$$

The condition $h = -1$ is therefore expressed by the equation

$$0 = 2a_1 a_2 x'x'' + 2b_1 b_2 y'y'' + 2c_1 c_2 z'z'' + (b_1 c_2 + c_1 b_2)(y'z'' + z'y'')$$
$$+ (c_1 a_2 + a_1 c_2)(z'x'' + x'z'') + (a_1 b_2 + b_1 a_2)(x'y'' + y'x''),$$

which is, as it ought to be, symmetric relatively to the two points Q and Q', and will therefore allow of our rationally eliminating the quotients $x' : y' : z'$ and $x'' : y'' : z''$, and introducing in their stead, the coefficients a, b, c of the given equation of the conic.

246. If for this purpose, we first eliminate linearly z' and z'', and write

$$a_1 - c_1 = a_1', \quad b_1 - c_1 = b_1',$$
$$a_2 - c_2 = a_2', \quad b_2 - c_2 = b_2',$$

the equation $h = -1$ becomes

$$0 = (a_1' x' + b_1' y')(a_2' x'' + b_2' y'') + (a_1' x'' + b_1' y'')(a_2' x' + b_2' y')$$
$$= 2a_1' a_2' x' x'' + (a_1' b_2' + b_1' a_2')(x' y'' + y' x'') + 2b_1' b_2' y' y'';$$

that is, by the proportion (article 245)

$$0 = (b + c) a_1' a_2' - c(a_1' b_2' + b_1' a_2') + (a + c) b_1' b_2';$$

where the coefficient of c is

$$a_1' a_2' - a_1' b_2' - b_1' a_2' + b_1' b_2' = (a_1' - b_1')(a_2' - b_2') = (a_1 - b_1)(a_2 - b_2);$$

also

$$a_1 - b_1 = (x + y) z_1 - z(x_1 + y_1) = s z_1 - z s_1$$

and

$$a_2 - b_2 = sz_2 - zs_2,$$

if we make

$$s = z + y + z, \qquad s_1 = x_1 + y_1 + z_1, \qquad s_2 = x_2 + y_2 + z_2;$$

with other but quite similar reductions. Hence, finally, the sought condition comes out, under the following very simple and symmetric form, and one from which determinants have been eliminated:

$$0 = a(sx_1 - xs_1)(sx_2 - xs_2) + b(sy_1 - ys_1)(sy_2 - ys_2) + c(sz_1 - zs_1)(sz_2 - zs_2).$$

247. If this points P_1, P_2 happened to be *on* the unit-line we have then $s_1 = 0$, $s_2 = 0$, and the last equation becomes still more simple, namely –

$$ax_1 x_2 + by_1 y_2 + cz_1 z_2 = 0;$$

being indeed in this case, independent of the position of the point P, as it ought to be, because the *group* $P_1 QP_2 Q'$, having thus become *collinear*, has a value for its anharmonic function, which is independent of the position of the *vertex* P of the *pencil*. Conversely, we might have deduced, even more easily the last form first, and then have inferred the other from it. For the following formula for this *group* (supplied by article 188)

$$h = (P_1 QP_2 Q') = \begin{vmatrix} x_1, & y_1 \\ x', & y' \end{vmatrix} \begin{vmatrix} x_2, & y_2 \\ x'', & y'' \end{vmatrix} : \begin{vmatrix} x', & y' \\ x_2, & y_2 \end{vmatrix} \begin{vmatrix} x'', & y'' \\ x_1, & y_1 \end{vmatrix},$$

would have reduced the harmonic relation $h = -1$ to the form,

$$\begin{aligned} 0 &= (x_1 y' - y_1 x')(x_2 y'' - y_2 x'') + (x_1 y'' - y_1 x'')(x_2 y' - y_2 x') \\ &= 2x_1 x_2 y' y'' + 2y_1 y_2 x' x'' - (x_1 y_2 + y_1 x_2)(x' y'' + y' x''), \end{aligned}$$

or, by article 245, and by the equation of the unit-line,

$$\begin{aligned} 0 &= (a + c) x_1 x_2 + (b + c) y_1 y_2 + c(x_1 y_2 + y_1 x_2) \\ &= ax_1 x_2 + by_1 y_2 + cz_1 z_2, \end{aligned}$$

as above. And if we now return to the more general supposition (article 244), that P_1 and P_2 are *any* points in the given plane, we easily find that the *intersections* of the lines, PP_1, PP_2 with the unit-line are the points (X_1, Y_1, Z_1) and (X_2, Y_2, Z_2), where

$$\begin{aligned} X_1 &= sx_1 - xs_1, \qquad Y_1 = sy_1 - ys_1, \qquad Z_1 = sz_1 - zs_1, \\ X_2 &= sx_2 - xs_2, \qquad Y_2 = sy_2 - ys_2, \qquad Z_2 = sz_2 - zs_2; \end{aligned}$$

substituting therefore these values in the formula,

$$aX_1 X_2 + bY_1 Y_2 + cZ_1 Z_2 = 0,$$

we recover the general equation of the foregoing article – I am not sure whether I shall be disposed, before the close of this long Letter, to enter upon any comparison of that equation (article 246), with what I called in a much earlier article 46 a formula of *anharmonic perpendicularity*, and reduced (in article 47) to the form

$$l_1 l_1' + m_1 m_1' + n_1 n_1' = 0,$$

beyond observing that I then virtually supposed the constants a, b, c, to be *equal*, or employed the imaginary conic $x^2 + y^2 + z^2 = 0$. Instead of the slightly more general conic $ax^2 + by^2 + cz^2 = 0$, of article 244.

248. The two equations,

$$aX_1 X_2 + bY_1 Y_2 + cZ_1 Z_2 = 0, \qquad X_2 + Y_2 + Z_2 = 0,$$

allow us to write,

$$X_2 = bY_1 - cZ_1, \qquad Y_2 = cZ_1 - aX_1, \qquad Z_2 = aX_1 - bY_1;$$

or to assign any values proportional to these, to the coordinates of the point in which the unit-line is met by the line PP_2. And when the points P and P_2 are given, any one of the equations

$$\frac{sx_2 - xs_2}{X_2} = \frac{sy_2 - ys_2}{Y_2} = \frac{sz_2 - zs_2}{Z_2},$$

determines in general the position of the line PP_2, and consequently the point in which it meets any given line. For example, if we suppose $y_2 = 0$, and therefore $s_2 = x_2 + z_2$ we thus obtain the formula

$$\frac{Z_2 - X_2}{Z_2 + X_2} = \frac{(sz_2 - zs_2) - (sx_2 - xs_2)}{(sz_1 - zs_1) + (sx_2 - xs_2)} = \frac{x - z}{y} + \frac{s}{y} \cdot \frac{z_2 - x_2}{z_2 + x_2},$$

or

$$\frac{z_2 - x_2}{z_2 + x_2} = \frac{y}{s}\left(\frac{z - x}{y} + \frac{Z_2 - X_2}{Z_2 + X_2}\right);$$

and thus the intersection $(x_2, 0, z_2)$ of PP_2 with the side CA can be determined. Let this intersection be called B_2, the point P_1 being placed at the contact B_1 of the same side of the given triangle with the conic V; which latter point (by article 174) is $(\alpha^{-1}, 0, \gamma^{-1})$, or $(\gamma, 0, \alpha)$ so that we may write

$$x_1 = \gamma, \quad y_1 = 0, \quad z_1 = \alpha.$$

Let also the vertex P of the pencil be still the point of contact of the *two* conics (U) and (V), so that (by article 172) its coordinates are,

$$x = \alpha(\beta - \gamma)^2, \quad y = \beta(\gamma - \alpha)^2, \quad z = \gamma(\alpha - \beta)^2;$$

while the constants a, b, c may (by article 171) be thus expressed

$$a = \alpha(\alpha + \gamma) + \beta(\alpha - \gamma),$$

$$b = -\alpha\gamma + \beta(\alpha + \gamma) + \beta^2,$$

$$c = \gamma(\alpha + \gamma) + \beta(\gamma - \alpha).$$

We shall thus have all the requisite elements for the determination of the coordinates x_2 and z_2, or of their ratio, in terms of α, β, γ, which are here regarded as known constants; and are such that the three right lines, AA_1, BB_1, CC_1, drawn from the corners of the given triangle to the contacts of its sides, with the conic (V), concur (by article 174) in the point

$$(\alpha^{-1}, \beta^{-1}, \gamma^{-1}).$$

And thus, we can determine on the side CA the point B_2, which satisfies the harmonic relation (cf. article 244)

$$(P.B_1 QB_2 Q') = -1;$$

and which is therefore such that the *angle* B_1PB_2 *becomes right*, when we pass to the *limiting case*, of the *bisecting and inscribed circles.*

249. Dividing $sx_1 - xs_1$, $sy_1 - ys_1$, $sz_1 - zs_1$, by what is found to be their common factor $\beta(\alpha - \gamma)$, we obtain the following coordinates of the point in which the right line PP_1, or PB_1, intersects the unit-line:

$$X_1 = \gamma(3\alpha - \gamma) - \beta(\alpha + \gamma);$$

$$Y_1 = \gamma^2 - \alpha^2;$$

$$Z_1 = \alpha(\alpha - 3\gamma) + \beta(\alpha + \gamma);$$

of which, as a verification, the sum is zero; and in which it may be noticed, that Z_1 can be formed from X_1, by first interchanging α and γ, and then changing all the signs. We have next the three products

$$aX_1 = \alpha\gamma(3\alpha^2 + 2\alpha\gamma - \gamma^2) + \beta(-\alpha^3 + \alpha^2\gamma - 5\alpha\gamma^2 + \gamma^3) - \beta^2(\alpha^2 - \gamma^2);$$

$$bY_1 = \alpha\gamma(\alpha^2 - \gamma^2) + \beta(-\alpha^3 - \alpha^2\gamma + \alpha\gamma^2 + \gamma^3) - \beta^2(\alpha^2 - \gamma^2);$$

$$cZ_1 = \alpha\gamma(\alpha^2 - 2\alpha\gamma - 3\gamma^2) + \beta(-\alpha^3 + 5\alpha^2\gamma - \alpha\gamma^2 + \gamma^3) - \beta^2(\alpha^2 - \gamma^2);$$

where cZ_1 may be formed from aX_1, by the same rule as Z_1 from X_1. Dividing the three differences,

$$bY_1 - cZ_1, \quad cZ_1 - aX_1, \quad aX_1 - bY_1,$$

by $2\alpha\gamma$, we get the coordinates of the intersection of the line PP_2 (or PB_2) with the unit-line, as follows

$$X_2 = \gamma(\alpha + \gamma) + \beta(\gamma - 3\alpha);$$

$$Y_2 = (\alpha + \gamma)(2\beta - \alpha - \gamma);$$

$$Z_2 = \alpha(\alpha + \gamma) + \beta(\alpha - 3\gamma);$$

with the verification, $X_2 + Y_2 + Z_2 = 0$; whence

$$\frac{Z_2 - X_2}{Z_2 + X_2} = \frac{\alpha - \gamma}{\alpha + \gamma} \cdot \frac{\alpha + 4\beta + \gamma}{\alpha - 2\beta + \gamma}.$$

We have also $\dfrac{z - x}{y} = \dfrac{\alpha\gamma - \beta^2}{\beta(\alpha - \gamma)}$; and in adding the last two fractions, the numerator of the sum is found to be

$$(\alpha-\gamma)^2\beta(\alpha+4\beta+\gamma)+(\alpha+\gamma)(\alpha\gamma-\beta^2)(\alpha-2\beta+\gamma)$$
$$=(\alpha+2\beta+\gamma)\{\alpha(\beta-\gamma)^2+\beta(\gamma-\alpha)^2+\gamma(\alpha-\beta)^2\}$$
$$=(\alpha+2\beta+\gamma)\,s;$$

multiplying then this expression by $\frac{y}{s}$, and dividing by $\beta(\alpha-\gamma)$, we get $(\alpha-\gamma)(\alpha+2\beta+\gamma)$, as the quantity which is to be divided by Z_2+X_2, or by $(\alpha+\gamma)(\alpha-2\beta+\gamma)$, in forming an expression for the quotient $\frac{z_2-x_2}{z_2+x_2}$, by the process of article 248. Hence,

$$\frac{z_2-x_2}{z_2+x_2}=\frac{\alpha-\gamma}{\alpha+\gamma}\cdot\frac{\alpha+2\beta+\gamma}{\alpha-2\beta+\gamma};$$

that is

$$\frac{x_2}{z_2}=\frac{\gamma(\alpha+\gamma)-2\alpha\beta}{\alpha(\alpha+\gamma)-2\gamma\beta};$$

and therefore the required point B_2 may be denoted as follows:

$$B_2=(\alpha\gamma+\gamma^2-2\alpha\beta,\ 0,\ \alpha^2+\alpha\gamma-2\beta\gamma).$$

250. If now we compare the point B_2, thus found with the three points B_1, B'', B''', on the same side CA, for which (by articles 174, 170, and 181) we may employ the symbols $B_1=(\gamma,0,\alpha)$, $B''=(1,0,1)$, $B'''=(c,0,a)$, we soon perceive that while the three lines AA'', BB'', CC'' concurring in the point $(1,1,1)$ and the three lines AA''', BB''', CC''' in $\varpi=(a^{-1},b^{-1},c^{-1})$, as AA_1, BB_1, CC_1 concur (article 248) in $(\alpha^{-1},\beta^{-1},\gamma^{-1})$,

$$(\alpha+2\beta+\gamma)(B_1)=(B''')+\beta(\alpha+\gamma)(B''),$$

we have also

$$(B_2)=(B''')-\beta(\alpha+\gamma)(B'').$$

It follows that

$$(B_1B''B_2B''')=-1,$$

or that the points B_1 and B_2 are harmonically conjugate, with respect to the points B'' and A'''; whence, of course, we have the corresponding harmonic equation, for the *pencil* with its vertex at P,

$$(P.B_1P''B_2B''')=-1.$$

But we had determined the point B_2, so as to have (article 248) the analogous equation

$$(P.B_1QB_2Q')=-1;$$

it follows that PB_1 and PB_2 are the two *double rays* of an involution, whereof PB'', PB''' are *one pair* of conjugate rays, and PQ, PQ' are *another pair*. Passing therefore to the case of the *two circles*, since the rays PQ, PQ' are in *that* case directed to the *circular points at infinity*, we see that "*the line PB_1 is one of the bisectors of the angle $B''PB'''$*"; which was (article 242) for that case, the thing to be proved.

251. Of course, I do not pretend that the foregoing is the shortest, or the easiest mode, of proving, even in an extended form, and by *calculation* the correctness of that graphic determination of the point P; or of the statement in article 242 of a certain *property* of that point, mentioned by you to me. In fact, I think that I see – though perhaps I may not write down, – *another extension* of the same sort, but likely to lead to *simpler calculations,* which shall depend on the circumstance that "*the tangent to the conic* (U), *at the point where the chord* PB_1 *of the inscribed conic* (V) *meeting* (U) *again, passes through the point* B', *where the side CA of the given triangle ABC, is met by the given transversal* $A'B'C'$."

252. Whether the proof of this *last* extension shall, or shall not, be written out, we may already state this *Theorem*:-

"If the sides of a triangle ABC touch in the points A_1, B_1, C_1 a conic (V) and if a homothetic conic (U) be described so as to bisect those sides in A'', B'', C'', and in cutting them again in A''', B''', C''', then not only will these two conics *touch* each other (by article 177) in some point P but also *if the chords* PA'', PA''' of the conic (U) be treated as *conjugate rays* of a system of lines *in involution, two other conjugate rays* through the point of contact P of the two conics being *parallel* to the two (real or imaginary) asymptotes of either, then *the chord* PA_1 *of the conic* (V) *will be one of the double rays* of the system"; and similarly PB_1 will be a double ray of a second involution PC_1 of a third, these two other involutions being determined on a precisely similar plan.

253. In general, if P_0 be any given vertex (x_0, y_0, z_0) of a pencil, and if we write

$$X = s_0 x - s x_0, \quad Y = s_0 y - s y_0, \quad Z = s_0 z - s z_0, \quad s = x + y + z, \text{ \&c.},$$

the equation

$$aX^2 + bY^2 + cZ^2 = 0$$

represents *the pair of rays* from P_0 to the two real or imaginary points Q, Q', in which (as before) the conic

$$ax^2 + by^2 + cz^2 = 0$$

is cut by the unit line $z + y + z = 0$. Let P_1, P_2 be any two given points, and write

$$X_1 = s_0 x_1 - s_1 x_0, \text{ \&c.}, \quad X_2 = s_0 x_2 - s_2 x_0, \text{ \&c.},$$

$$a' = Y_1 Z_2 + Z_1 Y_2, \quad b' = Z_1 X_2 + X_1 Z_2, \quad c' = X_1 Y_2 + Y_1 X_2;$$

then the equation

$$a'X^2 + b'Y^2 + c'Z^2 = 0$$

will represent the new pair of rays, P_0P_1, P_0P_2, from the same given vertex as before. Let P_0P', P_0P'' be supposed to be the two double rays of the involution, which is determined by the two former pairs of rays, regarded as two pairs of conjugates therein; that is, let us suppose, that the two following harmonic relations hold good

$$(P_0.P'QP''Q') = -1, \quad (P_0.P'P_1P''P_2) = -1.$$

Then the first relation is expressed by the equation (cf. article 247),

$$aX'X'' + bY'Y'' + cZ'Z'' = 0;$$

and the second harmonic relation is expressed by the analogous equation

$$a'X'X'' + b'Y'Y'' + c'Z'Z'' = 0$$

but $X' + Y' + Z' = 0$, $X'' + Y'' + Z'' = 0$, and therefore (cf. article 248) we may write

$$X'' = bY' - cZ', \quad Y'' = cZ' - aX', \quad Z'' = aX' - bY',$$

or any expressions proportional to these, as consequences of the first harmonic equation; substituting then these values for X'', Y'', Z'', the second equation becomes

$$0 = a'X'(bY' - cZ') + b'Y'(cZ' - aX') + c'Z'(aX' - bY');$$

with an exactly similar result of the elimination of X', Y', Z'. We may therefore say that the equation

$$(bc' - cb')YZ + (ca' - ac')ZX + (ab' - ba')XY = 0,$$

represents the system of the two double rays, of the involution above described. And thus we have a sufficiently simple and perfectly *general solution* of the problem, which in the present method replaces (as an extension) that of *finding the two bisectors* (P_0P', P_0P'') *of a given angle* $(P_1P_0P_2)$.

254. It is not worth while to exemplify this general Solution, in the particular question which we were lately considering. But I think that the following little investigation may have some interest, as serving at once to prove, in my own way, the theorem (or extension) which was stated in article 251; and also to show how even for the *general case* of *two conics,* the point R where the chord PB_1 of (V) meets the conic (U) again, and which (by the theorem) is *one* point on the *polar* of B', may be *geometrically distinguished* from the *other point,* where the same conic (U) is met by the *same polar.*

255. Since the constants a, b, c, (not *here* the *sides* of the triangle) can be *rationally expressed,* as in articles 171 and 248, in terms of the other constants α, β, γ, but *not* vice versa, it seems convenient to *eliminate* the *former* set of constants from the first equation in article 170 of the conic (U), and to introduce the *latter* set in their stead. We have thus the equation

$$0 = U = (\alpha + \beta + \gamma)(\alpha x^2 + \beta y^2 + \gamma z^2) - \beta\gamma x^2 - \gamma\alpha y^2 - \alpha\beta z^2$$
$$- (\beta + \gamma)^2 yz - (\gamma + \alpha)^2 zx - (\alpha + \beta)^2 xy = (\text{say})\, U_{x,y,z}.$$

The analysis of article 172 showed us, that this equation is satisfied by the coordinates of the point P, namely,

$$x = \alpha(\beta - \gamma)^2, \quad y = \beta(\gamma - \alpha)^2, \quad z = \gamma(\alpha - \beta)^2.$$

Let x', y', z' be the coordinates of the point (R in Figure 29) where the line PB_1 meets, as above, the conic (U) again; the coordinates of B_1 being γ, 0, α, as in article 248.

We shall thus have expressions of the forms

$$x' = x + t\gamma, \quad y' = y, \quad z' = z + t\alpha,$$

which are to be made to satisfy the condition

$$U_{x',y',z'} = 0;$$

and which will do so, if the auxiliary coefficient t be determined by the linear equation

$$0 = (\gamma D_x + \alpha D_z)\, U_{x,y,z} + tU_{\gamma,0,\alpha}.$$

But for the point P of contact of the two conics, we had (in article 172)

$$D_x U = (\alpha+\beta)(\beta+\gamma)(\gamma+\alpha)\lambda, \ldots$$

$$D_z U = (\alpha+\beta)(\beta+\gamma)(\gamma+\alpha)\nu,$$

where (by article 171)

$$\lambda = (\gamma-\alpha)(\alpha-\beta), \quad \mu = (\alpha-\beta)(\beta-\gamma),$$

$$\nu = (\beta-\gamma)(\gamma-\alpha);$$

hence

$$\gamma\lambda + \alpha\nu = -\beta(\alpha-\gamma)^2,$$

and

$$(\gamma D_x + \alpha D_z)\, U_{x,y,z} = -\beta(\alpha-\gamma)^2(\alpha+\beta)(\beta+\gamma)(\gamma+\alpha).$$

Again, without going back beyond the present article, we have

$$U_{\gamma,0,\alpha} = \alpha\gamma(\alpha+\gamma)(\alpha+\beta+\gamma) - \beta(\alpha^3+\gamma^3) - \alpha\gamma(\alpha+\gamma)^2$$

$$= -\beta(\alpha+\gamma)(\alpha-\gamma)^2;$$

whence

$$t = -(\alpha+\beta)(\beta+\gamma).$$

The second intersection (R) of PB_1 with (U) has therefore the coordinates,

$$x' = \alpha(\beta-\gamma)^2 - \gamma(\alpha+\beta)(\beta+\gamma);$$

$$y' = \beta(\gamma-\alpha)^2;$$

$$z' = \gamma(\alpha-\beta)^2 - \alpha(\alpha+\beta)(\beta+\gamma);$$

or, dividing by β,

$$x' = \beta(\alpha-\gamma) - \gamma(3\alpha+\gamma);$$

$$y' = (\alpha-\gamma)^2;$$

$$z' = -\beta(\alpha-\gamma) - \alpha(\alpha+3\gamma).$$

256. If now we seek the point, say $R' = (x'', y'', z'')$ in which the same chord PB_1 of the conic (V) meets the polar of the point $B' = (-1, 0, 1)$ taken with respect to the conic (U) the equation of which polar is, in x'', y'', z'',

$$0 = (D_{x''} - D_{z''})\, U_{x'',y'',z''},$$

or more fully,

$$0 = a''x'' + b''y'' + c''z'',$$

when

$$a'' = 2\alpha(\alpha+\beta+\gamma) - 2\beta\gamma + (\alpha+\gamma)^2,$$
$$b'' = (\beta+\gamma)^2 - (\alpha+\beta)^2,$$
$$c'' = -2\gamma(\alpha+\beta+\gamma) + 2\alpha\beta - (\alpha+\gamma)^2,$$

we are to make

$$x'' = x + t'\gamma, \quad y'' = y, \quad z'' = z + t'\alpha,$$

and determine the value of the new coefficient t'. Substitution gives, after a few reductions,

$$a''x + b''y + c''z = (\gamma^2 - \alpha^2)(\alpha - 2\beta + \gamma)(\alpha+\beta)(\beta+\gamma),$$
$$a''\gamma + c''\alpha = (\gamma^2 - \alpha^2)(\alpha - 2\beta + \gamma);$$

whence

$$t' = -(\alpha+\beta)(\beta+\gamma) = t;$$

and therefore the point R' coincides with the point R. The point last named is therefore on the polar of B', with respect to (U); and consequently the tangent to the conic (U), at this point R, passes through the point B', as stated in article 251.

257. This circumstance enables us easily to calculate a simple expression for the *anharmonic* h' *of the group* of the four points A'', B'', C'', R, *on the conic* (U); hence we may write (cf. end of article 18)

$$h' = ((A''B''C''R)) = (R.A''B''C''B'),$$

and then employ the general formula of article 151; which gives here

$$h' = \begin{vmatrix} x', & y', & z' \\ 0, & 1, & 1 \\ 1, & 0, & 1 \end{vmatrix} \begin{vmatrix} x', & y', & z' \\ 1, & 1, & 0 \\ -1, & 0, & 1 \end{vmatrix} : \begin{vmatrix} x', & y', & z' \\ 1, & 0, & 1 \\ 1, & 1, & 0 \end{vmatrix} \begin{vmatrix} x', & y', & z' \\ -1, & 0, & 1 \\ 0, & 1, & 1 \end{vmatrix} = \frac{x' + y' - z'}{x' - y' - z'}$$

as the anharmonic of the *pencil* which has the point (x', y', z') for vertex, and of which the rays pass in order through A'', B'', C'', B' wherever in the plane that vertex may be placed; at least if it be not upon the line $C''A''B'$ in which case $h' = \frac{0}{0}$, because a suppressed factor then vanishes, since

$$\begin{vmatrix} x', & y', & z' \\ 1, & 1, & 0 \\ -1, & 0, & 1 \end{vmatrix} = \begin{vmatrix} x', & y', & z' \\ 0, & 1, & 1 \\ -1, & 0, & 1 \end{vmatrix} = x' - y' + z'.$$

For the point R, we have (by article 255) the coordinates, (before their division by β,),

$$x' = \alpha(\beta-\gamma)^2 - \gamma(\alpha+\beta)(\beta+\gamma), \quad y' = \beta(\gamma-\alpha)^2,$$
$$z' = \gamma(\alpha-\beta)^2 - \alpha(\alpha+\beta)(\beta+\gamma);$$

for this point, therefore

$$x' - z' = \beta(\alpha - \gamma)(\alpha + 2\beta + \gamma);$$

$$x' - z' + y' = 2\beta(\alpha - \gamma)(\alpha + \beta), \quad x' - z' - y' = 2\beta(\alpha - \gamma)(\beta + \gamma);$$

the anharmonic h' of the group A'', B'', C'', R on (U) has therefore this very simple value, (cf. article 241)

$$h' = ((A''B''C''R)) = \frac{\alpha + \beta}{\beta + \gamma} = (C_1 C'AB).(A_1 BCA').$$

For the case of two homothetic conics (articles 177, &c.), this becomes

$$h' = \frac{AB}{C_1 B} \cdot \frac{BA_1}{BC};$$

and when we pass to the *limit*, of a bisecting circle (U), and an *inscribed* circle (V), we have then simply $h' = \frac{c}{a}$, where c and a are again *sides* of the triangle, and are therefore proportional to the chords $A''B''$, $B''C''$. The two other chords $C''R$, RA'' are therefore in this case *equally long*, as was to be expected; and *because* $h' > 0$, *the two points* R and B'' *are situated at one common side of the chord* $C''A''$, as in my Figure 29, and (with other letters) in your figure also.

258. If we took that one of the three enscribed circles, which touches the *mean* side CA *itself*, but the two *extreme* sides *prolonged*, then the *new* point of contact, say (B_1) of the side CA with this new circle, should *still* be joined to the point R of Figure 29, in order that the new joining line $(B_1)R$, might furnish the new point of contact (P), of one circle with the other; as is exhibited below, in the new Figure, numbered 30*, which will serve to illustrate many other things. And this was to be expected from the analysis of the last article; because each of the two new quotients, $\frac{BA}{B(C_1)} \frac{B(A_1)}{BC}$, like the former quotient corresponding, is positive; and therefore the *new anharmonic* (h'), which is their product *is also positive*; so that the point in which the chord $(P)(B_1)$ of the new (or exscribed) circle, meets the bisecting circle again, and which by the theory *bisects one or other* of the two arcs of the last circle which have $A''C''$ for their common chord, is thus proved to be *at the same side as before* of that chord; and consequently to be (as in Figure 30) the former point R *itself*, and *not* the point diametrically *opposite* thereto.

259. But whereas, if we had in Figure 29, as we have done in Figure 30, bisected the "*little arc*" $A''A'''$ in a point Q, and joined that point to A_1, the joining line QA_1 (prolonged) would have passed through the sought point of contact P, and so have sufficed (instead of the line RB_1) to determine that sought point of contact of the bisecting circle K with the *inscribed* circle I; when, on the contrary, we seek to determine similarly the point (P), in which the same bisecting circle K is touched by that *exscribed* circle (I), represented in Figure 30, which touches, as above, the two extreme sides of the triangle prolonged, we are to join the point of contact (A_1) *not now* to the point Q *itself*, *but* to the diametrically *opposite* point (Q) upon the circle K, or to what may be called by contrast the bisecting point of the "*big arc*"

* A systematic description of this Figure 30 is given in articles 280, &c.

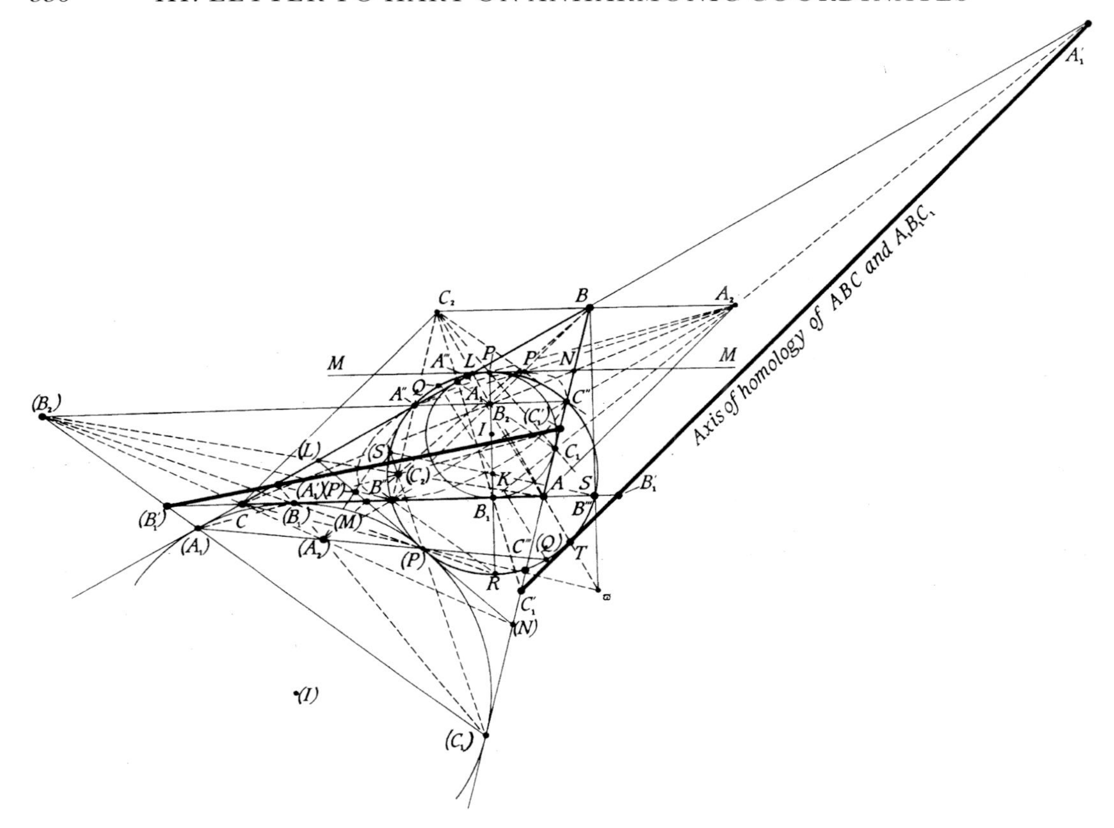

Fig. 30 [See article 280 for a description.]

$A''A'''$, in order that the joining line $(A_1)(Q)$ may pass through the new point (P), and so determine it on either circle. Yet here again, the geometrical result is in exact conformity with the algebraical analysis. For if, on the plan of article 257, we had sought the value of the anharmonic h of the group $C''A''B''Q$, upon the *conic* (U), *defining* Q to be the point in which the right line $A_1 P$ meets that conic again, as R was (in article 255) *defined* to be the point of second intersection of the line $B_1 P$ with that conic; a mere cyclical exchange of letters shows that we should have found, *generally*, and *without any distinction of cases*, the formula

$$h = ((C''A''B''Q)) = \frac{\gamma + \alpha}{\alpha + \beta} = (B_1 B'CA).(C_1 ABC');$$

or, for two homothetic conics, (the line $A'B'C'$ going off to infinity,)

$$h = \frac{AC}{AB_1} \cdot \frac{AC_1}{AB}.$$

Now if, as a limiting case, we suppose that the conics become *circles*, and that the one which touches the sides is *in*scribed, as in Figure 29, – to which, however, you need not take the trouble of referring, since all its parts have been incorporated, although upon a smaller scale, with the later and more complex Figure 30, – we have the values

$$\frac{AC}{AB_1} = \frac{b}{s-a}, \quad \frac{AC_1}{AB} = \frac{s-a}{c},$$

and therefore

$$h = +\frac{b}{c}$$

with the usual significations of a, b, c, s in trigonometry; the anharmonic of the group is therefore in this case *positive*, and the point Q lies *at the* same side of the chord $B''C''$ as the point A'': so that it bisects, according to your remark the *little* arc $A''A'''$ for that Q bisects *either that* arc, or the *big* one with the same chord, results from the numerical (or absolute) value of the anharmonic h, in virtue of which the *distances* of this point Q from C'' and B'' are *equal*; because the distances of A'' from those two points are proportional to the sides b and c, and the anharmonic of a *circular group* is $= \pm$ a product of quotients of *chords*. *But* when we suppose that the conic (V) becomes the *ex*scribed circle (I) of Figure 30, and when as in that Figure, we enclose, for distinction, in parentheses, the symbols of the *altered points*; then although we have *still* an expression of precisely the *same general form* for the anharmonic of the group, namely,

$$(h) = \frac{AC}{A(B_1)} \cdot \frac{A(C_1)}{AB},$$

yet we have now the new values of these two factors

$$\frac{AC}{A(B_1)} = \frac{b}{s-c}, \quad \frac{A(C_1)}{AB} = -\frac{s-c}{c},$$

which gives the *negative value* for their product, namely

$$h = -\frac{b}{c}.$$

The *distances* of the *new point* (Q) like those of the *old* point Q, from the given points C'' and B'', are therefore *equal*, so that this new point, like the old one must be situated at *one or other end* of that *diameter* through K which *bisects the chord* $B''C''$; but because the *anharmonic* of the group is now *negative*, the points (Q) and A'' must lie at *opposite sides* of that chord, and therefore (Q) must be, as in Figure 30, *diametrically opposite* to Q.

260. Another cyclical interchange of letters would give the *general* formula,

$$h'' = ((B''C''A''S)) = \frac{\beta + \gamma}{\gamma + \alpha} = (A_1 A'BC).(B_1 CAB');$$

or for two homothetic conics,

$$h'' = \frac{CB}{CA_1} \cdot \frac{CB_1}{CA};$$

S being defined to be the second intersection of the right line $C_1 P$ with the conic (U). For the case of the bisecting and *in*scribed circles

$$\frac{CB}{CA_1} = \frac{a}{s-c}, \qquad \frac{CB_1}{CA} = \frac{s-c}{b}, \qquad h = +\frac{a}{b};$$

the point S is therefore in this case not only *equally* distant from the ends of the chord $A''B''$, but also on the *same side* of that chord as the point C''; it therefore *bisects* what may be called the *little arc* $C''C'''$, so that it *happens to coincide* in Figures 29 and 30, with the point B'''; because (cf. article 239) in constructing both those figures the sides a, b, c were taken as proportional to the numbers 4, 3, 2, which has led to this and a few other undesigned peculiarities, such as the parallelism of the common tangent LMN at P to the side CA, not at all affecting however the general argument. But when we take the case of the *ex*scribed circle in Figure 30, and use parentheses for distinction as before, we have then the values,

$$\frac{CB}{C(A_1)} = \frac{-a}{s-a}, \qquad \frac{C(B_1)}{CA} = \frac{s-a}{b}, \qquad (h'') = -\frac{a}{b};$$

the right line $(C_1)P$ therefore now meets the bisecting circle K, *not* in the old point S, *but* in that other point (S) which is (as in Figure 30) diametrically *opposite* thereto, and may be said to *bisect the big arc* $C''C'''$. – Analogous remarks would of course apply, to the two other exscribed circles.

261. Comparing these results, it is not difficult to abstract from them the following ***RULE***, which meets *all the cases* of the *circles*, and is therefore *a little more complete* than the one that you *stated* to me: "According as the point of contact of the tangent circle *with a side* is *on that side itself*, or on the side *prolonged*, the right line connecting it with the point of contact of the *two circles* bisects *that one of the two arcs* of the bisecting circle, having their common chord coincident in position with the side, which is *remote from*, or is *near to*, the *vertex* opposite to that side of the triangle". – I daresay *you know* this completed *RULE* quite well, but I worked it out for myself as above.

262. Returning for a moment to the analysis of article 257, and comparing it with that of article 241, we see that the quantities which have been lately called h, h', h'' may be interpreted as anharmonics of *pencils*, as follows

$$h = (A''.B''CC''Q); \qquad h' = (B''.C''AA''R); \qquad h'' = (C''.A''BB''S);$$

where Q, R, S are still the second intersections of the sought lines, PA_1, PB_1, PC_1, with the conic (U). These three points of intersection can therefore be found *directly*, as the second intersections of the same conic with *known lines*, drawn from the points A'', B'', C'' of that curve. For example, the second formula gives immediately and without any original reference to *circles*

$$\frac{\sin A''B''R}{\sin RB''C''} = h' \frac{\sin A''B''A}{\sin AB''C''}$$

with

$$h' = \frac{\alpha+\beta}{\beta+\gamma} = (C_1C'AB).(A_1BCA'),$$

as before; and thus the *direction* of the *line* $B''R$, and thence the *position* of the *point* R, can *generally* be determined; after which by drawing the line $B_1 R$, the point P can be found, as the second intersection of this *latter* line with either of the two conics. When they are made *homothetic*, the points A'', B'', C'' *bisect* the sides BC, CA, AB; and (still abstaining from *circles*) we have, by mere *triangles*

$$\frac{\sin A''B''A}{\sin AB''C''} = -\frac{\sin A}{\sin C} = -\frac{a}{c};$$

also, as in article 257, we have then

$$h' = \frac{AB}{C_1 B} \cdot \frac{BA_1}{BC};$$

hence

$$\sin A''B''R : \sin C''B''R = \pm \overline{BA_1} : \overline{BC_1},$$

$\overline{BA_1}$ and $\overline{BC_1}$ denoting the *lengths* of the two *tangents* to the conic (V) from the point B, and the upper or the lower sign being taken, according as h' is positive or negative. For the *circles*, the tangents are of course *equally long*, and we have again *your graphic rule* (extended to meet all *cases*), under the form

$$\sin A''B''R = \pm \sin C''B''R;$$

but I do not see how, *by projective properties*, we could pass from this *last* equation, to the scarcely less simple *proportion*, written just above, between the sines of the corresponding *angles*, and the *lengths* of *tangents to a conic*. Perhaps, however, an *expert* geometer – such as I do not at all pretend to be – might see this at a glance.

263. On a little reflection, however, I *do* see how to interpret *each* of the two last formulæ geometrically, in such a manner as to show that they *can* be *projectively* extended, and interchanged. In fact, they *both* express, attention being paid to signs, that the chord $B''R$ of the conic (U) is *parallel* to the chord $C_1 A_1$ of the conic (V), when those two conics are *homothetic*; or more generally, and with more evident projectivity of relation, that "*these two chords* ($B''R$ *and* $C_1 A_1$) *meet upon that other common chord of the two conics, which contains those two of their intersection, that do not coalesce in the point of contact* P". Accordingly, the right line through the two points $A_1 = (0, \gamma, \beta)$, and $C_1 = (\beta, \alpha, 0)$, meets the unit-line in the point $(\beta + \gamma, \alpha - \gamma, -\alpha - \beta)$; the line joining this latter point to $B'' = (1, 0, 1)$ has for equation,

$$(\alpha - \gamma)(x - z) = (\alpha + 2\beta + \gamma)\, y;$$

and this last equation is satisfied (article 257), by the coordinates x', y', z' of the point R, determined in article 255.

264. The *parallelism* of $B''R$ to $A_1 C_1$, in Figures 29 and 30, did not occur to me when I drew those figures; and it is perhaps curious that I saw it, half an hour ago, by reasoning about *conics*, many minutes before I was able to verify it by any special argument from *circles*. Such verification, however, should certainly have offered itself sooner to any of your junior freshman friends. (I was of course a junior freshman in my time; and was never beaten in geometry – nor as it chanced, in any other subject in my division; but things have so much advanced generally in our University since *then*, that I might well meet with a different fate, in every department, if I could go in again *now*, as I have sometimes half wished to do, in order to be lectured by Salmon and Townsend and others.)

265. The *verification* by circles is, notwithstanding, *very easy*. Whether in Figure 29, or in Figure 30, the chord $B''R$ of the bisecting circle is the *external bisector* of the angle $A''B''C''$; and is therefore *parallel* to the external bisector of the angle B of the *triangle* ABC; to which latter bisector the chord $A_1 C_1$ of the inscribed circle is *also* parallel.

266. The recognition of this *parallelism of the two chords* as extending *to all cases* of the *two circles*, (in fact, it exists, as we have seen for the more general case of two *homothetic conics*, with a *projective extension* above indicated) throws some fresh light upon its construction. Thus $B''R$ being parallel, in Figure 30, to the chord $(A_1)(C_1)$ of the exscribed circle (I), as well as to the chord $A_1 C_1$ of the inscribed circle I, the lines $B_1 P$ and $(B_1)(P)$ both pass as before, through *some* point R of the bisecting circle K. But the chord $(A_1)(B_1)$ of the exscribed circle is *not* parallel, but *perpendicular*, to the corresponding chord $A_1 B_1$ of the inscribed; while then we might have found the point S upon the circle K, (accidentally coincident, as remarked in article 260 in the recent figures, with B''',) through which $C_1 P$ passes, by drawing $C''S$ parallel to the *latter* chord $A_1 B_1$, we must (on the same plan) determine the corresponding point (S) on the new line $(C_1)(P)$, by drawing from C'' a line *parallel* to the *former* chord $(A_1)(B_1)$ and therefore at right angles to $C''S$. In like manner, the chords $A''Q$ and $A''(Q)$ of the bisecting circle are indeed perpendicular *to each other*, but they are respectively *parallel* to the chords $B_1 C_1$ and $(B_1)(C_1)$, of the inscribed and exscribed circles.

267. Before I had thought of this, or of any other *rule* for *distinguishing* between the bisecting points of the *little* and *big* arcs in your figure; and for extending such distinction to the more general problem of the *conics*, so as to determine for example, *by some geometrical construction* "***which*** *intersection of the conic (U) with the polar of the point B′* (articles 251, 254, and 256) *was to be taken as the point above called R*": I reasoned in the following way. *Usually*, when any geometrical investigation conducts to the consideration of the intersections of a right line with a circle, or other conic, those *two* intersections enter *jointly* and *symmetrically* into the result, and are indeed *inseparably combined*, so that no reason can be assigned for *preferring*, *generally*, *one* of them to the other. *Calculation* conducts, in almost every mode of treating such a question by *algebra*, to a *quadratic equation*, of which the *roots* cannot usually be separately and *rationally expressed*, and of which therefore we must, *in general*, retain the *system*. (In fact, if I remember rightly, it is on this principle that Chasles founds, in his *Géométrie Supérieure*,* his theory of *geometrical imaginaries*.) Whenever, therefore, we *know*, *aliunde*, that *one point of intersection* has any *geometrical prerogative* above the *other*, so that it *can* be treated *separately*, we may be sure *algebraically* that the *quadratic equation breaks up into linear and rational factors*; and geometrically, that *some linear construction exists*, whereby, *the particular intersection required* can be determinately found, *without the aid of the conic*.

268. I therefore looked out for some mode of otherwise determining – or of dispensing with – the auxiliary point, by me called R, of your graphic construction; and in short, I sought

* [Michel Chasles, 1793–1880, *Traité géométrie superieure*, Paris, 1852.]

to find *some other auxiliary point, on the chord PB_1 of the inscribed circle,* or of the conic (V), (distinct from B_1 and from R,) which *could itself be graphically constructed,* or determined, *without my being obliged to use,* for that purpose, the bisecting circle, or the conic (U), at all. Nevertheless, as I allowed myself the fuller use of *anharmonic coordinates,* as *analytic instruments* in this inquiry, I collected into one view the coordinates which had been *already calculated,* of all the points *previously considered* upon that chord PB_1 without excluding the point R itself. These coordinates were the following

for P, $x = \alpha(\beta-\gamma)^2$, $y = \beta(\gamma-\alpha)^2$, $z = \gamma(\alpha-\beta)^2$; (article 172)
for B_1, $x_1 = \gamma$, $y_1 = 0$, $z_1 = \alpha$; (article 248)
for R, $x' = \beta(\alpha-\gamma) - \gamma(3\alpha+\gamma)$, $y' = (\alpha-\gamma)^2$, $z' = \beta(\gamma-\alpha) - \alpha(\alpha+3\gamma)$; (article 255)

and for the point in which the line PB_1 met the unit-line,

$$X_1 = \gamma(3\alpha-\gamma) - \beta(\alpha+\gamma), \quad Y_1 = \gamma^2 - \alpha^2, \quad Z_1 = \alpha(\alpha-3\gamma) + \beta(\alpha+\gamma). \quad \text{(article 249)}$$

If then, any one set of these coordinates be denoted (as above) by x', y', z', and any other set by x'', y'', z'', I was to form some such combination as the following

$$(x''', y''', z''') = t(x', y', z') + u(x'', y'', z'');$$

for the *new point*

$$P' = (x''', y''', z'''),$$

thus determined, would like the other, be *on the sought line* PB_1, and might happen to be simply *constructible,* if the *coefficients* t and u were judiciously chosen, and if it should be possible to *divide* the resulting coordinates by any *common factor.*

269. When this conception had been formed, there was no difficulty in putting it into immediate execution. Taking simply the sums and the differences of the coordinates of the two points last mentioned, I had, at once,

$$X_1 + x' = -2\gamma(\beta+\gamma), \quad Y_1 + y' = 2\gamma(\gamma-\alpha), \quad Z_1 + z' = 2\gamma(\beta-3\alpha);$$
$$X_1 - x' = +2\alpha(3\gamma-\beta), \quad Y_1 - y' = 2\alpha(\gamma-\alpha), \quad Z_1 - z' = 2\alpha(\alpha+\beta);$$

whence there were seen to be these *two new points* on the line PB_1, *each* having its coordinates expressible *linearly* in terms of α, β, γ:

$$P' = \left(\frac{X_1+x'}{-2\gamma}, \frac{Y_1+y'}{-2\gamma}, \frac{Z_1+z'}{-2\gamma}\right) = (\beta+\gamma, \alpha-\gamma, 3\alpha-\beta);$$

$$P'' = \left(\frac{X_1-x'}{2\alpha}, \frac{Y_1-y'}{2\alpha}, \frac{Z_1-z'}{2\alpha}\right) = (3\gamma-\beta, \gamma-\alpha, \alpha+\beta).$$

But the presence of the *numerical coefficient,* 3, made these expressions *not yet simple enough* for my purpose. But on taking the semidifferences of the two *new* sets of coordinates, or employing the formula,

$$2(P''') = (P'') - (P'),$$

I was led to determine a *third new point,* P''', on the line PB_1, namely,

$$P''' = (\gamma - \beta, \gamma - \alpha, \beta - \alpha),$$

of which the expression was sufficiently simple, to induce me to try to *construct* it.

270. As few people ever arrive at a simple thing in a simple way, I need not fear to confess that I assisted myself here by suggestions drawn from sources which might seem remote. Even *quaternions* were called into play, and really with very good effect: but it would interfere too much with what I hope has been the *homogeneity* of the present Letter, – whatever may have been its *discursiveness*, – if I were to say anything about quaternions at present. Nor shall I at this moment speak of a little investigation which I made, of the value of the anharmonic, $((A_1 B_1 C_1 P))$ of the *group* thus denoted *upon the conic* (V), or on the inscribed circle in Figure 29, though it led me to some rather pretty results, and supplied me with some suggestions. But I may at once mention that I soon came to suspect, that the *last auxiliary point* P''' of article 269, which I shall now denote by B_2, writing therefore

$$B_2 = (\gamma - \beta, \gamma - \alpha, \beta - \alpha),$$

(since we shall not have any further reason for the point B_2 of articles 248, 249, and 250) was the *intersection of the chords* $C''A''$ and $C_1 A_1$, *of the conics* (U) and (V).

271. And in fact I found that this suspicion was correct. For the *coordinates* of the extremities of these two chords being given by the expressions,

$$A'' = (0, 1, 1), \quad C'' = (1, 1, 0), \quad A_1 = (0, \gamma, \beta), \quad C_1 = (\beta, \alpha, 0),$$

and *equations* of the same chords are, for $A''C$,

$$x - y + z = 0,$$

and for $A_1 C_1$,

$$\alpha x - \beta y + \gamma z = 0;$$

and these are evidently both satisfied by the recently written coordinates of B_2. Hence, then, the following ***Construction*** results, *for graphically determining the point P*, of contact of the two conics, *without employing the conic* (U), or the bisecting circle, at all: "Find the point B_2, in which the chord $A_1 C_1$ of the conic (V) is intersected by the right line $A''C''$; the right line $B_1 B_2$ will then have the sought point P for its second intersection with that conic." It will be remembered that although the *line* $A''C''$, here used, has been *made*, by the construction in article 170, &c., a *chord* of the *other* conic (U), – as when we drew a circle or conic through the middle points of the sides of a given triangle, – yet it is *given* or known, *before* this circle K, or conic (U), is constructed; which justifies my statement, that in the process of the present article, *we determine the point P on one conic, without any assistance from the other.*

272. From the nature of the analysis, the point B_1, although it happens in Figure 29 when the point of contact of the *mean side CA* of the given triangle with the inscribed circle, has *no general prerogative* above the two other points of contact, A_1 and C_1. We may therefore enunciate this Theorem, or set of Theorems:

"If a triangle ABC be touched in the points A_1, B_1, C_1, by a conic (V), and if its sides be cut in the points A', B', C' by a right line; then, taking on those sides the harmonically conjugate points A'', B'', C'', and forming the two inscribed triangles $A_1B_1C_1$ and $A''B''C''$, if we denote the intersections of their corresponding sides as follows,

$$A_2 = B_1C_1\dot{}B''C'', \quad B_2 = C_1A_1\dot{}C''A'', \quad C_2 = A_1B_1\dot{}A''B'',$$

we shall have concurrences and contacts as below:

Ist the three right lines A_1A_2, B_1B_2, C_1C_2, will concur, in one common point P;
IInd this point will be *upon the conic* (V);
IIIrd it will be also *upon that other conic* (U), which passes through the three points A'', B'', C'', and through the two points in which the former conic (V) is intersected by the transversal $A'B'C'$; and
IVth the two conics, (V) and (U), will *touch* at this point P.''

273. Accordingly, among the many new points &c. which have been introduced in Figure 30, you may see A_2, B_2, C_2, as the respective intersections of the three pairs of lines, B_1C_1 and $B''C''$, C_1A_1 and $C''A''$, A_1B_1 and $A''B''$; and may observe the concurrence of the three lines A_1A_2, B_1B_2, C_1C_2 in the point P of the *in*scribed circle. For the *ex*scribed circle, in the same figure, I have been content to determine *one** auxiliary point, namely (A_2), as the intersection of the old line $B''C''$ with the new chord $(B_1)(C_1)$; but you may see that the new line $(A_1)(A_2)$ passes through the new point of contact (P), and would suffice to determine that point without any use of the bisecting circle.

274. In the same complex figure, you may observe that the three right lines B_2C_2, C_2A_2, A_2B_2, pass respectively through the three points A, B, C; so that *the new triangle* $A_2B_2C_2$ may be said to be *excribed to the given triangle ABC*. Nor is this result accidental. For while we have, by article 270,

$$B_2 = (\gamma - \beta, \gamma - \alpha, \beta - \alpha),$$

we have, in like manner,

$$C_2 = (\gamma - \beta, \alpha - \gamma, \alpha - \beta).$$

Since the coordinates of C_2 must satisfy the two equations, (cf. article 271,)

$$y - z + x = 0, \quad \text{and} \quad \beta y - \gamma z + \alpha x = 0;$$

so that it is not sufficient, in passing from B_2 to C_2, to change cyclically α, β, γ to β, γ, α, but we must also change x, y, z to y, z, x, or must write *first* the coordinate $\gamma - \beta$, which is obtained from the *last* coordinate $\beta - \alpha$ of B_2: and therefore

$$(B_2) + (C_2) = (2\gamma - 2\beta, 0, 0) = 2(\gamma - \beta)(1, 0, 0) = 2(\gamma - \beta)(A).$$

A little more simply, if we denote by α', β', γ' any quantities which satisfy the two equations,

$$\alpha' + \beta' + \gamma' = 0, \quad \alpha\alpha' + \beta\beta' + \gamma\gamma' = 0,$$

for example, the quantities,

* I have, however, since inserted the points (B_2) and (C_2), and drawn the lines $(B_1)(B_2)$ and $(C_1)(C_2)$, which are seen to pass likewise, through (P).

$$\alpha' = \beta - \gamma, \quad \beta' = \gamma - \alpha, \quad \gamma' = \alpha - \beta,$$

we may write

$$A_2 = (-\alpha', \beta', \gamma'), \quad B_2 = (\alpha', -\beta', \gamma'), \quad C_2 = (\alpha', \beta', -\gamma');$$

and under these forms, the combinations,

$$(B_2) + (C_2) = (2\alpha', 0, 0) = 2\alpha'(A),$$

$$(C_2) + (A_2) = (0, 2\beta', 0) = 2\beta'(B),$$

$$(A_2) + (B_2) = (0, 0, 2\gamma') = 2\gamma'(C),$$

exhibit at once the *three collineations,*

$$AB_2C_2, \quad BC_2A_2, \quad CA_2B_2,$$

which are, as above, exemplified in Figure 30.

275. The recent quantities α', β', γ' have interesting geometrical significations, which will lead us to some additional theorems, illustrated by the same figure. In fact, because their sum is zero, they are the coordinates of a point D upon the unit-line; and on account of the other linear relations which they have been made to satisfy, this point D must also be upon the right line which has for symbol $[\alpha, \beta, \gamma]$, or for equation

$$\alpha x + \beta y + \gamma z = 0.$$

Now this last line admits of being simply constructed. For in general the right line $[l, m, n]$ has for its *pole,* with respect to the unit-triangle, the point (l^{-1}, m^{-1}, n^{-1}); in the same sense as that in which the *unit-point* O has been called, in former articles, the *pole* of the *unit-line with respect to the same unit-triangle*; and in which the *mean point,* or *centre of gravity of any triangle,* is *the pole of the line at infinity.* (I *think* that this phraseology is a *received* one, but I forget, just now where to *refer* for it.) Now the pole $(\alpha^{-1}, \beta^{-1}, \gamma^{-1})$, of the recent line $[\alpha, \beta, \gamma]$, is precisely that point in which it was remarked in article 248 that the three right lines AA_1, BB_1, CC_1 concur. *The line* $[\alpha, \beta, \gamma]$ *is through the axis of homology of the two triangles* ABC *and* $A_1B_1C_1$; since *its pole* (with respect to *one* of those two triangles, and therefore also with respect to the *other*) *is their centre* $(\alpha^{-1}, \beta^{-1}, \gamma^{-1})$ *of homology.* It is therefore the line $A_1'B_1'C_1'$, (one of the two that are strongly marked in Figure 30,) if the points A_1', &c. be the *intersections of the corresponding sides* of the two triangles; namely,

$$A_1' = BC^{\cdot}B_1C_1 = (0, \gamma, -\beta),$$

$$B_1' = CA^{\cdot}C_1A_1 = (-\gamma, 0, \alpha),$$

$$C_1' = AB^{\cdot}A_1B_1 = (\beta, -\alpha, 0),$$

which in fact are seen to satisfy the equation

$$\alpha x + \beta y + \gamma z = 0,$$

and have of course harmonic properties, not necessary to be here written down. The new auxiliary point

$$D = (\alpha', \beta', \gamma'),$$

is therefore to be conceived to be, in Figure 30, *the point at infinity on the right line* $A_1' B_1' C_1'$; and in general, for the case of the *two homothetic conics,* it is the point at infinity on the line similarly determined. More generally still, since the unit-point is the *centre,* and the unit-line is the *axis,* of homology of the two triangles ABC and $A''B''C''$, we may say that "the point D *is the intersection of the two axes of homology, of the two inscribed triangles* $A''B''C''$ and $A_1 B_1 C_1$, *with the given triangle* ABC".

276. If we now resume the expressions of article 274 for the points A_2, B_2, C_2, and compare them with the recent expression for D, we find the symbolical relations,

$$\begin{aligned}(D) &= (A_2) + (B_2) + (C_2)\\ &= (A_2) + 2\alpha'(A) = (B_2) + 2\beta'(B) = (C_2) + 2\gamma'(C);\end{aligned}$$

whence follows this new Theorem, namely, that "the *three lines* AA_2, BB_2, CC_2, *concur in the point* D". (It was through *Quaternions,* cf. article 270, that I first perceived this Theorem, and I assure you, as the result of a very easy calculation, with which however I shall not at present trouble you.) When, therefore, – as for the case of the circles, or *homothetic* conics, – the points A'', B'', C'' are made to *bisect* the sides of the given triangle, "*these three lines,* AA_2, &c., *concur at infinity,* or *are parallel to the axis of homology* $A_1' B_1' C_1'$ *of the two triangles* ABC *and* $A_1 B_1 C_1$." Accordingly you may *see* the *parallelism of the four lines,* AA_2, BB_2, CC_2, $A_1' B_1' C_2'$, in Figure 30, for the case of the *in*scribed circle. For the *ex*scribed circle in that figure, I have drawn the new axis of homology $(A_1')(B_1')(C_1')$, and marked it strongly, like the other, but have been content to exhibit *one** of its parallels, namely the line $A(A_2)$, for this case. It ought to be observed that when A_2 is considered as the intersection of the lines AD, $B_1 C_1$, or (A_2) as the corresponding intersection of the lines $A(D)$, $(B_1)(C_1)$, (AD $A(D)$ being in the Figure, the *parallels* through A to the strong black lines,) we have only to draw the lines $A_1 A_2$, $(A_1)(A_2)$, as before, in order to obtain, by their second intersections with the circles I and (I), the points of contact P and (P); *which points can thus be graphically determined,* at least for the case of circles or *homothetic conics, without bisecting the sides of the given triangle at all*: such *conceived bisections* being now *replaced* by throwing off D and (D) to *infinity.*

277. Since the two triangles ABC and $A_2 B_2 C_2$ are *homologic,* and have the point D, or $(\alpha', \beta', \gamma')$, for their *centre* of homology, the right *line* of which the anharmonic *coordinates* are the respective *reciprocals* of the coordinates of that point; that is to say, the line of which the symbol is, $[\alpha'^{-1}, \beta'^{-1}, \gamma'^{-1}]$, or $[(\beta-\gamma)^{-1}, (\gamma-\alpha)^{-1}, (\alpha-\beta)^{-1}]$. But we had, in article 176, the proportion

$$\lambda : \mu : \nu = (\beta-\gamma)^{-1} : (\gamma-\alpha)^{-1} : (\alpha-\beta)^{-1};$$

in which λ, μ, ν were the *coordinates of the tangent at P,* to either of the two conics. We arrive then at this other new Theorem: "*The common tangent* $[\lambda, \mu, \nu]$ *to the two conics* (U) *and V, at the point P, is the axis of homology of the auxiliary triangle* $A_2 B_2 C_2$ *with the given triangle* ABC".

* Since this article was written, I have inserted the points (B_2), (C_2), and drawn the parallels $B(B_2)$, $C(C_2)$; article 260 being called into play.

278. This last theorem may be thus stated: "*the intersections L, M, N, of corresponding sides of the two triangles, ABC and $A_2B_2C_2$, are points upon the common tangent at P.*" It was only just now that I perceived the *theorem* thus enunciated: but I remember that when I was constructing, a few days ago, that complicated Figure 30, or a sketch of it on another sheet of paper, I was struck by the *parallelism* of C_2A_2 to CA, which answers to their meeting at infinity in M and by the tangent at P appearing to *concur*, even more exactly than in Figure 30 of this Letter, in one point N, with the lines AB and A_2B_2; while an analogous concurrence seemed to take place at L: but I then supposed that the concurrences might be *accidental*, whereas I now see that they were *necessary*. For I am now entitled to assert that the tangent at P is the line LMN, where

$$L = BC^{\cdot}B_2C_2, \quad M = CA^{\cdot}C_2A_2, \quad N = AB^{\cdot}A_2B_2;$$

or by article 274,

$$L = (0, \beta', -\gamma'), \quad M = (-\alpha', 0, \gamma'), \quad N = (\alpha', -\beta', 0);$$

which indeed may be inferred from the equation

$$\lambda x + \mu y + \nu z = 0$$

of that *common tangent* to the conics, because by articles 171 and 274, we have the values

$$\lambda = \beta'\gamma', \quad \mu = \gamma'\alpha', \quad \nu = \alpha'\beta'.$$

279. Quite enough may have been already said, about that Figure 30; yet, as I have yielded to the temptation of inserting some new points and lines therein, even since the two or three last articles were written, I may as well give here a systematic account of the Construction of that Diagram, before I put it out of my power, by posting the last sheets, to add or alter anything therein.

Descriptions of Figure 30, Article 258
(With a Recapitulation of several former Theorems and Constructions.)

280. ABC is a triangle, supposed to be given, and of which the sides a, b, c, have been here assumed proportional to the numbers 4, 3, 2; so that they may be equated to numbers, and we may write $a = 4$, $b = 3$, $c = 2$. Those three sides, BC, CA, AB, are conceived to be cut, by the transversal at infinity, in the (unexhibited) points A', B', C'.

The harmonic conjugates to those three points are marked as the points A'', B'', C'', which of course bisect the sides, and are joined by three right lines, forming an inscribed triangle $A''B''C''$.

From the three corners, A, B, C, of the given triangle, perpendiculars are let fall upon the sides a, b, c, meeting those sides in the points A''', B''', C''', and each other in a point.

A circle, called the *bisecting circle*, with centre K, (called also the circle K, and representing the conic (U) of article 170) is described so as to pass through the three points A'', B'', C''; and therefore also, (in virtue of the elementary theorem mentioned in a note of yours,) through the three other points, A''', B''', C'''.

The particular lengths assumed for the sides give for the segments made by the three perpendiculars, the following numerical values, marked as positive when they are measured in the same directions as the corresponding sides, measured from the corners first named for each, but as negative when they happen to have the respectively opposite directions:

$$AB''' = \frac{b^2 + c^2 - a^2}{2b} = -\frac{1}{2}; \quad BC''' = \frac{c^2 + a^2 - b^2}{2c} = +\frac{11}{4}; \quad CA''' = +\frac{21}{8};$$

$$CB''' = \frac{b^2 + a^2 - c^2}{2b} = +\frac{7}{2}; \quad AC''' = \frac{c^2 + b^2 - a^2}{2c} = -\frac{3}{4}; \quad BA''' = +\frac{11}{8}.$$

(In *quaternions*,

$$AB''' + B'''C = AC, \quad \text{or} \quad B'''A + CB''' = CA,$$

and not

$$AB''' + CB''' = AC, \quad \text{nor} = CA.$$

But it is *here* meant merely that

$$\frac{AB'''}{AC} = -\frac{1}{2b}, \text{ and that } \frac{CB'''}{CA} = +\frac{7}{2b} \text{\&c.; whence}$$

$$\frac{B'''A}{CA} = -\frac{1}{2b}, \text{ and } \frac{CB''' + B'''A}{CA} = \frac{7-1}{2b} = \frac{6}{2b} = 1,$$

as it should be.)

281. Continuing the description of the Figure, we have next an *inscribed circle*, with centre at I, (called also the circle I, and answering to the conic (V) of article 171,) which touches the sides of the triangle ABC at the points A_1, B_1, C_1; the segments formed by these three points, of contact being, $\left(\text{because } s = \frac{a+b+c}{2} = \frac{9}{2},\right)$

$$AB_1 = AC_1 = s - a = \frac{1}{2}; \quad BC_1 = BA_1 = s - b = \frac{3}{2}; \quad CA_1 = CB_1 = s - c = \frac{5}{2}.$$

And we have an *excribed circle*, with its centre at the point (I), (called also the circle (I), and being another representation of the conic (V),) which touches the same sides in the points (A_1), (B_1), (C_1); the segments thus formed being the following, with signs determined as above:

$$A(B_1) = s - c = +\frac{5}{2}; \quad B(C_1) = s = +\frac{9}{2}; \quad C(A_1) = a - s = -\frac{1}{2};$$

$$C(B_1) = s - a = +\frac{1}{2}; \quad A(C_1) = c - s = -\frac{5}{2}; \quad B(A_1) = s = +\frac{9}{2};$$

the algebraical *sign* of the *side b*, which is still touched *internally*, having changed in the passage from the one tangent circle to the other.

282. The sides of the triangle are cut by the two transversals, LMN, $(L)(M)(N)$, with the following ratios of segments (cf. articles 177, 178, 238, and 239):

$$\frac{LB}{CL} = \frac{a-b}{a-c} = +\frac{1}{2}; \quad \frac{MC}{AM} = \frac{b-c}{b-a} = -1; \quad \frac{NA}{BN} = \frac{c-a}{c-b} = +2;$$

$$\frac{(L)B}{C(L)} = \frac{a+b}{a-c} = +\frac{7}{2}; \quad \frac{(M)C}{A(M)} = \frac{b+c}{b+a} = +\frac{5}{7}; \quad \frac{(N)A}{B(N)} = \frac{c-a}{c+b} = -\frac{2}{5};$$

so that

$$BL = +\frac{4}{3}, \ CM = \pm\infty, \ AN = +\frac{4}{3}; \quad B(L) = +\frac{28}{9}, \ C(M) = +\frac{5}{4}, \ A(V) = -\frac{4}{3};$$

the first transversal thus happening to be parallel to the side *CA*, and to trisect each of the two others, with the lesser segments adjacent to *B*; while the second transversal cuts (cf. again article 178) the lesser of those two sides externally.

283. The transversal *LMN* touches (as by article 177 it ought to do) the bisecting circle *K*, in the point *P* where (by the theorem of article 169, of which you informed me) that circle touches the inscribed circle *I*; so that this point of contact *P* happens, in Figure 30, to be diametrically opposite on the inscribed circle, to the point B_1 of contact of that circle with the side *CA*. In like manner (as was to be expected from article 178), the transversal (L), (M), (N) touches the bisecting and the exscribed circles, in that common point (P) in which (by the same theorem of article 169) they touch one another. And this construction of mine is *sufficient* for the determination of those two points of contact; indeed, it would be sufficient to find this, by simple *ratios of segments* (article 282), *any one* of the three points *L*, *M*, *N*, and *any one* of the three other points (L), (M), (N), and then to draw from the *first* point so found, a *second tangent* to the *inscribed* circle, or from the *second* point a second tangent to the *exscribed* circle; the *side* of the triangle, on which the point is thus determined, being in each case a *first tangent.* But whereas *this* construction involves a *metric element,* (although a very simple one,) *graphic constructions* have subsequently been assigned, which require *no use of ratios*, and which will be described a little farther on.

284. In the same Figure 30, – of which, with copious recapitulation, it is thus proposed to give here a complete and systematic description, – the points *Q*, *R*, and *S* (which last happens to coincide with B''') bisect those three "*little arcs*" (article 259) of the circle *K*, which have the lines $A''A'''$, $B''B'''$, $C''C'''$ for their chords, and are respectively *remote* from the corners *A*, *B*, *C* of the given triangle; also (Q) and (S) are the points diametrically *opposite* to *Q* and *S* on the same circle *K*, so that they may be said to bisect those "*big arcs*" (article 259) of that circle, which have $A''A'''$ and $C''C'''$ for their chords. And we see that, in conformity with *your graphic construction* (article 240), *the three right lines* QA_1, RB_1, SC_1 *concur in the point of contact P* of the circles *K* and *I*; *any one* of them being thus sufficient to determine it. In like manner, with only the modification (no doubt well known to you) of using in certain cases the middle points of *big*, instead of those of *little* arcs, according to the *rule* assigned in article 261, the three other right lines, $(Q)(A_1)$, $R(B_1)$, $(S)(C_1)$, *concur* in the point of contact (P) of the circles *K* and (I); which point can therefore be determined by *any one* of those three lines.

285. Instead of thus distinguishing between *big* and *little* arcs, the *selection* of the points, *Q* and (Q), &c., may be made without *thinking* of those points as *bisectors*, by means of *another Rule*, or *Theorem*, in virtue of which the *chord* $B''R$ of the bisecting circle *K* is *parallel* (articles 263 and 266) at once to the *chord* C_1A_1 of the inscribed circle (I) as is exhibited in Figure 30.

The chords $A''Q$, $A''(Q)$, and $C''S$, $C''(S)$, of the circle K, are not actually drawn in that Figure; but it is easy to recognise, by the eye, that they are (as the theorem of article 263 requires, and as is remarked in article 266) respectively parallel to the chords B_1C_1, $(B_1)(C_1)$, and A_1B_1, $(A_1)(B_1)$, of the two other circles. And *any one* of the *three chords* $A''Q$, $B''R$, $C''S$ of the bisecting circle, thus drawn parallel to a known chord, B_1C_1 or C_1A_1, or A_1B_1, of the inscribed circle, would suffice to determine a point, Q or R or S, upon the former circle K, such that a right line drawn from it to a known point, A_1 or B_1 or C_1, of the latter circle I, would give the sought point P of contact of those two circles: with an analogous construction for the contact (P_1) of the bisecting and exscribed circles.

286. The same Figure 30, however, is designed to illustrate another, and a totally different process, for *graphically determining* those points of contact, P and (P); namely, by the use of certain *other auxiliary points*, A_2, &c., as follows, and without its being *necessary* to employ the *bisecting circle at all*. For if we denote, as in article 272, and in the Figure, by A_2, B_2, C_2 the intersections of the sides of the *inscribed and bisecting triangle* $A''B''C''$ with the corresponding sides of the *triangle of contacts* $A_1B_1C_1$, which *also* may be said to be *inscribed* in the given triangle ABC, then *the three lines* A_1A_2, B_1B_2, C_1C_2 *concur in the point* P; which point can thus be graphically determined, *upon the circle* I, *without any use of the circle* K, by *any one* of those three lines. In like manner, the figure exhibits the determination of three *other* auxiliary points, (A_2), (B_2), (C_2), as the intersections of corresponding sides of the two triangles $A''B''C''$ and $(A_1)(B_1)(C_1)$; and the *concurrence* of the *three lines* $(A_1)(A_2)$, $(B_1)(B_2)$, $(C_1)(C_2)$ in the point (P); which point can thus be found *on the exscribed circle*, by any one of these three last lines, without using the *bisecting circle*.

287. But, although that *circle* K has not been used, in any of the constructions described in the foregoing article, yet the *bisecting triangle* $A''B''C''$ has been employed in all of them. Even this *triangle* can, however, be dispensed with, and the points P and (P) on the circles I and (I) be *graphically* determined, *without bisecting the sides of the given triangle*, in the following way, which likewise is illustrated by the Figure. Considering only the *triangle of contacts* $A_1B_1C_1$, let its sides meet the corresponding points of the given triangle ABC in the points A_1', B_1', C_1'; those *three points* are necessarily *collinear*, by a well known property of any *conic* inscribed in a *triangle*, in virtue of which the three right lines AA_1, BB_1, CC_1 *concur*, in a *centre of homology* not marked in the figure; and the *axis of homology*, of the triangles ABC, $A_1B_1C_1$, is the line $A_1'B_1'C_1'$ thus determined, and strongly marked in the diagram. Then the auxiliary points A_2, B_2, C_2, upon the sides of $A_1B_1C_1$, can be found (in virtue of the theorem in article 276) by drawing the *parallels* AA_2, BB_2, CC_2 *to this axis* $A_1'B_1'C_1'$; after which the lines A_1A_2, B_1B_2, C_1C_2, or any one of them, will determine the point P as before. In like manner, $(A_1')(B_1')(C_1')$ with another strong black line in Figure 30, is the axis of homology of the triangles ABC and $(A_1)(B_1)(C_1)$; and the auxiliary points (A_2), (B_2), (C_2), upon the sides of this last triangle of contacts, can be found by drawing parallels to this last axis, from the corners of the given triangle ABC: after which the point (P) can be found, as before, upon the exscribed circle, by means of any one of the three lines $(A_1)(A_2)$, &c. – The near approach, in the Figure, to coincidence of the former axis $A_1'B_1'C_1'$ with the *tangent* to the circle K, at the second intersection T of that circle with the line AA''', is a purely accidental circumstance, arising from the assumed ratios of the sides a, b, c. The *latter axis* of homology,

$(A_1')(B_1')(C_1')$, as you see, does not even *approach* to contact, with any of the circles of the figure.

288. The same Diagram illustrates also a certain *graphic determination* of the *common tangent* at P to the two circles K and I, and of the corresponding tangent at (P) to the circles K and (I), whereby we may perhaps with advantage replace that earlier and *metric determination* (articles 177 and 178, or 282 and 283) of the position of those two tangents, which occurred to me very soon after you mentioned the theorem of contact between the circles: although I still think my *metric* method simple too, especially as it extended, with little modification, to homothetic and other conics; which, however, the *graphic* process can likewise easily be made to do. This later and graphic method of construction for the common tangent at P is (by article 278) the following: "Find the point L, as the intersection of the two right lines BC, B_2C_2; M as the intersection of CA, C_2A_2; and N as that of AB, A_2B_2; then these three points, L, M, N, are *collinear*, and the right line connecting them is the tangent sought." In short, this common tangent LMN at P is (article 277) the *axis of homology* of the *given* triangle ABC, and the *auxiliary* triangle $A_2B_2C_2$; the *centre* D, of homology of these two triangles, being (by articles 275 and 276) the *point at infinity on the axis of homology* $A_1'B_1'C_1'$ *of the other pair of triangles*, ABC, $A_1B_1C_1$. And in like manner the tangent $(L)(M)(N)$ at (P) is the axis of homology of the triangles ABC and $(A_2)(B_2)(C_2)$, of which the centre of homology (D) is the infinitely distant point on the axis of homology $(A_1')(B_1')(C_1')$, of the triangles ABC, $(A_1)(B_1)(C_1)$.

289. The only other points marked in the figure are the two new ones, P' and (P'), about which I have said nothing as yet: I must therefore mention what they are, and what suggested them to me. I consider, then, that so far as the *linear* element of all this construction was concerned, the triangles $A''B''C''$ and $A_1B_1C_1$ are really indistinguishable, neither having any prerogative over the other, at least when projective changes are allowed. They are simply *two triangles inscribed in the given triangle* ABC, and *having it for a common homologue.* Accordingly, *my auxiliary triangle* $A_2B_2C_2$ *is symmetrically related to them*; its *corners* having been originally determined (articles 272, and 286) as the *intersections of their corresponding sides*, so that it may be said to be *their common inscribed triangle*, as the *given* one was their *common exscribed.* But I have *proved*, – anharmonically, I admit, and not without calculation, – in article 272, with the help of article 271, that this common inscribed triangle $A_2B_2C_2$ is *homologous to one* of the two triangles, $A''B''C''$ and $A_1B_1C_1$ namely to the latter of them, since the lines A_1A_2, B_1B_2, C_1C_2 have been shown to *concur* in the point P; it must therefore, by the *symmetry* of relation above mentioned, be *homologous to the other* also, so that *the three new lines*, $A''A_2$, $B''B_2$, $C''C_2$ *must concur, in some new point* P', as the figure sufficiently exhibits, although the framework of it had been constructed, before it occurred to me to draw these and other additional lines. In like manner, *the two triangles* $A''B''C''$ and $(A_2)(B_2)(C_2)$ *are homologic*; the three lines $A''(A_2)$ &c., which connect their corresponding corners, *concurring* in *another new point* (P'), which is the *last* that I have inserted in the Figure.

290. A few words more, however, must be said, respecting these two new points, P' and (P'). You may observe that they fall very well, *in the Figure,* – which I congratulate Mr. Gill's

Engraver* on not having to take in charge, – *on the tangents* at P *and* (P). And this is *not accidental.* In fact, the former tangent, LMN, has been proved, – by anharmonic coordinates, I admit, but probably *you* can prove it otherwise, – in articles 277, and 278, to be (as mentioned in article 288) the *axis of homology* of the two triangles ABC and $A_2 B_2 C_2$. But these are *symmetrically related,* as we have seen, to the two triangles $A''B''C''$ and $A_1 B_1 C_1$; one being their common exscribed, and the other their common inscribed, and each being a common homologue to them. Since, then, the *axis* LMN of homology of the pair $A_1 B_1 C_1$, $A_2 B_2 C_2$, it must pass *also* through the centre P' of homology of the other pair $A''B''C''$, $A_2 B_2 C_2$: in other words, *this axis must coincide with the line of those two centres. The right line* PP' *is therefore* (as in the figure) *the common tangent to the circles* K and I; and in like manner, the line $(P)(P')$ is the common tangent to K and (I), as exhibited. – And this much may for the present suffice, as *Explanation*† *of the Diagram*: but it may be proper to devote another page or two to the consideration of some (perhaps) *new properties,* of conics and even of triangles, which are suggested by the whole construction.

291. Abstracting now from all the *details* of Figure 30, or using that Diagram merely in *illustration* of some general properties of triangles and conics, which were just now alluded to, I remark that, on the principle stated in article 289, we may infer from the Figure, or rather from the reasonings which led me to construct it:

I. "If any two triangles, t_0 and t_1, be *each* homologous to their *common exscribed* triangle, t, they are *also* each homologous to their *common inscribed triangle, say* t_2; and these two triangles, t and t_2, thus exscribed and inscribed to the other two, are homologous *to each other*". In the Figure, t, t_0, t_1, t_2 are exemplified by the four triangles, ABC, $A''B''C''$, $A_1 B_1 C_1$, $A_2 B_2 C_2$.

292. It may be noticed, that the fourth triangle t_2 is *ex*scribed to the first triangle t; or that the *given* triangle, t, or ABC, is conversely *in*scribed in t_2, or in $A_2 B_2 C_2$: which is no mere accident of the construction, as the calculations and reasonings, already set forth, are sufficient to show. Thus, whether we take the three triangles t, t_0, t_2, or the three others, t, t_1, t_2, they form in each case a ***Cycle of Triangles***, *each* of which is an *exscribed homologue* of that one which *follows* it, in the cyclical succession thus assigned. – A somewhat simpler and perhaps more general theorem of the same sort, may be mentioned a little later.

293. In the next place it may be asserted, under the same conditions (article 291), that:

II. "If a_0 and a_1 denote the axes of homology of the two pairs of triangles, t, t_0, and t, t_1, then the intersection $c = a_0 \cdot a_1$ of these two axes is the centre of homology of the two triangles, t, t_2." In the Figure, a_0 may be conceived to be the line at infinity, and a_1 is represented by the strong black line to the right; to which line, accordingly, AA_2, BB_2, CC_2 were found to be parallel. We might have taken that *other* strong black line, which is more to the *left* of the diagram, as representing the axis of homology (a_1), when the triangle (t_1) is

* [Michael Henry Gill, 1794–1879, was Printer at the Dublin University Press from 1842 until 1872; the engraver was William Oldham 1811–1885. See Vincent Kinane, *A History of the Dublin University Press 1734–1976*, Gill and Macmillan, Dublin: 1994.]

† That is, so far as *points, lines* and *circles* are concerned; for I must postpone my account of the *inscribed ellipse,* which indeed I have inserted since this sheet was written.

exemplified by the triangle of contacts $(A_1)(B_1)(C_1)$ with the exscribed circle; the centre (c) of homology of the two triangles t and (t_2) where (t_2) denotes the auxiliary triangle $(A_2)(B_2)(C_2)$, being the intersection $a_0{}^{\cdot}(a_1)$, that is, the point at infinity on that other axis (a_2); to which axis accordingly the three lines $A(A_2)$, $B(B_2)$, $C(C_2)$ are parallel.

294. Again, we have the Theorem:

III. "If c_0', c_1', be the centres of homology of the two pairs of triangles, t_0, t_2, and t_1, t_2, then their joining line, $a = \overline{c_0' c_1'}$, is the axis of homology of the two triangles t and t_2." In the Figure the centres c_0' and c_1' are represented by the points P' and P, if we attend to the inscribed circle; or by (P') and (P), if we employ the exscribed circle in its stead.

295. Proceeding to theorems which involve not only *triangles* but also *conics*, we may notice the two following, which are obviously connected with each other:

IV. "If a conic v_1 be inscribed in the triangle t, so as to touch its sides at the corners of t_1, this conic will pass through the centre of homology c_1' of t_1 and t_2, and will have for its tangent at that point the axis a of homology of the triangles t, t_2," or, by III., the line $\overline{c_1' c_0'}$. In the Figure, v_1 may be the inscribed circle, touched at P by the line PP'; or the theorem may be exemplified by the contact of the exscribed circle, at (P), with the line $(P)(P')$.

296. The obvious variation, above alluded to, of the Theorem IV, is the following:-

IV'. "If a conic v_0 be inscribed in t, touching its sides at the corners of t_0, it will be touched at c_0', by the same axis a of the pair t, t_2; which line $\overline{c_0' c_1'}$, will thus be a *common tangent* to the two conics v_0, v_1, although those conics do not touch *one another*". In the Figure, the conic v_0 is represented by the *inscribed ellipse* (see footnote to article 290) touching the given triangle ABC at the *middle points* of its sides; and touching also, at the point P', the line PP', which line is thus a *common tangent*, to this ellipse and to the *inscribed circle*. The same inscribed ellipse has also, in conformity with the same theorem IV'., the line $(P)(P')$ for a common tangent with the *exscribed circle*, as may be seen in the Figure: *all* the parts of which have now been fully (and perhaps more than fully) *described*, but which I shall still *use, for illustrations* of another general Theorem.

297. V. "If a conic u_1 pass through the corners of the triangle t_0, and be touched by the axis *a at the point* c_1', which *point* will thus count (by IV.) as *two intersections* of this last conic u_1 with the conic v_1, the *two other* intersections of these two conics are *on the axis* a_0 of homology of t and t_0; which axis may thus be said to be a *common chord* of theirs." In the Figure, the conics u_1, v_1 are the *bisecting* and *inscribed circles, touching* each other and the axis PP' at P, and having for their *other intersections* the *circular points at infinity*, which thus are situated (by article 293) upon the axis a. Or the Theorem may be exemplified by the contact at (P) of the bisecting and *exscribed* circles, with each other and with the axis $(P)(P')$ of homology of the two triangles ABC and $(A_2)(B_2)(C_2)$.

298. More generally we may infer, from the same recent Theorem V., what was stated in article 177 as an *extension* of the elementary proposition respecting such contacts of the *bisecting circle* with others, of which you have given so elegant an investigation based on principles of *geometry* alone; namely, that "if through the middle points of a plane triangle, exscribed to a given conic, a *homothetic conic* be made to pass, those two conics will *touch* one another". For there exists, by the Theorem V., a point c_1' upon the conic v_1 to which the triangle t is exscribed, such that if a second conic u_1 be made to touch the first conic v_1 at that point, and to bisect the sides of the triangle, the *line at infinity* a_0 will be a *common chord* of the two conics, which consequently are *homothetic*: but a conic, homothetic to a *given* one, is *determined* by three given points.

299. I have had the curiosity to *verify*, by *Cartesian Coordinates*, which may be *oblique* this *Theorem of the homethetic conics*, as follows. Let the given conic v_1 be the *hyperbola*, $xy = 1$, not necessarily equilateral. Let the three points of contact with the sides of the exscribed triangle t be the points $\left(a, \frac{1}{a}\right)$, $\left(a', \frac{1}{a'}\right)$, $\left(a'', \frac{1}{a''}\right)$; when the abscissæ a, a', a'' are the three roots of a given cubic equation;

$$x^3 - sx^2 + tx - v = 0.$$

Then, for the middle points of the sides of the triangle, I found expressions of the form,

$$\left(\frac{s+a}{t+a^2}, \frac{v+ta}{t+a^2}\right);$$

so that the coefficients k, l, m, n, in the equation of the *homothetic conic*, or hyperbola with *asymptotes parallel* to the axes of x and y, which is evidently of the form

$$kxy + lx + my + n = 0,$$

must be such as to allow the *biquadratic* equation,

$$k(s+x)(v+tx) + l(s+x)(t+x^2) + m(v+tx)(t+x^2) + n(t+x^2)^2 = 0,$$

to be *divisible by the cubic* equation

$$x^3 - sx^2 + tx - v = 0;$$

whence it follows that I might write,

$$k = st - v, \qquad l = -t^2, \qquad m = -s^2, \qquad n = st - v.$$

The *bisecting hyperbola*, with its *asymptotes* in the *given directions*, was thus found to have for its equation,

$$(v - st)(xy - 1) + t^2x + s^2y - 2st = 0;$$

so that it *touches the given hyperbola* (as by the theorem it ought to do), namely at the point $\left(\frac{s}{t}, \frac{t}{s}\right)$.

300. The Theorem V. may obviously be varied thus,

V′. "If a conic u_0 be exscribed to the triangle t_1 and touch the axis a, and therefore also (by IV′) the inscribed conic v_0 at the point c_0', its other (real or imaginary) intersections with that conic will be upon the axis a_1. In the Figure, I *might* have described a *second ellipse*, touching the *inscribed ellipse* v_0 (article 295) there drawn as the point P', and passing *through the points of contact* A_1, B_1, C_1, of the *inscribed circle* with the sides of the given triangle. This *second* ellipse, which I thought it better *not* to draw, would then have been the conic u_0 of the present theorem V′.; it would have lain *wholly outside* the *first* ellipse v_0, except just at their point of *contact* P'; but the *two ellipses* would have had the *same two imaginary intersections*, with the axis a_1 of homology of the two triangles ABC and $A_1B_1C_1$, that is with the strong black line to the right of the Figure. – I know that I mentioned, by anticipation, some or all of these theorems in a separate Note [not found] of Saturday last, (I write now on Monday, May 21st,) but it seemed convenient to incorporate them with the present Series of Sheets, to which indeed they naturally belong.

301. The analysis by which I was led to perceive the foregoing Theorems, was perhaps not altogether inelegant; but it must be admitted to have been very long, and to have involved a great number of reductions, some of which were rather formidable. If, however, I were to *begin again*, – which, happily for your patience, and even for my own, I am not going to do, – I see how the processes might, in many respects, have been made shorter, and more simple. But without returning on the question stated and solved by you, I think that it may be no waste of time and space, to devote a few pages to the statement, and to the proof by anharmonic coordinates, of a few theorems respecting triangles and conics, which are perhaps a little simpler, if not more general, than those that have been recently enunciated; because they involve the consideration of only *one inscribed triangle* t_1; in a *given* triangle t, instead of *two* such triangles, t_0 and t_1; although I shall still have to consider a *third* triangle t_2, which is (like $A_2B_2C_2$ in the Figure) at once *ex*scribed to t and *in*scribed in t_1, and is *homologous* to *each* separately, as they likewise are to each other. In short, I am disposed to say something about a *single cycle of triangles*, t, t_1, t_2, instead of the *double cycle* in article 292, – of which *each* is (as in that article) an *exscribed homologue* of the one which *follows* it in the succession; and is of course, at the same time, an *inscribed homologue* of the triangle which *precedes* it, in the same given ***Cycle***.

302. Dismissing then, all considerations of Figure 30, let us attend only to the simpler Figure 31, which is annexed; and in which ABC is, as above, any given triangle t, while $A_1B_1C_1$ is a triangle t_1, inscribed in it; and $A_2B_2C_2$ is another triangle t_2, which is at once *in*scribed in t_1, and *ex*scribed to t. Taking ABC as usual for the unit-triangle, let us at first suppose that the corners of t_1 are expressed as follows:

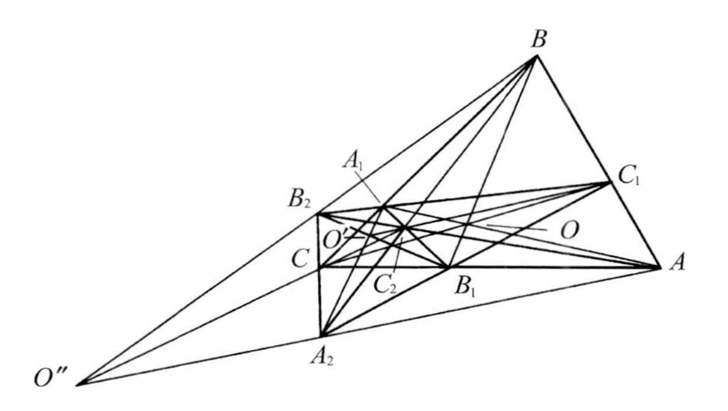

Fig. 31

$$A_1 = (0, 1, a); \quad B_1 = (1, 0, 1); \quad C_1 = (1, 1, 0);$$

where a is any constant. The equation of the line $B_1 C_1$ being then

$$-x + y + z = 0,$$

if we assume any point (α, β, γ) upon the unit-line, so that

$$\alpha + \beta + \gamma = 0,$$

we are at liberty to write

$$A_2 = (-\alpha, \beta, \gamma),$$

so far as the condition of *collinearity* of the three points A_2, B_1, C_1 is alone concerned, and conversely *every point* A_2, upon the *side* $B_1 C_1$ of t_1, is expressible by a symbol of this last *form*, with the recent relation between α, β, γ. If we next write,

$$B_2 = CA_2 \cdot C_1 A_1 = (x, y, z),$$

and

$$C_2 = A_2 B \cdot A_1 B_1 = (x', y', z'),$$

we shall have the equations,

$$0 = \begin{vmatrix} x, & y, & z \\ 0, & 0, & 1 \\ -\alpha, & \beta, & \gamma \end{vmatrix}, \quad 0 = \begin{vmatrix} x, & y, & z \\ 1, & 1, & 0 \\ 0, & 1, & a \end{vmatrix},$$

and

$$0 = \begin{vmatrix} x', & y', & z' \\ -\alpha, & \beta, & \gamma \\ 0, & 1, & 0 \end{vmatrix}, \quad 0 = \begin{vmatrix} x', & y', & z' \\ 0, & 1, & a \\ 1, & 0, & 1 \end{vmatrix};$$

that is,

$$0 = \beta x + \alpha y, \qquad 0 = ax - ay + z,$$
$$0 = -\gamma x' - \alpha z', \quad 0 = x' + ay' - z';$$

whence we may write,

$$B_2 = (\alpha, -\beta, a\gamma), \quad C_2 = (a\alpha, \beta, -a\gamma);$$

and the line $B_2 C_2$, thus determined, will pass through the point A, because we shall have the symbolical equation,

$$(B_2) + (C_2) = (1 + a)\alpha(A),$$

since $A = (1, 0, 0)$.

303. The *conditions of inscription* are therefore thus all satisfied; but we have made no use, as yet, of any *conditions of homology*. But if we now inquire, under what condition will the *third* triangle t_2 be homologous to the *first triangle* t, we have for the three lines AA_2, &c., connecting their corresponding corners, the equations,

$$\gamma y - \beta z = 0, \quad az - a\gamma x = 0, \quad \beta x - a\alpha y = 0;$$

if then these three lines *concur* in any common point, we must have the equation,

$$0 = \begin{vmatrix} 0, & \gamma, & -\beta \\ -a\gamma, & 0, & \alpha \\ \beta, & -a\alpha, & 0 \end{vmatrix} = \alpha\beta\gamma(1 - a^2),$$

of which the only useful factor is $1 - a = 0$; and such is also the *condition of concurrence* of AA_1, BB_1, CC_1. The first triangle will therefore be homologous to the *third*, if it be homologous to the *second*, but not otherwise; that is, other things remaining the same, if these three triangles admit of being expressed as follows:

$$t = ABC = (1, 0, 0)(0, 1, 0)(0, 0, 1);$$

$$t_1 = A_1 B_1 C_1 = (0, 1, 1)(1, 0, 1)(1, 1, 0);$$

$$t_2 = A_2 B_2 C_2 = (-\alpha, \beta, \gamma)(\alpha, -\beta, \gamma)(\alpha, \beta, -\gamma);$$

with the relation

$$\alpha + \beta + \gamma = 0,$$

as before.

304. With these coordinates of the corners, the centres of homology of the first triangle, t_1, with the two others separately taken, which *centres* in Figure 31 are the points O and O'', but which may also be expressively denoted by the symbols $c_{0,1}$, $c_{2,0}$, are evidently the following:

$$O = c_{0,1} = (1, 1, 1); \quad O'' = c_{2,0} = (\alpha, \beta, \gamma).$$

And because t has been assumed as the unit-triangle, we can already infer, from the general theory mentioned, or alluded to, in articles 275 and 277, that the *axes* of homology of the same two pairs of triangles, t_0, t_1, and t_2, t_0, may be represented by the following analogous symbols:

$$a_{0,1} = [1, 1, 1]; \quad a_{2,0} = [\alpha^{-1}, \beta^{-1}, \gamma^{-1}].$$

Accordingly, the intersections of corresponding sides of the pair t_0, t_1 are the three points,

$$BC\,\dot{}\,B_1 C_1 = (0, 1, -1), \quad CA\,\dot{}\,C_1 A_1 = (-1, 0, 1), \quad AB\,\dot{}\,A_1 B_1 = (1, -1, 0)$$

which are evidently on the unit line [1, 1, 1], by some of the earliest results of this whole calculus; and the intersections of corresponding sides, of the pair of triangles t_2, t_0, are the points,

$$B_2 C_2\,\dot{}\,BC = (0, \beta, -\gamma), \quad C_2 A_2\,\dot{}\,CA = (-\alpha, 0, \gamma), \quad A_2 B_2\,\dot{}\,AB = (\alpha, -\beta, 0),$$

which obviously all satisfy the common equation

$$\alpha^{-1} x + \beta^{-1} y + \gamma^{-1} z = 0,$$

and are therefore ranged upon one common axis of homology, of which the *symbol* is, as above,

$$a_{2,0} = [\alpha^{-1}, \beta^{-1}, \gamma^{-1}].$$

305. Already, simple and unlaborious as the recent calculations have been, we may derive from them a *theorem* as follows. Since the axis $a_{0,1}$, or $[1, 1, 1]$, passes through the centre $c_{2,0}$, or (α, β, γ), because $\alpha + \beta + \gamma = 0$; and since the three triangles t, t_1, t_2 of article 303 may represent *any cycle*, of the kind described in article 301, we are entitled to assert that "*In any cycle of three triangles, of which each is an exscribed homologue of the next, the axis of homology of first and second passes through the centre of homology of first and third*". (In Figure 30, this was exemplified by the passage of the axis of ABC, $A_1 B_1 C_1$ through the centre of ABC, $A_2 B_2 C_2$; or by that centre being at infinity, and therefore on the axis of homology of ABC, $A''B''C''$; &c.)

306. The *cycle of inscription* may be written as $t_2 t t_1$; t being inscribed in t_2, &c. But we have proved that in this *new* arrangement, the first triangle is homologous to the second; it must therefore, by article 303, be homologous also to the third; that is to say, the two triangles t_1, t_2 must have a centre of homology, $c_{1,2}$, or O', in which lines joining corresponding corners *concur*, as is exhibited in Figure 31. – to verify this geometrical result by calculation, and to determine the coordinates of this new centre O', we may proceed as follows. Let the equations of $A_1 A_2$, $B_1 B_2$, $C_1 C_2$ be for the moment, denoted and written thus:

$$lx + my + nz = 0, \qquad l'x + m'y + n'z = 0, \qquad l''x + m''n + n''z = 0.$$

Then by the values of the coordinates of the corners in article 303, the coordinates l, m, n of the joining line $A_1 A_2$ must satisfy the two equations,

$$0 = m + n, \qquad 0 = -\alpha l + \beta m + \gamma n;$$

whence

$$l = \beta - \gamma, \qquad m = \alpha, \qquad n = -\alpha.$$

In like manner,

$$0 = n' + l', \qquad 0 = -\beta m' + \gamma n' + \alpha l';$$

$$m' = \gamma - \alpha, \qquad n' = \beta, \qquad l' = -\beta;$$

and similarly,

$$n'' = \alpha - \beta, \qquad l'' = \gamma, \qquad m'' = -\gamma.$$

If then we merely want to *prove* that the three lines $A_1 A_2$, $B_1 B_2$, $C_1 C_2$, or $[l, m, n]$, $[l', m', n']$, $[l'', m'', n'']$, *concur somewhere*, it is sufficient to observe that the equation,

$$0 = \begin{vmatrix} l, & m, & n \\ l', & m', & n' \\ l'', & m'', & n'' \end{vmatrix} = \begin{vmatrix} \beta - \gamma, & \alpha, & -\alpha \\ -\beta, & \gamma - \alpha, & \beta \\ \gamma, & -\gamma, & \alpha - \beta \end{vmatrix}$$
$$= (\beta - \gamma)(\gamma - \alpha)(\alpha - \beta) + (\beta - \gamma)\beta\gamma + (\gamma - \alpha)\gamma\alpha + (\alpha - \beta)\alpha\beta + \alpha\beta\gamma - \alpha\beta\gamma,$$

is satisfied, independently of any relation between α, β, γ. But if we also wish to *assign the coordinates x, y, z* of the *point of concurrence*, we may write, further,

$$x = \begin{vmatrix} m, & n \\ m', & n' \end{vmatrix} = \begin{vmatrix} \alpha, & -\alpha \\ \gamma - \alpha, & \beta \end{vmatrix} = \alpha(\beta + \gamma) - \alpha^2;$$

$$y = \begin{vmatrix} n, & l \\ n', & l' \end{vmatrix} = \begin{vmatrix} -\alpha, & \beta - \gamma \\ \beta, & -\beta \end{vmatrix} = \beta(\gamma + \alpha) - \beta^2;$$

$$z = \begin{vmatrix} l, & m \\ l', & m' \end{vmatrix} = \begin{vmatrix} \beta - \gamma, & \alpha \\ -\beta, & \gamma - \alpha \end{vmatrix} = \gamma(\alpha + \beta) - \gamma^2;$$

no use having *yet* been made, in the present article, of the relation $\alpha + \beta + \gamma = 0$. We may, however, now with advantage introduce that relation; and thus, dividing each coordinate by -2, may write the following symbol for the remaining centre of homology,

$$O' = c_{1,2} = (\alpha^2, \beta^2, \gamma^2).$$

307. The theorem of article 305, applied to the cycle $t_2 t t_1$ of article 306, would have enabled us to foresee that this axis of homology $a_{2,0}$ of the two triangles t_2, t must pass through the centre of homology $c_{1,2}$ of t_1 and t_2. Accordingly we know now that the symbols of this axis and of this centre are, by articles 304 and 306,

$$a_{2,0} = [\alpha^{-1}, \beta^{-1}, \gamma^{-1}]; \quad c_{1,2} = (\alpha^2, \beta^2, \gamma^2);$$

and the relation $\alpha + \beta + \gamma = 0$ gives evidently the verification,

$$\alpha^{-1}.\alpha^2 + \beta^{-1}.\beta^2 + \gamma^{-1}.\gamma^2 = 0.$$

308. We have not yet investigated the axis $a_{1,2}$ of homology of the two triangles t_1 and t_2; which axis, as a new verification of the theory, ought to pass through the centre $c_{0,1}$ of homology of t and t_1 because the three triangles t_1, t_2, t form a cycle, of the kind considered in article 301: so that if we write

$$a_{1,2} = [l_{1,2}, m_{1,2}, n_{1,2}],$$

we ought to find the relation,

$$l_{1,2} + m_{1,2} + n_{1,2} = 0.$$

Accordingly, the equation of $B_1 C_1$ being still

$$-x + y + z = 0,$$

as in article 302, and the equation of $B_2 C_2$ being

$$\beta^{-1} y + \gamma^{-1} z = 0,$$

(by the value of the coordinates, as given in article 303,) we have, for the intersection of these two corresponding sides of the triangles t_1 and t_2, the symbol,

$$B_1 C_1 \dot{} B_2 C_2 = (\beta - \gamma, \beta, -\gamma),$$

and similarly, for the two other intersections of corresponding sides, we have,

$$C_1 A_1 \dot{} C_2 A_2 = (-\alpha, \gamma - \alpha, \gamma),$$

and

$$A_1 B_1 \cdot A_2 B_2 = (\alpha, -\beta, \alpha - \beta).$$

These three intersections are thus all ranged upon the one straight line,

$$\alpha x + \beta y + \gamma z = 0,$$

and we may write the symbol:

$$a_{1,2} = [\alpha, \beta, \gamma];$$

with the expected verification, that this axis $a_{1,2}$ passes through the centre $c_{0,1}$, or through the unit-point.

309. It may be convenient here to *tabulate* the symbols which have been thus obtained, for the centres and axes of homology, of the three pairs of triangles considered. They are, then, the following:

$$c_{0,1} = (1, 1, 1); \quad c_{1,2} = (\alpha^2, \beta^2, \gamma^2); \quad c_{2,0} = (\alpha, \beta, \gamma);$$
$$a_{0,1} = [1, 1, 1]; \quad a_{1,2} = [\alpha, \beta, \gamma]; \quad a_{2,0} = [\alpha^{-1}, \beta^{-1}, \gamma^{-1}].$$

310. The supposed relation of t_1 to t, as an inscribed and homologous triangle, allows us to conceive a *conic*, υ, which shall be at once *in*scribed to t, and *ex*scribed to t_1; and of which the equation is easily found to be, ([left blank in MS])

$$x^{\frac{1}{2}} + y^{\frac{1}{2}} + z^{\frac{1}{2}} = 0, \quad \text{or,} \quad x^2 + y^2 + z^2 - 2yz - 2zx - 2xy = 0.$$

But this equation is satisfied, when we suppose

$$x = \alpha^2, \ y = \beta^2, \ z = \gamma^2,$$

with the relation

$$\alpha \pm \beta \pm \gamma = 0;$$

the conic υ passes therefore through the centre $c_{1,2}$. We have therefore this Theorem: "If a conic υ be inscribed in a given triangle t, and if a triangle t_2 be constructed, which is at once exscribed to that given triangle t, and inscribed in the triangle of contacts t_1, this conic will pass through a point $c_{1,2}$, which will be a centre of homology of the two triangles t_1 and t_2." I need not repeat how this theorem is illustrated by Figure 30; but may just observe that it is ([Left blank in MS]) in some respects simpler than theorems stated before (IV., IV′.), as now involving the consideration of only *three triangles*.

311. The *tangent* to the conic υ, at the point $c_{1,2}$, is the line $[D_x\upsilon, D_y\upsilon, D_z\upsilon]$, [(Left blank in MS]) when

$$2\upsilon = x^2 + y^2 + z^2 - 2yz - 2zx - 2xy;$$

it is therefore the line

$$[x - y - z, \ y - z - x, \ z - x - y], \quad \text{or} \quad [\alpha^2 - \beta^2 - \gamma^2, \beta^2 - \gamma^2 - \alpha^2, \gamma^2 - \alpha^2 - \beta^2]$$

or (dividing by $2\alpha\beta\gamma$), it is $[\alpha^{-1}, \beta^{-1}, \gamma^{-1}]$; but this last has been seen in article 304 to be the symbol of the axis $a_{2,0}$. We have therefore thus the Theorem (which is again illustrated by Figure 30, and includes theorems already stated):

"The tangent to the inscribed conic v, at the point $c_{1,2}$, or at the centre of homology of its inscribed triangle t_1, and of the third or auxiliary triangle t_2, is the axis of homology $a_{2,0}$ of the same auxiliary triangle t_2, and of the exscribed triangle t." – In fact, the right line

$$\frac{x}{\alpha} + \frac{y}{\beta} + \frac{z}{\gamma} = 0,$$

of article 304, touches the conic

$$x^{\frac{1}{2}} + y^{\frac{1}{2}} + z^{\frac{1}{2}} = 0$$

of article 310, in the point $x = \alpha^2$, $y = \beta^2$, $z = \gamma^2$, if the condition $\alpha + \beta + \gamma = 0$ be satisfied.

312. Of course, by symmetry, we may in like manner conceive a *second conic*, v_1, inscribed in this triangle t_1, and exscribed to t_2; and then this new conic will be touched by the axis $a_{0,1}$ at the centre of homology $c_{2,0}$, of the pair of triangles indicated by these symbols. And similarly a *third conic* v_2 may be inscribed in t_2, and exscribed to t, which shall be touched by $a_{1,2}$ at $c_{0,1}$. Accordingly the equations of these two new conics are, by the conditions of their description,

$$0 = 2v_1 = \alpha^{-1}x^2 + \beta^{-1}y^2 + \gamma^{-1}z^2,$$

$$0 = v_2 = \alpha yz + \beta zx + \gamma xy;$$

the former passes through the point (α, β, γ), or $c_{2,0}$, and is touched there by the line $[1, 1, 1]$, or $a_{0,1}$, and the latter passes through $(1, 1, 1)$ or $c_{0,1}$, and has for its tangent at that point the line $[\alpha, \beta, \gamma]$, or $a_{1,2}$. – (In Figure 30, if we still suppose the triangles ABC and $A_2B_2C_2$ to be denoted by t and t_2, then according as we consider t_1 to represent the *triangle of contacts* $A_1B_1C_1$ or the *bisecting triangle*, $A''B''C''$, the conic v_1 will either be a *hyperbola*, touching the sides of the *former* triangle at the points $A_2B_2C_2$, and having the strong black line $A_1'B_1'C_1'$ for *one* of its *asymptotes*; or else a *parabola*, touching the sides of the *latter* triangle at the same points A_2, B_2, C_2, and having the same line $A_1'B_1'C_1'$ for a *parallel* to its *axis* of figure.)

313. *Another* system is, however, suggested by preceding investigations, respecting which new system it may be proper to say something here. Conceive then a *new triangle*, t_1', or $A_1'B_1'C_1'$, which shall (like t_1) be an *inscribed homologue* of the given triangle t, and therefore also (like t_1 again) an *exscribed homologue* of t_2, by the theorem of article 303.

Let the centre and axis of homology of the new pair of triangles t, t_1' be thus denoted in conformity with a theory already explained,

$$c_{0,1}' = (a, b, c), \quad a_{0,1}' = [a^{-1}, b^{-1}, c^{-1}];$$

then, because, by the theorem of article 305, this new axis must pass through the old centre $c_{2,0}$, or (α, β, γ), the new and old coordinates must be connected by the equation,

$$a^{-1}\alpha + b^{-1}\beta + c^{-1}\gamma = 0;$$

which, when combined with the old equation

$$\alpha + \beta + \gamma = 0,$$

allows us to write

$$\alpha = aa', \quad \beta = bb', \quad \gamma = cc',$$

or makes, for conciseness,

$$b - c = a', \quad c - a = b', \quad a - b = c';$$

the *ratios* only of the coordinates being important here.

Exscribe now, to the new triangle t_1', a new *conic*, u, which shall *touch* the conic v of article 310, and therefore also, by article 311, the line $a_{2,0}$, or $[\alpha^{-1}, \beta^{-1}, \gamma^{-1}]$, at the point $c_{1,2}$; the equation of the new conic will be, on account of the given *condition of contact*, of the form,

$$0 = 2u = 2v + (\alpha^{-1}x + \beta^{-1}y + \gamma^{-1}z)(\lambda x + \mu y + \nu z),$$

with the expression for the *function* $2v$ which was assigned in article 311, and with values for the three *constants* λ, μ, ν which are to be determined by the *conditions of description*; namely, in virtue of that form (article 311), of $2v$,

$$0 = (b - c)^2 + (\beta^{-1}b + \gamma^{-1}c)(\mu b + \nu c);$$

$$0 = (c - a)^2 + (\gamma^{-1}c + \alpha^{-1}a)(\nu c + \lambda a);$$

$$0 = (a - b)^2 + (\alpha^{-1}a + \beta^{-1}b)(\lambda a + \mu b);$$

which arise from our having

$$t_1' = (0, a, b)(a, 0, c)(a, b, 0).$$

But, by the present article,

$$\beta^{-1}b + \gamma^{-1}c = b'^{-1} + c'^{-1} = -a'b'^{-1}c'^{-1} = -a'/(b'c'),$$

because

$$a' + b' + c' = 0; \quad \text{and} \quad (b - c)^2 = a'^2,$$

with analogous reductions for the 2 other equations of exscription. Those three equations may therefore be thus written,

$$a'b'c' = \mu b + \nu c = \nu c + \lambda a = \lambda a + \mu b,$$

giving

$$\lambda a = \mu b = \nu c = \tfrac{1}{2}a'b'c';$$

the equation of the new conic u is consequently

$$0 = 4u = 4v + a'b'c'(\alpha^{-1}x + \beta^{-1}y + \gamma^{-1}z)(a^{-1}x + b^{-1}y + c^{-1}z).$$

314. Proceeding to *interpret* this result, we see that while the line

$$\alpha^{-1}x + \beta^{-1}y + \gamma^{-1}z = 0,$$

or $[\alpha^{-1}, \beta^{-1}, \gamma^{-1}]$ or the axis $a_{2,0}$ of homology of the pair t, t_2, is (by the *construction* to which the recent *analysis* answers) the *common tangent* to the two conics, u and v, at the point $c_{1,2}$, or

$(\alpha^2, \beta^2, \gamma^2)$, their *common chord* (connecting their *two intersections*, which are *distinct* from that point of *contact*) is the line

$$a^{-1}x + b^{-1}y + c^{-1}z = 0,$$

or

$$[a^{-1}, b^{-1}, c^{-1}],$$

or finally, by the foregoing article, the axis of homology, $a'_{0,1}$, of the new pair of triangles t, t'_1. We have therefore, in this new way a *Theorem*, which includes those above numbered as V. and V.′, in articles 297 and 300; namely, the following:-

"If a conic u be so exscribed to the triangle t'_1 as to *touch* the conic v *at the centre* $c_{1,2}$ of homology of the two triangles t_1, t_2 its *other intersections* (real or imaginary) with that conic will be situated *upon the axis* of homology $a'_{0,1}$ of the triangles t and t'_1.

315. It follows that if a conic u be made to pass through the three corners of the triangle t'_1, and through the two points, real or imaginary, in which the conic v is *cut* by the axis $a'_{0,1}$, these two conics will *touch* one another; namely at the centre $c_{1,2}$, their common tangent there being the axis $a_{2,0}$. Hence we derive, in a new way, the earlier Theorem (articles 177, 298), that *two homothetic conics* (and therefore in particular the two *circles*) *touch*, if *one* (as v) be *inscribed in a triangle* (as t, *including* of course what in elementary language is called *exscription* of a *circle*), and if the *other* (as u) pass through the *middle points* of the *sides* of the same triangle. – I need say nothing about the conics, u_1 and u_2, which touch the other conics v_1 or v_2, nor about the illustration of these Theorems of contacts of Conics which are supplied by Figure 30.

316. I may, however, before taking leave of the Subject once more, – and I hope this time *finally*, so far at least as the present Letter is concerned, – just *sum up* briefly, and *in general terms*, the leading *Theorems* to which my analysis has conducted me, respecting such *cycles* of triangles, &c.: though doubtless it must be possible to assign some much more *elementary proof* of them, derived from *geometry alone.*

317. ***Theorem*** A. – "If a first triangle (plane or spherical) be exscribed to a second, to which it is homologous, and if a third triangle be at once inscribed in the second, and exscribed to the first, this third triangle is homologous to each of them; so that they form a *cycle of triangles*, wherein *each* is an *exscribed homologue* of the one that *follows* it in the cycle, and an *inscribed homologue* of the one that *precedes* it, in the same cyclical succession."

318. ***Theorem*** B. – "Under the same conditions, the *axis* of homology of the *first* and *third* triangles passes through the *centre* of homology of the *second* and *third*; and touches there the *conic*, which is at once *in*scribed in the *first* triangle, and *ex*scribed to the *second*." Or we may say (articles 305 and 312) that the axis of homology of first and second passes through the centre of homology of first and third, and touches there the conic, which is inscribed in the second and exscribed to the third. Or, etc.. It may thus be seen that "the *triangle of centres* of homology is *inscribed in the triangle of axes*."

319. ***Theorem*** C. – "If a *fourth* triangle be (like the second) inscribed in the first, and exscribed to the third, in which case it will be (by A.) a *common homologue* of *theirs*, although *not* of the *second* triangle; and if a *second conic*, exscribed to this fourth triangle, pass through the intersections (real or imaginary) of the first conic with the axis of homology of the first and fourth triangles; *this second conic will touch the first*, at the point where (by B.) that former conic touches the axis of homology of the first and third triangles; which axis is thus a *common tangent* to these two conics, as the axis of the first and fourth triangles is (by construction) a *common chord* of theirs." – And so I *close my account of what I had to say upon this* subject: but must add a *little* more on *cubic curves* before I conclude this Letter.

320. May 28th, 1860. – (More last words.) Although I had fully intended not to devote another sheet of this Letter to those little Theorems and Constructions which your Question led me to perceive; yet as other things have prevented me hitherto from resuming Cubic Curves, – about which indeed I had not proposed to say much more, and as I happened to see, this morning, a very simple and *purely geometrical* proof of the Theorems which I lately marked as *A.* and *B.* in articles 317 and 318, I have amused myself just now by drawing the annexed Figure, which might indeed have been called 31 bis, although I thought the variations considerable enough to entitle it to a new number, as Figure 32.

321. Let ABC be any given triangle, t, and O any given point in its plane; draw AOA_1, &c. and so construct an *inscribed* and *homologous* triangle, t_1, or $A_1B_1C_1$; on the side B_1C_1 take any point A_2, and draw A_2C, A_2B; let B_2, C_2 be the points in which these last lines meet the other sides, C_1A_1, A_1B_1, of t_1; and join B_2C_2. We shall thus have constructed a third triangle, t_2 or $A_2B_2C_2$, which is evidently *in*scribed in the *second* triangle, t_1; but I say that it is *also ex*scribed to the *first* triangle, t; and that it is *homologous* to *each.*

322. I propose then to prove, *geometrically*: first, that side B_2C_2 of the third triangle passes *through the vertex* A of the first; second, that the three right lines A_1A_2, B_1B_2, C_1C_2 concur in

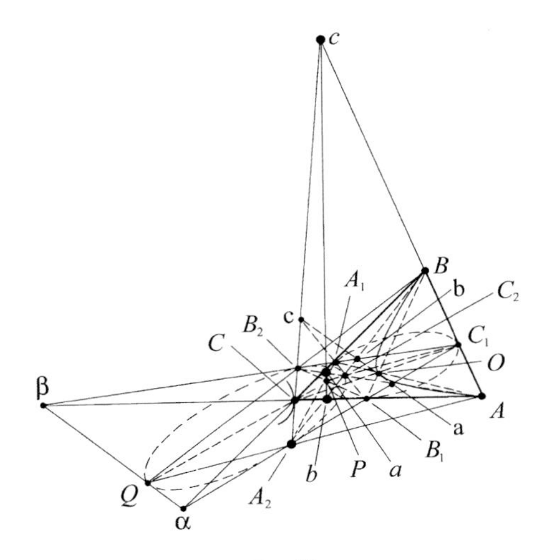

Fig. 32

one point or *centre* of *homology*, P; and third, that if a, b, c be the points in which the sides BC, CA, AB of the first triangle are intersected by the corresponding sides B_2C_2, C_2A_2, A_2B_2 of the third triangle, then these three points of intersection a, b, c are ranged *on one right line*, or *axis* of homology: from which it will of course follow that the three lines AA_2, BB_2, CC_2 *concur* in *another centre*, namely that which is marked as Q in the Figure. But I say farther, and fourth, that the *axis abc*, of homology of t, t_2, passes *through the centre* P, of homology of t_1t_2; and fifth that this axis *touches* there the conic v, or $A_1B_1C_1$, which is at once *inscribed* in the *first* triangle, t, and *exscribed* to the *second* triangle, t_1.

323. *Analysis* (*Geometrical*). – *Suppose it true*: and let the conic v be described, which thus touches, at the points A_1, B_1, C_1, the sides of the given triangle ABC. The centre of homology P of t_1t_2 will then be the second intersection of that conic with the right line A_1A_2; and the axis of homology abc of tt_2 will cut the sides of the given triangle in the same points as the tangent to the conic at P; also the three lines aA, bB, cC will coincide in position with the sides B_2C_2, C_2A_2, A_2B_2 of the triangle t_2. We shall thus have three quadrilaterals, $bcBC$, $caCA$, $abAB$, exscribed to the conic, with the three corresponding but inscribed quadrilaterals, $PC_1A_1B_1$, $PA_1B_1C_1$, $PB_1C_1A_1$; and the diagonals bB, cC, of the first exscribed quadrilateral, are thus found to *concur* (*if the theorem be true*) in one common point A_2, with the diagonals PA_1, B_1C_1 of the first inscribed: with analogous *concurrences of diagonals*, for the two other pairs of exscribed and inscribed quadrilaterals. But such concurrences are otherwise known to exist; the *analysis* therefore *succeeds*.

324. *Synthesis*. – To convert now the *geometrical analysis* of the foregoing article into a *synthesis*, let the point P be *defined* to be the second intersection of the line A_1A_2 with the conic $A_1B_1C_1$, inscribed as before in the given triangle; and let a, b, c be *defined* to be the points in which the sides BC &c., of that triangle are cut by the tangent to the conic at P. We have thus again the same three exscribed and the same three inscribed quadrilaterals as before; but we *now assume as known* the *concurrence* in one point of the *four diagonals* of any one such pair. Hence,

Ist the diagonals bB, cC, of the exscribed quadrilateral $bcBC$, pass each through the intersection A_2 of the diagonals PA_1, B_1C_1, of the inscribed quadrilateral $PC_1A_1B_1$;

IInd the lines aA, and PB_1, which are diagonals respectively of the two quadrilaterals $caCA$ and $PA_1B_1C_1$, pass through the intersection B_2, of the lines cC (or CA_2) and C_1A_1, which are the other diagonals of the same two quadrilaterals; and

IIIrd the diagonals aA and PC_1, of $abAB$ and $PB_1C_1A_1$, pass through the intersection C_2 of bB (or A_2B) and A_1B_1; so that the *sides* B_2C_2, C_2A_2, A_2B_2 of t_2 are now *proved* (as they were before in article 323, *supposed*) to *coincide* with the lines aA, bB, cC. Hence first, the side B_2C_2 passes through A and the *third triangle is exscribed to the first*; second, the lines A_1A_2, B_1B_2, C_1C_2 concur in the point P, and the *third triangle is homologous to the second*; third, this point of concurrence, or *centre of homology* P, is *on the inscribed conic*, v; fourth, the three points of intersection,

$$a = BC^{\cdot}B_2C_2, \quad b = CA^{\cdot}C_2A_2, \quad c = AB^{\cdot}A_2B_2$$

are *collinear*, and the *third triangle is homologous to the first*; fifth, the *axis of homology*, *abc*, of tt_2

passes *through the centre* of homology P of $t_1\,t_2$; and sixth, this axis *touches*, at this centre, the inscribed *conic*, v. Thus *all* the assertions of article 322 are justified, and the Theorems A. and B. are fully and *geometrically proved*: for it is clear that the third triangle cannot be (as supposed in A.) at once inscribed in the second triangle and exscribed to the first, *unless* its vertex A_2 be situated somewhere upon the side $B_1\,C_1$, and *unless* the two other vertices B_2 and C_2 admit of being *derived* from this one, by the constructions mentioned in article 321. – (The axis α, β, γ, of t, t_1, (γ not mentioned in the Figure) touches, at the centre Q of $t_2\,t_1$ the conic $A_2\,B_2\,C_2$, inscribed in t_1; and the axis abc of t_1, t_2 touches, at the centre O of t, t_1, the conic ABC inscribed in t_2.)

325. If the *point* A_2 *move* along the line $B_1\,C_1$, the *triangle* $A_2\,B_2\,C_2$, constructed as in article 322, will vary; but it will be *still* an *exscribed homologue* of the given triangle ABC; and their *varying axis* of homology, *abc*, will *still touch* the *conic* $A_1\,B_1\,C_1$, which conic is thus its *envelope*. Let a *second conic* v', or $A_1'\,B_1'\,C_1'$, be inscribed in the given triangle; so that the *triangle* t_1' or $A_1'\,B_1'\,C_1'$, is a second inscribed homologue of ABC. The *sides* of that given triangle t being thus *three common tangents* to the two conics v and v', there must be a *fourth* real and common tangent, which may be taken as the axis *abc* of homology of ABC or t, and of a triangle t_2, or $A_2\,B_2\,C_2$, *exscribed to* t as before; and this latter triangle t_2 will then be a *common inscribed homologue of the two inscribed triangles*, t_1 and t_1'. Again, this *axis abc* of t, t_2 will pass not only *through the old centre* P of homology of t, t_1, but also (for the same reason) *through the new centre* P', of homology of t and t_1'; it is therefore the *joining line* PP' of those *two centres*, and may be found as such. Again, if $\alpha = BC^{\cdot}B_1\,C_1$, and $\beta = CA^{\cdot}C_1\,A_1$, the line $\alpha\beta$, that is, the axis of homology of t, t_1, will pass (as in Figure 32) through the centre of homology Q of t_2, t, for the same reason that the axis *abc* of t_2, t passes through the centre P of t_1, t_2; and in like manner the axis $\alpha'\beta'$ (or $\alpha'\beta'\gamma'$) of homology of t, t_1' if it were drawn, would necessarily be found to pass through the same centre Q of t_2, t; which centre Q is therefore the *intersection* of those two *axes*, $\alpha\beta\gamma$ and $\alpha'\beta'\gamma'$. And thus it seems that the Theorems I., II., III., IV., IV'., of articles 291, 293, 294, 295, and 296, with the remark made in article 292, receive a sufficient proof, *independently of any calculation.* But I do not yet see any simple and purely *geometrical proof* of the Theorem V. of 297; nor (of course) of its variation v' in article 300 or corollary in article 298.

326. (June 2nd, 1860.) – I have this day read, and (as I conceive) understood, the geometrical investigation contained in your note of the 31st of May [not found], which happened to be brought to me rather late in the evening of yesterday:- having been entrusted by the postman to someone. Perhaps I followed the proof the more easily, because you had paid me the compliment of adopting several letters and accents from recent diagrams of mine – with which *yours*, in its simplicity, contrasts very advantageously for itself: but I wanted, wisely or not, to illustrate *many things at once* in one synoptical view. If either of us shall draw any more figures, we shall of course not be bound to retain the same *notations.*

327. The *only hint*, which you seem to have taken from this *Letter* of mine, consists (I think) in your recognition of that useful *triangle* $A_2\,B_2\,C_2$, which is the *co-inscribed* of the *bisecting* triangle $A''B''C''$, and of the triangle of contacts $A_1\,B_1\,C_1$. You prove (as I take it), with the help of Ptolemy's Theorem (*Almagest*, ôÕã. ê., òÍ—. Ò., . . . ŠÓššĘÕÿÚ Íj¼ρÓœÿÚ ÞÉŬ . . .)*, first, that

$$PA'' : PB'' : PC'' = A_1A'' : B_1B'' : C_1C'';$$

if P be *defined* to be the second intersection of the right line RB_2 with the circle $A''B''C''$; R being on that circle (as in Figure 30) the middle point of the *arc* $A''B''C''$; and B_2 being the intersection of the sides $C''A''$, C_1A_1 of the two triangles inscribed in the given triangle ABC.

328. You prove, second, with the help of the celebrated Theorem of Ptolemy, and by employing the quadrilateral $PA''B'''C''$, (where B''' is the foot of the perpendicular from B on CA, or the second intersection of that side CA with the bisecting circle,) in combination with the formerly employed quadrilateral $PA''B''C''$, that the point B_1 is on the line B_2R.

329. You prove, very simply and elegantly, third, that the lines which I called A_2Q and C_2S (in my Figure 30) meet in the same point P, because $P'B_2$ still *bisects the angle* $A''P'C_2$, if P' be determined &c.; and therefore P' coincides with P [it does not seem to be possible to reconcile this statement with Figure 30].

330. In the fourth place, you prove, with an elegant use of elementary theorems about the bisector of the vertical angle of a plane triangle, that the following proportion exists:

$$PA_1 : PB_1 : PC_1 = PQ : PR : PS.$$

And hence you infer, by known principles of homology (or even of similarity) of figures, that the circle which touches the bisecting circle at the point P (determined as in article 327), and which passes through the point B_1 passes also through the two other points, of the same kind, C_1 and A_1: or, in other words, that *the two circles,* $A''B''C''$ and $A_1B_1C_1$, *touch one another* (namely, at the point P). But this last is a form of the *Theorem,* which following the example of

* [Claudius Ptolemy (*c.* 100–178 AD) '*Almagest,* Book I, Chapter 9, . . . a most useful theorem . . .'. See: *Composition mathématique de Claude Ptolémée,* translated by l'Abbé Nicholas B. Halma (1755–1828) with notes by Jean-Baptiste Joseph Delambre (1749–1822), 2 Vols, Paris: 1813–16, Vol. I, p. 29; *Ptolemy's ALMAGEST,* translated and annotated by G. J. Toomer, pp. 50–51, Duckworth, London: 1984; D. Pedoe, *Geometry,* p. 90, Cambridge University Press, Cambridge: 1974.

In a separate letter to Hart, dated 13 June 1860 (Trinity College Dublin MS 1493/1171, f. 88–89), Hamilton writes:

I am happy to say that I wrote as far as page 215 [article 331, and following] of my *Letter,* about ten days ago – more accurately on the Saturday before last; intending then, as I still do intend to *finish that* "Letter" with page 216 [article 336]. But the sheet is not just now at my hand; it was laid aside, I think, for the purpose of inserting some *references* – one of them being to the Almagest of Ptolemy, a large part of which I have read in the Greek, and in which the great Author very justly speaks of what we are accustomed to call *his* Theorem as ŠÓššĘÕÿÚ Íj¼ρÓœÿÚ ÞÉÚ. I have never had patience with Delambre's depreciation of Ptolemy. He was, no doubt, in an important sense, a pupil, this love & study, of Hipparchus [*c.* 190–*c.* 120 BC]; who was perhaps the more original genius of the two: but how little would we now know of that

eÚÓρ —ÕŠÿÞÿÚÿŒ κÃÕ —ÕŠÃŠÓÒÓŒ

['diligent and upright man'] as Ptolemy delights to call Hipparchus, if the Almagest had not recorded the disinterested admiration of its author! (I fancy that Hipparchus was *far* inferior to Ptolemy as a geometer.)']

our friend Salmon, I must be permitted to call "***Hart's Theorem***", until it is *proved*, to *us*, that it had been anticipated abroad.

331. You *feel*, however, that it is natural to *wish* for a more purely *projective proof*; or for one which shall be more completely independent of all *ratios* of *lengths* of lines. Just now, I am not disposed to search for any *such* proof of my Theorem C., or V.; but I think that you could not wish me to suppress, before closing my remarks on this Subject, some short account of what, at this moment, appears to me to be a remarkable or at least a *curious* extension, of the geometrical fact of *your circle* being a *common tangent* to *four* assignable *circles*: which extension has been pointed out to me by my son William,* who was a pupil of Salmon's, and attended many lectures of Townsend, when in College.

332. My son observes, then, that if to denote, as in my Figures 29 and 30, the point of concourse of the perpendicularity AA''', BB''', CC''' of the *given triangle ABC*, *your circle* passes *through the feet of the perpendiculars* of each of the *three new triangles*, ϖBC, ϖCA, ϖAB: and therefore is the *bisecting circle* of *each* of them. *Admitting* then your Theorem, as proved by you, he *infers* that your circle $A''B''C''$, or $A'''B'''C'''$, *touches not only your four other circles, inscribed or exscribed* to the *given* triangle ABC, but *also* those *twelve other circles*, which are *in*scribed to any of these *three new triangles*, with the point ϖ for their common vertex. And I believe that he has made some drawings, to illustrate this result, which to me was totally unexpected.

333. I see, of course, that we are thus entitled to enunciate a *Theorem*, respecting a *conic*, *touching at once sixteen other conics*, which I shall not delay to write down. But as the point of concourse ϖ acquires thus a new interest, I shall just observe here that if we write,

$$e = abc(a + b + c), \qquad a'' = e - b^2c^2,$$

$$b'' = e - c^2a^2, \qquad c'' = e - a^2b^2,$$

a, b, c having the same signification as in article 313, this *centre* of homology ϖ, *of the two triangles* ABC, $A'''B'''C'''$, is found to be denoted as follows:

$$\varpi = \left(\frac{1}{a''}, \frac{1}{b''}, \frac{1}{c''}\right);$$

the *axis* of homology of those two triangles being therefore on my plan denoted by the symbol, $[a'', b'', c'']$.

334. The Subject is by no means exhausted, but I must bring this long *Letter* to a close. If I had resumed the subject of *Cubic Curves*, I might have sought to illustrate, in a new way, a Theorem to which some interest seems to attach, and which is indeed one of the main results of some articles, respecting the *geometrical distinction* which exists (according to me) between the *real positive*, and the *real negative* root, of the biquadratic in $\frac{\lambda}{\mu}$. And I had other connected projects, which the time warns me to abandon.

* [William Edwin Hamilton (1834–1902), was Hamilton's eldest son. He graduated in 1857.]

335. (June 28th, 1860.) My son has lately pointed out to me a still greater extension of your Theorem, as regards the number of the circles to which yours is a common tangent, and which can be simply enough assigned, in a natural and geometrical connexion with the original triangle. The *principle* of this new extension is, that a *series of triangles* may be obtained *from the original one*, each capable of replacing it in the theorem: namely either, first, by forming a first derived triangle, such that the *old feet* of perpendiculars from vertices on sides shall be the *new middles* of sides; with *other* derived triangles *ad infinitum*, each similarly related to the preceding; or, second, by forming from the given triangle a first derived triangle, and from it an *infinite succession*, so that the connecting construction at each stage may be the *converse* of those in [the first].

336. (November 16th, 1860.) This Sheet, having at last been found, is forwarded, to complete your Series, – if you have any fancy of stitching all the Sheets together. It completes what I called the ***Letter***, on Anharmonic Coordinates: the ***Postscript***, which I take up only at odd times, may almost be considered as finished too, although about a couple of sheets more may yet be forwarded.

In the meantime I remain, my dear Doctor Hart,
faithfully yours,
William Rowan Hamilton
Andrew Searle Hart, Esq., LL. D., F.T.C.D.

Postscript to Letter on Anharmonic Coordinates.

Observatory, August 28th, 1860.

My dear Dr. Hart

337. It is almost exactly six months since the First Sheet of my Letter on Anharmonic Coordinates was begun; and was handed to you at the Royal Irish Academy on the eve of their last Meeting in February: nor have I yet sent what was designed to be the *last sheet* because it chanced to be always missing, just when I wanted it for you. Indeed, I remember that the *last page* of it had not been written, when I laid the sheet aside, nearly three months ago; it was my project to have mentioned in that paper, some additional results of my son William, derived from the study of your Theorem: perhaps I may still fill up the page, and send the sheet, when I find it. Meanwhile I know that the last paragraph, or article, was to have been numbered 336; wherefore I number the present paragraph 337, as above: although I regard anything more, which I may write now, in this resumed Series of Sheets as merely a Postscript to the (theoretically) concluded Letter. To write, with your permission, such a Postscript, will be useful to myself: and though it may extend to several new pages, will yet serve rather as an economy of time to me than the reverse, by acting as a sort of safety valve, to carry off superfluous steam, and leave me freer in thought, to say little or nothing on the subject to the public.

338. (September 5th, 1860.) – I received from Salmon yesterday a proof slip of a Paper of his, for the Quarterly Journal,* on matters connected with the Theorem which I still persist

* [*The Quarterly Journal of Pure and Applied Mathematics.*]

in calling *yours,* until some definite evidence to the contrary is produced. I am getting the paper copied here, but hope to post it back to Salmon this evening as I know that he wishes to let you see it; and I am sure that you feel, without surprise, that he has sought to do justice to *all* parties: even to your imaginary (or at least unknown) rival.

339. – You are aware that I lately felt a wish to know whether the Theorem could be *extended,* to triangles and *circles* on a *sphere*: for, of course, lemmas of mine about *ellipses* &c., extend at once, by radial projection, to *spherical* conics. My first *vague impression* had been, that it was *likely* to admit of such extension: and on applying quaternions, I found that certain verifications were borne, (as was to be expected,) for the case of an equilateral spherical triangle, however large. But the verifications did not succeed, – as you have lately assigned to me reasons why they should (or might) have been expected *not* to do – when the spherical triangle was *scalene.*

340. Still, as the theorem is undoubtedly true for the plane, whatever the *shape* of the original triangle maybe, it was an object of interest to investigate the ***Law,*** – analogous to that of the *Spherical Excess,* – according to which, what is (in rigour) *false* for the *sphere,* passes into being *true* for the *plane.*

341. To state now definitely the question:-

Let ABC be a spherical and scalene triangle, inscribed in a small circle, of which the spherical centre (or pole) is K, and the arcual radius is c, so that

$$\overset{\frown}{KZ} = \overset{\frown}{KB} = \overset{\frown}{KC} = \mathsf{c}.$$

Let $A_1 B_1 C_1$ be another spherical triangle, of which the sides are bisected by the corners of the triangle ABC. Let A', B', C' be the points in which the sides of this second triangle are touched by its inscribed small circle, of which we shall call the spherical centre K', and the arcual radius c'; so that

$$\overset{\frown}{K'A'} = \overset{\frown}{K'B'} = \overset{\frown}{K'C'} = \mathsf{c}'.$$

And let the arcual distance between the two centres K and K', be denoted by d; so that

$$\mathsf{d} = \overset{\frown}{KK'}.$$

342. Then, *at the plane limit,* by your Theorem,* we have

$$\mathsf{c} - \mathsf{c}' = \mathsf{d};$$

or,

* [Andrew S. Hart, 'Extension of Terquem's theorem respecting the circle which bisects three sides of a triangle', *Quarterly Journal of Pure and Applied Mathematics,* Vol. IV, pp. 260–261, 1861 (communicated 15th December 1860). See also: John Casey (1820–1891), *A treatise on spherical trigonometry,* p. 82, Hodges, Figgis, & Co., Dublin: 1889; Isaac Todhunter (1820–1884) *Spherical trigonometry,* (revised by John Gaston Leathem (1871–)) pp. 162–170, Macmillan and Co. Limited, London: 1949; and footnotes to articles 360 and 382 of this letter.]

$$\mathsf{a} = \mathsf{c} - \mathsf{c}' - \mathsf{d} = 0;$$

for it is easy to prove that $\mathsf{c} > \mathsf{c}'$, in general, or that the *inscribed* circle touches the *bisecting* (or Hartian) circle *internally*: the former circle being *smaller* than the latter, except when (as for the *equilateral* case) they happen entirely to *coincide*.

343. But, *for the sphere*, if we retain the *definition*,

$$\mathsf{a} = \mathsf{c} - \mathsf{c}' - \mathsf{d},$$

the quantity a does *not*, in general, *vanish*. Is it, then, *positive*, or *negative*, – to begin with? Does the *smaller* circle fall *wholly within* the *larger* one, so as to produce what may be called an ***Annular Eclipse*** of the *Hartian*? or is there a *partial overlapping*? and can any *law* be discovered, according to which the one case or the other, (if it be true that sometimes one, and at other times the other exists,) can be *predicted* to be the *case of the question*?

344. Again, the quantity a is certainly *small*, by your Theorem, if the *sides* of the spherical triangle ABC be *small*, whatever the *shape* of that triangle may be: but of *what order* of smallness is a? If there is an *annular eclipse* (article 343), *about* what is the *least breadth* of the *annulus*? Or if there be an *overlapping* of the two circles, *about how far* do they overlap?

345. – Being absolutely unacquainted with any branch of Mathematical Science, except the Calculus of Quaternions, by which problems of this sort could be attacked, I lately "turned on the steam" – or, in short, applied my own Calculus.

346. But, although I wish for your permission to write to you soon on the subject of the Quaternions themselves, the plan of the Letter, to which this sheet forms part of a Postscript, requires that I should merely *state results*, derived from them, and *verify* them by other *methods*.

347. From *Quaternions*, then, I deduce the *approximate formula*, – substantially the same as that which I mentioned in a note of about a week ago, but more conveniently expressed, –

Least Breadth of Annulus =

$$\mathsf{a} = 8\mathsf{c}\left(\sin \mathsf{c}.\tan\theta.\sin\frac{B-C}{2}.\sin\frac{B-C}{2}.\sin\frac{A-B}{2}\right)^2;$$

where c is still the arcual radius of the small circle, in which the triangle ABC is inscribed: θ is an auxiliary angle, such that

$$\sin\theta = \left(8\sin\frac{A}{2}.\sin\frac{B}{2}.\sin\frac{C}{2}\right)^{\frac{1}{2}};$$

and A, B, C may represent either the angles of the *spherical* or (as a limit) of the *plane triangle*, ABC.

348. The *positive* character of a shows that your circle is *annularly* eclipsed; the *evanescence* of $\frac{\mathsf{a}}{\mathsf{c}}$ with c, is a *new proof of your Theorem*.

349. *As to quantity,* (considered independently of *sign,*) I have just finished a most satisfactory verification, by *spherical trigonometry,* – for I know no other *check*: but it has cost me three or four days of very hard work, on account of the long chain of triangles employed, and the minute accuracy which was required.

350. (September 6th, 1860) I assumed as my three data, the three following quantities

$$\mathsf{c} = \overset{\frown}{KA} = \overset{\frown}{KB} = \overset{\frown}{KC} = 10°; \qquad AKB = 58°; \qquad BKC = 93°;$$

(whence $AKC = 58° + 93° = 151°$;) and began by deducing from these values, supposed to be rigorous, the values of the sides and angles of the triangle *ABC,* as well as I could, to hundredths of seconds, with the help of Taylor's Tables of Logarithms*, using seven decimal places, and proportional parts throughout [Figure 33]. In fact, if this last precaution had been omitted, the results could not have been relied on, to the degree of accuracy at which I thought it necessary to aim: because the value of a can be *foreseen* to be *small.*

351. For, before we *begin* to use *spherical trigonometry* at all, my formula enables me to *estimate* the value of that sought *annular breadth,* a, as follows. Although we do not yet know the *spherical angles A, B, C,* we can already assign, without any logarithmic calculation, the angles of the *plane triangle ABC,* because we know the *angles* $AK^{\backprime}B$, $BK^{\backprime}C$, which the sides $\overline{AB}$, $\overline{BC}$ of *that* triangle subtend at the centre $K^{\backprime}$ of its own circumscribed circle; since these are precisely *equal* to the *spherical angles AKB, BKC,* or (by supposition) 58° and 93°, which the sides $\overset{\frown}{AB}$, $\overset{\frown}{BC}$ of the spherical triangle *ABC* subtend at the *spherical centre* (or pole) *K* of the same circle, considered as a small one on the sphere. Hence, for the *plane* triangle, we have, *exactly,*

$$A = \tfrac{1}{2}BK^{\backprime}C = \tfrac{1}{2}BKC = 46°\,30'; \qquad C = \tfrac{1}{2}AK^{\backprime}B = \tfrac{1}{2}AKB = 29°\,0';$$

and therefore

$$B = \mathsf{c} - C - A = 104°\,30';$$

and a can be *approximately* calculated, by substituting *these* values for *A, B, C,* in the formula which I deduced from Quaternions.

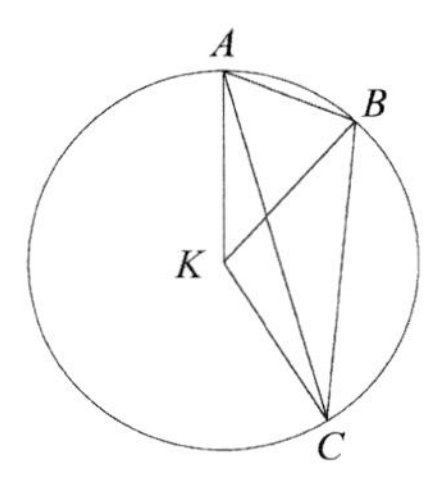

Fig. 33

* [Michael Taylor, (1756–1789), *Tables of logarithms,* London: 1792.]

352. For *this* little calculation, it will be found to be quite sufficient to use *minutes*, and *five-figure logarithms*:

And the whole work is as exhibited below.

353.

$$
\begin{array}{rr|rr}
\dfrac{A}{2} = 23^\circ\,15', & \log_{10}\sin\ldots\bar{1}{\cdot}596\,32; & \dfrac{B-C}{2} = 37^\circ\,45', & \log_{10}\sin\ldots\bar{1}{\cdot}786\,91 \\
\dfrac{B}{2} = 52^\circ\,15', & \log_{10}\sin\ldots\bar{1}{\cdot}898\,01; & \dfrac{A-C}{2} = 08^\circ\,45', & \log_{10}\sin\ldots\bar{1}{\cdot}182\,20 \\
\dfrac{C}{2} = 14^\circ\,30', & \log_{10}\sin\ldots\bar{1}{\cdot}398\,60; & \dfrac{B-A}{2} = 29^\circ\,00', & \log_{10}\sin\ldots\bar{1}{\cdot}685\,57 \\
 & \log_{10}8\ldots 0{\cdot}903\,09; & \mathsf{c} = 29^\circ\,00', & \log_{10}\sin\ldots\bar{1}{\cdot}239\,67 \\
 & (\div 2)\ldots\overline{\bar{1}{\cdot}796\,02} & & \\
\theta = 52^\circ\,15', & \log_{10}\sin\ldots\bar{1}{\cdot}898\,01 & \theta = 52^\circ\,15', & \log_{10}\tan\ldots 0{\cdot}111\,10; \\
\hline
 & & & \bar{2}{\cdot}005\,45 \\
 & & & \times 2 \\
 & & & \overline{\bar{4}{\cdot}010\,90} \\
 & & \dfrac{8\mathsf{c}}{1''} = 288\,000; & \log_{10}\ldots 5{\cdot}459\,39 \\
 & & \dfrac{\mathsf{a}}{1''} = 29{\cdot}53; & \log_{10}\ldots\overline{1{\cdot}470\,29}
\end{array}
$$

The value of a, thus estimated, amounts to no more than 29″·53; whence it becomes *necessary* to attend to *seconds*, and *prudent* to attend to *tenths* of seconds, in the subsequent calculations with spherical trigonometry. But, as already stated, I have chosen to employ even *hundredths* of seconds throughout, to prevent accumulation of errors.

354. Letting fall the arcual perpendicular, *KA*\`, *KB*\`, *KC*\`, from the spherical centre *K*, on the sides *BC*, *CA*, *AB* of the spherical triangle *ABC*, I form six right-angled triangles, equal to each other two by two [Figure 34]. The triangle *AKC*\` gives, by the usual rules, if α, β, γ denote the sides, and *A*, *B*, *C* the angles of the triangle *ABC*, considered as a spherical one:

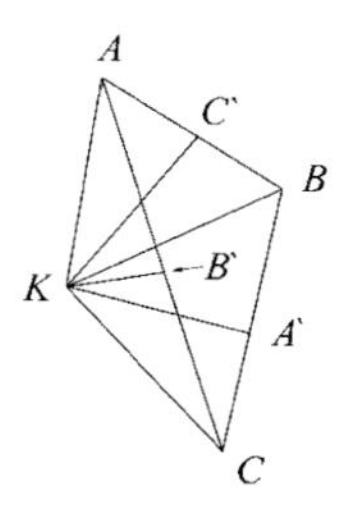

Fig. 34

$$\sin\frac{\gamma}{2} = \sin\frac{1}{2}AB = \sin AC' = \sin KA \sin AKC' = \sin \mathsf{c} \sin\frac{1}{2}AKB;$$

that is, here,

$$\sin\frac{\gamma}{2} = \sin 10^\circ.\sin 29^\circ = \sin 4^\circ\,49'\,45''\!\cdot\!25;$$

and

$$\cot C'AK = \cos KA.\tan AKC' = \cos \mathsf{c} \tan\frac{1}{2}AKB,$$

or,

$$\cot BAK = \cos 10^\circ.\tan 29^\circ = \cot 61^\circ\,22'\,13''\!\cdot\!46;$$

with the verification that

$$\tan AC' = \sec C'AK = \tan KA,$$

or here that

$$\tan 4^\circ\,49'\,45''\!\cdot\!25.\sec 61^\circ\,22'\,13''\!\cdot\!46 = \tan 10^\circ\,0'\,0''\!\cdot\!00.$$

355. In like manner, the right angled triangle BKA' gives,

$$\frac{\alpha}{2} = 7^\circ 14'\,10''\!\cdot\!29;\ KCB = CBK = 43^\circ 56'\,17''\!\cdot\!24;$$

and the triangle AKB' gives,

$$\frac{\beta}{2} = 9^\circ 40'\,42''\!\cdot\!10;\ KCA = CAK = 14^\circ 42'\,50''\!\cdot\!58;$$

while the triangle ABC' gave (article 354),

$$\frac{\gamma}{2} = 4^\circ 49'\,45''\!\cdot\!25;\ KBA = BAK = 61^\circ 22'\,13''\!\cdot\!46.$$

The sides and angles of the oblique spherical triangle ABC, which is thus inscribed in the small circle of 10° radius round K, are therefore:

$$\begin{cases} \alpha = 2\times 7^\circ\,14'\,10''\!\cdot\!29 = 14^\circ\,28'\,20''\!\cdot\!58; & \log\sin\alpha = \bar{1}\!\cdot\!397\,789\,4; \\ \beta = 2\times 9^\circ\,40'\,42''\!\cdot\!10 = 19^\circ\,21'\,24''\!\cdot\!20; & \log\sin\beta = \bar{1}\!\cdot\!520\,416\,2; \\ \gamma = 2\times 4^\circ\,49'\,45''\!\cdot\!25 = \ \ 9^\circ\,39'\,30''\!\cdot\!50; & \log\sin\gamma = \bar{1}\!\cdot\!224\,727\,0; \end{cases}$$

$$\begin{cases} A = BAK - CAK = 61^\circ\,22'\,13''\!\cdot\!46 - 14^\circ\,42'\,50''\!\cdot\!58 = 46^\circ\,39'\,22''\!\cdot\!88; & \sin\ \ldots\ \bar{1}\!\cdot\!861\,683\,8; \\ B = CBK + KBA = 43^\circ\,56'\,17''\!\cdot\!24 + 61^\circ\,22'\,13''\!\cdot\!46 = 105^\circ\,18'\,30''\!\cdot\!70; & \sin\ \ldots\ \bar{1}\!\cdot\!984\,310\,4; \\ C = KCB - KCA = 43^\circ\,56'\,17''\!\cdot\!24 - 14^\circ\,42'\,50''\!\cdot\!58 = 29^\circ\,13'\,26''\!\cdot\!66; & \sin\ \ldots\ \bar{1}\!\cdot\!688\,621\,2; \end{cases}$$

and although I do not *answer* for the *last figures* in these *logarithms*, nor for *last decimals* of the seconds, (which have however been *all* computed by myself,) yet that the results so far are very accurate may be judged by observing that they bear extremely well the verification:

$$\sin\beta\sin\gamma\sin A = \sin\gamma\sin\alpha\sin B = \sin\alpha\sin\beta\sin C = \text{(say)}\ \sin\varepsilon.$$

In fact, the three values thus found for log sin ε are,

$$\log\sin\varepsilon = \bar{2}\cdot 606\,827\,0; \quad = \bar{2}\cdot 606\,826\,8; \quad \text{and} \quad = \bar{2}\cdot 606\,826\,8;$$

which all agree, *to the hundredth of a second*, in giving

$$\varepsilon = 2°\,19'\,3''\cdot 93.$$

356. If we now again apply my formula for a, using the angles of the *spherical* triangle $A_1\,B_1\,C_1$ as more accurate than those of the plane triangle, but retaining the rigorous value $\mathsf{c} = 10°\,00'\,00''$, we have this new little calculation, which conducts to the new value, $\mathsf{a} = 31''\cdot 65$:-

$\frac{A}{2} = 23°\,19'\,41''$, $\log_{10}\sin \ldots \bar{1}\cdot 597\,69$;	$\frac{B-C}{2} = 38°\,02'\,32''$, $\log_{10}\sin \ldots \bar{1}\cdot 789\,75$
$\frac{B}{2} = 52°\,39'\,15''$, $\log_{10}\sin \ldots \bar{1}\cdot 900\,36$;	$\frac{A-C}{2} = 08°\,42'\,58''$, $\log_{10}\sin \ldots \bar{1}\cdot 180\,52$
$\frac{C}{2} = 14°\,36'\,43''$, $\log_{10}\sin \ldots \bar{1}\cdot 401\,87$;	$\frac{B-A}{2} = 29°\,19'\,34''$, $\log_{10}\sin \ldots \bar{1}\cdot 690\,00$
$\log_{10} 8 \ldots 0\cdot 903\,09$;	$\mathsf{c} = 29°\,00'$, $\log_{10}\sin \ldots \bar{1}\cdot 239\,67$
$(\div 2) \ldots \bar{1}\cdot 803\,01$	
$\theta = 52°\,51'\,07''$, $\log_{10}\sin \ldots \bar{1}\cdot 901\,50$;	$\theta = 52°\,51'\,07''$, $\log_{10}\tan \ldots 0\cdot 120\,55$
	$\bar{2}\cdot 020\,49$
	$\times 2$
	$\bar{4}\cdot 040\,98$
	$\frac{8\mathsf{c}}{1''} = 288\,000$; $\log_{10} \ldots 5\cdot 459\,39$
	$\frac{\mathsf{a}}{1''} = 31\cdot 65$; $\log_{10} \ldots 1\cdot 500\,37$
	$\mathsf{a} = 31''\cdot 65$

357. Although this last value for the least breadth of the annulus, namely,

$$\mathsf{a} = 31''\cdot 65,$$

computed as above from the formula which Quaternions had given me, is found to coincide, to the most astonishing degree of accuracy, with the result of a long chain of calculation with spherical triangles, as I intend in this Postscript to show: yet when I entered on that chain, I did not venture to expect more than that this *annular breadth* a would turn out, by spherical trigonometry, to be *about half a minute*, for the Example here selected. – I must now explain how I proceeded to determine the *bisected triangle* $A_1\,B_1\,C_1$ (article 341); and the spherical *centre* K', and the arcual *radius* c', of the *new small circle* $A'B'C'$, which is *inscribed* in that bisected triangle, as also the *distance of centres*, $\mathsf{d} = KK'$, which I *expected* to find connected with c and c', by the *approximate relation*,

$$\mathsf{a} = \mathsf{c} - \mathsf{c}' - \mathsf{d} = 30'', \quad \textit{nearly};$$

so that, because $\mathsf{c} = 10^\circ$, *exactly*, we ought to have

$$\mathsf{c}' + \mathsf{d} = 9^\circ\,59'\,30'', \quad \textit{nearly};$$

or, *more nearly*, if the *last* value of a be adopted,

$$\mathsf{c}' + \mathsf{d} = 9^\circ\,59'\,28''{\cdot}35.$$

In point of fact, my final results from spherical trigonometry, have been (for the present Example),

$$\mathsf{c}' = 6^\circ\,7'\,25''{\cdot}49;$$

and

$$\mathsf{d} = 3^\circ\,52'\,2''{\cdot}82;$$

giving the sum,

$$\mathsf{c}' + \mathsf{d} = 9^\circ\,59'\,28''{\cdot}31$$

the agreements of which with what I may call the *result for quaternions* would really appear suspicious to myself, if I did not know the perfect independence of the two methods, and were not conscious of the extreme care employed.

358.[*] The next step, – and the only one which can present any difficulty to a person who is accustomed to the known rules of solution of spherical triangles, – is the *inscription* of the *bisected triangle* $A_1\,B_1\,C_1$; or the determination of its *sides*, which I shall call $2\alpha_1$, $2\beta_1$, $2\gamma_1$, and its *angles* which I shall denote simply by A_1, B_1, C_1.

259. Quaternions, however, furnished me long ago with a mode of conquering this little difficulty: for they show with ease, that we have the rigorous equations,

$$\frac{\cos\alpha}{\cos\alpha_1} = \frac{\cos\beta}{\cos\beta_1} = \frac{\cos\gamma}{\cos\gamma_1} = \cos\varepsilon;$$

where ε is the auxiliary angle determined by the condition (see article 355)

$$\sin\varepsilon = \sin\beta\sin\gamma\sin A = \&\text{c}.$$

(I think, indeed, that in one of my printed "Lectures"[†], I proved this Theorem by spherical geometry; and, very probably it may be a known one otherwise.)

360. Applying it here, we have the values,

$$\log\cos\alpha_1 = \bar{1}{\cdot}986\,351\,1; \quad \log\cos\beta_1 = \bar{1}{\cdot}975\,085\,0; \quad \log\cos\gamma_1 = \bar{1}{\cdot}994\,155\,4;$$

whence, at least nearly,

$$\alpha_1 = 14^\circ\,17'\,22''{\cdot}20; \quad \beta_1 = 19^\circ\,13'\,21''{\cdot}88; \quad \gamma_1 = 9^\circ\,22'\,43''{\cdot}75.$$

[*] [Articles 358–366 are based solely on MS 1493/1171.]
[†] [W. R. Hamilton, *Lectures on Quaternions*, Hodges and Smith, Dublin: 1853.]

Arcs, however, which are *even so small* as these, are not well given by their *cosines*; and I adopted the modified formulæ (with certain auxiliary angles α', β', γ'):

$$\tan\alpha' = \tan^2\frac{\varepsilon}{2}\cot\alpha; \quad \tan\beta' = \&c.; \quad \tan\gamma' = \&c.;$$

$$\tan\frac{\alpha-\alpha_1}{2} = \tan^2\frac{\varepsilon}{2}\cot(\alpha-\alpha'); \quad \tan\frac{\beta-\beta_1}{2} = \&c.; \quad \tan\frac{\gamma-\gamma_1}{2} = \&c.;$$

which gave the slightly altered values, for the half sides of the bisected triangle:

$$\alpha_1 = 14°\,17'\,22''\cdot17; \quad \beta_1 = 19°\,13'\,21''\cdot89; \quad \gamma_1 = 9°\,22'\,43''\cdot97;$$

for the minute decimal fractions of which, notwithstanding, I by no means ventured confidently to accept.

(End of Sheet LVI. Sent to the post by Fitzpatrick on Friday 7 September 1860: with a memorandum to mention that "I posted a note this morning to tell you my son William finds the Theorem [article 342] to be Terquem's*".)

361. Calling for brevity, the *bisecting* triangle ABC the *inner* triangle, and the *bisected* triangle $A_1 B_1 C_1$ (and that one of which the *sides* are bisected by A, B, C) the *outer* one: while the three triangles $A_1 BC$, $AB_1 C$, ABC_1, which together with the inner make up the outer triangle, may be called the three *attached* triangles, the *angles* of these latter may be denoted as follows (see Figure 35):

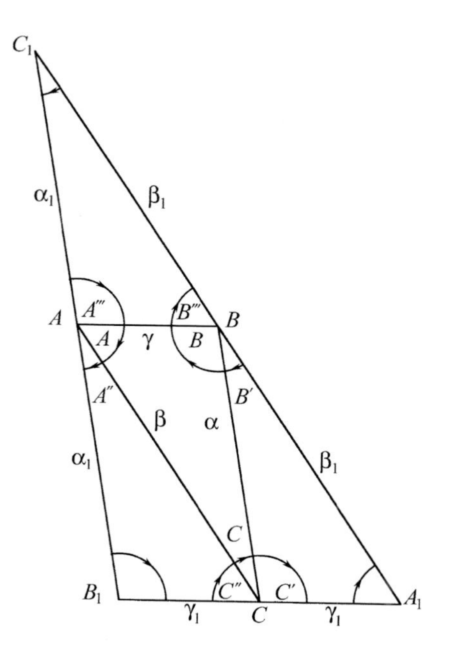

Fig. 35

* [Olry Terquem (1782–1862).]

$$CA_1B = A_1; \qquad A_1BC = B'; \qquad BCA_1 = C';$$

$$CAB_1 = A''; \qquad AB_1C = B_1; \qquad B_1CA = C'';$$

$$C_1AB = A'''; \qquad ABC_1 = B'''; \qquad BC_1A = C_1;$$

while the *sides* of the same three attached triangles, respectively opposite to these angles, are

$$\alpha, \gamma_1, \beta_1; \quad \gamma_1, \beta, \alpha_1; \quad \beta_1, \alpha_1, \gamma.$$

362. The values of the *nine angles* (article 361) were next deduced *from the sides*, in the three attached triangles, by the usual roots: and thus I found

$$A_1 = 46°\,55'\,21''\cdot4; \quad B' = 28°\,26'\,37''\cdot70; \quad C' = 105°\,47'\,7''\cdot48;$$

$$B_1 = 108°\,18'\,21''\cdot78; \quad A'' = 27°\,49'\,32''\cdot84; \quad C'' = 44°\,59'\,25''\cdot40$$

$$C_1 = 29°\,24'\,23''\cdot58; \quad A''' = 105°\,31'\,5''\cdot72; \quad B''' = 46°\,14'\,52''\cdot07;$$

while we had before (article 355), for the angles of the inner triangle,

$$A = 46°\,39'\,22''\cdot88; \quad B = 105°\,18'\,30''\cdot70; \quad C = 29°\,13'\,26''\cdot66;$$

so that the sines of the angles adjacent to the corners of the inner triangle ABC are:

$$A + A'' + A''' = 180°\,0'\,1''\cdot44;$$

$$B + B''' + B' = 180°\,0'\,0''\cdot47;$$

$$C + C' + C'' = 179°\,59'\,59''\cdot54;$$

and the sum of the angles of the outer triangle is,

$$A_1 + B_1 + C_1 = 184°\,38'\,6''\cdot77.$$

363. That the angles thus obtained are very nearly accurate, may be judged by the *four* following verifications: for each of the three points A_1, B_1, C_1 the sum of the three angles at that point is nearly *equal* to two right angles; and the sum of the three angles at A_1, B_1, C_1 *exceeds* two right angles by nearly the theoretical value of the *spherical excess* of the outer triangle; since such excess ought (by a theorem which I derived from Quaternions several years ago, and confirmed by spherical geometry, and which is probably known) to be *exactly double* of that *auxiliary angle* for the inner triangle, which has been above called ε: but (by article 355) we have

$$2\varepsilon = 4°\,38'\,7''\cdot86;$$

and (by article 362),

$$A_1 + B_1 + C_1 - \pi = 4°\,38'\,6''\cdot77.$$

364. In order that the *three first verifications* which appeared to be the simplest, might be *rigorously* borne by the adopted angles at the corners of the *inner* triangle, I applied to the six new angles at those points the following very small corrections:

$$\delta A'' = \delta A''' = -0''\cdot 72; \quad \delta B' = -0''\cdot 24; \quad \delta B''' = -0''\cdot 23;$$

$$\delta C' = +0''\cdot 24; \quad \delta C'' = +0''\cdot 22;$$

and then *recomputed* the three attached angles, assuming as *new data* the values:

for A_1BC, $BC = \alpha = 14^\circ 28' 20''\cdot 58$; $B' = 28^\circ 26' 37''\cdot 46$; $C' = 105^\circ 47' 7''\cdot 72$;

for B_1CA, $CA = \beta = 19^\circ 21' 24''\cdot 20$; $C'' = 44^\circ 59' 25''\cdot 62$; $A'' = 27^\circ 49' 22''\cdot 12$;

for C_1AB, $AB = \gamma = 9^\circ 39' 30''\cdot 50$; $A''' = 105^\circ 31' 5''\cdot 00$; $B''' = 46^\circ 14' 51''\cdot 84$;

employing for this purpose the *Equations of Gauss**, (which I have very often worked with, although I forget whether they are given in any usual English book,) namely,

$$\frac{\cos}{\sin}\frac{1}{2}(a+b).\sin\frac{1}{2}C = \cos\frac{1}{2}(A \pm B).\frac{\cos}{\sin}\frac{1}{2}c;$$

$$\frac{\cos}{\sin}\frac{1}{2}(a-b).\cos\frac{1}{2}C = \sin\frac{1}{2}(A \pm B).\frac{\cos}{\sin}\frac{1}{2}c;$$

which *include* the *Analogies of Napier*, but have several advantages over them.

365. In this manner I found, from the three attached triangles thus *separately* resolved anew, the values:

$$\begin{array}{lll} A_1 = 46^\circ 55' 21''\cdot 05; & & \beta_1 = 19^\circ 13' 21''\cdot 82; \quad \gamma_1 = 9^\circ 22' 43''\cdot 02; \\ B_1 = 108^\circ 18' 22''\cdot 30; & \alpha_1 = 14^\circ 17' 22''\cdot 25; & \qquad\qquad \gamma_1 = 9^\circ 22' 43''\cdot 75; \\ C_1 = 29^\circ 24' 24''\cdot 58; & \alpha_1 = 14^\circ 17' 21''\cdot 70; & \beta_1 = 19^\circ 13' 21''\cdot 38; \end{array}$$

in which the near agreement of the two values of α_1 shows that the *side* B_1C_1 of the outer triangle is very well *bisected*, and similarly for the *two other sides*; which the *spherical excess* of that triangle $A_1B_1C_1$ being, with these last values,

$$A_1 + B_1 + C_1 - \pi = 4^\circ 38' 7''\cdot 93,$$

agrees as well as we could desire, – and a little better than as recently constructed, with the *theoretical value* (article 363); namely,

$$2\varepsilon = 4^\circ 38' 7''\cdot 84.$$

In fact, it would be quite unreasonable to expect a closer agreement.

366. I therefore *adopted* as the elements of the *outer* triangle $A_1B_1C_1$, the sides and angles following:

$$\left\{\begin{array}{lll} 2\alpha_1 = 28^\circ 34' 43''\cdot 95; & 2\beta_1 = 38^\circ 26' 43''\cdot 20; & 2\gamma_1 = 18^\circ 45' 27''\cdot 57; \\ A_1 = 46^\circ 55' 21''\cdot 05; & B_1 = 108^\circ 18' 22''\cdot 30; & C_1 = 29^\circ 24' 24''\cdot 58: \end{array}\right.$$

while I *retain* the sides α, β, γ, and angles A, B, C, of the *inner* triangle ABC, which had been previously constructed (article 355): namely, – to have them in one view, –

* [See I. Todhunter, *Spherical trigonometry* (details in footnote to article 342), p. 36, eqns 43–46, who attributes them to J. B. J. Delambre.]

$$\begin{cases} \alpha = 14^\circ\, 28'\, 20''{\cdot}58; & \beta = 19^\circ\, 21'\, 24''{\cdot}20; & \gamma = 9^\circ\, 39'\, 30''{\cdot}50; \\ A = 46^\circ\, 39'\, 22''{\cdot}88; & B = 105^\circ\, 18'\, 30''{\cdot}70; & C = 29^\circ\, 13'\, 26''{\cdot}66: \end{cases}$$

And thus, these *two* triangles, – and the *three* which we have called the attached ones (article 361), – may be now considered to have had their elements numerically determined, by spherical trigonometry and logarithms, as well (perhaps) as Tables with Seven Decimal Figures will permit: while the arcual *radius* c, of the circle ABC, continues to be *exactly* $10^\circ\, 0'\, 0''{\cdot}00$ in virtue of our first *assumption*.

367. Continuing to spare no pains upon this *one* selected Example, and seeking rather to multiply verifications than to employ abridgements, I resolved, independently of each other, the three new spherical triangles,

$$K'B_1C_1, \quad K'C_1A_1, \quad K'A_1B_1;$$

in which K' is still (article 341) the centre of the small circle inscribed in the outer triangle $A_1B_1C_1$; and A', B', C' are still the three points of contact, or the feet of the perpendiculars from K' while ρ_1', ρ_2', ρ_3' may denote the distances of that centre K' from the corners A_1, B_1, C_1, see Figure 36.

368. In the triangle $K'B_1C_1$, I had the side, namely

$$B_1C_1 = 2\alpha_1 = 28^\circ\, 34'\, 43''{\cdot}95,$$

and the two adjacent angles,

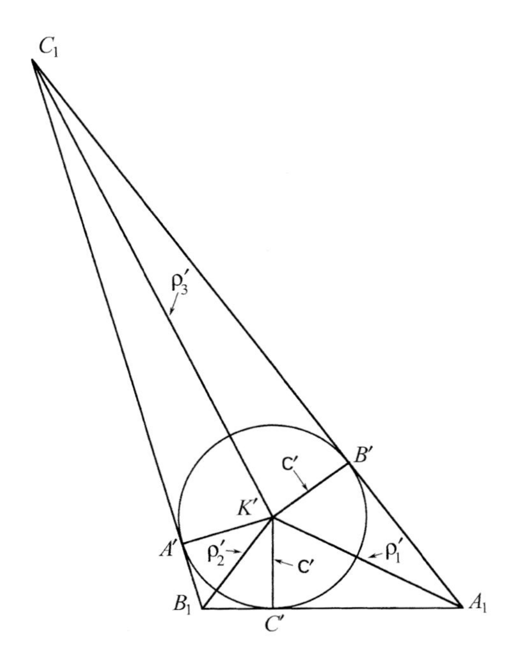

Fig. 36

$$C_1 B_1 K' = \tfrac{1}{2} B_1 = 59^\circ\, 9'\, 11''\cdot 15, \qquad K' C_1 B_1 = \tfrac{1}{2} C_1 = 14^\circ\, 42'\, 12''\cdot 29;$$

whence I deduced (still using Gauss's Formulæ – though ones more common might of course have been employed) the following values for the two other sides, and the remaining angle:

$$\rho_3' = K'C_1 = 24^\circ\, 51'\, 10''\cdot 54; \qquad \rho_2' = K'B_1 = 7^\circ\, 33'\, 44''\cdot 28;$$

$$B_1 K' C_1 = 112^\circ\, 41'\, 30''\cdot 06;$$

and for the altitude,

$$\sin \mathsf{c}' = \sin K'A' = \sin \rho_2' \sin \tfrac{1}{2} B_1 = \sin \rho_3' \sin \tfrac{1}{2} C_1 = \sin 6^\circ\, 7'\, 25''\cdot 51.$$

369. In like manner, the triangle $K' C_1 A_1$ gave:

$$\rho_1' = K'A_1 = 15^\circ\, 32'\, 31''\cdot 05; \qquad \rho_3' = K'C_1 = 24^\circ\, 51'\, 10''\cdot 51; \qquad C_1 K' A_1 = 143^\circ\, 54'\, 54''\cdot 58;$$

and

$$\mathsf{c}' = K'B' = 6^\circ\, 7'\, 25''\cdot 49;$$

while the triangle $KA_1 B_1$, – each being still treated independently of the rest, – gave,

$$\rho_1' = K'A_1 = 15^\circ\, 32'\, 30''\cdot 95; \qquad \rho_2' = K'B' = 7^\circ\, 33'\, 44''\cdot 21; \qquad A_1 K' B_1 = 103^\circ\, 23'\, 34''\cdot 98;$$

and

$$\mathsf{c}' = K'C' = 6^\circ\, 7'\, 25''\cdot 46.$$

370. The sum of the three angles at K', in these triangles is

$$112^\circ\, 41'\, 30''\cdot 06 + 143^\circ\, 54'\, 54''\cdot 58 + 103^\circ\, 23'\, 34''\cdot 98 = 359^\circ\, 59'\, 59''\cdot 62;$$

it therefore scarcely differs from 360°, with which theoretically it ought to coincide: and thus one verification is obtained. The near agreement of the two values found for each arc ρ' which is a common side of two triangles, supplies three other verifications; and we may safely *adopt*, for these arcs of distance, the arithmetic means,

$$\rho_1' = K'A_1 = 15^\circ\, 32'\, 31''\cdot 00; \qquad \rho_2' = K'B_1 = 7^\circ\, 33'\, 44''\cdot 25; \qquad \rho_3' = K'C_1 = 24^\circ\, 51'\, 10''\cdot 52,$$

of which the differences from the separate values are insensible. Finally, as a fifth verification, the three last triangles give almost exactly the same altitude for each; and the following value may therefore be now adopted, for the radius of the inscribed circle $A'B'C'$:

$$\mathsf{c}' = 6^\circ\, 7'\, 25''\cdot 49;$$

as was mentioned by anticipation in article 357:– No doubt, this value might have been more rapidly deduced from the three sides $2\alpha_1$, $2\beta_1$, $2\gamma_1$, of the outer triangle, $A_1 B_1 C_1$, by a well known formula: but besides choosing to have verifications, I wanted other arcs, determined as above. (In fact, the formula,

$$\tan \mathsf{c}' = \sqrt{\frac{\sin(\beta_1 + \gamma_1 - \alpha_1)\sin(\gamma_1 + \alpha_1 - \beta_1)\sin(\alpha_1 + \beta_1 - \gamma_1)}{\sin(\alpha_1 + \beta_1 + \gamma_1)}},$$

gives *precisely* the last value for c'.)

371. I proposed also to determine the arcs of *distance,* ρ_1, ρ_2, ρ_3, of the *other centre, K,* from the *same three corners,* of the outer triangle; and between those distances and certain other arcs: and I resolved a system of new triangles for the purpose, as follows.

372. In the triangle $A_1 BK$, we have two sides and the included angle, namely,

$$BK = \mathsf{c} = 10°; \quad BA_1 = \beta_1 = 19° 13' 21''{\cdot}60; \quad \text{and}$$

$$A_1 BK = A_1 BC + A^{\backprime} BK = B' + CBK,$$

that is, by articles 355, 364,

$$A_1 BK = 28° 26' 37''{\cdot}46 + 43° 56' 17''{\cdot}24 = 72° 22' 54''{\cdot}70;$$

whence the third side

$$\rho_1 = KA_1 = 18° 42' 3''{\cdot}05;$$

and the two remaining angles are,

$$KA_1 B = 31° 4' 36''{\cdot}66;$$

$$BKA_1 = 78° 9' 7''{\cdot}34.$$

[See Figure 37.]

373. Again, in the triangle KCA_1, we have the two sides,

$$CK = \mathsf{c} = 10°; \quad CA_1 = \gamma_1 = 9° 22' 43''{\cdot}78;$$

and the included angle,

$$KCA_1 = C' + KCB = 105° 47' 7''{\cdot}72 + 43° 56' 17''{\cdot}24 = 149° 43' 24''{\cdot}96;$$

whence

$$\rho_1 = KA_1 = 18° 42' 3''{\cdot}41; \quad CA_1 K = 15° 50' 45''{\cdot}69; \quad A_1 KC = 14° 50' 51''{\cdot}31;$$

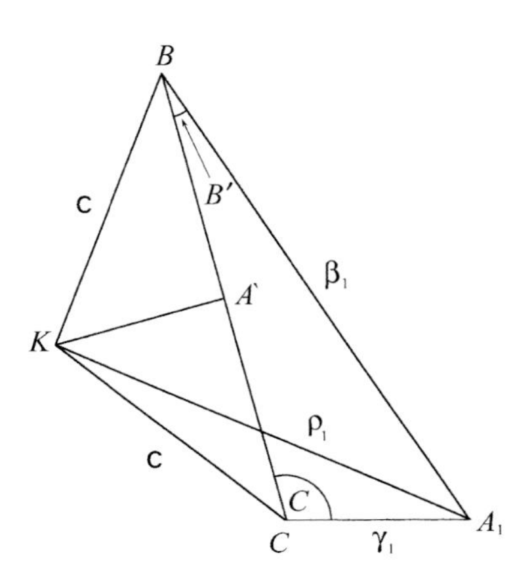

Fig. 37

374. As verifications, the two values of the distance ρ_1, considered as a *common side* of the two last triangles, are almost exactly equal; and we shall *adopt* their mean, namely

$$\rho_1 = KA_1 = 18^\circ\, 42'\, 3''{\cdot}23.$$

Again, the two angles at K had their sum $= 92^\circ\, 59'\, 58''{\cdot}65$; whereas in rigour it ought (article 350) to be, $BKC = 93^\circ\, 0'\, 0''{\cdot}00$; and I make the small difference disappear, by applying the corrections $+0''{\cdot}68$, and $+0''{\cdot}67$, or by *adopting* the values:

$$BKA_1 = 78^\circ\, 9'\, 8''{\cdot}02; \qquad A_1\, KC = 14^\circ\, 50'\, 51''{\cdot}98.$$

Finally, the sum of the two partial angles at A_1 exceeds, by $+1''{\cdot}30$, the value above adopted (article 366) for the total angle A_1, namely, $46^\circ\, 55'\, 21''{\cdot}05$; I therefore apply to each part the correction $-0''{\cdot}65$ and write,

$$KA_1\, B = 31^\circ\, 4'\, 36''{\cdot}01; \qquad CA_1\, K = 15^\circ\, 50'\, 45''{\cdot}04.$$

375. In like manner, from the two triangles, KAB_1, $KB_1\, C$, (see Figure 38) I infer that the following values are very nearly correct:

$$\rho_2 = KB_1 = 5^\circ\, 4'\, 25''{\cdot}85;$$

$$B_1\, KA = 140^\circ\, 43'\, 15''{\cdot}98; \qquad CKB_1 = 68^\circ\, 16'\, 44''{\cdot}02;$$

$$AB_1\, K = 26^\circ\, 26'\, 57''{\cdot}94; \qquad KB_1\, C = 81^\circ\, 51'\, 24''{\cdot}36;$$

the sum of the two former angles being

$$209^\circ = 360^\circ - 151^\circ = 2\pi - AKC;$$

(in which, 151° is the sum $58^\circ + 93^\circ = AKB + BKC$, as in article 350;) and the sum of the two latter angles is

$$B_1 = 108^\circ\, 18'\, 22''{\cdot}30 \quad \text{(as in article 366)}.$$

376. And the two triangles, KBC_1, $KC_1\, A$, (for which it is unnecessary to draw a new Figure,) give:

$$\rho_3 = KC_1 = 24^\circ\, 7'\, 59''{\cdot}98;$$

$$C_1\, KB = 50^\circ\, 7'\, 44''{\cdot}65; \qquad AKC_1 = 7^\circ\, 52'\, 15''{\cdot}35;$$

$$BC_1\, K = 23^\circ\, 52'\, 41''{\cdot}01; \qquad KC_1\, A = 5^\circ\, 31'\, 43''{\cdot}57;$$

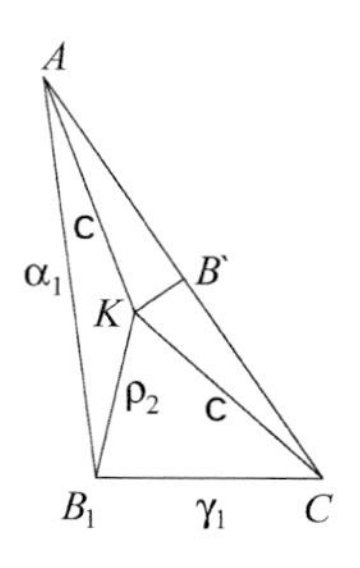

Fig. 38

the sum of the two former being $AKB = 58°$, as in article 350; and the sum of the two latter being $C_1 = BCA = 29° 24' 24''\cdot 58$, as in article 366.

377. We are now in possession of elements, which have been carefully collated and verified, and which enable us to determine the sought *distance of centres*, d or KK', in *three distinct ways*, as the *common base* of *three new spherical triangles*; which have their *vertices* at the points A_1, B_1, C_1. For we have determined, in the three last articles, the distances ρ of K from these three vertices, and in article 370 the distances ρ' of K' from the same; and the angles which the two arcs ρ_1, ρ_1', or ρ_2, ρ_2', or ρ_3, ρ_3' make with each other, may be found as the *semidifferences* of the angles (in the three last articles), into which the arcs ρ divide the angles A_1, B_1, C_1; because these last angles are *bisected* by the arcs ρ'. Thus,

$$\left\{\begin{aligned} KA_1K' &= \tfrac{1}{2}(KA_1B - CA_1K) = \tfrac{1}{2}(31° 4' 36''\cdot 01 - 15° 50' 45''\cdot 04) = 7° 36' 55''\cdot 48; \\ KB_1K' &= \tfrac{1}{2}(KB_1C - AB_1K) = \tfrac{1}{2}(81° 51' 24''\cdot 36 - 26° 26' 57''\cdot 94) = 27° 42' 13''\cdot 21; \\ KC_1K' &= \tfrac{1}{2}(KC_1A - BC_1K) = \tfrac{1}{2}(5° 31' 43''\cdot 57 - 23° 52' 41''\cdot 01) = -9° 10' 28''\cdot 72; \end{aligned}\right.$$

the *negative sign* implying here that the *rotation* round C_1 from K to K' has a *contrary character* to that of the other rotation; so that we may write,

$$K'C_1K = +9° 10' 28''\cdot 72.$$

378. Solving now the triangle KA_1K', with the data,

$$A_1K = \rho_1 = 18° 42' 3''\cdot 23; \quad A_1K' = \rho_1' = 15° 32' 31''\cdot 00; \quad KA_1K' = 7° 36' 55''\cdot 48;$$

I find,

$$A_1K'K = 140° 57' 7''\cdot 04; \quad K'KA_1 = 31° 45' 59''\cdot 18;$$

and

$$\mathsf{d} = KK' = 3° 52' 2''\cdot 7.9.$$

379. Solving the triangle KB_1K', with the three data,

$$B_1K = \rho_2 = 5° 4' 25''\cdot 85; \quad B_1K' = \rho_2' = 7° 3' 44''\cdot 25; \quad KB_1K' = 27° 42' 13''\cdot 21,$$

I obtain the values,

$$K'KB_1 = 114° 53' 33''\cdot 48; \quad B_1K'K = 37° 33' 33''\cdot 44; \quad \mathsf{d} = KK' = 3° 52' 2''\cdot 85.$$

380. And solving the third triangle $K'C_1K$, with

$$C_1K = \rho_3 = 24° 7' 59''\cdot 98; \quad C_1K' = \rho_3' = 24° 51' 10''\cdot 52; \quad K'C_1K = 9° 10' 28''\cdot 72,$$

there results,

$$C_1KK' = 96° 30' 51''\cdot 49; \quad KK'C_1 = 75° 8' 0''\cdot 37; \quad \mathsf{d} = KK' = 3° 52' 2''\cdot 80.$$

381. The spherical *angles*, found by these three last triangles, may easily be a second or two in error, notwithstanding all the pains that have been taken; yet they bear several verifications very well. But the three closely concurrent values of the *arc KK′*, which is their common *base*, seem likely to be *very* accurate: and I *adopt* their *mean*, namely,

$$\mathsf{d} = KK' = 3°\,52'\,2''{\cdot}81,$$

as mentioned (with a different hundredth) in article 357; whence the *least breadth of the annulus* is, in this Example,

$$\mathsf{a} = \mathsf{c} - \mathsf{c}' - \mathsf{d} = 10°\,0'\,0''{\cdot}0 - 6°\,7'\,25''{\cdot}49 - 3°\,52'\,2''{\cdot}81 = +31''{\cdot}70,$$

by spherical trigonometry; while it had been found (in article 356) to be *nearly*

$$\mathsf{a} = +31''{\cdot}65$$

by quaternions.

382.[*] (September 12th, 1860) In my desire not to exaggerate the degree of approximation thus obtained, I must admit that after all the pains which have been taken, an error of a second, or even of two seconds, may *possibly* exist in the value of the annular breadth a, as thus calculated by a long chain of spherical triangles: although I own that I *think* the error does not exceed *one* second, in one direction or the other. And again I admit that there was something arbitrary and conjectural, in the substitution (article 356) of the angles of the *spherical* triangle *ABC*, for those of the *plane* triangle, which latter had been more immediately *suggested* to me by the *quaternion formula*. So that, after all, I do not venture to assert more than this: that *quaternions* had led me to *expect* that the value of a would be found to be (article 357) *about half a minute*, in the Example selected; and that *this* expectation was fully borne out, by the result of the long calculation with *spherical trigonometry*.

383. It must however, be remembered, that one very important part of the *prediction*, derived from quaternions, was the *positive* chord of a, or of $\mathsf{c} - \mathsf{c}' - \mathsf{d}$; and that *this* also was confirmed by spherical trigonometry: whereas I have at present no notion how such a result could have been *foreseen* from Geometry, although I suppose that some way of doing it will be discovered – though not by me. How was one to know *beforehand* that the ***Eclipse*** of the ***Median Circle*** (for I fear that some such *name* must now be adopted, since the theorem of its *contacts* in the plane has been traced to a foreign[†] author) would be *Annular* (article 343)?

384. This may be a convenient place for mentioning something of the *nature* of my *Quaternion Investigation*[‡] so far at least as to show, more distinctly than I have yet done, the *character* of the *result* to which it led me.

385. I started with *three unit vectors*, α, β, γ, directed from the centre *O* of the sphere to the three points *A*, *B*, *C*,

[*] [Articles 382–391 are based solely on MS 1493/1171.]

[†] *Viz*.: Terquem [see footnote to article 360], who published and proved it in his [he was an editor] *Nouvelles Annales* [*de Mathématiques*], Vol. 1, p. 196, &c., Paris, 1842. – The place was shown me by my son William, as I stated in my note of September 7th.

[‡] [In addition to Hamilton's publications on quaternions, useful references are: *An elementary treatise on quaternions*, by P. G. Tait (1831–1901), Clarendon Press, Oxford: 1867; and Chapter X of *Vector and tensor analysis*, by L. Brand, Chapman & Hall, London: 1947.]

$$OA = \alpha, \quad OB = \beta, \quad OC = \gamma,$$

$$\alpha^2 = \beta^2 = \gamma^2 = -1, \quad T\alpha = T\beta = T\gamma = 1;$$

[$T\alpha$ denotes the magnitude of the vector α] which *vectors*, α, β, γ, became thus the fundamental *data* of the question.

386. Writing,

$$a = \tfrac{1}{2}T(\beta - \gamma), \quad b = \tfrac{1}{2}T(\gamma - \alpha), \quad c = \tfrac{1}{2}T(\alpha - \beta),$$

or

$$a = \sqrt{\tfrac{1}{2}(1 + S\beta\gamma)}, \quad b = \sqrt{\tfrac{1}{2}(1 + S\gamma\alpha)}, \quad c = \sqrt{\tfrac{1}{2}(1 + S\alpha\beta)},$$

and taking all these radicals positively, the *scalars*, a, b, c, represent, on my principles, the *sines of the half sides* of the spherical triangle ABC;

$$a = \sin\tfrac{1}{2}BC, \quad b = \sin\tfrac{1}{2}CA, \quad c = \sin\tfrac{1}{2}AB;$$

and *all other scalars* in the investigation must be considered as *functions of these three*, and may be developed according to their ascending powers and products, when the triangle is supposed to be *small*.

387. And because the *final scalars*, – such as those which have been called c, c', d, a, – are necessarily *Symmetric Functions* of them, – at least till we come to consider the circles which touch *externally*, – I conceived the *three fundamental scalars*, a, b, c, to be the *three roots* of a given *cubic equation*,

$$x^3 - sx^2 + tx - p = 0;$$

in other words, I made

$$s = a + b + c, \quad t = bc + ca + ab, \quad p = abc,$$

and sought to express other symmetric scalars, and especially and ultimately a, (which is *not assumed* to be *positive*,) as *rational functions of* s, t, p, at least through ascending series.

388. Introducing a new vector, κ, and two new scalars, e, h, defined by the equations,

$$\kappa = V(\beta\gamma + \gamma\alpha + \alpha\beta), \quad e = S\alpha\beta\gamma, \quad h = TV\alpha\beta\gamma,$$

a general theorem in quaternions gave me the expression,

$$\begin{aligned} h^2 = 1 - e^2 &= (S\beta\gamma)^2 + (S\gamma\alpha)^2 + (S\alpha\beta)^2 + 2S\beta\gamma S\gamma\alpha S\alpha\beta, \\ &= (1 - 2a^2)^2 + (1 - 2b^2)^2 + (1 - 2c^2)^2 - 2(1 - 2a^2)(1 - 2b^2)(1 - 2a^2), \end{aligned}$$

whence

$$e^2 = 4(\Delta^2 - 4p^2),$$

if we write

$$\Delta^2 = s(4st - s^3 - 8p)$$
$$= 2(b^2c^2 + c^2a^2 + a^2b^2) - (a^4 + b^4 + c^4)$$
$$= (a + b + c)(b + c - a)(c + a - b)(a + b - c);$$

so that Δ represents the *area* of the *plane* triangle ABC, of which the *sides* are $2a$, $2b$, $2c$; and

$$\frac{2p}{\Delta} = r = \sin \mathsf{c}$$

[which is equal to the] *linear radius* of the circle circumscribed about the plane triangle ABC, c being still the *arcual radius* of the *small circle ABC* on the *sphere*, that is, of the *Median Circle* (article 383), as it may be called with respect to that other spherical triangle $A_1 B_1 C_1$, which has not yet been constructed. And if we suppose the order of the *rotation ABC* to be so chosen as to render $e > 0$, as well as $\Delta > 0$, we have the very simple expressions,

$$e = 2\Delta \cos \mathsf{c}; \quad 4p = 2\Delta \sin \mathsf{c};$$

all the formulæ, so far, being *rigorous*. It may be noted that quaternions give also,*

$$e = S.\alpha V\beta\gamma = S.\beta V\gamma\alpha = S.\gamma V\alpha\beta$$
$$= \text{sine of base} \times \text{sine of altitude},$$

for the *spherical* triangle ABC; and therefore

$$e = \sin \varepsilon, \quad h = \cos \varepsilon,$$

if ε be the angle so denoted in former articles, namely the *semiexcess* of the bisected or *outer* triangle, $A_1 B_1 C_1$. Quaternions give also this other very simple relation which (like all the foregoing) is rigorous, –

$$T\kappa = 2\Delta.$$

389. Instead of immediately considering the *circle ABC*, you conceive that my method leads me rather to consider the *cone of revolution*, which has its *vertex* at the centre O of the sphere, and has that circle for its *base*. And I easily find that the *quaternion-equation* of this cone may thus be written:

$$(S\kappa\rho)^2 + e^2\rho^2 = 0.$$

This, then, may be said to be, in the present investigation, the *Equation of the Median Cone*: and if K be still the spherical centre of the *median-circle ABC*, I find that

$$OK = -U\kappa.^\dagger$$

390. *Any other cone* of revolution, with the same vertex O, may be represented by an equation of the *same form*, such as

$$(S\kappa'\rho)^2 + e'^2\rho^2 = 0;$$

* [The dot after the S is to show that it refers to the whole product that follows it.]

† [$U\kappa$ is the unit vector of κ.]

and the *condition of contact* for these two concentric cones to be expressed by the formula,

$$TV\kappa\kappa' = T(e\kappa' \pm e'\kappa);$$

or

$$(V\kappa\kappa')^2 = (e\kappa' \pm e'\kappa)^2;$$

or, by another quaternion transformation,

$$(S\kappa\kappa' \mp ee')^2 = (\kappa^2 + e^2)(\kappa'^2 + e'^2);$$

or,

$$S\kappa\kappa' \mp ee' = \pm'\sqrt{(T\kappa)^2 - e^2}\sqrt{(T\kappa')^2 - e'^2},$$

when the accent in $\pm'$ implies that the second ambiguous sign is independent of the first; or

$$e^{-1}S\kappa\kappa' \mp e' = \pm' 16p,$$

if we remember that (by article 388)

$$\sqrt{(T\kappa)^2 - e^2} = \sqrt{4\Delta^2 - e^2} = 4p,$$

and suppose p' to be a new scalar such that

$$\sqrt{(T\kappa')^2 - e'^2} = 4ep';$$

or

$$e^{-1}S\kappa\kappa' \mp' 16pp' = \pm e'.$$

391. For the case of internal contact, which exists at the plane limit, I found that I might take the *upper signs*; or writing

$$N = e^{-1}S\kappa\kappa' - 16pp' - e' = (\text{say})\, N' + N'' + N''',$$

in which κ' and e' and therefore p' are supposed to be adapted to that *second cone* which rests on the *inscribed circle* $A'B'C'$, I knew (by what you had told me) that N must vanish with a, b, c: and I sought to *develope* it, for the *sphere*, according to their *powers* and *products* (article 386); or at least to express its *principal part*, which I found to be of the *fifth dimension*, N_5, as a *rational* (although not necessarily an *integral*) *function of* s, t, p (article 387).

392. To exhibit the geometrical signification of each of the three parts N', N'', N''' of this expression for N, and to show *why* the whole should *vanish* for the case of *internal contact*, I observe that while the vector κ makes equal *obtuse* angles with the three unit vectors α, β, γ, because, by article 388,

$$S\alpha\kappa = S\beta\kappa = S\gamma\kappa = S\alpha\beta\gamma = e > 0,$$

so that $-\kappa$, and not $+\kappa$, is the *interior semiaxis* of the *half cone* which is *first* considered, the vector κ' will on the contrary be presently determined so as to make equal *acute* angles with three new vectors α', β', γ', answering to the *sides* of *contact* of the *second* (or *inscribed*) half cone with the *planes* $\beta_1\gamma_1$, $\gamma_1\alpha_1$, $\alpha_1\beta_1$; where

$$U\alpha_1 = OA_1, \quad U\beta_1 = OB_1, \quad U\gamma_1 = OC_1.$$

In other words, when we shall have *exscribed* to ABC a *new spherical triangle* $A_1 B_1 C_1$, such that the *sides* of the new shall be *bisected* by the corners of the old, if (as in former articles of this Postscript) we *inscribe* a *new small circle* $A'B'C'$ in this new triangle, and denote its spherical *centre* (or *interior pole*) by K', then while we had (article 389),

$$OK = -U\kappa,$$

we shall have

$$OK' = +U\kappa';$$

and therefore

$$SU\alpha'\kappa' = SU\beta'\kappa' = SU\gamma'\kappa' = -\cos \mathsf{c}'$$

if c' be the arcual radius of the new small circle,

$$\mathsf{c}' = K'A' = K'B' = K'C';$$

whereas

$$SU\alpha\kappa = SU\beta\kappa = SU\gamma\kappa = \frac{e}{T\kappa} = +\cos \mathsf{c},$$

where

$$\mathsf{c} = KA = KB = KC$$

is the arcual radius of the *median* (article 388), if K be still the centre (or inner pole) of *that* small circle. And according as ρ represents a side (or ray) OP of the *median* half cone, or of the *inscribed* one, which vector ρ may be supposed to terminate in a point P of one or other circle, the one or the other of these two equations will exist:

$$+SU\kappa\rho = \cos KP = \frac{e}{T\kappa} = \cos \mathsf{c}:$$

$$-SU\kappa'\rho = \cos K'P = \frac{e'}{T\kappa'} = \cos \mathsf{c}';$$

which accordingly agree with the equations, in articles 389 and 390, already assigned for the two cones and might (though with doubtful signs) have been inferred from them.

393. Without yet actually exscribing and inscribing as above, – or making the quaternion calculations which answer to those conceived geometrical processes, – we can already transform the first term of N in article 391, as follows. Denoting still by d the arc KK', of distance between the centres of the two small circles, or the angle KOK' between the interior semiaxes of the two half cones, we have, by quaternions,

$$\cos \mathsf{d} = SU\kappa\kappa';$$

and therefore,

$$N' = e^{-1}S\kappa\kappa' = e^{-1}T\kappa\kappa'.\cos \mathsf{d} = T\kappa' \sec \mathsf{c}.\cos \mathsf{d}.$$

394. As regards the two other terms of N, we have

$$-N''' = e' = T\kappa'.\cos \mathsf{c}';$$

and therefore by article 390,

$$4ep' = T\kappa'.\sin \mathsf{c}';$$

but, by article 388,

$$4e^{-1}p = \tan \mathsf{c};$$

therefore

$$16pp' = T\kappa'.\tan \mathsf{c}.\sin \mathsf{c}' = -N'';$$

consequently,

$$16pp' + e' = T\kappa'.\sec \mathsf{c}.\cos(\mathsf{c} - \mathsf{c}') = N' - N;$$

and the whole expression, in article 391, for N takes this new form:

$$N = T\kappa' \sec \mathsf{c}\{\cos \mathsf{d} - \cos(\mathsf{c} - \mathsf{c}')\};$$

in which *nothing* (so far) has been *neglected*. And we see that the *condition of interior contact*,

$$N = 0,$$

expresses simply, when thus *interpreted*, that the *difference of the radii*, $\mathsf{c} - \mathsf{c}'$, or $\mathsf{c}' - \mathsf{c}$, is equal to the *distance of centres*, d, upon the sphere.

395. It was, however, – as I may remark in passing, – in a less geometrical way, and by a process of a more general and (so to speak) *analytical* kind, that I obtained the forms, in article 390, for the condition of contact, internal *or* external, of the *two cones* of revolution,

$$(S\kappa\rho)^2 + e^2\rho^2 = 0, \quad (S\kappa'\rho)^2 + e'^2\rho^2 = 0;$$

which condition, when cleared of radicals, may be expressed as follows:

$$\begin{aligned} 0 =& (e^2\kappa'^2 + e'^2\kappa^2 - (V\kappa\kappa')^2 + 2ee'S\kappa\kappa') \\ & \times (e^2\kappa'^2 + e'^2\kappa^2 - (V\kappa\kappa')^2 - 2ee'S\kappa\kappa'), \end{aligned}$$

or,

$$0 = (e^2\kappa'^2 + e'^2\kappa^2 - (V\kappa\kappa')^2)^2 - 4e^2e'^2(S\kappa\kappa')^2.$$

396. *For the plane*, we have generally,

$$\mathsf{c} > \mathsf{c}', \quad \text{and} \quad \mathsf{c} - \mathsf{c}' = \mathsf{d};$$

or writing (as in former articles)

$$\mathsf{a} = \mathsf{c} - \mathsf{c}' - \mathsf{d},$$

we know that a *vanishes* for the *plane*. It may then be *expected* to be *small*, for a *small spherical triangle ABC*: and even to be small of an *order higher than the first*, if the *sines a, b, c,* of the half sides of that triangle (article 385) be supposed to be small of the *first* order. And *however large* that triangle ABC may be, *if it be equilateral*, the *two circles ABC* and $A'B'C'$ will *coincide*: so that we may *expect*, with full confidence, to find

$$\mathsf{c} = \mathsf{c}', \quad \mathsf{d} = 0, \quad \mathsf{a} = 0,$$

and finally

$$N = 0,$$

when

$$a = b = c.$$

But even if we have *only two* of these sines *equal*, or in other words if the triangle ABC be only *isosceles*, (without being *equilateral*,) a *contact* of *two* circles will evidently exist, for the *sphere* as well as for the *plane*; whence the equation of condition $N = 0$, must be satisfied, *if the cubic equation* in article 387 have *two equal roots*.

397. Introducing then, a *new scalar* δ, such that

$$\delta^2 = (a-b)^2(b-c)^2(c-a)^2 = t^2(s^2-4t) - 2ps(2s^2-9t) - 27p^2$$
$$= \tfrac{4}{3}(s^2-3t)(t^2-3ps) - \tfrac{1}{3}(st-9p)^2,$$

is the *discriminant* of the *cubic*, of which a, b, c are the *roots*, we may *foresee* that the quaternion expression (article 391) for N, when developed according to ascending powers and products of a, b, c (article 386), and transformed (article 387) to a *rational function of the coefficients s, t, p*, will be found, – if the calculations are properly conducted, – to involve this combination of them, or *this discriminant* δ^2, *as a factor*: which accordingly I have ascertained to be the case, so far as I have carried the developments.

398. But I have not shown how I determined the vectors $\alpha_1, \beta_1, \gamma_1, \alpha', \beta', \gamma', \kappa'$, and the e', p'. Quaternions lead me to write,

$$\alpha_1 = V\beta\alpha^{-1}\gamma, \quad \beta_1 = V\gamma\beta^{-1}\alpha, \quad \gamma_1 = V\alpha\gamma^{-1}\beta;$$

whence

$$T\alpha_1 = T\beta_1 = T\gamma_1 = TV\alpha\beta\gamma = h \text{ (article 388)};$$

so that these three new vectors $\alpha_1, \beta_1, \gamma_1$ are *equally long*. The expressions for $\alpha_1, \beta_1, \gamma_1$ may also be thus written:

$$\alpha_1 = V\gamma\alpha^{-1}\beta, \quad \beta_1 = V\alpha\beta^{-1}\gamma, \quad \gamma_1 = V\beta\gamma^{-1}\alpha;$$

but

$$\beta^{-1}\gamma + \gamma^{-1}\beta = -(\beta\gamma + \gamma\beta) = -2S\beta\gamma = 2(1-2a^2) = 2\cos BC;$$

hence

$$\beta_1 + \gamma_1 = 2\alpha(1-2a^2); \; \gamma_1 + \alpha_1 = 2\beta(1-2b^2); \; \alpha_1 + \beta_1 = 2\gamma(1-2c^2);$$

and as the *sides* of the *median triangle ABC* are here supposed to be each less than a quadrant, so that $\cos BC > 0$, &c., it follows that the vectors α, β, γ bisect respectively the angles between

$\beta_1, \gamma_1; \gamma_1, \alpha_1$; and α_1, β_1. Thus the three vectors, $\alpha_1, \beta_1, \gamma_1$ have the required *directions* of OA_1, OB_1, OC_1; and if they be divided by their *common length h*, the quotients are these last *unit vectors* themselves; so that we may write,

$$OA_1 = U\alpha_1 = h^{-1}\alpha_1;\ OB_1 = U\beta_1 = h^{-1}\beta_1;\ OC_1 = U\gamma_1 = h^{-1}\gamma_1; \quad \text{or}$$

$$OA_1 = UV\beta\alpha^{-1}\gamma, \quad OB_1 = UV\gamma\beta^{-1}\alpha, \quad OC_1 = UV\alpha\gamma^{-1}\beta;$$

and thus the *exscribed* (or *bisecand*) *triangle*, $A_1 B_1 C_1$, is determined, very simply, by quaternions.

399. The recent values for $\beta_1 + \gamma_1$, &c., give easily:

$$\alpha_1 = \beta(1 - 2b^2) + \gamma(1 - 2c^2) - \alpha(1 - 2a^2);$$

$$\beta_1 = \gamma(1 - 2c^2) + \alpha(1 - 2a^2) - \beta(1 - 2b^2);$$

$$\gamma_1 = \alpha(1 - 2a^2) + \beta(1 - 2b^2) - \gamma(1 - 2c^2);$$

or more concisely,

$$\alpha_1 = \alpha S\beta\gamma - \beta S\gamma\alpha - \gamma S\alpha\beta,$$

$$\beta_1 = \beta S\gamma\alpha - \gamma S\alpha\beta - \alpha S\beta\gamma,$$

$$\gamma_1 = \gamma S\alpha\beta - \alpha S\beta\gamma - \beta S\gamma\alpha;$$

expressions which accordingly result at once, by general quaternion transformations, from

$$\alpha_1 = -V\beta\alpha\gamma, \quad \beta_1 = -V\gamma\beta\alpha, \quad \gamma_1 = -V\alpha\gamma\beta.$$

400. The same expressions for $\alpha_1, \beta_1, \gamma_1$, give:

$$S\alpha\beta_1 = S\alpha\gamma_1 = S\beta\gamma; \quad S\beta\gamma_1 = S\beta\alpha_1 = S\gamma\alpha; \quad S\gamma\alpha_1 = S\gamma\beta_1 = S\alpha\beta;$$

but we have seen that

$$T\alpha_1 = T\beta_1 = T\gamma_1 = h = \cos\varepsilon;$$

whence the *proportion of cosines* results (compare article 359):

$$\frac{\cos BC}{\cos\frac{1}{2}B_1 C_1} = \frac{\cos CA}{\cos\frac{1}{2}C_1 A_1} = \frac{\cos AB}{\cos\frac{1}{2}A_1 B_1} = \cos\varepsilon;$$

ε being still the semiexcess, or semiarea of the triangle $A_1 B_1 C_1$, of which the sides are bisected by the corners of the triangle *ABC*. Quaternions gave me this theorem with great ease, as above, a good many years ago; and it is stated somewhere in my Lectures: but probably it was known* before, and in fact it can be otherwise proved, either by geometry, or by trigonometry.

* My son William remarks to me that something of this sort – perhaps the very thing – is given by Mulcahy† in his *Mod. Geometry*.

† [John Mulcahy (1810–1853), *Principles of modern geometry*, Hodges & Smith, Dublin: 1852.]

401. The expressions for $\alpha_1, \beta_1, \gamma_1$, give also,

$$h^2 \cos B_1 C_1 = -S\beta_1\gamma_1 = (\beta S\gamma\alpha - \gamma S\alpha\beta)^2 - (\alpha S\beta\gamma)^2$$
$$= (S\beta\gamma)^2 - (S\gamma\alpha)^2 - 2S\alpha\beta S\beta\gamma S\gamma\alpha$$
$$= 2(S\beta\gamma)^2 - h^2;$$

as might have been foreseen from the equation,

$$h \cos \tfrac{1}{2} B_1 C_1 = -S\alpha\beta_1 = -S\beta\gamma.$$

And we see, at the same time, that

$$S\beta_1\gamma_1 = h^2 - 2(S\beta\gamma)^2 = h^2 - 2(1 - 2a^2)^2; \quad S\gamma_1\alpha_1 = \&c.; \quad S\alpha_1\beta_1 = \&c.$$

402. We shall require also expressions for the three quantities,

$$a_1 = h^2 \sin B_1 C_1 = TV\beta_1\gamma_1, \quad b_1 = TV\gamma_1\alpha_1, \quad c_1 = TV\alpha_1\beta_1,$$

in terms of a, b, c. These new scalars, which are here considered as positive, are given by their *squares*, namely,

$$a_1^2 = h^4 - (S\beta_1\gamma_1)^2 = 4(1 - 2a^2)^2\{h^2 - (1 - 2a^2)^2\}$$
$$= 16(1 - 2a^2)^2\left(a^2 - a^4 - \frac{e^2}{4}\right)$$
$$= 16a^2(1 - 2a^2)^2\left(1 - a^2 - \frac{e^2}{4a^2}\right)$$
$$= 16a^2(1 - 2a^2)^2\left(1 - a^2 - \frac{\Delta^2}{a^2} + 4b^2c^2\right); \quad b_1^2 = \&c.; \quad c_1^2 = \&c.;$$

whence

$$a_1 = 4a(1 - 2a^2)\sqrt{\left(1 - a^2 - \frac{\Delta^2}{a^2} + 4b^2c^2\right)};$$

$$b_1 = 4b(1 - 2b^2)\sqrt{\left(1 - b^2 - \frac{\Delta^2}{b^2} + 4c^2a^2\right)};$$

$$c_1 = 4c(1 - 2c^2)\sqrt{\left(1 - c^2 - \frac{\Delta^2}{c^2} + 4a^2b^2\right)}.$$

The same expressions for a_1, b_1, c_1, or for $h^2 \sin B_1 C_1$, &c., may also be obtained by observing that

$$h \cos \tfrac{1}{2} B_1 C_1 = \cos BC = 1 - 2a^2, \ \&c.;$$

and that therefore

$$h\sin\tfrac{1}{2}B_1C_1 = \sqrt{\{h^2-(1-2a^2)^2\}}$$
$$= \sqrt{(4a^2-4a^4-e^2)}$$
$$= 2\sqrt{(a^2-a^4-\Delta^2+4p^2)}$$
$$= 2a\sqrt{(1-a^2-a^{-2}\Delta^2+4b^2c^2)} \text{ \&c.}$$

403. Without troubling you at present with any further details of *quaternion analysis,* I may just *state* that the remaining vectors and scalars of the question (article 398) are given by the equations:

$$\kappa' = a_1\alpha_1 + b_1\beta_1 + c_1\gamma_1;$$
$$p' = -S\beta\gamma S\gamma\alpha S\alpha\beta;$$
$$e' = +\sqrt{(T\kappa'^2 - 16e^2p'^2)};$$
$$e'\alpha' = \kappa' + 4ep'UV\beta_1\gamma_1; \quad e'\beta' = \kappa' + 4ep'UV\gamma_1\alpha_1; \quad e'\gamma' = \kappa' + 4ep'UV\alpha_1\beta_1;$$

whence follow the relations,

$$S.\kappa'UV\beta_1\gamma_1 = S.\kappa'UV\gamma_1\alpha_1 = S.\kappa'UV\alpha_1\beta_1 = S\alpha_1\beta_1\gamma_1 = 4\varepsilon p';$$
$$TV.\kappa'UV\beta_1\gamma_1 = TV.\kappa'UV\gamma_1\alpha_1 = TV.\kappa'UV\alpha_1\beta_1 = e';$$
$$S\alpha'\beta_1\gamma_1 = S\beta'\gamma_1\alpha_1 = S\gamma'\alpha_1\beta_1 = 0;$$
$$S.\alpha'\kappa'V\beta_1\gamma_1 = S.\beta'\kappa'V\gamma_1\alpha_1 = S.\gamma'\kappa'V\alpha_1\beta_1 = 0;$$
$$\alpha'^2 = \beta'^2 = \gamma'^2 = -1;$$
$$S\alpha'\kappa' = S\beta'\kappa' = S\gamma'\kappa' = -e' < 0;$$
$$\alpha' = OA', \quad \beta' = OB', \quad \gamma' = OC';$$
$$+U\kappa' = OK'; \quad K'A' = K'B' = K'C' = \mathsf{c}' < \frac{\pi}{2}; \quad \cos\mathsf{c}' = \frac{e'}{T\kappa'}.$$

404. It must be added that the recent expression (article 403),

$$\kappa' = a_1\alpha_1 + b_1\beta_1 + c_1\gamma_1,$$

for a vector which *coincides* in direction with the interior semiaxis OK' (article 392) of the second or inscribed half cone, gives easily, by the rules of quaternions, combined with the expression (article 388)

$$\kappa = V(\beta\gamma + \gamma\alpha + \alpha\beta),$$

for a vector which is in direction *opposite* to the interior semiaxis OK of the first or median half cone (article 392), and with the formulæ already written, the equations:

$$e^{-1}S\kappa\alpha_1 = S\beta\gamma - S\gamma\alpha - S\alpha\beta = 1 + 2(a^2-b^2-c^2) = 1 - 2s_2 + 4a^2,$$

where*

$$s_2 = a^2 + b^2 + c^2 = s^2 - 2t;$$

[see articles 387 and 388 for the definitions of s, p, t, and e] similarly

$$e^{-1}S\kappa\beta_1 = 1 - 2s_2 + 4b^2; \quad e^{-1}S\kappa\gamma_1 = 1 - 2s_2 + 4c^2;$$

and therefore, for the *first part* (article 391) of N, we have the value,

$$N' = e^{-1}S\kappa\kappa' = (1 - 2s_2)\sum a_1 + 4\sum a^2 a_1.$$

405. As regards the *second part* of the same expression for N, we have (by articles 403 and 386),

$$p' = (1 - 2a^2)(1 - 2b^2)(1 - 2c^2) = 1 - 2s_2 + 4s_{2,2} - 8p^2,$$

if

$$s_{2,2} = \sum b^2 c^2 = t^2 - 2ps;$$

so that

$$N'' = -16pp' = -16p + 32ps_2 - 64ps_{2,2} + 128p^3.$$

406. To express the *third part* N''' of N, let

$$k' = T\kappa'$$

be a new positive scalar, such that

$$\begin{aligned} k'^2 &= -\kappa'^2 = -(a_1\alpha_1 + b_1\beta_1 + c_1\gamma_1)^2 \\ &= a_1^2 T\alpha_1^2 + b_1^2 T\beta_1^2 + c_1^2 T\gamma_1^2 - 2b_1 c_1 S\beta_1\gamma_1 - 2c_1 a_1 S\gamma_1\alpha_1 - 2a_1 b_1 S\alpha_1\beta_1 \\ &= h^2\left(\sum a_1^2 - 2\sum b_1 c_1\right) + 4\sum(1 - 2a^2)^2 b_1 c_1 \\ &= \left(\sum a_1\right)^2 - 16\sum a^2 b_1 c_1 + 16\sum a^4 b_1 c_1 + e^2\sum(2b_1 c_1 - a_1^2); \end{aligned}$$

then

$$e' = \sqrt{(k'^2 - 16e^2 p'^2)} = (\text{say})\sum a_1 - 8e'';$$

and

$$N''' = -\sum a_1 + 8e'',$$

where

* [Hamilton uses the following notation:

$$s_m = \sum a^m = a^m + b^m + c^m,$$

$$s_{n,m} = s_{m,n} = \sum a^m b^n = a^m(b^n + c^n) + b^m(a^n + c^n) + c^m(a^n + b^n);$$

given on page 162 of his Notebook E (1860) (Trinity College Dublin, MS 1492/159) which contains details of some of the quaternion calculations used in this letter.]

$$e''\sum a_1 = \sum a^2 b_1 c_1 - \sum a^4 b_1 c_1 + \frac{e^2}{16}\sum(a_1^2 - 2b_1 c_1) + e^2 p'^2 + 4e''^2.$$

407. Collecting the three parts, we have thus the expression:

$$N = \left(-2s_2\sum a_1 + 4\sum a^2 a_1\right) + 8e'' - 16pp';$$

in which *nothing* (so far) has been *neglected*, or even (necessarily) considered as *small*.

408. But if we *now* introduce the supposition, above alluded to (articles 386 and 396), that the quantities a, b, c are *small* of the *first order*, we have

$$a_1 = 4a; \quad b_1 = 4b; \quad c_1 = 4c, \text{ nearly};$$

which appears at once from the expressions (article 402) for a_1, b_1, c_1, and might have been foreseen from the very simple geometrical consideration, that the sines of the *whole* sides of the *outer* triangle, $A_1 B_1 C_1$, are *ultimately* the *quadruples* of the sines of the *half* sides of the inner triangle ABC. Hence

$$\sum a_1 = 4\sum a = 4s, \text{ nearly},$$

$$\sum a^2 a_1 = 4\sum a^3 = 4s_3 = 4(s^3 - 3st + 3p), \text{ nearly};$$

and

$$e'' = \frac{16\sum a^2 bc + e^2 p'^2}{4s} = 4p + \frac{\Delta^2}{s} = 4st - s^3 - 4p, \text{ nearly};$$

so that *if we neglect terms of the fifth dimension, the four parts* of the *last* expression (article 407) for N become

$$-2s_2\sum a_1 = -8ss_2 = -8s^3 + 16st;$$

$$4\sum a^2 a_1 = 16s_3 = +16s^3 - 48st + 48p;$$

$$8e'' = 8\left(4p + \frac{\Delta^2}{s}\right) = -8s^3 + 32st - 32p;$$

$$-16pp' = -16p;$$

and therefore their *sum* is

$$N = 0;$$

a result which furnishes *at the plane limit*, a *new proof of the* ***Median Theorem***, at least for the case of the *inner inscribed circle* $A'B'C'$; but (with a little reflexion on the nature of the analysis employed) for the *outer inscribed*, or (as it has been proposed to call them) for the *ex-inscribed** *circles also*: since the passage from one to the other will be found to be performed (as in more elementary questions), by simply *changing the sign* of one of the three sides.

* It appears that L'Huillier† proposed (at least for the *plane* case) the name of "cercles ex-inscrits" for those which are commonly called, less elegantly, *ex*scribed circles, but which are obviously a *sort* of *in*scribed ones.

† [Simon-Antoine-Jean L'Huiller, 1750–1840.]

409. If, however, I had *only* wanted to *prove* the theorem of *contacts* for the *plane*, I should not have thought of employing so refined and perhaps complex *apparatus*; although really the *labour* of the *calculations* so far, has been extremely trifling. The important thing is, that I thus reduce the calculation of the least *Annular breadth*, λ, for the *sphere*, (article 347), to the determination of the *part*, or *term*, N_5 in the development of N, which is (as stated in article 391) of the *fifth dimension*.

410. In calculating this part N_5, we may neglect all terms of the *seventh* and higher orders, in the expressions which present themselves; we may therefore write simply

$$N = N_5, \quad \text{and} \quad N_5 = N_5' + N_5'' + N_5''',$$

this *last* equation being *rigorous*, although the one *before* it is only *approximate*: if we denote by N_5, N_5', N_5'', N_5''', the *parts* of the developments of the four functions, N, $N' - \sum a_1$, N'', $N''' + \sum a_1$, which are homogeneous of the *third dimension*, with respect to a, b, c.

411. Denoting by $a_{1,3}$, $b_{1,3}$, $c_{1,3}$ the parts of the developments of a_1, b_1, c_1, which are homogeneous of the *third* dimension, we have, at sight, from the complete but irrational expressions in article 402,

$$a_{1,3} = -10a^3 - 2a^{-1}\Delta^2 = -10a^3 - 2p^{-1}\Delta^2 bc;$$

$$b_{1,3} = -10b^3 - 2b^{-1}\Delta^2 = -10b^3 - 2p^{-1}\Delta^2 ca;$$

$$c_{1,3} = -10c^3 - 2c^{-1}\Delta^2 = -10c^3 - 2p^{-1}\Delta^2 ab;$$

if

$$s_5 = \sum a^5 = 5p(s^2 - t) + s(s^4 - 5s^2 t + 5t^2);$$

whence

$$\sum a_{1,3} = -10s_3 - 2p^{-1}t\Delta^2; \quad \sum a^2 a_{1,3} = 10s_5 - 2s\Delta^2;$$

and therefore, by articles 404 and 410,

$$N_5' = -2s_2 \sum a_{1,3} + 4\sum a^2 a_{1,3}.$$

$$= \text{a known symmetric function of } a,\ b,\ c,$$

$$= \text{a rational function of the coefficients } s,\ t,\ p;$$

namely (by a little *algebraic* and very *easy* work)

$$N_5' = 4p(-19s^2 + 20t) + 4s(-3s^4 + 9s^2 t - 4t^2) - 4p^{-1}s^2 t(s^4 - 6st^2 t + 8t^2).$$

412. It is even more easy to express N_5'' as a rational function of s, t, p: namely by articles 405 and 410,

$$N_5'' = 32ps_2 = 32p(s^2 - 2t).$$

413. As regards N_5''', we have, in the first place,

$$N_5''' = 8e_5''',$$

by articles 406 and 410, if

$$e'' = e_3'' + e_5'' + e_7'' + \&c.,$$

whence

$$e'' = -4p - s(s^2 - 4t),$$

by article 408; it remains then to calculate this term of fifth dimension, e_5'', with the help of the last equation of article 406, on a plan exactly analogous to that of extracting the *square root* of a development.

414. Attending only to terms of the *sixth* order, in the *product* $e''\sum a_1$, the equation just cited gives:

$$4se_5'' = -e_3''\sum a_{1,3} + 4e_3''^2 + 4\sum a^2bc_{1,3} - 16ps_3 - 4\Delta^2(3s^2 - 4t) - 16p^2;$$

because

$$e^2 = 4\Delta^2 - 16p^2, \quad p'^2 = 1 - 4s_2, \ldots, \sum(a_1^2 - 2b_1c_1) = 16(s_2 - 2t) \ldots;$$

and therefore [to the same order],

$$\frac{e^2}{16}\sum(a_1^2 - 2b_1c_1) + e^2p'^2 - 4\Delta^2 = 4\Delta^2(s^2 - 4t) - 16\Delta^2(s^2 - 2t) - 16p^2$$
$$= -12s^2\Delta^2 + 16t\Delta^2 - 16p^2;$$

also

$$\sum a^2bc_{1,3} = -10ps_{1,2} - 2p^{-1}\Delta^2s_{2,3},$$

if

$$s_{1,2} = \sum ab^2 = -3p + st, \quad \text{and} \quad s_{2,3} = \sum a^2b^3 = -p(2s^2 + t) + st^2.$$

Hence, by [the following*] algebraic reductions, of a sufficiently simple character.

Auxiliary calculations, connected with N_5' and N_5'''

By article 410

$$N_5' = \text{part of fifth dimension in } N' - 2\sum a_1;$$

by article 404

$$N_5' = \text{part of fifth dimension in } -2s_2\sum a_1 + 4\sum a^2a_1;$$

therefore as in article 411

$$N_5' = -2s_2\sum a_{1,3} + 4\sum a^2a_{1,3}; \quad (s_2 = s^2 - 2t;)$$

but, from article 411,

$$\sum a_{1,3} = -10s_3 - 2p^{-1}t\Delta^2, \quad \sum a^2a_{1,3} = 10s_5 - 2s\Delta^2;$$

* [Taken from p. 34 of Trinity College Dublin MS 1493/1171, dated Tuesday, 18 September 1860.]

from articles 408 and 411 [see Notebook *E*60, p. 162 (footnote to article 404)]

$$s_3 = 3p + s(s^2 - 3t), \qquad s_5 = 5p(s^2 - t) + s(s^4 - 5s^2t + 5t^2);$$

and from article 388 [see Notebook *E*60, p. 163]

$$\Delta^2 = -8ps - s^2(s^2 - 4t);$$

therefore

$$\begin{aligned}\sum a_{1,3} &= -30p - 10s(s^2 - 3t) + 16st + 2p^{-1}s^2t(s^2 - 4t)\\ &= -30p - 2s(5s^2 - 23t) + 2p^{-1}s^2t(s^2 - 4t);\end{aligned}$$

hence

$$\begin{cases}-2s_2\sum a_{1,3} = p(60s^2 - 120t) + 4s(5s^4 - 33s^2t + 46t^2)\\ \qquad\qquad - 4p^{-1}s^2t(s^4 - 6s^2t + 8t^2);\\ 4\sum a^2a_{1,3} = -4s_5 - 8s\Delta^2 = -200p(s^2 - t) - 40s(s^4 - 5s^2t + 5t^2)\\ \qquad\qquad + 64ps^2 + 8s^3(s^2 - 4t)\\ \qquad\qquad = p(-136s^2 + 200t) + 4s(-8s^4 + 42s^2t - 50t^2);\end{cases}$$

therefore

$$\begin{aligned}N_5' &= p(-76s^2 + 80t) + 4s(-3s^4 + 9s^2t - 4t^2) - 4p^{-1}s^2t(s^4 - 6s^2t + 8t^2)\\ &= 4p(-19s^2 + 20t) + 4s(-3s^4 + 9s^2t - 4t^2) - 4p^{-1}s^2t(s^2 - 2t)(s^2 - 4t).\end{aligned}$$

In the expression for $4se_5''$ [at the beginning of] article 414, the first term is $-e_3''\sum a_{1,3}$, now, from article 413

$$-e_3'' = 4p + s(s^2 - 4t),$$

and with the expression above for $\sum a_{1,3}$ we get

$$-e_3''\sum a_{1,3} = -120p^2 - 2ps(35s^2 - 152t) - 2s^2(s^2 - 4t)(5s^2 - 27t) + 2p^{-1}s^3t(s^2 - 4t)^2. \tag{I}$$

And

$$4e_3''^2 = +64p^2 + 32ps(s^2 - 4t) + 4s^2(s^2 - 4t)^2. \tag{II}$$

[From Notebook *E*60, p. 162] we have

$$4\sum a^2bc_{1,3} = -40ps_{1,2} - 8p^{-1}\Delta^2s_{2,3};$$

$$s_{1,2} = -3p + st; \qquad s_{2,3} = -p(2s^2 + t) + st^2;$$

$$-8p^{-1}\Delta^2 = 64s + 8p^{-1}s^2(s^2 - 4t);$$

therefore

$$-40ps_{1,2} = +120p^2 - 40pst;$$

and

$$-8p^{-1}\Delta^2 s_{2,3} = -64ps(2s^2+t)+64s^2t^2-8s^2(2s^4-7s^2t-4t^2)$$
$$+64p^{-1}s^3t^2(s^2-4t);$$

which gives

$$4\sum a^2bc_{1,3} = +120p^2-8ps(16s^2+13t)+8s^2(-2s^4+7s^2t+12t^2)$$
$$+8p^{-1}s^3t^2(s^2-4t). \qquad \text{(III)}$$

Also

$$-16ps_3 = -48p^2-16ps(s^2-3t), \qquad \text{(IV)}$$
$$-4\Delta^2(3s^2-4t) = 32ps(3s^2-4t)+4s^2(3s^2-4t)(s^2-4t)$$
$$= 32ps(3s^2-4t)+4s^2(3s^4-16s^2t+16t^2), \qquad \text{(V)}$$

and

$$-16p^2 = -16p^2. \qquad \text{(VI)}$$

The sum of these six parts is $4se_5'' = \frac{1}{2}sN_5'''$; and in this sum the coefficient of p^2 is $-120+64+120-48-16=0$; of ps^3, $-70+32-128-16+96=-86$; of pst, $+304-128-104+48-128=-8$; of s^6, $-10+4-16+12=-10$; of s^4t, $94-32+56-64=+54$; of s^2t^2, $-216+64+96+64=+8$; of t^3, 0; and of $2p^{-1}(s^2-4t)s^3t$, $s^2-4t+4t=s^2$.

Hence

$$N_5''' = -4p(43s^2+4t)+4s(-5s^4+27s^2t+4t^2)+4p^{-1}s^4t(s^2-4t),$$

as in article 414.

Hence

$$N_5' + N_5''' = 8p(-31s^2+8t)+16s^3(-2s^2+9t)+8p^{-1}s^2t^2(s^2-4t).$$

But, by article 412,

$$N_5'' = 32ps_2 = 8p(4s^2-8t).$$

415. Collecting the three parts (article 410) of the sought term N_5 of N, we have (by articles 411, 412, and 414) the following expression for that term

$$N_5 = N_5' + N_5'' + N_5'''$$
$$= -216ps^2+16s^3(-2s^2+9t)+8p^{-1}s^2t^2(s^2-4t);$$

or finally (by article 397)

$$N_5 = 8\frac{s^2\delta^2}{p};$$

a simple and remarkable result, which is *rigorous*, so far as it goes, – as an expression for the term N_5, though *not* for the *function* N, which involves also other terms, N_7 &c., of the *seventh* and *higher* dimensions.

416. I arrived at the same result, a few weeks ago, by two processes which were distinct, in their details, from the recent process, and from each other; and which were much less simple, because it had not then occurred to me to isolate the terms $\pm\sum a_1$, in the developments of N' and N'', which (as we see) destroy each other in the sum. Accordingly, I was obliged, in those former calculations, to take account of terms $a_{1,5}$, $b_{1,5}$, $c_{1,5}$, of the *fifth* dimension, and of their combinations with other terms: which introduced the *inverse square* and *inverse cube* of p although I found, after troublesome reductions, that the terms involving *those* negative powers neutralised each other. – The process *above* described may fairly claim (I think) to be regarded as *simple*, when the difficulty of the question is taken into account. The work with *quaternions* is really not laborious, to a person familiar with the *principles* of my Calculus: Tait, I am sure, would *follow* the preceding pages (say articles 385–401) *at sight*: though I confess that it tasked somewhat my invention, to strike out the *plan* of conducting the inquiry. The *labour*, – such as it is – and incomparably inferior is it to that of the complete solution of two or three spherical triangles, when such great numerical accuracy is aimed at, as in earlier articles of this Postscript, – does not *begin*, until ***Quaternions*** (with their best bow) hand over the task to be *finished* by their elder sister, the ***Algebra of Scalars***.

417. *At the limit*, – or if you choose *just after or before* the limit (corresponding to the *plane*), – the factor, $T\kappa'.\sec\mathsf{c}$, in the rigorous expression (article 394) for N, may be treated as [equal to] $T\kappa' = k'$ (article 406); or ultimately as [equal to] $\sum a_1$, or [to] $4s$; all these quantities bearing to each other a limiting ratio of equality, as N does to its principal term N_5. Hence, if we introduce a *new auxiliary angle*, e, such that

$$\sin\mathsf{e} = \frac{1}{2}\{\cos\mathsf{d} - \cos(\mathsf{c} - \mathsf{c}')\} = \sin\frac{\mathsf{a}}{2}\sin\left(\mathsf{d} + \frac{\mathsf{a}}{2}\right),$$

we must have, as an approximation,

$$\sin\mathsf{e} = \frac{N\cos\mathsf{c}}{2T\kappa'} = \frac{N_5}{8s} = \frac{s\delta^2}{p};$$

s, p, δ^2, d being *small* of the first, third, *sixth*, and first orders, and [therefore] e, a being small of the *fourth* and *third*: and as a ***Theorem***, true in all mathematical rigour, –

$$\lim_{\mathsf{c}=0}.\frac{\mathsf{e}p}{s\delta^2} = 1;$$

which may be said to be the *result* of my Investigation, at least for the case of the *approximate internal contact* between the *Median* and *inscribed small circles*, ABC, $A'B'C'$: or for the *Annular Eclipse* (article 343) of the four circles of the latter, on the sphere. – That the eclipse *is* annular, is shown by the *positive character of* e, or of a (article 348).

418. It may be remembered that (by articles 387 and 397),

$$s = a + b + c; \quad p = abc; \quad \delta^2 = (a-b)^2(b-c)^2(c-a)^2;$$

where (by article 386)

$$a = \sin\tfrac{1}{2}\widehat{BC}; \quad b = \sin\tfrac{1}{2}\widehat{CA}; \quad c = \sin\tfrac{1}{2}\widehat{AB};$$

and therefore,

$$a = r\sin 2\theta, \quad b = r\sin 2\varphi, \quad c = r\sin 2\psi,$$

if 2θ, 2φ, 2ψ denote the *angles of the plane triangle ABC*, so that

$$\theta + \varphi + \psi = \frac{\pi}{2};$$

while $r = \sin \mathsf{c}$ (article 388), if c [is equal to the] *arcual radius of the median circle ABC*, as before.

419. Hence,

$$\frac{s}{2r} = \frac{\sin 2\theta + \sin 2\varphi}{2} + \frac{\sin 2\psi}{2} = \sin(\theta+\varphi)\cos(\theta-\varphi) + \sin\psi\cos\psi$$
$$= \cos\psi\{\cos(\theta-\varphi) + \cos(\theta+\varphi)\}$$
$$= 2\cos\theta\cos\varphi\cos\psi;$$

that is,

$$s = 4r\cos\theta\cos\varphi\cos\psi,$$

rigorously.

420. Also,

$$p = 8r^3\sin\theta\sin\varphi\sin\psi\cos\theta\cos\varphi\cos\psi;$$
$$\frac{p}{s} = 2r^2\sin\theta\sin\varphi\sin\psi.$$

421. Again,

$$\frac{a-b}{2r} = \frac{\sin 2\theta - \sin 2\varphi}{2} = \sin\psi\sin(\theta-\varphi), \quad \&c.$$
$$\delta = (a-b)(b-c)(c-a)$$
$$= 8r^3\sin\theta\sin\varphi\sin(\theta-\varphi)\sin(\varphi-\psi)\sin(\psi-\theta),$$

and

$$\frac{N_5}{8s} = \frac{s\delta^2}{p} = \sin \mathsf{e}_0,$$

rigorously if e_0 be any new auxiliary angle, *defined* by the equation [see Notebook *E*, p. 236]:

$$\sin \mathsf{e}_0 = 32r^4\sin\theta\sin\varphi\sin\psi\sin^2(\theta-\varphi)\sin^2(\varphi-\psi)\sin^2(\psi-\theta).$$

422. The *Theorem* of article 417 may now be enunciated:

"*The angles* e, e_0 *are ultimately equal*",

or in symbols,

$$\lim_{c=0} \frac{\mathsf{e}}{\mathsf{e}_0} = 1.$$

423. In the *Example* above selected (article 350, &c.), I *assumed*

$$r = \sin 10°;$$

and (compare article 351),

$$\theta = 23° \, 15'; \quad \varphi = 52° \, 15'; \quad \psi = 14° \, 30';$$

whence the last formula for sin e_0 gives *at once, without spherical trigonometry*, the value,

$$\mathsf{e} = +0''{\cdot}956.$$

And the result (article 381) of a *long train of spherical triangles* gave:

$$\mathsf{a} = +31''{\cdot}70; \quad \mathsf{d} = 3° \, 52' \, 2''{\cdot}81;$$

whence, by the first formula (article 417) for sin e

$$\mathsf{e} = +1''{\cdot}070.$$

Besides the *grand point* (article 383) of the *agreement in sign*, thus exhibited, I consider this degree of *agreement in value* between the results of the processes so different, to be *sufficiently close*; especially as some of the *triangles* employed are really *not very small*: and consequently the *omitted terms*, N_7, &c., must have *begun* to produce sensibly an effect, which is *at first* extremely minute, but *increases very rapidly* with r: namely, for the auxiliary angle e, as the sixth power of that radius: and for the annular breadth a, as the *fifth power*.

424. (September 18th, 1860) – Although I propose to return on the investigation just described, and especially to show how it adapts itself to the case of an *Ex-inscribed Circle* (article 408), – yet I am now desirous to turn for a while to a quite different subject, which yet has a closer connexion with the *title* of this Postscript: I mean, the *extension* of Anharmonic Coordinates, from the *plane* to *space*; to which it is necessary, just now, that my own thoughts should be recalled, because the Abstract of my last Communication to the Royal Irish Academy, respecting such extension, has not yet been finally revised for the Proceedings.*

425. Still, – lest anything should interfere with my resuming the account of the investigation about the circles, – I may as well give here some brief *sketch* of my results, obtained by *quaternions*, and of their approximate *verification* by *spherical trigonometry*, for the *non-contacts* of the *median* and with the *ex-inscribed* circles.

426. At first, I feared that nearly the whole of the calculations, – which, *as formerly conducted* (article 416) were certainly very troublesome, – would have to be gone over anew, for passing to a circle of the *latter* kind: and what was worse, that I should have *no verification of the final formula*, when obtained, so simple and searching as the existence of the *discriminant*, δ^2, as a *factor* (article 397) in the result. In fact, it is clear that the sines a, b, c cannot enter *symmetrically* into any result, in which *one* exinscribed circle is *distinguished* from the *other two*.

* [See p. 507 of this volume.]

And I actually *began* an entirely new calculation, in which *two* of those *three sines a, b, c* were treated as the *roots* of a *quadratic* equation, while the *other* remained isolated from them.

427. No doubt, with patience, I might thus have at last succeeded, in resolving what had the air of a *new problem*, and was about to be treated by an almost enticing *new method*: but I perceived that the work might be greatly abridged, or rather that *no new labour* need be expended, if a simple *principle* were introduced.

428. Let K'' be the spherical centre of that ex-inscribed small circle, which touches the side $B_1 C_1$ *itself* in a point A'', but the *prolongations* of the sides $A_1 C_1$, $A_1 B_1$, of the triangle $A_1 B_1 C_1$, in points B'' and C''. Let the arcual radius of this new circle be c'', so that

$$K''A'' = K''B'' = K''C'' = \mathsf{c}'';$$

and let d' be the new distance of centres,

$$KK'' = \mathsf{d}'.$$

Finally, let

$$\mathsf{a}' = \mathsf{d}' - (\mathsf{c} + \mathsf{c}'').$$

429. Then $\frac{\mathsf{a}'}{\mathsf{c}}$ *vanishes at the plane limit*; or more fully, a' is a *small quantity*, of *an order higher than the first*, if the triangles and circles on the *sphere* be *small*, and if we still treat *a, b, c*, and therefore c, as small of the *first* order: whence it was easy to foresee that this new *interval of circumferences* a' would turn out to be a quantity of the *third order*; but this alone gave no clue to its algebraic *sign*. Was I to expect that the two circles would *partially overlap* (compare article 343), and so present the case ($\mathsf{a}' < 0$) of a *Partial Eclipse*, however trifling, of the *Median*; or that the contrary case ($\mathsf{a}' > 0$) would exist, and that thus the two circles would be *wholly external to each other*? Or might there be *sometimes the one result*, and at *other times* the *other*, according to the *ratios* of the *sines, a, b, c*, or the *ultimate shape* of the *median triangle* ABC?

430. The only *anticipation*, which I at first supposed myself at liberty to form, with any confidence, was that because, when $B_1 C_1$ is the base of an *isosceles* triangle (compare article 396), the median and the exinscribed circles *touch*, at the *middle point A* of that base, with which point A' and A'' here coincide, – therefore the general expression for a', or for the connected quantity (compare article 417),

$$\sin \mathsf{e}' = \frac{1}{2}(\cos(\mathsf{c} + \mathsf{c}'') - \cos \mathsf{d}') = \sin \frac{\mathsf{a}'}{2} \sin\left(\mathsf{d}' - \frac{\mathsf{a}'}{2}\right),$$

in which e' is a new auxiliary angle, *defined* by this equation, would be found to *contain* $(b - c)^2$ *as a factor*, the *other* factor being *some symmetric function of b and c;* though this expression would *not* be a symmetric function of the *three* quantities, *a, b, c*, and therefore *not a rational function of s, t, p.*

431. I own, however, that I *did expect*, with a rather strong *feeling* of *confidence*, derived from *analogy*, although at first without any *proof*, that as the *inner inscribed circle* $A'B'C'$ was *wholly* *within* the *median ABC*, so the *exinscribed circle $A''B''C''$* would be found to be *wholly without* that median, or bisecting circle.

432. Applying the quaternion calculus, I easily proved that if we write (compare article 403),

$$\kappa'' = -a_1\alpha_1 + b_1\beta_1 + c_1\gamma_1, \quad \text{and} \quad e'' = +\sqrt{(T\kappa''^2 - 16e^2p'^2)},$$

when the vectors α_1, β_1, γ_1, and the scalars a_1, b_1, c_1, e, p', have the same values as before, then the equation (compare article 390),

$$(S\kappa''\rho)^2 + e''^2\rho^2 = 0,$$

represents that *ex-inscribed half cone,* which has its vertex at the centre O of the sphere, and rests on the new small circle $A''B''C''$; while (compare articles 393 and 394) we have also the relations,

$$e'' = T\kappa''.\cos \mathsf{c}'', \quad 4ep' = T\kappa''.\sin \mathsf{c}'', \quad OK'' = +U\kappa'',$$

and

$$\cos \mathsf{d}' = SU\kappa\kappa''.$$

433. Hence, if we make for abridgement (compare article 391),

$$N^{(1)} = e'' - 16\,pp' - e^{-1}S\kappa\kappa'',$$

we have rigorously (compare articles 394 and 417),

$$N^{(1)} = T\kappa''.\sec \mathsf{c}.\{\cos(\mathsf{c} + \mathsf{c}'') - \cos \mathsf{d}\} = 2T\kappa''.\sec \mathsf{c}.\sin \mathsf{e}';$$

and therefore, *nearly*,

$$\mathsf{e}' = \frac{N^{(1)}}{2T\kappa''}, \quad \text{and} \quad \mathsf{a}' = \frac{N^{(1)}}{T\kappa''.\mathsf{d}'};$$

so that I was led to calculate the *principal term of $N^{(1)}$*.

434. I *foresaw* that this would be a term $N_5^{(1)}$, of the *fifth order* (compare article 409); and that it would be *symmetric* with respect to b and c, and involve $(b - c)^2$ as a *factor* (compare article 430); but *at first*, I *foresaw nothing more*, and *commenced* a *new calculation.*

435. I soon perceived, however, that I was thus operating on $-a$, $+b$, $+c$, precisely as I *had* operated on a, b, c; and therefore that if I supposed $-a$, b, and c to be the *roots of a new cubic,*

$$x^3 - s'x^2 + t'x + p = 0, \quad \text{(compare article 387)}$$

where $s' = b + c - a$, and $t' = bc - a(b + c)$, but $p = abc$, as before, *then the new function,*

$$-N^{(1)} = e^{-1}S\kappa\kappa'' + 16pp' - e'',$$

in which, as before (article 405),

$$p' = (1 - 2a^2)(1 - 2b^2)(1 - 2a^2),$$

must admit of being *formed from the old function,*

$$+N^{(1)} = e^{-1}S\kappa\kappa'' - 16pp' - e',$$

by simply *changing* s, t, p to s', t', $-p$.

436. I had therefore, *at once, without any new trouble, the new formula:*

$$N_5^{(1)} = \frac{8s'^2\delta'^2}{p};$$

in which

$$\delta'^2 = (a+b)^2(a+c)^2(b-c)^2,$$

so that the *factor* $(b-c)^2$ enters as was expected (article 430), and the whole is symmetric (as it ought to be) with respect to b and c. And because $N_5^{(1)} > 0$, and therefore $\mathsf{a}' > 0$, it followed, as I had thought *likely* from *analogy* (article 431), – that (compare article 429) "*the ex-inscribed circle is wholly exterior to the median*"; at least if these circles be *not too large*: although I *believe* that this last restriction is unnecessary, for I *think* that the *omitted terms*, $N_7^{(1)}$, &c., of $N^{(1)}$, *never become so great* as to *change the character* of the result.

437. The approximate value of $T\kappa''$ is $4s'$, as that of $T\kappa'$ was $4s$ (article 417); hence the auxiliary angle e' (article 430) is *nearly* [equal to]

$$\frac{N_5^{(1)}}{8s'} = \frac{s'\delta'^2}{p}$$

(by article 436). Transforming this last expression, $\frac{s'\delta'^2}{p}$, as the corresponding expression $\frac{s\delta^2}{p}$ was transformed before, and equating it to the sine of a new auxiliary angle e'_0 we have, as the *definition* of this new angle, the equation (compare article 419):

$$\sin \mathsf{e}'_0 = 32r^4 \sin\theta \cos\varphi \cos\psi \sin^2(\varphi-\psi)\cos^2(\psi-\theta)\cos^2(\theta-\psi);$$

where r $(= \sin \mathsf{c})$, and θ, φ, ψ have the same values as before. And the *new Theorem*, derived from *Quaternions*, is (compare article 422) that "*the angles* e' *and* e'_0 *are ultimately equal*"; or,

$$\lim_{\mathsf{c}=0} \frac{\mathsf{e}'}{\mathsf{e}'_0} = 1.$$

438. I have compared these angles, e' and e'_0, for each of the three ex-inscribed circles, retaining the numerical data of our Example, but being obliged to resolve several new spherical triangles. At present I shall only say, that the *agreements in value* were as close as could fairly be demanded, considering that *larger arcs than before* were introduced. The *essential test of algebraic sign* was borne in *each* of the three cases. In other words, *each of the three exinscribed circles when found, by spherical trigonometry, to be wholly external to the median*:- as you are aware that I had foreseen (article 436).

439. September 21st 1860. – Resuming now the Anharmonic Coordinates themselves, which have been much out of my thoughts for a good while, I think that I shall not inflict on you any long Addition to what was said in the "Letter" itself: which *Letter* if you have ever read through, your patience must have been greater than mine is with the Copy of it, that has been preserved for me. However I know that to *parts* of it you were so good as to lend your attention: and I may perhaps *refer* to parts, in what I am about to write.

440. (September 29th) To use a vulgar, but expressive phrase, I must confess that I have been, for the last two or three months, *out of conceit* with the whole System of "Anharmonic Coordinates," at least as applied to *Space*: because I seem to myself to have (rather lately) invented a much *better System*, or at least one which, for many purposes, and especially for the study of *Geometrical Nets in Space*, is such.

441. In applications of this sort, I have been conducted to use *Quinary Symbols*, and *Quinary Types*, for *Points, Lines*, and *Planes*; which can easily be *reduced* to *Quaternary* (or *Anharmonic*) Symbols and Types: but not without some sort of *Symmetry*, and consequently of facility of conception, and of reasoning.

442. I am not sure that I shall enter *at all* on the subject of ***Nets***, in the present *Postscript*, miscellaneous as its nature may be. And as they (the nets) supply me with the chief *motive*, at present, for preferring *quinary* to *quaternary* symbols for *space*, I shall endeavour to *compel* myself to wind up with some account of the *anharmonic extension*.

443. As regards the mere *definition* of the "Anharmonic Coordinates of a Point in Space", it is nearly enough to say that they are *four* positive or negative *numbers*, x, y, z, w, of which the *ratios*, or the *quotients*, $\frac{x}{w}$, $\frac{y}{w}$, $\frac{z}{w}$, express the *anharmonic functions* of certain *Pencils of Planes*, which *determine the position* of the *point*,

$$P = (x,\ y,\ z,\ w).$$

444. But to state this conception a little more in detail. Let A, B, C, D, E, F be any six points of space; and let us select any *two* of them, suppose E and F, to determine a right *line* EF, through which *four planes* are drawn,

$$AEF,\ BEF,\ CEF,\ DEF,$$

so as to pass also in succession through the *four other points*. We have thus a *Pencil of Planes*, which it seems natural to denote by the Symbol, $EF.ABCD$; (*without* parentheses, as yet:) and if any arbitrary rectilinear Transversal, suppose GH, be drawn across this Pencil, of *four collinear planes*, (for such appears to be a convenient designation of their relation to each other,) and we denote its intersections with the four planes by A', B', C', D', so that

$$A' = GH^{\cdot}AEF,\ B' = GH^{\cdot}BEF,\ C' = AG^{\cdot}CEF,\ D' = GH^{\cdot}DEF;$$

then the *Anharmonic* $(A',\ B',\ C',\ D')$, of the *Group* of the Four Points of intersection, is obviously independent of the position and direction of the Transversal GH; and may therefore be naturally said to be the "Anharmonic of the Pencil* of Planes", and *denoted*, as such, by the new Symbol, (in which I *now* introduce parentheses)

$$(EF.ABCD).$$

We have therefore thus the *equation*, – which is however (as you see) merely a mode of stating a *definition*, –

$$(EF.ABCD) = (A'B'C'D');$$

if

$$A' = GH^{\cdot}AEF,$$

&c., as above, and if the symbol $(A'B'C'D')$, for the Anharmonic of a Group of Four Collinear Points, be still interpreted as in the early article 5 of this Letter, to which the present Postscript is appended: namely by the formula,

$$(A'B'C'D') = \frac{A'B'}{B'C'} \cdot \frac{C'D'}{D'A'}.$$

445. So much being laid down, – or admitted, – respecting the *general signification* of a symbol of the *form* $(EF.ABCD)$, I next proceed to consider the *particular* (or at all events, the *less general*) *symbols*:

$$(BC.AEDP), \quad (CA.BEDP), \quad (AB.CEDP);$$

in which A, B, C, D, E denote *any five given points of space*, whereof *no four are coplanar*, and P is *any sixth point*. I denote the three anharmonics of pencils by the algebraic *quotients*,

$$\frac{x}{w}, \frac{y}{w}, \frac{z}{w};$$

and I call (compare article 443) x, y, z, w the *Anharmonic Coordinates* of the *Point P*, or (x, y, z, w), taken with respect to the *given system of the five points A, . . . , E.*

446. If the point P be in the plane BCD. but not on the line BC, then BCP is a determined plane, namely the *fourth plane* of the first pencil (article 445), and it coincides with the given plane BCD, which is the *third plane* of that pencil; hence, by the definition adopted, the anharmonic of the pencil vanishes, and thus we may write

$x = 0$, as the *anharmonic equation* of the *given plane BCD*.

$y = 0$ is the equation of the plane CAD;

and

$z = 0$ is the equation of the plane ABD.

* *Chasles* has a definition introducing *sines* of *angles*, when first treating of *pencils*, of *lines*, and of *planes*: but this it would be foreign to my plan to employ. *Möbius* was quite familiar in 1827 with what he calls the "Ratio bissectionalis", for pencils as well as for groups, which we now call *anharmonic ratio*. [See footnotes to articles 1 and 5.]

447. If P be in the plane ABC, but not on any one of the lines BC, CA, AB, the *fourth plane* of each pencil is still *determined*, but it now coincides with the *first* plane of that pencil; the anharmonic of each pencil is therefore in this case *infinite*, and we have thus

$$w = 0,$$

as the equation of the given plane ABC.

448. If P be on the line BC, which is the *axis* of the first pencil, – or the line *common* to all its planes, – then the fourth plane of *that* pencil becomes *indeterminate*, and its anharmonic $\frac{z}{w}$ takes the form $\frac{0}{0}$; while the second and third anharmonics are *infinite*, because *their* fourth planes, CAP and ABP, coincide now with their common first plane ABC. For these reasons, we are led to write

$$x = 0, \quad w = 0,$$

as the joint *equations of the line BC*; and in fact we have just seen that these two equations, taken *separately*, represent the *two planes BCD, ABC*, of which that line is the *intersection*. In like manner,

the line CA is represented by the two equations,	$y = 0,\ w = 0;$
" " AB " " " " " "	$z = 0,\ w = 0;$
" " DA " " " " " "	$y = 0,\ z = 0;$
" " DB " " " " " "	$z = 0,\ x = 0;$
" " DC " " " " " "	$x = 0,\ y = 0.$

449. For the *point A* we may write

$$x = 1, \quad y = 0, \quad z = 0, \quad w = 0,$$

because the first of the three anharmonics becomes infinite, and each of the others is indeterminate; and accordingly, this point may be considered as the intersection of the line DA, of which the equations are $y = 0$, $z = 0$, with the plane ABC, of which the equation is $w = 0$: the value, $+1$, for x being merely assumed as the simplest different from 0. Hence we may write,

$$A = (1, 0, 0, 0),$$

or sometimes more concisely

$$A = (1\,0\,0\,0);$$

and in like manner

$$B = (0, 1, 0, 0), \quad C = (0, 0, 1, 0), \quad D = (0, 0, 0, 1),$$

or briefly,

$$B = (0\,1\,0\,0), \quad C = (0\,0\,1\,0), \quad D = (0\,0\,0\,1).$$

And on account of these very simple forms for the *four corners* of the *pyramid ABCD*, when expressed thus by their *anharmonic coordinates*, I call this pyramid $ABCD$ the ***Unit-Pyramid***.

450. The equations of the three planes *BCE, CAE, ABE* are

$$x = w, \quad y = w, \quad z = w;$$

hence the four coordinates of the point *E* are *equal* to each other, and may be supposed each [equal to] +1; I write, therefore,

$$E = (1, 1, 1, 1), \quad \text{or} \quad E = (1\,1\,1\,1);$$

and I call this *fifth given point E* the ***Unit-Point.***

451. If we write,

$$A' = BC^{\cdot} ADE,$$

that is, if we denote by A' the point in which the line *BC* is intersected by the plane *ADE*, we are conducted to the expression,

$$A' = (0, 1, 1, 0), \quad \text{or} \quad A' = (0\,1\,1\,0);$$

with the analogous expressions,

$$B' = CA^{\cdot} BDE = (1\,0\,1\,0), \quad C' = AB^{\cdot} CDE = (1\,1\,0\,0).$$

In like manner, if

$$A_1 = EA^{\cdot} BCD, \quad B_1 = EB^{\cdot} CAD, \quad C_1 = EC^{\cdot} ABD,$$
$$A_2 = DA^{\cdot} BCE, \quad B_2 = DB^{\cdot} CAE, \quad C_2 = DC^{\cdot} ABE,$$

and

$$D_1 = DE^{\cdot} ABC,$$

it is found that

$$A_1 = (0\,1\,1\,1), \quad B_1 = (1\,0\,1\,1), \quad C_1 = (1\,1\,0\,1), \quad D_1 = (1\,1\,1\,0),$$
$$A_2 = (1\,0\,0\,1), \quad B_2 = (0\,1\,0\,1), \quad C_2 = (0\,0\,1\,1);$$

and I may just remark in passing, that these *ten derived points,* of which I have thus assigned (without proof) the *anharmonic symbols,* are what I call the *points of first construction,* for the *Geometrical Net in Space,* which is determined by the *five given points A,* ..., *E*; whereof it is still supposed that *no four are situated in any common plane.*

452. More generally, we have the formulæ, – which, like several others, to save time, I offer here without demonstration, although you may be sure that I *have* proofs for them, –

$$(AD.BECP) = \frac{y}{z}; \quad (BD.CEAP) = \frac{z}{x}; \quad (CD.AEBP) = \frac{x}{y};$$

with the following immediate transformations,

$$(AD.CEBP) = \frac{z}{y}; \quad (BD.AECP) = \frac{x}{z}; \quad (CD.BEAP) = \frac{y}{x}.$$

Hence every plane through the line *AD* has an equation of the form: $\frac{y}{z} =$ constant; and the particular plane *ADE* has for equation,

$$y = z;$$

while the planes *BDE, CDE* have for their equations,

$$z = x, \quad \text{and} \quad x = y.$$

If this be supposed to be known, the expressions (article 451) for the points A', B', C', in which these three last planes are cut by the lines *BC, CA, AB,* are at once seen to be correct; and it may be remarked that the three other planes, *BCE, CAE, ABE,* which are drawn through the unit-point E (article 450), and through the three other edges of the unit pyramid *ABCD* (article 449), are represented by the three analogous equations,

$$x = w, \quad y = w, \quad z = w;$$

which may be employed to deduce the expressions (article 451) for the points A_2, B_2, C_2. As regards the four other derived points, $A_1, \ldots, D_1$, it may be noted that the four lines, *EA, EB, EC, ED,* drawn from the unit-point to the corners of the unit-pyramid, are represented by the four following pairs of equations (which immediately follow from those above assigned);

$$y = z = w; \quad z = x = w; \quad x = y = w; \quad x = y = z.$$

453. From the foregoing examples, it is natural to suppose that a *plane, generally,* is *anharmonically represented* by a *homogeneous and linear equation,* connecting the *four anharmonic coordinates* of a *variable point* on the plane: and such I find to be the fact. Let *Q, R, S* denote the points in which the plane Π intersects the three edges of the unit-pyramid, which meet in what we may call its *vertex, D;* so that

$$Q = DA^{\cdot}\Pi, \quad R = DB^{\cdot}\Pi, \quad S = DC^{\cdot}\Pi;$$

and let A_2', B_2', C_2' denote the points which, on the same three edges, are harmonically conjugate to the points A_2, B_2, C_2 (article 451), so that

$$(DA_2 AA_2') = (DB_2 BB_2') = (DC_2 CC_2') = -1;$$

then, writing

$$\frac{l}{r} = (DA_2' AQ), \quad \frac{m}{r} = (DB_2' BR), \quad \frac{n}{r} = (DC_2' CS),$$

I *define* that *l, m, n, r,* or any numbers proportional to them, are the *Anharmonic Coordinates of the Plane* Π, which I *denote* accordingly by the *Symbol,*

$$\Pi = [l, m, n, r],$$

or sometimes briefly by

$$\Pi = [l\, m\, n\, r],$$

and I *prove* (*quaternions* and *quines* having *assisted* me, although of course the question may be treated by *geometry alone*) that "*if* $P = (x\, y\, z\, w)$ *be any point on this plane* $\Pi = [l\, m\, n\, r]$, *then the coordinates of point and plane are connected by the equation,*

$$lx + my + nz + rw = 0;\text{"}$$

a *Theorem* which is analogous to that stated and proved in the Letter (articles 10, 61, and 80), for *points* and *lines* in the *plane*; and which you can at once see to be of great, and even of fundamental importance, in the application of this *Anharmonic Method* to *Space.*

454. One immediate consequence of this Theorem is, that "*if any four points P, P_1, P_2, P_3 be complanar*, then *their sixteen coordinates are connected by the equation,*

$$\begin{vmatrix} x & y & z & w \\ x_1 & y_1 & z_1 & w_1 \\ x_2 & y_2 & z_2 & w_2 \\ x_3 & y_3 & z_3 & w_3 \end{vmatrix} = 0;\text{''}$$

which may be regarded as one form of the *local equation of the plane $P_1 P_2 P_3$*, or of the equation of that plane considered as the *locus Π of a variable point P*. And similarly "*if any four planes Π, Π_1, Π_2, Π_3 be concurrent,* that is, if they meet in any one *common point P*, then *their* coordinates must satisfy the analogous condition,

$$\begin{vmatrix} l & m & n & r \\ l_1 & m_1 & n_1 & r_1 \\ l_2 & m_2 & n_2 & r_2 \\ l_3 & m_3 & n_3 & r_3 \end{vmatrix} = 0;\text{''}$$

which is a form for what may be called (article 10) the *tangential equation of the point P*, considered as a *pivot*, round which a *variable plane Π may turn*, being obliged *always to pass through that point P*, whatever *direction* it may assume in space.

455. And the very general and very useful *Theorem*, or rather *pair of theorems*, connected with each other by the *Principle of Geometrical Duality*, is the following.

I. Every Quotient of two given homogeneous and linear Functions, of the Coordinates of a variable Point may be expressed as the Anharmonic of a Pencil of Planes; whereof three are given and collinear, (compare article 444) while the fourth passes through their common line, and through the variable point.

II. Every Quotient of two given homogeneous and linear Functions, of the Coordinates of a variable Plane, may be expressed as the Anharmonic of a Group of Points: whereof three are given and collinear, while the fourth is the intersection of their common line, with the variable plane.

456. More fully, I., if $f(x\,y\,z\,w)$ and $f_1(x\,y\,z\,w)$, or briefly f and f_1, be any two given homogeneous and linear functions of x, y, z, w; and if we determine three planes, Π_1, Π_2, Π_3, by the local equations,

$$f = 0, \quad f_1 = f, \quad f_1 = 0;$$

if also through the fixed line Λ which is common (by their form) to these three planes and through the variable point $P = (x\,y\,z\,w)$, we draw a variable plane $\Pi = \Lambda P$: then we shall have the equation,

$$\frac{f_1}{f} = (\Pi_1\,\Pi_2\,\Pi_3\,\Pi).$$

457. And in like manner, II., if $F(l\,m\,n\,r)$ and $F_1(l\,m\,n\,r)$, or briefly F and F_1, be any two given homogeneous and linear functions of l, m, n, r; if we determine three points, P_1, P_2, P_3, by the tangential equations,

$$F = 0, \quad F_1 = F, \quad F_1 = 0,$$

which points will always be ranged on one common and given line Λ; and if we denote by P the point $\Lambda\dot{}\Pi$ in which this fixed line intersects the variable plane $\Pi = [l\,m\,n\,r]$; then we shall have the equation,

$$\frac{F_1}{F} = (P_1\,P_2\,P_3\,P).$$

458. To illustrate these two Theorems by Examples, let us consider the two quotients, $\frac{z}{y}$, and $\frac{n}{m}$. For the first we have the values,

$$f = y, \quad f_1 = z;$$

and for the second,

$$F = m, \quad F_1 = n.$$

Hence the planes Π_1, Π_2, Π_3, are, for the first example,

$$y = 0, \quad y = z, \quad z = 0,$$

or ACD, ADE, ABD; (articles 446, and 452), their common line Λ is therefore in this case AD; and the variable plane is $\Pi = ADP$. We thus recover the equation,

$$\frac{z}{y} = (AD.CEBP);$$

which had been stated in article 452.

459. Again,

$$m = 0, \quad n = m, \quad n = 0,$$

are the tangential equations (article 454) of the three collinear points,

$$B = (0\,1\,0\,0), \quad A'' = (0\,1\,\bar{1}\,0), \quad C = (0\,0\,1\,0);$$

(in which, to save commas, I write $\bar{1}$ for -1;) because they express the condition requisite for a plane Π, or $[l,\ m,\ n,\ r]$, passing through the three points respectively: if then we denote by L the point in which a variable plane $\Pi = [l\,m\,n\,r]$ intersects the fixed line BC, we have, by the Theorem in article 457, the equation,

$$\frac{n}{m} = (B\,A''\,C\,L);$$

when A'' is, on BC, the harmonic conjugate of A', so that $(B\,A'\,C\,A'') = -1$.

460. Similarly,

$$\frac{l}{n} = (C\,B''\,A\,M), \quad \text{and} \quad \frac{m}{l} = (A\,C''\,B\,N),$$

if $B'' = (\bar{1}\,0\,1\,0)$, $C'' = (1\,\bar{1}\,0\,0)$, so that $(C\,B'\,A\,B'') = (A\,C'\,B\,C'') = -1$, and if M and N be the points in which the same plane Π intersects the given lines CA and AB: with the verification that these three last expressions for $\frac{n}{m}, \frac{l}{m}, \frac{m}{l}$, give the product,

$$(B\,A''\,C\,L).(C\,B''\,A\,M).(A\,C''\,B\,N) = +1;$$

which answers to the circumstance that $A''B''C''$ and LMN are two rectilinear transversals of the triangle ABC as in Figure 39, so that we have the two equations,

$$\frac{BA''}{A''C}\cdot\frac{CB''}{B''A}\cdot\frac{AC''}{C''B} = -1,$$

$$\frac{BL}{LC}\cdot\frac{CM}{MA}\cdot\frac{AN}{NB} = -1;$$

on dividing the former of which by the latter, the recent product of anharmonics is obtained.

461. The points A'', B'', C'' here correspond to those which were marked as A', B', C' in earlier articles, for example in article 79, and Figure 21; and in fact we had *there* the equations,

$$\frac{n}{m} = (C\,L\,B\,A')\text{\&c.},$$

which might have been written as

$$\frac{n}{m} = (B\,A'\,C\,L),\ \text{\&c.}$$

Conversely, I have thought it convenient, in recent articles, to write A', B', C' instead of the old A'', B'', C''; the point O, in which the lines AA'', BB'', CC'', of article 77, and Figure 18, concurred, being now replaced by the point of concourse of the three lines AA', BB', CC', which point is now D_1: namely the intersection (article 451) of the line DE with the plane ABC. But it is unnecessary to draw a new Figure, to illustrate these small changes of notation which have been suggested in part by the passage from the plane to space.

462. As another verification, and at the same time as a new illustration of the theory, it may be remarked that we can *recover* the *equations of definition*, – if we suppose ourselves to have *forgotten* them, – for the quotients $\frac{x}{w}, \frac{y}{w}, \frac{z}{w}$, (in article 445), and for the quotients $\frac{l}{r}, \frac{m}{r}, \frac{n}{r}$, (in article 453) by applying the theorems of articles 456 and 457: which may be otherwise enunciated as follows.

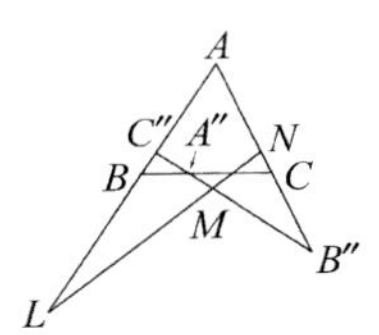

Fig. 39

I. If $\varphi(x\,y\,z\,w)$ be any *Rational Fraction*, expressed as the Quotient of two given homogeneous and Linear Functions of the form variables, x, y, z, w; and if we determine a line Λ and three planes Π_1, Π_2, Π_3 through that line, by the equations,

$$\varphi = \frac{0}{0}, \quad \varphi = \infty, \quad \varphi = 1, \quad \varphi = 0,$$

the Fractional Function φ may then be expressed as the Anharmonic of a Pencil of Planes as follows:

$$\varphi(x\,y\,z\,w) = (\Pi_1\,\Pi_2\,\Pi_3\,\Pi);$$

where Π is the plane which passes through the fixed line Λ, and through the variable point $P = (x\,y\,z\,w)$.

II. If $\phi(l\,m\,n\,r)$ be, in like manner, a *Fractional Function* of l, m, n, r, expressed as the Quotient of two given homogeneous and linear functions of those four variables; and if we determine a line Λ, and three points P_1, P_2, P_3 on that line, by the equations

$$\phi = \frac{0}{0}, \quad \phi = \infty, \quad \phi = 1, \quad \phi = 0;$$

we shall then be able to express the Fraction or Function ϕ as the Anharmonic of a Group of Points, as follows:

$$\phi(l\,m\,n\,r) = (P_1\,P_2\,P_3\,P);$$

where P is the intersection of the fixed line Λ with the variable plane, $\Pi = [l\,m\,n\,r]$.

463. When $\varphi = \frac{x}{w}$, the three fixed planes are (by articles 446, 452, and 447), *ABC*, *BCE*, *BCD*, and their common line is *BC*; we therefore recover the equation of definition (article 445),

$$\frac{x}{w} = (BC.AEDP),$$

as one which is *included* in the Theorem I., of articles 455, 456, and 462; and in like manner, when $\phi = \frac{l}{r}$, the three fixed points are D, A_2', A, so that if Q be (as before) the point of intersection of the fixed line DA with the variable plane Π, we are conducted anew to the equation (article 453),

$$\frac{l}{r} = (DA_2'AQ);$$

with (of course) similar expressions for $\frac{y}{w}, \frac{z}{w}, \frac{m}{r}, \frac{n}{r}$. But since *these* equations had been *defined* to be true, our thus arriving at them *again* can only be regarded as a *verification* (article 462) of the *consistency* of this whole theory; and as *an illustration* of the *meaning* of the theorems.

464. You may perhaps remember, that so early in the Letter as in Article 11 I introduced, and very often afterwards employed, a Theorem for the Plane, which indeed was stated without proof: namely that if there be any four points,

$$P_0 = (x_0\,y_0\,z_0), \quad P_1 = (x_1\,y_1\,z_1), \quad P_2 = \&c., \quad P_3 = \&c.,$$

related so that

$$x_2 = tx_0 + ux_1, \qquad y_2 = ty_0 + uy_1, \qquad z_2 = tz_0 + uz_1,$$

$$x_3 = t'x_0 + u'x_1, \qquad y_2' = t'y_0 + u'y_1, \qquad z_2' = t'z_0 + u'z_1,$$

– under which conditions those four points must be *collinear*, – then the Anharmonic of their Group may be expressed as follows:

$$(P_0\, P_2\, P_1\, P_3) = \frac{t'u}{tu'}.$$

The demonstration by *quaternions* of this Theorem is very easy, and I dashed it off one day on the back of a note, while sitting in the Council Room of the Academy, for the purpose of proving by it to Graves* the equation

$$(O\, U\, C\, U') = -1, \qquad \text{(article11)}$$

which in connexion with Figure 6 (of the same article) expresses the harmonic relation of pole and polar for a plane conic: but I really forget whether I ever proved the theorem by *geometry* alone, my conviction of its truth being so complete from the first (through quaternions), and applications of it springing up so abundantly. I have now repeated the enunciation, for the purpose of stating, in connexion with it, an analogous *Theorem for Space*, which like this former has very many applications.

465. This new Theorem is, that "if the anharmonic symbols, $(x_0\, y_0\, z_0\, w_0)$, &c., or briefly $(P_0)\,(P_1)\,(P_2)\,(P_3)$, of any four points of *space*, be connected by the two symbolic equations,

$$(P_1) = t(P_0) + u(P_2), \qquad (P_3) = t'(P_0') + u'(P_2'),$$

then not only are those four points collinear (compare articles 453, and 454), but the anharmonic of their group has the value

$$(P_0\, P_1\, P_2\, P_3) = \frac{ut'}{tu'}."$$

466. Another Theorem, – which may be said (compare article 455) to be the *Dual* of this last, – exists, and may be thus enunciated:-

"If the anharmonic symbols, $[l_0\, m_0\, n_0\, r_0], \ldots, [l_3\, m_3\, n_3\, r_3]$, of any four planes, be connected by two equations of the forms,

$$[\Pi_1] = t[\Pi_0] + u[\Pi_2], \; [\Pi_3] = t'[\Pi_0] + u'[\Pi_2],$$

then not only (compare article 454) are these four planes collinear, but also the anharmonic of their pencil has the value,

$$(\Pi_0\, \Pi_1\, \Pi_2\, \Pi_3) = \frac{ut'}{tu'}."$$

* [Charles Graves, 1812–1899, at the time Erasmus Smith's Professor of Mathematics at Trinity College, Dublin.]

467. It is understood that the symbolic equation,

$$[\Pi_1] = t(\Pi_0) + u[\Pi_2],$$

is designed to represent the system of the *four* ordinary equations,

$$l_1 = tl_0 + ul_2, \ldots, r_1 = tr_0 + ur_2;$$

and in like manner that the *four* equations,

$$x_1 = tx_0 + ux_2, \ldots, w_1 = tw_0 + uw_2,$$

are briefly represented by the *one* symbolic formula,

$$(P_1) = t(P_0) + u(P_2).$$

Analogous abridgments had occurred before, in the Letter; for example, in articles 74, 150, &c. And it is important to remember, that any *one* formula, such as

$$(P_1) = t(P_0) + u(P_2),$$

or

$$[\Pi_1] = t[\Pi_0] + u[\Pi_2],$$

expresses nothing more than the *graphical relation* of *collinearity*, between the *three points*, P_0, P_1, P_2, or the *three planes* Π_0, Π_1, Π_2: and that *no metric* (or *anharmonic*) *relation* is expressed, until we have, for points or planes (compare article 161), a system of *two* such formulæ.

468. (October 3rd 1860) – I think that the foregoing are the *most* (or *among* the most) general and useful things which I made out, not very recently, about *Anharmonic Coordinates in Space*: but which have been since *absorbed* (article 441) in a Theory of *Quinary Coefficients, Symbols,* and *Types*, whereon, however, I may as well abstain from entering here. But it seems that I ought to add something respecting the *Anharmonic Method,* as applied to *Curved Surfaces*: although, in the mere *mechanism* of calculation, the fundamental theorem,

$$lx + my + nz + rw = 0, \quad \text{(article 453)}$$

causes that method to differ as little usually, in its mere *working*, from the known method of *Quadriplanar Coordinates*, as Anharmonics differ, *within* the *plane*, from *Trilinears* (article 144).

469. Let it then be required to determine the *anharmonic and local equation* of the *surface of the second order*, which passes through the *nine points*, denoted by

$$A, \quad B, \quad C, \quad D, \quad E, \quad A', \quad A_2, \quad C', \quad C_2,$$

in articles 445, and 451. In the first place, this *equation*,

$$f(x\,y\,z\,w) = 0,$$

must be *homogeneous*, because only *ratios* enter; and of the *second dimension*, because if we establish any *two linear equations* between the *four variable coordinates*, to express that the point $x\,y\,z\,w$ of the locus is *on a given right line*, and then *eliminate two* of those four variables, the resulting homogeneous equation between the *other two* must have *two roots* (real or imaginary), and *not more* than two. But, by article 449, we have

$$A = (1, 0, 0, 0), \ldots, D = (0, 0, 0, 1);$$

and the surface is to pass through these four points (the corners of the unit-pyramid); its equation must therefore *not contain the squares* of the variables, and one may write,

$$f = ayz + bzx + (ex + gy + hz)w;$$

in which it remains to determine the ratios of the six coefficients, by the condition that the surface shall pass also through the five other given points.

470. One of these points is (by article 451)

$$A' = (0, 1, 1, 0);$$

the coefficient of the product yz therefore vanishes, and the equation at this stage takes the form,

$$f = x(bz + cy) + w(ex + gy + hz) = 0;$$

it is therefore satisfied by every point for which z and w vanish; that is (article 448), *by every point of the right line BC*. The same conclusion would have followed, if we had not supposed the coefficients of x^2, w^2, that is, if we had not obliged the locus to pass through A and D; in fact the *three* points B, C, A' are *collinear*, and any surface of the second order which passes through *them* must therefore contain the *whole indefinite right line* upon which they are situated.

471. Obliging our locus to contain also the points A_2, C', C_2, or $(1, 0, 0, 1)$, $(1, 1, 0, 0)$, $(0, 0, 1, 1)$, we are led to suppress also the coefficients of xw, xy, zw; the equation therefore reduces itself to the binomial form,

$$f = bxz + gwy = 0:$$

and represents *any hyperboloid* on which the *gauche** *quadrilateral ABCD* is *superscribed*; or which has AB, BC, CD, DA for four of its generating lines. And when we make it pass through E, we have, finally

$$b + g = 0;$$

so that the equation of the sought locus, or surface, becomes simply,

$$0 = f(x\, y\, z\, w) = xz - yw.$$

– Of course, the *same nine sets* of *coordinates* would have given the *same equation*, in the *quadriplanar method* also (article 468).

472. But here comes into play a difference of *interpretation*; for while, in quadriplanar coordinates, the equation expresses that *one product of two perpendiculars on planes is equal to another such product*, my method supposes *no use* (nor even knowledge) of *perpendiculars*, or any thing else connected with the *sphere*: and accordingly, I interpret the equation as

* [Not planar.]

expressing, in either of two ways, an *equality between the anharmonic functions of two pencils of planes.* In fact, if we write it as

$$\frac{x}{w} = \frac{y}{z},$$

it expresses (by articles 445 and 452) the equality,

$$(BC.AEDP) = (DA.BECP);$$

and if we write it as

$$\frac{z}{w} = \frac{y}{x},$$

it signifies the other equality,

$$(AB.CEDP) = (CD.BEAP);$$

whence can be easily derived the *double generation,* and the *anharmonic properties* of *any ruled hyperboloid.*

473. (November 9th, 1860.) I perceive, by a memorandum, that it is just a month since I posted that last sheet of this "Postscript" but I have since been otherwise fully employed, and of course I do not flatter myself that you have regretted the delay. *This* time, however, – that is, within a day or two, – I trust that I shall really conclude, though writing only at intervals. We had (article 472), the anharmonic equations,

$$(BC.AEDP) = (DA.BECP);$$
$$(AB.CEDP) = (CD.BEAP),$$

as two forms of the equation of the hyperboloid, on which the gauche quadrilateral $ABCD$ is superscribed, and which passes through the point E; see Figure 40. And we saw that, in anharmonic coordinates, the second of the two last written equations corresponded to the following:

$$\frac{z}{w} = \frac{y}{x}; \text{ or } xz - yw = 0.$$

It is now to be shown that this equation can be *interpreted,* as signifying (without any reference to *perpendiculars*), that the hyperboloid in question is the *locus* of a *point P,* such that if we

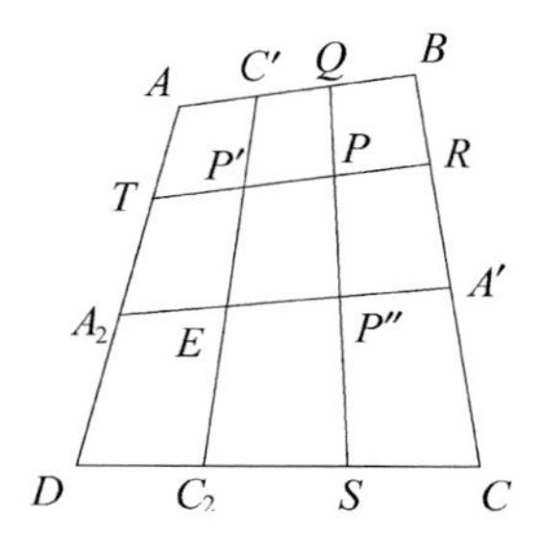

Fig. 40

draw through it a right line *QPS*, which *cuts the two opposite sides*, *AB* and *DC*, of the quadrilateral, *this line divides those two sides homographically*: or more fully, – or at least more symbolically, – that we shall have the anharmonic equation,

$$(A\,C'\;B\,Q) = (D\,C_2\;C\,S),$$

$$\text{if } Q = AB^{\cdot}DCP, \quad \text{and} \quad S = DC^{\cdot}ABP,$$

$$\text{as } C' = AB^{\cdot}DCE, \quad \text{and} \quad C_2 = DC^{\cdot}ABE,$$

by the notations of a recent article (451). But in fact this last anharmonic equation, under the form $(C\,C_2\;D\,S) = (B\,C'\;A\,Q)$, is an immediate consequence of the equation previously established, namely,

$$(AB.CEDP) = (CD.BEAP);$$

in deriving which consequence, we have only to conceive that the pencil through *AB* is cut by *CD* as a transversal; and that the pencil through *CD* is cut by *BA* as a transversal.

474. It is clear that in like manner, the other equation (article 472) between anharmonics of *pencils of planes*,

$$(BC.AEDP) = (DA.BECP),$$

when we cut the one pencil by *AD*, and the other by *BC*, taking the form of an equation between anharmonics of *groups of points*, as follows:

$$(A\,A_2\;D\,T) = (B\,A'\;C\,R);$$

when

$$A' = BC^{\cdot}ADE, \text{ and } A_2 = AD^{\cdot}BCE,$$

as in article 451; and

$$R = BC^{\cdot}ADP, \; T = AD^{\cdot}BCP,$$

as is partially represented in the recent Figure 40. It follows that the right line *RPT*, which is drawn through *P across the other pair of opposite sides BC, AD*, of the same gauche quadrilateral, *divides that pair also* homographically:- as is known. (Compare [Michel] Chasles, [*Traité de*] *Géometrie Supérieure*, page 298: Paris, 1852. You probably have the book – a copy was sent me by the author.)

475. So far we have only been deriving (known) *properties* of the *surface of the second order*, which is *represented* by the *equation*,

$$xz = yw,$$

and which is here considered as *determined* thereby. In *that* view, the shortest way of *proving* that the two systems of homographic divisions *exist*, might be to remark, that with our last significations of *Q*, *R*, *S*, *T*, as intersections of lines with planes, we have the anharmonic symbols,

$$Q = (x\,y\,0\,0); \quad R = (0\,y\,z\,0); \quad S = (0\,0\,z\,w); \quad T = (x\,0\,0\,w).$$

Nothing more is *here* supposed, than that through the point $P = (x\,y\,z\,w)$, anywhere situated in space, are drawn two right lines, QPS and RPT, so as to intersect the lines AB, BC, CD, DA in the points Q, R, S, T. Hence, with this signification of these four last letters, and with A' &c. signifying as before, we have the following *general interpretations* of the *ratios* of the four anharmonic coordinates of a point P *in space*:

$$\frac{x}{y} = (A\,C'\,B\,Q); \quad \frac{y}{z} = (B\,A'\,C\,R); \quad \frac{z}{w} = (C\,C_2\,D\,S); \quad \frac{w}{x} = (D\,A_2\,A\,T);$$

which are quite *independent of the conception of any curved surface* whatever, and *might* have been assumed as *definitions of anharmonic coordinates* (of a point). As an easy verification, the four last equations give,

$$1 = (A\,C'\,B\,Q)(B\,A'\,C\,R)(C\,C_2\,D\,S)(D\,A_2\,A\,T),$$

$$\frac{AC'}{C'B}\cdot\frac{BA'}{A'C}\cdot\frac{CC_2}{C_2D}\cdot\frac{DA_2}{A_2A} = \frac{AQ}{QB}\cdot\frac{BR}{RC}\cdot\frac{CS}{SD}\cdot\frac{DT}{TA},$$

for *any gauche quadrilateral*, cut by *any two transversal planes*, in the points $C'\,A'\,C_2\,A_2$, and $Q\,R\,S\,T$; and in fact each of the two last *products of quotients of segments of sides* is known to be equal to positive unity, by a theorem which (I think) is due to Carnot*, but at all events is stated by Chasles. Admitting, then, these *general* interpretations of $\frac{x}{y}$, &c., we see at once that if a curved surface be represented by the equation,

$$xz = yw, \quad \text{or} \quad \frac{x}{y}.\frac{z}{w} = 1,$$

it must have the *anharmonic property*,

$$(A\,C'\,B\,Q)(C\,C_2\,D\,S) = 1, \quad \text{or} \quad (A\,C'\,B\,Q) = (D\,C_2\,C\,S),$$

as before; so that it is the *locus of a right line* QS, which *divides* AB and DC *homographically*: AD, $C'C_2$ and BC being three of its positions. And because we should have with equal ease this other anharmonic equation,

$$\frac{x}{w} = \frac{y}{z}, \quad \text{or,} \quad (A\,A_2\,D\,T) = (B\,A'\,C\,R),$$

as before, we should see that the *same surface of the second order* may be *generated* in the (well known) *second mode*, as the *locus of another right line* TR, which divides AD and BC homographically, having AB, A_2A', and DC for three of its positions.

476. We have here been *interpreting an equation*, of the second order: but we might have proposed to *find a locus*, namely the *locus of the right line* which divides two given right lines in space homographically; or (more definitely) the *locus of a variable point* P, on the variable line QS, when the condition

$$(A\,C'\,B\,Q) = (D\,C_2\,C\,S)$$

is satisfied, as before. For this purpose, the following method appears natural. Since Q is on AB, its anharmonic symbol must be of the form $(t\,u\,0\,0)$; and since S is its homologue on DC, C_2 being the homologue of C', we must then have $(0\,0\,u\,t)$ as the corresponding symbol for

* [Lazare-Nicolas-Marguerite Carnot, 1753–1823.]

S. But *P* is collinear with *Q*, *S*; its symbol $(x\,y\,z\,w)$ must therefore be of the form $q(Q) + s(S)$, or we must have

$$(x\,y\,z\,w) = q(t\,u\,0\,0) + s(0\,0\,u\,t);$$

(compare articles 465, and 467) whence follow the four separate equations,

$$x = qt, \quad y = qu, \quad z = su, \quad w = st;$$

and by elimination of auxiliary quantities, the required *equation* of *the locus*; of the *point P*, or of the *line QS*, is found again to be,

$$xz = yw.$$

477. I have been avoiding throughout this Letter and Postscript, to make much use of *Quaternions*, or even of *Vectors*; but cannot resist the temptation to write here a *vector-formula* for a ruled hyperboloid, which has a natural connexion with what has been said respecting the anharmonic coordinates, although it may be arrived at, and interpreted, separately. Let then *OA*, *OB*, *OC*, *OD*, or α, β, γ, δ, be four given and coinitial vectors not terminating on any common plane; and let *OP* or ρ be a variable vector. Let a, b, c, d be four given or constant scalars, and let s, t, u, v be four variable scalars, of which however only the two ratios, $s : u$, and $t : v$ are important. Then if we suppose that

$$\rho = \frac{sta\alpha + tub\beta + uvc\gamma + vsd\delta}{sta + tub + uvc + vsd},$$

the locus of *P* is a single-sheeted hyperboloid, having the four sides of the quadrilateral *ABCD* for generating lines, and passing through the point *E* of which the vector is,

$$E = \frac{a\alpha + b\beta + c\gamma + d\delta}{a + b + c + d}.$$

478. (November 23rd, 1860.) – I find that another week has passed away, – although not unemployed, – since I posted my last packet to you, including the missing Sheet LIV, which completed the "Letter" itself. Perhaps I was fearful of going too far, for consistency with my plan, – into the application of *Vectors*, which I had sketched in the last Paragraph, to the discussion of the Ruled Hyperboloid. Let me add, however, that with the same vector-formula,

$$\rho = \frac{sta\alpha + tub\beta + uvc\gamma + vsd\delta}{sta + tub + uvc + vsd},$$

– in which a, b, c, d are constant scalars, and s, t, u, v are variable scalars, while α, β, γ, δ are constant vectors, but ρ is a variable vector, namely that of a variable point *P* on the hyperboloid, – the vector κ of the *centre K* is, by my method, found to be:

$$\kappa = \frac{ac(\alpha + \gamma) - bd(\beta + \delta)}{2(ac - bd)};$$

for this among other reasons, that if we make

$$s' = bt + cv, \quad t' = cu + ds, \quad u' = dv + at, \quad v' = as + bu,$$

and

$$\rho' = \frac{s't'a\alpha - t'u'b\beta + u'v'c\gamma - v's'd\delta}{s't'a - t'u'b + u'v'c - v's'd},$$

so that ρ' is the vector of *another point* of the *same surface*, then

$$\rho + \rho' = 2\kappa, \quad \text{or} \quad \rho' = 2\kappa - \rho;$$

and thus *every chord PP'* of the hyperboloid, which passes through the *fixed point K*, is *bisected* in that point. The theorem that the centre K of the surface is on the right line which *bisects the two diagonals, AC* and *BD*, is also much in evidence by the foregoing expression for the vector κ.

479. When $ac = bd$, the *centre* is *infinitely distant*; and the surface becomes a *hyperbolic paraboloid. In this case,* since $sta.uvc = tub.vsd$, no generality is lost by *suppressing* the constant factors a, b, c, d, and writing simply,

$$\rho = \frac{st\alpha + tu\beta + uv\gamma + vs\alpha}{st + tu + uv + vs};$$

respecting which I shall only here observe, that it shows that the point E, of which the vector is now reduced to the simple form,

$$\varepsilon = \tfrac{1}{4}(\alpha + \beta + \gamma + \delta),$$

so that it is the *mean point* (or centre of gravity) of the *pyramid ABCD*, – is still a point of the *surface*. And conversely, if E be such a mean point, the surface of the second order which passes through it, and has the quadrilateral *ABCD* superscribed upon it, is thus found to be a ruled *paraboloid*:- a theorem which is probably known.

480. Resuming, for a short time, the *anharmonic coordinates*, I shall merely *use* the recent results with *vectors*, so far as to observe that they enable us to infer, – what can be otherwise proved, – that the harmonic conjugate of the fixed point,

$$K = (c, \ -d, \ a, \ -b),$$

taken with respect to the chord PP' which passes through that point K, and of which the extremities are denoted by the anharmonic symbols,

$$P = (st, \ tu, \ uv, \ vs),$$
$$P' = (s't, \ -t'u', \ u'v', \ -v's'),$$

when

$$s' = bt + cv, \quad t' = cu + ds, \quad u' = dv + at, \quad v' = as + bu,$$

as in article 478, is a point P'' upon the plane

$$\Pi = [a, \ b, \ c, \ d];$$

that is, on the *plane at infinity*. In fact, *without vectors*, if we write for abridgment,

$$ac - bd = e, \quad ast + btu + cuv + dvs = w',$$

these values of s', t', u', v' give,

$$s't' + est = cw'; \quad -t'u' + etu = -dw'; \quad u'v' + euv = aw'; \quad -v's' + evs = -bw';$$

therefore

$$(P') + e(P) = w'(K),$$

a formula which exhibits the asserted *collinearity* of the *three points, P, K, P'*. Again, as regards the harmonic *conjugation,* if we suppose that it *exists, or that* P'' is determined so as to satisfy the condition,

$$(P \quad P' \quad P'') = -1,$$

we may then write,

$$-(P') + e(P) = (P'');$$

or

$$(P'') = 2e(P) - w'(K);$$

or

$$P'' = (2est - cw', 2etu + dw', 2euv - aw', 2evs + bw');$$

if then

$$P'' = (x, y, z, w),$$

we have

$$\begin{aligned} & ax + by + cz + dw \\ &= a(2est - cw') + b(2etu + dw') + c(2euv - aw') + d(2evs + bw') \\ &= 2e(ast + btu + cuv + dvw) - 2(ac - bd)w' \\ &= 2ew' - 2ew' = 0; \end{aligned}$$

so that the point P'' is on the plane $[a, b, c, d]$, as asserted.

481. If we write, with a different signification of P'',

$$P' = (y, y, z, z), \qquad P'' = (x, y, y, x),$$

then the point P' is at once collinear with $(y, y, 0, 0)$ and $(0, 0, z, z)$, or with $(1, 1, 0, 0)$ and $(0, 0, 1, 1)$, or with C' and C_2, and also with $(0, y, z, 0)$ or R, and with $(y, 0, 0, z)$ or $(x, 0, 0, w)$ or T; hence we may write (compare Figure 40),

$$P' = C'C_2 \cdot RT,$$

so that the *variable generatrix RT* of *one* system *intersects* (as is well known) the *three fixed generatrices AD, BC,* $C'C_2$ of the other system. And similarly the point last called P'' is the intersection,

$$P'' = A'A_2 \cdot QS.$$

482. We may write

$$(E) = (C') + (C_2), \quad (P') = y(C') + z(C_2);$$

whence results the anharmonic,

$$(C'\, E\, C_2\, P') = \frac{y}{z} = (B\, C\, A'\, R) = (A\, D\, A_2\, T);$$

so that the variable generatrix RT not only *intersects* the *three* generatrices AD, BC, $C'C_2$, as before, but also *divides* them *homographically*: another known property of the hyperboloid. We must of course expect to find that RT divides QS *also* homographically, with AD and BC; or that

$$(Q\, P''\, S\, P) = (B\, A'\, C\, R) = \frac{y}{z}.$$

Accordingly, if we write

$$Q = (x,\ y,\ 0,\ 0), \quad S = (0,\ 0,\ z,\ w),$$

as before, we have

$$(P) = (Q) + (S), \quad z(P'') = (zx,\ zy,\ zy,\ yw) = z(Q) + y(S);$$

whence

$$(S\, P\, Q\, P'') = \frac{y}{z},$$

as was expected.

483. Let $[l,\ m,\ n,\ r]$ be the anharmonic symbol for the plane $QRST$ of the two generatrices QS, RT through P. Then

$$0 = lx + my = my + nz = nz + rw = vw + lx;$$

whence

$$l : m : n : r = x^{-1} : -y^{-1} : z^{-1} : -w^{-1} = z : -w : x : -y;$$

the plane $[z,\ -w,\ x,\ -y]$ is therefore the *tangent plane* to the hyperboloid at the point P, or $(x,\ y,\ z,\ w)$. And as the *local equation* of that surface is:

$$0 = f(x\, y\, z\, w) = xz - yw,$$

so the *tangential equation* of the same surface is

$$0 = F(l\, m\, n\, r) = lr - mn.$$

(When $ac = bd$, this last equation is satisfied by $[a,\ b,\ c,\ d]$: or *the ruled paraboloid touches* (as is known) *the plane at infinity*.)

484. More generally, the *point* $(x,\ y,\ z,\ w)$ and the *plane* $[l,\ m,\ n,\ r]$ are related as *pole* and *polar*, with respect to the hyperboloid in question,

$$l = z, \quad m = -w, \quad n = x, \quad r = -y;$$

whence (*without vectors*) the *centre* K of that surface, or the *pole* of the *plane at infinity*, namely of $[a, b, c, d]$, is seen again to be the point $(c, -d, a, -b)$.

485. In general, if $f = 0$ be the *local equation* of *any curved surface*, I find that $[D_x f, D_y f, D_z f, D_w f]$ is the anharmonic *symbol* of its *tangent plane* at the *point* $(x\ y\ z\ w)$; and that in like manner, if $F = 0$ be the *tangential equation* of *any surface*, then $(D_l F, D_m F, D_n F, D_r F)$ is the anharmonic symbol of the *point of contact* of *the plane* $[l\ m\ n\ r]$ with its *envelope*: whence it is easy to conceive how problems of *reciprocal surfaces* may be treated, by a *mechanism* which is in fact the *same* as that used in the known method of quadriplanar coordinates. (Compare article 468.)

486. As bearing on the general determination of a *tangent plane*, let $P = (x, y, z, w)$ be any given point, and let

$$P' = (x + \Delta x,\ y + \Delta y,\ z + \Delta z,\ w + \Delta w),$$

and

$$P'' = (x + \Delta' x,\ y + \Delta' y,\ z + \Delta' z,\ w + \Delta' w),$$

be any two other points, on a surface given by its *local equation*, $f = 0$; where f is any given and homogeneous function of x, y, z, w. Let $[L, M, N, R]$ be the symbol of the *secant plane* Π, or $P P' P''$, which passes through these three points; we shall then have the three equations,

$$Lx + My + Nz + Rw = 0,$$

$$L\Delta x + M\Delta y + N\Delta z + R\Delta w = 0,$$

$$L\Delta' x + M\Delta' y + N\Delta' z + R\Delta' w = 0,$$

whereby the *ratios* of L, M, N, R, or the *position* of this secant plane, may generally be determined, when x, y, z, w, and $\Delta x, \ldots, \Delta w, \Delta' x, \ldots, \Delta' w$, are known. Let a fixed right line PP_1 be conceived to be drawn through the point P, so as to be normal, or at least not tangential, to the surface: and let the points P', P'' be conceived to *approach* indefinitely to P, along the respective *sections*, made by the two planes, $PP_1 P'$ and $PP_1 P''$. The *differences*, $\Delta x, \ldots$, and $\Delta' x, \ldots$, will then *diminish* indefinitely, and will tend to *vanish* together; but they will, in general, have certain *limiting ratios*, and will be *ultimately proportional*, for each of the two points P' and P'', to certain *finite quantities*: which, in the spirit of the Newtonian Doctrine of *Fluxions*, and not without support from modern authority, may be called the *differentials* of the coordinates $x, \ldots$, and may be denoted as such by $\mathrm{d}x, \ldots, \mathrm{d}w$, and $\mathrm{d}'x, \ldots, \mathrm{d}'w$; the latter sets being here supposed to be *distinct* from the former. Then the coordinates L, M, N, R of the *secant plane* will also vary, but will *tend* to acquire certain *determined ratios*, namely those of the coordinates l, m, n, r of the sought tangent *plane* at P; which latter plane must therefore satisfy the three conditions,

$$0 = lx + my + nz + rw, \qquad 0 = l\mathrm{d}x + m\mathrm{d}y + n\mathrm{d}z + r\mathrm{d}w,$$

$$0 = l\mathrm{d}'x + m\mathrm{d}'y + n\mathrm{d}'z + r\mathrm{d}'w;$$

when $\mathrm{d}x, \ldots$ and $\mathrm{d}'x, \ldots$ are only obliged to satisfy the two linear equations,

$$0 = \mathrm{d}f = D_x f.\mathrm{d}x + D_y f.\mathrm{d}y + D_z f.\mathrm{d}z + D_w f.\mathrm{d}w,$$

$$0 = \mathrm{d}'f = D_x f.\mathrm{d}'x + D_y f.\mathrm{d}'y + D_z f.\mathrm{d}'z + D_w f.\mathrm{d}'w,$$

$D_x f, \ldots, D_w f$ being *partial derivatives*. But, by the supposed *homogeneity* of f, we have $0 = xD_x f + yD_y f + zD_z f + wD_w f$; hence

$$l : m : n : r = D_x f : D_y f : D_z f : D_w f;$$

and therefore, as asserted in article 485,

$$[D_x f, D_y f, D_z f, D_w f]$$

may be taken as the *anharmonic symbol of the tangent plane* to the surface $f = 0$, at the point (x, y, z, w).

487. In like manner, let $F(l\,m\,n\,r) = 0$, when F is any homogeneous function, be the given *tangential equation* of a surface, considered as the *envelope* of a *variable plane* $[l, m, n, r]$, or Π. Let

$$\Pi' = [l + \Delta l, \ldots, r + \Delta r], \quad \textit{and} \quad \Pi'' = [l + \Delta' l, \ldots, r + \Delta' r],$$

denote any two other planes of the same system; and let the common *intersection* of *these planes* be the point,

$$P = \Pi^{\cdot}\Pi'^{\cdot}\Pi'' = (X, Y, Z, W).$$

Then the position of this point P, or the ratios of its coordinates $X, \ldots,$ may generally be determined by the three equations,

$$0 = Xl + Ym + Zn + Wr;$$

$$0 = X\Delta l + Y\Delta m + Z\Delta n + W\Delta r;$$

$$0 = X\Delta' l + V\Delta' m + Z\Delta' n + W\Delta' r.$$

At the *limit*, when P becomes a point $(x\,y\,z\,w)$ of the *envelope* by the three planes tending to *coincidence*, the three last equations become,

$$0 = xl + ym + zn + wr,$$

$$0 = x\mathrm{d}l + y\mathrm{d}m + z\mathrm{d}n + w\mathrm{d}r,$$

$$0 = x\mathrm{d}'l + y\mathrm{d}'m + z\mathrm{d}'n + w\mathrm{d}'r;$$

where $\mathrm{d}l, \ldots,$ and $d'l, \ldots,$ are only subject to the two equations

$$0 = \mathrm{d}F = D_l F.\mathrm{d}l + \cdots + D_r F.\mathrm{d}r, \quad 0 = \mathrm{d}'F = D_l F.\mathrm{d}'l + \cdots + D_r F.\mathrm{d}'r.$$

But by the supposed homogeneity of F, we have also,

$$0 = lD_l F + mD_m F + nD_n F + rD_r F;$$

hence the *coordinates of contact* $x, y, z, w,$ must be such that

$$x : y : z : w = D_l F : D_m F : D_n F : D_r F;$$

and $(D_lF,\ D_mF,\ D_nF,\ D_rF)$ is therefore, as in article 485, the *anharmonic symbol of the point of contact* of the enveloped surface, $F=0$, with the plane $[l,\ m,\ n,\ r]$.

488. In general, if we differentiate the equation

$$lx+my+nz+rw=0,$$

we get

$$l\mathrm{d}x+m\mathrm{d}y+n\mathrm{d}z+r\mathrm{d}w=-(x\mathrm{d}l+y\mathrm{d}m+z\mathrm{d}n+w\mathrm{d}r);$$

if then we suppose it *known* that $[D_xf,\ D_yf,\ D_zf,\ D_wf]$ is the symbol of the *tangent plane* $[l,\ m,\ n,\ r]$ to a surface given by the *local equation* $f=0$, or that

$$l\mathrm{d}x+m\mathrm{d}y+n\mathrm{d}z+r\mathrm{d}w=0,$$

for *all* values of the differentials $\mathrm{d}x,\ \ldots,\ \mathrm{d}w$, which are consistent with the differential equation $\mathrm{d}f=0$, we can *infer* (without any new appeal to *geometry*) that

$$x\mathrm{d}l+y\mathrm{d}m+z\mathrm{d}n+w\mathrm{d}r=0,$$

for all values of $\mathrm{d}l,\ \ldots,\ \mathrm{d}r$ which are consistent with this *other differential* equation, $\mathrm{d}F=0$, when $F=0$ is the *tangential equation* of the same surface; or that

$$x:y:z:w=D_lF:D_mF:D_nF:D_rF;$$

or finally that $(D_lF,\ D_mF,\ D_nF,\ D_rF)$ is, as above, the symbol of the *point of contact.* And in like manner we can *return* from this last symbol, to the symbol for the tangent plane. – It is scarcely necessary to remark that in the present method, as in that of quadriplanar coordinates, the *degree* of the *local equation,* $f=0$, (when rational and integral,) determines the *order of the surface*; and that the *degree* of the *tangential equation,* $F=0$, determines its *class.*

489. (November 24th, 1860.) But it is really time to *conclude this* "Postscript," and therefore I shall not venture to *look back* lest I should be tempted to add anything more. For as regards the question of the *bisecting circle* on a *sphere,* I remember that I pursued, with some success, a couple of months ago, beyond the stage described in former sheets, the investigation of that question by quaternions. I had also something more to say, – but shall not say it, – with reference to *cubic curves.*

490. In conclusion, then, let me thank you for your patience and good humour, shown in allowing me to write to you at such great length, and at such irregular intervals, whenever I fancied that I could clear up *my own thoughts,* by so *expressing* them. Not more than a *tenth part,* as I suppose, if so much, has been transferred to *printed* sheets: including a sketch of the *Anharmonic Coordinates* for the *Proceedings** *of the Royal Irish Academy.* – I do not venture to propose to begin a new Letter, on the subject of *Quaternions*: but should be glad if you were

* [The short account, which will be found on pp. 507–515 of this volume, was published in 1860 in part XI of volume VII (publication of this volume was not completed until 1862) of the *Proceedings of the Royal Irish Academy,* pp. 286–289 (com. Apr. 9 1860), 329 (com. May 28, 1860), 350–354 (com. Jun. 25, 1860). It was also published, in 1860, in *The Natural History Review,* Vol. 7, pp. 242–246, 325, 505–509. *The Natural History Review* was a quarterly journal containing the proceedings of a number of Irish scientific bodies, including those of the Royal Irish Academy.]

to look over part of the manuscript, or at least some of the proof sheets, or slips, of the *Second Book* of my "*Elements*", when I get them.* The First Book is nearly printed: and a pretty large stock of *copy* is ready for the Second, although perhaps it may be all rewritten for the press. You would find, I hope, that *Second Book* quite intelligible, without its having been *necessary* to read more of the *First*, than that *initial part* which was selected by yourself some months ago, as what might be recommended for the perusal of a student, when entering on the subject. The progress of the printing has been hitherto slow: and consequently the *Grant* has not been so much encroached on, as you might apprehend, from the early sheets.† Enough, I think remains, to bring out in a satisfactory manner, the theory of the *Quaternions themselves*: although I cannot consider as foreign, or irrelevant, the previous Sections on *Vectors*.

I am, my dear Dr. Hart, once more, faithfully yours,
William Rowan Hamilton.
Andrew Searle Hart Esq., LL. D., &c.

* [Hamilton's major treatise, *The Elements of Quaternions* (London: Longmans, Green, & Co.) was not published until 1866, the year after his death in 1865.]

† [See Thomas Leroy Hankins, *Sir William Rowan Hamilton*, pp. 359–365, John Hopkins University Press, Baltimore, Maryland, and London: 1980.]

Part II
GEOMETRY

IV.

ON SYMBOLICAL GEOMETRY (1846–49)

[*Cambridge and Dublin Mathematical Journal,* **1** 45–57, 137–154, 256–263 (1846); **2** 47–52, 130–133, 204–209 (1847); **3** 68–84, 220–225 (1848); **4** 84–89, 105–118 (1849).]

Introductory Remarks

The present paper is an attempt towards constructing a symbolical geometry, analogous in several important respects to what is known as symbolical algebra, but not identical therewith; since it starts from other suggestions, and employs, in many cases, other rules of combination of symbols. One object aimed at by the writer has been (he confesses) to illustrate, and to exhibit under a new point of view, his own theory, which has in part been elsewhere published, of algebraic quaternions. Another object, which interests even him much more, and will probably be regarded by the readers of this Journal as being much less unimportant, has been to furnish some new materials towards judging of the general applicability and usefulness of some of those principles respecting symbolical language which have been put forward in modern times. In connexion with this latter object he would gladly receive from his readers some indulgence, while offering the few following remarks.

An opinion has been formerly published* by the writer of the present paper, that it is possible to regard Algebra as a *science,* (or more precisely speaking) as a *contemplation,* in some degree *analogous to Geometry,* although not to be confounded therewith; and to separate it, as such, in our conception, from its own *rules of art* and *systems of expression*: and that when so regarded, and so separated, its ultimate subject-matter is found in what a great metaphysician has called the inner intuition of *time.* On which account, the writer ventured to characterise Algebra as being the *Science of Pure Time*; a phrase which he also expanded into this other: that it is (ultimately) the Science of *Order and Progression.* Without having as yet seen cause to abandon that former view, however obscurely expressed and imperfectly developed it may have been, he hopes that he has since profited by a study, frequently resumed, of some of the works of Professor Ohm,† Dr. Peacock,‡ Mr. Gregory,§ and some other authors; and imagines that he has come to seize their meaning, and appreciate their value, more fully than he was prepared to do, at the date of that former publication of his own to which he has referred. The whole theory of the laws and logic of symbols is indeed one of no small subtlety; insomuch that (as is well known to the readers of the *Cambridge Mathematical Journal,* in which periodical many papers of great interest and importance on this very subject have appeared) it requires a close and long-continued attention, in order to be able to form a judgment of

* *Trans. Royal Irish Acad.,* vol. XVII. Dublin, 1835. [See Vol. III, p 3; and this Vol. p. 762.]

† [Martin Ohm (1792–1872)].

‡ [George Peacock (1791–1858)].

§ [Duncan Farquharson Gregory (1813–1844)].

any value respecting it: nor does the present writer venture to regard his own opinions on this head as being by any means sufficiently matured; much less does he desire to provoke a controversy with any of those who may perceive that he has not yet been able to adopt, in all respects, their views. That he has adopted *some* of the views of the authors above referred to, though in a way which does not seem to himself to be contradictory to the results of his former reflexions; and especially that he feels himself to be under important obligations to the works of Dr. Peacock upon Symbolical Algebra, are things which he desires to record, or mark, in some degree, by the very *title* of the present communication; in the course of which there will occur opportunities for acknowledging part of what he owes to other works, particularly to Mr. Warren's Treatise* on the Geometrical Representation of the Square Roots of Negative Quantities.

Observatory of Trinity College, Dublin, Oct. 16, 1845.

Uniliteral and Biliteral Symbols

1. In the following pages of an attempt towards constructing a symbolical geometry, it is proposed to employ (as usual) the roman capital letters A, B, &c., with or without accents, as symbols of *points* in space; and to make use (at first) of binary combinations of those letters, as symbols of straight *lines*: the symbol of the beginning of the line being written (for the sake of some analogies†) towards the right hand, and the symbol of the end towards the left. Thus BA will denote the line *to* B *from* A; and is not to be confounded with the symbol AB, which denotes a line having indeed the same extremities, but drawn in the opposite direction. A biliteral symbol, of which the two component letters denote determined and different points, will thus denote a finite straight line, having a determined length, direction, and situation in space. But a biliteral symbol of the particular form AA may be said to be a *null* line, regarded as the limit to which a line tends, when its extremities tend to coincide: the conception or at least the name and symbol of such a line being required for symbolic generality. All lines BA which are not null, may be called by contrast *actual*; and the two lines AB and BA may be said to be the *opposites* of each other. It will then follow that a null line is its own opposite, but that the opposites of two actual lines are always to be distinguished from each other.

On the Mark =.

2. An *equation* such as

$$\mathrm{B} = \mathrm{A}, \tag{1}$$

between two uniliteral symbols, may be interpreted as denoting that A and B are *two names for one common point*; or that a point B, determined by one geometrical process, coincides with a point A determined by another process. When a formula of the kind (1) holds good, in any calculation, it is allowed to *substitute*, in any other part of that calculation, either of the two

* [John Warren (1796–1852) *A treatise on the geometrical representation of the square roots of negative quantities*, Cambridge: 1828.]

† The writer regards the line to B from A as being in some sense an interpretation or construction of the symbol $\mathrm{B} - \mathrm{A}$; and the evident possibility of reaching the point B, by going along that line from the point A, may, as he thinks, be symbolized by the formula $\mathrm{B} - \mathrm{A} + \mathrm{A} = \mathrm{B}$.

equated symbols for the other; and every other equation between two symbols of one common class must be interpreted so as to allow a similar substitution. We shall not violate this principle of symbolical language by interpreting, as we shall interpret, an equation such as

$$DC = BA, \tag{2}$$

between two biliteral symbols, as denoting that the two lines,* of which the symbols are equated, have *equal lengths and similar directions*, though they may have different situations in space: for if we call such lines *symbolically equal*, it will be allowed, in *this* sense of equality, which has indeed been already proposed by Mr. Warren, Dr. Peacock, and probably by some of the foreign writers referred to in Dr. Peacock's Report, as well as in that narrower sense which relates to magnitudes only, and for lines in space as well as for those which are in one plane, to assert that lines *equal* to the same line are equal to each other. (Compare *Euclid*, XI. 9.) It will also be true, that

$$D = B, \quad \text{if} \quad DA = BA, \tag{3}$$

or in words, that the ends of two symbolically equal lines coincide if the beginnings do so; a consequence which it is very desirable and almost necessary that we should be able to draw, for the purposes of symbolical geometry, but which would not have followed, if an equation of the form (2) had been interpreted so as to denote *only* equality of lengths, or *only* similarity of directions. The opposites of equal lines are equal in the sense above explained; therefore the equation (2) gives also this *inverse* equation,

$$CD = AB. \tag{4}$$

Lines joining the similar extremities of symbolically equal lines are themselves symbolically equal (*Euc.* I. 33); therefore the equation (2) gives also this *alternate* equation,

$$DB = CA. \tag{5}$$

The *identity* BA = BA gives, as its alternate equation,

$$BB = AA, \tag{6}$$

which symbolic result may be expressed in words by saying that any two null lines are to be regarded as equal to each other. Lines equal to opposite lines may be said to be themselves opposite lines.

* The writer regards the relation between two lines, mentioned in the text, as a sort of interpretation of the following symbolic equation, $D - C = B - A$; which may also denote that the point D is ordinally related (in space) to the point C as B is to A, and may in that view be also expressed by writing the *ordinal analogy*, D..C : : B..A; which admits of *inversion* and *alternation*. The same relation between four points may, as he thinks, be thus symbolically expressed, $D = B - A + C$. But by writing it as an equation between lines, he deviates less from received notation.

On the Mark +.

3. The equation*

$$CB + BA = CA \tag{7}$$

is true in the most elementary sense of the notation, when B is any point upon the finite straight line CA; but we propose now to *remove this restriction for the purposes of symbolical geometry*, and to regard the formula (7) as being universally *valid, by definition, whatever three points of space may be denoted* by the three letters ABC. The equation (7) will then *express nothing about those points*, but will serve to *fix the interpretation of the mark + when inserted between any two symbols of lines*; for if we meet any symbol formed by such insertion, suppose the symbol HG + FE, we have only to draw, or conceive drawn, from any assumed point A, a line BA = FE, and from the end B of the line so drawn, a new line CB = HG; and then the proposed symbol HG + FE will be interpreted by (7) as denoting the line CA, or at least a line equal thereto. In like manner, by defining that

$$DC + CB + BA = DA, \tag{8}$$

we shall be able to interpret any symbol of the form

$$KI + HG + FE,$$

as denoting a determined (actual or null) line; at least if we now regard a line as *determined* when it is *equal* to a determined line: and similarly for any number of biliteral symbols, connected by marks + interposed. Calling *this* act of connection of symbols, the operation of *addition*; the added symbols, *summands*; and the resulting symbol, a *sum*; we may therefore now say, that the sum of any number of symbols of given lines is itself a symbol of a determined line; and that this symbolic sum of lines represents the *total* (or final) *effect* of all those successive rectilinear *motions*, or translations of a point in space, which are represented by the several summands. This *interpretation of a symbolic sum of lines* agrees with the conclusions already published by the authors above alluded to; though the modes of symbolically obtaining and expressing it, here given, may possibly be found to be new. The same interpretation satisifies, as it ought to do, the condition that the sums of equals shall be equal (compare the demonstration of *Euclid*, XI. 10); and also this other condition, almost as much required for the advantageous employment of symbolical language, that those lines which, when added to equal lines, give equal sums, shall be themselves equal lines: or that

$$FE = DC, \quad \text{if} \quad FE + BA = DC + BA. \tag{9}$$

It shows too that the sum of two opposite lines, and generally that the sum of all the successive sides of any closed polygon, or of lines respectively equal to those sides, is a null line: thus

* On the plan mentioned in former notes, this equation would be written as follows:

$$(C - B) + (B - A) = C - A.$$

It might also be thus expressed: the ordinal relation of the point C to the point A is compounded of the relations of C to B and of B to A.

$$\mathrm{AA} = \mathrm{AB} + \mathrm{BA} = \mathrm{AC} + \mathrm{CB} + \mathrm{BA} = \&\mathrm{c}. \tag{10}$$

The symbolic sum of any two lines is found to be *independent of their order*, in virtue of the same interpretation; so that the equation

$$\mathrm{FE} + \mathrm{HG} = \mathrm{HG} + \mathrm{FE}, \tag{11}$$

is true, in the present system, *not as an independent definition*, but rather as one of the modes of *symbolically expressing that elementary theorem of geometry*, (*Euclid*, I. 33), on which was founded the rule for deducing, from any equation (2) between lines, the *alternate* equation (5). For if we assume, as we may, that three points A, B, C, have been so chosen as to satisfy the equations FE = BA, HG = CA; and that a fourth point D is chosen so as to satisfy the equation DC = BA; the same points will then, by the theorem just referred to, satisfy also the equation DB = CA; and the truth of the formula (11) will be proved, by observing that each of the two symbols which are equated in that formula is equal to the symbol DA, in virtue of the definition (7) of +, without any new definition: since

$$\mathrm{FE} + \mathrm{HG} = \mathrm{DC} + \mathrm{CA} = \mathrm{DA} = \mathrm{DB} + \mathrm{BA} = \mathrm{HG} + \mathrm{FE}.$$

A like result is easily shown to hold good, for any number of summands; thus

$$\mathrm{FE} + \mathrm{HG} + \mathrm{KI} = \mathrm{KI} + \mathrm{HG} + \mathrm{FE}; \tag{12}$$

since the first member of this last equation may be put successively under the forms

$$(\mathrm{FE} + \mathrm{HG}) + \mathrm{KI}, \quad \mathrm{KI} + (\mathrm{FE} + \mathrm{HG}), \quad \mathrm{KI} + (\mathrm{HG} + \mathrm{FE}),$$

and finally under the form of the second member; the stages of this successive transformation of symbols admitting easily of geometrical interpretations: and similarly in the cases. *Addition of lines in space* is therefore generally (as Mr. Warren has shown it to be for lines in a single plane) a *commutative operation*; in the sense that the summands may interchange their places, without the sum being changed. It is also an *associative* operation, in the sense that any number of successive summands may be associated into one group, and collected into one partial sum (denoted by enclosing these summands in parentheses); and that then this partial sum may be added, as a single summand, to the rest: thus

$$(\mathrm{KI} + \mathrm{HG}) + \mathrm{FE} = \mathrm{KI} + (\mathrm{HG} + \mathrm{FE}) = \mathrm{KI} + \mathrm{HG} + \mathrm{FE}. \tag{13}$$

On the Mark −.

4. The equation*

$$\mathrm{CA} - \mathrm{BA} = \mathrm{CB} \tag{14}$$

is true, in the most elementary sense of the notation, when B is on CA; but we may remove this restriction by a *definitional extension* of the formula (14), for the purposes of symbolical geometry, as has been done in the foregoing article with respect to the formula (7); and then the equation (14), so extended, will express *nothing about the points* A, B, C, but will serve to fix

* On the plan mentioned in some former notes, this equation would take the form

$$(\mathrm{C} - \mathrm{A}) - (\mathrm{B} - \mathrm{A}) = \mathrm{C} - \mathrm{B}.$$

the *interpretation of any symbol*, such as KI − FE, formed by *inserting the mark − between the symbols of any two lines.* This general meaning of the effect of the mark −, so inserted, is consistent with the particular interpretation which suggested the formula (14); it is also consistent with the usual symbolical opposition between the effects of + and −; since the comparison of (14) with (7) gives the equations

$$(\mathrm{CA} - \mathrm{BA}) + \mathrm{BA} = \mathrm{CA}, \tag{15}$$

and

$$(\mathrm{CB} + \mathrm{BA}) - \mathrm{BA} = \mathrm{CB}, \tag{16}$$

either of which two equations, if regarded as a general formula, and combined with the formula (7), would include, reciprocally, the definition (14) of −, and might be substituted for it.

Symbolical *subtraction* of one line from another is thus equivalent to the *decomposition* of a given rectilinear *motion* (CA) into two others, of which one (BA) is given; or to the *addition of the opposite* (AB) of the line which was to be subtracted: so that we may write the symbolical equation

$$-\mathrm{BA} = +\mathrm{AB}, \tag{17}$$

because the second member of (14) may be changed by (7) to CA + AB. These conclusions respecting symbolical subtraction of lines, differ only in their notation, and in the manner of arriving at them, from the results of the authors already referred to, so far as the present writer is acquainted with them. In the present notation, when an isolated biliteral symbol is preceded by + or −, we may still interpret it as denoting a line, if we agree to prefix to it, for the purpose of such interpretation, the symbol of a null line; thus we may write

$$+\mathrm{AB} = \mathrm{AA} + \mathrm{AB} = \mathrm{AB}, \quad -\mathrm{AB} = \mathrm{BB} - \mathrm{AB} = \mathrm{BA}; \tag{18}$$

+AB will, therefore, on this plan, be another symbol for the line AB itself, and −AB will be a symbol for the opposite line BA.

Abridged Symbols for Lines

5. Some of the foregoing formulæ may be presented more concisely, and also in a way more resembling ordinary Algebra, by using now some new *uniliteral* symbols, such as the small roman letters a, b, &c., with or without accents, as symbols of lines, instead of binary combinations of the roman capitals, in cases where the lines which are compared are not supposed to have necessarily any common point, and generally when the *situations* of lines are disregarded, but not their lengths nor their directions. Thus we shall have, instead of (11) and (12), (13), (15) and (16), these other formulæ of the present Symbolical Geometry, which agree in all respect with those used in Symbolical Algebra:

$$\mathrm{a} + \mathrm{b} = \mathrm{b} + \mathrm{a}, \quad \mathrm{a} + \mathrm{b} + \mathrm{c} = \mathrm{c} + \mathrm{b} + \mathrm{a}; \tag{19}$$

$$(\mathrm{c} + \mathrm{b}) + \mathrm{a} = \mathrm{c} + (\mathrm{b} + \mathrm{a}) = \mathrm{c} + \mathrm{b} + \mathrm{a}; \tag{20}$$

$$(\mathrm{b} - \mathrm{a}) + \mathrm{a} = \mathrm{b}, \quad (\mathrm{b} + \mathrm{a}) - \mathrm{a} = \mathrm{b}; \tag{21}$$

and because the isolated but *affected* symbols +a, −a, may denote, by (18), the line a itself, and the opposite of that line, we have also here the usual *rule of the signs*,

$$+(+\mathrm{a}) = -(-\mathrm{a}) = +\mathrm{a}, \quad +(-\mathrm{a}) = -(+\mathrm{a}) = -\mathrm{a}. \tag{22}$$

Introduction of the Marks × and ÷.

6. Continuing to denote lines by letters, the formula

$$(\mathrm{b} \div \mathrm{a}) \times \mathrm{a} = \mathrm{b}, \tag{23}$$

which is, for the relation between multiplication and division, what the first of the two formulæ (21) is for the relation between addition and subtraction, will be true, in the most elementary sense of the multiplication of a length by a number, for the case when the line b is the sum of several summands, each equal to the line a, and when the number of those summands is denoted by the quotient b ÷ a. And we shall now, for the purposes of symbolical generality, *extend* this formula (23), so as to make it be valid, *by definition, whatever two lines* may be denoted by a and b. The formula will then *express nothing respecting those lines* themselves, which can serve to distinguish them from any other lines in space; but will furnish a *symbolic condition,* which we must satisfy by the *general interpretation* of a *geometrical quotient,* and of the *operation of multiplying a line* by such a quotient.

To make such general interpretation consistent with the particular case where a quotient becomes a *quotity,* we are led to write

$$\mathrm{a} \div \mathrm{a} = 1, \quad (\mathrm{a} + \mathrm{a}) \div \mathrm{a} = 2, \ \&\mathrm{c}., \tag{24}$$

and conversely

$$1 \times \mathrm{a} = \mathrm{a}, \quad 2 \times \mathrm{a} = \mathrm{a} + \mathrm{a}, \ \&\mathrm{c}.; \tag{25}$$

and because, when quotients can be thus interpreted as quotities, the four equations

$$(\mathrm{c} \div \mathrm{a}) + (\mathrm{b} \div \mathrm{a}) = (\mathrm{c} + \mathrm{b}) \div \mathrm{a}, \tag{26}$$

$$(\mathrm{c} \div \mathrm{a}) - (\mathrm{b} \div \mathrm{a}) = (\mathrm{c} - \mathrm{b}) \div \mathrm{a}, \tag{27}$$

$$(\mathrm{c} \div \mathrm{a}) \times (\mathrm{a} \div \mathrm{b}) = \mathrm{c} \div \mathrm{b}, \tag{28}$$

$$(\mathrm{c} \div \mathrm{a}) \div (\mathrm{b} \div \mathrm{a}) = \mathrm{c} \div \mathrm{b}, \tag{29}$$

are true in the most elementary sense of arithmetical operations on whole numbers, we shall now *define* that these four equations are valid, *whatever three lines* may be denoted by a, b, c; and thus shall have conditions for the general *interpretations of the four operations* + − × ÷ *performed on geometrical quotients.*

We shall in this way be led to interpret a quotient of which the divisor is an actual line, but the dividend a null one, as being equivalent to the symbol 1 − 1, or *zero*; so that

$$(\mathrm{a} - \mathrm{a}) \div \mathrm{a} = 0, \quad 0 \times \mathrm{a} = \mathrm{a} - \mathrm{a}. \tag{30}$$

Negative numbers will present themselves in the consideration of such quotients and products as

$$(-\mathrm{a}) \div \mathrm{a} = 0 - 1 = -1, \quad (-1) \times \mathrm{a} = -\mathrm{a}, \quad \&\mathrm{c}.; \tag{31}$$

fractional numbers in such formulæ as

$$\mathrm{a} \div (\mathrm{a} + \mathrm{a}) = 1 \div 2 = \tfrac{1}{2}, \quad \tfrac{1}{2} \times (\mathrm{a} + \mathrm{a}) = \mathrm{a}, \quad \&\mathrm{c.}; \tag{32}$$

and *incommensurable* numbers, by the conception of the connected *limits* of quotients and products, and by the formula, which symbolical language leads us to assume,

$$\left(\lim \frac{n}{m}\right) \times \mathrm{a} = \lim \left(\frac{n}{m} \times \mathrm{a}\right). \tag{33}$$

If then we give the name of SCALARS to all numbers of the kind called usually *real*, because they are all contained on the one *scale* of progression of number from negative to positive infinity; and if we agree, for the present, to denote such numbers generally by small italic letters, *a, b, c,* &c.; and to insert the mark $\|$ between the symbols of two lines when we wish to express that the directions of those lines are either exactly similar or exactly opposite to each other, in each of which two cases the lines may be said to be *symbolically parallel*; we shall have generally two equations of the forms

$$\mathrm{b} \div \mathrm{a} = a, \quad a \times \mathrm{a} = \mathrm{b}, \quad \text{when } \mathrm{b} \,\|\, \mathrm{a}. \tag{34}$$

That is to say, the *quotient of two parallel lines* is generally a *scalar number*; and, conversely, to multiply a given line (a) by a given scalar (or real) number a, is to determine a new line (b) parallel to the given line (a), the direction of the one being similar or opposite to that of the other, according as the number is positive or negative, while the length of the new line bears to the length of the given line a ratio which is marked by the same given number. So that if $\mathrm{A}_0\ \mathrm{A}_1\ \mathrm{A}_a$ denote any three points on one common axis of rectilinear progression, which are related to each other, upon that axis, as to their order and their intervals, in the same manner as the three scalar numbers 0, 1, a, regarded as ordinals, are related to each other on the scale of numerical progression from $-\infty$ to $+\infty$, then the equations

$$\mathrm{A}_a\mathrm{A}_0 \div \mathrm{A}_1\mathrm{A}_0 = a, \quad a \times \mathrm{A}_1\mathrm{A}_0 = \mathrm{A}_a\mathrm{A}_0, \tag{35}$$

will be true by the foregoing interpretations.

It is easy to see that this mode of interpreting a quotient of parallel lines renders the formulæ (26) (27) (28) (29) consistent with the received rules for performing the operations $+ - \times \div$ on what are called real numbers, whether they be positive or negative, and whether commensurable or incommensurable; or rather reproduces those rules as consequences of those formulæ.

On Vectors, and Geometrical Quotients in General

7. The other chief relation of directions of lines in space, besides parallelism, is perpendicularity; which it is not unusual to denote by writing the mark $\perp$ between the symbols of two perpendicular lines. And the other chief class of geometrical quotients which it is important to study, as preparatory to a general theory of such quotients, is the class in which the dividend is a line perpendicular to the divisor. A quotient of this latter class we shall call a VECTOR, to mark its connection (which is closer than that of a *scalar*) with the conception of *space*, and for other reasons which will afterwards appear: and if we agree to denote, for the present, such vector quotients (of perpendicular lines) by small Greek letters, in contrast to the scalar class of quotients (of parallel lines) which we have proposed to denote by small italic letters, we shall then have generally two equations of the forms

$$\mathrm{c} \div \mathrm{a} = \alpha, \quad \mathrm{c} = \alpha \times \mathrm{a}, \quad \text{if } \mathrm{c} \perp \mathrm{a}. \tag{36}$$

Any line e may be put under the form $\mathrm{c} + \mathrm{b}$, in which $\mathrm{b} \parallel \mathrm{a}$, and $\mathrm{c} \perp \mathrm{a}$; a *general geometrical quotient* may therefore, by (26) (34) (36), be considered as the *symbolic sum of a scalar and a vector,* zero being regarded as a common limit of quotients of these two classes; and consequently, if we adopt the notation just now mentioned, we have generally an equation of the form

$$\mathrm{e} \div \mathrm{a} = \alpha + a. \tag{37}$$

This *separation of the scalar and vector parts* of a general geometrical quotient corresponds (as we see) to the decomposition, by *two separate projections,* of the dividend line into two other lines of which it is the symbolic sum, and of which one is parallel to the divisor line, while the other is perpendicular thereto. To be able to mark on some occasions more distinctly, in writing, than by the use of two different alphabets, the conception of such separation, we shall here introduce two new symbols of operation, namely the abridged words Scal. and Vect., which, where no confusion seems likely to arise from such farther abridgment, we shall also denote more shortly still by the letters S and V, prefixing them to the symbol of a general geometrical quotient in order to form separate symbols of its scalar and vector parts: so that we shall now write generally, for any two lines a and e,

$$\mathrm{e} \div \mathrm{a} = \mathrm{Vect.}(\mathrm{e} \div \mathrm{a}) + \mathrm{Scal.}(\mathrm{e} \div \mathrm{a}); \tag{38}$$

or more concisely,

$$\mathrm{e} \div \mathrm{a} = \mathrm{V}(\mathrm{e} \div \mathrm{a}) + \mathrm{S}(\mathrm{e} \div \mathrm{a}); \tag{39}$$

in which expression the order of the two summands may be changed, in virtue of the definition (26) of addition of geometrical quotients, because the order of the two partial dividends may be changed without preventing the dividend line e from being still their symbolic sum. A scalar cannot become equal to a vector, except by each becoming zero; for if the divisor of the vector quotient be multiplied separately by the scalar and the vector, the products of these two multiplications will be (by what has been already shown) respectively lines parallel and perpendicular to that divisor, and therefore not symbolically equal to each other, except it be at the limit where both become null lines, and are on that account regarded as equal. A scalar quotient $\mathrm{b} \div \mathrm{a} = a$, $(\mathrm{b} \parallel \mathrm{a})$, has been seen to denote the relative length and relative direction (as similar or opposite) of two parallel lines a, b: and in like manner a vector quotient $\mathrm{c} \div \mathrm{a} = \alpha$, $(\mathrm{c} \perp \mathrm{a})$, may be regarded as denoting the *relative length and relative direction* (depending on *plane* and *hand*) *of two perpendicular lines* a, c; or as indicating at once *in what ratio* the length of one line a must be altered (if at all) in order to become equal to the length of another line c, and also *round what axis,* perpendicular to both these two rectangular lines, the direction of the divisor line a must be caused or conceived to turn, right-handedly, through a right angle, in order to attain the original direction of the dividend line c. A line drawn in the direction of this *axis of* (what is here regarded as) *positive rotation,* and having its length in the same ratio to some assumed *unit* of length as the length of the dividend to that of the divisor, may be called the INDEX of the vector. We shall thus be led to substitute, for any equation between two vector quotients, an equation between two lines, namely between their indices; for if we define that two vector quotients, such as $c \div a$ and $c' \div a'$ if $c \perp a$ and $c' \perp a'$, are *equal* when they have *equal indices*, we shall satisfy all conditions of symbolical equality, of the kinds already considered in connection with other definitions; we shall also be able to say that in every case of two such equal quotients, the two dividend lines (c and c′) bear to their own divisor lines (a and a′), respectively, one common ratio of lengths, and one common relation of directions. We shall thus also, by (23), be able to *interpret the multiplication* of any given line a′ by any given vector $c \div a$, *provided that the one is perpendicular to the index of the other*, as the operation of deducing from a′ another line c′, by altering (generally) its length in a given ratio, and by turning (always) its direction round a given axis of rotation, namely round the index of the vector, right-handedly, through a right angle. And we can now *interpret an equation between two general geometrical quotients*, such as

$$e' \div a' = e \div a, \tag{40}$$

as being equivalent to a *system of two separate equations*, one between the scalar and another between the vector parts, namely the two following:

$$S(e' \div a') = S(e \div a); \quad V(e' \div a') = V(e \div a); \tag{41}$$

of which each separately is to be interpreted on the principles already laid down; and which are easily seen (by considerations of similar triangles) to imply, when taken jointly, that the length of e′ is to that of a′ in the same ratio as the length of e to that of a; and also that the same rotation, round the index of either of the two equal vectors, which would cause the direction of a to attain the original direction of e, would also bring the direction of a′ into that originally occupied by e′. At the same time we see how to interpret the operation of multiplying any given line a′ by any given geometrical quotient $e \div a$ of two other lines, *whenever the three given lines* a, e, a′, *are parallel to one common plane*; namely as being the complex operation of altering (generally) a given length in a given ratio, and of turning a given line round a given axis, through a given amount of right-handed rotation, in order to obtain a certain new line e′, which may be thus denoted, in conformity with the definition (23),

$$e' = (e \div a) \times a'. \tag{42}$$

The relation between the four lines a, e, a′, e′, may also be called a *symbolic analogy*, and may be thus denoted:

$$e' : a' :: e : a; \tag{43}$$

a′ and e being the *means*, and e′ and a the *extremes* of the analogy. An analogy or equation of this sort admits (as it is easy to prove) of *inversion* and *alternation*; thus (43) or (42) gives, *inversely*,

$$a' : e' :: a : e, \quad a' \div e' = a \div e, \tag{44}$$

and *alternately*,

$$e' : e :: a' : a, \quad e' \div e = a' \div a. \tag{45}$$

These results respecting analogies between *co-planar lines*, that is, between lines which are in or parallel to one common plane, agree with, and were suggested by, the results of Mr. Warren. But it will be necessary to introduce other principles, or at least to pursue farther the track already entered on, before we can arrive at an interpretation of a *fourth proportional to three lines which are not parallel to any common plane*: or can interpret the multiplication of a line by a quotient of two others, when it is not perpendicular to what has been lately called the index of the vector part of that quotient.

Determinateness of the First Four Operations on Geometrical Fractions (or Quotients)

8. Meanwhile the principles and definitions which have been already laid down, are sufficient to conduct to clear and determinate interpretations of all operations of combining geometrical quotients among themselves, by any number of additions, subtractions, multiplications, and divisions: each *quotient* of the kind here mentioned being regarded, by what has been already shown, as the *mark of a certain complex relation between two straight lines in space*, depending not only on their *relative lengths*, but also on their *relative directions*. If we denote now by a symbol of fractional form, such as $\frac{\mathrm{b}}{\mathrm{a}}$, the quotient thus obtained by dividing one line b by another line a, when directions as well as lengths are attended to, the definitional equations (26), (27), (28), (29), will take these somewhat shorter forms:*

$$\frac{\mathrm{c}}{\mathrm{a}}+\frac{\mathrm{b}}{\mathrm{a}}=\frac{\mathrm{c}+\mathrm{b}}{\mathrm{a}}; \quad \frac{\mathrm{c}}{\mathrm{a}}-\frac{\mathrm{b}}{\mathrm{a}}=\frac{\mathrm{c}-\mathrm{b}}{\mathrm{a}}; \tag{46}$$

$$\frac{\mathrm{c}}{\mathrm{a}}\times\frac{\mathrm{a}}{\mathrm{b}}=\frac{\mathrm{c}}{\mathrm{b}}; \quad \frac{\mathrm{c}}{\mathrm{a}}\div\frac{\mathrm{b}}{\mathrm{a}}=\frac{\mathrm{c}}{\mathrm{b}}; \tag{47}$$

* On the principles alluded to in former notes, the formulæ for the addition, subtraction, multiplication, and division, of any two geometrical fractions, might be thus written:

$$\frac{\mathrm{D}-\mathrm{C}}{\mathrm{B}-\mathrm{A}}+\frac{\mathrm{C}-\mathrm{A}}{\mathrm{B}-\mathrm{A}}=\frac{\mathrm{D}-\mathrm{A}}{\mathrm{B}-\mathrm{A}},$$
$$\frac{\mathrm{D}-\mathrm{A}}{\mathrm{B}-\mathrm{A}}-\frac{\mathrm{C}-\mathrm{A}}{\mathrm{B}-\mathrm{A}}=\frac{\mathrm{D}-\mathrm{C}}{\mathrm{B}-\mathrm{A}},$$
$$\frac{\mathrm{D}-\mathrm{A}}{\mathrm{C}-\mathrm{A}}\times\frac{\mathrm{C}-\mathrm{A}}{\mathrm{B}-\mathrm{A}}=\frac{\mathrm{D}-\mathrm{A}}{\mathrm{B}-\mathrm{A}},$$
$$\frac{\mathrm{D}-\mathrm{A}}{\mathrm{B}-\mathrm{A}}\div\frac{\mathrm{C}-\mathrm{A}}{\mathrm{B}-\mathrm{A}}=\frac{\mathrm{D}-\mathrm{A}}{\mathrm{C}-\mathrm{A}};$$

A, B, C, D being symbols of any four points of space, and B − A being a symbol of the straight line drawn to B from A. If we denote this line by the biliteral symbol BA, we obtain the following somewhat shorter forms, which do not however all agree so closely with the forms of ordinary algebra:

$$\frac{\mathrm{DC}}{\mathrm{BA}}+\frac{\mathrm{CA}}{\mathrm{BA}}=\frac{\mathrm{DA}}{\mathrm{BA}},$$
$$\frac{\mathrm{DA}}{\mathrm{BA}}-\frac{\mathrm{CA}}{\mathrm{BA}}=\frac{\mathrm{DC}}{\mathrm{BA}},$$
$$\frac{\mathrm{DA}}{\mathrm{CA}}\times\frac{\mathrm{CA}}{\mathrm{BA}}=\frac{\mathrm{DA}}{\mathrm{BA}},$$
$$\frac{\mathrm{DA}}{\mathrm{BA}}\div\frac{\mathrm{CA}}{\mathrm{BA}}=\frac{\mathrm{DA}}{\mathrm{CA}}.$$

which agree in all respects with the corresponding formulæ of ordinary algebra, and serve to fix, in the present system, the meanings of the operations +, −, ×, ÷, on what may be called *geometrical fractions*. These FRACTIONS being only other forms for what we have called *geometrical quotients* in earlier articles of this paper, we may now write the identity,

$$\frac{\mathrm{b}}{\mathrm{a}} = \mathrm{b} \div \mathrm{a}. \tag{48}$$

For the same reason, an *equation between any two such fractions*, for example the following,

$$\frac{\mathrm{f}}{\mathrm{e}} = \frac{\mathrm{b}}{\mathrm{a}}, \tag{49}$$

is to be understood as signifying, 1st, that the *length* of the one *numerator* line f is to the length of its own *denominator* line e *in the same ratio* as the length of the other numerator line b to the length of the other denominator line a; 2nd, that these four lines are *co-planar*, that is to say, in or parallel to one common plane; and 3rd, that the *same amount and direction of rotation*, round an axis perpendicular to this common plane, which would bring the line a into the direction originally occupied by b, would also bring the line e into the original direction of f. The same complex relation between the same four lines may also (by what has been already seen) be expressed by the *inverse* equation

$$\frac{\mathrm{e}}{\mathrm{f}} = \frac{\mathrm{a}}{\mathrm{b}}, \tag{50}$$

or by the *alternate* form

$$\frac{\mathrm{f}}{\mathrm{b}} = \frac{\mathrm{e}}{\mathrm{a}}. \tag{51}$$

Two fractions which are, in this sense, *equal* to the same third fraction, are also equal to each other; and the *value* of such a fraction is not altered by altering the lengths of its numerator and denominator in any common ratio; nor by causing both to run together through any common amount of rotation, in a common direction, round an axis perpendicular to both; nor by transporting either or both, without rotation, to any other positions in space. When the lengths and directions of any three co-planar lines, a, b, e, are given, it is always possible to determine the length and direction of a fourth line f, which shall be co-planar with them, and shall satisfy an equation between fractions, of the form (49). It is therefore possible to *reduce any two geometrical fractions to a common denominator*; or to satisfy not only the equation (49), but also this other equation,

$$\frac{\mathrm{h}}{\mathrm{g}} = \frac{\mathrm{c}}{\mathrm{a}}, \tag{52}$$

by a suitable choice of the three lines a, b, c, when the four lines e, f, g, h, are given; since, whatever may be the given directions of these four lines, it is always possible to find (or to conceive as found) a fifth line a, which shall be at once co-planar with the pair e, f, and also with the pair g, h. For a similar reason it is always possible to transform two given geometrical fractions into two others equivalent to them, in such a manner, that the new denominator of one shall be equal to the new numerator of the other; or to satisfy the two equations

$$\frac{h}{g} = \frac{c'}{a'}, \quad \frac{f}{e} = \frac{a'}{b'}, \tag{53}$$

by a suitable choice of the three lines a′, b′, c′, whatever the four given lines e, f, g, h, may be. Making then for abridgment

$$c + b + d, \quad c - b = d', \tag{54}$$

and interpreting a sum or difference of lines as has been done in former articles, we see that it is always possible to choose eight lines a, b, c, d, a′, b′, c′, d′, so as to satisfy the conditions (49), (52), (53), (54); and thus, by (46) and (47), to interpret the sum, the difference, the product, and the quotient of *any two* given geometrical fractions, $\frac{f}{e}$ and $\frac{h}{g}$, as being each equal to *another given fraction* of the same sort, as follows:

$$\frac{h}{g} + \frac{f}{e} = \frac{d}{a}, \quad \frac{h}{g} - \frac{f}{e} = \frac{d'}{a}, \tag{55}$$

$$\frac{h}{g} \times \frac{f}{e} = \frac{c'}{b'}, \quad \frac{h}{g} \div \frac{f}{e} = \frac{c}{b}, \tag{56}$$

any variations in the new numerators and denominators, which are consistent with the foregoing conditions, being easily seen to make no changes in the values of the fractions which result. The *intepretations* of those four symbolic combinations, which are the first members of the four equations (55) and (56), are thus entirely *fixed*: and we are *no longer at liberty, in the present system,* to introduce arbitrarily any *new meanings* for those symbolic forms, or to subject them to any *new laws* of combination among themselves, without examining whether such meanings or such laws are consistent with the principles and definitions which it has been thought right to establish already, as appearing to be more simple and primitive, and more intimately connected with the application of symbolical language to geometry, or at least with the plan on which it is here attempted to make that application, than any of those other laws or meanings. If, for example, it shall be found that, in virtue of the foregoing principles, the *successive addition* of any number of geometrical fractions gives a result which is independent of their order, this consequence will be, for us, a *theorem,* and not a definition. And if, on the contrary, the same principles shall lead us to regard the *multiplication* of geometrical fractions as being in general a *non-commutative* operation, or as giving a result which is *not* independent of the order of the factors, we shall be obliged to accept this conclusion also, that we may preserve consistency of system.

Separation of the Scalar and Vector parts of Sums and Differences of Geometrical Fractions

9. To develope the geometrical meaning of the first equation (46), we may conceive each of the two numerator lines b, c, and also their sum d, to be orthogonally projected, first on the common denominator line a itself, and secondly on a plane perpendicular to that denominator. The former projections may be called b_1, c_1, d_1; the latter, b_2, c_2, d_2; and thus we shall have the nine relations,

$$\left.\begin{array}{lll} b_2 + b_1 = b, & b_1 \parallel a, & b_2 \perp a, \\ c_2 + c_1 = c, & c_1 \parallel a, & c_2 \perp a, \\ d_2 + d_1 = d, & d_1 \parallel a, & d_2 \perp a, \end{array}\right\} \tag{57}$$

together with the three equations

$$c + b = d, \quad c_1 + b_1 = d_1, \quad c_2 + b_2 = d_2; \tag{58}$$

of which the two last are deducible from the first, by the geometrical properties of projections. We have, therefore, by (46),

$$\frac{c}{a} + \frac{b}{a} = \frac{d}{a} = \frac{d_2}{a} + \frac{d_1}{a}, \tag{59}$$

$$\frac{d_1}{a} = \frac{c_1}{a} + \frac{b_1}{a}, \quad \frac{d_2}{a} = \frac{c_2}{a} + \frac{b_2}{a}. \tag{60}$$

Since the three projections b_1, c_1, d_1, are parallel to a (in that sense of the word *parallel* which does not exclude coincidence), the three quotients in the first equation (60) are what we have already named *scalars*; that is, they are what are commonly called real numbers, positive, negative, or zero: they are also the scalar parts of the three quotients in the first equation (59), so that we may write

$$\frac{b_1}{a} = S\frac{b}{a}, \quad \frac{c_1}{a} = S\frac{c}{a}, \quad \frac{d_1}{a} = S\frac{d}{a}, \tag{61}$$

using the letter S here, as in a former article, for the characteristic of the operation of *taking the scalar part* of any geometrical quotient, or fraction. (If any confusion should be apprehended, on other occasions, from this use of the letter S, and if the abridged word Scal. should be thought too long, the sign S might be employed.) Eliminating the four symbols b_1, c_1, d_1, d, between the first equation (59), the first equation (60), and the three equations (61), we obtain the result

$$S\left(\frac{c}{a} + \frac{b}{a}\right) = S\frac{c}{a} + S\frac{b}{a}; \tag{62}$$

in which, by the foregoing article, $\frac{b}{a}$ and $\frac{c}{a}$ may represent any two geometrical fractions: so that we may write generally

$$S\left(\frac{h}{g} + \frac{f}{e}\right) = S\frac{h}{g} + S\frac{f}{e}, \tag{63}$$

and may enunciate in words the same result by saying, that the *scalar of the sum* of any two such fractions is equal to the *sum of the scalars*. In like manner, the three other projections b_2, c_2, d_2, being each perpendicular to a, the three other partial quotients, which enter into the second equation (60), are what we have already called *vectors* in this paper, or more fully they are the vector parts of the three quotients in the first equation (59); so that we may write

$$\frac{b_2}{a} = V\frac{b}{a}, \quad \frac{c_2}{a} = V\frac{c}{a}, \quad \frac{d_2}{a} = V\frac{d}{a}, \tag{64}$$

V being here used, as in a former article, for the characteristic of the operation of *taking the vector part*; we have, therefore,

$$\mathrm{V}\left(\frac{\mathrm{c}}{\mathrm{a}}+\frac{\mathrm{b}}{\mathrm{a}}\right)=\mathrm{V}\frac{\mathrm{c}}{\mathrm{a}}+\mathrm{V}\frac{\mathrm{b}}{\mathrm{a}}, \tag{65}$$

$$\mathrm{V}\left(\frac{\mathrm{h}}{\mathrm{g}}+\frac{\mathrm{f}}{\mathrm{e}}\right)=\mathrm{V}\frac{\mathrm{h}}{\mathrm{g}}+\mathrm{V}\frac{\mathrm{f}}{\mathrm{e}}, \tag{66}$$

and may assert that the *vector of the sum* of any two geometrical fractions is equal to the *sum of the vectors.* These formulæ (63) and (66) are important in the present system; they are however, as we see, only symbolical expressions of those very simple geometrical principles from which they have been derived, through the medium of the equations (58); namely, the principles that, *whether on a line or on a plane,* the *projection of a sum* of lines is equal to the *sum of the projections,* if the word *sum* be suitably interpreted. The analogous interpretation of a *difference* of lines, combined with similar considerations, gives in like manner the formulæ

$$\mathrm{S}\left(\frac{\mathrm{h}}{\mathrm{g}}-\frac{\mathrm{f}}{\mathrm{e}}\right)=\mathrm{S}\frac{\mathrm{h}}{\mathrm{g}}-\mathrm{S}\frac{\mathrm{f}}{\mathrm{e}}; \tag{67}$$

$$\mathrm{V}\left(\frac{\mathrm{h}}{\mathrm{g}}-\frac{\mathrm{f}}{\mathrm{e}}\right)=\mathrm{V}\frac{\mathrm{h}}{\mathrm{g}}-\mathrm{V}\frac{\mathrm{f}}{\mathrm{e}}; \tag{68}$$

that is to say, the *scalar and vector of the difference* of any two geometrical fractions are respectively equal to the *differences of the scalars and of the vectors* of those fractions; precisely as, and because, the *projection of a difference* of two lines, whether on a line or on a plane, is equal to the *difference of the projections.*

Addition and Subtraction of Vectors by their Indices

10. We see, then, that in order to combine by addition or subtraction any two geometrical fractions, it is sufficient to combine separately their scalar and their vector parts. The former parts, namely the scalars, are simply *numbers,* of the kind called commonly real; and are to be added or subtracted among themselves according to the usual rules of algebra. But for effecting with convenience the combination of the latter parts among themselves, namely the vectors, which have been shown in a former article to be of a kind essentially distinct from all stages of the progression of real number from negative to positive infinity (and therefore to be rather *extra-positives* than either positive or *contra*-positive numbers), it is necessary to establish other rules: and it will be found useful for this purpose to employ the consideration of certain connected *lines,* namely the *indices,* of which each is determined by, and in its turn completely characterises, that vector quotient or fraction to which it corresponds, according to the construction assigned in the 7th article. If we apply the rules of that construction to determine the indices of the vector parts of any two fractions and of their sum, we may first, as in recent articles, reduce the two fractions to a common denominator; and may, for simplicity, take this denominator line a of a length equal to that assumed unit of length which is to be employed in the determination of the indices. Then, having projected, as in the last article, the new numerators b and c, and their sum d, on a plane perpendicular to a, and having called these projections b_2, c_2, d_2, as before; we may conceive a right-handed rotation of each of these three projected lines, through a right angle, round the line a as a common axis, which shall transport them without altering their lengths or relative directions, and therefore without affecting their mutual relation as summands and sum, into coincidence with three other lines b_3, c_3, d_3, such that

$$d_3 = c_3 + b_3; \tag{69}$$

and these three new lines will be the three indices required. For a right-handed rotation through a right angle, round the line b_3 as an axis, would bring the line a into the direction originally occupied by b_2; and the length of b_2 is to the length of a in the same ratio as the length of b_3 to the assumed unit of length; therefore b_3 is, in the sense of the 7th article, the index of the vector quotient $\frac{b_2}{a}$, that is, the index of the vector part of the fraction $\frac{b}{a}$, or $\frac{f}{e}$; and similarly for the indices of the two other fractions, in the first equation (59). We may therefore write, as consequences of the construction lately assigned, and of the equations (49) and (52),

$$b_3 = I\frac{f}{e}; \quad c_3 = I\frac{h}{g}; \quad d_3 = I\left(\frac{h}{g} + \frac{f}{e}\right); \tag{70}$$

if we agree for the present to prefix the letter I to the symbol of a geometrical fraction, as the characteristic of the operation of *taking the index of the vector part*. Eliminating now the three symbols b_3, c_3, d_3 between the four equations (69) and (70), we obtain this general formula:

$$I\left(\frac{h}{g} + \frac{f}{e}\right) = I\frac{h}{g} + I\frac{f}{e}, \tag{71}$$

which may be thus enunciated: the *index of the vector part of the sum* of any two geometrical fractions is equal to the *sum of the indices* of the vector parts of the summands. Combining this result with the formula (63), which expresses that the scalar of the sum is the sum of the scalars, we see that the complex *operation of adding any two geometrical fractions*, of which each is determined by its scalar and by the index of its vector part, may be in general *decomposed into two* very simple but *essentially distinct operations*; namely, *first*, the operation of adding together *two numbers*, positive or negative or null, so as to obtain a third number for their sum, according to the usual rules of elementary algebra; and *second*, the operation of adding together *two lines* in space, so as to obtain a third line, according to the geometrical rules of the composition of motions, or by drawing the diagonal of a parallelogram. In like manner the operation of *taking the difference* of two fractions may be decomposed into the two operations of taking separately the difference of two numbers, and the difference of two lines; for we can easily prove that

$$I\left(\frac{h}{g} - \frac{f}{e}\right) = I\frac{h}{g} - I\frac{f}{e}; \tag{72}$$

or, in words, that the *index* (of the vector part) *of the difference* of any two fractions is equal to the *difference of the indices*. And because it has been seen that not only for numbers but also for lines, considered among themselves, any number of summands may be in any manner grouped or transposed without altering the sum; and that the sum of a scalar and a vector is equal to the sum of the same vector and the same scalar, combined in a contrary order; it follows that the *addition* of any number of geometrical fractions is an *associative* and also a *commutative* operation: in such a manner that we may now write

$$\frac{\mathrm{h}}{\mathrm{g}}+\frac{\mathrm{f}}{\mathrm{e}}=\frac{\mathrm{f}}{\mathrm{e}}+\frac{\mathrm{h}}{\mathrm{g}}; \quad \frac{\mathrm{k}}{\mathrm{i}}+\left(\frac{\mathrm{h}}{\mathrm{g}}+\frac{\mathrm{f}}{\mathrm{e}}\right)=\left(\frac{\mathrm{k}}{\mathrm{i}}+\frac{\mathrm{h}}{\mathrm{g}}\right)+\frac{\mathrm{f}}{\mathrm{e}}=\frac{\mathrm{f}}{\mathrm{e}}+\frac{\mathrm{h}}{\mathrm{g}}+\frac{\mathrm{k}}{\mathrm{i}}, \quad \text{\&c.,} \tag{73}$$

whatever straight lines in space may be denoted by e, f, g, h, i, k, &c. We may also write, concisely,

$$\mathrm{S}\Sigma=\Sigma\mathrm{S}; \quad \mathrm{V}\Sigma=\Sigma\mathrm{V}; \quad \mathrm{I}\Sigma=\Sigma\mathrm{I}; \tag{74}$$

$$\mathrm{S}\Delta=\Delta\mathrm{S}; \quad \mathrm{V}\Delta=\Delta\mathrm{V}; \quad \mathrm{I}\Delta=\Delta\mathrm{I}; \tag{75}$$

using Σ, Δ as the characteristics of sum and difference, while S, V, I are still the signs of scalar, vector, index.

Separation of the Scalar and Vector Parts of the Product of any Two Geometrical Fractions

11. The definitions (46), (47) of addition and multiplication of fractions, namely

$$\frac{\mathrm{c}}{\mathrm{a}}+\frac{\mathrm{b}}{\mathrm{a}}=\frac{\mathrm{c}+\mathrm{b}}{\mathrm{a}}, \quad \frac{\mathrm{c}}{\mathrm{a}}\times\frac{\mathrm{a}}{\mathrm{b}}=\frac{\mathrm{c}}{\mathrm{b}},$$

give obviously, for any 4 straight lines a, b, c, a′, the formula

$$\left(\frac{\mathrm{c}}{\mathrm{a}}+\frac{\mathrm{b}}{\mathrm{a}}\right)\times\frac{\mathrm{a}}{\mathrm{a}'}=\frac{\mathrm{c}+\mathrm{b}}{\mathrm{a}'}=\left(\frac{\mathrm{c}}{\mathrm{a}}\times\frac{\mathrm{a}}{\mathrm{a}'}\right)+\left(\frac{\mathrm{b}}{\mathrm{a}}\times\frac{\mathrm{a}}{\mathrm{a}'}\right), \tag{76}$$

and this other formula of the same kind,

$$\frac{\mathrm{a}'}{\mathrm{a}}\times\left(\frac{\mathrm{c}}{\mathrm{a}}+\frac{\mathrm{b}}{\mathrm{a}}\right)=\frac{\mathrm{a}'}{\dfrac{\mathrm{a}}{\mathrm{c}+\mathrm{b}}\times\mathrm{a}}=\left(\frac{\mathrm{a}'}{\mathrm{a}}\times\frac{\mathrm{c}}{\mathrm{a}}\right)+\left(\frac{\mathrm{a}'}{\mathrm{a}}\times\frac{\mathrm{b}}{\mathrm{a}}\right), \tag{77}$$

may be proved without difficulty to be a consequence of the same definitions; the operation of multiplying a line, by the quotient of two others with which it is co-planar, being interpreted by the definition (23), so as to give, in the present notation,

$$\frac{\mathrm{e}}{\mathrm{a}}\times\mathrm{a}=\mathrm{e}. \tag{78}$$

In fact, if we assume, as we may, seven new lines, db′c′d′b″c″d″, so as to satisfy the seven conditions

$$\left.\begin{array}{llll}\mathrm{c}+\mathrm{b}=\mathrm{d}, & \dfrac{\mathrm{b}}{\mathrm{a}}=\dfrac{\mathrm{a}}{\mathrm{b}'}, & \dfrac{\mathrm{c}}{\mathrm{a}}=\dfrac{\mathrm{a}}{\mathrm{c}'}, & \dfrac{\mathrm{d}}{\mathrm{a}}=\dfrac{\mathrm{a}}{\mathrm{d}'}, \\ & \dfrac{\mathrm{b}''}{\mathrm{a}'}=\dfrac{\mathrm{a}'}{\mathrm{b}'}, & \dfrac{\mathrm{c}''}{\mathrm{a}'}=\dfrac{\mathrm{a}'}{\mathrm{c}'}, & \dfrac{\mathrm{d}''}{\mathrm{a}'}=\dfrac{\mathrm{a}'}{\mathrm{d}'},\end{array}\right\} \tag{79}$$

we shall have the first member of the formula (77) equal to $\frac{\mathrm{a}'}{\mathrm{a}}\times\frac{\mathrm{a}}{\mathrm{d}'}=\frac{\mathrm{a}'}{\mathrm{d}'}=$ the second member of that formula; it will therefore be equal to $\frac{\mathrm{d}''}{\mathrm{a}'}$, and consequently will be shown to be $=\frac{\mathrm{c}''}{\mathrm{a}'}+\frac{\mathrm{b}''}{\mathrm{a}'}=\frac{\mathrm{a}'}{\mathrm{c}'}+\frac{\mathrm{a}'}{\mathrm{b}'}=$ the third member of that formula, if we can show that the conditions (79) give the relation

$$\mathrm{d}'' = \mathrm{c}'' + \mathrm{b}''. \tag{80}$$

Now those conditions show that the line a is common to the planes of b, b′ and c, c′, and that it bisects the angle between b and b′, and also the angle between c and c′; therefore the mutual inclination of the lines b′ and c′ is equal to the mutual inclination of b and c; while the lengths of the two former lines are, by the same conditions, inversely proportional to those of the two latter. And on pursuing this geometrical reasoning, in combination with the definitional meanings of the symbolic equations (79), it appears easily that the mutual inclinations of the lines b″, c″, d″, are equal to those of b′, c′, d′, and therefore to those of b, c, d; while the lengths of b″, c″, d″ are inversely proportional to those of b′, c′, d′, and therefore directly proportional to the lengths of b, c, d: since then the line d is the symbolic sum of b and c, or the diagonal of a parallelogram described with those two lines as adjacent sides, it follows that the line d″ is similarly related to b″ and c″, or that the relation (80) holds good. The formula (77) is therefore shown to be true: and although we have not *yet* proved that the multiplication of two geometrical fractions is *always* a *distributive* operation, we see at least that either factor may be distributed into two partial factors, and that the sum of the two partial products will give the total product, whenever either total factor and the two parts of the other factor are *co-linear*, that is, whenever the planes of these three fractions are *parallel to any common line*, such as the line a in the formulæ (76) (77): the *plane* of a geometrical fraction being one which contains or is parallel to the numerator and denominator thereof. A *scalar* fraction, being the quotient of two parallel lines, of which either may be transported without altering its direction to any other position in space while both may revolve together, may be regarded as having an entirely *indeterminate plane*, which may thus be rendered parallel to any arbitrary line; we shall therefore always satisfy the condition of *co-linearity*, by distributing either or both of two factors into their scalar and vector parts, and may consequently write,

$$\begin{aligned}\frac{\mathrm{h}}{\mathrm{g}}\times\frac{\mathrm{f}}{\mathrm{e}} &= \left(\mathrm{V}\frac{\mathrm{h}}{\mathrm{g}}\times\frac{\mathrm{f}}{\mathrm{e}}\right)+\left(\mathrm{S}\frac{\mathrm{h}}{\mathrm{g}}\times\frac{\mathrm{f}}{\mathrm{e}}\right)\\ &= \left(\frac{\mathrm{h}}{\mathrm{g}}\times\mathrm{V}\frac{\mathrm{f}}{\mathrm{e}}\right)+\left(\frac{\mathrm{h}}{\mathrm{g}}\times\mathrm{S}\frac{\mathrm{f}}{\mathrm{e}}\right)\\ &= \left(\mathrm{V}\frac{\mathrm{h}}{\mathrm{g}}\times\mathrm{V}\frac{\mathrm{f}}{\mathrm{e}}\right)+\left(\mathrm{V}\frac{\mathrm{h}}{\mathrm{g}}\times\mathrm{S}\frac{\mathrm{f}}{\mathrm{e}}\right)+\left(\mathrm{S}\frac{\mathrm{h}}{\mathrm{g}}\times\mathrm{V}\frac{\mathrm{f}}{\mathrm{e}}\right)+\left(\mathrm{S}\frac{\mathrm{h}}{\mathrm{g}}\times\mathrm{S}\frac{\mathrm{f}}{\mathrm{e}}\right);\end{aligned} \tag{81}$$

or more concisely,

$$(\beta + b)(\alpha + a) = \beta\alpha + \beta a + b\alpha + ba, \tag{82}$$

if we denote, as in a former article, vectors by greek and scalars by italic letters, and omit the mark of multiplication between any two successive letters of these two kinds, or between sums of such letters, when those sums are enclosed in parentheses. But the multiplication of scalars is effected, as we have seen, by the ordinary rules of algebra; and to multiply a vector by a scalar, or a scalar by a vector, is easily shown, by the definitions already laid down, to be equivalent to multiplying by the scalar, on the plan of the sixth article, either the index or the numerator of the vector, without altering the denominator of that vector: thus, in the second member of (82), the term ba is a known scalar, and the terms $b\alpha$, βa are known vectors, if the partial factors a, b, α, β be known; in order therefore to apply the equation (82), which in its form agrees with ordinary algebra, to any question of multiplication of any two geometrical fractions, it is sufficient to know how to interpret generally the remaining term βa, or the product of one vector by another. For this purpose we may always conceive the index $\mathrm{I}\beta$ of the vector β to be the sum of two other indices, which shall be respectively parallel and perpendicular to the index $\mathrm{I}\alpha$ of the other vector α, as follows:

$$\mathrm{I}\beta' \parallel \mathrm{I}\alpha, \quad \mathrm{I}\beta'' \perp \mathrm{I}\alpha, \quad \mathrm{I}\beta'' + \mathrm{I}\beta' = \mathrm{I}\beta; \tag{83}$$

and then the vector β itself will be, by the last article, the sum of the two new vectors β' and β'', and the planes of these two new vector fractions will be respectively parallel and perpendicular to the plane of the vector reaction α; consequently, the three fractions β', β'', α will be co-linear, and we shall have, by the principle (76),

$$\beta\alpha = (\beta'' + \beta')\alpha = \beta''\alpha + \beta'\alpha. \tag{84}$$

The problem of the multiplication of *any two* vectors is thus decomposed into the two simpler problems, of multiplying first *two parallel*, and secondly *two rectangular, vectors* together. If then we merely wish to separate the scalar and the vector parts, it is sufficient to observe that if, in the general formula (47), for the multiplication of any two fractions, we suppose the factors to be parallel vectors, then the line a is perpendicular to both b and c, and is also co-planar with them, so that they are necessarily parallel to each other, and the product $\frac{\mathrm{c}}{\mathrm{b}}$ is a scalar; but if, in the same general formula, we suppose the factors to be rectangular vectors, then the three lines a, b, c are themselves mutually rectangular, and the product of the fractions is a vector. Thus, in the formula (84), the partial product $\beta'\alpha$ is a scalar, but the other partial product $\beta''\alpha$ is a vector: and we may write

$$\mathrm{S}.\beta\alpha = \beta'\alpha; \quad \mathrm{V}.\beta\alpha = \beta''\alpha. \tag{85}$$

We may therefore, more generally, under the conditions (83), decompose the formula of multiplication (82) into the two following equations:

$$\left.\begin{aligned} \mathrm{S}.(\beta + b)(\alpha + a) &= \beta'\alpha + ba; \\ \mathrm{V}.(\beta + b)(\alpha + a) &= \beta''\alpha + \beta a + ba \end{aligned}\right\} \tag{86}$$

Or we may write, for abridgment,

$$c = \beta'\alpha + ba; \quad \gamma = \beta''\alpha + \beta a + ba; \tag{87}$$

and then we shall have this other equation of multiplication,

$$\gamma + c = (\beta + b)(\alpha + a). \tag{88}$$

And thus the general *separation of the scalar and vector parts* of the product of any two geometrical fractions may be effected. But it seems proper to examine more closely into the separate meanings of the two partial products of vectors, denoted here by the two terms $\beta'\alpha$ and $\beta''\alpha$; which will be done in the two following articles.

Products of Two Parallel Vectors; Geometrical Representations of the Square Roots of Negative Scalars

12. It was shown, in the last article, that the product of any two parallel vectors, such as α and β', that is, the product of any two vectors of which the planes or the indices are parallel, is equal to a scalar. By pursuing the reasoning of that article, it is easy to show, farther, that this *scalar product of two parallel vectors* is equal to the *product of the numbers* which express the lengths of the two parallel indices; this numerical product being taken with a *negative* or with a *positive* sign, according as these indices are *similar* or *opposite* in direction. In fact, in the general formula $\frac{\mathrm{c}}{\mathrm{a}} \times \frac{\mathrm{a}}{\mathrm{b}} = \frac{\mathrm{c}}{\mathrm{b}}$, we have now $\mathrm{b} \perp \mathrm{a}$, $\mathrm{c} \parallel \mathrm{b}$; the length of c is to the length of b, in a ratio compounded of the ratio of the length of c to that of a, and of the ratio of the length of a to that of b; and the direction of c is opposite or similar to that of b, according as the two quadrantal rotations in one common plane, from b to a, and from a to c, are performed right-handedly round the same index, or round opposite indices.

We know then perfectly how to interpret the product of any two parallel vectors; and, as a case of such interpretation, if we agree to say that the product of any two equal fractions is the *square* of either, and to write

$$\frac{\mathrm{b}}{\mathrm{a}} \times \frac{\mathrm{b}}{\mathrm{a}} = \left(\frac{\mathrm{b}}{\mathrm{a}}\right)^2, \tag{89}$$

whatever two lines may be denoted by a and b, we see that, in the present system, the *square of a vector is always a negative scalar,* namely the negative of the square of the number which denotes the length of the index of the vector; in such a manner that, for any vector α, we shall have the equation

$$\alpha^2 = -\overline{\alpha}^2, \tag{90}$$

if we agree to denote by the symbol $\overline{\alpha}$ that positive or absolute number which expresses the *length of the index* Iα. We have then, reciprocally,

$$\overline{\mathrm{a}}^2 = -\alpha^2; \tag{91}$$

and may therefore write

$$\overline{\alpha} = \sqrt{(-\alpha^2)}, \tag{92}$$

$-\alpha^2$ being here a positive number (because α^2 is negative), and $\sqrt{(-\alpha^2)}$ being its positive or absolute *square root,* which is an entirely *determined* (and real) *number,* when the vector α, or even when the length of its index, is determined. But although we might be led to write, in like manner, from (90), the equation

$$\alpha = \sqrt{(-\overline{\alpha}^2)}, \tag{93}$$

yet the same principles prove that this expression, which may denote generally any *square root of a negative number,* by a suitable choice of the positive number $\overline{\alpha}$, is equal to a *vector* α, of which the index Iα has indeed a *determined length,* but has an entirely *undetermined direction;* the symbol in the second member of the equation (93) may therefore receive (in the present system) infinitely many different geometrical representations, or constructions, though they have all one common character: and it will be a little more consistent with the analogies of ordinary algebra to write the equation under the form

$$\alpha = (-\overline{\alpha}^2)^{\frac{1}{2}}, \tag{94}$$

using a fractional exponent which suggests a certain degree of indeterminateness, rather than a radical sign which it is often convenient to restrict to one determined value. Thus, for example, the symbol $(-1)^{\frac{1}{2}}$, or the *square root of negative unity*, will, in the present system, denote, or be geometrically constructed by, *any vector of which the index is equal to the unit of length*; that is, any geometrical fraction of which the numerator and the denominator are lines equal to each other in length, but perpendicular to each other in direction. And we see that the geometrical principle, on which this conclusion ultimately depends, is simply this: that *two successive and similar quadrantal rotations, in any arbitrary plane, reverse the direction* of any straight line in that plane. Mr. Warren, confining himself to the consideration of lines in *one fixed plane*, has been led to attribute to his geometrical representations of the square roots of negative numbers, *one fixed direction*, or rather axis, perpendicular to that other axis on which he represents square roots of positive numbers. And other authors, both before and since the publication of Mr. Warren's work,* seem to have been in like manner disposed to represent positive or negative numbers by lines in some one direction, or in the direction opposite, but symbols of the form $a\sqrt{(-1)}$ by lines perpendicular thereto. Such is at least the impression on the mind of the present writer, produced perhaps by an insufficient acquaintance with the works of those who have already written on this class of subjects. It will however be attempted to show, in a future article of this paper, that the geometrical fractions which have been called *vectors*, in the present and in former articles, may be symbolically equated to their own indices; and that thus *every straight line having direction in space* may properly be looked upon *in the present system* as *a geometrical representation of a square root of a negative number*; while positive and negative numbers are in the same system regarded indeed as belonging to one common *scale* of progression, from $-\infty$ to $+\infty$, but to a scale which is not to be considered as having any one direction rather than any other, in tridimensional space.

Products of Two Rectangular Vectors; Non-commutativeness of the Factors, in the General Multiplication of Two Geometrical Fractions

13. The reasoning by which it was shown, in the 11th article, that the *product* $\beta''\alpha$ of *any two rectangular vectors*, α and β'', is *itself a vector*, may be continued so as to show that the number expressing the length of the index of this vector product is the product of the numbers which express the lengths of the indices of the factors; or that, in a notation similar to one employed in the last article,

$$\overline{\beta''\alpha} = \overline{\beta}''\overline{\alpha}, \quad \text{when } \mathrm{I}\beta'' \perp \mathrm{I}\alpha; \tag{95}$$

and therefore that, by the principle (92), for the same case of *rectangular vectors*, we have the formula

$$\sqrt{\{-(\beta''\alpha)^2\}} = \sqrt{(-\beta''^2)}\sqrt{(-\alpha^2)}. \tag{96}$$

Also in the general formula of multiplication $\frac{\mathrm{c}}{\mathrm{a}} \times \frac{\mathrm{a}}{\mathrm{b}} = \frac{\mathrm{c}}{\mathrm{b}}$, the three lines a, b, c compose here a rectangular system; and therefore the *index of the product* is parallel to the line a, and is consequently *perpendicular to the indices of the two factors*; $\mathrm{I}.\beta''\alpha$ is therefore perpendicular to

* *Treatise on the Geometrical Representation of the Square Roots of Negative Quantities*, by the Rev. John Warren, Cambridge, 1828. See also Dr. Peacock's Treatises on Algebra [*A Treatise on Algebra*, Cambridge: 1836; *A Treatise on Algebra*, 2 Vols, Cambridge: 1842–5], and his Report to the British Association, containing references to other works.

both $\mathrm{I}\beta''$ and $\mathrm{I}\alpha$; a conclusion which may be extended by (83) and (85) to the multiplication of *any two vectors*, so that we may write generally,

$$\mathrm{I}.\beta\alpha \perp \mathrm{I}\beta; \quad \mathrm{I}.\beta\alpha \perp \mathrm{I}\alpha. \tag{97}$$

Again, we are allowed to suppose, in applying the same general formula of multiplication to the same case of rectangular vectors, that the index $\mathrm{I}\alpha$ of the multiplicand $\frac{\mathrm{a}}{\mathrm{b}}$ is not only parallel to the line c, but similar (and not opposite) in direction to that line; in such a manner that the rotation round c from b to a is positive: and then the rotation round b from a to c is positive, and so is the rotation round a from c to b, and also that round −a from b to c; therefore the index $\mathrm{I}\beta''$ of the multiplier is similar in direction to +b, and the index $\mathrm{I}.\beta''\alpha$ of the product is similar in direction to −a; consequently *the rotation round the index of the product, from the index of the multiplier to that of the multiplicand, is positive.* And although this last result has only been proved here for the case of two rectangular vectors, yet it may easily be shown, by the principles of the 11th article, to extend to the multiplication of two general geometrical fractions. For, in the notation of that article, γ denoting the vector part of the product of any two such fractions, we have, by (87),

$$\mathrm{I}\gamma = \mathrm{I}.\beta''\alpha + a\mathrm{I}\beta + b\mathrm{I}\alpha; \tag{98}$$

$\mathrm{I}\gamma$ is therefore the symbolic sum of $\mathrm{I}.\beta''\alpha$ and of two other lines which are respectively parallel to the indices of the vector parts of the two factors, and which consequently have their sum co-planar with those indices, and therefore also co-planar, by (83), with $\mathrm{I}\beta''$ and $\mathrm{I}\alpha$; consequently $\mathrm{I}\gamma$ and $\mathrm{I}.\beta''\alpha$ both lie at the same side of the plane of $\mathrm{I}\alpha$ and $\mathrm{I}\beta''$; and therefore the rotation round $\mathrm{I}\gamma$, like that round $\mathrm{I}.\beta''\alpha$, from $\mathrm{I}\beta''$ to $\mathrm{I}\alpha$, and consequently from $\mathrm{I}\beta$ to $\mathrm{I}\alpha$, is positive. Hence also the rotation round $\mathrm{I}\beta$ from $\mathrm{I}\alpha$ to $\mathrm{I}\gamma$ is positive; that is to say, in the multiplication of two general geometrical fractions, *the rotation round the index of the vector part of the multiplier, from that of the multiplicand to that of the product, is positive*; from which may immediately be deduced a remarkable consequence, already alluded to by anticipation in the 8th article, namely—that the *multiplication of two general geometrical fractions is not a commutative operation*, or that the *order of the factors is not in general indifferent*; since the index of the vector part of the product lies at one or at the other side of the plane of the indices of the vector parts of the two factors, according as those factors are taken in one or in the other order. We have, for example, by the present article, a relation of *opposition* of signs between the products of two *rectangular* vectors, taken in two opposite orders; which relation may be expressed by the following *equation of perpendicularity*,

$$\alpha\beta'' = -\beta''\alpha, \quad \text{when } \mathrm{I}\beta'' \perp \mathrm{I}\alpha. \tag{99}$$

But in the case where the indices of the vector parts α and β of two fractional factors are *parallel* (which includes the case where either of those indices vanishes, the corresponding factor becoming then a scalar), the part β'' of the vector β vanishes, and the latter vector reduces itself by (83) to its other part β'; so that in *this* case, by the results of the last article, the order of the factors is indifferent, and the operation of multiplication is commutative: and thus we may write, as the *equation of parallelism* between two vectors,

$$\alpha\beta' = \beta'\alpha, \quad \text{when } \mathrm{I}\beta' \parallel \mathrm{I}\alpha. \tag{100}$$

It is easy to infer hence, by (84) and (77), that in the more general case of the multiplication of any two vectors α and β, we may write, instead of (85), the following formulæ for the separation of the scalar and vector parts of the product:

$$\left.\begin{aligned} \mathrm{S}.\beta\alpha &= \tfrac{1}{2}(\beta\alpha + \alpha\beta) = \mathrm{S}.\alpha\beta \\ \mathrm{V}.\beta\alpha &= \tfrac{1}{2}(\beta\alpha - \alpha\beta) = -\mathrm{V}.\alpha\beta \end{aligned}\right\} \tag{101}$$

with corresponding formulæ instead of (86), which give

$$(\beta + b)(\alpha + a) - (\alpha + a)(\beta + b) = \beta\alpha - \alpha\beta, \tag{102}$$

the second member of this last equation being a vector different from 0, unless it happen that the planes (or the indices) of the vectors α and β are parallel to each other. Finally, we may here observe that in virtue of the principles and definitions already laid down, *the length of the index* ($\mathrm{I}.\beta\alpha$) *of the vector part of the product of any two vectors bears to the unit of length the same ratio which the area of the parallelogram under the indices* ($\mathrm{I}\beta$ and $\mathrm{I}\alpha$) *of the factors bears to the unit of area*; the *direction* of this index of the product being also (as we have seen) *perpendicular to the plane* of the indices of the factors, and therefore to the plane of the parallelogram under them; and being changed to its own *opposite* when the order of the factors is inverted, which *inversion* of their order may be considered as corresponding to a *reversal of the face* of the parallelogram: for all which reasons, there appears to be a propriety in considered this index of the vector part of a product of any two vectors as a symbolical representation of this parallelogram under the indices of the factors, and in writing the symbolical equation

$$\mathrm{I}.\beta\alpha = \square(\mathrm{I}\beta,\ \mathrm{I}\alpha). \tag{103}$$

It will be remembered that the indices $\mathrm{I}(\beta + \alpha)$, $\mathrm{I}(\beta - \alpha)$, of the sum and difference of the same two vectors, are symbolically equal to two different diagonals of the same parallelogram, by former articles of this paper.

On the Distributive Character of the Operation of Multiplication, as Performed Generally on Geometrical Fractions

14. We are now prepared to extend the formulæ (76), (77), respecting the multiplication of sums of geometrical fractions; and to show that similar results hold good, even when the condition of colinearity, assumed in those two formulæ, is no longer supposed to be satisfied. That is, the two equations

$$\left(\frac{\mathrm{h}}{\mathrm{g}} + \frac{\mathrm{f}}{\mathrm{e}}\right) \times \frac{\mathrm{k}}{\mathrm{i}} = \left(\frac{\mathrm{h}}{\mathrm{g}} \times \frac{\mathrm{k}}{\mathrm{i}}\right) + \left(\frac{\mathrm{f}}{\mathrm{e}} \times \frac{\mathrm{k}}{\mathrm{i}}\right), \tag{104}$$

$$\frac{\mathrm{k}}{\mathrm{i}} \times \left(\frac{\mathrm{h}}{\mathrm{g}} + \frac{\mathrm{f}}{\mathrm{e}}\right) = \left(\frac{\mathrm{k}}{\mathrm{i}} \times \frac{\mathrm{h}}{\mathrm{g}}\right) + \left(\frac{\mathrm{k}}{\mathrm{i}} \times \frac{\mathrm{f}}{\mathrm{e}}\right), \tag{105}$$

can both be shown to be true, whatever may be the lengths and directions of the six lines e, f, g, h, i, k; although, by the general non-commutativeness of geometrical fractions as factors, which was pointed out in the last article, the expressions contained in these two equations are not to be confounded with each other.

Making for this purpose

$$\left.\begin{array}{llll} \dfrac{\mathrm{f}}{\mathrm{e}} = \beta_1 + b_1, & \dfrac{\mathrm{h}}{\mathrm{g}} = \beta_2 + b_2, & \dfrac{\mathrm{k}}{\mathrm{i}} = \alpha + a, & \\ \mathrm{I}\beta_1' \parallel \mathrm{I}\alpha, & \mathrm{I}\beta_1'' \perp \mathrm{I}\alpha, & \mathrm{I}\beta_1'' + \mathrm{I}\beta_1' = \mathrm{I}\beta_1, & \\ \mathrm{I}\beta_2' \parallel \mathrm{I}\alpha, & \mathrm{I}\beta_2'' \perp \mathrm{I}\alpha, & \mathrm{I}\beta_2'' + \mathrm{I}\beta_2' = \mathrm{I}\beta_2, & \\ \beta_2' + \beta_1' = \beta', & \beta_2'' + \beta_1'' = \beta'', & \beta_2 + \beta_1 = \beta, & b_2 + b_1 = b, \end{array}\right\} \quad (106)$$

the conditions (83) will be satisfied; and if we still assign to γ and c the meanings (87), the equation (88) will hold good, and $\gamma + c$ will be an expression for the first member of (104). Making also, in imitation of (87),

$$\left.\begin{array}{ll} c_1 = \beta_1'\alpha + b_1 a, & \gamma_1 = \beta_1''\alpha + \beta_1 a + b_1 \alpha, \\ c_2 = \beta_2'\alpha + b_2 a, & \gamma_2 = \beta_2''\alpha + \beta_2 a + b_2 \alpha, \end{array}\right\} \quad (107)$$

the second member of the same equation (104) becomes, by the principles of the 11th article, $(\gamma_2 + c_2) + (\gamma_1 + c_1)$; and the equation resolves itself into the two following,

$$c = c_2 + c_1, \quad \gamma = \gamma_2 + \gamma_1; \quad (108)$$

which are easily seen to reduce themselves to these two,

$$(\beta_2' + \beta_1')\alpha = \beta_2'\alpha + \beta_1'\alpha; \quad (\beta_2'' + \beta_1'')\alpha = \beta_2''\alpha + \beta_1''\alpha; \quad (109)$$

the one being an equation between scalars, and the other between vectors. In like manner the equation (105) may be shown to depend on the two following equations, less general than itself, but of the same form,

$$\alpha(\beta_2' + \beta_1') = \alpha\beta_2' + \alpha\beta_1'; \quad \alpha(\beta_2'' + \beta_1'') = \alpha\beta_2'' + \alpha\beta_1''. \quad (110)$$

And since, by (101), the three scalar products in the equations (110) are respectively equal, and the three vector products are respectively opposite (in their signs) to the corresponding products in the equations (109), it is sufficient to prove either of these two pairs of equations; for example, the pair (110). Now the first equation of this pair is true, because the scalars denoted by the three products $\alpha\beta_1'$, $\alpha\beta_2'$, $\alpha(\beta_2' + \beta_1')$, are proportional, both in their magnitudes and in their signs, to the indices of the three parallel vectors β_1', β_2', $\beta_2' + \beta_1'$; and the second equation of the same pair is true, because the indices of the vectors denoted by the three other products $\alpha\beta_1''$, $\alpha\beta_2''$, $\alpha(\beta_2'' + \beta_1'')$ are formed from the indices of the three coplanar vectors β_1'', β_2'', $\beta_2'' + \beta_1''$, by causing the three latter indices to revolve together, as one system, in their common plane, round the index $\mathrm{I}\alpha$, their lengths being at the same time changed (if at all) in one common ratio, namely, in that of $\overline{\alpha}$ to 1. The formulæ (104) (105) are therefore proved to be true; and the same reasoning shows, that in any multiplication of two geometrical fractions, either of the factors may be *distributed* into *any number* of parts, and that the sum of the partial products will be equal to the total product: so that we may write, generally,

$$\left(\Sigma\frac{\mathrm{k}}{\mathrm{i}}\right) \times \left(\Sigma\frac{\mathrm{f}}{\mathrm{e}}\right) = \Sigma\left(\frac{\mathrm{k}}{\mathrm{i}} \times \frac{\mathrm{f}}{\mathrm{e}}\right). \quad (111)$$

The *multiplication of geometrical fractions* is therefore *a distributive operation*; although it has been shown to be *not*, in general, a *commutative* one.

On the Associative Property of the Multiplication of Geometrical Fractions

15. Proceeding now, with the help of the distributive property established in the last article, and of the principle that a product is multiplied by a scalar when any one of its factors is multiplied thereby, to prove that the multiplication of geometrical fractions is generally an *associative* operation, or that the formula

$$\frac{\mathrm{k}}{\mathrm{i}} \times \left(\frac{\mathrm{h}}{\mathrm{g}} \times \frac{\mathrm{f}}{\mathrm{e}}\right) = \left(\frac{\mathrm{k}}{\mathrm{i}} \times \frac{\mathrm{h}}{\mathrm{g}}\right) \times \frac{\mathrm{f}}{\mathrm{e}}, \tag{112}$$

holds good for *any three fractions* (with other formulæ of the same sort for more fractional factors than three), it will be sufficient to prove that the formula is true for *any three vectors*; or that we may write generally

$$\gamma \times \beta\alpha = \gamma\beta \times \alpha, \tag{113}$$

the vector γ being not here obliged to satisfy the equation (87); we may even content ourselves with proving that the equation (113) is true in the two following cases, namely first, when any two of the three vectors are parallel; and secondly, when all three are rectangular to each other. The first case may be expressed by the three following equations as its types—

$$\beta \times \beta\alpha = \beta\beta \times \alpha, \tag{114}$$

$$\beta \times \alpha\beta = \beta\alpha \times \beta, \tag{115}$$

$$\alpha \times \beta\beta = \alpha\beta \times \beta; \tag{116}$$

and the second case may be expressed by the equation

$$\alpha\beta \times \beta\alpha = (\alpha\beta \times \beta) \times \alpha, \quad \text{when } \beta \perp \alpha; \tag{117}$$

because, under this last condition, $\alpha\beta$ is, by Art. 13, a vector, rectangular to both α and β. Under the same condition we may, by (99), change $\alpha\beta$ to $-\beta\alpha$; therefore the first member of the equation (117) may be equated to $-(\beta\alpha)^2$, and consequently, by (96), to $(-\beta^2) \times (-\alpha^2) = \beta^2 \times \alpha^2 = \beta^2\alpha \times \alpha = (\alpha \times \beta\beta) \times \alpha$, because β^2 or $\beta\beta$ is, by Art. 12, a scalar; thus we may make (117) depend on (116), which again depends on (114), and on the following equation,

$$\beta \times \beta\alpha = \alpha\beta \times \beta. \tag{118}$$

Equations (118) and (115) may both be proved by observing that, by Art. 13, whatever two vectors may be denoted by α and β, we have the expressions

$$\left.\begin{aligned} \beta\alpha &= \mathrm{S}.\beta\alpha + \mathrm{V}.\beta\alpha, \\ \alpha\beta &= \mathrm{S}.\beta\alpha - \mathrm{V}.\beta\alpha, \end{aligned}\right\} \tag{119}$$

with the relations

$$\left.\begin{aligned} \beta \times \mathrm{S}.\beta\alpha - \mathrm{S}.\beta\alpha \times \beta = 0, \\ \beta \times \mathrm{V}.\beta\alpha + \mathrm{V}.\beta\alpha \times \beta = 0, \end{aligned}\right\} \tag{120}$$

It remains then to prove the equation (114); and it is sufficient to prove this for the case where α and β are two rectangular vectors. But, in this case, $\beta\alpha$ is a vector formed from α by causing its index $\mathrm{I}\alpha$ to revolve right-handedly through a right angle round the index $\mathrm{I}\beta$, to which it is perpendicular, changing at the same time in general the length of this revolving index from $\overline{\alpha}$ to $\overline{\beta}\times\overline{\alpha}$; and the repetition of this process, directed by the symbol $\beta\times\beta\alpha$, conducts to a new vector, of which the index is in direction opposite to the original direction of $\mathrm{I}\alpha$, and in length equal to $\overline{\beta}^2\times\overline{\alpha}$: this new vector may therefore be otherwise denoted by $-\overline{\beta}^2\times\alpha$, or by $\beta^2\times\alpha$, and the equation (114) is true. The equations (113) and (112) are therefore also true; and since the latter formula may easily be extended to any number of fractional factors, we are now entitled to conclude what it was at the beginning of the present article proposed to prove; namely, that the *multiplication of geometrical fractions is always an associative operation*: as the addition of fractions, and the addition of lines, have in former articles been shown to be. In other words, any number of successive fractional factors may be *associated* or grouped together by multiplication (without altering their order) into a single product, and this product substituted as a single factor in their stead; a result which constitutes a new agreement (the more valuable on account of the absence of identity in some other important respects), between the *rules of operation* on ordinary algebra, and those of the present Symbolical Geometry.

Other forms of the Associative Principle of Multiplication

16. By the principles already established respecting the transformation of geometrical fractions, any three such fractions, $\frac{\mathrm{f}}{\mathrm{e}}, \frac{\mathrm{h}}{\mathrm{g}}, \frac{\mathrm{k}}{\mathrm{i}}$, may be so prepared that the numerator of the first shall be in the plane of the second, and that the numerator of the second shall coincide with the denominator of the third; we may, therefore, without diminishing the generality of the theorem expressed by the formula (112), suppose that the line i is equal to h, and that the fourth proportional to g, h, f, is a new line l; and with this preparation the associative principle of multiplication, established in the foregoing article, may be put under the following form, in which the mark of multiplication between two fractional factors is omitted for the sake of conciseness:

$$\text{if } \frac{\mathrm{h}}{\mathrm{g}}=\frac{\mathrm{l}}{\mathrm{f}}, \quad \text{then } \frac{\mathrm{k}}{\mathrm{g}}\frac{\mathrm{l}}{\mathrm{e}}=\frac{\mathrm{k}}{\mathrm{g}}\frac{\mathrm{f}}{\mathrm{e}}; \tag{121}$$

that is to say, *the product of any two geometrical fractions will remain unaltered* in value, or will still continue to represent the same third fraction, *if the denominator of the multiplier and the numerator of the multiplicand be changed to any two new lines to which they are proportional*, or with which they form a *symbolic analogy*, including a relation between *directions* as well as a proportion of lengths, of the kind considered in Mr. Warren's work, (and earlier by Argand and Français,) and in the seventh article of this paper. Reciprocally, by the associative principle, the former of the two equations (121) is in general a consequence of the latter; that is, if the product of two geometrical fractions be equal to the product of two other fractions of the same sort, and if the multipliers have a common numerator, and the multiplicands a common denominator, then the numerators of the two multiplicands and the denominators of the two multipliers are the antecedents and consequents of a symbolical proportion or analogy, of the kind considered in the seventh article: for we may write

$$\frac{h}{g} = \frac{h}{k}\left(\frac{k}{g}\frac{f}{e}\right)\frac{e}{f}, \quad \frac{h}{k}\left(\frac{k}{h}\frac{l}{e}\right)\frac{e}{f} = \frac{l}{f};$$

so that the first equation (121) may be obtained from the second, by suitably grouping or associating the factors.

Again, the same associative principle shows that

$$\text{if } \frac{c}{c'} = \frac{b'}{b}\frac{a'}{a}, \quad \text{then } \frac{c}{b'} = \frac{c'}{a}\frac{a'}{b}; \tag{122}$$

for the first equation (122) may be replaced by the system of the three following equations,

$$\frac{a'}{a} = \frac{b''}{a''}, \quad \frac{b'}{b} = \frac{c''}{b''}, \quad \frac{c'}{c} = \frac{a''}{c''}; \tag{123}$$

of which the two last give, for the first member of the second equation (122), the expression

$$\frac{c}{b'} = \frac{c'}{a''}\frac{b''}{b},$$

which is equal to the second member of the same second equation (122), by the first of the three equations (123), and by the theorem (121): whenever, therefore, we meet an equation between one geometrical fraction and the product of two others, we are at liberty to *interchange the denominator of the product and the numerator of the multiplier*, provided that we at the same time *interchange the denominators of the two factors*; no change being made in the numerators of the product and the multiplicand. Conversely, this assertion respecting the liberty to make these interchanges, and the formula (122), to which the assertion corresponds, are modes of expressing the associative principle of multiplication; for by introducing the equations (123) we find that the theorem (122) conducts to the following relation, or *identity between the two ternary products of three fractions*, associated in two different ways, but with one common order of arrangement,

$$\frac{c'}{a''}\left(\frac{a''}{a}\frac{a'}{b}\right) = \left(\frac{c'}{a''}\frac{a''}{a}\right)\frac{a'}{b}; \tag{124}$$

in which last form, as in (112), the three factors multiplied together may represent any three geometrical fractions. We may also present the same principle under the form of the following theorem—

$$\text{if } \frac{c'}{c}\frac{b'}{b}\frac{a'}{a} = 1, \quad \text{then } \frac{c'}{a}\frac{a'}{b}\frac{b'}{c} = 1; \tag{125}$$

and may derive from it, with the help of (123), the following value of a certain product of six fractional factors,

$$\frac{a''}{c''}\frac{c}{a}\frac{b''}{a''}\frac{a}{b}\frac{c''}{b''}\frac{b}{c} = 1: \tag{126}$$

which must hold good whenever the three lines a, b, c are respectively coplanar with the three pairs a″b″, b″c″, c″a″. Finally, it may be stated here, as a theorem essentially equivalent to the associative principle of multiplication, although not expressly involving any product of two or more fractions, that *in the system of the six equations* of which those marked (123) are three, and of which the others are the three following analogous equations,

$$\frac{a}{c'} = \frac{a'''}{c'''}, \quad \frac{b}{a'} = \frac{b'''}{a'''}, \quad \frac{c}{b'} = \frac{c'''}{b'''}; \tag{127}$$

any five equations of the system include the sixth.

Geometrical Interpretation of the Associative Principle: Symbolic Equations between Arcs upon a Sphere: Theorem of the two Spherical Hexagons

17. If we attended only to the *lengths* of the various lines compared, the associative principle of multiplication, under all the foregoing forms, would be nothing more than an easy and known consequence of a few elementary theorems respecting compositions of ratios of magnitudes. On the other hand it is permitted, in the present symbolical geometry, to assume at pleasure the *situations* of straight lines denoted by small roman letters, provided that the lengths and directions are preserved. The general theorem or property of multiplication, which has been expressed in various ways in the two foregoing articles, may therefore be regarded as being essentially a *relation, or system of relations, between the directions of certain lines in space.*

In this view of the subject no essential loss of generality (or at least none which cannot easily be supplied by known and elementary principles) will be sustained by supposing all the straight lines abc, a′b′c′, a″b″c″, a‴b‴c‴, efghikl, of the two last articles to be *radii of one sphere,* setting out from one *common origin* or centre O, and terminating in points upon one *common spheric surface,* which may be denoted respectively by the symbols ABC, A′B′C′, A″B″C″, A‴B‴C‴, EFGHIKL. In order more conveniently to study and express relations between points so situated, we may agree to say that two *arcs upon one sphere,* such as those from G to H and from F to L, are *symbolically equal,* when they are *equally long and similarly directed portions of the circumference of one great circle*; and may denote this *symbolical equality between arcs,* so called for the sake of suggesting that (like the symbolical equality between straight lines considered in the second article) it involves a relation of *identity of directions,* as well as a relation of equality of lengths, by writing any one of the three formulæ,

$$\left.\begin{aligned} \frown \mathrm{LF} &= \frown \mathrm{HG}, \\ \frown \mathrm{FL} &= \frown \mathrm{GH}, \\ \frown \mathrm{LH} &= \frown \mathrm{FG}, \end{aligned}\right\} \tag{128}$$

of which the second may be called the *inverse,* and the third the *alternate* of the first. Any one of these three formulæ (128) will thus express the *same relation between the directions of four coplanar radii,* namely, the four lines fghl, as that expressed by the first equation (121), or by its inverse, or its alternate equation; that is, by any one of the three following equations between geometrical fractions,

$$\frac{l}{f} = \frac{h}{g}, \quad \frac{f}{l} = \frac{g}{h}, \quad \frac{l}{h} = \frac{f}{g}. \tag{129}$$

The formulæ (128) express also the same relation between the same four directions, as that which would be expressed in a notation of a former article, by any one of the three following *symbolic analogies* between the same four lines,

$$\mathrm{l} : \mathrm{f} : : \mathrm{h} : \mathrm{g}, \quad \mathrm{f} : \mathrm{l} : : \mathrm{g} : \mathrm{h}, \quad \mathrm{l} : \mathrm{h} : : \mathrm{f} : \mathrm{g}; \tag{130}$$

although it must not be forgotten that any one of the six latter formulæ, (129) and (130), expresses at the same time a proportion between the lengths of four straight lines, not generally equal to each other, which is not expressed by any one of the three former symbolical equations (128), between pairs of arcs upon a sphere. In this notation (128), the last form of the associative principle of multiplication which was assigned in the foregoing article, so far as it relates to directions only, may be expressed by saying that *any one of the six following symbolical equations between arcs is a consequence of the other five,*

$$\left.\begin{aligned} \frown \mathrm{A'A} &= \frown \mathrm{B''A''}, \\ \frown \mathrm{B'B} &= \frown \mathrm{C''B''}, \\ \frown \mathrm{C'C} &= \frown \mathrm{A''C''}, \end{aligned}\right\} \tag{131}$$

$$\left.\begin{aligned} \frown \mathrm{BA'} &= \frown \mathrm{B'''A'''}, \\ \frown \mathrm{CB'} &= \frown \mathrm{C'''B'''}, \\ \frown \mathrm{AC'} &= \frown \mathrm{A'''C'''}. \end{aligned}\right\} \tag{132}$$

Regarding *any six points* upon a spheric surface, in *any one order* of succession, as the *six corners of a spherical hexagon* (which may have re-entrant angles, and of which two or more sides may cross each other without being prolonged), we may speak of the arcs joining *successive corners* as the *sides*; those joining *alternate* corners, as the *diagonals*; and those joining *opposite* corners, as the *diameters* of this hexagon: the first side, first diagonal, and first diameter, respectively, being those three arcs which are drawn from the first corner to the second, third, and fourth corners of the figure. With this phraseology, the form just now obtained for the result of the two foregoing articles may be expressed as a relation between two spherical hexagons, AA′BB′CC′, A″A‴B″B‴C″C‴, and may be enunciated in words as follows: *If five successive sides of one spherical hexagon be respectively and symbolically equal to five successive diagonals of another spherical hexagon, the sixth side of the first hexagon will be symbolically equal to the sixth diagonal of the second hexagon.* This theorem of spherical geometry, which may be called, for the sake of reference, the *theorem of the two hexagons*, is therefore a consequence, and may be regarded as an interpretation of the associative principle of multiplication: and conversely, in all applications to spherical geometry, and generally in all investigations respecting relations between the directions of straight lines in space, the associative principle of multiplication may be replaced by the theorem of the two spherical hexagons.

Other Interpretation of the Associative Principle of Multiplication: Theorem of the two Conjugate Transversals of a Spherical Quadrilateral (which are the Cyclic Arcs of a circumscribed Spherical Conic)

18. The theorem of the two hexagons gives also the following theorem: If upon each of the four sides of a spherical quadrilateral, or on that side prolonged, a portion be taken commedial with the side (two arcs being said to be *commedial* when they have one common point of bisection); and if four extreme points of the four portions thus obtained be ranged on one transversal arc of a great circle, in such a manner that the part of this arc comprised between the first and third sides is commedial with the part comprised between the second and fourth: then the four other extremities of the same four portions will be ranged on another great circle; and the parts of this second or *conjugate* transversal, which are intercepted respectively by the same two pairs of opposite sides of the quadrilateral, will be in like manner commedial with each other.

For let the corners of the quadrilateral be denoted by the letters A, B, C, D, and let the side from A to B be cut in two points A′ and B″, while the three other sides are cut in three other pairs of points, which may be called B′ and C″, C′ and D″, and D′ and A″ respectively. Then, if the arcs from A′ to C′ and from B′ to D′ be commedial portions of one common great circle, or of a first transversal arc, the arcs from A′ to B′ and from D′ to C′ will be *symbolically equal arcs*, in the sense of the preceding article; and therefore, in the notation of that article, we may now write the equation

$$\frown \mathrm{B'A'} = \frown \mathrm{C'D'}. \tag{133}$$

In like manner the conditions, that the four portions of the sides of the quadrilateral shall be commedial with the sides themselves, give the four other equations of the same kind,

$$\left.\begin{array}{ll} \frown \mathrm{A'A} = \frown \mathrm{BB''}; & \frown \mathrm{B'B} = \frown \mathrm{CC''}; \\ \frown \mathrm{C'C} = \frown \mathrm{DD''}; & \frown \mathrm{D'D} = \mathrm{AA''}. \end{array}\right\} \tag{134}$$

Hence, by alternation and inversion, we find that the five successive sides

$$\frown \mathrm{AB''}, \quad \frown \mathrm{D'A}, \quad \frown \mathrm{C'D'}, \quad \frown \mathrm{CC'}, \quad \frown \mathrm{C''C},$$

of the spherical hexagon B″AD′C′CC″ are respectively and symbolically equal to the five successive diagonals

$$\frown \mathrm{A'B}, \quad \frown \mathrm{DA''}, \quad \frown \mathrm{B'A'}, \quad \frown \mathrm{D''D}, \quad \frown \mathrm{BB'},$$

of the other hexagon BA″A′DB′D″; and therefore, by the theorem of the two hexagons, the sixth side of the former figure must be symbolically equal to the sixth diagonal of the latter; that is, we may write the symbolical equation,

$$\frown \mathrm{B''C''} = \frown \mathrm{A''D''}. \tag{135}$$

But this expresses a relation equivalent to the following, that the two arcs from A″ to C″ and from B″ to D″ are commedial portions of one common great circle, or second transversal arc, which was the thing to be proved.

Reciprocally, the associative principle of geometrical multiplication, in so far as it relates to the directions of straight lines in space, may be expressed by the assertion that the symbolical equation between arcs (135) is a consequence of the five other equations of the same kind (133) and (134); this principle of symbolical geometry may therefore be so interpreted as to coincide with the foregoing *theorem of the two conjugate transversals* of a spherical quadrilateral, instead of the theorem of the two spherical hexagons. It is easy to see that to a given quadrilateral correspond infinitely many such pairs of conjugate transversal arcs; and those readers who are familiar with the theory of *spherical conics** will recognise in these conjugate

* The plane of the first side of the quadrilateral, or the plane of OAB, if O denote the centre of the sphere, is cut by the plane of the first transversal arc in the radius A′O, and by the plane of the second transversal arc in the radius B″O. Thus the four plane faces of the tetrahedral angle, of which the four edges are the four radii from O to the four corners A, B, C, D of the quadrilateral, are cut by any secant

transversals, A′B′C″D′, A″B″C″D″, the two *cyclic arcs* of such a conic, circumscribed about the proposed quadrilateral ABCD; but it suits better the plan of this communication on symbolical geometry to pass at present to another view of the subject.

It may however be noticed here, that in the first of the two hexagons already mentioned, *any two pairs of opposite sides intercept commedial portions on either of the two sides remaining*; and that the associative principle asserts that *if* a spherical hexagon have *five* of its sides thus *cut commedially*, the *sixth* side also will be cut in the same way. Or, because the two sets of alternate diagonals of the second hexagon are sides of two triangles, which have for their corners the alternate corners of this hexagon, we may in another way eliminate this second hexagon, and may express the same principle of spherical geometry by saying, that *if one set of alternate sides of a* (first) *spherical hexagon*, taken in their order (as first, third, and fifth), *be respectively and symbolically equal to the three successive sides of a triangle*, then the *other set of alternate sides of the same hexagon will be* in like manner *symbolically equal to the sides of another triangle*. This last interpretation of the associative principle is even more immediately suggested than any other, by the forms of the equations (131) (132); in the notation of the present article, *the two triangles* are BA′B′ and A″DD″, which may be considered as having their *bases* A′B′ and A″D″ *on the two cyclic arcs* above alluded to, while their *vertical angles* at B and D may be said to be *angles in the same segment* (or in alternate segments) *of the spherical conic*: since, by (134), the two arcual sides BA′, BB′ of the one angle intersect respectively the two sides DA″, DD″ of the other angle, in the points A and C, which points of intersection, as well as the vertices B and D, are corners of the quadrilateral inscribed in that spherical conic.

Symbolical Addition of Arcs upon a Sphere; Associative and Non-commutative Properties of such Addition

19. The foregoing geometrical interpretations of the associative principle or property of the multiplication of geometrical fractions, may assist us in forming and applying the conception of the symbolical addition of arcs of great circles upon a sphere, and in establishing and interpreting an analogous principle or property of such symbolical addition.

As it has been already proposed in the third article of this paper, and also in the works of other writers on subjects connected with the present, to adopt, for the *addition of straight lines having direction*, a rule expressed by the formula

plane parallel to the plane of the first transversal arc in four indefinite straight lines, which are respectively parallel to the four other radii A′O, B′O, C′O, D′O of the sphere; and consequently, in virtue of the equation (133), between the arcs which these last radii include, these four new lines in one common secant plane have the angular relation required for their being the (prolonged) sides of a (plane) quadrilateral inscribed in a circle; therefore the four edges of the same tetrahedral angle are cut by the same secant plane in points which are on the circumference of a circle; therefore they are edges or sides of a cone which has this circle for its base, and has its vertex at the centre of the sphere. But the intersection of such a cone with such a concentric sphere is called a *spherical conic*; a plane through its vertex, parellel to its circular base, is called a *cyclic plane*; and the intersection of this latter plane with the sphere has received the designation of a *cyclic arc*. Therefore the first transversal arc A′B′C″D′ is (as asserted in the text) a cyclic arc of a spherical conic circumscribed about the quadrilateral ABCD: and by a reasoning of exactly the same kind it may be proved, that the second transversal A″B″C″D″ is another cyclic arc of the same conic, or that its plane is a second cyclic plane, being parallel to the plane of another (or *subcontrary*) circular section.

$$\mathrm{CB} + \mathrm{BA} = \mathrm{CA}, \tag{7}$$

in whatever manner the three points ABC may be situated or related to each other; so it seems natural to adopt now, for the analogous *addition of arcs upon a sphere*, when directions as well as lengths are attended to, the corresponding formula,

$$\frown \mathrm{CB} + \frown \mathrm{BA} = \frown \mathrm{CA}. \tag{136}$$

Admitting this latter formula as *the definition of the effect of the sign + when inserted between two such symbols of arcs*, and granting also that it is permitted, in any such formula, to substitute for any arcual symbol another which is *equal* thereto, we shall have, by the two first and two last equations (134) respectively, the two following other equations,

$$\left.\begin{array}{l} \frown \mathrm{B''C''} = \frown \mathrm{AA'} + \frown \mathrm{B'C} \\ \frown \mathrm{A''D''} = \frown \mathrm{AD'} + \frown \mathrm{C'C} \end{array}\right\} \tag{137}$$

The two sums in these second members will therefore be symbolically equal if we have the equation

$$\frown \mathrm{A'D'} = \frown \mathrm{B'C'}, \tag{138}$$

because (135) has been seen to follow from (133) and (134). But by (136) and (138), we have the expression

$$\frown \mathrm{B'C} = \frown \mathrm{A'D'} + \frown \mathrm{C'C} \tag{139}$$

consequently the associative principle of multiplication, considered in several recent articles, when combined with the *formula of arcual addition* (136), conducts to the following formula,

$$\frown \mathrm{AA'} + (\frown \mathrm{A'D'} + \frown \mathrm{C'C}) = (\frown \mathrm{AA'} + \frown \mathrm{A'D'}) + \frown \mathrm{C'C}, \tag{140}$$

or, as it may be more concisely written,

$$\frown''' + (\frown'' + \frown') = (\frown''' + \frown'') + \frown': \tag{141}$$

which in its form agrees with ordinary algebra, and may be said to express the *associative principle of the symbolical addition of arcs*; since the three arcs added in (140) or (141) may be any three arcs of great circles upon one common spheric surface. It is remarkable that so much geometrical meaning should be contained in so simple and elementary a form; for this form (141), which is *apparently an algebraic truism*, and has been here deduced from the associative principle of multiplication of geometrical fractions, may reciprocally be substituted for it, and therefore includes in its interpretation, *if we adopt the symbolical definition* (136) of the effect of + between two symbols of arcs, all those theorems respecting spherical great circles, triangles, quadrilaterals, hexagons, and conics, which have been deduced or mentioned as geometrical results of the associative principle in the two foregoing articles. And this encouragment to adopt the foregoing very simple defintion (136) of the meaning of a symbol such as $\frown'' + \frown'$, is the more worthy of attention, because the *same definition* conducts to a *departure from the ordinary rules of symbolical addition* in *another* important point; since, when combined with the *definition of symbolical equality between arcs* assigned in the 17th article, it shows that *addition of arcs is in general a non-commutative operation.* For if we conceive two arcs of different great circles on one sphere, from A to B and from C to D, to bisect each other in a point E, we shall then have the two symbolical equations

$$\frown AE = \frown EB, \qquad \frown CE = \frown ED; \tag{142}$$

and therefore, whereas by (136),

$$\frown AE + \frown ED = \frown AD, \tag{143}$$

the result of the addition of the same two arcs, taken in a different order, will be

$$\frown ED + \frown AE = \frown CB. \tag{144}$$

And although the two *sum-arcs*, $\frown$ AD and $\frown$ CB, thus obtained, connecting two opposite pairs of extremities of the two commedial arcs $\frown$ AB and $\frown$ CD, are *equally long*, yet they are in general *parts of different great circles*, and therefore *not symbolically equal* in the sense of the 17th article. This result, which may at first sight seem a paradox, illustrates and is intimately connected with the analogous result obtained in the 13th article, respecting the general non-commutativeness of geometrical multiplication; for we shall find that there exists a species of *logarithmic connexion* between arcs situated in different great circles on a sphere and fractional factors belonging to different planes, which is analogous to, and includes as a limiting case, the known connexion between ordinary imaginary logarithms and angles in a single plane. It may be here remarked, that with the same definition (136) *in any symbolical addition of three successive arcs, the two partial sum-arcs,*

$$\frown'' + \frown' \quad \text{and} \quad \frown''' + \frown'', \tag{145}$$

are portions of the cyclic arcs of a certain spherical conic, circumscribed about a quadrilateral which has

$$\frown', \quad \frown'', \quad \frown''', \quad \text{and} \quad \frown''' + \frown'' + \frown', \tag{146}$$

that is, *the three proposed summand-arcs and their total sum-arc, for portions of its four sides*, or of those sides prolonged; as will appear by supposing that the three summands, $\frown'$, $\frown''$, $\frown'''$, coincide respectively with the arcs $\frown$ CC′, $\frown$ B′C, $\frown$ AA′, in the notation of the preceding article.

Symbolical Expressions for a Cyclic Cone; Relations of such a Cone, and of its Cyclic Planes, to a Product of two Geometrical Fractions

20. It is evidently a determinate* problem to construct a *cyclic cone*, that is, a cone with circular base (called usually a cone of the second degree), when three of the *sides* (or generating straight lines) of the cone are given in position, and when the plane of the base is parallel to a given *cyclic plane*, which passes through the vertex. To treat this problem, which may be regarded as a fundamental one in the theory of such cones, by a method derived from the principles of foregoing articles, let the three given sides be denoted by the letters a, b, c;

* The evident and known determinateness of this problem, corresponding to that of the elementary problem of circumscribing a circle about a given plane triangle, was tacitly assumed, but might with advantage have been expressly referred to, in the outline of a demonstration which was given in the note to Art. (18). The reasoning, towards the end of that note, would then stand thus:– If D be any fourth point on the determined spherical conic, which passes through the three points A, B, C, and has the arc A′B′ for a cyclic arc, it is also a fourth point on the determined spherical conic which passes through the same three points and has the arc B″C″ for a cyclic arc; therefore the two conics, determined by these two sets of conditions, coincide one with the other: or, in other words, the arc B″C″ is a *second* cyclic arc of the *same* spherical conic, of which the arc A′B′ is a *first* cyclic arc.

and let the two known lines, in which the given cyclic plane is cut by the planes of the two pairs, ab and bc, be denoted by a′ and b′; also let d denote any fourth side of the sought cyclic cone, and c′, d′ the lines of intersection of the given cyclic plane with the variable planes of cd and da; then, if suitable lengths be assigned to these straight lines, of which the relative *directions* in space are the chief object of the present investigation, the following equality between two products of certain geometrical fractions will exist, and may be regarded as a form of the *equation of the cone*:

$$\frac{\mathrm{c}}{\mathrm{b}'}\frac{\mathrm{a}'}{\mathrm{a}} = \frac{\mathrm{c}}{\mathrm{c}'}\frac{\mathrm{d}'}{\mathrm{a}}. \tag{147}$$

That is to say, when this equation is satisfied, the two lines which are the respective intersections of the planes of the fractional factors of these two equal products, namely the intersection b of the planes aa′ and b′c, and the intersection d of the planes ad′ and c′c, are two sides of a cyclic cone, which has for two other sides the lines a and c, and which has for one cyclic plane the common plane of the four lines a′, b′, c′, and d′; these eight lines, a, b, c, d, a′, b′, c′, d′, being here supposed to diverge from one common origin, namely the vertex (or centre) of the cone. This may easily be shown to be a consequence of what has been already established, respecting the connexion of the cyclic arcs of a spherical conic with the symbolic sums of certain other arcs. Or, without introducing any sphere, we may observe that, by (121) and its converse, the equation (147) may be abridged to the following:

$$\frac{\mathrm{a}'}{\mathrm{b}'} = \frac{\mathrm{d}'}{\mathrm{c}'}; \quad \text{or,} \quad \frac{\mathrm{a}'}{\mathrm{b}'}\frac{\mathrm{c}'}{\mathrm{d}'} = 1; \tag{148}$$

which shows, in virtue of the notation here employed, that besides a certain proportionality of lengths, not necessary now to be considered, there exists an equality between the angles of rotation, in one common plane, which would transport the lines b′ and c′, respectively, into the directions of a′ and d′. But the four lines a′, b′, c′, d′ are respectively parallel to the four symbolic differences, b − a, c − b, d − c, a − d, or to the four straight lines BA, CB, DC, AD, that is to the successive sides of the plane quadrilateral ABCD, if we now suppose the lines a, b, c, d to terminate, in the points A, B, C, D, on a transversal plane parallel to the plane of a′ b′ c′ d′. We may therefore present the relation (148) under either of the two forms:

$$\frac{\mathrm{b}-\mathrm{a}}{\mathrm{c}-\mathrm{b}}\frac{\mathrm{d}-\mathrm{c}}{\mathrm{a}-\mathrm{d}} = x; \quad \text{or} \quad \frac{\mathrm{BA}}{\mathrm{CB}}\frac{\mathrm{DC}}{\mathrm{AD}} = x; \tag{149}$$

in which x is a positive or negative scalar; or, using the characteristic V of the operation of taking the vector part, we may write;

$$\mathrm{V}.\frac{\mathrm{b}-\mathrm{a}}{\mathrm{c}-\mathrm{b}}\frac{\mathrm{d}-\mathrm{c}}{\mathrm{a}-\mathrm{d}} = 0; \quad \text{or } \mathrm{V}.\frac{\mathrm{BA}}{\mathrm{CB}}\frac{\mathrm{DC}}{\mathrm{AD}} = 0. \tag{150}$$

When the scalar x is positive, then, by considering the two rotations above mentioned, we easily perceive that the two points B and D are at one common side of the straight line AC, and that this line subtends equal angles at those two points; being in one common plane with them, as indeed the second equation (149) sufficiently expresses, since it gives

$$\mathrm{V}\frac{\mathrm{BA}}{\mathrm{CB}} = x\mathrm{V}\frac{\mathrm{DA}}{\mathrm{CD}}; \tag{151}$$

so that the two triangles ABC, ADC, on the common base AC, have one common perpendicular to their planes, which must therefore coincide with each other. In the contrary case, namely when x is negative, the equation (151) still shows that the four points are (as above) coplanar with each other; and while the points B and D are now at opposite sides of the line AC, the angles which this line subtends at those two points are now not equal but supplementary. In each case, therefore, the four points ABCD are on the circumference of one common circle; the four lines a, b, c, d are consequently sides of a cyclic cone; and the plane of the four other lines a′, b′, c′, d′ is a cyclic plane of that cone.

21. In the foregoing article, the coplanarity of each of the four sets of three lines, a′ab, b′bc, c′cd, d′da, allows us to suppose that four other lines b″, c″, d″, a″, in the same four planes respectively, and all, like the eight former lines, diverging from the vertex of the cone, are determined so as to satisfy the four equations:

$$\frac{\mathrm{b}''}{\mathrm{b}} = \frac{\mathrm{a}}{\mathrm{a}'}; \quad \frac{\mathrm{c}''}{\mathrm{c}} = \frac{\mathrm{b}}{\mathrm{b}'}; \quad \frac{\mathrm{d}''}{\mathrm{d}} = \frac{\mathrm{c}}{\mathrm{c}'}; \quad \frac{\mathrm{a}''}{\mathrm{a}} = \frac{\mathrm{d}}{\mathrm{d}'}; \tag{152}$$

and then, since these equations, combined with (148), give, by the associative property of the multiplication of geometrical fractions, this other equation,

$$\frac{\mathrm{b}''}{\mathrm{c}''} = \frac{\mathrm{a}''}{\mathrm{d}''}, \tag{153}$$

it follows that these four new lines are in one common plane; and also that the rotations in that plane, from b″ and c″ to a″ and d″, respectively, are equal. And this new plane is evidently a *second** *cyclic plane of the same cone*; for we may now write, instead of (147), the analogous equation:

$$\frac{\mathrm{c}}{\mathrm{c}''}\frac{\mathrm{b}''}{\mathrm{a}} = \frac{\mathrm{c}}{\mathrm{d}''}\frac{\mathrm{a}''}{\mathrm{a}}; \tag{154}$$

the two members being here equal respectively to the reciprocals of the two members of the first equation (148): nor is it necessary to retain the restriction that the lines a, b, c, d should terminate in one common plane. In like manner, the two members of the equation (147) are respectively equal to the reciprocals of the two members of the equation (153); a geometrical (like an arithmetical) fraction being said to be changed to its *reciprocal*, when the numerator and denominator are interchanged. We have therefore this theorem :– *A cyclic cone is the locus of the intersection of the planes of two geometrical fractions, of which the product is a constant fraction, while the numerator of the multiplier and the denominator of the multiplicand are constant lines. These two lines are two fixed sides of the cone; the plane of the two other and variable lines*, which enter as denominator and numerator into the expressions of the same two fractional factors, *is one cyclic plane of that cone*; and *the plane of the constant product is the other cyclic plane*. The investigation in the last article shows also that the condition for four points ABCD being *concircular* or *homocyclic*, that is, for their being corners of a quadrilateral inscribed in a circle, is expressed by the second equation (150); which may therefore be called the *equation of homocyclicism*. The same investigation shows that if we only know that ABCD are four points on

* See the remarks made in the note to the foregoing article.

one common plane, we may still write an equation of the form (151); which may for that reason be said to be a *formula of coplanarity*.

Symbolical Expressions and Investigations of some Properties of Cyclic Cones, with reference to their Tangent Planes

22. If the side b of the cyclic cone be conceived to approach to the side a, and ultimately to coincide with it, the first equation (152) will take this limiting form:

$$\frac{\mathrm{b}''}{\mathrm{a}} = \frac{\mathrm{a}}{\mathrm{a}'}; \tag{155}$$

which expresses the known theorem that the side of contact a bisects the angle between the traces a′ and b″ of the tangent plane on the two cyclic planes; bisecting also the vertically opposite angle between the traces −a′ and −b″, but being pependicular to the bisector of either of the two other angles, which are supplementary to the two already mentioned, namely the angle between the traces a′ and −b″, and that between −a′ and b″. And if in like manner we conceive the side d to approach indefinitely to the side c, the plane of these two sides will tend to become another tangent plane to the cone; of which plane the traces c′ and d″ on the two cyclic planes will satisfy an equation of the same form as that last written, namely the following, which is the limiting form of the third equation (152):

$$\frac{\mathrm{d}''}{\mathrm{c}} = \frac{\mathrm{c}}{\mathrm{c}'}. \tag{156}$$

At the same time, the two secant planes bc and da will tend to coalesce in one secant plane, containing the two sides of contact a and c, with which the two other sides b and d tend to coincide; so that the traces d′ and a″ of the latter secant plane, on the two cyclic planes, will ultimately coincide with the traces b′ and c″ of the former secant plane on the same two cyclic planes; and the equations (148) (153) become:

$$\frac{\mathrm{a}'}{\mathrm{b}'} = \frac{\mathrm{b}'}{\mathrm{c}'}; \quad \frac{\mathrm{b}''}{\mathrm{c}''} = \frac{\mathrm{c}''}{\mathrm{d}''}; \tag{157}$$

which express that the traces b′ and c″ of the one remaining secant plane bisect respectively the angles between the pairs of traces, a′, c′, and b″, d″, of the two tangent planes, on the two cyclic planes. And the two remaining equations (152) concur in giving this other equation:

$$\frac{\mathrm{c}''}{\mathrm{c}} = \frac{\mathrm{a}}{\mathrm{b}'}: \tag{158}$$

expressing that the rotations in the secant plane from b′ to a and from c to c″, that is to say from one trace to one side, and from the other side to the other trace, are equal in amount, and similarly directed; in such a manner that these two traces b′ and c″, of the secant plane on the two cyclic planes, are equally inclined to the straight line which bisects the angle between these two sides a and c, along which the plane cuts the cone: all which agrees with the known properties of cones of the second degree.

23. The eight straight lines a, c, a′, b′, c′, b″, c″, d″, being supposed to be equally long, the first of them, which has been seen to coincide in direction with the bisector of the angle between the third and sixth, can differ only by a scalar (or real and numerical) coefficient from their symbolic sum; because the diagonals of a plane and equilateral quadrilateral figure (or rhombus) bisect the angles of that figure. We have therefore, by (155), and by the supposition of the equal lengths of the eight lines,

$$\mathrm{a}' + \mathrm{b}'' \parallel \mathrm{a}; \quad \text{or,} \quad \mathrm{a}' + \mathrm{b}'' = l\mathrm{a}, \tag{159}$$

l being a numerical coefficient, and the sign of parallelism being designed to include the case of coincidence.

In like manner, by (156), we have

$$\mathrm{d}'' + \mathrm{c}' \parallel \mathrm{c}; \quad \text{or,} \quad \mathrm{d}'' + \mathrm{c}' = l'\mathrm{c}, \tag{160}$$

l' being another scalar coefficient. Again, by (157),

$$\left.\begin{array}{ll} \mathrm{c}' + \mathrm{a}' \parallel \mathrm{b}'; & \mathrm{c}' + \mathrm{a}' = m\mathrm{b}'; \\ \mathrm{b}'' + \mathrm{d}'' \parallel \mathrm{c}''; & \mathrm{b}'' + \mathrm{d}'' = m'\mathrm{c}''; \end{array}\right\} \tag{161}$$

m and m' being two other scalars. But, by (158),

$$\frac{\mathrm{c}''}{\mathrm{c}}\frac{\mathrm{b}'}{\mathrm{a}} = 1; \tag{162}$$

therefore

$$\frac{\mathrm{b}'' + \mathrm{d}''}{\mathrm{d}'' + \mathrm{c}'}\frac{\mathrm{c}' + \mathrm{a}'}{\mathrm{a}' + \mathrm{b}''} = \frac{m'}{l'}\frac{m}{l} = V^{-1}0; \tag{163}$$

this symbol $V^{-1}0$ denoting generally, in the present system, *any geometrical fraction of which the vector part is zero,* and therefore any positive or negative number (including zero). (Compare the definition and remarks in the 7th article).

By comparing this equation (163) with the first form (150), we see that the four straight lines,

$$-\mathrm{b}'', \quad \mathrm{d}'', \quad -\mathrm{c}', \quad \mathrm{a}', \tag{164}$$

which have been supposed to diverge from one common origin, namely the vertex of the cone, have their terminations on the circumference of one common circle. But these four lines, by supposition, are also equally long; they must therefore be four sides of a new cone, which is not only cyclic, as having a circular base, but is also a *cone of revolution.* The axis of revolution of this new cone is perpendicular to the plane of the circle in which the four lines (164) terminate; and this plane is parallel to the plane of the symbolic differences of those four lines, namely, the following,

$$\mathrm{d}'' + \mathrm{b}'', \quad -\mathrm{c}' - \mathrm{d}'', \quad \mathrm{a}' + \mathrm{c}', \quad -\mathrm{b}'' - \mathrm{a}'; \tag{165}$$

but these have been seen to be parallel respectively to the four lines c″, c, b′, a, which are contained in the secant plane of the former cone; consequently the axis of revolution of the new cone is perpendicular to this secant plane. We arrive therefore, by this symbolical process, at a new proof of the known theorem, discovered by M. Chasles,* that two planes,

* See the Translation of Two Geometrical Memoirs by M. Chasles, on the General Properties of Cones of the Second Degree, and on the Spherical Conics; which Translation was published, with an Appendix, by the Rev. Charles Graves, in Dublin, 1841.

touching a cyclic cone along any two sides, intersect the two cyclic planes in four right lines, which are sides of one common cone of revolution, whose axis of revolution is perpendicular to the plane of the two sides of contact of the former cone.

24. If we conceive the first and fourth of the sides (164) of the cone of revolution to tend to coincide with each other, then the fourth of the sides (165) of the plane quadrilateral inscribed in the circular base of that cone will tend to vanish; consequently the direction of this last mentioned side $-b'' - a'$, or the opposite direction of $a' + b''$, will become at last tangential to this circular base; and the plane of the two sides previously mentioned, namely $-b''$ and a', which plane has been seen to touch the cyclic cone along the side a, will become ultimately tangential also to the cone of revolution, touching it along the line a', which becomes one trace of the second cyclic plane on the first cyclic plane; the opposite line, $-a'$, being of course also situated in the intersection of those two planes, so that it may be regarded as the opposite trace of one cyclic plane on the other. Thus, at the limit here considered, the equation (155) and the second equation (157) are replaced by the equations

$$\frac{-a'}{a} = \frac{a}{a'}, \quad \frac{-a'}{c''} = \frac{c''}{d''}; \tag{166}$$

of which the first expresses that the side a is equally inclined to the two opposite traces, a' and $-a'$; while the numerical coefficient l vanishes, and the formula (163) is replaced by this other,

$$V.\frac{d'' - a'}{d'' + c'}\frac{c' + a'}{a} = 0. \tag{167}$$

We see also that the two rectangular but equally long lines a, a', of which the former is a side of the cyclic cone, while the latter is part of the line of intersection of the two cyclic planes of that cone, are such that their plane is a common tangent to both the cyclic cone and the cone of revolution; which latter cone has also, as sides of the same sheet with a', the two other of the four lines (164), namely the lines $-c'$ and d''. Indeed, the formula (167) is sufficient to show, by comparison with the first formula (150), that if the three straight lines a', d'', $-c'$ be still supposed to diverge from one common origin, the circle passing through the three points in which they terminate is touched, at the termination of the line a', by a straight line parallel to the line a; and therefore that the cone of revolution, having these three equally long lines a', $-c'$, d'' for sides of one common sheet, is touched along the side a' by the plane which contains the two rectangular lines aa'; so that we may regard this formula (167) as containing the symbolical solution of the problem, to draw a tangent plane, along any proposed side, to the cone of revolution which passes through that side and through two other sides also given, and belonging to the same sheet as the former. Now if three such sides be connected by three planes, forming three faces of a triangular pyramid, inscribed in a single sheet of a cone of revolution, and having its vertex at the vertex of that cone, while the sheet is touched by a fourth plane along one edge of the pyramid, it follows from the most elementary principles of solid geometry, that the difference between the two exterior angles which the faces meeting at that edge make with the tangent plane to the cone is equal to the difference of the two interior angles which the same two faces make with the third face of the pyramid; the greater exterior angle being the one which is the more remote from the greater interior angle; as may be shown by conceiving three planes to pass through the three edges respectively, and through the axis of revolution of the cone. The same equality between the differences of these two pairs of angles between planes, will become still more evident if, without making use of any formula of spherical trigonometry, we consider a spherical triangle inscribed in a small circle on the sphere, which small circle is touched at one corner of the triangle by a great circle, while arcs are drawn to that and to the two other corners from a pole of the small circle; the only principles required being these: that the base angles of a spherical isosceles triangle are equal, and that the arcs from the pole of a small circle are all perpendicular to its perimeter. If then we denote by the symbol $\angle(\mathrm{a, b, c})$ the acute or right or obtuse dihedral or spherical angle, at the edge b, between the planes ab and bc, in such a manner as to write, generally,

$$\angle(\mathrm{a, b, c}) = \angle(\mathrm{c, b, a}) = \angle(-\mathrm{a, b, -c}) = \angle(\mathrm{a, -b, c}) = \pi - \angle(\mathrm{a, b, -c}); \tag{168}$$

π being the symbol for two right angles, we shall have, in the present question, the equation

$$\angle(\mathrm{a', d'', -c'}) - \angle(\mathrm{a', -c', d''}) = \angle(-\mathrm{a, a', -c'}) - \angle(\mathrm{a, a', d''}); \tag{169}$$

and therefore, by subtracting both members from π,

$$\angle(\mathrm{a', d'', c'}) + \angle(\mathrm{a', c', d''}) = \angle(-\mathrm{a, a', c'}) + \angle(\mathrm{a, a', d''}). \tag{170}$$

We have also here the relation

$$\angle(\mathrm{c', a', d''}) = \angle(\mathrm{a, a', c'}) + \angle(\mathrm{a, a', d''}); \tag{171}$$

because the plane aa′ is intermediate between the planes a′c′ and a′d″, or lies *within* the dihedral angle (c′, a′, d″) itself, and not within either of the two angles which are exterior and supplementary thereto; which again depends on the circumstance that both the cyclic planes are necessarily exterior to each sheet of the cyclic cone. Adding therefore the equations (170) and (171), member to member, and subtracting π on both sides of the result, we find for the *spherical excess* of the new triangular pyramid (a′, c′, d″), or for the excess of the sum of the mutual inclinations of its three faces a′c′, a′d″, c′d″, above two right angles, the expression:

$$\angle(\mathrm{a', d'', c'}) + \angle(\mathrm{a', c', d''}) + \angle(\mathrm{c', a', d''}) - \pi = 2\angle(\mathrm{a, a', d''}). \tag{172}$$

This spherical excess therefore remains unchanged, while the two lines c′, d″, move together on the two cyclic planes, in such a manner that their plane, always passing through the vertex of the cone, continues to touch that cyclic cone; a′ being still a line situated in the intersection of the two cyclic planes, and a being still a side of contact of the cone with a plane drawn through that intersection. And hence, or more immediately from the equation (170), the known property of a cyclic cone is proved anew, that the sum of the inclinations (suitably measured) of its variable tangent plane to its two fixed cyclic planes is constant.

Condition of Concircularity, resumed. New Equation of a Cyclic Cone

25. The equation (150) of *homocyclicism*, or of *concircularity*, which was assigned in the 20th article, and which expresses the condition requisite in order that four straight lines in space, a, b, c, d, diverging from one common point O, as from an origin, may terminate in four other points A, B, C, D, which shall all be contained on the circumference of one common circle, may also, by (149), be put under the form

$$\frac{\mathrm{b}-\mathrm{a}}{\mathrm{c}-\mathrm{b}} = x\frac{\mathrm{d}-\mathrm{a}}{\mathrm{c}-\mathrm{d}}, \tag{173}$$

where x is a scalar coefficient. It gives therefore the two following separate equations, one between scalars, and the other between vectors:

$$\mathrm{S}\frac{\mathrm{b}-\mathrm{a}}{\mathrm{c}-\mathrm{b}} = x\mathrm{S}\frac{\mathrm{d}-\mathrm{a}}{\mathrm{c}-\mathrm{d}}; \quad \mathrm{V}\frac{\mathrm{b}-\mathrm{a}}{\mathrm{c}-\mathrm{b}} = x\mathrm{V}\frac{\mathrm{d}-\mathrm{a}}{\mathrm{c}-\mathrm{d}}, \tag{174}$$

of which the latter is only another way of writing the equation (151). If then we agree to use, for conciseness, a new characteristic of operation, $\frac{\mathrm{V}}{\mathrm{S}}$, of which the effect on any geometrical fraction, to the symbol of which it is prefixed, shall be defined by the formula

$$\frac{\mathrm{V}}{\mathrm{S}}\cdot\frac{\mathrm{b}}{\mathrm{a}} = \mathrm{V}\frac{\mathrm{b}}{\mathrm{a}} \div \mathrm{S}\frac{\mathrm{b}}{\mathrm{a}}; \tag{175}$$

so that this new characteristic $\frac{\mathrm{V}}{\mathrm{S}}$, which (it must be observed) *is not a distributive symbol*, is to be considered as directing to *divide the vector by the scalar part* of the geometrical fraction on which it operates; we shall then have, as a consequence of (173), this other form of the *equation of concircularity*:

$$\frac{\mathrm{V}}{\mathrm{S}}\cdot\frac{\mathrm{b}-\mathrm{a}}{\mathrm{c}-\mathrm{b}} = \frac{\mathrm{V}}{\mathrm{S}}\cdot\frac{\mathrm{d}-\mathrm{a}}{\mathrm{c}-\mathrm{d}}. \tag{176}$$

Conversely we can return from this latter form (176) to the equation (173); for if we observe that, in the present system of symbolical geometry, *every geometrical fraction is equal to the sum of its own scalar and vector parts*, so that we may write generally (see article 7),

$$\mathrm{S}\frac{\mathrm{b}}{\mathrm{a}} + \mathrm{V}\frac{\mathrm{b}}{\mathrm{a}} = \mathrm{V}\frac{\mathrm{b}}{\mathrm{a}} + \mathrm{S}\frac{\mathrm{b}}{\mathrm{a}} = \frac{\mathrm{b}}{\mathrm{a}}, \tag{177}$$

or more concisely,

$$\mathrm{S} + \mathrm{V} = \mathrm{V} + \mathrm{S} = 1; \tag{178}$$

and, if we add the identity,

$$\mathrm{S}\frac{\mathrm{b}-\mathrm{a}}{\mathrm{c}-\mathrm{b}} \div \mathrm{S}\frac{\mathrm{b}-\mathrm{a}}{\mathrm{c}-\mathrm{b}} = \mathrm{S}\frac{\mathrm{d}-\mathrm{a}}{\mathrm{c}-\mathrm{d}} \div \mathrm{S}\frac{\mathrm{d}-\mathrm{a}}{\mathrm{c}-\mathrm{d}}, \tag{179}$$

of which each member is equal to unity, to the equation (176), attending to the definition (175) of the new characteristic lately introduced, we are conducted to this other formula,

$$\frac{\mathrm{b}-\mathrm{a}}{\mathrm{c}-\mathrm{b}} \div \mathrm{S}\frac{\mathrm{b}-\mathrm{a}}{\mathrm{c}-\mathrm{b}} = \frac{\mathrm{d}-\mathrm{a}}{\mathrm{c}-\mathrm{d}} \div \mathrm{S}\frac{\mathrm{d}-\mathrm{a}}{\mathrm{c}-\mathrm{d}}; \tag{180}$$

which allows us to write also

$$\frac{\mathrm{b}-\mathrm{a}}{\mathrm{c}-\mathrm{b}} \div \frac{\mathrm{d}-\mathrm{a}}{\mathrm{c}-\mathrm{d}} = \mathrm{S}\frac{\mathrm{b}-\mathrm{a}}{\mathrm{c}-\mathrm{b}} \div \mathrm{S}\frac{\mathrm{d}-\mathrm{a}}{\mathrm{c}-\mathrm{d}}, \tag{181}$$

where the second member, being the quotient of two scalars, is itself another scalar, which may be denoted by x; and thus the equation (173) may be obtained anew, as a consequence of the equation (176). We may therefore also deduce from the last-mentioned equation the following form,

$$\frac{\mathrm{c}-\mathrm{d}}{\mathrm{d}-\mathrm{a}} = x\frac{\mathrm{c}-\mathrm{b}}{\mathrm{b}-\mathrm{a}}; \tag{182}$$

and thence also, by a new elimination of the scalar coefficient x, performed in the same manner as before, may derive this other form,

$$\frac{\mathrm{V}}{\mathrm{S}}\cdot\frac{\mathrm{c}-\mathrm{d}}{\mathrm{d}-\mathrm{a}} = \frac{\mathrm{V}}{\mathrm{S}}\cdot\frac{\mathrm{c}-\mathrm{b}}{\mathrm{b}-\mathrm{a}}. \tag{183}$$

Indeed, the geometrical signification of the condition (176) shows easily that we may in any manner transpose, in that condition, the symbols a, b, c, d; since if, *before* such a transposition, those symbols denoted four diverging straight lines (not generally in one common plane), which terminate on the circumference of one common circle, then *after* this transposition they must still denote four such diverging lines. We may therefore interchange the symbols a and c, in the condition (176), which will thus become

$$\frac{\mathrm{V}}{\mathrm{S}}\cdot\frac{\mathrm{b}-\mathrm{c}}{\mathrm{a}-\mathrm{b}} = \frac{\mathrm{V}}{\mathrm{S}}\cdot\frac{\mathrm{d}-\mathrm{c}}{\mathrm{a}-\mathrm{d}}; \tag{184}$$

but also, as in ordinary algebra, we have here,

$$\frac{\mathrm{b}-\mathrm{c}}{\mathrm{a}-\mathrm{b}} = \frac{\mathrm{c}-\mathrm{b}}{\mathrm{b}-\mathrm{a}}; \quad \frac{\mathrm{d}-\mathrm{c}}{\mathrm{a}-\mathrm{d}} = \frac{\mathrm{c}-\mathrm{d}}{\mathrm{d}-\mathrm{a}}; \tag{185}$$

this equation (183) might therefore have been in this other way deduced from the equation (176), as another form of the same condition of concircularity: and it is obvious that several other forms of the same condition may be obtained in a similar way.

26. From the fundamental importance of the *circle* in geometry, it is easy to foresee that these various forms of the condition of concircularity must admit of a great number of geometrical applications, besides those which have already been given in some of the preceding articles of this essay on Symbolical Geometry. For example, we may derive in a new way a solution of the problem proposed at the beginning of the 20th article, by conceiving that the symbols a, b, c denote three given sides of a *cyclic cone*, extending from the vertex to some given plane which is parallel to that one of the two *cyclic planes* which in the problem is supposed to be given; for then the equation (183) may be employed to express that the variable line d is a fourth side of the same cyclic cone, drawn from the same vertex as an origin, and bounded by the same given plane, or terminating on the same circumference, or circular base of the cone, as the three given sides, a, b, c. Or we may change the symbol d to another symbol of the form xx, and may conceive that x denotes a variable side of the cone, still drawn as before from the vertex, but not now necessarily terminating on any one fixed plane, nor otherwise restricted as to its length; while x shall denote a scalar coefficient, or multiplier, so varying with the side or line x as to render the product-line xx a side of which the extremity is (like that of d) concircular with the given extremities of a, b, c; and we may express these conceptions and conditions by writing as the equation of the cone the following:

$$\frac{\mathrm{V}}{\mathrm{S}}.\frac{\mathrm{c}-x\mathrm{x}}{x\mathrm{x}-\mathrm{a}}=\beta; \tag{186}$$

where β is a given geometrical fraction of the *vector* class, namely, that vector which is determined by the equation

$$\frac{\mathrm{V}}{\mathrm{S}}.\frac{\mathrm{c}-\mathrm{b}}{\mathrm{b}-\mathrm{a}}=\beta. \tag{187}$$

The *index* $\mathrm{I}\beta$ of this vector β is such that

$$\mathrm{I}\beta \parallel \mathrm{I}\frac{\mathrm{c}-\mathrm{b}}{\mathrm{b}-\mathrm{a}}, \tag{188}$$

it is therefore (by the principles of articles 7 and 10) a line perpendicular to each of the two lines represented by the two symbolical differences c − b, b − a, and therefore also perpendicular to the line denoted by their symbolic sum, c − a; so that we may establish the three formulæ,

$$\mathrm{I}\beta \perp \mathrm{c}-\mathrm{b};\quad \mathrm{I}\beta \perp \mathrm{b}-\mathrm{a};\quad \mathrm{I}\beta \perp \mathrm{c}-\mathrm{a}, \tag{189}$$

and may say that $\mathrm{I}\beta$ is a line *perpendicular to the plane* in which the three lines a, b, c all terminate. This constant index $\mathrm{I}\beta$, connected with the equation (186) of the cyclic cone just now determined, as being the *index of the constant vector fraction* β, to which the first member of that equation is equal, is therefore perpendicular also to the given cyclic plane of the same cone, and may be regarded as a symbol for one of the two *cyclic normals* of that conical locus of the variable line x lately considered. In the particular case when the three given lines a, b, c are all *equally long*, so that the cyclic cone (186) becomes a *cone of revolution*, then the index $\mathrm{I}\beta$, which has been generally a symbol for a cyclic normal, becomes a symbol for the *axis of revolution* of the cone. Other forms of equations of such cyclic and other cones will offer themselves when the principles of the present system of symbolical geometry shall have been more completely unfolded; but the forms just given will be found to be sufficient, when combined with some of the equations assigned in previous articles, to conduct to the solution of some interesting geometrical problems: to which class it will perhaps be permitted to refer the general determination of the *curvature of a spherical conic*, or the construction of the cone of revolution which *osculates* along a given side to a given cyclic cone.

Curvature of a Spherical Conic, or of a Cyclic Cone

27. To treat this problem by a method which shall harmonise with the investigations of recent articles of this paper, let the symbols a′, c′, d″, be employed with the same significations as in article 24, so as to denote three equally long straight lines, of which a′ is a trace of one cyclic plane on the other, while c′ and d″ are the traces of a tangent plane on those two cyclic planes; and let c (still bisecting the angle between c′ and d″) be still the equally long side of contact of that tangent plane with the given cyclic cone. We shall then have, by (156), the symbolic analogy,

$$\mathrm{d}'' : \mathrm{c} : : \mathrm{c} : \mathrm{c}', \tag{190}$$

which, on account of the supposed equality of the lengths of the lines c, c′, d″, gives also the two following formulæ, of parallelism and perpendicularity,

$$\mathrm{d}'' + \mathrm{c}' \parallel \mathrm{c}; \quad \mathrm{d}'' - \mathrm{c}' \perp \mathrm{c}; \tag{191}$$

of which indeed the former has been given already, as the first of the two formulæ (160). Conceive next that through the side of contact c we draw two secant planes, cutting the same sheet of the cone again in two known sides, c_1, c_2, and having for their known traces on the first cyclic plane (which contains the trace c' of the tangent plane) the lines c'_1, c'_2, but for their traces on the second cyclic plane (or on that which contains d'') the lines d''_1, d''_2; these lines, $\mathrm{c}\mathrm{c}_1\mathrm{c}_2\mathrm{c}'_1\mathrm{c}'_2\mathrm{d}''_1\mathrm{d}''_2$, being supposed to be all equally long. We may then write (in virtue of what has been shown in former articles) at once the two new symbolic *analogies*,

$$\mathrm{d}''_1 : \mathrm{c}_1 :: \mathrm{c} : \mathrm{c}'_1; \quad \mathrm{d}''_2 : \mathrm{c}_2 :: \mathrm{c} : \mathrm{c}'_2; \tag{192}$$

the two new *parallelisms*,

$$\mathrm{d}''_1 + \mathrm{c}'_1 \parallel \mathrm{c}_1 + \mathrm{c}; \quad \mathrm{d}''_2 + \mathrm{c}'_2 \parallel \mathrm{c}_2 + \mathrm{c}; \tag{193}$$

and the two new *perpendicularities*,

$$\mathrm{d}''_1 - \mathrm{c}'_1 \perp \mathrm{c}_1 + \mathrm{c}; \quad \mathrm{d}''_2 - \mathrm{c}'_2 \perp \mathrm{c}_2 + \mathrm{c}: \tag{194}$$

we shall have also these two other formulæ of parallelism,

$$\mathrm{d}''_1 - \mathrm{c}'_1 \parallel \mathrm{c}_1 - \mathrm{c}; \quad \mathrm{d}''_2 - \mathrm{c}'_2 \parallel \mathrm{c}_2 - \mathrm{c}. \tag{195}$$

Now if we conceive a cone of revolution to contain upon one sheet the three equally long lines c, c_1, c_2, which are also (by the construction) three sides of one sheet of the given cyclic cone, we may (by the last article) represent a line in the direction of the *axis* of this cone of revolution by the symbol,

$$\mathrm{I}\frac{\mathrm{c}_2 - \mathrm{c}}{\mathrm{c} - \mathrm{c}_1}; \tag{196}$$

or by this other symbol, which denotes indeed a line having an opposite direction, but still one contained upon the indefinite axis of the same cone of revolution, if drawn from a point on that axis,

$$\mathrm{I}\frac{\mathrm{c}_2 - \mathrm{c}}{\mathrm{c}_1 - \mathrm{c}}. \tag{197}$$

On account of the parallelisms (195) we may substitute for the last symbol (197) this other of the same kind,

$$\mathrm{I}\frac{\mathrm{d}''_2 - \mathrm{c}'_2}{\mathrm{d}''_1 - \mathrm{c}'_1}; \tag{198}$$

which expression, when we add to it another, which is a symbol of a null line (because in general the index of a scalar vanishes), namely the following,

$$\mathrm{I}\frac{\mathrm{c}'_1 - \mathrm{d}''_1}{\mathrm{d}''_1 - \mathrm{c}'_1} = 0, \tag{199}$$

takes easily this other form,

$$\mathrm{I}\frac{\mathrm{d}''_2 - \mathrm{c}'_2}{\mathrm{d}''_1 - \mathrm{c}'_1} = \mathrm{I}\frac{\mathrm{d}''_2 - \mathrm{d}''_1}{\mathrm{d}''_1 - \mathrm{c}'_1} + \mathrm{I}\frac{\mathrm{c}'_1 - \mathrm{c}'_2}{\mathrm{d}''_1 - \mathrm{c}'_1}. \tag{200}$$

The sought axis of the cone of revolution through the sides $\mathrm{c}\mathrm{c}_1\mathrm{c}_2$ of the cyclic cone, or a line in the direction of this axis, is therefore thus given, by the expression (200), as the symbolic *sum* of two other lines; which two new lines, by comparison of their expressions with the form (188), are seen to be in the directions of the axes of revolution of two new or *auxiliary cones* of revolution; one of these auxiliary cones containing, upon a single sheet, the three lines

$$\mathrm{c}_1', \quad \mathrm{d}_1'', \quad \mathrm{d}_2'', \tag{201}$$

so that it may be briefly called the cone of revolution $\mathrm{c}_1', \mathrm{d}_1'', \mathrm{d}_2''$; while the other auxiliary cone of revolution, which may be called in like manner the cone $\mathrm{c}_2', \mathrm{c}_1', \mathrm{d}_1''$, contains on one sheet this other system of three straight lines,

$$\mathrm{c}_1', \quad \mathrm{d}_1'', \quad \mathrm{c}_2'. \tag{202}$$

The symbolic *difference* of the same two lines, namely, that of the lines denoted by the symbols

$$\mathrm{I}\frac{\mathrm{d}_2''-\mathrm{d}_1''}{\mathrm{d}_1''-\mathrm{c}_1'}, \quad \mathrm{I}\frac{\mathrm{c}_1'-\mathrm{c}_2'}{\mathrm{d}_1''-\mathrm{c}_1'}, \tag{203}$$

which lines are thus in the directions of the axes of these two new cones of revolution, may easily be expressed under the form

$$\mathrm{I}\frac{\frac{1}{2}(\mathrm{d}_2''+\mathrm{c}_2')-\frac{1}{2}(\mathrm{d}_1''+\mathrm{c}_1')}{\frac{1}{2}(\mathrm{d}_1''+\mathrm{c}_1')-\mathrm{c}_1'}; \tag{204}$$

it is therefore (by the same last article) a line perpendicular to the plane in which the three following straight lines terminate, if drawn from one common point, such as the common vertex of the four cones,

$$\mathrm{c}_1', \quad \tfrac{1}{2}(\mathrm{d}_1''+\mathrm{c}_1'), \quad \tfrac{1}{2}(\mathrm{d}_2''+\mathrm{c}_2'). \tag{205}$$

This plane contains also the termination of the line d_1'', if that line be still drawn from the same vertex; because, in general, whatever may be the value of the scalar x, the three straight lines denoted by the symbols

$$\mathrm{c}_1', \quad (1-x)\mathrm{d}_1''+x\mathrm{c}_1', \quad \mathrm{d}_1'', \tag{206}$$

all terminate on one straight line, if they be drawn from one common origin; and this last straight line is situated in the first secant plane, and connects the extremities of the two equally long lines $\mathrm{c}_1', \mathrm{d}_1''$, which are the traces of that secant plane on the two cyclic planes. The remaining line, $\frac{1}{2}(\mathrm{d}_2''+\mathrm{c}_2')$, of the system (205), if still drawn from the same vertex as before, bisects that other straight line, situated in the second secant plane, which connects the extremities of the two equally long traces $\mathrm{c}_2', \mathrm{d}_2''$, of that other secant plane on the same two cyclic planes. And these two connecting lines, thus situated respectively in the first and second secant planes do not generally intersect each other; because they cut the line of mutual intersection of those two secant planes, namely the side c of the given cyclic cone, in points which are in general situated at different distances from the vertex. It is therefore in general a determinate problem, to draw through the first of these two connecting lines a plane which shall bisect the second: and we see that the plane so drawn, being that in which the three lines (205) terminate, is perpendicular to the line (204), that is to the symbolic difference,

$$\mathrm{I}\frac{d''_2 - d''_1}{d''_1 - c'_1} - \mathrm{I}\frac{c'_1 - c'_2}{d''_1 - c'_1}, \tag{207}$$

of the two lines (203), of which the symbolic sum (200) has been seen to be a line in the direction of the axis (197) of the first cone of revolution considered in the present article; while the two lines (203), of which we have thus taken the symbolic sum and difference, have been perceived to be in the directions of the axes of the two other and auxiliary cones of revolution, which we have also had occasion to consider. But in general, by one of those fundamental principles which the present system of symbolical geometry has in *common* with other systems, the symbolical sum and difference of two adjacent and coinitial sides of a parallelogram may be represented or constructed geometrically by the two diagonals of that figure; namely the sum by that diagonal which is intermediate between the two sides, and the difference by that other diagonal which is transversal to those sides: and every other transversal straight line, which is drawn across the same two sides in the same direction as the second diagonal, is bisected by the first diagonal, because the two diagonals themselves bisect each other. We may therefore enunciate this theorem:–*If across the axes* (203) *of the two auxiliary cones of revolution, which contain respectively the two systems of straight lines* (201) *and* (202), (each system of three straight lines being contained upon a single sheet), *we draw a rectilinear transversal, perpendicular to the plane which contains the first and bisects the second of the two connecting lines, drawn as before in the two secant planes; and if we then bisect this transversal by a straight line drawn from the common vertex of the cones: this bisecting line will be situated on the axis of revolution* (197) *of that other cone of revolution, which contains upon one sheet the three given sides of the given cyclic cone.* (The drawing of this transversal is possible, because the preceding investigation shows that the plane of the axes of revolution of the two auxiliary cones is perpendicular to that other plane which is described in the construction.)

28. Since, generally, in the present system of symbolical geometry, the vector part of the quotient of any two parallel lines, and the scalar part of the quotient of any two perpendicular lines, are respectively equal to zero, we may express that *three* straight lines, a, b, c, if drawn from a common origin, all *terminate on one common straight line*, by writing the equation

$$\mathrm{V}\frac{c - a}{b - a} = 0; \tag{208}$$

and may express that *two* straight lines, a, c, are *equally long*, or that they are fit to be made adjacent sides of a rhombus (of which the two diagonals are mutually rectangular), by this other formula:

$$\mathrm{S}\frac{c + a}{c - a}. \tag{209}$$

If then we combine these two conditions, which will give

$$\mathrm{S}\frac{c + a}{b - a} = 0, \tag{210}$$

and therefore

$$\mathrm{S}\frac{\mathrm{c}}{\mathrm{b}-\mathrm{a}} = -\mathrm{S}\frac{\mathrm{a}}{\mathrm{b}-\mathrm{a}}, \quad \mathrm{V}\frac{\mathrm{c}}{\mathrm{b}-\mathrm{a}} = \mathrm{V}\frac{\mathrm{a}}{\mathrm{b}-\mathrm{a}}, \tag{211}$$

we shall thereby express that the chord or secant of a circle or sphere, which passes through the extremity of one given radius a, and also through the extremity of another given and coinitial straight line b, meets the circumference of the same circle or the surface of the same sphere again at the extremity of the other straight line denoted by c, which will thus be another radius. But with the same mode of abridgment as that employed in the formula (178), we have, by (211),

$$(\mathrm{V}+\mathrm{S})\frac{\mathrm{c}}{\mathrm{b}-\mathrm{a}} = (\mathrm{V}-\mathrm{S})\frac{\mathrm{a}}{\mathrm{b}-\mathrm{a}}, \tag{212}$$

and therefore

$$\mathrm{c} = (\mathrm{V}-\mathrm{S})\frac{\mathrm{a}}{\mathrm{b}-\mathrm{a}}.(\mathrm{b}-\mathrm{a}). \tag{213}$$

This last is consequently an expression for the second radius c, in terms of the first radius a, and of the other given line b from the same centre, which terminates at some given point upon the common chord or secant, connecting the extremities of the two radii. If therefore we write for abridgment

$$\mathrm{m} = \tfrac{1}{2}(\mathrm{d}_2'' + \mathrm{c}_2'), \tag{214}$$

so that m shall be a symbol for the last of the three lines (205); and if we employ the two following expressions, formed on the plan (213),

$$\left.\begin{aligned} \mathrm{m}' &= (\mathrm{V}-\mathrm{S})\frac{\mathrm{c}_1'}{\mathrm{m}-\mathrm{c}_1'}.(\mathrm{m}-\mathrm{c}_1') \\ \mathrm{m}' &= (\mathrm{V}-\mathrm{S})\frac{\mathrm{d}_1''}{\mathrm{m}-\mathrm{d}_1''}.(\mathrm{m}-\mathrm{d}_1'') \end{aligned}\right\} \tag{215}$$

the symbols c_1', d_1'' retaining their recent meanings; then the four straight lines,

$$\mathrm{c}_1', \quad \mathrm{d}_1'', \quad \mathrm{m}', \quad \mathrm{m}'', \tag{216}$$

all drawn from the given vertex of the cones, will be equally long, and will terminate in four concircular points; or, in other words, their extremities will be the four corners of a certain quadrilateral inscribed in a circle: of which plane quadrilateral the two diagonals, connecting respectively the ends of c_1', m′, and of d_1'', m″, will intersect each other at the extremity of the line m, which is drawn from the same vertex as before. It may also be observed respecting this line m, that in virtue of its definition (214), and of the second parallelism (193), it bisects the angle between the two equally long sides c, c_2, of the given cyclic cone. Thus *the four lines* (216) *are four sides of one common sheet of a new cone of revolution, of which the axis is perpendicular to the plane described in the construction of the foregoing article*; because these four equally long lines (216) terminate on the same plane as the three lines (205), that is on a plane perpendicular to the line (204) or (207), which latter line has thus the direction of the axis of revolution of the new auxiliary cone. It is usual to say that four diverging straight lines are *rays of an harmonic pencil*, or simply that they are *harmonicals*, when a rectilinear transversal, parallel to the fourth, and bounded by the first and third, is bisected by the second of these lines: so that, in general, any four diverging straight lines which can be represented by the four symbols

$$a, \quad a+b, \quad b, \quad a-b,$$

or by symbols which are obtained from these by giving them any scalar coefficients, have the *directions* of four such harmonicals. We are then entitled to assert that *the fourth harmonical to the axes of the three cones of revolution*

$$(c_1' d_1'' d_2''), \quad (c c_1 c_2), \quad (c_2' c_1' d_1''), \tag{217}$$

which three axes have been already seen to be all situated in one common plane, *is the axis of that new or fourth cone of revolution* $(c_1' d_1'' m' m'')$, *which contains on one sheet the four straight lines* (216). And if we regard the four last-mentioned lines as *edges of a tetrahedral angle*, inscribed in this new cone of revolution, we see that *the two diagonal planes* of this tetrahedral angle *intersect each other along a straight line* m, *which bisects the plane angle* (c, c_2) *between two of the edges of the trihedral angle* $(c c_1 c_2)$; which latter angle is at once inscribed in the given cyclic cone, and also in that cone of revolution which it was originally proposed to construct.

29. Conceive now that this original cone of revolution $(c c_1 c_2)$ comes to *touch* the given cyclic cone along the side c, as a consequence of a gradual and unlimited approach of the second secant plane $(c c_2)$, to coincidence with the given tangent plane $(c' c d'')$, which touches the given cone along that side; or in virtue of a gradual and indefinite tendency of the side c_2 to coincide with the given side c. The line m, bisecting always the angle between these two sides c, c_2, will thus itself also tend to coincide with c; and the diagonal planes of the tetrahedral angle $(c_1' d_1'' m' m'')$, which planes still intersect each other in m, will tend at the same time to contain the same given side. But that side c is (by the construction) a line in the plane of one face of that tetrahedral angle, namely in the plane of c_1' and d_1'', which was the first secant plane of the cyclic cone; consequently the tetrahedral angle itself, and its circumscribed cone of revolution, tend generally to flatten together into coincidence with this secant plane, as c_2 thus approaches to c: and the axis of the cone $(c_1' d_1'' m' m'')$ coincides ultimately with the normal to the first secant plane $(c_1' d_1'')$. At the same time the traces c_2' and d_2'', of the second secant plane on the two cyclic planes, tend to coincide with the traces c' and d'' of the given tangent plane thereupon. We have therefore this new theorem, which is however only a limiting form of that enunciated in article 27:– If through a given side (c) of a given cyclic cone, we draw a tangent plane $(c' c d'')$, and a secant plane $(c_1' c c_1 d_1'')$; and if we then describe three cones of revolution, the first of these three cones containing on one sheet the two traces (c_1', d_1'') of the secant plane, and one trace (d'') of the tangent plane; the second cone of revolution touching the cyclic cone along the side of contact (c), and cutting it along the side of section (c_1); and the third cone of revolution containing the same two traces (c_1', d_1'') of the secant plane, and the other trace (c') of the tangent plane: *the fourth harmonical to the axes of revolution of these three cones will be perpendicular to the secant plane.*

30. Finally, conceive that the remaining secant plane $c_1' d_1''$ tends likewise to coincide with the tangent plane $c' d''$; the cone of revolution which lately *touched* the given cyclic cone along the given side c, will now come to *osculate* to that cone along that side: and because a line in the direction of the mutual intersection of the two cyclic planes has been already denoted by a′, therefore the first and third of the three last-mentioned cones of revolution tend now to touch the planes a′d″ and a′c′, respectively, along the lines d″ and c′. The theorem of article 27, at the limit here considered, takes therefore this new form:– *If three cones of revolution be described, the first cone cutting the first cyclic plane* (a′c′) *along the first trace* (c′) *of a given tangent plane* (c′cd′) *to a given cyclic cone, and touching the second cyclic plane* (a′d″) *along the second trace* (d″) *of the same tangent plane; the second cone of revolution osculating to the same cyclic cone, along the given side of contact* (c); *and the third cone of revolution touching the first cyclic plane and cutting the second cyclic plane, along the same two traces as before: then the fourth harmonical to the axes of revolution of these three cones will be the normal to the plane* (c′d″) *which touches at once the given cyclic cone, and the sought osculating cone, along the side* (c) *of contact or of osculation.*

31. To deduce from this last theorem an *expression* for a line e in the direction of the axis of the osculating cone of revolution, by the processes of this symbolical geometry, we may remark in the first place, that when any two straight lines a, b, are equally long, we have the three equations following:

$$\mathrm{S}\frac{\mathrm{a}}{\mathrm{b}} = \mathrm{S}\frac{\mathrm{b}}{\mathrm{a}}, \quad \mathrm{V}\frac{\mathrm{a}}{\mathrm{b}} = -\mathrm{V}\frac{\mathrm{b}}{\mathrm{a}}, \quad \mathrm{I}\frac{\mathrm{a}}{\mathrm{b}} = -\mathrm{I}\frac{\mathrm{b}}{\mathrm{a}}, \tag{218}$$

from the two former of which it may be inferred that the relation

$$\frac{\mathrm{V}}{\mathrm{S}}\cdot\frac{\mathrm{a}}{\mathrm{b}} = -\frac{\mathrm{V}}{\mathrm{S}}\cdot\frac{\mathrm{b}}{\mathrm{a}} \tag{219}$$

holds good, not only when the two lines a, b are thus equal in length, but generally for any two lines: because if we multiply or divide either of them by any scalar coefficient, we only change thereby in one common (scalar) ratio both the scalar and vector parts of their quotient, and so do not affect that other quotient which is obtained by dividing the latter of these two parts by the former. We may also obtain the equation (219), as one which holds good for any two straight lines a, b, under the form

$$\mathrm{S}\frac{\mathrm{b}}{\mathrm{a}}\mathrm{V}\frac{\mathrm{a}}{\mathrm{b}} + \mathrm{V}\frac{\mathrm{b}}{\mathrm{a}}\mathrm{S}\frac{\mathrm{a}}{\mathrm{b}} = 0, \tag{220}$$

by operating with the characteristic V on the identity,

$$\mathrm{S}\frac{\mathrm{b}}{\mathrm{a}}\cdot\frac{\mathrm{a}}{\mathrm{b}} + \mathrm{V}\frac{\mathrm{b}}{\mathrm{a}}\cdot\frac{\mathrm{a}}{\mathrm{b}} = \frac{\mathrm{b}}{\mathrm{b}} = 1; \tag{221}$$

while, if we operate on the same identity (221) by the characteristic S, we obtain this other general formula, which likewise holds good for any two straight lines a, b, whether equal or unequal in length, and will be useful to us on future occasions,

$$\mathrm{S}\frac{\mathrm{b}}{\mathrm{a}}\mathrm{S}\frac{\mathrm{a}}{\mathrm{b}} + \mathrm{V}\frac{\mathrm{b}}{\mathrm{a}}\mathrm{V}\frac{\mathrm{a}}{\mathrm{b}} = 1. \tag{222}$$

Again, if there be three equally long lines, a, b, c, then since the principle contained in the third equation (218) gives

$$\mathrm{I}\frac{\mathrm{b}-\mathrm{a}}{\mathrm{c}} = \mathrm{I}\frac{\mathrm{b}}{\mathrm{c}} - \mathrm{I}\frac{\mathrm{a}}{\mathrm{c}} = \mathrm{I}\frac{\mathrm{c}}{\mathrm{a}} - \mathrm{I}\frac{\mathrm{c}}{\mathrm{b}}, \tag{223}$$

which last expression is only multiplied by a scalar when the line c is multiplied thereby; while the index of a geometrical fraction is (among other properties) a line perpendicular to both the numerator and denominator of the fraction; we see that the symbol $\mathrm{I}\frac{\mathrm{c}}{\mathrm{a}} - \mathrm{I}\frac{\mathrm{c}}{\mathrm{b}}$ denotes generally a line perpendicular to both c and b − a, if only the two lines a and b have their own lengths equal to each other, without any restriction being thereby laid on the length of c: this symbol denotes therefore, under this single condition, a straight line contained in a plane perpendicular to c, and having equal inclinations to a and b. Thus, under the same condition, the same symbol $\mathrm{I}\frac{\mathrm{c}}{\mathrm{a}} - \mathrm{I}\frac{\mathrm{c}}{\mathrm{b}}$ may represent the axis d of a cone of revolution, which contains upon one sheet the two equally long lines a and b, while the third line c is in or parallel to the *single* cyclic plane of this *monocyclic cone*, or the plane of its circular base, or of one of its circular sections; or coincides with or is parallel to some tangent to such circular base or section. If then we know any other line a′, contained in the plane which touches this monocyclic cone along the side a, we may substitute for c, in this symbol $\mathrm{I}\frac{\mathrm{c}}{\mathrm{a}} - \mathrm{I}\frac{\mathrm{c}}{\mathrm{b}}$, that part or component of this new line a′ which is perpendicular to the side of contact a; and therefore may write with this view,

$$\mathrm{c} = \mathrm{V}\frac{\mathrm{a}'}{\mathrm{a}}.\mathrm{a} = \mathrm{a}' - \mathrm{S}\frac{\mathrm{a}'}{\mathrm{a}}.\mathrm{a}, \tag{224}$$

which will give

$$\mathrm{d} = \mathrm{I}\frac{\mathrm{a}'}{\mathrm{a}} - \mathrm{I}\frac{\mathrm{a}'}{\mathrm{b}} + \mathrm{S}\frac{\mathrm{a}'}{\mathrm{a}}\mathrm{I}\frac{\mathrm{a}}{\mathrm{b}}, \tag{225}$$

as a general expression for a line d in the direction of the axis of a cone of revolution which is touched by the plane aa′ along the side of contact a, and contains on the same sheet the equally long side b. We may also remark that because the normal plane to a cone of revolution, drawn along any side of that cone, contains the axis of revolution, so that the plane containing the axis and the side is perpendicular to the tangent plane, we have a relation between the three directions of a, a′, d, which does not involve the direction of b, and may be expressed by any one of the three following formulæ:–

$$\angle(\mathrm{a}', \mathrm{a}, \mathrm{d}) = \frac{\pi}{2}, \quad \mathrm{d} \perp \mathrm{V}\frac{\mathrm{a}'}{\mathrm{a}}.\mathrm{a}, \quad \mathrm{S}\frac{\mathrm{a}'}{\mathrm{d}} = \mathrm{S}\frac{\mathrm{a}'}{\mathrm{a}}\mathrm{S}\frac{\mathrm{a}}{\mathrm{d}}; \tag{226}$$

in each of which it is allowed to reverse the direction of d, or to change d to −d. (Compare the formulæ (168), for the notation of dihedral angles.) It may indeed be easily proved, without the consideration of any cone, that any one of these three formulæ (226) involves the other two; but we see also by the recent reasoning, that they may all be deduced when an expression of the form (225) for d is given; or when this line d can be expressed in terms of a, a′, and of another line b which is supposed to have the same length as a, by any symbol which differs only from the form (225) through the introduction of a scalar coefficient.

These things being premised, if we change a, b, d, in this form (225), to c′, d″, n′, we find

$$\mathrm{n}' = \mathrm{I}\frac{\mathrm{a}'}{\mathrm{c}'} - \mathrm{I}\frac{\mathrm{a}'}{\mathrm{d}''} + \mathrm{S}\frac{\mathrm{a}'}{\mathrm{c}'}\mathrm{I}\frac{\mathrm{c}'}{\mathrm{d}''}, \tag{227}$$

as an expression for a line n' in the direction of the axis of revolution of the cone which touches the first cyclic plane $a'c'$ along the first trace c' of the tangent plane, and cuts the second cyclic plane $a'd''$ along the second trace d' of the same tangent plane; that is to say, in the direction of the axis of the third cone of revolution, described in the enunciation of the theorem of article 30. Again, if we change a, b, d, in the same general formula (225), to d'', c', $-n''$, and attend to the third equation (218), we find

$$n'' = I\frac{a'}{c'} - I\frac{a'}{d''} + S\frac{a'}{d''}I\frac{c'}{d''}, \tag{228}$$

as an expression for another line n'', in the direction of the axis of another cone of revolution, which cuts the first cyclic plane $a'c'$ along the trace c', and touches the second cyclic plane $a'd''$ along the other trace d'' of the tangent plane; that is, in the direction of the axis of revolution of the first of the three cones, described in the enunciation of the same theorem of article 30. And since these expressions give

$$n'' - n' = \left(S\frac{a'}{d''} - S\frac{a'}{c'}\right)I\frac{c'}{d''}, \tag{229}$$

we have the two perpendicularities

$$n'' - n' \perp c', \quad n'' - n' \perp d''; \tag{230}$$

so that a transversal drawn across the two axes of revolution last determined, in the direction of this symbolic difference $n'' - n'$, is perpendicular to both the traces of the tangent plane $c'd''$, and therefore has the direction of the normal to that plane, or to the cyclic cone; or, in other words, this transversal has the direction of the fourth harmonical mentioned in the theorem. But the lines n'' and n', of which the symbolic *difference* has thus been taken, have been seen to be in the directions of the first and third of the same four harmonicals; and the axis of the osculating cone, which axis we have denoted by e, has (by the theorem) the direction of the second harmonical: it has therefore the direction of the symbolical *sum* of the same two lines n'', n', because it bisects their transversal drawn as above. Thus by conceiving the bisector to terminate on the transversal, we find, as an expression for this sought axis e, the following,

$$e = \tfrac{1}{2}(n'' + n') = I\frac{a'}{c'} - I\frac{a'}{d''} + \tfrac{1}{2}\left(S\frac{a'}{c'} + S\frac{a'}{d''}\right)I\frac{c'}{d''}. \tag{231}$$

32. This symbolical expression for e contains, under a not very complex form, the solution of the problem on which we have been engaged; namely, *to find the axis of the cone of revolution, which osculates along a given side to a given cyclic cone.* It may however be a little simplified, and its geometrical interpretation made easier, by resolving the line a' into two others, which shall be respectively parallel and perpendicular to the *lateral normal plane*, as follows:

$$a' = a^{\backprime} + a^{\backprime\backprime}; \quad a^{\backprime} \perp d'' - c'; \quad a^{\backprime\backprime} \parallel d'' - c'; \tag{232}$$

so that

$$a^{\backprime} = V\frac{a'}{d'' - c'}.(d'' - c'); \quad a^{\backprime} = S\frac{a'}{d'' - c'}.(d'' - c'); \tag{233}$$

which will give, by (191) and (218), because $a^{\backprime\backprime} \perp d'' + c'$,

$$S\frac{d''}{a^{\backprime\backprime}} + S\frac{c'}{a^{\backprime\backprime}} = 0; \quad S\frac{a^{\backprime\backprime}}{c'} + S\frac{a^{\backprime\backprime}}{d''} = 0; \tag{234}$$

also

$$I\frac{d''}{a^{\backprime\backprime}} - I\frac{c'}{a^{\backprime\backprime}} = 0; \quad I\frac{a^{\backprime\backprime}}{c'} - I\frac{a^{\backprime\backprime}}{d''} = 0; \tag{235}$$

and

$$S\frac{d''}{a^{\backprime}} - S\frac{c'}{a^{\backprime}} = 0; \quad S\frac{a^{\backprime}}{c'} - S\frac{a^{\backprime}}{d''} = 0. \tag{236}$$

For by thus resolving a', in (231), into the two components $a^{\backprime}$ and $a^{\backprime\backprime}$, it is at once seen, by (234) (235), that the latter component $a^{\backprime\backprime}$ disappears from the result, which reduces itself by (236) to the following simplified form,

$$e = I\frac{a^{\backprime}}{c'} - I\frac{a^{\backprime}}{d''} + S\frac{a^{\backprime}}{c'}I\frac{c'}{d''}; \tag{237}$$

and this gives, by comparison with the forms (225) and (226), a remarkable relation of rectangularity between two planes, of which one contains the axis e of the osculating cone, namely the planes $a^{\backprime}c'$ and $c'e$; which relation is expressed by the formula,

$$\angle(a^{\backprime}, c', e) = \frac{\pi}{2}. \tag{238}$$

In like manner, from the same expression (231), by the same decomposition of a', we may easily deduce, instead of (237), this other expression for the axis of the osculating cone,

$$e = I\frac{a^{\backprime}}{c'} - I\frac{a^{\backprime}}{d''} - S\frac{a^{\backprime}}{d''}I\frac{d''}{c'}; \tag{239}$$

and may derive from it this other relation, of rectangularity between two other planes, namely the planes $a^{\backprime}d''$ and $d''e$,

$$\angle(a^{\backprime}, d'', e) = \frac{\pi}{2}. \tag{240}$$

Hence follows immediately this theorem, which furnishes a remarkably simple *construction with planes*, for determining generally a line in the required direction of the axis of the osculating cone:– *If we project the line* a' *of mutual intersection of the two cyclic planes* $a'c'$, $a'd''$, *of any given cyclic cone, on the lateral normal plane which is drawn along any given side* c; *if we next draw two planes,* $a^{\backprime}c'$, $a^{\backprime}d''$, *through the projection* $a^{\backprime}$ *thus obtained, and through the two traces,* c', d'', *of the tangent plane on the two cyclic planes; and if we then draw two new planes,* $c'e$, $d''e$, *through the same two traces of the tangent plane, perpendicular respectively to the two planes* $a^{\backprime}c'$, $a^{\backprime}d''$, *last drawn: these two new planes will intersect each other along the axis* e *of the cone of revolution, which osculates along the given side* c *to the given cyclic cone.*

And by considering, instead of these cones and planes, their intersections with a spheric surface described about the common vertex, we arrive at the following *spherographic* *construction,** for finding the *spherical centre of curvature of a given spherical conic* at a given point, or the pole of the small circle which osculates at that point to that conic:– *From one of the two points of mutual intersection of the two cyclic arcs let fall a perpendicular upon the normal arc to the conic, which latter arc is drawn through the given point of osculation; connect the foot of this (arcual) perpendicular by two other arcs of great circles, with those two known points, equidistant from the point upon the conic, where the tangent arc meets the two conic arcs; draw through the same two points two new arcs of great circles, perpendicular respectively to the two connecting arcs: these two new arcs will cross each other on the normal arc, in the pole of the osculating circle, or in the spherical centre of curvature of the spherical conic,* which centre it was required to determine.

On Elliptic Cones, and on their Osculating Cones of Revolution

33. With the same significations of a′, a‵, c, c′, d″, and e, as symbols of certain straight lines, connected with a given cyclic cone, as in the last article of this Essay; and with the same use of the sign I, as the characteristic of the *index* of the vector part of any geometrical fraction in general; if we now write

$$\mathrm{f} = \mathrm{I}\frac{\mathrm{a}'}{\mathrm{c}'}; \quad \mathrm{g} = \mathrm{I}\frac{\mathrm{d}''}{\mathrm{a}'}; \quad \mathrm{h} = \mathrm{I}\frac{\mathrm{d}''}{\mathrm{c}} = \mathrm{I}\frac{\mathrm{c}}{\mathrm{c}'}; \tag{241}$$

$$\mathrm{i} = \mathrm{I}\frac{\mathrm{a}^{\backprime}}{\mathrm{c}'}; \quad \mathrm{k} = \mathrm{I}\frac{\mathrm{d}''}{\mathrm{a}^{\backprime}}; \quad \mathrm{l} = \frac{\mathrm{h}}{\mathrm{c}}\mathrm{a}^{\backprime} \parallel \mathrm{I}\frac{\mathrm{a}'}{\mathrm{a}^{\backprime}}; \tag{242}$$

we shall thus form symbols for certain other straight lines, f, g, h, and i, k, l, which may be conceived to be all drawn from the same common origin as the former lines, namely from the vertex of the cyclic cone. And these new lines will be found to be connected with *another* cone, which may be called an *elliptic*§ *cone*; namely the cone which is *normal, supplementary*, or *reciprocal* to the former *cyclic* cone. They may also be employed to assist in the determination of the *cone of revolution*, which *osculates* along a given side to this new or elliptic cone; as will be seen by the following investigation.

* This construction was communicated to the Royal Irish Academy (see *Proceedings*)†, at its meeting of November 30th, 1847, along with a simple geometrical construction for generating a system of two reciprocal ellipsoids by means of a moving sphere, as new applications of the author's Calculus of Quaternions to Surfaces of the Second Order. With that Calculus, of which the fundamental principles and formulæ were communicated to the same Academy‡ on the 13th of November, 1843, it will be found that the present System of Symbolical Geometry is connected by very intimate relations, although the subject is approached, in the two methods, from two quite different points of view: the *algebraical quaternion* of the one method being *ultimately* the same as the *geometrical fraction* of the other.

† [See Vol. III, pp. 378–380.]

‡ [See Vol. III, pp. 111–116.]

§ The methods of the present Symbolical Geometry might here be employed to prove that the *normal cone*, here called *elliptic*, from its connexion with its two focal lines, is itself *another cyclic cone*; being cut in circles by two sets of planes, which are perpendicular respectively to the two focal lines of the former cone. But it may be sufficient thus to have alluded to this well-known theorem, which it is not necessary for our present purpose to employ. There is even a convenience in retaining, for awhile, the two contrasted designations of *cyclic* and *elliptic*, for these two reciprocal cones, to mark more strongly the difference of the modes in which they here present themselves to our view.

34. The lines f and g being, as is shown by their expressions (241), perpendicular respectively to the planes a′c′ and a′d″, which were the two cyclic planes of the former or cyclic cone, are themselves the two *cyclic normals* of that cone; and because the line h is, by the same system of expressions (241), perpendicular to the plane c′d″ which touches that cyclic cone along the side c, it is the variable normal of that former cone: or this new line h is the *side* of the new or normal cone, which *corresponds* to that old side c. The inclinations of h to f and g, respectively, are given by the following equations, which are consequences of the same expressions (241):

$$\left.\begin{aligned}\angle(\mathrm{f},\mathrm{h}) &= \angle(\mathrm{a}',\mathrm{c}',\mathrm{c}) = \angle(\mathrm{a}',\mathrm{c}',\mathrm{d}'')\\ \angle(\mathrm{h},\mathrm{g}) &= \angle(\mathrm{a}',\mathrm{d}'',\mathrm{c}) = \angle(\mathrm{a}',\mathrm{d}'',\mathrm{c}')\end{aligned}\right\} \tag{243}$$

and we have seen, in article 24, that for the cyclic cone an equation which may now be thus written holds good:

$$\angle(\mathrm{a}',\mathrm{c}',\mathrm{d}'') + \angle(\mathrm{a}',\mathrm{d}'',\mathrm{c}') = 2a; \tag{244}$$

where a is a constant angle: therefore for the cone of normals to that cyclic cone, the following other equation is satisfied:

$$\angle(\mathrm{f},\mathrm{h}) + \angle(\mathrm{h},\mathrm{g}) = 2a; \tag{245}$$

a being here the same constant as before. The sum of the inclinations of the variable side h of the new or *elliptic* cone to the two fixed lines f and g is therefore constant; in consequence of which known property, these two fixed lines are called the *focal lines* of the elliptic cone. And we see that these two *focal lines* f, g, of the *normal* cone, coincide, respectively, in their directions, with the two *cyclic normals* (or with the normals to the two cyclic planes) of the *original* cone: which is otherwise known to be true.

35. Another important and well-known property of the elliptic cone may be proved anew by observing that the expressions (241) give

$$\left.\begin{aligned}\angle(\mathrm{f},\mathrm{h},\mathrm{c}) &= \angle(\mathrm{f},\mathrm{h},\mathrm{c}') - \angle(\mathrm{c},\mathrm{h},\mathrm{c}') = \tfrac{1}{2}\pi - \angle(\mathrm{c},\mathrm{c}')\\ \angle(\mathrm{c},\mathrm{h},\mathrm{g}) &= \angle(\mathrm{d}'',\mathrm{h},\mathrm{g}) - \angle(\mathrm{d}'',\mathrm{h},\mathrm{c}) = \tfrac{1}{2}\pi - \angle(\mathrm{d}'',\mathrm{c})\end{aligned}\right\} \tag{246}$$

and that we have, by (190),

$$\angle(\mathrm{d}'',\mathrm{c}) = \angle(\mathrm{c},\mathrm{c}'); \tag{247}$$

for thus we see that

$$\angle(\mathrm{f},\mathrm{h},\mathrm{c}) = \angle(\mathrm{c},\mathrm{h},\mathrm{g}); \tag{248}$$

that is to say, the lateral normal plane hc to the reciprocal or elliptic cone (which is at the same time the lateral normal plane of the original or cyclic cone) bisects the dihedral angle $\angle(\mathrm{f},\mathrm{h},\mathrm{g})$, comprised between the two *vector planes*, fh, hg, which connect the side h of the elliptic cone with the two focal lines f and g.

Or because the expressions (241) show that these two vector planes, fh, hg, of the elliptic cone, are perpendicular respectively to the two *traces* c′ and d″ of the tangent plane to the cyclic cone, on the two cyclic planes of that cone; which traces are, as the formula (247) expresses, inclined equally to the side of contact c of the original or cyclic cone, while that side or line c is also the normal to the reciprocal or elliptic cone; we might hence infer that the tangent plane to the latter cone is equally inclined to the two vector planes: which is another form of the same known relation.

36. The expressions (242), combined with (241), show that the two new lines i and k, as being perpendicular respectively to the two traces c′ and d″, are contained respectively in the two vector planes, fh and hg. But each of the same two new lines, i, k, is also perpendicular to the line a‵, to which the remaining new line l is also perpendicular, as the same expressions show; they show too that a‵ is a line in the common lateral and normal plane ch of the two cones, while l is also contained in that plane: the plane ik therefore cuts the plane ch perpendicularly in the line l. This latter line l is also, by the same expressions, perpendicular to the line a‵ (that is to the intersection of the two cyclic planes of the cyclic cone), which is perpendicular to both f and g; and therefore l can be determined, as the intersection of the common normal plane ch with the plane of the two focal lines fg; after which, by drawing through the line l, thus found, a plane ik perpendicular to ch, the lines i and k may be obtained, as the respective intersections of this last perpendicular plane with the two vector planes, fh, hg. And we see that these three new lines, i, k, l, introduced by the expressions (242), are such as to satisfy the following conditions of dihedral perpendicularity:

$$\tfrac{1}{2}\pi = \angle(\mathrm{h, l, i}) = \angle(\mathrm{k, l, h}); \tag{249}$$

$$\tfrac{1}{2}\pi = \angle(\mathrm{h, i, c'}) = \angle(\mathrm{d'', k, h}); \tag{250}$$

$$\tfrac{1}{2}\pi = \angle(\mathrm{a`, c', i}) = \angle(\mathrm{a`, d'', k}); \tag{251}$$

with which we may combine the following relations:

$$\angle(\mathrm{f, h, i}) = \angle(\mathrm{k, h, g}) = 0; \quad \angle(\mathrm{f, l, g}) = \pi; \quad \angle(\mathrm{f, h, l}) = \angle(\mathrm{l, h, g}). \tag{252}$$

37. The positions of these three lines i, k, l, being thus fully known, by means of the expressions (242), or of the corollaries which have been deduced from those expressions, let us now consider, in connexion with them, the two formulæ of dihedral perpendicularity, (238), (240), which were given in article 32, to determine the axis e of a cone of revolution, which osculates along the side c to the given cyclic cone, and which formulæ may be thus collected:

$$\tfrac{1}{2}\pi = \angle(\mathrm{a`, c', e}) = \angle(\mathrm{a`, d'', e}). \tag{253}$$

The comparison of (253) with (251) shows that the planes c′e, d″e, must coincide respectively with the planes c′i, d″k; because they are drawn like them respectively through the lines c′, d″, and are like them perpendicular respectively to the planes a‵c′, a‵d″; the line e must therefore be the intersection of the two planes c′i, d″k, which contain respectively the two lines i, k, and are, by (250), perpendicular to the two planes ih, kh, or (by what has been seen in the last article) to the two vector planes fh, gh. We can therefore construct the line e as the intersection of the two planes ie, ke, which are thus drawn through the lately determined lines i, k, at right angles to the two vector planes; and we may write, instead of (253), the formulæ

$$\tfrac{1}{2}\pi = \angle(\mathrm{h, i, e}) = \angle(\mathrm{h, k, e}). \tag{254}$$

38. Again, because this line e is (by Art. 32) the axis of a cone of revolution which *osculates* to the given cyclic cone, or which touches that cone not only along the side c itself but also along another side infinitely near thereto; while, in general, the lateral normal planes of a cone of revolution all cross each other along the axis of that cone; it is clear that e must be the line along which the common lateral and normal plane ch of the two reciprocal cones is intersected by an infinitely near normal and lateral plane to the first or cyclic cone, which is also at the same time a lateral and normal plane to the second or elliptic cone; consequently *the two cones of revolution which osculate to these two reciprocal cones, along these two corresponding sides,* c and h, *have one common axis,* e. And it is evident that a similar result for a similar reason holds good, in the more general case of *any two reciprocal cones,* which have a common vertex, and of which each contains upon its surface all the normals to the other cone, *however arbitrary the form* of either cone may be; any two such cones having always *one common system of lateral and normal planes,* and *one common conical envelope of all those normal planes*: which common envelope is thus the *common conical surface of centres of curvature,* for the two reciprocal cones.

Eliminating therefore what belongs, in the present question, to the original or cyclic cone, or confining ourselves to the formulæ (245), (249), (252), (254), we are conducted to the following *construction, for determining the axis* e *of that new cone of revolution, which osculates along a given side* h *to a given elliptic cone*; this latter cone having f and g for its given focal lines, or being represented by an equation of the form (245):– *Draw, through the given side,* h, *the normal plane* hl, *bisecting the angle between the two vector planes,* fh, gh, *and meeting in the line* l *the plane* fg *of the two given focal lines; through the same line* l *draw another plane* ik, *perpendicular to the normal plane* hl, *and cutting the vector planes in two new lines,* i *and* k; *through these new lines draw two new planes,* ie, ke, *perpendicular respectively to the two vector planes,* fi, gk, *or* fh, gh: *these new planes will cross each other on the normal plane, in the sought axis* e *of the osculating cone of revolution.*

39. Or if we prefer to consider, instead of cones and planes, their intersections with a spheric surface described about the common vertex, as its centre; we then arrive at the following *spherographic construction, for finding the spherical centre of curvature of a given spherical ellipse,* at any given point of that curve, which may be regarded as being the *reciprocal* of the construction assigned at the end of the 32nd article of this essay:– Draw, from the given point H, of the ellipse, the normal arc HL, bisecting the spherical angle FHG between the two vector arcs FH, GH, and terminated at L by the arc FG which connects the two given foci, F and G; through L draw an arc of a great circle IK, perpendicular to the normal arc HL, and cutting one of the two vector arcs HF, HG, and the other of those two vector arcs prolonged, in two new points, I and K; through these two new points draw two new arcs of great circles, IE, KE, perpendicular respectively to the two vector arcs, or to the arcs HI, HK: *the two new arcs so drawn will cross each other on the normal arc (prolonged), in a point* E, *which will be the spherical centre of curvature sought,* or the pole of the small circle which osculates at the given point H to the given spherical ellipse.

And since it is obvious (on account of the spherical right angles HIE, HKE, in the construction), that the points I, K are the respective middle points of those portions of the vector arcs, or of those arcs prolonged, which are comprised within this osculating circle; so that the arc IK, which has been seen to pass through the point L, and which crosses at that point L the arcual major axis of the ellipse (because that axis passes through both foci), is the *common bisector* of these two intercepted portions of the vector arcs, which intercepted arcs of great circles may be called (on the sphere) the two *focal chords of curvature* of the spherical ellipse; we are therefore permitted to enunciate the following *theorem*,* which is in general sufficient for the determination of the spherical centre of curvature, or pole of the osculating small circle, at any proposed point of any such ellipse:– *The great circle which bisects the two focal (and arcual) chords of curvature of any spherical ellipse, for any point of osculation, intersects the (arcual) axis major in the same point in which that axis is cut by the (arcual) normal to the ellipse, drawn at the point of osculation.*

On the Tensor of a Geometrical Quotient

40. The equations (218) (222), of Art. 31, show that for any two equally long straight lines, a, b, the following relation holds good,

$$\left(\mathrm{S}\frac{\mathrm{b}}{\mathrm{a}}\right)^2 - \left(\mathrm{V}\frac{\mathrm{b}}{\mathrm{a}}\right)^2 = 1; \tag{255}$$

or, more concisely, that

$$\mathrm{T}\frac{\mathrm{b}}{\mathrm{a}} = 1, \tag{256}$$

if we introduce a new characteristic T of operation on a geometrical quotient, defined by the general formula,

$$\mathrm{T}\frac{\mathrm{b}}{\mathrm{a}} = \sqrt{\left\{\left(\mathrm{S}\frac{\mathrm{b}}{\mathrm{a}}\right)^2 - \left(\mathrm{V}\frac{\mathrm{b}}{\mathrm{a}}\right)^2\right\}}; \tag{257}$$

where it is to be observed, that the expression of which the square root is taken is essentially a positive scalar, because the square of *every* scalar is *positive*, while the square of *every* vector is on the contrary a *negative* scalar, by the principles of the 12th article. Hence, generally, for *any two* straight lines a, b, of which the lengths are denoted by $\bar{\mathrm{a}}$, $\bar{\mathrm{b}}$, we have the equation,

$$\mathrm{T}\frac{\mathrm{b}}{\mathrm{a}} = \bar{\mathrm{b}} \div \bar{\mathrm{a}}; \tag{258}$$

because the expression (257) is doubled, tripled, or multiplied by any positive number, when the line b is multiplied by the same number, whatever it be, while the line a remains unchanged. This geometrical signification of the expression $\mathrm{T}\frac{\mathrm{b}}{\mathrm{a}}$, may induce us to name that

* This theorem was proposed by the present writer, in June 1846, at the Examination for Bishop Law's Mathematical Premium, in Trinity College, Dublin; and it was shown by him in a series of Questions on that occasion, which have since been printed in the Dublin University Calendar for 1847, (see p. LXX), among the University Examination Papers for the preceding year, that this theorem, and several others connected therewith (for example, that the trigonometric tangent of the focal half chord of curvature is the harmonic mean between the tangents of the two focal vector arcs), might be deduced by *spherical trigonometry*, from the known constancy of the sum of the two vector arcs, or focal distances, for any one spherical ellipse. But in the method employed in the present essay, no use whatever has hitherto been made of any formula of spherical or even plane trigonometry, any more than of the doctrine of coordinates.

expression the TENSOR of the geometrical quotient $\frac{\mathrm{b}}{\mathrm{a}}$, on which the characteristic T has operated; because this *tensor* is a number which directs us how to *extend* (directly or inversely, that is, in what ratio to lengthen or shorten) the denominator line a, in order to render it *as long* as the numerator line b: and it appears to the writer, that there are other advantages in adopting this name "tensor", with the signification defined by the formula (257). Adopting it, then, we might at once be led to see, by (258), from considerations of compositions of ratios between the lengths of lines, that in any multiplication of geometrical quotients among themselves, "the tensor of the product is equal to the product of the tensors." But to establish this important principle otherwise, we may observe that by the equations (87), (88), (99), (100), of Arts. 11, 13, the vector part γ of the product $c+\gamma$ of any two geometrical quotients, represented by the binomial forms $b+\beta$, $a+\alpha$, is changed to its own opposite, $-\gamma$, while the scalar part c of the same product remains unchanged, when we change the signs of the vector parts β, α, of the two factors, without changing their scalar parts b, a, and also *invert*, at the same time, the *order* of those factors; in such a manner that either of the two following *conjugate equations* includes the other:

$$\left.\begin{aligned} c+\gamma &= (b+\beta)(a+\alpha) \\ c-\gamma &= (a-\alpha)(b-\beta) \end{aligned}\right\} \tag{259}$$

and these two conjugate equations give, by multiplication,

$$c^2-\gamma^2=(b^2-\beta^2)(a^2-\alpha^2), \tag{260}$$

because the product $(a+\alpha)(a-\alpha)=a^2-\alpha^2$ is scalar, so that

$$(a^2-\alpha^2)(b-\beta)=(b-\beta)(a^2-\alpha^2).$$

This product, $a^2-\alpha^2$, of the two *conjugate expressions*, or *conjugate geometrical quotients*, denoted here by

$$a+\alpha, \qquad a-\alpha, \tag{261}$$

is not only scalar, but is also *positive*; because we have, by the principles of the 12th article, the two inequalities,

$$a^2>0, \qquad \alpha^2<0. \tag{262}$$

Making then, in conformity with (257),

$$\mathrm{T}(a+\alpha)=\mathrm{T}(a-\alpha)=\sqrt{(a^2-\alpha^2)}, \tag{263}$$

we see that either of the two conjugate equations (259) gives, by (260),

$$\mathrm{T}(c+\gamma)=\mathrm{T}(b+\beta).\mathrm{T}(a+\alpha); \tag{264}$$

or eliminating $c+\gamma$,

$$\mathrm{T}(b+\beta)(a+\alpha)=\mathrm{T}(b+\beta).\mathrm{T}(a+\alpha). \tag{265}$$

It is easy to extend this result to any number of geometrical quotients, considered as factors in a multiplication; and thus to conclude generally that, as already stated, *the tensor of the product is equal to the product of the tensors*; a theorem which may be concisely expressed by the formula,

$$\mathrm{T\Pi} = \mathrm{\Pi T}. \tag{266}$$

On Conjugate Geometrical Quotients

41. It will be found convenient here to introduce a new characteristic, K, to denote the operation of passing from any geometrical quotient to its *conjugate*, by preserving the scalar part unchanged, but changing the sign of the vector part; with which new characteristic of operation K, we shall have, generally,

$$\mathrm{K}\frac{\mathrm{b}}{\mathrm{a}} = \mathrm{S}\frac{\mathrm{b}}{\mathrm{a}} - \mathrm{V}\frac{\mathrm{b}}{\mathrm{a}}; \tag{267}$$

or,

$$\mathrm{K}(a + \alpha) = a - \alpha, \tag{268}$$

if a be still understood to denote a scalar, but α a vector quotient. The *tensors of two conjugate quotients are equal to each other*, by (263); so that we may write

$$\mathrm{TK}\frac{\mathrm{b}}{\mathrm{a}} = \mathrm{T}\frac{\mathrm{b}}{\mathrm{a}}, \quad \text{or briefly,} \quad \mathrm{TK} = \mathrm{T}; \tag{269}$$

and *the product of any two such conjugate quotients is equal to the square of their common tensor*,

$$\frac{\mathrm{b}}{\mathrm{a}}\mathrm{K}\frac{\mathrm{b}}{\mathrm{a}} = \left(\mathrm{T}\frac{\mathrm{b}}{\mathrm{a}}\right)^2. \tag{270}$$

By separation of symbols, we may write, instead of (267),

$$\mathrm{K} = \mathrm{S} - \mathrm{V}, \tag{271}$$

and the characteristic K is a *distributive symbol*, because S and V have been already seen to be such: so that the equations (74) (75), of Art. 10, give now the analogous equations,

$$\mathrm{K\Sigma} = \mathrm{\Sigma K}, \quad \mathrm{K\Delta} = \mathrm{\Delta K}, \tag{272}$$

or in words, *the conjugate of a sum* (of any number of geometrical quotients) *is the sum of the conjugates*; and in like manner, the conjugate of a difference is equal to the difference of the conjugates. But also we have seen, in (178), that

$$1 = \mathrm{S} + \mathrm{V},$$

because a geometrical quotient is always equal to the sum of its own scalar and vector parts; we may therefore now form the following *symbolical expressions for our two old characteristics of operation, in terms of the new characteristic* K,

$$\left.\begin{aligned} \mathrm{S} &= \tfrac{1}{2}(1 + \mathrm{K}) \\ \mathrm{V} &= \tfrac{1}{2}(1 - \mathrm{K}) \end{aligned}\right\}. \tag{273}$$

We may also observe that

$$\mathrm{KK}\frac{\mathrm{b}}{\mathrm{a}} = \frac{\mathrm{b}}{\mathrm{a}}, \quad \text{or} \quad \mathrm{K}^2 = 1; \tag{274}$$

the *conjugate of the conjugate* of any geometrical quotient being equal to that quotient itself.

Combining (273), (274), we find, by an easy symbolical process, which the formulæ (272) show to be a legitimate one,

$$\left.\begin{aligned} \mathrm{KS} &= \tfrac{1}{2}(\mathrm{K} + \mathrm{K}^2) = \tfrac{1}{2}(\mathrm{K} + 1) = +\mathrm{S} \\ \mathrm{KV} &= \tfrac{1}{2}(\mathrm{K} - \mathrm{K}^2) = \tfrac{1}{2}(\mathrm{K} - 1) = -\mathrm{V} \end{aligned}\right\} \tag{275}$$

and accordingly the operation of *taking the conjugate* has been defined to consist in changing the sign of the vector part, without making any change in the scalar part, of the quotient on which the operation is performed. From (273), (274), we may also infer the symbolical equations,

$$\left.\begin{aligned} \mathrm{S}^2 &= \tfrac{1}{4}(1 + \mathrm{K})^2 = \tfrac{1}{2}(1 + \mathrm{K}) = \mathrm{S} \\ \mathrm{V}^2 &= \tfrac{1}{4}(1 - \mathrm{K})^2 = \tfrac{1}{2}(1 - \mathrm{K}) = \mathrm{V} \\ \mathrm{SV} &= \mathrm{VS} = \tfrac{1}{4}(1 - \mathrm{K}^2) = 0 \end{aligned}\right\} \tag{276}$$

and in fact, after once separating the scalar and vector parts of any proposed geometrical quotient, no farther separation of the same kind is possible; so that the operation denoted by the characteristic S, if it be again performed, makes no change in the scalar part first found, but reduces the vector part to zero; and, in like manner, the operation V reduces the scalar part to zero, while it leaves unchanged the vector part of the first or proposed quotient. We may note here that the same formulæ give these other symbolical results, which also can easily be verified:

$$\mathrm{KS} = \mathrm{SK}; \quad \mathrm{KV} = \mathrm{VK}; \tag{277}$$

and

$$(\mathrm{S} + \mathrm{V})^n = \mathrm{S}^n + \mathrm{V}^n = \mathrm{S} + \mathrm{V} = 1; \tag{278}$$

at least if the exponent n be any positive whole number, so as to allow a finite and integral development of the symbolic power

$$(\mathrm{S} + \mathrm{V})^n = 1^n. \tag{279}$$

With respect to the *geometrical signification* of the relation between conjugate quotients, we may easily see that if c and d denote any two equally long straight lines, and x any scalar coefficient or multiplier, then the two quotients

$$\frac{x\mathrm{c}}{\mathrm{c} + \mathrm{d}}, \quad \frac{x\mathrm{d}}{\mathrm{c} + \mathrm{d}} \tag{280}$$

will be, in the foregoing sense, *conjugate*; because their *sum* will be a *scalar*, namely x, but their *difference* will be a *vector*, on account of the mutual perpendicularity of the lines $\mathrm{c} - \mathrm{d}$ and $\mathrm{c} \div \mathrm{d}$, which are there the diagonals of a rhombus, and of which the latter bisects the angle between the sides c and d of that rhombus. (Compare (209).)

Conversely, if the relation

$$\frac{\mathrm{b}'}{\mathrm{a}} = \mathrm{K}\frac{\mathrm{b}}{\mathrm{a}}, \tag{281}$$

be given, we shall have, by the definition (267) of K,

$$0 = \mathrm{V}\frac{\mathrm{b}' + \mathrm{b}}{\mathrm{a}} = \mathrm{S}\frac{\mathrm{b}' - \mathrm{b}}{\mathrm{a}}; \tag{282}$$

whence it is easy to infer that, *if two conjugate geometrical quotients or fractions be so prepared as to have a common denominator* (a), *their numerators* (b, b′) *will be equally long, and will be equally inclined to the denominator, at opposite sides thereof, but in one common plane with it*; in such a manner that the line a (or −a) *bisects the angle* between the lines b and b′, if these three straight lines be supposed to have all one common origin. We are then conducted, in this way, to a very simple and useful *expression, for* (what may be called) *the reflexion* (b′) *of a straight line* (b), *with respect to another straight line* (a), namely the following:

$$\mathrm{b}' = \mathrm{K}\frac{\mathrm{b}}{\mathrm{a}} \times \mathrm{a}. \tag{283}$$

And whenever we meet with an expression of this form, we shall know that the two lines b and b′ are equally long; and also that if they have a common origin, the angle between them is bisected there by one of the two opposite lines ± a, or by a parallel thereto.

Finally, we may here note that, by the principles of the present article, and of the foregoing one, we have the following expressions, which hold good for any pair of straight lines, a and b:

$$\left.\begin{aligned} \frac{\mathrm{a}}{\mathrm{b}} &= \left(\mathrm{T}\frac{\mathrm{a}}{\mathrm{b}}\right)^2 \mathrm{K}\frac{\mathrm{b}}{\mathrm{a}} \\ \mathrm{S}\frac{\mathrm{a}}{\mathrm{b}} &= \left(\mathrm{T}\frac{\mathrm{a}}{\mathrm{b}}\right)^2 \mathrm{S}\frac{\mathrm{b}}{\mathrm{a}} \\ \mathrm{V}\frac{\mathrm{a}}{\mathrm{b}} &= -\left(\mathrm{T}\frac{\mathrm{a}}{\mathrm{b}}\right)^2 \mathrm{V}\frac{\mathrm{b}}{\mathrm{a}} \end{aligned}\right\} \tag{284}$$

42. The equation

$$\mathrm{S}\frac{\mathrm{r}}{\mathrm{a}} = 0, \tag{285}$$

signifies, by what has been already shown, that the straight line r is perpendicular to a; it is therefore the equation of a *plane* perpendicular to this latter line, and passing through some fixed origin of lines, if r be regarded as a variable line, but a as a fixed line from that origin. The equation

$$\mathrm{S}\frac{\mathrm{r} - \mathrm{a}}{\mathrm{a}} = 0, \quad \text{or} \quad \mathrm{S}\frac{\mathrm{r}}{\mathrm{a}} = 1, \tag{286}$$

expresses, for a similar reason, that the variable line r terminates on another plane, parallel to the former plane, and having the line a for the perpendicular let fall upon it from the origin. If b denote the perpendicular let fall from the same origin upon a third plane, the equation of this third plane will of course be, in like manner,

$$\mathrm{S}\frac{\mathrm{r}}{\mathrm{b}} = 1; \tag{287}$$

and it is not difficult to prove, with the help of the transformations (284), that this other equation,

$$S\frac{r}{b} = S\frac{r}{a}, \tag{288}$$

represents a fourth plane, which passes through the intersection of the second and third planes just now mentioned, namely, the planes (286), (287), and through the origin.

In general, the equation

$$S\left(\frac{r}{a} + \frac{r}{a'} + \frac{r}{a''} + \&c.\right) = a, \tag{289}$$

expresses that r terminates on a fixed plane, if it be drawn from a fixed origin, and if the lines a, a′, a″, &c., and the number *a* be given. It may also be noted here that the equation of the plane which perpendicularly bisects the straight line connecting the extremities of two given lines, a and b, may be thus written:

$$T\frac{r-b}{r-a} = 1. \tag{290}$$

43. On the other hand, the equation

$$S\frac{r-b}{r-a} = 0, \tag{291}$$

expresses that the lines from the extremities of a and b to the extremity of r are perpendicular to each other; or that the line r terminates upon a *spheric surface*, in two diametrically opposite points of which surface the lines a and b respectively terminate: and this diameter itself, from the end of a to the end of b, regarded as a *rectilinear locus*, is represented by the equation

$$V\frac{r-b}{r-a} = 0; \tag{292}$$

which may however be put under other forms. A transformation of the equation (291) is the following:

$$T\left(\frac{2r-b-a}{b-a}\right) = 1; \tag{293}$$

which expresses that the variable radius $r - \frac{1}{2}(b+a)$ has the same length as the fixed radius $\frac{1}{2}(b-a)$. For example, by changing $-a$ to $+b$, in this last equation of the sphere, we find

$$T\frac{r}{b} = 1, \quad \text{or} \left(S\frac{r}{b}\right) - \left(V\frac{r}{b}\right)^2 = 1 \tag{294}$$

as the equation of a spheric surface described about the origin of lines, as centre, with the line b for one of its radii, so as to touch, at the end of this line b, the plane (287). (Comp. (255)).

And a small *circle* of this sphere (294), if it be situated on a secant plane, parallel to this tangent plane (287), which new plane will thus have for its equation,

$$S\frac{r}{b} = x, \tag{295}$$

where x is a scalar, numerically less than unity, and constant for each particular circle, will also be situated on a certain corresponding cylinder of revolution, which will have for its equation

$$\left(\mathrm{V}\frac{\mathrm{r}}{\mathrm{b}}\right)^2 = x^2 - 1; \tag{296}$$

where $x^2 - 1$ is negative, as it ought to be, by the 12th article, being equal to the square of a vector. The sphere may be regarded as the locus of these small circles; and its equation (294) may be supposed to be obtained by the elimination of the scalar x between the equations of the plane (295), and of the cylinder (296).

44. Conceive now that instead of cutting the cylinder (296) *perpendicularly* in a *circle,* we cut it *obliquely,* in an *ellipse,* by the plane having for equation

$$\mathrm{S}\frac{\mathrm{r}}{\mathrm{a}} = x, \tag{297}$$

where x is the same scalar as before; so that this new plane is parallel to the fixed plane (286), and cuts the plane of the circle (295) in a straight line situated on that other fixed plane (288), which has been seen to contain also the intersection of the same fixed plane (286) with the tangent plane (287). The *locus of the elliptic sections,* obtained from the circular cylinders by this construction, will be an *ellipsoid;* and conversely, an ellipsoid may in general be regarded as such a locus. The equation of the *ellipsoid,* thus found, by eliminating x between the equations (296), (297), is the following;

$$\left(\mathrm{S}\frac{\mathrm{r}}{\mathrm{a}}\right)^2 - \left(\mathrm{V}\frac{\mathrm{r}}{\mathrm{b}}\right)^2 = 1; \tag{298}$$

and by some easy modifications of the process, it may be shown that a *hyperboloid,* regarded as a certain other locus of ellipses, may in general be represented by an equation of the form

$$\left(\mathrm{S}\frac{\mathrm{r}}{\mathrm{a}}\right)^2 + \left(\mathrm{V}\frac{\mathrm{r}}{\mathrm{b}}\right)^2 = \mp 1. \tag{299}$$

The upper sign belongs to a hyperboloid of *one sheet,* but the lower sign to a hyperboloid of *two sheets;* while the *common asymptotic cone* of these two (conjugate) hyperboloids (299) is the locus of a certain other system of ellipses, and is represented by the analogous but intermediate equation,

$$\left(\mathrm{S}\frac{\mathrm{r}}{\mathrm{a}}\right)^2 + \left(\mathrm{V}\frac{\mathrm{r}}{\mathrm{b}}\right)^2 = 0. \tag{300}$$

These equations admit of several instructive transformations, to some of which we shall proceed in the following article.

On some Transformations and Constructions of the Equation of the Ellipsoid

45. The equation (298) of the ellipsoid resolves itself into factors, as follows:

$$\left(\mathrm{S}\frac{\mathrm{r}}{\mathrm{a}} + \mathrm{V}\frac{\mathrm{r}}{\mathrm{b}}\right)\left(\mathrm{S}\frac{\mathrm{r}}{\mathrm{a}} - \mathrm{V}\frac{\mathrm{r}}{\mathrm{b}}\right) = 1 \tag{301}$$

where the sum and difference, which when thus multiplied together give unity for their product, are *conjugate expressions* (in the sense of recent articles); they have therefore a *common tensor*, which must itself be equal to unity; and consequently we may write the equation of the ellipsoid thus,

$$\mathrm{T}\left(\mathrm{S}\frac{\mathrm{r}}{\mathrm{a}} + \mathrm{V}\frac{\mathrm{r}}{\mathrm{b}}\right) = 1, \tag{302}$$

where the sign of the vector may be changed. Substituting for the characteristics of operation, S and V, their symbolical values (273), we are led to introduce two new fixed lines g and h, depending on the two former fixed lines a and b, and determined by the equations

$$\frac{\mathrm{r}}{2\mathrm{a}} + \frac{\mathrm{r}}{2\mathrm{b}} = \frac{\mathrm{r}}{\mathrm{g}}; \quad \frac{\mathrm{r}}{2\mathrm{a}} - \frac{\mathrm{r}}{2\mathrm{b}} = \frac{\mathrm{r}}{\mathrm{h}}; \tag{303}$$

and thus the equation of the ellipsoid may be changed from (302) to this other form,

$$\mathrm{T}\left(\frac{\mathrm{r}}{\mathrm{g}} + \mathrm{K}\frac{\mathrm{r}}{\mathrm{h}}\right) = 1; \tag{304}$$

which, by the principles (269), (272), (274), may also be thus written,

$$\mathrm{T}\left(\frac{\mathrm{r}}{\mathrm{h}} + \mathrm{K}\frac{\mathrm{r}}{\mathrm{g}}\right) = 1; \tag{305}$$

so that the symbols, g and h, may be interchanged in either of the two last forms of the equation of the ellipsoid.

46. Let $\bar{\mathrm{r}}$, $\bar{\mathrm{g}}$, $\bar{\mathrm{h}}$ be conceived to be numerical symbols, denoting respectively the lengths of the three lines r, g, h; and make, for conciseness,

$$\mathrm{r} \div \bar{\mathrm{r}}^2 = \mathrm{r}'; \quad \mathrm{g} \div \bar{\mathrm{g}}^2 = \mathrm{g}'; \quad \mathrm{h} \div \bar{\mathrm{h}}^2 = \mathrm{h}'; \tag{306}$$

so that the symbols r′, g′, h′ shall denote three new lines, having the *same directions* as the three former lines r, g, h, but having their *lengths* respectively *reciprocals* of the lengths of those three former lines. Then, by the properties of conjugate quotients already established, we shall have the transformations

$$\frac{\mathrm{r}}{\mathrm{g}} = \mathrm{K}\frac{\mathrm{g}'}{\mathrm{r}'}; \quad \mathrm{K}\frac{\mathrm{r}}{\mathrm{h}} = \frac{\mathrm{h}'}{\mathrm{r}'}; \tag{307}$$

whereby the equation (304) of the ellipsoid becomes

$$\mathrm{T}\left(\frac{\mathrm{h}'}{\mathrm{r}'} + \mathrm{K}\frac{\mathrm{g}'}{\mathrm{r}'}\right) = 1. \tag{308}$$

Let g″ be a new line, not fixed but variable, and determined for each variable direction of r′ or of r by the formula

$$\mathrm{g}'' = \mathrm{K}\frac{\mathrm{g}'}{\mathrm{r}'} \times \mathrm{r}'; \quad \text{or } \mathrm{g}'' = \mathrm{K}\frac{\mathrm{g}'}{\mathrm{r}} \times \mathrm{r}; \tag{309}$$

so that this new and variable line g″ is, by what was shown respecting the expression (283), the *reflexion* of the fixed line g′ with respect to a line having the variable direction just mentioned, of r′ or of r: we may then write the equation (308) of the ellipsoid as follows,

$$\mathrm{T}\frac{\mathrm{h}' + \mathrm{g}''}{\mathrm{r}'} = 1. \tag{310}$$

And by comparing this with the formula (256), we see that the length of the line r′, or *the reciprocal of the length* r *of the variable semidiameter* r *of the ellipsoid, is equal to the length of the line* h′ + g″; which latter line is the symbolical sum of one fixed line, h′, and of the variable reflexion, g″, of another fixed line, g′; this *reflexion* having been already seen to be performed with respect to the variable radius vector or semidiameter, r, of the ellipsoid, *of which semidiameter the dependence of the length on the direction admits of being thus represented, or constructed, by a very simple geometrical rule.*

47. To make more clear the conception of this geometrical rule, let A denote the centre of the ellipsoid, which centre is the origin of the variable line r; and let two other fixed points, B and C, be determined by the symbolical equations

$$\mathrm{g}' = \mathrm{A} - \mathrm{C} = \mathrm{AC}; \quad \mathrm{h}' = \mathrm{B} - \mathrm{C} = \mathrm{BC}: \tag{311}$$

these two notations, AC and A − C, (of which one has been already used in the *text* of the first article* of this Essay on Symbolical Geometry, while the other was suggested in a *note* to the same early article,) being each designed to denote or signify a straight line drawn *to* the point A from the point C. Let D be a new or fourth point, not fixed but variable, and determined by the analogous equation

$$\mathrm{g}'' = \mathrm{C} - \mathrm{D} = \mathrm{CD}: \tag{312}$$

then because, in virtue of the relation (309), the lines g′, g″ are equally long, it follows that the variable point D is situated on the surface of that fixed and *diacentric sphere*, which we may conceive to be described *round* the fixed point C as centre, so as to pass *through* the centre A of the ellipsoid as through a given superficial point of this diacentric sphere. Again, in virtue of the same relation (309), or of the geometrical reflexion which the second formula so marked expresses, the symbolic sum of the two lines, g′, g″, has the direction of the line r, or the exactly contrary direction; in fact, that relation (309) conducts to the following scalar quotient,

$$\frac{\mathrm{g}' + \mathrm{g}''}{\mathrm{r}} = \frac{\mathrm{g}'}{\mathrm{r}} + \mathrm{K}\frac{\mathrm{g}'}{\mathrm{r}} = 2\mathrm{S}\frac{\mathrm{g}'}{\mathrm{r}}; \tag{313}$$

and this symbolic sum, g′ + g″, may also, by (311) (312), be thus expressed,

$$\mathrm{g}' + \mathrm{g}'' = (\mathrm{A} - \mathrm{C}) + (\mathrm{C} - \mathrm{D}) = \mathrm{A} - \mathrm{D} = \mathrm{AD}. \tag{314}$$

If then we denote by E that variable point on the surface of the ellipsoid at which the line r terminates, so that

$$\mathrm{r} = \mathrm{E} - \mathrm{A} = \mathrm{EA}, \tag{315}$$

* It was for the sake of making easier the transition to the notation B − A, which appears to the present writer an expressive one, for the straight line drawn *to* the point B *from* the point A, that he proposed to use, with the *same* geometrical signification, the symbol BA, instead of AB: although it is certainly more ususal, and perhaps also more natural, when *direction* is attended to, to employ the latter symbol AB, and not the former BA, to denote the line thus drawn from A to B.

we shall have the relation

$$\frac{\text{A}-\text{D}}{\text{E}-\text{A}}=\frac{\text{AD}}{\text{EA}}=2\text{S}\frac{\text{g}'}{\text{r}}=\text{V}^{-1}0, \tag{316}$$

which requires that the three points, A, D, E, should be situated on one common straight line. We know then the geometrical position of the auxiliary and variable point D, or have a simple construction for determining this variable point D, as corresponding to any particular point E on the surface of the ellipsoid, when the centre A, and the two other fixed points, B and C, are given; for we see that we have merely to seek the *second intersection* of the semi-diameter E − A (or EA) of the ellipsoid, with the surface of the diacentric sphere, the *first* intersection being the centre A itself; since this second point of intersection will be the required point D.

48. But also, by (311) (312), we have

$$\text{h}'+\text{g}''=(\text{B}-\text{C})+(\text{C}-\text{D})=\text{B}-\text{D}=\text{BD}: \tag{317}$$

this line BD has therefore, by (310), the length of the line r′; which length is, by (306), (315) the reciprocal of the length r of the semidiameter EA of the ellipsoid. The lines g, h have generally unequal lengths; and because, by (304) (305), their symbols may be interchanged, we may choose them so that the former shall be the longer of the two, or that the inequality

$$\overline{\text{g}}>\overline{\text{h}} \tag{318}$$

shall be satisfied; and then, by (306), the line g′ will, on the contrary, be shorter than the line h′, or the fixed point B will be *exterior* to the fixed diacentric sphere. Drawing, then, from this external point B, a tangent to this diacentric sphere, and taking the length of the tangent so drawn for the unit of length, the reciprocal of the length of the line BD, which is considered in (317), will be the length of that other line BD′, which has the same direction as BD, but terminates at another variable point D′ on the surface of the diacentric sphere; in such a manner that this new variable point D′, without generally coinciding with the point D, shall satisfy the two equations,

$$\frac{\text{D}'-\text{B}}{\text{D}-\text{B}}=\text{V}^{-1}0;\quad \frac{\text{D}'-\text{C}}{\text{D}-\text{C}}=\text{T}^{-1}1; \tag{319}$$

for then the two lines D′ − B, D − B (or D′B, DB) will be, in this or in the opposite order, the whole secant and external part, while the length of the tangent to the sphere has been above assumed as unity. Under these conditions, then, *the lengths of the lines* D′B *and* EA *will be equal*, because they will have the length of the line DB for their common reciprocal; so that we shall have the equation

$$\frac{\text{E}-\text{A}}{\text{D}'-\text{B}}=\text{T}^{-1}1; \tag{320}$$

or, in a more familiar notation,

$$\overline{\text{AE}}=\overline{\text{BD}'}. \tag{321}$$

It may be noted here that the new radius D′ − C of the diacentric sphere admits (compare the formula (213)) of being symbolically expressed as follows,

$$\mathrm{D}' - \mathrm{C} = \mathrm{K}\frac{\mathrm{g}''}{\mathrm{h}' + \mathrm{g}''}.(\mathrm{h}' + \mathrm{g}''); \tag{322}$$

and, accordingly, this last expression satisfies the two conditions (319), because it gives

$$\frac{\mathrm{D}' - \mathrm{B}}{\mathrm{D} - \mathrm{B}} = \mathrm{S}\frac{\mathrm{h}' - \mathrm{g}''}{\mathrm{h}' + \mathrm{g}''}, \tag{323}$$

and

$$\frac{\mathrm{D}' - \mathrm{C}}{\mathrm{D} - \mathrm{C}} = \mathrm{K}\frac{\mathrm{g}''}{\mathrm{h}' + \mathrm{g}''}.\frac{\mathrm{h}' + \mathrm{g}''}{-\mathrm{g}''}, \tag{324}$$

of which latter expression the tensor is unity.

49. The remarkably simple formula,

$$\overline{\mathrm{AE}} = \overline{\mathrm{BD}'}, \tag{321}$$

to which we have thus been conducted for the ellipsoid, admits of being easily translated into the following rule for constructing that important surface; which rule for the *construction of the ellipsoid* does not seem to have been known to mathematicians, until it was communicated by the present writer to the Royal Irish Academy in 1846,* as a result of his Calculus of Quaternions, between which and the present Symbolical Geometry a very close affinity exists.

From a fixed point A, *on the surface of a given sphere, draw a variable chord of that sphere,* DA; *let* D′ *be the second point of intersection of the spheric surface with the secant* DB, *which connects the variable extremity* D *of this chord* DA *with a fixed external point* B; *and take the radius vector* EA *equal in length to* D′B, *and in direction either coincident with, or opposite to, the chord* DA: *the locus of the point* E, *thus constructed, will be an ellipsoid, which will have its centre at the fixed point* A, *and will pass through the fixed point* B.

The fixed sphere through A, in this construction of the ellipsoid, is the *diacentric sphere* of recent articles; it may also be called a *guide-sphere,* from the manner in which it assists to mark or to *represent the direction,* and at the same time serves to *construct the length* of a variable semidiameter of the ellipsoid; while, for a similar reason, the points D and D′ upon the surface of this sphere may be said to be *conjugate guide-points*; and the chords DA and D′A may receive the appellation of *conjugate guide-chords.* In fact, while either of these two guide-chords of the sphere, for instance (as above) the chord DA, coincides in *direction* with a semidiameter EA of the ellipsoid, the distance $\overline{\mathrm{D}'\mathrm{B}}$ of the extremity D′ of the other or conjugate guide-chord, D′A, from the fixed external point B, represents, as we have seen, the *length* of that semidiameter. And that the fixed point B, although exterior to the diacentric sphere, is a superficial point of the ellipsoid, appears from the construction, by conceiving the conjugate guide-point D′ to approach to coincidence with A; for E will then tend to coincide either with the point B itself, or with another point diametrically opposite thereto, upon the surface of the ellipsoid.

50. Some persons may prefer the following mode of stating the same geometrical construction, or the same fundamental property, of the ellipsoid: which other mode also was

* [See Vol. III, p. 375.]

communicated by the present writer to the Royal Irish Academy in 1846.* *If, of a rectilinear quadrilateral* ABED′, *of which one side* AB *is given in length and in position the two diagonals* AE, BD′ *be equal to each other in length, and intersect* (in D) *on the surface of a given sphere* (with centre C), *of which sphere a chord* AD′ *is a side of the quadrilateral adjacent to the given side* AB, *then the other side* BE, *adjacent to the same given side* AB, *is a chord of a given ellipsoid.*

Thus, denoting still the centre of the sphere by C, while A is still the centre of the ellipsoid, we see that the form, magnitude, and position, of this latter surface are made by the foregoing construction to depend, according to very simple geometrical rules, on the positions of the three points A, B, C; or on the form, magnitude, and position of what may (for this reason) be named the *generating triangle* ABC. Two of the sides of this triangle, namely BC and CA, are perpendicular, as it is not difficult to show from the construction, to the two *planes of circular section* of the ellipsoid; and the third side AB is perpendicular to one of the two *planes of circular projection* of the same ellipsoid: this third side AB being the axis of revolution of a circumscribed circular cylinder; which also may be proved, without difficulty, from the construction assigned above. (See Articles 52, 53.) The length $\overline{\text{BC}}$ of the side BC of the triangle, is (by the construction) the semisum of the lengths of the greatest and least semidiameters of the ellipsoid; and the length $\overline{\text{CA}}$ of the side CA is the semidifference of the lengths of those extreme semidiameters, or principal semiaxes, of the same ellipsoid: while (by the same construction) these greatest and least *semiaxes*, or their prolongations, intersect the surface of the diacentric sphere in points which are situated, respectively, on the finite side CB of the triangle ABC itself, and on the side CB prolonged through C. The *mean* semiaxis of the ellipsoid, or the semidiameter perpendicular to the greatest and least semiaxes, is (by the construction) equal in length (as indeed it is otherwise known to be) to the radius of the enveloping cylinder of revolution, or to the radius of either of the two diametral and circular sections: the length of this mean semiaxis is also constructed by the portion $\overline{\text{BG}}$ of the axis of the enveloping cylinder, or of the side BA of the generating triangle, if G be the point, distinct from A, in which this side BA meets the surface of the diacentric sphere. And hence we may derive a simple geometrical signification, or property, of this remaining side BA of the triangle ABC, as respects its length $\overline{\text{BA}}$; namely, that this length is a fourth proportional to the three semiaxes of the ellipsoid, that is to say, to the mean, the least, and the greatest, or to the mean, the greatest, and the least of those three principal and rectangular semiaxes.

On the Law of the Variation of the Difference of the Squares of the Reciprocals of the Semiaxes of a Diametral Section

51. To give a specimen of the facility with which the foregoing construction serves to establish some important properties of the ellipsoid, we shall here employ it to investigate anew the known and important law, according to which the difference of the squares of the reciprocals of the greatest and least semidiameters, of any plane and diametral section, varies in passing from one such section to another. Conceive then that the ellipsoid itself, and the

* [See Vol. III. p. 375.]

auxiliary or diacentric sphere which was employed in the foregoing construction, are both cut by a plane AB′C′, passing through the centre A of the ellipsoid, and having B′ and C′ for the orthogonal projections, upon this secant plane, of the fixed points B and C. The auxiliary or guide-point D comes thus to be regarded as moving on the circumference of a circle, which passes through A, and has its centre at C′: and since the semidiameter EA of the ellipsoid, as being equal in length to D′B, by the formula (321) of Art. 48, (or because these are the two equally long diagonals of the quadrilateral ABED′ of Art. 50), must vary inversely as DB (by an elementary property of the sphere), we are led to seek the difference of the squares of the greatest and least values of DB, or of DB′, since the square of the perpendicular B′B is constant for the section. But the shortest and longest straight lines, D_1B', D_2B', which can thus be drawn to the circumference of the auxiliary circle round C′ (namely the section of the diacentric sphere), from the fixed point B′ in its plane, are those drawn to the extremities D_1, D_2 of that diameter $D_1C'D_2$ which passes through, or tends towards this point B′; in such a manner that the four points $B'D_1C'D_2$ are situated on one straight line. Hence the difference of the squares of D_1B', D_2B', is equal to four times the rectangle under D_1C', or AC′, and B′C′; that is to say, under the projections of the sides AC, BC, of the generating triangle, on the plane of the diametral section. *It is, then, to this rectangle, under these two projections of two fixed lines, on any variable plane through the centre of the ellipsoid, that the difference of the squares of the reciprocals of the extreme semidiameters of the section is proportional.* Hence, in the language of trigonometry, this difference of squares is proportional (as indeed it is well known to be) to the product of the sines of the inclinations of the cutting plane to two fixed planes of circular section; which latter planes are at the same time seen to be perpendicular to the two fixed sides AC, BC, of the generating triangle in the construction.

It seems worth noting here, that the foregoing process proves at the same time this other well-known property of the ellipsoid, that the greatest and least semidiameters of a plane section through the centre are perpendicular to each other; and also gives an easy geometrical rule for *constructing the semiaxes of any proposed diametral section*; for it shows that these semiaxes have the *directions* of the two rectangular guide-chords D_1A, D_2A; while their *lengths* are equal, respectively, to those of the lines $D_1'B$, $D_2'B$.

On the Planes of Circular Section and Circular Projection

52. It may not be uninstructive to state briefly here some simple geometrical reasonings, by which the line BG of Art. 50 may be shown to have its length equal to that of the radius of an enveloping cylinder of revolution, as was asserted in that article; and also to the radius of either of the two diametral and circular sections of the ellipsoid. First, then, as to the cylinder: the equation $\overline{AE} = \overline{BD'}$ shows that the rectangle under the two lines AE and BD is constant for the ellipsoid, because the rectangle under BD′ and BD is constant for the sphere; and the point D has been seen to be situated upon the straight line AE (prolonged if necessary). Hence the double area of the triangle ABE, or the rectangle under the fixed line AB, and the perpendicular let fall thereon from the variable point E of the ellipsoid, is always less than the lately mentioned constant rectangle; or than the square of the tangent to the diacentric sphere from B; or, finally, than the rectangle under the same fixed line AB and its constant part GB: except at the limit where the angle ADB is right, at which limit the double area of the triangle ABE becomes equal to the last mentioned rectangle. The ellipsoid is therefore entirely enveloped by that cylinder of revolution which has AB for axis, and $\overline{\text{GB}}$ for radius; being situated entirely *within* this cylinder, except for a certain limiting curve or system of points, which are *on* (but not outside) the cylinder, and are determined by the condition that ADB shall be a right angle. This limiting condition determines a *second spherical locus* for the guide-point D, besides the diacentric sphere; it serves therefore to assign a *circular locus* for that point, which circle passes through the centre A of the ellipsoid, because this centre is situated on each of the two spherical loci. And hence by the construction we obtain an *elliptic locus* for the point E, namely the ellipse of contact of the ellipsoid and cylinder; which ellipse presents itself here as the intersection of that enveloping cylinder of revolution with the plane of the circle which has been seen to be the locus of D.– It may also be shown geometrically, by pursuing the same construction into its consequences, that the ellipsoid is enveloped by *another* (equal) cylinder of revolution, giving a *second* diametral *plane of circular projection*; the first such plane being (by what precedes) perpendicular to the line AB: and that the axis of this second circular cylinder, or the normal to this second plane of circular projection of the ellipsoid, is parallel to the straight line which touches, at the centre C of the diacentric sphere, the circle circumscribed about the generating triangle ABC.

53. Again, with respect to the diametral and circular sections of the ellipsoid, considered as results of the construction: if we conceive that the guide-point D, in that construction, approaches in any direction, on the surface of the diacentric sphere, to the centre A of the ellipsoid, the conjugate guide-point D′ must then approach to the point G, because this is the second point of intersection of the side BA of the triangle with the surface of the diacentric sphere, if the point A itself be regarded as the first point of such intersection. Thus, during this approach of D to A, the semidiameter EA of the ellipsoid, having always (by the construction) the direction of ± DA, and the length of D′B, must tend to touch the diacentric sphere at A, and to have the same fixed length as the line BG, or as the radius of the cylinder. And in this way the construction offers to our notice a *circle* on the ellipsoid, whose radius $= \overline{\text{BG}}$, and whose plane is perpendicular to the side AC of the generating triangle; which side is thus seen to be a *cyclic normal* of the ellipsoid, by this process as well as by that of the 51st article.

Finally, with respect to that *other* cyclic plane which is perpendicular to the side BC of the triangle ABC, it is sufficient to observe that if we conceive the point D′ to revolve in a small circle on the surface of the diacentric sphere, from G to G again, preserving a constant distance from the fixed external point B, then the semidiameter EA of the ellipsoid will retain, by the construction, during this revolution of D′, a constant length $= \overline{\text{BG}}$; while, by the same construction, the guide-chord DA, and the semidiameter EA of the ellipsoid, will at the same time revolve together in a diametral plane perpendicular to BC: in which *second cyclic plane*, therefore, the point E will thus trace out a *second circle* on the ellipsoid, with a radius equal to the radius of the former circle; or to that of the *mean sphere* (constructed on the mean axis as diameter, and containing both the circles hitherto considered); or to the radius of either of the two enveloping cylinders of revolution.— It is evident that if the guide-point D describe any other circle on the diacentric sphere, parallel to this second cyclic plane, the conjugate guide-point D′ will describe another parallel circle, leaving the length $\overline{\text{BD}'} = \overline{\text{EA}}$ unaltered; whence the known theorem flows at once, that if the ellipsoid be cut by a concentric sphere, the section is a spherical ellipse;* and also that the concentric cyclic cone which rests thereon (being the cone described by the guide-chord DA in the construction) has its two cyclic planes coincident with the two cyclic planes of the ellipsoid.

* This easy mode of deducing, from the author's construction of the ellipsoid, the known spherical ellipses on that surface, was pointed out to him in 1846, by a friend to whom he had communicated that construction, namely by the Rev. J. W. Stubbs, Fellow of Trinity College, Dublin.† Several investigations, by the present author, connected with the same construction of the ellipsoid, have appeared in the *Proceedings* of the Royal Irish Academy, (see in particular those for July 1846); and also in various numbers of the (London, Edinburgh, and Dublin) *Philosophical Magazine*, in which magazine several articles on Quaternions have been already published by the writer,‡ and are likely to be hereafter continued, which may on some points be perhaps usefully compared with the present Essay on Symbolical Geometry.

† [See Vol. III, p. 375, footnote.]

‡ [See Vol. III.]

V.

ON SOME NEW APPLICATIONS OF QUATERNIONS TO GEOMETRY (1849)

[*British Association Report* 1849, Part II., p. 1.]

VI.

ON POLYGONS INSCRIBED ON A SURFACE OF THE SECOND ORDER (1850)

[*British Association Report* 1850, Part II., p. 2.]

VII.

SYMBOLICAL EXTENSIONS OF QUATERNIONS;
AND
GEOMETRICAL APPLICATIONS OF QUATERNIONS (1855)

[Communicated June 11, 1855]
[*Proceedings of the Royal Irish Academy*, **6** 250 (1858).]

SIR WILLIAM R. HAMILTON commenced the reading of a Paper on some symbolical extensions of quaternions, and especially on a theory of associative quines.

2. Sir William R. Hamilton also commenced an account of some geometrical applications of his theories, especially as founded on the notion of the anharmonic quaternion, and as leading to an enlarged conception of involution, not merely in one plane, but on a sphere, and generally in space.

[Communicated June 25, 1855]
[*Proceedings of the Royal Irish Academy*, **6** 260 (1858).]

Sir William R. Hamilton read a continuation of his Paper on some new geometrical applications of quaternions.

[Communicated February 25 1856]
[*Proceedings of the Royal Irish Academy*, **6** 311 (1858).]

Sir W. R. Hamilton, LL. D., read a paper on a geometrical extension of the Calculus of Quaternions, as concerns its fundamental interpretations.

VIII.

ON A GENERAL EXPRESSION BY QUATERNIONS FOR CONES OF THE THIRD ORDER (1857)

[Communicated May 11, 1857]
[*Proceedings of the Royal Irish Academy*, **6** 506 (1858).]

The President read the following note:-

"Sir William Rowan Hamilton wished to hand in a memorandum of the following 'General Expression by Quaternions, for Cones of the Third Order,' which he hoped to be allowed to develope and illustrate at some subsequent meeting of the Academy during the present Session. The equation in question is,

$$Sq\rho q'\rho q''\rho = 0; \qquad \text{(A)}$$

where ρ is the variable vector (or side) of the cone of the third order, drawn from its vertex as the origin; while q, q', q'', are three arbitrary but constant quaternions, which may be regarded as fixed parameters of the surface."

[Communicated May 25, 1857]
[*Proceedings of the Royal Irish Academy*, **6** 512 (1858).]

Sir W. R. Hamilton, LL. D., read some remarks on the General Equation in Quaternions for Cones of the Third Order.

IX.

ON A CERTAIN HARMONIC PROPERTY OF THE ENVELOPE OF THE CHORD CONNECTING TWO CORRESPONDING POINTS OF THE HESSIAN OF A CUBIC CONE (1857)

[Communicated June 22, 1857]
[*Proceedings of the Royal Irish Academy*, **6** 524 (1858).]

Sir William R. Hamilton read a paper on a certain harmonic property of the envelope of the chord connecting two corresponding points of the Hessian of a cubic cone.

X.

ON SOME APPLICATIONS OF QUATERNIONS TO CONES OF THE THIRD DEGREE (1857)

[*British Association Report* 1857, Part II., p. 3.]

XI.

ON ANHARMONIC CO-ORDINATES (1860)

[Communicated April 9, 1860; May 28, 1860; June 25, 1860]
[*Proceedings of the Royal Irish Academy*, **7**, 1862, pp. 286–289, 329, 350–354
and
Natural History Review and Journal of Science, **7**, 1860, pp. 242–246, 325–327, 506–509.]

---

1. Let ABC be any given triangle; and let O, P be any two points in its plane, whereof O shall be supposed to be given or constant, but P variable. Then, by a well-known theorem, respecting the six segments into which the sides are cut by right lines drawn from the vertices of a triangle to any common point the three following *anharmonics of pencils* have a product equal to positive unity:-

$$(\text{A.PCOB}).(\text{B.PAOC}).(\text{C.PBOA}) = +1.$$

It is, therefore, allowed to establish the following system of three equations, of which any one is a consequence of the other two:-

$$\frac{y}{z} = (\text{A.PCOB}); \quad \frac{z}{x} = (\text{B.PAOC}); \quad \frac{x}{y} = (\text{C.PBOA});$$

and, when this is done, I call the three quantities x, y, z, or any quantities proportional to them, the *Anharmonic Co-ordinates of the Point* P, with respect to the *given triangle* ABC, and to the *given point* O. And I *denote* that point P by the *symbol*,

$$\text{P} = (x,\ y,\ z); \quad \text{or}, \quad \text{P} = (tx,\ ty,\ tz); \quad \&\text{c}.$$

2. When the variable point P takes the given position O, the three anharmonics of pencils above mentioned become each equal to unity; so that we may write then,

$$x = y = z = 1.$$

The given point O is therefore denoted by the symbol,

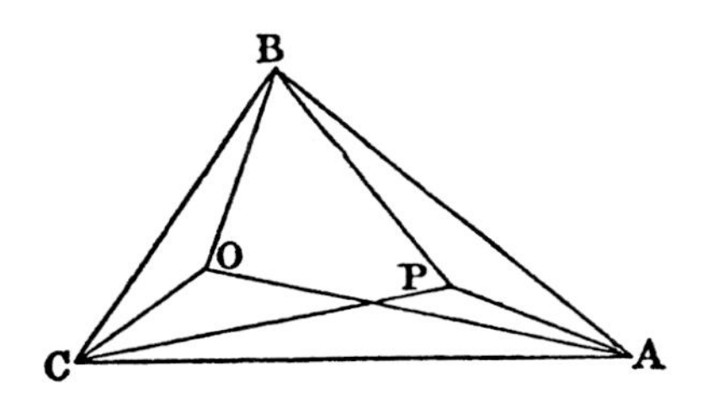

Fig. 1

$$O = (1, 1, 1);$$

on which account I call it the *Unit-Point.*

3. When the variable point P comes to coincide with the given point A, so as to be at the vertex of the first pencil, but on the second ray of the second pencil, and on the fourth ray of the third, without being at the vertex of either of the two latter pencils, then the first anharmonic becomes indeterminate, but the second is equal to zero, and the third is infinite. We are, therefore, to consider y and z, but not x, as vanishing for this position of P; and consequently may write,

$$A = (1, 0, 0).$$

In like manner,

$$B = (0, 1, 0), \quad \text{and } C = (0, 0, 1);$$

and on account of these simple representations of its three corners, I call the given triangle ABC the *Unit-Triangle.*

4. Again, let the sides of this given triangle ABC be cut by a given transversal A′B′C′, and by a variable transversal LMN. Then, by another very well known theorem respecting segments, we shall have the relation,

$$(\mathrm{LBA'C}).(\mathrm{MCB'A}).(\mathrm{NAC'B}) = +1;$$

it is therefore permitted to establish the three equations,

$$\frac{m}{n} = (\mathrm{LBA'C}), \quad \frac{n}{l} = (\mathrm{MCB'A}), \quad \frac{l}{m} = (\mathrm{NAC'B});$$

where l, m, n, or any quantities proportional to them, are what I call the *Anharmonic Co-ordinates of the Line* LMN, with respect to the given triangle ABC, and the given transversal A′B′C′. And I *denote* the *line* LMN by the *symbol,*

$$\overline{\mathrm{LMN}} = [l, m, n].$$

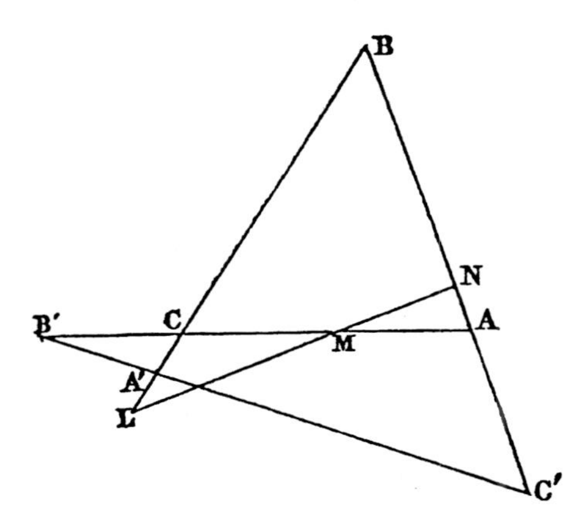

Fig. 2

For example, if this variable line come to coincide with the given line A′B′C′, then

$$l = m = n;$$

so that this given line may be thus denoted,

$$\overline{\mathrm{A'B'C'}} = [1, 1, 1];$$

on which account I call the *given transversal* A′B′C′ the *Unit-Line* of the Figure. The *sides*, BC, &c., of the given triangle ABC, take on this plan the symbols [1, 0, 0], [0, 1, 0], [0, 0, 1].

5. Suppose now that the *unit-point* and *unit-line* are related to each other, as being (in a known sense) *pole* and *polar*, with respect to the given or *unit-triangle*; or, in other words, let the lines OA, OB, OC be supposed to meet the sides BC, CA, AB of that given triangle, in points A″, B″, C″, which are, with respect to those sides, the harmonic conjugates of the points A′, B′, C′, in which the same sides are cut by the given transversal A′B′C′. Also, let the variable *point* P be situated upon the variable *line* LMN; and let Q, R, S be the intersections of AP, BP, CP with BC, CA, AB. Then, because

$$(\mathrm{BA'CA''}) = (\mathrm{CB'AB''}) = (\mathrm{AC'BC''}) = -1,$$

we have

$$\begin{cases} -\dfrac{m}{n} = (\mathrm{LBA''C}), & -\dfrac{n}{l} = (\mathrm{MCB''A}), & -\dfrac{l}{m} = (\mathrm{NAC''B}) \\ -\dfrac{n}{m} = (\mathrm{LCA''B}), & -\dfrac{l}{n} = (\mathrm{MAB''C}), & -\dfrac{m}{l} = (\mathrm{NBC''A}); \end{cases}$$

as well as

$$\begin{cases} \dfrac{y}{z} = (\mathrm{QCA''B}), & \dfrac{z}{x} = (\mathrm{RAB''C}), & \dfrac{x}{y} = (\mathrm{SBC''A}), \\ \dfrac{z}{y} = (\mathrm{QBA''C}), & \dfrac{x}{z} = (\mathrm{RCB''A}), & \dfrac{y}{x} = (\mathrm{SAC''B}); \end{cases}$$

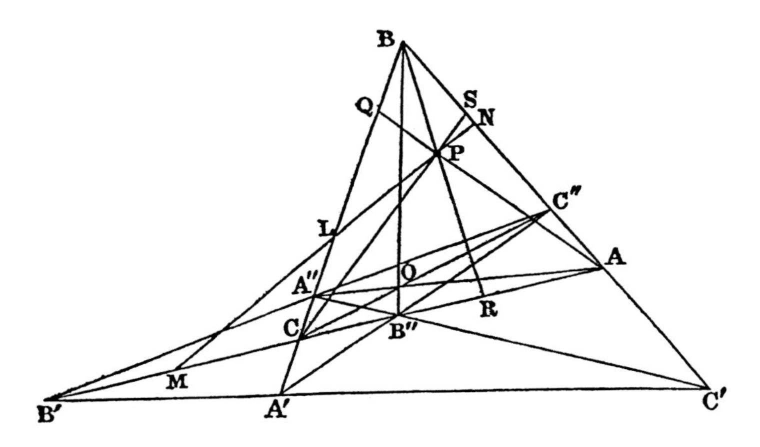

Fig. 3

and therefore,

$$\frac{-lx}{nz} = (\text{MARC}); \quad \frac{-my}{nz} = (\text{LBQC}).$$

But, by the pencil through P,

$$(\text{MARC}) = (\text{LQBC});$$

and by the *definition* of the symbol (ABCD), for any four collinear points,

$$(\text{ABCD}) = \frac{\text{AB}}{\text{BC}} \cdot \frac{\text{CD}}{\text{DA}},$$

which is here throughout adopted, we have the *identity*,

$$(\text{ABCD}) + (\text{ACBD}) = 1;$$

therefore

$$(\text{MARC}) + (\text{LBQC}) = 1,$$

or,

$$lx + my + nz = 0.$$

6. We arrive then at the following *Theorem*, which is of fundamental importance in the present system of Anharmonic Co-ordinates:-

"If the unit-point O be the pole of the unit-line A′B′C′, with respect to the unit-triangle ABC, and if a variable point P, or (x, y, z), be situated anywhere on a variable right line LMN, or $[l, m, n]$, then the sum of the products of the corresponding co-ordinates of point and line is zero."

7. It may already be considered as an evident consequence of this Theorem, that any *homogeneous equation* of the pth *dimension*,

$$f_p(x, y, z) = 0,$$

represents a *curve of the* p^{th} *order*, considered as the *locus* of the variable *point* P; and that any homogeneous equation of the qth dimension, of the form

$$F_q(l, m, n) = 0,$$

may in like manner be considered as the *tangential equation* of a *curve of the* q^{th} *class*, which is the *envelope* of the variable *line* LMN. But any examples of such applications must be reserved for a future communication. Meantime, I may just mention that I have been, for some time back, in possession of an analogous method for treating Points, Lines, Planes, Curves, and Surfaces in Space, by a system of Anharmonic Co-ordinates.

8. As regards the *advantages* of the Method which has been thus briefly sketched, the *first* may be said to be its geometrical *interpretability*, in a manner *unaffected by perspective.* The *relations*, whether between *variables* or between *constants*, which enter into the formula of this method, are *all projective*; because they *all* depend upon, and are referred to, *anharmonic functions*, of groups or of pencils.

9. In the *second* place, we may remark that the great principle of *geometrical Duality* is recognised from the very outset. Confining ourselves, for the moment (as in the foregoing articles), to figures in a *given plane*, we have seen that the *anharmonic co-ordinates* of a *point*, and those of a *right line*, are deduced by processes absolutely *similar*, the one from a system of *four given points*, and the other from a system of *four given right lines*. And the *fundamental equation* ($lx + my + nz = 0$) which has been found to *connect* these *two systems* of co-ordinates, is evidently one of the most perfect *symmetry*, as regards *points* and *lines*. An analogous symmetry will show itself afterwards, in relation to points and planes.

10. The *third advantage* of the anharmonic method may be stated to consist in its possessing an *increased number of disposable constants*. Thus, within the plane, *trilinear* co-ordinates give us *only six* such constants, corresponding to the *three* disposable *positions* of the *sides* of that assumed *triangle*, to the perpendicular distances from which the co-ordinates are supposed to be proportional; but *anharmonics*, by admitting an *arbitrary unit-point*, enable us to treat *two other constants* as disposable, the *number* of such constants being thus raised from *six* to *eight*. Again, in *space*, whereas *quadriplanar* co-ordinates, considered as the *ratios of the distances* from *four assumed planes*, allow of only *twelve* disposable constants, corresponding to the possible selection of the *four planes of reference*, *anharmonic co-ordinates*, on the contrary, which admit either *five planes* or *five points* as *data*, and which might, therefore, be called *quinquiplanar* or *quinquipunctual*, permit us to dispose of no fewer than *fifteen constants as arbitrary*, in the general treatment of *surfaces*.

11. To myself it naturally appears as a *fourth advantage* of the anharmonic method, that it is found to harmonize well with the method of *quaternions*, and was in fact *suggested* thereby; though not without suggestions from other methods previously known.

12. Thus, if α, β, γ denote three given vectors, OA, OB, OC, from a given origin O, while a, b, c are three given and constant scalars, but t, u, v are three variable scalars, subject to the condition that their sum is zero,

$$t + u + v = 0;$$

then the equation,

$$\text{OP} = \rho = \frac{t^r a\alpha + u^r b\beta + v^r c\gamma}{t^r a + u^r b + v^r c},$$

in which r is any positive and whole exponent, expresses generally that the *locus* of the point P is a *curve of the* r^{th} *order*, in the given plane of ABC; which curve has the property, that it is met in r coincident points, by any one of the three sides of the given triangle ABC. But the coefficients $t^r\ u^r\ v^r$ are examples here of what have been above called anharmonic co-ordinates.

13. Proceeding to *space*, let A, B, C, D be the four corners of a given triangular pyramid, and let E be any fifth given point, which is not on any one of the four faces of that pyramid. Let P be any sixth point of space; and let $xyzw$ be four positive or negative numbers, such that

$$\frac{x}{w} = (\text{BC.AEDP}), \quad \frac{y}{w} = (\text{CA.BEDP}), \quad \frac{z}{w} = (\text{AB.CEDP});$$

the right-hand member of these equations representing *anharmonics of pencils of planes*, in a way which is easily understood, with the help of the definition (5) of the symbol (ABCD). Then I call x, y, z, w (or any numbers proportional to them), the *Anharmonic Co-ordinates* of the *Point* P, with respect to what may be said to be the *Unit-Pyramid*, ABCD, because its corners may (on the present plan) be thus denoted,

$$\text{A} = (1, 0, 0, 0); \quad \text{B} = (0, 1, 0, 0); \quad \text{C} = (0, 0, 1, 0); \quad \text{D} = (0, 0, 0, 1);$$

and with respect to that fifth given point E, which may be called the *Unit-Point*, because its symbol, in the present system, may be thus written:-

$$\text{E} = (1, 1, 1, 1).$$

And I denote the general or variable point by the symbol,

$$\text{P} = (x, y, z, w).$$

14. When we have thus five given points, A ... E, of which no four are situated in any common plane, we can connect any two of them by a right line, and the three others by a plane, and determine the point in which these last intersect each other, deriving in this way a system of ten lines, ten planes, and ten points, whereof the latter may be thus denoted:-

$$\begin{aligned} &\text{A}' = \text{BC}^{\cdot}\text{ADE} = (0, 1, 1, 0), \quad \text{B}' = \&\text{c.}, \quad \text{C}' = \&\text{c.}; \\ &\text{A}_1 = \text{AE}^{\cdot}\text{BCD} = (0, 1, 1, 1), \quad \text{B}_1 = \&\text{c.}, \quad \text{C}_1 = \&\text{c.}; \\ &\text{A}_2 = \text{AD}^{\cdot}\text{BCE} = (1, 0, 0, 1), \quad \text{B}_2 = \&\text{c.}, \quad \text{C}_2 = \&\text{c.}; \\ &\text{D}_1 = \text{DE}^{\cdot}\text{ABC} = (1, 1, 1, 0), \quad ; \end{aligned}$$

and the harmonic conjugates of these last points, with respect to the ten given lines on which they are situated, may on the same plan be represented by the following symbols:-

$$\begin{aligned} &\text{A}'' = (0, 1, -1, 0), \quad \text{B}'' = \&\text{c.} \quad \text{C}'' = \&\text{c.} \\ &\text{A}_1' = (2, 1, \ \ 1, 1), \quad \text{B}_1' = \&\text{c.} \quad \text{C}_1' = \&\text{c.} \\ &\text{A}_2' = (1, 0, 0, -1), \quad \text{B}_2' = \&\text{c.} \quad \text{C}_2' = \&\text{c.} \\ &\text{D}_1' = (1, 1, 1, \ \ 2); \end{aligned}$$

so that

$$(\text{BA}'\text{CA}'') = \ldots = (\text{EA}_1\text{AA}_1') = \ldots = (\text{DA}_2\text{AA}_2') = \ldots = (\text{ED}_1\text{DD}_1') = -1.$$

15. Let any plane Π intersect the three given lines DA, DB, DC in points Q, R, S; and let $lmnr$ be any positive or negative numbers, such that

$$\frac{l}{r} = (\text{DA}_2'\text{AQ}), \quad \frac{m}{r} = (\text{DB}_2'\text{BR}), \quad \frac{n}{r} = (\text{DC}_2'\text{CS});$$

then I call l, m, n, r, or any numbers proportional to them, the *Anharmonic Coordinates* of the *Plane* Π; which *plane* I also denote by the *Symbol*,

$$\Pi = [l,\ m,\ n,\ r].$$

In particular the four faces of the unit pyramid come thus to be denoted by the symbols,

$$\text{BCD} = [1,\ 0,\ 0,\ 0], \quad \text{CAD} = [0,\ 1,\ 0,\ 0], \quad \text{ABD} = [0,\ 0,\ 1,\ 0], \quad \text{ABC} = [0,\ 0,\ 0,\ 1];$$

and the six planes through its edges, and through the unit point, are denoted thus:-

$$\text{BCE} = [1,\ 0,\ 0,\ -1]; \quad \text{CAE} = [0,\ 1,\ 0,\ -1]; \quad \text{ABE} = [0,\ 0,\ 1,\ -1];$$

$$\text{ADE} = [0,\ 1,\ -1,\ 0]; \quad \text{BDE} = [-1,\ 0,\ 1,\ 0]; \quad \text{CDE} = [1,\ -1,\ 0,\ 0];$$

in connexion with which last planes it may be remarked that we have, generally, as a consequence of the foregoing definitions, the formulæ,

$$\frac{n}{m} = (\text{BA}''\text{CL}), \quad \frac{l}{n} = (\text{CB}''\text{AM}), \quad \frac{m}{l} = (\text{AC}''\text{BN}),$$

if L, M, N be the points in which a variable plane Π intersects the sides BC, &c., of the given triangle ABC: as we have also, generally,

$$\frac{z}{y} = (\text{AD.CEBP}), \quad \frac{x}{z} = (\text{BD.AECP}), \quad = \frac{y}{x}(\text{CD.BEAP}).$$

16. If a *point*, $\text{P} = (xyzw)$, be situated *on a plane*, $\Pi = [lmnr]$, then I find that the following relation between their co-ordinates exists, which is entirely analogous to that already assigned (6) for the case of a *point* and *line* in a given plane, and is of fundamental importance in the application of the present *Anharmonic Method* to *space*:

$$lx + my + nz + rw = 0;$$

or in words, "*the sum of the products of corresponding co-ordinates*, of point and plane, *is zero.*"

For example, all planes through the unit point (1, 1, 1, 1) are subject to the condition,

$$l + m + n + r = 0,$$

as may be seen for the six planes (15) already drawn through that point E; and the six points $\text{A}''\text{B}''\text{C}''\text{A}_2'\text{B}_2'\text{C}_2'$ (14), in which the six edges BC, CA, AB, DA, DB, DC, of the given or unit pyramid ABCD, intersect the six corresponding edges of the inscribed and homologous pyramid $\text{A}_1\text{B}_1\text{C}_1\text{D}_1$, with the unit point E for their centre of homology, are all ranged on one common plane of homology, of which the *equation* and the *symbol* may be thus written,

$$x + y + r + w = 0, \quad [\text{E}] = [1,\ 1,\ 1,\ 1],$$

and which may be called (comp. 4) the *Unit-Plane.*

17. Any four collinear points, P_0, P_1, P_2, P_3, have their anharmonic symbols connected by two equations of the forms,

$$(\text{P}_1) = t(\text{P}_0) + u(\text{P}_2), \quad \text{P}_3 = t'(\text{P}_0) + u'(\text{P}_2)$$

each including four ordinary linear equations between the co-ordinates of the four points, such as

$$x_1 = tx_0 + ux_2, \quad y_1 = ty_0 + uy_2, \quad \text{\&c.};$$

and the anharmonic of their *group* is then found to be expressed by the formula,

$$(P_0, P_1, P_2, P_3) = \frac{ut'}{tu'}.$$

And similarly, if any four planes $\Pi_0 . . \Pi_3$ be collinear (that is, if they have any one right line common to them all), their symbols satisfy two linear equations of the corresponding forms,

$$[\Pi_1] = t[\Pi_0] + u[\Pi_2], \qquad [\Pi_3] = t'[\Pi_0] = u'[\Pi_2];$$

and the anharmonic of the *pencil* is,

$$(\Pi_0 \Pi_1 \Pi_2 \Pi_3) = \frac{ut'}{tu'}.$$

18. If $\phi(xyzw)$ be any *rational fraction*, the numerator and denominator of which are any two given homogeneous and linear functions of the co-ordinates of a *variable point*; and if we determine a *line* Λ, and *three planes* Π_0, Π_1, Π_2 through that line, by the four *local* equations,

$$\phi = \frac{0}{0}, \quad \phi = \infty, \quad \phi = 1, \quad \phi = 0;$$

then I find that the function ϕ may be expressed as the anharmonic of a *pencil of planes*, as follows:-

$$\phi(xyzw) = (\Pi_0 \Pi_1 \Pi_2 \Pi);$$

where Π is the variable plane ΛP, which passes through the fixed line Λ, and through the variable point $P = (xyzw)$.

19. And in like manner, as the *geometrical dual* (9) of this last theorem (18), if $\Phi(lmnr)$ be any rational fraction, of which the numerator and denominator are any two given functions, homogeneous and linear, of the co-ordinates of a *variable plane*; and if we determine a *line* Λ, and *three points* P_0, P_1, P_2 on that line, by the four *tangential* equations,

$$\Phi = \frac{0}{0}, \quad \Phi = \infty, \quad \Phi = 1, \quad \Phi = 0;$$

I find that the proposed function Φ may then be thus expressed as the anharmonic of a *group of points*.

$$\Phi(lmnr) = (P_0 P_1 P_2 P);$$

P here denoting the variable point $\Lambda \cdot \Pi$, in which the fixed line Λ intersects the variable plane $\Pi = [lmnr]$.

20. All problems respecting *intersections of lines with planes*, &c., are resolved, with the help of the Fundamental Theorem (16) respecting the relation which exists between the anharmonic co-ordinates of point and plane, as easily by the present method, as by the known method of *quadriplanar* co-ordinates (10); and indeed, by the very *same mechanism*, of which it is therefore unnecessary here to speak.

But it may be proper to say a few words respecting the application of the anharmonic method to *Surfaces* (7); although here again the known mechanism of *calculation* may in great part be preserved unchanged, and only the *interpretations* need be *new*.

21. In general, it is easy to see (comp. 7) that, in the present method, as in older ones, the *order* of a curved surface is denoted by the *degree* of its *local equation*, $f(xyzw) = 0$; and that the *class* of the same surface is expressed, in like manner, by the degree of its *tangential equation*, $F(lmnr) = 0$: because the *former* degree (or dimension) determines the *number of points* (distinct or coincident, and real or imaginary), in which the surface, considered as a locus, is *intersected* by an arbitrary right *line*; while the *latter* degree determines the *number of planes* which can be drawn *through* an arbitrary right line, so as to touch the same surface, considered as an *envelope*. It may be added, that I find the *partial derivatives of each* of these two functions, f and F, to be proportional to the *co-ordinates* which enter as variables into the *other*; thus we may write

$$[D_x f,\ D_y f,\ D_z f,\ D_w f],$$

as the symbol (15) of the *tangent plane* to the *locus* f, at the point $(xyzw)$; and

$$(D_l F,\ D_m F,\ D_n F,\ D_r F),$$

as a symbol for the *point of contact* of the *envelope* F, with the plane $[lmnr]$: whence it is easy to conceive how problems respecting the *polar reciprocals* of *surfaces* are to be treated.

22. As a very simple *example*, the surface of the *second order* which passes through the *nine points*, above called ABCDEA$'$A$_2$C$'$C$_2$, is easily found to have for its *local* equation, $0 = f = xz - yw$; whence the co-ordinates of its tangent plane are, $l = D_x f = z$, $m = D_y f = -w$, $n = D_z f = x$, $r = D_w f = -y$, and its *tangential* equation is, therefore, $0 = F = ln - mr$, so that it is also a surface of the *second class*. In fact it is the *hyperboloid* on which the gauche quadrilateral ABCD is superscribed, and which passes also through the point E; and the known *double generation*, and *anharmonic properties*, of this surface, may easily be deduced from either of the foregoing forms of its anharmonic equation, whereof the first may (by 13, 15) be expressed as an equality between the anharmonic functions of two *pencils of planes*, in either of the two following ways:-

$$(\mathrm{BC.AEDP}) = (\mathrm{DA.BECP}); \quad (\mathrm{AB.CEDP}) = (\mathrm{CD.BEAP}).$$

XII.

ON GEOMETRICAL NETS IN SPACE

[Communicated June 24, 1861]
[*Proceedings of the Royal Irish Academy*, **7**, 532–582 (1862).]

June 24, 1861.

THE VERY REV. DEAN GRAVES, President, in the Chair.

The following communication was read:-

"On Geometrical Nets in Space." By SIR WM. R. HAMILTON, LL. D., Astronomer Royal of Ireland, and Andrews' Professor of Astronomy.

1. When any five points of space, ABCDE, are given, whereof no four are supposed to be complanar, we can connect any two of them by a right line, and the three others by a plane, and determine the point in which these last intersect each other: *deriving* thus a system of *ten lines* Λ_1, *ten planes* Π_1, and *ten points* P_1, from the *given* system of *five points* P_0, by what may be called a *First Construction.*

We may next propose to determine all the new and distinct lines Λ_2, and planes Π_2, which connect the ten derived points P_1, with the five given points P_0, and with each other; and may then inquire what new and distinct points P_2 arise, as intersections* $\Lambda{\cdot}\Pi$ of lines and planes already obtained: all *such* new lines, planes, and points being said to belong to a *Second Construction.* And then we might proceed, on the same plan, to a *Third* Construction, and to indefinitely many others following: building up thus what Professor *Möbius*, in his *Barycentric Calculus*,† has proposed to call a *Geometrical Net in Space.*

2. In general, if n denote five or any greater number of *independent* points of space, the number of the derived points of the *form* $\Lambda{\cdot}\Pi$, or AB·CDE, which can be obtained by what is relatively to *them* a First Construction, of the kind just now described, is easily seen to be the function,

$$f(n) = \frac{n(n-1)}{2} \cdot \frac{(n-2)(n-3)(n-4)}{2.3};$$

so that $f(5) = 10$, as above, but $f(15) = 30030$. If then the *fifteen points* P_0, P_1 were thus *independent*, or *unconnected* with each other, we might expect to find that the number of points

* Intersections $\Lambda{\cdot}\Lambda$ of *line with line* (when complanar) are *included* in this class $\Lambda.\Pi$; and intersections $\Pi{\cdot}\Pi{\cdot}\Pi$ of *three distinct planes*, when *not* included at this stage, may be reserved for a *subsequent construction*, in which they naturally offer themselves, as of the standard form $\Lambda{\cdot}\Pi$.

† Der Calcul Barycentrische, Leipzig, 1827, p. 291. Some first results connected with the subject were given, according to the writer's recollection, in a Memoir by *Carnot* on *Transversals*, to which he cannot at present refer.

P_2 *derived* from them, at the next stage, should *exceed thirty thousand.* And although it was obvious that many *reductions* of this number must occur, on account of the *dependence* of the ten points P_1 on the five points P_0, yet when I happened to feel a curiosity, some time ago, to determine the precise *number* of those which have been above called *Points of Second Construction,* and to assign their chief geometrical relations to each other, and to the fifteen former points, it must be confessed that I thought myself about to undertake the solution of a rather formidable Problem. But the motive which had led me to attack that problem, namely the desire to try the efficiency of a certain system of *Quinary Symbols,* for points, lines, and planes in space, which the *Method of Vectors* had led me to invent, inspired me with a hope, which I trust that the result of the attempt has not altogether failed to justify. And, in the present communication, I wish first to present some outline of what may be called perhaps a *Quinary Calculus,* before proceeding to give, in the second place, some sketch of the results of its application to the geometrical *Net in Space.*

Part I. On a Quinary Calculus for Space

3. Let ABCDE be (as in **1.**) any five given points of space, whereof no four are situated in any common plane; then, by decomposing ED in the directions of EA, EB, EC, we can always obtain an equation of the form,

$$a.\mathrm{EA} + b.\mathrm{EB} + c.\mathrm{EC} + d.\mathrm{ED} = 0, \tag{1}$$

in which the coefficients *abcd* have determined ratios. And if we next introduce a fifth coefficient *e*, such that

$$a + b + c + d + e = 0, \tag{2}$$

and add to (1) the identity

$$(a + b + c + d + e)\mathrm{OE} = 0, \tag{3}$$

in which O is any arbitrary point (or origin of vectors), we arrive at the following equivalent but more symmetric form,

$$a.\mathrm{OA} + b.\mathrm{OB} + c.\mathrm{OC} + d.\mathrm{OD} + e.\mathrm{OE} = 0, \tag{4}$$

in which *abcde* may be called the *five* (*numerical*) *constants* of the given system of *five points*, A ... E, although only their *ratios* are important, and (as above) their *sum* is *zero.*

4. Let P be any other point of space, and let *xyzwv* be coefficients satisfying the equation,

$$(x - v)\,a.\mathrm{PA} + (y - v)\,b.\mathrm{PB} + (z - v)\,c.\mathrm{PC} + (w - v)\,d.\mathrm{PD} = 0; \tag{5}$$

then, adding the identity,

$$v(a.\mathrm{PA} + b.\mathrm{PB} + c.\mathrm{PC} + d.\mathrm{PD} + e.\mathrm{PE}) = 0, \tag{6}$$

which results from (4), we obtain this other symmetric formula,

$$xa.\mathrm{PA} + yb.\mathrm{PB} + zc.\mathrm{PC} + wd.\mathrm{PD} + ve.\mathrm{PE} = 0, \tag{7}$$

which may also be thus written,

$$\text{OP} = \frac{xa.\text{OA} + yb.\text{OB} + zc.\text{OC} + wd.\text{OD} + ve.\text{OE}}{xa \quad + yb \quad + zc \quad + wd \quad + ve}, \tag{8}$$

O being again an arbitrary origin; and the *five new and variable coefficients*, *xyzwv*, whereof the *ratios of the differences* determine the *position of the point* P, when the five points A . . E are given, may be called the *Quinary Coordinates of that Point* P, with respect to the given system of five points.

5. Under these conditions, we may agree to write, briefly,

$$\text{P} = (x,\ y,\ z,\ w,\ v), \text{ or even } \text{P} = (xyzwv), \tag{9}$$

whenever it seems that the omission of the commas will not give rise to any confusion; and may call this form a *Quinary Symbol of the Point* P. But because (as above) only the ratios of the differences of the coefficients or coordinates are important, we may establish the following *Formula of Quinary Congruence*, between two *equivalent Symbols* of one *common point*,

$$(x'\,y'\,z'\,w'v') \equiv (xyzwv), \tag{10}$$

$$\text{if } x' - v' : y' - v' : z' - v' : w' - v' = x - v : y - v : w - v; \tag{11}$$

reserving the *Quinary Equation*,

$$(x'\,y'\,z'\,w'v') = (xyzwv), \tag{12}$$

to imply the coexistence of the *five* separate and ordinary equations,

$$x' = x,\ y' = y,\ z' = z,\ w' = w,\ v' = v. \tag{13}$$

We shall also adopt, as abridgments of notation, the formulæ,

$$t(x,\ y,\ z,\ w,\ v) = (tx,\ ty,\ tz,\ tw,\ tv); \tag{14}$$

$$(x' \,.\,.\, v') \pm (x \,.\,.\, v) = (x' \pm x,\ \,.\,.\, v' \pm v); \tag{15}$$

and shall find it convenient to employ occasionally what may be called the *Quinary Unit Symbol*,

$$\text{U} = (11111); \tag{16}$$

although *this* symbol represents *no determined point*, because both the denominator and numerator of the expression (8) vanish, by (2) and (4), when the five coefficients *xyzwv* become each equal to unity.

6. With these notations, if Q and Q′ be any *other* quinary symbols, and t and u any two coefficients, we shall have the congruence,

$$\text{Q}' \equiv \text{Q}, \quad \text{if } \text{Q}' = t\text{Q} + u\text{U}; \tag{17}$$

the *two points* P and P′, which are denoted by these *two symbols*, in this case *coinciding*. Again the equation,

$$\text{Q}'' = t\text{Q} + t'\text{Q}' + u\text{U}, \tag{18}$$

is found to express that Q, Q′, Q″ are symbols of *three collinear points*; and the *complanarity of four points*, of which the symbols are Q, Q′, Q″, Q‴, is expressed by this other equation of the same form,

$$Q''' = tQ + t'Q' + t''Q'' + uU. \tag{19}$$

7. If then a *variable point* P be thus *complanar* with *three given points*, P_0, P_1, P_2, its coordinates
4. must be connected with theirs, by five equations of the form,

$$x = t_0 x_0 + t_1 x_1 + t_2 x_2 + u; \;..\; v = t_0 v_0 + t_1 v_1 + t_2 v_2 + u; \tag{20}$$

whence, by elimination of the four arbitrary coefficients $t_0 t_1 t_2 u$, a *linear equation* is obtained, of the form

$$lx + my + nz + rw + sv = 0, \tag{21}$$

with the general relation

$$l + m + n + r + s = 0 \tag{22}$$

between its coefficients; and this equation (21) may be said to be the *Quinary Equation of the Plane* $P_0P_1P_2$. The five new coefficients *lmnrs* may be called the *Quinary Coordinates of that Plane*; and the plane itself may be denoted by the *Quinary Symbol*,

$$\Pi = [l, m, n, r, s], \text{ or briefly, } \Pi = [lmnrs], \tag{23}$$

when the commas can be omitted without confusion.
R, R′, .. be symbols of this form, for planes Π, Π′, .., then the equation

$$R' = tR, \tag{24}$$

in which t is an arbitrary coefficient, expresses that the *two planes* Π, Π′ *coincide*; the equation

$$R'' = tR + t'R' \tag{25}$$

expresses that the *three planes* Π, Π′, Π″ are *collinear*, or that the *third* passes *through the line of intersection* of the *other two*; and the equation

$$R''' = tR + t'R' + t''R'' \tag{26}$$

expresses that the *four planes* Π, Π′, Π″, Π‴ are *compunctual* (or *concurrent*), or that the *fourth* passes *through the point of intersection* of the *other three*.

8. It is easy to conceive how problems respecting *intersections of lines and planes* can be resolved, on the foregoing principles. And if we define that a point P, or plane Π, is a *Rational Point*, or a *Rational Plane*, of the *System* determined by the *five given Points* A .. E, or that it is *rationally related* to those five points, when its *coordinates* are equal (or proportional) to *whole numbers*, it is obvious, from the nature of the *eliminations* employed, that a *plane* which is determined as containing *three rational points*, or a *point* which is determined as the intersection of *three rational planes*, is itself, in the above sense, *rational*. We may also say that a *right line* Λ is a *Rational Line*, when it is the line P·P which *connects* two rational points, or the *intersection* Π·Π of two rational planes: and then the intersection of a rational line with a rational plane, or of two complanar and rational lines with each other, will be a rational point.

9. When any two points, P, P′, or any two planes, Π, Π', have symbols which differ only by the *arrangement* (or *order*) of the five coefficients or coordinates in each, those points, or those planes, may then be said to have one *common type*; or briefly, to be *syntypical*. For example, the five *given* points are thus syntypical, because (omitting commas, as in **5.**) their symbols are,

$$\mathrm{A} = (10000), \quad \mathrm{B} = (01000), \quad \mathrm{C} = (00100), \quad \mathrm{D} = (00010), \quad \mathrm{E} = (00001). \tag{27}$$

In general, any two syntypical points, or planes, admit of being *derived* from the five given points, by precisely *similar processes of construction*, the *order* only of the *data* being *varied*; and in the *most general* case, a *single type* includes 120 *distinct points*, or *distinct planes*, although this *number* may happen to be diminished, even when the coordinates are all unequal: for example, the type (12345) includes only *sixty* distinct points, because, by (17), we have in this case the congruence,

$$(12345) \equiv (54321). \tag{28}$$

10. The *anharmonic function* of any group of four collinear points ABCD being denoted by the symbol (ABCD), and defined by the equation,

$$(\mathrm{ABCD}) = \frac{\mathrm{AB}}{\mathrm{BC}} \cdot \frac{\mathrm{CD}}{\mathrm{DA}} = \frac{\mathrm{AB}}{\mathrm{CB}} : \frac{\mathrm{AD}}{\mathrm{CD}}, \tag{29}$$

it will be found that if $\mathrm{P}_0 \,..\, \mathrm{P}_3$ be thus *any four collinear points*, of which therefore, by (18), the quinary symbols $Q_0 \,..\, Q_3$ are connected by two linear relations, of the forms,

$$Q_1 = t_0 Q_0 + t_2 Q_2 + uU, \qquad Q_3 = t_0' Q_0 + t_2' Q_2 + u'\mathrm{U}, \tag{30}$$

then the *anharmonic of this group of points* is given by the formula,

$$(\mathrm{P}_0\mathrm{P}_1\mathrm{P}_2\mathrm{P}_3) = \frac{t_2\, t_0'}{t_0\, t_2'}, \tag{31}$$

of which the applications are numerous and important.

And in like manner, if $\Pi_0 \,..\, \Pi_3$ be *any four collinear planes*, of which consequently, by (25), the symbols $R_0 \,..\, R_3$ are connected by two other linear relations, such as

$$R_1 = t_0 R_0 + t_2 R_2, \qquad R_3 = t_0' R_0 + t_2' R_2, \tag{32}$$

we have then this other very useful formula of the same kind, for the *anharmonic of this pencil of planes*,

$$(\Pi_0\Pi_1\Pi_2\Pi_3) = \frac{t_2\, t_0'}{t_0\, t_2'}; \tag{33}$$

it being understood that the anharmonic function of such a *pencil* is the same as that of the *groups* of *points*, in which its *planes* are *cut* by any rectilinear *transversal*: so that we may write generally, for *any six points* A .. F, the formula,

$$(\mathrm{EF.ABCD}) = (\mathrm{A'B'C'D'}), \tag{34}$$

if any transversal GH cut the four planes EFA, .. EFD in the four points A′, .. D′; or in symbols, if

$$\mathrm{A'} = \mathrm{GH{\cdot}EFA}, \; ..\, \mathrm{D'} = \mathrm{GH{\cdot}EFD}. \tag{35}$$

11. The expression of fractional form,

$$\varphi(xyzwv) = \frac{l'x + m'y + n'z + r'w + s'v}{lx + my + nz + rw + sv} = \frac{f'}{f}, \tag{36}$$

in which the ten coefficients, $l \,..\, s$ and $l' \,..\, s'$, are supposed to be given, and to be such (comp. (22)) that

$$l + .. + s = 0, \quad \text{and} \quad l' + .. + s' = 0, \tag{37}$$

may represent the quotient of any two linear and homogeneous functions, f and f', of the coordinates $x \,..\, v$ of a variable point P, or rather of the *differences* of those coordinates (comp. 5.); and if we assign any *particular* or *constant value,* such as k, to this *quotient,* or *fractional function,* φ, the equation so obtained will represent (comp. (21)) a *plane locus* for that point, which *plane* Π will always pass *through a given line* Λ, determined by equating separately the denominator and numerator of φ to zero. Hence the *four equations,*

$$f = 0, \quad f' = f, \quad f' = 0, \quad f' = kf, \tag{38}$$

which answer to the four values,

$$\varphi = \infty, \quad \varphi = 1, \quad \varphi = 0, \quad \varphi = k, \tag{39}$$

represent a *pencil of four planes* $\Pi_0 \,..\, \Pi_3$, of which the quinary symbols (23) may be thus written:-

$$R_0 = [lmnrs]; \quad R_2 = [l'm'n'r's']; \quad R_1 = R_2 - R_0; \quad R_3 = R_2 - kR_0; \tag{40}$$

and of which the *anharmonic* is consequently, by (33), the same *quotient,*

$$(\Pi_0\Pi_1\Pi_2\Pi_3) = (k = \varphi =)\frac{f'}{f}, \tag{41}$$

as before. We have therefore this *Theorem*:-

"*The Quotient of any two given homogeneous and linear Functions, of the Differences of the Quinary Coordinates of a variable Point in Space, can always be expressed as the Anharmonic of a Pencil of Planes, whereof three are given, while the fourth passes through the variable Point, and through a given Right Line, which is common to the three former Planes.*"

12. For example, we find thus that

$$\frac{x - v}{w - v} = (\text{BC.AEDP}); \quad \frac{y - v}{w - v} = (\text{CA.BEDP}); \quad \frac{z - v}{w - v} = (\text{AB.CEDP}); \tag{42}$$

and that

$$\frac{x - v}{y - v} = (\text{CD.AEBP}); \quad \frac{y - v}{z - v} = (\text{AD.BECP}); \quad \frac{z - v}{x - v} = (\text{BD.CEAP}); \tag{43}$$

the product of these three last anharmonics of pencils being therefore equal to positive unity, so that we have, for *any six points of space,* ABCDEF, the general equation,

$$(\text{AD.BECF}).(\text{BD.CEAF}).(\text{CD.AEBF}) = 1. \tag{44}$$

If then we *suppress the fifth coefficient,* v, *in the quinary symbol* (9) *of a point* P, which comes to first substituting, as the congruence (10) permits, the differences $x - v$, $y - v$, $z - v$, $w - v$, and $v - v$ or 0, for x, y, z, w, and v, and then writing simply $x, .. w$ instead of $x - v$, $.. w - v$, and omitting the final zero, whereby the quinary symbol (00001) for the fifth given point E (27) becomes first (−1, −1, −1, −1, 0) or (11110), and is then reduced to the *quaternary unit symbol* (1111), we shall *fall back on that system of anharmonic coordinates in space*, of which some account was given in a former communication* to this Academy: the *anharmonic* (or *quaternary*) *symbol of a plane* Π being, in like manner, *derived from the quinary symbol* (23), by simply *suppressing the fifth coefficient*, or coordinate, *s*. *Anharmonic coordinates*, whether for *point* or for *plane*, are therefore *included in quinary ones*; but although they have some advantages of *simplicity*, it appears that their *less perfect symmetry*, of reference to the *five given points* A . . E, renders them less adapted to investigations respecting the *Geometrical Net in Space*, which is constructed with those *five* points as data: and that therefore they are less fit than *quinary* coordinates for the purposes of the present paper.

13. Retaining then the *quinary form*, we may next observe that although, *when the five coefficients* $l \,..\, s$ *are given*, as in **7.**, and the *coordinates* $x \,..\, v$ of a *point* P are *variable*, the *linear equation* $lx + \,..\, + sv = 0$ (21) may be said to be the *Local Equation of a Plane*, namely of the *plane* $[l \,..\, s]$, considered as the *locus of the point* $(x \,..\, v)$; yet if, on the contrary, we *now* regard $x \,..\, v$ as *given*, and $l \,..\, s$ as *variable*, the *same linear equation* (21) expresses the *condition* necessary, in order that a *variable plane* $[l \,..\, s]$ may pass *through a given point* $(x \,..\, v)$; and *in this view*, the formula (21) may be considered to be the *Tangential Equation of that given Point*. Thus the very simple equation,

$$l = 0, \tag{45}$$

expresses the condition requisite for the plane $[l \,..\, s]$ passing through the given point (10000), or A (27); and it is, in that sense, the tangential *equation of that point*: while $m = 0$ is, in like manner, the equation of B, &c. This being understood, if we suppose that F and F′ denote two given, linear, and homogeneous functions of the coordinates $l \,..\, s$ of a variable plane Π, we may consider the four equations,

$$\mathrm{F} = 0, \quad \mathrm{F}' = \mathrm{F}, \quad \mathrm{F}' = 0, \quad \mathrm{F}' = k\mathrm{F}, \tag{46}$$

as the tangential equations of *four collinear points*, P_0, P_1, P_2, P_3, whereof the three first are entirely given, but the fourth varies with the value of the coefficient k, although always remaining on the line Λ of the other three; and then it is easy to deduce, from the formula (31), by reasonings analogous to those employed in **11.**, the following *anharmonic of the group*:

$$(\mathrm{P}_0\mathrm{P}_1\mathrm{P}_2\mathrm{P}_3) = k = \frac{\mathrm{F}_1}{\mathrm{F}}. \tag{47}$$

We have therefore this new *Theorem*, analogous to one lately stated:-

"*The Quotient of any two given, homogeneous, and linear Functions, of the Quinary Coordinates of a variable Plane, may always be expressed as the Anharmonic of a Group of Points; whereof three are given and collinear, while the fourth is the Intersection of the variable Plane with the given Line on which the other three are situated.*"

14. For example, if we wish in this way to *interpret the quotient* $m : n$, of these two coordinates of a *variable plane* Π, or $[lmnrs]$ (23), as denoting the *anharmonic of a group of points*, the three first points P_0, P_1, P_2 of that group (47) have here for their tangential equations,

* See the Proceedings for the Session of 1859–60.

$$n = 0, \quad m - n = 0, \quad m = 0, \tag{48}$$

whereof the *third* has recently been seen **13.** to represent the given point B, and the *first* represents in like manner another given point, namely C, of the initial system: while the *second* represents the point $(0, 1, -1, 0, 0)$, or briefly $(01\overline{1}00)$, if, to save commas, we write $\overline{1}$ for -1. To *construct* this last point, let us write

$$\text{A}' = (01100) \equiv (10011), \quad \text{and} \quad \text{A}'' = (01\overline{1}00); \tag{49}$$

then, by (18), these two new points A′ and A″ are each *collinear* with B, C, or are on the *line* BC; and they are, with respect to that line (or to its extreme points) *harmonically conjugate* to each other, because the formula (31) gives easily, by the *first* symbol for A′, the *harmonic equation*,

$$(\text{BA}'\text{CA}'') = -1; \tag{50}$$

but also the *second* (or congruent) symbol for A′ shows, by (19), that A′ is in the *plane* ADE; we may therefore write the *formula of intersection*,

$$\text{A}' = \text{BC}^{\cdot}\text{ADE}, \tag{51}$$

whereby this point A′ is entirely determined; and then the point A″, as being its harmonic conjugate with respect to B and C, or as satisfying the equation (50), is to be considered as being itself a known point. We have thus assigned the three first points P_0, P_1, P_2, of the *group* (47), namely the points C, A″, B; and if we denote by L the point $\text{BC}^{\cdot}\Pi$ in which the variable plane Π, or $[l\,..\,s]$, intersects the given line BC, so that

$$\text{L} = (0,\ n,\ -m,\ 0,\ 0), \text{ or briefly, } \text{L} = (0\,n\,\overline{m}00), \tag{52}$$

writing $\overline{m}$ for $-m$, then the fourth point P_3 is L; and the required *formula of interpretation* for the quotient $m:n$ becomes,

$$\frac{m}{n} = (\text{CA}''\text{BL}). \tag{53}$$

In like manner, if we write

$$\text{B}' = (10100), \quad \text{C}' = (11000), \quad \text{B}'' = (\overline{1}0100), \quad \text{C}'' = (1\overline{1}000), \tag{54}$$

and

$$\text{M} = (\overline{n}0\,l00), \quad \text{N} = (m\overline{l}000), \tag{55}$$

in which $\overline{n} = -n$, and $\overline{l} = -l$, so that $\text{M} = \text{CA}^{\cdot}\Pi$, $\text{N} = \text{AB}^{\cdot}\Pi$, and

$$\text{B}' = \text{CA}^{\cdot}\text{BDE}, \quad \text{C}' = \text{AB}^{\cdot}\text{CDE}, \quad (\text{CB}'\text{AB}'') = (\text{AC}'\text{BC}'') = -1, \tag{56}$$

we shall have these two other formulæ of interpretation, analogous to (53),

$$\frac{n}{l} = (\text{AB}''\text{CM}), \quad \frac{l}{m} = (\text{BC}''\text{AN}); \tag{57}$$

and therefore,

$$(\text{AB}''\text{CM}).(\text{BC}''\text{AN}).(\text{CA}''\text{BL}) = 1. \tag{58}$$

15. Again, if we denote by Q, R, S the intersections DA·Π, DB·Π, DC·Π, so that

$$\text{Q} = (\bar{r}00\,l0), \quad \text{R} = (0\bar{r}0\,m0), \quad \text{S} = (00\bar{r}\,n0), \tag{59}$$

where $\bar{r} = -r$; if also we introduce seven new points syntypical **9.** with the three points A′B′C′, and seven others syntypical with A″B″C″, as follows:

$$\text{A}_1 = (10001), \quad \text{B}_1 = (01001), \quad \text{C}_1 = (00101), \quad \text{D}_1 = (00011); \tag{60}$$

$$\text{A}_2 = (10010), \quad \text{B}_2 = (01010), \quad \text{C}_2 = (00110); \tag{61}$$

$$\text{A}_1' = (1000\bar{1}), \quad \text{B}_1' = (0100\bar{1}), \quad \text{C}_1' = (0010\bar{1}); \quad \text{D}_1' = (0001\bar{1}); \tag{62}$$

$$\text{A}_2' = (100\bar{1}0), \quad \text{B}_2' = (010\bar{1}0), \quad \text{C}_2' = (001\bar{1}0); \tag{63}$$

so that, by principles already established, we shall have the seven relations of intersection,

$$\text{A}_1 = \text{EA}\cdot\text{BCD}, \quad \text{B}_1 = \text{EB}\cdot\text{CAD}, \quad \text{C}_1 = \text{EC}\cdot\text{ABD}, \quad \text{D}_1 = \text{ED}\cdot\text{ABC}, \tag{64}$$

$$\text{A}_2 = \text{DA}\cdot\text{BCE}, \quad \text{B}_2 = \text{DB}\cdot\text{CAE}, \quad \text{C}_2 = \text{DC}\cdot\text{ABE}, \tag{65}$$

and the seven harmonic relations,

$$(\text{EA}_1\text{AA}_1') = (\text{EB}_1\text{BB}_1') = (\text{EC}_1\text{CC}_1') = (\text{ED}_1\text{DD}_1') = -1, \tag{66}$$

$$(\text{DA}_2\text{AA}_2') = (\text{DB}_2\text{BB}_2') = (\text{DC}_2\text{CC}_2') = -1, \tag{67}$$

by means of which 14 last relations these 14 new points can all be geometrically constructed; we shall then be able to interpret, on the recent plan **13.**, the three new quotients, $l : r$, $m : r$, $n : r$, as anharmonics of groups, as follows:

$$\frac{l}{r} = (\text{DA}_2'\text{AQ}); \quad \frac{m}{r} = (\text{DB}_2'\text{BR}); \quad \frac{n}{r} = (\text{DC}_2'\text{CS}); \tag{68}$$

with the analogous interpretations,

$$\frac{l}{s} = (\text{EA}_1'\text{AX}); \quad \frac{m}{s} = (\text{EB}_1'\text{BY}); \quad \frac{n}{s} = (\text{EC}_1'\text{CZ}); \quad \frac{r}{s} = (\text{ED}_1'\text{DW}), \tag{69}$$

if X, Y, Z, W denote the intersections EA·Π, EB·Π, EC·Π, ED·Π, so that

$$\text{X} = (\bar{s}000\,l), \quad \text{Y} = (0\bar{s}00\,m), \quad \text{Z} = (00\bar{s}0\,n), \quad \text{W} = (000\bar{s}r), \text{ where } \bar{s} = -s. \tag{70}$$

16. As regards the *notations* employed, it may be observed that although we have often, as in (9) or (27), &c., *equated a point,* or rather its *literal symbol,* A or P, &c., to the *corresponding quinary symbol* (10000) or (*xyzwv*), &c., of that point, yet in some formulæ, such as (17) (18) (19), in which we had occasion to treat of *linear combinations* of such quinary symbols, we substituted *new letters,* such as Q, Q′, for P, P′, &c., in order to avoid the apparent strangeness of writing such expressions* as $t\text{P} + t'\text{P}'$ &c. To *economise symbols,* however, we may agree to *retain the literal symbols first used,* for any system of given or derived points, but to *enclose them in parentheses,* when we wish to employ them as *denoting quinary symbols in combination with each other*; writing, at the same time, for the sake of uniformity, (U) instead of *U*, as the *quinary unit*

* Expressions of this *form* occur continually in the *Barycentric Calculus of Moebius* but with significations entirely different from those here proposed.

symbol (16). And thus, if we agree also that an *equation* between *two unenclosed* and *literal symbols of points*, P and P′, shall be understood as expressing that the two points so denoted *coincide*, we may write anew those formulæ (17) (18) (19) as follows:

$$\text{P}' = \text{P}, \text{ if } (\text{P}') = t(\text{P}) + u(\text{U}); \tag{71}$$

$$\text{P}'' \textit{ on line } \text{PP}', \text{ if } (\text{P}'') = t(\text{P}) + t'(\text{P}') + u(\text{U}); \tag{72}$$

$$\text{P}''' \textit{ in plane } \text{PP}'\text{P}'', \text{ if } (\text{P}''') = t(\text{P}) + t(\text{P}') + t''(\text{P}'') + u(\text{U}). \tag{73}$$

17. We may also occasionally denote a point *in the given plane* of A, B, C by the *ternary symbol,*

$$(x,\ y,\ z), \quad \text{or} \quad (xyz), \tag{74}$$

considered here as an *abridgment* of the *quinary* symbol ($xyz00$); and the *right line* which is the *trace on that plane,* of any *other plane* Π, or [$lmnrs$] (23), may be denoted by this *other ternary symbol,*

$$[l,\ m,\ n], \quad \text{or} \quad [lmn]; \tag{75}$$

these two last ternary symbols being *connected* by the relation,

$$lx + my + nz = 0, \tag{76}$$

if the *point* (xyz) be *on the line* [lmn]. And the *point* P in which any *other* line Λ, *not* situated in the plane ABC, *intersects* that *plane,* may be said to be the *trace* of that *line.*

18. For example, the *point* D_1 is, by (64), the *trace of the line* DE; and if we write,

$$\text{A}_0 = (\bar{1}11), \quad \text{B}_0 = (1\bar{1}1), \quad \text{C}_0 = (11\bar{1}), \tag{77}$$

then these three points are the respective traces of the three lines A_1A_2, B_1B_2, C_1C_2; because they are, by the notation (74), in the given plane, and we have, by (60) and (61), the three following symbolical equations of the form (72),

$$(\text{A}_0) + (\text{A}_1) + (\text{A}_2) = (\text{B}_0) + (\text{B}_1) + (\text{B}_2) = (\text{C}_0) + (\text{C}_1) + (\text{C}_2) = (\text{U}), \tag{78}$$

which express the three collineations, $\text{A}_0\text{A}_1\text{A}_2$, $\text{B}_0\text{B}_1\text{B}_2$, $\text{C}_0\text{C}_1\text{C}_2$.

We have also the three other collineations, $\text{AD}_1\text{A}'$, $\text{BD}_1\text{B}'$, $\text{CD}_1\text{C}'$, because the quinary symbols (27) (49) (54) (60) give the equations,

$$(\text{A}) + (\text{A}') + (\text{D}_1) = (\text{B}) + (\text{B}') + (\text{D}_1) = (\text{C}) + (\text{C}') + (\text{D}_1) = (\text{U}); \tag{79}$$

and these *three lines,* $\text{AA}'\text{D}_1$, &c., are the *traces of the three planes* ADE, BDE, CDE, of which *planes* the respective *equations* (21), and *quinary symbols* (23), are

$$y - z = 0, \quad z - x = 0, \quad x - y = 0, \tag{80}$$

and

$$[01\bar{1}00], \quad [\bar{1}0100], \quad [1\bar{1}000]; \tag{81}$$

so that the *ternary symbols* of the three last *lines,* regarded as their *traces,* are simply, by (75),

$$[01\bar{1}], \quad [\bar{1}01], \quad [1\bar{1}0]. \tag{82}$$

Accordingly, whether we consider the point A = (100), or A′ = (011), or D_1 = (111), (this *ternary* symbol of D_1 being *congruent* to the former *quinary* symbol (00011) for that point (60),) we have in each case the relation $y - z = 0$ between its coordinates; and similarly for the two other lines.

19. As other examples, the *four planes,*

$$A_1B_1C_1, \quad A_2B_2C_2, \quad A_1'B_1'C_1', \quad A_2'B_2'B_2', \tag{83}$$

have for their quinary equations,

$$x+y+z=2w+v, \quad x+y+z=w+2v, \quad x+y+z+v=4w, \quad x+y+z+w=4v, \tag{84}$$

and for their quinary symbols,

$$[111\overline{2}\overline{1}], \quad [111\overline{1}\overline{2}], \quad [111\overline{4}1], \quad [1111\overline{4}]; \tag{85}$$

they have therefore a *common trace,* namely the line

$$[111], \text{ or } A''B''C'', \tag{86}$$

because, by (49) and (54), we may now write,

$$A'' = (01\overline{1}), \quad B'' = (\overline{1}01), \quad C'' = (1\overline{1}0), \tag{87}$$

and the coordinates of each of these three last points satisfy the equation,

$$x+y+z=0. \tag{88}$$

Accordingly, because we have, by (60) (61) (62) (63), the three following sets of symbolical equations of the form (72),

$$\left.\begin{aligned}(A'') &= (B_1)-(C_1) = (B_2)-(C_2) = (B_1')-(C_1') = (B_2')-(C_2'),\\ (B'') &= (C_1)-(A_1) = (C_2)-(A_2) = (C_1')-(A_1') = (C_2')-(A_2'),\\ (C'') &= (A_1)-(B_1) = (A_2)-(B_2) = (A_1')-(B_1') = (A_2')-(B_2'),\end{aligned}\right\} \tag{89}$$

we see that the *point* A'' is the *common trace* of the *four lines,* B_1C_1, B_2C_2, $B_1'C_1'$, $B_2'C_2'$; B'' of C_1A_1, C_2A_2, $C_1'A_1'$, $C_2'A_2'$; and C'' of A_1B_1, A_2B_2, $A_1'B_1'$, $A_2'B_2'$.

20. In all such cases as these, in which we have to consider a *set of three points* P, or a *set of three planes* Π, of which the *first* is *geometrically derived* from ABCDE according to the *same rule of construction,* as that according to which the *second* is derived from BCADE, and the *third* from CABDE, we can *symbolically derive the second from the first,* and in like manner the *third* from the *second,* (or again the first from the third,) by writing, in each case, the *third, first,* and *second coefficients,* or coordinates, in the places of the *first, second,* and *third,* respectively. In symbols, we may express this *law of successive derivation,* of certain *syntypical* points or planes **9.** from one another, by the formulæ,

$$\text{if } P(ABC) = (xyzwv), \text{ then } P(BCA) = (zxywv), \text{ and } P(CAB) = (yzxwv); \tag{90}$$

and if

$$\Pi(ABC) = [lmnrs], \text{ then } \Pi(BCA) = [nlmrs], \text{ and } \Pi(CAB) = [mnlrs]; \tag{91}$$

as has been already exemplified in the systems (27), (60), (61), (62), (63), (77), (81), (87), for *points* or *planes*, and in (82) for *lines*, considered as *traces* of planes. In all these cases, therefore, we can, with perfect clearness and *definiteness* of signification, *abridge the notation*, by writing *only the first* (or indeed *any one*) of the *three* equations (90) or (91), and then appending an "&c."; for the *law* which has been just stated will always enable us to *recover* (or deduce) *the other two*. We may therefore briefly but sufficiently express several of the foregoing results, by writing,

$$\left.\begin{array}{l} \mathrm{A} = (100),\ \&\text{c.};\ \mathrm{A}' = (011),\ \&\text{c.};\ \mathrm{A}'' = (01\overline{1}),\ \&\text{c.};\ \mathrm{A}_0 = (\overline{1}11),\ \&\text{c.}; \\ \mathrm{A}_1 = (10001),\ \&\text{c.};\ \mathrm{A}_2 = (10010),\ \&\text{c.};\ \mathrm{A}_1' = (1000\overline{1}),\ \&\text{c.};\ \mathrm{A}_2' = (100\overline{1}0),\ \&\text{c.}; \end{array}\right\} \tag{92}$$

$$\textit{Plane}\ \mathrm{ADE} = [01\overline{1}00],\ \&\text{c.};\ \textit{Line}\ \mathrm{AD_1A}' = [01\overline{1}],\ \&\text{c.}; \tag{93}$$

to which we may add these other symbols of planes and lines, each supposed to be followed by an "&c.":

$$\text{plane BCD} = [1000\overline{1}];\ \text{BCE} = [100\overline{1}0];\ \text{trace} = \text{BC} = [100]; \tag{94}$$

$$\text{plane } \mathrm{DB'B_1C'C_1} = [\overline{1}110\overline{1}];\ \mathrm{EB'B_2C'C_2} = [\overline{1}11\overline{1}0];\ \text{trace} = \mathrm{B'C'A''} = [\overline{1}11] \tag{95}$$

$$\text{plane } \mathrm{AB_1C_2C_1B_2} = [011\overline{1}\overline{1}];\ \text{trace} = \mathrm{AA''} = [011]; \tag{96}$$

this line $\mathrm{AA''}$ passing also, by (77), through the two points $\mathrm{B_0}$ and $\mathrm{C_0}$;

$$\text{plane } \mathrm{B_1C_1D_1} = [\overline{2}111\overline{1}];\ \mathrm{B_2C_2D_1} = [\overline{2}111\overline{1}];\ \text{trace} = \mathrm{D_1A''} = [2\overline{1}1]; \tag{97}$$

$$\left.\begin{array}{ll} \text{plane } \mathrm{A'B_1B_2} = [\overline{2}\,\overline{1}111]; & \text{trace} = \mathrm{A'B_0} = [\overline{2}\,\overline{1}1]; \\ \text{plane } \mathrm{A'C_1C_2} = [\overline{2}1\overline{1}11]; & \text{trace} = \mathrm{A'C_0} = [\overline{2}1\overline{1}]; \end{array}\right\} \tag{98}$$

where it may be noticed that the symbol for $\mathrm{A'C_1C_2}$, or for $\mathrm{A'C_0}$, may be deduced from that for $\mathrm{A'B_1B_2}$ or for $\mathrm{A'B_0}$, by simply interchanging the second and third coefficients, or coordinates. It is easy to see that the quinary symbol for the plane ABC itself is on the same plan $[0001\overline{1}]$, the equation of that plane being $w = v$; and it will be remembered that, by **18.**, the ternary symbol for the point $\mathrm{D_1}$ in that plane is (111).

21. A *right Line* Λ *in Space* may be regarded in two principal views, as follows. Ist, it may be considered as the *locus of a variable point* P, *collinear with two given points* $\mathrm{P_0}$, $\mathrm{P_1}$; and in this view, the *symbol*

$$t_0(\mathrm{P_0}) + t_1(\mathrm{P_1}), \qquad \text{(comp. (72),)}$$

for the variable *point* upon the line, may be regarded as a *Local Symbol* (or *Point-Symbol*) *of the Line* Λ *itself.* Thus

$$(0\,tt'),\ \text{or}\ (0\,yz), \tag{99}$$

may either represent *an arbitrary point on the line* BC; or, *as a local symbol*, that *line itself.* Or IInd, we may consider a line Λ as a *hinge, round which a plane* Π *turns*, so as to be always *collinear* **7.** *with two given planes* Π_0, Π_1 through the line; and then a symbol of the form

$$t_0[\Pi_0] + t_1[\Pi_1], \qquad \text{(comp. (25),)}$$

which represents immediately the *variable plane* Π, may be regarded as being *also a Tangential Symbol* (or *Plane-Symbol*) *for the Line* Λ. For example, the line BC may thus be represented, not only by the *local* symbol (99), but also by the *tangential* symbol,

$$[\bar{\sigma}00\,tu],\ \text{if } \sigma = t + u, \text{ and } \bar{\sigma} = -\sigma. \tag{100}$$

In fact, this last symbol can be derived, by linear combinations, from the symbols (94) for the two planes BCD, BCE, which intersect in the line BC; and if any particular value be assigned to the ratio $t:u$, a particular *plane through that line* results. But it is time to apply these general principles to the *Geometrical Net in Space.*

Part II. Applications to the Net in Space: Enumeration and Classification of the Lines, Planes, and Points of that Net, to the end of the Second Construction

22. The *data* of the *Geometrical Net* are, by **1.**, the *five points* ABCDE, or P_0; of which the *quinary symbols* (27) have been assigned, and shown to be syntypical **9.**; and also the *ternary* symbols (92) of the three first of them. Of these the symbol

$$\text{A} = (100)$$

may be taken as the *type*; and the point A itself may be said to be a *First Typical Point.*

23. The *derived lines* Λ_1, of *First Construction* **1.**, are the *ten* following,

BC, &c.; DA, &c.; EA, and DE;

the "&c." being interpreted as in **20.**; and each line Λ_1 connecting, by its construction, *two* points P_0. Among these the line BC may be selected, as a *First Typical Line*; and its *symbols* **21.**, namely,

$$(0\,yz), \text{ and } [\bar{\sigma}00\,tu],$$

whereof the former represents this line BC considered as the *locus* of a *variable point,* while the latter represents the same line considered as the *hinge* of a *variable plane,* may be taken as *types* (the *point-type* and the *plane-type*) of the *group* of the *ten lines* Λ_1.

24. The *derived planes* Π_1 of *first* construction are in like manner *ten*; namely,

ADE, &c.; BCE, &c.; BCD, &c.; and ABC,

each obtained by connecting *three* points P_0. Of these the last has, by **20.** the quinary *symbol,*

$$\text{ABC} = [0001\bar{1}],$$

which may be taken as a *type* of the *group* Π_1; and the plane ABC itself may be called a *First Typical Plane.* As a verification, we see that when we make $\sigma = t + u = 0$, in the second symbol **23.**, and divide by t, we are led to the recent symbol for ABC, as one of the planes which pass through the line BC.

25. The *derived points* P_1, of the same *first* construction, which are all, by **1.**, of the *form* $\Lambda_1 \cdot \Pi_1$, are in like manner *ten*; namely the intersections,

BC˙ADE, &c.; DA˙BCE, &c.; EA˙BCD, &c.; and DE˙ABC,

which have been denoted in **14.** and **15.** by the letters, or *literal symbols,*

$$A', \text{ \&c.; } A_2, \text{ \&c.; } A_1, \text{ \&c.; and } D_1,$$

and for which *quinary symbols* (49) (54) (60) (61) have been assigned. Of these ten points *four,* namely A', B', C', D_1, are situated *in the plane* ABC, and have accordingly been represented **20.** by *ternary symbols* also: and we may take the particular symbol of this sort,

$$A' = (011),$$

as a *type* of this *group* P_1; understanding, however, that the *full* or *quinary type* is to be recovered from this *ternary type,* by *restoring the two omitted zeros*; so that we have, more fully,

$$A' = (01100) \equiv (10011).$$

And the *point* A' itself may be considered as a *Second Typical Point.*

26. We have thus denoted, by *literal* and by *quinary symbols,* whereof some have been *abridged* to *ternary* ones **17.**, and have been also represented by *types* **9.**, not only the *five given points* P_0, but all the *ten lines* Λ_1, *ten planes* Π_1, and *ten points* P_1, of what has been called, in **1.**, the *First Construction.* And it is evident that we have, at this stage, *ten triangles* T_1, namely the ten,

ADE, &c.; BCE, &c.; BCD, &c.; and ABC,

whereof each is contained in a plane Π_1; and also *five pyramids* R_1, each bounded by *four* of these *triangles,* namely the pyramids,

BCDE, CADE, ABDE, ABCE, ABCD,

which may be called the pyramids A, B, C, D, E; each being marked by the literal symbol of *that one* of the five points P_0, which is *not a corner* of the pyramid.

27. It may be remarked, that *ten arbitrary lines* in space *intersect* generally, *ten arbitrary planes,* in *one hundred points*; but that this *number* of intersections $\Lambda_1 \cdot \Pi_1$ is *here* reduced to *fifteen,* whereof only *ten* are *new*; because *each* of the *five points* P_0 *counts* as *twelve,* since in each of those points *four lines cut* (each) *three planes,* while *each* of the *ten planes contains three lines*; so that *thirty binary combinations* are *not cases of intersection,* and *sixty* such cases conduct only to the five *old* (or given) points. This sort of *arithmetical verification* of the accuracy of an *enumeration of derived points,* or lines, or planes, will be found useful in more complex cases, although it was not necessary here.

28. Proceeding to a *Second Construction* **1.**, we may begin by determining the *lines* Λ_2, whereof each connects some *two* (at least) of the *fifteen points* P_0, P_1, but *not* any two of the *five* points P_0, since otherwise it would be a line Λ_1. If the 15 points to be connected were *independent,* they would give, generally, by their binary combinations, 105 lines; but the *ten collineations of construction,*

$$BCA', \text{ \&c.; } DAA_2, \text{ \&c.; } EAA_1, \text{ \&c.; and } EDD_1,$$

show that 30 of these *combinations* are to be rejected, as giving only the ten old lines. The remaining number, 75, is still farther reduced by the consideration that we have (comp. (79)) the *fifteen derived collineations,*

$$\text{AA}'\text{D}_1,\ \&\text{c.};\ \text{AB}_1\text{C}_2,\ \&\text{c.};\ \text{AC}_1\text{B}_2,\ \&\text{c.};\ \text{DA}'\text{A}_1,\ \&\text{c.};\ \text{EA}'\text{A}_2,\ \&\text{c.};$$

which represent only *fifteen new lines,* of a *group* which we shall denote by $\Lambda_{2,1}$, but *count* (comp. **27.**) as 45 binary combinations of the 15 points. There remain therefore only 30 such combinations to be considered; and these give in fact a *second group,* $\Lambda_{2,2}$, consisting of *thirty lines of second construction*: namely, the *thirty edges* of the *five new pyramids* R_2,

$$\text{C}'\text{B}'\text{A}_2\text{A}_1, \quad \text{A}'\text{C}'\text{B}_2\text{B}_1, \quad \text{B}'\text{A}'\text{C}_2\text{C}_1, \quad \text{A}_2\text{B}_2\text{C}_2\text{D}_1, \quad \text{A}_1\text{B}_1\text{C}_1\text{D}_1,$$

which are respectively *inscribed* in the five former pyramids R_1 **26.**, and are *homologous* to them, the five given points A . . E being the respective *centres of homology*; for example, $\text{C}' = \text{AB}^{\cdot}\text{CDE}$, &c. The corresponding *planes of homology* will present themselves somewhat later, in connexion with the points P_2.

29. On the whole, then, there are only *forty-five distinct lines of second construction* Λ_2; and these naturally divide themselves into *two groups,* of 15 lines $\Lambda_{2,1}$, and 30 lines $\Lambda_{2,2}$, as above. *Each* line of the *first group* $\Lambda_{2,1}$ connects *one* point P_0 with *two* points P_1; as each line Λ_1 had connected *one* point P_1 with *two* points P_0; but *no* line of the *second group* $\Lambda_{2,2}$ connects, at this stage of the construction, more than *two* points, which are *both* points P_1. Through *no* point P_0, therefore, can we draw *any line* $\Lambda_{2,2}$; but through *each* point P_0 we can draw *three lines* $\Lambda_{2,1}$; and each of these is determined as the *intersection of two planes* Π_1 through that point, or as *crossing two opposite edges* of that *pyramid* R_1, which has *not* the point P_0 for a corner (comp. **26.**): for example, $\text{AA}'\text{D}_1$ is the intersection of ABC, ADE, and crosses the lines BC, DE. And besides being, as in **28.**, the *edges* of certain *other* and *inscribed* pyramids R_2, the 30 lines $\Lambda_{2,2}$ are also the *sides* of *ten new triangles* T_2, namely,

$$\text{D}_1\text{A}_1\text{A}_2,\ \&\text{c.}; \quad \text{C}_1\text{B}_1\text{A}',\ \&\text{c.}; \quad \text{C}_2\text{B}_2\text{A}',\ \&\text{c.}; \text{ and } \text{A}'\text{B}'\text{C}',$$

situated *in the ten planes* Π_1, and *inscribed* in the *ten old triangles* T_1, to which also they are *homologous*; the corresponding *centres of homology* being the ten points P_1, in the same order,

$$\text{A}',\ \&\text{c.}; \quad \text{A}_2,\ \&\text{c.}; \quad \text{A}_1,\ \&\text{c.}; \quad \text{and} \quad \text{D}_1, \text{ as before.}$$

The *axes of homology* of these *ten pairs of triangles* T_1, T_2, will offer themselves a little later, in connexion with points P_2.

30. All this may be considered as evident from *geometry* alone, at least with the assistance of *literal symbols,* such as those used above. But to deduce the same things by *calculation,* with *quinary symbols* and *types,* on the plan of the present Paper, we may observe that the symbolical equation,

$$(10000) + (01100) + (00011) = (11111),$$

considered as a type of all equations of the same form, proves by (18) or (72) that each point P_1 can, in three different ways, be combined with another point P_1, so that their joining line shall pass through a point P_0; and that thus the *group* of the 15 lines $\Lambda_{2,1}$ arises, of which the line $\text{AA}'\text{D}_1$ is a specimen, and may be called a *Second Typical Line* (the *first* such line having been BC, by **23.**). The *complete* quinary symbol of a *point* on this line is $(tuuvv)$, which is however congruent to one of the form $(tuu00)$, and may therefore be abridged to the ternary symbol (tuu), or (xyy); and the quinary symbol of a *plane* through the same line is of the form $[0\,m\overline{m}\,r\overline{r}]$, or $[0\,t\overline{t}\,u\overline{u}]$; we may therefore, by **21.** (comp. **23.**) consider the two expressions,

$$(xyy), \text{ and } [0\,t\overline{t}\,u\overline{u}],$$

as being not only *local and tangential symbols* for the *particular* (or *typical*) *line* $\mathrm{AA'D_1}$ *itself*, but also *local and tangential types* for the *group* $\Lambda_{2,1}$; or as the *point-type*, and the *plane-type*, of that group.

31. The two points $\mathrm{P_1}$, of which the quinary symbols have been thus combined in **30.**, had *no common coordinate different from zero*; but there remains to be considered the case, in which two points of that group *have* such a coordinate: for example, when the points have for their symbols,

$$(10100) \text{ and } (11000), \text{ or } (101) \text{ and } (110).$$

The *point-symbol* and *plane-symbol* of the *line* Λ_2 connecting *these* two points $\mathrm{P_1}$ are easily seen to be (with the same significations of σ and $\overline{\sigma}$ as before),

$$(\sigma\,tu00), \text{ or } (\sigma\,tu), \text{ and } [\overline{t}ttu\overline{\sigma}];$$

but no choice of the arbitrary *ratio*, $t : u$, with $\sigma = t + u$, will reduce the symbol $(\sigma\,tu)$ to denote *any one* of the 15 points $\mathrm{P_0}$, $\mathrm{P_1}$, except the *two* points $\mathrm{P_1}$ (in this example, $\mathrm{B'}$ and $\mathrm{C'}$), by joining which the line is obtained; considering therefore the two last *symbols* as *types*, we see that they represent a *second group*, consisting of *thirty lines* $\Lambda_{2,2}$; but that there can be *no third group*, of *lines* Λ_2 of *second construction.* The *particular line* $\mathrm{B'C'}$, which the symbols in the present paragraph represent, may be taken as *typical* of this *second group*; and may be called (comp. **23.** and **30.**) a *Third Typical Line* of the System, or *Net*, determined by the five given points A . . E. And the *pyramids* $\mathrm{R_1}$, $\mathrm{R_2}$, and *triangles* $\mathrm{T_1}$, $\mathrm{T_2}$, of first and second constructions, of which the *literal symbols* have been assigned in **26. 28. 29.**, might also have easily been suggested and studied, by *quinary* symbols and types alone.

32. As regards the *Planes* Π_2 of *Second Construction* **1.**, it is easily seen that no such plane contains any *two* points $\mathrm{P_0}$, or any *one* line Λ_1; for example, the *first typical line* BC **23.** *contains* the point $\mathrm{A'}$; and if we *connect* it with any one of the four points A, $\mathrm{B'}$, $\mathrm{C'}$, $\mathrm{D_1}$, we only get a plane Π_1, namely ABC; if with D, $\mathrm{A_1}$, $\mathrm{B_2}$, or $\mathrm{C_2}$, we get another plane Π_1, namely BCD; and if with any one of the four remaining points E, $\mathrm{A_2}$, $\mathrm{B_1}$, $\mathrm{C_1}$, the plane BCE is obtained. Accordingly, the general symbol $[\overline{\sigma}00\,tu]$, in **23.**, for a plane through the line BC, gives $\sigma = 0$, or $t = 0$, or $u = 0$, when we seek to particularize it, by the first, the second, or the third of these three sets of conditions respectively.

33. But if we take the symbol $[0\,t\overline{t}\,u\overline{u}]$, in **30.**, for a plane through the *second typical line* $\mathrm{AA'D_1}$, and seek to particularize *this* symbol by the condition of passing through some one of the eight points $\mathrm{P_1}$ which are not situated upon it, we are conducted to the following results. The points $\mathrm{B'}$, $\mathrm{C'}$ give $t = 0$, and the points $\mathrm{A_1}$, $\mathrm{A_2}$ give $u = 0$; these points therefore give only two planes Π_1, namely the two planes ABC and ADE, of which the line $\Lambda_{2,1}$ is the intersection. But the points B_1, C_2 give $t = u$, and the points C_1, B_2 give $t = -u$; *these points* therefore give *two planes* of a *new group*, $\Pi_{2,1}$, namely (comp. **20.**) the two following:

$$\text{plane } AA'D_1B_1C_2 = [01\bar{1}1\bar{1}]; \quad \text{plane } AA'D_1C_1B_2 = [01\overline{11}1];$$

which are of the same *type* as the plane (96), namely,

$$\text{plane } AB_1C_2C_1B_2 = [011\overline{11}].$$

There are *fifteen* such *planes* $\Pi_{2,1}$, as the type sufficiently shows; each passes through *one point* P_0, and contains *two lines* $\Lambda_{2,1}$, containing also four lines $\Lambda_{2,2}$; as, for instance, the last-mentioned plane $AB_1C_2C_1B_2$, which we shall call (comp. **24.**) the *Second Typical Plane*, contains the *two* lines AB_1C_2, AC_1B_2 **28.**, and the *four* lines B_1C_1, C_1C_2, C_2B_2, B_2B_1; that is to say, the *two diagonals* and the *four sides* of the *quadrilateral* $B_1C_1C_2B_2$, of which the *plane* $\Pi_{2,1}$ passes through A.

34. We have now exhausted all the planes Π_2 which contain any point P_0; but there exists a *second group of planes*, $\Pi_{2,2}$, each of which is determined as connecting *three* points P_1, although passing through *no* point P_0. Thus if we take the *third typical line* $B'C'$ **31.**, and the symbol $[\bar{t}ttu\bar{\sigma}]$ for a plane through it, we get indeed $t = 0$, or a plane Π_1, namely, ABC, if we oblige the plane through $B'C'$ to contain A, or B, or C, or A', or D_1; and we get $u = 0$, or $[\bar{1}110\bar{1}]$, or a plane $\Pi_{2,1}$, namely $DB'B_1C'C_1$, as in (95), if we oblige it to contain D, or B_1, or C_1; while we get $\sigma = 0$, or $[\bar{1}11\bar{1}0]$, or $EB'B_2C'C_2$, again as in (95), if we oblige it to contain E, or B_2, or C_2. But there remain the two points A_1 and A_2, determining the two new planes $B'C'A_1$ and $B'C'A_2$, for the former of which we have $t + \sigma = 0$, or $u = -2t$, $\sigma = -t$, and therefore have the symbol $[\bar{1}11\bar{2}1]$; while for the latter we have $u = t$, $\sigma = 2t$, and therefore the syntypical symbol $[\bar{1}111\bar{2}]$. There are *twenty planes* of this *group* $\Pi_{2,2}$, as may be at once concluded from inspection of the *type*; among which (comp. **19.**) we shall select the following,

$$\text{plane } A_1B_1C_1 = [11\overline{21}],$$

and call this a *Third Typical Plane.* And it is evident that these 20 planes $\Pi_{2,2}$ are the *twenty faces* of the *five inscribed pyramids* R_2 **28.**, of which the *edges* have been seen to be the *thirty lines* $\Lambda_{2,2}$. On the whole, then, there are only *thirty-five planes* Π_2 of *second construction*; which thus divide themselves into *two groups*, of *fifteen* and *twenty*, respectively.

35. To *verify arithmetically* (comp **27. 28.**) the *completeness* of the foregoing *enumeration* of the *planes* Π_2, we may proceed as follows. In general, *fifteen independent points* would determine 455 planes, by their *ternary combinations*; but the *25 collineations* **28.**, which give only the *lines* Λ_1, $\Lambda_{2,1}$, *account* for 25 such combinations, leaving only 430 to be accounted for, by so many *triangles.* Now each plane Π_1 contains three points P_0, and four points P_1, connected by six collineations; it contains therefore 29 $(= 35 - 6)$ triangles, and thus the ten planes Π_1 account for 290 triangles, leaving only 140, situated in planes Π_2. But each of the 15 planes $\Pi_{2,1}$ contains one point P_0, and four points P_1, connected by two collineations; it contains therefore 8 $(= 10 - 2)$ triangles, and thus 120 are accounted for, leaving only 20 ternary combinations to be represented, by triangles in other planes Π_2. And these accordingly have presented themselves, as the twenty faces $\Pi_{2,2}$ of the five inscribed pyramids R_2. It must be mentioned, that the *enumeration* and *classification* of the foregoing *lines* and *planes* had been completely performed by MÖBIUS, although with an entirely different *notation* and *analysis.*

36. It is much more difficult, however, or at least without the aid of *types* it *would* be so, to *enumerate* and *classify* what we have called in **1.** the *Points* P_2 of *Second Construction*; and to assign their chief *geometrical relations,* to each other, and to the *five given* and *ten* (formerly) *derived* points, P_0 and P_1. In fact, it is obvious that these *new points* P_2, being (by their definition) *all the intersections* of lines Λ_1 or Λ_2 with planes Π_1 or Π_2, which have *not already occurred,* as points P_0 or P_1, may be expected to be (comp. **2.**) considerably *more numerous,* than either the *lines* or the *planes* themselves.

37. The *total number* of *derived lines and planes,* so far, is exactly *one hundred*; namely, 55 lines Λ, and 45 planes Π, of first and second constructions. Their *binary combinations,* of the *form* $\Lambda\Pi$, are therefore 2475 in number; but as it is not difficult to prove that there are 240 distinct cases of *coincidence* of line with plane (or of a plane *containing* a line), we must subtract this from the former number, and thus there remain only 2235 cases of *intersection,* of the kind which we have proposed to consider. *Every one,* however, of these 2235 cases, must be accounted for, either as a *given point* P_0, or as a *derived point* P_1 of *first* construction, or finally as one of those *new points* P_2, of which we have proposed to accomplish the *enumeration,* and to determine the natural *groups,* as represented by their respective *types.*

38. We saw, in **27.**, that each point P_0, as for instance the point A, represents *twelve intersections* of the form $\Lambda_1 \cdot \Pi_1$; and it is easy to prove that the same point P_0 represents *twelve other* intersections of the form $\Lambda_1 \cdot \Pi_{2,1}$; *twelve,* of the form $\Lambda_{2,1} \cdot \Pi_1$; and *three,* of the form $\Lambda_{2,1} \cdot \Pi_{2,1}$; but none of any other form. It represents therefore, on the whole, a system of 39 *intersections,* included in the *general form* $\Lambda \cdot \Pi$; and we must, for this reason, subtract 195 $(= 5 \times 39)$ from 2235, leaving 2040 *other* cases of intersection of line with plane, to be accounted for by the old and new *derived points,* P_1 and P_2.

39. An analysis of the same kind shows, that each of the *ten points* of *first construction,* as for example the *typical point* A′ **25.**, represents *one* intersection of the form $\Lambda_1 \cdot \Pi_1$; *six,* of the form $\Lambda_1 \cdot \Pi_{2,1}$; *six,* of the form $\Lambda_1 \cdot \Pi_{2,2}$; *six,* of the form $\Lambda_{2,1} \cdot \Pi_1$; *twelve,* of the form $\Lambda_{2,1} \cdot \Pi_{2,1}$; *eighteen,* of the form $\Lambda_{2,1} \cdot \Pi_{2,2}$; *eighteen,* of the form $\Lambda_{2,2} \cdot \Pi_1$; *twenty-four* of the form $\Lambda_{2,2} \cdot \Pi_{2,1}$; and *twenty-four* others, of the remaining form $\Lambda_{2,2} \cdot \Pi_{2,2}$. It represents, therefore, in all, 115 intersections $\Lambda \cdot \Pi$; and there remain only 890 $(= 2040 - 1150)$ cases of intersection to be accounted for, or represented, by the points P_2 of which we are in search. But all these 890 cases of intersection *must* be accounted for, by *such new points,* if the investigation is to be considered as *complete.*

40. A *first,* but important, and well-known *group* of such points P_2, consists of the *ten points* (already considered in Part I. of this Paper),

$$A'', \&c.; A'_2, \&c.; A'_1, \&c.; \text{and } D'_1,$$

namely, the *harmonic conjugates* of the *ten points* P_1, with respect to the *ten lines* Λ_1, which we shall call collectively the points, or the group, $P_{2,1}$; and among which we shall select the point

$$\text{A}'' = (01\bar{1}),$$

as a *Third Typical Point* of the *Net.* In fact, it is what we have called a point P_2, because, without belonging to either of the two former groups, P_0, P_1, it is an *intersection* $\Lambda_1 \cdot \Pi_{2,2}$; or rather, it represents *six* such intersections, of the line BC with planes of second construction, and of the second group: namely, with two such through $\text{B}'\text{C}'$, two through B_2C_2, and two through B_1C_1, being pairs of faces **28.** of three pyramids B_2, inscribed in those three pyramids R_1, which have been distinguished, in **26.**, by the letters A, D, E. The same point A″ is also the intersection of the same line BC with *three* planes $\Pi_{2,1}$; namely, with the three which connect, two by two, the three lines $\text{B}'\text{C}'$, B_2C_2, B_1C_1, and contain the three points A, D, E. It is also, in *six* ways, the intersection of one or other of these three last lines $\Lambda_{2,2}$, with a plane Π_1; in *three* ways, with a plane $\Pi_{2,1}$; and in *twelve* ways, with a plane $\Pi_{2,2}$; so that a *single point* $\text{P}_{2,1}$ represents *thirty intersections* of the form $\Lambda \cdot \Pi$; and the *group* of the *ten* such points represents 300 such intersections. We have therefore only to account for 590 (= 890 − 300) intersections $\Lambda \cdot \Pi$, by *other groups* $\text{P}_{2,2}$, &c., of points of *second construction.*

41. A *second group,* $\text{P}_{2,2}$, of such points P_2, has already presented itself, in the case of the *traces* A_0, B_0, C_0 **18.**, of the *lines* A_1A_2, B_1B_2, C_1C_2, on the plane ABC. The *ternary* symbol of the point A_0 has been found (77) (92) to be $(\bar{1}11)$; its *quinary* symbol is therefore $(\bar{1}1100)$, which is *congruent* (10) with (20011); hence in the *full,* or *quinary sense* **9.**, this point A_0 is *syntypical* with the following *other point, in the same plane* ABC,

$$\text{A}''' = (211),$$

which we shall call a *Fourth Typical Point,* and shall consider as representing the *group* $\text{P}_{2,2}$; this group consisting of *thirty* such *points,* namely of two on each of the 15 lines $\text{A}_{2,1}$.

42. Each of these thirty points $\text{P}_{2,2}$ represents *seven intersections* of line with plane; namely, two of each of the three forms, $\Lambda_{2,1} \cdot \Pi_{2,1}$, $\Lambda_{2,1} \cdot \Pi_{2,2}$, $\Lambda_{2,2} \cdot \Pi_{2,1}$, and one of the form $\Lambda_{2,2} \cdot \Pi_1$. For example, the typical point A‴, which is the intersection of the *two lines* $\text{AA}'\text{D}_1$ and $\text{B}'\text{C}'$, is at the same time the intersection of the former line $\Lambda_{2,1}$ with each of four planes Π_2 which contain the latter line $\Lambda_{2,2}$; being also the intersection of this last line $\text{B}'\text{C}'$ with a plane Π_1, namely ADE, and with two planes $\Pi_{2,1}$ which contain the first line $\text{AA}'\text{D}_1$. The *group* $\text{P}_{2,2}$ represents therefore 210 intersections $\Lambda \cdot \Pi$; and there remain only 380 (= 590 − 210) intersections of this standard form, to be accounted for by *other groups* of *second construction,* such as $\text{P}_{2,3}$, &c.

43. In investigating such *groups,* we need only seek for *typical points*; and because every such *point* is on a *line* of one of the *three forms,* Λ_1, $\Lambda_{2,1}$, $\Lambda_{2,2}$, we may confine ourselves to the *three typical lines,*

$$\text{BC},\ \text{AA}'\text{D}_1,\ \text{B}'\text{C}'; \quad \text{or} \quad (0tu),\ (tuu),\ (\sigma tu);$$

in which, as before, $\sigma = t + u$, and in which the ratio of t to u is to be determined. And because a line in the plane ABC intersects any *other plane* in the point in which it intersects the *line* which is the *trace* of the latter plane upon the former, we need only, for the present purpose, consider these lines, or traces: whereof there are, by what has been already seen, *seven* distinct *ternary types,* namely the following:

$$[100],\ [01\bar{1}],\ [\bar{1}11],\ [111],\ [011],\ [\bar{2}11],\ [\bar{2}1\bar{1}];$$

which answer to the *seven typical traces* of planes,

$$\text{BC},\ \text{AA}'\text{D}_1,\ \text{B}'\text{C}',\ \text{A}''\text{B}''\text{C}'',\ \text{AA}'',\ \text{D}_1\text{A}'',\ \text{A}'\text{C}_0.$$

There are 22 $(= 3+3+3+1+3+3+6)$ such *lines*, answering to 44 $(= 3.2+3.3+3.4+1.2+3.1+3.2+6.1)$ *planes*; namely to *all* the 45 planes Π_1, Π_2, *except* the particular plane ABC, on which the *traces* are thus taken. And we have now to *combine* these *seven types of lines*, with the *three symbols of points*, $(0tu)$, (tuu) (σtu), according to the general law, $lx + my + nz = 0$ (76).

44. The line BC is itself one of the three traces of the first type; and it intersects the twelve other traces, of the five first types, only in points which have been already considered. The line $\text{AA}'\text{D}_1$ is, in like manner, a trace of the second type; and it gives no new point, by its intersections with the eight other traces, of the three first types; but its intersection with the common trace $\text{A}''\text{B}''\text{C}''$, of the two planes $\text{A}_1\text{B}_1\text{C}_1$ and $\text{A}_2\text{B}_2\text{C}_2$ **19.**, which is the only line of the fourth type, gives what we shall call a *Fifth Typical Point*, namely the following:

$$\text{A}^{iv} = (\bar{2}11);\ \text{or more fully},\ \text{A}^{iv} = (\bar{2}1100) \equiv (30011).$$

This last quinary symbol shows that the point A^{iv} is syntypical with this other point in the plane ABC,

$$\text{A}_1^{iv} = (31100) = (311);$$

so that this *plane* contains *six points* $\text{P}_{2,3}$, which (in the *quinary* sense) belong to one *common group*, although their two *ternary types* are *different*. In fact, the point A_1^{iv} is the common intersection of the line $\text{AA}'\text{D}_1$ with the two planes $[1\bar{2}\bar{1}11]$ and $[1\bar{1}\bar{2}11]$, or $\text{B}'\text{C}_1\text{C}_2$ and $\text{C}'\text{B}_1\text{B}_2$, as the point A^{iv} is the common intersection of the same line with the two planes $[111\bar{2}\bar{1}]$ and $[111\bar{1}\bar{2}]$, or $\text{A}_1\text{B}_1\text{C}_1$ and $\text{A}_2\text{B}_2\text{C}_2$, as above.

45. There are *thirty* distinct points $\text{P}_{2,3}$, of this *third group* of *second construction*; and *each* represents *two* (but only two) intersections, which are both of the form $\Lambda_{2,1}{\cdot}\Pi_{2,2}$. The *group* therefore represents a system of 60 intersections $\Lambda{\cdot}\Pi$; and there remain only 320 $(= 380 - 60)$ such intersections to be accounted for by *other* points, or groups, such as $\text{P}_{2,4}$, &c. It will be found that we have now exhausted all the points, or groups, of *second* construction, which are situated on lines $\Lambda_{2,1}$; but that two other groups of points P_2 may be determined on lines Λ_1, by combining the typical line BC with the two last sets of traces **43.** as follows.

46. Combining thus BC with $\text{D}_1\text{C}''$ and $\text{D}_1\text{B}''$, or with the traces $[11\bar{2}]$ and $[1\bar{2}1]$, we get the two following points, of a *fourth group of second construction*,

$$\text{A}^{v} = (021);\quad \text{A}_1^{v} = (012);$$

whereof the former may be taken as a *Sixth Typical Point*. There are *twenty points* of this *group* $\text{P}_{2,4}$, whereof each represents *three* intersections, of the form $\Lambda_1{\cdot}\Pi_{2,2}$; for example, the typical point A^{v} is the common intersection of the line BC with the three planes $\text{C}'\text{A}_1\text{A}_2$, $\text{D}_1\text{A}_1\text{B}_1$,

$D_1A_2B_2$; the group therefore represents *sixty* intersections $\Lambda \cdot \Pi$, and there remain 260 (= 320 − 60) to be accounted for.

47. Again, combining BC with $C'B_0$, and with $B'C_0$, or with $[11\bar{2}]$ and $[12\bar{1}]$, we get the two following other points, belonging to a *fifth group* of *second construction*,

$$A^{vi} = (02\bar{1}); \quad A_1^{vi} = (0\bar{1}2);$$

whereof the first may be said to be a *Seventh Typical Point.* There are *twenty* points of this new group $P_{2,5}$, whereof each represents only *one* intersection, of the form $\Lambda_1 \cdot \Pi_{2,2}$; for example, $A^{vi} = BC \cdot C'B_1B_2$. We are therefore to subtract 20 from the recent number 260; and thus there remain still 240 intersections to be accounted for, by new points P_2 upon the lines $\Lambda_{2,2}$: since the lines Λ_1 as well as $\Lambda_{2,1}$ have been exhausted, as on examination will easily appear.

48. The line $B'C'$ intersects the traces BB'' and CC'' of the *fifth type* **43.** in the two following points, of a *sixth group* of *second construction*,

$$A^{vii} = (12\bar{1}); \quad A_1^{vii} = (1\bar{1}2);$$

whereof the former may be called an *Eighth Typical Point.* There are *sixty* points of this new group, $P_{2,6}$, whereof each represents *one* intersection, of the form $\Lambda_{2,2} \cdot \Pi_{2,1}$; for example, A^{vii} is the intersection of the line $B'C'$ with the plane $BC_1A_2A_1C_2$; there remain therefore 180 (= 240 − 60) intersections $\Lambda \cdot \Pi$ to be still accounted for, by other points P_2, on the same set of lines $\Lambda_{2,2}$.

49. The traces D_1B'', D_1C'', which belong to the *sixth type* **43.** intersect the line $B'C'$ in two new points, namely,

$$A^{vii} = (321); \quad A_1^{vii} = (312);$$

which belong to a *seventh group* $P_{2,7}$, of *second construction*, and of which the former may be regarded as a *Ninth Typical Point.* There are *sixty* points of this group, namely two on each of the 30 lines $\Lambda_{2,2}$; and each is the intersection of *one* such line with *two* distinct planes $\Pi_{2,2}$; their *group* therefore represents a system of 120 such intersections; and only 60 (= 180 − 120) intersections *remain* to be accounted for, by *other* points of this last *form*, $\Lambda_{2,2} \cdot \Pi_{2,2}$.

50. Accordingly, when we combine the line $B'C'$ with the traces $A'C_0$, $A'B_0$, which are of the *seventh type* **43.**, we obtain, for the intersections of that line $\Lambda_{2,2}$ with two new planes $\Pi_{2,2}$, namely with $A'C_1C_2$ and $A'B_1B_2$ (98), two new points, belonging to a new or *eighth group* $P_{2,8}$ of *second construction*, namely,

$$A^{ix} = (23\bar{1}); \quad A_1^{ix} = (2\bar{1}3);$$

whereof the former may be selected, as a *Tenth* (and, for our purpose, *last*) *Typical Point*: for the *sixty* points of this last group represent each *one* intersection, and thus account for *all* the intersections which lately *remained* **49.**, after all the preceding groups had been exhausted.

51. We are now therefore enabled to assert that the proposed *Enumeration of the Points* P_2 *of Second Construction*, and the proposed *Classification of such Points in Groups*, have both been completely effected. For the *number* of such *groups* $P_{2,1}, \ldots P_{2,8}$ has been seen to be *eight*, represented by the 8 *typical points*, $A'' \ldots A^{ix}$; which, along with the *first given point* A, and the *first derived point* A', make up a system of *ten types*, as follows:

$$A = (100); \quad A' = (011); \quad A'' = (01\bar{1}); \quad A''' = (211); \quad A^{iv} = (\bar{2}11);$$
$$A^{v} = (021); \quad A^{vi} = (02\bar{1}); \quad A^{vii} = (12\bar{1}); \quad A^{viii} = (321); \quad A^{ix} = (23\bar{1});$$

and the *number* of the *points* P_2 is $(10 + 30 + 30 + 20 + 20 + 60 + 60 + 60 =)290$; so that, when combined with the points P_1, they make up a system of exactly *three hundred points*, P_1, P_2, *derived from the five points* P_0.

52. It is to be remembered that the three other *ternary types*,

$$D_1 = (111), \quad A_0 = (\bar{1}11), \quad A_1^{iv} = (311),$$

have been seen to represent points which are, in the *quinary* sense, *syntypical* with A', A''', A^{iv}, and therefore belong to the same three groups, P_1, $P_{2,2}$, $P_{2,3}$; all these three points being in the plane ABC, and on the line $AA'D_1$. And it is evident that the five other points,

$$A_1^{v} = (012); \quad A_1^{vi} = (0\bar{1}2); \quad A_1^{vii} = (1\bar{1}2); \quad A_1^{viii} = (312); \quad A_1^{ix} = (2\bar{1}3),$$

belong (as has been seen) to the same five last groups, $P_{2,4}, \ldots P_{2,8}$, as the five points above selected as typical thereof, namely the points $A^{v} \ldots A^{ix}$, and are situated on the same two typical lines, BC and $B'C'$. The transition from A' to B', C', or from A'' to B'', C'', &c., is very easily made, by a rule already stated **20.**; and therefore it is unnecessary to write down here the symbols for *these* derived points, B', B'', &c., or C', C'', &c. But we must now proceed, in the remainder of this Paper, to investigate some of the chief *Geometrical Relations* which connect the points, lines, and planes of the *Net*, so far as they have been hitherto determined: namely, to the end of the *Second Construction*.

Part III. Applications to the Net, continued: Enumeration and Classification of the Collineations of the Fifty-Two Points in a Plane of First Construction

53. The plane ABC has been seen to contain, besides the three points P_0 which determine it, four points P_1, namely A', B', C', and D_1; and it contains forty-five points P_2 namely the three points A'', B'', C'' of the group $P_{2,1}$, and six points of each of the seven remaining groups of second construction. This *plane* Π_1 contains therefore *fifty-two points* P_0, P_1, P_2; and we propose to examine, in the first place, the various *relations of collinearity* which connect these different points among themselves: intending afterwards to investigate their principal *harmonic and involutionary relations.*

54. The points on the *first typical line* BC **23.** are, in number, *eight*; their literal symbols being, by what precedes,

$$B,\ C,\ A',\ A'',\ A^{v},\ A_1^{v},\ A^{vi},\ A_1^{vi};$$

the ternary symbols corresponding to which have been shown to be,

$$(010),\quad (001),\quad (011),\quad (01\bar{1}),\quad (021),\quad (012),\quad (02\bar{1}),\quad (0\bar{1}2).$$

In fact, that these eight points are all on the line BC, is evident on mere inspection of their *symbols*, which are all of the common *form*,

$$(0yz)\quad [\mathbf{23.}]$$

55. The points on the *second typical line*, AA′ **30.**, are in number *seven*: their literal symbols being,

$$\mathrm{A},\ \mathrm{A}',\ \mathrm{D}_1,\ \mathrm{A}''',\ \mathrm{A}_0,\ \mathrm{A}^{iv},\ \mathrm{A}_1^{iv};$$

and their ternary symbols being,

$$(100),\quad (011),\quad (111),\quad (211),\quad (\bar{1}11),\quad (\bar{2}11),\quad (311).$$

In fact, each of these seven symbols is evidently of the form (tuu), or (xyy) **30.**.

56. The points on the *third typical line*, B′C′ **31.**, are in number *ten*; namely the points,

$$\mathrm{B}_1',\quad \mathrm{C}',\quad \mathrm{A}'',\quad \mathrm{A}''',\quad \mathrm{A}_1^{vii},\quad \mathrm{A}_1^{viii},\quad \mathrm{A}^{ix},\quad \mathrm{A}_1^{ix},$$

of which the ternary symbols are,

$$(101),\quad (110),\quad (01\bar{1}),\quad (211),\quad (12\bar{1}),\quad (1\bar{1}2),\quad (321),\quad (312),\quad (23\bar{1}),\quad (2\bar{1}3);$$

each of these ten symbols being of the form (σtu) **31.**, with $\sigma = t + u$, as before.

57. These *three* typical *lines*, in the plane ABC, which may be denoted by the ternary symbols, $[100]$, $[0\bar{1}1]$, $[\bar{1}11]$, and represent a system of *nine lines* Λ_1, Λ_2 in that plane Π_1, are also three typical *traces* **43.** of *other* planes thereon; and the remaining traces of such planes are in number *thirteen*, represented by *four* other lines, as *types*: of which lines, considered as such traces, the ternary symbols have been found **43.** to be,

$$[111],\quad [011],\quad [\bar{2}11],\quad [\bar{2}1\bar{1}];$$

answering to the literal symbols,

$$\mathrm{A''B''C''},\quad \mathrm{AA''},\quad \mathrm{D_1A''},\quad \mathrm{A'C_0},$$

and serving as abridged expressions for the four *equations* of ternary form,

$$x + y + z = 0,\quad y + z = 0,\quad 2x = y + z,\quad 2x = y - z.$$

58. Each of these four last lines passes through *six* points; thus the trace $[111]$ passes through the points $(01\bar{1})$ $(\bar{1}01)$ $(1\bar{1}0)$ $(\bar{2}11)$ (121) $(11\bar{2})$, or through $\mathrm{A''B''C''}\ \mathrm{A}^{iv}\ \mathrm{B}^{iv}\ \mathrm{C}^{iv}$; $[011]$ through (100) $(01\bar{1})$ $(\bar{1}11)$ $(11\bar{1})$ $(2\bar{1}1)$ $(21\bar{1})$, or $\mathrm{AA''B_0C_0C}^{vii}\mathrm{B}_1^{vii}$; $[\bar{2}11]$ through (111) $(01\bar{1})$ (102) (120) (213) (231), or $\mathrm{D_1A''B}^{v}\mathrm{C}_1^{v}\mathrm{C}^{viii}\mathrm{B}_1^{viii}$; and $[\bar{2}1\bar{1}]$ through (011) $(11\bar{1})$ (131) (120) $(\bar{1}02)$ $(23\bar{1})$, or $\mathrm{A'C_0B}_1^{iv}\ \mathrm{C}_1^{v}\mathrm{B}^{vi}\mathrm{A}^{ix}$; the correctness of the *ternary symbols* being evident on inspection, if the law $lx + my + nz = 0$ (76) be remembered: and the *literal symbols* being thence at once deduced, by **51.** and **52.**.

59. *So far*, then, that is when we attend only to the *twenty-two traces* **43.** of planes Π_1, Π_2 on the plane ABC, we find a system of three collineations of eight points; three of seven points;

three of ten points; and thirteen of six points each. Each collineation of the first of these four systems *counts* as 28 binary combinations of the 52 points in the plane **53.**; each of the second system counts as 21 such combinations; each of the third system as 45; and each of the fourth as 15. We therefore account, in this way, for $84 + 63 + 135 + 195 = 477$ binary combinations: but the total number is $26.51 = 1326$; there remain then 849 to be accounted for, by lines Λ_3 which are *not traces*, of any one of the foregoing groups.

60. In seeking for such new lines, it is natural to consider first those which pass through one or other of the three given points A, B, C; and the *types* of such are found to be the five following, each representing a new group of six lines Λ_3:

$$[021]; \quad [02\bar{1}]; \quad [03\bar{1}]; \quad [03\bar{2}]; \quad [031].$$

As *symbols*, these answer respectively to the five new *lines*:

$$(100)(11\bar{2})(0\bar{1}2)(1\bar{1}2)(3\bar{1}2), \quad \text{or } AC^{iv}A_1^{iv}A_1^{vii}C^{ix};$$

$$(100)(112)(012)(\bar{1}12)(312), \quad \text{or } AC'''A_1^{v}B^{vii}A^{viii};$$

$$(100)(113)(213), \quad \text{or } AC_1^{iv}C^{viii};$$

$$(100)(123)(\bar{1}23), \quad \text{or } AC_1^{viii}B^{ix};$$

$$(100)(2\bar{1}3), \quad \text{or } AA_1^{ix}.$$

We have thus *twelve* lines Λ_3, each connecting a point P_0, with *four* points P_2, and counting as *ten* binary combinations; *twelve* other lines, each connecting a point P_0 with *two* points P_2, and counting as *three* such combinations; and *six* lines, each of which connects a point P_0 with *one* point P_2, and counts as only *one* combination. In this manner, then, we account for $120 + 36 + 6 = 162$, out of the 849 which had remained in **59.**; but there still remain 687 combinations to be accounted for, by new lines of third construction, which pass through no given point.

61. Considering next the new lines which connect a point of the *first* construction, with one or more points of the *second*, we find these five new types,

$$[31\bar{1}]; \quad [12\bar{2}]; \quad [12\bar{3}]; \quad [13\bar{3}]; \quad \text{and} \quad [13\bar{4}];$$

which as *symbols* denote the five lines,

$$\left.\begin{array}{r}(011)(1\bar{2}1)(1\bar{1}2); \quad (011)(201)(2\bar{1}0); \quad (111)(2\bar{1}0)(\bar{1}21);\\ (011)(312); \quad (111)(\bar{1}32);\end{array}\right\}$$

$$\text{or } A'B^{iv}A_1^{vii}; \quad A'B_1^{v}C^{vi}; \quad D_1C^{vi}C_1^{vii}; \quad A'A_1^{viii}; \quad \text{and } D_1C_1^{ix};$$

but as *types* represent each a *group* of *six lines*. We thus get 18 new lines, each passing through 1 point P_1, and 2 points P_2; and 12 other lines, each connecting a point P_1 with only *one* point P_2. And these thirty lines Λ_3 account for $54 + 12 = 66$ binary combinations of points: leaving however 621 such combinations to be accounted for, by new lines Λ_3, of which each must connect at least two points P_2, without passing through any point P_0 or P_1, and without being any one of the traces already considered.

62. The *symbol* $[\bar{2}33]$, which denotes a line passing through *two* points P_2, namely, $(01\bar{1})$ and (311), or A'' and A_1^{iv}, but through *no other* point, represents, when considered as a *type*, a group of *three* such lines; and 40 *other types*, as for example $[1\bar{3}4]$, which as a symbol denotes the line $(\bar{1}11)$ (132), or $\text{A}_0\text{B}^{viii}$, are found to exist, representing each a group of *six* lines, whereof each connects in like manner *two* points P_2, but *only* those two points. We have thus a system of 243 new lines, which represent only so many binary combinations: and there remain 378 such combinations to be accounted for, by new lines Λ_3, whereof each must connect *at least three points* P_2.

63. For lines connecting *three* such points, and *no more*, it is found that there are *twenty types*; whereof *eight*, as for instance the type $[\bar{3}11]$, which as a symbol denotes the line $(01\bar{1})$ (121) (112), or $\text{A}''\text{B}'''\text{C}'''$, represent each a group of *three* such lines; while each of the *twelve others*, like $[1\bar{2}3]$, which as a symbol denotes the line $(\bar{1}11)$ (121) (210), or $\text{A}_0\text{B}'''\text{C}^v$, represents a group of *six* lines. We have thus 96 new lines, counting as 288 binary combinations: but we must still account for 90 *other* combinations, by new lines Λ_3, connecting each *more than three points* P_2.

64. Accordingly, we find *three new types of lines*, which *alone remain*, when all those which have been above exhibited, or alluded* to, are set aside: namely

$$[1\bar{2}4]; \quad [\bar{1}24]; \quad [112].$$

And these represent, respectively, groups of *six*, of *six*, and of *three* new lines, and therefore on the whole a system of *fifteen* new lines, each passing *through four points* P_2, and consequently counting as *six* combinations; for example, as *symbols*, they denote the three following lines:

$$(210)(\bar{2}11)(021)(231), \quad \text{or } \text{C}^v\text{A}^{iv}\text{A}^v\text{B}_1^{viii};$$

$$(210)(2\bar{1}1)(02\bar{1})(23\bar{1}), \quad \text{or } \text{C}^v\text{C}^{vii}\text{A}^{vi}\text{A}^{ix};$$

$$(20\bar{1})(1\bar{1}0)(02\bar{1})(11\bar{1}), \quad \text{or } \text{B}_1^{vi}\text{C}''\text{A}^{vi}\text{C}_0.$$

But $6.15 = 90$; we are therefore entitled to say, that *all the* 1326 *binary combinations* **59.**, *of the* 52 *points* P_0, P_1, P_2 **53.** *in the plane* ABC, *have now been fully accounted for.*

65. Collecting the results, respecting the *collineations in the plane* ABC, it has been found that there are 261 *lines* Λ_3, whereof each *connects two*, but *only two*, of the 52 *points* in that plane; and that *these lines*, which at the present stage of the construction are not properly *cases of collinearity* at all, are represented by a system of 44 *ternary types.*

* It has been thought that it could not be interesting to set down *all* the *types of lines*, above referred to; especially as those which relate to lines *not* passing through *at least four points* give rise, at the present stage of the construction, to no *theorems* of *harmonic* (or *anharmonic*) *ratio.*

66. There are 126 *other lines* Λ_3, each connecting *three* (but *only three*) *points*; they are represented by a system of 25 *types*; and account for 378 binary *combinations*.

67. There are 15 lines Λ_3, each connecting *four points* P_2; they are represented by a system of 3 types, and account for 90 combinations.

68. There are 12 lines Λ_3, each connecting *one* point P_0 with *four* points P_2; they are represented by 2 types, and represent 120 combinations.

69. There are 13 other lines Λ_3, namely the *traces of planes* Π_1 or Π_2, whereof each connects *six points*, namely a point P_0 or P_1 with five points P_2, or else six points P_2 with each other; they are represented by 4 types, and account for 195 combinations.

70. There are 3 lines $\Lambda_{2,2}$, each connecting *two* points P_1 with *eight* points P_2; they have one common type, and represent 135 combinations.

71. There are, in like manner, 3 lines $\Lambda_{2,1}$, each connecting *one point* P_0 with *two points* P_1, and with *four points* P_2, but having only *one* common type; and they represent 63 combinations.

72. Finally, there are (in the same plane) 3 *lines* Λ_1, each connecting *two points* P_0 with *one point* P_1, and with *five points* P_2; these lines also have all *one type*; and they account for 84 *combinations*: with the *arithmetical verification*, that

$$261 + 378 + 90 + 120 + 195 + 135 + 63 + 84 = 1326 = 26.51;$$

which proves that the *enumeration* is *complete.*

73. The *total number of distinct lines,* above obtained, is $261 + 126 + 15 + 12 + 13 + 3 + 3 + 3 = 436$; and the total number of their *ternary types* is 81. But *if we set aside* (as conducting to *no general metric relations*) *all those lines which contain fewer than four points,* there *then remain only forty-nine lines,* and *only twelve types,* to be discussed, with reference to *harmonic* (or *anharmonic*) *relations,* of the points upon those lines.

74. For the purpose of studying completely all *such* relations, it will therefore be permitted to confine ourselves to the *three first typical lines,* BC, AA$'$, B$'$C$'$, or [100], [011], [$\bar{1}$11]; the *four other typical traces,* A$''$B$''$C$''$, AA$''$, D_1A$''$, A$'C_0$, or [111], [011], [$\bar{2}$11], [$\bar{2}$1$\bar{1}$]; and *five new typical lines* Λ_3, connecting each *at least four points*: namely the *two lines,* [021] and [02$\bar{1}$], of **60.**, whereof each connects the given point A with *four* points P_2; and the *three lines* [1$\bar{2}$4], [$\bar{1}$24], [112], of **64.**, of which each connects *four* other points P_2 among themselves, but does not pass through any point P_0, or P_1.

Part IV. Applications to the Net, continued: Harmonic and Involutionary Relations, of the Points situated on the Twelve Typical Lines, in a Plane of First Construction

75. Commencing here with the examination of the last typical lines, because they contain only *four* points each, let us adopt, as temporary symbols, of the *literal* kind, the ten following:

$$a = (210), \quad b = (\bar{2}11), \quad c = (021), \quad d = (231);$$
$$b' = (2\bar{1}1), \quad c' = (02\bar{1}), \quad d' = (23\bar{1});$$
$$a'' = (20\bar{1}), \quad b'' = (1\bar{1}0), \quad d'' = (11\bar{1});$$

instead of the more systematic but less simple symbols, $\mathrm{C}^{v}\mathrm{A}^{iv}\mathrm{A}^{v}\mathrm{B}_1^{viii}\mathrm{C}^{vii}\mathrm{A}^{vi}\mathrm{A}^{ix}\mathrm{B}_1^{vi}\mathrm{C}''\mathrm{C}_0$.

76. The three lines referred to 64., are then the three following:

$$abcd; \quad ab'c'd'; \quad a''b''c'd''.$$

And because we have (comp. **16.**) the six symbolical relations,

$$(c) - (a) = (b); \quad (c) + (a) = (d);$$
$$(a) - (c') = (b'); \quad (a) + (c') = (d');$$
$$(a'') - (c') = 2(b''); \quad (a'') + (c') = 2(d''),$$

it results (31) that the three *harmonic equations* exist:

$$(abcd) = (ab'c'd') = (a''b''c'd'') = -1.$$

We have therefore this *Theorem*:-

"*Each of the* 150 *lines* Λ_3, *which connect four points* P_2, *in any one of the ten planes* Π_1, *and pass through no other of the* 305 *points* P_0, P_1, P_2, *is harmonically divided*."

77. As verifications, the three right lines bb', cc', dd' concur in the point C; bd', cc', db', in B; aa'', $b'b''$, $d'd''$, in A′; and aa'', $b'd''$, $d'b''$, in a point P_3, namely in $(41\bar{1})$: the existence of which four *concurrences* of lines was to be expected, from a known principle of *homography*, as consequences of the harmonic relations **76.**. It is worth noticing, however, how simply these concurrences are here *expressed*, by the *ternary symbols* of the *points*, according to the *law* (18); or, if we choose, by the corresponding symbols of the *lines*, with the analogous law (25): for example, the three last concurrent lines ad'', &c., have for their respective symbols, $[1\bar{2}2]$, $[011]$, and $[115] = [1\bar{2}2] + [033]$.

78. To examine, in like manner, the analogous relations of arrangement, on the two new typical lines **60.**, namely [021] and $[02\bar{1}]$, whereof each connects the given point A with four points of second construction, let us write as eight new temporary symbols of the literal kind, more convenient than the former symbols, $\mathrm{C}^{iv}\,\mathrm{A}_1^{vi}\,\mathrm{A}_1^{vii}\,\mathrm{C}^{ix}\,\mathrm{B}^{vii}\,\mathrm{A}_1^{v}\,\mathrm{C}'''\,\mathrm{A}_1^{viii}$, the following:

$$b = (11\bar{2}), \quad c = (0\bar{1}2), \quad d = (1\bar{1}2), \quad e = (3\bar{1}2);$$
$$\beta = (\bar{1}12), \quad \gamma = (012), \quad \delta = (112), \quad \epsilon = (312);$$

so that the two lines in question are,

$$\mathrm{A}bcde, \text{ and } \mathrm{A}\beta\gamma\delta\epsilon.$$

We have thus the eight following new symbolical relations, A being still = (100):

$$(\mathrm{A}) - (c) = (b), \quad (\mathrm{A}) + (c) = (d); \quad (e) - (b) = 2(d), \quad (e) + (b) = 4(\mathrm{A});$$

$$(\gamma) - (\text{A}) = (\beta), \quad (\gamma) + (\text{A}) = (\delta); \quad (\epsilon) + (\beta) = 2(\delta), \quad (\epsilon) - (\beta) = 4(\text{A});$$

whence result at once the *four harmonic relations*,

$$(\text{A}bcd) = (\text{A}bde) = (\text{A}\beta\gamma\delta) = (\text{A}\beta\delta\epsilon) = -1.$$

These *two* lines from A are therefore *homographically divided*, the point A corresponding to *itself*, and b to β, &c.; and accordingly the *four right lines*, $b\beta$, $c\gamma$, $d\delta$, $e\epsilon$, *which connect corresponding points*, concur in one common point, which is easily found to be B. And other *verifications*, by such *concurrences*, can be assigned with little trouble.

79. It may assist the conception of the *common law of arrangement*, of the *five points* on each of the *two typical lines* last considered, to suppose that the joining line $b\beta$ is *thrown off*, by projection, *to infinity*: or, what comes to the same thing, that the *two points* b and β, themselves, are thus made infinitely distant. For thus the harmonic equations **78.** will simply express that, *in this projected state of the figure*, the *four points*, d, e, δ, ϵ, *bisect* respectively the *four intervals*, Ac, Ad, Aγ, Aδ; whence it is easy to construct a diagram, not necessary here to be exhibited. The consideration of the *two other lines* through the same given point A, which have $[012]$ $[0\bar{1}2]$ for their symbols, and belong to the same two types as the two last, would offer to our notice a *pencil of four rays*, which has some interesting properties, especially as regards its *intersections* with *other* pencils, but which we cannot here delay to describe.

80. It may, however, be worth while to state here, as a consequence from the preceding discussion, this other *Theorem*:-

"*The* 120 *lines* Λ_3, *in the ten planes* Π_1, *whereof each connects a point* P_0 *with four points* P_2, *and with no other of the* 305 *points, although not all syntypical, are all homographically divided.*"

81. Proceeding to consider the arrangements of those six typical lines **58.** which contain each *six points*, we find that whether we write, as new temporary and literal symbols,

$$a = (01\bar{1}), \quad b = (\bar{1}01), \quad c = (1\bar{1}0), \quad a' = (\bar{2}11), \quad b' = (1\bar{2}1), \quad c' = (11\bar{2}),$$

or

$$a = (011), \quad b = (11\bar{1}), \quad c = (120), \quad a' = (23\bar{1}), \quad b' = (131), \quad c' = (\bar{1}02),$$

the six points $abca'b'c'$ being in the one case on the line $[111]$, and in the other case on the line $[\bar{2}1\bar{1}]$, we have in each case the three harmonic equations:

$$(caba') = (abcb') = (bcac') = -1.$$

We may then at once infer this *Theorem*:

"*The* 70 *lines* Λ_3, *in the ten planes* Π_1, *which are represented by the fourth and seventh typical traces of planes on the plane* ABC, *although not all syntypical* (or generated by similar processes of construction), *are all homographically divided.*"

82. This *common mode* of their *division* may deserve, however, a somewhat closer examination, its consequences being not without interest. When any six collinear points, $a \ldots c'$, are connected by the three equations **81.**, we are permitted to suppose that their symbols are so *prepared* (if necessary), by *coefficients*,* as to give,

$$(a) + (b) + (c) = 0;$$

$$(a') = (b) - (c), \quad (b') = (c) - (a), \quad (c') = (a) - (b);$$

and therefore,

$$(a') + (b') + (c') = 0,$$

$$3(a) = (c') - (b'), \quad 3(b) = (a') - (c'), \quad 3(c) = (b') - (a').$$

Whenever, then, the three harmonic equations **81.** *exist, for a system of six collinear points,* $a\,..\,c'$, *the three other harmonic equations*, formed by interchanging accented and unaccented letters,

$$(c'\,a'\,b'\,a) = (a'\,b'\,c'\,b) = (b'\,c'\,a'\,c) = -1,$$

are also satisfied; and *the three pairs* (or *segments*),

$$aa',\ bb',\ cc',$$

which connect corresponding points, compose an involution.†

83. Under the same conditions, the two points a and a' are harmonically conjugate to each other, not only with respect to b and c, but also with respect to b' and c'; they are therefore the *double points* (or *foci*) of that *other involution* which is determined by the *two pairs* of points, bc, $b'c'$. In like manner, b, b' are the double points of the involution, determined by the two pairs, or segments, ca, $c'a'$; and c, c' are the double points of the involution determined by ab, $a'b'$.

84. From any one of the three last involutions **83.**, we could *return*, by known principles, to the involution **82.**; we can also infer from them that the *three new pairs of points* (or *segments of the common line*), aa', bc', cb'; the three pairs, or segments, bb', ca', ac'; and the three others, cc', ab', ba', form *three other involutions*, making *seven distinct involutions of the six points*, so far: in *three* of which, as we have seen in **83.** *two* of those *six points* are *their own conjugates.*

85. For these and other reasons we propose to say, that *when any three collinear points* (as a, b, c) *are assumed* (or *given*), *and three other points on the same line are derived from them, by the condition that each shall be the harmonic conjugate of one, with respect to the other two, then these two sets of points are two Triads of Points in Involution.* And it is easy to extend this definition so as to include cases of two *triads* of complanar and co-initial *lines*, or of collinear *planes*, which shall be, in the same general but (as it is supposed) *new sense*, in *involution* with each other: every such *involution of triads* including, by what precedes, a *system of seven involutions* of the *old* or *usual* kind.

* For example, in the second case **81.**, we should change the symbols for c and b' to their negatives, before employing the formulæ of **82.**

† Compare p. 127 of the *Géométrie Supérieure* (Paris, 1852). In general, the reader is supposed to be acquainted with the chapter (chap. ix.) of that excellent work of M. *Chasles*, which treats of *Involution*.

86. For example, because the two *triads of points*, A″B″C″ and $\text{A}^{iv}\text{B}^{iv}\text{C}^{iv}$, are thus in involution, by the equations **81.** applied to the fourth typical trace 43., it follows that the *two pencils*, each of *three rays*,

$$\text{D}_1.\text{A}''\text{B}''\text{C}'', \quad \text{and} \quad \text{D}_1.\text{ABC},$$

are *triads of lines, in involution* with each other; and that, for a similar reason, the *two triads of planes*, all passing through the line DE,

$$\text{DEA}, \quad \text{DEB}, \quad \text{DEC}, \quad \text{and} \quad \text{DEA}'', \quad \text{DEB}'', \quad \text{DEC}'',$$

are, in the sense above explained, in *involution*. In fact, when the point D_1 is thus taken as a *vertex of the pencils* in the plane ABC, the three harmonic equations of the first case **81.**, namely,

$$(\text{C}''\text{A}''\text{B}''\text{A}^{iv}) = (\text{A}''\text{B}''\text{C}''\text{B}^{iv}) = (\text{B}''\text{C}''\text{A}''\text{C}^{iv}) = -1,$$

or rather the three reciprocal equations (comp. **82.**),

$$(\text{C}^{iv}\text{A}^{iv}\text{B}^{iv}\text{A}'') = (\text{A}^{iv}\text{B}^{iv}\text{C}^{iv}\text{B}'') = (\text{B}^{iv}\text{C}^{iv}\text{A}^{iv}\text{C}'') = -1,$$

correspond simply to the elementary equations, (50), (56),

$$(\text{CA}'\text{BA}'') = (\text{AB}'\text{CB}'') = (\text{BC}'\text{AC}'') = -1,$$

which may be employed to *define* the three important points A″, B″, C″, (87), of the *first group* of *second construction* **40.**, as being the (well known) *harmonic conjugates* of the points A′, B′, C′ of *first* construction, with respect to the three *lines* of the same first construction, BC, CA, AB, on which those points are situated.

87. The equations **82.**, which connect the *symbols* $(a) \ldots (c')$ of the *six points*, give, by easy eliminations, these other equations of the same kind:

$$(b') = (b) + 2(c); \quad -(c') = 2(b) + (c);$$

we have therefore, by (31), the following *anharmonic of the group* b, b', c, c':

$$(bb'cc') = +4;$$

and other easy calculations of the same sort given, in like manner, the equal anharmonics,

$$(cc'aa') = +4; \quad (aa'bb') = +4.$$

But in general, for any four collinear points, a, b, c, d, the *definition* (29) of the *symbol* $(abcd)$ gives easily the relation,

$$(abcd) + (acbd) = 1;$$

and hence, or immediately by calculations such as those recently used, we have this *other* set of anharmonics, with a *new* common value:

$$(bcb'c') = (cac'a') = (aba'b') = -3;$$

the *negative* character of which shows, by the same definition (29), that the segment (or interval) aa', for example, is cut *internally* by *one* of the two points b, b', or by one of the two points c, c', and *externally* by the *other*: with similar results for each of the two other segments, bb', cc'.

88. We may then say that *each of the three segments, aa', bb', cc', overlaps each of the two others,* in the sense that *any two* of them have a *common part,* and also *parts not common*: whence it immediately follows that the *involution* **82.**, *to which these three segments belong, has its double points imaginary*: whereas it may be proved, on the same plan, that each of the three involutions of segments mentioned in **84.**, namely aa', bc', cb'; bb', ca', ac'; cc', ab', ba', has *real* double points*; and the double points of the three other involutions, determined by the three *pairs* of segments, bc, $b'c'$; ca, $c'a'$; ab, $a'b'$, are likewise *real,* and have been assigned **83.**; namely, in each of these three last cases, the two remaining points of the system.

89. Now, in general, when the *foci* (or double points) of an involution of collinear segments, aa', bb', ... are *imaginary,* so that *conjugate points, a, a',* or b, b', &c., fall at *opposite sides* of the *central point* O, it is known, and may indeed be considered as evident, that if an *ordinate* OP be erected, equal to the constant *geometrical mean* between the two distances Oa, Oa', or Ob, Ob', &c., then, *all the segments aa', bb', &c., subtend right angles, at the extremity* P *of this ordinate.* It follows, then, by what has been proved in **82.** and **88.**, and by the *first case* of **81.**, that *each of the three segments* $\mathrm{A}''\ \mathrm{A}^{iv}$, $\mathrm{B}''\ \mathrm{B}^{iv}$, $\mathrm{C}''\ \mathrm{C}^{iv}$, *of the fourth typical trace* 43., *subtends a right angle at some one point,* P, *in the plane* ABC, or rather generally at *each* of *two* such points: and in like manner, by the *second case* **81.**, that each of the *three other segments,* $\mathrm{A}'\ \mathrm{A}^{ix}$, $\mathrm{C}_0\ \mathrm{B}_1^{iv}$, $\mathrm{C}_1^{v}\ \mathrm{B}^{vi}$, of the *seventh typical trace,* subtends a *right angle,* at each of two *other* points, P, P$'$, in the same plane.

90. *These* results, by their nature, like *all the foregoing results* of the present Paper, are quite *independent of the assumed arrangement of the five given* (or *initial*) *points of space* A..E, and are *unaffected by projection,* or *perspective.* In saying this, it is not meant, of course, that one *right angle* will generally be *projected* into *another*; or that the *new point* P, at which the *three new segments* $\mathrm{A}''\mathrm{A}^{iv}$, $\mathrm{B}''\mathrm{B}^{iv}$, $\mathrm{C}''\mathrm{C}^{iv}$, or $\mathrm{A}'\mathrm{A}^{ix}$, $\mathrm{C}_0\mathrm{B}_1^{iv}$, $\mathrm{C}_1^{v}\mathrm{B}^{vi}$, *subtend* right angles, will be itself (what may be called) the *projection* of the *old point* P **89.**, which was so related to the three *old segments,* denoted by the same literal symbols, when the *arrangement* (or *configuration*) of the five *initial* points is *varied,* by a process *analogous* to projection. We only assert that there will *always,* in *every state* of the Figure, or of the *Net,* be *some point* P, possessing the above-mentioned property: or rather that there will be a *circle* of such points *in space,* having for its *axis* the *line* to which the three segments belong.

91. To fix a little more definitely the conceptions, let A, B, C, D be supposed, for a moment, to be the *corners* of a *regular pyramid,* with E for its *mean point,* or *centre of gravity.* With this arrangement of the five *given* points P_0, *six* of the *derived* points P_1, namely A$'$, B$'$, C$'$, A_2, B_2, C_2, *bisect* the *six edges,* BC, CA, AB, DA, DB, DC, of the given pyramid; and the *four* other points P_1, namely A_1, B_1, C_1, D_1, are the *mean points* of the *four faces,* opposite to A, B, C, D. *Six* of the ten points $\mathrm{P}_{2,1}$, namely A$''$, B$''$, C$''$, A_2', B_2', C_2', are now *infinitely distant*; and the *line* $\mathrm{A}''\mathrm{B}''\mathrm{C}''\ \mathrm{A}^{iv}\mathrm{B}^{iv}\mathrm{C}^{iv}$ to which three of the lately mentioned *segments* belong, becomes the *line at infinity* in the plane ABC: which might seem, at first sight, to render difficult, with respect at least to *them,* the verification of a recent theorem **89.**. That theorem is, however, verified in a very

* The determination of these double points gives rise naturally to some new theorems, which cannot conveniently be stated here.

simple manner, by observing that, with the arrangement here conceived, *the three angles* $A''D_1A^{iv}$, $B''D_1B^{iv}$, $C''D_1C^{iv}$, which those *infinite* and *infinitely distant segments* may be imagined to *subtend* at the point D_1, *are all right angles*; D_1A'', for example, being *parallel* to the *side* BC of the *triangle* ABC, which is now an *equilateral* one; while D_1A^{iv} is *perpendicular* to the same side, because it is drawn from the *mean point* D_1, and passes through the *opposite corner*, A. As another verification of the theorem **89.**, it will be found that, with the arrangement here supposed, the *segments* $A'A^{ix}$, $C_0B_1^{iv}$, $C_1^vB^{vi}$, *of the seventh trace* **43.**, *subtend right angles at the given point* B.

92. the *involution of the three segments* **82.** is only *one* of the consequences of the *three harmonic equations* **81.**, or of what we have called in **85.** the *Involution of the two Triads, abc,* and $a'b'c'$. We can therefore *infer more*, respecting the *geometrical relations* of the *six points*, even in the *general state* of the whole *Figure*, or NET, than merely that those three segments subtend *right angles*, as above, *at every point of one real circle*, which has its *centre on the common line*, and its *plane perpendicular thereto*. The *order of succession* of the six points being supposed to be the following, $ac'ba'cb'$, from which it can only differ, if at all, by changes not important to the argument, let P be, as in **90.**, a point such that the angles aPa', bPb', cPc', are *right*. Then, because the *three pencils*,

$$P.ac'bc, \quad P.c'ba'b', \quad \text{and} \quad P.ba'ca,$$

are all *harmonic* pencils by **81.**, it follows that (with the supposed *order* of the points) the lines Pc' and Pc are respectively the *internal* and *external bisectors* of the *angle* aPb; Pb and Pb', of the angle $c'Pa'$; and Pa', Pa, of bPc: the line Pc bisecting also the angle $a'Pb'$ internally. Hence it is easy to infer the following *continued equation between angles* (which is supposed to be new):

$$aPc' = c'Pb = bPa' = a'Pc = cPb' = \frac{\pi}{6};$$

and therefore we may enunciate this *Theorem*:- "*When six collinear points form a system of two triads in involution, their five successive intervals subtend angles each equal to the third part of a right angle, at every point of a certain circle, of which the axis is their common line.*"

For example, with the particular arrangement **91.** of the five initial points A . . E, it is found that the five successive portions, C_0A^{ix}, $A^{ix}C_1^v$, $C_1^vB_1^{iv}$, $B_1^{iv}A'$, $A'B^{vi}$, of the seventh trace, subtend each an angle of *thirty degrees*, at the given point B; and the six lines D_1A'', D_1C^{iv}, D_1B'', D_1^{iv}, D_1C'', D_1B^{iv}, if suitably distinguished from their own opposites, succeed each other at angular intervals, of the same common amount.

93. In general, if *three equally inclined diameters* of a circle, forming a regular and *six-rayed star*, be taken as a *given triad of lines* **85.**, the *triad in involution* therewith is represented by that *other star* of the same kind, of which the diameters *bisect the angles* between those of the former star: so that if we consider any *six successive rays* of the *compound* or *twelve-rayed star*, which results from the combination of these *two*, their *successive angles* are evidently each equal to thirty degrees. But we now see further, that if a *star* of this last kind be *cut in six points* by an *arbitrary transversal* in its plane; and if these six points of section be in any manner put into perspective, by any *new* pencil and transversal: the *six new points*, thus obtained, as forming still *two triads* *in involution,* must admit of having their *five successive intervals seen,* from *every point* of some *new circle,* under *angles still equal each* to the same *third part of a right angle.*

94. We have not yet considered the arrangement of the six points on either the *fifth* or the *sixth* typical *trace* **43.**; but it is easy to do this as follows. Let $abc\alpha\beta\gamma$ denote, as new temporary symbols, either the six points of the fifth trace (comp. **58.**),

$$\text{I.}\quad a=(100),\quad b=(1\bar{1}1),\quad c=(11\bar{1}),\quad \alpha=(01\bar{1}),\quad \beta=(21\bar{1}),\quad \gamma=(2\bar{1}1);$$

or these six other points, belonging to the sixth trace,

$$\text{II.}\quad a=(111),\quad b=(102),\quad c=(120),\quad \alpha=(01\bar{1}),\quad \beta=(231),\quad \gamma=(213);$$

we shall then have, in each case, the three harmonic equations,

$$(bac\alpha) = (c\beta a\alpha) = (a\gamma b\alpha) = -1.$$

In *each* case, therefore, we may consider ourselves at first deriving from three points a fourth, as the harmonic conjugate of the first with respect to the other two; and then deriving a fifth point, and a sixth, as the harmonic conjugates of that fourth point, with respect, on the one hand, to the third and first points; and on the other hand, to the first and second points of the system.

95. Having regard merely to this *common law,* we may enunciate (comp. **80. 81.**) this theorem:-

"*The sixty lines, in the ten planes of first construction, represented by the fourth and fifth typical traces of the planes on the plane* ABC, *although not all syntypical, are all homographically divided.*"

And this *common mode* of their *division* is such, that if the fourth point be thrown off to infinity, the first point bisects the interval between the second and third; the fifth point bisects the interval between third and first; and the sixth point bisects the interval between first and second: so that, on the whole, we have a *finite line, bc quadrisected* in the points γ, a, β, and cut at infinity in α; whereas if, on either the *fourth* or the *seventh* trace, *one* of the six points, but only one, had been thus made *infinitely distant,* the *five others* would have presented the figure of a *finite right line, bisected* and *trisected.* With the equations **94.**, if a, instead of α, be projected to infinity, it is then the line $\beta\gamma$ which is quadrisected, namely, in the points c, α, b. In general, with these last equations, the *first set* of three points, abc, can be *derived* from the *second set,* $\alpha\beta\gamma$, by the *same rule* **94.**, as that by which the second set has been derived from the first: so that there is a sense in which *these two sets* may be said to be *reciprocal triads,* although they are *not triads in involution,* according to the definition **85.**.

96. It may be added that, on either the *fifth* or the *sixth* trace, the two points which we have called *first* and *fourth,* are the *double points* of a new *involution,* determined by the *two pairs, second* and *third, fifth* and *sixth*; or, with the recent notations **94.**, that $a\alpha$ are the foci of the involution bc, $\beta\gamma$; because the three last harmonic equations conduct to this fourth equation,

$$(\beta a\gamma\alpha) = -1.$$

97. And, as regards the *homography* of the divisions on the same two traces, if we denote, for the sake of distinction, the six points on the sixth trace by $a' \,..\, \gamma'$, then (because $\alpha' = \alpha$) the *five lines* aa', bb', a', $\beta\beta'$, $\gamma\gamma'$, or (comp. **58.**) the five lines,

$$\mathrm{AD}_1, \quad \mathrm{B}_0\mathrm{B}^{v}, \quad \mathrm{C}_0\mathrm{C}_1^{v}, \quad \mathrm{B}_1^{vii}\mathrm{B}_1^{viii}, \quad \mathrm{C}^{vii}\mathrm{C}^{viii},$$

ought to *concur* in some *one point*: which accordingly it is easy to see that they do, namely in the point A'; in fact, with the recent signification of $a, ..$ and $a', ..,$ we have the symbolic equations,

$$(a') - (a) = (b') - (b) = (c') - (c) = (011) = (\mathrm{A}');$$

and

$$(\beta') - (\beta) = (\gamma') - (\gamma) = (022) = 2(\mathrm{A}').$$

98. The *two sets of six points*, on these two traces, with one point common, are thus the points in which a certain *six-rayed pencil*, with A' for vertex, is *cut* by the two traces as transversals; the *symbols* of the six *rays* being the following:

$$\mathrm{A}'\mathrm{AD}_1 = [01\bar{1}]; \quad \mathrm{A}'\mathrm{B}_0\mathrm{B}^{v} = [\bar{2}\bar{1}1]; \quad \mathrm{A}'\mathrm{C}_0\mathrm{C}_1^{v} = [\bar{2}1\bar{1}];$$

$$\mathrm{A}'\mathrm{A}'' = [100]; \quad \mathrm{A}'\mathrm{B}_1^{vii}\mathrm{B}_1^{viii} = [1\bar{1}1]; \quad \mathrm{A}'\mathrm{C}^{vii}\mathrm{C}^{viii} = [11\bar{1}].$$

And from a mere inspection of these symbols, we can infer (comp. (33)) that the *first* and *fourth rays* are the *common harmonic conjugates* of the *two pairs, second* and *third, fifth* and *sixth*; or that they are the *double rays* of the *involution*, which those two *pairs of rays* determine: the theorem **96.** being thus, in a new way, confirmed.

99. We have now discussed the arrangements of the *points* on those *nine* typical *lines* Λ_3, whereof each passes through not less than *four*, nor more than *six*, of the 52 points in the plane ABC; but we have still *three other typical lines* to consider, namely the lines Λ_1 and Λ_2, of which each passes through *at least seven points*. Taking first, for this purpose, the typical line $\Lambda_{2,1}$, namely, AA', which contains *only seven* points, whereof the ternary symbols have been assigned in **55.**, and the literal symbols there given may be retained, we shall, for the moment, reserve the consideration of the two points $\mathrm{P}_{2,3}$; but shall introduce a new and auxiliary point $\mathrm{P}_{3,1}$ on the same line, which may be thus denoted:

$$\mathrm{A}^{x} = (122) = \mathrm{AA}'\cdot\mathrm{BC}'''\cdot\mathrm{CB}''';$$

and which may be said to *represent*, or *typify*, a *first group of third construction*, containing *fifteen points*, *one* on each of the *fifteen lines* $\Lambda_{2,1}$; although, in the present Paper, we can only *allude* to *such* new points P_3, and cannot *here* attempt to *enumerate*, or even to *classify* them.

100. We have thus again *six* points, at this stage, to consider, namely the points A, A', D_1, A''', A_0, A^{x}; and their symbols easily show that they are connected by the *three* following harmonic equations,

$$(\mathrm{AA}'\mathrm{D}_1\mathrm{A}''') = (\mathrm{AD}_1\mathrm{A}'\mathrm{A}_0) = (\mathrm{A}'\mathrm{AD}_1\mathrm{A}^{x}) = -1;$$

from which it follows, by **85.**, that the *two triads of points*,

$$\text{AA}'\text{D}_1 \text{ and } \text{A}^x\text{A}'''\text{A}_0,$$

are triads in involution: with, of course, all the properties which have been proved, in recent paragraphs of this Paper, to belong generally to *any two* such triads. As a verification, it may be mentioned that, with the particular arrangement **91.** of the five initial points A .. E, if we determine two new points P, P′, of *third* construction, by the formulæ,

$$\text{P} = (214) = \text{BC}'''\cdot\text{CA}''', \quad \text{P}' = (241) = \text{CB}'''\cdot\text{BA}''',$$

it can be proved that each of the five successive intervals (comp. 92.) between the six points,

$$\text{A}, \quad \text{A}''', \quad \text{D}_1, \quad \text{A}^x, \quad \text{A}', \quad \text{A}_0,$$

subtends the third part of a right angle at each of these two new auxiliary points, P and P′. But with *other* initial configurations, the *coordinates* of these two *new vertices* would be different, because they are connected with *angles*, which are not generally *projective* **90.**; although, as has been already remarked, there would always be *some* new points P, or rather a *circle* of such, possessing the property in question.

101. We may however enunciate generally, and without reference to any such particular *arrangement* of the five initial points, this *Theorem*:-

"*On any one of the fifteen lines* $\Lambda_{2,1}$, *of second construction, and first group, the given point* P_0, *and the two derived points of first construction* P_1, *compose a triad, the triad in involution to which* **85.** *consists of the point* $\text{P}_{3,1}$, *of third construction and first group, and of the two points* $\text{P}_{2,2}$, *of second construction and second group, upon that line*;" with *seven involutions of segments* (comp. **84.**) *included* under this general relation.

For example, on the line AA′, the *three segments* AA^x, $\text{A}'\text{A}'''$, D_1A_0 form always an *involution* of the *ordinary* kind, with its *double points imaginary*; the *three other sets* of segments, AA^x, $\text{A}'\text{A}_0$, $\text{D}_1\text{A}'''$; $\text{A}'\text{A}'''$, AA_0, D_1A^x; and D_1A_0, AA''', $\text{A}'\text{A}^x$, form *each* an involution, with *real double points*; the points A, A^x are the *real foci* of a *fifth* involution, determined by the *two pairs* of segments $\text{A}'\text{D}_1$, and $\text{A}'''\text{A}_0$; the points A′, A‴ are, in like manner, the real double points of that *sixth* involution, which the *two other pairs*, A, D_1, and A_0, A^x, determine: and finally, D_1 and A_0 are such points, for the *seventh* involution, determined by AA′, $\text{A}'''\text{A}^x$.

102. Introducing now the consideration of the two lately *reserved points* $\text{P}_{2,3}$ **99.**, of *second construction* and *third group* **45.**, upon the typical line $\Lambda_{2,1}$, we may derive them from the point P_0, the two points P_1, and the two points $\text{P}_{2,2}$, upon that line AA′, by the two following harmonic equations:

$$(\text{AA}'''\text{A}'\text{A}^{iv}) = (\text{AA}_0\text{D}_1\text{A}_1^{iv}) = -1;$$

or by these two others,

$$(\text{AA}'\text{A}_0\text{A}^{iv}) = (\text{AD}_1\text{A}'''\text{A}_1^{iv}) = -1,$$

which may indeed be inferred from the two former, with the help of the relations between the six points previously considered: for, in general, if *abc*, *a′b′c′* be collinear triads in involution, and if *d* and *d′* be the harmonic conjugates of *b′* and *c′*, with respect to the two pairs, *ab*, *ac*, they are also the harmonic conjugates of *b* and *c*, with respect to the two *other* pairs, *ac′*, *ab′*; or in symbols,

$$(abc'd) = (acb'd') = -1, \text{ if } (ab'bd) = (ac'cd') = -1,$$

when the three harmonic equations **81.** exist. We have also, generally, under these conditions, the equation

$$(ada'd') = -1;$$

for example, on the line AA′, we have

$$(\mathrm{AA}^{iv}\mathrm{A}^{x}\mathrm{A}_{1}^{iv}) = -1.$$

103. It is scarcely worth while to remark that the 15 lines $\Lambda_{2,1}$ of the net, as being all *syntypical*, are all *homographically divided*; although it may just be noticed, as a verification, that the six lines,

$$\mathrm{BC}, \quad \mathrm{B'C'}, \quad \mathrm{B'''C'''}, \quad \mathrm{B_0C_0}, \quad \mathrm{B^{iv}C^{iv}}, \quad \mathrm{B_1^{iv}C_1^{iv}},$$

which connect corresponding points on the two other lines of the same group in the given plane, namely $\mathrm{BB'D_1}$ and $\mathrm{CC'D_1}$, *concur* in one point A″. But it may not be without interest to observe, that A^x is the *common harmonic conjugate* of A, with respect to *each* of the three pairs, $\mathrm{A'D_1}$, $\mathrm{A'''A_0}$, $\mathrm{A^{iv}A_1^{iv}}$; which *three pairs,** or segments, form thus an *involution*, with A and A^x for its *double points*. We have therefore this *Theorem*:-

"*On each of the fifteen lines* $\Lambda_{2,1}$, *the three pairs of derived points, of first and second constructions,* namely the *pair* P_1, *the pair* $\mathrm{P}_{2,2}$, *and the pair* $\mathrm{P}_{2,3}$, *compose an involution, one double point of which is the given point* P_0; *the other double point being the point* $\mathrm{P}_{3,1}$, *of third construction and first group, upon the line.*"

104. We have thus discussed the arrangements of the points P_0, P_1, P_2, on each of the *ten* typical lines which connect not *fewer* than *four*, and not *more* than *seven* of them; but there are still *two other* typical lines to be considered, belonging to the groups Λ_1 and $\Lambda_{2,2}$; whereof one, as BC, passes through *eight* points **54.**; and the other, as B′C′, has *ten* points upon it **56.**. Beginning with the first, we easily find that the two sets of points, A′BC and $\mathrm{A''A_1^{v}A^{v}}$, are *triads in involution* **85.**; the latter set being thus deducible from the former: while the two other points upon the line may be determined by the condition that they satisfy this *other involution of two triads*, A″BC, $\mathrm{A'A_1^{vi}A^{vi}}$. With the *initial arrangement* **91.**, the line $\mathrm{A^{vi}A_1^{vi}}$ is *trisected* in B and C, and its *middle part* BC is likewise *trisected* in A^{v} and A_1^{v}; while *each* line is *bisected* in A′, and *cut at infinity* in A″. And in general we may enunciate these *two Theorems*:-

I. "*On every line of first construction, the point* P_1 *and the two points* P_0 *form a triad, the triad in involution with which consists of the point* $\mathrm{P}_{2,1}$, *and the two points* $\mathrm{P}_{2,4}$."

II. "*On every such line* Λ_1, *the triad formed by the point* $\mathrm{P}_{2,1}$, *and the two points* P_0, *is in involution with a triad which consists of the point* P_1 *and the two points* $\mathrm{P}_{2,5}$."

105. Besides these *two involutions of triads*, we have *two distinct involutions* of the *ordinary* kind, into *each* of which *all the eight points enter*; *two* being *double points* in each. For we have these *two*

* That the *two first* of these three pairs belong to an involution, with those two double points, was seen in **101**.

other Theorems, deducible, indeed, from the two former, but perhaps deserving to be separately stated:-

III. "*On every line of first construction, the two given points are foci of an involution of six points, in which the points* P_1, $P_{2,1}$, *are one pair of conjugates, while the two other pairs are of the common form*, $P_{2,4}$, $P_{2,5}$." For example, A^{v}, A^{vi} are such a pair, on the line BC.

IV. "*On every such line* Λ_1, *the points* P_1, $P_{2,1}$, *are the double points of a second involution of six points, obtained by pairing the two points of each of the three other groups.*"

106. Finally, as regards the *remaining typical line* B′C′, which connects *two points* P_1, and passes through *eight points* P_2, if we reserve for a moment the consideration of the *last pair*, $P_{2,8}$, or A^{ix} and A_1^{ix}, we have a *system of eight points upon that line, homographic with the recent system of eight points on the line* BC; being indeed the *intersections* of the line B′C′ with the *eight-rayed pencil*, $A.A'BCA''A_1^{v}A^{v}A_1^{vi}A^{vi}$, when taken in the order $A'''C'B'A''\,A_1^{viii}A^{viii}A_1^{vii}A^{vii}$. No description of the arrangement of these latter *points* is therefore at this stage required: but as regards the *pencil*, it may be remarked that, by **104.**, the 1st, 2nd and 3rd *rays* form a *triad of lines, in involution* **85.** *with the triad* formed by the 4th, 5th and 6th; and that the *triad* of the 2nd, 3rd and 4th rays is, in the same new sense, in *involution* with the *triad* of the 7th, 8th, and 1st: from which *double involutions of triads*, the *five last rays* may be *derived*, if the *three first* are *given*. We have also by **105.** a *double involution of the rays*, considered as *paired* with *each other*, or with *themselves*: thus the second and third rays are the *double rays* of an involution (of the *usual* kind), in which the first is conjugate to the fourth, the fifth to the seventh, and the sixth to the eighth; while the first and fourth rays are the double rays of *another* involution, in which the second and third, the fifth and sixth, and the seventh and eighth are conjugate.

107. It only remains to assign the arrangement of the *two last points of second construction*, $P_{2,8}$, with respect to the *other points*, P_1, P_2, on a line $\Lambda_{2,2}$, or to some *three* of them; or to show how A^{ix} and A_1^{ix} can be *derived*,* for example, from B′, C′, and A″: which derivation may easily be effected, on the plan already described for the fifth and sixth typical traces. In fact, if we denote the six points $A''C'B'A'''A_1^{ix}A^{ix}$ by $abc\alpha\beta\gamma$, we have the three harmonic equations of **94.**; and if, by one of the modes of *perspective*, or *projection*, mentioned in **95.**, which answers to the initial arrangement **91.**, we throw off the first point A″ to *infinity*, the finite line $A^{ix}A_1^{ix}$ is then *quadrisected*: being *itself bisected* at A‴, while C′ and B′ *bisect its halves.* In general, we shall have again the equations **94.**, if we otherwise represent the six lately mentioned points on B′C′ by $\alpha\beta\gamma\,abc$; and thus it is seen that *those six points* are *always homographic, in every state* of the figure, or *net*, with the six points $A''B_1^{vii}C^{vii}AB_0C_0$ on the *fifth trace* AA″, and with the six points $A''B_1^{viii}C^{viii}D_1B^{v}C_1^{v}$ on the *sixth trace* D_1A''; in fact they are, if taken in a suitable order, the points in which the *six-rayed pencil* **98.**, with A′, for vertex, is cut by the line B′C′.

108. We have thus shown for each of the *twelve typical lines* **74.**, in the plane ABC, how *all the points but three*, upon that line, may be derived *from those three* by a *system of harmonic equations*,

* This point A^{ix} may also, by **81.**, be determined on the *seventh trace*, or *seventh typical line* **74.**, as the *harmonic conjugate* of A′, with respect to C_0 and C_1^{v}.

not *necessarily* employing any point P_3, or other *foreign** or merely *auxiliary point*: although it appeared that something was gained, in respect to elegance and clearness, by introducing, on the line AA', such a point $A^{\times}$ **99.**; or by considering generally, on any one of the fifteen lines $\Lambda_{2,1}$, a point $P_{3,1}$ of *third construction*, belonging to what may perhaps deserve to be regraded as a *first group* **103.** of the points P_3, in any future *extension* **1.** of the results of the present Paper.

Part V. Applications to the Net, continued: Distribution of the Given or Derived Points, in a Plane of Second Construction, and of First or Second Group

109. It will be necessary to be much more concise, in our remarks on the distribution of the *net-points* in *planes* of *second construction*; but a few general remarks may here be offered, from which it will appear that each plane $\Pi_{2,1}$ contains *forty-seven* of the 305 points P_0, P_1, P_2; and that each plane $\Pi_{2,2}$ contains *forty-three* of those points; with many cases of *collineation* for each.

110. We saw in **33.**, that each plane $\Pi_{2,1}$ contains two lines $\Lambda_{2,1}$, which intersect in a point P_0, and may be regarded as the diagonals of a quadrilateral, of which the four sides are lines $\Lambda_{2,2}$. It contains, therefore, as has been seen, one point P_0, and four points P_1; but it is found to contain also 42 points P_2, arranged in *six groups*, as follows.

111. There are 2 points $P_{2,1}$, namely the intersections of opposite sides of the quadrilateral; thus, in what we have called the *second typical plane* **33.**, *the sides* B_1C_1, C_2B_2 *intersect in the point* A''; *and the sides* C_1C_2, B_2B_1 *in* D_1' *(62).*

112. The plane contains also 8 points $P_{2,2}$; namely, *two* on each of the *two diagonals*, and *one* on each of the *four sides*; and it contains 4 points $P_{2,3}$, namely two on each diagonal: but it contains *no* point of either of the two groups, $P_{2,4}$, $P_{2,5}$, as a comparison of their *types* sufficiently proves, or as may be inferred from the *laws* of their construction **46. 47.**

113. The same plane contains 12 points $P_{2,6}$; namely two on each side of the quadrilateral; and four others, in which the plane is intersected by four lines $\Lambda_{2,2}$; as the *types* sufficiently prove. But to show, geometrically, *why* there should be *only four such intersections*, conducting thus to new points $P_{2,6}$ in the plane, let the five inscribed pyramids **28.** be denoted by the symbols A' .. E'; then the six edges of the pyramid A' are found to intersect the present plane $\Pi_{2,1}$ in points already considered, namely in the two points $P_{2,1}$, of *meetings of opposite sides*, and in those four points $P_{2,2}$, which are situated *on the diagonals* of the quadrilateral; they give therefore *no new points*. Also, each *side* of the same quadrilateral is an *edge* of one of the *four other pyramids*, B' .. E'; but there remains, for each such pyramid, an *opposite edge*; and these are

* This *non-requirement* of *foreign points* is the only remarkable thing here: for the *anharmonic function* of *every group* of *four collinear net-points* is necessarily *rational*; and whenever $(abcd) =$ any positive or negative quotient of *whole numbers*, it is *always possible* to deduce the *fourth point* d from the *three* points a, b, c, by *some system of auxiliary points*, derived successively from them through *some system of harmonic equations*.

the *four lines, out of the plane,* which *intersect it* in the *four points* $P_{2,6}$, additional to the *eight points* $P_{2,6}$, which are ranged, two by two, *upon the sides.* There are thus *twelve points* of the *group* $P_{2,6}$, in any one plane $\Pi_{2,1}$; and we have now exhausted the intersections of that plane with lines $\Lambda_{2,2}$; and also, as it will be found, with the lines $\Lambda_{2,1}$, and Λ_1.

114. But there remain *eight* points $P_{2,7}$, and *eight* points $P_{2,8}$, in the plane now considered; namely *two* of *each group,* on each of the *four sides* of the quadrilateral. There are, therefore, 16 such points; which, with the 12 points $P_{2,6}$ the 4 points $P_{2,3}$; the 8 points $P_{2,2}$; the 2 points $P_{2,1}$; the 4 points P_1; and the one point P_0, make up (as has been said in **109.**) a system of 47 points, *given or derived,* in any one of the fifteen planes $\Pi_{2,1}$.

It may be remarked that with the initial arrangement **91.** of the five given points, the four points $B'C'B_2C_2$, in a new plane $\Pi_{2,1}$, are corners of a *square,* which has the point E for its *centre*; and that thus the Figure, of the 47 points in such a plane, may be thrown into a clear and elegant perspective.

115. As regards the distribution in a plane $\Pi_{2,2}$, such as the *Third Typical Plane* **34.**, it may here be sufficient to observe, that besides containing *three lines* $\Lambda_{2,2}$, namely the *sides of a triangular face* **34.** of one of the *five inscribed pyramids* **28.**, and *three points* P_1, which are the *corners* of that *triangle,* and serve to *determine the plane* **1.**, it contains also *forty points* P_2, which are arranged in *groups,* as follows. *Each* of the *four first groups,* of *second construction,* $P_{2,1} \,..\, P_{2,4}$, gives *three points* to the plane; the *fifth group,* P_2 furnishes only *one point*; and the *sixth, seventh,* and *eighth* groups, $P_{2,6}, .. P_{2,8}$, supply *six, twelve,* and *nine* points, respectively, Of these 40 points P_2, *twenty-four* are ranged, eight by eight, *on the three sides* of the triangle, as was to be expected from 56.; and the existence of *at least* 27 *points,* P_1, P_2, in a plane $\Pi_{2,2}$, might thus have been at once foreseen. But we have also to consider the *traces,* on that plane, of the 52 *lines,* Λ_1, Λ_2, which are not contained therein. Of these lines, it is found that 36 *intersect the sides* of the triangle, and give therefore *no new points.* But the *sixteen other lines* intersect the *plane,* in so many *new* and *distinct points*; and thus the *total number* **109.**, of *forty-three derived points,* P_1, P_2, in a plane $\Pi_{2,2}$, which contains *no given point* P_0, is made up.

116. Without attempting here to enumerate the cases of *collineations,* in either of the two typical planes Π_2, we may just remark, that while the traces of four of the planes Π_1 on the typical plane $\Pi_{2,1}$ are the four sides, and the traces of four others are the diagonals, of the quadrilateral already mentioned, the trace of a ninth plane Π_1, namely ABC, on that plane $\Pi_{2,1}$, has been already considered, as the trace AA″ of the latter on the former; but that the trace of the *tenth plane* Π_1, namely ADE, or $[01\bar{1}00]$, on $AB_1C_2C_1B_2$, or on $[011\bar{1}\bar{1}]$, is a *new line,* $AD_1{}'$; which passes thus through one point P_0 and one point $P_{2,1}$, and also through two points $P_{2,2}$, namely (01120) and (01102), and through two points $P_{2,6}$, namely $(2001\bar{1})$ and $(200\bar{1}1)$: being, however, *syntypical* with the formerly considered trace AA″, and therefore leading to no new harmonic or anharmonic relations.

117. As a specimen of a case of collineation which conducts to such *new relations,* let us take the four following points P_2, in the second typical plane,

$$a = (01120), \quad b = (00211), \quad c = (0203\bar{1}), \quad d = (0\bar{1}302),$$

whereof the two first are points $P_{2,2}$, and the two last are points $P_{2,8}$; and of which the symbols satisfy the equations,

$$(c) = 2(a) - (b), \quad (d) = -(a) + 2(b); \quad \text{whence } (adbc) = 4.$$

These four points, therefore, with which it is found that *no other* given or derived point of the system P_0, P_1, P_2 is *collinear*, do *not* form a *harmonic group*; and consequently we *cannot construct the fourth point, d*, when the *three other* points, *a, b, c* are *given*, by means of *harmonic relations alone* (comp. **108.**), unless we introduce some *auxiliary point*, or points, *e*, . . , which shall be at lowest of the *third construction*. But if we write

$$e = (12020) \equiv (01\bar{1}1\bar{1}), \quad f = (\bar{1}0220) \equiv (01331),$$

so that *e* is a point $P_{3,1}$ **99.**, while *f* may be said to be a point $P_{3,2}$, we find that these two *new* or *auxiliary* points, *e, f*, are the *double points* of the *involution*, determined by the *two pairs, ab, cd*; because we have the two harmonic equations,

$$(aebf) = (cedf) = -1.$$

And because we have also,

$$(cabe) = (abde) = -1,$$

we need only employ the *one* auxiliary point *e*, considered as the harmonic conjugate of *a*, with respect to *b* and *c*; and then determine the fourth point *d*, as the harmonic conjugate of *a*, with respect to *b* and *e*. It may be added that *abe* and *dcf* are *triads in involution* **85.**; so that if *e* be projected to infinity, the finite line *cd* is *trisected* at *a* and *b*.

Part VI. On some other Relations of Complanarity, Collinearity, Concurrence, or Homology, for Geometrical Nets in Space

118. Although we have not proposed, in the present Paper, to enumerate, or even to *classify*, any points, lines, or planes, beyond what we have called the *Second Construction* **1.**, yet *some* such points, lines, and planes have offered themselves naturally to our consideration: and we intend, in this *Sixth Part*, to consider a few others, chiefly in connexion with relations of *homology*, of triangles or pyramids which have been already mentioned.

119. It was remarked in **29.**, that the thirty lines $\Lambda_{2,2}$ are the sides of *ten triangles* T_2, of *second construction*, which are certain *inscribed homologues* of ten *other* triangles T_1, of *first* construction **26.**; the *ten* corresponding *centres* of homology being the ten points P_1. For example, the triangle $A'B'C'$ is inscribed in ABC, and is homologous thereto, the point D_1 being their *centre* of homology; because we have the three relations of *intersection*,

$$A' = D_1A \cdot BC, \text{ \&c.};$$

or because, A' being a point on BC, &c., the *three joining lines* AA', &c., *concur* in the point D_1.

120. Proceeding to determine the *axis* of this homology, or the right line which is the locus of the points of intersection of corresponding sides, we easily see that it is the line $A''B''C''$; because we had $A'' = BC \cdot B'C'$, &c. And because an analogous result must take place in *each* of the *ten planes* Π_1, we see that *the ten points* $P_{2,1}$, *are ranged, three by three, on ten lines* $\Lambda_{3,1}$, *in the* *ten planes* Π_1; namely on the *axes of homology* of the *ten pairs of triangles,* T_1, T_2, in those ten planes: which axes are the lines,

$$D_1'A_1'A_2', \text{ \&c.; } C_1'B_1'A'', C_2'B_2'A'', \text{ \&c.; and } A''B''C'';$$

each point $P_{2,1}$ being thus *common* to *three* of them, because it is common to those *three planes* Π_1, which contain the line Λ_1 whereupon it is situated. Each point $P_{2,1}$ is also the *common intersection* of this last line with *three lines* $\Lambda_{2,2}$; we have for example, the *formulæ of concurrence,*

$$A'' = BC\cdot B'C'\cdot B_1C_1\cdot B_2C_2.$$

121. The line $A''B''C''$ was seen to be the *common trace of two planes* $\Pi_{2,2}$, namely of $A_1B_1C_1$ and $A_2B_2C_2$, on the plane Π_1, namely ABC, in which it is situated; and a similar result must evidently hold good for *each* of the *ten lines* $\Lambda_{3,1}$. But we may add that the *three triangles* ABC, $A_1B_1C_1$, $A_2B_2C_2$, in the plane of *each* of which the line $A''B''C''$ is contained, are *homologous, two by two,* and have this line for the *common axis of homology* of each of their *three pairs*; having however *three distinct centres* of homology, namely D_1' for second and third, D for third and first, and E for first and second: with (as we need not again repeat) analogous results for the *other* lines $\Lambda_{3,1}$, of which *group* we here take the line $A''B''C''$ as *typical.* It may be remarked that the *four centres,* recently determined, are *collinear,* and compose an *harmonic group*; and that the *inscribed* triangle $A'B'C'$ is also *homologous* with *each* of the two triangles $A_1B_1C_1$, $A_2B_2C_2$, although not *complanar* with *either*; the line $A''B''C''$ being *still* the *common axis* of homology; while the *two centres,* of these two last homologies, are the two given points, D and E.

122. The *six points* $P_{2,2}$, in the plane ABC, have been seen to range themselves, according to their *two ternary types* **41.**, into *two sets of three,* which are the *corners* of *two new triangles*; *one* of these, namely $A'''B'''C'''$, being an *inscribed homologue* of $A'B'C'$; while the *other,* namely $A_0B_0C_0$, is an *exscribed homologue* of ABC; and these two new triangles are also homologous to *each other*: the *line* $A''B''C''$ being still the *common axis,* and the *point* D_1 being the *common centre* of homology. And the same thing holds good for any one of these four triangles, $A_0B_0C_0$, ABC, $A'B'C'$, $A'''B'''C'''$, in the plane Π_1 here considered, as compared with the triangle $A_1^{iv}B_1^{iv}C_1^{iv}$, whereof the corners are those three points $P_{2,3}$, which are *not* ranged on the line $A''B''C''$, as the three *other* points $P_{2,3}$, namely A^{iv}, B^{iv}, C^{iv}, have been seen to be.

123. It was remarked in **28.**, that each of the *five pyramids* R_2 is not only *inscribed* in the corresponding pyramid R_1 **26.**, but is also *homologous* therewith; the *centre* of their homology being a point P_0: thus the point E is such a centre, for the two pyramids $ABCD$ and $A_1B_1C_1D_1$, or for those which we have lettered as E and E' **26. 113.**. The *planes* BCD, $B_1C_1D_1$, of two corresponding *faces,* intersect in the *line* $C_2'B_2'A''$; the planes CAD, $C_1A_1D_1$ in $A'_2C'_2B''$; the planes ABD, $A_1B_1D_1$ in $B_2'A_2'A''$; and the planes ABC, $A_1B_1C_1$ in $A''B''C''$. Hence it is easy to infer that *these six points* $P_{2,1}$, namely

$$A'', \quad B'', \quad C'', \quad A_2', \quad B_2', \quad C_2',$$

are all situated *in one plane,* which is the *plane of homology* of the *two pyramids* E and E', and which we shall denote by $[E]$; its *quinary symbol* being

$$[E] = [1111\bar{4}],$$

which may also serve as a *type* of the *group* [A] .. [E]. And in fact, the quinary symbols of the six points all satisfy the *equation* (comp. **19.**),

$$x + y + z + w = 4v.$$

124. It may be noted that the *two planes* of homology, [D] and [E], have the *line* A″B″C″ for their *common trace* on the plane ABC; and that the traces of the *three other planes* of the same group, [A], [B], [C], which have

$$[\bar{4}11], \quad [1\bar{4}1], \quad [11\bar{4}],$$

for their *ternary symbols,* pass respectively through the points A^x, B^x, C^x, (comp. **99.**), and coincide with the lines $B_1^{iv}C_1^{iv}$, &c., or with the *sides* of the last mentioned *triangle* **122.**. And it follows from **123.**, that *the ten points* $P_{2,1}$ are ranged *six by six,* and that the *ten lines* $\Lambda_{3,1}$ are ranged *four by four,* in *five planes* $\Pi_{3,1}$; namely, in the five planes [A] .. [E] of *homology of pyramids.* But *these last laws* of arrangement, of points and lines, must be considered as included in results which have been comparatively long known, respecting *transversal** *lines and planes in space.*

125. Instead of *inscribing* a pyramid E′ in the pyramid E, we may propose to *exscribe* to the latter a new pyramid A‵B‵C‵D‵, or E‵, which shall be *homologous* with it, the given point E being still the *centre* of homology. In other words, the *four new planes* B‵C‵D‵, ., A‵B‵C‵, or E_A, E_B, E_C, E_D, are to pass *through the four given points* A, B, C, D; and the *four new lines* AA‵, BB‵, CC‵, DD‵ are to *concur,* in the *fifth given point* E. The solution of this problem is found to be expressed by the following quinary symbols for the four sought planes:

$$[E_A] = [0111\bar{3}], \ .. [E_D] = [1110\bar{3}].$$

In fact, the pyramid E‵, with these four planes for *faces* is evidently *exscribed* to the pyramid ABCD, or E; and because its *corners* may be represented by these other quinary symbols,

$$A‵ = (30001), \ .. D‵ = (00031),$$

the condition of *concurrence* is satisfied. We may remark that the plane [E] of **123.** is the plane of homology of the two last pyramids E and E‵; and that this *exscribed pyramid* E‵ is homologous also to the *inscribed* pyramid E′, the point E being still the *centre,* and the plane [E] the *plane* of their homology.

126. It may be remarked that the *common trace* of the two planes E_D and D_E, on the plane ABC, is the line A″B″C″; to *construct,* then, the *exscribed pyramid* E‵, we may construct the plane E_D of *one* of its *faces,* by connecting the *point* D with the line A″B″C″; and similarly for the rest. Or if we wish to determine separately the *new point,* or corner, D‵, which *corresponds* to the given point D, we may do so, by the *anharmonic equation,*

$$(DD_1ED‵) = 3;$$

for which may be substituted† the system of the *two* following *harmonic* equations:

* Compare the second note to **1.** † Compare the note to **108.**.

$$(\mathrm{DD_1EF}) = (\mathrm{DD'D_1F}) = -1;$$

where F is an auxiliary point, namely $\mathrm{D}_1{}'$.

Part VII. On the Homography and Rationality of Nets in Space; and on a Connexion of such Nets with Surfaces of the Second Order

127. In general, *all geometric nets in space* are *homographic figures*; *corresponding points, lines,* and *planes,* being those which have the *same* (or *congruent*) *quinary symbols,* in whatever manner we may pass from one to another system of *five initial points,* A . . E; whereof it is still supposed that *no four are complanar. All* points, lines, and planes of any such *Net* are evidently *rational,* in the sense **8.** already defined, with respect to the initial system; and conversely it is not difficult to prove that every *rational point, line,* or *plane,* in space, *is a net-point, net-line,* or *net-plane,* whatever that initial system of five points may be. It follows that although *no irrational point, line,* or *plane,* can possibly *belong* to the *net,* with respect to which it *is* thus irrational, yet it can be *indefinitely approached to,* by points, lines, or planes which *do* so belong: a remarkable and interesting theorem, which appears to have been first discovered by *Möbius*;* to whom indeed, as has been already said, the *conception of the net* is due, but whose *analysis* differs essentially from that employed in the present Paper.

128. As regards the *passage from one net in space to another,* let the quinary symbols of some five given points, $\mathrm{P}_1 \,.\,.\, \mathrm{P}_5$, whereof no four are in one plane, be with respect to the *given* initial system A . . E the following:-

$$\mathrm{P}_1 = (x_1 \,.\,.\, v_1), \quad .\,.\, \mathrm{P}_5 = (x_5 \,.\,.\, v_5);$$

and let $a' \,.\,.\, e'$ and u' be six coefficients, determined so as to satisfy the *quinary equations* **5.**,

$$a'(\mathrm{P}_1) + b'(\mathrm{P}_2) + c'(\mathrm{P}_3) + d'(\mathrm{P}_4) + e'(\mathrm{P}_5) = -u'(\mathrm{U}),$$

or the five ordinary equations which it includes, namely,

$$a'x_1 + .\,.\, + e'x_5 = \ldots = a'v_1 + .\,.\, + e'v_5 = -u'.$$

Let P' be any sixth point of space, such that

$$(\mathrm{P}') = xa'(\mathrm{P}_1) + yb'(\mathrm{P}_2) + zc'(\mathrm{P}_3) + wd'(\mathrm{P}_4) + ve'(\mathrm{P}_5) + u(\mathrm{U});$$

then *this sixth point* P' *can be derived from the five points* $\mathrm{P}_1 \,.\,.\, \mathrm{P}_5$, *by the same constructions, as those by which the point* $\mathrm{P} = (xyzwv)$ *is derived from the five given points* ABCDE. For example, if we take the five points,

$$\mathrm{A}_1 = (10001), \quad \mathrm{B}_1 = (01001), \quad \mathrm{C}_1 = (00101), \quad \mathrm{D}_1 = (00011), \quad \mathrm{E} = (00001),$$

we have the symbolic equation,

* See page 295 of the *Barycentric Calculus.* As regards the theory of *homographic figures,* chapter XXV, of the *Géométrie Supérieure* of M. Chasles may be consulted with advantage. But with respect to *anharmonic ratio,* generally, it must be remarked that Professor *Möbius* was thoroughly familiar with its theory and practice, when he published in 1827; although he called it by the longer but perhaps more expressive name of Doppelschnittsverhältniss (*ratio bissectionalis*). It may be added that he denotes by (A, C, B, D), what I write as (ABCD).

$$(A_1) + (B_1) + (C_1) + (D_1) - 3(E) = (U);$$

if then we write $v' = x + y + z + w - 3v$, the point $(xyzwv')$ is derived from $A_1B_1C_1D_1E$, by the same constructions as $(xyzwv)$ from ABCDE. In particular, D is related to $A_1B_1C_1D_1E$, as the point $P = (00031)$ is related to ABCDE; but this point P satisfies the anharmonic equation, $(DD_1EP) = +3$; if then $E_1 = D_1E \cdot A_1B_1C_1 = (000\bar{1}2)$, we must have the corresponding equation $(D_1E_1ED) = +3$: which is accordingly found to exist, and furnishes a *construction* for *exscribing a pyramid* ABCD *to a given pyramid* $A_1B_1C_1D_1$, with which it is to be *homologous*, and to have a *given point* E for the *centre* of their homology, agreeing with the construction assigned in **126.** for a similar problem of *exscription*. And in general, *from any five given points of a net* whereof no four are complanar, we can (as was first shown by *Möbius*) *return, by linear constructions, to the five initial points* A . . E; and therefore can, in this way, *reconstruct the net.*

129. If we content ourselves *with quarternary* (or *anharmonic*) *coordinates* **12.**, or suppose (as we may) that $v = o$, the *equation of a surface of the second order* takes the form,

$$0 = f(xyzw) = ax^2 + \beta y^2 + \gamma z^2 + \delta w^2 + 2(\epsilon yz + \zeta zx + \eta xy) + 2w(\theta x + \iota y + \kappa z);$$

and if the ten coefficients $\alpha \ldots \kappa$, or their ratios, be determined by the condition that the surface shall pass *through nine given net-points*, those *coefficients* may then be replaced by *whole numbers*, and the *surface* may be said to be *rationally related to the given net*, or to the *initial system* A . . E, or briefly to be (comp. **8.**) a *Rational Surface.* For example, if the nine points be ABCDEC′A′C_2A_2, so that, besides passing through E, the surface has the gauche quadrilateral ABCD superscribed upon it, the equation is

$$\text{I} \ldots 0 = f = xz - yw;$$

and if they be A, B, A′, B′, A_2, B_2, A_1, $A^{vii} = (12\bar{1}0)$, and $F = (120\bar{1})$, so that this new point F, like A^{vii}, belong to the group $P_{2,6}$, the equation of the surface is then found to be,

$$\text{II} \ldots 0 = f = w^2 + z^2 - (w + z)(x + y) - 2xy.$$

130. In general, whether the surface of the second order be *rational* or not, it results from the principles of a former communication that any point $P = (xyzw)$ of space is the *pole* of the *plane* $\Pi = [XYZW]$, if $X \ldots W$ be the *derivatives*,

$$X = D_x f, \quad Y = D_y f, \quad Z = D_z f, \quad W = D_w f;$$

hence, in particular, the *pole of the plane* [E] *of homology* of the three pyramids E, E′, E‵, **26. 113. 125.**, of which plane the *quaternary symbol* **12.** is [1111], is the point K determined by the equations,

$$X = Y = Z = W, \text{ or } D_x f = D_y f = D_z f = D_w f;$$

and if the point E be the mean point of the pyramid ABCD, the *plane* [E] is then *infinitely distant*, and this *point* K is the *centre of the surface.*

131. For example, in the case of the Ist surface **129.**, this *pole* K is the point $(1\bar{1}1\bar{1}) \equiv (20201)$, which belongs to the group $P_{3,1}$; and because it is *on* the plane [E], that plane *touches* the surface in that point: so that when the point E is the *mean* point of the pyramid ABCD, the surface becomes a ruled *paraboloid*. In the case of the IInd surface **129.**, the pole K of [E] is always the point (1100), or C′; this point C′ becomes therefore the *centre* of the surface, when E is the *mean* point of the pyramid; and the five following lines,

$$\text{AB}, \quad \text{A}'\text{B}_1^{vii}, \quad \text{B}'\text{A}^{vii}, \quad \text{A}_2\text{F}, \quad \text{and } \text{B}_2\text{G},$$

where G is the new point $(210\overline{1})$ of the group $\text{P}_{2,6}$, which are *always chords through* C′, become in that case *diameters*. It may be added that, with the initial arrangement **91.**, the surface last considered becomes the *sphere*, which is described with AB for diameter; and that it *always* passes through the auxiliary point P, of *third* construction, which was mentioned in **100.**.

132. We have then here an *example*, of a surface of the second order, which was *determined* so as to pass **129.** through *nine net-points*

$$\text{A}, \quad \text{B}, \quad \text{A}', \quad \text{B}', \quad \text{A}_2, \quad \text{B}_2, \quad \text{A}_1, \quad \text{A}^{vii}, \quad \text{and F},$$

but which has been subsequently *found* to pass *also* through at least *four other points of the net*, namely

$$\text{B}_1, \quad \text{B}_1^{vii}, \quad \text{G}, \quad \text{and P}.$$

This is, however, only a very particular *case* of a much more general *Theorem*, with the enunciation of which I shall conclude the present Paper, regretting sincerely that it has already extended to a length, so much exceeding the usual limits of communications designed for the *Proceedings** of the Academy, but hoping that some at least of its processes and results will be thought not wholly uninteresting:-

"If a Surface of the Second Order be determined by the condition of passing through nine given points of a Geometrical Net in Space, it passes also through indefinitely many others: and every Point upon the Surface, which is not a point of the Net, can be included within a Geodetic Triangle on that surface, of which the corners are net-points, and of which the sides can be made as small as we may desire."

In fact, the *surface* is a *rational* one **129.**, or the coefficients of its equation may be made whole numbers; and therefore *every rational line* **8.**, from any *one net point*, or rational point, upon it, if not happening to *touch* the surface, is easily proved to meet it *again*, in *another rational point*: whence, with the aid of a lately mentioned principle **127.**, the theorem evidently follows.

* *Some* of the early formulæ of this Paper are unavoidably repeated from a communication of the preceding Session (1859–60), but with extended significations, as connected now with a *quinary calculus*. And in a not yet published volume, entitled "*Elements of Quaternions*," the subject of *Nets in Space* is incidentally discussed, as an illustration of the *Method of Vectors*. But it will be found that the present Paper is far from being a mere reprint of the Section on Nets, in the unpublished work thus referred to: many new *theorems* having been introduced, and the *plan* of treatment generally being different, although the *notations* have, on the whole, been retained. Besides it was thought that Members of the Academy might like to see the subject treated in their Proceedings without any express reference to *quaternions*: with which indeed the *nets* have not any *necessary* connexion.

XIII.

ON GEOMETRICAL NETS IN SPACE (1861)

[*British Association Report* 1861, Part II., p. 4.]

XIV.

ELEMENTARY PROOF THAT EIGHT PERIMETERS OF THE REGULAR INSCRIBED POLYGON OF TWENTY SIDES EXCEED TWENTY-FIVE DIAMETERS OF THE CIRCLE* (1862)

[*Philosophical Magazine* (4S), **23**, 267–269 (1862).]

It was proved by Archimedes that 71 perimeters, of a regular polygon of 96 sides inscribed in a circle, exceed 223 diameters; whence follows easily the well-known theorem, that eight circumferences of a circle exceed twenty-five diameters, or that $8\pi > 25$. Yet the following elementary proof, that eight perimeters of the regular inscribed polygon of *twenty sides* are greater than twenty-five diameters, has not perhaps hitherto appeared in any scientific† work or periodical; and if a page of the Philosophical Magazine can be spared for its insertion, some readers may find it interesting from its extreme simplicity. In fact, for completely understanding it, no preparation is required beyond the four first Books of Euclid, and the few first Rules of Arithmetic, together with some rudimentary knowledge of the connexion between arithmetic and geometry.

1. It follows from the Fourth Book of Euclid's 'Elements,' that the rectangle under the side of the regular decagon inscribed in a circle, and the same side increased by the radius, is equal to the square of the radius. But the product of the two numbers, 791 and 2071, whereof the latter is equal to the former increased by 1280, is less than the square of 1280 (because 1638161 is less than 1638400). If then the radius be divided into 1280 equal parts, the side of the inscribed decagon must be greater than a line which consists of 791 such parts; or briefly, if the radius be equal to 1280, the side of the decagon exceeds 791.

2. When a diameter of a circle bisects a chord, the square of the chord is equal, by the Third Book, to the rectangle under the doubled segments of that diameter. But the product of the two numbers, 125 and 4995, which together make up 5120, or the double of the double of 1280, is less than the square of 791 (because 624375 is less than 625681). If then the radius be still represented by 1280, and therefore the doubled diameter by 5120, and if the bisected chord be a side of the regular decagon, and therefore greater (by what has just been proved) than 791, the lesser segment of the diameter is greater than the line represented by 125.

* Communicated by the Author.

† A sketch of the proof was published, at the request of a friend, in an eminent literary journal last summer, but in a connexion not likely to attract the attention of mathematical readers in general. At all events, it pretends to no merit but that of brevity, and the simplicity of the principles on which it rests.

3. The rectangle under this doubled segment and the radius, is equal to the square of the side of the regular inscribed polygon of twenty sides. But the product of 125 and 1280 is equal to the square of 400; and if the radius be still 1280, it has been proved that the doubled segment exceeds 125; with this representation of the radius, the side of the inscribed polygon of twenty sides exceeds therefore the line represented by 400; and the perimeter of that polygon is consequently greater than 8000.

4. Dividing then the numbers 1280 and 8000 by their greatest common measure 320, we find that if the radius be now represented by the number 4, or the diameter by 8, the perimeter of the polygon will be greater than the line represented by 25; or in other words, that *eight perimeters of the regular inscribed polygon of twenty sides* (and by still stronger reason, *eight circumferences of the circle* itself) *exceed twenty-five diameters.*

Observatory, March 7, 1862.

XV.

ON THE LOCUS OF THE OSCULATING CIRCLE TO A CURVE IN SPACE (1863)

[Communicated June 22, 1863]
[*Proceedings of the Royal Irish Academy* **8**, 394 (1864).]

Sir W. R. Hamilton read a communication "On the Locus of the Osculating Circle to a Curve in Space."

XVI.

ON THE EIGHT IMAGINARY UMBILICAL GENERATRICES OF A CENTRAL SURFACE OF THE SECOND ORDER (1864)

[Communicated January 11, 1864]
[*Proceedings of the Royal Irish Academy* **8**, 471 (1864).]

[Sir W. R. Hamilton] stated that he had been lately led, by quaternions, to perceive that the twelve known umbilics of such a surface are ranged on *eight imaginary right lines*, of which he has assigned the vector equations, and deduced a variety of properties.

XVII.

ON RÖBER'S CONSTRUCTION OF THE HEPTAGON* (1864)

[*Philosophical Magazine* (4S), **27**, 124–132 (1864).]

---

1. In a recent Number of the Philosophical Magazine, observations were made on some approximate constructions of the regular heptagon, which have recalled my attention to a very remarkable construction of that kind, invented by a deceased professor of architecture at Dresden, Friedrich Gottlob Röber†, who came to conceive, however, that it had been known to the Egyptians, and employed by them in the building of the temple at Edfu. Röber, indeed, was of opinion that the connected triangle, in which each angle at the base is triple of the angle at the vertex, bears very important relations to the plan of the human skeleton, and to other parts of nature. But without pretending to follow him in such speculations, attractive as they may be to many readers, I may be permitted to examine here the accuracy of the proposed geometrical construction, of such an isosceles triangle, or of the heptagon which depends upon it. The closeness of the approximation, although short of mathematical rigour, will be found to be very surprising.

2. Röber's diagram is not very complex, and may even be considered to be elegant; but the essential parts of the construction are sufficiently expressed by the following formulæ: in which p denotes a side of a regular pentagon; r, r' the radii of its inscribed and circumscribed circles; r'' the radius of a third circle, concentric with but exterior to both; p' a segment of the side p; and q, s, t, u, v five other derived lines. The result is, that in the right-angled triangle of which the inner diameter $2r$ is the hypotenuse, and u, v supplementary chords, the former chord (u) is *very nearly* equal to a side of a regular heptagon, inscribed in the interior circle; while the latter chord (v) makes with the diameter ($2r$) an angle ϕ, which is *very nearly* equal to the vertical angle of an isosceles triangle, whereof each angle at the base is triple of the angle at the vertex. In symbols, if we write

$$u = 2r \sin \phi, \quad v = 2r \cos \phi,$$

* Communicated by the Author.

† The construction appears to have been first given in pages 15, 16 of a quarto work by his son, Friedrich Röber, published at Dresden in 1854, and entitled *Beiträge zur Erforschung der geometrischen Grundformen in den alten Tempeln Aegyptens, und deren Beziekung zur alten Naturerkenntniss.* It is repeated in page 20 of a posthumous work, or collection of papers, edited by the younger Röber, and published at Leipzig in 1861, entitled *Elementar-Beiträge zur Bestimmung des Naturgesetzes der Gestaltung und des Widerstandes, und Anwendung dieser Beiträge auf Natur und alte Kunstgestaltung,* von Friedrich Gottlob Röber, ehemaligen Königlich-Sächsischen Professor der Baukunst und Land-Baumeister. Both works, and a third upon the pyramids, to which I cannot at present refer, are replete with the most curious speculations, into which however I have above declined to enter.

then ϕ is found to be very nearly $= \frac{\pi}{7}$. It will be seen that the equations can all be easily constructed by right lines and circles alone, having in fact been formed as the expression of such a construction; and that the numerical ratios of the lines, including the numerical values of the sine and cosine of ϕ, can all be arithmetically computed*, with a few extractions of square roots.

$$
(A)\quad \begin{cases}
(r+r')^2 = 5r^2 & \dfrac{r'}{r} = 1{\cdot}2360680 \\[2ex]
p^2 = 4(r'^2 - r^2) & \dfrac{p}{r} = 1{\cdot}4530851 \\[2ex]
\dfrac{p'}{p} = \dfrac{r + \frac{1}{2}r'}{r+r'} & \dfrac{p'}{r} = 1{\cdot}0514622 \\[2ex]
q^2 = p^2 - p'^2 & \dfrac{q}{r} = 1{\cdot}0029374 \\[2ex]
s^2 + ps = (p - q + r)^2 & \dfrac{s}{r} = 0{\cdot}8954292 \\[2ex]
r''^2 = r^2 + s^2 & \dfrac{r''}{r} = 1{\cdot}3423090 \\[2ex]
t^2 = \left(\dfrac{r'r''}{r}\right)^2 - (r'' - r)^2 & \dfrac{t}{r} = 1{\cdot}6234901 \\[2ex]
u^2 = 2r(2r - t) & \dfrac{u}{r} = 0{\cdot}8677672 \\[2ex]
v^2 = 2rt & \dfrac{v}{r} = 1{\cdot}8019379 \\[2ex]
u = 2r\sin\phi & \sin\phi = 0{\cdot}4338836 \\[1ex]
v = 2r\cos\phi & \cos\phi = 0{\cdot}9009689
\end{cases}
$$

3. On the other hand, the true septisection of the circle may be made to depend on the solution of the cubic equation,

$$8x^3 + 4x^2 - 4x - 1 = 0,$$

of which the roots are $\cos\frac{2\pi}{7}$, $\cos\frac{4\psi}{7}$, $\cos\frac{6\pi}{7}$. Calculating then, by known methods†, to eight decimals, the positive root of this equation, and thence deducing to seven decimals, by square roots, the sine and cosine of $\frac{\pi}{7}$, we find, without tables, the values:

* The computations have all been carried to several decimal places beyond what are here set down. Results of analogous calculations have been given by Röber, and are found in page 16 of the first-cited publication of his son, with the assumption $p = \sqrt{3}$, and with one place fewer of decimals.

† Among these the best by far appears to be Horner's method,—for practically arranging the figures in the use of which method, a very compact and convenient form or scheme was obligingly communicated to me by Professor De Morgan, some time ago. We arrived independently at the following value, to 22 decimals, of the positive root of the cubic mentioned above:

$$\cos\frac{2\pi}{7} = 0{\cdot}62348\,98018\,58733\,53052\,50.$$

$$\cos\frac{2\pi}{7} = x = 0{\cdot}62348980;$$

$$\sin\frac{\pi}{7} = \sqrt{\frac{1-x}{2}} = 0{\cdot}4338837;$$

$$\cos\frac{\pi}{7} = \sqrt{\frac{1+x}{2}} = 0{\cdot}9009689;$$

and these last agree so nearly with the values (A) of $\sin\phi$ and $\cos\phi$, that at this stage a doubt may be felt, *in which direction does the construction err*. In fact, Röber appears to have believed that the construction above described was geometrically rigorous, and had been known and prized as such from a very remote antiquity, although preserved as a secret doctrine, entrusted only to the initiated, and recorded only in stone.

4. The following is an easier way, for a reader who may not like so much arithmetic, to satisfy himself of the extreme closeness of the approximation, by formulæ adapted to logarithms, but rigorously derived from the construction. It being evident that

$$r' = r\sec\frac{\pi}{5}, \quad \text{and } p = 2r\tan\frac{\pi}{5},$$

let $\phi_1 \ldots \phi_6$ be six auxiliary angles, such that

$$r' = 2r\tan\phi_1, \quad p' = p\sin 2\phi_2, \quad p - q = r\tan^2\phi_3,$$
$$p - q + r = \tfrac{1}{2}p\tan 2\phi_4, \quad s = r\tan 2\phi_5, \quad r(r'' - r) = r'r''\sin\phi_6;$$

we shall then have the following system of equations, to which are annexed the angular values, deduced by interpolation from Taylor's seven-figure logarithms, only eleven openings of which are required, if the logarithms of two and four be remembered, as they cannot fail to be by every calculator.

$$(\text{B})\left\{\begin{array}{lll}
\cot\phi_1 = 2\cos\dfrac{\pi}{5} & & \phi_1 = 31^\circ 43'\ 2''{\cdot}91 \\
\sin 2\phi_2 = \cos^2\phi_1 & & \phi_2 = 23\ 10\ 35{\cdot}52 \\
\tan^2\phi_3 = 4\sin^2\phi_2\tan\dfrac{\pi}{5} & & \phi_3 = 33\ 51\ 31{\cdot}90 \\
\cot 2\phi_4 = \cos^2\phi_3\tan\dfrac{\pi}{5} & & \phi_4 = 31\ 41\ 39{\cdot}37 \\
\tan 2\phi_5 = 2\sin^2\phi_4\sec 2\phi_4\tan\dfrac{\pi}{5} & & \phi_5 = 20\ 55\ 15{\cdot}93 \\
\sin\phi_6 = \sin^2\phi_5\cot\phi_1 & & \phi_6 = 11\ 54\ 22{\cdot}60 \\
\cos^2\phi = \cos\phi_6\sec 2\phi_5\tan\phi_1 & & \phi = 25\ 42\ 51{\cdot}4.
\end{array}\right.$$

I had however found, by trials, before using Horner's method, the following approximate value:

$$\cos\frac{2\pi}{7} = 0{\cdot}62348\,98018\,587;$$

which was more than sufficiently exact for comparison with Röber's construction.

It is useless to attempt to estimate hundredths of seconds in this last value, because the difference for a second, in the last logarithmic cosine, amounts only to ten units in the seventh place of decimals, or to one in the sixth place. But if we thus confine ourselves to tenths of seconds, a simple division gives immediately that final value, under the form

$$\frac{\pi}{7} = \frac{180°}{7} = 25°42'51''\cdot 4;$$

it appears therefore to be difficult, if it be possible, to decide by Taylor's tables, whether the equations (B), deduced from Röber's construction, give a value of the angle ϕ, which is *greater* or *less* than the *seventh part of two right angles.* (It may be noted that $\tan 2\phi_1 = 2$; but that to take out ϕ_1 by this equation would require another opening of the tables.)

5. To fix then decisively the *direction* of the *error* of the approximation, and to form with any exactness an estimate of its *amount*, or even to prove quite satisfactorily by *calculation* that any such error *exists*, it becomes necessary to fall back upon arithmetic; and to carry at least the first extractions to several more places of decimals,—although fewer than those which have been actually used in the resumed computation might have sufficed, except for the extreme accuracy aimed at in the resulting values. For this purpose, it has been thought convenient to introduce eight auxiliary numbers, $a \ldots h$, which can all be calculated by square roots, and are defined with reference to the recent equations (B), as follows:

$$a = 1 + 2\tan\phi_1; \quad b = 4\cos 2\phi_2 \cot\frac{\pi}{5}; \quad c = 2\cos 2\phi_2 - \cot\frac{\pi}{5};$$

$$d = \sec 2\phi_4; \quad e = \sec 2\phi_5; \quad f = 2\cos^2\phi; \quad g = 2\cos\phi; \quad h = 2\cos\frac{\phi}{2};$$

or thus with reference to the earlier equations (A):

$$a = \frac{r + r'}{r}; \quad b = \frac{8qr}{p^2}; \quad c = \frac{2(q-r)}{p}; \quad d = \frac{2s + p}{p}; \quad e = \frac{r''}{r};$$

$$f = \frac{t}{r}; \quad g = \frac{v}{r}; \quad h^2 = \frac{2r + v}{r};$$

and respecting which it is to be observed that c, like the rest, is positive, because it may be put under the form

$$c = \sqrt{\frac{14 - 2\sqrt{5}}{5}} - \sqrt{\frac{5 + 2\sqrt{5}}{5}},$$

and $14 - 2\sqrt{5} > 5 + 2\sqrt{5}$, because $9 > 4\sqrt{5}$, or $9^2 > 4^2 5$. With these definitions, then, of the numbers $a \ldots h$, and with the help of the following among other identities,

$$\cos\frac{7\phi}{2}\sec\frac{\phi}{2} = 2\cos 3\phi - 2\cos 2\phi + 2\cos\phi - 1$$

$$= 2(2\cos\phi - 1)\cos 2\phi - 1,$$

I form *without tables* a system of values as below, the early numbers of which have been computed to several decimals more than are set down.

$$
\text{(C)}\quad\begin{cases}
a^2 = 5 & a = \quad 2{\cdot}23606\,79774\,99789\,6964\\
b^2 = 8 + \dfrac{72a}{25} & b = \quad 3{\cdot}79998\,36545\,96345\,0138\\
c^2 = \dfrac{19}{5} - b & c = \quad 0{\cdot}00404\,29449\,23565\,7641\\
d^2 = 1 + (2-c)^2 & d = \quad 2{\cdot}23245\,25898\,01044\,7849\\
e^2 = 1 + (5-2a)(d-1)^2 & e = \quad 1{\cdot}34230\,90137\,74792\,5831\\
f^2 = (5-2a)\,e^2 + 2e - 1 & f = \quad 1{\cdot}62349\,00759\,24105\,2470\\
g^2 = 2f & g = \quad 1{\cdot}80193\,78878\,99638\,5912\\
h^2 = 2 + g & h = \quad 1{\cdot}94985\,58633\,65197\,2049\\
\sin\dfrac{\pi - 7\phi}{2} = h\left((f-1)(g-1) - \dfrac{1}{2}\right) & = +0{\cdot}00000\,06134\,49929\,1683.
\end{cases}
$$

Admitting then the known value,

$$\pi = 3{\cdot}14159\,26535\,89793 \quad \ldots,$$

or the deduced expression,

$$1'' = \frac{\pi}{648000} = 0{\cdot}00000\,48481\,36811\,095 \quad \ldots,$$

I infer as follows:

$$
\text{(D)}\quad\begin{cases}
\dfrac{\pi - 7\phi}{2} = +0''{\cdot}12653\,31307\,822,\\
\dfrac{\pi}{7} - \phi = +0''{\cdot}03615\,23230\,806,\\
\phi = 25^\circ\ 42'\ 51''{\cdot}39241\,91054\,91,
\end{cases}
$$

and think that these twelve decimals of a second, in the value of the angle ϕ, may all be relied on, from the care which has been taken in the calculations.

6. The following is a quite different way, as regards the few last steps, of deducing the same ultimate numerical results. Admitting (comp. Art. 3) the value*,

$$2\cos\frac{2\pi}{7} = z = 1{\cdot}24697\,96037\,17467\,06105,$$

as the positive root, computed by Horner's method, of the cubic equation

$$z^3 + z^2 - 2z - 1 = 0,$$

and employing the lately calculated value f of $1 + \cos 2\phi$, I find by square roots the following sines and cosines, with the same resulting error of the angle ϕ as before:

* Compare a preceding note.

$$
(\mathrm{E})\quad\begin{cases}
\sin\dfrac{\pi}{7}=\dfrac{1}{2}\sqrt{2-z} & = 0{\cdot}43388\,37391\,17558\,1205;\\[2ex]
\cos\dfrac{\pi}{7}=\dfrac{1}{2}\sqrt{2+z} & = 0{\cdot}90096\,88679\,02419\,1262;\\[2ex]
\sin\phi=\dfrac{1}{2}\sqrt{4-2f} & = 0{\cdot}43388\,35812\,03469\,1138;\\[2ex]
\cos\phi=\dfrac{1}{2}\sqrt{2f}=\dfrac{1}{2}g & = 0{\cdot}90096\,89439\,49819\,2956\\[2ex]
\sin\left(\dfrac{\pi}{7}-\phi\right) & = +0{\cdot}00000\,01752\,71408\,3339;\\[2ex]
\dfrac{\pi}{7}-\phi & = +0''{\cdot}03615\,23230\,806.
\end{cases}
$$

7. If we continue the construction, as Röber did, so as to form an isosceles triangle, say ABC, with ϕ for its vertical angle, and if we content ourselves with thousandths of seconds, the angle of this triangle will be as follows:

$$
(\mathrm{F})\quad\begin{cases}
\mathrm{A}=\phi & = 25^\circ 42' 51''{\cdot}392;\\[1ex]
\mathrm{B}=\dfrac{\pi-\phi}{2} & = 77^\circ\ 8' 34''{\cdot}304;\\[1ex]
\mathrm{C}=\mathrm{B} & = 77^\circ\ 8' 34''{\cdot}304;
\end{cases}
$$

and we see that each base-angle exceeds the triple of the vertical by only about an eighth part of a second, namely by that small angle which occurs first in the system (D), and of which the sine is the last number in the preceding system (C). And if we compare a base-angle of the triangle thus constructed, with the base-angle $\frac{3\pi}{7}=77^\circ 8' 34''{\cdot}2857\ldots$ of the true triangle, in which each angle at the base is triple of the angle at the vertex, we find an error in excess equal nearly to $0''{\cdot}018$, or, more exactly,

$$
\mathrm{B}-\frac{3\pi}{7}=\frac{\pi}{14}-\frac{\phi}{2}=+0''{\cdot}01807\,61615\,403,
$$

which amounts to *less* than a *fifty-fifth part* of a *second*, but of which I conceive that all the thirteen decimals here assigned are correct. And I suppose that no artist would undertake to construct a triangle which should more perfectly, or so perfectly, fulfil the conditions proposed. The *problem*, therefore, of constructing *such* a *triangle*, and with it the *regular heptagon*, by *right lines and circles only*, has been *practically solved* by that process which Röber believed to have been *known* to the ancient Egyptians, and to have been *employed* by them in the architecture of some of their temples—some *hints*, as he judged, being intentionally preserved in the details of the workmanship, for the purpose of being *recognized*, by the initiated of the time, or by men of a later age.

8. Another way of rendering conceivable the extreme smallness of the practical error of that process, is to imagine a series of seven successive chords inscribed in a circle, according to the construction in question, and to inquire *how near* to the initial point the final point would be.

The answer is, that the *last* point would fall *behind* the *first*, but only by about *half a second* (more exactly by $0''\cdot506$). If then we suppose, for illustration, that these chords are *seven successive tunnels*, drawn *eastward* from station to station of the *equator of the earth*, the last tunnel would emerge to the *west* of the first station, but only by about *fifty feet*.

9. My own studies have not been such as to entitle me to express an opinion whether the architectural and geometrical drawings of Röber in connexion with the plan of the temple at Edfu, and his comparisons of the numbers deduced from the *details* of his construction with French measurements previously made, are sufficient to bear out his conclusion, that the process had been anciently used: but I wish that some qualified person would take up the inquiry, which appears to me one of great interest, especially as I see no antecedent improbability in the supposition that the construction in question may have been invented in a very distant age. The geometry which it employs is in no degree more difficult than that of the Fourth Book of Euclid*; and although I have no conjecture to offer as to what may have *suggested* the particular process employed, yet it seems to me quite as likely to have been discovered thousands of years ago, perhaps after centuries of tentation, as to have been first found in our own time, which does not generally attach so much importance to the heptagon as a former age may have done, and which perhaps enjoys no special facilities in the search after such a *construction*, although it supplies means of proving, as above, that the proposed solution of the problem is *not mathematically perfect*.

10. If Röber's professional skill as an architect, combined with the circumstance stated of his having previously invented the construction for himself, did really lead him to a correct interpretation† of the plan of the temple at Edfu, which he believed to embody a tradition much older than itself, we are thus admitted to a very curious glimpse, or even a partial view, of the nature and extent, but at the same time the imperfection, of that knowledge of geometry which was possessed, but kept secret, by the ancient priests of Egypt. I say the *imperfection*, on the supposition that the above described construction of the *heptagon*, if known to them at all, was thought by them to be *equal in rigour*, as the elder Röber appears to have thought it to be, to that construction of the *pentagon* which Euclid *may* have learned from them, *rejecting* perhaps, at the same time, the *other* construction, as being not based on demonstration, and not by him demonstrable, although Euclid may not have *known* it to be, in its result, imperfect. The interest of the speculation stretches indeed back to a still earlier age, and may be connected in imagination with what we read of the "wisdom of the Egyptians." But I trust that I shall be found to have abstained, as I was bound to do, from any expression which could imply an acquaintance of my own with the archæology of Egypt, and that I may at least be pardoned, if not thanked, for having thus submitted, to those who may

* The segment p' of the side p of the pentagon, and the fourth proportional $\frac{r'r''}{r}$ to the three radii, which enter into the equations (A), and of which the latter is the greater segment of the third diameter, $2r''$, if this last be cut in extreme and mean ratio, may at first appear to depend on the Sixth Book of Euclid, but will be found to be easily constructible without going beyond the Fourth Book.

† It ought in fairness to be stated that Röber's *interpretation* of Egyptian antiquities included a vast deal more than what is here described, and that he probably considered the *geometrical part* of it to be the *least interesting*, although still, in his view, an essential and *primary element*.

be disposed to study the subject, a purely mathematical* discussion, although connected with a question of other than mathematical interest.

W. R. H.
Observatory of Trinity College, Dublin,
December 22, 1863.

* *Note added during printing.*—Some friends of the writer may be glad to know that these long arithmetical calculations have been to him rather a relaxation than a distraction from his more habitual studies, and that there are already in type 672 octavo pages of the 'Elements of Quaternions', a work which (as he hopes) is rapidly approaching to the stage at which it may be announced for publication.
Observatory, January 19, 1864

XVIII.

ON A NEW SYSTEM OF TWO GENERAL EQUATIONS OF CURVATURE (1865)

[Communicated June 26, 1865]
[*Proceedings of the Royal Irish Academy* **9**, 302–305 (1867).]

The President read the following paper by the late Sir William R. Hamilton:—

ON A NEW SYSTEM OF TWO GENERAL EQUATIONS OF CURVATURE,

Including as easy consequences a new form of the Joint Differential Equation of the Two Lines of Curvature, with a new Proof of their General Rectangularity; and also a new Quadratic for the Joint Determination of the Two Radii of Curvature: all deduced by Gauss's Second Method, for discussing generally the Properties of a Surface; and the latter being verified by a Comparison of Expressions, for what is called by him the Measure of Curvature.

1. Notwithstanding the great beauty and importance of the investigations of the illustrious Gauss, contained in his *Disquisitiones Generales circà Superficies Curvas*, a Memoir* which was communicated to the *Royal Society of Göttingen* in October, 1827, and was printed in Tom. vi. of the *Commentationes Recentiores*, but of which a Latin reprint has been since very judiciously given, near the beginning of the Second Part (Deuxième Partie, Paris, 1850) of LIOUVILLE'S *Edition*†‡ *of* MONGE, it still appears that there is room for some not useless Additions to the Theory of *Lines* and *Radii* of *Curvature*, for *any given Curved Surface*, when treated by what Gauss calls the *Second Method* of discussing the *General Properties of Surfaces*. In fact, the *Method* here alluded to, and which consists chiefly in treating the *three* co-ordinates of the *surface* as being so many *functions* of *two* independent variables, does not seem to have been used *at all* by Gauss, for the determination of the *Directions of the Lines of Curvature*; and as regards the *Radii of Curvature* of the *Normal Sections* which *touch* those *Lines* of Curvature, he appears to have employed the *Method, only for the Product*, and *not also* for the *Sum*, of the *Reciprocals*, of those *Two Radii*.

2. As regards the *notations*, let x, y, z be the rectangular co-ordinates of a point P upon a surface (S), considered as *three* functions of *two* independent variables, t and u; and let the 15 partial derivatives, or 15 partial differential coefficients, of x, y, z taken with respect to t and u, be given by the nine differential expressions.§

* [Reprinted in: Carl Friedrich Gauss (1777–1855), *Werke*, Vol. 4, pp. 217–258, Göttingen: 1863.]
† The foregoing dates, or references, are taken from a note to page 505 of that Edition.
‡ [Gaspard Monge, comte de Péluse (1741–1818), *Application de l'analyse à la géométrie*, 5th edn, edited by Joseph Liouville (1809–1882), Paris: 1850.]
§ [$x' = dx/dt$; $x_, = dx/du$.]

$$\left\{\begin{array}{lll} dx = x'dt + x_, du; & dx' = x''dt + x_,'du; & dx_, = x_,'dt + x_{,,} du; \\ dy = y'dt + y_, du; & dy' = y''dt + y_,'du; & dy_, = y_,'dt + y_{,,} du; \\ dz = z'dt + z_, du; & dz' = z''dt + z_,'du; & dz_, = z_,'dt + z_{,,} du. \end{array}\right. \tag{a}$$

3. Writing also, for abridgment,

$$e = x'^2 + y'^2 + z'^2; \quad e' = x'x_, + y'y_, + z'z_,; \quad e'' = x_,^2 + y_,^2 + z_,^2 \tag{b}$$

we shall have

$$ee'' - e'^2 = K^2, \tag{c}$$

if

$$.\,.\; K^2 = L^2 + M^2 + N^2, \tag{d}$$

and

$$L = y'z_, - z'y_,; \quad M = z'x_, - x'z_,; \quad N = x'y_, - y'x_,; \tag{e}$$

so that

$$Lx' + My' + Nz' = 0, \qquad Lx_, + My_, + Nz_, = 0. \tag{f}$$

Hence $K^{-1}L$, $K^{-1}M$, $K^{-1}N$ are the *direction-cosines* of the *normal* to the surface (S) at P; and if x, y, z be the co-ordinates of any *other* point Q of the same normal, we shall have the equations,

$$K(X - x) = LR; \quad K(Y - y) = MR; \quad K(Z - z) = NR; \tag{g}$$

with

$$R^2 = (X - x)^2 + (Y - y)^2 + (Z - z)^2; \tag{h}$$

where R denotes the normal line PQ, considered as changing sign in passing through zero.

4. The following, however, is for some purposes a more convenient *form* (comp. (f)) of the *Equations of the Normal*;

$$(X - x)x' + (Y - y)y' + (Z - z)z' = 0; \tag{i}$$

$$(X - x)x_, + (Y - y)y_, + (Z - z)z_, = 0. \tag{j}$$

Differentiating these, as if X, Y, Z were constant, that is, treating the point Q as an intersection of two consecutive normals, we obtain these two other equations,

$$\left\{\begin{array}{l} (X - x)dx' + (Y - y)dy' + (Z - x)dz' = x'dx + y'dy + z'dz; \\ (X - x)dx_, + (Y - y)dy_, + (Z - x)dz_, = x_,dx + y_,dy + z_,dz. \end{array}\right. \tag{k}$$

If, then, we write, for abridgment,

$$\left\{\begin{array}{ll} v = du : dt; & E = Lx'' + My'' + Nz''; \\ E' = Lx_,' + My_,' + Nz_,'; & E'' = Lx_{,,} + My_{,,} + Nz_{,,}; \end{array}\right. \tag{l}$$

we shall have, by (a) (b) (g), the two important formulæ:

$$R(E+E'v)=K(e+e'v); \quad R(E'+E''v)=K(e'+e''v); \tag{m}$$

which we propose to call the two general *Equations of Curvature.*

5. In fact, by elimination of R, these equations (m) conduct to a *quadratic in* v, of which the roots may be denoted by v_1 and v_2, which first presents itself under the form,

$$(e+e'v)(E'+E''v)=(e'+e''v)(E+E'v), \tag{n}$$

but may easily be thus transformed,

$$\begin{cases} Av^2-Bv+C=0, \text{ or } Adu^2-Bdtdu+Cdt^2=0, \\ \text{with } A=e'E''-e''E', \; B=e''E-eE'', \; C=eE''-e'E; \end{cases} \tag{o}$$

so that we have the following *general relation,*

$$eA+e'B+e''C=0, \tag{p}$$

(of which we shall shortly see the geometrical signification), between the *coefficients, A, B, C,* of the *joint differential equation* of the system of the two *Lines of Curvature* on the surface.

6. The root v_1 of the quadratic (o) determines the *direction* of what may be called the *First Line of Curvature,* through the point P of that surface; and the *first Radius of Curvature,* for the same point P, or the Radius R_1 of curvature of the *normal section* of the surface which *touches* that *first line,* may be obtained from *either* of the two equations (m), as the value of R which corresponds in that equation to the value v_1 of v. And in like manner, the *Second Radius of Curvature* of the same surface at the same point has the value R_2, which answers to the value v_2 of v, in each of the same two *Equations of Curvature* (m). We see, then, that this *name* for those two equations is justified by observing that when the two independent variables t and u are given or known; and therefore also the seven functions of them, above denoted by e, e', e'', E, $E'E''$, and K. The equations (m) are satisfied by *two* (but *only two*) *systems of values,* v_1, R_1, and v_2, R_2, of (I.) the *differential quotient* v, or $\frac{du}{dt}$, which determines the *direction* of a *line of curvature* on the surface; and (II.) the symbol R, which determines (comp. No. 4) at once the *length* and the *direction,* of the *radius of curvature* corresponding to that *line.*

7. Instead of eliminating R between the two equations (m), we may *begin* by eliminating v; a process which gives the following quadratic in R^{-1} (the curvature):–

$$(eR^{-1}-eK^{-1})(e''R^{-1}-e''K^{-1})=(e'R^{-1}-eK^{-1})^2; \tag{q}$$

or

$$R^{-2}-FR^{-1}+G; \text{ where (because } ee''-e'^2=K^2), \tag{r}$$

$$F=R_1{}^{-1}+R_2{}^{-1}=(eE''-2e'E'+e''E)K^{-3}, \text{ and} \tag{s}$$

$$G=R_1{}^{-1}R_2{}^{-1}=(EE''-E'^2)K^{-4}. \tag{t}$$

We ought, therefore, as a *First General Verification,* to find that this last expression, which may be also thus written,

$$G = R_1^{-1} R_2^{-1} = \frac{EE'' - E'E'}{(L^2 + M^2 + N^2)^2}, \tag{u}$$

agrees with that reprinted in page 521 of Liouville's Monge, for what Gauss calls the *Measure of Curvature* (k) of a *Surface*; namely,

$$k = \frac{DD'' - D'D'}{(AA + BB + CC)^2}; \tag{v}$$

which accordingly it evidently does, because our symbols $L\ M\ N\ A\ B\ C$ represent the combinations which he denotes by ABCD D′D″.

8. As a *Second General Verification*, we may observe that if I be the *inclination* of any *linear element*, $du = v\,dt$, to the *element* $du = 0$, at the point P, then

$$\tan I = \frac{Kv}{e + e'v}; \tag{w}$$

and therefore, that if H be the *angle* at which the *second crosses the first* of *any two lines* represented *jointly* by such an equation as

$$Av^2 - Bv + C = 0, \text{ with } v_1 \text{ and } v_2 \text{ for roots, then} \tag{x}$$

$$\tan H = \tan(I_2 - I_1) = \frac{K(B^2 - 4AC)^{1/2}}{eA + e'B + e''c}; \tag{y}$$

so that the *Condition of Rectangularity* ($\cos H = 0$), for any *two* such lines, may be thus written:

$$eA + e'B + e''C = 0. \tag{z}$$

But this *condition* (z) had already occurred in No. 5, as an equation (p) which is satisfied generally by the *Lines of Curvature*; we see therefore anew, by this analysis, that those *lines* on *any surface* are in general (as is indeed well known) *orthogonal* to each other.

9. Finally, as a *Third General Verification*, we may assume x and y *themselves* (instead of t and u), as the two independent variables of the problem, and then, if we use *Monge's Notation* of p, q, r, s, t, we shall easily recover all his leading results respecting *Curvatures of Surfaces*, but by transformations on which we cannot here delay.

Part III
ANALYSIS

XIX.

ON THE ERROR OF A RECEIVED PRINCIPLE OF ANALYSIS, RESPECTING FUNCTIONS WHICH VANISH WITH THEIR VARIABLES (1830)

[Read January 25, 1830]
[*Transactions of the Royal Irish Academy* **16**, 63–64 (1830).]

It appears to be a received principle of analysis, that if a real function of a positive variable (x) approaches to zero with the variable, and vanishes along with it, then that function can be developed in a real series of the form

$$Ax^{\alpha} + Bx^{\beta} + Cx^{\gamma} + \&c.,$$

the exponents α, β, γ, .. being constant and positive, and the coefficients A, B, C, . . . being constant, and all these constant exponents and coefficients being finite and different from zero. This principle has been made the foundation of important theories, and has not ever, so far as I know, been questioned; but I believe that the following example of exception, which it would be easy to put in a more general form, will sufficiently prove it to be erroneous; since if the principle be true, it is by its nature universal.

The real function $e^{-x^{-2}}$, in which e is the base of the neperian logarithms, approaches to zero along with x and vanishes along with it. Yet if we could develope this function in a series of the kind described, we should have

$$x^{-\alpha}e^{-x^{-2}} = A + Bx^{\beta-\alpha} + Cx^{\gamma-\alpha} + \&c.$$

in which we might suppose α the least exponent; and then, while x approached to 0, the second member would tend to the limit A, which by hypothesis is different from 0; and yet, from the nature of exponential functions, the limit of the first member is zero.

We conclude, therefore, that the function $e^{-x^{-2}}$ cannot be developed in a series of the kind assumed, although it vanishes with its variable; and consequently that, if we only know this property of a function, that it vanishes when its variable vanishes, we cannot correctly assume that it may be developed in such a series.

If any doubt should be felt respecting the truth of the remark, that the function $x^{-\alpha}e^{-x^{-2}}$ tends to zero along with x, when α is any positive constant, this doubt will be removed by observing that the function $x^{\alpha}e^{x^{-2}}$, which is the reciprocal of the former, increases without limit while x decreases to zero. For we may develope this latter function, by the known theorems, in the essentially converging series,

$$x^{a} e^{x^{-2}} = y^{-a} e^{y^2} = y^{-a} + y^{2-a} + \frac{1}{2} y^{4-a} + \frac{1}{2.3} y^{6-a} + \frac{1}{2.3.4} y^{8-a} + \&c.,$$

y being the reciprocal of x; and while y tends to $+\infty$, the terms of this series remain all positive, and all after a certain constant number increase indefinitely.

XX.

NOTE ON A PAPER ON THE ERROR OF A RECEIVED PRINCIPLE OF ANALYSIS (1831)

[Read April 18, 1831]

[*Transactions of the Royal Irish Academy* **16**, 129–130 (1830).]

THE Royal Irish Academy having done me the honour to publish, in the First Part of the Sixteenth Volume of their Transactions, a short Paper, in which I brought forward a certain exponential function as an example of the Error of a received Principle respecting Developments, I am desirous to mention that I have since seen (within the last few days) an earlier Memoir by a profound French Mathematician, in which the same function is employed to prove the fallacy of another usual principle. In the French Memoir, (tom. v. of the Royal Academy of Sciences, at page 13, of the History of the Academy,) the exponential (e^{-1/x^2}) is given by M. CAUCHY, as an example of the vanishing of a function and of all its differential coefficients, for a particular value of the variable (x), without the function vanishing for other values of the variable. In my Paper the same exponential is given as an example of a function, which vanishes with its variable, and yet cannot be represented by any development according to powers of that variable, with constant positive exponents, integer or fractional. There is therefore a difference between the purposes for which this function has been employed in the two Memoirs, although there is also a sufficient resemblance to induce me to wish, that at the time of publishing my Paper, I had been acquainted with the earlier remarks of M. CAUCHY, in order to have noticed their existence.

OBSERVATORY,
April 16, 1831.

XXI.

ON FLUCTUATING FUNCTIONS (1840)

[Communicated June 22, 1841]
[*Proceedings of the Royal Irish Academy* **1**, 475–477 (1841).]

The President gave an account of some investigations respecting *Fluctuating Functions*, from which the following are extracts:-

"Let P_x denote any real function x, continuous or discontinuous, but such that its first and second integrals,

$$\int_0^x dx\, \mathrm{P}_x, \quad \text{and} \quad \left(\int_0^x dx\right)^2 \mathrm{P}_x,$$

are always comprised between given finite limits. Let also the equation

$$\left(\int_0^x dx\right)^2 \mathrm{P}_x = \mu,$$

in which μ is some given constant, have infinitely many real roots, both positive and negative, which are not themselves comprised between any finite limits, but are such that the interval between any one and the next greater root is never greater than some given finite interval. Then,

$$\lim_{t=\infty} \int_a^b dx \int_0^{tx} dy \mathrm{P}_y \mathrm{F}_x = 0, \tag{A}$$

if a and b are any finite values of x, between which the function F_x is finite.

"Again, the same things being supposed, let the arbitrary function F_x vary gradually between the same values of x; and let P_x be finite and vary gradually when x is infinitely small; then

$$\mathrm{F}_y = \varpi^{-1} \int_0^\infty dt \int_a^b dx\, \mathrm{P}_{tx-ty} \mathrm{F}_x, \quad \left(y \begin{matrix} > a \\ < b \end{matrix}\right), \tag{B}$$

in which

$$\varpi = \int_{-\infty}^\infty dx \int_0^1 dt\, \mathrm{P}_{tx}.$$

"For the case $y = a$, we must change ϖ, in (B), to

$$\varpi' = \int_0^\infty dx \int_0^1 dt\, \mathrm{P}_{tx};$$

and for the case $y = b$, we must change it to

$$\varpi'' = \int_{-\infty}^{0} dx \int_{0}^{1} dt \, \mathrm{P}_{tx}.$$

"For values of $y > b$, or $< a$, the second member of the formula (B) vanishes.

"If F_x, although finite, were to receive any sudden change for some particular value of y between a and b, so as to pass suddenly from the value F'' to the value F', we should then have, for this value of y,

$$\int_{0}^{\infty} dt \int_{a}^{b} dx \, \mathrm{P}_{tx-ty} \, \mathrm{F}_x = \varpi'\mathrm{F}' + \varpi''\mathrm{F}''.$$

By changing P_x to $\cos x$, we obtain from (B) the celebrated theorem of Fourier. Indeed, that great mathematician appears to have possessed a clear conception of the *principles* of fluctuating functions, although he is not known to have deduced from them consequences so general as the above.

"Again, another celebrated theorem is comprised in the following:-

$$\mathrm{F}_y = \varpi^{-1} \mathrm{P}_0 \left(\int_{a}^{b} dx \, \mathrm{F}_x + \sum_{(n)1}^{\infty} \int_{a}^{b} dx \, \mathrm{Q}_{x-y,n} \, \mathrm{F}_x \right), \tag{C}$$

in which, the function Q is defined by the conditions

$$\mathrm{Q}_{x,n} \int_{0}^{x} dx \, \mathrm{P}_x = \int_{2nx-x}^{2nx+x} dx \, \mathrm{P}_x;$$

y is $> a$, $< b$; and no real root of the equation

$$\int_{0}^{x} dx \, \mathrm{P}_x = 0,$$

except the root 0, is included between the negative number $a - y$ and the positive number $b - y$, nor are those numbers themselves supposed to be roots of that equation. When these conditions are not satisfied, the theorem (C) takes other forms, which, with other analogous results, may be deduced from the same principles."

XXII.

ON FLUCTUATING FUNCTIONS (1840)

[Read June 22, 1840]
[*Transactions of the Royal Irish Academy* **19**, 264–321 (1843).]

The paper now submitted to the Royal Irish Academy is designed chiefly to invite attention to some consequences of a very fertile principle, of which indications may be found in FOURIER'S Theory of Heat, but which appears to have hitherto attracted little notice, and in particular seems to have been overlooked by POISSON. This principle, which may be called the *Principle of Fluctuation,* asserts (when put under its simplest form) the evanescence of the integral, taken between any finite limits, of the product formed by multiplying together any two finite functions, of which one, like the sine or cosine of an infinite multiple of an arc, changes sign infinitely often within a finite extent of the variable on which it depends, and has for its mean value zero; from which it follows, that if the other function, instead of being always finite, becomes infinite for some particular values of its variable, the integral of the product is to be found by attending only to the immediate neighbourhood of those particular values. The writer is of opinion that it is only requisite to develope the foregoing principle, in order to give a new clearness, and even a new extension, to the existing theory of the transformations of arbitrary functions through functions of determined forms. Such is, at least, the object aimed at in the following pages; to which will be found appended a few general observations on this interesting part of our knowledge.

1. The theorem, discovered by FOURIER, that between any finite limits, a and b, of any real variable x, any arbitrary but finite and determinate function of that variable, of which the value varies gradually, may be represented thus,

$$fx = \frac{1}{\pi}\int_a^b d\alpha \int_0^\infty d\beta \cos(\beta\alpha - \beta x)\, f\alpha, \tag{a}$$

with many other analogous theorems, is included in the following form:

$$fx = \int_a^b d\alpha \int_0^\infty d\beta\phi(x, \alpha, \beta)\, f\alpha; \tag{b}$$

the function ϕ being, in each case, suitably chosen. We propose to consider some of the conditions under which a transformation of the kind (b) is valid.

2. If we make, for abridgment,

$$\psi(x, \alpha, \beta) = \int_0^\beta d\beta\phi(x, \alpha, \beta), \tag{c}$$

the equation (b) may be thus written:

$$fx = \int_a^b d\alpha \psi(x, \alpha, \infty) f\alpha. \tag{d}$$

This equation, if true, will hold good, after the change of $f\alpha$, in the second member, to $f\alpha + \mathrm{F}\alpha$; provided that, for the particular value $\alpha = x$, the additional function $\mathrm{F}\alpha$ vanishes; being also, for other values of α, between the limits a and b, determined and finite, and gradually varying in value. Let then this function F vanish, from $\alpha = a$ to $\alpha = \lambda$, and from $\alpha = \mu$ to $\alpha = b$; λ and μ being included, either between a and x, or between x and b; so that x is not included between λ and μ, though it is included between a and b. We shall have, under these conditions,

$$0 = \int_\lambda^\mu d\alpha \psi(x, \alpha, \infty) \mathrm{F}\alpha; \tag{e}$$

the function F, and the limits λ and μ, being arbitrary, except so far as has been above defined. Consequently, unless the function of α, denoted here by $\psi(x, \alpha, \infty)$, be itself $= 0$, it must change sign at least once between the limits $\alpha = \lambda$, $\alpha = \mu$, however close those limits may be; and therefore must change sign indefinitely often, between the limits a and x, or x and b. A function which thus changes sign indefinitely often, within a finite range of a variable on which it depends, may be called a *fluctuating function.* We shall consider now a class of cases, in which such a function may present itself.

3. Let N_α be a real function of α, continuous or discontinuous in value, but always comprised between some finite limits, so as never to be numerically greater than $\pm c$, in which c is a finite constant; let

$$\mathrm{M}_\alpha = \int_0^\alpha d\alpha \mathrm{N}_\alpha; \tag{f}$$

and let the equation

$$\mathrm{M}_\alpha = \mathrm{a}, \tag{g}$$

in which a is some finite constant, have infinitely many real roots, extending from $-\infty$ to $+\infty$, and such that the interval $\alpha_{n+1} - \alpha_n$, between any one root α_n and the next succeeding α_{n+1}, is never greater than some finite constant, b. Then,

$$0 = \mathrm{M}_{\alpha_{n+1}} - \mathrm{M}_{\alpha_n} = \int_{\alpha_n}^{\alpha_{n+1}} d\alpha \mathrm{N}_\alpha; \tag{h}$$

and consequently the function N_α must change sign at least once between the limits $\alpha = \alpha_n$ and $\alpha = \alpha_{n+1}$; and therefore at least m times between the limits $\alpha = \alpha_n$ and $\alpha = \alpha_{n+m}$, this latter limit being supposed, according to the analogy of this notation, to be the mth root of the equation (g), after the root α_n. Hence the function $\mathrm{N}_{\beta\alpha}$, formed from N_α by multiplying α by β, changes sign at least m times between the limits $\alpha = \lambda$, $\alpha = \mu$, if*

* These notations $\ntriangleright$ and $\ntriangleleft$ are designed to signify the contradictories of $>$ and $<$; so that "$a \ntriangleright b$" is equivalent to "a not $> b$", and "$a \ntriangleleft b$" is equivalent to "a not $< b$."

$$\lambda \not> \beta^{-1} a_n, \ \mu \not< \beta^{-1} a_{n+m};$$

the interval $\mu - \lambda$ between these limits being less than $\beta^{-1}(m+2)\mathrm{b}$, if

$$\lambda > \beta^{-1} a_{n-1}, \ \mu < \beta^{-1} a_{n+m+1};$$

so that, under these conditions, (β being > 0,) we have

$$m > -2 + \beta \mathrm{b}^{-1} (\mu - \lambda).$$

However small, therefore, the interval $\mu - \lambda$ may be, provided that it be greater than 0, the number of changes of sign of the function $\mathrm{N}_{\beta\alpha}$, within this range of the variable α, will increase indefinitely with β. Passing then to the extreme or limiting supposition, $\beta = \infty$, we may say that the function $\mathrm{N}_{\infty\alpha}$ *changes sign infinitely often* within a finite range of the variable α on which it depends; and consequently that it is, in the sense of the last article, a FLUCTUATING FUNCTION. We shall next consider the integral of the product formed by multiplying together two functions of α, of which one is $\mathrm{N}_{\infty\alpha}$, and the other is arbitrary, but finite, and shall see that this integral vanishes.

4. It has been seen that the function N_α changes sign at least once between the limits $\alpha = a_n$, $\alpha = a_{n+1}$. Let it then change sign k times between those limits, and let the k corresponding values of α be denoted by $a_{n,1}, a_{n,2}, \ldots a_{n,k}$. Since the function N_α may be discontinuous in value, it will not necessarily vanish for these k values of α; but at least it will have one constant sign, being throughout not < 0, or else throughout not > 0, in the interval from $\alpha = a_n$ to $\alpha = a_{n,1}$; it will be, on the contrary, throughout not > 0, or throughout not < 0, from $a_{n,1}$ to $a_{n,2}$; again, not < 0, or not > 0, from $a_{n,2}$ to $a_{n,3}$; and so on. Let then N_α be never < 0 throughout the whole of the interval from $a_{n,i}$ to $a_{n,i+1}$; and let it be > 0 for at least some finite part of that interval; i being some integer number between the limits 0 and k, or even one of those limits themselves, provided that the symbols $a_{n,0}$, $a_{n,k+1}$ are understood to denote the same quantities as a_n, a_{n+1}. Let F_α be a finite function of α, which receives no sudden change of value, at least for that extent of the variable α, for which this function is to be employed; and let us consider the integral

$$\int_{a_{n,i}}^{a_{n,i+1}} d\alpha \mathrm{N}_\alpha \mathrm{F}_\alpha. \tag{i}$$

Let F' be the algebraically least, and F'' the algebraically greatest value of the function F_α, between the limits of integration; so that, for every value of α between these limits, we shall have

$$\mathrm{F}_\alpha - \mathrm{F}' \not< 0, \ \mathrm{F}'' - \mathrm{F}_\alpha \not< 0;$$

these values F' and F'', of the function F_α, corresponding to some values $\alpha'_{n,i}$ and $\alpha''_{n,i}$ of the variable α, which are not outside the limits $a_{n,i}$ and $a_{n,i+1}$. Then, since, between these latter limits, we have also

$$\mathrm{N}_\alpha \not< 0,$$

we shall have

$$\left.\begin{aligned}&\int_{\alpha_{n,i}}^{\alpha_{n,i+1}} d\alpha \mathrm{N}_\alpha (\mathrm{F}_\alpha - \mathrm{F}^{\backprime}) \not< 0;\\ &\int_{\alpha_{n,i}}^{\alpha_{n,i+1}} d\alpha \mathrm{N}_\alpha (\mathrm{F}'' - \mathrm{F}_\alpha) \not< 0;\end{aligned}\right\} \tag{k}$$

the integral (i) will therefore be not $< s_{n,i}\mathrm{F}^{\backprime}$, and not $> s_{n,i}\mathrm{F}''$, if we put, for abridgment,

$$s_{n,i} = \int_{\alpha_{n,i}}^{\alpha_{n,i+1}} d\alpha \mathrm{N}_\alpha; \tag{l}$$

and consequently this integral (i) may be represented by $s_{n,i}\mathrm{F}'$, in which

$$\mathrm{F}' \not< \mathrm{F}^{\backprime},\ \mathrm{F}' \not> \mathrm{F}'',$$

because, with the suppositions already made, $s_{n,i} > 0$. We may even write

$$\mathrm{F}' > \mathrm{F}^{\backprime},\ \mathrm{F}' < \mathrm{F}'',$$

unless it happen that the function F_α has a constant value through the whole extent of the integration; or else that it is equal to one of its extreme values, $\mathrm{F}^{\backprime}$ or F'', throughout a finite part of that extent, while, for the remaining part of the same extent, that is, for all other values of α between the same limits, the factor N_α vanishes. In all these cases, F' may be considered as a value of the function F_α, corresponding to a value $\alpha'_{n,i}$ of the variable α which is included between the limits of integration; so that we may express the integral (i) as follows:

$$\int_{\alpha_{n,i}}^{\alpha_{n,i+1}} d\alpha \mathrm{N}_\alpha \mathrm{F}_\alpha = s_{n,i}\mathrm{F}_{\alpha'_{n,i}}; \tag{m}$$

in which

$$\alpha'_{n,i} > \alpha_{n,i},\ < \alpha_{n,i+1}. \tag{n}$$

In like manner, the expression (m), with the inequalities (n), may be proved to hold good, if N_α be never > 0, and sometimes < 0, within the extent of the integration, the integral $s_{n,i}$ being in this case < 0; we have, therefore, rigorously,

$$\int_{\alpha_n}^{\alpha_{n+1}} d\alpha \mathrm{N}_\alpha \mathrm{F}_\alpha = s_{n,0}\mathrm{F}_{\alpha'_{n,0}} + s_{n,1}\mathrm{F}_{\alpha'_{n,1}} + \cdots + s_{n,k}\mathrm{F}_{\alpha'_{n,k}}. \tag{o}$$

But also, we have, by (h)

$$0 = s_{n,0} + s_{n,1} + \cdots + s_{n,k}; \tag{p}$$

the integral in (o) may therefore be thus expressed, without any loss of rigour:

$$\int_{\alpha_n}^{\alpha_{n+1}} d\alpha \mathrm{N}_\alpha \mathrm{F}_\alpha = s_{n,0}\Delta_{n,0} + \cdots + s_{n,k}\Delta_{n,k}, \tag{q}$$

in which

$$\Delta_{n,i} = \mathrm{F}_{\alpha'_{n,i}} - \mathrm{F}_{\alpha_n}; \tag{r}$$

so that $\Delta_{n,i}$ is a finite difference of the function F_α, corresponding to the finite difference $\alpha'_{n,i} - \alpha_n$ of the variable α, which latter difference is less than $\alpha_{n+1} - \alpha_n$, and therefore less than the finite constant b of the last article. The theorem (q) conducts immediately to the following,

$$\int_{\beta^{-1}\alpha_n}^{\beta^{-1}\alpha_{n+1}} d\alpha \mathrm{N}_{\beta\alpha}\mathrm{F}_\alpha = \beta^{-1}(s_{n,o}\delta_{n,o} + \cdots + s_{n,k}\delta_{n,k}), \tag{s}$$

in which

$$\delta_{n,i} = \mathrm{F}_{\beta^{-1}\alpha'_{n,i}} - \mathrm{F}_{\beta^{-1}\alpha_n}; \tag{t}$$

so that, if β be large, $\delta_{n,i}$ is small, being the difference of the function F_α corresponding to a difference of the variable α, which latter difference is less than β^{-1}b. Let $\pm\delta_n$ be the greatest of the $k+1$ differences $\delta_{n,o}, \ldots \delta_{n,k}$, or let it be equal to one of those differences and not exceeded by any other, abstraction being made of sign; then, since the $k+1$ factors $s_{n,o}, \ldots s_{n,k}$ are alternately positive and negative, or negative and positive, the numerical value of the integral (s) cannot exceed that of the expression

$$\pm\beta^{-1}(s_{n,o} - s_{n,1} + s_{n,2} - \cdots + (-1)^k s_{n,k})\delta_n. \tag{u}$$

But, by the definition (1) of $s_{n,i}$, and by the limits $\pm$c of value of the finite function N_α, we have

$$\pm s_{n,i} \ngtr (\alpha_{n,i+1} - \alpha_{n,i})\mathrm{c}; \tag{v}$$

therefore

$$\pm(s_{n,o} - s_{n,1} + \cdots + (-1)^k s_{n,k}) \ngtr (\alpha_{n+1} - \alpha_n)\mathrm{c}; \tag{w}$$

and the following rigorous expression for the integral (s) results:

$$\int_{\beta^{-1}\alpha_n}^{\beta^{-1}\alpha_{n+1}} d\alpha \mathrm{N}_{\beta\alpha}\mathrm{F}_\alpha = \theta_n\beta^{-1}(\alpha_{n+1} - \alpha_n)\mathrm{c}\delta_n; \tag{x}$$

θ_n being a factor which cannot exceed the limits ± 1. Hence, if we change successively n to $n+1$, $n+2$, $\ldots$ $n+m-1$, and add together all the results, we obtain this other rigorous expression, for the integral of the product $\mathrm{N}_{\beta\alpha}\mathrm{F}_\alpha$, extended from $\alpha = \beta^{-1}\alpha_n$ to $\alpha = \beta^{-1}\alpha_{n+m}$:

$$\int_{\beta^{-1}\alpha_n}^{\beta^{-1}\alpha_{n+m}} d\alpha \mathrm{N}_{\beta\alpha}\mathrm{F}_\alpha = \theta\beta^{-1}(\alpha_{n+m} - \alpha_n)\mathrm{c}\delta; \tag{y}$$

in which δ is the greatest of the m quantities δ_n, $\delta_{n+1}, \ldots$, or is equal to one of those quantities, and is not exceeded by any other; and θ cannot exceed ± 1. By taking β sufficiently large, and suitably choosing the indices n and $n+m$, we may make the limits of integration in the formula (y) approach as nearly as we please to any given finite values, a and b; while, in the second member of that formula, the factor $\beta^{-1}(\alpha_{n+m} - \alpha_n)$ will tend to become the finite quantity $b - a$, and θc cannot exceed the finite limits $\pm$c; but the remaining factor δ will tend indefinitely to 0, as β increases without limit, because it is the difference between two values of the function F_α, corresponding to two values of the variable α of which the difference diminishes indefinitely. Passing then to the limit $\beta = \infty$, we have, with the same rigour as before:

$$\int_a^b d\alpha \mathrm{N}_{\infty\alpha}\mathrm{F}_\alpha = 0; \tag{z}$$

which is the theorem that was announced at the end of the preceding article. And although it has been here supposed that the function F_α receives no sudden change of value, between the limits of integration; yet we see that if this function receive any finite number of such sudden changes between those limits, but vary gradually in value between any two such changes, the foregoing demonstration may be applied to each interval of gradual variation of value separately; and the theorem (z) will still hold good.

5. This theorem (z) may be thus written:

$$\lim_{\beta=\infty} \int_a^b d\alpha \mathrm{N}_{\beta\alpha}\mathrm{F}_\alpha = 0; \tag{a$'$}$$

and we may easily deduce from it the following:

$$\lim_{\beta=\infty} \int_a^b d_\alpha \mathrm{N}_{\beta(\alpha-x)}\mathrm{F}_\alpha = 0; \tag{b$'$}$$

the function F_α being here also finite, within the extent of the integration, and x being independent of α and β. For the reasonings of the last article may easily be adapted to this case; or we may see, from the definitions in article **3.**, that if the function N_α have the properties there supposed, then $\mathrm{N}_{\alpha-x}$ will also have those properties. In fact, if N_α be always comprised between given finite limits, then $\mathrm{N}_{\alpha-x}$ will be so too; and we shall have, by (f),

$$\int_0^\alpha d\alpha \mathrm{N}_{\alpha-x} = \int_{-x}^{\alpha-x} d\alpha \mathrm{N}_\alpha = \mathrm{M}_{\alpha-x} - \mathrm{M}_{-x}; \tag{c$'$}$$

in which M_{-x} is finite, because the suppositions of the third article oblige M_α to be always comprised between the limits $\mathrm{a} \pm \mathrm{bc}$; so that the equation

$$\int_0^\alpha d\alpha \mathrm{N}_{\alpha-x} = \mathrm{a} - \mathrm{M}_{-x}, \tag{d$'$}$$

which is of the form (g), has infinitely many real roots, of the form

$$\alpha = x + \alpha_n, \tag{e$'$}$$

and therefore of the kind assumed in the two last articles. Let us now examine what happens, when, in the first member of the formula (b$'$), we substitute, instead of the finite factor F_α, an expression such as $(\alpha - x)^{-1} f_\alpha$, which becomes infinite between the limits of integration, the value of x being supposed to be comprised between those limits, and the function f_α being finite between them. That is, let us inquire whether the integral

$$\int_a^b d\alpha \mathrm{N}_{\beta(\alpha-x)}(\alpha - x)^{-1} f_\alpha, \tag{f$'$}$$

(in which $x > a$, $< b$), tends to any and to what finite and determined limit, as β tends to become infinite.

In this inquiry, the theorem (b$'$) shows that we need only attend to those values of α which are extremely near to x, and are for example comprised between the limits $x \mp \epsilon$, the quantity

ϵ being small. To simplify the question, we shall suppose that for such values of α, the function f_α varies gradually in value; we shall also suppose that $\mathrm{N}_0 = 0$, and that $\mathrm{N}_\alpha \alpha^{-1}$ tends to a finite limit as α tends to 0, whether this be by decreasing or by increasing; although the limit thus obtained, for the case of infinitely small and positive values of α, may possibly differ from that which corresponds to the case of infinitely small and negative values of that variable, on account of the discontinuity which the function N_α may have. We are then to investigate, with the help of these suppositions, the value of the double limit:

$$\lim_{\epsilon=0}.\lim_{\beta=\infty}.\int_{x-\epsilon}^{x+\epsilon} d\alpha \mathrm{N}_{\beta(\alpha-x)} (\alpha - x)^{-1} f_\alpha; \tag{g'}$$

this notation being designed to suggest, that we are first to assume a small but not evanescent value of ϵ, and a large but not infinite value of β, and to effect the integration, or conceive it effected, with these assumptions; then, retaining the same value of ϵ, make β larger and larger without limit; and then at last suppose ϵ to tend to 0, unless the result corresponding to an infinite value of β shall be found to be independent of ϵ. Or, introducing two new quantities y and η, determined by the definitions

$$y = \beta(\alpha - x), \qquad \eta = \beta\epsilon, \tag{h'}$$

and eliminating α and β by means of these, we are led to seek the value of the double limit following:

$$\lim_{\epsilon=0}.\lim_{\eta=\infty}.\int_{-\eta}^{\eta} dy \mathrm{N}_y y^{-1} f_{x+\epsilon\eta^{-1}y}; \tag{i'}$$

in which η tends to ∞, before ϵ tends to 0. It is natural to conclude that since the sought limit (g′) can be expressed under the form (i′), it must be equivalent to the product

$$f_x \times \int_{-\infty}^{\infty} dy \mathrm{N}_y y^{-1}; \tag{k'}$$

and in fact it will be found that this equivalence holds good; but before finally adopting this conclusion, it is proper to consider in detail some difficulties which may present themselves.

6. Decomposing the function $f_{x+\epsilon\eta^{-1}y}$ into two parts, of which one is independent of y, and is $= f_x$, while the other part varies with y, although slowly, and vanishes with that variable; it is clear that the formula (i′) will be decomposed into two corresponding parts, of which the first conducts immediately to the expression (k′); and we are now to inquire whether the integral in this expression has a finite and determinate value. Admitting the suppositions made in the last article, the integral

$$\int_{-\zeta}^{\zeta} dy \mathrm{N}_y y^{-1}$$

will have a finite and determinate value, if ζ be finite and determinate; we are therefore conducted to inquire whether the integrals

$$\int_{-\infty}^{-\zeta} dy \mathrm{N}_y y^{-1}, \qquad \int_{\zeta}^{\infty} dy \mathrm{N}_y y^{-1},$$

are also finite and determinate. The reasonings which we shall employ for the second of these integrals, will also apply to the first; and, to generalize a little the question to which we are thus conducted, we shall consider the integral

$$\int_a^\infty d\alpha \mathrm{N}_\alpha \mathrm{F}_\alpha; \tag{l'}$$

F_α being here supposed to denote any function of α which remains always positive and finite, but decreases continually and gradually in value, and tends indefinitely towards 0, while α increases indefinitely from some given finite value which is not greater than a. Applying to this integral (l′) the principles of the fourth article, and observing that we have now $\mathrm{F}_{\alpha'_{n,i}} < \mathrm{F}_{\alpha_n}$, $\alpha'_{n,i}$ being $> \alpha_n$, and α_n being assumed $\not< a$; and also that

$$\pm(s_{n,0} + s_{n,2} + \cdots) = \mp(s_{n,1} + s_{n,3} + \cdots) \not> \tfrac{1}{2}\mathrm{bc}; \tag{m'}$$

we find

$$\pm\int_{\alpha_n}^{\alpha_{n+1}} d\alpha \mathrm{N}_\alpha \mathrm{F}_\alpha < \tfrac{1}{2}\mathrm{bc}(\mathrm{F}_{\alpha_n} - \mathrm{F}_{\alpha_{n+1}}); \tag{n'}$$

and consequently

$$\pm\int_{\alpha_n}^{\alpha_{n+m}} d\alpha \mathrm{N}_\alpha \mathrm{F}_\alpha < \tfrac{1}{2}\mathrm{bc}(\mathrm{F}_{\alpha_n} - \mathrm{F}_{\alpha_{n+m}}). \tag{o'}$$

This latter integral is therefore finite and numerically less than $\frac{1}{2}\mathrm{bc}\,\mathrm{F}_{\alpha_n}$, however great the upper limit α_{n+m} may be; it tends also to a determined value as m increases indefinitely, because the part which corresponds to values of α between any given value of the form α_{n+m} and any other of the form α_{n+m+p} is included between the limits $\pm\frac{1}{2}\mathrm{bc}\,\mathrm{F}_{\alpha_{n+m}}$, which limits approach indefinitely to each other and to 0, as m increases indefinitely. And in the integral (l′), if we suppose the lower limit a to lie between α_{n-1} and α_n, while the upper limit, instead of being infinite, is at first assumed to be a large but finite quantity b, lying between α_{n+m} and α_{n+m+1}, we shall only thereby add to the integral (o′) two parts, an initial and a final, of which the first is evidently finite and determinate, while the second is easily proved to tend indefinitely to 0 as m increases without limit. The integral (l′) is therefore itself finite and determined, under the conditions above supposed, which are satisfied, for example, by the function $\mathrm{F}_\alpha = \alpha^{-1}$, if a be > 0. And since the suppositions of the last article render also the integral

$$\int_0^a d\alpha \mathrm{N}_\alpha \alpha^{-1}$$

determined and finite, if the value of a be such, we see that with these suppositions we may write

$$\varpi` = \int_0^\infty d\alpha \mathrm{N}_\alpha \alpha^{-1}, \tag{p'}$$

$\varpi`$ being itself a finite and determined quantity. By reasonings almost the same we are led to the analogous formula

$$\varpi'' = \int_{-\infty}^{0} d\alpha \mathrm{N}_\alpha \alpha^{-1}; \tag{q'}$$

and finally to the result

$$\varpi = \varpi' + \varpi'' = \int_{-\infty}^{\infty} d\alpha \mathrm{N}_\alpha \alpha^{-1}; \tag{r'}$$

in which ϖ'' and ϖ are also finite and determined. The product (k′) is therefore itself determinate and finite, and may be represented by ϖf_x.

7. We are next to introduce, in (i′), the variable part of the function f, namely,

$$f_{x+\epsilon\eta^{-1}y} - f_x,$$

which varies from $f_{x-\epsilon}$ to $f_{x+\epsilon}$, while y varies from $-\eta$ to $+\eta$, and in which ϵ may be any quantity > 0. And since it is clear, that under the conditions assumed in the fifth article,

$$\lim_{\epsilon=0}.\lim_{\eta=\infty}.\int_{-\zeta}^{\zeta} dy \mathrm{N}_y y^{-1}(f_{x+\epsilon\eta^{-1}y} - f_x) = 0, \tag{s'}$$

if ζ be any finite and determined quantity, however large, we are conducted to examine whether this double limit vanishes when the integration is made to extend from $y = \zeta$ to $y = \eta$. It is permitted to suppose that f_α continually increases, or continually decreases, from $\alpha = x$ to $\alpha = x + \epsilon$; let us therefore consider the integral

$$\int_{\zeta}^{\eta} d\alpha \mathrm{N}_\alpha \mathrm{F}_\alpha \mathrm{G}_\alpha, \tag{t'}$$

in which the function F_α decreases, while G_α increases, but both are positive and finite, within the extent of the integration.

By reasonings similar to those of the fourth article, we find under these conditions,

$$\pm\int_{\alpha_n}^{\alpha_{n+1}} d\alpha \mathrm{N}_\alpha \mathrm{F}_\alpha \mathrm{G}_\alpha < \mathrm{bc}(\mathrm{F}_{\alpha_n}\mathrm{G}_{\alpha_{n+1}} - \mathrm{F}_{\alpha_{n+1}}\mathrm{G}_{\alpha_n}); \tag{u'}$$

and therefore

$$\left.\begin{aligned} \pm\frac{1}{\mathrm{bc}}\int_{\alpha_n}^{\alpha_{n+m}} d\alpha \mathrm{N}_\alpha \mathrm{F}_\alpha \mathrm{G}_\alpha &< \mathrm{F}_{\alpha_{n+m-1}}\mathrm{G}_{\alpha_{n+m}} - \mathrm{F}_{\alpha_{n+1}}\mathrm{G}_{\alpha_n} \\ &+ (\mathrm{F}_{\alpha_n} - \mathrm{F}_{\alpha_{n+2}})\mathrm{G}_{\alpha_{n+1}} + (\mathrm{F}_{\alpha_{n+2}} - \mathrm{F}_{\alpha_{n+4}})\mathrm{G}_{\alpha_{n+3}} + \&\mathrm{c.} \\ &+ (\mathrm{F}_{\alpha_{n+1}} - \mathrm{F}_{\alpha_{n+3}})\mathrm{G}_{\alpha_{n+2}} + (\mathrm{F}_{\alpha_{n+3}} - \mathrm{F}_{\alpha_{n+5}})\mathrm{G}_{\alpha_{n+4}} + \&\mathrm{c.} \end{aligned}\right\} \tag{v'}$$

This inequality will still subsist, if we increase the second member by changing, in the positive products on the second and third lines, the factors G to their greatest value $\mathrm{G}_{\alpha_{n+m}}$; and, after adding the results, suppress the three negative terms which remain in the three lines of the expression, and change the functions F, in the first and third lines, to their greatest value F_{α_n}. Hence,

$$\pm\int_{\alpha_n}^{\alpha_{n+m}} d\alpha \mathrm{N}_\alpha \mathrm{F}_\alpha \mathrm{G}_\alpha < 3\mathrm{bc}\,\mathrm{F}_{\alpha_n}\mathrm{G}_{\alpha_{n+m}}; \tag{w'}$$

this integral will therefore ultimately vanish, if the product of the greatest values of the functions F and G tend to the limit 0. Thus, if we make

$$\mathrm{F}_\alpha = \alpha^{-1}, \quad \mathrm{G}_\alpha = \pm(f_{x+\epsilon\eta^{-1}\alpha} - f_x),$$

the upper sign being taken when f_α increases from $\alpha = x$ to $\alpha = x + \epsilon$; and if we suppose that ζ and η are of the forms α_n and α_{n+m}; we see that the integral (t′) is numerically less than $3\mathrm{bc}\,\alpha_n^{-1}(f_{x+\epsilon} - f_x)$, and therefore that it vanishes at the limit $\epsilon = 0$. It is easy to see that the same conclusion holds good, when we suppose that η does not coincide with any quantity of the form α_{n+m}, and when the limits of the integration are changed to $-\eta$ and $-\zeta$. We have therefore, rigorously,

$$\lim_{\epsilon=0}.\lim_{\eta=\infty}.\int_{-\eta}^{\eta} dy \mathrm{N}_y y^{-1}(f_{x+\epsilon\eta^{-1}y} - f_x) = 0, \tag{x′}$$

notwithstanding the great and ultimately infinite extent over which the integration is conducted. The variable part of the function f may therefore be suppressed in the double limit (i′), without any loss of accuracy; and that limit is found to be exactly equal to the expression (k′); that is, by the last article, to the determined product ϖf_x. Such, therefore, is the value of the limit (g′), from which (i′) was derived by the transformation (h′); and such finally is the limit of the integral (f′), proposed for investigation in the fifth article. We have, then, proved that under the conditions of that article,

$$\lim_{\beta=\infty}.\int_a^b d\alpha \mathrm{N}_{\beta(\alpha-x)}(\alpha - x)^{-1} f_\alpha = \varpi f_x; \tag{y′}$$

and consequently that the arbitrary but finite and gradually varying function f_x, between the limits $x = a$, $x = b$, may be transformed as follows:

$$f_x = \varpi^{-1}\int_a^b d\alpha \mathrm{N}_{\infty(\alpha-x)}(\alpha - x)^{-1} f_\alpha; \tag{z′}$$

which is a result of the kind denoted by (d) in the second article, and includes the theorem (a) of FOURIER. For all the suppositions made in the foregoing articles, respecting the form of the function N, are satisfied by assuming this function to be the sine of the variable on which it depends; and then the constant ϖ, determined by the formula (r′), becomes coincident with π, that is, with the ratio of the circumference to the diameter of a circle, or with the least positive root of the equation

$$\frac{\sin x}{x} = 0.$$

8. The known theorem just alluded to, namely, that the definite integral (r′) becomes $= \pi$, when $\mathrm{N}_\alpha = \sin\alpha$, may be demonstrated in the following manner. Let

$$\mathrm{A} = \int_0^\infty d\alpha \frac{\sin\beta\alpha}{\alpha};$$

$$\mathrm{B} = \int_0^\infty d\alpha \frac{\cos\beta\alpha}{1+\alpha^2};$$

then these two definite integrals are connected with each other by the relation

$$\text{A} = \left(\int_0^\beta d\beta - \frac{d}{d\beta} \right) \text{B},$$

because

$$\int_0^\beta d\beta \text{B} = \int_0 d\alpha \frac{\sin \beta\alpha}{\alpha(1+\alpha^2)},$$

$$-\frac{d}{d\beta}\text{B} = \int_0^\infty d\alpha \frac{\alpha \sin \beta\alpha}{1+a^2};$$

and all these integrals, by the principles of the foregoing articles, receive determined and finite (that is, not infinite) values, whatever finite or infinite value may be assigned to β. But for all values of $\beta > 0$, the value of A is constant; therefore, for all such values of β, the relation between A and B gives, by integration,

$$e^{-\beta} \left\{ \left(\int_0^\beta d\beta + 1 \right) \text{B} - \text{A} \right\} = \text{const.};$$

and this constant must be $= 0$, because the factor of $e^{-\beta}$ does not tend to become infinite with β. That factor is therefore itself $= 0$, so that we have

$$\text{A} = \left(\int_0^\beta d\beta + 1 \right) \text{B}, \text{ if } \beta > 0.$$

Comparing the two expressions for A, we find

$$\text{B} + \frac{d}{d\beta}\text{B} = 0, \text{ if } \beta > 0;$$

and therefore, for all such values of β,

$$\text{B}\, e^{\beta} = \text{const.}$$

The constant in this last result is easily proved to be equal to the quantity A, by either of the two expressions already established for that quantity; we have therefore

$$\text{B} = \text{A} e^{-\beta},$$

however little the value of β may exceed 0; and because B tends to the limit $\frac{\pi}{2}$ as β tends to 0, we find finally, for all values of β greater than 0,

$$\text{A} = \frac{\pi}{2}, \ \text{B} = \frac{\pi}{2} e^{-\beta}.$$

These values, and the result

$$\int_{-\infty}^{\infty} d\alpha \frac{\sin \alpha}{\alpha} = \pi,$$

to which they immediately conduct, have long been known; and the first relation, above mentioned, between the integrals A and B, has been employed by LEGENDRE to deduce the former integral from the latter; but it seemed worth while to indicate a process by which that relation may be made to conduct to the values of both those integrals, without the necessity of expressly considering the second differential coefficient of B relative to β, which coefficient presents itself at first under an indeterminate form.

9. The connexion of the formula (z′) with FOURIER'S theorem (a), will be more distinctly seen, if we introduce a new function P_α defined by the condition

$$\mathrm{N}_\alpha = \int_0^\alpha d\alpha\, \mathrm{P}_\alpha, \tag{a''}$$

which is consistent with the suppositions already made respecting the function N_α. According to those suppositions the new function P_α is not necessarily continuous, nor even always finite, since its integral N_α may be discontinuous; but P_α is supposed to be finite for small values of α, in order that N_α may vary gradually for such values, and may bear a finite ratio to α. The value of the first integral of P_α is supposed to be always comprised between given finite limits, so as never to be numerically greater than $\pm$c; and the second integral,

$$\mathrm{M}_\alpha = \left(\int_0^\alpha d\alpha\right)^2 \mathrm{P}_\alpha, \tag{b''}$$

becomes infinitely often equal to a given constant, a, for values of α which extend from negative to positive infinity, and are such that the interval between any one and the next following is never greater than a given finite constant, b. With these suppositions respecting the otherwise arbitrary function P_α, the theorems (z) and (z′) may be expressed as follows:

$$\lim_{\beta=\infty} . \int_a^b d\alpha \left(\int_0^{\beta\alpha} d\gamma \mathrm{P}_\gamma\right) f_\alpha = 0; \tag{A}$$

and

$$f_x = \varpi^{-1} \int_a^b d\alpha \int_0^\infty d\beta\, \mathrm{P}_{\beta(\alpha-x)} f_\alpha; \ (x > a, < b) \tag{B}$$

ϖ being determined by the equation

$$\varpi = \int_{-\infty}^\infty d\alpha \int_0^1 d\beta\, \mathrm{P}_{\beta\alpha}. \tag{c''}$$

Now, by making

$$\mathrm{P}_\alpha = \cos\alpha,$$

(a supposition which satisfies all the conditions above assumed), we find, as before,

$$\varpi = \pi,$$

and the theorem (B) reduces itself to the less general formula (a), so that it includes the theorem of FOURIER.

10. If we suppose that x coincides with one of the limits, a or b, instead of being included between them, we find easily, by the foregoing analysis,

$$f_a = \varpi'^{-1} \int_a^b d\alpha \int_0^\infty d\beta \, \mathrm{P}_{\beta(\alpha-a)} f_\alpha; \tag{d''}$$

$$f_b = \varpi''^{-1} \int_a^b d\alpha \int_0^\infty d\beta \, \mathrm{P}_{\beta(\alpha-b)} f_\alpha; \tag{e''}$$

in which

$$\varpi' = \int_0^\infty d\alpha \int_0^1 d\beta \, \mathrm{P}_{\beta\alpha}; \tag{f''}$$

$$\varpi'' = \int_{-\infty}^0 d\alpha \int_0^1 d\beta \, \mathrm{P}_{\beta\alpha}; \tag{g''}$$

so that, as before,

$$\varpi = \varpi' + \varpi''.$$

Finally, when x is outside the limits a and b, the double integral in (B) vanishes; so that

$$0 = \int_a^b d\alpha \int_0^\infty d\beta \, \mathrm{P}_{\beta(\alpha-x)} f_\alpha, \text{ if } x < a, \text{ or} > b. \tag{h''}$$

And the foregoing theorems will still hold good, if the function f_α receive any number of sudden changes of value, between the limits of integration, provided that it remain finite between them; except that for those very values α' of the variable α, for which the finite function f_α receives any such sudden variation, so as to become $= f'$ for values of α infinitely little greater than α', after having been $= f''$ for values infinitely little less than α', we shall have, instead of (B), the formula

$$\varpi' f' + \varpi'' f'' = \int_a^b d\alpha \int_0^\infty d\beta \, \mathrm{P}_{\beta(\alpha-\alpha')} f_\alpha. \tag{i''}$$

11. If P_α be not only finite for small values of α, but also vary gradually for such values, then, whether α be positive or negative, we shall have

$$\lim_{\alpha=0} . \mathrm{N}_\alpha \alpha^{-1} = \mathrm{P}_0; \tag{k''}$$

and if the equation

$$\mathrm{N}_{\alpha-x} = 0 \tag{l''}$$

have no real root α, except the root $\alpha = x$, between the limits a and b, nor any which coincides with either of those limits, then we may change f_α to $\frac{(\alpha - x)\mathrm{P}_0}{\mathrm{N}_{\alpha-x}} f_\alpha$, in the formula (z′), and we shall have the expression:

$$f_x = \varpi^{-1} \mathrm{P}_0 \int_a^b d\alpha \, \mathrm{N}_{\infty(\alpha-x)} \mathrm{N}_{\alpha-x}^{-1} f_\alpha. \tag{m''}$$

Instead of the infinite factor in the index, we may substitute any large number, for example, an uneven integer, and take the limit with respect to it; we may, therefore, write

$$f_x = \varpi^{-1} \mathrm{P}_0 \lim_{n=\infty} \int_a^b d\alpha \frac{\int_0^{(2n+1)(\alpha-x)} d\alpha\, \mathrm{P}_\alpha}{\int_0^{\alpha-x} d\alpha\, \mathrm{P}_\alpha} f_\alpha. \tag{n''}$$

Let

$$\int_{(2n-1)\alpha}^{(2n+1)\alpha} d\alpha\, \mathrm{P}_\alpha = \mathrm{Q}_{\alpha,n} \int_0^\alpha d\alpha\, \mathrm{P}_\alpha; \tag{o''}$$

then

$$1 + \mathrm{Q}_{\alpha,1} + \mathrm{Q}_{\alpha,2} + \cdots + \mathrm{Q}_{\alpha,n} = \frac{\int_0^{(2n+1)\alpha} d\alpha\, \mathrm{P}_\alpha}{\int_0^\alpha d\alpha\, \mathrm{P}_\alpha}, \tag{p''}$$

and the formula (n″) becomes

$$f_x = \varpi^{-1} \mathrm{P}_0 \left(\int_a^b d\alpha\, f_\alpha + \textstyle\sum_{(n)1}^{\infty} \int_a^b d\alpha\, \mathrm{Q}_{\alpha-x,n} f_\alpha \right); \tag{c}$$

in which development, the terms corresponding to large values of n are small. For example, when $\mathrm{P}_\alpha = \cos\alpha$, then

$$\varpi = \pi,\ \mathrm{P}_0 = 1,\ \mathrm{Q}_{\alpha,n} = 2\cos 2n\alpha,$$

and the theorem (c) reduces itself to the following known result:

$$f_x = \pi^{-1} \left(\int_a^b d\alpha\, f_\alpha + 2\textstyle\sum_{(n)1}^{\infty} \int_a^b d\alpha \cos(2n\alpha - 2nx)\, f_\alpha \right); \tag{q''}$$

in which it is supposed that $x > a$, $x < b$, and that $b - a \ntriangleright \pi$, in order that $\alpha - x$ may be comprised between the limits $\pm\pi$, for the whole extent of the integration; and the function f_α is supposed to remain finite within the same extent, and to vary gradually in value, at least for values of the variable α which are extremely near to x. The result (q″) may also be thus written:

$$f_x = \pi^{-1} \textstyle\sum_{(n)-\infty}^{\infty} \int_a^b d\alpha \cos(2n\alpha - 2nx)\, f_\alpha; \tag{r''}$$

and if we write

$$\alpha = \frac{\beta}{2},\ x = \frac{y}{2},\ f_{\frac{y}{2}} = \phi_y,$$

it becomes

$$\phi_y = \frac{1}{2\pi} \textstyle\sum_{(n)-\infty}^{\infty} \int_{2a}^{2b} d\beta \cos(n\beta - ny)\phi_\beta, \tag{s''}$$

the interval between the limits of integration relatively to β being now not greater than 2π, and the value of y being included between those limits. For example, we may assume

$$2a = -\pi,\ 2b = \pi,$$

and then we shall have, by writing α, x, and f, instead of β, y, and ϕ,

$$f_x = \frac{1}{2\pi} \sum\nolimits_{(n)-\infty}^{\infty} \int_{-\pi}^{\pi} d\alpha \cos(n\alpha - nx) f_\alpha, \tag{t''}$$

in which $x > -\pi$, $x < \pi$. It is permitted to assume the function f_α such as to vanish when $\alpha < 0$, $> -\pi$; and then the formula (t″) resolves itself into the two following, which (with a slightly different notation) occur often in the writings of POISSON, as does also the formula (t″):

$$\tfrac{1}{2} \int_0^\pi d\alpha f_\alpha + \sum\nolimits_{(n)1}^{\infty} \int_0^\pi d\alpha \cos(n\alpha - nx) f_\alpha = \pi f_x; \tag{u''}$$

$$\tfrac{1}{2} \int_0^\pi d\alpha f_\alpha + \sum\nolimits_{(n)1}^{\infty} \int_0^\pi d\alpha \cos(n\alpha + nx) f_\alpha = 0; \tag{v''}$$

x being here supposed > 0, but $< \pi$; and the function f_α being arbitrary, but finite, and varying gradually, from $\alpha = 0$ to $\alpha = \pi$, or at least not receiving any sudden change of value for any value x of the variable α, to which the formula (u″) is to be applied. It is evident that the limits of integration in (t″) may be made to become $\mp l$, l being any finite quantity, by merely multiplying $n\alpha - nx$ under the sign cos., by $\frac{\pi}{l}$, and changing the external factor $\frac{1}{2\pi}$ to $\frac{1}{2l}$; and it is under this latter form that the theorem (t″) is usually presented by POISSON: who has also remarked, that the difference of the two series (u″) and (v″) conducts to the expression first assigned by LAGRANGE, for developing an arbitrary function between finite limits, in a series of sines of multiples of the variable on which it depends.

12. In general, in the formula (m″), from which the theorem (c) was derived, in order that x may be susceptible of receiving all values $> a$ and $< b$ (or at least all for which the function f_x receives no sudden change of value), it is necessary, by the remark made at the beginning of the last article, that the equation

$$\int_0^\alpha d\alpha\, \mathrm{P}_\alpha = 0, \tag{w''}$$

should have no real root α different from 0, between the limits $\mp(b - a)$. But it is permitted to suppose, consistently with this restriction, that a is < 0, and that b is > 0, while both are finite and determined; and then the formula (m″), or (c) which is a consequence of it, may be transformed so as to receive new limits of integration, which shall approach as nearly as may be desired to negative and positive infinity. In fact, by changing α to $\lambda\alpha$, x to λx, and $f_{\lambda x}$ to f_x, the formula (c) becomes

$$f_x = \lambda \varpi^{-1} \mathrm{P}_0 \left(\int_{\lambda^{-1}a}^{\lambda^{-1}b} d\alpha\, f_\alpha + \sum\nolimits_{(n)1}^{\infty} \int_{\lambda^{-1}a}^{\lambda^{-1}b} d\alpha\, \mathrm{Q}_{\lambda\alpha - \lambda x, n} f_\alpha \right); \tag{x''}$$

in which $\lambda^{-1}a$ will be large and negative, while $\lambda^{-1}b$ will be large and positive, if λ be small and positive, because we have supposed that a is negative, and b positive; and the new variable x is only obliged to be $> \lambda^{-1}a$, and $< \lambda^{-1}b$, if the new function f_x be finite and vary gradually between these new and enlarged limits. At the same time, the definition (o″) shows that $\mathrm{P}_0\,\mathrm{Q}_{\lambda\alpha-\lambda x,n}$ will tend indefinitely to become equal to $2\mathrm{P}_{2n\lambda(\alpha-x)}$; in such a manner that

$$\lim_{\lambda=0}.\frac{\mathrm{P}_0\mathrm{Q}_{\lambda\alpha-\lambda x,n}}{2\,\mathrm{P}_{2n\lambda(\alpha-x)}}=1, \tag{y″}$$

at least if the function P be finite and vary gradually. Admitting then that we may adopt the following ultimate transformation of a sum into an integral, at least under the sign $\int_{-\infty}^{\infty} d\alpha$,

$$\lim_{\lambda=0}.2\lambda\left(\tfrac{1}{2}\mathrm{P}_0+\textstyle\sum_{(n)1}^{\infty}\mathrm{P}_{2n\lambda(\alpha-x)}\right)=\int_0^{\infty}d\beta\,\mathrm{P}_{\beta(\alpha-x)}, \tag{z″}$$

we shall have, as the limit of (x″), this formula:

$$f_x=\varpi^{-1}\int_{-\infty}^{\infty}d\alpha\int_0^{\infty}d\beta\,\mathrm{P}_{\beta(\alpha-x)}\,f_\alpha; \tag{D}$$

which holds good for all real values of the variable x, at least under the conditions lately supposed, and may be regarded as an extension of the theorem (B), from finite to infinite limits. For example, by making P a cosine, the theorem (D) becomes

$$f_x=\pi^{-1}\int_{-\infty}^{\infty}d\alpha\int_0^{\infty}d\beta\cos(\beta\alpha-\beta x)\,f_\alpha, \tag{a‴}$$

which is a more usual form than (a) for the theorem of FOURIER. In general, the deduction in the present article, of the theorem (D) from (C), may be regarded as a verification of the analysis employed in this paper, because (D) may also be obtained from (B), by making the limits of integration infinite; but the demonstration of the theorem (B) itself, in former articles, was perhaps more completely satisfactory, besides that it involved fewer suppositions; and it seems proper to regard the formula (D) as only a limiting form of (B).

13. This formula (D) may also be considered as a limit in another way, by introducing, under the sign of integration relatively to β, a factor $\mathrm{F}_{k\beta}$ such that

$$\mathrm{F}_0=1,\ \mathrm{F}_\infty=0, \tag{b‴}$$

in which k is supposed positive but small, and the limit taken with respect to it, as follows:

$$f_x=\lim_{k=0}.\varpi^{-1}\int_{-\infty}^{\infty}d\alpha\left(\int_0^{\infty}d\beta\,\mathrm{P}_{\beta(\alpha-x)}\mathrm{F}_{k\beta}\right)f_\alpha. \tag{E}$$

It is permitted to suppose that the function F decreases continually and gradually, at a finite and decreasing rate, from 1 to 0, while the variable on which it depends increases from 0 to ∞; the first differential coefficient F′ being thus constantly finite and negative, but constantly tending to 0, while the variable is positive and tends to ∞. Then, by the suppositions already made respecting the function P, if $\alpha-x$ and k be each different from 0, we shall have

$$\int_0^{\beta}d\beta\mathrm{P}_{\beta(\alpha-x)}\mathrm{F}_{k\beta}=\mathrm{F}_{k\beta}\mathrm{N}_{\beta(\alpha-x)}(\alpha-x)^{-1}-k(\alpha-x)^{-1}\int_0^{\beta}d\beta\mathrm{N}_{\beta(\alpha-x)}\mathrm{F}'_{k\beta}; \tag{c‴}$$

and therefore, because $\mathrm{F}_\infty=0$, while N is always finite, the integral relative to β in the formula (E) may be thus expressed:

$$\int_0^\infty d\beta\, \mathrm{P}_{\beta(\alpha-x)}\, \mathrm{F}_{k\beta} = (\alpha - x)^{-1} \psi_{k^{-1}(\alpha-x)}, \tag{d'''}$$

the function ψ being assigned by the equation

$$\psi_\lambda = -\int_0^\infty d\gamma\, \mathrm{N}_{\lambda\gamma}\, \mathrm{F}'_\gamma. \tag{e'''}$$

For any given value of λ, the value of this function ψ is finite and determinate, by the principles of the sixth article; and as λ tends to ∞, the function ψ tends to 0, on account of the fluctuation of N, and because F' tends to 0, while γ tends to ∞; the integral (d‴) therefore tends to vanish with k, if α be different from x; so that

$$\lim_{k=0} . \int_0^\infty d\beta \mathrm{P}_{\beta(\alpha-x)} \mathrm{F}_{k\beta} = 0, \text{ if } \alpha \gtrless x. \tag{f'''}$$

On the other hand, if $\alpha = x$, that integral tends to become infinite, because we have, by (b‴),

$$\lim_{k=0} . \mathrm{P}_0 \int_0^\infty d\beta\, \mathrm{F}_{k\beta} = \infty. \tag{g'''}$$

Thus, while the formula (d‴) shows that the integral relative to β in (E) is a homogeneous function of $\alpha - x$ and k, of which the dimension is negative unity, we see also, by (f‴) and (g‴), that this function is such as to vanish or become infinite at the limit $k = 0$, according as $\alpha - x$ is different from or equal to zero. When the difference between α and x, whether positive or negative, is very small and of the same order as k, the value of the last mentioned integral (relative to β) varies vary rapidly with α; and in this way of considering the subject, the proof of the formula (E) is made to depend on the verification of the equation

$$\varpi^{-1} \int_{-\infty}^\infty d\lambda \psi_\lambda \lambda^{-1} = 1. \tag{h'''}$$

But this last verification is easily effected; for when we substitute the expression (e‴) for ψ_λ, and integrate first relatively to λ, we find, by (r′),

$$\int_{-\infty}^\infty d\lambda\, \mathrm{N}_{\lambda\gamma} \lambda^{-1} = \varpi; \tag{i'''}$$

it remains then to show that

$$-\int_0^\infty d\gamma\, \mathrm{F}'_\gamma = 1; \tag{k'''}$$

and this follows immediately from the conditions (b‴). For example, when P is a cosine, and F a negative neperian exponential, so that

$$\mathrm{P}_\alpha = \cos \alpha,\ \mathrm{F}_\alpha = e^{-\alpha},$$

then, making $\lambda = k^{-1}(\alpha - x)$, we have

$$\int_0^\infty d\beta\, e^{-k\beta} \cos(\beta\alpha - \beta x) = (\alpha - x)^{-1} \psi_\lambda;$$

$$\psi_\lambda = \int_0^\infty d\gamma e^{-\gamma} \sin \lambda\gamma = \frac{\lambda}{1 + \lambda^2};$$

and

$$\varpi^{-1}\int_{-\infty}^{\infty} d\lambda\psi_\lambda\lambda^{-1} = \pi^{-1}\int_{-\infty}^{\infty}\frac{d\lambda}{1+\lambda^2} = 1.$$

It is nearly thus that POISSON has, in some of his writings, demonstrated the theorem of FOURIER, after putting it under a form which differs only slightly from the following:

$$f_x = \pi^{-1}\lim_{k=0}\int_{-\infty}^{\infty} d\alpha\int_0^{\infty} d\beta e^{-k\beta}\cos(\beta\alpha - \beta x) f_\alpha; \tag{1'''}$$

namely, by substituting for the integral relative to β its value

$$\frac{k}{k^2 + (\alpha - x)^2};$$

and then observing that, if k be very small, this value is itself very small, unless α be extremely near to x, so that f_α may be changed to f_x; while, making $\alpha = x + k\lambda$, and integrating relatively to λ between limits indefinitely great, the factor by which this function f_x is multiplied in the second member of (1'''), is found to reduce itself to unity.

14. Again, the function F_α retaining the same properties as in the last article for positive values of α, and being further supposed to satisfy the condition

$$\mathrm{F}_{-\alpha} = \mathrm{F}_\alpha, \tag{m'''}$$

while k is still supposed to be positive and small, the formula (D) may be presented in this other way, as the limit of the result of two integrations, of which the first is to be effected with respect to the variable α:

$$f_x = \lim_{k=0}.\varpi^{-1}\int_0^{\infty} d\beta\int_{-\infty}^{\infty} d\alpha\, \mathrm{F}_{k\alpha}\, \mathrm{P}_{\beta(\alpha-x)} f_\alpha. \tag{F}$$

Now it often happens that if the function f_α be obliged to satisfy conditions which determine all its values by means of the arbitrary values which it may have for a given finite range, from $\alpha = a$ to $\alpha = b$, the integral relative to α in the formula (F) can be shown to vanish at the limit $k = 0$, for all real and positive values of β, except those which are roots of a certain equation

$$\Omega_\rho = 0; \tag{G}$$

while the same integral is, on the contrary, infinite, for these particular values of β; and then the integration relatively to β will in general change itself into a summation relatively to the real and positive roots ρ of the equation (G), which is to be combined with an integration relatively to α between the given limits a and b; the resulting expression being of the form

$$f_x = \textstyle\sum_\rho \displaystyle\int_a^b d\alpha\, \phi_{x,\alpha,\rho} f_\alpha. \tag{H}$$

For example, in the case where P is a cosine, and F a negative exponential, if the conditions relative to the function f be supposed such as to conduct to expressions of the forms

$$\int_0^{\infty} d\alpha e^{-h\alpha} f_\alpha = \frac{\psi(h)}{\phi(h)}, \qquad (\text{n}''')$$

$$\int_0^{-\infty} d\alpha e^{h\alpha} f_\alpha = \frac{\psi(-h)}{\phi(-h)}, \qquad (\text{o}''')$$

in which h is any real or imaginary quantity, independent of α, and having its real part positive; it will follow that

$$\int_{-\infty}^{\infty} d\alpha e^{-k\sqrt{\alpha^2}} (\cos\beta\alpha - \sqrt{-1}\sin\beta\alpha) f_\alpha = \frac{\psi(\beta\sqrt{-1}+k)}{\phi(\beta\sqrt{-1}+k)} - \frac{\psi(\beta\sqrt{-1}-k)}{\phi(\beta\sqrt{-1}-k)}, \qquad (\text{p}''')$$

in which $\sqrt{\alpha^2}$ is $= \alpha$ or $= -\alpha$, according as α is $>$ or < 0, and the quantities β and k are real, and k is positive. The integral in (p‴), and consequently also that relative to α in (F), in which, now,

$$\text{P}_\alpha = \cos\alpha, \ \text{F}_\alpha = e^{-k\sqrt{\alpha^2}},$$

will therefore, under these conditions, tend to vanish with k, unless β be a root ρ of the equation

$$\phi(\rho\sqrt{-1}) = 0, \qquad (\text{q}''')$$

which here corresponds to (G); but the same integral will on the contrary tend to become infinite, as k tends to 0, if β be a root of the equation (q‴). Making therefore $\beta = \rho + k\lambda$, and supposing $k\lambda$ to be small, while ρ is a real and positive root of (q‴), the integral (p‴) becomes

$$\frac{k^{-1}}{1+\lambda^2}(\text{A}_\rho - \sqrt{-1}\text{B}_\rho), \qquad (\text{r}''')$$

in which A_ρ and B_ρ are real, namely,

$$\left.\begin{aligned} \text{A}_\rho &= \frac{\psi(\rho\sqrt{-1})}{\phi'(\rho\sqrt{-1})} + \frac{\psi(-\rho\sqrt{-1})}{\phi'(-\rho\sqrt{-1})}, \\ \text{B}_\rho &= \sqrt{-1}\left(\frac{\psi(\rho\sqrt{-1})}{\phi'(\rho\sqrt{-1})} - \frac{\psi(-\rho\sqrt{-1})}{\phi'(-\rho\sqrt{-1})}\right); \end{aligned}\right\} \qquad (\text{s}''')$$

ϕ' being the differential coefficient of the function ϕ. Multiplying the expression (r‴) by $\pi^{-1}\, d\beta(\cos\beta x + \sqrt{-1}\sin\beta x)$, which may be changed to $\pi^{-1}\, kd\lambda(\cos\rho x + \sqrt{-1}\sin\rho x)$; integrating relatively to λ between indefinitely great limits, negative and positive; taking the real part of the result, and summing it relatively to ρ; there results,

$$f_x = \textstyle\sum_\rho (\text{A}_\rho \cos\rho x + \text{B}_\rho \sin\rho x); \qquad (\text{t}''')$$

a development which has been deduced nearly as above, by POISSON and LIOUVILLE, from the suppositions (n‴), (o‴), and from the theorem of FOURIER presented under a form equivalent to the following:

$$f_x = \lim_{k=0} .\pi^{-1} \int_0^{\infty} d\beta \int_{-\infty}^{\infty} d\alpha e^{-k\sqrt{\alpha^2}} \cos(\beta\alpha - \beta x) f_\alpha; \qquad (\text{u}''')$$

and in which it is to be remembered that if 0 be a root of the equation (q‴), the corresponding terms in the development of f_x must in general be modified by the circumstance, that in calculating these terms, the integration relatively to λ extends only from 0 to ∞.

For example, when the function f is obliged to satisfy the conditions

$$f_{-\alpha} = f_\alpha, \ f_{l-\alpha} = -f_{l+\alpha}, \qquad \text{(v''')}$$

the suppositions (n''') (o''') are satisfied; the functions ϕ and ψ being here such that

$$\phi(h) = e^{hl} + e^{-hl},$$

$$\psi(h) = \int_0^l d\alpha (e^{h(l-\alpha)} - e^{h(\alpha-l)}) f_\alpha;$$

therefore the equation (q''') becomes in this case

$$\cos \rho l = 0, \qquad \text{(w''')}$$

and the expressions (s''') for the coefficients of the development (t''') reduce themselves to the following:

$$\mathrm{A}_\rho = \frac{2}{l} \int_0^l d\alpha \cos \rho_\alpha f_\alpha; \quad \mathrm{B}_\rho = 0; \qquad \text{(x''')}$$

so that the method conducts to the following expression for the function f, which satisfies the conditions (v'''),

$$f_x = \frac{2}{l} \sum\nolimits_{(n)1}^{\infty} \cos \frac{(2n-1)\pi x}{2l} \int_0^l d\alpha \cos \frac{(2n-1)\pi\alpha}{2l} f_\alpha; \qquad \text{(y''')}$$

in which f_α is arbitrary from $\alpha = 0$ to $\alpha = l$, except that f_l must vanish. The same method has been applied, by the authors already cited, to other and more difficult questions; but it will harmonize better with the principles of the present paper to treat the subject in another way, to which we shall now proceed.

15. Instead of introducing, as in (E) and (F), a factor which has unity for its limit, we may often remove the apparent indeterminateness of the formula (D) in another way, by the principles of fluctuating functions. For if we integrate first relatively to α between indefinitely great limits, negative and positive, then, under the conditions which conduct to developments of the form (H), we shall find that the resulting function of β is usually a fluctuating one, of which the integral vanishes, except in the immediate neighbourhood of certain particular values determined by an equation such as (G); and then, by integrating only in such immediate neighbourhood, and afterwards summing the results, the development (H) is obtained. For example, when P is a cosine, and when the conditions (v''') are satisfied by the function f, it is not difficult to prove that

$$\int_{-2ml-l}^{2ml+l} d\alpha \cos(\beta\alpha - \beta x) f_\alpha = \frac{2\cos(2m\beta l + \beta l + m\pi)}{\cos \beta l} \cos \beta x \int_0^l d\alpha \cos \beta\alpha f_\alpha; \qquad \text{(z''')}$$

m being here an integer number, which is to be supposed large, and ultimately infinite. The equation (G) becomes therefore, in the present question and by the present method, as well as by that of the last article,

$$\cos \rho l = 0;$$

and if we make $\beta = \rho + \gamma$, ρ being a root of this equation, we may neglect γ in the second member of (z‴), except in the denominator

$$\cos \beta l = -\sin \rho l \sin \gamma l,$$

and in the fluctuating factor of the numerator

$$\cos(2m\beta l + \beta l + m\pi) = -\sin \rho l \sin(2m\gamma l + \gamma l);$$

consequently, multiplying by $\pi^{-1} d\gamma$, integrating relatively to γ between any two small limits of the forms $\mp\epsilon$, and observing that

$$\lim_{m=\infty} . \frac{2}{\pi}\int_{-\epsilon}^{\epsilon} d\gamma \frac{\sin(2ml\gamma + l\gamma)}{\sin l\gamma} = \frac{2}{l},$$

the development

$$f_x = \frac{2}{l}\sum\nolimits_\rho \cos \rho x \int_0^l d\alpha \cos \rho\alpha\, f_\alpha,$$

which coincides with (y‴), and is of the form (H), is obtained.

16. A more important application of the method of the last article is suggested by the expression which FOURIER has given for the arbitrary initial temperature of a solid sphere, on the supposition that this temperature is the same for all points at the same distance from the centre. Denoting the radius of the sphere by l, and that of any layer or shell of it by α, while the initial temperature of the same layer is denoted by $\alpha^{-1} f_\alpha$, we have the equations

$$f_0 = 0,\ f'_l + \nu f_l = 0, \tag{a^{IV}}$$

which permit us to suppose

$$f_\alpha + f_{-\alpha} = 0,\ f'_{l+\alpha} + f'_{l-\alpha} + \nu(f_{l+\alpha} + f_{l-\alpha}) = 0; \tag{b^{IV}}$$

ν being here a constant quantity not less than $-l^{-1}$, and f' being the first differential coefficient of the function f, which function remains arbitrary for all values of α greater than 0, but not greater than l. The equations (b^{IV}) give

$$(\beta \cos \beta l + \nu \sin \beta l)\int_{l-\alpha}^{l+\alpha} d\alpha \sin \beta\alpha\, f_\alpha = \tag{c^{IV}}$$

$$(\beta \sin \beta l - \nu \cos \beta l)\int_{\alpha-l}^{\alpha+l} d\alpha \cos \beta\alpha\, f_\alpha - \cos \beta\alpha (f_{\alpha+l} + f_{\alpha-l});$$

so that

$$(\rho \sin \rho l - \nu \cos \rho l)\int_{\alpha-l}^{\alpha+l} d\alpha \cos \rho\alpha\, f_\alpha = \cos \rho\alpha(f_{\alpha+l} + f_{\alpha-l}), \tag{d^{IV}}$$

if ρ be a root of the equation

$$\rho \cos \rho l + \nu \sin \rho l = 0. \tag{e^{IV}}$$

This latter equation is that which here corresponds to (G); and when we change β to $\rho+\gamma$, γ being very small, we may write, in the first member of (c^{IV}),

$$\beta\cos\beta l+\nu\sin\beta l=\gamma\{(+\nu l)\cos\rho l-\rho l\sin\rho l\}, \tag{f^{IV}}$$

and change β to ρ in all the terms of the second member, except in the fluctuating factor $\cos\beta\alpha$, in which α is to be made extremely large. Also, after making $\cos\beta\alpha=\cos\rho\alpha\cos\gamma\alpha-\sin\rho\alpha\sin\gamma\alpha$, we may suppress $\cos\gamma\alpha$ in the second member of (c^{IV}), before integrating with respect to γ, because by (d^{IV}) the terms involving $\cos\gamma\alpha$ tend to vanish with γ, and because $\gamma^{-1}\cos\gamma\alpha$ changes sign with γ. On the other hand, the integral of $\dfrac{d\gamma\sin\gamma\alpha}{\gamma}$ is to be replaced by π, though it be taken only for very small values, negative and positive, of γ, because α is here indefinitely large and positive. Thus in the present question, the formula

$$f_x=\frac{1}{\pi}.\lim_{\alpha=\infty}.\int_0^{\infty}d\beta\sin\beta x\int_{l-\alpha}^{l+\alpha}d\alpha\sin\beta\alpha f_\alpha, \tag{g^{IV}}$$

(which is obtained from (a‴) by suppressing the terms which involve $\cos\beta x$, on account of the first condition (b^{IV}),) may be replaced by a sum relative to the real and positive roots of the equation (e^{IV}); the term corresponding to any one such root being

$$\frac{R_\rho\sin\rho x}{(+\nu l)\cos\rho l-\rho l\sin\rho l}, \tag{h^{IV}}$$

if we suppose $\rho>0$, and make for abridgment

$$R_\rho=(\nu\cos\rho l-\rho\sin\rho l)\int_{\alpha-l}^{\alpha+l}d\alpha\sin\rho\alpha f_\alpha+\sin\rho\alpha(f_{\alpha+l}+f_{\alpha-l}). \tag{i^{IV}}$$

The equations (b^{IV}) show that the quantity R_ρ does not vary with α, and therefore that it may be rigorously thus expressed:

$$R_\rho=2(\nu\cos\rho l-\rho\sin\rho l)\int_0^l d\alpha\sin\rho\alpha f_\alpha; \tag{k^{IV}}$$

we have also, by (e^{IV}), ρ being >0,

$$\frac{2(\nu\cos\rho l-\rho\sin\rho l)}{\cos\rho l+l(\nu\cos\rho l-\rho\sin\rho l)}=\frac{2\rho}{\rho l-\sin\rho l\cos\rho l}. \tag{l^{IV}}$$

And if we set aside the particular case where

$$\nu l+1=0, \tag{m^{IV}}$$

the term corresponding to the root

$$\rho=0, \tag{n^{IV}}$$

of the equation (e^{IV}), vanishes in the development of f_x; because this term is, by (g^{IV}),

$$\frac{x}{\pi}\int_0^{\beta}d\beta\left(\beta\int_{l-\alpha}^{l+\alpha}d\alpha\sin\beta\alpha f_\alpha\right), \tag{o^{IV}}$$

α being very large, and β small, but both being positive; and unless the condition (m^{IV}) be satisfied, the equation (c^{IV}) shows that the quantity to be integrated in (o^{IV}), with respect to

β, is a finite and fluctuating function of that variable, so that its integral vanishes, at the limit $\alpha = \infty$. Setting aside then the case (m^{IV}), which corresponds physically to the absence of exterior radiation, we see that the function f_x, which represents the initial temperature of any layer of the sphere multiplied by the distance x of that layer from the centre, and which is arbitrary between the limits $x = 0$, $x = l$, that is, between the centre and the surface, (though it is obliged to satisfy at those limits the conditions (a^{IV})), may be developed in the following series, which was discovered by FOURIER, and is of the form (H):

$$f_x = \sum\nolimits_\rho \frac{2\rho \sin \rho x \int_0^l d\alpha \sin \rho\alpha f_\alpha}{\rho l - \sin \rho l \cos \rho l}; \tag{p^{IV}}$$

the sum extending only to those roots of the equation (e^{IV}) which are greater than 0. In the particular case (m^{IV}), in which the root (n^{IV}) of the equation (e^{IV}) must be employed, the term (o^{IV}) becomes, by (c^{IV}) and (d^{IV}),

$$\frac{3x}{\pi l^3}\left\{\int_{\alpha-l}^{\alpha+l} d\alpha f_\alpha \alpha \mathrm{c} - l(f_{\alpha+l} + f_{\alpha-l})\alpha \mathrm{c}\right\}, \tag{q^{IV}}$$

in which, at the limit here considered,

$$\mathrm{c} = \int_0^\infty d\theta \frac{\operatorname{vers}\theta}{\theta^2} = \frac{\pi}{2}; \tag{r^{IV}}$$

but also, by the equations (b^{IV}), (m^{IV}),

$$\int_{\alpha-l}^{\alpha+l} d\alpha f_\alpha \alpha - l(f_{\alpha+l} + f_{\alpha-l})\alpha = 2\int_0^l d\alpha f_\alpha \alpha; \tag{s^{IV}}$$

the sought term of f_x becomes, therefore, in the present case,

$$\frac{3x}{l^3}\int_0^l d\alpha f_\alpha \alpha, \tag{t^{IV}}$$

and the corresponding term in the expression of the temperature $x^{-1} f_x$ is equal to the mean initial temperature of the sphere; a result which has been otherwise obtained by POISSON, for the case of no exterior radiation, and which might have been anticipated from physical considerations. The supposition

$$\nu l + 1 < 0, \tag{u^{IV}}$$

which is inconsistent with the physical conditions of the question, and in which FOURIER'S development (p^{IV}) may fail, is excluded in the foregoing analysis.

17. When a converging series of the form (H) is arrived at, in which the coefficients ϕ of the arbitrary function f, under the sign of integration, do not tend to vanish as they correspond to larger and larger roots ρ of the equation (G); then those coefficients $\phi_{x,\alpha,\rho}$ must in general tend to become fluctuating functions of α, as ρ becomes larger and larger. And the sum of those coefficients, which may be thus denoted,

$$\sum\nolimits_\rho \phi_{x,\alpha,\rho} = \psi_{x,\alpha,\rho}, \tag{I}$$

and which is here supposed to be extended to all real and positive roots of the equation (G), as far as some given root ρ, must tend to become a fluctuating function of α, and to have its mean value equal to zero, as ρ tends to become infinite, for all values of α and x which are different from each other, and are both comprised between the limits of the integration relative to α; in such a manner as to satisfy the equation

$$\int_{\lambda}^{\mu} d\alpha \psi_{x,\alpha,\infty} f_{\alpha} = 0, \tag{K}$$

which is of the form (e), referred to in the second article; provided that the arbitrary function f is finite, and that the quantities λ, μ, x, α are all comprised between the limits a and b, which enter into the formula (H); while α is, but x is not, comprised also between the new limits λ and μ. But when $\alpha = x$, the sum (I) tends to become infinite with ρ, so that we have

$$\psi_{x,x,\infty} = \infty, \tag{L}$$

and

$$\int_{x-\epsilon}^{x+\epsilon} d\alpha \psi_{x,\alpha,\infty} f_{\alpha} = f_x, \tag{M}$$

ϵ being here a quantity indefinitely small. For example, in the particular question which conducts to the development (y'''), we have

$$\phi_{x,\alpha,\rho} = \frac{2}{l} \cos \rho x \cos \rho \alpha, \tag{v^{IV}}$$

and

$$\rho = \frac{(2n-1)\pi}{2l}; \tag{w^{IV}}$$

therefore, summing relatively to ρ, or to n, from $n = 1$ to any given positive value of the integer number n, we have, by (I),

$$\psi_{x,\alpha,\rho} = \frac{\sin \dfrac{n\pi(\alpha - x)}{l}}{2l \sin \dfrac{\pi(\alpha - x)}{2l}} + \frac{\sin \dfrac{n\pi(\alpha + x)}{l}}{2l \sin \dfrac{\pi(\alpha + x)}{2l}}; \tag{x^{IV}}$$

and it is evident that this sum tends to become a fluctuating function of α, and to satisfy the equation (K), as ρ, or n, tends to become infinite, while α and x are different from each other, and are both comprised between the limits 0 and l. On the other hand, when α becomes equal to x, the first part of the expression (x^{IV}) becomes $= \frac{n}{l}$, and therefore tends to become infinite with n, so that the equation (L) is true. And the equation (M) is verified by observing, that if $x > 0, < l$, we may omit the second part of the sum (x^{IV}), as disappearing in the integral through fluctuation, while the first part gives, at the limit,

$$\lim_{n=\infty} \int_{x-\epsilon}^{x+\epsilon} d\alpha \frac{\sin \dfrac{n\pi(\alpha - x)}{l}}{2l \sin \dfrac{\pi(\alpha - x)}{2l}} f_{\alpha} = f_x. \tag{y^{IV}}$$

If x be equal to 0, the integral is to be taken only from 0 to ϵ, and the result is only half as great, namely,

$$\lim_{n=\infty} . \int_0^\epsilon d\alpha \frac{\sin \dfrac{n\pi\alpha}{l}}{2l \sin \dfrac{\pi\alpha}{2l}} f_\alpha = \tfrac{1}{2} f_0; \qquad (\text{z}^{IV})$$

but, in this case, the other part of the sum (x^{IV}) contributes an equal term, and the whole result is f_0. If $x = l$, the integral is to be taken from $l - \epsilon$ to l, and the two parts of the expression (x^{IV}) contribute the two terms $\frac{1}{2} f_l$ and $-\frac{1}{2} f_l$, which neutralize each other. We may therefore in this way prove, *à posteriori*, by the consideration of fluctuating functions, the truth of the development (y‴) for any arbitrary but finite function f_x, and for all values of the real variable x from $x = 0$ to $x = l$, the function being supposed to vanish at the latter limit; observing only that if this function f_x undergo any sudden change of value, for any value x' of the variable between the limits 0 and l, and if x be made equal to x' in the development (y‴), the process shows that this development then represents the semisum of the two values which the function f receives, immediately before and after it undergoes this sudden change.

18. The same mode of *à posteriori* proof, through the consideration of fluctuating functions, may be applied to a great variety of other analogous developments, as has indeed been indicated by FOURIER, in a passage of his Theory of Heat. The spirit of POISSON'S method, when applied to the establishment, *à posteriori*, of developments of the form (H), would lead us to multiply, before the summation, each coefficient $\phi_{x,\alpha,\rho}$ by a factor $\text{F}_{k,\rho}$ which tends to unity as k tends to 0, but tends to vanish as ρ tends to ∞; and then instead of a *generally fluctuating sum* (I), there results a *generally evanescent sum* (k being evanescent), namely,

$$\textstyle\sum_\rho \text{F}_{k,\rho} \phi_{x,\alpha,\rho} = \chi_{x,\alpha,k,\rho}, \qquad (\text{N})$$

which conducts to equations analogous to (K) (L) (M), namely,

$$\lim_{k=0} \int_\lambda^\mu d\alpha \chi_{x,\alpha,k,\infty} f_\alpha = 0; \qquad (\text{O})$$

$$\lim_{k=0} \chi_{x,x,k,\infty} = \infty; \qquad (\text{P})$$

$$\lim_{k=0} \int_{x-\epsilon}^{x+\epsilon} d\alpha \chi_{x,\alpha,k,\infty} f_\alpha = f_x. \qquad (\text{Q})$$

It would be interesting to inquire what form the generally evanescent function χ would take immediately before its vanishing, when

$$\text{F}_{k,\rho} = \epsilon^{-k\rho},$$

and

$$\phi_{x,\alpha,\rho} = \frac{2\rho \sin \rho x \sin \rho\alpha}{\rho l - \sin \rho l \cos \rho l},$$

ρ being a root of the equation

$$\rho l \operatorname{cotan} \rho l = \text{const.},$$

and the constant in the second member being supposed not greater than unity.

19. The development (C), which, like (H), expresses an arbitrary function, at least between given limits, by a combination of summation and integration, was deduced from the expression (m″) of the eleventh article, which conducts also to many other analogous developments, according to the various ways in which the factor with the infinite index, $\mathrm{N}_{\infty(\alpha-x)}$, may be replaced by an infinite sum, or other equivalent form. Thus, if, instead of (o″), we establish the following equation,

$$\int_{(2n-2)\alpha}^{2n\alpha} d\alpha \, \mathrm{P}_\alpha = \mathrm{R}_{\alpha,n} \int_0^\alpha d\alpha \, \mathrm{P}_\alpha, \qquad (\mathrm{a}^V)$$

we shall have, instead of (C), the development:

$$f_x = \varpi^{-1} \mathrm{P}_0 \textstyle\sum_{(n)1}^{\infty} \displaystyle\int_a^b d\alpha \, \mathrm{R}_{\alpha-x,n} f_\alpha; \qquad (\mathrm{R})$$

which, when P is a cosine, reduces itself to the form,

$$f_x = \frac{2}{\pi} \textstyle\sum_{(n)1}^{\infty} \displaystyle\int_a^b d\alpha \cos(\overline{2n-1}.\overline{\alpha - x}) \, f_\alpha, \qquad (\mathrm{b}^V)$$

x being $> a$, $< b$, and $b - a$ being not $> \pi$; and easily conducts to the known expression

$$f_x = \frac{1}{l} \textstyle\sum_{(n)1}^{\infty} \displaystyle\int_{-l}^{l} d\alpha \cos \frac{(2n-1)\pi(\alpha - x)}{2l} f_\alpha, \qquad (\mathrm{c}^V)$$

which holds good for all values of x between $-l$ and $+l$. By supposing $f_\alpha = f_{-\alpha}$, we are conducted to the expression (y‴); and by supposing $f_\alpha = -f_{-\alpha}$, we are conducted to this other known expression,

$$f_x = \frac{2}{l} \textstyle\sum_{(n)1}^{\infty} \sin \dfrac{(2n-1)\pi x}{2l} \displaystyle\int_0^l d\alpha \sin \frac{(2n-1)\pi\alpha}{2l} f_\alpha; \qquad (\mathrm{d}^V)$$

which holds good even at the limit $x = l$, by the principles of the seventeenth article, and therefore offers the following transformation for the arbitrary function f_l:

$$f_l = -\frac{2}{l} \textstyle\sum_{(n)1}^{\infty} (-1)^n \displaystyle\int_0^l d\alpha \sin \frac{(2n-1)\pi\alpha}{2l} f_\alpha. \qquad (\mathrm{e}^V)$$

For example, by making $f_\alpha = \alpha^i$, and supposing i to be an uneven integer number; effecting the integration indicated in (e^V), and dividing both members by l^i, we find the following relation between the sums of the reciprocals of even powers of odd whole numbers:

$$1 = [i]^1 \omega_2 - [i]^3 \omega_4 + [i]^5 \omega_6 - \ldots; \qquad (\mathrm{f}^V)$$

in which

$$[i]^k = i(i-1)(i-2) \ldots (i-k+1); \qquad (\mathrm{g}^V)$$

and

$$\omega_{2k} = 2\left(\frac{2}{\pi}\right)^{2k} \textstyle\sum_{(n)1}^{\infty} (2n-1)^{-2k}; \qquad (\mathrm{h}^V)$$

thus

$$1 = \omega_2 = 3\omega_2 - 3.2.1.\omega_4 = 5\omega_2 - 5.4.3\omega_4 + 5.4.3.2.1\omega_6, \qquad (\mathrm{i}^V)$$

so that

$$\omega_2 = 1,\ \omega_4 = \tfrac{1}{3},\ \omega_6 = \tfrac{2}{15}. \qquad (\mathrm{k}^V)$$

Again, by making $f_\alpha = \alpha^i$, but supposing $i =$ an uneven number $2k$, we get the following additional term in the second member of the equation (f^V),

$$(-1)^k[2k]^{2k}\omega_{2k+1}, \qquad (\mathrm{l}^V)$$

in which

$$\omega_{2k+1} = -2\left(\frac{2}{\pi}\right)^{2k+1} \Sigma_{(n)1}^{\ \infty}(-1)^n(2n-1)^{-2k-1}; \qquad (\mathrm{m}^V)$$

thus

$$1 = \omega_1 = 2\omega_2 - 2.1\omega_3 = 4\omega_2 - 4.3.2\omega_4 + 4.3.2.1\omega_5, \qquad (\mathrm{n}^V)$$

so that

$$\omega_1 = 1,\ \omega_3 = \tfrac{1}{2},\ \omega_5 = \tfrac{5}{24}. \qquad (\mathrm{o}^V)$$

Accordingly, if we multiply the values (k^V) by $\frac{\pi^2}{8}, \frac{\pi^4}{32}, \frac{\pi^6}{128}$, we get the known values for the sums of the reciprocals of the squares, fourth powers, and sixth powers of the odd whole numbers; and if we multiply the values (o^V) by $\frac{\pi}{4}, \frac{\pi^3}{16}, \frac{\pi^5}{64}$, we get the known values for the sums of the reciprocals of the first, third, and fifth powers of the same odd numbers, taken however with alternately positive and negative signs. Again, if we make $f_\alpha = \sin\alpha$, in (e^V), and divide both members of the resulting equation by $\cos l$, we get this known expression for a tangent,

$$\tan l = \Sigma_{(n)-\infty}^{\ \infty} \frac{2}{(2n-1)\pi - 2l}; \qquad (\mathrm{p}^V)$$

which shows that, with the notation (h^V),

$$\tan l = \omega_2 l^1 + \omega_4 l^3 + \omega_6 l^5 + \cdots; \qquad (\mathrm{q}^V)$$

so that the coefficients of the ascending powers of the arc in the development of its tangent are connected with each other by the relations (f^V), which may be briefly represented thus:

$$\sqrt{-1} = (+\sqrt{-1}\mathrm{D}_0)^{2k-1} \tan 0; \qquad (\mathrm{r}^V)$$

the second member of this symbolic equation being supposed to be developed, and $\mathrm{D}_0^i \tan 0$ being understood to denote the value which the ith differential coefficient of the tangent of α, taken with respect to α, acquires when $\alpha = 0$; thus,

$$\left.\begin{aligned} 1 &= \mathrm{D}_0 \tan 0 = 3\mathrm{D}_0 \tan 0 - \mathrm{D}_0^3 \tan 0 \\ &= 5\mathrm{D}_0 \tan 0 - 10\mathrm{D}_0^3 \tan 0 + \mathrm{D}_0^5 \tan 0. \end{aligned}\right\} \qquad (\mathrm{s}^V)$$

Finally, if we make $f_\alpha = \cos\alpha$, and attend to the expression (p^V), we obtain, for the secant of an arc l, the known expression:

$$\sec l = \sum_{(n)-\infty}^{\infty} \frac{2(-1)^{n+1}}{(2n-1)\pi - 2l}; \tag{t^V}$$

which shows that, with the notation (m^V),

$$\sec l = \omega_1 l^0 + \omega_3 l^2 + \omega_5 l^4 + \cdots, \tag{u^V}$$

and therefore, by the relations of the form (n^V),

$$\sqrt{-1}(-(\sqrt{-1}\,\mathrm{D}_0)^{2k} \sec 0) = (+\sqrt{-1}\,\mathrm{D}_0)^{2k} \tan 0; \tag{v^V}$$

thus

$$\left.\begin{aligned} 1 = \sec 0 &= 2\mathrm{D}_0 \tan 0 - \mathrm{D}_0^2 \sec 0 \\ &= 4\mathrm{D}_0 \tan 0 - 4\mathrm{D}_0^3 \tan 0 + \mathrm{D}_0^4 \sec 0. \end{aligned}\right\} \tag{w^V}$$

Though several of the results above deduced are known, the writer does not remember to have elsewhere seen the symbolic equations (r^V), (v^V), as expressions for the laws of the coefficients of the developments of the tangent and secant, according to ascending powers of the arc.

20. In the last article, the symbol R was such, that

$$\sum_{(n)1}^{n} \mathrm{R}_{\alpha,n} = \mathrm{N}_{2n\alpha}\,\mathrm{N}_\alpha^{-1}; \tag{x^V}$$

and in article **11.**, we had

$$1 + \sum_{(n)1}^{n} \mathrm{Q}_{\alpha,n} = \mathrm{N}_{2n\alpha+\alpha}\,\mathrm{N}_\alpha^{-1}. \tag{y^V}$$

Assume, now, more generally,

$$\nabla_\beta \mathrm{S}_{\alpha,\beta} = \mathrm{N}_{\beta\alpha}\,\mathrm{N}_\alpha^{-1}; \tag{z^V}$$

and let the operation ∇_β admit of being effected after, instead of before, the integration relatively to α; the expression (m″) will then acquire this very general form:

$$f_x = \varpi^{-1}\mathrm{P}_0 \nabla_\infty \int_a^b d\alpha\, \mathrm{S}_{\alpha-x,\beta} f_\alpha; \tag{S}$$

which includes the transformations (C) and (R), and in which the notation ∇_∞ is designed to indicate that after performing the operation ∇_β we are to make the variable β infinite, according to some given law of increase, connected with the form of the operation denoted by ∇.

21. In order to deduce the theorems (C), (R), (S), we have hitherto supposed (as was stated in the twelfth article), that the equation $\mathrm{N}_\alpha = 0$ has no real root different from 0 between the limits $\mp(b-a)$, in which a and b are the limits of the integration relative to α, between which latter limits it is also supposed that the variable x is comprised. If these conditions be not satisfied, the factor $\mathrm{N}_{\alpha-x}^{-1}$, in the formula (m″), may become infinite within the proposed extent of integration, for values of α and x which are not equal to each other; and it will then be necessary to change the first member of each of the equations (m″), (C), (R), (S), to a function different from f_x, but to be determined by similar principles. To simplify the question, let it be supposed that the function N_α receives no sudden change of value, and that the equation

$$\mathrm{N}_\alpha = 0, \qquad (\mathrm{a}^{VI})$$

which coincides with (w″), has all its real roots unequal. These roots must here coincide with the quantities $\alpha_{n,i}$ of the fourth and other articles, for which the function N_α changes sign; but as the double index is now unnecessary, while the notation α_n has been appropriated to the roots of the equation (g), we shall denote the roots of the equation (a^{VI}), in their order, by the symbols

$$\nu_{-\infty}, \ldots \nu_{-1}, \nu_0, \nu_1, \ldots \nu_\infty; \qquad (\mathrm{b}^{VI})$$

and choosing ν_0 for that root of (a^{VI}) which has already been supposed to vanish, we shall have

$$\nu_0 = 0, \qquad (\mathrm{c}^{VI})$$

while the other roots will be $>$ or <0, according as their indices are positive or negative. If the differential coefficient P_α be also supposed to remain always finite, and to receive no sudden change of value in the immediate neighbourhood of any root ν of (a^{VI}), we shall have, for values of α in that neighbourhood, the limiting equation:

$$\lim_{\alpha=\nu} .\mathrm{N}_\alpha (\alpha - \nu)^{-1} = \mathrm{P}_\nu; \qquad (\mathrm{d}^{VI})$$

and P_ν will be different from 0, because the real roots of the equation (a^{VI}) have been supposed unequal. Conceive also that the integral

$$\int_{-\infty}^{\infty} d\alpha \, \mathrm{N}_{\alpha+\beta\nu} \alpha^{-1} = \varpi_{\nu,\beta} \qquad (\mathrm{e}^{VI})$$

tends to some finite and determined limit, which may perhaps be different for different roots ν, and therefore may be thus denoted,

$$\varpi_{\nu,\infty} = \varpi_\nu, \qquad (\mathrm{f}^{VI})$$

as β tends to ∞, after the given law referred to at the end of the last article. Then, by writing

$$\alpha = x + \nu + \beta^{-1} y, \qquad (\mathrm{g}^{VI})$$

and supposing β to be very large, we easily see, by reasoning as in former articles, that the part of the integral

$$\int_a^b d\alpha \, \mathrm{N}_{\beta(\alpha-x)} \mathrm{N}_{\alpha-x}^{-1} f_\alpha, \qquad (\mathrm{h}^{VI})$$

which corresponds to values of $\alpha - x$ in the neighbourhood of the root ν, is very nearly expressed by

$$\varpi_\nu \mathrm{P}_\nu^{-1} f_{x+\nu}; \qquad (\mathrm{i}^{VI})$$

and that this expression is accurate at the limit. Instead of the equation (s), we have therefore now this other equation:

$$\textstyle\sum_\nu \varpi_\nu \mathrm{P}_\nu^{-1} f_{x+\nu} = \nabla_\infty \int_a^b d\alpha\, \mathrm{S}_{\alpha-x,\beta} f_\alpha; \tag{T}$$

the sum in the first member being extended to all those roots ν of the equation (a^{VI}), which satisfy the conditions

$$x + \nu > a, < b. \tag{k^{VI}}$$

If one of the roots ν should happen to satisfy the condition

$$x + \nu = a, \tag{l^{VI}}$$

the corresponding term in the first member of (T) would be, by the same principles,

$$\varpi'_\nu \mathrm{P}_\nu^{-1} f_a, \tag{m^{VI}}$$

in which

$$\varpi'_\nu = \lim_{\beta=\infty} \int_0^\infty d\alpha\, \mathrm{N}_{\alpha+\beta\nu} \alpha^{-1}. \tag{n^{VI}}$$

And if a root ν of (a^{VI}) should satisfy the condition

$$x + \nu = b, \tag{o^{VI}}$$

the corresponding term in the first member of (T) would then be

$$\varpi''_\nu \mathrm{P}_\nu^{-1} f_b, \tag{p^{VI}}$$

in which

$$\varpi''_\nu = \lim_{\beta=\infty} \int_{-\infty}^0 d\alpha\, \mathrm{N}_{\alpha+\beta\nu} \alpha^{-1}. \tag{q^{VI}}$$

Finally, if a value of $x + \nu$ satisfy the conditions (k^{VI}), and if the function f undergo a sudden change of value for this particular value of the variable on which that function depends, so that $f = f''$ immediately before, and $f = f'$ immediately after the change, then the corresponding part of the first member of the formula (T) is

$$\mathrm{P}_\nu^{-1} (\varpi'_\nu f' + \varpi''_\nu f''). \tag{r^{VI}}$$

And in the formulæ for ϖ_ν, ϖ'_ν, ϖ''_ν, it is permitted to write

$$\mathrm{N}_{\alpha+\beta\nu} \alpha^{-1} = \int_0^1 dt\, \mathrm{P}_{t\alpha+\beta\nu}. \tag{s^{VI}}$$

22. One of the simplest ways of rendering the integral (e^{VI}) determinate at its limit, is to suppose that the function P_α is of the periodical form which satisfies the two following equations,

$$\mathrm{P}_{-\alpha} = \mathrm{P}_\alpha, \ \mathrm{P}_{\alpha+p} = -\mathrm{P}_\alpha; \tag{t^{VI}}$$

p being some given positive constant. Multiplying these equations by $d\alpha$, and integrating from $\alpha = 0$, we find, by (a″),

$$\mathrm{N}_{-\alpha} + \mathrm{N}_\alpha = 0, \ \mathrm{N}_{\alpha+p} + \mathrm{N}_\alpha = \mathrm{N}_p; \tag{u^{VI}}$$

therefore

$$\mathrm{N}_p = \mathrm{N}_{\frac{p}{2}} + \mathrm{N}_{-\frac{p}{2}} = 0, \tag{v^{VI}}$$

and

$$\mathrm{N}_{\alpha+p} = -\mathrm{N}_\alpha,\ \mathrm{N}_{\alpha+2p} = \mathrm{N}_\alpha,\ \&\mathrm{c}. \tag{w^{VI}}$$

Consequently, if the equations (t^{VI}) be satisfied, the multiples (by whole numbers) of p will all be roots of the equation (a^{VI}); and reciprocally that equation will have no other real roots, if we suppose that the function P_α, which vanishes when α is any odd multiple of $\frac{p}{2}$, preserves one constant sign between any one such multiple and the next following, or simply between $\alpha = 0$ and $\alpha = \frac{p}{2}$. We may then, under these conditions, write

$$\nu_i = ip, \tag{x^{VI}}$$

i being any integer number, positive or negative, and ν_i denoting generally, as in (b^{VI}), any root of the equation (a^{VI}). And we shall have

$$\int_{-\infty}^{\infty} d\alpha\, \mathrm{N}_{\alpha+kp}\alpha^{-1} = (-1)^k \varpi, \tag{y^{VI}}$$

k being any integer number, and ϖ still retaining the same meaning as in the former articles. Also, for any integer value of k,

$$\mathrm{P}_{kp} = (-1)^k \mathrm{P}_0. \tag{z^{VI}}$$

These things being laid down, let us resume the integral (e^{VI}), and let us suppose that the law by which β increases to ∞ is that of coinciding successively with the several uneven integer numbers 1, 3, 5, &c., as was supposed in deducing the formula (C). Then $\beta\nu$ in (e^{VI}) will be an odd or even multiple of p, according as ν is the one or the other, so that we shall have by (x^{VI}), (y^{VI}), the following determined expression for the sought limit (f^{VI}):

$$\varpi_{\nu_i} = (-1)^i \varpi; \tag{a^{VII}}$$

but also, by (x^{VI}), (z^{VI}),

$$\mathrm{P}_{\nu_i} = (-1)^i \mathrm{P}_0; \tag{b^{VII}}$$

therefore

$$\varpi_\nu \mathrm{P}_\nu^{-1} = \varpi \mathrm{P}_0^{-1}, \tag{c^{VII}}$$

the value of this expression being thus the same for all the roots of (a^{VI}). At the same time, in (i^{VI}),

$$f_{x+\nu} = f_{x+ip}; \tag{d^{VII}}$$

the equation (T) becomes therefore now

$$\textstyle\sum_i f_{x+ip} = \varpi^{-1} \mathrm{P}_0\, \nabla_\infty \displaystyle\int_a^b d\alpha\, \mathrm{S}_{\alpha-x,\beta} f_\alpha, \tag{U}$$

β tending to infinity by passing through the successive positive odd numbers, and i receiving all integer values which allow $x + ip$ to be comprised between the limits a and b. If any integer value of i render $x + ip$ equal to either of these limits, the corresponding term of the sum in the first member of (U) is to be $\frac{1}{2}f_a$, or $\frac{1}{2}f_b$; and if the function f receive any sudden change of value between the same limits of integration, corresponding to a value of the variable which is of the form $x + ip$, the term introduced thereby will be of the form $\frac{1}{2}f' + \frac{1}{2}f''$.

For example, when

$$\text{P}_\alpha = \cos\alpha,\ \varpi = \pi,\ p = \pi, \tag{e^{VII}}$$

we obtain the following known formula, instead of (r″),

$$\sum_i f_{x+i\pi} = \pi^{-1}\sum_{(n)-\infty}^{\infty}\int_a^b d\alpha\cos(2n\alpha - 2nx)\,f_\alpha; \tag{f^{VII}}$$

which may be transformed in various ways, by changing the limits of integration, and in which halves of functions are to be introduced in extreme cases, as above.

On the other hand, if the law of increase of β be, as in (R), that of coinciding successively with larger and larger even numbers, then

$$\varpi_\nu = \varpi,\ \text{P}_\nu = \mp\text{P}_0, \tag{g^{VII}}$$

and the equation (T) becomes

$$\sum_i (-1)^i f_{x+i\pi} = \varpi^{-1}\text{P}_0\nabla_\infty\int_a^b d\alpha\,\text{S}_{\alpha-x,\beta}f_\alpha. \tag{V}$$

For example, in the case (e^{VII}), we obtain this extension of the formula (b^{V}),

$$\sum_i (-1)^i f_{x+i\pi} = \pi^{-1}\sum_{(n)-\infty}^{\infty}\int_a^b d\alpha\cos(\overline{2n-1}.\overline{\alpha - x})\,f_\alpha. \tag{h^{VII}}$$

We may verify the equations (f^{VII}) (h^{VII}) by remarking that both members of the former equation remain unchanged, and that both members of the latter are changed in sign, when x is increased by π. A similar verification of the equations (U) and (V) requires that in general the expression

$$\nabla_\infty\int_a^b d\alpha\,\text{S}_{\alpha-x,\beta}f_\alpha \tag{i^{VII}}$$

should either receive no change, or simply change its sign, when x is increased by p, according as β tends to ∞ by coinciding with large and odd or with large and even numbers.

23. In all the examples hitherto given to illustrate the general formulæ of this paper, it has been supposed for the sake of simplicity, that the function P is a cosine; and this supposition has been sufficient to deduce, as we have seen, a great variety of known results. But it is evident that this function P may receive many other forms, consistently with the suppositions made in deducing those general formulæ; and many new results may thus be obtained by the method of the foregoing articles.

For instance, it is permitted to suppose

$$\text{P}_\alpha = 1,\ \text{if}\ \alpha^2 < 1; \tag{k^{VII}}$$

$$\text{P}_1 = 0; \tag{1^{VII}}$$

$$\text{P}_{\alpha+2} = -\text{P}_\alpha; \tag{m^{VII}}$$

and then the equations (t^{VI}) of the last article, with all that were deduced from them, will still hold good. We shall now have

$$p = 2; \tag{n^{VII}}$$

and the definite integral denoted by ϖ, and defined by the equation (r′), may now be computed as follows. Because the function N_α changes sign with α, we have

$$\varpi = 2\int_0^\infty d\alpha\, \text{N}_\alpha\, \alpha^{-1}; \tag{o^{VII}}$$

but

$$\left.\begin{array}{llll} \text{N}_\alpha = \alpha, \text{ from } \alpha = 0 & \text{to } \alpha = 1; \\ \ldots 2 - \alpha, \quad \ldots 1 & \ldots \quad 3; \\ \ldots \alpha - 4, \quad \ldots 3 & \ldots \quad 4; \end{array}\right\} \tag{p^{VII}}$$

and

$$\text{N}_{\alpha+4} = \text{N}_\alpha. \tag{q^{VII}}$$

Hence

$$\int_0^4 d\alpha\, \text{N}_\alpha\, \alpha^{-1} = 6\log 3 - 4\log 4, \tag{r^{VII}}$$

the logarithms being Napierian; and generally, if m be any positive integer number, or zero,

$$\begin{aligned} \int_{4m}^{4m+4} d\alpha\, \text{N}_\alpha\, \alpha^{-1} &= \int_0^4 d\alpha\, N_\alpha(\alpha + 4m)^{-1} \\ &= 4m\log(4m) - (8m+2)\log(4m+1) + (8m+6)\log(4m+3) \\ &\quad - (4m+4)\log(4m+4) \\ &= \textstyle\sum_{(k)1}^{\infty} \dfrac{1 - 2^{-2k}}{k(k+\frac{1}{2})}(2m+1)^{-2k}. \end{aligned} \tag{s^{VII}}$$

But, by (h^{V}),

$$\textstyle\sum_{(m)0}^{\infty} (2m+1)^{-2k} = \frac{1}{2}\left(\dfrac{\pi}{2}\right)^{2k} \omega_{2k}, \tag{t^{VII}}$$

if k be any integer number > 0; therefore

$$\varpi = \textstyle\sum_{(k)1}^{\infty} \dfrac{1 - 2^{-2k}}{k(k+\frac{1}{2})}\left(\dfrac{\pi}{2}\right)^{2k} \omega_{2k}; \tag{u^{VII}}$$

ω_{2k} being by (q^{V}) the coefficient of x^{2k-1} in the development of $\tan x$. From this last property, we have

$$\sum_{(k)1}^{\infty} \frac{\omega_{2k} x^{2k}}{k(k+\frac{1}{2})} = \frac{4}{x}\left(\int_0^x dx\right)^2 \tan x = \frac{4}{x}\int_0^x dx \log \sec x; \qquad (\mathrm{v}^{VII})$$

therefore, substituting successively the values $x = \frac{\pi}{2}$ and $x = \frac{\pi}{4}$, and subtracting the result of the latter substitution from that of the former, we find, by (u^{VII}),

$$\begin{aligned} \varpi &= \frac{8}{\pi}\left(\int_{\frac{\pi}{4}}^{\frac{\pi}{2}} - \int_0^{\frac{\pi}{4}}\right) dx \log \sec x \\ &= \frac{8}{\pi}\int_{\frac{\pi}{4}}^{\frac{\pi}{2}} dx \log \tan x \\ &= \frac{8}{\pi}\int_0^{\frac{\pi}{4}} dx \log \operatorname{cotan} x. \end{aligned} \qquad (\mathrm{w}^{VII})$$

Such, in the present question, is an expression for the constant ϖ; its numerical value may be approximately calculated by multiplying the Napierian logarithm of ten by the double of the average of the ordinary logarithms of the cotangents of the middles of any large number of equal parts into which the first octant may be divided; thus, if we take the ninetieth part of the sum of the logarithms of the cotangents of the ninety angles $\frac{1^0}{4}, \frac{3^0}{4}, \frac{5^0}{4}, \ldots \frac{177^0}{4}, \frac{179^0}{4}$, as given by the ordinary tables, we obtain nearly, as the average of these ninety logarithms, the number 0,5048; of which the double, being multiplied by the Napierian logarithm of ten, gives, nearly, the number 2,325, as an approximate value of the constant ϖ. But a much more accurate value may be obtained with little more trouble, by computing separately the doubles of the part (r^{VII}), and of the sum of (s^{VII}) taken from $m = 1$ to $m = \infty$; for thus we obtain the expression

$$\begin{aligned} \varpi = {} & 12 \log 3 - 8 \log 4 \\ & + 2\sum_{(k)1}^{\infty} \frac{1 - 2^{-2k}}{k(k+\frac{1}{2})} \sum_{(m)1}^{\infty} (2m+1)^{-2k}, \end{aligned} \qquad (\mathrm{x}^{VII})$$

in which each sum relative to m can be obtained from known results, and the sum relative to k converges tolerably fast; so that the second line of the expression (x^{VII}) is thus found to be nearly $=$ 0,239495, while the first line is nearly $=$ 2,092992; and the whole value of the expression (x^{VII}) is nearly

$$\varpi = 2,332487. \qquad (\mathrm{y}^{VII})$$

There is even an advantage in summing the double of the expression (s^{VII}) only from $m = 2$ to $m = \infty$, because the series relative to k converges then more rapidly; and having thus found $2\int_8^{\infty} d\alpha\, \mathrm{N}_\alpha\, \alpha^{-1}$, it is only necessary to add thereto the expression

$$2\int_0^8 d\alpha\, \mathrm{N}_\alpha\, \alpha^{-1} = 12 \log 3 - 20 \log 5 + 28 \log 7 - 16 \log 8. \qquad (\mathrm{z}^{VII})$$

The form of the function P and the value of the constant ϖ being determined as in the present article, it is permitted to substitute them in the general equations of this paper; and thus to deduce new transformations for portions of arbitrary functions, which might have been employed instead of those given by FOURIER and POISSON, if the discontinuous function P, which receives alternately the values 1, 0, and −1, had been considered simpler in its properties than the trigonometrical function cosine.

24. Indeed, when the conditions (t^{VI}) are satisfied, the function P_x can be developed according to cosines of the odd multiples of $\frac{\pi x}{p}$, by means of the formula (y'''), which here becomes, by changing l to $\frac{p}{2}$, and f to P,

$$P_x = \sum_{(n)1}^{\infty} A_{2n-1} \cos \frac{(2n-1)\pi x}{p}, \qquad (a^{VIII})$$

in which

$$A_{2n-1} = \frac{4}{p}\int_0^{\frac{p}{2}} d\alpha \cos \frac{(2n-1)\pi\alpha}{p} P_\alpha; \qquad (b^{VIII})$$

the function N_x at the same time admitting a development according to sines of the same odd multiples, namely,

$$N_x = \frac{p}{\pi}\sum_{(n)1}^{\infty} \frac{A_{2n-1}}{2n-1} \sin \frac{(2n-1)\pi x}{p}; \qquad (c^{VIII})$$

and the constant ϖ being equal to the following series,

$$\varpi = p\sum_{(n)1}^{\infty} \frac{A_{2n-1}}{2n-1}. \qquad (d^{VIII})$$

Thus, in the case of the last article, where $p = 2$, and $P_\alpha = 1$ from $\alpha = 0$ to $\alpha = 1$, we have

$$A_{2n-1} = \frac{4}{\pi}\frac{(-1)^{n+1}}{2n-1}; \qquad (e^{VIII})$$

$$P_x = \frac{4}{\pi}\left(\cos\frac{\pi x}{2} - 3^{-1}\cos\frac{3\pi x}{2} + 5^{-1}\cos\frac{5\pi x}{2} - \cdots\right); \qquad (f^{VIII})$$

$$N_x = \frac{8}{\pi^2}\left(\sin\frac{\pi x}{2} - 3^{-2}\sin\frac{3\pi x}{2} + 5^{-2}\sin\frac{5\pi x}{2} - \cdots\right); \qquad (g^{VIII})$$

$$\varpi = \frac{8}{\pi}(1^{-2} - 3^{-2} + 5^{-2} - 7^{-2} + \cdots); \qquad (h^{VIII})$$

so that, from the comparison of (w^{VII}) and (h^{VIII}), the following relation results:

$$\int_0^{\frac{\pi}{4}} dx \log \cot x = \sum_{(n)0}^{\infty} (-1)^n (2n+1)^{-2}. \qquad (i^{VIII})$$

But most of the suppositions made in former articles may be satisfied, without assuming for the function P the periodical form assigned by the conditions (t^{VI}). For example, we might assume

$$P_\alpha = \frac{4}{\pi}\int_0^{\pi} d\theta \sin\theta^2 \cos(2\alpha\sin\theta); \qquad (k^{VIII})$$

which would give, by (a''), and (b''),

$$\mathrm{N}_\alpha = \frac{2}{\pi}\int_0^\pi d\theta \sin\theta \sin(2\alpha \sin\theta); \qquad (\mathrm{l}^{VIII})$$

$$\mathrm{M}_\alpha = \frac{1}{\pi}\int_0^\pi d\theta \operatorname{vers}(2\alpha \sin\theta); \qquad (\mathrm{m}^{VIII})$$

and finally, by (r′),

$$\varpi = 2\int_0^\pi d\theta \sin\theta = 4. \qquad (\mathrm{n}^{VIII})$$

This expression (k^{VIII}) for P_α satisfies all the conditions of the ninth article; for it is clear that it gives a value to N_α which is always numerically less than $\frac{4}{\pi}$; and the equation

$$\mathrm{M}_\alpha = 1, \qquad (\mathrm{o}^{VIII})$$

which is of the form (g), is satisfied by all the infinitely many real and unequal roots of the equation

$$\int_0^\pi d\theta \cos(2\alpha \sin\theta) = 0, \qquad (\mathrm{p}^{VIII})$$

which extend from $\alpha = -\infty$ to $\alpha = \infty$, and of which the interval between any one and the next following is never greater than π, nor even so great; because (as it is not difficult to prove) these several roots are contained in alternate or even octants, in such a manner that we may write

$$\alpha_n > \frac{n\pi}{2} - \frac{\pi}{4}, < \frac{n\pi}{2}. \qquad (\mathrm{q}^{VIII})$$

We may, therefore substitute the expression (k^{VIII}) for P, in the formulæ (A), (B), (C), &c.; and we find, by (B), if $x > a, < b$,

$$f_x = \pi^{-1}\int_a^b d\alpha \int_0^\infty d\beta \int_0^\pi d\theta \sin\theta^2 \cos\{2\beta(\alpha - x)\sin\theta\} f_\alpha; \qquad (\mathrm{r}^{VIII})$$

that is,

$$f_x = \frac{1}{2\pi}\lim_{\beta=\infty}\int_0^\pi d\theta \sin\theta \int_a^b d\alpha \sin\{2\beta(\alpha - x)\sin\theta\}(\alpha - x)^{-1} f_\alpha; \qquad (\mathrm{s}^{VIII})$$

a theorem which may be easily proved *à posteriori*, by the principles of fluctuating functions, because those principles show, that (if x be comprised between the limits of integration) the limit relative to β of the integral relative to α, in (s^{VIII}), is equal to πf_x. In like manner, the theorem (C), when applied to the present form of the function P, gives the following other expression for the arbitrary function f_x:

$$f_x = \tfrac{1}{2}\int_a^b d\alpha f_\alpha + \Sigma_{(n)1}^{\ \infty}\int_a^b d\alpha f_\alpha \frac{\int_0^\pi d\theta \sin\theta \sin(2(\alpha - x)\sin\theta)\cos(4n(\alpha - x)\sin\theta)}{\int_0^\pi d\theta \sin\theta \sin(2(\alpha - x)\sin\theta)}; \qquad (\mathrm{t}^{VIII})$$

x being between a and b, and $b - a$ being not greater than the least positive root ν of the equation

$$\frac{1}{\nu}\int_0^\pi d\theta \sin\theta \sin(2\nu \sin\theta) = 0. \qquad (\mathrm{u}^{VIII})$$

And if we wish to prove, *à posteriori*, this theorem of transformation (t^{VIII}), by the same principles of fluctuating functions, we have only to observe that

$$1 + 2\textstyle\sum_{(n)1}^{n} \cos 2ny = \frac{\sin(2ny + y)}{\sin y}, \qquad (\mathrm{v}^{VIII})$$

and therefore that the second member of (t^{VIII}) may be put under the form

$$\lim_{n=\infty} \int_a^b d\alpha\, f_\alpha \frac{\int_0^\pi d\theta \sin\theta \sin((4n+2)(\alpha - x)\sin\theta)}{2\int_0^\pi d\theta \sin\theta \sin(2(\alpha - x)\sin\theta)}; \qquad (\mathrm{w}^{VIII})$$

in which the presence of the fluctuating factor

$$\sin((4n+2)(\alpha - x)\sin\theta),$$

combined with the condition that $\alpha - x$ is numerically less than the least root of the equation (u^{VIII}), shows that we need only attend to values of α indefinitely near to x, and may therefore write in the denominator,

$$\int_0^\pi d\theta \sin\theta \sin(2(\alpha - x)\sin\theta) = \pi(\alpha - x); \qquad (\mathrm{x}^{VIII})$$

for thus, by inverting the order of the two remaining integrations, that is by writing

$$\int_a^b d\alpha \int_0^\pi d\theta \ldots = \int_0^\pi d\theta \int_a^b d\alpha \ldots, \qquad (\mathrm{y}^{VIII})$$

we find first

$$\lim_{n=\infty} \int_a^b d\alpha\, f_\alpha \frac{\sin((4n+2)(\alpha - x)\sin\theta)}{2\pi(\alpha - x)} = \tfrac{1}{2} f_x, \qquad (\mathrm{z}^{VIII})$$

for every value of θ between 0 and π, and of x between a and b; and finally,

$$\tfrac{1}{2} f_x \int_0^\pi d\theta \sin\theta = f_x.$$

25. The results of the foregoing articles may be extended by introducing, under the functional signs N, P, a product such as $\beta\gamma$, instead of $\beta\alpha$, γ being an arbitrary function of α; and by considering the integral

$$\int_a^b d\alpha\, \mathrm{N}_{\beta\gamma}\, \mathrm{F}_\alpha, \qquad (\mathrm{a}^{IX})$$

in which F is any function which remains finite between the limits of integration. Since γ is a function of α, it may be denoted by γ_α, and α will be reciprocally a function of γ, which may be denoted thus:

$$\alpha = \phi_{\gamma_\alpha}. \qquad (\mathrm{b}^{IX})$$

While α increases from a to b, we shall suppose, at first, that the function γ_a increases constantly and continuously from γ_a to γ_b, in such a manner as to give always, within this extent of variation, a finite and determined and positive value to the differential coefficient of the function ϕ, namely,

$$\frac{d\alpha}{d\gamma} = \phi'_\gamma. \qquad (\mathrm{c}^{IX})$$

We shall also express, for abridgment, the product of this coefficient and of the function F by another function of γ, as follows,

$$\phi'_\gamma \, \mathrm{F}_\alpha = \psi. \qquad (\mathrm{d}^{IX})$$

Then the integral (a^{IX}) becomes

$$\int_{\gamma_a}^{\gamma_b} d\gamma \, \mathrm{N}_{\beta\gamma} \, \psi_\gamma; \qquad (\mathrm{e}^{IX})$$

and a rigorous expression for it may be obtained by the process of the fourth article, namely

$$\left(\int_{\gamma_a}^{\beta^{-1}\alpha_n} + \int_{\beta^{-1}\alpha_{n+m}}^{\gamma_b} \right) d\gamma \, \mathrm{N}_{\beta\gamma} \, \psi_\gamma + \theta\beta^{-1}(\alpha_{n+m} - \alpha_n)\mathrm{c}\delta; \qquad (\mathrm{f}^{IX})$$

in which, as before, α_n, α_{n+m} are suitably chosen roots of the equation (g); c is a finite constant; θ is included between the limits ± 1; and δ is the difference between two values of the function ψ_γ, corresponding to two values of the variable γ of which the difference is less than β^{-1}b, b being another finite constant. The integral (a^{IX}) therefore diminishes indefinitely when β increases indefinitely; and thus, or simply by the theorem (z) combined with the expression (e^{IX}), we have, rigorously, at the limit, without supposing here that N_0 vanishes,

$$\int_a^b d\alpha \, \mathrm{N}_{\infty\gamma} \, \mathrm{F}_\alpha = 0. \qquad (\mathrm{W})$$

The same conclusion is easily obtained, by reasonings almost the same, for the case where γ continually decreases from γ_a to γ_b, in such a manner as to give, within this extent of variation, a finite and determined and negative value to the differential coefficient (c^{IX}). And with respect to the case where the function γ is for a moment stationary in value, so that its differential coefficient vanishes between the limits of integration, it is sufficient to observe that although ψ in (e^{IX}) becomes then infinite, yet F in (a^{IX}) remains finite, and the integral of the finite product $d\alpha \, \mathrm{N}_{\beta\gamma} \, \mathrm{F}_\alpha$, taken between infinitely near limits, is zero. Thus, generally, the theorem (W), which is an extension of the theorem (z), holds good between any finite limits a and b, if the function F be finite between those limits, and if, between the same limits of integration, the function γ never remain unchanged throughout the whole extent of any finite change of α.

26. It may be noticed here, that if β be only very large, instead of being infinite, an approximate expression for the integral (a^{IX}) may be obtained, on the same principles, by attending only to values of α which differ very little from those which render the coefficient

(c^{IX}) infinite. For example, if we wish to find an approximate expression for a large root of the equation (p^{VIII}), or to express approximately the function

$$f_\beta = \frac{1}{\pi}\int_0^\pi d\alpha \cos(2\beta \sin\alpha), \tag{g^{IX}}$$

when β is a large positive quantity, we need only attend to values of α which differ little from $\frac{\pi}{2}$; making then

$$\sin\alpha = 1 - y^2,\ d\alpha = \frac{2\,dy}{\sqrt{2-y^2}}, \tag{h^{IX}}$$

and neglecting y^2 in the denominator of this last expression, the integral (g^{IX}) becomes

$$f_\beta = \mathrm{A}_\beta \cos 2\beta + \mathrm{B}_\beta \sin 2\beta, \tag{i^{IX}}$$

in which, nearly,

$$\left.\begin{aligned} \mathrm{A}_\beta &= \frac{\sqrt{2}}{\pi}\int_{-\infty}^{\infty} dy \cos(2\beta y^2) = \frac{1}{\sqrt{2\pi\beta}}; \\ \mathrm{B}_\beta &= \frac{\sqrt{2}}{\pi}\int_{-\infty}^{\infty} dy \sin(2\beta y^2) = \frac{1}{\sqrt{2\pi\beta}}; \end{aligned}\right\} \tag{k^{IX}}$$

so that the large values of β which make the function (g^{IX}) vanish are nearly of the form

$$\frac{n\pi}{2} - \frac{\pi}{8}, \tag{l^{IX}}$$

n being an integer number; and such is therefore the approximate form of the large roots α_n of the equation (p^{VIII}): results which agree with the relations (q^{VIII}), and to which POISSON has been conducted, in connexion with another subject, and by an entirely different analysis.

The theory of fluctuating functions may also be employed to obtain a more close approximation; for instance, it may be shown, by reasonings of the kind lately employed, that the definite integral (g^{IX}) admits of being expressed (more accurately as β is greater) by the following semiconvergent series, of which the first terms have been assigned by POISSON:

$$f_\beta = \frac{1}{\sqrt{\pi\beta}}\Sigma_{(i)0}^{\infty}[0]^{-i}([-\tfrac{1}{2}]^i)^2(4\beta)^{-i}\cos\left(2\beta - \frac{\pi}{4} - \frac{i\pi}{2}\right); \tag{m^{IX}}$$

and in which, according to a known notation of factorials,

$$\left.\begin{aligned} [0]^{-i} &= 1^{-1}.2^{-1}.3^{-1}.\dots i^{-1}; \\ [-\tfrac{1}{2}]^i &= \frac{-1}{2}.\frac{-3}{2}.\frac{-5}{2}\cdots\frac{1-2i}{2}. \end{aligned}\right\} \tag{n^{IX}}$$

For the value $\beta = 20$, the 3 first terms of the series (m^{IX}) give

$$\left.\begin{aligned} f_{20} &= \left(1 - \frac{9}{204800}\right)\frac{\cos 86°49'52''}{\sqrt{20\pi}} + \frac{1}{320}\frac{\sin 86°49'52''}{\sqrt{20\pi}} \\ &= 0{,}0069736 + 0{,}0003936 = +0{,}0073672. \end{aligned}\right\} \tag{o^{IX}}$$

For the same value of β, the sum of the first sixty terms of the ultimately convergent series

$$f_\beta = \textstyle\sum_{(i)0}^{\infty} ([0]^{-i})^2(-\beta^2)^i \tag{p^{IX}}$$

gives

$$\left.\begin{aligned} f_{20} = +7\ 447\ 387\ 396\ 709\ 949,9657957 \\ -7\ 447\ 387\ 396\ 709\ 949,9584289 \\ = +0,0073668. \end{aligned}\right\} \tag{q^{IX}}$$

The two expressions (m^{IX}) (p^{IX}) therefore agree, and we may conclude that the following numerical value is very nearly correct:

$$\frac{1}{\pi}\int_0^\pi d\alpha \cos(40 \sin \alpha) = +0,007367. \tag{r^{IX}}$$

27. Resuming the rigorous equation (w), and observing that

$$\int_0^\infty d\beta \mathrm{P}_{\beta\gamma} = \lim_{\beta=\infty} .\mathrm{N}_{\beta\gamma}\, \gamma_\alpha^{-1}, \tag{s^{IX}}$$

we easily see that in calculating the definite integral

$$\int_a^b d\alpha \int_0^\infty d\beta \mathrm{P}_{\beta\gamma}\, f_\alpha, \tag{t^{IX}}$$

in which the function f is finite, it is sufficient to attend to those values of α which are not only between the limits a and b, but are also very nearly equal to real roots x of the equation

$$\gamma_x = 0. \tag{u^{IX}}$$

The part of the integral (t^{IX}), corresponding to values of a in the neighbourhood of any one such root x, between the above-mentioned limits, is equal to the product

$$\frac{f_x}{\gamma'_x} \times \int_{-\infty}^{\infty} d\alpha \frac{\mathrm{N}_{\beta\gamma'_x(\alpha-x)}}{\alpha - x}, \tag{v^{IX}}$$

in which β is indefinitely large and positive, and the differential coefficient γ'_x of the function γ is supposed to be finite, and different from 0. A little consideration shows that the integral in this last expression is $= \pm\varpi$, ϖ being the same constant as in former articles, and the upper or lower sign being taken according as γ'_x is positive or negative. Denoting then by $\sqrt{\gamma'^2_x}$ the positive quantity, which is $= +\gamma'_x$ or $= -\gamma'_x$, according as γ'_x is >0 or <0, the part (v^{IX}) of the integral (t^{IX}) is

$$\frac{\varpi f_x}{\sqrt{\gamma'^2_x}}; \tag{w^{IX}}$$

and we have the expression

$$\int_a^b d\alpha \int_0^\infty d\beta \mathrm{P}_{\beta\gamma} f_\alpha = \varpi \textstyle\sum_x \dfrac{f_x}{\sqrt{\gamma'^2_x}}, \tag{x^{IX}}$$

the sum being extended to all those roots x of the equation (u^{IX}) which are $> a$ but $< b$. If any root of that equation should coincide with either of these limits a or b, the value of α in its neighbourhood would introduce, into the second member of the expression (x^{IX}), one or other of the terms

$$\frac{\varpi' f_a}{\gamma'_a}, \frac{-\varpi'' f_a}{\gamma'_a}, \frac{\varpi'' f_b}{\gamma'_b}, \frac{-\varpi' f_b}{\gamma'_b}; \tag{y^{IX}}$$

the first to be taken when $\gamma_a = 0$, $\gamma'_a > 0$; the second when $\gamma_a = 0$, $\gamma'_a < 0$; the third when $\gamma_b = 0$, $\gamma'_b > 0$; and the fourth when $\gamma_b = 0$, $\gamma'_b < 0$. If, then, we suppose for simplicity, that neither γ_a nor γ_b vanishes, the expression (x^{IX}) conducts to the theorem

$$\sum_x f_x = \varpi^{-1} \int_a^b d\alpha \int_0^\infty d\beta \, \text{P}_{\beta\gamma} f_\alpha \sqrt{\gamma_\alpha'^2}; \tag{X}$$

and the sign of summation may be omitted, if the equation $\gamma_x = 0$ have only one real root between the limits a and b. For example, that one root itself may then be expressed as follows:

$$x = \varpi^{-1} \int_a^b d\alpha \int_0^\infty d\beta \, \text{P}_{\beta\gamma} \alpha \sqrt{\gamma_\alpha'^2}. \tag{z^{IX}}$$

The theorem (X) includes some analogous results which have been obtained by CAUCHY, for the case when P is a cosine.

28. It is also possible to extend the foregoing theorem in other ways; and especially by applying similar reasonings to functions of several variables. Thus, if γ, $\gamma^{(1)}$ ... be each a function of several real variables α, $\alpha^{(1)}$, ...; if P and N be still respectively functions of the kinds supposed in former articles, while $\text{P}^{(1)}$, $\text{N}^{(1)}$, ... are other functions of the same kinds; then the theorem (W) may be extended as follows:

$$\int_a^b d\alpha \int_{a^{(1)}}^{b^{(1)}} d\alpha^{(1)} \dots \text{N}_{\infty\gamma} \text{N}^{(1)}_{\infty\gamma^{(1)}} \dots \text{F}_{\alpha,\alpha^{(1)},\dots} = 0, \tag{Y}$$

the function F being finite for all values of the variables α, $\alpha^{(1)}$, ..., within the extent of the integrations; and the theorem (X) may be thus extended:

$$\left.\begin{aligned} \sum f_{x,x^{(1)},\dots} = \varpi^{-1}\varpi^{(1)-1} \dots \int_a^b d\alpha \int_{a^{(1)}}^{b^{(1)}} d\alpha^{(1)} \dots \int_0^\infty d\beta \int_0^\infty d\beta^{(1)} \dots \text{P}_{\beta\gamma} \text{P}^{(1)}_{\beta^{(1)}\gamma^{(1)}} \dots \\ \dots f_{\alpha,\alpha^{(1)},\dots} \sqrt{\text{L}^2}; \end{aligned}\right\} \tag{Z}$$

in which, according to the analogy of the foregoing notation,

$$\varpi^{(i)} = \int_{-\infty}^\infty d\alpha \int_0^1 d\beta \, \text{P}^{(i)}_{\beta\alpha}; \tag{a^X}$$

and L is the coefficient which enters into the expression, supplied by the principles of the transformation of multiple integrals,

$$\text{L} d\alpha \, d\alpha^{(1)} \dots = d\gamma \, d\gamma^{(1)} \dots; \tag{b^X}$$

while the summation in the first member is to be extended to all those values of x, $x^{(1)}$, ...

which, being respectively between the respective limits of integration relatively to the variables $\alpha, \alpha^{(1)}, \ldots$ are values of those variables satisfying the system of equations

$$\gamma_{x,x^{(1)},\ldots} = 0,\ \gamma^{(1)}_{x,x^{(1)},\ldots} = 0,\ \ldots. \qquad (c^X)$$

And thus may other remarkable results of CAUCHY be presented under a generalized form. But the theory of such extensions appears likely to suggest itself easily enough to any one who may have considered with attention the remarks already made; and it is time to conclude the present paper by submitting a few general observations on the nature and the history of this important branch of analysis.

LAGRANGE appears to have been the first who was led (in connexion with the celebrated problem of vibrating cords) to assign, as the result of a species of interpolation, an expression for an arbitrary function, continuous or discontinuous in form, between any finite limits, by a series of sines of multiples, in which the coefficients are definite integrals. Analogous expressions, for a particular class of rational and integral functions, were derived by DANIEL BERNOUILLI, through successive integrations, from the results of certain trigonometric summations, which he had characterized in a former memoir as being *incongruously true.* No farther step of importance towards the improvement of this theory seems to have been made, till FOURIER, in his researches on Heat, was led to the discovery of his well known theorem, by which any arbitrary function of any real variable is expressed, between finite or infinite limits, by a double definite integral. POISSON and CAUCHY have treated the same subject since, and enriched it with new views and applications; and through the labours of these and, perhaps, of other writers, the theory of the development or transformation of arbitrary functions, through functions of determined forms, has become one of the most important and interesting departments of modern algebra.

It must, however, be owned that some obscurity seems still to hang over the subject, and that a farther examination of its principles may not be useless or unnecessary. The very existence of such transformations as in this theory are sought for and obtained, appears at first sight paradoxical; it is difficult at first to conceive the possibility of expressing a perfectly arbitrary function through any series of sines or cosines; the variable being thus made the subject of known and series of sines or cosines; the variable being thus made the subject of known and determined operations, whereas it had offered itself originally as the subject of operations unknown and undetermined. And even after this first feeling of paradox is removed, or relieved, by the consideration that the number of the operations of known form is infinite, and that the operation of arbitrary form reappears in another part of the expression, as performed on an auxiliary variable; it still requires attentive consideration to see clearly how it is possible that none of the values of this new variable should have any influence on the final result, except those which are extremely nearly equal to the variable originally proposed. This latter difficulty has not, perhaps, been removed to the complete satisfaction of those who desire to examine the question with all the diligence its importance deserves, by any of the published works upon the subject. A conviction, doubtless, may be attained, that the results are true, but something is, perhaps, felt to be still wanting for the full rigour of mathematical demonstration. Such has, at least, been the impression left on the mind of the present writer, after an attentive study of the reasonings usually employed, respecting the transformations of arbitrary functions.

POISSON, for example, in treating this subject, sets out, most commonly, with a series of cosines of multiple arcs; and because the sum is generally indeterminate, when continued to infinity, he alters the series by multiplying each term by the corresponding power of an auxiliary quantity which he assumes to be less than unity, in order that its powers may diminish, and at last vanish; but, in order that the new series may tend indefinitely to coincide with the old one, he conceives, after effecting its summation, that the auxiliary quantity tends to become unity. The limit thus obtained is generally zero, but becomes on the contrary infinite when the arc and its multiples vanish; from which it is inferred by POISSON, that if this arc be the difference of two variables, an original and an auxiliary, and if the series be multiplied by any arbitrary function of the latter variable, and integrated with resepct thereto, the effect of all the values of that variable will disappear from the result, except the effect of those which are extremely nearly equal to the variable originally proposed.

POISSON has made, with consummate skill, a great number of applications of this method; yet it appears to present, on close consideration, some difficulties of the kind above alluded to. In fact, the introduction of the system of factors, which tend to vanish before the integration, as their indices increase, but tend to unity, after the integration, for all finite values of those indices, seems somewhat to change the nature of the question, by the introduction of a foreign element. Nor is it perhaps manifest that the original series, of which the sum is indeterminate, may be replaced by the convergent series with determined sum, which results from multiplying its terms by the powers of a factor infinitely little less than unity; while it is held that to multiply by the powers of a factor infinitely little greater than unity would give an useless or even false result. Besides there is something unsatisfactory in employing an apparently arbitrary contrivance for annulling the effect of those terms of the proposed series which are situated at a great distance from the origin, but which do not themselves originally tend to vanish as they become more distant therefrom. Nor is this difficulty entirely removed, when integration by parts is had recourse to, in order to show that the effect of these distant terms is insensible in the ultimate result; because it then becomes necessary to differentiate the arbitrary function; but to treat its differential coefficient as always finite, is to diminish the generality of the inquiry.

Many other processes and proofs are subject to similar or different difficulties; but there is one method of demonstration employed by FOURIER, in his separate Treatise on Heat, which has, in the opinion of the present writer, received less notice than it deserves, and of which it is proper here to speak. The principle of the method here alluded to may be called the *Principle of Fluctuation*, and is the same which was enunciated under that title in the remarks prefixed to this paper. In virtue of this principle (which may thus be considered as having been indicated by FOURIER, although not expressly stated by him), if any function, such as the sine or cosine of an infinite multiple of an arc, changes sign infinitely often within a finite extent of the variable on which it depends, and has for its mean value zero; and if this, which may be called a *fluctuating function*, be multiplied by any arbitrary but finite function of the same variable, and afterwards integrated between any finite limits; the integral of the product will be zero, on account of the mutual destruction or neutralization of all its elements.

It follows immediately from this principle, that if the factor by which the fluctuating function is multiplied, instead of remaining always finite, becomes infinite between the limits of integration, for one or more particular values of the variable on which it depends; it is then only necessary to attend to values in the immediate neighbourhood of these, in order to obtain the value of the integral. And in this way FOURIER has given what seems to be the most satisfactory published proof, and (so to speak) the most natural explanation of the theorem called by his name; since it exhibits the actual process, one might almost say the interior mechanism, which, in the expression assigned by him, destroys the effect of all those values of the auxiliary variable which are not required for the result. So clear, indeed, is this conception, that it admits of being easily translated into geometrical constructions, which have accordingly been used by FOURIER for that purpose.

There are, however, some remaining difficulties connected with this mode of demonstration, which may perhaps account for the circumstance that it seems never to be mentioned, nor alluded to, in any of the historical notices which POISSON has given on the subject of these transformations. For example, although FOURIER, in the proof just referred to, of the theorem called by his name, shows clearly that in integrating the product of an arbitrary but finite function, and the sine or cosine of an infinite multiple, each successive positive portion of the integral is destroyed by the negative portion which follows it, if infinitely small quantities be neglected, yet he omits to show that the infinitely small outstanding difference of values of these positive and negative portions, corresponding to the single period of the trigonometric function introduced, is of the second order; and, therefore, a doubt may arise whether the infinite number of such infinitely small periods, contained in any finite interval, may not produce, by their accumulation, a finite result. It is also desirable to be able to state the argument in the language of limits, rather than in that of infinitesimals; and to exhibit, by appropriate definitions and notations, what was evidently foreseen by FOURIER, that the result depends rather on the *fluctuating* than on the *trigonometric* character of the auxiliary function employed.

The same view of the question had occurred to the present writer, before he was aware that indications of it were to be found among the published works of FOURIER; and he still conceives that the details of the demonstration to which he was thus led may be not devoid of interest and utility, as tending to give greater rigour and clearness to the proof and the conception of a widely applicable and highly remarkable theorem.

Yet, if he did not suppose that the present paper contains something more than a mere expansion or improvement of a known proof of a known result, the Author would scarcely have ventured to offer it to the Transactions* of the Royal Irish Academy. It aims not merely to give a more perfectly satisfactory demonstration of FOURIER's celebrated theorem than any which the writer has elsewhere seen, but also to present that theorem, and many others analogous thereto, under a greatly generalized form, deduced from the principle of fluctuation. Functions more general than sines or cosines, yet having some correspondent properties, are introduced throughout; and constants, distinct from the ratio of the circumference to the diameter of a circle, present themselves in connexion therewith. And thus, if the

* The Author is desirous to acknowledge, that since the time of his first communicating the present paper to the Royal Irish Academy, in June, 1840, he has had an opportunity of entirely rewriting it, and that the last sheet is only now passing through the press, in June, 1842. Yet it may be proper to mention also that the theorems (A) (B) (C), which sufficiently express the character of the communication, were printed (with some slight differences of notation) in the year 1840,† as part of the *Proceedings* of the Academy for the date prefixed to this paper.

† [See this volume, p. 583.]

intention of the writer have been in any degree accomplished, it will have been shown, according to the opinion expressed in the remarks prefixed to this paper, that the development of the important principle above referred to gives not only a new clearness, but also (in some respects) a new extension, to this department of science.

XXIII.

SUPPLEMENTARY REMARKS ON FLUCTUATING FUNCTIONS (1842)

[Communicated February 28, 1842]
[*Proceedings of the Royal Irish Academy* **2**, 232–238 (1844).
(Identical with pp. 317–321 of No. 4.)]

XXIV.

NEW DEMONSTRATION OF FOURIER'S THEOREM (1841)

[Communicated June 28, 1841]
[*Proceedings of the Royal Irish Academy* **2**, 129 (1844).]

The President communicated a new demonstration of Fourier's theorem.

XXV.

ON CERTAIN DISCONTINUOUS INTEGRALS CONNECTED WITH THE DEVELOPMENT OF THE RADICAL WHICH REPRESENTS THE RECIPROCAL DISTANCE BETWEEN TWO POINTS* (1842)

[*Philosophical Magazine* (3S) **20**, 288–294 (1842).]

---

1. It is well known that the radical

$$(1 - 2xp + x^2)^{-\frac{1}{2}}, \tag{1}$$

in which x and 1 may represent the radii vectors of two points, while p represents the cosine of the angle between those radii, and the radical represents therefore the reciprocal of the distance of the one point from the other, may be developed in a series of the form

$$\mathrm{P}_0 + \mathrm{P}_1 x + \mathrm{P}_2 x^2 + \ldots + \mathrm{P}_n x^n + \ldots; \tag{2}$$

the coefficients P_n being functions of p, and possessing many known properties, among which we shall here employ the following only,

$$\mathrm{P}_n = [0]^{-n} \left(\frac{d}{dp}\right)^n \left(\frac{p^2 - 1}{2}\right)^n; \tag{3}$$

the known notation of factorials being here used, according to which

$$[0]^{-n} = \frac{1}{1} \cdot \frac{1}{2} \cdot \frac{1}{3} \cdots \frac{1}{n}. \tag{4}$$

It is proposed to express the sum of the first n terms of the development (2.), which may be thus denoted,

$$\textstyle\sum_{(n)0}^{n-1} \mathrm{P}_n x^n = \mathrm{P}_0 + \mathrm{P}_1 x + \mathrm{P}_2 x^2 + \ldots + \mathrm{P}_{n-1} x^{n-1}. \tag{5}$$

2. In general, by Taylor's theorem,

$$f(p + q) = \textstyle\sum_{(n)0}^{\infty} [0]^{-n} q^n \left(\frac{d}{dp}\right)^n f(p); \tag{6}$$

hence, by the property (3.), P_n is the coefficient of q^n in the development of

$$\left(\frac{(p + q)^2 - 1}{2}\right)^n; \tag{7}$$

it is therefore also the coefficient of q^0 in the development of

* Communicated by the Author.

$$\left(\frac{p^2-1}{2q}+p+\frac{q}{2}\right)^n. \tag{8}$$

If then we make, for abridgment,

$$\vartheta = p+\frac{p^2}{2}\cos\theta+\sqrt{-1}\left(1-\frac{p^2}{2}\right)\sin\theta, \tag{9}$$

we shall have the following expression, which perhaps is new, for P_n:

$$P_n = \frac{1}{2\pi}\int_{-\pi}^{\pi}\vartheta^n\,d\theta; \tag{10}$$

and hence, immediately, the required sum (5.) may be expressed as follows:

$$\textstyle\sum_{(n)0}^{n-1} P_n x^n = \frac{1}{2\pi}\int_{-\pi}^{\pi} d\theta\,\frac{1-\vartheta^n x^n}{1-\vartheta x}; \tag{11}$$

in which it is to be observed that x may be any quantity, real or imaginary.

3. We have therefore, rigorously, for the sum of the n first terms of the series

$$P_0+P_1+P_2+\dots, \tag{12}$$

the expression

$$\textstyle\sum_{(n)0}^{n-1} P_n = \frac{1}{2\pi}\int_{-\pi}^{\pi} d\theta\,\frac{1-\vartheta^n}{1-\vartheta}; \tag{13}$$

of which we propose to consider now the part independent of n, namely,

$$F(p) = \frac{1}{2\pi}\int_{-\pi}^{\pi}\frac{d\theta}{1-\vartheta}, \tag{14}$$

and to examine the form of this function F of p, at least between the limits $p=-1$, $p=1$.

4. A little attention shows that the denominator $1-\vartheta$ may be decomposed into factors, as follows:

$$1-\vartheta = -\tfrac{1}{2}(\alpha+e^{\theta\sqrt{-1}})(1-\beta e^{-\theta\sqrt{-1}}); \tag{15}$$

in which,

$$\alpha = 2s(1-s),\ \beta = 2s(1+s), \tag{16}$$

and

$$p = 1-2s^2; \tag{17}$$

so that s may be supposed not to exceed the limits 0 and 1, since p is supposed not to exceed the limits -1 and 1. Hence

$$\frac{1}{1-\vartheta} = \frac{-2(\alpha+e^{-\theta}\sqrt{-1})(1-\beta e^{\theta}\sqrt{-1})}{(1+2\alpha\cos\theta+\alpha^2)(1-2\beta\cos\theta+\beta^2)}; \tag{18}$$

of which the real part may be put under the form

$$\frac{\lambda}{1+2\alpha\cos\theta+\alpha^2}+\frac{\mu}{1-2\beta\cos\theta+\beta^2}, \tag{19}$$

if λ and μ be so chosen as to satisfy the conditions

$$\lambda(1+\beta^2)+\mu(1+\alpha^2)=2(\beta-\alpha), \tag{20}$$

$$\lambda\beta-\mu\alpha=1-\alpha\beta, \tag{21}$$

which give

$$\lambda=\frac{1-\alpha^2}{\alpha+\beta}, \quad \mu=\frac{\beta^2-1}{\alpha+\beta}. \tag{22}$$

The imaginary part of the expression (18) changes sign with θ, and disappears in the integral (14.); that integral therefore reduces itself to the sum of the two following:

$$\mathrm{F}(p)=\frac{1}{4s\pi}\int_0^\pi\frac{(1-\alpha^2)\,d\theta}{1+2\alpha\cos\theta+\alpha^2}+\frac{1}{4s\pi}\int_0\frac{(\beta^2-1)\,d\theta}{1-2\beta\cos\theta+\beta^2}; \tag{23}$$

in which, by (16), $\alpha+\beta$ has been changed to $4s$. But, in general if $a^2>b^2$,

$$\int_0^\pi\frac{d\theta}{a+b\cos\theta}=\frac{\pi}{\sqrt{a^2-b^2}}, \tag{24}$$

the radical being a positive quantity if a be such; therefore, in the formula (23),

$$\int_0^\pi\frac{(1-\alpha^2)\,d\theta}{1+2\alpha\cos\theta+\alpha^2}=\pi, \tag{25}$$

because, by (16.), α cannot exceed the limits 0 and $\frac{1}{2}$, s being supposed not to exceed the limits 0 and 1, so that $1-\alpha^2$ is positive. On the other hand, β varies from 0 to 4, while s varies from 0 to 1; and β^2-1 will be positive or negative, according as s is greater or less than the positive root of the equation

$$s^2+s=\tfrac{1}{2}. \tag{26}$$

Hence, in (23), we must make

$$\int_0^\pi\frac{(\beta^2-1)\,d\theta}{1-2\beta\cos\theta+\beta^2}=\pi,\ \text{or}=-\pi, \tag{27}$$

according as

$$s>\text{or}<\frac{\sqrt{3}-1}{2}; \tag{28}$$

and thus we find, under the same alternative,

$$\mathrm{F}(p)=\frac{1}{4s}(1\pm1), \tag{29}$$

that is,

$$\mathrm{F}(p)=\frac{1}{2s},\ \text{or}=0. \tag{30}$$

But, by (17),

$$s = \sqrt{\frac{1-p}{2}}; \tag{31}$$

the function $F(p)$, or the definite integral (14), receives therefore a sudden change of form when p, in varying from -1 to 1, passes through the critical value

$$p = \sqrt{3} - 1; \tag{32}$$

in such a manner that we have

$$F(p) = (2 - 2p)^{-\frac{1}{2}}, \quad \text{if } p < \sqrt{3} - 1; \tag{33}$$

and, on the other hand,

$$F(p) = 0, \quad \text{if } p > \sqrt{3} - 1. \tag{34}$$

For the critical value (32) itself, we have

$$s = \frac{\sqrt{3} - 1}{2}, \quad \alpha = 2\sqrt{3} - 3, \quad \beta = 1, \tag{35}$$

and the real part of (18) becomes

$$\frac{1 - \alpha}{1 + 2\alpha \cos\theta + \alpha^2}; \tag{36}$$

multiplying therefore by $d\theta$, integrating from $\theta = 0$ to $\theta = \pi$, and dividing by π, we find, by (25) and (14), this formula instead of (29),

$$F(p) = \frac{1}{1 + \alpha} = \frac{1}{4s}, \tag{37}$$

that is,

$$F(p) = \tfrac{1}{2}(2 - 2p)^{-\frac{1}{2}}, \quad \text{if } p = \sqrt{3} - 1. \tag{38}$$

The value of the discontinuous function F is therefore, in this case, equal to the semisum of the two different values which that function receives, immediately before and after the variable p attains its critical value, as usually happens in other similar cases of discontinuity.

5. As verifications of the results (33), (34), we may consider the particular values $p = 0$, $p = 1$, which ought to give

$$F(0) = 2^{-\frac{1}{2}}, \; F(1) = 0. \tag{39}$$

Accordingly, when $p = 0$, the definitions (9) and (14) give

$$\vartheta = \sqrt{-1} \sin\theta, \tag{40}$$

$$F(0) = \frac{1}{2\pi} \int_{-\pi}^{\pi} \frac{d\theta}{1 - \sqrt{-1}\sin\theta} = \frac{1}{\pi} \int_0^{\pi} \frac{d\theta}{1 + \sin\theta^2}; \tag{41}$$

which easily gives, by (24),

$$F(0) = \frac{2}{\pi} \int_0^{\pi} \frac{d\theta}{3 - \cos 2\theta} = \frac{1}{\pi} \int_0^{2\pi} \frac{d\theta}{3 - \cos\theta} = 2^{-\frac{1}{2}}. \tag{42}$$

And when $p = 1$, we have

$$1 - \vartheta = -\tfrac{1}{2}(\cos\theta + \sqrt{-1}\sin\theta), \tag{43}$$

$$\frac{1}{2\pi}\frac{d\theta}{1-\vartheta} = -\pi^{-1}(\cos\theta - \sqrt{-1}\sin\theta)\,d\theta, \tag{44}$$

of which the integral, taken from $\theta = -\pi$ to $\theta = \pi$, is $\mathrm{F}(1) = 0$.

6. Let us consider now this other integral,

$$\mathrm{G}(p) = \frac{1}{2\pi}\int_{-\pi}^{\pi}\frac{\vartheta^n\,d\theta}{\vartheta - 1}. \tag{45}$$

The expression (13) gives

$$\textstyle\sum_{(n)0}^{n-1}\mathrm{P}_n = \mathrm{F}(p) + \mathrm{G}(p); \tag{46}$$

therefore, by (34), we shall have

$$\mathrm{G}(p) = \textstyle\sum_{(n)0}^{n-1}\mathrm{P}_n, \quad \text{if } p > \sqrt{3} - 1. \tag{47}$$

For instance, let $p = 1$; then multiplying the expression (44) by

$$-\vartheta^n = -(1 + \tfrac{1}{2}e^{\theta\sqrt{-1}})^n, \tag{48}$$

the only term which does not vanish when integrated is $\frac{1}{2}n\pi^{-1}\,d\theta$, and this term gives the result

$$\mathrm{G}(1) = n, \tag{49}$$

which evidently agrees with the formula (47), because it is well known that

$$\mathrm{P}_n = 1 \text{ when } p = 1, \tag{50}$$

the series (2) becoming then the development of $(1 - x)^{-1}$.

7. On the other hand, let p be $< \sqrt{3} - 1$; then, observing that, by (33),

$$\mathrm{F}(p) = (2 - 2p)^{-\frac{1}{2}} = \textstyle\sum_{(n)0}^{\infty}\mathrm{P}_n, \tag{51}$$

we find, by the relation (46) between the functions F and G,

$$\left.\begin{aligned}\mathrm{G}(p) &= -\textstyle\sum_{(n)n}^{\infty}\mathrm{P}_n \\ &= -(\mathrm{P}_n + \mathrm{P}_{n+1} + \mathrm{P}_{n+2} + \ldots).\end{aligned}\right\} \tag{52}$$

For instance, let $p = 0$; then, by (40) and (45),

$$\mathrm{G}(0) = \frac{-(\sqrt{-1})^n}{2\pi}\int_{-\pi}^{\pi}\frac{d\theta(\sin\theta)^n}{1 - \sqrt{-1}\sin\theta}; \tag{53}$$

that is,

$$\mathrm{G}(0) = \frac{(-1)^{i+1}}{\pi}\int_0^{\pi}\frac{d\theta\sin\theta^{2i}}{1 + \sin\theta^2}; \tag{54}$$

if n be either $=2i-1$, or $=2i$. Now, when $p=0$, P_n is the coefficient of x_n in the development of $(1+x^2)^{-\frac{1}{2}}$; therefore,

$$\mathrm{P}_{2i-1}=0, \quad \text{when } p=0, \tag{55}$$

and, in the notation of factorials,

$$\mathrm{P}_{2i}=[0]^{-i}[-\tfrac{1}{2}]^{i}=(-1)^{i}\pi^{-1}\int_0^{\pi} d\theta \sin\theta^{2i}; \tag{56}$$

so that, by (54),

$$\mathrm{G}(0)=-(\mathrm{P}_{2i}+\mathrm{P}_{2i+2}+\ldots), \tag{57}$$

when $p=0$, and when n is either $2i$ or $2i-1$.

8. For the critical value $p=\sqrt{3}-1$, we have, by (38),

$$\mathrm{F}(p)=\tfrac{1}{2}\textstyle\sum_{(n)0}^{\infty}\mathrm{P}_n; \tag{58}$$

therefore, for the same value of p, by (46),

$$\begin{aligned}\mathrm{G}(p)&=\tfrac{1}{2}\textstyle\sum_{(n)0}^{n-1}\mathrm{P}_n-\tfrac{1}{2}\sum_{(n)\,n}^{\infty}\mathrm{P}_n\\ &=\tfrac{1}{2}(\mathrm{P}_0+\mathrm{P}_1+\ldots+\mathrm{P}_{n-1}-\mathrm{P}_n-\mathrm{P}_{n+1}-\ldots);\end{aligned} \tag{59}$$

so that the discontinuous function G, like F, acquires, for the critical value of p, a value which is the semisum of those which it receives immediately before and afterwards.

9. We have seen that the sum of these two discontinuous integrals, F and G, is always equal to the sum of the first n terms of the series (12), so that

$$\mathrm{F}(p)+\mathrm{G}(p)=\mathrm{P}_0+\mathrm{P}_1+\ldots+\mathrm{P}_{n-1}; \tag{60}$$

and it may not be irrelevant to remark that this sum may be developed under this other form:

$$\frac{1}{2\pi}\int_{-\pi}^{\pi} d\theta\frac{\vartheta^{n}-1}{\vartheta-1}=\textstyle\sum_{(k)1}^{n}[n]^{k}[0]^{-k}\mathrm{Q}_{k-1}; \tag{61}$$

in which the factorial expression $[n]^k[0]^{-k}$ denotes the coefficient of x^k in the development of $(1+x)^n$; and

$$\mathrm{Q}_k=\frac{1}{2\pi}\int_{-\pi}^{\pi} d\theta(\vartheta-1)^k. \tag{62}$$

Thus

$$\left.\begin{aligned}&\mathrm{P}_0=\mathrm{Q}_0;\\ &\mathrm{P}_0+\mathrm{P}_1=2\mathrm{Q}_0+\mathrm{Q}_1;\\ &\mathrm{P}_0+\mathrm{P}_1+\mathrm{P}_2=3\mathrm{Q}_0+3\mathrm{Q}_1+\mathrm{Q}_2;\\ &\&\mathrm{c}.;\end{aligned}\right\} \tag{63}$$

and consequently

$$\left. \begin{aligned} P_0 &= Q_0; \\ P_1 &= Q_0 + Q_1; \\ P_2 &= Q_0 + 2Q_1 + Q_2; \\ &\&c.; \end{aligned} \right\} \tag{64}$$

which last expressions, indeed, follow immediately from the formula (10).

10. With respect to the calculation of Q_0, Q_1, &c. as functions of p, it may be noted, in conclusion, that, by (15) and (62), Q_k is the term independent of θ in the development of

$$2^{-k}(\alpha + e^{\theta}\sqrt{-1})^k(1 - \beta e^{-\theta}\sqrt{-1})^k; \tag{65}$$

thus

$$\left. \begin{aligned} Q_0 &= 1, \\ Q_1 &= 2^{-1}(\alpha - \beta), \\ Q_2 &= 2^{-2}(\alpha^2 - 4\alpha\beta + \beta^2), \\ Q_3 &= 2^{-3}(\alpha^3 - 9\alpha^2\beta + 9\alpha\beta^2 - \beta^3), \\ &\&c. \end{aligned} \right\} \tag{66}$$

in which the law of formation is evident. It remains to substitute for α, β, their values (16) as functions of s, and then to eliminate s^2 by (17); and thus we find, for example,

$$\left. \begin{aligned} Q_1 &= p - 1; \\ Q_2 &= \tfrac{1}{2}(p-1)(3p-1); \\ Q_3 &= \tfrac{1}{2}(p-1)^2(5p+1); \\ Q_4 &= \tfrac{1}{8}(p-1)^2(35p^2 - 10p - 13). \end{aligned} \right\} \tag{67}$$

This, then, is at least one way, though perhaps not the easiest, of computing the initial values of the successive differences of the function P_n, that is, the quantities

$$\left. \begin{aligned} Q_0 &= \Delta^0 P_0 = P_0, \\ Q_1 &= \Delta^1 P_0 = P_1 - P_0, \\ Q_2 &= \Delta^2 P_0 = P_2 - 2P_1 + P_0, \\ &\&c. \end{aligned} \right\} \tag{68}$$

And we see that it is permitted to express generally those differences, as follows:

$$\Delta^k P_0 = s^k \textstyle\sum_{(i)0}^{k} (-1)^i ([k]^i [0]^{-i})^2 (1+s)^i (1-s)^{k-i}; \tag{69}$$

in which

$$s^2 = \tfrac{1}{2}(1 - p). \tag{70}$$

Observatory of Trinity College, Dublin,
Feb. 12, 1842.

XXVI.

ON A MODE OF EXPRESSING FLUCTUATING OR ARBITRARY FUNCTIONS BY MATHEMATICAL FORMULÆ (1842)

[*British Association Report* 1842, Part II., p. 10.]

XXVII.

ON AN EXPRESSION FOR THE NUMBERS OF BERNOULLI BY MEANS OF A DEFINITE INTEGRAL, AND ON SOME CONNECTED PROCESSES OF SUMMATION AND INTEGRATION (1843)

[*Philosophical Magazine* (3S) **13**, 360–367 (1843).]

The following analysis, extracted from a paper which has been in part communicated to the Royal Irish Academy, but has not yet been printed, may interest some readers of the Philosophical Magazine.

1. Let us consider the function of two real variables, m and n, represented by the definite integral

$$y_{m,n} = \int_0^\infty dx \left(\frac{\sin x}{x}\right)^m \cos nx; \tag{1}$$

in which we shall suppose that m is greater than zero; and which gives evidently the general relation

$$y_{m,-n} = y_{m,n}.$$

By changing m to $m+1$; integrating first the factor $x^{-m-1}dx$, and observing that $x^{-m}\sin x^{m+1}\cos nx$ vanishes both when $x=0$, and when $x=\infty$; and then putting the differential coefficient $\frac{d}{dx}(\sin x^{m+1}\cos nx)$ under the form

$$\tfrac{1}{2}\sin x^m\{(m+1+n)\cos(nx+x)+(m+1-n)\cos(nx-x)\};$$

we are conducted to the following equation, in finite and partial differences,

$$2\,my_{m+1,n} = (m+1+n)\,y_{m,n+1} + (m+1-n)\,y_{m,n-1}; \tag{2}$$

and if we suppose that the difference between the two variables on which y depends is an even integer number, this equation takes the form

$$my_{m+1,m+1-2k} = (m+1-k)\,y_{m,m+2-2k} + ky_{m,m-2k}. \tag{3}$$

The same equation in differences (2) shows easily that

$$y_{m+1,n} = 0, \quad \text{when } n = \text{or} > m+1,$$
$$\text{if } y_{m,n-1} = 0, \quad \text{when } n-1 > m;$$

but, by a well-known theorem, which in the present notation becomes

$$y_{1,0} = \frac{\pi}{2}, \tag{4}$$

it is easy to prove, not only that

$$y_{1,1} = y_{1,-1} = \frac{\pi}{4}, \tag{5}$$

but also that

$$y_{1,n} = 0, \quad \text{if } n^2 > 1; \tag{6}$$

we have therefore, generally, for all whole values of $m > 1$, and for all real values of n,

$$y_{m,n} = 0, \quad \text{unless } n^2 < m^2. \tag{7}$$

2. If then we make

$$\mathrm{T}_m = \sum y_{m,m-2k}(-t)^k, \tag{8}$$

the sign $\sum$ indicating a summation which may be extended from as large a negative to as large a positive whole value of k as we think fit, but which extends at least from $k = 0$ to $k = m$, m being here a positive whole number; this sum will in general (namely when $m > 1$) include only $m - 1$ terms different from 0, namely those which correspond to $k = 1, 2, \ldots m - 1$; but in the particular case $m = 1$, the sum will have two such terms, instead of none, namely those answering to $k = 0$ and $k = 1$, so that we shall have

$$\mathrm{T}_1 = y_{1,1} - y_{1,-1}\,t = \frac{\pi}{4}(1 - t). \tag{9}$$

Multiplying the first member of the equation in differences (3) by $(-t)^k$, and summing with respect to k, we obtain $m\mathrm{T}_{m+1}$, m being here any whole number > 0. Multiplying and summing in like manner the second member of the same equation (3), the term $my_{m,m+2-2k}$ of that member gives $-mt\mathrm{T}_m$, because we may change k to $k+1$ before effecting the indefinite summation; $ky_{m,m-2k}$ gives $t\dfrac{d}{dt}\mathrm{T}_m$; and $(1 - k)\,y_{m,m+2-2k}$ gives $t^2\dfrac{d}{dt}\mathrm{T}_m$; but

$$-mt.\mathrm{T}_m + (t + t^2)\frac{d}{dt}\mathrm{T}_m = (1 + t)^{m+1}\frac{td}{dt}(1 + t)^{-m}\mathrm{T}_m;$$

therefore

$$m(1 + t)^{-m-1}\mathrm{T}_{m+1} = \frac{d}{d\log t}(1 + t)^{-m}\mathrm{T}_m. \tag{10}$$

This equation in mixed differences gives, by (9),

$$\mathrm{T}_m = \frac{\pi}{4}\frac{(1 + t)^m}{1.2.3\ldots(m - 1)}\left(\frac{d}{d\log t}\right)^{m-1}\frac{1 - t}{1 + t}; \tag{11}$$

the factorial denominator being considered as $= 1$, when $m = 1$, as well as when $m = 2$. If $m > 1$, we may change $\dfrac{1 - t}{1 + t}$ to $\dfrac{2}{1 + t}$, from which it only differs by a constant; and then by changing also t to e^h, and multiplying by $\dfrac{2}{\pi}$, we obtain the formula:

$$\frac{(e^h + 1)^m}{1.2.3\ldots(m - 1)}\left(\frac{d}{dh}\right)^{m-1}(e^h + 1)^{-1} = \frac{2}{\pi}\sum\nolimits_{(k)1}^{m-1}\int_0^\infty dx\left(\frac{\sin x}{x}\right)^m(-e^h)^k\cos(mx - 2kx); \tag{12}$$

which conducts to many interesting consequences. A few of them shall be here mentioned.

3. The summation indicated in the second member of this formula can easily be effected in general; but we shall here consider only the two cases in which m is an odd or an even whole number greater than unity, while h becomes $=0$ after the $m-1$ differentiations of $(e^h+1)^{-1}$, which are directed in the first member.

When m is odd (and greater than one), each power, such as $(-e^h)^k$ in the second member, is accompanied by another, namely $(-e^h)^{m-k}$, which is multiplied by the cosine of the same multiple of x; and these two powers destroy each other, when added, if $h=0$: we arrive therefore in this manner at the known result, that

$$\left(\frac{d}{dh}\right)^{2p}(e^h+1)^{-1}=0,\ \text{when}\ h=0,\ \text{if}\ p>0. \tag{13}$$

On the contrary, when m is even, and $h=0$, the powers $(-e^h)^k$ and $(-c^h)^{m-k}$ are equal, and their sum is double of either; and because

$$(-1)^p\{1-2\cos 2x+2\cos 4x-\cdots+(-1)^{p-1}2\cos(2px-2x)\}=-\frac{\cos(2px-x)}{\cos x},$$

by making $m=2p$ we arrive at this other result, which perhaps is new, that (if $p>0$ and $h=0$)

$$\left(\frac{d}{dh}\right)^{2p-1}(e^h+1)^{-1}=\frac{-1.2.3\ldots(2p-1)}{2^{2p-1}\pi}\int_0^\infty dx\left(\frac{\sin x}{x}\right)^{2p}\frac{\cos(2px-x)}{\cos x}. \tag{14}$$

Developing therefore $(e^h+1)^{-1}$ according to ascending powers of h; subtracting the development from $\frac{1}{2}$, multiplying by h, and changing h to $2h$; we obtain

$$h\frac{e^h-e^{-h}}{e^h+e^{-h}}=\frac{2}{\pi}\int_0^\infty\frac{dx}{\cos x}\Sigma_{(p)1}^\infty\left(\frac{h\sin x}{x}\right)^{2p}\cos(2px-x); \tag{15}$$

that is, effecting the summation, and dividing by h^2,

$$\frac{1}{h}\frac{e^h-e^{-h}}{e^h+e^{-h}}=\frac{2}{\pi}\int_0^\infty\frac{dxx^{-2}\sin x^2(1-h^2x^{-2}\sin x^2)}{1-2h^2x^{-2}\sin x^2\cos 2x+h^4x^{-4}\sin x^4}; \tag{16}$$

or, integrating both members with respect to h,

$$\int_0^h\frac{dh}{h}\frac{e^h-e^{-h}}{e^h+e^{-h}}=\frac{1}{\pi}\int_0^\infty\frac{dx}{x}\tan x\log\sqrt{\frac{1+hx^{-1}\sin 2x+h^2x^{-2}\sin x^2}{1-hx^{-1}\sin 2x+h^2x^{-2}\sin x^2}}. \tag{17}$$

It might seem, at first sight, from this equation, that the integral in the first member ought to vanish, when taken from $h=0$ to $h=\infty$; because, if we set about to develope the second member, according to descending powers of h, we see that the coefficient of h^0 vanishes; but when we find that, on the same plan, the coefficient of h^{-1} is infinite, being $=\frac{2}{\pi}\int_0^\infty dx$, we perceive that this mode of development is here inappropriate: and in fact, it is clear that the first member of the formula (17) increases continually with h, while h increases from 0.

4. Again, since

$$\frac{-h}{e^h+1}=\psi(2h)-\psi(h), \quad \text{if } \psi(h)=\frac{h}{e^h-1}, \tag{18}$$

we shall have (for $p>0$) the expression

$$A_{2p}=\frac{2^{1-2p}\pi^{-1}}{2^{2p}-1}\int_0^\infty dx\left(\frac{\sin x}{x}\right)^{2p}\frac{\cos(2px-x)}{\cos x}, \tag{19}$$

if, according to a known form of development, which the foregoing reasonings would suffice to justify, we write

$$\frac{h}{e^h-1}+\frac{h}{2}=1+A_2h^2+A_4h^4+A_6h^6+\&c. \tag{20}$$

If p be a large number, the rapid and repeated changes of sign of the numerator of the fraction $\frac{\cos(2px-x)}{\cos x}$ produce nearly a mutual destruction of the successive elements of the integral (19), except in the neighbourhood of those values of x which cause the denominator of the same fraction to vanish; namely those values which are odd positive multiples of $\frac{\pi}{2}$ (the integral itself being not extended so as to include any negative values of x). Making therefore

$$x=(2i-1)\frac{\pi}{2}+\omega, \tag{21}$$

in which i is a whole number >0, and ω is positive or negative, but nearly equal to 0; we shall have

$$\cos(2px-x)=(-1)^{p+i-1}\sin(2p\omega-\omega),$$

exactly, and $\cos x=(-1)^i\omega$, nearly; changing also $\left(\frac{\sin x}{x}\right)^{2p}$ to the value which it has when $\omega=0$, namely $\left(\frac{2}{\pi}\right)^{2p}(2i-1)^{-2p}$; and observing that

$$\int_{-\omega}^{\omega}d\omega\frac{\sin(2p\omega-\omega)}{\omega}=\pi, \text{ nearly}, \tag{22}$$

even though the extreme values of ω may be small, if p be very large; we find that the part of A_{2p}, corresponding to any one value of the number i, is, at least nearly, represented by the expression

$$\frac{(-1)^{p-1}2(2i-1)^{-2p}}{(2^{2p}-1)\pi^{2p}}; \tag{23}$$

which is now to be summed, with reference to i, from $i=1$ to $i=\infty$. But this summation gives rigorously the relation

$$\textstyle\sum_{(i)1}^{\infty}(2i-1)^{-2p}=(1-2^{-2p})\sum_{(i)1}^{\infty}i^{-2p}; \tag{24}$$

we are conducted, therefore, to the expression

$$A_{2p}=(-1)^{p-1}2(2\pi)^{-2p}\textstyle\sum_{(i)1}^{\infty}i^{-2p}, \tag{25}$$

as *at least* approximately true, when the number p is large. But in fact the expression (25) is *rigorous* for all whole values of p greater than 0; as we shall see by deducing from it an analogous expression for a Bernoullian number, and comparing this with known results.

5. The development

$$\frac{1}{e^h - 1} + \frac{1}{2} = h^{-1} + \mathrm{B}_1 \frac{h}{1.2} - \mathrm{B}_3 \frac{h^3}{1.2.3.4} + \&c., \tag{26}$$

being compared with that marked (20), gives, for the pth Bernoullian number, the known expression

$$\mathrm{B}_{2p-1} = (-1)^{p-1} 1.2.3.4 \ldots 2p \mathrm{A}_{2p}; \tag{27}$$

and therefore, rigorously, by the equation (19) of the present paper,

$$\mathrm{B}_{2p-1} = \frac{(-1)^{p-1} 1.2 \ldots 2p}{2^{2p-1}(2^{2p}-1)\pi} \int_0^\infty dx \left(\frac{\sin x}{x}\right)^{2p} \frac{\cos(2px - x)}{\cos x}; \tag{28}$$

a formula which is believed to be new. Treating the definite integral which it involves by the process just now used, we necessarily obtain the same result as if we combine at once the equations (25) and (27). We find, therefore, in this manner, that the equation

$$\frac{2^{2p-1}\pi^{2p}\mathrm{B}_{2p-1}}{1.2.3.4 \ldots 2p} = \Sigma_{(i)1}^{\infty} i^{-2p}, \tag{29}$$

(in which, by the notation here employed,

$$\Sigma_{(i)1}^{\infty} i^{-2p} = 1^{-2p} + 2^{-2p} + 3^{-2p} + \&c.)$$

is *at least nearly true*, when p is a large number; but Euler has shown, in his *Institutiones Calculi Differentialis* (vol. i. cap. v. p. 339. ed. 1787)*, that this equation (29) is *rigorous*, each member being the coefficient of u^{2p} in the development of $\frac{1}{2}(1 - \pi u \cot \pi u)$. [See also Professor De Morgan's Treatise on the Diff. and Int. Calc., 'Library of Useful Knowledge,' part xix. p. 581.]†. The analysis of the present paper is therefore not only verified generally, but also the modifications which were made in the form of that definite integral which entered into our rigorous expressions (19) and (28) for A_{2p} and B_{2p-1}, by the process of the last article, (on the ground that the parts omitted or introduced thereby must at least nearly destroy each other, through what may be called the "principle of fluctuation,") are now seen to have produced no ultimate error at all, their mutual compensation being perfect; a result which may tend to give increased confidence in applying a similar process of approximation, or transformation, to the treatment of other similar integrals; although the logic of this process may deserve to be more closely scrutinized. Some assistance towards such a scrutiny may be derived from the essay on "Fluctuating Functions," which has been published by the present writer in the second part of the nineteenth volume of the Transactions of the Royal Irish Academy.‡

* [See Euler, L., *Opera Omnia*, Series 1, Vol. 10, p. 327, Leipzig and Berlin: 1913.]
† [De Morgan, A., *The differential and integral calculus*, London: 1842.]
‡ [See this volume, p. 585.]

6. It may be worth while to notice, in conclusion, that the property marked (7) of the definite integral (1), enables us to change $\frac{\cos(2px - x)}{\cos x}$ to $\sin 2px \tan x$, in the equations (14), (15), (19), (28); so that the pth Bernoullian number may rigorously be expressed as follows:-

$$\mathrm{B}_{2p-1} = \frac{(-1)^{p-1}.1.2\ldots 2p}{2^{2p-1}(2^{2p}-1)\pi}\int_0^\infty dx\left(\frac{\sin x}{x}\right)^{2p}\sin 2px\tan x; \tag{30}$$

under which form the preceding deduction of its transformation (29) admits of being slightly simplified. The same modification of the foregoing expressions conducts easily to the equation

$$\log\frac{e^h + e^{-h}}{2} = \frac{1}{\pi}\int_0^\infty dx \tan x \tan^{-1}\frac{h^2\sin x^2 \sin 2x}{x^2 - h^2 \sin x^2 \cos 2x}; \tag{31}$$

in which $\tan^{-1}$ is a characteristic equivalent to arc tang., and which may be made an expression for $\log\sec x$, by merely changing the sign of h^2 in the last denominator; and from this equation (31) it would be easy to return to an expression for the coefficients in the development of $\frac{e^h - e^{-h}}{e^h + e^{-h}}$, or in that of $\tan h$, and therefore to the numbers of Bernoulli. Those numbers might thus be deduced from the following very simple equation:

$$\pi\log\sec h = \int_0^\infty dx\ y\tan x; \tag{32}$$

in which y is connected with x and h by the relation

$$\frac{\sin y}{\sin(2x - y)} = \left(\frac{h\sin x}{x}\right)^2. \tag{33}$$

Observatory of Trinity College, Dublin,
October 6, 1843.

XXVIII.

ON THE INTEGRATIONS OF CERTAIN EQUATIONS (1854)

[Communicated February 27, 1854]

[*Proceedings of the Royal Irish Academy* **6**, 62–63 (1858).]

Rev. Dr. Graves read a note from Sir W. R. Hamilton, in which he stated that he had lately arrived at a variety of results respecting the integrations of certain equations, which might not be unworthy of the acceptance of the Academy, and the investigation of which had been suggested to him by Mr. Carmichael's* printed Paper, and by a manuscript which he had lent Sir W. Hamilton, who writes, – "In our conclusions we do not quite agree, but I am happy to acknowledge my obligations to his writings for the suggestions above alluded to, as I shall hereafter more fully express.

"So long ago as 1846, I communicated to the Royal Irish Academy a transformation which may be written thus (see the Proceedings for the July of that year)†:

$$D_x^2 + D_y^2 + D_z^2 = -(iD_x + jD_y + kD_z)^2; \tag{1}$$

and which was obviously connected with the celebrated equation of Laplace.

"But it had quite escaped my notice that the principles of quaternions allow also this other transformation, which Mr. Carmichael was the first to point out:

$$D_z^2 + D_x^2 + D_y^2 = (D_z - iD_x - jD_y)(D_z + iD_x + jD_y). \tag{2}$$

And therefore I had, of course, not seen, what Mr. Carmichael has since shown, that the integration of Laplace's equation of the *second* order may be made to depend on the integrations of *two linear* and conjugate equations, of which one is

$$(D_z - iD_x - jD_y)V = 0. \tag{3}$$

"I am disposed, for the sake of reference, to call this '*Carmichael's Equation*;' and have had the pleasure of recently finding its integral, under a form, or rather forms, so general as to extend even to *bi*quaternions.

"One of those forms is the following:‡

$$V_{xyz} = e^{z(iD_x + jD_y)} V_{xy0}. \tag{4}$$

* [Robert Bell Carmichael, 1828–1861, Fellow of Trinity College, Dublin.]

† [See Vol. III, pp. 376–7.]

‡ "*Note, added during printing*.—Since writing the above, I have convinced myself that Mr. Carmichael had been in full possession of the exponential form of the integral, and probably also of my chief transformations thereof; although he seems to have chosen to put forward more prominently certain other forms, to which I have found objections, arising out of the non-commutative character of the symbols *ijk* as factors, and on which forms I believe that he does not now insist.—W. R. H."

"Another is

$$V_{xyz} = (D_z + iD_x + jD_y)\int_0^z \cos\{z(D_x^2 + D_y^2)^{\frac{1}{2}}\} V_{xy0}\, dz; \tag{5}$$

where V_{xy0} is generally an *initial biquaternion*; and where the *single* definite integral admits of being usefully put under the form of a *double definite integral*, exactly analogous to, and (when we proceed to Laplace's equation) reproducing, a well known expression of Poisson's, to which Mr. Carmichael has referred.

"These specimens may serve to show to the Academy that I have been aiming to collect materials for future communications to their Transactions."

XXIX.

ON THE SOLUTION OF THE EQUATION OF LAPLACE'S FUNCTIONS (1855)

[Communicated February 26, 1855]
[*Proceedings of the Royal Irish Academy* **6**, 181–185 (1858).]

Rev. Professor Graves communicated the following extract from a letter addressed to him (under date of January 26th, 1855) by Sir William R. Hamilton:-

"MY DEAR GRAVES, – You may like, perhaps, to see a way in which I have to-day, for my own satisfaction, confirmed (not that they required confirmation) some of the results announced by you to the Academy on Monday evening last.

"Let us then consider the function (suggested by you),

$$\sum i^l j^m k^n = (l,\ m,\ n)\, i^l j^m k^n; \tag{1}$$

where l, m, n are positive and integer exponents (0 included); the summation $\sum$ refers to all the possible arrangements of the $l + m + n$ factors, whereof the number is

$$N_{l,m,n} = \frac{(l + m + n)!}{l!\, m!\, n!}; \tag{2}$$

each of these N arrangements gives (by the rules of ijk) a product $= \pm 1 . i^l j^m k^n$; and the sum of all these positive or negative unit-coefficients, ± 1, thus obtained, is the numerical coefficient denoted by $(l,\ m,\ n)$.

"Since each arrangement must have i or j or k to the left, we may write,

$$\sum i^l j^m k^n = i \sum i^{l-1} j^m k^n + j \sum i^l j^{m-1} k^n + k \sum i^l j^m k^{n-1}; \tag{3}$$

and it is easy to see that the coefficient $(l,\ m,\ n)$, or the sum $\sum(\pm 1)$, vanishes, if *more than one* of the exponents, l, m, n, be *odd*. Assume, therefore, as a new notation,

$$(2\lambda,\ 2\mu,\ 2\nu) = \{\lambda,\ \mu,\ \nu\}; \tag{4}$$

which will give, by (3), and by the principle last mentioned respecting odd exponents,

$$\begin{aligned} (2\lambda + 1,\ 2\mu,\ 2\nu) &= \{\lambda,\ \mu,\ \nu\}; \\ (2\lambda - 1,\ 2\mu,\ 2\nu) &= \{\lambda - 1,\ \mu,\ \nu\}. \end{aligned} \tag{5}$$

We shall then have, by the mere notation,

$$\sum i^{2\lambda} j^{2\mu} k^{2\nu} = \{\lambda,\ \mu,\ \nu\}\, i^{2\lambda} j^{2\mu} k^{2\nu}; \tag{6}$$

and, by treating this equation on the plan of (3),

$$\{\lambda,\ \mu,\ \nu\} = \{\lambda - 1,\ \mu,\ \nu\} + \{\lambda,\ \mu - 1,\ \nu\} + \{\lambda,\ \mu,\ \nu - 1\}. \tag{7}$$

By a precisely similar reasoning, attending only to j and k, or making $\lambda = 0$, we have an expression of the form,

$$\sum j^{2\mu} k^{2\nu} = \{\mu, \nu\} j^{2\mu} k^{2\nu}, \tag{8}$$

where the coefficients $\{\mu, \nu\}$ must satisfy the analogous equation in differences,

$$\{\mu, \nu\} = \{\mu - 1, \nu\} + \{\mu, \nu - 1\}, \tag{9}$$

together with the initial conditions,

$$\{\mu, 0\} = 1, \{0, \nu\} = 1. \tag{10}$$

Hence, it is easy to infer that

$$\{\mu, \nu\} = \frac{(\mu + \nu)!}{\mu!\nu!}; \tag{11}$$

one way of obtaining which result is, to observe that the generating function has the form,

$$\sum \{\mu, \nu\} u^{\mu} v^{\nu} = (1 - u - v)^{-1}. \tag{12}$$

In like manner, if we combine the equation in differences (7), with the initial conditions derived from the foregoing solution of a less complex problem, namely, with

$$\{0, \mu, \nu\} = \{\mu, \nu\}, \quad \{\lambda, 0, \nu\} = \{\lambda, \nu\}, \quad \{\lambda, \mu, 0\} = \{\lambda, \mu\}, \tag{13}$$

when the second members are interpreted as in (11), we find that the (slightly) more complex generating function sought is,

$$\sum \{\lambda, \mu, \nu\} t^{\lambda} u^{\mu} v^{\nu} = (1 - t - u - v)^{-1}; \tag{14}$$

and therefore that the required form of the coefficient is,

$$\{\lambda, \mu, \nu\} = \frac{(\lambda + \mu + \nu)!}{\lambda!\mu!\nu!}; \tag{15}$$

as, I have no doubt, you had determined it to be.

"With the same signification of { }, we have, by (2),

$$N_{l,m,n} = \{l, m, n\}; \tag{16}$$

therefore, dividing $\sum$ by N, or the *sum* by the *number*, we obtain, as an expression for what you happily call the MEAN VALUE of the product $i^{2\lambda} j^{2\mu} k^{2\nu}$, the following:

$$M i^{2\lambda} j^{2\mu} k^{2\nu} = \frac{\{\lambda, \mu, \nu\}}{\{2\lambda, 2\mu, 2\nu\}} i^{2\lambda} j^{2\mu} k^{2\nu}; \tag{17}$$

or, substituting for { } its value (15), and writing for abridgment

$$\kappa = \lambda + \mu + \nu, \tag{18}$$

$$M i^{2\lambda} j^{2\mu} k^{2\nu} = \frac{(-1)^{\kappa} \kappa! (2\lambda)! (2\mu)! (2\nu)!}{(2\kappa)! \, \lambda! \, \mu! \, \nu!}. \tag{19}$$

In like manner,

$$Mi^{2\lambda+1}j^{2\mu}k^{2\nu} = \frac{i(-1)^{\kappa}\kappa!(2\lambda+1)!(2\mu)!(2\nu)!}{(2\kappa+1)!\,\lambda!\,\mu!\,\nu!}. \tag{20}$$

"The whole theory of what you call the *mean values*, of *products* of positive and integer *powers* of *ijk*, being contained in the foregoing remarks, let us next apply it to the determination of the mean value of a *function* of $x+iw$, $y+jw$, $z+kw$; or, in other words, let us investigate the equivalent for your

$$Mf(x+iw,\ y+jw,\ z+kw): \tag{21}$$

by developing this function f according to ascending powers of w, and by substituting, for every product of powers of *ijk*, its *mean* value determined as above. Writing, as you propose,

$$\frac{d}{dw} = D, \quad \frac{d}{dx} = D_1, \quad \frac{d}{dy} = D_2, \quad \frac{d}{dz} = D_3, \tag{22}$$

we are to calculate and to sum the general term of (21), namely,

$$Mi^l j^m k^n \times \frac{w^{l+m+n}}{l!\,m!\,n!} D_1^l D_2^m D_3^n f(x,\ y,\ z). \tag{23}$$

One only of the exponents, l, m, n, can usefully be *odd*, by properties of the *mean* function, which have been already stated. If *all* be *even*, and if we make

$$l = 2\lambda, \qquad m = 2\mu, \qquad n = 2\nu, \tag{24}$$

the corresponding part of the general term of Mf, namely, the part independent of *ijk*, is by (15), (18), (19),

$$\frac{(-w^2)^{\kappa}}{(2\kappa)!}\{\lambda,\ \mu,\ \nu\} D_1^{2\lambda} D_2^{2\mu} D_3^{2\nu} f(x,\ y,\ z); \tag{25}$$

whereof the sum, relatively to λ, μ, ν, when *their* sum κ is given, is,

$$\frac{(-w^2)^{\kappa}}{(2\kappa)!}(D_1^2 + D_2^3 + D_3^2)^{\kappa} f(x,\ y,\ z) = \frac{(w\triangleleft)^{2\kappa}}{(2\kappa)!} f(x,\ y,\ z), \tag{26}$$

if my signification of $\triangleleft$ be adopted, so that

$$\triangleleft = iD_1 + jD_2 + kD_3; \tag{27}$$

and another summation, performed on (26), with respect to κ, gives, for the part of Mf which is independent of *ijk*, the expression,

$$\tfrac{1}{2}(e^{w\triangleleft} + e^{-w\triangleleft}) f(x,\ y,\ z). \tag{28}$$

"Again, by supposing, in (23),

$$l = 2\lambda + 1, \qquad m = 2\mu, \qquad n = 2\nu, \tag{29}$$

and by attending to (20), we obtain the term,

$$\frac{wiD_1(-w^2)}{(2\kappa+1)!}\{\lambda,\ \mu,\ \nu\} D_1^{2\lambda} D_2^{2\mu} D_3^{2\nu} f(x,\ y,\ z). \tag{30}$$

Adding the two other general terms correspondent, in which iD_1 is replaced by jD_2 and by kD_3, we change iD_1 to $\triangleleft$; and obtain, by a first summation, the term

$$\frac{(w\triangleleft)^{2\kappa+1}}{(2\kappa+1)!}f(x,\ y,\ z); \tag{31}$$

and, by a second summation, we obtain

$$\tfrac{1}{2}(e^{w\triangleleft} - e^{-w\triangleleft})f(x,\ y,\ z), \tag{32}$$

as the *part* of the mean function Mf, which involves expressly ijk. Adding the two parts, (28) and (32), we are conducted finally to the very simple and remarkable transformation of the MEAN FUNCTION Mf, of which the discovery is due to you:

$$Mf(x+iw,\ y+jw,\ z+kz) = e^{w\triangleleft}f(x,\ y,\ z). \tag{33}$$

In like manner,

$$M\phi(x-iw,\ y-jw,\ z-kz) = e^{-w\triangleleft}\phi(x,\ y,\ z). \tag{34}$$

Each of these two means of arbitrary functions, and therefore also their sum, is thus a value of the expression

$$(D^2 - \triangleleft^2)^{-1}0; \tag{35}$$

that is, the partial differential equation,

$$(D^2 + D_1^2 + D_2^2 + D_3^2)V = 0, \tag{36}$$

has its general integral, with two arbitrary functions, f and ϕ, expressible as follows:

$$V = Mf(x+iw,\ y+jw,\ z+kw) + M\phi(x-iw,\ y-jw,\ z-kw); \tag{37}$$

which is another of your important results. You remarked that if the second member of the equation (36) had been U, the expression for V would contain the additional term,

$$e^{w\triangleleft}D^{-1}e^{-2w\triangleleft}D^{-1}e^{w\triangleleft}U. \tag{38}$$

In fact,

$$D+\triangleleft = e^{-w\triangleleft}De^{w\triangleleft}, \qquad D-\triangleleft = e^{w\triangleleft}De^{-w\triangleleft}, \tag{39}$$

and therefore,

$$(D-\triangleleft)^{-1}(D+\triangleleft)^{-1} = e^{w\triangleleft}D^{-1}e^{-2w\triangleleft}D^{-1}e^{w\triangleleft}. \tag{40}$$

"Most of this letter is merely a repetition of your remarks, but the analysis employed may perhaps not be in all respects identical with yours: a point on which I shall be glad to be informed.

"I remain faithfully yours,
"WILLIAM ROWAN HAMILTON.
"*The Rev. Charles Graves, D. D.*"

XXX.

ON THE CALCULATION OF NUMERICAL VALUES OF A CERTAIN CLASS OF MULTIPLE AND DEFINITE INTEGRALS* (1857)

[*Philosophical Magazine* (4S) **14**, 375–382 (1857).]

Section I

1. The results, in part numerical, of which a sketch is here to be given, may serve to illustrate some points in the theory of functions of large numbers, and in that of definite and multiple integrals. In stating them, it will be convenient to employ a notation which I have formerly published, and have often found to be useful; namely the following,

$$\mathrm{I}_t = \int_0^t dt; \tag{1}$$

or more fully,

$$\mathrm{I}_t ft = \int_0^t ft\,dt; \tag*{(1)$'$}$$

with which I am now disposed to combine this other symbol,

$$\mathrm{J}_t = \int_t^\infty dt; \tag{2}$$

in such a manner as to write,

$$\mathrm{J}_t ft = \int_t^\infty ft\,dt, \tag*{(2)$'$}$$

and therefore

$$\mathrm{I}_t + \mathrm{J}_t = \int_0^\infty dt. \tag{3}$$

I shall also retain, for the present, the known notation of Vandermonde for factorials, which has been described and used by Lacroix, and in which, for any positive whole value of n,

$$[x]^n = x(x-1)(x-2)\ldots(x-n+1); \tag{4}$$

so that there are the transformations,

$$[x]^n = [x]^m[x-m]^{n-m} = [x]^{n+m} : [x-n]^m, \text{ \&c.}; \tag*{(4)$'$}$$

* Communicated by the Author.

which are extended by definition to the case of null and negative indices, and give, in particular,

$$[0]^{-n} = \frac{1}{[n]^{n}} = \frac{1}{1.2.3 \ldots n}. \tag{4)''$$

For example,

$$(1+x)^{n} = \sum_{m=0}^{m=\infty} [n]^{m}[0]^{-m}x^{m}. \tag{5}$$

It is easy, if it be desired, to translate, these into other known notations of factorials, but they may suffice on the present occasion.

2. With the notations above described, it is evident that

$$\mathrm{I}_t^n 1 = [0]^{-n} t^n; \tag{6}$$

and more generally, that

$$\mathrm{I}_t^n t^m = \frac{t^{m+n}}{[m+n]^{n}} = [m]^{-n} t^{m+n} \tag{6)'$$

Hence results the series,

$$(1 + \mathrm{I}_t + \mathrm{I}_t^2 + \ldots)1 = (1 - \mathrm{I}_t)^{-1}1 = e^t; \tag{7}$$

and accordingly, we have the finite relation,

$$\mathrm{I}_t e^t = e^t - 1. \tag{7)'$$

The imaginary equation,

$$e^{t\sqrt{-1}} = (1 - \sqrt{-1}\mathrm{I}_t)^{-1}1, \tag{8}$$

breaks up into the two real expressions,

$$\cos t = (1 + \mathrm{I}_t^2)^{-1}1, \tag{8)'$$

$$\sin t = \mathrm{I}_t(1 + \mathrm{I}_t^2)^{-1}1. \tag{8)''$$

The series of Taylor may be concisely denoted by the formula,

$$f(x+t) = (1 - \mathrm{I}_t D_x)^{-1} fx; \tag{9}$$

and accordingly,

$$\mathrm{I}_t D_x f(x+t) = \mathrm{I}_t f'(x+t) = f(x+t) - fx. \tag{9)'$$

And other elementary applications of the symbol I_t may easily be assigned, whereof some have been elsewhere indicated.

3. The following investigations relate chiefly to the function,

$$\mathrm{F}_{n,r} t = \mathrm{I}_t^n (1 + 4\mathrm{I}_t^2)^{-r-\frac{1}{2}}1; \tag{10}$$

or

$$\mathrm{F}_{n,r}\,t = \mathrm{I}_t^n (1 + 4\mathrm{I}_t^2)^{-r} ft, \tag{10}'$$

where

$$ft = \mathrm{F}_{0,0}\,t = (1 + 4\mathrm{I}_t^2)^{-\frac{1}{2}} 1. \tag{11}$$

Developing by (5) and (6), and observing that

$$2^{2m}[-\tfrac{1}{2}]^m = (-1)^m [2m]^m, \tag{12}$$

and that therefore

$$2^{2m}[-\tfrac{1}{2}]^m [0]^{-m} [0]^{-2m} = (-1)^m ([0]^{-m})^2, \tag{12}'$$

we find the well-known series,

$$ft = 1 - \left(\frac{t}{1}\right)^2 + \left(\frac{t^2}{1.2}\right)^2 - \left(\frac{t^3}{1.2.3}\right)^2 + \&c., \tag{13}$$

which admits of being summed as follows,

$$ft = \frac{2}{\pi} \int_0^{\frac{\pi}{2}} d\omega \cos(2t \cos\omega); \tag{13}'$$

the function ft being thus equal to a celebrated definite integral, which is important in the mathematical theory of heat, and has been treated by Fourier and by Poisson.

4. It was pointed out* by the great analyst last named, that if there were written the equation,

$$y = \int_0^{\pi} \cos(k \cos\omega)\, d\omega, \tag{14}$$

so that, in our recent notation,

$$y = \pi f\left(\frac{k}{2}\right), \tag{14}'$$

then for large, real, and positive values of k, the function $y\sqrt{k}$ might be developed in a series of the form,

$$y\sqrt{k} = \left(\mathrm{A} + \frac{\mathrm{A}'}{k} + \frac{\mathrm{A}''}{k^2} + \&c.\right)\cos k + \left(\mathrm{B} + \frac{\mathrm{B}'}{k} + \frac{\mathrm{B}''}{k^2} + \&c.\right)\sin k; \tag{15}$$

where a certain differential equation of the second order, which $y\sqrt{k}$ was obliged to satisfy, was proved to be sufficient for the successive deduction of as many of the other constant coefficients, A′, A″, . . and B′, B″, . . of the series, as might be desired, through an assigned system of equations of condition, after the two first constants, A and B, were determined; and certain processes of definite integration gave for them the following values,

$$\mathrm{A} = \mathrm{B} = \sqrt{\pi}; \tag{15}'$$

* In his Second Memoir on the Distribution of Heat in Solid Bodies, *Journal de l'Ecole Polytechnique*, tome xii. cahier 19, Paris, 1823, pages 349, &c.

so that when k is very large, we have nearly, as Poisson showed,

$$y\sqrt{k} = (\cos k + \sin k)\sqrt{\pi}. \tag{15)''$$

5. In my own paper on Fluctuating Functions*, I suggested a different process for arriving at this important formula of approximation, (15)″, which, with some slight variation, may be briefly stated as follows. Introducing the two definite integrals,

$$\left.\begin{aligned} A_t &= \frac{2}{\pi}\int_0^{\frac{\pi}{2}} d\omega \cos(2t \operatorname{vers}\omega), \\ B_t &= \frac{2}{\pi}\int_0^{\frac{\pi}{2}} d\omega \sin(2t \operatorname{vers}\omega), \end{aligned}\right\} \tag{16}$$

which give the following *rigorous transformation* of the expression (13)′, or of the function ft,

$$ft = A_t \cos 2t + B_t \sin 2t; \tag{16)'$$

and employing the *limiting values*,

$$\left.\begin{aligned} \lim_{t=\infty} . t^{\frac{1}{2}} A_t &= \frac{2}{\pi}\int_0^{\infty} dx \cos(x^2) = (2\pi)^{-\frac{1}{2}}, \\ \lim_{t=\infty} . t^{\frac{1}{2}} B_t &= \frac{2}{\pi}\int_0^{\infty} dx \sin(x^2) = (2\pi)^{-\frac{1}{2}}; \end{aligned}\right\} \tag{16)''$$

(which two last and well-known integrals have indeed been used by Poisson also,) I obtained (and, as I thought, more rapidly than by his method) the following *approximate expression*, equivalent to that lately marked as (15)″, for large, real, and positive values of t:

$$ft = (\pi t)^{-\frac{1}{2}} \sin\left(2t + \frac{\pi}{4}\right); \tag{17}$$

which is sufficient to show that the *large and positive roots* of the transcendental equation,

$$\int_0^{\frac{\pi}{2}} d\omega \cos(2t\cos\omega) = 0, \tag{17)'$$

are (as is known)† very nearly of the form,

$$t = \frac{n\pi}{2} - \frac{\pi}{8}, \tag{17)''$$

where n is a large whole number.

* In the Transactions of the Royal Irish Academy, vol. xix. part 2, p. 313; Dublin, 1843. Several copies of the paper alluded to were distributed at Manchester in 1842, during the Meeting of the British Association for that year: one was accepted by the great Jacobi.

† It must, I think, be a misprint, by which, in p. 353 of Poisson's memoir, the expression $k = i\pi + \frac{\pi}{4}$, is given, instead of $i\pi - \frac{\pi}{4}$, for the large roots of the transcendental equation $y = 0$. It is remarkable, however, that this error of sign, if it be such, does not appear to have had any influence on the correctness of the physical conclusions of the memoir: which, no doubt, arises from the circumstance that the number i is treated as infinite, in the applications.

6. Poisson does not appear to have required, for the applications which he wished to make, any more than the *two* constants, which he called A and B, of his descending series (15); although (as has been said) he showed how all the *other* constants of that series *could* be *successively* computed, from them, if it had been thought necessary or desirable to do so. In other words, he seems to have been content with assigning the values (15)′, and the formula (15)″, as sufficient for the purpose which he had in view. In my own paper, already cited, I gave the *general term of the descending series* for *ft*, by assigning a formula, which (with one or two unimportant differences of notation) was the following:

$$(\pi t)^{\frac{1}{2}} ft = \sum_{m=0}^{m=\infty} [0]^{-m}([-\tfrac{1}{2}]^{m})^{2}(4t)^{-m}\cos\left(2t-\frac{\pi}{4}-\frac{m\pi}{2}\right). \tag{18}$$

As an example of the numerical approximation attainable hereby, when t was a moderately large number, (not necessarily whole,) I assumed $t=20$; and found that *sixty terms* of the ultimately convergent, but initially divergent series (13), gave

$$\begin{aligned} f(20) &= \frac{2}{\pi}\int_0^{\frac{\pi}{2}} d\omega \cos(40\cos\omega) \\ &= +7\,447\,387\,396\,709\,949{\cdot}\,965\,795\,7 - 7\,447\,387\,396\,709\,949{\cdot}\,958\,428\,9 \\ &= +0{\cdot}007\,366\,8; \end{aligned} \tag{19}$$

while only *three terms* of the ultimately divergent, but initially convergent series (18), sufficed to give almost exactly the same result, under the form,

$$\begin{aligned} f(20) &= \left(1-\frac{9}{204800}\right)\frac{\cos 86^{\circ}\,49'\,52''}{\sqrt{20\pi}} + \frac{1}{320}\frac{\sin 86^{\circ}\,49'\,52''}{\sqrt{20\pi}} \\ &= 0{\cdot}0069736 + 0{\cdot}0003936 = +0{\cdot}0073672. \end{aligned} \tag{19$'$}$$

7. The function ft becomes infinitely small, when t becomes infinitely great, on account of the indefinite fluctuation which $\cos(2t\cos\omega)$ then undergoes, under the sign of integration in (13)′; so that we may write

$$\mathrm{F}_{0,0}\infty = f\infty = 0. \tag{20}$$

But it is by no means true that the value of this *other* series,

$$\mathrm{F}_{1,0}\,t = \mathrm{I}_t ft = \frac{t}{1} - \frac{t}{3}\left(\frac{t}{1}\right)^2 + \frac{t}{5}\left(\frac{t^2}{1.2}\right)^2 - \&\text{c.}, \tag{21}$$

which may be expressed by the definite integral,

$$\mathrm{F}_{1,0}\,t = \frac{1}{\pi}\int_0^{\frac{\pi}{2}} d\omega \sec\omega \sin(2t\cos\omega), \tag{21$'$}$$

is infinitesimal when t is infinite. On the contrary, by making

$$2t\cos\omega = x, \qquad d\omega\sec\omega = -\frac{dx}{x}\left(1-\frac{x^2}{4t^2}\right)^{-\frac{1}{2}}, \tag{22}$$

the integral (21)′ becomes, at the limit in question,

$$F_{1,0}\infty = \frac{1}{\pi}\int_0^\infty \frac{dx}{x}\sin x = \tfrac{1}{2}. \qquad (21)''$$

Accordingly I verified, many years ago, that the series (21) takes *nearly* this constant value, $\frac{1}{2}$, when t is a large and positive number. But I have lately been led to inquire what is the *correction* to be applied to this approximate value, in order to obtain a more accurate numerical estimate of the function $F_{1,0}\,t$, or of the integral $I_t ft$, when t is large. In other words, having here, by (3) and (21)″, the *rigorous* relation,

$$F_{1,0}\,t = I_t ft = \tfrac{1}{2} - J_t ft, \qquad (23)$$

I wished to evaluate, at least *approximately*, this *other* definite integral, $-J_t ft$, for large and positive values of t. And the result to which I have arrived may be considered to be a very simple one; namely, that

$$-J_t ft = D_t^{-1} ft; \qquad (24)$$

where $D_t^{-1} ft$ is a development analogous to the series (18), and reproduces that series, when the operation D_t is performed.

8. As an example, it may be sufficient here to observe that if we thus operate by D_t on the function,

$$f`t = \left(1 - \frac{129}{2^9 t^2}\right)\frac{\sin\left(2t - \frac{\pi}{4}\right)}{2\sqrt{\pi t}} - \frac{5\cos\left(2t - \frac{\pi}{4}\right)}{2^5 t\sqrt{\pi t}}, \qquad (25)$$

and suppress $t^{-\frac{1}{2}}$ in the result, we are led to this other function of t,

$$D_t f`t = \left(1 - \frac{9}{2^9 t^2}\right)\frac{\cos\left(2t - \frac{\pi}{4}\right)}{\sqrt{\pi t}} + \frac{\sin\left(2t - \frac{\pi}{4}\right)}{2^4 t\sqrt{\pi t}}; \qquad (25)'$$

which coincides, so far as it has been developed, with the expression (18) for ft: so that we may write, as at least approximately true, the equation

$$f`t = D_t^{-1} ft. \qquad (25)''$$

Substituting the value 20 for t, in order to obtain an arithmetical comparison of results, we find,

$$\begin{aligned} f`(20) &= \left(1 - \frac{129}{204800}\right)\frac{\sin 86°\,49'\,52''}{\sqrt{20\pi}} - \frac{\cos 86°\,49'\,52''}{128\sqrt{20\pi}} \\ &= +0{\cdot}062942 - 0{\cdot}000054 = +0{\cdot}062888; \end{aligned} \qquad (26)$$

which ought, if the present theory be correct, to be nearly equal to the definite integral, $-J_t ft$, for the case where $t = 20$. In other words, I am thus led to expect, after adding the constant term $\frac{1}{2}$, that the value of the connected integral,

$$I_t ft = \pi^{-1}\int_0^{\frac{\pi}{2}} d\omega \sec\omega \sin(40\cos\omega), \qquad (26)'$$

must be nearly equal to the following number,

$$+0{\cdot}562888. \qquad (26)''$$

And accordingly, when this last integral (26)′ is developed by means of the *ascending* series (21), I find that the sum of the first sixty terms (beyond which it would be useless for the present purpose to go) gives, as the small difference of two large but nearly equal numbers, (which are *themselves* of interest, as representing certain *other* definite integrals,) the value:

$$\begin{aligned}\pi^{-1}\int_0^{\frac{\pi}{2}} d\omega \sec\omega \sin(40\cos\omega) &= +3\,772\,428\,770\,679\,800{\cdot}537\,705\,8 \\ &\quad -3\,772\,428\,770\,679\,799{\cdot}974\,817\,7 \\ &= +0{\cdot}562\,888\,1; \qquad (26)'''\end{aligned}$$

which can scarcely (as I estimate) be wrong in its last figure, the calculation having been pushed to more decimals than are here set down; and which exhibits as close an agreement as could be desired with the result (26)″ of an entirely different method.

9. It must however be stated, that in extending the method thus exemplified to higher orders of integrals, the development denoted by $D_t^{-n} ft$, or the definite and multiple integral $(-J_t)^n ft$, to which it is equivalent, comes to be *corrected*, in passing to the *other* integral $I_t^n ft$, not by a *constant term*, such as $\frac{1}{2}$, but by a *finite algebraical function*, which I shall here call $f_n t$, and of which I happened to perceive the existence and the law, while pursuing some unpublished researches respecting vibration, a considerable time ago. Lest anything should prevent me from soon submitting a continuation of the present little paper, (for I wish to write on one or two other subjects,) let me at least be permitted now to mention, that the spirit of the process alluded to, for determining this finite and algebraical *correction**,

$$I_t^n ft - (-J_t)^n ft = I_t^n ft - D_t^{-n} ft = f_n t, \qquad (27)$$

(where $\cdot D_t^{-n} ft$ still denotes a descending and periodical series, analogous to and including those above marked (18) and (25),) consists in *developing the algebraical expression* (10), (for the case $r = 0$, but with a corresponding development for the more general case,) *according to descending powers of the symbol* I_t, and *retaining only those terms in which the exponent of that symbol is positive or zero*: which process gives the formula,

$$f_n t = \tfrac{1}{2} I_t^{n-1}(1 + 2^{-2} I_t^{-2})^{-\frac{1}{2}} 1 = (\tfrac{1}{2} I_t^{n-1} - \tfrac{1}{16} I_t^{n-3} + \tfrac{3}{236} I_t^{n-5} - \,.\,.) 1; \qquad (28)$$

that is, by (5) and (6),

* Although this *algebraical part*, $f_n t$, of the multiple integral $I_t^n ft$, is *here* spoken of as a *correction* of the *periodical part*, denoted above by $D_t^{-n} ft$, yet for *large* and *positive values* of t it is, *arithmetically* speaking, by much the *most important portion* of the whole: and accordingly I perceived (although I did not publish) it long ago, whereas it is only very lately that I have been led to *combine* with it the *trigonometrical series*, deduced by a sort of extension of Poisson's analysis. – When I thus venture to speak of any results on this subject as being my own, it is with every deference to the superior knowledge of other Correspondents of this Magazine, who may be able to point out many anticipations of which I am not yet informed. The formulæ (27) (28) are perhaps those which have the best chance of being new.

$$\mathrm{f}_n t = \sum_{m=0}^{m=\infty} 2^{-2m-1}[-\tfrac{1}{2}]^m [0]^{-m} [0]^{-(n-2m-1)} t^{n-2m-1}; \tag{28)'$$

where the series may be written as if it were an infinite one, but the terms involving negative powers of t have each a null coefficient, and are in this question to be suppressed.

For instance, I have arithmetically verified, at least for the case $t = 10$, that the two finite algebraical functions,

$$\mathrm{f}_6 t = \frac{t^5}{240} - \frac{t^3}{96} + \frac{3t}{256}, \qquad (28)''$$

$$\mathrm{f}_7 t = \frac{t^6}{1440} - \frac{t^4}{384} + \frac{3t^2}{512} - \frac{5}{2048}, \qquad (28)'''$$

express the values of the two following sums or differences of integrals,

$$\mathrm{f}_6 t = \mathrm{I}_t^6 ft - \mathrm{J}_t^6 ft, \qquad (27)'$$

$$\mathrm{f}_7 t = \mathrm{I}_t^7 ft + \mathrm{J}_t^7 ft; \qquad (27)''$$

the calculations having been carried to several places of decimals, and the integrals $\mathrm{I}_t^6 ft$, $\mathrm{I}_t^7 ft$ having each been found as the difference of two large numbers.

Observatory of Trinity College, Dublin,
September 29, 1857.

[To be continued.]

Part IV
ASTRONOMY

XXXI.

INTRODUCTORY LECTURE ON ASTRONOMY (1832)

[Delivered November 8, 1832]
[*Dublin University Review* **1**, 72–85 (1833).]

The time has returned when, according to the provisions of this our University, we are to join our thoughts together, and direct them in concert to astronomy—the parent of all the sciences, and the most perfect and beautiful of all. And easily and gladly could I now expatiate on the dignity and interest of astronomy, but the very assurance of your complete and perfect sympathy renders needless any attempt at excitement. I must not and cannot suppose that any of those who are assembled here this day, are insensible to the inward impulses, and unconscious of the high aspirations, by which the stars, from their thrones of glory and of mystery, excite and win toward themselves the heart of man; that the golden chain has been let down in vain; and that celestial beauty and celestial power have offered themselves in vain to human view. And if I could suppose that this were so—that any here had been till now untouched by the majesty and loveliness with which astronomy communes—still less could I persuade myself that in the mind of such a person my words could do what the heavens had failed to effect. The heart, because it is human, say rather because it is not wholly not divine, lifts itself up in aspiration, and claims to mingle with the lights of heaven; and joyfully receives into itself the skyey influences, and feels that it is no stranger in the courts of the moon and the stars. Though between us and the nearest of those stars there be a great gulph fixed, yet beyond that mighty gulph (oh, far beyond!) fly, on illimitable pinions, the thoughts and affections of man, and tell us that there, too, are beings, akin to us—members of one great family—beings animated, thoughtful, loving—susceptible of joy and hope, of pain and fear—able to adore God, or to rebel against him—able to admire and speculate upon that goodly array of worlds with which they also are surrounded. And often this deep instinct of affection, to the wide family of being, to the children of God thus scattered throughout all worlds, has stirred within human bosoms; often have men, tired of petty cares and petty pleasures, fretting within this narrow world of ours, seeking for other suns and ampler ether, gone forth as it were colonists from earth, and become naturalized and denizens in heaven. Not of one youthful enthusiast alone, are the words of a great living poet* true, that,

> "Thus, before his eighteenth year was told,
> Accumulated feelings pressed his heart
> With still increasing weight; he was o'erpower'd
> By nature—by the turbulence subdued

* Wordsworth.

Of his own mind—by mystery and hope,
And the first virgin passion of a soul
Communing with the glorious universe."

I must not and do not doubt, that many, let me rather say that all, of those whom I now address, have, from time to time, been stirred by such visitations, and been conscious of such aspirings; and that you need not me to inform you, that astronomy, though a science, and an eminent one, is yet more than a science,—that it is a chain woven of feeling as well as thought—an influence pervading not the mind only, but the soul of man. Thus much, therefore, it may suffice to have indulged in the preliminary and general expression of these our common aspirations; and I now may pass to the execution of my particular duty, my appointed and pleasant task, and fulfil, so far as in me lies, the intentions and wishes of the heads of our University; who, in fixing the order of your studies, directed first your attention to the sciences of the pure reason—the logical, the metaphysical, and the mathematical—and call you now to those in which the reason is combined with experience; and who have judged it expedient, among all the physical sciences, to propose astronomy the first, as a favourable introduction to the rest, and a specimen and type of the whole.

It is, then, my office, this day, to present to you astronomy as itself a part, and as an introduction to the other parts, of physical science in general—and thus to greet you at the first steps and vestibule of that majestic edifice which patient intellect has been rearing up through many a past generation; and which, with changes doubtless, but such as rather improve than destroy the unity of the whole, shall remain, as we trust, for the exercise, the contemplation, and the delight, of many a generation yet unborn. It were difficult for any one, and it is impossible for me, to do full justice to so vast a subject; but I shall hope for a renewal of that indulgent attention with which I have more than once before been favoured upon similar occasions, while, in pursuit and illustration of the subject, I touch briefly, and as it were by allusion only, on the following points:—the distinction between the physical and the purely mathematical sciences—the end which should be considered as proposed in physical science in general—and the means which are to be employed for the attainment of this end—the objections, utilitarian and metaphysical, which are sometimes expressed, and perhaps oftener felt, against the study of physical science—the existence of a scientific faculty analogous to poetical imagination, and the analogies of other kinds between the scientific and the poetical spirit.

I have said that I design to speak briefly of the end proposed, and the means employed, in the physical sciences on which you are entering; and of the distinction between them and the pure mathematics, in which you have lately been engaged. It seems necessary, or at least useful, for this purpose, to remind you of the nature and spirit of these your recent studies—the sciences of geometry and algebra. In all the mathematical sciences we consider and compare relations. The relations of geometry are evidently those of space; the relations of algebra resemble rather those of time. For geometry is the science of figure and extent; algebra, of order and succession. The relations considered in geometry are between points, and lines, and surfaces; the relations of algebra, at least those primary ones, from the comparison of which others of higher kinds are obtained, are relations between successive thoughts, viewed as successive and related states of one more general and regularly changing thought. Thus algebra, it appears, is more refined, more general, than geometry; and has its foundation deeper in the very nature of man; since the ideas of order and succession appear to be less foreign, less separable from us, than those of figure and extent. But, partly from its very refinement and generality, algebra is more easily and often misconceived; more easily and often degraded to a mere exercise of memory—a mere application of rules—a mere legerdemain of symbols: and thus, except in the hands of a very skilful and philosophical teacher, it is likely to be a less instructive discipline to the mind of a beginner in science.

Motion, although its causes and effects belong to physical science, yet furnishes, by its conception and by its properties, a remarkable application of each of these two great divisions of the pure mathematics: of geometry, by its connexion with space; of algebra, by its connexion with time. Indeed, the thought of position, whether in space or time, as varied in the conception of motion, is an eminent instance of that passage of one general and regularly changing thought, through successive and related states, which has been spoken of as suggesting to the mind the primary relations of algebra. We may add, that this instance, motion, is also a type of such passage; and that the phrases which originally belong to and betoken motion, are transferred by an expressive figure to every other unbroken transition. For with time and space we connect all continuous change; and by symbols of time and space we reason on and realise progression. Our marks of temporal and local site, our *then* and *there*, are at once signs and instruments of that transformation by which thoughts become things, and spirit puts on body, and the act and passion of mind are clothed with an outward existence, and we behold ourselves from afar.

These purely mathematical sciences of algebra and geometry, are sciences of the pure reason, deriving no weight and no assistance from experiment, and isolated, or at least isolable, from all outward and accidental phenomena. The idea of order, with its subordinate ideas of number and of figure, we must not indeed call innate ideas, if that phrase be defined to imply that all men must possess them with equal clearness and fulness; they are, however, ideas which seem to be so far born with us, that the possession of them, in any conceivable degree, appears to be only the development of our original powers, the unfolding of our proper humanity. Foreign, in so far that they touch not the will, nor otherwise than indirectly influence our moral being, they yet compose the scenery of an inner world, which depends not for its existence on the fleeting things of sense, and in which the reason, and even the affections may at times find a home and a refuge. The mathematician, dwelling in that inner world, has hopes, and fears, and vicissitudes of feeling of his own; and even if he be not disturbed by anxious yearnings for an immortality of fame, yet has he often joy, and pain, and ardour: the ardour of successful research, the pain of disappointed conjecture, and the joy that is felt in the dawning of a new idea. And when, as on this earth of ours must sometimes happen, he has sent forth his wishes and hopes from that lonely ark, and they return to him, having found no resting-place: while he drifts along the turbulent current of passion, and is tossed about by the storm and agony of grief, some sunny bursts may visit him, some moments of delightful calm may be his, when his old habits of thought recur, and the "charm severe" of lines and numbers is felt at intervals again.

It has been said, that in all the mathematical sciences we consider and compare relations. But the relations of the pure mathematics are relations between our own thoughts themselves; while the relations of mixed or applied mathematical science are relations between our thoughts and phenomena. To discover laws of nature, which to us are links between reason and experience—to explain appearances, not merely by comparing them with other appearances, simpler or more familiar, but by showing an analogy between them on the one hand, and our own laws and forms of thought on the other, "darting our being through earth, sea, and air"*—such seems to me the great design and office of genuine physical science, in that highest and most philosophical view in which also it is most imaginative. But, to fulfil this design—to execute this office—to discover the secret unity and constancy of nature amid its seeming diversity and mutability—to construct, at least in part, a history and a prophecy of the outward world adapted to the understanding of man—to account for past, and to predict future phenomena—new forms and new manifestations of patience and of genius become requisite, for which no occasion had been in the pursuits of the pure mathematics. Induction must be exercised; probability must be weighed. In the sphere of the pure and inward reason, probability finds no place; and if induction ever enter, it is but tolerated as a mode of accelerating and assisting discovery, never rested in as the ground of belief, or testimony of that truth, which yet it may have helped to suggest. But in the physical sciences, we can conclude nothing, can know nothing without induction. Two elements there are in these, the outward and the inward; and if the latter, though higher in dignity, usurp the place which of right pertains to the former, there ensues only a specious show, a bare imagination, and not a genuine product of the imaginative faculty, exerting itself in due manner and measure on materials which nature supplies. Here, then, in the use and need of induction and probability, we have a great and cardinal distinction between the mixed and the pure mathematics.

Does any, then, demand what this induction is, which has been called the groundwork of the physical sciences, the key to the interpretation of nature? To answer this demand, I must resume my former statement of the main design and office of physical science in general. I said, that this design was to explain and account for phenomena, by discovering links between reason and experience. Now the essence of genuine induction appears to me to consist in this, that in seeking for such links we allow to experience its due influence, and to reason not more than its due—that we guard against false impressions from the mechanism and habits of our own understandings—and submit ourselves teachably to facts; not that we may ultimately abide in mere facts, and sensations, and arranged recollections of sensation, but from the deep and sublime conviction, that the author, and sustainer, and perpetual mover of nature has provided in nature a school, in which the human understanding may advance ever more and more, and discipline itself with continual improvement. We must not conclude a law from facts too small in number, or observed with too little care; or if the scientific imagination, impatient of restraint, press onward at once to the goal, and divine from the falling of an apple the law of gravitation, and in the trivial and every day changes which are witnessed around us on this earth perceive the indications of a mighty power, extending through all space, and compelling to their proper orbits the "planets struggling fierce towards heaven's free wilderness;"† yet must such divinations be long received, even by the favoured discoverer himself, if he be of the true inductive school, with candid diffidence and philosophic doubt, until they have been confirmed by new appeals to other, and more

* Shakespeare.

† "As the sun rules, even with a tyrant's gaze,
The unquiet republic of the maze
Of planets struggling fierce towards heaven's free wilderness." — *Shelley*.

remote, and more varied phenomena. If, as in this case of gravitation, the law, concluded or anticipated from the first few facts, admit of a mathematical enunciation, and consequently can be made a basis of mathematical reasoning, then it is consistent with, and required by, the spirit of induction, that the law should be made such a basis. We may and ought to employ *à priori* reasoning here, and consider what consequences must happen if the law supposed be a true one. These consequences ought to be mathematically developed, and a detailed prediction made of the yet unobserved phenomena which the law includes, and with which it must stand or fall, the truth of the one and of the other being connected by an indissoluble tie. New and more careful observations must then be made, to render closer and more firm the connexion between thoughts and things. For,* in order to derive from phenomena the instruction which they are fitted to afford, we must not content ourselves with the first vague perceptions, and obvious and common appearances. We must discriminate the similar from the same—must vary, must measure, must combine—until, by the application of reason and of the scientific imagination to carefully recorded facts, we ascend to an hypothesis, a theory, a law, which includes the particular appearances, and enables them to be accounted for and foreseen. Then, when the passive of our being has been so far made subject to the active, and sensation absorbed or sublimed into reason, the philosopher reverses the process, and asks how far the conceptions of his mind are realised in the outward world. By the deductive process following up induction, he seeks to make his theory more than a concise expression of the facts on which it first was founded; he seeks to deduce from it some new appearances which ought to be observed if the theory be co-extensive with nature. He then again consults sensation and experience, and often their answer is favourable; but often, too, they speak an unexpected language. Yet, undismayed by the repulse, and emboldened by partial success, he frames, upon the ruins of the former, some new and more general theory, which equally with the former accounts for the old appearances, while it includes within its ampler verge the results of more recent observation. Nor can this struggle ever end between the active and the passive of our being—between the imagination of the theorist and the patience of the observer—until the time, if such a time can ever come, when the mind of man shall grasp the infinity of nature, and comprehend all the scope, and character, and habits of those innumerable energies which to our understanding compose the material universe. Meanwhile, this struggle, with its alternate victories and defeats, its discoveries of laws and exceptions, forms an appointed discipline for the mind, and its history is justly interesting. Nor can we see without admiring sympathy, the triumph of astronomy and Newton; Newton, who in astronomy, by one great stride of thought, placed theory at once so far in advance of observation, that the latter has not even yet overtaken the former, nor has the law of gravitation, in all its wide dominion, yet met with one rebellious fact in successful revolt against its authority. Yet, haply, those are right, who, seeing that Newton himself had sat at the feet of another master, and had deeply drunk from the fountain of a still more comprehending intellect, have thought it just to divide the glory, and award more than half to Bacon. He, more than any other man, of ancient or of modern times, appears to have been penetrated with the desire, and to have conceived and shown the possibility, of uniting the mind to things, say rather of drawing things into the mind. Deeply

* Some of the following remarks on physical science were published in the Dublin Literary Gazette in 1830.

he felt, and eloquently and stirringly he spake. In far prophetic vision he foresaw, and in language as of inspiration he gave utterance to the vision, of the progress and triumphs of the times then future—nay more, of times which even now we do but look for. And thus, by highest suffrage, and almost unanimous consent, the name of Bacon has been enrolled as eminent high-priest in the spousal temple* of man's mind and of the universe. And if, impressed with the greatness of his task and importance of his office, and burning to free mankind from those intellectual letters in which the injudicious manner of their admiration of the philosophers of Greece had bound them, he appears to have been sometimes blind to the real merit of those great philosophers, and uttered harsh words, and words seeming to imply a spirit which (we will trust) was not the habitual spirit of Bacon; let us pardon this weakness of our great intellectual parent, let us reverently pardon, but let us not imitate it. For, I cannot suppress my fear, that the signal success, which since the time, and in the country, and by the method of Bacon, has attended the inductive research into the phenomena of the material universe, has injuriously drawn off the intellect from the study of itself and its own nature; and that while we know more than Plato did of the outward and visible world, we know less, far less, of the inward and ideal. But not now will I dwell on this high theme, fearing to desecrate and degrade by feeble and unworthy utterance those deep ideal truths which in the old Athenian days the eloquent philosopher poured forth.

I have now touched on some of the points which at the beginning of this lecture I proposed. I have stated my view of the great aim and design of physical science in general—the explanation of appearances, by linking of experience to reason; an aim which is itself subordinate to another higher end, but to an end too high and too transcendent to come within the sphere of science, till science shall attain its bright consummation in wisdom—the end of restoring and preserving harmony between the various elements of our own being; a harmony which can be perfect only when it includes reconciliation with our God. I have stated the chief means which since the time of Bacon are generally admitted as fit and necessary for the just explanation of appearances—the alternate use of induction and deduction, and the judicious appreciation of probabilities; and have shown how by this use of induction and probability, an essential difference is established between the physical sciences—among which astronomy ranks so high—and the sciences of the pure mathematics; and as an example of successful induction, have referred you to the discovery of gravitation. Many other examples will occur in the course of the subsequent lectures, in which I shall have occasion to speak of ancient as well as of modern discoveries, and to show you from the Almagest of Ptolemy, what the state of astronomy was in his time and the time of Hipparchus. You will, I think, accompany and share the interest which I have felt in a review of the science of a time so ancient. The contemplations, like the objects, of astronomy, are not all of modern growth. Not to us first do Arcturus, and Orion, and the Pleiades glide on in the still heaven. The Bear, forbidden here and now to bathe in ocean, circled the Pole in that unceasing round, three thousand years ago, and its portraiture was imagined by Homer as an ornament for the shield of Achilles. And if that old array of "cycle and epicycle, orb in orb," with which the Greek astronomer had filled the planetary spaces, have now departed with its principle of uniform and circular motion, yet the memory of it will long remain, as of a

* And thus, by the divine assistance, we shall have prepared and decked the nuptial chamber of the mind and of the universe. — *Bacon.*

mighty work of mind, and (for the time) a good explanation of phenomena. The principle itself has in a subtler form revived, and seems likely to remain for ever, as a conviction that some discoverable unity exists, some mathematical harmony in the frame of earth and heaven. We live under no despotism of caprice, are tossed about in no tempest and whirlwind of anarchy; what is law and nature in one age, is not repealed and unnatural in the next; the acquisitions of former generations are not all obsolete and valueless in ours, nor is ours to transmit nothing which the generations that are to come shall prize: our life, the life of the human race, is no life of perpetual disappointment and chaotic doubt, nor doomed to end in blank despondence, it is a life of hope and progress, of building on foundations laid, and of laying the foundations for other and yet greater buildings. And thus are distant generations knit together in one celestial chain, by one undying instinct: while, yielding to kindred impulses, our fathers, ourselves, and our children all seek and find, in the phenomenal and outward world, the projection of our own inward being, of the image of God within us. Astronomy is to man an old and ancestral possession. Through a long line of kings of mind, the sceptre of Astronomy has come down, and its annals are enshrined among the records of the royalty of genius. Its influence has passed, with silent but resistless progress, from simple shepherds watching their flocks by night, to the rulers of ancient empires, and the giants of modern thought. When we thus trace its history, and change of habitation, from the first rude pastoral and patriarchal tents of Asia, to some old palace roof of Araby or Egypt, or to the courts of that unforgotten king of China, who, noting in his garden the shadows of summer and of winter, left a record by which we measure after three thousand years the changes that the seasons have undergone; and passing from these imperial abodes of the East to dwellings not less worthy, when we see astronomy shrined in the observatories and studies of Europe, and nation vying with nation, and man with man, which shall produce the worthier temple, and yield the more acceptable homage; when we review the long line of scientific ancestry, from Hipparchus and Ptolemy, to Copernicus and Galileo, from Tycho and Kepler, to Bradley, Herschel, and Brinkley; or call before us those astronomical mathematicians, who, little provided with instruments and outward means of observing, while they seemed in the silence of their closets to have abandoned human affairs, and to live abstracted and apart, have shown that genius in the very solitude of its meditations is yet essentially sympathetic, and must rule the minds of men by the instinct of its natural regality, and have filled the intervals of the great succession, from Archimedes to Newton, from Newton to Lagrange: when the imagination is crowded and possessed by all these old and recent associations, must we not then, if self be not quite forgotten, if our own individuality be not all merged in this extended and exalted sympathy, this wide and high communion, yet long to bow for a while, and veil ourselves, as before superior spirits, and think it were a lot too happy, if we might but follow in the train, and serve under the direction of this immortal band!

In such a mood, can we discuss with patience, can we hear without indignation the utilitarian objection, "of what *use* is Astronomy?" meaning thereby, what money will it make?—what sensual pleasure will it procure?

Against astronomy, indeed, the objection from utility is singularly infelicitous, and almost ludicrously inapplicable: astronomy, which binding in so close connexion the earth with the visible heaven, and mapping the one in the other, has guided through wastes, which else were trackless, the fleet and the caravan, and made a path over the desert and the deep. But suppose it otherwise, or take some other science which has not yet been so successfully applied. What, then; and is the whole of life to be bound down to the exchange and the market-place? Are there no desires, no pleasures, but the sensual—no wants, and no enjoyments, but of the outward and visible kind? Are we placed here only to eat, and drink, and die? Some less magnificent stage, methinks, might have sufficed for that. It was not needed, surely, for such a race of sorry animals—so void themselves of power and beauty within, so incapable and so undesigned for the contemplation of power and beauty without—that they should have been placed in this world of power and beauty; and the evermoving universe commanded to roll before our view, "making days and equal years, an all-sufficing harmony,"* that the heavens should declare the glory of God, and the firmament show his handywork. I am almost ashamed to have dwelt so long, here, amid these influences, and before such an audience, on objections of a class and character so quite unworthy of your consideration. More important is it that I should endeavour to answer another class of objections, founded on the misapprehension and misapplication of deep, and inward, and important truth, and of a nature fitted to captivate and carry away the young and ardent spirit.

It is, then, sometimes said, and, perhaps, oftener felt, that astronomy itself is too unrefined—too material a thing;—that the mind ought to dwell within its own sphere of reason and imagination, and not be drawn down into the world of phenomena and experience. Now, with respect to the pure Reason, I will grant that this objection would assume a force, which I cannot now concede to it, if it were indeed possible for man, on that etherial element alone, to feed and live. But if this be not so—if we must quit at all the sphere of the pure reason, and descend at all into the world of experience, as surely we must sometimes do—why narrow our intercourse with experience to the smallest possible range? why tread, with delicate step, this common earth of ours, and not rather wander freely through all her heights and depths, and gaze upon the wonders and beauties that are her own, and store our minds and memories with truths of fact, were it only to have them ready, as materials and implements, for the exercise of that transforming and transmuting power, which is gradually to draw those truths into its own high sphere, and to prepare them for the ultimate beholding of pure and inward intuition? And as to the imagination, it results, I think, from the analysis which I have offered of the design and nature of physical science, that into such science generally, and eminently into astronomy, imagination enters as an essential element: if that power be imagination, which "darts our being through earth, sea, and air;" and if I rightly transferred this profound line of our great dramatist to the faculty which constructs dynamical and other physical theories, by seeking for analogies in the laws of outward phenomena to our own inward laws and forms of thought. Be not startled at this, as if in truth there were no beauty, and in beauty no truth; as if these two great poles of love and contemplation were separated by a diametral space, impassable to the mind of man, and no connecting influences could radiate from their common centre. Be not surprised that there should exist an analogy, and that not faint nor distant, between the workings of the

* "And bade the ever-moving universe
Roll round us, making days and equal years,
An all-sufficing harmony."

From a Manuscript Poem, by A. de. V. [Aubrey de Vere]

poetical and of the scientific imagination; and that those are kindred thrones whereon the spirits of Milton and Newton have been placed by the admiration and gratitude of man. With all the real differences between Poetry and Science, there exists, notwithstanding, a strong resemblance between them; in the power which both possess to lift the mind above the stir of earth, and win it from low-thoughted care; in the enthusiasm which both can inspire, and the fond aspirations after fame which both have a tendency to enkindle; in the magic by which each can transport her votaries into a world of her own creating; and perhaps, in the consequent unfitness for the bustle and the turmoil of real life, which both have a disposition to engender. Doubtless there are enthusiasts here this day, whom, without knowing, I affectionately sympathise with: who bear upon them that character of all good and genuine enthusiasm, highly to conceive, intensely to admire, and ardently to aspire after excellence. If any such have chosen poetry for its own sake, and with a hope of adding to the literature of his country; aware of the greatness of the task and responsibility of the office, knowing that the poet should be no pandar to sensual pleasure, no trifler upon frivolous themes, but an interpreter between the heart and beauty, an utterer of divine and of eternal oracles; and if no more imperious duty interfere, I do not seek to dissuade him: but if he have only been repelled from science by its seeming to possess no power of similar excitement, I would not that, so far as in me lay, he should be unaware of the kindred enthusiasm. In science, as in poetry, there are enthusiasts, who, fixing their gaze upon the monuments which kindred genius has reared, press on to those pyramids in the desert, forgetting the space between. And when I think that among the new hearers whom a new year has brought, it is likely that some, perhaps many, are conscious of such aspirations; that some may go forth from this room to-day, whom after-times shall hail with love and reverence, as worthy children and champions of their college and their country; and that I, in however small a degree, may have influenced and confirmed their purpose: I feel, I own, "a presence that disturbs me with the joy of elevated thoughts,"* a sublime and kindling sense of the unseen majesty of mind. Doubtless in that period of generous ardour to which in part the philosophic poet† alluded, when, mourning over the too frequent degeneracy that attends the cares and temptations of manhood, the loss of enthusiasm without the gain of wisdom, or with the acquisition only of "that half-wisdom half-experience gives," he framed that magnificent stanza—

"Not in entire forgetfulness,
And not in utter nakedness,
But trailing clouds of glory do we come
From God who is our home;
Heaven lies about us in our infancy;
Shades of the prison-house begin to close
Upon the growing boy,
But he beholds the light, and whence it flows,
He sees it in his joy;
The youth, who daily farther from the east
Must travel, still is nature's priest,
And by the vision splendid

* Wordsworth. † Ibid.

Is on his way attended;
At length the man beholds it die away,
And fade into the light of common day:''

doubtless, (I was about to say,) in this period of youthful ardour, there are many vague and some determined aspirations after excellence, among those whom I now address; and some assuredly there are, who, burning to consecrate themselves to the service of truth and goodness, and ideal beauty, and wedding themselves in imagination to the spirit of the human race, feed on the hope of future and perpetual fame, and fondly look for that pure ideal recompense, and long to barter ease, and health, and life itself for that influence, surviving life, that power and sympathy, which has been attained by the few, who, after long years of thought, produce some immortal work, a Paradise Lost, or a Principia, and win their sublime reward of praise and wonder;* who do not wholly die, but through all time continue to influence the minds and hearts of men; who leave behind them some enduring monument, which, while it shall be claimed as the honour of their age and nation, bears also their own name engraven on it in imperishable characters, like Phidias on the statue of Minerva. Of such emotions I will not risk the weakening, by dwelling now on a conceivable superior state, in which perfection should be sought for its own sake, and as independent even of this fine unmercenary reward: and the spirit, purified even from this ''last infirmity of noble minds,''† feel, in the words of one who has attained the earthly and (we will trust) the heavenly fame, the words of the immortal Milton, that

Fame is no plant that grows on mortal soil,
Nor in the glistering foil,
Set off to the world, nor in broad rumour lies;
But lives, and spreads abroad, by those pure eyes,
And perfect witness of alljudging God:
As He pronounces lastly of each deed,
Of so much fame in Heaven expect thy meed.

* And win he knows not what sublime reward
Of praise and wonder.—AKENSIDE.

† Milton.

XXXII.

THE COMET

[*Dublin Penny Journal* 207, 208 (Dec. 1832); and 223, 224 (Jan. 1833).]

The little comet which at present excites so much the interest of astronomers and of the public, was first seen by Montaigne, at Limoges, in March 1772. It was soon afterwards observed by Messier, and on the 3rd of April in the same year, a small telescopic star was seen by him shining through it, and was mistaken for a nucleus, or bright and solid body of the comet, such as many are found to possess; a more attentive examination, however, confirmed the suspicion suggested by its first appearance, that it had neither nucleus nor tail. In its appearance it resembled perfectly those faint nebulæ, or little clouds of light, which are seen in great numbers among the stars by the help of telescopes, in every part of the sky, and could only be distinguished from a nebula by observing that it shifted its place among the stars, and passed from constellation to constellation. In 1806, a little comet was observed, which passed the perihelion of its orbit, that is the point nearest to the sun, on the 2nd of January in that year; it was not, however, at that time recognised to be the same comet which had been seen in 1772, nor was this identity perceived until the comet was re-discovered in 1826. On the evening of the 27th of February in that year, it was perceived by Biela, whose name it now bears, at Josephstadt, in Bohemia, as a small round nebulosity, which on the following evening had advanced about a degree towards the east, and had a little increased in size and brightness. Biela continued to observe it for some time. Gambart, at Marseilles, discovered it independently, on the 9th of March following, and from his observations, he concluded that it passed its perihelion on the 18th of the same month. The comet was observed soon afterwards, at Göttingen, by Harding, and at Altona, by Clausen; it disappeared about the beginning of the following May. The calculations of Gambart and of Clausen established the identity of the comet, thus discovered by Biela in 1826, with that of 1772, and with that of 1806, and shewed that this little body revolves about the sun in an ellipse or oval curve, the sun being out of the centre of the oval by about three quarters of the half length of the oval; that is, as it is technically expressed, the linear eccentricity being about three quarters of the mean distance. The time of revolution was found to be about six years and three quarters; so that, although through its faintness it happened not to be perceived, it must have passed its perihelion five times in the interval from 1772 to 1806, and twice in the interval from 1806 to 1826. It was also an inference from the time of revolution thus found, that the comet would pass its perihelion again towards the end of 1832; and a more accurate calculation since made by Baron Damoiseau, in which the attractions of the planets, especially of Jupiter, were allowed for, conducted to the expectation that this perihelion passage would take place on the 27th of November, 1832. Whether this prediction has been verified to the very letter, it is difficult as yet to say, for the faintness of the comet has scarcely suffered it to be seen; it has, however, been perceived in the excellent telescopes of Herschel, and in a part of the sky but little differing from that expected.

In our former number, we stated that the perihelion passage of the comet was predicted to take place on the 27th of November, 1832. It was also predicted that it would pass near the orbit of the earth. These predictions, which there is no reason to think inaccurate, excited however a very disproportionate interest in the minds of persons unacquainted with astronomy, who thought that nobody would take all this trouble about an almost invisible thing, and that the expected and talked-of comet must be some enormous and terrible visitant. Accordingly, a few months ago, the newspapers abounded with accounts of persons, who, without any incumbrance of telescopes, (and for ought we know in the day time) had seen this monstrous comet: there were even detailed descriptions of *comet hurricanes*, which regularly set in, every night at the very time of the comet's passing the meridian. In reality, however, the comet is so extremely faint an object that we have not received any authentic account of its being seen at all, except by Sir John Herschel. He, indeed, inheriting the immense reflectors, and not degenerating from the fame of his illustrious father, was able to detect this little wanderer in a place not much differing from that which theory had assigned: and the sublime delight was experienced, which attends the fulfilment of scientific prediction, the realization of scientific idea.

It is this periodical return and consequent fitness for frequent comparison with theory, that invests with so great an interest, in the minds of astronomers, a body which, from its smallness and faintness, would otherwise be utterly insignificant. Revolving about the sun in less than seven years, it seems to belong to our own system, to our own solar family. We can compare its motions with those of our own sister planets, and trace in the one, as in the other, the influence of the sun's attraction, and the fulfilment of the laws of Kepler: an influence and a fulfilment which can indeed be also traced in the orbits of all other comets, for example in that of the great comet of 1811, but only through very small portions of those enormous orbits, the rest being invisible by distance. Only a very few, out of the many comets that have been seen, are known as yet to revolve in moderate periods about the sun; and the comet of Biela was hailed with interest and delight, as an accession to this little band. Astronomers expect that near the end of 1835 another of these periodic comets will return to its perihelion, the celebrated comet of Halley, which was last seen about the time of the accession [1760] of George the Third, and which, at intervals of about three quarters of a century, had several times before attracted the notice of Europe.

The periodicity of Biela's comet has been assigned as a reason for its interesting astronomers *notwithstanding* its smallness. But there is a view, in which its *very smallness* gives to it an interest that it would not otherwise possess. To explain this other source of interest, we must be allowed to make some remarks on another little periodic comet, the comet of Encke's, which passed its perihelion last May, though from its faintness and southern position it was not seen this year in Europe, and was only detected at the observatory of the Cape of Good Hope, by Mr. Henderson, who has the charge of that establishment. This comet, also, though never easily visible, had been long watched by astronomers with interest from the rapidity of its revolution about the sun. But this interest has of late been greatly increased, by the detection, in its motion, of a little irregularity, which has been successfully accounted for on the supposition of a resisting ether diffused through the planetary spaces, while it does not seem to admit of any other explanation. No effect, indeed, of such resistance, has been yet detected in the motions of the planets: but this objection to an ether is not formidable, much less decisive. For there is abundant evidence that the planets are bodies far more dense and massy than any ordinary comets, and especially than those little comets now in question; and we know that a feather is greatly resisted by an atmosphere through which a stone makes its way without any sensible hindrance. And as, in order to observe the effects of the resistance of our atmosphere, we do not use the densest but the rarest moveables, a feather not a stone, so astronomers are glad when they find themselves furnished in the heavens with a new celestial feather, if we may call it so, wherewith to prove the existence and watch the effects of that fine ether, through which the old cannon balls of planets held on so free a way. Thus then the very smallness of Biela's (as of Encke's) comet, is favourable to the inquiries of astronomers on an important question. But if this smallness were not combined with periodical return, the comet would disappear for ever, before it could be long enough observed to give any decisive testimony on this question of a resisting ether, and then it would be only one of a crowd, to which every few years are adding, and which have little other interest than the possibility that a remote posterity may one day detect, in the celestial phenomena of their age, some unexpected connexions with our then ancient records.

We have not spoken of the near approach of the orbit of Biela's comet to the orbit of the earth, although this has often been popularly put forward as a ground of alarming interest. There is indeed a very near approach of the two oval orbits to each other at their nearest points, an approach within about twenty thousand miles, according to the elements of Baron Damoiseau; so that the comet, moving in its own oval path, came, about the end of last October, according to calculations founded on those elements, to be only twenty thousand miles from the nearest point of the oval path of the earth: and, therefore, if the earth had happened to be at the same time in that nearest point, the comet would have been much nearer than the moon, and would only have been removed from the earth's surface by about two diameters of the earth. But the earth did not come up to that nearest point till about the end of November; and this difference of a month in time had so great an influence on the mutual distance of the two bodies, that the comet in its late approach came no nearer than within about fifty millions of miles, and is now receding from us. It is little likely that the earth and comet will ever happen to arrive together at the nearest points of their respective paths; still less that the changes which the attractions of the planets are perpetually making will ever so alter the comet's oval as to make the two paths exactly intersect, or approach so near as to render a collision possible: but even a collision with a body so light and cloudlike as the comet, would not be attended with those disastrous effects which some have amused themselves with imagining, and we need not be much concerned on *this* head for the fate of our great great great great grand children.

H.

XXXIII.

REVIEW OF ARAGO'S WORK, "*THE COMET*", TRANSLATED BY COLONEL CHARLES GOLD, LONDON, 1833

[*Dublin University Review* **1** 365–372 (1833).]

This entertaining little Treatise of M. Arago* will, we think, be welcomed in its English form by an extensive class of readers, who might not have happened to meet with it in the pages of the French Annuaire. The original appears to have been drawn up officially, at the request of the French Government,† to prevent the danger of a return of one of those *comet-panics*, which so lately as the time of Lalande produced the most serious effects in Paris. Lalande, we believe, had published a calculation of the injuries which *might* be occasioned to the earth, *if* a Comet sufficiently large were to come sufficiently near; and Lalande being known to the Parisian public as one who used to give notice of eclipses, and compile astronomical almanacs, a report went abroad that he had given notice of an *actually coming Comet*, which was *in fact to do* what he had only spoken of as possible. In vain did the Astronomer protest against this interpretation, and refer to his printed reasonings; the alarmists answered, that in order to pacify people he had suppressed the original work, and substituted another in its place. In short, there was a comet-fever in Paris, and many died in consequence. On the recent occasion of the approach of Biela's Comet, to which chiefly the work before us relates, fears as unreasonable, though less intense, prevailed. It had been predicted that the Comet would *cross the orbit of the Earth;* and in the minds of many persons, unfamiliar with astronomical language, and uninformed in astronomical theory, the announcement excited a fear of approaching collision, which was not always removed by the assurance of astronomers, that at the predicted time of crossing, the Earth would be a month in arrear. "Who knows," said the alarmists, 'whether we can depend on these Astronomers for this month of odds? They own that they do not know every thing about the nature of Comets, and that their methods of cometary calculation are less perfect than their planetary rules; perhaps the truth is, that they know nothing about the matter, and only use fine words to disguise their real ignorance." Such thoughts, perhaps, were at least silently entertained by many persons; and it was judged useful to lay before the public, in the instructive and amusing Treatise upon Comets which we have now the pleasure of noticing, a clear and frank avowal of the remaining degree of astronomical ignorance on the one hand, and a plain and popular sketch of the extent of astronomical knowledge on the other.

Respecting the chemical constitution of Comets, our ignorance may be said to be complete; but respecting the manner of their motion, the knowledge of Astronomers is almost perfect. When we think of a transparent mass, exceeding the earth in size, and,

* [Dominique François Jean Arago, 1786–1853.] † Translator's Preface

perhaps, the air in rarity, undergoing rapid and enormous changes in shape and magnitude, but such that it seems to contract with heat and dilate with cold, and when we compare it with any of the known terrestrial or planetary bodies, we are compelled to account it a most striking and singular novelty. But when we find this novel object pursuing tranquilly its regular ellipse, and harmonising with the ancient planets by the long known laws of Kepler; obedient to the same sun as centre of attraction, and subject to the same mutual influences, according to the same theory of Newton; disturbed only in a greater, though still in a slight degree, by the resistance of a celestial ether, which opposed no sensible obstruction to its elder and stronger brethren: we recover the serenity of contemplation, and merge the wonder of novelty in the admiration of harmony and law.

In the Treatise before us, M. Arago having aimed to correct some prejudices, and to circulate some accurate but simple notions respecting the present state of cometary science, has judiciously begun by giving some popular and preliminary illustrations of the elliptic and parabolic orbits in which the Planets and the Comets move. This part of the work is made still more clear and popular by an interesting diagram, which the Translator has procured from M. Pond. The diagram contains an outline of the nearly circular orbits of the Planets, as far as Jupiter, and of the much more oval orbit of that Comet of last year, which forms the chief subject of the Treatise, and is usually called Biela's Comet, from the name of the Bohemian Astronomer who first perceived it in its last foregoing approach to the Sun, at which time the shortness of its period (not quite seven years) was discovered; although the Comet itself had been seen in 1772 by Montaigne and Messier, and afterwards at an intermediate passage in 1806. The successive positions of this Comet for some months, before and after its perihelion passage of last November, and for the first days of several years to come, are marked in the diagram, in a manner easily understood, and likely to interest the public.

After the preliminary illustrations already mentioned, of cometary orbits in general, the author proceeds to lay before his readers a sketch of the course and history of some of the most remarkable periodical Comets yet known, namely those which are usually called after the names of Halley, Lexell, Encke, and Biela. Of these the Comet of Halley is remarkable, as that of which first, in the history of science, a prediction of the return was made and realised; thus manifesting, even to the unlearned, the power of theory, and existence of order, in a region which had before seemed strewed only with vague alarms and meteoric symbols of uncertainty: a tranquil ministry, imposed upon that very wanderer, which but three centuries before had conspired with the Turks to alarm invaded Christendom, and "from its horrid hair" seemed to "shake pestilence and war." The Comet of Lexell is noticed as having perplexed *astronomers* at least, (though not *monarchs*,) by its singular changes of course, until it was found to have had its orbit, by one approach to Jupiter, inflected towards the earth, and brought within reach of our vision—but, at its next approach to the same great disturbing planet, deflected anew to regions beyond our ken. Next comes in review the faint cloud-like Comet of Encke, scarce visible except in powerful telescopes and favouring skies—but interesting to astronomers, as having the shortest period of all, and as that which first has given proof, within our own times, of the resistance of the celestial ether: And last the Comet of Biela, already mentioned as having lately crossed the orbit of the earth, and as having, by the expectation of that event, excited some ignorant alarm.

It fell within the plan of the treatise to state the times when these different Comets might be expected, according to astronomical calculation, to return next to their respective perihelia, and become visible to us again. Of the four great Comets just mentioned, that of Lexell indeed is not likely to be ever seen again, since its last great deflection from the disturbance of Jupiter; but when the work was written the three other Comets were approaching to the sun and to the earth, though two have since passed perihelion. That of Halley, last seen about the time of its last perihelion passage, in 1759, is announced, on the authority of Baron Damoiseau, as expected to return to perihelion on the 16th of November, 1835. We may add, that from the extreme intricacy of the calculations respecting a body which moves in so long and so eccentric an orbit, withdrawing from the sun in one part of its course, to a distance almost double that of the Herschel, and approaching in another part, within a distance less than that of Venus, astronomers differ a little in their predictions of the time of its next return, though only by a few days out of more than seventy-six years. Baron Damoiseau himself, by a later calculation, in which he allowed for the attraction of the earth, was led to alter his former result from the 16th to the 4th of November; and M. Pontecoulant, in a recent and elaborate memoir, which has received the prize of the French Academy of Sciences, as the memoir of Damoiseau did the prize of the Academy of Turin, has assigned the 7th of the same month (November, 1835) as the time of the next passage. The Comet will, no doubt, be seen before that passage, but it seems to have been gradually wasting away, at least diminishing in brightness, from period to period—so that we need not hope to see, at its next return, appearances so terrible or splendid as those which history records of it. Astronomers, however, will still regard it with deep interest, as a fresh trial of their theories on a large scale, and as a new witness (probably) respecting the resistance of the ether, and perhaps the disturbance of some planet too distant to be visible to us. The time, no doubt, will come, when the few *days* of present uncertainty, in the predicted epochs of its secular returns, will, by the progress of science, be reduced to a few *hours*, and these again to narrower limits.

The perihelion passage of the Comet of Encke was predicted for the 4th of May, 1832; but it was added, that the position of this faint Comet would be unfavourable for observations in the northern hemisphere. Accordingly, in the treatise before us, we find it remarked by M. Arago, as it had been by Professor Encke, that the chief hope of seeing this Comet must rest with the southern observatories of Paramatta and the Cape of Good Hope. These expectations have since been realized: the Comet, though not (we believe) perceived by any one in Europe, was seen by Mr. Henderson, at the Cape, and by M. Mossotti, at Buenos Ayres, and its position in the heavens agreed remarkably with the predictions of Encke, founded on the combined theories of attraction and etherial resistance.

Having spoken of Encke's result respecting this etherial resistance, as a recent and remarkable discovery, we must add, that the *possibility* of an ether, diffused through the planetary spaces, had been conjectured by Newton; who, in the celebrated *Queries*, while he attacks the hypothesis of a *dense* celestial fluid, as incompatible with the observed freeness and continuance of the planetary motions, yet thinks it probable that an etherial medium exists, far rarer and more elastic than our air. Thus, in the 22d query, he asks, "May not Planets and Comets perform their motions more freely and with less resistance, in this etherial medium, than in any fluid which fills space adequately, without leaving any pores? And may not its resistance be so small as to be inconsiderable? For instance, if this ether (for so I will call it) should be supposed seven hundred thousand times more elastic than our air, and above seven hundred thousand times more rare; its resistance would be above six hundred million times less than that of water. And so small a resistance would scarce make any sensible alterations in the motions of the planets in ten thousand years." Again, in the 28th query—"And therefore to make way for the regular and lasting motions of the Planets and Comets, it is necessary to empty the heavens of all matter, except perhaps some very thin vapours, steams, or effluvia, arising from the atmospheres of the Earth, Planets, and Comets, and from such an exceedingly rare ethereal medium as we described above." But, notwithstanding these early conjectures of one endowed with a sagacity so uncommon, the existence of an ether, or in other words, the non-existence of a vacuum between planet and planet, remained long doubtful and doubted, or rather disbelieved; and the establishment of the fact on a mathematical discussion of astronomical phenomena, was reserved for our own days and for Professor Encke.

The Comet of Encke has another title to attention, as appearing to have established another very remarkable fact in celestial physics—which seemed indeed to result from an old observation of Hevelius, but was little noticed or believed till lately: the fact which we already mentioned of a Comet's contracting with heat and dilating with cold, at least in its visible dimensions. The treatise of M. Arago presents us with a little table of the observed magnitudes of Encke's Comet, when approaching to the sun in 1828: from which it appears, that on the 24th of December, in that year, the Comet was nearly three times nearer to the sun than on the 28th of October; and yet that its volume, or bulk, when we may thus presume it to have been exposed to a *nearly ninefold increase of heat,* was *diminished to about the sixteen thousandth** *part* of what it originally was.

M. Valz, an accomplished Astronomer, of Nismes, has endeavoured to account for this extraordinary circumstance, by supposing that the ether, which in Encke's theory is considered to be denser near the Sun, exerts in that neighbourhood a more powerful compression on the Comet; as a bladder, which seems full of air when on the top of a mountain, contracts and falls in when brought down to a denser atmosphere. But against this explanation it has been urged, that a compression of this kind requires an impenetrable envelope; the bladder, we know, is *air-tight,* but where is the *skin* of the Comet? Sir John Herschel, if we are rightly informed, has offered lately two other explanations of the phenomenon. He thinks that the Comet of Encke, being little else than a fog or light cloud of vapour, may have its particles so slightly connected, that, without any sensible cohesion among themselves, they describe each its own parabola, or rather ellipse, about the Sun; the whole family of these elliptic paths separating never far from each other, but more so as they recede from the sun—a supposition which does not seem to contradict any known law or principle, and which, if admitted, would account for a *real decrease of dimension of the groupe of cometary particles,* when approaching independently to the Sun. But, secondly, it is suggested by Herschel, that this decrease may perhaps *not be real,* but *only an optical illusion,* arising from increased transparency. In fact, if the Comet were *perfectly transparent,* we could *not see it at all,* though we might, perhaps, conclude by calculation, that it was interposed between us and a star, from some observed refraction. If, then, the increase of heat should quite remove the slight defect of transparency which before existed, in a large part of the cometary vapour, it is plain that a real increase of volume might be accompanied with an apparent diminution; and that thus the observed optical fact respecting this singular body might be reconciled with terrestrial analogies.

* Not sixteen *millionth,* as in the Translation. A few other strange mistakes occur in the English version, though it seems to be generally faithful, and is written in a spirited and pleasant style.

As we have mentioned the observed return of this Comet of Encke, in the summer of last year, our readers may be glad to know that Biela's Comet also—the last of those spoken of above—which was expected to pass its perihelion on the 27th of November (1832), has been seen, though only with powerful telescopes, and has been found to follow nearly the path and time which calculation had assigned. It was perceived first by Sir John Herschel, on the 23d of last September, and afterwards (though with great difficulty, from its faintness,) by other Astronomers. Professor Nicolai, of Manheim, has concluded from his observations, that it passed its perihelion on the 26th of November, that is, about a day sooner than the time predicted by Baron Damoiseau. Even this small difference between theory and observation—this anticipation of a day in an orbit of nearly seven years—will (there can be little doubt) disappear hereafter, when instruments and methods shall be improved, and especially when allowance shall be made, in the case of this as of Encke's Comet, for the accelerating effect of the ether already mentioned. For, strange as it must appear on the first view, that a Comet should come *sooner* to its perihelion because it is *resisted*, yet Astronomers are unanimous that such must be the effect of an ether. The difficulty of understanding how *resistance* can ever produce *acceleration*, arises from not attending to another *antagonist power* (in this case the attraction of the Sun,) which is, as it were, provoked into stronger action. The *immediate* and *direct* effect of the resisting ether is indeed to *retard*; and *if* the Comet were confined to one particular path, like a bead strung upon a wire, there would then be nothing to compensate for the retardation, and the arrival at perihelion would be delayed; *but*, in reality, the Comet being free to obey the solar attraction, which has not now so great a centrifugal force to overcome, the *orbit is diminished* in its dimensions, and the *perihelion is changed in place*; and, at the same time, the *increase of attraction* arising from diminished distance produces an *indirect increase* in the velocity, which is more than enough to make up for the former direct diminution. The subject may be illustrated by the motions of pendulum clocks, which are found to go faster and gain, when they are about to stop for want of oil; the direct retarding effect of the clogging of the wheels being more than compensated by the indirect acceleration produced by the shortening of the swing.

Besides the Comets of Encke and Biela, another faint telescopic Comet was seen in 1832. It was discovered on the 19th of July, by Gambart, at Marseilles, and passed its perihelion in September. Its orbit appears to be parabolic, or nearly so, and, consequently, it can never return, or not till after many ages. Encke's Comet will return in 1835, and Biela's in 1839.

We regret that our limits will not permit us to extract from M. Arago's work, any of his humerous pleadings *in favour of Comets*, that is for their utter innocence and insignificance in the great physical changes which the earth has undergone, and even in the more modern and less important changes of weather and of temperature. Under this latter head, of atmospheric changes, which by some have been attributed to Comets, the treatise contains many curious facts respecting the great European fogs of 1783 and 1831; and also respecting the *harmattan*, a periodical wind which blows from Africa to the Atlantic, remarkable for its extreme dryness, and for disinfecting qualities; which, as the Translator ingeniously remarks, might, perhaps, be practically applied to the prevention or cure of many contagious diseases, if the *harmattan* were chemically analysed. But we think that enough has been said to convince our readers of the entertainment and information to be found in this little volume, and that they will join with us in thanking Colonel Gold for introducing the work to the notice of the British public.

XXXIV.

ON THE DAY OF THE VERNAL EQUINOX AT THE TIME OF THE COUNCIL OF NICE (1842)

[Communicated May 9, 1842]:
[*Proceedings of the Royal Irish Academy* **2**, 249–250 (1844).]

The President made some remarks on the day of the Vernal Equinox at the time of the Council of Nice [Nicaea].

It has been stated by some eminent writers on astronomy, for example by Brinkley and Biot, and seems to be generally supposed, that the vernal equinox in the year 325, A.D. fell on the 21st of March. But Sir W. Hamilton finds that Vince's Solar Tables (or Delambre's, from which those are formed) conduct to about $2\frac{1}{4}$ hours before the Greenwich mean noon of the 20th of March, as the true date of the equinox in that year; which thus appears to have been assigned to a wrong day, by some erroneous computation or report, perhaps as long ago as the time of the phenomenon in question.

As this result is curious, Sir W. Hamilton conceives that it may not be uninteresting to confirm it by a very simple process of calculation, derived from the Gregorian Calendar. According to that calendar, 400 years contain 146097 days, being a number less by 3 than that of the days in four Julian centuries; and if the farther refinement be adopted, which some have suggested, of suppressing the intercalary day in each of the years, 4000, 8000, &c., then, in the calendar thus improved, 4000 civil years will contain 1460969 solar days. Assuming then, as a sufficiently near approximation, that such is the real length of 4000 tropical years; multiplying by 3, and dividing by 8, we find that 1500 tropical years are equivalent to 547863 days and a fraction; which fraction of a day, according to this simple arithmetic, would be equivalent to 9 hours. But 1500 Julian years contain 547875 days, that is, 12 more than the number last determined; and these 12 days are precisely the difference of new and old styles in the present century. If, then, we neglect the fraction, the new-style date of an equinox in any year of the nineteenth century ought to be the same with the old-style date of the same equinox in the corresponding year of the fourth century; and in particular the vernal equinox of 325 ought to have fallen on the 20th of March, because that of 1825 fell on the day so named: while the fraction of a day above referred to, though not entirely to be relied on, renders this result a little more exact, by throwing back the equinox from the evening to a time more near to noon.

XXXV.

ON APPROXIMATING TO THE CALCULATION OF ECLIPSES (1844)

[Communicated May 27, 1844]
[*Proceedings of the Royal Irish Academy* **2**, 597 (1844).]

The President communicated a method of mentally approximating to the calculation of ancient eclipses, and applied it to the eclipse of the moon recorded by Tacitus as having happened soon after the death of Augustus.

XXXVI.

ON THE NEW PLANET, *METIS* (1848)

[Communicated May 22, 1848]
[*Proceedings of the Royal Irish Academy* **4**, 169 (1850).]

Sir William R. Hamilton handed in the following diagram, representing (rudely) the manner in which the planet* Metis was seen on April 28, 1848, in an inverting telescope:

On April 30, 1848, the other seven stars, *a*, *b*, *c*, *d*, *e*, *f*, *g*, of this group, retained their respective positions; but the planet *Metis* had withdrawn from the position *p*, and had left the (circular) *field* indicated above, in the direction of the arrow.

The planet was thus seen at the Observatory of Trinity College, Dublin, in consequence of information from the discoverer, Mr. Graham†, principal assistant to E. J. Cooper, Esq‡.

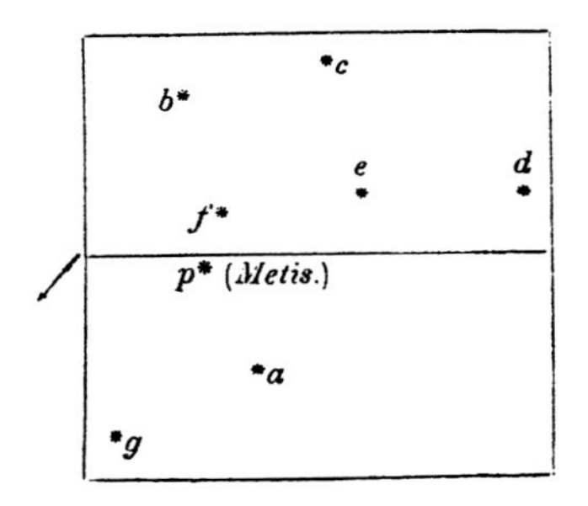

Fig. 1

* [In modern parlance: Asteroid 9 Metis.]
† [Andrew Graham, 1815– .]
‡ [Edward Joshua Cooper, 1798–1863.]

XXXVII.

REMARKS, CHIEFLY ASTRONOMICAL, ON WHAT IS KNOWN AS THE PROBLEM OF HIPPARCHUS (1855)

[*Hermathena* (*Dublin*) **4**, 480–506 (1883).]

To the Editor of 'HERMATHENA.'

MY DEAR SIR,

Among the papers of Sir W. Rowan Hamilton which are in my hands, I find two bearing titles connecting them with *The Problem of Hipparchus.* They are incomplete; but I send them for your perusal, thinking that they may be deemed by you suited to publication in the pages of 'HERMATHENA'.

In sequence with them is a manuscript book containing an extended discussion of a problem of Snellius, 'having much affinity to' that of Hipparchus, but which, in contradistinction to the astronomical character of the latter, is called by Sir William Hamilton a *geodetical* problem; the distances from the observer to the things observed being what was sought, instead of the central point of the excentric or epicycle.

Sir William Hamilton took much delight in studying the *Almagest* of Ptolemy, and expressed admiration of his mathematical powers, and of the justice done by him to the astronomical discoveries of his predecessor Hipparchus. The papers I send may be regarded as evidencing the Author's respect for these ancient men of science, and also possess an interest as exhibiting Hamilton at work upon matter strictly astronomical, instead of on the higher mathematics, which more generally occupied his attention. I may remark, however, that he considered himself, by his *Essay on Dynamics*, as well as by other scientific memoirs, to have earned a place in the history of Physical Astronomy, and not to have been, as by some he has been thought to be, an originator only in the region of pure mathematics.

I am able to report that Dr. Ball, our present Royal Astronomer, is of opinion that value, both intrinsic and personal, attaches to these papers of Hamilton; and I may add my belief, that an examination by competent persons of the scientific manuscripts of Hamilton in my hands and in the Library of Trinity College, would not improbably be rewarded by the discovery of other unpublished work that ought to be rescued from oblivion. I have heard with great satisfaction that such an examination is contemplated by the authorities of the University: it would give me pleasure to contribute towards it every facility in my power.

I remain,
Very faithfully yours,
R. P. GRAVES.
1, WINTON-ROAD, DUBLIN,
7th July, 1883.

Remarks, Chiefly Astronomical on What is Known as the Problem of Hipparchus

§ *I.—On the Mean Motions of the Sun and Moon, as Determined by Hipparchus, and Recorded in the Almagest of Ptolemy*

1. Hipparchus* estimated the length of the tropical year as falling short, by one day in three hundred years, of the old approximate amount, best known to us as the Julian, of 365 days and a-quarter. He therefore conceived that 300 revolutions of the sun in longitude occupied

$$300 \times 365 + 75 - 1 = 109574 \text{ days;}$$

or that 150 tropical years were = 54787 days; whence the mean daily motion of the sun in longitude is

$$\frac{150 \times 360^\circ}{54,787} = 0^\circ\, 59'\, 8''\, 17'''\, 13^{\text{iv}}\, 12^{\text{v}}\, 31^{\text{vi}},$$

as, in fact, he found† it to be, though probably by a very different arithmetic.

2. Astronomers before Hipparchus—ÿÕ^ ƒÕ ÞÃŠÃÕĮρÿÕ, as described by Ptolemy‡—namely, the Chaldeans, as is commonly supposed, or the Greeks, perhaps Meton, as is conjectured by Delambre,§ who, in the absence of evidence in their favour, is not disposed to concede even so much of accurate and mathematical knowledge to the Chaldeans as would be involved in their understanding what goes to the name of their own *Saros*,‖ had conceived that in a certain *Period* of a little more than 18 years, but more exactly in one of $6585\frac{1}{3}$ days,¶ the following lunar revolutions were accomplished. The moon had gained (it was supposed) 223 circumferences on the sun, or had performed 223 complete lunations, by having described 241 circumferences of the ecliptic, with a surplus of about 10° 40′, as compared with the *fixed*** *stars* (ÞρNŒ ÿOŒ ÃÞŠÃÚĨĩŒ ÃœïþÃŒ), while the sun had described only 18 circumferences, with the same sidereal surplus. In this interval it was thought that the moon had gone 239 times through her known†† alternations of slow and quick motion in longitude, or had described 239 revolutions with respect to its (variable) place of slower motion, as referred to the ecliptic; or had accomplished 239 *restitutions of anomaly*.

It was believed also by some astronomers *before* Hipparchus, that in the same *period*, called the Chaldean, and which, in fact, the Chaldean observers are likely enough to have *recognized* at Babylon, as approximately bringing back *eclipses* of the moon (whether they knew, or even *suspected*, anything of all these *mathematical* conceptions of the lunar motions, more probably due to the Greeks), that the moon accomplished 242 *restitutions of latitude*,‡‡ or (as we should

* See Halma's edition of the Almagest, Paris, tome i. page 164, to which work other references shall be made, sometimes by tome and page, at other times by book and chapter, or by both.

† Almagest, tome i. p. 166, Book iii. chap. ii. (šÿÕρῶÛ ō ÚÒ′ Ó″ ÕŒ‴ ÕÐ⁗ Õã′′′′′ ŠÃ′′′′′′ fÐÐÕÊÃ).

‡ Alm., tom. i. p. 215, book iv., chap. ii.

§ Hist. of Anct. Astry., by Delambre, tome ii., Paris, 1817, p. 144. This volume will for the present be referred to simply as 'Delambre.'

‖ Brinkley's Astronomy, &c. ¶ Ptolemy, vol. i. p. 216. ** *Ibid.*

†† Delambre doubts whether the Chaldeans had ever noticed these variations in the longitudinal motion of the moon at all; and indeed the discovery of them, but still more any theory to account for them, seems more likely to be due to the Greeks (Delambre ii. p. 143).

‡‡ Ptolemy, vol. i. p. 216.

now express it) performed 242 complete revolutions with respect to its ascending *node*; the node (as we should now say) having thus *regressed once* (242 − 241 = 1) in the course of the Chaldean Period.

3. It could not, of course, have been with any hope of improving the astronomical *accuracy*, but only of increasing the arithmetical *convenience* of such determinations as these, that some astronomer (perhaps Meton) before Hipparchus (one of the 'yet more ancients' already referred to, as mentioned by Ptolemy in the Almagest) proposed to *triple** *all* the numbers, for the sake of escaping from fractions. It had thus been collected, as a mere numerical inference from former results, that in a certain longer period, called 'éŸÍŠÕÐšŒ (= Evolution), consisting of 19756 days, the moon performed 669 lunations, or gained 669 circumferences on the sun; the moon describing 723 times the ecliptic, but the sun only 54 times, with a common *sidereal* surplus of 32°. The moon was also calculated to make, in the same tripled period, or '*Exeligm*,' of 19756 days, 717 restitutions of anomaly, and 726 restitutions of latitude. But all these estimates, though originally not quite useless, came to be set aside by Hipparchus, in his deeper study of the subject.

4. In their stead, Hipparchus introduced into the theory of the moon's mean motions *two other principal periods:* one relating chiefly to the 'restitution of the anomaly,' but involving also means for a more accurate determination of the mean returns of the moon to the sun, or of the length of a mean lunation; and the other referring to what was called (as above mentioned) the 'restitution of the latitude.' The *first*, or *anomalistic period*, of Hipparchus consisted, according to Ptolemy, of 126,007 days and *about* an hour.† Ptolemy does not even insert, in his mention of it, as regards the surplus *hour*, the usual fÐÐÕœ̧Ã, but I find myself obliged to believe that Hipparchus had designed this period to consist, *very exactly*, of 126,007 days, one hour, five minutes, and three and one-third seconds, such as those by which we now count time; and that Ptolemy suppressed the five minutes odd, as being of very slight astronomical importance in a period of about 345 tropical years; though I find myself obliged to *restore* this trifling surplus, in so long an interval, as what appears to me a *correction* of the *lost text* of Hipparchus, on this particular subject, as reported in the Almagest; because thus, and thus only, I can recover by calculation the *mean motions* there recorded, to the very *sixth* of the sexagesimal division. Be that as it may, from the comparison of ancient Babylonian eclipses of the moon with others observed, some centuries later, by himself, Hipparchus inferred that in his first period, of about 126,007 days, the moon accomplished 4573 restitutions of anomaly, and 4267 complete lunations; having described the ecliptic 4612 times, with a defect of about $7\frac{1}{2}^\circ$, as compared with the fixed stars, inclination being *here* neglected; while the sun had only moved through 345 circumferences, with the same sidereal deficiency. His *second*, or *latitudinal* period, after which the lunar eclipses were observed to return with nearly the same *magnitudes* as before, consisted of 5458 lunations,‡ wherein the moon was found by him to accomplish 5923 restitutions of latitude, or revolutions with regard to its regressing node.

5. If we divide the period of 126,007 days and an hour by the number, namely 4267, of lunations which it was found to contain, we shall obtain a quotient which exceeds

$$29^d\ 12^h\ 44^m\ 3^s,$$

* *Ibid.* † Ptolemy, vol. i. p. 216. ‡ Ptolemy, vol. i. p. 216.

by somewhat more than a quarter of a second of time. But whether it were that Hipparchus suspected that the surplus *hour* of his first period required to be a little increased, or merely that he wished to combine, with a sufficient accuracy of determination, a more manageable expression, he adopted (what with our division of the day is equivalent to) the *exact third part of a second*, as what was to be added to the approximate quotient above mentioned. His adopted value of the *mean lunation* was then, in our notation, exactly

$$29^{\mathrm{d}}\,12^{\mathrm{h}}\,44^{\mathrm{m}}\,3\tfrac{1}{3}^{\mathrm{s}};$$

or in his more *purely sexagesimal** *division of the day*,

$$29^{\mathrm{d}}+\frac{31^{\mathrm{d}}}{60}+\frac{50^{\mathrm{d}}}{60^2}+\frac{8^{\mathrm{d}}}{60^3}+\frac{20^{\mathrm{d}}}{60^4};$$

the *hour* of the period coming thus to be increased, as I have already said, by $5^{\mathrm{m}}\,3\tfrac{1}{3}^{\mathrm{s}}$, in above 345 years. In vulgar fractions, each expression becomes,

$$\text{Mean lunation} = 29^{\mathrm{d}}\frac{13754}{25920}=\frac{765434^{\mathrm{d}}}{12^3.15}.$$

6. Dividing the whole circumference by this number of days (and fractions of a day) in the mean lunation, I find:

Mean daily motion of the moon, in *elongation* from the sun,

$$=\frac{25920\times 360^\circ}{765433}=12^\circ\,11'\,26''\,41'''\,20^{\mathrm{iv}}\,17^{\mathrm{v}}\,59^{\mathrm{vi}}$$

$$=568771489079^{\mathrm{vi}}.$$

And adding to this quantity, the

$$59'\,8''\,17'''\,13^{\mathrm{iv}}\,12^{\mathrm{v}}\,31^{\mathrm{vi}},$$

which were found, in Art. 1, to be the mean daily motion of the sun, we obtain,

Mean daily motion of the moon in *longitude*,

$$=13^\circ\,10'\,34''\,58'''\,33^{\mathrm{iv}}\,30^{\mathrm{v}}\,30^{\mathrm{vi}}.$$

And such are *precisely* the results of Hipparchus, respecting the mean motions of the moon in elongation and in longitude, as recorded in the Almagest† of Ptolemy. (YŸÿšÍÚ ÃÞÿ¼ῆŒ šïœÿÚ Ό̂ šÍρßœÕÿÚ κÔÚÓšÃ, šÿÔþÃŒ Õβ̄ ÕÃ′ κŒ″ šÃ‴ κ⁗ ÕÌ′′′′′ ÚÒ′′′′′′ RŸÿšÍÚ Ό̂ šÍρßœÕÿÚ šïœÿÚ κÔÚÓšÃ šßκÿ̌Œ, šÿÔþÃŒ Õγ̄ Õ′ Šð″ ÚÓ‴ ŠÐ⁗ Š′′′′′ Š′′′′′′ fÐÐÕœ̧Ã).

7. In the anomalistic period of 126,007 days, Hipparchus had found (Art. 4) that there were 4573 restitutions of the moon's anomaly, but only 4267 returns to opposition with the sun: the mean motion in anomaly was, therefore, concluded to be more rapid than the mean motion in elongation, in the ratio of 4573 to 4267, or of 269 to 251. Increasing, therefore, the last-mentioned motion in this ratio, I find:

Mean daily motion of the moon in *anomaly*,

* Ptolemy, vol. i. p. 223. † Ptolemy, vol. i. p. 223.

$$= \frac{269}{251} \times 568,\ 771,\ 489,\ 079^{\text{vi}} = 609,\ 559,\ 882,\ 718^{\text{vi}}$$

$$= 13^\circ\, 3'\, 53''\, 56'''\, 29^{\text{iv}}\, 38^{\text{v}}\, 38^{\text{vi}};$$

and such, to the very *sixth*, is the result* of Hipparchus (RŸÿšÍÚ κÃM ĄÚªšÃŠÔÃŒ Ó̂ šÍþßœÕÿÚ šïœÿÚ κÔÚÓšÃ, šÿÔρÃŒ Õγ̄ Đ′ ÚĐ″ ÚŒ‴ κÒ⁗ ŠÓ′′′′′ ŠÓ′′′′′′). And by subtracting this mean motion in *anomaly* from the mean motion in *longitude*, namely, from $13^\circ\, 10'\, 34''\, 58'''\, 33^{\text{iv}}\, 30^{\text{v}}\, 30^{\text{vi}}$, we obtain:

Mean daily *progression* of the moon's *apogee*,

$$= 0^\circ\, 6'\, 41''\, 2'''\, 3^{\text{iv}}\, 51^{\text{v}}\, 52^{\text{vi}};$$

the word 'apogee' being here used unhypothetically, to express merely that variable and progressive point of the ecliptic where the moon was observed to move most slowly in longitude.

8. Finally, the latitudinal period of 5458 lunations, with 5923 restitutions of latitude, gives, in exact agreement with what Hipparchus determined it to be,†

Mean daily motion of the moon with respect to its ascending *node*,

$$= \frac{5923}{5458} \times 568,\ 771,\ 489,\ 079^{\text{vi}} = 617,\ 228,\ 569,\ 039^{\text{vi}}$$

$$= 13^\circ\, 13'\, 45''\, 39'''\, 40^{\text{iv}}\, 17^{\text{v}}\, 19^{\text{vi}}.$$

(RŸÿšÍÚ κÃM ÞŠĘÿŒ Ó̂ šÍþßœÕÿÚ šïœÿÚ κÔÚÓšÃ, šÿÔρÃŒ Õγ̄ ÕĐ′ šÍ″ ŠÒ‴ š⁗ ÕÌ′′′′′ ÕÒ′′′′′′).

Whence, by subtracting the mean motion of the moon in longitude, it may be inferred, as a consequence of the foregoing data, that

Mean daily *regression* of the moon's *node*,

$$= 0^\circ\, 3'\, 10''\, 41'''\, 6^{\text{iv}}\, 46^{\text{v}}\, 49^{\text{vi}}.$$

But of course all these results have merely an *arithmetical* accuracy, as being consistent among themselves, and cannot be relied on as astronomically correct, to anything like the extent to which they have been developed.

9. As regards the *node*, I may remark that Hipparchus's ratio of 5923 to 5458 is very nearly the same as that of 777 to 716, which many years ago occurred to me, from more modern data, as approximately expressing the rate of the moon's mean gain upon its ascending node, as compared with its mean gain upon the sun, and which I have often found useful in the mental or approximate calculation of the returns of eclipses of the moon. Let the arc $\frac{\pi}{358}$, which is little more than half a degree, be called, for conciseness, a *moon-breadth* (or sometimes simply a 'moon'); then, in *one* mean lunation, one satellite, on an average, overtakes a given (say the ascending node), and passes it by 61 'moon-breadths'; that is, by about a sign of the zodiac, rendering thus the return of a lunar eclipse impossible, after so short an interval. After *six* lunations, supposed, for simplicity, to commence with the moment of one central and total eclipse in the ascending node, the moon has gained $6 \times 61 = 366$

* Ptolemy, vol. i. p. 223. † *Ibid.*

moonbreadths on that node, or has passed the *opposite* (the descending node), by 366 − 358 = 8 such parts, = about 4 degrees, rendering thus the return of an eclipse *certain.* After 12 lunations, or one *lunar year,* the moon has passed the original node by 16 moon-breadths (of the kind above described), = about 8 degrees, and an eclipse must again take place. After 18 lunations the opposite node is passed by 24 such spans: after 24 lunations the original node is passed by 32 moon-breadths, and no eclipse can take place. After 48 lunations the excess on the first node amounts to 64 such parts; and therefore (subtracting 61), after 47 lunations, the excess is only 3 parts (moon-breadths), and a great eclipse is certain to recur. After 141 = 3 × 47 lunations the surplus on the original node amounts to 3 × 3 = 9 parts; and, therefore, subtracting 6 from the number of lunations, and 8 from the number of parts, and changing the node, the moon is found to be only 1 moon-breadth advanced beyond the opposite node, after 135 lunations; which interval is therefore a pretty good *period* of eclipses of the moon, so far as mere *nodations* are concerned. The (so-called) Chaldean Saros is a sort of *complement* of this little period; for in 358 lunations there are (according to the approximations here adopted) 777 *semi-nodations,* bringing thus the moon to the opposite node; subtracting, therefore, 135 from 358, and 1 from 0, and again reversing the node, we find that in 223 lunations the mean moon *falls short* of returning to its *original* node by about *one* moon-breadth. All this I have occasionally lectured on.

§ *II.—On Hipparchus's Hypothesis of the Excentric*

10. Such being the chief *mean motions* (ŷ šÃŠÃM κÕÚßœÍÕŒ) of the sun and moon, as determined by Hipparchus, we have next to consider the *hypotheses* by which he sought to account for, and reduce to calculation, the apparent inequalities (ĄÚªšÃŠÔÃÕ) of the observed motions of those two bodies. Plato is reported by Delambre* to have laid down the principle, that the object of mathematicians ought to be to represent all the celestial phenomena by uniform and circular motions. (... Ptolémée, pour suivre le principe de Platon, que l'objet des mathématiciens doit être de representer tous les phénomènes célestes par des mouvements circulaires et uniformes, ...). The *principle* is worthy of Plato, and I agree with Moebius† and with Laplace in considering that it is only in *appearance* obsolete; but I regret to be obliged to confess that I know not *where,* in Plato's works, the enunciation of it is to be found. In the Almagest it is thus laid down, as having been at least *adopted* by Ptolemy‡:—ÞρÒÍœÕÚ šKÚ κÃM œκÿÞNÚ Ó Ðÿ»šÍÒÃ ðÍĩÚ^ ÞÉρ¼ÍÕÚ̧ Ꞡ šÃÒÓšĄÕκꞠ ðÍĩŸÃÕ̧ J —ÃÕÚšÍÚÃ ÍÚ̧ Ꞡ ŷþÃÚꞠ ÞÉŲÃ, ðÕ \ ŷ šÃŠῶÚ κÃM ÍÐκκŠÔªÚ κÕÚßœÍªÚ ĄÞÿÍŠÿ»šÍÚÃ, ... The author of the Almagest had, however, the advantage of the example of his great master, Hipparchus, whom he is never weary of praising as 'a labour-loving and truth-loving man,' ÃÚÓρ —ÕŠÿÞÿÚÿŒ κÃM —ÕŠÃŠÓÒÓŒ; and whom we too must reverence (in this nineteenth century of Christ), as the true founder of modern astronomy: ancient, indeed, if 2000 years can make him such, but not less *modern,* in a deeper sense, than Thucydides. Between Hipparchus and all known predecessors of his in astronomy the difference is one of *kind* rather than of degree. Compared with *him,* the Chaldeans, for instance, remind one of those children at play on the woody banks of the Orinoko, who were

* Delambre, vol. ii. p. 113. † Die Mechanik des Himmels. ‡ Ptolemy, vol. i. p. 165.

found by Humboldt rubbing the dry, flat, and shining seeds of a creeping leguminous plant (he thought it might be the negretia) until they attracted fibres of cottonwool and chips of the bamboo, and thus *exemplified* 'electricity by friction,' without having even *begun* to *theorize* upon the subject.

11. Hipparchus selected the very simple and natural conception of which he seems to have been, in the strictest sense, the author, and which Ptolemy scarcely improved by a modification proposed by himself—that the sun and moon moved each in a certain '*excentric circle*' of its own, and that each described its own excentric *uniformly*. More precisely (see earlier articles of this little Paper), he conceived that the *sun's* excentric circle was *fixed* with respect to the equinoctial points; for he had failed, and so did Ptolemy, to detect any progression of the sun's apogee in longitude, which is no way to be wondered at, though the precession of the equinoxes was one of the many discoveries of Hipparchus; and he regarded the sun as describing its excentric at the uniform rate of 59′ 8″ 17‴ ... for each mean solar day. The *moon's* excentric was conceived by him to be described also uniformly, but at the greater mean daily rate of about 13° 3′ 54″ (Art. 7), from apogee to apogee; while the lunar apogee itself, or rather the projection of the apogee diameter of the moon's orbit on the plane of the ecliptic, had a mean daily progression of about 6′ 41″ in longitude. Hipparchus thus regarded the apparent *anomalies* of the observed motions of the sun and moon as phenomena purely *optical*; and doubtless it was right to *try this* mode of explanation before seeking for any more refined one. That it is insufficient *we*, with the help of telescopes, can very easily establish. For it would have given, in the theory of the sun, the equality which we *now* know *not* to exist:

$$\text{Spring plus autumn} = \text{summer plus winter;}$$

if by the word 'spring,' as denoting an *interval*, be understood the time elapsed between the vernal equinox and the summer solstice; and similarly in the other cases. Hipparchus, however, supposed the equality to hold good; for he estimated the spring quarter of the year as $= 94\frac{1}{2}$ days; the summer quarter as $92\frac{1}{2}$ days; and deduced by a *calculation* to be soon explained:

Autumn quarter = nearly $88\frac{8}{60}$ days (ÍÚ Ó̂ šïþÃÕŒ ÞÓ κÃM Ó′)

and

Winter quarter = nearly $90\frac{8}{60}$ days (ÍÚ Ó̂ šïρÃÕŒ Œ κÃM Ó′)

the favourite 'à peu près,' or mĐĐÕŒÃ, being added. Indeed Hipparchus can have only considered these results as rough *approximations* to the truth; for they would have given the length of the tropical year

$$= 94\tfrac{1}{2} + 92\tfrac{1}{2} + 88\tfrac{8}{60} + 90\tfrac{8}{60} \text{ days} = 365\tfrac{4}{15} \text{ days;}$$

whereas he was well aware (see Art. 1) that the surplus of the year was *less* than a *quarter* of a day.

12. To determine the *details* of the sun's apparently *anomalous* motion (ÞÍρM̨ η̃Œ̨ y̌ Ó̂ ŠÔy̌ —ÃÕÚy̌šïÚÓŒ ÃÚªšÃŠÔÃŒ), Hipparchus rested his whole weight on those *two* determinations, which he had made with all the care in his power, and which Ptolemy, after more than two centuries, found himself unable to improve, of the lengths (in days) of the *two* intervals called lately 'spring' and 'summer.' Very rudely determined, no doubt, those intervals were;

the observation of a solstice, even of the summer one, being still ruder in that age, and indeed essentially more difficult still, than that of an equinox. It would, therefore, be merely pedantic to attempt to improve, by a new calculation, on the numbers, $93^\circ 9'$, and $91^\circ 11'$, which Hipparchus estimated as expressing the arcs of *mean anomaly* (or of mean longitude) described by the sun during the spring and summer quarters of the year. The *first case*, therefore, of the

'Problem of Hipparchus,'

as actually and *historically* proposed and resolved by that great and venerable astronomer, consists in finding the *position of the excentric point* (the earth's centre) *at which two given consecutive arcs,* $93^\circ 9'$, and $91^\circ 11'$, *of a given circle* (the sun's excentric) *shall subtend angles of* 90° *each,* with the same apparent directions of motion.

13. In a question of so great historical interest, I may be permitted to transcribe the diagram, which is preserved in the Almagest of Ptolemy.

In this diagram Figure 1 A, B, Γ, Δ represent the four points of the ecliptic (supposed to be *homocentric* with the earth), to which the observer at the point E refers the four places of the sun, at the moments of spring, summer, autumn, and winter; their longitudes being 0°, 90°, 180°, and 270°. The centre of the sun's excentric circle is supposed to be Z; the apogee, as referred to the ecliptic, is H, and the perigee, so referred, is I; the true positions of the sun in its excentric, at the moments of spring, summer, and autumn, are conceived to be Φ, K, Λ; the diameters NTZΞO, and ΠΦZΣ, of the excentric are drawn parallel to the equinoctial and solstitial diameters of the eclipse, namely, AΓ and BΔ; and to these are drawn parallel the chords of the excentric, XΦK and ϒTΘ.

14. Hipparchus (as it has been said) adopted the values $93^\circ 9'$ and $91^\circ 11''$ for the arcs, ΘK and KΛ, of the excentric circle of the sun, as computed from the observed times elapsed, namely, as the arcs described between the observed spring equinox, summer solstice, and autumn equinox, in virtue of the known (or assumed) rate of mean daily motion of the sun, which is a consequence of the assumed length of the tropical year. He had thus—

$$\text{(arc) } \Theta\Upsilon = \Theta N + O\Lambda = \Theta K + K\Lambda - NO = 93^\circ 9' + 91^\circ 11' - 180 = 4^\circ 20';$$

$$\text{(arc) } XK = \Theta K - \Theta X = \Theta K - K\Lambda = 93^\circ 9' - 91^\circ 11' = \mathrm{I}^\circ 58'.$$

Hence, if the radius,

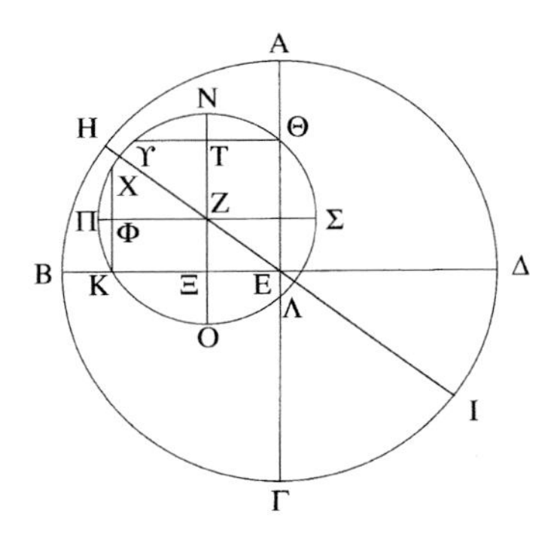

Fig. 1

$$Z\Theta = ZN = ZX = Z\Xi = Z\Pi = ZK = ZO = Z\Lambda = Z\Sigma,$$

of the sun's excentric be taken for unity, we have nearly the numbers adopted by Hipparchus—

$$E\Xi = \Theta T = \tfrac{1}{2}\text{ chord } \Theta N\Upsilon = \tfrac{1}{2}\text{ chord } 4^\circ 20' = \frac{2}{60} + \frac{16}{60^2};$$

$$Z\Xi = \Phi K = \tfrac{1}{2}\text{ chord } X\Pi K = \tfrac{1}{2}\text{ chord } 1^\circ 58' = \frac{1}{60} + \frac{2}{60^2};$$

which last numbers might, however, have been more accurately determined, at least by Ptolemy, from his Table of Chords in the Almagest, to have been, respectively—

$$E\Xi = \sin 2^\circ 10' = \frac{2}{60} + \frac{16}{60^2} + \frac{6}{60^3};$$

$$Z\Xi = \sin 0^\circ 59' = \frac{1}{60} + \frac{1}{60^2} + \frac{47}{60^3}.$$

15. As neither Hipparchus, nor even Ptolemy, possessed any Table of *Tangents*, the right-angled triangle was resolved by finding its *hypotenuse* on the famous Pythagorean principle. It having been estimated that (with $Z\Theta = \&c. = 1$),

$$3600.E\Xi = 136, \qquad = 3600 \sin 2^\circ 10',$$

and

$$3600.Z\Xi = 62, \qquad = 3600 \sin 59',$$

it was inferred that

$$3600.EZ = \sqrt{136^2 + 62^2} = 149\tfrac{1}{2}, \text{ very nearly}, = \text{nearly } 150;$$

whence $EZ = \frac{1}{24}$, nearly. In other words, he judged that the distance between the centre of the earth (E) and the centre of the sun's excentric (Z) equalled very nearly the 24th part of the radius of that excentric circle: whence he inferred that the greatest equation of the centre, or the greatest difference between mean and apparent longitude, was $(\sin^{-1}\frac{1}{24}) = 2^\circ 23'$. *This* result would not have been materially improved by using more accurate expressions for the sides about the right angle, EΞ and ZΞ, in the triangle. But the slightly too large (assumed) value for the side EΞ, and the slightly too small (assumed) value for the other side ZΞ, led Hipparchus to an expression somewhat too small for the longitude of the apogee of the sun, as determined from his own data. He judged that longitude, ΞZH, to be only 65° 30′; because he found that

$$Z\Xi : EZ = 62 : 149\tfrac{1}{2} = \frac{24}{60} + \frac{53}{60^2} = \tfrac{1}{2}\left(\frac{49}{60} + \frac{46}{60^2}\right)$$

$$= \tfrac{1}{2}\text{ chord } 49^\circ 0' = \sin 24^\circ 30';$$

and therefore judged the apogee to *precede* the summer solstice by 24 degrees and a-half.

16. The more accurate values for the sides about the right angle, which may be taken out (as above) from the Table of Chords in the Almagest, would have given, however,

$$216000.\mathrm{E\Xi} = 216000 \sin 2^\circ 10' = 8166;$$

$$216000.\mathrm{E\Xi} = 216000 \sin 59' = 3707;$$

$$\therefore 216000.\mathrm{EZ} = \sqrt{8166^2 + 3707^2} = 8968, \text{ nearly;}$$

sine of greatest equation of centre

$$= \frac{8968}{216000} = \frac{2}{60} + \frac{29}{60^2} + \frac{28}{60^3};$$

chord of same angle

$$= \frac{4}{60} + \frac{58}{60^2} + \frac{56}{60^3} = \text{nearly } \frac{4}{60} + \frac{59}{60^2} = \text{chord } 4^\circ 46',$$

and greatest equation of centre = nearly 2° 23′, as before;

$$\text{but } 2(\mathrm{Z\Xi}: \mathrm{EZ}) = \frac{2 \times 3707}{8968} = \frac{49}{60} + \frac{36}{60^2} + \frac{11}{60^3} = \text{chord } 48^\circ 50',$$

nearly, by the Table in the Almagest; so that (with the supposed data) the apogee *preceded* the solstice of summer (that is, had *less* longitude) by only about 24° 25′, instead of 24° 30′; or the longitude of the apogee ought to have been inferred to be, not 65° 30′, but 65° 35′. Hipparchus, however, was probably well aware that it was idle, astronomically, to insist on an accuracy of *minutes*, in a result of calculation from observed data, which might easily permit a final error of *degrees*.

17. Our modern tables of logarithmic sines and tangents give easily the same *greatest equation*, 2° 23′, answering to the same approximate *excentricity*, $\frac{1}{24}$; and also the same (corrected) *longitude of the apogee*, 65° 35′, as above. Let $e = \sin \varepsilon$ = excentricity = distance EZ of centres, divided by the radius ZΘ of the excentric; let ω = longitude of apogee, = AEH = ΞZE = 90° − ZEΞ; and, finally, let $90^\circ + \iota$ = ΘK = mean motion in spring quarter, and $90^\circ + \iota'$ = KΛ = mean motion of the sun in the summer quarter of the year: these latter arcs being computed from the observed number of days elapsed, with the help of the known mean daily motion, derived from the length of the year. Then the equations for this *case* of the 'Problem of Hipparchus' are simply these two:—

$$e \sin \omega = \sin \frac{\iota + \iota'}{2}; \; e \cos \omega = \sin \frac{\iota - \iota'}{2};$$

giving, of course, $\tan \omega = \sin \frac{\iota + \iota'}{2} : \sin \frac{\iota - \iota'}{2}$.

Hipparchus assumed $\iota = 3^\circ 9'$; $\iota' = 1^\circ 11'$; whence

$$\frac{\iota + \iota'}{2} = 2^\circ 10'; \; \frac{\iota - \iota'}{2} = 59';$$

$$\tan \omega = \frac{\sin 2^\circ 10'}{\sin 59'} = \tan 65^\circ 35';$$

$$e = \sin \varepsilon = \frac{\sin 2^\circ 10'}{\sin 65^\circ 35'} = \frac{\sin 59'}{\cos 65^\circ 35'} = \frac{1}{24};$$

$$\varepsilon = 2^\circ 23': \text{ all as before.}$$

$$\begin{aligned} \log\sin\ 2^\circ 10' &= 8.57757 - 10; \\ \log\sin\ 59' &= 8.23456 - 10; \\ \hline \log\tan 65^\circ 35' &= 10.34301 - 10; \end{aligned}$$

taking out this angle to the nearest minute.

$$\begin{aligned} \log\sin 2^\circ 10' &= 8.57757 - 10; \\ \log\sin 65^\circ 35' &= 9.95931 - 10; \\ &\underline{8.61826 - 10} \\ (\text{Mean})\log\sin(\varepsilon = 2^\circ 23') &\ldots 8.61824 - 10. \\ \log\sin 59' &= 8.23456 - 10; \\ \log\cos 65^\circ 35' &= 9.61634 - 10; \\ &\underline{8.61822 - 10} \end{aligned}$$

$$(\text{Compt.})\, e^{-1} = 24.086 \ldots 1.38176.$$

§ *III.—On the Problem of Hipparchus*

18. The Problem, which was resolved two thousand years ago by Hipparchus, and which is at this day known by his name, may be thus stated:—

'Given, in degrees, &c., the angles which two successive arcs, AB, BC, of a given circle ABC, subtend at an excentric point D, as well as at the centre E of the circle, directions of rotation being included: to find the position of the excentric point'.

More fully, if we write [see Figure 2]

$$ADB = \theta,\ BDC = \theta',\ AEB = \iota + \theta,\ BEC = \iota' + \theta',$$

$$DBE = \kappa,\ EDB = \nu,\ ED : DB = e,$$

the four angles θ, θ', ι, ι' are given, with their respective algebraic signs; and the two angles κ, ν, and the ratio e, are sought. In the astronomical applications, if we retain Hipparchus's *own* hypothesis of the *excentric*, θ, θ' are the two observed or geocentric motions in longitude, in two successive intervals of time, diminished (for the moon) by the computed progressions of the apogee, in order to render the two extreme observations (at A and C) comparable with the middle observation (at B), by allowing for the supposed progressive motion of the excentric; ι, ι', are the computed mean motions in longitude, minus the observed motions in longitude; whence also $\iota + \theta$, $\iota' + \theta'$, are equal to the computed mean motions in anomaly, since these are the mean motions in longitude, minus the progressions of the apogee, in the intervals between the three observations (from first to second, and from second to third); thus the four angles θ, θ', and ι, ι', with their algebraic signs, are known, *without trigonometry*, from observation and arithmetic: and as regards the three sought quantities, κ, ν, e, the first, namely, κ, is the correction for excentricity (or equation of the centre), at the time of the middle observation, to be algebraically added to the observed longitude, or to the angle. $\lambda = YDB$, in order to obtain the *mean longitude*, $\kappa + \lambda = Y'EB$ (where Y' may be confounded with Y), the second sought angle, ν, is the apparent or *geocentric anomaly*, EDB or FDB, at the time of the same middle observation; so that

$$\lambda - \nu = \omega = YDF = Y'EF = \text{longitude of apogee at that time;}$$

also

$$\kappa + \nu = FEB = \textit{mean anomaly} \text{ of the body at the middle time;}$$

so that if m denote the mean daily motion in anomaly, and t the middle time, expressed in days, and counted from some fixed era, $t - \dfrac{\kappa+\nu}{m}$ is the date of the *apogean passage*; finally, e may be called the *numerical excentricity* of the orbit; and in Hipparchus's *own* hypothesis (though *not* in Ptolemy's *modification* of it, by *epicycles*) this excentricity was *less than unity* (the earth being *interior* to the excentric, although *exterior* to the epicycle), so that we may write (as in former articles) $e = \sin\varepsilon$, where ε = greatest equation of the centre.

19. It is evident that $EDA = \nu - \theta$, and $EDC = \nu + \theta'$; and it is not difficult to prove that $DAE = \kappa - \iota$, $DCE = \kappa + \iota'$; in fact, it is clear that ι, ι', as being the excesses of the mean over the observed motions in longitude, must be the (algebraical) increments of the correction (κ) for excentricity (applied as above), in the intervals between the three observations. Since then we had also $EDB = \nu$, $DBE = \kappa$, and $DE = e = \sin\varepsilon$, if $EA = EB = EC = 1$, the three equations of the Problem are as follows:—

$$\text{(I).}\quad \begin{cases} \sin(\kappa - \iota) = \sin\varepsilon\sin(\nu - \theta); & (1)\\ \sin\kappa = \sin\varepsilon\sin\nu; & (2)\\ \sin(\kappa + \iota') = \sin\varepsilon\sin(\nu + \theta'); & (3)\end{cases}$$

which are now to be resolved by trigonometry, so as to deduce the 3 angles κ, ν, ε, from the 4 angles θ, θ', ι, ι'. And to render the question still more definite, it is permitted to assume that ε is positive and acute, and that κ lies between the limits $\pm\varepsilon$.

20. In the *solar example* of Hipparchus, $\theta = \theta' = 90°$; also $\lambda = 90°$, $\omega = 90° - \nu$; $\sin(\iota - \kappa) = \sin(\iota' + \kappa) = \sin\varepsilon\cos\nu = \sin\varepsilon\sin\omega$; $\kappa = \dfrac{\iota - \iota'}{2}$; and the *general equations* (I.) take (as in Art. 17) the simplified forms,

$$\sin\varepsilon\sin\omega = \sin\frac{\iota + \iota'}{2},\quad \sin\varepsilon\cos\omega = \sin\frac{\iota - \iota'}{2},$$

giving $\omega = 65°35'$, $\varepsilon = 2°23'$, as before, if $\iota = 3°9'$, $\iota' = 1°11'$. But, of course, the *general solution* cannot be expected to be so simple.

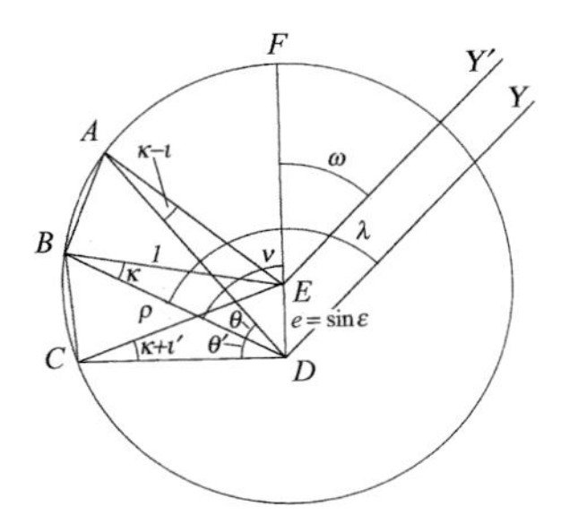

Fig. 2

21. In general, the identity,

$$\sin\theta' \sin(\nu - \theta) + \sin\theta \sin(\nu + \theta') = \sin(\theta + \theta') \sin\nu, \tag{4}$$

makes it easy to eliminate ν and ε between the three equations (I.), and gives,

$$\sin\theta' \sin(\kappa - \iota) + \sin\theta \sin(\kappa + \iota') = \sin(\theta + \theta') \sin\kappa, \ldots \tag{5}$$

whence

$$D.\cot\kappa = \sin\theta' \cos\iota + \sin\theta \cos\iota' - \sin(\theta + \theta'), \ldots \tag{6}$$

if

$$D = \sin\theta' \sin\iota - \sin\theta \sin\iota', \ldots \tag{7}$$

and this result, though not in the received and technical sense, *adapted to logarithms*, I have found not ill-suited to calculations with the usual logarithmic tables. We have also,

$$D.\tan(\nu + 90^\circ) = \cos\theta' \sin\iota + \cos\theta \sin\iota' - \sin(\iota + \iota'), \ldots \tag{8}$$

with the same denominator, D; but when κ has once been computed, by the foregoing or by a better method, to be presently explained, we may more conveniently deduce ν from it, by any one of the three following formulæ, which follow easily from the original system (I.):—

$$\text{(II.)} \quad \begin{cases} \tan\left(\nu - \dfrac{\theta}{2}\right) = \tan\dfrac{\theta}{2} \cot\dfrac{\iota}{2} \tan\left(\kappa - \dfrac{\iota}{2}\right), \ldots & (9) \\[2ex] \tan\left(\nu + \dfrac{\theta'}{2}\right) = \tan\dfrac{\theta'}{2} \cot\dfrac{\iota'}{2} \tan\left(\kappa + \dfrac{\iota'}{2}\right), \ldots & (10) \\[2ex] \tan\left(\nu + \dfrac{\theta' - \theta}{2}\right) = \tan\dfrac{\theta' + \theta}{2} \cot\dfrac{\iota' + \iota}{2} \tan\left(\kappa + \dfrac{\iota' - \iota}{2}\right), \ldots & (11) \end{cases}$$

and may serve as verifications of each other. With tables of *natural sines*, the angle κ might be easily computed from the formula, derived from the equations (6) and (7),

$$\cot\kappa = \frac{\sin(\theta' - \iota) + \sin(\theta' + \iota) + \sin(\theta + \iota') + \sin(\theta - \iota') - 2\sin(\theta + \theta')}{\cos(\theta' - \iota) - \cos(\theta' + \iota) + \cos(\theta + \iota') - \cos(\theta - \iota')} \tag{12}$$

And there would be an analogous expression for $\tan(\nu + 90^\circ)$. When κ and ν have been found, the original equations give three distinct expressions for $\sin\varepsilon$, which may be used as checks on each other. If we dispense with all such verifications, and have no natural sines at hand, the system (6 and 7) requires *six openings* of a table of logarithmic sines ($\sin\iota$ and $\cos\iota$ being taken out at one opening, and $\sin\iota'$, $\cos\iota'$ at another); it requires also 7 openings, of a table of logarithms of numbers, or (in all) 13 openings of tables, in order to compute the angle κ: after which 4 openings will give ν, by (9); and then 3 more openings will enable us to compute ε by (2). Instead of these 20 openings of logarithmic tables, Delambre has supplied a method which requires essentially only 17 different logarithms; but his method appears to me to be much embarrassed by *constructions*, which render it difficult to adapt the process to new varieties of the *figure*, without recommencing the reasonings. I shall, therefore, mention here a second method of my own, in which *only* 8 *logarithms* are required for the calculation of κ from the equation (5), and therefore *only* 15 *logarithms* in all; unless (as will always be prudent) we choose to employ formulæ of verification, which my method also furnishes. No reference whatever to a *figure* need be made, if once the algebraical *signs* of the given angles $\theta, \theta', \iota, \iota'$ have been determined, as already explained.

22. Writing the equation (5) under the form,

$$\sin\theta'\{\sin\kappa\cos\theta + \sin(\iota - \kappa)\} = \sin\theta\{\sin(\iota' + \kappa) - \sin\kappa\cos\theta'\}, \tag{13}$$

and adding, on both sides, $\cos\kappa\sin\theta\sin\theta'$, we find

$$\sin\theta'\{\sin(\theta + \kappa) + \sin(\iota - \kappa)\} = \sin\theta\{\sin(\iota' + \kappa) + \sin(\theta' - \kappa)\}, \tag{14}$$

that is,

$$\sin\theta'\sin\frac{\theta + \iota}{2}\cos\left(\kappa + \frac{\theta - \iota}{2}\right) = \sin\theta\sin\frac{\theta' + \iota'}{2}\cos\left(\kappa - \frac{\theta' - \iota'}{2}\right). \tag{15}$$

Hence

$$\text{(III.)}\quad\begin{cases} \tan\left(\kappa + \dfrac{\theta - \iota - \theta' + \iota'}{4}\right) = \tan(\xi - 45^\circ)\cot\dfrac{\theta - \iota - \theta' - \iota'}{4}, & (16) \\[2ex] \text{if } \tan\xi = \dfrac{\sin\theta'\sin\dfrac{\theta + \iota}{2}}{\sin\theta\sin\dfrac{\theta' + \iota'}{2}}; & (17) \end{cases}$$

and we see that *this* system (III.) requires only 8 different logarithms, for the calculation of κ. Any one of the 3 equations (II.) will then give ν, by 4 other logarithms; and all the 3 equations of that system may be used as checks on each other. And, finally, any one of the 3 original equations (I.) will give ε, by 3 logarithms; only 15 *logarithms* being thus essentially required, in this Second Method of mine. If, however, for any reason, we wish to calculate, though not really required for the astronomical purpose of Hipparchus, the *distance*, $\rho = DB$, *from the earth to the body observed*, at the middle time, it is easily found that we may do so by either of the two following formulæ:—

$$\text{(IV.)}\quad\begin{cases} DB = \rho = \dfrac{2\sin\dfrac{\theta + \iota}{2}}{\sin\theta}\cos\left(\kappa + \dfrac{\theta - \iota}{2}\right); & (18) \\[2ex] \text{or,}\quad \rho = \dfrac{2\sin\dfrac{\theta' + \iota'}{2}}{\sin\theta'}\cos\left(\kappa - \dfrac{\theta' - \iota'}{2}\right); & (19) \end{cases}$$

and thus shall introduce 2 new logarithms, raising the total to Delambre's number of 17. In fact, it will be found that

$$\angle BAD = \frac{\pi}{2} - \frac{\theta + \iota}{2} - (\kappa - \iota) = \angle BAE - \angle DAE, \tag{20}$$

and

$$\angle DCB = \frac{\pi}{2} - \frac{\theta' + \iota'}{2} + (\kappa + \iota') = \angle ECB + \angle DCE; \tag{21}$$

while the arcs AB, BC, of the excentric are $\theta + \iota$, and $\theta' + \iota'$, respectively.

NOTES AND EXAMPLES CONNECTED WITH THE PROBLEM OF HIPPARCHUS.

In the original Problem of Hipparchus, the points here represented by A, B, C were three positions of the sun or moon, in the fixed or moveable excentric but circular orbit of that body, reduced to one common date, by allowing when necessary for the progression of the apogee; *D* was the fixed centre of the earth, and *E* was the centre of the excentric circle, supposed to be fixed (in longitude) for the sun, but to advance uniformly in longitude at a mean daily rate of about $6' 41''$ for the moon; while *F* (in the present figure [Fig. 3]) denotes what was supposed to be the fixed or revolving apogean point of the circle. The two angles *AEB*, *BEC*, were known by computation from the two observed intervals of time between three observations of longitude of the body (equinoxes and solstices were selected for the sun, eclipses for the moon), combined with the known (or supposed) mean daily motion in anomaly ($59' 8''$ for the sun, $13° 3' 54''$ for the moon); these angles being conceived by Hipparchus to be described uniformly about the fixed or revolving centre *E*, in respect of the apogee *F*, according to the order of the signs, and therefore from right to left as seen as in a northern latitude, whereas the present figure exhibits the contrary rotation.

The two other angles, *ADB*, *BDC*, subtended by the same two chords of the excentric circle, not at its own centre, but at the centre of the earth, were the observed motions in geocentric longitude, diminished (in the case of the moon) by the before-mentioned progressions of the apogee, in order to make the observations comparable, by reduction of them to a common date. Thus, the problem which was (really and historically) proposed and solved by Hipparchus was this:—From the four known angles, subtended by the two chords, *AB*, *BC*, of the (reduced) excentric, at the centre *E* of that circle, and at the centre *D* of the earth (directions of rotation being included), to find the position of *D*, with respect to the excentric. Or more fully, to find the *angles* of the triangle *BDE*, at the time of the middle observation; and also the *ratio*, $ED : EB$, of the distance between the centres of the earth and the excentric, to the radius of that excentric circle: for Hipparchus had no particular motive for investigating the *other* ratio, $DB : EB$, of sides of that triangle, since he had no mode of observing the angular diameters of sun or moon, and was but rudely acquainted with their parallaxes. Ptolemy's discovery of the Lunar Evection gave him a motive and an excuse for substituting, in the stead of Hipparchus's Hypothesis of the *Excentric*, another hypothesis of the *Epicycle*; but of this I need not speak at present, because Ptolemy himself took pains to prove that, *mathematically* considered, the *one* view was equivalent to the *other* (nor does he seem to have attached any the slightest physical *reality* to either of them); and because, in fact, the only geometrical modification made by him, in the problem of Hipparchus, consists in his having placed the sought point *D* *outside* an *epicycle*, instead of placing it *inside* an *excentric*.

SIR W. R. HAMILTON.

OBSERVATORY, *Dec.*, 1855.

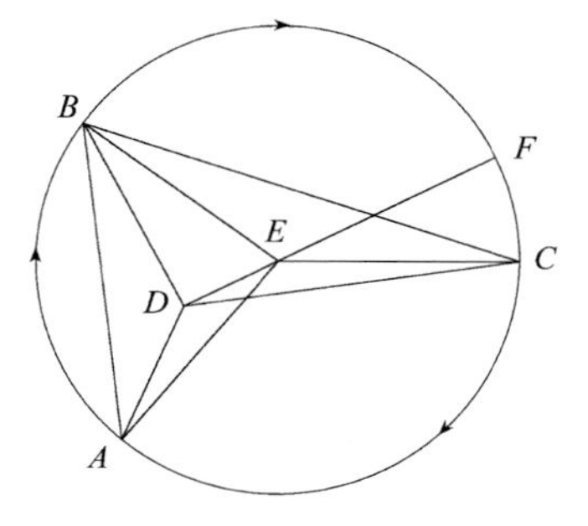

Fig. 3

Part V

PROBABILITY AND FINITE DIFFERENCES

XXXVIII.

ON DIFFERENCES AND DIFFERENTIALS OF FUNCTIONS OF ZERO (1831)

[Read June 13, 1831]
[*Transactions of the Royal Irish Academy* **17**, 235–236 (1830).
Translation in Quetelet's *Correspondance Mathématique et Physique* **8**, 235–237 (1834).]

The first important researches on the differences of powers of zero, appear to be those which Dr. BRINKLEY published in the Philosophical Transactions for the year 1807.* The subject was resumed by Mr. HERSCHEL in the Philosophical Transactions for 1816†; and in a collection of Examples on the Calculus of Finite Differences, published a few years afterwards at Cambridge. In the latter work, a remarkable theorem is given, for the development of any function of a neperian exponential, by means of differences of powers of zero. In meditating upon this theorem of Mr. HERSCHEL, I have been led to one more general, which is now submitted to the Academy. It contains three arbitrary functions, by making one of which a power and another a neperian exponential, the theorem of Mr. HERSCHEL may be obtained.

Mr. HERSCHEL'S Theorem is the following:

$$f(e^t) = f(1) + tf(1+\Delta)\,o^1 + \frac{t^2}{1.2} f(1+\Delta)\,o^2 + \&c. \tag{A}$$

$f(1+\Delta)$ denoting any function which admits of being developed according to positive integer powers of Δ, and every product of the form $\Delta^m o^n$ being interpreted, as in Dr. BRINKLEY'S notation, as a difference of a power of zero.

The theorem which I offer as a more general one may be thus written:

$$\phi(1+\Delta)\, f\psi(o) = f(1+\Delta')\phi(1+\Delta)(\psi(o))^{o'}; \tag{B}$$

or thus

$$F(D)\, f\psi(o) = f(1+\Delta')\, F(D)\,(\psi(o))^{o'}. \tag{C}$$

In these equations, f, ϕ, F, ψ, are arbitrary functions, such however that $f(1+\Delta')$, $\phi(1+\Delta)$, $F(D)$, can be developed according to positive integer powers of Δ' Δ D; and after this development, Δ' Δ are considered as marks of differencing, referred to the variables o' o, which vanish after the operations, and D as a mark of derivation by differentials, referred to

* [Brinkley, J., 'An investigation of the general term of an important series in the inverse method of finite differences' *Philosophical Transactions of the Royal Society of London*, **97**, 114–132 (1807).]

† [Herschel, J. F. W., 'On the development of exponential functions together with several new theorems relating to finite differences' *Philosophical Transactions of the Royal Society of London*, **106**, 25–45 (1816).]

the variable o. And if in the form (C) we particularise the functions F, ψ, by making F a power, and ψ a neperian exponential, we deduce the following corollary:

$$D^x f(e^o) = f(1+\Delta')\, D^x e^{o'} = f(1+\Delta')\, o'^x;$$

that is, the coefficient of $\dfrac{t^x}{1.2\,.\,.\,x}$ in the development of $f(e^t)$ may be represented by $f(1+\Delta)\, o^x$; which is the theorem (A) of Mr. HERSCHEL.

June 13, 1831.

ADDITION

The two forms (B) (C) may be included in the following:

$$\nabla' f\psi(o') = f(1+\Delta)\nabla'(\psi(o'))^o. \tag{D}$$

To explain and prove this equation, I observe that in MACLAURIN'S series,

$$f(x) = f(o) + \frac{Df(o)}{1}x + \frac{D^2 f(o)}{1.2}x^2 + .\,.\,+ \frac{D^n f(o)}{1.2\,.\,.\,n}x^n + .\,.$$

we may put $x = (1+\Delta)\, x^o$ and therefore may put the series itself under the form

$$f(x) = f(o) + \frac{Df(o)}{1}.(1+\Delta)\, x^o + \frac{D^2 f(o)}{1.2}.(1+\Delta)^2 x^o + \&c.$$

or more concisely thus

$$f(x) = f(1+\Delta)\, x^o: \tag{E}$$

which latter expression is true even when MACLAURIN'S series fails, and which gives, by considering x as a function ψ of a new variable o' and performing any operation ∇' with reference to the latter variable,

$$\nabla' f\psi(o') = \nabla' f(1+\Delta)\,(\psi(o'))^o. \tag{F}$$

If now the operation ∇' consist in any combination of differencings and differentiatings, as in the equations (B) and (C), and generally if we may transpose the symbols of operation ∇' and $f(1+\Delta)$, which happens for an infinite variety of forms of ∇', we obtain the theorem (D). It is evident that this theorem may be extended to functions of several variables.

June 20, 1831.

XXXIX.

ON A THEOREM IN THE CALCULUS OF DIFFERENCES (1843)

[*British Association Report* 1843, Part II., pp. 2–3.]

It is a curious and may be considered as an important problem in the Calculus of Differences to assign an expression for the sum of the series

$$\mathrm{X} = u_n(x+n)^n - u_{n-1}.\frac{n}{1}.(x+n-1)^n + u_{n-2}.\frac{n(n-1)}{1.2}.(x+n-2)^n - \&\mathrm{c}.; \qquad (1)$$

which differs from the series for $\Delta^n x^n$ only by its introducing the coefficients u, determined by the conditions that

$$u_i = +1,\ 0,\ \text{or}\ -1,\ \text{according as}\ x+i>0,\ =0,\ \text{or}\ <0. \qquad (2)$$

These conditions may be expressed by the formula

$$u_i = \frac{2}{\pi}\int_0^\infty \frac{dt}{t}\sin(xt+it); \qquad (3)$$

and if we observe that

$$\frac{d}{dt}\sin(at+b) = a\sin\left(at+b+\frac{\pi}{2}\right),$$

$$\left(\frac{d}{dt}\right)^n \sin(at+b) = a^n \sin\left(at+b+\frac{n\pi}{2}\right),$$

we shall see that the series (1.) may be put under the form

$$\mathrm{X} = \frac{2}{\pi}\int_0^\infty \frac{dt}{t}\left(\frac{d}{dt}\right)^n \Delta^n \sin\left(xt-\frac{n\pi}{2}\right); \qquad (4)$$

the characteristic Δ of difference being referred to x. But

$$\Delta\sin(2\alpha x+\beta) = 2\sin\alpha\sin\left(2\alpha x+\beta+\alpha+\frac{\pi}{2}\right),$$

$$\Delta^n\sin(2\alpha x+\beta) = (2\sin\alpha)^n\sin\left(2\alpha x+\beta+n\alpha+\frac{n\pi}{2}\right);$$

therefore, changing t, in (4.), to 2α, we find

$$\mathrm{X} = \int_0^\infty \frac{d\alpha}{\alpha}\frac{d^n\Lambda}{d\alpha^n}, \qquad (5)$$

if we make, for abridgement,

$$\Lambda = \frac{2}{\pi} \sin \alpha^n \sin(2x\alpha + n\alpha). \tag{6}$$

Again, the process of integration by parts gives

$$\int_0^\infty \frac{d\alpha}{\alpha^i} \frac{d^{n-i+1}\Lambda}{d\alpha^{n-i+1}} = i \int_0^\infty \frac{d\alpha}{\alpha^{i+1}} \frac{d^{n-i}\Lambda}{d\alpha^{n-i}},$$

provided that the function

$$\frac{1}{\alpha^i} \frac{d^{n-i}\Lambda}{d\alpha^{n-i}}$$

vanishes both when $\alpha = 0$ and when $\alpha = \infty$, and does not become infinite for any intermediate value of α, conditions which are satisfied here; we have, therefore, finally,

$$\mathrm{X} = 1.2.3 \ldots n \int_0^\infty d\alpha \frac{\Lambda}{\alpha^{n+1}}. \tag{7}$$

Hence, if we make

$$\mathrm{P} = \frac{\mathrm{X}}{1.2.3 \ldots n}, \text{ and } c = 2x + n, \tag{8}$$

we shall have the expression

$$\mathrm{P} = \frac{2}{\pi} \int_0^\infty d\alpha \left(\frac{\sin \alpha}{\alpha}\right)^n \frac{\sin c\alpha}{\alpha}, \tag{9}$$

as a transformation of the formula

$$\left.\begin{aligned} \mathrm{P} = \frac{1}{1.2.3 \ldots n.2^n} \Big\{ & (n+c)^n - \frac{n}{1}(n+c-2)^n + \frac{n(n-1)}{1.2}(n+c-4)^n - \&\text{c.} \\ & -(n-c)^n + \frac{n}{1}(n-c-2)^n - \frac{n(n-1)}{1.2}(n-c-4)^n + \&\text{c.} \Big\}; \end{aligned}\right\} \tag{10}$$

each partial series being continued only as far as the quantities raised to the nth power are positive. Laplace has arrived at an equivalent transformation, but by a much less simple analysis.

XL.

ON THE CALCULUS OF PROBABILITIES (1843)

[Communicated July 31, 1843]
[*Proceedings of the Royal Irish Academy* **2**, 420–422 (1844).]

Dr. Lloyd having taken the Chair, the President gave an account of some researches in the Calculus of Probabilities.

Many questions in the mathematical theory of probabilities conduct to approximate expressions of the form

$$p = \frac{2}{\sqrt{\pi}} \int_o^t dt e^{-t^2};$$

that is,

$$p = \theta(t),$$

θ being the characteristic of a certain function which has been tabulated by Encke in a memoir* on the Method of Least Squares, translated from the Berlin Ephemeris, in vol. ii. part 7 of Taylor's Scientific Memoirs; p being the probability sought, and t an auxiliary variable.

Sir William Hamilton proposes to treat the equation

$$p = \theta(t)$$

as being in all cases rigorous, by suitably determining the auxiliary variable t, which variable he proposes to call the *argument of probability*, because it is the argument with which Encke's Table should be entered, in order to obtain from that Table the value of the probability p. He shows how to improve several of Laplace's approximate expressions for the argument t, and uses in many such questions a transformation of a certain double definite integral, of the form,

$$\frac{4s^{\frac{1}{2}}}{\pi} \int_0^r dr \int_0^\infty du\, e^{-su^2}\, \mathrm{U} \cos(2s^{\frac{1}{2}} ru\, \mathrm{V}) = \theta(r(1 + \nu_1 s^{-1} + \nu_2 s^{-2} + \ldots));$$

in which

$$\mathrm{U} = 1 + \alpha_1 u^2 + \alpha_2 u^4 + \cdots$$

$$\mathrm{V} = 1 + \beta_1 u^2 + \beta_2 u^4 + \cdots$$

while $\nu_1, \nu_2, \ldots$ depend on $\alpha_1, \ldots \beta_1, \ldots$ and on r; thus

* [J. F. Encke (1791–1865), 'On the method of least squares', pp. 317–370, *Taylor's Scientific Memoirs*, Vol. II, London: 1841.]

$$\nu_1 = \tfrac{1}{2}\alpha_1 - \beta_1 r^2.$$

The function θ has the same form as before, so that if, for sufficiently large values of the quantity s (which represents, in many questions, the number of observations or events to be combined), a probability p can be expressed, exactly or nearly, by the foregoing double definite integral, then the argument t, of this probability p, will be expressed nearly by the formula,

$$t = r(1 + \nu_1 s^{-1} + \nu_2 s^{-2}).$$

Numerical examples were given, in which the approximations thus obtained appeared to be very close. For instance, if a common die (supposed to be perfectly fair) be thrown six times, the probability that the sum of the six numbers which turn up in these six throws shall not be less than 18, nor more than 24, is represented rigorously by the integral

$$p = \frac{2}{\pi}\int_0^{\frac{\pi}{2}} dx \frac{\sin 7x}{\sin x}\left(\frac{\sin 6x}{6 \sin x}\right)^6, \text{ or by the fraction } \tfrac{27448}{46656};$$

while the approximate formula deduced by the foregoing method gives 27449 for the numerator of this fraction, or for the product $6^6 p$; the error of the resulting probability being therefore in this case only 6^{-6}. The advantage of the method is that the quantity which has here been called the argument of probability, depends in general more simply than does the probability itself on the conditions of a question; while the introduction of this new conception and nomenclature allows some of the most important known results respecting the mean results of many observations to be enunciated in a simple and elegant manner.

XLI.

ON SOME INVESTIGATIONS CONNECTED WITH THE CALCULUS OF PROBABILITIES (1843)

[*British Association Report* 1843, Part II., pp. 3–4.]

Identical with previous paper.

Part VI
MISCELLANEOUS

XLII.

ON THE DOUBLE REFRACTION OF LIGHT ACCORDING TO THE PRINCIPLES OF FRESNEL* (REVIEW OF TWO SCIENTIFIC MEMOIRS OF JAMES MACCULLAGH B.A.) (1830)

[*National Magazine* **1**, 145–149. Dublin (August 1830).]

It was noticed, in the last number of this magazine, under the head of proceedings of the Royal Irish Academy, that two papers of Mr. M'Cullagh, (a scholar of Trinity College, Dublin, and candidate for a fellowship there,)† were to have been read at the June meeting of the academy, and were only prevented from being so, by the adjournment which took place, in consequence of the death of the king. These papers relate, one to the double refraction of light, and the other to the rectification of the conic sections; they had been before referred, in the council of the academy, to a committee, consisting of Dr. Lloyd, Dr. Sadleir, and Professor Hamilton; and in consequence of the favourable report of the committee, they have since been ordered to be printed, and will appear in the forthcoming volume of the Irish Transactions. Although this volume will very soon be published, yet as we have been favoured with a sight of Mr. M'Cullagh's papers, and think them creditable to the university of which he is a member, we shall endeavour to present our readers with a sketch of their subject and plan.

The first of the two papers relates, as we have already said, to the double refraction of light, which is one of the most curious and interesting subjects in mathematical and physical optics. The refraction produced by water has long been known; for Virgil alludes to the *oar broken in the Sicilian wave;* although the discovery of the law of this breaking, was reserved for Snellius and Descartes. But the double refraction produced by crystals, seems to have been unknown to the ancients; the earliest observations on the subject having been published in 1669, by Erasmus Bartholinus, professor of geometry at Copenhagen. Bartholinus found that when a piece of the transparent crystal, technically called carbonate of lime, and more familiarly known by the name of Iceland spar, was laid upon any object, the object was seen *double*; each line or letter, for example, of a printed book, being changed into two such lines or letters, both equally distinct. Huyghens, the great contemporary, and almost rival of Newton, made new experiments upon the subject, and deduced a remarkable law of the phenomena, which he connected, by elegant geometrical reasonings, with his theory of the undulations of a luminous ether. So decidedly was Newton opposed to this theory of luminous undulations,

* Mr. M'Cullagh's Papers. (Transactions of the Royal Irish Academy Vol. xvi. *Part Second.*)

† [See: Scaife, B. K. P., 'James MacCullagh, M. R. I. A., F. R. S., 1809–1847' *Proceedings of the Royal Irish Academy* **90C**, 67–106 (1990).]

that it seems to have prevented him from giving sufficient attention to any researches which were connected therewith; although he was always ready to admit the great merit of Huyghens, as a geometer and as a physician. At least, this bias against the theory of waves appears to be the only mode of accounting for Newton's having proposed, in one of the celebrated Queries annexed to the third book of his Optics, an inaccurate law of double refraction, instead of that which Huyghens had given; although he had seen the treatise of his contemporary, and alludes to it in another of his Queries. The authority of Newton seems to have prejudiced those who succeeded him, until the experiments of Wollaston confirmed the law of Huyghens, and revived it after a century of oblivion. Herschell, in his treatise upon Light, (Encyclopædia Metropolitana,) observes, that since this time, a new impulse has been given to this department of optics; and the successive labours of Laplace, Malus, Brewster, Biot, Arago, and Fresnel, present a picture of emulous and successful research, than which nothing prouder has adorned the annals of physical science, since the development of the true system of the universe. "To enter, however, (Mr. Herschell continues,) into the history of these discoveries, or to assign the share of honour which each illustrious labourer has reaped in this ample field, forms no parts of our plan. Of the splendid constellation of great names just enumerated, we admire the living and revere the dead, far too warmly and too deeply, to suffer us to sit in judgment on their respective claims to priority, in this or that particular discovery, to balance the mathematical skill of one against the experimental dexterity of another, or the philosophical acumen of a third. So long as one star differs from another in glory—so long as there shall exist varieties, or even incompatibilities of excellence—so long will the admiration of mankind be found sufficient for all who truly merit it." In the sentiments thus expressed by Mr. Herschell, we fully concur; and think ourselves, by still stronger reason, excused and forbidden from attempting to decide on the respective eminence of these illustrious names. We shall, however, endeavour to give such brief and popular account of the facts and hypotheses respecting double refraction, as may explain the value of that theory of Fresnel, which Mr. M'Cullagh, in the paper before us, has sought to illustrate and simplify.

We have already mentioned the obvious and fundamental fact, of objects appearing double, when seen through certain crystals. But in order to derive from the phenomena of nature, the instruction and intellectual exercise which they are fitted to afford, we must not content ourselves with such general and obvious perceptions. We must discriminate the similar from the same; must vary, must measure, must combine; until, by the application of reasoning and invention to carefully recorded facts, we ascend to an hypothesis, a law, a theory, which includes the particular appearances, and enables them to be accounted for and foreseen. Then, when the passive of our being has been so far made subject to the active, and sensation absorbed or sublimed into reason, the philosopher reverses the process, and asks how far the conceptions of his intellect are realised in the outward world. He seeks to extend his hypothesis beyond the facts on which it was founded, and to deduce from it some new appearance which ought to be observed, if his theory be coextensive with nature. He then consults sensation and experiment, and often their answer is favourable; but often too, they speak an unexpected language. Yet undismayed by the repulse, and emboldened by partial success, he frames upon the ruins of his former, some new and more general theory; which, equally with the former, accounts for the old appearances, while it includes, within its ampler verge, the results of more recent observation. Nor can this struggle ever end, between the active and the passive of our being, between the ingenuity of the theorist and the patience of the observer, until the mind of man shall grasp the infinity of nature; and comprehend all the scope and character and habits of those innumerable energies which compose the material universe. Meanwhile this warfare, with its alternate triumphs and defeats, its discoveries of laws and exceptions, forms an appointed discipline for the human mind, and its history is justly interesting. And in no part of physical inquiry, has the struggle been more vigorous and various, than in the research of the phenomena of light. In astronomy, Newton, by one great stride of thought, respecting the law according to which the heavenly bodies act upon each other, placed theory at once so far in advance of observation, that the latter has not even yet overtaken the former, nor has the law of gravitation, in all its wide dominions, yet met with one rebellious fact in successful revolt against its authority. But in optics, no theory sits thus upon an old ancestral throne. The very nature of light, and the mode of its propagation, have been the subjects of a lengthened controversy. The theory that light, like sound, diffuses itself in waves, which had been maintained by Huyghens, and combated by Newton, is now, after a long interval of neglect, again acquiring defenders, and gaining ground in scientific opinion. And to return to the more immediate subject of the present article, the opinions respecting double refraction have undergone a great and recent change, in consequence of the researches of Fresnel.

To give an idea of this revolution of opinion, we must distinguish between the phenomena of *single-axed* and *double-axed* crystals; and for this purpose we shall borrow Mr. Herschell's illustration of the meaning of axes of a crystal. "Suppose a mass of brick-work, or masonry, of great magnitude, built of bricks, all laid parallel to each other. Its exterior form may be what we please; a cube, a pyramid, or any other figure. We may cut it (when hardened into a compact mass,) into any shape—a sphere, a cone, a cylinder; but the edges of the bricks within it will still lie parallel to each other; and their directions, as well as those of the diagonals of their surfaces, or of their solid figures, may all be regarded as so many *axes*, that is, lines having (so long as the mass remains at rest,) a determinate position, or rather *direction* in space, no way related to the exterior surfaces, or linear boundaries of the mass, which may cut across the edges of the bricks in any angles we please. Whenever then we speak of fixed lines, or *axes of*, or *within*, a crystal, we always mean directions in space parallel to each of a system of lines drawn in the several elementary molecules of the crystal, according to given geometrical laws, and related in a given manner to the sides and angles of the molecules themselves. We must conceive the axis, then, of a crystallized mass, not as a single line having a given place, but as any line whatever having a given *direction* in space, that is, parallel to the *axis of each molecule*, which is a line having a determinate place and position within it." We may add that the directions of those axes of a crystal which have important optical properties, are usually connected in a regular manner with the planes in which the crystal is found to cleave most easily, or with the other fixed planes or lines which the facts of crystallography indicate. For example, in the case of Iceland spar, the natural or primitive form appears to be that of a regular rhomboid, with six plane faces, and eight corners, of which two are blunter than the rest, and are opposite to one another; and the line that may be conceived to go within the crystal, from one of these corners to the other, is called by mineralogists the primitive axis of the crystal. And the same line is said to be the optical axis also, for the reason which we are going to mention.

When we look perpendicularly down upon any object, through a piece of common glass with two parallel horizontal faces, as for instance, when we lay upon a book a part of a window pane, or a cube of glass, we not only see the object single, but we see it exactly under us. But when we look in the same manner at an object, through a piece of Iceland spar, *one of the two images* which (as was before remarked) we *almost always* see, is not exactly under us; and thus the light which forms this image has undergone an *extraordinary refraction*, since it has come obliquely through the crystal to the eye, while in the ordinary mode of refraction (such as takes place with ordinary refracting substances,) it would have come perpendicularly through. However there is *one case* in which this extraordinary refraction *disappears*, namely, when the two plane faces of the crystal, through which we are supposed to look, have been artificially obtained, by cutting off the two bluntest corners of the primitive form, in directions perpendicular to the line that joins those corners; for then, that is when we look perpendicularly down through the crystal in the direction of its primitive axis, we see but *one image* of the object below, and that one *perpendicularly under us*.

Analogous phenomena are observed in other crystals. In each there is at least *one optical axis*, distinguished by the property that along it there is *no double refraction*. To all such crystals the law of Huyghens applies. But there are other crystals, more recently discovered, which possess *two optical axes*, distinguished by the same property. To these latter crystals the law of Huyghens is inapplicable, and many attempts had been made to find another which should replace it. In the course of these attempts, many valuable theories were discovered, respecting the connexions between the two separate rays produced by the double refraction. But it was always *taken for granted* that *one* of the two rays followed that law of *ordinary refraction* which Snellius and Descartes had long since established for the passage of light through water and glass, and which Huyghens had found to extend to one of the two rays produced by a single-axed crystal. Fresnel, however, has *overthrown this supposition*, by his experiments upon prisms of topaz, and by other experiments and reasonings; and has thus produced that great revolution of opinion to which we have before alluded.

Nor is this change of opinion confined to the rejection of a law which had been extended by a too hasty induction from single-axed to double-axed crystals. The whole theory of the motion of light has undergone a change, in consequence of the researches and speculations of Fresnel. On the ruins of old hypotheses, he has erected a new one, which may indeed itself be overthrown or changed hereafter, but which will, probably, long continue to guide the thoughts of mathematicians, and the experiments of the physical observer.

The sketch (or at least the notice) which we have thus endeavoured to give, of the optical labours of Fresnel, is indeed a rapid and imperfect one; yet it may serve to give an idea of the nature and importance of those labours, and so may explain the *subject* of that memoir of Mr. M'Cullagh, in which the latter mathematician has chosen to be a commentator on the former. With regard to the *plan* of the commentary, we can only mention here that Mr. M. has aimed, and (we think) successfully, to simplify, by the use of geometrical reasoning, and by several elegant lemmas upon the properties of ellipsoids, the proof of some theorems respecting the shape of the luminous waves, and the positions of the planes of polarization, in the double refraction of light, to which Fresnel had arrived by long analytic processes.

We must be still more brief in our notice of Mr. M'Cullagh's other paper, which relates to the rectification of the conic sections, and contains a geometrical demonstration of a celebrated theorem of Landen, respecting the connexion of hyperbolic and elliptic arcs, which had before been proved by means of definite integrals, in an entirely different manner.

It is evident that Mr. M. wields, with no common skill, the ancient armour of geometry. Those who have compared the synthetic and the analytic methods, have usually concluded that the latter is fitter for the discovery, and the former for the communication of truth. Perhaps, however, as algebra is now employed so often as a mode of elementary exposition, the powers of geometry may not have been exhausted as an instrument of high investigation. In the papers which form the subject of our foregoing notice, Mr. M'Cullagh has, on the whole, sought rather to illustrate known theorems by elegant geometrical demonstrations, than to open for himself a field entirely new; since, though the method which he uses, and most of the steps, are his own, yet the grand results had before been obtained by others. But he has given a favourable omen that he may hereafter apply his facility of geometrical conception, not only to illustrate but to discover. Let us, therefore, hope that he will aim at this brilliant union; and we, though ourselves devoted analysts, shall attend him with the sincerest good wishes, and give him our most cordial applause.

XLIII.

ADDRESS AS SECRETARY OF THE DUBLIN MEETING OF THE BRITISH ASSOCIATION (1835)

[*British Association Report* 1835, pp. xli–lvi.]

It has fallen to my lot, Gentlemen, as one of your Secretaries for the year, to address you on the present occasion. The duty would, indeed, have been much better discharged had it been undertaken by my brother secretary; but so many other duties of our secretaryship had been performed almost entirely by him, that I could not refuse to attempt the execution of this particular office, though conscious of its difficulty and its importance. For if we may regard it as a thing established now by precedent and custom that an annual address should be delivered, it is not, therefore, yet, and I trust that it will never be, an office of mere cold routine, a filling up of a vacant hour, on the ground that the hour must be some way or other got rid of. You have not left your homes–you have not adjourned from your several and special businesses—you have not gathered here, to have your time thus frittered away in an idle and unmeaning ceremonial. There ought to be, and there is, a reason that some such thing should be done; that from year to year, at every successive reassembling, an officer of your body should lay before you such an address; and in remembering what this reason is, we shall be reminded also of the spirit in which the duty should be performed. The reason is the fitness and almost the necessity of providing, so far as an address can provide, for the permanence and progression of the body, by informing the new members, and reminding the old, of the objects and nature of the Association, or by giving utterance to at least a few of those reflections which at such a season present themselves respecting its progress and its prospects; and it is a valid reason, and deserves to be acted upon now, however little may have been left unsaid in the addresses of my predecessors in this office. For if even amongst the members who have attended former meetings, and have heard those eloquent addresses delivered by former secretaries, it is possible that some may have been so dazzled by the splendour of the spectacle, and so rapt away by the enthusiasm of the time, as to have given but little thought to the purport and the use, the meaning and the function of the whole; much more may it be presumed that of the several hundred persons who have lately joined themselves as new members to this mighty body, there are some, and even many, who have reflected little as yet upon its characteristic and essential properties, and who have but little knowledge of what it has been, and what it is, and what it may be expected to become. First, then, the object of the Association is contained in its title; it is the advancement of science. Our object is not literature, though we have many literary associates, and though we hail and love as brethren those who are engaged in expressly literary pursuits, and who are either themselves the living ornaments of our land's language, or else make known to us the literary treasures of other languages, and lands, and times. Our object is not religion in any special sense, though respect for religious things, and religious men, has always marked these meetings, and though we are all bound together by that great tie of brotherhood which unites the whole human family as children of one Father who is in heaven. Still less is our object politics, though we are not mere citizens of the world, but are essentially a British Association of fellow-subjects and of fellow-countrymen, who give, however, glad and cordial welcome to those our visitors who come to us from foreign countries, and thankfully accept their aid to accomplish our common purpose. That common purpose, that object for which Englishmen, and Scotchmen, and Irishmen have banded themselves together in this colossal Association, to which the eyes of the whole world have not disdained to turn, and to see which, and to raise it higher still, illustrious men from foreign lands have come, is SCIENCE; the acceleration of scientific discoveries, and the diffusion of scientific influences. And if it be inquired how is this aim to be accomplished, and through what means, and by what instruments and process we as a body hope to forward science—the answer briefly is, that this great thing is to be done by us through the agency of the social spirit, and through the means, and instruments, and process which are contained in the operation of that spirit. We meet, we speak, we feel *together now*, that we may *afterwards* the better think and act and feel *alone*. The excitement with which this air is filled will not pass at once away; the influences that are now among us will not (we trust) be transient, but abiding; those influences will be with us long—let us hope that they will never leave us; they will cheer, they will animate us still, when this brilliant week is over; they will go with us to our separate abodes, will attend us on our separate journeys; and whether the mathematician's study, or the astronomer's observatory, or the chemist's laboratory, or some rich distant meadow unexplored as yet by botanist, or some untrodden mountain-top, or any of the other haunts and homes and oracular places of science, be our allotted place of labour till we meet together again, I am persuaded that those influences will operate upon us all, that we shall all remember this our present meeting, and look forward with joyful expectation to our next reassembling, and by the recollection, and by the hope, be stimulated and supported. It is true, that it is the individual man who thinks and who discovers; not any aggregate or mass of men. Each mathematician for himself, and not any one for any other, not even all for one, must tread that more than royal road which leads to the palace and sanctuary of mathematical truth. Each, for himself, in his own personal being, must awaken and call forth to mental view the original intuitions of time and space; must meditate himself on those eternal forms, and follow for himself that linked chain of thought which leads, from principles inherent in the child and the peasant, from the simplest notions and marks of temporal and local site, from the questions when and where, to results so varied, so remote, and seemingly so inaccessible, that the mathematical intellect of full-grown and fully cultivated man cannot reach and pass them without wonder, and something of awe. Astronomers, again, if they would be more than mere artizans, must be more or less mathematicians, and must separately study the mathematical grounds of their science; and although in this as in every other physical science, in every science which rests partly on the observation of nature, and not solely on the mind of man, a faith in testimony is required, that the human race may not be stationary, and that the accumulated treasures of one man or of one generation of men may not be lost to another; yet even here, too, the individual must act, and must stamp on his own mental possessions the impress of his own individuality. The humblest student of astronomy, or of any other physical science, if he is to profit at all by his study, must in some degree go over for himself, in his own mind, if not in part with the aid of his own observation and experiment, that process of induction which leads from familiar facts to obvious laws, then to the observation of facts more remote, and to the discovery of laws of higher orders. And if even this *study* be a personal act, much more must that *discovery* have been individual. Individual energy, individual patience, individual genius, have all been needed, to tear fold after fold away, which hung before the shrine of nature; to penetrate, gloom after gloom, into those Delphic depths, and force the reluctant Sibyl to utter her oracular responses. Or if we look from nature up to nature's God, we may remember that it is written—"Great are the works of the Lord, sought out of all those who have pleasure therein." But recognising in the fullest manner the necessity for private exertion, and the ultimate connexion of every human act and human thought with the personal being of man, we must never forget that the social feelings make up a large and powerful part of that complex and multiform being. The affections act upon the intellect, the heart upon the head. In the very silence and solitude of its meditations, still genius is essentially sympathetic; is sensitive to influences from without, and fain would spread itself abroad, and embrace the whole circle of humanity, with the strength of a world-grasping love. For fame, it has been truly said, is love disguised. The desire of fame is a form of the yearning after love; and the admiration which rewards that desire, is a glorified form of that familiar and every-day love which joins us in common life to the friends whom we esteem. And if we can imagine a desire of excellence for its own sake, and can so raise ourselves *above* (Well if we do not in the effort sink ourselves *below*) the common level of humanity, as to account the aspiration after fame only "the last infirmity of noble minds," it will still be true that in the greatest number of cases, and of the highest quality,

> Fame is the spur that the clear spirit doth raise,
> To scorn delights, and live laborious days.

That mysterious joy—incomprehensible if man were wholly mortal—which accompanies the hope of influencing unborn generations; that rapture, solemn and sublime, with which a human mind, possessing or possessed by some great truth, sees in prophetic vision that truth acknowledged by mankind, and itself long ages afterward remembered and associated therewith, as its interpreter and minister, and sharing in the offering duly paid of honour and of love, till it becomes a power upon the earth, and fills the world with felt or hidden influence; that joy which thrills most deeply the minds the most contemptuous of mere ephemeral reputation, and men who care the least for common marks of popular applause or outward dignity—does it not show, by the revival, in another form, of an instinct seemingly extinguished, how deeply man desires, in intellectual things themselves, the sympathy of man? If then the *ascetics* of science—if those who seem to shut themselves up in their own separate cells, and to disdain or to deny themselves the ordinary commerce of humanity—are found, after all, to be thus influenced by the *social spirit,* we can have little hesitation in pronouncing that to the operation of this spirit must largely be ascribed the labours of ordinary minds; of those who do not even affect or seem to shun the commerce of their kind; who accept gladly, and with acknowledged joy, all present and outward marks of admiration or of sympathy, and who are willing, and confess themselves to be so, to do much for immediate reward, or speedy though perishing reputation. Look where we will, from the highest and most solitary sage who ever desired "the propagation of his own memory," and committed his lonely labours to the world, in full assurance that an age would come, when that memory would not willingly be led to die, down to the humblest labourer who was ever content to cooperate outwardly and subordinately with others, and hoped for nothing more than present and visible recompense, we still perceive the operation of that social spirit, that deep instinctive yearning after sympathy, to use the power, and (if it may be done) to guide the influences of which, this British Association was framed. Thus much I thought that I might properly premise, on the social spirit in general, and its influence upon the intellect of man; since that is the very bond, the great and ultimate reason, of this and of all other similar associations and companies of studious men. But you may well expect that in the short remaining time which your leisure this evening can spare, I should speak more especially, and more definitely, of this British Association in particular. And here it may be right to adopt in part a more technical style, and to enter more minutely into detail, than I could yet persuade myself to do, till I had eased myself in some degree of those overflowing emotions, which on such an occasion as this could hardly be altogether suppressed. Presuming, therefore, that some one now demands, how this Association differs from its fellows, and what peculiar means it has of awakening and directing to scientific purposes the power of the social spirit; or why, when there were so many old and new societies for science, it was thought necessary or expedient to call this society also into being: I proceed to speak of some of the characteristic and essential circumstances of this British Association, which contain the answer to that reasonable demand. First, then, it differs in its magnitude and universality from all lesser and more local societies. So evidently true is this, that you might justly blame me if I were to occupy your time by attempting any formal proof of it. What other societies do upon a small scale, this does upon a large; what others do for London, or Edinburgh, or Dublin, this does for the whole triple realm of England, Scotland, and Ireland. Its gigantic arms stretch even to America and India, insomuch that it is commensurate with the magnitude and the majesty of the British empire, on which the sun never sets; and that we hail with pleasure, but without surprise, the enrolment of him among our members who represents the sovereign here, and is to us the visible image of the head of that vast empire; and the joy with which we welcome to our assemblies and to our hospitality those eminent strangers who have come to us from foreign lands, rises almost above the sphere of private friendship, and partakes of the dignity of a compact between all the nations of the earth. Forgive me that I have not yet been able to speak calmly in such a presence, and on such a theme. But it is not merely in its magnitude and universality, and consequently higher power of stimulating intellect through sympathy, that this Association differs from others. It differs also from them in its constitution and details; in the migratory character of its meetings, which visit, for a week each year, place after place in succession, so as to indulge and stimulate all, without wearying or burdening any; in encouraging oral discussion, throughout its several separate sections, as the principal medium of making known among members the opinions, views, and discoveries of each other; in calling upon eminent men to prepare reports upon the existing state of knowledge in the principal departments of science; and in publishing only abstracts or notices of all those other contributions which it has not as a body called for; in short, in attempting to induce men of science to work more together than they do elsewhere, to establish a system of more strict cooperation between the labourers in one common field, and thus to effect, more fully than other societies can do, the combination of intellectual exertions. In other societies, the constitution and practice are such, that the labours of the several members are comparatively unconnected, and few attempts are systematically made to combine and harmonize them together; so that if we except that general and useful action of the social spirit upon the intellect of which I have already spoken, and the occasional incitement to specific research, by the previous proposal of prizes, there remains little beyond the publication of Transactions, whereby they seek as bodies to cooperate in the work of science. In them an author, of his own accord, hands in a paper; the title and subject are announced; it is referred to a Committee for examination, and if it be approved of, it is published at the expense of the society. This is a very great and real good, because the most valuable papers are seldom the most attractive to common purchasers, and because the authors of these papers are rarely able to defray from their own funds the cost of an expensive publication. There is no doubt that if it had not been for this resource, many essays of the greatest value must have been altogether suppressed, for want of pecuniary means. Besides, the approbation of a body of scientific men, which is at least partially implied in their undertaking to publish a paper, however limited and guarded it may be by their disclaimer of corporate responsibility, cannot fail to be accounted a high and honourable reward; and one, of which the hope must much assist to cheer and support the author in his toils, by virtue of the principle of sympathy. It is known, and (I believe) was mentioned in an address to this Association, at one of the former meetings, that the Principia and Optics of Newton were published at the request of the Royal Society of London. Newton, indeed, might well have thought that those works did not need that sanction, if the meekness of his high faculties had permitted him to judge of himself as all other men have judged of him; but our gratitude is not therefore due the less to the Society whose request prevailed over his own modest reluctance, and procured those treasures for that and for every age. It must be added that the Royal and Astronomical Societies print abstracts of their communications, for speedy circulation among their members, which is a useful addition to the service done in publishing the papers themselves, and is an example well worthy of being followed by all similar institutions; and that the Royal Society has even gone so far as to procure and print, in at least one recent instance (I mean in the case of a paper of Mr. Lubbock's), and perhaps also in some other instances, a report from some of its members, on a memoir presented by another, thus imitating an excellent practice of the Institute of France, which has probably contributed much to the high state of science in that country. This last procedure, and doubtless other acts of some other scientific societies, such as the discussions in the Geological Society, the lending of instruments by the Astronomical Society to its members, and the occasional exhibition of models and experiments by members to the body, in the Irish and other institutions, are examples of direct co-operation; and perhaps there is nothing to prevent such cases being greatly multiplied hereafter. But admitting freely these and other claims of the several societies and academies of the empire to our gratitude for their services to science, and accounting it a very valuable privilege to belong, as most of us do, to one or other of those bodies, and acknowledging that there is much work to be done which can only be done by them, we must still turn to this British Association, as the body which is *cooperative* by eminence.—The *discussions* in its sections are more animated, comprehensive and instructive, and make minds which were strangers, more intimately acquainted with each other, than can be supposed to be the case in any less general body; the *general meetings* bring together the cultivators of all different departments of science; and even the less formal *conversations*, which take place in its halls of assembly during every pause of business, are themselves the working together of mind with mind, and not only excite but *are* co-operation. Express requests also are systematically made to individuals and bodies of men, to cooperate in the execution of particular tasks in science, and these requests have often been complied with. But more perhaps than all the rest, the reports which it has called forth on the existing state of the several branches of knowledge are astonishing examples of industry and zeal exerted in the spirit and for the purpose of cooperation. No other society, I believe, has yet ventured to call on any of its members for any such report, and indeed it would be a difficult, perhaps an invidious thing, for any one of the other societies or academies so to do. For such a report should contain a large and comprehensive view of the treasures of all the academies; and would it not be difficult for a zealous member of any one of them, undertaking the task at the request of his own body, to form and to express that view with all the impartiality requisite? Would there not be some danger of a bias, in some things to palliate the defects of his own particular society, and in other things to exalt beyond what was strictly just, its true and genuine merits? But a body like the British Association which receives indeed all communications, but publishes (except by abstract) none, save only those very reports which it had previously and specially called for,—a body such as this, and governed by such regulations, may hope, that standing in one common relation to all the existing academies, and not belonging to the same great class of societies publishing papers, the members whom it has selected for the task may come before it to report what has resulted from the labours of all those different societies, without any excessive depression or any undue exultation, and in a more unbiassed mood of mind than would be possible under other circumstances. Accordingly the reports already presented by those eminent men who were selected for the office, (and rightly so selected, because a comprehensive mind was not less needed than industry,) appear to have been drawn up with as much impartiality as diligence; they comprise a very extensive and perfect view of the existing state of science in most of its great departments: and if in any case they do not quite bring down the history of science to this day (as certainly they go near to do), they furnish some of the best and most authentic materials to the future writer of such history. But we should not only underrate the value of those reports, but even quite mistake the character of that value if we were to refer it all to its connexion with distant researches, and some unborn generation. They will, indeed, assist the future historian of science; but it was not solely, nor even chiefly for that purpose they were designed, nor is it solely or chiefly for that purpose which they will answer. They belong to our own age; they are the property of ourselves as well as of our children. To stimulate the living, not less than to leave a record to the unborn, was hoped for, and will be attained, through those novel and important productions. In holding up to us a view of the existing state of science, and of all that has been done already, they show us that much is still to be done, and they rouse our zeal to do it. Can any person look unmoved on the tablet which they present of the brilliant discoveries of this century, in any one of the regions of science? Can he see how much has been achieved, what large and orderly structures have been in part already built up, and are still in process of building, without feeling himself excited to give his own aid also in the work, and to be enrolled among the architects, or at least among the workmen? or can any person have his attention guided to the many wants that remain, can he look on the gaps which are still unfilled, even in the most rich and costly of those edifices (like the unfinished window that we read of in the palace of eastern story), without longing to see those wants supplied, that palace raised to a still more complete perfection; without burning to draw forth all his own old treasures of thought, and to elaborate them all into one new and precious offering?—The volume containing the reports which were presented at the last meeting of the Association has been published so very recently, that it is perhaps scarcely yet in the hands of more than a few of the members; some notice of its contents may therefore be expected from me now, though the notice which I can give must of necessity be brief and inadequate. I shall speak first of two reports, which may in a certain sense be said to be on foreign science. Science, indeed, as has been well remarked, is not properly of any country; but men of science are, and in studying the works of their brethren of foreign nations, they at once increase their own stock of knowledge, and cultivate those kindly feelings of general good will, which are among the very best results of all our studies, and of all our assemblings together. The first report of the volume is that which Professor Rogers, of Philadelphia, has presented, at the request of the British Association, upon American Geology. The kindness of an eminent British Geologist, whose name would command attention if I thought myself at liberty to mention it, and whom I had requested to state to me in writing his opinion on this report, enables me to furnish you with a notice respecting its nature, which I shall accordingly read, instead of presuming to substitute any remarks of my own on the subject.

"The object proposed by Professor Rogers was to convey a clear summary of what had been ascertained concerning the geology of America, whether the knowledge acquired had been communicated to the public or not. This is not very different from the object contemplated by other reporters; but in the execution of the report it is found that a marked peculiarity arises. For the far greater portion of the report contains the result of Mr. Rogers's own reasonings on data, many of which appear for the first time in his essay. It has therefore more the character of a memoir than of an ordinary report. Were any one to adopt this plan in treating of the state of European geology he might be blamed, because the value of such a report would consist in the discussion of a vast mass of published data, and in the comparison of theoretical notions proposed by persons of high reputation. But in treating of America this was not the case; because, first, little authentic was known in Europe on the subject—second, there are few American authors of high repute in geology. This character of originality is certainly well supported by the author's own researches, and it is not surprising if his work contains some errors, still less remarkable that it should have excited some opposition at home. But the writer of the report has really taken much pains, has exhibited much patience, and has brought to his task a competent knowledge of European geology. It has certainly cleared our notions of the general features of American geology, and particularly augmented our positive knowledge of the more recent deposits, as regards organic remains, mineral characters, and geographical features. It is to be continued."

The other report which I alluded to, as almost entitled to be called a report on foreign science, is the report of the Rev. Mr. Challis on the theory of capillary attraction, which is a sequel to that presented at Cambridge on the common theory of fluids, and which the author proposes to follow up hereafter by another report on the propagation of motion as affected by the development of heat. Mr. Challis remarks, that while many questions in physics are to be resolved by unfolding through deductive reasoning the consequences of facts actually observed, there is also another class of questions in physical science, in which the facts that are to be reasoned from are not phænomena; for example, the fact of universal gravitation for which the evidence is inductive indeed, but yet essentially mathematical, the fact not coming itself under the cognisance of any of our senses, although its mathematical consequences are abundantly attested by observations. Mr. Challis goes on to say—"The great problem of universal gravitation, which is the only one of this class that can be looked upon as satisfactorily solved, relates to the large masses of the universe, to the dependence of their forms on their own gravitation, and the motions resulting from their actions on one another. The progress of science seems to tend towards the solution of another of a more comprehensive nature, regarding the elementary constitution of bodies and the forces by which their constituent elements are arranged and held together. Various departments of science appear to be connected together by the relation they have to this problem. The theories of light, heat, electricity, chemistry, mineralogy, crystallography, all bear upon it. A review, therefore, of the solutions that have been proposed of all such questions as cannot be handled without some hypotheses respecting the physical condition of the constituent elements of bodies, would probably conduce by a comparison of the hypotheses towards reaching that generalization to which the known connexion of the sciences seems to point." The author finally remarks, that "questions of this kind have of late largely engaged the attention of some French mathematicians, and the nature of their theories, and the results of the calculations founded on them, deserve to be brought as much as possible into notice." Acting upon these just views, Mr. Challis has accordingly performed, for the British Association and for the British public, the important office of reviewing and reporting upon those researches of Laplace, Poisson, and Gauss, respecting the connexion of molecular attraction, and of the repulsion of heat, with the ascent of fluids in tubes, which give to his report so much of that foreign character which I have already ventured to ascribe to it; yet, it is just to add, and, indeed, Mr. Challis does so, that as Newton first resolved the mathematical problem of gravitation, in its bearings on the motion of a planet about the sun, and went far to resolve the same extensive problem in its details of perturbation also; he likewise first resolved a problem of molecular forces, and clearly foresaw and foretold the extensive and almost universal application of such forces to the mathematical explanation of the most varied classes of phænomena; and that the theory of capillary attraction, in particular, has received some very valuable illustrations in England from the late Dr. Thomas Young. I ought to mention that a very interesting report, on the foreign mathematical theories of electricity and magnetism was read in part this morning to the mathematical and physical section, by the Rev. Mr. Whewell.

The next report after that of Mr. Challis in the volume, is the report I have already alluded to, by Professor Lloyd, on the progress and present state of physical optics; respecting which I should have much to say, if I did not fear to offend the modesty of the author, and were not restrained by the recollection that he is a member of the same University with myself, and a countryman and friend of my own. I shall therefore simply express my belief, that no person who shall hereafter set about to form an opinion of his own on the question between the two theories of light, will think himself at liberty to dispense with the study of this report. I may add that it also, as well as that of Mr. Challis, draws largely from foreign stores; but if Huygens was the first inventor, and Fresnel the finest unfolder, and Cauchy the profoundest mathematical dynamician, of the theory of the propagation of light by waves; and if the names of Malus, and Biot, and Arago, and Mitscherlich, and other eminent foreigners are familiar words in the annals of physical optics, we also can refer, among our own illustrious dead, to names enshrined in the history of this science—to the names of Newton, and Wollaston, and Young—and among our living fellow-countrymen and fellow-members of this Association, (unhappily not present here,) we have Brewster and Airy to glory of. It should be mentioned that the author of the report has himself made contributions to the science of light, more valuable than any one could collect from the statements in the report itself, and that important communications in that science are expected to be made during the present week, by Professor Powell, to a general meeting, and by Mr. MacCullagh to the physical section.

(The Secretary here read a notice, which he had procured from a scientific friend, of the report by Professor Jenyns on zoology; and afterwards continued his own remarks, as follows:)

The remaining reports in the new volume are those by Mr. Rennie on hydraulics; by Dr. Henry of Manchester, on the laws of contagion; and by Professor Clark of Cambridge, on animal physiology, and especially on our knowledge respecting the blood. Mr. Rennie's report contains, I believe, new facts from the manuscripts of his father, and is in other ways a valuable statement, industriously drawn up, of the recent improvements in the practice of hydraulics, to the theory of which science it is to be lamented that so little has lately been added: and without pretending to judge myself of the merits of the two other reports, I may mention them as compositions which I know to have interested persons, with whose professional and habitual pursuits they have no close connexion, and therefore, as an instance of the accomplishment of one great end proposed by our Association, that of drawing together different minds, and exciting intellectual sympathy. The other contents of the volume are accounts of researches undertaken at the request of the Association, notices in answers to queries and recommendations of the same body, and miscellaneous communications. Of these, it is of course impossible to speak now; your time would not permit it. Yet, perhaps, I ought not to pass over the mention of one particular recommendation which has happened to become the subject of remarks elsewhere—I mean that recommendation which advised an application to the Lords of the Treasury for a grant of money, to be used in the reduction of certain Greenwich observations, the result of which recommendation is noticed in the volume before us. In all that I have hitherto said respecting this Association, I have spoken almost solely of its internal effects, or those which it produces on the minds and acts of its own members. But it is manifest that such a society cannot fail to have also effects which are external, and that its influence must extend even beyond its own wide circle of members. It not only helps to diffuse through the community at large a respect and interest for the pursuits of scientific men, but ventures even to approach the throne, and to lay before the King the expression of the wishes of this his Parliament of science, on whatever subject of national importance belongs to science only, and is unconnected with the predominance in the state of any one political party. It was judged that the reduction of the astronomical observations on the sun and moon, and planets, which had been accumulating under the care of Bradley and his successors, at the Royal and national Observatory of Greenwich, since the middle of the last century, but which, except so far as foreign astronomers might use them, had lain idle and useless till now, to the great obstruction of the advance of practical as well as theoretical science, was a subject of that national importance, and worthy of such an approach to the highest functionaries of the state. It happened that I was not present when the propriety of making this application was discussed, so that I do not know whether the authority of Bessel was quoted. That authority has not at least been mentioned, to my knowledge, in any printed remarks upon the question, but as it bears directly and powerfully thereupon, you will permit me, perhaps, to occupy a few moments by citing it.

Professor Bessel of Koenigsberg, who, for consummate union of theory and practice, must be placed in the very foremost rank, may be placed perhaps at the head of astronomers now living and now working, published not long ago that classical and useful volume, the *Tabulæ Regiomontanæ,* which I now hold in my hand. In the introduction to this volume of tables, Bessel remarks, that "the present knowledge of the solar system has not made all the progress which might have been expected from the great number and goodness of the observations made on the sun, and moon, and planets, from the times of Bradley down. It may, indeed, be said with truth, that astronomical tables do not err now by so much as whole minutes from the heavens; but if those tables differ by more than five seconds now, after using all the present means of accurate reduction, from a well-observed opposition of a planet (for example), their error is as manifest and certain now as an error exceeding a minute was, in a former state of astronomy—and the discrepancies between the present tables and observations are not uncommonly outside that limit. The cause is doubtful. Errors of observation to such amount they cannot be; and therefore they can only arise from some wrong method of reduction, or wrongly assumed elliptic elements or masses of the planets, or insufficiently developed formulæ of perturbation, or else they point to some disturbing cause, which still remains obscure, and has not yet been reached by the light of theory. But it ought surely to be deemed the *highest problem of astronomy,* to examine with the utmost diligence into that which has been often said, but not as yet in every case sufficiently established, whether theory and experience do really always agree. When the solution of this weighty problem shall have been most studiously made trial of, in all its parts, then either will the theory of Newton be perfectly and absolutely confirmed, or else it will be known beyond all doubt that in certain cases it does not suffice without some little change, or that besides the known disturbing bodies there exist some causes of disturbance still obscure." And then after some technical remarks, less connected with our present subject, Bessel goes on to say, "To me, considering all these things together, it appears to be of the *highest moment* (*plurimum valere*) towards our future progress in the knowledge of the solar system, to reduce into catalogues as diligently as can be done, according to one common system of elements, *the places of all the planets observed since* 1750, than which labour, I believe that no other now will be of greater use to astronomy" (.. *quo labore nullum credo nunc majorem utilitatem Astronomiæ allaturum esse*). Such is the opinion of Bessel; but such is not the opinion of an anonymous censor, who has written of us in a certain popular review. To *him* it seems a matter of little moment that old observations should be reduced. Nothing good, *he* imagines, can come from the study of those obsolete records. It may be very well that thousands of pounds should continue to be spent by the nation, year after year, in keeping up the observatory at Greenwich; but as to the spending 500*l.* in turning to some scientific profit the accumulated treasures there, *that* is a waste of public money, and an instance of *misdirected influence* on the part of the British Association. For you, gentlemen, will rejoice to hear, if any of you have not already heard it, and those who have heard it already will not grudge to hear it again, that through the influence of this Association, what Bessel wished, rather than hoped, is now in process of accomplishment: and that, under the care of the man who in England has done most to show how much may be done with an observatory, that national disgrace is to be removed, of ignorance or indifference about those scientific treasures which England has almost unconsciously been long amassing, and which concern her as the country of Newton and the maritime nation of the world. For the spirit of exactness is diffusive, and so is the spirit of negligence. The closeness, indeed, of the existing agreement between the tables and the observations of astronomers is so great, that it cannot easily be conceived by persons unfamiliar with that science. No theory has ever had so brilliant a fortune, or ever so outrun experience, as the theory of gravitation has done. But if astronomers ever grow weary, and faintly turn back from the task which science and nature command, of constantly continuing to test even this great theory by observation, if they put any limit to the search, which nature has not put, or are content to leave any difference unaccounted for between the testimony of sense and the results of mathematical deduction, then will they not only become gradually negligent in the discharge of their other and more practical duties; and their observations themselves, and their nautical almanacs, will then degenerate instead of improving, to the peril of navies and of honour; but also they will have done what in them lay, to mutilate outward nature, and to rob the mind of its heritage. For, be we well assured that no such search as this, were it only after the smallest of those treasures which wave after wave may dash up on the shore of the ocean of truth, is ever unrewarded. And small as those five seconds may appear, which stir the mind of Bessel, and are to him a prophecy of some knowledge undiscovered, perhaps unimagined by man, we may remember that when Kepler was "feeling" as he said, "the walls of ignorance, ere yet he reached the brilliant gate of truth," he thus expressed himself respecting discrepancies which were not larger for the science of his time:—"These eight minutes of difference, which cannot be attributed to the errors of so exact an observer as Tycho, are about to give us the means of reforming the whole of astronomy." We indeed cannot dream that gravitation shall ever become obsolete; perhaps it is about to receive some new and striking confirmation; but Newton never held that the law of the inverse square was the only law of the action of body upon body; and the question is, whether some other law or mode of action, coexisting with this great and principal one, may not manifest some sensible effect in the heavens to the delicacy of modern observation, and especially of modern reduction. It was worthy of the British Association to interest themselves in such a subject: it was worthy of British rulers to accede promptly to such a request.

I have been drawn into too much length by the consideration of this instance of the external effects of our Association, to be able to do more than allude to the kindred instance of the publication of the observations on the tides in the port of Brest, which has, I am informed, been ordered by the French Government, at the request of M. Arago and the French Board of Longitudes, who were stimulated to make that request by a recommendation of the British Association at Edinburgh. Many other topics, also, connected with your progress and prospects, I must pass over, having occupied your time so long; and in particular I must waive what, indeed, is properly a subject for your general committee—the consideration whether anything can be done, or left undone, to increase still more the usefulness of this Association, and the respect and good will with which it is already regarded by the other institutions of this and of other countries. As an Irishman, and a native of Dublin, I may be suffered in conclusion to add my own to the many voices which welcome this goodly company of English, and Scottish, and foreign visitors to Ireland and to Dublin. We cannot, indeed, avoid regretting that many eminent persons, whose presence we should much enjoy, are not in this assembly; though not, we trust, in any case, from want of their good will or good opinion. Especially we must regret the absence of Sir David Brewster, who took so active a part in forming this association: but I am authorized, by a letter from himself, to mention that his absence proceeds entirely from private causes, and that they form the only reason why he is not here. Herschel, too, is absent; he has borne with him to another hemisphere his father's fame and his own; perhaps, from numbering the nebulæ invisible to northern eyes, he turns even now away to gaze upon some star, which we, too, can behold, and to be in spirit among us. And other names we miss; but great names, too, are here: enough to give assurance that in brilliance and useful effect, this Dublin meeting of the Association will not be inferior to former assemblings, but will realize our hopes and wishes, and not only give a new impulse to science, but also cement the kindly feeling which binds us all together already.

XLIV.

ON A NEW THEORY OF LOGOLOGUES:

ALSO

ON A NEW THEORY OF VARYING ORBITS:

AND

EXPLANATION OF THE METHOD OF INVESTIGATION PURSUED BY MR G. B. JERRARD FOR ACCOMPLISHING THE SOLUTION OF EQUATIONS OF THE FIFTH AND OF HIGHER DEGREE (1835)

[*British Association Report* 1835, Part 11., p. 7.]

XLV.

NEW APPLICATION OF THE CALCULUS OF PRINCIPAL RELATIONS; *AND* EXPOSITION OF MR. TURNER'S THEOREM RESPECTING THE SERIES OF ODD NUMBERS, AND THE CUBES AND OTHER POWERS OF THE NATURAL NUMBERS (1837)

[*British Association Report* 1837, Part II., p. 1.]

XLVI.

INAUGURAL ADDRESS AS PRESIDENT OF THE ROYAL IRISH ACADEMY (1838)

[January 8, 1838]

[*Proceedings of the Royal Irish Academy* **1**, pp. 107–120 (1841).]

My Lords and Gentlemen of the Royal Irish Academy,

The position in which your kindness has placed me, entitles me, perhaps, to address to you a few remarks. Called by your choice to fill a chair, which Charlemont, and Kirwan, and others, not less illustrious, have occupied, I cannot suffer this first occasion of publicly accepting that high trust to pass in silence by, as if it were to me a thing of course. Nor ought I to forego this natural opportunity of submitting to you some views respecting the objects and prospects of this Academy, which, if they shall be held to have no other interest, may yet be properly put forward now, as views, by the spirit at least of which I hope that my own conduct will be regulated, so long as your continuing approbation shall confirm your recent choice, and shall retain me in the office of your President.

First, then, you will permit me to thank you for having conferred on me an honour, to my feelings the most agreeable of any that could have been conferred, by the unsolicited suffrages of any body of men. Gladly indeed do I acknowledge a belief, which it would pain me not to entertain, that friendship had, in influencing your decision, a voice as potent as esteem. An Irishman, and attached from boyhood to this Academy of Ireland, I see with pleasure in your choice a mark of affection returned. But knowing that the elective act partakes of a judicial character, and that the exercise of friendship has its limits, I must suppose that the same long attachment to your body, which had won for me your personal regard, appeared also to you a pledge, more strong than promises could be, that if any exertions of mine could prevent the interests of the Academy from suffering through your generous confidence, those exertions should not be withheld; and that you thought they might not be entirely unavailing. After every deduction for kindness, there remains a manifestation of esteem, than which I can desire no higher honour, and for which I hope that my conduct will thank you better than my words.

And yet, Gentlemen, it is to me a painful thought, that the opportunity for your so soon bestowing this mark of confidence and esteem has arisen out of the deaths, too rapidly succeeding each other, of the two last Presidents of our body, who, while they are on public grounds deplored, and for their private worth were honoured and beloved by all of us, must ever be remembered by me with peculiar love and honour:—Brinkley, who introduced me to your notice, by laying on your table long ago my first mathematical paper; and Lloyd, whose works, addressed to the University of Dublin, first opened to me that new world of mind, the application of algebra to geometry. But of these personal feelings, the occasion has betrayed me into speaking perhaps too much already. Into that fault, I trust, I shall not often fall again. I pass to the exposition of views respecting the objects and prospects of our Society.

The Royal Irish Academy was incorporated (as you know) in 1786, having been founded a short time before, for the promotion generally, but particularly in Ireland, of Science, Polite Literature, and Antiquities. Its objects were to be the *True*, the *Beautiful*, and the *Old*: with which ideas, of the True and Beautiful, is intimately connected the coordinate (and perhaps diviner) idea of the *Good*. So comprehensive, therefore, was the original plan of this Academy, that it was designed to include nearly every object of human contemplation, and might almost be said to adapt itself to all conceivable varieties of study; insomuch that scarce any meditation or inquiry is directly and necessarily excluded from a place among our pleasant labours: and precedents may accordingly be found, among our records, for almost every kind of contribution. If only a diligence and patient zeal be shown, such as befit the high aims of our body; and if due care be taken, that the spirit of love be not violated, nor brother offend brother in anything; no strict nor narrow rules prevent us from receiving whatever may be offered to our notice, with an indulgent and joyful welcome. And though we meet only as studious, meditative men, and abstain from including among our objects any measures of immediate, outward, practical utility, such as improvements in agriculture, or other useful arts,—a field which had been occupied, in this metropolis, by another and elder society, before the institution of our own; yet no philosopher nor statesman, who has reflected sufficiently on the well-known connexion between theory and practice, or on the refining and softening tendencies of quiet study, will think that therefore we must necessarily be useless or unimportant as a body, to Ireland, or to the Empire.

The *object* of this Academy being thus seen to be the encouragement of STUDY, we have next to consider the *means* by which we are to accomplish, or to tend towards accomplishing that object. Those means are of many kinds, but they may all be arranged under the two great heads of *inward* and *outward* encouragement; or, in other words, *stimuli* and *assistances*; in short, SPURS and HELPS to study. The encouragement that is given may act as supplying a motive, or as removing a hindrance; it may be indirect, or it may be direct; invisible or visible; mental or material. Not that these two great kinds of good and useful action are altogether separated from each other. On the contrary, they are usually combined; and what gives a stimulus, gives commonly a facility too. In our *meetings*, for example, the *stimulating* principle prevails; yet in them we are not only caused to feel an increased *interest* in study generally, through the operation of that social spirit, or spirit of sympathy, of which I spoke so largely, in the presence of most of you, at the meeting of the British Association* in this city; but also are directly *assisted* in pursuing our own particular studies, by having the results of other studious persons early laid before us, and commented upon, by themselves and by others, in a fresh familiar way. We are not only spurred but helped to study, by mixing freely with other students.—A *library*, again, is designed rather to *assist* than to stimulate; and yet it is impossible for a person of ardent mind to contemplate a well selected assemblage of books, containing what Milton has described as "the precious life-blood of a master-spirit, embalmed and treasured up on purpose to a life beyond life," without feeling a deep desire to add, to

* See the Address printed in the Fifth Report of the British Association for the Advancement of Science,—*Note by* PRESIDENT. [See this Vol. p. 716.]

the store already accumulated, some newer treasure of his own. Our library, then, spurs as well as helps.—The *prizes* which from time to time we award for successful exertion in the various departments of study, might seem to be *stimulants* only; yet if we were to act sufficiently upon the spirit of precedents, of which we have several among our past proceedings, and which allow us to make our awards in part pecuniary, as well as honorary, they might become important *assistances*, and not merely *excitements* to study; they might serve, for instance, to enrich the private libraries of the authors on whom they were conferred. Why might we not, for example, instead of giving one gold *medal*, which can (according to the custom of this country) only be gazed at for a while and then shut up, allow the author who has been thought worthy of a prize to select any *books* for himself, which he might think most useful for his future researches, within a certain specified limit of expense; and then not only purchase those books for him out of our own prize funds, but also stamp them with the arms of the Academy, or otherwise testify that they were given to him by us as a reward? Or might not some such presentation of books be at least combined with the presentation of medals? But the whole system of prizes will deserve an attentive reconsideration, for which this is not the proper time nor place; and anything that I may now have said, or may yet say on that subject, in this address, is to be looked upon as merely intended to *illustrate* a few general *views* and principles, and not as any *proposal* of *measures* for your adoption; since, upon measures of detail, I have not as yet even made my own mind up; and am aware that, by the constitution of our Society, all measures of that kind must first be matured in the Council, before they are submitted to the Academy at large for final sanction or rejection.

The publication of our *Transactions* is another field of action for our body, and perhaps the most important of all; in which it is not easy to determine whether the stimulating or the assisting principle prevails; so much both of inducement and of facility do they give to study and to its communication. It is indeed a high reward for past, and inducement to future labours, to know that whatever of value may be elicited by the studies of any members of this body, (nor are we to be thought to wish to *confine* the advantage to *them*,) is likely or rather is sure to be adopted by the Society at large, and published to the world, at least to the learned world, in the name and by the order of the whole:—the responsibility for any errors of detail, and the credit for any merit of originality, remaining still in each case with the author, while the Academy exercises only a right of preliminary or *primâ facie* examination, and a superintendence of a general kind. Nay, the more rigorous this preliminary examination is, and the more strict this general superintendence, the greater is the compliment paid to the writer whose productions stand the test; and the more honourable does it become to any particular essay, to be admitted among the memoirs of a Society, in proportion as those memoirs are made more select, and expected and required to be more high. But besides this honorary stimulus, which we should all in our several spheres exert ourselves to make more effective, by each endeavouring, according to his powers, to contribute, or to judge, or to diffuse, there is also a powerful and direct *assistance* given to study, by the publishing of profound intellectual works at the expense of a corporate body, rather than at the expense of individuals; a course which spares the private funds of authors and of readers; and thus procures, for the collections of learned and studious men, many works of value, which otherwise might never have appeared. Indeed, the publication of Transactions has long been regarded by me as the most direct and palpable advantage resulting from the institution of scientific and literary societies like our own; and, I believe, that I expressed myself accordingly, on the occasion* to which I lately alluded. But having *then* to deal with science only, I felt that it was unnecessary, and would have been improper for me to have introduced any view of the connexion and contrast between science and other studies, which are, not less than science, included among the objects of this Academy, and may therefore be fitly, if briefly, brought now before your notice. The union of all studies is indeed that at which we aim; but the three great departments, which our founders distinguished without dividing, may now also with advantage be distinctly considered, and separated, that they may be re-combined; a clearness of conception being likely to be thus attained, without any sacrifice of unity.

Directing our attention, therefore, first to science, or the study of the True,—

Inter sylvas Academi quærere verum,—

we find that, even when thus narrowed, the field to be examined is still so wide as to make necessary a minuter distinction; whether we would inquire, however briefly, what has been already done by this Academy, or what may fitly be desired and hopefully proposed to be done. Were we to rush into this inquiry without any previous survey of its limits, and, as were natural, allowed ourselves to begin by considering the actual and possible relation of our studies to the primal science, or First Philosophy, the Science of the Mind itself; we might easily be drawn, by the consideration of this one topic, into a discussion, interesting indeed, and (it might be) not uninstructive, but of such vast extent as to leave no room for other topics, which ought even less to be omitted, because they have hitherto come, and are likely to come hereafter, more often than it before our notice, in actual contributions to our Transactions. Indeed I think it prudent at this moment to resist altogether the temptation of expatiating on this attractive theme, of Philosophy, eminently so called; and to content myself with remarking, that as metaphysical investigation has more than once already found place among the scientific labours of this Academy, so ought it to take rank among them still, and to reappear in that character, from time to time, in our pages.

Confining ourselves, therefore, at present to Science, in the usual acceptation of the term, and inquiring what are its chief divisions, in relation mainly to the connected distribution or classification of scientific essays in our Transactions, we soon perceive that three such parts of science may conveniently be distinguished from each other, and marked out for separate consideration; namely those three, which, with some latitude of language, are not uncommonly spoken of as Mathematics, Physics, and Physiology. The first, or *mathematical* part, being understood to include not only the pure but the mixed mathematics; not only the results of our original intuitions of time and space, but also the results of the combination of those intuitions with the not less original notion of cause, and with the observed laws of nature, so far and no farther than that ever-widening sphere extends, within which observation is subordinate to reasoning; in short, all those deductive studies, in which Algebra and Geometry are dominant, though the dynamical and the physical may enter as elements also. The second, or *physical* part of science, embracing all those inductive studies respecting unliving or unorganized bodies, which proceed mainly through outward observation or experiment, and can as yet make little progress in "the high *priori* road." And finally, the

* See Address, already cited, p. 720.—*Note by* PRESIDENT.

third, or *physiological* part, including all studies of an equally inductive kind, respecting living or organized bodies. (I do not pretend that this arrangement is the most philosophical that can be imagined, but it may suffice for our present purpose.)

In all these divisions of science, and in several subdivisions of each, our published Transactions contain many valuable essays; and there seems to be no cause for apprehension that in *this* respect, at least, (if indeed in any other,) the Academy is likely to lose character. Death has, it is true, removed some mighty names from among us—elders and chiefs of our society: but the stimulus and instruction of their example have not been thrown away: an ardent band of followers has been raised up by themselves to succeed them. To keep the trust thus handed down, is an arduous, but noble charge, from which it is not to be thought that any here will shrink, whatever his share of that charge may be.

And yet, while Mathematics and Physics seem likely not to be neglected here, or rather certain to be ardently pursued, it may be pardoned me if I express a fear and a regret, that Physiology, or more precisely, the study of the phenomena and laws of life, and living bodies, has not been represented lately in the published Transactions of our Academy, to a degree correspondent with the eminence of the existing School of physiological study in Dublin. Our medical men and anatomists, our zoologists and botanists also, will take, I hope, this little hint in good part. They know how far I am from pretending to criticize their productions, and that I only wish to have more of their results brought forward here, for the instruction of myself and of others. *That* is not, I think, too much to ask from gentlemen who have subscribed the obligation which is signed by every member of this body, and who are qualified, by intellect and education, to take an enlarged yet not exaggerated view of the importance of a central society. I know that many other, and indeed more appropriate outlets exist, for the publication of curious, isolated, or semi-isolated facts: but it is not so much remarkable *facts*, as remarkable *views*, that I wish to see communicated to us, and through us to the world; although such views ought, of course, to be illustrated and confirmed by facts.

It seems possible, that in each of the three great divisions of science already enumerated, our Transactions may be enriched in future, through a judicious system of rewards, (of the kinds to which I lately alluded,) intended to encourage contributions of a more elaborate kind than usual, from strangers as well as from members of our body. It has appeared, for example, to some members of your Council, and to me, that for each of those three divisions of science a *triennial prize* might be given; these three triennial prizes succeeding each other in such rotation, for mathematics, physics, and physiology, that a prize should be awarded every year, on some one principal class of scientific subjects, for the best essay which had been communicated for publication, on any subject of that class, whether by a member or by a stranger, during the three preceding years. A plan of this sort has been lately tried, and (it would seem) with advantage, in the distribution of the Royal Medals entrusted by the late King* to the Royal Society of London; and the principle is not unsanctioned by you, that a greater range of investigation may sometimes be allowed to the authors of prize-essays, than the terms of an ordinary prize-question would allow. So that it only remains for your Council to consider and report to you, as they are likely soon to do, to what extent this principle may

* And continued by her present Majesty: whose gracious intention of becoming Patroness of the Royal Irish Academy has been made known since the delivery of this Address.—*Note by* PRESIDENT.

advantageously be pushed, and by what regulations it may conveniently be carried into effect. In saying this, I do not presume to pronounce that it is expedient to give up entirely the system of proposing occasionally prize-questions, of a much more definite kind than those to which I have been referring as desirable; but thus much I may venture to lay down, that original genius in inquirers ought to be as far indulged as it is possible to indulge it, both in respect of subject and of time; and that due time ought also to be allowed to those members of a Scientific Society, on whom is put the important and delicate office of pronouncing an award in its name.

The length at which I have spoken of our relations to Science, as a Society publishing Transactions, though far from exhausting that subject, leaves me but little room, in this address, to speak of our relations to Literature and Antiquities; subjects to which, indeed, I am still less able to do justice, than to that former theme. But the spirit of many of my recent remarks applies to these other subjects also; and you will easily make the application, without any formal commentary from me. A word or two, however, must be said on some points of distinction and connexion between the one set of subjects and the other.

As, in Science, or the study of the *True*, the highest rank must be assigned to the science of the investigating Mind itself, and to the study of those Faculties by which we become cognizant of truth; so, in Literature, or the study of the *Beautiful*, the highest place belongs to the relation of Beauty to the mind, and the study of those essential Forms, or innate laws of taste, in and by which, alone, man is capable of beholding the beautiful. Above all particular fair things is the Idea of Beauty general: which in proportion as a man has suffered to possess his spirit, and has, as it were, won down from heaven to earth, to irradiate him with inward glory, in the same proportion does he become fitted to be a minister of the spirit of beauty, in the poetry of life, or of language, or of the sculptor's, or the painter's art. The mathematician himself may be inspired by this in-dwelling beauty, while he seeks to behold not only truth but harmony; and thus the profoundest work of a Lagrange may become a scientific poem. And though I am aware that little can be communicated by expressions so general (and some will say so vague) as these, and check myself accordingly, to introduce some remarks more specific and definite; yet I will not regret that I have thus for a moment attempted to give words to that form of emotion, which many here will join with me in acknowledging to be the ultimate spring of all genuine and genial criticism, in literature and in all the fine arts. For we, in so far as we are an Academy of Literature, are also a Court of Criticism;—Criticism which is to Beauty, what Science is to Nature. Between the divine of genius and the human of enjoyment, we hold a kind of middle place; creating not, nor merely feeling, but aspiring to understand: and yet incapable of rightly understanding, unless we at the same time sympathize.

To express myself then in colder and more technical terms, I should wish that metaphysico-ethical and metaphysico-æsthetical essays,—those which treat generally of the beautiful in action and in art, and are connected rather with the study of the beauty-loving mind itself, than of the particular products or objects which that mind may generate or contemplate,—should be considered as entitled to the foremost place among our literary memoirs. After these *à priori* inquiries into the PRINCIPLES of beauty, which are rather *preparatory* to criticism than criticism itself, or which, at least, deserve to be called *criticism universal*, should be ranked, I think, that important but *à posteriori* and inductive species of criticism, which, from the study of some actual master-pieces, collects certain great RULES as *valid*, without deducing them as *necessary* from any higher principles. And last, yet still deserving of high honour, I would rank those researches of DETAIL, those particulars, and helps, and applications of criticism, which, if they be, in a large philosophical view, subordinate and subsidiary to principles, and to rules of universal validity, yet form perhaps the larger part of the habitual and ordinary studies of men of erudition; such as the differences and affinities of languages, and the explication of obscure passages in ancient authors. Whatever metaphysical preference I may feel for inquiries of the two former kinds, no one, I hope, will misconceive me as speaking of this last class of researches with any other feelings than those of profound respect, and of desire and hope to see them cultivated here; nor as presenting other than hearty congratulations to the Academy on the fact, that whereas no single paper on Literature appeared in our last volume, two memoirs, interesting and erudite, have been presented to us, and probably are by this time printed, to be in readiness for our next publication;—one, on the Punic Passage in Plautus, by a near and dear relative of my own; and the other, on the Sanscrit Language, by a gentleman of great attainments and of high station in our national University: from which seat of learning, it seems not too much to hope, that we shall soon receive many other contributions in the department of Polite Literature, as well as in other departments. It is, of course, understood that the awarding of prizes is not to be confined to scientific papers, but is to be extended, as indeed it has always been, under some convenient regulations, to literary and antiquarian papers also.

I was to say a few words respecting that other department of our Transactions, namely, Antiquities, or the study of the Old; and if, at this stage of my address, those words must be very few, I regret this circumstance the less, because I know that the study is deservedly a favourite here, and that I am surrounded by persons who are, beyond all comparison, more familiar with the subject than myself.

In general, I may say, that whether the study of Antiquities be regarded in its highest aspect, as the guardian of the purity of history,—the history of nations and of mankind; or as ministering to literature, by recovering from the wreck of time the fragments of ancient compositions; or as indulging a natural and almost filial curiosity to know the details of the private life of eminent men of old, and to gaze upon those relics which invest the past with reality, as the palæontologist from his fossils reconstructs lost forms of life: in all these various aspects, the study is worthy to interest any body of learned men, and to occupy a considerable part of the Transactions of any society so comprehensive as our own. The historian of the Peloponnesian war was also himself an antiquarian; and prefaced that work which was to be "a possession for ever," by an inquiry into the antiquities of Greece. And while he complained of the ḅªŒ ĄÃŠÃÔÞªρÿŒ̧ ÿĩŒ ÞÿŠŠÿĩŒ ῆ ÌßÓœÕŒ̧ ῆŒ ĄŠÓÒÍÔÃŒ, that easy search after truth which cost the multitude nothing; he also claimed to have arrived at an ÎŸῆŒ̧ ÍõšÓρÔÿÚ, a linked chain of antiquarian proof, by which he could establish his correction of their errors. Indeed, the uninitiated are apt to doubt,—perhaps too they may sometimes smile,—when they observe the earnest confidence which the zealous Antiquary reposes in results deduced from arguments which seem to them to be but slight; nor dare I say that I have never yielded to that sort of sceptical temptation. But I remember a fact which ought to have given me a lesson, on the danger of hastily rejecting conclusions which have been maturely considered by others. A learned Chancellor of Ireland, now no more, assured me often and earnestly, that he gave no faith to the inductions of astronomers respecting the distances and sizes of the sun and moon; and hinted that he disliked our year, for containing the odd fraction of a day. Yet this was a man, not only of great private worth, but of great intellectual power, and eminent in his profession as in the state. Astronomers and mathematicians, it may be, look sometimes on other inductions with a not less unfounded incredulity. It is one of the advantages of an Academy, so constituted as ours is, that it brings together persons of the most different tastes and the most varied mental habits, and teaches them an intellectual toleration, which may ripen into intellectual comprehension. Thus, while the antiquary catches from the scientific man his ardent desire for progression, and for that clearer light which is future, the man of science imbibes something in return, of the antiquarian reverence for that which remains from the past. The literary man and the antiquary, again, re-act upon each other, through the connexion of the Beautiful and the Old, which in conception are distinct, but in existence are often united. And finally, the scientific man learns elegance of method from the man of literature, and teaches him precision in return.

Before I leave the subject of Transactions, I may remark that their value, both as stimulants and as assistants to study, must much depend on the rapidity and extent of their circulation, and on the care that is taken to put them as soon as possible into the hands or within the reach of studious men abroad. Reciprocally it is of importance that measures should be taken for obtaining speedy information here of what is doing by such men in other countries. On both these points, some reforms have lately been made, but others still are needed, and will soon be submitted to your Council. On these and all questions of improvement, I rely upon receiving the assistance of all those gentlemen who are in authority among us; but especially am encouraged by the hope of the cordial co-operation of your excellent Vice-President, Professor Lloyd, who has done so much already for this Academy, in these and in other respects.

It may deserve consideration, as connected with the last-mentioned point, whether Reports upon some foreign memoirs of eminent merit, accompanied by extracts, and, perhaps, translations, might not sometimes be advantageously called for. There is, I think, among our early records, some hint that the Academy had once a paid Translator. It may or it may not be expedient to revive the institution of such an office; or to give direct encouragement to the exertions of those,* who, without any express reference to our own body, work in this way for us, while working for the public; but no one can doubt that it is desirable to diminish the too great isolatedness which at present exists among the various learned bodies of the world. The Reports of the British Association on the actual state of science in each of its leading subdivisions, do not exactly meet the want to which I have alluded; because, upon the whole, they aim rather at condensing into one view the ultimate *conclusions* of scientific men in general, than at diffusing the fame and light of individual scientific genius, by selecting some few great foreign works, and making known at home their *method* as well as well as their results. Besides we must remember that far as that colossal Association exceeds the body to which we belong, in numbers, wealth, and influence, yet in plan it is less comprehensive; since it restricts itself to science exclusively, while we aspire, as I have said, to comprehend

* For instance, Mr. Richard Taylor [1781–1858], of London, F. S. A., &c., who lately began to publish *Scientific Memoirs*, selected and translated from the Transactions of Foreign Academies of Science, and other foreign sources; which valuable publication is now suspended for want of sufficient support from the public.—*Note by* PRESIDENT.

nearly the whole sphere of thought,—at least of thought as applied to merely human things: in making which last reservation, I shall not, I hope, be supposed wanting in reverence for things more sacred and divine.

With that powerful and good Association, however, we should endeavour to continue always on our present, or if possible, on closer terms of amicable relation. I need not say that we should also aim to preserve and improve our friendly relation with all the other Scientific, Literary, and Antiquarian Societies, of these and of foreign countries. Especially we ought to regard, with a kind of filial feeling of respect and love, the Royal Society of London—that central and parent institution, from which so many others have sprung; over which Newton once presided; and in which our own Brinkley wrote. While feelings of this sort are vigilantly guarded, and public and private jealousies excluded vigilantly, a vast and almost irresistible moral weight belongs to companies like these, of studious men; and, amid the waves of civil affairs, the gentle voice of mind makes itself heard at last. Societies such as ours, if they do their duty well, and fulfil, so far as in them lies, their own high purpose, become entitled to be regarded as being, on all purely intellectual and unpolitical questions, hereditary counsellors of crown and nation. The British Association has already made applications to government with success, for the accomplishment of scientific objects; and I am not without hopes that our own recent memorial, for the printing, at the public expense, of some valuable manuscripts in our possession, adapted to throw light on history, and interesting in an especial degree to us as Irishmen, will receive a favourable consideration.

On the present occasion, which to me is solemn, and to you not unimportant, I may be pardoned for expressing, in conclusion, the pleasure which it gives me to believe, that while we cautiously abstain from introducing polemics or politics, or whatever else might cause an angry feeling in this peaceful and happy society, some great and fundamental principles, of duty to heaven and to the state, are universally recognized amongst us. Admitted at an early age to join your body, I now have known you long, and hope to know you longer; but have never seen the day, and trust that I shall never see it, when piety to God, or loyalty to the Sovereign, shall be out of fashion here.

XLVII.

ADDRESS AS PRESIDENT OF THE ROYAL IRISH ACADEMY ON PROFESSOR MACCULLAGH'S PAPER ON THE LAWS OF CRYSTALLINE REFLEXION AND REFRACTION (1838)

[June 25, 1838]
[*Proceedings of the Royal Irish Academy* **1**, 212–221 (1841).]

The time has now arrived for terminating the present session; and it will, no doubt, be gratifying to you, as it is to me, that our closing act should be the public presentation of a Medal to one of our most distinguished Members; that Medal being the first which has been awarded by your Council in the exercise of the new and fuller power confided by you lately to them, and in execution of the plan which was announced to you at the time when you gave them that enlarged discretion, with respect to the bestowal of honorary rewards.

That plan, as you may remember, differs little from the scheme suggested by me in the inaugural address which I had the honor to deliver on the occasion of first taking the chair of this Academy: the only difference, indeed, so far as science is concerned, being the subsequent adoption of a suggestion of Professor Lloyd, respecting a change of distribution of those subjects which were included by me under the two great heads of Physics and Physiology, but by him under those of Experimental and Observational Science, or Physics and Natural History. The time for acting upon this modification has not, however, as yet arrived; and before the suffrages of your Council were collected, at its last meeting, on the question of the absolute and relative merits of the various communications which have lately been made to our Transactions, it was resolved to postpone, till after the recess, the consideration of all scientific or other awards, except only that which should be made for the most important paper in pure or mixed mathematics, communicated during the three years which ended in March 1837, and already actually printed. The papers coming within this definition were few; the authors of them were only two, Professor MacCullagh and myself. The decision, which in theory is a decision of the President and Council, and which did in fact receive my cordial and previously expressed concurence, was in favour of Mr. MacCullagh's paper "On the Laws of Crystalline Reflexion and Refraction," contained in the just published part of the eighteenth volume of the Transactions of this Academy.

It may happen that upon future occasions of this sort, if it shall again become my duty to present from this Chair those Medals which may hereafter be awarded, for papers of other triennial cycles, and upon other subjects, I may not think it necessary or expedient to occupy your time by any but the briefest statement of the grounds on which those future awards may have been made. But on the present occasion, which is (to me at least, and in relation to our new plan) the first occasion of its kind; while the subject is one of a class to which my own inquiries have been much directed, and upon which, therefore, I may speak with a less risk of impropriety than upon many others; and while we, as an Academy, by extra hours and extra nights of attendance, during that busy session which is now about to close, have earned for ourselves a little leisure, on this last night of meeting, without interfering (as we hope) with the rights, or even with the convenience of authors; I think myself allowed to enter more at large into the merits of the award, and to lay before you some of the thoughts which the perusal of the present prize essay has suggested to my own mind.

When ordinary light is reflected at the common boundary of two transparent and uncrystallized media, as when we see (for example) for reflexion of the sun in water, the reflected light differs from the incident in both direction and intensity, according to laws which were known to Euclid in so far as they regard *direction*, but of which the discovery, in so far as *intensity* is concerned, was reserved for the sagacity of Fresnel. In general, the laws which regulate the changes of the direction of light have been found easier of discovery than those which regulate its changes of intensity; the laws of the reflexions and refractions of the lines along which light is propagated, than the laws of the accompanying determinations or alterations of its planes of polarisation; or, to express the same distinction in the language of the theory of indulations, it has been found easier to assign the form of the waves which spread from any origin of disturbance through any given portion of the elastic luminiferous ether, than to assign the directions and relative magnitudes of the vibrations which constitute those waves, and the laws which regulate the changes of such vibrations, in the passage from one medium to another.

The laws which regulate such *changes of vibration*, produced by reflexion and refraction, at the boundaries of crystallized media, have been the special object of Mr. MacCullagh's investigations, in the paper now before us. But in investigating them, he has been obliged to consider also the laws which regulate the vibrations of the ether, in the *interior* of a crystallized body, and not at its *surface* only; the laws of the *propagation* as well as those of the *reflexion* and *refraction* of light. His researches are therefore connected intimately with a wide range of optical phenomena; and the hypotheses on which his formulæ are founded, and which seem to have their own correctness proved by the experiments of many kinds with which they have been successfully compared, though liable, of course, like every physical induction, to be modified in some degree by future observation, appear to be entitled to assume henceforth a very high rank among the principles of physical optics.

The method which Mr. MacCullagh has adopted may be said to be in general the method of *mathematical induction*, as distinguished from *dynamical deduction*. He has not sought to deduce, from any pre-supposed attractions or repulsions, and arrangements of the molecules of the ether, any conclusions respecting the vibrations in the interior or at the boundaries of a medium, as necessary consequences of those dynamical principles or assumptions. But he has sought to gather from phenomena a system of mathematical laws by which those phenomena might be expressed and grouped together, be conceived in connexion with each other, and receive an inductive unity. He has sought to arrive at laws which might bear somewhat the same relation to the optical observations already made, as the laws of Kepler did to the astronomical observations of his predecessor Tycho Brahe, without seeking yet to deduce these laws, as Newton did the laws of Kepler, from any higher and dynamic principle. And though, no doubt, it is to such deduction that science must continually tend; and though, in optics, some progress has been actually made, by Cauchy and by others, to a dynamical theory of light, as a system of vibrations regulated by forces of attraction and repulsion; yet it may well be judged a matter of congratulation when minds are found endowed with talents so high as those which Mr. MacCullagh possesses, and willing to apply them to the preparatory but important task of discovering, from the phenomena themselves, the mathematical laws which connect and represent those phenomena, and are in a manner intermediate between facts and principles, between appearances and causes.

It was thus, that, in a former paper, Mr. MacCullagh proposed, as mathematical expressions for the phenomena of Quartz, a system of differential equations, which are indeed simple in themselves, and seem to agree well with observation, but have not yet been shown to be consistent with dynamic views. And in that later memoir for which the present prize is awarded, he has, in like manner, adopted some hypotheses, and rejected others, without apparently regarding whether and how far it may seem possible at present to reconcile such adoption or such rejection with received opinions respecting the mechanism of light; exhibiting thus, a kind of intellectual courage, in admiring which I am fortified by the opinion of Sir John Herschel, who lately, in a conversation and a letter, expressed himself thus to me: "The perusal of Mr. MacCullagh's paper on the Laws of Reflexion and Polarisation in Crystals, has, although cursory, produced a very strong impression on my mind that the theory of light is on the eve of some considerable improvement, and that by abandoning for a while the *à priori* or deductive path, and searching among phenomena for laws simple in their geometrical enunciation, and of more or less wide applicability, *without (for a while) much troubling ourselves how far those laws may be in apparent accordance with any preconceived notions, or even with what we are used to consider as general principles in dynamics,* it may be possible to unite scattered fragments of knowledge into such groups and masses as shall afford glimpses of their fitness to combine into a regular edifice."

The hypotheses which are the bases of Mr. MacCullagh's theory of Crystalline Reflexion and Refraction are the following. He supposes that the form of the wave surface in a doubly-refracting crystal is that which was assigned by Fresnel, and that the vibrations are tangential to this surface, but that they are perpendicular to the ray, and consequently *parallel* to the plane of polarisation; whereas Fresnel supposed them to coincide with the projection of the ray upon the wave, and consequently to be perpendicular to the plane of polarisation. Professor MacCullagh supposes also, with Fresnel, that the *vis viva* is preserved, or in other words, that the reflected and refracted lights are together equal to the incident; but in applying this principle to investigate the refracted vibrations, he supposes, in opposition to Fresnel, that the density of the ether is *not changed* in passing from one body to another. And he supposes, finally, that *the vibrations in two contiguous transparent media are equivalent*; or, in other words, that the resultant of the incident and reflected vibrations is the same, both in length and direction, as the resultant of the refracted vibrations; whereas Fresnel had supposed only that the vibrations parallel to the separating surface, but not that the vibrations perpendicular to the same surface were equivalent.

And here I may be permitted to state, what indeed cannot fail to be remembered by many here, that when the British Association for the Advancement of Science met in this city, about three years ago, (in August, 1835), a communication was made by Mr. MacCullagh to the Mathematical and Physical section, "on the Laws of Reflexion and Refraction at the Surface of Crystals," which embodied nearly all the principles or hypotheses that I have now recited, and of which an abstract was printed in the London and Edinburgh Philosophical Magazine for October, 1835, having indeed been published even earlier (in September, 1835) by Mr.

Hardy here. The only supposition, which was not either formally stated or clearly indicated in this abstract, was that of the preservation of the *vis viva*; instead of which principle of Fresnel, Mr. MacCullagh was, at one time, inclined to employ a relation between pressures, proposed by M. Cauchy. Since, therefore, the leading principles of the new theory of Reflexion and Refraction were all made known by Mr. MacCullagh so early as the August of 1835, were printed in Dublin in the September of that year, and in London in the October following, it will not, perhaps, be attributed solely to national partiality if we claim for him the priority of discovery on this curious and important question, notwithstanding that a very valuable and elaborate memoir on the same subject, embodying the same results, was communicated, in December, 1835, to the Academy of Sciences at Berlin, by M. Neumann, and was published in 1837, before the publication (though after the reading) of that essay of Mr. MacCullagh, to which the present prize is awarded.

It is, however, an interesting circumstance, and one which is adapted to increase our confidence in these new laws of light, that they should have been independently and almost simultaneously discovered in these and in foreign countries; and it will not, I trust, be supposed that I desire to depreciate M. Neumann's admirable essay, if having recalled some facts and dates which bear upon the question of priority, I proceed to point out a few of the features of Mr. MacCullagh's briefer paper, which have appeared to me to deserve a peculiar and special attention. I mean the geometrical elegance of the principal enunciations, and the philosophical character of the interspersed remarks.

As a specimen of the former, I shall select the theorem of the *polar plane*. When light in air is incident on a doubly-refracting crystal, it may be polarised in such a plane, that one of the two refracted rays shall disappear; and then the one refracted vibration which corresponds to the one remaining refracted ray, must (by the hypotheses or laws already mentioned) be the resultant of the one incident and one reflected vibration; and consequently these three vibrations must be contained in one common plane, which plane it is therefore an object of interest to assign a simple rule for constructing. In fact, the refracted vibration is known, in direction, from the laws of propagation of light in the crystal, and the hypotheses already mentioned; if, then, we know how to draw through its direction the plane just now referred to, we should only have to examine in what lines this plane intersected the incident and reflected waves, in order to obtain the direction of the incident and reflected vibrations, and afterwards (by the rules of statical composition) the relative magnitudes of all the three vibrations, or the relative intensities of the incident, reflected, and refracted lights. Now Mr. MacCullagh shows, that the desired construction can be deduced from the properties of the doubly refracting medium or wave, as follows: Let OT, OP represent in length and in direction the velocity of the refracted ray, and the slowness of the refracted wave; so that, by what has been before supposed, the refracted vibration OV is perpendicular to the plane TOP; then, if a plane be drawn *through the vibration* OV, *parallel to the line* TP, this plane, which Mr. MacCullagh calls the *polar plane* of the ray OT, will be the plane desired; that is, it will contain the incident and the reflected vibrations, if these be uniradial, or, in other words, if they have such directions, or correspond to such polarisations, as to cause one of the two refracted rays in the crystal to disappear.

Many elegant geometrical corollaries are drawn, in the Essay, from this theorem of the polar plane; but I shall only mention one, (which includes, as a particular case, the remarkable law for determining the angle of polarisation of light reflected at the surface of an ordinary medium, discovered by Sir David Brewster,) namely, that when the light reflected from the surface of a doubly refracting crystal is completely polarised, or, in other words, when the reflected vibration has a determined direction, independent of the direction of the incident vibration, then *the reflected ray is perpendicular to the intersection of the polar planes of the two different refracted rays.*

In this and other applications of the theorem of the polar plane to the case where the incident light is polarised so as to undergo a double refraction, the obvious manner of proceeding is to decompose its *one biradial* vibration into *two uniradial* vibrations, and to treat these separately, by applying to each the construction above described. Yet Mr. MacCullagh remarks, that it requires proof that the reflected and refracted intensities, thus determined, will have their sum exactly equal to the intensity of the incident light; or, in other words, that the law of the *vis viva* will hold good for the resultant vibrations, though we know, by the construction, that it holds good for each system of uniradial components taken separately. In fact, if the two separate incident vibrations, which correspond to the two separate refracted vibrations, be inclined at an acute angle to each other, they will generate by their superposition (according to the law of interference) a compound incident light, of which the intensity exceeds, by a determined amount, the sum of the two separate or component intensities; and it requires proof that the two separate reflected vibrations will in like manner be inclined to each other at that precise acuteness of angle which will allow the intensity of the compound reflected light to exceed, by precisely the same determined amount, the sum of the two separate intensities, corresponding to the two separate reflected vibrations: (or that the same sort of equality of differences between incident and reflected resultants and sums will take place, when the angles are obtuse and not acute;) the two refracted vibrations being not in general (in either case) superposed upon each other. Professor MacCullagh has arrived at an equation of condition, as necessary for the foregoing agreement, which expresses a property of the laws of propagation deduced from the laws of reflexion and refraction, however singular it may appear that the latter laws should give any information respecting the former; and he states that he has found this equation to express rigorously a property of Fresnel's wave. His demonstration of this latter property having not yet been published, I have been induced to investigate one for myself; and have thus been conducted to a construction of the condition in question, so simple that it may perhaps be mentioned here. Let R and W denote the planes VOT and VOP in the figure before referred to, which may also be called the planes of ray-polarisation and of wave-polarisation, for the ray OT, or for the corresponding wave; and let P', T', R', W' be analogous to P, T, R, W, but referred to any other ray or wave; then the following is the relation to be satisfied:

$$\mathrm{OT}.\mathrm{OP}'.\cos \mathrm{RW}' = \mathrm{OT}'.\mathrm{OP}.\cos \mathrm{R}'\mathrm{W};$$

RW' and $\mathrm{R}'\mathrm{W}$ denoting here diedral angles. Under this form, it is easily proved that Fresnel's wave surface possesses rigorously the property in question. Mr. MacCullagh's equation has been otherwise obtained by M. Neumann, namely, as a condition for the possibility of depressing the equation of the *vis viva* to the first from the second degree.

On this and many other points of the investigation, Mr. MacCullagh (as I have already said) has thrown out many interesting and philosophical remarks; for instance, that the perfect adaptation which thus appears to exist between the laws of the propagation and those of the reflexion and refraction of light, is a strong indication that these two sets of laws are derived from some one common source, in other and more intimate laws not yet discovered; and that it is allowed to hope that the next step in physical optics will lead us to those higher and more elementary principles by which the laws of reflexion and propagation are linked together as parts of the same system. His remarks on the probable connexion between the theories of metallic and crystalline reflexion, and on the hopefulness of ascending to a true theory of light by the method of mathematical induction from phenomena, (exemplified, as has been seen, in his own papers,) rather than by attempting prematurely to make deductions from dynamical principles, are also well worthy of attention, though my own habits of thought lead me to feel an even stronger interest in dynamic and deductive researches.

But I have suffered myself to speak at greater length than has been usually occupied by others before, or is likely to be occupied by me hereafter on other similar occasions, and certainly at greater length than was required to justify the award of your Council. The reasons which I pleaded at the commencement of this address may, perhaps, serve partly as my excuse for having occupied your time so long; and some additional indulgence may have been thought due by those who remember that many years ago, both here and elsewhere, in public and in private, I expressed strongly my admiration of the talents of him to whom I have now the gratifying office of presenting this first public mark of honour from his scientific brethren and cotemporaries.

[*The President then, delivering the Medal to Professor MacCullagh, addressed him as follows:—*]

Professor MacCullagh,

I present to you this medal, awarded to you by the President and Council of the Royal Irish Academy. Accept it as a mark of the interest and intellectual sympathy with which we regard your researches; of the pleasure with which we have received the communications wherewith you have already favoured us; and of our hope to be favoured with other communications hereafter. And when your genius shall have filled a wider sphere of fame than that which (though already recognized, and not here only) it has yet come to occupy, let *this* attest, that minds were found which could appreciate and admire you early in this your native country.

XLVIII.

NOTICE OF A SINGULAR APPEARANCE OF THE CLOUDS OBSERVED ON THE 16TH OF DECEMBER, 1838 (1839)

[Communicated January 14, 1839]
[*Proceedings of the Royal Irish Academy* **1**, 249 (1841).]

The President gave an account of a singular appearance of the clouds, observed on the 16th December, 1838, at the Observatory of Trinity College, Dunsink. They appeared, for at least the last four hours of day light, to be arranged in arches which converged very exactly to the N. E. and S. W. points of the horizon; while the breaks or joints in these arches were directed, though with less exactness, to two other horizontal points, which seemed to be always opposite to each other, but ranged from N. W. and S. E. to N. and S. Conjectures were offered with respect to the cause of this appearance.

XLIX.

ADDRESS AS PRESIDENT OF THE ROYAL IRISH ACADEMY ON DR. APJOHN'S RESEARCHES ON THE SPECIFIC HEAT OF GASES (1839)

[February 25, 1839]

[*Proceedings of the Royal Irish Academy* **1**, 276–284 (1841).]

The President delivered the following Address to the Academy.

I have now the honour to inform you, that your Council, in the exercise of the discretion entrusted to them by you, have taken into their consideration, since the commencement of the present session, the various papers which had been for a few years past communicated to our Transactions, on several different subjects, in order to determine whether any and which of those papers should be distinguished by the award of a Cunningham Medal: and that the medal for the most important Paper in Physics, communicated to us during the three years ending in March, 1838, has been adjudged to Dr. Apjohn, for his Essay on a New Method of investigating the Specific Heats of the Gases, published in the First Part of the Eighteenth Volume of the Transactions of this Academy.

The importance of the study of what are called the imponderable agents, is known to all physical inquirers. Indeed it would appear, that as the scientific history of Newton, and of his successors during the century which followed the publication of his Principles of Natural Philosophy, is connected mainly with the establishment of the law of universal gravitation, and with the deduction of its chief consequences; so are the mathematical and physical researches of the present age likely to be associated, for the most part, with the study of light and heat and electricity, and of their causes, effects, and connexions. Whatever, then, whether on the practical or on the theoretical side, in the inductive or the deductive way, may serve to extend or to improve the knowledge of these powerful and subtle agents or states of body, which are always and everywhere present, but always and everywhere varying, and which seem to be concerned in all the phenomena of the whole material world, must be received by scientific men as a welcome and valuable acquisition.

Among researches upon heat, the highest rank is, (I suppose,) by common assent, assigned to such works as those of Fourier and Poisson, which bring this part of physics within the domain of mathematical analysis. That such reduction, and to such extent, is possible, is itself a high fact in the intellectual history of man; and from the contemplation of this fact, combined with that of the analogous success which it was allowed to Newton to attain in the study of universal gravitation, we derive a new encouragement to adopt the sublime belief, that all physical phenomena could be contemplated by a sufficiently high intelligence as consequences of one harmonious system of intelligible laws, ordained by the Author and Upholder of the universe; perhaps as the manifold results of one such mathematical law.

But if those profound and abstract works, in which so large a part is occupied by purely mathematical reasoning, suggest more immediately the thought of that great intellectual consummation, we must not therefore overlook the claims of experimental and practical inquirers, nor forget that they also have an important office to perform in the progress of human knowledge; and that the materials must be supplied by them, though others may arrange and refine them. Especially does it become important to call in the aid of experimental research, when facts of a primary and (so to speak) a central character require to be established; above all, if the establishment of such facts has been attempted in vain, or with only doubtful success, by eminent experimentalists already. Now, in the theory of heat, the research of the specific heats of the gases is one not far removed from such primary or central position, being no mere question of detail, but intimately connected with the inquiry into the nature of heat itself; it is also one which has been agitated by eminent men, and results have been obtained by some, and disputed by others, of which it is interesting, in a high degree, to examine the correctness or invalidity. For a new examination of this kind, conducted by new methods of experiment, the present award has been made. Of the nature and grounds of this award, I now proceed briefly to speak; and first, it may be proper that I should remind the Academy of the meaning of this phrase *specific heats*, and of the phenomena which suggest the name and the conception.

When any two equal volumes of water at any two unequal temperatures are mixed together, the mixture acquires, in general, a temperature which is either exactly or at least very nearly intermediate between the two original temperatures, being as many degrees of the thermometer below the one, as it is higher than the other. But if a pint of mercury at 60° and a pint of water at 80° be brought in contact and acquire thereby a common temperature, it is found that this last is not so low as 70°; and that thus, this passage of heat, from the warmer water to the colder mercury, has cooled the former less than it has warmed the latter, as indicated by the degrees of a thermometer. Phenomena of this kind suggest the conception, that only a part of the heat contained at any one time, in any particular body, affects the senses or the thermometer; and that the remainder of the heat is insensible, latent, or hidden: so that water, for example, absorbs or hides more heat than the same bulk of mercury at any temperature common to both, and that for any given increase of that temperature (measured by the thermometer) the former absorbs or renders latent more than the latter, while, on the contrary, in cooling through any given number of degrees, it sets a greater quantity free. Many other phenomena are made intelligible by such a conception, and even more immediately suggest it. Thus, if we put a pound of freshly frozen ice in contact with a pound of water, which is warmer than it by about 140° of Fahrenheit's thermometer, the result will be two pounds of water, not at an intermediate, but at the lower temperature; the excess of heat of the originally warmer water having been all employed in the mere act of melting the ice, or having all become insensible or latent, in the new water formed by melting it. And the principle that heat is absorbed or rendered latent in the production of steam from water, but is given out or set free again when the former is condensed into the latter, is part of the theory of the steam-engine. But because this phraseology suggests a view of the intimate nature of heat, which is at most hypothetical only, it has by many persons been thought better to use the word *specific*, instead of *latent*; and to speak of the specific heats of bodies in a sense analogous to that in which we speak of their specific gravities, to express only certain known and measurable properties of these bodies, in relation to the unknown principle of heat. And thus we say, that water has a greater *specific heat* than mercury, implying only that, whatever be the reason, any given bulk or weight of water produces a more powerful heating effect than is produced by the same bulk or weight of mercury, when both are cooled through the same number of degrees, by contact with a body of a lower temperature.

The specific heats of solids and of liquids are comparatively easy of determination; but the great rarity or lightness of the gases renders the measure of their specific heats more difficult. The former may be investigated with much accuracy, by the aid of Laplace's calorimeter: which is an instrument for measuring (by weight) the quantity of ice that is melted by the heat produced or set free in the cooling of a given weight of the proposed solid or liquid body through a given range of temperature. But in applying the same method to the latter question, that is to the inquiry into the heats of the gases, it appears to be difficult to disentangle the small effect of this sort produced by the cooling of any moderate bulk of gas from the effect produced by the cooling of the envelope in which that gas is contained. Several other methods also of inquiry into this delicate subject, however ingeniously devised and carefully executed, by men of deservedly high reputation, have been considered liable to the same or to other objections, and have failed to inspire any general confidence in their results. It seems, however, that the problem has been at length, to a great extent, resolved, by the employment of that other method, which was invented a few years ago by Dr. Apjohn here, and elsewhere by Dr. Suerman;* and which may be said to consist in determining, (indirectly,) through the help of a thermometer with moistened bulb, the weight of gas which is required for the conversion (at a known temperature and under a known pressure) of a known weight of water into vapour, by cooling through a number of degrees which is known from observation of another thermometer.

The general theory of the evaporation hygrometer, or the manner of employing a thermometer with moistened bulb, to discover the amount of moisture which is contained at any given time in the atmosphere, was very well and clearly set forth by Mr. (now Sir James) Ivory, in Tilloch's Philosophical Magazine for August, 1822. The same theory was also discovered by M. August of Berlin, with the date of whose work upon the subject I am unacquainted, having only seen the extracts made from it in M. Kupffer's Meterological and Magnetical Observations, (published at St. Petersburgh in 1837,) and in a recent volume of M. Quetelet's Correspondence. It appears, indeed, that M. Gay Lussac had prepared the way for this discovery, by his researches on the cold of evaporation; and the laws of the elastic force of vapour, and of its mixture with the gases, without which the theory could not have been constructed, are due to the venerable Dalton. Notwithstanding all that had thus been done, the subject seems to have attracted little general notice in these countries, until it was recommended to the attention of scientific men at the first meeting of the British Association; and Dr. Apjohn, who was thus led to examine it anew,† was not aware of the results that had been already obtained. He thus arrived at a new and independent solution, of which he had the satisfaction of testing the correctness, by several different series of experiments; and this success encouraged him to extend the research, and to apply the same principles and

* Dissertatio Physica Inauguralis de Calore Fluidorum Elasticorum Specifico; auctore A. C. G. Suerman: Trajecti ad Rhenum, 1836. An excellent work, to which every student of this subject must refer.

† It appears that another Member of the Academy, Dr. Henry Hudson, was also led, by this recommendation, to consider this interesting subject.

methods to other gases, and not to atmospheric air alone. He perceived that whatever the gas* might be, in a current of which was placed the thermometer with moistened bulb, the minimum or stationary temperature of that thermometer must be attained when just enough of heat was given out in cooling, by each new portion of gas, to cause the evaporation of that new portion of moisture with which this gas was at the same time saturated; and that thus the amount of depression would vary inversely as the specific heat of the gas, all other circumstances being the same. He investigated, however, the allowances that should be made for variations in such other circumstances, and took all other precautions which his experience pointed out to be important. The consequence has been a new determination of the specific heats of several different gases, on which it seems that much reliance may be placed, from the nature of the method, and from the agreement of the partial results with each other, and with those of Dr. Suerman, though some of these results differ widely from those obtained by methods previously employed; the specific heat of hydrogen, for instance, being found by Apjohn and Suerman, to be, under equal volumes, greater than that of atmospheric air in the ratio nearly of seven to five; whereas some former experimenters had supposed it to be equal or inferior. And by such results the law which had been thought to be obtained by a former eminent observer, namely, that all the simple gases have, under equal volumes, the same specific heat, appears to be overthrown. It is impossible not to feel some degree of regret, when we are thus compelled to abandon a view which had recommended itself by its simplicity, and had been found to be in at least partial accordance with facts; but besides that the search after truth is the primary duty of science, the whole tenor of scientific history assures us, that each new seeming complexity, or apparent anomaly, which the study of nature presents, is adapted ultimately to lead to the discovery of some new and higher simplicity.

A somewhat more distinct conception than the foregoing remarks may have given, of the nature of Dr. Apjohn's method, may be attained by a short study of that first experiment described by him, in which it was found that in a stream of dry hydrogen gas, in which a thermometer with a dry bulb stood at 68°, the one that had the moistened bulb was cooled to 48° of Fahrenheit; the barometer indicating at the same time an atmospheric pressure of 30.114 inches. From the stationary state to which the second of these two thermometers had been reduced, it is clear that the continual supply of heat, required for the continuing evaporation of moisture from the bulb, was supplied neither from the water with which that bulb was moistened, nor from the mercury which it contained, but only from the stream of warmer gas which continued to pass along it; the small effect of radiation from surrounding bodies being neglected in comparison herewith. Each new portion of the current of hydrogen, in cooling from 68° to 48°, must therefore have given out very nearly the precise amount of heat absorbed by that new portion of moisture, which passed at the same time from the state of water to the state of vapour, at the temperature of 48°. It is also assumed, apparently upon good grounds, that after the moist bulb attains its stationary temperature, the whole (or almost the whole) of the new gas, in becoming *fully cooled,* becomes at the same time *fully moistened,* or *saturated* with the new vapour; this vapour being intimately mixed with

* Dr. Suerman states, that M. Gay Lussac perceived that the specific heat of any gas must be connected with the degree of cold produced by the evaporation of a liquid placed therein; but the remark appears to have been merely made in passing, and to have been afterwards neglected and forgotten.

the gas which had assisted to form it; and every cubic inch of this mixture containing exactly (or almost exactly) as much moisture as a cubic inch *could* contain, in the form of vapour, at its own temperature: a quantity which is known from the results of Dalton, respecting the elastic force of vapour. From those results it follows, that in the present case, the temperature of the vapour being 48°, its elastic force must have been such that it could by itself have supported the pressure of a column of mercury, 35 hundredths of an inch in height; but the pressure upon the mixture was equivalent to a column 30 inches and 11 hundredths high; therefore the pressure which could have been supported by the hydrogen alone, at the same temperature of 48°, was equivalent to 29 inches and 76 hundredths: so that, by the known proportionality between density and pressure, the weight of the gas which was contained in the whole or in any part of this mixture would have exceeded the weight of the vapour in the ratio of 2976 to 35, or in the ratio nearly of 85 to 1, if the weight of a cubic inch of hydrogen gas were as great as that of watery vapour, under a common pressure, and at a common temperature. But under such circumstances, a cubic inch of vapour weighs about nine times as much as a cubic inch of hydrogen; we must therefore divide the number 85 by 9, and we find that in the present case the mixture contained only about $9\frac{1}{2}$ grains of hydrogen for every grain of vapour; and thus we learn, from this experiment, that the heat required for the evaporation of a grain of water at the temperature of 48° might be (and was in fact) supplied by the cooling of about $9\frac{1}{2}$ grains of hydrogen from 68° to 48°. But in order to produce the same amount of evaporation by the heat which water would give out, in cooling through the same range of temperature, it is known from other experiments that it would be necessary to employ about 56 grains; therefore $9\frac{1}{2}$ grains of hydrogen have nearly as much heating power as 56 grains of water, or one grain of the former contains almost as much specific heat as six grains of the latter. All this is stated in round numbers, and with the omission of all lesser corrections, for the sake merely of such members as may not have attended to the subject, and yet may wish to have a clear, though general notion of it. Those who desire a more exact account will, of course, turn to the Essay itself.*

* The formula given by Dr. Apjohn for the general solution of the problem of the moist bulb hygrometer, in any gaseous atmosphere, is,

$$f'' = f' - \frac{48ad}{e} \times \frac{p}{30};$$

in which e is the caloric of elasticity of vapour, at the temperature t' of the hygrometer; p is the atmospheric pressure; d is the difference between the temperatures of the dry and wet thermometers: f' and f'' are the elastic forces of the vapour of water, at the temperature of the hygrometer, and at that other temperature at which dew would begin to be deposited; and a is the specific heat of the gas, compared with that of an equal weight of water, and multiplied by the specific gravity of the same gas, compared with that of atmospheric air. For the case of a current of dry gas $f'' = o$, and

$$a = \frac{ef'}{48d} \times \frac{30}{p};$$

in which, as also in the other formula, it would be a little more exact to write $p - f'$ instead of p. A correction is given for the case of a mixture of gas with air; and the influence of other corrections also is taken into account. When a is divided by the known number 0,267 the quotient is the specific heat of the gas compared with that of an equal volume of atmospheric air: and the sensible inequality of the specific heats so found, for different simple gases, is the chief physical conclusion of the paper.

With respect to those independent, but analogous researches of Dr. Suerman,* to which allusion has been made, they seem (as has been said) to confirm as closely as could be expected, under the differing circumstances of the experiments, the results of Dr. Apjohn; of whose labours, indeed, that eminent foreigner has spoken in the most handsome terms, and in favour of whom he has freely waived, upon this subject, all contest for priority. But even if among the many persons who now are cultivating science in many distant countries, and whose results are sometimes long in coming to the knowledge of each other, it should be found that some one has anticipated our countryman and brother academician in the publication or invention of the method which I have endeavoured briefly to describe to you, or if, on the other hand, his own future reflections and experiments, or those of any other person, shall indicate hereafter the necessity of any new improvement, your Council still will have no cause to regret that they have adjudged the present distinction to a paper which contains so much of independent thought, and so much of positive merit.

[*The President then delivered the Medal to Doctor Apjohn, addressing him as follows.*]

Doctor Apjohn,

In the name of the Royal Irish Academy, I present to you this Medal, for your investigations respecting the specific heats of the gases; hoping that it will be received and valued by you, as attesting our sense of the services which you have already rendered to that important and delicate department of physical research; and that it will also be to you a stimulus and an encouragement to pursue the same inquiry further still, so as to improve still more the results already obtained, and to establish other new ones; and thus to connect, more and more closely, your name and our Transactions with the history of this part of Science.

* It is proper to remember that Dr. Suerman published his Dissertation without having seen the last and most correct results of Dr. Apjohn, contained in the present prize Essay. This remark applies particularly to the specific heat of hydrogen.

L.

ADDRESS AS PRESIDENT OF THE ROYAL IRISH ACADEMY ON MR. PETRIE'S PAPER ON THE HISTORY AND ANTIQUITIES OF TARA HILL (1839)

[June 24, 1839]
[*Proceedings of the Royal Irish Academy* **1**, 350–354 (1841).]

The President delivered the following Address:

Before the present session closes, as it is now about to do, I am to inform you, that your Council have continued to consider the expediency of awarding any medal or medals, from the resources of the Cunningham Fund, to any of the papers which had been communicated to us for publication, within the last few years, and which had not previously been so distinguished; adopting still the same plan of triennial cycles, and the same principles connected with that plan, which have been announced to you on former occasions; and thinking themselves bound to lean rather to the side of caution, than to that of indulgence, in deliberating on questions of this kind. The award of a medal, in the name of a learned body, is attended with a grave responsibility. It does not indeed pronounce, in the name of the Society, on the rigorous accuracy, or perfect novelty, of the paper which is thus marked out; but it at least offers the peculiar thanks of that Society to the author of that paper, and expresses a desire, on the part of the body, to be connected, to a peculiar degree, in present observation and in future history, with the communication for which the honour is awarded. The with-holding of a medal is, for the converse reason, no expression of unfavourable opinion, nor any denial of the existence of a large share of positive merit in the paper or papers which it is thus forborne to distinguish: even when the principle of competition does not happen to come into play, and when no other essay, of the same class and cycle, is adjudged to have superior pretensions. It has, however, appeared to your Council, that they were authorized and bound to award a medal to Mr. Petrie, for his Paper on the History and Antiquities of Tara Hill, printed in the Second Part of the Eighteenth Volume of the Transactions of this Academy; as being, in their opinion, the most important of those which were communicated to us, during the three years ending with December, 1838, in the departments of Polite Literature and Antiquities; and as possessing also such amount of positive merit and interest as to entitle it to this mark of distinction. Having attended the discussions which took place in the committees on the merits of the various papers, and on Mr. Petrie's Essay in particular, I shall venture now to lay before you, in the briefest possible manner, a few of the grounds of this award; without attempting to offer a complete statement of those grounds, or anything approaching to a full analysis of the memoir itself, which memoir indeed will very soon be in your hands.

Mr. Petrie's Essay may be considered as consisting of two principal parts: the first containing an account of Events connected with Tara, compiled from Irish manuscripts and illustrative of the History of Ireland; and the second part being devoted to an identification of the existing Remains, including an examination of the various descriptive notices also contained in ancient Irish manuscripts. The documents brought forward, possess a great degree of curiosity and interest; many of them, also, are now for the first time published; and (which is of importance to observe) are given in an entire, unmutilated form; accompanied with literal translations, and with philological and other notes, adapted to increase their value to the student of the ancient literature and history of Ireland. And what gives to these literary relics a value and an interest perhaps greater than, or at least different from, what might attach to them if considered merely as curious fragments, illustrative of the mode of thinking and feeling in times long passed away, is the circumstance that the accuracy of their topographical descriptions has been tested by recent and careful examination. The resources of the Ordnance Survey have been called in, to check or to confirm, by appeal to existing vestiges, the statements still preserved of the writers of former centuries, respecting the relics of what was even then an ancient and almost forgotten greatness; the time-worn traces have been measured, and compared with those old descriptions; and an agreement has been found, which establishes as well the truly wonderful antiquity of the remains still to be found at Tara, on what was once, and for so many centuries, the royal hill of Ireland, as the correctness and authenticity of documents, which it has been little the fashion to esteem.

It is this clear establishment of the authenticity of what had been commonly thought doubtful, this employment of a manifestly rigorous method of inquiry in what had seemed to many persons a region of fancy and of fable, in a word this evident approach to the character of scientific proof, which has made (I own) a stronger impression on my own mind, and (I believe) on the minds of others too, than even the literary and antiquarian interest of those curious and valuable details (such as the Hymn of Patrick, and the particulars respecting the Lia Fail, or ancient Coronation Stone of Ireland,) brought forward in the present Essay. I shall not venture here to give utterance to any opinion respecting the extent to which the once common and still lingering prejudice against the value and authenticity of Irish Manuscripts, almost against the very existence of any ancient History or Literature of Ireland, may have been removed or exposed before, by the labours of other antiquaries. But it may be allowed me to express a conviction, that it is only by pursuing some such plan as that exemplified in Mr. Petrie's Essay, namely, by a diligent examination of existing Irish Manuscripts, and of existing Irish Remains, and by an unreserved publication of all which may be found in the one and in the other, that full historic certainty can be attained, respecting the ancient state of Ireland. And that if, on the other hand, this diligent search be made, and this full and free publication, they will not fail to produce a clearness and convergence of opinions, among all who attend to these subjects; and will throw such a steady light, not on Irish History alone, but on other cognate histories, as will repay the labour and expense required for such an enterprize.

The Royal Irish Academy has already, from its limited means, contributed much to accomplish this object, or to prepare materials for accomplishing it. By purchase or transcription, we have gradually collected originals, or carefully collated copies, of many of the most valuable manuscripts which are extant, in the ancient Irish language. At a no slight expense, our volumes of Transactions have been and still continue open to receive such fruits of diligent and judicious research, in this department of study, as are contained in the paper on

Tara. The sum which, by a recent vote, has been placed at the disposal of the Council, will enable them to push on with vigour the printing and engraving of that other elaborate work of the same author, which was honoured with the award of a medal here some years ago,—the Essay (by Mr. Petrie) on the Round Towers of Ireland. And the liberality of Members concurs with that of extern Subscribers to place, from time to time, upon our table, such splendid donations of ancient Irish Relics, as the Cross and the Torques of this evening.

It is, however, to the resources of the Nation that we must look, to aid us in accomplishing what is truly a national object. As it was long ago pronounced to be a symptom of the health of a State, and an element in its well-being, that all should interest themselves in the weal of each, and that if one member suffer, the whole body should suffer with it; in order that thus whatever injury was offered to a part might be repelled by the energy of the whole, and that every limb might be animated by one pervading vigour: so too it is another fruit and sign of the dignity and happiness of brotherhood, another opposite and contrast to the misery of savage isolation, when not the present only of a nation's life, but the past and future also are regarded with a vivid interest; and, caring for posterity, men care for their ancestors likewise. Each people owes it to the human race, to do what in it lies for preserving its own separate history, and guarding its own annals from decay: and each, according to its power, should cheer and help the rest in their exertions to accomplish this, which is an object common to all. Ireland is rich in records of an ancient civilization; and looks with a just hope to Britain for assistance towards rescuing those records from oblivion, and from the risk of perishing obscurely. Though this Academy possesses many manuscripts, and although many are contained in the Library of our national University, enough has not been done until they have been placed beyond all danger of destruction, and made accessible to students every where, by printing and by publishing them, with notes and with translations, such as can be supplied by some of the few persons who are now versed in the ancient Irish Language. For doing all this well, opportunities can now be had, which the lapse of a generation may almost remove, which the casualties of each year may diminish.

We have had more occasions than one to hear, this evening, of the assistance recently and wisely given by Government to Science. Nor ought (I think) the presence of the representative of our Sovereign and Patron, to restrain me from avowing the hope, in which you all will join, that our desire, long since expressed, for the publication of our Irish Records, may after no long time be granted; and that the State may soon resolve to undertake, or to assist in undertaking, a task for which the materials and the labourers are ready, but of which the expense, though to a Nation trifling, is too great for an Academy to bear.

Of the possibility of accomplishing that task, and of the fruit which may be expected from so doing, if a proof and specimen sought, they may be found in that Essay, on the History and Antiquities of Tara Hill, for which I now, in the name of this Academy, present this Medal to its Author.

The President then delivered the Gold Medal to George Petrie, Esq., R.H.A., M.R.I.A., and the Academy adjourned to November.

LI.

ON DR. ROBINSON'S TABLE OF MEAN REFRACTIONS (1843)

[Communicated May 22, 1843]
[*Proceedings of the Royal Irish Academy* **2**, 400–401 (1844).]

Sir William Hamilton remarked that Dr. Robinson's mean refractions, published in the second Part of the Nineteenth Volume of the Transactions of the Academy, might be represented nearly by the formula,

$$\text{R} = 57,546 \tan(\theta - 4'' \times \text{R}); \tag{1}$$

or by this other formula,

$$\text{R} \cot \theta + \text{R}^2 \sin 3'',8 = 57,346; \tag{2}$$

R being the number of seconds in the refraction corresponding to the apparent zenith distance θ, when the thermometer is 50°, and the barometer 29,60 inches.

The first formula seems to give a maximum positive deviation from Dr. Robinson's Table, of about a quarter of a second, at about 80° of zenith distance; it agrees with the Table at about 83° 10′; is deficient by a second at about 84° 30′; and by $\frac{4}{5}''$ at 85°.

The second formula, which may be reduced to logarithmic calculation by the equations,

$$\left.\begin{aligned} \log \tan 2\rho &= \log \tan \theta + \bar{2},81296, \\ \log \text{R} &= \log \tan \rho + 3,24657, \end{aligned}\right\} \tag{3}$$

does not agree quite so closely with Dr. Robinson's Table, in the earlier part of it; but the error, positive or negative, seems never to exceed half a second, within the extent of the Table, that is, as far as 85°.

It appeared to Sir W. H. worth noticing, that the results of such (necessarily) long and complex calculations, as those which Dr. R. had made, could be so nearly represented by formulæ so simple: of which, indeed, the first is evidently analogous to Bradley's well known form, but differs in its coefficients. The second form is more unusual, and gives (approximately) the mean refraction as a root of a quadratic equation. It has been used (with other logarithms) by Brinkley, for low altitudes.

LII.

ADDRESS AS PRESIDENT OF THE ROYAL IRISH ACADEMY ON DR. KANE'S RESEARCHES ON THE NATURE OF AMMONIA (1839)

[June 24, 1839]

[*Proceedings of the Royal Irish Academy* **2**, 411–419 (1844).]

The President, on presenting to Dr. Kane the Cunningham Medal, awarded to him for his Researches on the Nature of Ammonia, gave an account of the progress of his discoveries.

It is now my duty to inform you, that a Cunningham Medal has been awarded by the Council to Dr. Robert Kane, for his Researches on the Nature and Constitution of the Compounds of Ammonia, published in the First Part of the Nineteenth Volume of the Transactions of this Academy. It would, indeed, have been much more satisfactory to myself, and doubtless to you also, if one of your Vice-Presidents, who is himself eminent in Chemistry, had undertaken the task which thus devolves upon me, of laying before you a sketch of the grounds of this award; but at least, my experience of your kindness encourages me to hope, that while thus called upon officially to attempt the discharge of a duty, for which I cannot pretend to possess any personal fitness, or any professional preparation, I shall meet with all that indulgence of which I feel myself to stand so much in need.

Although, in consequence of the variety of departments of thought and study which are cultivated in this Academy, and the impossibility of any one mind's fully grasping all, it is likely that many of its members are unacquainted with the details of chemistry, yet it has become matter of even popular knowledge, that in general the chemist aims to determine the constitution or composition of the bodies with which we are surrounded, by discovering the natures and proportions of their elements. Few need, for instance, to be told that water, which was once regarded as itself a simple element, and which seems to be so unlike to air, or fire, or earth, has been found to result from the intimate union of two different airs or gases, known by the names of oxygen and hydrogen, of which the one is also, under other circumstances, the chief supporter of combustion, is an ingredient of the atmosphere we breathe, and is closely connected with the continuance and healthful action of our own vital processes, by assisting to purify the blood, and to maintain the animal heat; this same gas combining also, at other times, with some metals to form rusts, with others acids, with others again alkalies and earths, entering largely into the composition of marble and of limestone, and, in short, insinuating itself, with a more than Protean ease and variety, into almost every bodily thing around us or within us; while the other gas which contributes to compose water, though endowed with quite different properties, is also extensively met with in nature, especially in organized bodies, and in particular occurs as an element in that important substance, on the confines of the mineral and organic kingdoms, to which the Researches of Doctor Kane relate; ammonia being, as all chemists admit, a compound of hydrogen and nitrogen, which last-named gas is well known as being the other chief ingredient (besides oxygen) of atmospheric air.

Again, it is generally known, to those who take an interest in physical science, as a truth which is almost the foundation of modern chemistry, that the elements of bodies of well-marked and definite constitutions, such as pure (distilled) water, or dry (anhydrous) ammonia, are combined, not in arbitrary, but in fixed and determined proportions; for example, the oxygen contained in any quantity of pure water weighs exactly, or almost exactly, eight times as much as the hydrogen contained in the same quantity, but occupies (when collected and measured) a space or volume only half as great; and the nitrogen contained in any given amount of dry ammoniacal gas, is to the hydrogen with which it is combined, by weight as 14 to 3, and by volume in the proportion, equally fixed, of 1 to 3.

Yet such results as these, respecting the constitution of compound bodies, however numerous and accurate they may be, are still not sufficient to satisfy the curiosity, or to terminate the researches of chemists. They aspire to understand, if possible, not only the *ultimate* constitution of bodies, or the elements of which they are composed, and the proportions of those elements, but also the *proximate* constitution of the same bodies, or the manner in which they arise from other intermediate and less complex compounds. Water, for instance, is believed to enter, in many cases, into composition with other bodies, *as water*, not *as* oxygen and hydrogen. Has ammonia any such component, which itself is composite? It is admitted to consist of one volume of nitrogen, combined with three of hydrogen. Can any order be discovered in this combination, any proximate constituent, any simpler and earlier product, from which the ammonia is afterwards produced? Until experiments decide, it appears not impossible, may seem even not unlikely, that nitrogen *may* combine (more intimately than by mere mixture) not only with thrice but with twice or once its own volume of hydrogen, and that thus other substances *may* be formed, from which, by the addition of new hydrogen, ammonia *may* result. It is interesting, therefore, to inquire whether either of these conceived possibilities is actually realized in nature; whether these two important gases do ever actually combine with each other in either of these two proportions. In the symbolic language of chemists, as usually written in these countries, the compound NH_3 is well known, being no other than ammonia; but does NH* or does NH_2 exist?

An eminent French chemist, M. Dumas, in examining a substance, which he called oxamide, and which was one of the results of the action of oxalic acid on ammonia, was led to the conclusion, that the last mentioned compound of nitrogen and hydrogen, namely NH_2, does really exist in nature, and he proposed for it the name of *amide*. The same chemist considered it also to exist in the substance formed by heating potassium in ammoniacal gas; and the same combination, amide, had been (I believe) regarded as a proximate constituent of certain other compound bodies, such as urea, sulphamide, and carbamide, before Dr. Kane's researches on the White Precipitate of Mercury. Yet is has been judged by Berzelius, that the investigations of Dr. Kane have assisted in an important degree to establish the actual existence (der wirklichen existenz) of amide, or of *amidogene* (as Kane prefers to call it, from its analogy with oxygen and cyanogen), and have thrown much light upon its chemical history and relations.

* The compound NH, or as it is otherwise better written, HN, has been suspected to exist, as one of the proximate elements of melamine and of some connected bodies. See Gregory's edition of Turner's Chemistry, 1840, page 757.

In fact, the body *oxamide*, which seems to have first led Dumas to infer the existence of amide, was one of those organic compounds, respecting which it has often been found difficult, by chemical inquirers, to pass with confidence from the *empirical* to the *rational* formula; from the knowledge of the *ultimate* elements (or of those which are at present to be viewed as such), and of the proportions in which they combine, to a satisfactory view respecting the *proximate* elements, or intermediate and less complex combinations on which the final result depends. Oxamide may be, and was considered to be, *probably* composed of amide and carbonic oxide (in the foregoing notation, $NH_2 + C_2O_2$); but it was perceived to admit also* of being *possibly* compounded of nitric oxide and a certain combination of carbon and hydrogen ($NO_2 + C_2H_2$); or of cyanogen and water ($C_2N + H_2O_2$). And even the amidides of potassium (KNH_2) and of sodium ($NaNH_2$), have, from the energetic affinities of those metallic bases, been thought to prove less decisively the existence of amidogene itself, than the amidide of mercury ($HgNH_2$) discovered by Dr. Kane, in his analysis of the white precipitate of the last mentioned metal. (Trans. R.I.A., vol. xviii. part iii.)

Although this precipitate had been long known, and often analyzed, erroneous views (as they are now regarded) were entertained respecting its composition, and it had, for instance, been supposed to contain oxygen, till Kane pointed out the absence of this element, and showed, with a high degree of probability, that the proximate elements were the chloride and the amidide of mercury; white precipitate being thus a chlor-amidide of that metal ($HgCl + HgNH_2$, if the Berzelian equivalent of mercury be adopted, instead of its double). Ullgren, a friend of Berzelius, obtained the chemical prize from the Swedish Academy of Sciences, for the year 1836, for a paper in which, having with great care repeated and varied the experiments, he confirmed this and other connected results of our countryman; and Berzelius himself, in his Report read to the above-mentioned Academy in 1837, on the recent progress of the Physical Sciences in Europe (to which Report allusion has been made above), expressed his opinion that these researches of Kane were among the most important of the preceding year.†

In the essay for which your Council have awarded the present prize, Dr. Kane has pursued his researches on ammonia, and has shown, with apparently a high probability, that there exist amidides (though not yet insulated) of other‡ metals besides mercury, especially of silver and copper; that is, combinations of these metals with the proximate element amide or amidogene. He has also given, in great detail, a series of analyses performed by him on a large number of compound bodies, of which some had been imperfectly examined before, while others were discovered by himself. But as it would lead into far too great length, and too minute detail, if any attempt were made at present to review these laborious processes of analytical chemistry, and as indeed they derive their chief philosophical interest from the

* L'Oxamide peut donc, à volonté, être considérée comme un composé de cyanogène et d'eau, ou bien comme un composé de deutoxide d'azote et d'hydrogène bicarboné, ou bien enfin comme un composé d'oxide de carbone et d'un azoture d'hydrogène différent de l'ammoniaque.—Dumas, sur l'Oxamide, &c. Annales de Chimie et de Physique, tome xliv. page 142.

† Diese Untersuchungen von Kane gehören meiner Ansicht nach zu den wichtigeren des verflossenen Jahres.—Wöhler's German Translation of Berzelius's Report, Jahres-Bericht über die Fortschritte der physischen Wissenschaften, 17th year, page 179. (Tübingen, 1838).

‡ Dr. Kane has since made it probable that there exist amidides of palladium and platinum also. (Phil. Trans. 1842, part ii.)

views with which they have been associated, it may be proper to attempt no more than a very brief (I fear that it will also be a very inadequate) sketch of those views.

Dr. Kane considers that in ammonia, which, in the usual language of chemists, is said to consist of one atom of nitrogen and three atoms of hydrogen, one of these atoms of hydrogen is more loosely combined than the two others with the nitrogen, so as to be capable of a comparatively easy replacement, by many, perhaps by all, of the metals, as well as by organic radicals; the other two atoms of hydrogen being *already*, in the ammonia itself, and not merely in the products of such replacement of hydrogen by metals, combined in a particular way with the one atom of nitrogen, so as to form that substance named amide or amidogene, which was detected by Dumas (as has been mentioned) in performing the analysis of oxamide. From Dr. Kane's own study of the combinations of this substance amidogene (H_2N), with metals, he infers it to be a compound radical of feebly electro-negative energy, analogous to that important one cyanogen (C_2N), of which the discovery by Gay-Lussac has exercised so powerful an influence on modern chemistry. He considers this radical, amidogene, as existing ready formed, in combination with hydrogen, in ammonia; which latter substance is thus, according to him, to be regarded as *amidide of hydrogen*; and as, in this respect, analogous to water, and to the hydrocyanic, hydro-sulphuric, and muriatic acids, that is, to the oxide, cyanide, sulphuret, and chloride of hydrogen; from all of which bodies it is possible, as from ammonia, to expel an atom of hydrogen, and to replace it by an atom of metal,—if indeed hydrogen be not (as there seems to be a tendency to believe it to be) itself of metallic nature, notwithstanding its highly rarefied form. By developing this view of the constitution and function of ammonia, Dr. Kane has offered explanations of a large number of replacements of that substance by others, some of which replacements (I believe) were known before, while many have been discovered by himself.

One of the most remarkable points in Dr. Kane's views is the way in which he considers the ordinary salts of ammonia. Many of these are known to contain an atom of water, the existence of which led to the proposition of the very remarkable theory by Berzelius, of the existence in them of a compound metal *ammonium*, which has not indeed been insulated, but has been found to form, in combination with mercury, a certain metallic amalgam. Dr. Kane looks upon these salts as double salts of hydrogen. He considers them to contain ammonia ready formed, united with a hydrated acid or with a hydrogen acid. He seeks to establish the similarity of the common ammoniacal salts to those complex metallic amidides, whose nature he has developed by analysis.

Thus, for example, the well-known body, sal-ammoniac, is, in the Berzelian view, regarded as chloride of ammonium; but, in the view put forward by Dr. Kane, it is chlor-amidide of hydrogen. The former view supposes that the ammonia robs the hydrochloric acid of its hydrogen, to form, by a combination with it, a metallic base, NH_4, with which the chlorine unites; as this last element combines with the metal sodium, in the formation of common salt. The latter view supposes that in the action between hydrochloric acid and dry ammoniacal gas, there is no separation of the chlorine from the hydrogen,—no breaking up of a previously existing union,—no overcoming of the affinity which these two elements (chlorine and hydrogen) have for each other; but an exemplification of a general tendency of chlorides, oxides, and amidides of the same or similar radicals, to unite, and form chlor-oxides, chlor-amidides, or oxamidides. Sal-ammoniac is, according to Kane, a double haloid salt; he looks upon it as being a compound exactly analogous to the white mercurial precipitate, which was first accurately analyzed by himself; the one being HCl + HAd (if Ad be the symbol of amidogene), while the other is HgCl + HgAd, so that the mercury in the latter takes the place of the hydrogen in the former.

It was, however, in the oxysalts, such as the sulphate of ammonia, that the presence of an atom, or equivalent, of water, or at least of the elements required for the composition of such an equivalent, appears to have suggested to Berzelius the theory, that what seemed to be hydrate of ammonia ($NH_3 + HO$) was really oxide of ammonium ($NH_4 + O$). There are, undoubtedly, many temptations to adopt this view, besides the high reputation of its propounder. One is, that it assimilates the constitution of sulphate of ammonia to what seems to be regarded by the greater number of modern chemists, as the probable constitution of other sulphates, nitrates, &c., for example, the sulphate of iron. When green vitriol is to be formed by the action of sulphuric acid upon iron, it is requisite to dilute the acid with water, before the action will take place. The hydrogen of the water then bubbles off, but what becomes of the oxygen which had been combined with it? Does it combine immediately, and as it were in the first instance, with the iron, to form oxide of iron, on which the anhydrous sulphuric acid may act, to produce *sulphate of oxide* of iron, according to the view which seems, till lately, to have been adopted: or does this oxygen, from the water, combine rather with the sulphuric acid to produce a sort of oxide thereof, and does this *sulphat-oxygen* act on the pure metallic iron to form with it a *sulphat-oxide,* as many eminent chemists now appear to think? Whatever may be the final judgment of those who are entitled to form opinions on questions such as these, it cannot, I conceive, be justly said, that the questions themselves are unimportant. They touch on points connected with the philosophy of chemistry, are essentially connected with its theory, and cannot always be without an influence upon its practice.

Now according to the Berzelian view of sulphate of ammonia, that is the salt produced by the mutual action of sulphuric acid, water, and ammonia, this salt is properly a sulphat-oxide of the compound metal ammonium ($NH_4 + SO_4$), in the same way as green vitriol, on the view last mentioned, is sulphat-oxide of iron ($Fe + SO_4$), or as sulphate of potash is sulphat-oxide of potassium ($K + SO_4$), and this analogy is doubtless pleasing to contemplate.

Dr. Kane does not entirely reject this Berzelian theory of ammonium; he acknowledges that the substance NH_4, which he regards as *subamidide of hydrogen,* and compares to some suboxides, possesses metallic properties, and is a proximate constituent of certain compounds, especially of the ammoniacal amalgam; but he conceives that the evidence for the existence of ammonia itself, in many of the ammoniacal salts, is too strong to be resisted: and he looks upon the hydrated ammonia, which is found to combine with sulphuric and other oxacids, as being not, in general, oxide of ammonium, but *oxyamidide of hydrogen*; the sulphate of ammonia being thus a bibasic compound, of which one base is ammonia, while the other base is water.

Between the conflicting opinions of such men, supported each by powerful arguments and analogies,—and it will easily be conceived that in so short a sketch as this, and upon such a subject, it has been found impossible by me to mention even the names of all the eminent chemists whose experiments and writings should be studied, by persons inquiring for themselves,—not only do I not venture to express any judgment of mine, but I conceive also that your Council did not desire to express on their part any decision. To justify the present award, it was, I believe, deemed by them sufficient, that great research and great talents had been brought, in the investigations of the author to whom that award has been made, to bear on an important subject, which has derived, from those investigations, an additional degree of importance. Whatever may be the final and unappealable judgment of those persons who shall, at some future time, be competent and disposed to pronounce it, we need not fear that the honour of this Academy shall have been compromised by the recognition which the Council have thought it right on the present occasion to make, of that combination of genius and industry, which has already caused the researches of Kane to influence in no slight degree the progress of chemical science, and has won for him an European reputation.

The President then presented the Gold Medal to Dr. Kane, and the Academy adjourned for the summer.

LIII.

SIR W. ROWAN HAMILTON ON THE ELEMENTARY CONCEPTIONS OF MATHEMATICS

(Seven letters to Viscount Adare (March and April, 1835))
[*Hermathena* (*Dublin*) **3**, 469–489 (1879).]

To the Editor of "HERMATHENA."

SIR,

The accompanying letters from Sir William Rowan Hamilton to his pupil and friend Lord Adare,* afterwards Earl of Dunraven, were written in the early part of the year 1835. Forming a series of six letters, with a fragment of a seventh they were, nevertheless, scattered through the great mass of papers which after his death were placed in my hands by his representatives; and I have thought myself fortunate in being able to link them together, for they seem to me to contain an exposition of the elementary conceptions of Mathematics valuable at once for characteristic depth and comprehensiveness, and for clear development. The series, though continuous, stops far short of its intended completion, for it will be seen that, towards the close of the sixth, a seventh letter is promised, in which was to be *commenced* the application to Algebra of the principles previously laid down. Of such continuation only a draft of the beginning of the seventh letter has been discovered, but the fragment is an important one, and I have reason, from other parts of his correspondence, to infer that this seventh letter expanded into that *Essay on Algebra as the Science of Pure Time*, which was presented by him on the 1st of June, 1835, to the Royal Irish Academy, and published† in the seventeenth volume of the Academy's "Transactions", as preliminary to the treatise on *Conjugate Functions* which had been communicated by him to the same Body on the 4th of November, 1833.

A comparison of the *Introductory Remarks* to that Essay with these letters will show that in the former he resumes from the latter his distinction between the several schools of Algebraists, and that the very first of the letters anticipates the definition of Algebra which, through the title of his Essay, has since become famous.

I offer these letters for insertion in "HERMATHENA" in the belief that their contents will opportunely appear at a time when the fundamental ideas of Algebra are engaging much attention, and I desire that they may serve as earnest of more from the same source likely ere long to see the light; for I may take this occasion to state, that at length some hopeful progress has been made by me in the task of arranging for publication the papers of Sir W. R. Hamilton in connexion with a biographical memoir. The task has been unavoidably a very

* [Edwin Richard Windham Wyndham-Quin (1812–1871), Viscount Adare, later third Earl of Dunraven and Mount-Earl; BA Trinity College Dublin 1833.]

† [See Vol. III, p. 3.]

prolonged one, in consequence of the great number and the disorder of the papers, and of other causes needless here to mention.

I remain, Sir,
Faithfully yours,
R. P. GRAVES.
1, WINTON-ROAD,
October 1*st*, 1878.

I.

OBSERVATORY,
March, *4*, 1835.

MY DEAR ADARE,

I have often intended to try to revive, in an easy, but systematic form, my own and your recollections of early conversations upon Algebra, especially with regard to the spirit and philosophy of that science: and will not longer defer the attempt, though I may be only making a beginning now, which may remain for a long time, and perhaps for ever, an uncompleted sketch.

When you commenced your studies with me, I did not assume any knowledge on your part of Algebra, nor of anything beyond the first elements of Arithmetic. I was anxious to begin at the beginning, and to initiate you by a method which should suppose no previous attainment. And I was glad that you had, in fact, read no Algebraical work, though you were a good and expert arithmetician.

In *Arithmetic*, properly so-called, Number is considered as an answer to the question *How many*, and as constituting a Science of *Multitude*, founded on the relation of *more and fewer*, or ultimately of the *many*, and the *one*. In a more complex Science, of *Magnitude* and *Measure*, which may perhaps be called *Metrology* (though often classed as a higher part of Arithmetic) Number is the answer to the question *How much*, and the fundamental relation is that of *greater and less*, or of *whole and part*. But in *Algebra* I taught that Number answers the question *How placed in a succession*, the guiding relation being that of *before and after* (or of positive, negative, and zero); and the Science itself being one of *Order and Progression*, or, as it might be called concisely, of PURE TIME. To *count*, *to measure*, *to order*, are three different, although connected, acts of thought, and belong to these three different, although closely connected, Sciences, of *Arithmetic*, *Metrology*, and *Algebra*. Groups *as Counted*, magnitudes *as measured*, positions or states *as ordered*; and, therefore, finally the *relations* of the counted to the counter, of the measured to the measurer, of the ordered to the orderer—such are the ultimate objects of these three acts of thought, and the ultimate or elementary conceptions of these three Sciences.

To dwell a little longer on this distinction. In Arithmetic we consider and compare groups of individuals, with reference, not to the nature, but merely to the multitude, of those individuals; regarding, for example, a pair of stars and a pair of men as similar, in so far as both are pairs, and denoting both for this reason by a common name, by the cardinal or counting number "Two". In Metrology, we consider and compare such magnitudes as lengths or times, or any other measurable magnitudes, with reference to their measures merely; regarding, for example, a yard and an interval of three weeks as similar, if measured by a foot and by a week respectively, and denoting them then by one common name of measure, the quantitative or measuring number "Triple". In Algebra, we consider and compare positions of states of the same or of different progressions, with reference only to their arrangements in that or in those progressions; regarding, for example, *to-day* as similar to a point upon a line, if to-day be referred to yesterday and to-morrow as standards of arrangement in time, and if the point upon the line be referred in like manner to any two other points thereon, between which it is supposed to be also exactly intermediate: and we use in this case one common ordinal or ordering number, or name of arrangement [such as "Halfway"], for these two similar states, as will be afterwards more fully explained. Equinumerous groups in the first science, proportional magnitudes in the second, and corresponding positions in the third, are considered, for the purposes of these sciences, as entirely coinciding with each other, group with group, magnitude with magnitude, position with position; and the name of any one such group, or magnitude, or position, is extended to every other which in this view coincides with it. By forming such general thoughts, and marking them with such general names, it becomes possible to construct and to use a quotitative, quantitative, or ordinal language; and so to propose and resolve (at least in part) this widely comprehensive problem, including perhaps all others in these sciences—"To name every thought, and to interpret every name, of multitude, magnitude, or succession."

Let this suffice at present, from your affectionate friend,
WILLIAM R. HAMILTON.

II.

OBSERVATORY,
March 13, 1835.

MY DEAR ADARE,

I attempted, in the former letter, to distinguish Algebra, as the Science of Order, from Arithmetic, as the Science of Multitude, and from an intermediate Science of Magnitude, which I proposed to call Metrology. And having shown the possibility and advantage of establishing general names, or names of relation, quotitative, quantitative, and ordinal, which should belong, some to groups as counted, others to magnitudes as measured, and others to states of a succession as ordered, or ultimately to the *relations* of multitude, magnitude, and order—the comprehensive problem was proposed, "*To name* every thought, and to *interpret* every name, of multitude, magnitude, and succession"; or, "*To construct* and to *use* a quotitative, a quantitative, and an ordinal language."

Now, whether we look to the Arithmetical or to the Metrological, or to the Algebraical part of this great problem, we find that in each it is possible to adopt three principal views, and thus to impress on the research any one of three different characters. Whether Arithmetic, or Metrology, or Algebra be our study (and the remark extends to other studies also), we may belong to one or other of three great schools, which I shall call the *Theoretical*, the *Philological*, and the *Practical*, according as we chiefly aim at clearness of thought, or symmetry of expression, or ease of operation; — according as *Intuition*, or *Language*, or *Rule*, the *sapere*, or the *fari*, or the *agere*, is eminently prized and sought for; — according as obscurity, or inelegance, or tediousness is most dreaded and guarded against. You know enough of my habits and inclinations to determine without difficulty the school to which I belong, and to place me at once in the class of the *theoretical*, as seeking more a clear and lively intuition, by whatever cost of meditation or mental discipline to be attained, than language, however perfect in its structure, or rules, however easy of application. But you also know how willingly I admit the utility of those more *practical* persons, who study to improve such rules and such applications; and how highly I respect those algebraical grammarians or *philologists* who, pursuing Algebra as a language, care chiefly for removing its anomalies, and would reduce it to an elegant and symmetrical system of words and signs. For to this philological school belong very many of the best modern writers upon Algebra; and especially Woodhouse and Peacock, of Cambridge, and Professor Ohm, of Berlin, men of learning, patience, and originality, to whom we may add the brilliant name of Lagrange; though Fourier and Cauchy lean more to the Theoretical School, in the sense in which I have defined it, as elevating Intuition above Language.

Indeed it must be owned that if the Theoretical Algebraists be rightly possessed by the idea that Algebra is *more than a Language*, yet the language is so beautiful in its kind, so wonderful as an organ, so necessary and so prominent in the study of the science itself, that reasonably and naturally has it received a large share of attention, and not only now, but from the earliest times, has the philological spirit directed or influenced the progress of Algebraic discovery. And, therefore, though I have professed myself as belonging to the Theoretical School, and make it my chief aim to imbue and, as it were, impregnate the whole of Algebra with Intuition, yet in subordination to this I desire to cultivate its Philology too; and in now seeking to revive your remembrance of our old conversations, I shall perhaps make the language of the science — its symmetrical system of signs, with their logical rules of combination — hold even a prominent place; or at least shall treat this department of study with a deserved and sincere respect. But for this entrance on the *Language*, along with the *Science, of Succession*, — for the beginning of the solution of that great two-fold problem of Algebra, already mentioned, "To name every thought, and to interpret every name of order" — I must refer you to a future letter, and remain in the meanwhile,

Affectionately yours,
WILLIAM R. HAMILTON.

III.

OBSERVATORY,
March 14, 1835

MY DEAR ADARE,

In writing a third letter of this series — for a series it seems to be growing, and no doubt an extensive series would be needed, to do justice to Algebra as a subject — let me indulge myself a little longer with generalities before we proceed to details. It is the less improper to do so because you are not actually beginning, but only reviewing the study; and you, as well as I, must exert a sort of imaginative, and as it were dramatic power, in throwing ourselves back into that state and time in which you made your earliest steps in Algebra, while I had the pleasure to assist. Having, therefore, devoted the first of these Letters to the distinguishing of Algebra from other kindred SCIENCES; and the second Letter to the distinction between the Theoretical and other SCHOOLS; I shall now make a few general remarks on the connexion and the contrast of the Analytic and Synthetic processes, or FORMS of thought, as applicable to every Science and to every School.

The two Greek words, *Analysis* and *Synthesis* (ÃÚÃŠœÕŒ, œ̆ÚÒÍœÕŒ), are used by Mathematicians and Metaphysicians in many senses, which seem however to have all some reference to the etymology of those two words, or of the cognate Latin forms, Resolution and Composition; as if we said, in a more English style, Putting asunder and Putting together — Taking to pieces, and Making up — Loosing and Binding — Decomposing a thing or thought into its simpler elements, and Compounding these elements again, so as to produce that thing or thought. Thus, in Chemistry, there is an *Analysis* performed, when Water is decomposed into Oxygen and Hydrogen; and there is a *Synthesis*, when these elements are combined in such a manner that Water results. In Dynamical Astronomy, it *was Analysis*, when Newton extracted from the complex phenomena of the motions of the planets and satellites the elementary laws of motion and of attraction; it was *Synthesis*, when he proceeded to combine these elementary laws, and to deduce from them the planetary and lunar motions. The general process of reasoning, itself, was *analysed* by Aristotle into the principles and rules of Logic; and, consciously or unconsciously, those principles and rules are applied to, and combined *synthetically* with, the premisses of every argument, when anyone reasons correctly. Assertions, propositions, judgments, are divided by Kant into *Analytic* and *Synthetic*, by an analogous and subtle distinction. He holds that an assertion or *Judgment* is *Analytic*, when the agreement of the Predicate with the Subject of the assertion is an identical and purely logical truth, deduced from a mere analysis (or examination) of the *meanings* already supposed to belong to the two signs compared (without any new and foreign connexion between the two thoughts themselves, whether established by Experience, or by purely mental Intuition): as, for example, the assertion, *All Bodies are extended*, which is indeed an useful assertion, but only useful for purposes of language, or, at most, for clearness only, and not for enlargement of thought, belonging to the class Erläuterungsurtheile, but not to the class Erweiterungsurtheile; since it results from a mere analysis or examination of the meaning previously attached to the word *Bodies*. On the other hand, Kant gives the assertion, *All Bodies are heavy*, as an example of a *Synthetic Judgment*, deduced not from the mere taking to pieces of the meanings of subject and predicate, but rather from the putting of these meanings together, through the cement of Experience and Induction. The thought of Body, such as we usually form it, and such as in these assertions we intend it to be formed, *includes* the thought of Extension. But *not* the thought of Heaviness; it gives therefore Extension by *Analysis*, but cannot be connected with Heaviness, except through a foreign and (in this case) empirical *Synthesis*. Following out the same principle of distinction, Kant holds that *all* judgments of Experience, and *nearly all* those of Pure Mathematics, are *Synthetic*, not *Analytic*; the judgments of Experience pronouncing never that two thoughts are inseparable in the mind, but that two properties go together in nature; and those of Pure Mathematics, pronouncing indeed of two thoughts, that they *ought* to accompany each other, but not that the one *does* in fact contain the other, by logical, as well as scientific comprehension, if we except some few elementary axioms, such as "the Whole is greater than the Part." Thus, in geometry, he remarks that the thought of *straight* does not *contain* the thought of *short*, though we *see*, through the intuition of Space, that the shortest line is the straight one. And in Arithmetic, the thoughts of *Five* and *Seven* and *Sum* do not *include* the thought of *Twelve*; but we *find*, by a mental trial, by calling up an Intuition of counted things, that the sum of Seven and Five *is*

Twelve. I am disposed to agree with these remarks of Kant; but since they relate to the classification of Judgments, Assertions, Propositions, Sentences, they leave free for a separate classification the *combinations*, and the processes of *discovery*, of such judgments or propositions; and do not tend to weaken the received distinction between the *Analytic* and *Synthetic* METHODS, or processes, in mathematical researches generally, whether of a geometrical or algebraical kind. On *this* distinction, especially in its reference to arithmetic, I shall touch in the following Letter.

Meanwhile remaining
Your affectionate friend,
WILLIAM ROWAN HAMILTON.
THE VISCOUNT ADARE.

IV.

OBSERVATORY,
March 16, 1835.

MY DEAR ADARE

The remarks in my last letter on *Analysis* and *Synthesis* in general may have prepared you to recollect the distinction between the Analytic and Synthetic METHODS of arriving at any truth;—building it up;—the one by analysing the truth into its simpler elements, the other by constructing it from those;—the analytic method by going from the unknown to the known, the synthetic method by proceeding from the known to the unknown. If I were fonder than I am of coining words, or giving them new meanings I might be tempted to call these two the *Centripetal* and *Centrifugal*, or the *Centre-seeking* and *Centre-flying* Methods. For the Analytic tends always inwards from the obscure surrounding space to some central home of light already known and won; while the Synthetic Method radiates from that centre, and is ever aiming outwards to the unconquered world around. But Examples, better than Poetry, may make the contrast clear, and I shall take first the arithmetical example already given from Kant, as an instance of a *Synthetic Judgment*, which, although truly such, may yet be *discovered* by the *Analytic*, as well as by the Synthetic Method.

I admitted that the Assertion, "The Sum of Seven and Five is Twelve," is rightly called by Kant a *Synthetic* assertion, in the sense that we might fully understand the meanings of *Seven* and *Five* and *Sum*, and therefore the meaning of the "Sum of Seven and Five," as a number which is to be determined by the process of summing from the given numbers Seven and Five, and yet we might not have actually determined this new number, so as to know it to be the number "Twelve"; whereas the assertion, "Eight is the Sum of Seven and One," may fairly be called by contrast an *Analytic* assertion, because there is no better way of explaining the word, or of forming the notion of *Eight*, than by defining it to be (in Arithmetic) the mark of that multitude which is next greater than, or *exceeds by one*, the multitude named *Seven*; so that in thinking of *Eight*, we have *already* thought of the sum of Seven and One, while in merely thinking of *Twelve* (as the sum of Eleven and One, or, if you like that other definition better, as the Sum of Ten and Two) we have not *yet actually* thought of the Sum of Seven and Five; nor in merely thinking of this latter Sum have we *actually thought* of Twelve, though Arithmetical Trial or Proof may effect a *Synthetical connection* between these *two different complex thoughts*, and show that they belong or *conduct to one common intuition of multitude.* When we say that Eight is, the Sum of Seven and One, we take to pieces the complex meaning which we had already given to the subject of the proposition, we *analyse* the thought already formed of Eight; but when we say, "The Sum of Seven and Five is Twelve," we pass to a *new* thought, by a mental *pulling-together.* A simpler instance of what Kant would call an Analytic proposition is the following:—"Seven is Seven"; and a simpler instance of a Synthetic proposition is this, "*Five more than Seven* is also *Seven more than Five*": for this last assertion appears to me (as it has also done to several other thinkers) to require a *proof*, however easily such proof may be supplied.

But what is the process of mind by which we *discover* the Synthetical connexion above asserted to exist between "The Sum of Seven and Five," as one thought, and the number "Twelve," or the "Sum of Ten and Two," as the other? *This Process may* certainly be Synthetical; and for a full and formal *proof*, it *must* be so; but we may, instead, adopt an analytical method, so far at least as *discovery* is concerned. We may proceed synthetically from the known to the unknown, or analytically from the unknown to the known. It is a synthetical method, when setting out with the known number "Seven," we pass successively through the five other known numbers, "Eight, Nine, Ten, Eleven, Twelve," and rest in the last as the result. It is an analytical method, when we set out from the unknown and sought result, *the Sum, whatever it may be,* of Seven and Five, and go backwards by successive decomposition through the stages "Seven and Four and One," "Seven., and Three and One and One," till we are conducted to something known, such as we may here suppose that the sum of "Seven and Three" already is. But it is highly important to observe (as a check on the usurpations of Analysis, and in vindication of the rights of Synthesis), that even when something known has thus been reached, by this *centre-seeking* process, the problem is not yet fully and formally solved, the communication between the Subject and Predicate of the theorem is not yet entirely established, until we have returned upon our steps, or at least made sure that starting from the known we *can*, by the alternate way, the *centre-flying* method, arrive again at the unknown, and effect a Synthetic connexion. Thus, in the last example, if we suppose it known that the "Sum of Seven and Three" is "Ten," we have analysed the "Sum of Seven and Five" by successive steps into this other phrase, "The Sum of Ten and One and One"; but the arithmetical problem is not yet entirely resolved, nor is the theorem that was proposed entirely and formally demonstrated, until a Synthesis, though short, has compounded these last elements of "Ten and One and One" into "Eleven and One," or "Ten and Two," that is, ultimately into the complex thought which we have agreed to mark by the name "Twelve": besides that other Synthesis, supposed to have been before effected, which showed that the sum of Seven and Three was Ten.

This little example may suffice as a preliminary specimen of the connexion and contrast between the Analytical and Synthetical Methods; and I remain, &c., &c.,

W. R. H.

V.

OBSERVATORY,
March 17, 1835.

MY DEAR ADARE,

I do not know whether you may not think the remarks of the last two letters too subtle or too vague to be connected in any useful way with the problem of the two first;

but perhaps a little consideration will convince you that the distinction between *Analysis* and *Synthesis*, however obscurely or vaguely it may have been set forth by me, has really a very clear and close connexion with the distinction between the two parts of that problem: *the Naming of* thoughts, and the *Interpretation* of names, of multitude, magnitude, or order. To *name* a thought or an intuition, is either to pass from that Thought, or from that Intuition, to its own proper sign and mark in a system of signs *already* established and known; or else it is to fix *now* for the first time on some new sign, which sign, however, by the very act, we adopt as thenceforth to become a permitted point of reference, and which we now by definition attach to the thought or intuition in question, admitting thus *that* thought or that intuition into the household of the known and the familiar, and assigning to it, in that household, a place of its own for ever: and both these, modes of naming, the mediate and the immediate, seem plainly to be modes of *analysis*, as passages from the unknown to the known, from the *thought* regarded as obscure or new, to the *name* which is, or is to become familiar. But in the converse process of *interpreting* a name, we do not analyse but *construct* a thought, by rules already established, or from elements already given. The sign to be interpreted is here the starting post and centre, from which known point we go, by a known path of *Synthesis*, until we reach the thought or intuition signified. It appears then that on the whole, the *Naming* is an Analytic and the *Interpreting* is a Synthetic office; at least in that *Theoretical School* described in my Second Letter, in which the mind aspires to thoughts rather than words, though it must reach the former through the latter, and views things signified, through a transparent veil of signs. In the *Philological School*, on the contrary, in which Names are regarded not as media or *instruments* of vision, but as actually constituting of themselves the principal *objects* of research, the foregoing distinction would be reversed, and instead of Analysing or Constructing a *Thought* by Names, we should have to speak of Constructing or Analysing a *Name* by Thoughts; if indeed this School in its consistency and ultimate rigour should not reject *these* operations altogether, and substitute for them a mere logical and symbolical composition and decomposition of signs, according to rules of language, in which nothing but symmetry is required, and the absence of express contradiction: while the *meaning* or *signification* of the signs, — their reference to thoughts or intuitions, — is put aside as a foreign thing from this merely symbolical science. But in that other and (I think) higher School, the Theoretical — which looks beyond and through the sign to the thing signified, — the three great acts of mind described in my First Letter, *to count, to measure, and to order*, are evidently *analytic* acts, and the *naming* of things or thoughts through them is essentially an *analytic* process: while the converse office is the *synthetic*, which *interprets* a name proposed, and by the rule embodied in that name (of multitude, or magnitude, or succession) constructs the thought or the intuition (of the group, or quantity, or state), in Arithmetic, Metrology, or Algebra.

It may illustrate these general remarks to apply them to the Roman Numeration: for example to the symbol IX, which is, we know, equivalent to our English "Nine." This *written name*, or sign of multitude, IX, bears obvious traces of the analysis by which it has been formed. Hold up to view the fingers of both hands, counting both thumbs as fingers; you will have the intuition of a multitude or group, which from its natural connexion with our bodies and our wants has received in perhaps all spoken languages a simple or special name, and has been treated as a known or simple thought, to which analysis may tend, and from which synthesis may begin. The English call it "Ten": the Romans wrote it X. Hide now one thumb, and you will have* another intuition; you will see a different group, a lesser multitude: the English name it "Nine," a word as simple as the former, but the Romans marked it by a complex written sign, IX. Our *Ten* and *Nine*, and their *Decem* and *Novem*, seem all alike to be simple and arbitrary and uncompounded names, and they may be supposed to have been immediately given to the respective groups, without any derivation of the one English word from the other, or of the one Latin word from the other; as if these two different multitudes had, been named in an arbitrary order, and both the intuitions been regarded as equally simple and elementary — both equally fit to be the close of any future analysis, or the starting-post in a process of synthesis. But in the forming of the Roman written sign IX, there is an obvious reference to the sign X already chosen: we see that the group *ten*, composed of all the fingers, was regarded as more familiar (although more numerous) than the group *nine*, with one thumb away: the less familiar and (thus far) less simple intuition was compared with the more familiar, *it was analysed that it might be named*, and the result was the sign IX, recording this distinctive property of *Nine* that it has been found by *analysis* to be *one less than Ten*, — not indeed by a separation of the multitude Nine into its parts, but by a resolution of the *complex thought* into its simpler connected conceptions. I need not dwell upon the converse process of *Synthesis* by which we return from the familiar thoughts of One and Ten in this particular mode of combination, and *interpret* the written *name* IX, till it yields up the intuition of Nine: for you will easily apply these remarks to all the past arithmetical examples, and will see their bearing on those questions of Algebra to which we shall next proceed.

W. R. H.

VI.

OBSERVATORY,
March 18, 1835.

MY DEAR ADARE,

For the reasons given in my last letter I hold that in Algebra, as in other sciences (if studied in that Theoretical school which uses Names as subordinate to Thoughts), *to name* is an *analytic* act, and every *spoken or written name* (whether word or other sign) *is the result or record of an analysis*: declaring, if *immediately* and arbitrarily imposed (as in the Roman Arithmetic the word Decem or the sign X), that the signified thing or thought either *must be* or at least *is chosen to be* regarded as a *simple* element, insusceptible of further analysis, or at least not analysed yet into any simpler or more familiar components; or else recording, if the name be *mediately* given, and by rules of composition of signs (like the Roman numeral IX discussed in the foregoing letter), that the signified thought *has already* been analysed into others, and recording likewise the special process of that analysis. But since the use of a sign is to signify, and since every analysis supposes a previous possible synthesis — because a mark should be the mark of something and a thing or thought which may be analysed may be compounded also, — therefore I hold that every Name is not only *formed* as the result or record of an analysis, but may be likewise *used* as the beginning or the rule of a *synthesis*, and is accordingly so used when we *interpret* it *To name* is a verb active; it

* (Construct) — thus added to MS. by W. R. H., without striking out "have".

supposes some *object* to be named, some thing or thought or intuition, which either is or might be given and present. Now the having of such an *intuition*, the contemplating of such an *object*, is the natural and necessary antecedent of the *earliest* act of naming; although analogies of language and of thought may afterwards suggest *derived* and complex names, for which the objects and answering intuitions remain to be discovered or called forth. Such seems to be the process of the mind in the actual formation of Science; and such (I think) should be the process in study and instruction also. Among our *conscious* processes and acts, Analysis precedes and Synthesis must follow. The determination of the relation from the related thing must be the first act; and not till after this can we hopefully or reasonably attempt to determine consciously the related thing from the relation. For the conscious knowledge of a relation seems never in the growth of the human mind to precede the perception of a related thing, as given by thought or intuition; though relations themselves have all their seats *à priori* in the constitution of that very mind, and are so far innate ideas, or more correctly *innate forms*. The *form* precedes the *matter* in order of nature and of dignity, but in order of time and of visible progressive development the perception of the matter precedes the consciousness of the form. And therefore, in instruction, the *naming of a given object* should be earlier taught than the *interpreting of a given name*. The teacher ought first to present to the learner the *matter*, the complete intuition, and then assist him to attain from this to the consciousness of the *form* within him. But after lie has thus analysed he may construct; after he has acquired an *inductive method* of relating a given object to others more simple or familiar, he may be led to a converse method of constructing or applying this relation, so as to discover or make *new objects*, by *deduction*. Having first turned inward on himself from the obscure manifold around him, he may then look forth with cleared and strengthened sight, and choose his own path of progress.

It may seem inconsistent with these views, respecting the natural order of instruction, that I have dwelt so long on generalities and preliminary principles, instead of proceeding sooner, or perhaps immediately, to the actual objects or intuitions in Algebra to which they are designed to apply. But you remember that, at the outset of these letters, I announced it to be my chief purpose to revive your recollections, and to make them more clear and systematic, respecting the views which I had put forward in our old conversations, on the spirit and philosophy of the science. And since, as was remarked in an intermediate letter of the series, you are not now actually beginning but only reviewing the study, it is permitted and almost required that I should adopt a different method now from that which I employed when I wished to introduce you for the first time to a class of conceptions then new to you. I may now address you more as a man, and as one in whom the consciousness of the processes of his mind has been already developed by exercise — though there are none of us to whom it is not useful to have this consciousness developed more and more. Yet, it seems time that I should come to particulars, and apply to Algebra, as the Science of Order, those general remarks which I have been making on all scientific processes, and which I have only exemplified as yet in the simplest questions of Arithmetic. And this application to Algebra I think that I shall really *commence*, in my next letter, without any intervening digression: though I will not answer for my long keeping close to the details, or that I may not soon leave them for a while in pursuit of some new generality. Meantime, I remain, &c.,

W. R. H.
April 7.

Kant holds that in the application of our thoughts to things there is a Synthesis *pervious* to all Analysis, and that indeed we never analyse but what we had ourselves compounded: — and this, which is a fundamental article of Kant's system, may seem to contradict what I have said in the Sixth Letter; but you will observe that in dating (there) Analysis as earlier than Synthesis, I spoke of *conscious* acts, while that first Synthesis assumed or proved by Kant is confessed to be a blind unconscious working of the Mind reacting upon Sensible Objects, and giving them their unity, and even their existence as Objects definite and specific, by casting them in its own innate forms or moulds of sense and thought.

VII.

April 16*th*, 1835.

MY DEAR ADARE,

I have often expressed in conversation and in these letters my opinion* that Algebra is the Science of Order and Progression, or, more concisely, of PURE TIME; that is, the mathematical Science of Time as disengaged on the one hand from the dynamical notion of Cause and Effect, and on the other hand from all empirical marks and measures suggested by particular phenomena; just as Geometry is the Science of Pure Space, or the mathematical Science of Space as disengaged from all dynamical notions of localised force, and from all empirical knowledge of actual shapes and sizes, and positions and motions of bodies. I must now add that the *Science of Pure Time* according to this conception contains *not only Algebra*, as contrasted in these letters with Arithmetic and with Metrology, but *Arithmetic and Metrology also*; that is, it contains the Sciences of Quotity and Quantity, or of number and measure, as well as the Science of Order. We use familiarly the phrases "How many times?" "For how long a time?" "At what time?" — and these familiar forms of inquiry contain within themselves the outlines and principles of construction of the three last named Sciences. If we can answer, under all sufficient conditions, the question *How often*, or *How many times and can* interpret or understand every sufficient answer to such a question, then are we perfect masters of Arithmetic as the Science of Quotity or Number. If, whenever the conditions are sufficient, we can enunciate and can interpret an answer to the question *How long*. or *For how long a time*, we are acquainted with the whole of Metrology, as the abstract Science of Quantity and Measure. And if, in every case of appropriate and sufficient conditions, we can assign and interpret a date, so as to form and to use in each case the answer to the question *When*, or *At what time*, we possess [(as it appears to me)] all that Algebra can ask or teach as the Science of Order and Progression. And these three questions Quoties, Quamdiu, Quando, — the How often, [the] How long, and [the] When, — with their respective answers So often, So long, or Then, — the Toties, Tamdiu, and Tunc, — while they contain those three separate Sciences of Arithmetic, Metrology, and Algebra, are plainly subordinate to the general conception of Time; and a Science of Pure Time, to be complete, must comprehend them all.

* Without striking out this word, the word "conviction" is written over it by W.R.H.

LIV.

ADDRESS TO ACADEMY OF SIR W. R. HAMILTON ON RETIRING FROM THE PRESIDENCY (1846)

[March 16, 1846]

[*Proceedings of the Royal Irish Academy* **3**, 199–201 (1846).]

The President delivered the following Address:

"My Lords and Gentlemen of the Royal Irish Academy,—Although it is, I believe, well known to most, perhaps to all, of you, that it has been for a considerable time my wish and intention to retire this evening from the Chair to which, in 1837, your kindness called me, on the still lamented event of the death of my distinguished predecessor, the late admirable Doctor Lloyd, and in which your continuing confidence has since replaced me on eight successive occasions, yet a few parting words from me may be allowed, perhaps expected; and I should wish to offer them, were it only to guard against the possibility of any one's supposing that I look upon my thus retiring from your Chair as a step unimportant to myself, or as one which might be taken by me with indifference, or without deliberation. It was under no hasty impulse that I resolved to retire from the office of your President into the ranks of your private members, nor was it lightly that I determined to lay down the highest honour of my life.

"My reasons have been stated in an Address delivered in another place, at a meeting of some members of your body. They are, briefly, these: that after the expiration of several years, I have found the duties of the office press too heavily upon my energies, indeed, of late, upon my health, when combined with other duties; and that I have felt the anxieties of a concentrated responsibility—exaggerated, perhaps, by an ardent or excitable temperament—tend more to distract my thoughts from the calm pursuits of study, than I can judge to be desirable or right in itself, or consistent with the full redeeming of those pledges which I may be considered to have long since given, as an early Contributor to your Transactions.

"When I look back on the aspirations with which first I entered on that office from which I am now about to retire, it humbles me to reflect how far short I have come of realizing my own ideal; but it cheers me to remember how greatly beyond what I could then have ventured to anticipate, the Academy itself has flourished. Of this result I may speak with little fear, because little is attributable to myself. Gladly do I acknowledge that it has been my good fortune, rather than my merit, to have presided over your body during a period in which, through the exertions of others much more than through my own (though mine, too, have not been withheld), the Academy is generally felt to have prospered in all its departments. The original papers which have been read; the volumes of Transactions which have been published; the closer communication which has been established with kindred societies of our own and of foreign countries; the enhanced value of our Library and Museum, which have been, at least, as much enriched in the quality as in the quantity of their contents; the improved state (as it is represented to me) of our finances, combined with an increased strength of our claims on public and parliamentary support; the heightened interest of members and visiters in our meetings, which have been honoured on four occasions, during my presidency, by the presence of representatives of Royalty; even the convenience and appropriate adornment of the rooms in which we assemble;—all these are things, and others might be named, in which, however small may have been the share of him who now addresses you, the progress of the Academy has not been small, and of which the recollection tends to console one who may, at least, be allowed to call himself an attached member of the body, under the sense, very deeply felt by him, of his own personal and official deficiencies.

"Whoever may be the member elected by your suffrages, this evening, to occupy that important and honourable post which I am now about to resign, it will, of course, become my duty to give to that future President my faithful and cordial support, by any means within the compass of my humble power. But if it be true, as I collect it to be, that your unanimous choice will fall upon the very member whom, out of all others, I should have myself selected, if it could have been mine to make the selection—with whom I have been long connected by the closest ties of college friendship, strengthened by the earnest sympathy which we have felt in our aspirations for the welfare of this Academy which has already benefited by his exertions in many and important ways—then will that course, which would have been in any event my *duty*, be in an eminent degree my *pleasure* also.

"And now, my Lords and Gentlemen, understanding that an old and respected member is prepared to propose for your votes, as my successor, the friend to whom I have ventured to allude—very inadequately, as regards my opinion of his merits, yet, perhaps, more pointedly than his modesty will entirely forgive or approve of,—I shall detain you no longer from that stage of the proceedings of the evening which must be the most interesting to all of us, but shall conclude these words of farewell from this Chair, by expressing a hope that my future exertions, though in a less conspicuous position, shall manifest, at least in some degree, that grateful and affectionate sense which I must ever retain of the constant confidence and favour which you have, at all times, shewn towards me."

RESOLVED,—That the thanks of the Academy be given to Sir William R. Hamilton, and that the Academy desire to express their entire sense of the value of his services as President, of his high and impartial bearing in the Chair, and of his untiring efforts to advance the interests of the body; and they also wish to record their satisfaction that he has determined to remain in the Council of the Academy.

LV.

OBITUARY NOTICE OF SIR W. R. HAMILTON (1865)

[*Proceedings of the Royal Irish Academy* **9**, 383 (1867).]

We have lost by death, within the year, two Honorary Members, viz.:—

1. SIR W. JACKSON HOOKER, K. H., F. R. S.; elected June 27, 1825;
2. THE REV. WILLIAM WHEWELL, D. D., F. R. S., Master of Trinity College, Cambridge; elected January 25, 1836.

And nine Ordinary Members, viz.:—

1. REV. W. H. DRUMMOND, D. D.; elected November 29, 1817.
2. RIGHT HON. THE EARL OF DONOUGHMORE; elected January 11, 1864.
3. SIR W. R. HAMILTON, LL. D.; elected October 22, 1827.
4. EDWARD HUTTON, M. D.; elected April 27, 1835.
5. THOMAS HUTTON, Esq., D. L.; elected February 10, 1840.
6. GEORGE A. KENNEDY, M. D.; elected November 30, 1835.
7. GEORGE PETRIE, LL. D.; elected February 25, 1828.
8. SIR THOMAS STAPLES, BART.; elected June 13, 1842.
9. RIGHT HON. JOHN WYNNE; elected April 10, 1843.

In this list are included two names of European celebrity—names which stand in the foremost ranks of Mathematical Science, and Archæological Learning, respectively. After the able and eloquent tributes lately paid by our President to the memories of Sir William Hamilton and Dr. Petrie, it is unnecessary for us to enlarge on the loss sustained by our country, and by the republic of letters, in the deaths of these distinguished men. But we may be permitted to observe how closely the labours and the renown of both were associated with this institution. Almost all the important researches of Hamilton were published in our "Transactions" or "Proceedings," from the "Essay on Systems of Rays," which first established his reputation, and for which he obtained our Cunningham Medal, to the "Theory of Quaternions," which was the latest product of his genius. The first hints of this Theory were given in communications read before the Academy; and up to the close of his life, he continued to bring before us, from time to time, the newest developments and applications of the method.

LIST OF THE WORKS OF SIR WILLIAM ROWAN HAMILTON IN APPROXIMATE CHRONOLOGICAL ORDER

1. On Caustics. Part First. Unpublished manuscript. (1824)
   I. 345.

2. Theory of Systems of Rays. Part First. (1827)*
   Transactions of the Royal Irish Academy **15**, 69–174 (1828).
   I. 1.†

3. Theory of Systems of Rays. Part Second. Unpublished manuscript.
   I. 88.

4. Systems of Rays. Part Third. Unpublished manuscript.
   IV. 3.

5. Supplement to an Essay on the Theory of Systems of Rays. (1830)
   Transactions of the Royal Irish Academy **16**, 1–61 (1830).
   I. 107.

6. On the Error of a received principle of Analysis, respecting Functions which vanish with their Variables. (1830)
   Transactions of the Royal Irish Academy **16**, 63–64 (1830).
   IV. 580.

7. Note on a Paper on the Error of a received Principle of Analysis. (1831)
   Transactions of the Royal Irish Academy **16**, 129–130 (1830).
   IV. 582.

* In the case of papers communicated, the date (or dates) immediately following the title is the date of communication.
† The volume and page numbers, where the item will be found, are given at the end of each entry.

8. On the double refraction of light according to the principles of Fresnel (Review of two scientific memoirs of James MacCullagh, B.A.). (1830)
National Magazine (*Dublin*) **1**, 145–149 (1830).
IV. 711.

9. Second Supplement to an Essay on the Theory of Systems of Rays. (1830)
Transactions of the Royal Irish Academy **16**, 93–125 (1831).
I. 145.

10. Optical Investigations. Unpublished manuscript. (1831)
I. 364.

11. Third Supplement to an Essay on the Theory of Systems of Rays. (1832)
Transactions of the Royal Irish Academy **17**, 1–144 (1837).
I. 164.

12. On the Differences and Differentials of Functions of Zero. (1831)
Transactions of the Royal Irish Academy **17**, 235–236 (1837).
(French translation in: *Correspondance Mathématique et Physique* **8**, 235–237 (1834).)
IV. 701.

13. On a View of Mathematical Optics. (1832)
*British Association Report for 1st Meeting, York 1831, and 2nd Meeting, Cambridge 1832**, pp. 545–547.
I. 295.

14. The Comet. (1832)
Dublin Penny Journal, pp. 207, 208 (December 1832) and pp. 223, 224 (January 1833).
IV. 673.

15. Introductory Lecture on Astronomy. (1832)
Dublin University Review, pp. 72–85 (January 1833).
IV. 663.

16. Review of Arago's work, "*The Comet*", translated by Colonel Charles Gold, London, 1833. (1833)
Dublin University Review, pp. 365–372 (April 1833).
IV. 676.

17. On some Results of the View of a Characteristic Function in Optics. (1833)
British Association Report for 3rd Meeting, Cambridge 1833, pp. 360–370.
I. 297.

* The date indicated for Reports of Meetings of the *British Association for the Advancement of Science* is that of the meeting; publication was in London and usually occurred in the following year.

18. On a New Method of investigating the relations of Surfaces to their Normals, with results respecting the Curvature of Ellipsoids. (1833)
Dublin University Review, pp. 653–654 (July 1833).
I. 304.

19. On the Effect of Aberration in prismatic Interference. (1833)
Philosophical Magazine (3S) **2**, 191–194 (1833).
(German translation in *Annalen der Physik und Chemie, Leipzig*, **29**, 316–319 (1833).)
I. 305.

20. On the undulatory Time of Passage of Light through a Prism. (1833)
Philosophical Magazine (3S) **2**, 284–287 (1833).
(German translation in *Annalen der Physik und Chemie, Leipzig*, **29**, 323–328 (1833).)
I. 307.

21. Note on Mr Potter's Reply respecting his experiment of prismatic Interference. (1833)
Philosophical Magazine (3S) **2**, 371 (1833).
(German translation in *Annalen der Physik und Chemie, Leipzig*, **29**, 328–329 (1833).)
I. 310.

22. On a general Method of expressing the Paths of Light and of the Planets, by the Coefficients of a Characteristic Function. (1833)
Dublin University Review, pp. 795–826 (November 1833).
(French translation, with a condensed introduction, in: *Correspondance Mathématique et Physique* **8**, 69–89, 200–211 (1834).)
I. 311.

23. Remarques de M. Hamilton, directeur de l'observatoire de Dublin, sur un mémoire de M. Plana inséré dans le tome VII de la Correspondance Math. (extrait d'une lettre). (1833)
Correspondance Mathématique et Physique **8**, 27–30 (1834).
I. 333.

24. Theory of Conjugate Functions, or Algebraic Couples; with a Preliminary and Elementary Essay on Algebra as the Science of Pure Time. (1833, 1835)
Transactions of the Royal Irish Academy **17**, 293–422 (1837).
III. 3.

25. The auxiliary function *T* for a telescope, when the axis of eyepiece is not coincident with, but parallel to, that of object glass. Unpublished manuscript. (1833)
I. 367.

26. The auxiliary function *T* for two thin lenses close together in vacuo, and for a single thin lens in vacuo. Unpublished manuscript. (1833)
I. 369.

27. The aberrations of an optical instrument of revolution. Unpublished manuscript. (1833)
I. 376.

28. Problem of Three Bodies by my Characteristic Function. Unpublished manuscript. (1833)
II. 1.

29. On the Application to Dynamics of a General Mathematical Method previously applied to Optics. (1834)
British Association Report for 4th Meeting, Edinburgh 1834, pp. 513–518.
II. 212.

30. On Conjugate Functions, or Algebraic Couples, as tending to illustrate generally the Doctrine of Imaginary Quantities and as confirming the Results of Mr Graves respecting the Existence of Two independent Integers in the complete expression of an Imaginary Logarithm. (1834)
British Association Report for 4th Meeting, Edinburgh 1834, pp. 519–523.
III. 97.

31. On a General Method in Dynamics, by which the Study of the Motions of all free Systems of attracting or repelling Points is reduced to the Search and Differentiation of one central Relation, or Characteristic Function. (1834)
Philosophical Transactions of the Royal Society, London, Part II, 247–308 (1834).
II. 103.

32. Sir W. Rowan Hamilton on the Elementary Conceptions of Mathematics (Seven Letters to Viscount Adare (March and April 1835).) (1879)
Hermathena (*Dublin*) **3**, 469–489 (1879).
IV. 762.

33. Address as Secretary of the Dublin Meeting of the British Association. (1835)
British Association Report for 5th Meeting, Dublin 1835, pp. xli–lvi.
IV. 716.

34. On a new theory of Logologues. (1835)
British Association Report for 5th Meeting, Dublin 1835, part II, p. 7 (title only).
IV. 728.

35. On a new theory of varying Orbits. (1835)
British Association Report for 5th Meeting, Dublin 1835, part II, p. 7 (title only).
IV. 728.

36. Explanation of the method of investigation pursued by Mr G. B. Jerrard for accomplishing the solution of equations of the fifth or of higher degrees.
British Association Report for 5th Meeting, Dublin 1835, part II, p. 7 (title only).
IV. 728.

37. Second Essay on a General Method in Dynamics. (1835)
Philosophical Transactions of the Royal Society, London, Part I, 95–144 (1835).
II. 162.

38. On the Propagation of Light in Crystals. Unpublished manuscript. (1835–1838)
II. 413.

39. Correspondence. Unpublished manuscripts. (1835–1839)
II. 583.

40. On nearly Circular Orbits. Unpublished manuscript. (1836)
II. 217.

41. Calculus of Principal Relations. Unpublished manuscript. (1836)
II. 297.

42. Calculus of Principal Relations. Unpublished manuscript. (1836)
II. 332.

43. Calculus of Principal Relations. A new series of investigations. Unpublished manuscript.
II. 358.

44. Integration of Partial Differential Equations by the Calculus of Variations. Unpublished manuscript. (1836)
II. 391.

45. Calculus of Principal Relations. (1836)
British Association Report for 6th Meeting, Bristol 1836, part II, pp. 41–44.
II. 408.

46. Theorems respecting Algebraic Elimination, connected with the Question of the Possibility of resolving in finite Terms the general Equation of the Fifth Degree. (1836)
Philosophical Magazine (3S) **8**, 538–543 (1836), and **9**, 28–32 (1836).
III. 471.

47. Inquiry into the validity of a Method recently proposed by George B. Jerrard, Esq., for Transforming and Resolving Equations of Elevated Degrees. (1836)
British Association Report for 6th Meeting, Bristol 1836, pp. 295–348.
III. 481.

48. Theory of the Moon. Unpublished manuscript. (1837)
II. 238.

49. Correspondence with J. W. Lubbock. Unpublished manuscript. (1837)
II. 249.

50. Exposition of the Arguments of Abel, respecting Equations of the Fifth Degree. (1837)
British Association Report for 7th Meeting, Liverpool 1837, part II, p. 1 (title only). See Nos. 51 and 54, below.
Not included in Vol. III.

51. On the Argument of Abel, respecting the Impossibility of expressing a Root of any General Equation above the Fourth Degree, by any finite Combination of Radicals and Rational Functions. (1837)
Transactions of the Royal Irish Academy **18**, 171–259 (1839).
III. 517.

52. New Application of the Calculus of Principal Relations. (1837)
British Association Report for 7th Meeting, Liverpool 1837, part II, p. 1.
IV. 729.

53. Exposition of Mr Turner's Theorem respecting the Series of Odd Numbers and the Cubes and other powers of the Natural Numbers. (1837)
British Association Report 7th Meeting, Liverpool 1837, part II, p. 1 (title only).
IV. 729.

54. Investigations respecting equations of the Fifth Degree. (1837)
Proceedings of the Royal Irish Academy **1**, 76–80 (1841).
III. 478.

55. Inaugural Address as President of the Royal Irish Academy. (1838)
Proceedings of the Royal Irish Academy **1**, 107–120 (1841).
IV. 730.

56. Address as President of the Royal Irish Academy on Professor MacCullagh's Paper on the Laws of Crystalline Reflexion and Refraction. (1838)
Proceedings of the Royal Irish Academy **1**, 212–221 (1841).
IV. 739.

57. On the Propagation of Light in vacuo. (1838)
British Association Report for 8th Meeting, Newcastle 1838, part II, pp. 2–6.
II. 446.

58. On the Propagation of Light in Crystals. (1838)
British Association Report for 8th Meeting, Newcastle 1838, part II, p. 6.
II. 450.

59. Researches respecting Vibration connected with the Theory of Light. Unpublished manuscript. (1839)
II. 451.

60. Propagation of Motion in Elastic Medium — Discrete Molecules. Unpublished manuscript. (1839)
II. 527.

61. Notice of a singular appearance of the clouds observed on the 16th December, 1838. (1839)
Proceedings of the Royal Irish Academy **1**, 249 (1841).
IV. 745.

62. Researches on the Dynamics of Light. (1839)
Proceedings of the Royal Irish Academy **1**, 267–270 (1841).
II. 576.

63. Address as President of the Royal Irish Academy on Dr Apjohn's researches on the Specific Heat of Gases. (1839)
Proceedings of the Royal Irish Academy **1**, 276–284 (1841).
IV. 746.

64. Researches respecting Vibration connected with the Theory of Light. (1839)
Proceedings of the Royal Irish Academy **1**. 341–349 (1841).
II. 578.

65. Address as President of the Royal Irish Academy on Mr Petrie's Paper on the History and Antiquities of Tara Hill. (1839)
Proceedings of the Royal Irish Academy **1**, 350–354 (1841).
IV. 752.

66. On Fluctuating Functions. (1840)
Proceedings of the Royal Irish Academy **1**, 475–477 (1841).
IV. 583.

67. On Fluctuating Functions. (1840)
Transactions of the Royal Irish Academy **19**, 264–321 (1843).
IV. 585.

68. New Demonstration of Fourier's theorem. (1841)
Proceedings of the Royal Irish Academy **2**, 129 (1844).
IV. 631.

69. On the Focal Lengths and Aberrations of a thin Lens of Uniaxal Crystal, bounded by Surfaces which are of Revolution about its axis. (1841)
Philosophical Magazine (3S) **19**, 289–294 (1841).
I. 336.

70. On a mode of deducing the equation of Fresnel's Wave. (1841)
Philosophical Magazine (3S) **19**, 381–383 (1841).
I. 341.

71. On the Composition of Forces. (1841)
Proceedings of the Royal Irish Academy **2**, 166–168 (1844).
II. 284.

72. Supplementary Remarks on Fluctuating Functions. (1842)
Proceedings of the Royal Irish Academy **2**, 232–238 (1844).
Identical with pp. 317–321 of No. 67 above.
IV. P. 630.

73. On the Day of the Vernal Equinox at the time of the Council of Nice. (1842)
Proceedings of the Royal Irish Academy **2**, 249–250 (1844).
IV. 681.

74. On certain discontinuous Integrals connected with the Development of the Radical which represents the Reciprocal Distance between two Points. (1842)
Philosophical Magazine (3S) **20**, 288–294 (1842).
IV. 632.

75. On a mode of expressing Fluctuating or Arbitrary Functions by mathematical formulæ. (1842)
British Association Report for 12th Meeting, Manchester 1842, part II, p. 10.
IV. 639.

76. On a Method proposed by Professor Badano for the solution of Algebraic Equations. (1842)
Proceedings of the Royal Irish Academy **2**, 275–276 (1844).
III. 570.

77. On Equations of the Fifth Degree and especially on a certain System of Expressions connected with these Equations, which Professor Badano has lately proposed. (1842)
Transactions of the Royal Irish Academy **19**, 329–376 (1843).
III. 572.

78. Two Letters to Professor Phillips on the construction of object glasses. Unpublished manuscript. (1843–1844)
I. 383.

79. On a Theorem in the Calculus of Differences. (1843)
British Association Report for 13th Meeting, Cork 1843, part II, pp. 2–3.
IV. 703.

80. On some investigations connected with the Calculus of Probabilities. (1843)
British Association Report for 13th Meeting, Cork 1843, part II, pp. 3–4.
IV. 707.

81. On an Expression for the Numbers of Bernoulli by means of a Definite Integral, and on some connected Processes of Summation and Integration. (1843)
Philosophical Magazine (3S) **23**, 360–367 (1843).
IV. 640.

82. On Dr Robinson's Table of Mean Refraction. (1843)
Proceedings of the Royal Irish Academy **2**, 400–401 (1844).
IV. 755.

83. Address as President of the Royal Irish Academy on Dr Kane's Researches on the Nature of Ammonia. (1843)
Proceedings of the Royal Irish Academy **2**, 411–419 (1844).
IV. 756.

84. On the Calculus of Probabilities. (1843)
Proceedings of the Royal Irish Academy **2**, 420–422 (1844).
IV. 705.

85. Quaternions. Unpublished manuscript. (1843)
III. 104.

86. On Quaternions, or on a new System of Imaginaries in Algebra. (A letter to John T. Graves, dated 17 October 1843).
Philosophical Magazine (3S) **25**, 489–495 (1844).
III. 106.

87. On a new species of Imaginary Quantities connected with the Theory of Quaternions. (1843)
Proceedings of the Royal Irish Academy **2**, 424–434 (1844).
III. 111.

88. Researches respecting Quaternions. First Series. (1843)
Transactions of the Royal Irish Academy **21**, 199–296 (1848).
III. 159.

89. On the improvement of the double achromatic object glass. Unpublished manuscript. (1844)
I. 387.

90. On approximating to the calculations of Eclipses. (1844)
Proceedings of the Royal Irish Academy **2**, 597 (1844).
IV. 682.

91. On Quaterntions; or on a new System of Imaginaries in Algebra. (1844–1850)
Philosophical Magazine (3S) **25**, 10–13, 241–246 (1844); **26**, 220–224 (1845); **29**. 26–31, 113–122, 326–328 (1846); **30**, 458–461 (1847); **31**, 214–219, 278–293, 511–519 (1847); **32**, 367–374 (1848); **33**, 58–60 (1848); **34**, 294–297, 340–343, 425–439 (1849); **35**, 133–137, 200–204 (1849); **36**, 305–306 (1850).
III. 227.

92. On Quaternions. (1844)
British Association Report for 14th Meeting, York 1844, part II, p. 2 (brief abstract).
Not included in Vol. III.

93. On Quaternions, or on a New system of Imaginaries in Algebra; with some Geometrical Illustrations. (1844)
Proceedings of the Royal Irish Academy **3**, 1–16 (1847).
III. 355.

94. A Theorem on Spherical Quadrilaterals and Spherical Conics. (1845)
Proceedings of the Royal Irish Academy **3**, 109 (1847).
III. 363.

95. On Quaternions. (1845)
British Association Report for 15th Meeting, Cambridge 1845, part II, p. 3 (Brief summary of an exposition of basic ideas. Hamilton poses the question: ‘Is there not an analogy between the fundamental pair of equations $ij = k$, $ji = -k$, and the facts of opposite currents of electricity corresponding to opposite rotations?’).
Not included in Vol. III.

96. Illustrations from Geometry of the Theory of Algebraic Quaternions. (1845)
Proceedings of the Royal Irish Academy **3**, Appendix, pp. xxxi–xxxvi (1847).
III. 364.

97. On the application of the Method of Quaternions to some Dynamical Questions, (1845)

Proceedings of the Royal Irish Academy **3**, Appendix xxxvi–l (1847).
III. 441.

98. Additional applications of the Theory of Algebraic Quaternions to Dynamical Questions. (1845)
Proceedings of the Royal Irish Academy **3**, Appendix, pp. li–lx (1847).
III. 449.

99. On Symbolical Geometry. (1846–1849)
Cambridge and Dublin Mathematical Journal **1**, 45–57, 137–154, 256–263, (1846); **2**, 47–52, 130–133, 204–209 (1847); **3**, 68–84, 220–225 (1848); **4**, 84–89, 105–118 (1849).
IV. 431.

100. On a Proof of Pascal's Theorem by means of Quaternions; and on some other connected subjects. (1846)
Proceedings of the Royal Irish Academy **3**, 273–292 (1847).
III. 367.

101. On Theorems of Central Forces. (1846)
Proceedings of the Royal Irish Academy **3**, 308–309 (1847).
II. 286.

102. The Hodograph, or a new method of expressing in symbolical language the Newtonian Law of Attraction. (1846)
Proceedings of the Royal Irish Academy **3**, 344–353 (1847).
II. 287.

103. On Theorems of Hodographic and Anthodographic Isochronism. (1847)
Proceedings of the Royal Irish Academy **3**, 417, 465–466 (1847).
II. 293.

104. Address to Academy of Sir W. R. Hamilton on Retiring from the Presidency. (1846)
Proceedings of the Royal Irish Academy **3**, 199–201 (1847).
IV. 773.

105. On the application of the Calculus of Quaternions to the Theory of the Moon. (1847)
Proceedings of the Royal Irish Academy **3**, 507–520 (1847).
III. 455.

106. On some applications of the calculus of quaternions to the theory of the moon. (1847)
British Association Report for 17th Meeting, Oxford 1847, part II, p. 4 (title only).
Not included in Vol. III.

107. Example of an Isoperimetrical Problem treated by the Calculus of Quaternions. (1847)
British Association Report for 17th Meeting, Oxford 1847, part II, p. 4 (title only).
Not included in Vol. III.

108. On additional applications of Quaternions to Surfaces of the Second Order. (1847)
Proceedings of the Royal Irish Academy **4**, 14–19 (1850).
III. 378.

109. On Quarternions and the Rotation of a Solid Body. (1848)
Proceedings of the Royal Irish Academy **4**, 38–56 (1850).
III. 381.

110. On Quaternions and the determination of the Distances of any recently discovered Comet or Planet from the Earth. (1848)
Proceedings of the Royal Irish Academy **4**, 75 (1850).
III. 464.

111. On the New Planet *Metis*. (1848)
Proceedings of the Royal Irish Academy **4**, 169 (1850).
IV. 683.

112. On the Double Mode of Generation of an Ellipsoid. (1848)
Proceedings of the Royal Irish Academy **4**, 173 (1850).
III. 392.

113. Additional Theorems respecting certain reciprocal Surfaces. (1848)
Proceedings of the Royal Irish Academy **4**, 192–193 (1850).
III. 393.

114. On Quaternions applied to Problems respecting the construction of a Circle touching three given Circles on a Sphere; and of a Sphere touching four given Spheres. (1848)
Proceedings of the Royal Irish Academy **4**, 255 (1850).
III. 394.

115. On Theorems relating to Surfaces, obtained by the method of Quaternions. (1849)
Proceedings of the Royal Irish Academy **4**, 306–308 (1850).
III. 395.

116. On an Equation of the Ellipsoid. (1849)
Proceedings of the Royal Irish Academy **4**, 324–325 (1850).
III. 397.

117. On the inscription of certain 'gauche' Polygons in Surfaces of the Second Degree. (1849)

Proceedings of the Royal Irish Academy **4**, 325–326 (1850).
III. 398.

118. On the Construction of the Ellipsoid by two sliding Spheres. (1849)
Proceedings of the Royal Irish Academy **4**, 341–342 (1850).
III. 399.

119. On a Theorem respecting Ellipsoids, obtained by the method of Quaternions. (1849).
Proceedings of the Royal Irish Academy **4**, 349–350 (1850).
III. 401.

120. On some results obtained by the Quaternion analysis respecting the inscription of 'gauche' Polygons in Surfaces of the Second Order. (1849)
Proceedings of the Royal Irish Academy **4**, 380–387 (1850).
III. 403.

121. On some new applications of Quaternions to Geometry. (1849)
British Association Report for 19th Meeting, Birmingham 1849, part II, p. 1 (title only).
IV. 501.

122. Exercises in Quaternions. (1849)
Cambridge and Dublin Mathematical Journal **4**, 161–168 (1849).
III. 298.

123. On 'gauche' Polygons in Central Surfaces of the Second Order. (1850)
Proceedings of the Royal Irish Academy **4**, 541–557 (1850).
III. 407.

124. On Polygons inscribed on a Surface of the Second Order. (1850)
British Association Report for 20th Meeting, Edinburgh 1850, part II, p. 2 (title only).
IV. 502.

125. On a Proof from Quaternions of the celebrated Theorem of Joachimsthal. (1851)
Proceedings of the Royal Irish Academy **5**, 71 (1853).
III. 416.

126. A generalization of Pascal's Theorem. (1851)
Proceedings of the Royal Irish Academy **5**, 100–101 (1853).
III. 417.

127. Of the nature and properties of the Aconic Function of six Vectors. (1851)
Proceedings of the Royal Irish Academy **5**, 177–186 (1853).
III. 418.

128. On the connexion of Quaternions with Continued Fractions and Quadratic Equations. (1851–1852)
Proceedings of the Royal Irish Academy **5**, 219–221, 299–301 (1853).
III. 304.

129. On continued Fractions in Quaternions. (1852–1853)
Philosophical Magazine (4S) **3**, 371–373 (1852); **4**, 303 (1852); **5**, 117–118, 236–238, 321–326 (1853).
III. 307.

130. On Biquaternions. (1852)
British Association Report for 22nd Meeting, Belfast 1852, part II, p. 2 (brief abstract).
Not included in Vol. III.

131. On the Geometrical Interpretation of some results obtained by calculation with Biquaternions. (1853)
Proceedings of the Royal Irish Academy **5**, 388–390 (1853).
III. 424.

132. Sur les Quarternions. (1853)
Nouvelles Annales de Mathématiques **12**, 275–283 (1853).
An abbreviated version, in French, of the first paper in the series published in the *Philosophical Magazine*, starting in 1844. See No. 91 above and the footnote on p. 233 of Vol. III.

133. On the Geometrical Demonstration of some Theorems obtained by means of the Quaternion analysis. (1853)
Proceedings of the Royal Irish Academy **5**, 407–415 (1853).
III. 426.

134. A Theorem concerning Polygonic Syngraphy. (1853)
Proceedings of the Royal Irish Academy **5**, 474–475 (1853).
III. 431.

135. *Lectures on Quaternions*, i–viii, 64, ix–lxxii, 1–736, (Hodges and Smith, Dublin; Whittaker & Co., London; Macmillan & Co., Cambridge: 1853).
The preface (pp. 1–64) is reprinted in III. 117–158.

136. On the Integrations of certain Equations. (1854)
Proceedings of the Royal Irish Academy **6**, 62–63 (1858).
IV. 646.

137. On the celebrated Theorem of Dupin. (1854)
Proceedings of the Royal Irish Academy **6**, 86–88 (1858).
III. 432.

138. On some Extensions of Quaternions. (1854)
Proceedings of the Royal Irish Academy **6**, 114–115 (1858).
III. 316.

139. On some Extensions of Quaternions. (1854–1855)
Philosophical Magazine (4S) **7**, 492–499 (1854); **8**, 125–137, 261–269 (1854); **9**, 46–51, 280–290 (1855).
III. 317.

140. On an Extension of Quaternions. (1854)
British Association Report for 24th Meeting, Liverpool 1854, part II, p. 1 (brief abstract).
Not included in Vol. III.

141. On the Solution of the Equation of Laplace's Functions. (1855)
Proceedings of the Royal Irish Academy **6**, 181–185 (1858).
IV. 648.

142. Remarks, chiefly Astronomical, on what is known as the Problem of Hipparchus. (1855)
Hermathena (Dublin) **4**, 480–506 (1883).
IV. 684.

143. Symbolical Extensions of Quaternions; *and* Geometrical Applications of Quaternions. (1855)
Proceedings of the Royal Irish Academy **6**, 250, 260, 311 (1858).
Notices only, no details.
IV. 503.

144. On the conception of the Anharmonic Quaternion, and on its application to the Theory of Involution in space. (1855)
British Association Report for 25th Meeting, Glasgow 1855, part II, p. 7 (title only).
Not included in Vol. III.

145. Account of the Icosian Calculus. (1856)
Proceedings of the Royal Irish Academy **6**, 415–416, 462 (1858).
III. 609.

146. Memorandum respecting a New System of Roots of Unity. (1856)
Philosophical Magazine (4S) **12**, 446 (1856).
III. 610.

147. Memorandum for John T. Graves. Unpublished manuscript. (1856)
III. 611.

148. Letter to John T. Graves on the Icosian. Unpublished manuscript. (1856)
III. 612.

149. On a General Expression by Quaternions for Cones of the Third Order. (1857)
Proceedings of the Royal Irish Academy **6**, 506, 512 (1858).
IV. 504.

150. On a certain harmonic property of the envelope of the chord connecting two corresponding points of the Hessian of a Cubic Cone. (1857)
Proceedings of the Royal Irish Academy **6**, 524 (1858).
IV. 505.

151. On some applications of Quaternions to Cones of the Third Degree. (1857)
British Association Report for 27th Meeting, Dublin 1857, part II, p. 3 (title only).
IV. 506.

152. On the Icosian Calculus. (1857)
British Association Report for 27th Meeting, Dublin 1857, part II, p. 3 (brief abstract).
Not included in Vol. III.

153. On the Calculation of Numerical Values of a certain class of Multiple and Definite Integrals. (1857)
Philosophical Magazine (4S) **14**, 375–382 (1857).
IV. 652.

154. Two Letters to Augustus De Morgan. Unpublished manuscripts. (1858)
IV. 34.

155. On some Quaternion Equations connected with Fresnel's wave-surface for biaxal crystals. (1859)
Proceedings of the Royal Irish Academy **7**, 122–124, 163 (1862).
III. 465.

156. On some Quaternion Equations connected with Fresnel's wave-surface for biaxal crystals. (1859)
British Association Report for 29th Meeting, Aberdeen 1859, part II, p. 248.
A précis of No. 155 above; not included in Vol. III.

157. On some Quaternion Equations connected with Fresnel's wave-surface for biaxal crystals. (1859)
Natural History Review **6**, 240–242, 365 (1859).
Identical with No. 155 above.

158. Letter to Hart on Anharmonic Coordinates. Unpublished manuscript. (1860)
IV. 179.

159. On Anharmonic Co-ordinates. (1860)
Proceedings of the Royal Irish Academy **7**, 286–289, 329, 350–354 (1862).
IV. P. XI.

160. On Anharmonic Co-ordinates. (1860)
Natural History Review **7**, 242–246, 325–327, 506–509 (1860).
Identical with No. 159 above.

161. On Geometrical Nets in Space. (1861)
Proceedings of the Royal Irish Academy **7**, 532–582 (1862)
IV. 516.

162. On Geometrical Nets in Space. (1861)
British Association Report for 31st Meeting, Manchester 1861, part II, p. 4 (title only).
IV. 561.

163. Quaternion Proof of a Theorem of Reciprocity of Curves in Space. (1862)
British Association Reporty for 32nd Meeting, Cambridge 1862, part II, p. 4.
III. 434.

164. Elementary Proof that Eight Perimeters of the Regular inscribed Polygon of Twenty Sides exceed Twenty-five Diameters of the Circle. (1862)
Philosophical Magazine (4S) **23**, 267–269 (1862).
IV. 562.

165. On a New and General Method of Inverting a Linear and Quaternion Function of a Quaternion. (1862)
Proceedings of the Royal Irish Academy **8**, 182–183 (1864).
III. 348.

166. On the Existence of a Symbolic and Biquadratic Equation, which is satisfied by the Symbol of Linear Operation in Quaternions. (1862)
Proceedings of the Royal Irish Academy **8**, 190–191 (1864).
III. 350.

167. On the Existence of a Symbolic and Biquadratic Equation, which is satisfied by the Symbol of Linear Operation in Quaternions. (1862)
Philosophical Magazine (4S) **24**, 127–128 (1862).
Identical with No. 166 above.

168. On 'Gauche' Curves of the Third Degree. (1863)
Proceedings of the Royal Irish Academy **8**, 331–334 (1864).
III. 435.

169. On a General Centre of Applied Forces. (1863)
Proceedings of the Royal Irish Academy **8**, 394 (1864).
III. 468.

170. On the Locus of the Osculating Circle to a Curve in Space. (1863)
Proceedings of the Royal Irish Academy **8**, 394 (1864).
IV. 564.

171. On the Eight Imaginary Umbilical Generatrices of a Central Surface of the Second Order. (1864)
Proceedings of the Royal Irish Academy **8**, 471 (1864).
IV. 565.

172. On Röber's Construction of the Heptagon. (1864)
Philosophical Magazine (4S) **27**, 124–132 (1864).
IV. 566.

173. On a Theorem relating to the Binomial Coefficients. (1865)
Proceedings of the Royal Irish Academy **9**, 297–302 (1867).
III. 603.

174. On a New System of Two General Equations of Curvature. (1865)
Proceedings of the Royal Irish Academy **9**, 302–305 (1867).
IV. 574.

175. Obituary notice of Sir W. R. Hamilton. (1865)
Proceedings of the Royal Irish Academy **9**, 383 (1867).
IV. 775.

176. *Elements of Quaternions*, ed. W E Hamilton*, i–lix , 1– 762, (Longmans, Green, & Co., London: 1866).

177. *Elemente der Quaternionen*, trans. P. Glan†, 2 Vols, Vol.1 Theorie der Quaternionen, Vol. 2 Anwendungen, (Johann Ambrosius Barth, Leipzig: 1882–1884).

178. *Elements of Quaternions*, 2nd edn, ed. C. J. Joly‡, 2 Vols (Longmans, Green & Co., London, New York, and Bombay: 1899, 1901).

* [William Edwin Hamilton (1834–1902) was Hamilton's son. See foot note on p. 367 of this volume.]
† [Paul Glan (1846–1898).]
‡ [Charles Jasper Joly (1864–1906) was Director of Dunsink Observatory from 1897 until his death. See P. A. Wayman, *Dunsink Observatory, 1785–1985* (Dublin Institute for Advanced Studies and Royal Dublin Society, Dublin: 1987).]

INDEX

COMBINED INDEX FOR ALL FOUR VOLUMES

* The expression "aberration coefficients" is used by the Editors [of Vol. I.] to denote the coefficients of type Q in the expansion for T (I., p. 298), instead of Hamilton's expression "constants of aberration".

Printed in the United States
by Baker & Taylor Publisher Services